Fundamentals of Finite Element Analysis

Fundamentals of Finite Element Analysis

Linear Finite Element Analysis

Ioannis Koutromanos
Department of Civil and Environmental Engineering
Virginia Polytechnic Intitute and State University
Blacksburg, VA, United States

With single-chapter contributions from:

James McClure
Advanced Research Computing
Virginia Polytechnic Institute and State University
Blacksburg, VA, United States

Christopher Roy
Department of Aerospace and Ocean Engineering
Virginia Polytechnic Institute and State University
Blacksburg, VA, United States

This edition first published 2018

Registered Office(s)
John Wiley & Sons, Inc., 111 River Street, Hoboken, NJ 07030, USA
John Wiley & Sons Ltd, The Atrium, Southern Gate, Chichester, West Sussex, PO19 8SQ, UK

Editorial Office
The Atrium, Southern Gate, Chichester, West Sussex, PO19 8SQ, UK

For details of our global editorial offices, customer services, and more information about Wiley products visit us at www.wiley.com.

Wiley also publishes its books in a variety of electronic formats and by print-on-demand. Some content that appears in standard print versions of this book may not be available in other formats.

Library of Congress Cataloging-in-Publication Data

Names: Koutromanos, Ioannis, 1982– author. | McClure, James, 1981– contributor. | Roy, Christopher J., contributor.
Title: Fundamentals of finite element analysis: linear finite element analysis / by Ioannis Koutromanos ; With single-chapter contributions from James McClure, Christopher Roy.
Description: Hoboken, NJ : John Wiley & Sons, 2017. | Includes index. | Identifiers: LCCN 2017030699 (print) | LCCN 2017043821 (ebook) | ISBN 9781119260141 (pdf) | ISBN 9781119260127 (epub) | ISBN 9781119260080 (pbk.)
Subjects: LCSH: Finite element method.
Classification: LCC TA347.F5 (ebook) | LCC TA347.F5 K68 2017 (print) | DDC 518/.25–dc23
LC record available at https://lccn.loc.gov/2017030699

Cover Design: Wiley
Cover Image: Courtesy of Ioannis Koutromanos

Set in 10/12pt Warnock by SPi Global, Pondicherry, India

Printed in the UK

To my family

Contents

About the Companion Website

Don't forget to visit the companion website for this book:

www.wiley.com/go/koutromanos/linear

The companion website of this text contains a standalone, open-source, finite element analysis program, called VTFEA, for one-dimensional and multidimensional elasticity and heat conduction problems. The source code of the program (written in Fortran) and brief tutorials and sample input files are also provided in the website. These tutorials include the use of a graphical post-processor for generating plots of the obtained results. A brief tutorial on Fortran programming is also provided, to facilitate understanding of the source code.

Scan this QR code to visit the companion website.

1

Introduction

1.1 Physical Processes and Mathematical Models

In all disciplines of science and engineering, the focus is on determining what will happen when a physical system of interest (e.g., a building, a bridge, an aircraft, a collection of molecules, or even a living organism) is subjected to the effects of the environment, which may in turn lead to a *response* from the system. For example, if we have a building subjected to forces due to the weight of occupants or due to an earthquake, the building structural components will respond through deformations and stresses. Similarly, if a hot piece of metal is thrown into a large pool with cool water, the metal piece will quickly become cooler.

The aforementioned examples are only two special cases of *physical processes* in systems of matter—that is, changes that happen into a piece of matter that we isolate and examine—due to interaction of that piece with its surroundings. Any physical process can be described mathematically, in the sense that we can formulate a set of mathematical expressions describing the process and allowing us to determine how the process takes place in space and time. The cornerstone of mathematical descriptions of systems are quantities called the *state variables*, describing the state of the system for a phenomenon of interest. For example, when we have the case of the piece of metal immersed in cool water, the most appropriate state variable is the temperature. The state variables are not generally constant in space or time. Thus, the mathematical description of most physical processes is synonymous with the need to determine the spatial variation and temporal evolution of state variables.

The set of mathematical equations allowing the determination of state variables and, in turn, the analytical investigation of any physical process, is sometimes referred to as the *mathematical model* of a process. In most physical problems of interest, the mathematical models typically correspond to *differential equations*. This means that we can formulate a set of equations involving the derivatives of functions with respect to spatial variables and/or time. One example differential equation, for the case that we want to find the spatial distribution of a state variable $T(x)$, is the following:

$$\frac{dT}{dx} + s(x) = 0 \tag{1.1.1}$$

where $s(x)$ is some given function.

Fundamentals of Finite Element Analysis: Linear Finite Element Analysis, First Edition.
Ioannis Koutromanos, James McClure, and Christopher Roy.

Companion website: www.wiley.com/go/koutromanos/linear

In problems with differential equations, we want to find functions which satisfy a given equation involving their derivatives. Thus, for the system described by the differential Equation (1.1.1), the goal would be to determine the function $T(x)$. But what is the *physical meaning* of differential equations? In most cases, differential equations are mathematical statements of the principle of *conservation of a fundamental physical quantity*. That is, there are several fundamental physical quantities, which, by principle, must be conserved in any physical process. Example physical quantities that are conserved in nature are:

- mass
- linear momentum (i.e., the "product of velocity times the mass of a material point")
- energy
- electric charge

The physical problems that will be discussed in the present textbook involve conservation of the first three quantities just listed. The conservation of linear momentum will not be explicitly invoked; instead, we will include an equivalent statement, as described in the following Remark.

Remark 1.1.1: A special mention is deemed necessary for the conservation of linear momentum, which is the principle that governs problems in solid and structural mechanics. We usually prefer to stipulate the principle in an equivalent and more meaningful (for mechanicians and engineers) version—that is, *force equilibrium*. We will demonstrate the equivalence of the two statements (conservation of linear momentum and force equilibrium) here. For a material point with mass m, the linear momentum is simply equal to the product $m \cdot \vec{v}$, where $\vec{v}$ is the *velocity vector* of the material point. The conservation of linear momentum can simply be stated as Newton's second law, that is, "the rate of change (with time t) of the linear momentum of a material point is equal to the sum of the forces acting on the point." Thus, for the specific example of the material point, the conservation of momentum principle can be mathematically stated as follows:

$$\frac{d(m \cdot \vec{v})}{dt} = \sum \vec{F} \tag{1.1.2}$$

where $\sum \vec{F}$ is the total (resultant) force vector acting on the material point. If we further assume that the mass m in Equation (1.1.2) remains constant with time, we can write:

$$m \cdot \vec{a} = \sum \vec{F} \tag{1.1.3}$$

where $\vec{a} = \frac{d\vec{v}}{dt}$ is the *acceleration vector* of the material point. For the special case of very slow loading (where the change with time of velocity is negligible and, consequently, the acceleration is practically equal to zero), Equation (1.1.3) reduces to a *force equilibrium equation*:

$$\sum \vec{F} = \vec{0} \tag{1.1.4}$$

Even if we do have significant accelerations, we can still cast equation (1.1.3) into a force equilibrium-like form, by defining the *inertial (or D'Alembert) force,* $\vec{F}_I$, as follows.

$$\vec{F}_I = -m \cdot \vec{a} \tag{1.1.5}$$

Then, Equation (1.1.3) can be cast in the form:

$$\sum \vec{F} + \vec{F}_I = 0 \tag{1.1.6}$$

That is, the sum of the forces and inertial forces acting on the material point must equal zero. The bottom line of the considerations presented herein, is that, whenever we use force equilibrium to describe a physical process, we are indirectly accounting for the conservation of linear momentum. This text will use force equilibrium, without any special mentioning that this has stemmed from conservation of linear momentum. ▮

The governing differential equations of any physical problem can be written in an abstract, generic mathematical form:

$$L[f(x)] + s(x) = 0 \tag{1.1.7}$$

where $f(x)$ is the function that we are trying to determine, $s(x)$ is a given expression that does not depend on $f(x)$, and $L[\ ]$ is a *differential operator*—that is, a mathematical entity that receives (as input) some function and returns (as output) an expression involving derivatives of the function. For example, Equation (1.1.1) can be written as $L[T(x)] + s(x) = 0$, where $L[\ ] = \frac{d}{dx}$. The function $f(x)$, which depends on the spatial coordinates x, will be called a *field function.* We can imagine that the term x in Equation (1.1.7) symbolically expresses the set of coordinates that are necessary to uniquely identify the location of a point in space, for a problem of interest. If we have a one-dimensional problem, we need a single variable to describe the spatial location, if we have a two-dimensional problem, we need a pair of variables, and so on. Additionally, for multidimensional problems, the field function $f(x)$ may be scalar-valued or vector-valued. We may thus speak in our discussions of scalar-field or vector-field problems. Toward the end of this text, we will consider time-dependent scalar- and vector-field problems, wherein the field functions have an additional dependence on time.

1.2 Approximation, Error, and Convergence

The use of computational simulation is integrally connected to the concept of *approximation.* The term implies that the computational representation of a physical process is governed by "somehow modified" versions of the governing mathematical equations. This "modification" leads to the fact that computational simulation provides *approximate values* for quantities of interest. How accurate our approximate values are will depend on how closely the computational "modification" matches the original mathematical expressions.

The notion of approximation entails the need to discuss three additional concepts—namely, that of the *error*, the *approximation parameter*, and the *convergence*. Whenever we approximate a quantity, the *error* in our approximation is defined as the difference of the exact value of the quantity from the approximate value that we have obtained. In other words, we can write:

$$(error) = (approximate\ value) - (exact\ value)$$

The error can be thought of as a quantitative measure of how "well" an approximation matches the exact quantity. In computational simulation, our approximation is usually characterized by means of an *approximation parameter*. This parameter is a quantity indicating how "crude" or "refined" our approximation is expected to be. One example is shown in Figure 1.1a, which presents the values of a function $f(x)$, defined over the interval between $x = 0$ and $x = 20$. We try to approximately represent the function through a set of points, which are connected to each other with straight line segments (the dashed lines in Figure 1.1a). In the specific example, we can define the *spacing* Δx between consecutive points as the approximation parameter. Figure 1.1a indicates that if we use a spacing $\Delta x = 5$, the approximate representation of the curve is not very satisfactory. In other words, one would expect the error between the approximate representation and the exact curve to be large. On the other hand, if we reduce the spacing to $\Delta x = 0.5$, the approximate representation of the curve $f(x)$ is very close to the exact curve, as shown in Figure 1.1b. In other words, as the size of the approximation parameter reduces, the error is expected to reduce. One could verify that, if an even smaller spacing (e.g., $\Delta x = 0.1$) was selected, then the approximate representation would become practically identical to the exact curve. The property of an approximation to become identical to the exact value when an approximation parameter is sufficiently small is called *convergence*. It is obvious that an approximate method for computational simulation is to be used only if it ensures convergence.

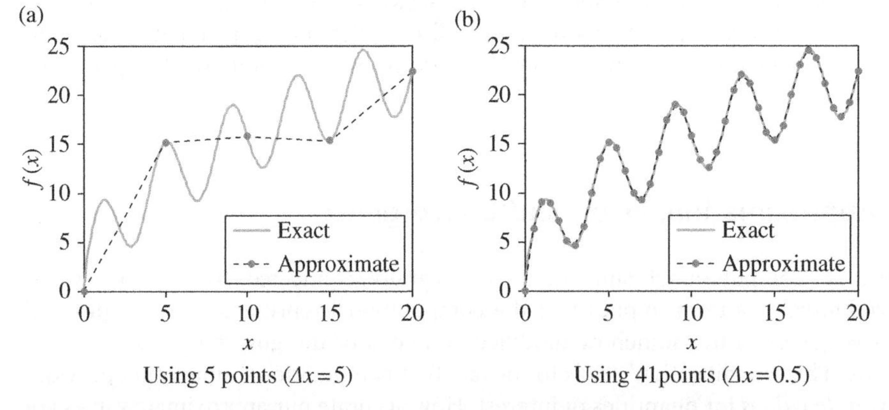

Figure 1.1 Approximate representation of a function $f(x)$, using points connected with straight-line segments.

Remark 1.2.1: The example provided in Figure 1.1 does not constitute a very rigorous treatment of approximation and convergence, but it hopefully provides an effective, qualitative means to facilitate introduction to the specific notions. This book will provide a more specific treatment of convergence in several chapters. ▮

1.3 Approximate Solution of Differential Equations and the Finite Element Method

Given the need for computational simulation of physical processes described by differential equations, one needs to establish methods for the approximate solution of such equations. Perhaps the most natural method for approximation—especially for students introduced to the topic—is the *finite difference method* (*FDM*). The salient feature of the FDM is that the differentiation operation is approximated with a difference equation, involving the values of the function at specific locations. These locations are called the *gridpoints*, and the set of all gridpoints comprises the *finite difference grid* for a domain of interest. For example, if we have the differential Equation (1.1.1), we could establish a *central difference approximation* of the derivative of the function. First, we need to create a finite difference approximation for the domain of values of the variable x, as shown in Figure 1.2. The finite difference domain will consist of multiple *gridpoints*, which in the specific figure are equally spaced at a distance Δx. For each location of a gridpoint i, $x = x_i$, we could use the two neighboring points, $x = x_{i-1}$ and $x = x_{i+1}$, to write the following approximation of the derivative at $x = x_i$.

$$\left.\frac{dT}{dx}\right|_{x=x_i} \approx \frac{T_{i+1}-T_{i-1}}{x_{i+1}-x_{i-1}} = \frac{T_{i+1}-T_{i-1}}{2\Delta x} \tag{1.3.1}$$

where $T_{i+1} = T(x = x_{i+1})$ and $T_{i-1} = T(x = x_{i-1})$. We can now write the following approximate expression for the differential equation (1.1.1), at the location $x = x_i$:

$$\frac{T_{i+1}-T_{i-1}}{2\Delta x} + s_i = 0 \tag{1.3.2}$$

where $s_i = s(x = x_i)$. A similar finite difference approximation of the differential equation can be written for all the gridpoint locations. Eventually, we can combine the gridpoint expressions into a system of equations. This system would ultimately allow to calculate the values of the unknown function T_i at the locations of the gridpoints. The approximation parameter for finite difference methods is simply the spacing of the gridpoints, Δx. The smaller the spacing, the closer the finite difference solution is to the exact solution.

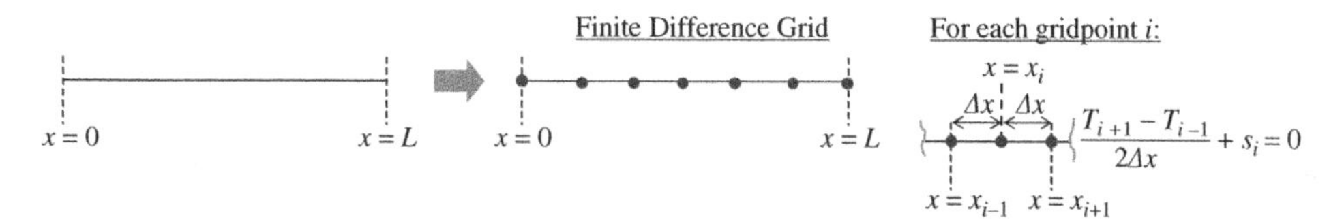

Figure 1.2 Summary of finite difference method for one-dimensional problems.

Instead of pursuing an approximation of the differentiation operation, the finite element method (FEM) takes a different approach for obtaining approximate solutions. First, we need to subdivide (*discretize*) each physical geometric domain into smaller pieces (*subdomains*), called finite elements. This process is schematically shown in Figures 1.3a and 1.3b, for a two-dimensional and a three-dimensional body, respectively. We then stipulate that the field function of interest will vary in each element in accordance with a given (usually polynomial) expression, which will only include some constant parameters to be determined. The full conceptual procedure, which is followed to obtain a finite element solution, is outlined in Section 1.7. A major advantage of the FEM compared to the FDM is the ease by which complicated geometries can be treated through subdivision into elements, and accurate results can be obtained for such geometries. The FDM may be at a disadvantage in complex geometries and multiple dimensions, due to the need to have a *topologically orthogonal network* of gridlines.

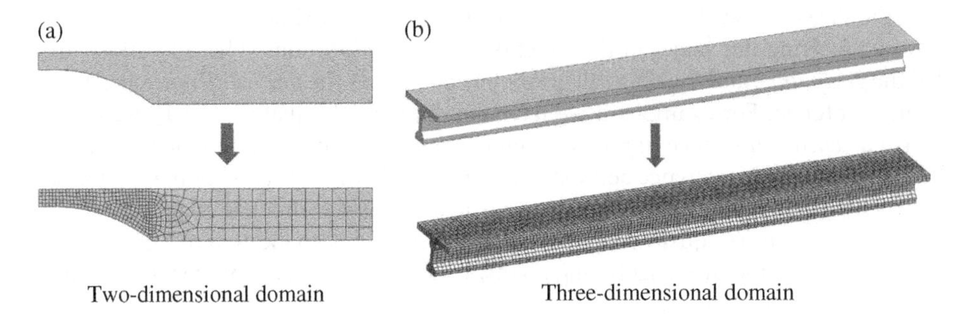

Figure 1.3 Example finite element discretization.

1.4 Brief History of the Finite Element Method

This section presents a brief history of key developments in the FEM and finite element analysis. For a complete, detailed account of the history of the finite element method, see Felippa (2000). Even today, there is some debate on who could be called the "creator" of the method. Still, it is universally accepted that the method originally stemmed from aerospace engineering research in the 1950s. The two key initial contributors of the method are Professors Jon Turner (in the United States) and John Argyris (in Europe). Professor Ray Clough should also be mentioned, as he was one of Turner's key collaborators. Perhaps the major motivation for developing novel computation methods at the time was the need to create new aircraft with swept-back wings. Fruitless efforts to obtain accurate results with the (very popular, back then) flexibility-based methods and beam or lattice modeling approaches instigated research for improved analysis tools.

In 1953, Turner came up with the idea of pursuing improved models of aircraft wings through representation of the skin with triangular elements, having constant stress. His thinking process was a beautiful example of engineering intuition, enabling users to satisfy conditions that were later found to be mathematically necessary for ensuring accurate results and the convergence of the method. For example, Turner understood that his approach allowed the sides of the triangular elements to remain straight lines during deformation, and this prevented the creation of "holes" between adjacent elements. In mathematical terms, this meant that *continuity of the displacement field* was ensured across inter-element boundaries—a condition that is a prerequisite for convergence of the finite element method. For a number of reasons described by Clough (1990), official publication of Turner's early work (as a paper) was delayed until 1956 (Turner et al. 1956). The specific method to analyze wings with triangular elements had been originally named the *direct stiffness method.*

The early contributions of Argyris were published in a series of articles in 1954 and 1955, providing a complete formulation of matrix structural analysis and also demonstrating the similarity in computational processes using the flexibility (force) method and the stiffness (displacement) method of analysis. These papers were finally combined and published as a book (Argyris 1960). Interestingly, there are references (e.g., Papadrakakis et al. 2006) stating that Argyris conducted analysis of swept-back wings with triangular elements as early as 1943, but there is apparently no official publication supporting this claim.

Several years after the initial efforts by Turner et al., Clough realized that the term *direct stiffness method* has a wider context (since it implies the process by which the matrices of an entire structure can be *assembled* from contributions of individual members, as described in more detail in Appendix B). Instead, he realized that a better name for the method would be *the finite element method*, and he published the first paper using this term in 1960 (Clough 1960). Subsequent work by Clough and his student, Ed Wilson, led to the creation of the first finite element program for stress analysis (in more specific terms, for elastostatic problems).

Major early contributions on finite element analysis also came from Professor Olgierd Zienkiewicz at the University of Swansea, Wales, and his collaborators. One of these collaborators, Bruce Irons, contributed the so-called isoparametric concept, an approach explained in detail in Chapter 6. The author considers the isoparametric concept as one of the most beautiful uses of applied math to solve challenging problems, such as the treatment of complicated, multi-dimensional geometries. Zienkiewicz (1967) also authored one of the first textbooks on the FEM, which has been reprinted in multiple editions (Professor R. Taylor of UC Berkeley has been a co-author in later editions of the book).

Despite the initial popularity of the method for applied structural analysis, many members of the scientific community were skeptical about its mathematical rigor. This instigated an effort to formally prove convergence for linear analysis. A classical textbook by Strang and Fix (2008), originally published in 1973, provides a mathematical treatment of the method and the conditions that ensure convergence. After the mathematical rigorous verification of the validity of the FEM, the method became vastly more popular, primarily thanks to research conducted at UC Berkeley by Taylor and Wilson. A major contribution by Wilson was the creation of the finite element program SAP (for Structural Analysis Program), the algorithmic structure of which is encountered in almost all analysis programs today.

A factor that created difficulties for the extensive use of the FEM was the need for computational resources, which were not available in many universities at the time. The use of the method has vastly increased thanks to the improvement in computer hardware capabilities. For example, a (reasonably priced) desktop workstation of today allows FEM-based investigations, which were impossible to envision three or four decades ago. Furthermore, the advent of high-performance computing and supercomputer facilities in many universities and research institutes has further allowed large-scale simulations, such as investigations of entire urban regions.

1.5 Finite Element Software

Since the FEM is a numerical/computational analysis method, it should come as no surprise that its popularity has motivated the development of multiple finite element programs, both commercial and research-oriented. Commercial software is typically marketed by private companies and accompanied by appropriate graphical user interfaces (GUI) to facilitate use. The first commercial program was *NASTRAN*, which stemmed from an effort by NASA to create a computer code for analysis of aerospace structures. The original effort began in the late 1960s by the MacNeal-Schwendler Corporation (MSC).

Another popular program is *ANSYS*, which was also one of the first programs to allow nonlinear analysis. Great impetus for software development was provided by efforts of E. Wilson at UC Berkeley, which spawned two types of programs. The first type evolved from the SAP code, and it ultimately led to a family of software solutions currently distributed by *Computers and Structures, Inc.*, a company based at Berkeley. The main programs of this family, which are used by the civil structural engineering community and emphasize seismic design, are *SAP 2000*, *ETABS* and *Perform3D* (the latter being the product of work by another Berkeley faculty member, G. Powell). The second type stemmed from work by K.-J. Bathe, a student of Wilson's and currently a faculty member at MIT, who created the commercial program *ADINA*.

Two popular commercial programs with industry are *ABAQUS*, originally created by David Hibbitt and currently marketed by Simulia, and *LS-DYNA*, originally created by John Hallquist at Livermore National Laboratories and currently marketed by Livermore Software and Technology Corporation (LSTC). Both of these programs have a very wide range of capabilities, including nonlinear analysis of solids and structures. They are also very attractive for researchers, by allowing the incorporation of user-defined code and ensuring compatibility with high-performance, parallel computing hardware.

Besides commercial software, several research groups have contributed open-source finite element codes, primarily aimed for researchers and educators. Access to the open source of an entire analysis program—which, for obvious reasons, is not possible for commercial software—is invaluable for understanding the method and the associated numerical analysis algorithms. Two very interesting and popular (at least with the civil structural engineering community) open-source programs are *FEAP* (created by Taylor at Berkeley) and *OpenSees* (created by a group of researchers of the Pacific Earthquake Engineering Research Center, also based at Berkeley, led by Professor Gregory Fenves). *OpenSees* may be thought of as a software of the modern, web era, in the sense that it emphasizes *object-oriented programming* and includes a *wiki*-like manual webpage, where various entries are constantly being updated and new information is added as the program capabilities are extended.

1.6 Significance of Finite Element Analysis for Engineering

It is hard to overstate the importance that the FEM has for design and analysis, in various disciplines of science and engineering. Several illustrative example applications of the FEM are provided in this section. The structural engineering discipline has extensively used the method for the simulation of civil and aerospace structures. Specific examples provided here are the simulation of building components and systems under extreme loading events, such as earthquakes causing structural damage and even collapse, as shown in Figure 1.4. The analysis in Figure 1.4a has been conducted by the author's research group and has used three-dimensional solid elements (described in Chapter 9) for a four-story, reinforced concrete (RC) building under seismic loads. The analysis in Figure 1.4b has been conducted for a 20-story RC building designed with modern standards. It uses (among others) beam elements, which are discussed in Chapter 13. Finite element analyses are instrumental for forensic structural engineering studies, as shown for the model in Figure 1.4c, which has primarily relied on shell

(a)

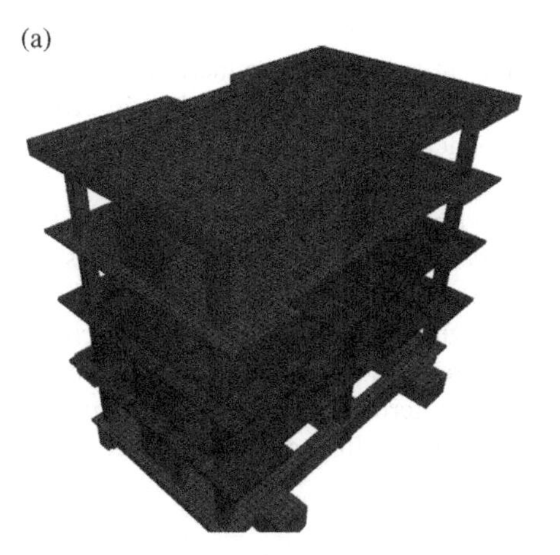

Analysis of three-dimensional, reinforced concrete building system under earthquake loading, using *LS-DYNA*

(b)

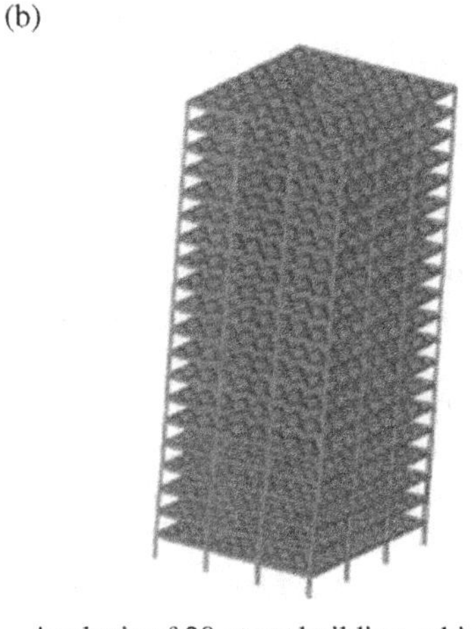

Analysis of 20-story building subjected to earthquake ground motion, using *OpenSees* (courtesy of Dr. Marios Panagiotou)

(c)

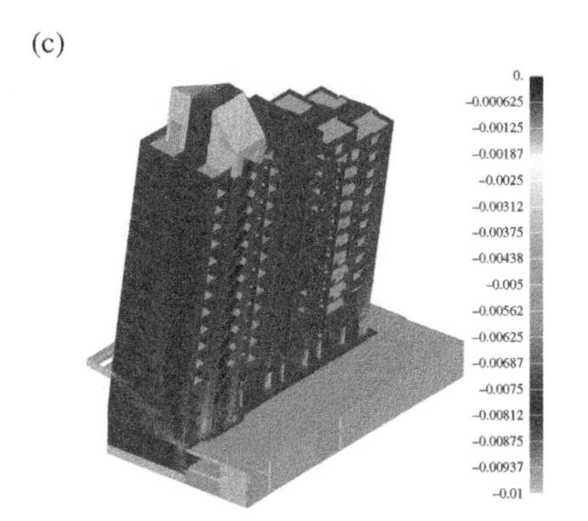

Analysis of multi-story reinforced concrete building that collapsed during the 2010 Chile earthquake, using *ABAQUS* (courtesy of Professor Jose I. Restrepo)

(d)

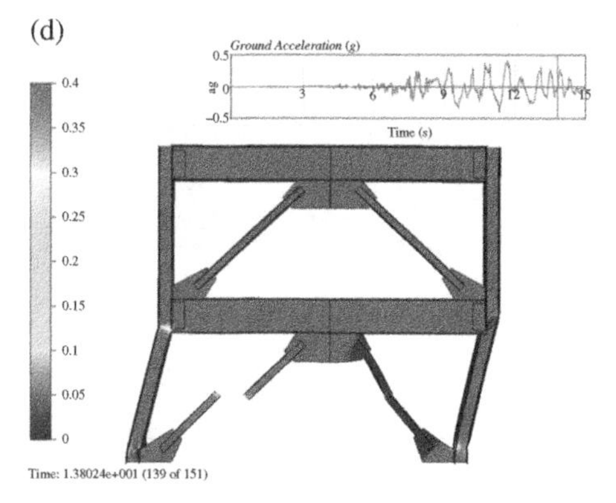

Analysis of damage and failure in concentrically braced steel frame under earthquake loads, using *LS-DYNA* (courtesy of Professor Stephen Mahin and Dr. Yuli Huang)

Figure 1.4 Example applications of finite element analysis for structural engineering.

elements (described in Chapter 14). The FEM can also be used for other types of construction, such as steel structures. An example case involving the high-fidelity analysis of a concentrically braced steel frame under earthquake loads is presented in Figure 1.4d. The analysis uses appropriate shell element formulations, material models, and solution algorithms, which can capture the damage in the form of brace buckling and rupture.

The automotive industry has also benefited by the FEM, thanks to the capability to replace (to an extent) expensive experimental crash tests with finite element analyses, as shown for an *LS-DYNA* analysis in Figure 1.5. It is interesting to mention that one of the reasons that motivated the early development and marketing of *LS-DYNA* was the need for software enabling crash-worthiness studies. Other uses of the FEM that elucidate our understanding of manufacturing processes involve *metal forming*, wherein a thin sheet of metal is subjected to pressure from a rigid *die* and a rigid *punch*, to create metal parts and objects through mechanical deformation. An example simulation of metal forming is provided in Figure 1.6. One may imagine that simulations such as car crashes, metal forming, and building failure and collapse require research to improve the mathematical description of, for example, how each piece of material deforms and gets stressed during the process at hand.

The FEM can also allow investigations of *micromechanics*, wherein the finer-scale characteristics of a material need to be taken into account, to elucidate the mechanisms which might ultimately undermine the macroscopically observed resistance. Applications of this type involve investigations associated with defense and security, such as the damage of materials to projectile penetration, as shown in Figure 1.7a. Other example applications for micromechanics are the formation of microscopic cracks of inhomogeneous, composite materials subjected to specific types of loading, as shown for example in Figure 1.7b.

Figure 1.5 Simulation of vehicle side-impact collision using *LS-DYNA* (courtesy of Ed Helwig, LSTC).

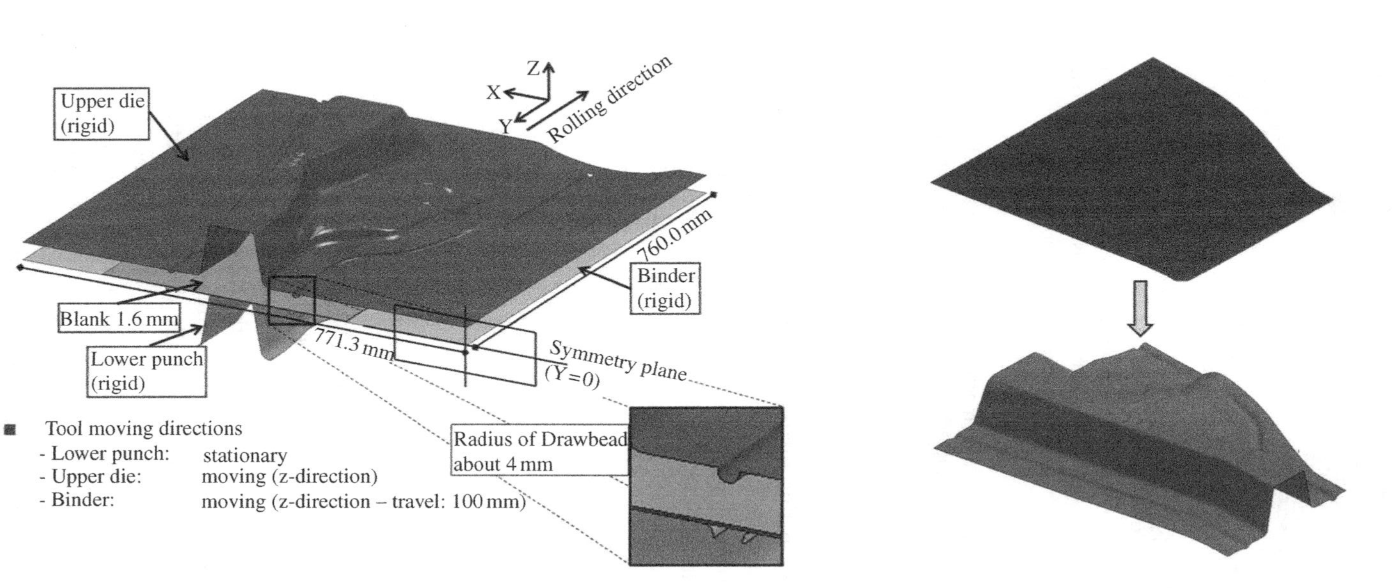

Figure 1.6 Simulation of metal forming using *LS-DYNA* (courtesy of Stefan Hartmann, Dynamore).

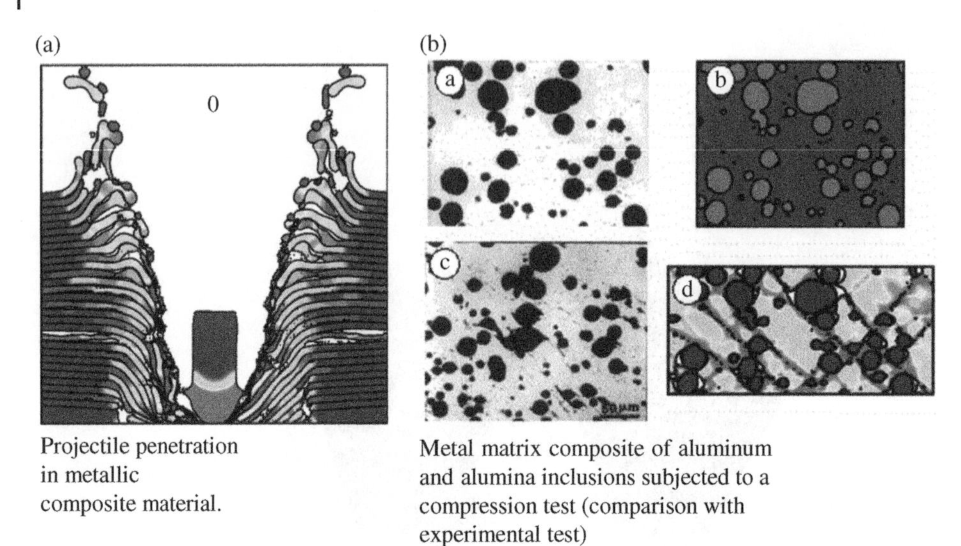

Figure 1.7 Example applications for micromechanics (figures courtesy of Professor David Benson—the specific simulations have been conducted with his research code *Raven*).

The cases presented in this section constitute a mere subset of physical problems that can be solved with the FEM. Further improvements in computational hardware guarantee a bright future for the method as a means of supplementing theoretical derivations and experimental testing, improving the understanding of physical processes, and enabling new concepts for materials, structural systems, machines and electronic devices.

1.7 Typical Process for Obtaining a Finite Element Solution for a Physical Problem

This section examines the general principles of linear finite element analysis. To facilitate understanding of the method, the various concepts will be presented with a focus on specific physical phenomena, such as mechanics of linearly elastic solids and structures and heat conduction in one or multiple dimensions. For any problem, the use of the FEM can be assumed to consist of four basic conceptual steps, summarized in Figure 1.8. The first step is to formulate the differential equations that describe the problem. As mentioned in Section 1.1, these differential equations express the governing physics of the problem at hand and usually comprise a mathematical statement of the conservation principle for a quantity such as mass, energy, or momentum. The differential equations for any problem are supplemented by *boundary conditions*, which merely provide statements about what happens at the boundary of the domain of the problem. For example, if we have a two-dimensional body, the boundary conditions will provide information on what happens on the bounding curve of the body.

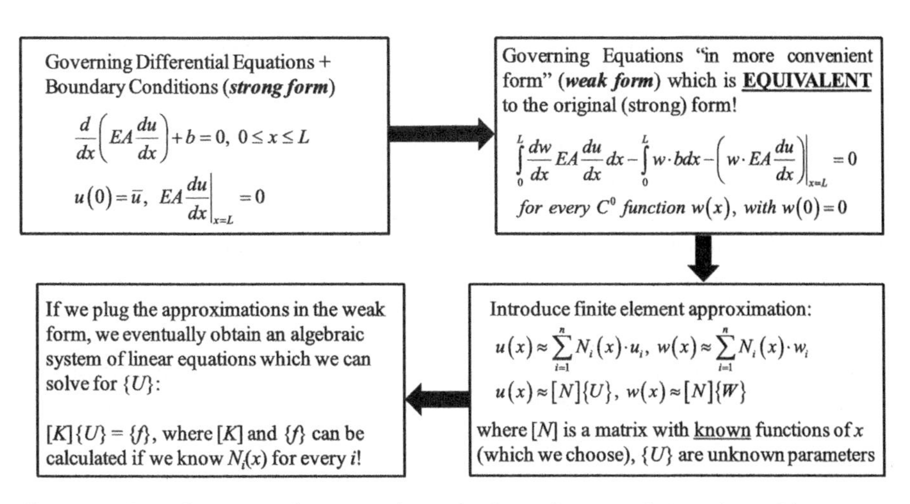

Figure 1.8 Typical conceptual steps to obtain the finite element solution of a problem.

The differential equations and the boundary conditions collectively comprise the *strong form* of a physical problem. In several (simple) cases, we may be able to solve the differential equation and obtain the *exact solution* to the strong form. However, in most cases (especially for multidimensional problems), it is not possible to analytically solve the differential equation; this, in turn, necessitates the use of numerical methods to obtain approximate solutions. While finite difference methods establish an approximation to the differentiation operation, the use of the FEM relies on a different approach, which requires the transformation of the strong form into an equivalent set of expressions called the *weak form*. The weak form of any problem typically involves integral equations (i.e., equations which have integral terms), as shown in Figure 1.8. The reason why we obtain the weak form is that it is "more convenient" for approximate methods such as the FEM for reasons explained throughout this textbook. After we obtain the weak form, we introduce our finite element approximation. Eventually, we will plug the finite element approximation into the weak form, which will allow us to transform the problem from an integral equation, to a system of linear equations for a set of unknown constant quantities, collectively placed in a vector $\{U\}$ in Figure 1.8. Knowledge of $\{U\}$ completely defines the finite element approximate solution.

Remark 1.7.1: The reader should not try (at this stage) to understand the exact meaning of the mathematical equations in Figure 1.8, which are a special case of *one-dimensional elasticity*, described in Chapters 2 and 3. All that the reader needs to understand is the conceptual content of the various steps in the figure. In summary, the strong form contains the governing differential equations of the problem, the weak form is an equivalent, integral version of the strong form, which is "more convenient" for the finite element approximation, and then a finite element approximation is introduced into the weak form, allowing to transform the weak form into a linear system of algebraic equations. Ideally, the reader should be able to completely understand the specific meaning of the mathematical expressions of Figure 1.8 after reading Chapters 2 and 3. ▮

1.8 A Note on Linearity and the Principle of Superposition

This textbook focuses on linear problems, where the response of a system due a disturbance from the environment is linearly proportional to that disturbance. For example, if we have a load applied on a linearly elastic structure, then doubling the magnitude of the load will lead to twice as large values of stresses, strains and displacements in the interior of the structure. Linear systems are the simplest type of systems to tackle, since their numerical solution with finite elements simply requires the solution of a linear algebraic system of equations. In fact, the assumption for linearity in the response of a system is often very accurate.

Linear systems are described by *linear differential equations*. In the context of the generic Equation (1.1.7), we will have linear differential operators $L[\ ]$. This means that, if we are given two functions $f_1(x)$ and $f_2(x)$, on which we can define the effect of $L[\ ]$, then we will satisfy:

$$L[c_1 \cdot f_1(x) + c_2 \cdot f_2(x)] = c_1 \cdot L[f_1(x)] + c_2 \cdot L[f_2(x)] \tag{1.8.1}$$

for any pair of constant numbers c_1, c_2.

A direct consequence of the linearity of differential equations is the satisfaction of the so-called *superposition principle*, which we will describe in mathematical terms. Let us imagine that we have a problem governed by the generic Equation (1.1.7). Let us also imagine that we have two given functions $s_1(x)$ and $s_2(x)$. Let us further assume that we have two functions $f_1(x)$ and $f_2(x)$, such that:

- $f_1(x)$ is the solution of the problem:

$$L[f(x)] + s_1(x) = 0 \tag{1.8.2a}$$

- $f_2(x)$ is the solution of the problem:

$$L[f(x)] + s_2(x) = 0 \tag{1.8.2b}$$

Then, the function $c_1 \cdot f_1(x) + c_2 \cdot f_2(x)$ for, any pair of constant numbers c_1, c_2, is the solution to the problem:

$$L[f(x)] + c_1 \cdot s_1(x) + c_2 \cdot s_2(x) \tag{1.8.3}$$

A direct consequence of the principle of superposition is that, if we have solved the problems governed by Equations (1.8.2a,b), then we immediately know the solution of the problem described by Equation (1.8.3). The proof of the principle of superposition for linear systems is straightforward and left as an exercise.

The most general case of linear relation between a quantity y and another quantity x can be written in the following form:

$$y = a \cdot x + b \tag{1.8.4}$$

where a and b are constant numbers. Here, we will almost exclusively study linear relations where $b = 0$.

Any nonlinear relation of the form $y = f(x)$ can be approximated by a linear one, through a process termed *linearization*. Specifically, the value of y can be approximated by a *first-order Taylor series expansion* about some value $x = x_o$, in accordance with the following expression.

$$f(x) \approx f(x_o) + \left.\frac{df}{dx}\right|_{x=x_o} \cdot (x - x_o) \tag{1.8.5}$$

In most cases, the value x_o is taken as $x_o = 0$. Still, when we decide to describe a problem as linear, it is important to first determine whether we know the expected *range* of values of x. For example, if we want to approximate the function $f(x) = e^x$ and we know that x is expected to attain values between 0 and 1, it may be better to select x_o as the midpoint of the range of values that x can attain, that is, $x_o = 0.5$. In this case, the linearized dependence of f on x can be written as follows.

$$f(x) \approx f(x_o) + \left.\frac{df}{dx}\right|_{x=x_o} \cdot (x - x_o) = f(0.5) + \left.\frac{df}{dx}\right|_{x=0.5} \cdot (x - 0.5) = e^{0.5} + e^x\big|_{x=0.5} \cdot (x - 0.5)$$
$$= e^{0.5}(x + 0.5) \tag{1.8.6}$$

The plots of the example nonlinear function and its linearized approximation over the range of values of x are compared in Figure 1.9. It is worth noticing that, for the specific example, the linearized approximation of the function gives a very "good" approximation of the nonlinear function; this is not always the case. Additionally, one may see from Figure 1.9 that the approximate linear representation coincides with the exact solution at $x = 0.5$, because we took the Taylor series expansion about this specific point. In other words, the linearized approximation and the actual nonlinear function are bound to coincide at the center of the Taylor series expansion of Equation (1.8.5). Furthermore, the disagreement between the nonlinear function and its linearized version grows as we move further from $x = 0.5$. This happens because, the further away we move from x_o (the center of the Taylor series expansion), the more significant will be the impact of *higher-order terms* in the series expansion, which have been neglected in Equation (1.8.5).

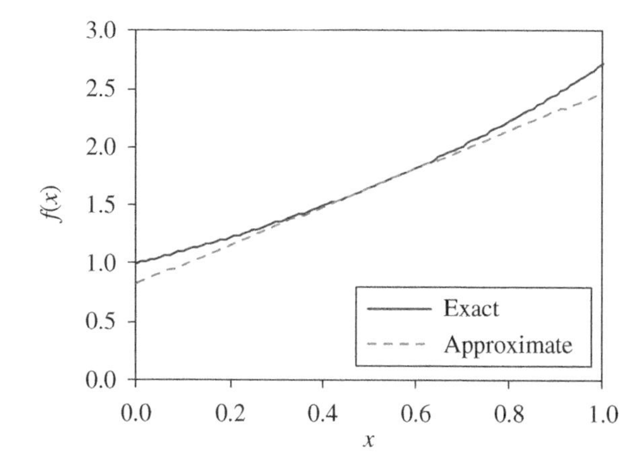

Figure 1.9 Example linearization of a nonlinear function over the range of interest of x.

References

Argyris, J. (1960). *Energy Theorems and Structural Analysis*. London: Butterworth Scientific Publications.

Clough, R. (1960). "The finite element method in plane stress analysis." *Proceedings of the 2nd ASCE Conference on Electronic Computation*.

Clough, R. (1990). "Original formulation of the finite element method." *Finite Elements in Analysis and Design*, 7, 89–101.

Felippa, C. (2000). "A Historical Outline of Matrix Structural Analysis: a Play in Three Acts." *Report No. CU-CAS-00-13*. Boulder: Center for Aerospace Structures, University of Colorado.

Papadrakakis, M., Hughes, T. J. R., and Oden, J. T. (2006). "In memoriam to Professor John H. Argyris." *Computer Methods in Applied Mechanics and Engineering*, 195, v–vii.

Strang, G., and Fix, G. (2008). *An Analysis of the Finite Element Method*, 2nd ed. Cambridge, MA: Wellesley.

Turner, M., Clough, R., Martin, H., and Topp, L. (1956). "Stiffness and deflection analysis of complex structures." *Journal of Aeronautical Science*, 23(9), 805–823.

Zienkiewicz, O. (1967). *The Finite Element Method in Structural and Continuum Mechanics*. New York: McGraw-Hill.

2

Strong and Weak Form for One-Dimensional Problems

As mentioned in Section 1.7, the finite element solution for a physical problem consists of several basic conceptual steps. In accordance with Figure 1.8, the two first steps that we need to take are to establish the *strong form* and the *weak form* of a problem. The strong form consists of mathematical equations physically expressing the conditions that the exact solution of a problem must satisfy at *any point* inside the system under consideration. The weak form is an alternative, integral version of the mathematical conditions governing a problem. This chapter provides an introduction to the derivation of the strong form and weak form for the simplest case of *one-dimensional problems*, wherein we have a single (scalar) field function that depends only on a single spatial coordinate, x. We will consider two specific types of problems—namely, one-dimensional elasticity and one-dimensional heat conduction. Besides allowing a nice introduction to the concept of the strong and weak forms and the proof of their equivalence, the examination of these problems will facilitate their extension in multiple dimensions, which is provided in Chapters 5 and 7.

This Chapter assumes that the reader is familiar with the concepts of differentiation and integration in one-dimensional problems, and also has a basic understanding of one-dimensional boundary value problems. Any textbook on calculus (e.g., Stewart 2015) and on differential equations (e.g., Boyce and DiPrima 2012) can serve as a reference for review of these concepts.

2.1 Strong Form for One-Dimensional Elasticity Problems

We will begin our discussion by focusing on one-dimensional elasticity. Let us consider an example structure, shown in Figure 2.1. The structure is a one-dimensional truss member (bar experiencing loads and deformations only along its axial direction), and we want to establish the variation of the axial displacement, $u(x)$, along the length. The function $u(x)$ is called the *axial displacement field*, (per our discussion in Section 1.1, a field function is simply a function of spatial coordinates, which for one-dimensional problems means a function of x). The bar is subjected to a *body force* $b(x)$—that is, a distributed axial loading along its length, which has units of force per unit length of the bar. The material of the structure is linearly elastic. We are also given that we apply a prescribed stress $\bar{t}$ at the left end of the bar, and that we also prescribe the

Fundamentals of Finite Element Analysis: Linear Finite Element Analysis, First Edition.
Ioannis Koutromanos, James McClure, and Christopher Roy.

Companion website: www.wiley.com/go/koutromanos/linear

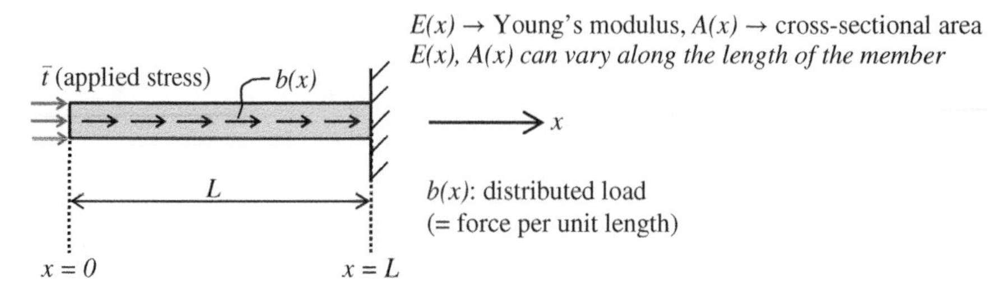

Figure 2.1 Example of one-dimensional truss structure.

displacement at the right end to be equal to zero. As shown in Figure 2.1, we assume that we know the value of Young's modulus, $E(x)$ and of cross-sectional area $A(x)$ for any point x. We will examine the most general case where these two parameters are not constant, but vary along the length of the member.

To formulate the strong form, we will establish mathematical expressions of the following conditions, which must always be satisfied in solid and structural mechanics: (i) equilibrium, (ii) compatibility, (iii) stress-strain law (constitutive law) of the material.

The location of any point in the structure can be described by means of the spatial coordinate, x, as shown in Figure 2.1. Notice that, based on the figure, we have the *range* $0 \leq x \leq L$. The range of all possible values of the spatial coordinates in a problem will be called the *domain* of the problem, and it will be denoted by the letter Ω.

For all the physical problems that we will consider, we will employ the procedure in Box 2.1.1.

Box 2.1.1 Obtaining the strong form

The key to obtaining the governing differential equation of any problem is to isolate a piece with *small* dimensions, establish the equations describing the governing physics and then consider the limit situation where the "small" dimensions tend to zero.

In accordance with Box 2.1.1, let us cut a small segment (i.e., piece) with length equal to Δx from the interior of the truss, as shown in Figure 2.2, and draw all the forces applied on this segment. Obviously, we will be having axial forces at both locations where we have taken sections ("cuts") and also we will be having an effect from the distributed loading on the segment. Since the length of the piece is small, we can assume

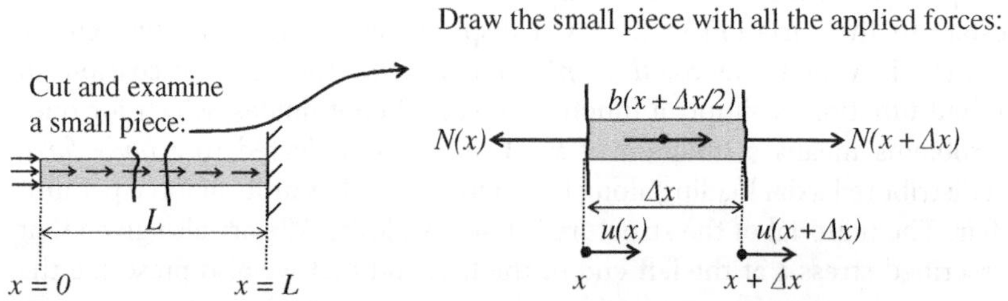

Figure 2.2 Free-body diagram from a small piece cut from the interior of the structure.

that the distributed load has a constant value, equal to the value at the midpoint of the piece, $b(x + \Delta x/2)$.

We can now establish the equation describing the governing physics of the problem, that is, the *force equilibrium equation* in the axial direction:

$$N(x+\Delta x)-N(x)+b\left(x+\frac{\Delta x}{2}\right)\cdot\Delta x=0\rightarrow\frac{N(x+\Delta x)-N(x)}{\Delta x}+b\left(x+\frac{\Delta x}{2}\right)=0 \quad (2.1.1)$$

If we take the limit as Δx tends to zero, Equation (2.1.1) yields:

$$\frac{dN}{dx}+b(x)=0 \quad (2.1.2)$$

Equation (2.1.2) *is the differential equation of equilibrium for one-dimensional truss structures*. The axial force, $N(x)$, can be written in terms of the axial stress, $\sigma(x)$, as:

$$N(x)=A(x)\cdot\sigma(x) \quad (2.1.3)$$

Compatibility is mathematically enforced through the strain-displacement equation, which can be obtained by establishing the strain of the piece of the structure that we have isolated in Figure 2.2:

$$\varepsilon(x)=\frac{(elongation)}{(initial\ length)}=\frac{u(x+\Delta x)-u(x)}{\Delta x} \quad (2.1.4)$$

If we now take the limit of Equation (2.1.4) as the size of the piece, Δx, tends to zero, Equation (2.1.4) becomes:

$$\varepsilon(x)=\frac{du}{dx} \quad (2.1.5)$$

Equation (2.1.5) is the *local strain-displacement relation for one-dimensional elasticity*.

Remark 2.1.1: As we will also see in Chapter 7, the strains are always obtained through differentiation of the displacement field. ∎

We now move on to the *local* relation between stresses and strains. The term "local" means that the stress at a specific location of our structure will depend on the strain of the same exact location! For a linearly elastic material, the uniaxial stress-strain law (= constitutive law) is written as follows:

$$\sigma(x)=E(x)\cdot\varepsilon(x) \quad (2.1.6)$$

If we combine equations (2.1.3) and (2.1.6), we obtain:

$$N(x)=E(x)\cdot A(x)\cdot\varepsilon(x) \quad (2.1.7)$$

and if we plug Equation (2.1.5) into (2.1.7), we have:

$$N(x)=E(x)\cdot A(x)\cdot\frac{du}{dx} \quad (2.1.8)$$

Finally, if we plug Equation (2.1.8) into Equation (2.1.2), we obtain the governing differential equation for one-dimensional elasticity (i.e., an axially loaded bar):

$$\frac{d}{dx}\left(EA\frac{du}{dx}\right)+b(x)=0 \tag{2.1.9}$$

Remark 2.1.2: The differential Equation (2.1.2) and Equation (2.1.3) apply for both linear and nonlinear materials. Equation (2.1.9) is a special case of (2.1.2) for linear elasticity. ∎

The obtained differential equation does *not fully* describe the problem at hand. There are two more pieces of information for which we have not accounted, namely the *prescribed stress* at $x = 0$ and the *support (restraint)* at $x = L$. These pieces of information correspond to the two ends of the member, which are called the *boundaries* for a one-dimensional problem. In one-dimensional problems, the boundaries are the two end points of the "line" system under consideration! A general consideration that we can keep in mind for boundaries and the domain of any problem is provided in Box 2.1.2.

Box 2.1.2 Dimensions of Boundary and of Domain
The "dimension of the boundary" is obtained if we reduce the dimension of the domain by 1. For one-dimensional problems, the dimension of the domain is 1 (we only measure lengths), and the dimension of the boundary is 0 (which means the boundary is mere points, which have dimension 0 in the sense that we do not have a "dimension to measure points"!).

Let us examine how to mathematically express the two conditions at the boundaries, which for obvious reasons are referred to as *the boundary conditions* of the problem.

We start with the left end, $x = 0$, where we have an applied stress. Prescribed stresses at boundaries of solids and structures are commonly referred to as *tractions*. To mathematically express traction boundary conditions, we need to cut a very small segment of the structure, as shown in Figure 2.3, with tiny length dx_1, in the vicinity of the boundary where we have the prescribed traction, and then establish the equilibrium equations for the segment. Notice that we can imagine that dx_1 is extremely small, for example, equal to 10^{-16} inches! Thus, every product of a quantity multiplied by dx_1 can be

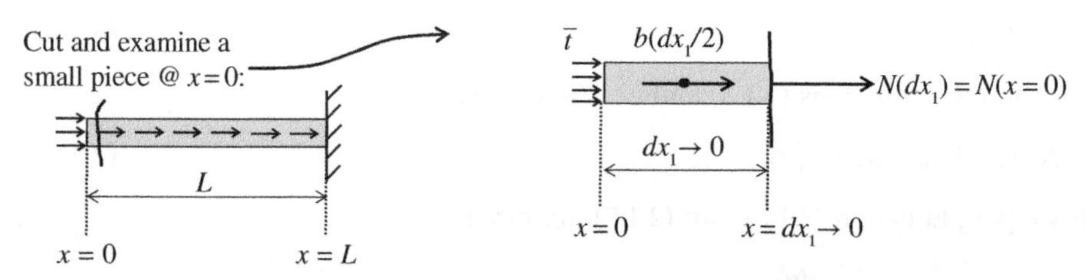

Figure 2.3 Free-body diagram from a small piece cut from the boundary with prescribed traction.

neglected, since it is practically equal to zero. Given the very tiny length of dx_1, we can also imagine that the body force for the tiny piece of the bar is constant and equal to the value of b at the midpoint of the tiny piece, as shown in Figure 2.3. The force equilibrium equation is:

$$\overset{+}{\rightarrow} \sum F_x = 0 = N(0) + \bar{t} \cdot A(0) + b\left(\frac{dx_1}{2}\right) \cdot dx_1 = 0$$

$= 0$, because $dx_1 \rightarrow 0$!

If we now account for Equation (2.1.8):

$$\left[EA\frac{du}{dx}\right]\Bigg|_{x=0} + \bar{t} \cdot A\big|_{x=0} = 0 \rightarrow \left[E\frac{du}{dx}\right]\Bigg|_{x=0} + \bar{t} = 0$$

Thus, we obtain:

$$\left[E\frac{du}{dx}\right]\Bigg|_{x=0} = -\bar{t} \tag{2.1.10}$$

Boundary conditions like Equation (2.1.10), involving derivatives of the field function $u(x)$, are referred to as *natural boundary conditions*. For elasticity problems, natural boundary conditions are also referred to as traction boundary conditions.

We can now continue by writing the mathematical expression for the other boundary condition, which simply states that the displacement at $x = L$ has a prescribed zero value.

$$u(x = L) = 0 \tag{2.1.11}$$

Boundary conditions involving prescribed values of the field function are referred to as *essential boundary conditions*. For solid mechanics problems such as one-dimensional elasticity, essential boundary conditions are also referred to as *geometric boundary conditions*.

Now, the governing differential equation combined with the boundary conditions define the *strong form* of the problem under consideration. So, for the problem in Figure 2.1, the strong form is written as shown in Box 2.1.3.

Box 2.1.3 Strong Form for One-Dimensional Elasticity Problem (with boundary conditions in Figure 2.1)

$$\frac{d}{dx}\left(EA\frac{du}{dx}\right) + b(x) = 0, \quad 0 < x < L$$

$$u(x = L) = 0$$

$$E\frac{du}{dx}\bigg|_{x=0} = -\bar{t}$$

Remark 2.1.3: We can schematically write:
(strong form) = (differential equation) + (boundary conditions). ■

In several (simple) situations, we can solve the strong form of a problem. Such a case is presented in the following example.

Example 2.1: Solution of strong form for one-dimensional elasticity
We want to determine the variation of axial deformation, $u(x)$, and of axial force, $N(x)$, for the structure shown in Figure 2.4.

We have constant E and A, and a zero $b(x)$, thus the differential equation of equilibrium becomes:

$$EA\frac{d^2u}{dx^2} = 0 \xrightarrow{E,A\neq 0} \frac{d^2u}{dx^2} = 0$$

This differential equation can be solved by directly integrating twice with respect to x:

$$\frac{d^2u}{dx^2} = 0 \rightarrow \frac{du}{dx} = c_1 \rightarrow u(x) = c_1 x + c_2$$

If we determine the values of the constants c_1 and c_2, then we will have obtained *the exact solution* (also called the *closed-form* solution) for the axial displacement field. To this end, we use the two boundary conditions for the structure:

- The boundary condition at the left end ($x = 0$) is an *essential* one, since the displacement at that end is prescribed: $u(x=0) = 0 \rightarrow \boxed{c_2 = 0}$
- The boundary condition at the right end, $x = 10$ ft, is a *natural* one, since there is no prescribed displacement at that end and we have a prescribed force (= prescribed traction), and we need to take the equilibrium of a small piece of material, in the vicinity of the right end, as shown in Figure 2.5.

We have the following equation for the equilibrium of the free-body in Figure 2.5.

$$-N(x = 10\,\text{ft}) + 10\,\text{kip} = 0 \rightarrow -EA\left.\frac{du}{dx}\right|_{x=120in} + 10\,\text{kip} = 0 \rightarrow -EA\cdot c_1|_{x=120in}$$
$$+10\,\text{kip} = 0 \rightarrow c_1 = \frac{10\,\text{kip}}{EA} \rightarrow c_1 = \frac{10\,\text{kip}}{20000\,\text{ksi}\cdot 5\,\text{in.}^2} \rightarrow \boxed{c_1 = 10^{-4}}$$

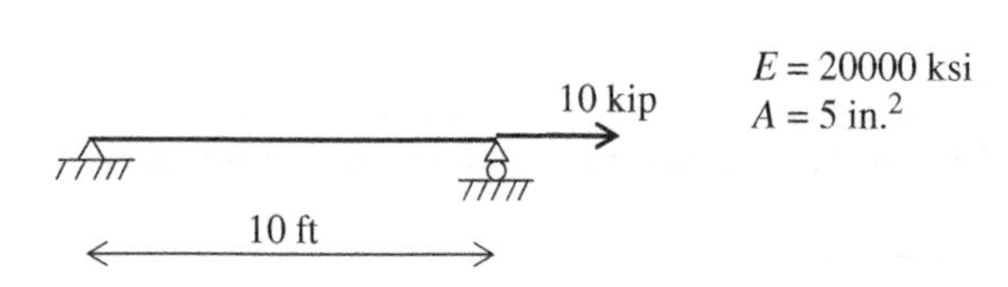

Figure 2.4 Example structure for solution of strong form of one-dimensional elasticity.

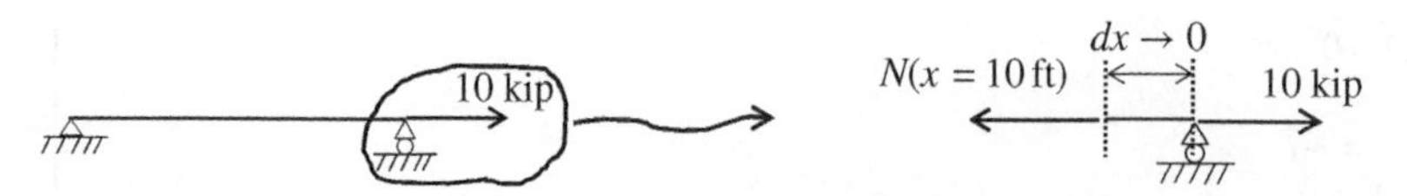

Figure 2.5 Isolation of small piece for natural boundary condition.

Thus, we have obtained $u(x) = 10^{-4} \cdot x$, and we can also obtain the axial force as a function of x, using Equation (2.1.8).

$$N(x) = E(x) \cdot A(x) \cdot \frac{du}{dx} = 20000\,\text{ksi} \cdot 5\,\text{in.}^2 \cdot 10^{-4} = 10\,\text{kip}$$

We can now create plots with the displacement field, $u(x)$ and the axial force field, $N(x)$, as shown in Figure 2.6. ∎

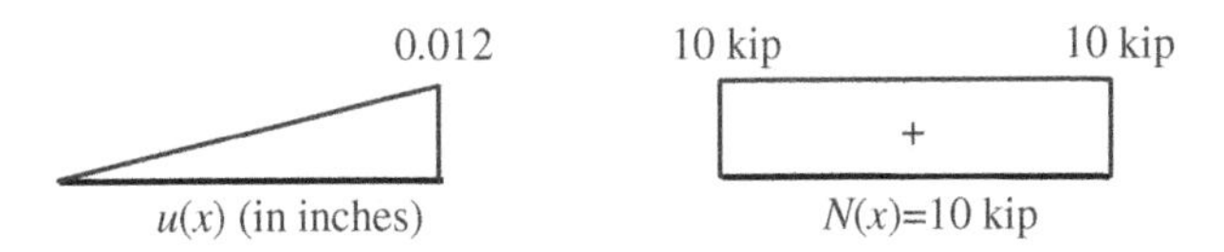

Figure 2.6 Axial displacement and axial force field for example structure.

2.2 General Expressions for Essential and Natural B.C. in One-Dimensional Elasticity Problems

We can generalize the expressions for the essential and natural boundary conditions (B.C.) for a one-dimensional elasticity problem. That is, the displacement at either of the two ends can have a prescribed value, or we may have a prescribed traction at either of the two ends. The two boundaries of a one-dimensional truss structure can be written as $x_i = 0$ (for the left end) or $x_i = L$ (for the right end). We will now distinguish the following two cases:

- For essential B.C. (*prescribed displacement*):

$$u(x_i) = \bar{u} \tag{2.2.1}$$

where $\bar{u}$ is a given value.

- For natural B.C. (*prescribed traction*):

$$n \cdot \sigma(x_i) = n_i \cdot \left[E\frac{du}{dx}\right]_{x=x_i} = \bar{t} \tag{2.2.2}$$

where $\bar{t}$ is a given value.

The quantity n in Equation (2.2.2) is *the unit outward vector normal to the boundary*, $x = x_i$, to which we prescribe the tractions. For one-dimensional problems we only have a single direction, so the "vector" n is a single value denoting whether we are pointing toward the positive or negative x-axis, as shown in Figure 2.7.

The value of the prescribed traction $\bar{t}$ in Equation (2.2.2) is positive when the traction is pointing towards the positive x-direction. As mentioned in Section 2.1, the domain of a

Figure 2.7 Definition of unit outward normal vector n, for boundaries of one-dimensional problems.

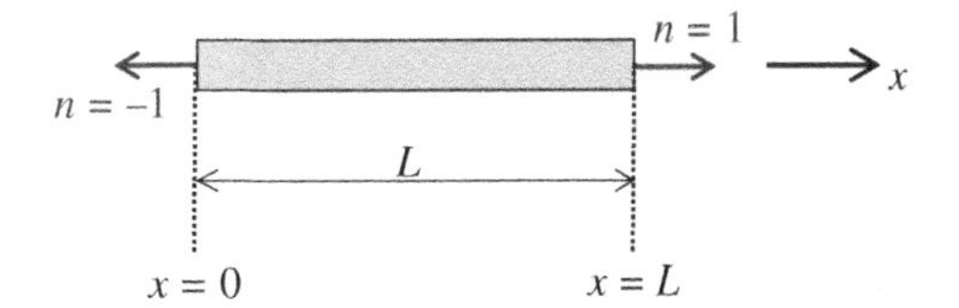

problem under consideration, especially in mathematical textbooks, is commonly denoted by Ω. We know that the differential equation of equilibrium must be satisfied anywhere inside the domain, so we can write:

$$\frac{d}{dx}\left(EA\frac{du}{dx}\right)+b(x)=0 \text{ in } \Omega \text{ or for } x\in\Omega \text{ (the latter expression means "}x\text{ belongs to }\Omega\text{").}$$

The boundaries of a domain under consideration are denoted by Γ. A further distinction of the boundary for solid and structural mechanics is in the *essential boundary*, Γ_u, and the *natural boundary*, Γ_t. So, for the structure in Figure 2.1, Γ_u corresponds to the location $x = L$ (which is where we had prescribed displacement) and Γ_t corresponds to the location $x = 0$ (which is where we had prescribed traction). In accordance with the above considerations, we can write the general version of the strong form for one-dimensional elasticity, provided in Box. 2.2.1.

Box 2.2.1 Strong Form for General One-Dimensional Elasticity Problem

$$\frac{d}{dx}\left(EA\frac{du}{dx}\right)+b(x)=0, \quad \text{for } x\in\Omega \ (0<x<L)$$

$$u=\bar{u} \text{ at } \Gamma_u \text{ (essential boundary condition)}$$

$$n\cdot\left[E\frac{du}{dx}\right]_{\Gamma_t}=\bar{t} \text{ at } \Gamma_t \text{ (natural boundary condition)}$$

2.3 Weak Form for One-Dimensional Elasticity Problems

We are now going to derive an alternative version of the one-dimensional elasticity problem called the *weak form*. Let us focus on the initial example in Figure 2.1, for which we obtained the strong form provided in Box 2.1.3.

If we multiply the differential equation in Box 2.1.3 by an *arbitrary* function $w(x)$ we obtain:

$$w(x)\left[\frac{d}{dx}\left(EA\frac{du}{dx}\right)+b(x)\right]=0, \quad \text{for any arbitrary } w(x) \tag{2.3.1}$$

The term *arbitrary function* will mean "any function that we can come up with" (provided that it does have some properties, such as continuity and an additional condition that we shall see in this section!).

Next, let us integrate the expression we obtained over the entire domain of the problem. The domain of the one-dimensional problem is simply the region with $0 \le x \le L$:

$$\int_0^L w(x)\left[\frac{d}{dx}\left(EA\frac{du}{dx}\right)+b(x)\right]dx=0, \quad \forall w(x) \tag{2.3.2}$$

The term "$\forall w(x)$" in Equation (2.3.2) means "for any arbitrary $w(x)$."

It is convenient to establish ***w*(*x*)** *such that it vanishes on the essential boundary* (= the region at which we have a prescribed value of u!). For the structure in Figure 2.1, the essential boundary is simply the point $x = L$. Thus, the expression "$\forall\, w(x)$" regarding $w(x)$ is replaced by "$\forall\, w(x)$, with $w(L) = 0$."

We can separate the two terms in the left-hand side of Equation (2.3.2) to obtain:

$$\int_0^L w(x)\left[\frac{d}{dx}\left(EA\frac{du}{dx}\right)\right]dx + \int_0^L w(x)\cdot b(x)\,dx = 0,\ \forall\, w(x) \text{ with } w(x=L) = w(L) = 0 \quad (2.3.3)$$

Notice that the first integral in Equation (2.3.3) lacks "symmetry" in terms of $w(x)$ and $u(x)$: This integral includes $w(x)$ and the second derivative of $u(x)$. It is preferable to have the *same order of derivative* for $w(x)$ and $u(x)$ (if this is possible), because this leads to symmetric coefficient ("stiffness") matrices in the finite element equations, as we will see in subsequent chapters. The way to transform the first term into one that has the same order of derivatives for $w(x)$ and $u(x)$ is to *use integration by parts*. Let us first provide the specific operation as a lemma.

Lemma 2.3.1: Integration by parts
For any two differentiable functions $f(x)$, $g(x)$, defined over the domain $0 \le x \le L$, the following equality applies.

$$\int_0^L f\cdot\frac{dg}{dx}dx = [f\cdot g]_0^L - \int_0^L \frac{df}{dx}\cdot g\,dx$$

where

$$[f\cdot g]_0^L = [f\cdot g]_{x=L} - [f\cdot g]_{x=0}$$

∎

We can now use Lemma 2.3.1 for $f(x) = w(x)$, $g(x) = EA\dfrac{du}{dx}$:

$$\int_0^L w(x)\frac{d}{dx}\left(EA\frac{du}{dx}\right)dx = \left[w\cdot EA\frac{du}{dx}\right]_0^L - \int_0^L \frac{dw}{dx}EA\frac{du}{dx}dx \quad (2.3.4)$$

If we now plug the constitutive law, that is, Equation (2.1.6), into the first term of the right-hand side of Equation (2.3.4), we have:

$$\left[w\cdot EA\frac{du}{dx}\right]_0^L = [w\cdot A\cdot\sigma]_0^L = [w\cdot A\cdot\sigma]_{x=L} - [w\cdot A\cdot\sigma]_{x=0} = 0 - [w\cdot A\cdot\sigma]_{x=0} \quad (2.3.5)$$

where we have accounted for the fact that $w(L) = 0$.

If we account for Equation (2.1.10), i.e. the natural boundary condition (B.C.) at the left end of the bar, $x = 0$, we obtain:

$$\left[E\cdot\frac{du}{dx}\right]_{x=0} = \sigma(x=0) = -\bar{t} \quad (2.3.6)$$

Thus, plugging Equation (2.3.6) into Equation (2.3.5) gives:

$$\left[w \cdot EA\frac{du}{dx}\right]_0^L = [w \cdot A]_{x=0} \cdot [-\sigma(0)] = [w \cdot A]_{x=0} \cdot \bar{t} \tag{2.3.7}$$

Eventually, if we plug Equations (2.3.4) and (2.3.7) into Equation (2.3.3), we have:

$$-\int_0^L \frac{dw}{dx} EA \frac{du}{dx} dx + [w \cdot A]_{x=0} \cdot \bar{t} + \int_0^L w \cdot b dx = 0 \tag{2.3.8}$$

If we rearrange the terms, and then also include the essential B.C. (since we did not account for this expression anywhere), we obtain the expression presented in Box 2.3.1, which is called the *weak form* of the one-dimensional elasticity problem of Figure 2.1. The same exact conceptual steps can be used to obtain the weak form for any general one-dimensional elasticity problem, whose strong form is given by Box 2.2.1. To this end, we can write the natural-boundary term $[w \cdot A]_{x=0} \cdot \bar{t}$ in the more generic form $[w \cdot A \cdot \bar{t}]_{\Gamma_t}$. This allows us to obtain the general weak form for one-dimensional elasticity, which is provided in Box 2.3.2.

Box 2.3.1 Weak Form for One-Dimensional Elasticity Problem (with boundary conditions in Figure 2.1)

$$\int_0^L \frac{dw}{dx} EA \frac{du}{dx} dx = \int_0^L w \cdot b dx + [w \cdot A]_{x=0} \cdot \bar{t} \quad \forall w(x) \text{ with } w(L) = 0$$

$u(x = L) = 0$ (Essential Boundary Condition)

Box 2.3.2 Weak Form for One-Dimensional Elasticity

$$\int_0^L \frac{dw}{dx} EA \frac{du}{dx} dx = \int_0^L w \cdot b dx + [w \cdot A \cdot \bar{t}]_{\Gamma_t}$$

$\forall w(x)$ with $w = 0$ at Γ_u (w vanishes at essential boundary)

$u = \bar{u}$ at Γ_u (essential boundary condition)

The function $w(x)$ in the weak form is called the *weight function*.

The weak form is *perfectly equivalent to the strong form*! In other words, the weak form itself is not an approximation to the strong form. The finite element approximation is introduced to the weak form, because the weak form is more convenient. Specifically:

- The weak form poses *lower continuity requirements* for the approximate solution (notice that in the weak form we have only first derivatives, while in the strong form we had second derivatives for u which would require that the first derivative of u is continuous).

- For many problems, the weak form leads to symmetric coefficient matrices (this is only the case when we reach a weak form wherein we have symmetry of terms including the field function and the weight function w).

It is worth noticing that the weak form includes conditions that $u(x)$ and $w(x)$ must satisfy. More specifically, $u(x)$ must satisfy the essential boundary conditions, while $w(x)$ must vanish at the essential boundary. Furthermore, both functions must be continuous, so that we can define their first derivative in the weak form. Functions $u(x)$ and $w(x)$ which satisfy these requirements are called *admissible*. The functions *must be admissible if a finite element approximate solution of the weak form is pursued.* In theory, however, $w(x)$ needs not be admissible. This course focuses on finite element analysis; thus, for the remainder of this course, whenever we have the weak form of a problem we will be stipulating that the functions u and w are admissible. When we introduce the finite element approximation in the weak form, then the approximate functions $u(x)$ are called *trial solutions*. Approximations will also be introduced for $w(x)$, and the approximate $w(x)$ in the weak form will be called the *test functions*.

The weak form for a problem usually has a specific name, depending on the physical context. For solid mechanics, the weak form of the equations of equilibrium is the *Principle of Virtual Work*: If we have a statically admissible set 1 of forces and stresses (statically admissible means that it satisfies all the pertinent equilibrium equations) and if we also have a *kinematically admissible* set 2 of deformation (a set of displacements and strains that satisfy the strain-displacement equation)—notice that sets 1 and 2 are generally *not* related to each other!—then the work done by the stresses of set 1 to the strains of set 2 (called the *internal virtual work*) is equal to the work done by the external forces of set 1 to the corresponding displacements of set 2 (this is called the *external virtual work*). In other words, if set 1 has a stress distribution $\sigma_1(x)$, an applied distributed axial force, $b_1(x)$, and an applied traction $\bar{t}_1$ at one of the member ends, and if set 2 has a strain $\varepsilon_2(x)$, a displacement field $u_2(x)$ and *a zero displacement at the essential boundary*, then we have the expression provided in Figure 2.8. The left-hand side of this expression corresponds to the internal virtual work, $\bar{W}_{\text{int}}$, and the right-hand side corresponds to the external virtual work, $\bar{W}_{ext}$.

Now, if we identify the statically admissible set 1 with the actual stress distribution and the actual traction and distributed load for the truss structure that we examined above, and then identify the kinematically admissible field with the weight function $w(x)$ (and of course $\varepsilon_2(x)$ is the first derivative of $w(x)$), we can verify indeed that the **weak form that we obtained for the one-dimensional elasticity problem is the principle of virtual work:**

$$\sigma_1 \equiv \sigma = E\frac{du}{dx},\ b_1 \equiv b,\ \bar{t}_1 \equiv \bar{t},\ u_2 \equiv w,\ \varepsilon_2 \equiv \frac{dw}{dx}$$

$$\underbrace{\int_0^L \sigma_1(x)\cdot A(x)\cdot \varepsilon_2(x)dx}_{\bar{W}_{int}\ \text{(internal virtual work)}} = \underbrace{\int_0^L b_1(x)\cdot u_2(x)dx + [\bar{t}_1\cdot A\cdot u_2]}_{\bar{W}_{ext}\ \text{(external virtual work)}}{}_{@\Gamma_t}$$

Figure 2.8 Principle of virtual work.

That is, the stipulation $\bar{W}_{int} = \bar{W}_{ext}$ yields the integral expression of Box 2.3.2.

It should now be clearer why the functions $w(x)$ need not vanish at the essential boundary. The principle of virtual work is perfectly valid even for displacements $w(x)$ that do not vanish at the supports of a structure (the supports are essential boundaries, since we have prescribed kinematic quantities, e.g., displacements).

Remark 2.3.1: After we have established the weak form, we can assume an approximation for $u(x)$ and $w(x)$. It is preferable to have the same type of approximation for $u(x)$ and $w(x)$. For example, if the approximate $u(x)$ is a quadratic function of x, it is preferable for the approximate $w(x)$ to also be a quadratic function of x! An assumed approximation introduces constant parameters to be determined (e.g., if we assume the approximation $u(x) = a_0 + a_1 x + a_2 x^2$, we need to determine the parameters a_0, a_1, and a_2). We plug the approximate expressions for $u(x)$ and $w(x)$ in the weak form, and we transform the vague weak form to an algebraic system of equations for the unknown parameters. ▮

Remark 2.3.2: A large number of textbooks and papers in the literature establish the weak form on the basis of *variational principles*. While we will not emphasize this approach in the present text, it is deemed necessary to present a basic treatment of variational principles, as applied to one-dimensional elasticity. This treatment is provided in Appendix C. ▮

2.4 Equivalence of Weak Form and Strong Form

So far, we have stipulated (without any mathematical proof) that the weak form is equivalent to the strong form. The proof of this equivalence for one-elasticity problems will be the goal of this section. Since we have shown that we can obtain the weak form from the strong form, we can establish the equivalence if we start from the weak form and obtain the strong form. The equivalence will be established for the example in Figure 2.1, whose strong form is given in Box 2.1.3.

In the previous section, we saw that, if we start from the strong form of the elasticity equations and appropriately manipulate various expressions, we obtain the weak form provided in Box 2.3.1. We will now obtain the strong form from the weak form. To this end, let us take the integral expression of Box 2.3.1:

$$\int_0^L \frac{dw}{dx} EA \frac{du}{dx} dx = [w \cdot A]_{x=0} \cdot \bar{t} + \int_0^L w \cdot b dx, \quad \forall w(x) \; with \, w(L) = 0$$

We first conduct integration by parts in the left-hand side:

$$\int_0^L \frac{dw}{dx} EA \frac{du}{dx} dx = \left[w \cdot EA \frac{du}{dx} \right]_0^L - \int_0^L w \frac{d}{dx}\left(EA \frac{du}{dx} \right) dx \tag{2.4.1}$$

Thus, the weak form expression becomes:

$$\left[w \cdot EA\frac{du}{dx}\right]_0^L - \int_0^L w\frac{d}{dx}\left(EA\frac{du}{dx}\right)dx = [w \cdot A \cdot \bar{t}]_{x=0} + \int_0^L w \cdot b dx, \quad \forall w(x) \ with\, w(L) = 0 \qquad (2.4.2)$$

If we bring all the terms of Equation (2.4.2) to the left-hand side and multiply both sides by −1, we have:

$$\int_0^L w\frac{d}{dx}\left(EA\frac{du}{dx}\right)dx - \left[w \cdot EA\frac{du}{dx}\right]_0^L + [w \cdot A \cdot \bar{t}]_{x=0} + \int_0^L w \cdot b dx = 0, \quad \forall w(x) \ with\, w(L) = 0$$

(2.4.3)

If we group the terms involving integration, we obtain:

$$\int_0^L w\left[\frac{d}{dx}\left(EA\frac{du}{dx}\right) + b\right]dx - \left[w \cdot EA\frac{du}{dx}\right]_0^L + [w \cdot A \cdot \bar{t}]_{x=0} = 0, \quad \forall w(x) \ with\, w(L) = 0 \qquad (2.4.4)$$

If we now examine the second term in the left-hand side of Equation (2.4.4), we have:

$$-\left[w \cdot EA\frac{du}{dx}\right]_0^L = -\left[w \cdot EA\frac{du}{dx}\right]_{x=L} + \left[w \cdot EA\frac{du}{dx}\right]_{x=0} \qquad (2.4.5)$$

We know that $w(L) = 0$, thus, the first term in the right-hand side of Equation (2.4.5) vanishes:

$$-\left[w \cdot EA\frac{du}{dx}\right]_0^L = \left[w \cdot EA\frac{du}{dx}\right]_{x=0} \qquad (2.4.6)$$

Plugging Equation (2.4.6) into Equation (2.4.4) yields:

$$\int_0^L w\left[\frac{d}{dx}\left(EA\frac{du}{dx}\right) + b\right]dx + \left[w \cdot A \cdot E\frac{du}{dx}\right]_{x=0} + [w \cdot A \cdot \bar{t}]_{x=0} = 0$$

And if we group together the two boundary terms corresponding to the location $x = 0$, we finally obtain:

$$\int_0^L w\left[\frac{d}{dx}\left(EA\frac{du}{dx}\right) + b\right]dx + \left[w \cdot A \cdot \left(E\frac{du}{dx} + \bar{t}\right)\right]_{x=0} = 0, \quad \forall w(x) \text{ with } w(L) = 0 \qquad (2.4.7)$$

We can now define the *equilibrium equation residual*, $r(x)$, and the *natural boundary condition residual*, r_{Γ_t}, as follows.

$$r(x) = \frac{d}{dx}\left(EA\frac{du}{dx}\right) + b(x) \qquad (2.4.8a)$$

$$r_{\Gamma_t} = r(0) = \left[A \cdot \left(E\frac{du}{dx} + \bar{t}\right)\right]_{x=0} \tag{2.4.8b}$$

Remark 2.4.1: We can provide a general definition of the natural boundary residual for one-dimensional elasticity:

$$r_{\Gamma_t} = \left[A \cdot \left(-E\frac{du}{dx} \cdot n + \bar{t}\right)\right]_{\Gamma_t}, \text{ where } n = -1 \text{ for } x = 0,\ n = 1 \text{ for } x = L \tag{2.4.9}$$ ■

In light of Equations (2.4.8a) and (2.4.8b), Equation (2.4.7) can be written in the following form:

$$\int_0^L w \cdot r(x) dx + [w \cdot r_{\Gamma_t}]_{\Gamma_t\ (i.e.,\ x=0)} = 0, \text{ for every } w(x) \text{ with } w(L) = 0 \tag{2.4.10}$$

Now, there is a key point: we know that the weak form applies *for any arbitrary function* $w(x)$ (which of course vanishes on the essential boundary, $w(L) = 0$).

Let us now consider functions of the form:

$$w(x) = \psi(x) \cdot \left[\frac{d}{dx}\left(EA\frac{du}{dx}\right) + b\right], \text{ with } \psi(x) > 0 \text{ for } 0 < x < L \text{ and } \psi(0) = 0,\ \psi(L) = 0 \tag{2.4.11}$$

We can verify that functions given by Equation (2.4.11) are a special case of the arbitrary function in the weak form (because they *do* vanish at the essential boundary!). Thus, since the weak form is satisfied by *any* arbitrary $w(x)$ that vanishes at the essential boundary, it must also be satisfied for $w(x)$ given by (2.4.11):

$$\int_0^L \psi\left[\frac{d}{dx}\left(EA\frac{du}{dx}\right) + b\right]^2 dx + \left[\psi\left[\frac{d}{dx}\left(EA\frac{du}{dx}\right) + b\right] \cdot A \cdot \left(E\frac{du}{dx} + \bar{t}\right)\right]_{x=0} = 0$$

(second term) $= 0$, because $\psi(0) = 0$!

Since we have $r(x) = \frac{d}{dx}\left(EA\frac{du}{dx}\right) + b(x)$, we can write the obtained expression as follows:

$$\int_0^L \underbrace{\psi}_{>0} \cdot \underbrace{[r(x)]^2}_{\geq 0} dx = 0 \tag{2.4.12}$$

Now, we notice that the integrand in Equation (2.4.12) is the product of a positive quantity, $\psi(x)$, and of a non-negative quantity, $[r(x)]^2$. The product of these quantities must also be non-negative, that is, greater than or equal to zero. From basic calculus, the only way for the integral of a non-negative function to be zero is for the function to be equal to zero for all values of x between the bounds in the integral. Thus, the only way to satisfy the weak form for any arbitrary $w(x)$ with $w(L) = 0$, is to have:

$$r(x) = \frac{d}{dx}\left(EA\frac{du}{dx}\right) + b(x) = 0, \; for \; 0 < x < L \tag{2.4.13}$$

If Equation (2.4.13) is not satisfied, then we cannot satisfy the weak form for the special case given by Equation (2.4.11), which would contradict our starting assumption that the weak form is satisfied for any arbitrary $w(x)$!

Now, if we account for Equation (2.4.13), Equation (2.4.7) becomes:

$$\left[w \cdot A \cdot \left(E\frac{du}{dx} + \bar{t}\right)\right]_{x=0} = 0, \; \forall \, w(x) \; with \, w(L) = 0 \tag{2.4.14}$$

This expression to be satisfied for *any* possible arbitrary function $w(x)$, which is zero at $x = L$. For example, it must be satisfied for the following special cases (all of which vanish at $x = L$):

$$w(x) = 1 - \frac{x}{L}, \; w(x) = e^{x/L} - e, \; w(x) = \cos\left(\frac{\pi \cdot x}{2L}\right)$$

It may be apparent that the only way for Equation (2.4.14) to always give zero left-hand side, is to satisfy:

$$\left[E\frac{du}{dx}\right]_{x=0} + \bar{t} = 0 \tag{2.4.15}$$

which is the natural boundary condition!

Thus, starting with the weak form, we eventually obtain the following expressions:

$$\frac{d}{dx}\left(EA\frac{du}{dx}\right) + b(x) = 0, \; 0 < x < L$$

$u(L) = 0$ (We had this expression in the weak form of Box 2.3.1, so we must carry it here!)

$$\left[E\frac{du}{dx}\right]_{x=0} + \bar{t} = 0$$

One can easily verify that the above three expressions constitute the differential equation and boundary conditions of the original problem—that is, the strong form provided in Box 2.1.3!

In summary, we have proven (in the previous section) that the strong form gives the weak form and (in this section) that the weak form gives the strong form; thus, the strong form is equivalent to the weak form. Of course, if we introduce approximations for the trial function, $u(x)$, and for the weight function, $w(x)$, we will obtain solutions from the weak form which do not exactly satisfy the strong form. This non-satisfaction of the strong form is the result of the function approximation and not of the use of the weak form!

Remark 2.4.2: The strong form can also be expressed as: $r(x) = 0$ and $r_{\Gamma t} = 0$. Establishing the terms $r(x)$ and $r_{\Gamma t}$ has a great practical significance: These two quantities will not be exactly equal to zero for an approximate solution, thus, they can be measures of how "good" an approximate solution is. If we have an approximate solution for which $r(x) = 0$ over the entire domain, and $r_{\Gamma t} = 0$, then this approximate solution will be the exact solution to the problem!

Thus, in summary, the approximate solutions will *not satisfy exactly* the differential equations and the *natural* boundary conditions! They will exactly satisfy the *essential* boundary conditions. ■

Remark 2.4.3: Notice that, what we obtained by starting with the weak form and working "backward" was:

$$\int_0^L w \cdot r(x) dx + [w \cdot r_{\Gamma_t}]_{\Gamma_t} = 0 \tag{2.4.16}$$

Equation (2.4.16) can be seen as a weighted average of the residual over the domain and natural boundary, using $w(x)$ as the weighting function. Thus, the weak form of a problem can also be obtained using Equation (2.4.16), which mathematically expresses the so-called *method of weighted residuals*, attributed to B. Galerkin. Thus, textbooks that refer to the *weighted residual method* or *the Galerkin method* actually refer to the weak form of the governing equations. This book will avoid the use of terms such as "weighted residual method," because the word *method* may convey the false impression that we have some form of "approximation," while the term *weak form* conveys the correct message—that is, that the weak form is an alternative version of the governing equations for a problem. ■

2.5 Strong Form for One-Dimensional Heat Conduction

We are now going to derive the strong form for one-dimensional (1D), steady-state, heat conduction. A *steady-state problem* is one for which the various field quantities do not change with time. In the steady problem that we will examine, we will formulate the governing equations in rate form (meaning we will examine the variation of various quantities *per unit time*). The only quantity that will not appear in rate form will be the temperature, Θ (because, based on what was already mentioned, the temperature in a steady-state heat conduction problem will only vary with x, not with time!). Although we will see that the strong and weak form are mathematically identical to those for 1D elasticity, examination of 1D heat conduction provides insight on the underlying physics. This will prove useful for establishing the strong and weak form of multidimensional heat conduction (which differs from the multidimensional elasticity, as we shall see in Chapters 5 and 7).

Let us consider the steady-state heat conduction problem of a one-dimensional system, namely a bar, whose cross-sectional area, $A(x)$, can vary over the length, as shown in Figure 2.9. The fundamental quantities in heat conduction are the heat (or thermal energy) and the temperature $\Theta(x)$. Related to heat is the *heat flux*, $q_x(x)$, which gives how much heat is transferred through a specific cross-section of the system, per unit sectional area and per unit time. The heat flux q_x *in the interior* of a one-dimensional body is always considered positive when it is "pointing" along the positive x-direction. Additionally, we will define the quantity $s(x)$, corresponding to the rate of heat (=energy) that is added per unit length of the system from a distant source (in a very simplistic context, we can imagine that the sun is such a source). The quantity $s(x)$ is commonly referred to as the *heat source* and—for 1D problems—it has units of energy per unit length per unit time.

To obtain the strong form for 1D heat conduction, we follow the exact same steps as for one-dimensional elasticity: We cut a small piece of the bar with length equal to Δx, as shown in Figure 2.9, and examine the pertinent equations. Notice that, since the piece is small, we imagine that the heat source term is constant along the piece, and equal to the value at the midpoint of the small piece. Since heat conduction is related to "flow" of energy in the interior of the body, the equation that we need to establish will be based on the conservation of energy principle: *The amount of energy (per unit time) that is added to the piece is equal to the amount of energy (per unit time) that leaves the piece.*

Let us define as Q_x the amount of heat per unit time that passes through a cross-section of the bar. We can write:

$$Q_x(x) = q_x(x) \cdot A(x) \tag{2.5.1}$$

We then have two terms with heat entering or leaving through the two cross-sections. Now, if we take the conservation of energy principle, we can write an equation by stating that the sum of the energy terms that are *added* to the small piece in Figure 2.9 per unit time—that is, the heat source term and the flux across the left section of the piece—minus the energy that leaves the piece per unit time (i.e., the flux across the right section of the piece), must equal zero:

$$\begin{aligned} &q_x(x) \cdot A(x) + s\left(x + \frac{\Delta x}{2}\right) \cdot \Delta x - q_x(x + \Delta x) \cdot A(x + \Delta x) = 0 \\ &\rightarrow Q_x(x) + s\left(x + \frac{\Delta x}{2}\right) \cdot \Delta x - Q_x(x + \Delta x) = 0 \end{aligned}$$

We divide both sides of our equation by the size of the piece, Δx:

$$\frac{-Q_x(x + \Delta x) + Q_x(x)}{\Delta x} + s\left(x + \frac{\Delta x}{2}\right) = 0 \tag{2.5.2}$$

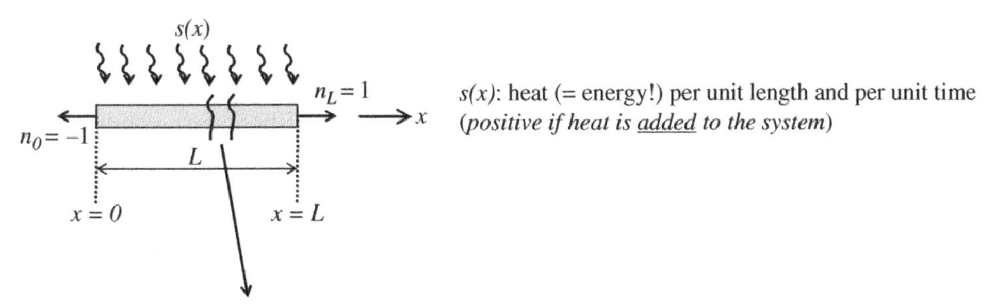

Cut, draw and examine a small piece with energy inflow and outflow:

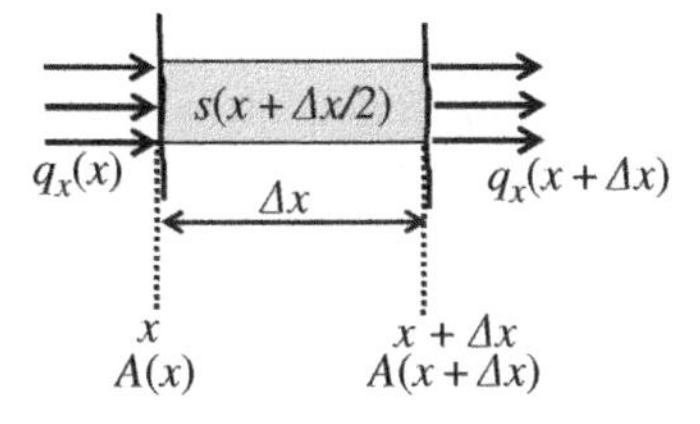

Figure 2.9 One-dimensional heat conduction problem.

If we finally take the limit of Equation (2.5.2) as the size Δx tends to zero, $\lim_{\Delta x \to 0}$, we obtain:

$$-\frac{dQ_x}{dx} + s(x) = 0,$$

and using the definition we established for Q_x in Equation (2.5.1), we can eventually write:

$$-\frac{d(q_x A)}{dx} + s(x) = 0 \tag{2.5.3}$$

The constitutive law for steady-state heat conduction is *Fourier's law*, which states that the heat flux q_x is linearly proportional to the *opposite* of the derivative of the temperature with x. It is a mathematically local[1] version of the statement that heat flows from regions of high temperature to regions of low temperature:

$$q_x = -\kappa \frac{d\Theta}{dx} \tag{2.5.4}$$

The material constant κ in Equation (2.5.4) is called the *thermal conductivity* of the material. The value of κ will generally vary over the length of the bar (just like the Young's modulus will generally vary along the length for one-dimensional elasticity).

If we plug the constitutive Equation (2.5.4) in the differential equation (2.5.3), we eventually obtain the governing *differential equation for 1D heat conduction*:

$$\frac{d}{dx}\left(\kappa A \frac{d\Theta}{dx}\right) + s(x) = 0 \tag{2.5.5}$$

The differential equation is expressed with respect to the *temperature field function*, *$\Theta(x)$*, which is what we will be trying to solve for in heat conduction problems.

Just like for one-dimensional elasticity, the strong form also includes the boundary conditions. There are two types of boundary conditions, namely essential, corresponding to prescribed values of temperature, and natural, corresponding to prescribed quantities related to the derivative of $\Theta(x)$. *The natural boundary conditions for heat conduction correspond to prescribed heat outflow*, $\bar{q}$, across the element natural boundary, which is denoted by Γ_q. The mathematical equation for an essential boundary condition is straightforward:

$$\Theta = \bar{\Theta} \text{ at } \Gamma_\Theta \tag{2.5.6}$$

where $\bar{\Theta}$ is a given quantity and Γ_Θ is the essential boundary for heat conduction (for the simple bar we are examining, it can be at $x = 0$ or $x = L$).

The natural boundary conditions can be written in a generic form by cutting a small piece in the vicinity of the prescribed outflow and taking the limit as the length of the piece tends to zero. We need to note that a prescribed heat outflow $\bar{q}$ is taken *positive* when heat is *leaving* the system under consideration. Thus, we can distinguish two cases, shown in Figure 2.10, with the expression for the natural boundary condition depending

---

1 In mathematics, the local version of a physical law is one that applies for a "differential" piece of the system under consideration (a piece whose dimensions tend to zero). Usually, a local version ends up involving derivatives. The strain-displacement equation and the constitutive equation that were examined for one-dimensional Elasticity in Section 2.1 are also examples of local expressions in physical systems.

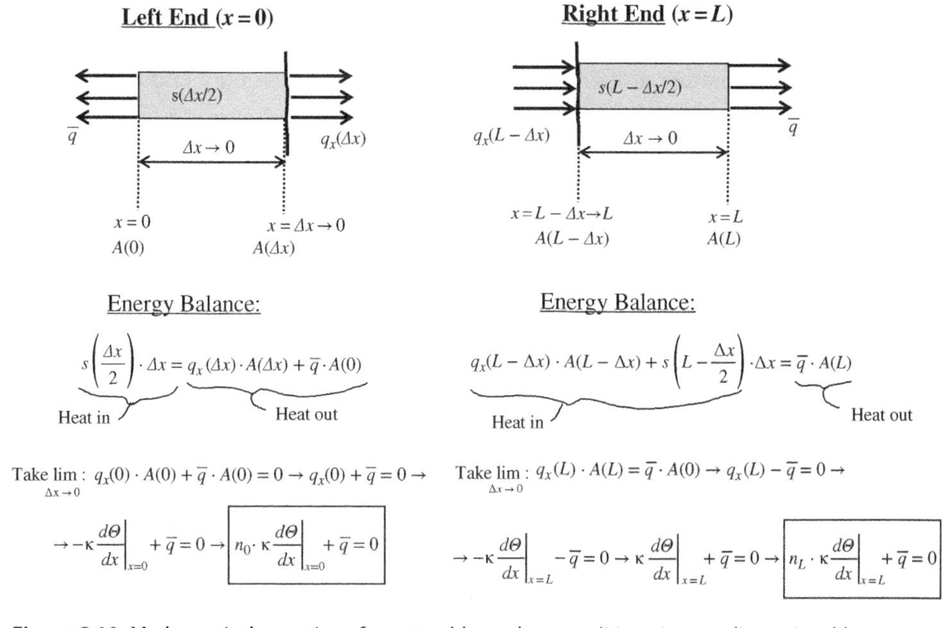

Figure 2.10 Mathematical equations for natural boundary conditions in one-dimensional heat conduction.

on whether the prescribed flow is at the left ($x = 0$) or right ($x = L$) end of the system. Note that these expressions include the unit normal outward vector to the boundary, which was schematically defined in Figure 2.7.

Based on what we obtain in Figure 2.10, we can write a general expression for the natural (=flux) boundary condition for heat conduction:

$$\left[n\cdot\kappa\frac{d\Theta}{dx}\right]_{\Gamma_q} = -\bar{q} \quad \text{OR} \quad \left[n\cdot q_x\right]_{\Gamma_q} = \bar{q} \tag{2.5.7}$$

We can finally combine the governing differential Equation (2.5.5), and the boundary conditions (Equations 2.5.6 and 2.5.7) to obtain the strong form for one-dimensional heat conduction, presented in Box. 2.5.1.

Box 2.5.1 Strong Form for One-Dimensional Heat Conduction

$$\frac{d}{dx}\left(A\cdot\kappa\frac{d\Theta}{dx}\right) + s(x) = 0,\ 0 < x < L$$

$$\Theta = \bar{\Theta} \text{ at essential boundary, } \Gamma_\Theta$$

$$-\left[n\cdot\kappa\frac{d\Theta}{dx}\right]_{\Gamma_q} = \bar{q} \text{ at natural boundary, } \Gamma_q$$

For one-dimensional heat conduction, we can have three cases of boundary conditions, as summarized in Box 2.5.2.

Box 2.5.2 Alternative Cases of Different Boundary Conditions for 1D Heat Conduction
1) Prescribed temperature at the left end $x = 0$, prescribed flux at right end $x = L$: Γ_Θ is at $x = 0$, Γ_q is at $x = L$ 2) Prescribed temperature at $x = L$, prescribed flux at $x = L$: Γ_Θ is at $x = L$, Γ_q is at $x = 0$ 3) Prescribed temperature BOTH at $x = 0$ AND at $x = L$: Γ_Θ includes BOTH $x = 0$ AND at $x = L$. There is no Γ_q (or we can write $\Gamma_q = \emptyset$)!

Remark 2.5.1: The third case in Box 2.5.2 corresponds to the situation that both end-points are essential boundary locations. In this case, we can write $\Gamma_q = \emptyset$, to imply that the natural boundary is the *empty set*, i.e. there are no points constituting natural boundary locations. ∎

Example 2.2: Solution of strong form for 1D heat conduction
In this example, we will find the temperature distribution for the system shown in Figure 2.11.

$$\frac{d}{dx}\left(A \cdot \kappa \frac{d\Theta}{dx}\right) + s(x) = 0 \xrightarrow{A,\kappa \text{ constant!}} A \cdot \kappa \frac{d^2\Theta}{dx^2} + s(x) = 0 \rightarrow 0.5 \cdot 2 \frac{d^2\Theta}{dx^2} + 5\frac{x^2}{5^2} = 0 \rightarrow$$

$$\rightarrow \frac{d^2\Theta}{dx^2} = -\frac{x^2}{5} \rightarrow \frac{d\Theta}{dx} = -\frac{x^3}{15} + C_1 \rightarrow \boxed{\Theta(x) = -\frac{x^4}{60} + C_1 \cdot x + C_2}$$

Now, we specify the value of the constants of integration C_1 and C_2, using the boundary conditions of the problem:

$$\Theta(x = 0) = 0 \rightarrow C_2 = 0$$

At $x = 5$ m, we have a *natural* boundary (since we do not have prescribed temperature, thus we do not have an essential boundary), and we have a zero prescribed flux:

$$-\left[\kappa \frac{d\Theta}{dx}\right]_{x=5m} = 0 \xrightarrow{\kappa \neq 0} \left[\frac{d\Theta}{dx}\right]_{x=5m} = 0 \rightarrow \left[-\frac{x^3}{15} + C_1\right]_{x=5m}$$

$$= 0 \rightarrow -\frac{5^3}{15} + C_1 = 0 \rightarrow \boxed{C_1 = \frac{25}{3}}$$

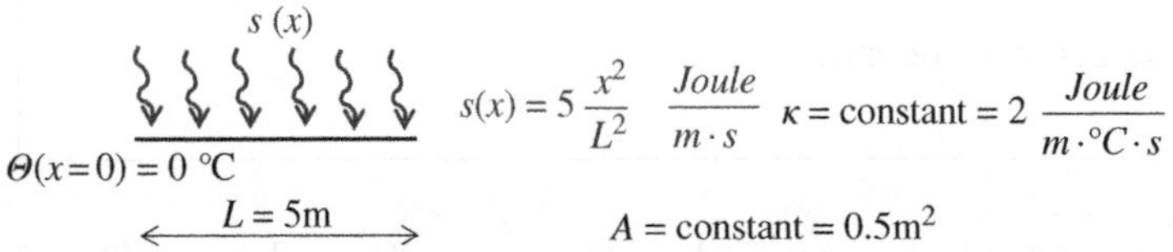

Figure 2.11 Example system for solving the strong form of heat conduction.

Thus, we finally obtain: $\boxed{\Theta(x) = -\frac{x^4}{60} + \frac{25}{3}x}$, and we can plot the function $\Theta(x)$, as shown in Figure 2.12. ∎

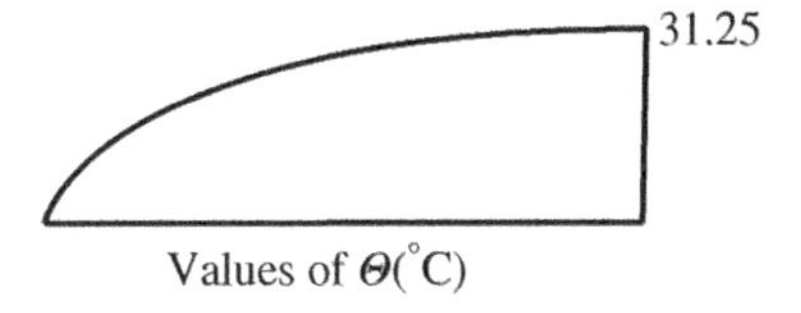

Figure 2.12 Plot of temperature field $\Theta(x)$ for bar in Figure 2.11.

2.6 Weak Form for One-Dimensional Heat Conduction

In the previous section, we obtained the strong form for one-dimensional heat conduction, which is presented in Box 2.5.1. To obtain the weak form, we once again follow the same procedure that we followed in Section 2.3 for one-dimensional elasticity. Thus, we multiply the governing differential equation of Box 2.5.1 by an arbitrary function, $w(x)$, which must *vanish at the essential boundary*. We then obtain:

$$w(x)\left[\frac{d}{dx}\left(\kappa A\frac{d\Theta}{dx}\right)+s(x)\right]=0, \text{ for any arbitrary } w(x) \text{ with } w=0 \text{ at } \Gamma_\Theta \tag{2.6.1}$$

Next, let us integrate Equation (2.6.1) over the entire domain of the problem. The domain of the one-dimensional problem is simply the range of values for x, that is, the region with $0 \leq x \leq L$. We conduct the integration and separate into two integral terms:

$$\int_0^L w(x)\left[\frac{d}{dx}\left(\kappa A\frac{d\Theta}{dx}\right)\right]dx+\int_0^L w(x)\cdot s(x)dx=0, \forall w(x) \text{ with } w=0 \text{ at } \Gamma_\Theta \tag{2.6.2}$$

Just like we did in Section 2.3, we use *integration by parts* in the first term of Equation (2.6.2) to render it symmetric with respect to $\Theta(x)$ and $w(x)$ (i.e., we make sure we have the same order of differentiation for the two functions):

$$\int_0^L w(x)\frac{d}{dx}\left(\kappa A\frac{d\Theta}{dx}\right)dx=\left[w\cdot\kappa A\frac{d\Theta}{dx}\right]_0^L-\int_0^L\frac{dw}{dx}\kappa A\frac{d\Theta}{dx}dx \tag{2.6.3}$$

The boundary term in the right-hand side of Equation (2.6.3) can be separated into two parts, one corresponding to the essential boundary, Γ_Θ, and the other corresponding to the natural boundary, Γ_q. We will also account for the fact that the function $w(x)$ must vanish at Γ_Θ:

$$\left[w\cdot\kappa A\frac{d\Theta}{dx}\right]_0^L=\left[n\cdot w\cdot\kappa A\frac{d\Theta}{dx}\right]_\Gamma=\left[n\cdot w\cdot\kappa A\frac{d\Theta}{dx}\right]_{\Gamma_\Theta}+\left[n\cdot w\cdot\kappa A\frac{d\Theta}{dx}\right]_{\Gamma_q}$$

$= 0$, because $w = 0$ at Γ_Θ!

Thus, we have:

$$\left[w \cdot \kappa A \frac{d\Theta}{dx}\right]_0^L = \left[n \cdot w \cdot \kappa A \frac{d\Theta}{dx}\right]_{\Gamma_q} \tag{2.6.4}$$

Since at the natural boundary for heat conduction we have satisfaction of the equation $\left[n \cdot \kappa \frac{d\Theta}{dx}\right]_{\Gamma_q} = -\bar{q}$, the natural boundary term in Equation (2.6.4) becomes:

$$\left[n \cdot w \cdot \kappa A \frac{d\Theta}{dx}\right]_{\Gamma_q} = \left[w \cdot A \cdot n \cdot \kappa \frac{d\Theta}{dx}\right]_{\Gamma_q} = -\left[w \cdot A \cdot \bar{q}\right]_{\Gamma_q} \tag{2.6.5}$$

If we plug Equations (2.6.4) and (2.6.5) into Equation (2.6.3), we obtain:

$$-\int_0^L \frac{dw}{dx} \kappa A \frac{d\Theta}{dx} dx - \left[w \cdot A \cdot \bar{q}\right]_{\Gamma_q} + \int_0^L w \cdot s dx = 0 \tag{2.6.6}$$

Finally, if we rearrange the terms, we obtain the *weak form for the one-dimensional heat conduction problem*, shown in Box 2.6.1.

Box 2.6.1 Weak Form for One-Dimensional Heat Conduction

$$\int_0^L \frac{dw}{dx} \kappa A \frac{d\Theta}{dx} dx = \int_0^L w \cdot s dx - \left[w \cdot A \cdot \bar{q}\right]_{\Gamma_q}$$

$\forall w(x)$ with $w = 0$ at Γ_Θ (w vanishes at essential boundary)

$\Theta = \bar{\Theta}$ at essential boundary, Γ_Θ

It can be proven (in exactly the same way as for 1D elasticity in Section 2.4, where we exploited the arbitrary nature of $w(x)$) that the weak form is perfectly equivalent to the strong form.

Remark 2.6.1: The key to obtaining the generic natural boundary term in the weak form for heat conduction was Equation (2.6.4), wherein we separated the boundary evaluation of a function into an essential-boundary part and a natural-boundary part. This approach also allows us to obtain the general weak form for one-dimensional elasticity, which is provided in Box 2.3.2 ■

Example 2.3: Weak form for heat conduction

Establish the weak form for the bar that was examined in Example 2.2.

For this problem, the natural boundary is at $x = 5$ m, where we have $\bar{q} = 0$. We thus have:

$$\int_0^L \frac{dw}{dx} \kappa A \frac{d\Theta}{dx} dx = \int_0^L w \cdot 5 \frac{x^2}{25} dx - \left[w \cdot A \cdot 0\right]_{x=5}$$

$= 0$, because $\bar{q} = 0$ at Γ_q!

$\forall\, w(x)$ with $w(0) = 0$

We must also remember to include the essential boundary condition of the problem, $\Theta(0) = 0$. Thus, the weak form for the specific example is the following:

$$\int_0^L \frac{dw}{dx} 2 \cdot 0.5 \frac{d\Theta}{dx} dx = \int_0^L w \cdot \frac{x^2}{5} dx$$

$$\forall w(x) \text{ with } w = 0 \text{ at } \Gamma_\Theta \text{ (}w\text{ vanishes at essential boundary)}$$

$$T(0) = 0$$

■

Example 2.4: Approximate solution of weak form

Let us now examine an *approximate solution to the weak form*, for the case that we considered in Examples 2.2 and 2.3. More specifically, we want to find an approximate solution, based on the assumption that both the trial solution and the weight function are *quadratic:*

$$\tilde{\Theta}(x) = a_0 + a_1 \cdot x + a_2 \cdot x^2, \quad \tilde{w}(x) = b_0 + b_1 \cdot x + b_2 \cdot x^2$$

■

Remark 2.6.2: In accordance with Remark 2.3.1, we will always be assuming the same type of approximation for the trial solution and the weight function (e.g., that both are linear functions of x or that both are cubic functions of x etc.). ■

We must enforce the admissibility conditions for the trial solutions and weighting functions, that is, that the trial solution must satisfy the essential boundary conditions and the weight function must vanish at the essential boundary. Thus, $\tilde{\Theta}(x)$ must satisfy the essential boundary condition:

$$\tilde{\Theta}(x=0) = 0 \rightarrow a_0 + a_1 \cdot 0 + a_2 \cdot 0^2 = 0 \rightarrow \boxed{a_0 = 0}$$

and $\tilde{w}(x)$ must vanish at the essential boundary, which is at $x = 0$:

$$\tilde{w}(x=0) = 0 \rightarrow b_0 + b_1 \cdot 0 + b_2 \cdot 0^2 = 0 \rightarrow \boxed{b_0 = 0}$$

Thus, after enforcing the admissibility conditions, we have:

$$\tilde{\Theta}(x) = a_1 \cdot x + a_2 \cdot x^2, \quad \tilde{w}(x) = b_1 \cdot x + b_2 \cdot x^2$$

We can also establish expressions for the first derivatives of the trial solution and weight function, which are needed for the weak form:

$$\frac{d\tilde{\Theta}}{dx} = a_1 + 2a_2 \cdot x, \quad \frac{d\tilde{w}}{dx} = b_1 + 2b_2 \cdot x$$

All that remains to be done is plug the assumed expressions for Θ and w in the weak form:

$$\int_0^5 (b_1 + 2b_2 \cdot x)\kappa A(a_1 + 2a_2 \cdot x)dx = \int_0^5 \left(b_1 \cdot x + b_2 \cdot x^2\right) \cdot \frac{x^2}{5} dx$$

$$\rightarrow \int_0^5 (b_1 + 2b_2 \cdot x)\kappa A(a_1 + 2a_2 \cdot x)dx - \int_0^5 \left(b_1 \cdot x + b_2 \cdot x^2\right) \cdot \frac{x^2}{5} dx = 0$$

$$\rightarrow b_1 \int_0^5 2 \cdot 0.5(a_1 + 2a_2 \cdot x)dx + b_2 \int_0^5 2 \cdot 2 \cdot 0.5\left(a_1 x + 2a_2 \cdot x^2\right)dx - b_1 \int_0^5 \frac{x^3}{5} dx - b_2 \int_0^5 \frac{x^4}{5} dx = 0$$

$$\rightarrow b_1 [(a_1 \cdot x + a_2 \cdot x^2)]_0^5 - b_1 \left[\frac{x^4}{20}\right]_0^5 + b_2 \left[\left(a_1 \cdot x^2 + 4a_2 \cdot \frac{x^3}{3}\right)\right]_0^5 - b_2 \left[\frac{x^5}{25}\right]_0^5 = 0$$

$$\rightarrow b_1(5a_1 + 25a_2) - b_1 \cdot 31.25 + b_2\left(25a_1 + \frac{500}{3} a_2\right) - b_2 \cdot 125 = 0$$

$$\rightarrow b_1(5a_1 + 25a_2 - 31.25) + b_2\left(25a_1 + \frac{500}{3} a_2 - 125\right) = 0 \quad \forall\ \tilde{w}(x)$$

Since we have already stipulated that $\tilde{w}(x) = b_1 \cdot x + b_2 \cdot x^2$, the only way to obtain different functions $\tilde{w}(x) = b_1 \cdot x + b_2 \cdot x^2$ is to arbitrarily vary the constant parameters b_1 and b_2. Thus, the statement "$\forall\ \tilde{w}(x)$" is translated to "$\forall\ b_1, b_2$"! Thus, we now have:

$$b_1(5a_1 + 25a_2 - 31.25) + b_2\left(25a_1 + \frac{500}{3} a_2 - 125\right) = 0 \quad \forall\ b_1, b_2$$

The only way for the obtained expression to *always* give zero result for any arbitrary combination of b_1 and b_2 is to stipulate that the two brackets that multiply b_1 and b_2 are equal to zero:

$$\left.\begin{array}{l} 5a_1 + 25a_2 - 31.25 = 0 \\ 25a_1 + \frac{500}{3} a_2 - 125 = 0 \end{array}\right\} \rightarrow \left.\begin{array}{l} 5a_1 + 25a_2 = 31.25 \\ 25a_1 + \frac{500}{3} a_2 = 125 \end{array}\right\} \rightarrow \begin{bmatrix} 5 & 25 \\ 25 & \frac{500}{3} \end{bmatrix} \begin{Bmatrix} a_1 \\ a_2 \end{Bmatrix} = \begin{Bmatrix} 31.25 \\ 125 \end{Bmatrix}$$

This system of two equations for two parameters a_1 and a_2, can be solved to yield the values of these two parameters. This computation is conducted in the next example, allowing us to also introduce an efficient method for solving systems of linear equations.

Example 2.5: Determination of approximate solution and comparison with exact solution

In this example, we will solve the system of equations of Example 2.4 to obtain an approximate solution of the weak form. We will initially deviate from the main focus of the chapter, to provide a first discussion of how linear systems of equations, $[A]\{x\} = \{b\}$, where [A], {b} are known and vector {x} is unknown, are solved in the computer. One would assume that the computer inverts the coefficient matrix $[A]$ in the left-hand side, and then obtain $\{x\}$ by premultiplying the right-hand-side vector $\{b\}$ by the inverse of $[A]$. It turns out that this procedure is computationally expensive. For this reason, we can use another method, called *Gauss elimination*. We will introduce Gauss elimination by using it for the solution of the system of equations obtained in Example 2.4. What we do first is write an *expanded* matrix, which is obtained if we attach the right-hand-side vector of the system as an additional column of the coefficient array:

$$\left[\begin{array}{cc|c} 5 & 25 & 31.25 \\ 25 & \frac{500}{3} & 125 \end{array}\right]$$

We then conduct appropriate operations, taking linear combinations of the rows of the expanded array, so that the coefficient matrix (in our case, the 2 × 2 matrix) becomes *upper triangular*. For the specific example, we multiply the first row by 5, then subtract it from the second row, and we substitute the second row by the resulting row from the subtraction:

$$\left[\begin{array}{cc|c} 5 & 25 & 31.25 \\ 25 & \frac{500}{3} & 125 \end{array}\right] \xrightarrow{\text{row } 2 \leftarrow \text{row } 2-5\times(\text{row } 1)} \left[\begin{array}{cc|c} 5 & 25 & 31.25 \\ 0 & \frac{125}{3} & -31.25 \end{array}\right]$$

$$\left[\begin{array}{cc|c} 25 & \frac{500}{3} & 125 \end{array}\right]$$
$$-5\left[\begin{array}{cc|c} 5 & 25 & 31.25 \end{array}\right]$$

The operation we have done is equivalent to multiplying the first equation of the system by 5, and subtracting it from the second equation:

$$\left.\begin{array}{l} 25a_1 + \frac{500}{3}a_2 - 125 = 0 \\ -5(5a_1 + 25a_2 - 31.25 = 0) \end{array}\right\} \rightarrow 25a_1 + \frac{500}{3}a_2 - 125 - (25a_1 + 125a_2 - 156.25) = 0$$

$$\rightarrow \frac{125}{3}a_2 = -31.25$$

Now, if we replace the second equation of the system by the equation we obtained after the subtraction, we have:

$$\left.\begin{array}{ll} 5a_1 + 25a_2 - 31.25 = 0 & (1) \\ \frac{125}{3}a_2 = -31.25 & (2) \end{array}\right\}$$

If we have rendered the coefficient matrix upper triangular, we are able to solve without matrix inversion! We can see that the last equation of the system, which is equation (2), can be directly solved for a_2:

$$\frac{125}{3}a_2 = -31.25 \rightarrow a_2 = -0.75$$

Now, we can substitute the value of a_2 in equation (1), and calculate a_1:

$$5a_1 + 25a_2 - 31.25 = 0 \rightarrow 5a_1 + 25(-0.75) - 31.25 = 0 \rightarrow a_1 = 10$$

Notice that, *after rendering the coefficient matrix upper triangular*, we used the last equation to calculate the last unknown (a_2) first, then we plugged the a_2-value in the previous equation to find the previous unknown (a_1). This procedure is called *back-substitution*. The procedure of rendering the coefficient matrix upper triangular is called *forward elimination*.

Thus, we have obtained the approximate solution $\tilde{\Theta}(x) = 10 \cdot x - 0.75 \cdot x^2$

The corresponding approximate flux field is:

$$\tilde{q}(x) = -\kappa \frac{d\tilde{\Theta}}{dx} = -20 + 3x$$

At the natural boundary, $x = 5$, we have: $\tilde{q}_x(x = 5) = -5 \frac{J}{m^2 s} \neq 0$. As expected from Remark 2.4.2, the approximate solution does not exactly satisfy the *natural* boundary condition! It can also be verified that the approximate solution does not exactly satisfy the governing differential equation of the strong form.

We can now assess the overall "accuracy" of the assumed quadratic trial solution for the temperature field by plotting the variation of the approximate temperatures and fluxes, and comparing them to the values of the exact solution obtained in Example 2.2. The comparison is presented in Figure 2.13.

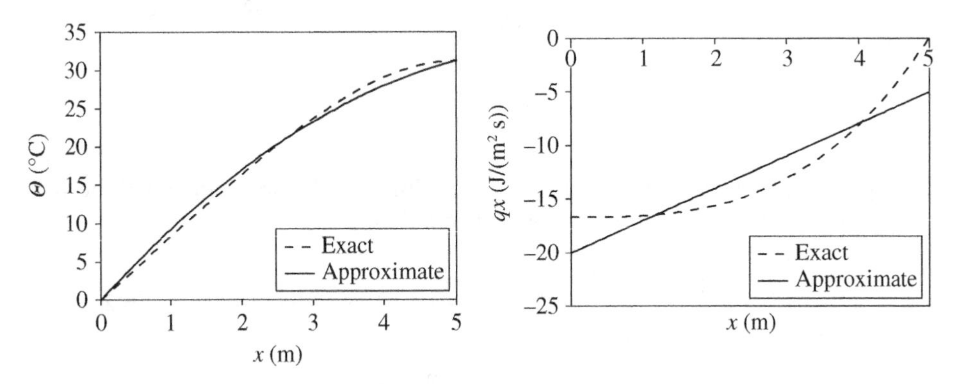

Figure 2.13 Comparison between approximate quadratic solution and exact solution, for the system shown in Figure 2.11.

It is now deemed necessary to also discuss another example with Gaussian elimination for solving systems of equations. This time, we will work with a larger system—a system of three equations for three unknowns.

Example 2.6: Gaussian Elimination

In this example, we will further examine the use of Gaussian elimination, by solving the following linear system of equations.

$$2x_1 + x_2 + x_3 = 2$$

$$-x_1 + 2x_2 + x_3 = 0$$

$$x_1 + x_2 + x_3 = 1$$

First of all, we cast the system in matrix form:

$$\begin{bmatrix} 2 & 1 & 1 \\ -1 & 2 & 1 \\ 1 & 1 & 1 \end{bmatrix} \begin{Bmatrix} x_1 \\ x_2 \\ x_3 \end{Bmatrix} = \begin{Bmatrix} 2 \\ 0 \\ 1 \end{Bmatrix}$$

We then write the expanded matrix, where we simply append to the coefficient matrix (as an extra column) the right-hand-side vector. We also use shading for the entities of the expanded matrix that correspond to the original coefficient array:

$$\left[\begin{array}{ccc:c} 2 & 1 & 1 & 2 \\ -1 & 2 & 1 & 0 \\ 1 & 1 & 1 & 1 \end{array}\right]$$

Now, we begin a series of manipulations (taking linear combinations of row vectors) on the expanded array, until we render the shaded area *upper triangular*. First, we set the components of the first column for rows 2 and 3 equal to zero:

$$\left[\begin{array}{ccc:c} 2 & 1 & 1 & 2 \\ -1 & 2 & 1 & 0 \\ 1 & 1 & 1 & 1 \end{array}\right]$$

$$\Downarrow \quad \text{row 2} \leftarrow \text{row 2} + 0.5 \times (\text{row 1})$$

$$\left[\begin{array}{ccc:c} 2 & 1 & 1 & 2 \\ 0 & 2.5 & 1.5 & 1 \\ 1 & 1 & 1 & 1 \end{array}\right]$$

$$\Downarrow \quad \text{row 3} \leftarrow \text{row 3} - 0.5 \times (\text{row 1})$$

$$\left[\begin{array}{ccc:c} 2 & 1 & 1 & 2 \\ 0 & 2.5 & 1.5 & 1 \\ 0 & 0.5 & 0.5 & 0 \end{array}\right]$$

The next step is to set, for all the rows below the second row, the components of the second column equal to zero. In our case, we only have one row, namely row 3, below the second row:

$$\left[\begin{array}{ccc:c} 2 & 1 & 1 & 2 \\ 0 & 2.5 & 1.5 & 1 \\ 0 & 0.5 & 0.5 & 0 \end{array}\right]$$

$$\Downarrow \quad \text{row 3} \leftarrow \text{row 3} - 0.2 \times (\text{row 2})$$

$$\left[\begin{array}{ccc:c} 2 & 1 & 1 & 2 \\ 0 & 2.5 & 1.5 & 1 \\ 0 & 0 & 0.2 & -0.2 \end{array}\right]$$

Now, we can understand that we have rendered the shaded area (= the coefficient array) *upper triangular*. What we have eventually obtained is the following system:

$$\begin{bmatrix} 2 & 1 & 1 \\ 0 & 2.5 & 1.5 \\ 0 & 0 & 0.2 \end{bmatrix} \begin{Bmatrix} x_1 \\ x_2 \\ x_3 \end{Bmatrix} = \begin{Bmatrix} 2 \\ 1 \\ -0.2 \end{Bmatrix} \rightarrow \left.\begin{aligned} 2x_1 + x_2 + x_3 &= 2 \\ 2.5x_2 + 1.5x_3 &= 1 \\ 0.2x_3 &= -0.2 \end{aligned}\right\}$$

We can now conduct *back-substitution*—that is, solve the third equation for x_3, plug the obtained value of x_3 into the second equation and obtain x_2, and then plug the obtained value of x_2 and x_3 into the first equation to obtain x_1:

$$\underline{3rd\,Equation:}\quad 0.2x_3 = -0.2 \rightarrow x_3 = -1$$

$$\underline{2nd\,Equation:}\quad 2.5x_2 + 1.5x_3 = 1 \rightarrow 2.5x_2 + 1.5(-1) = 1 \rightarrow x_2 = 1$$

$$\underline{1st\,Equation:}\quad 2x_1 + x_2 + x_3 = 2 \rightarrow 2x_1 + 1 + (-1) = 2 \rightarrow x_1 = 1$$

A general, rigorous description of Gaussian Elimination for linear systems of equations is provided in Section 19.2.1. ■

Problems

2.1 We are given the one-dimensional, linearly elastic truss member shown in Figure P2.1. The axial rigidity of the member, EA, is a linear function of x, as shown in the same figure. Knowing the strong form for 1D elasticity, find the stiffness matrix of the member. (*Hint:* What is the meaning of each term, k_{ij}, of the stiffness matrix of a member?).

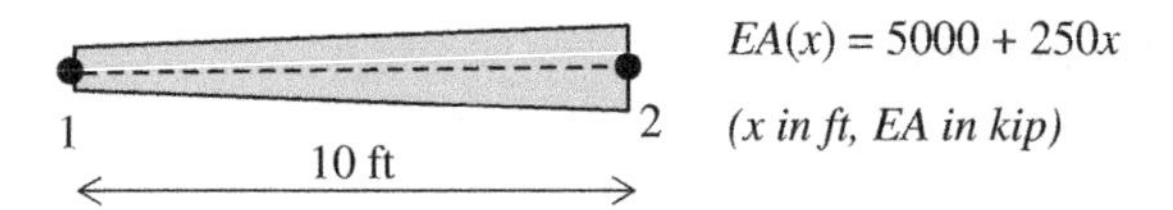

Figure P2.1

2.2 We are given the structure in Figure P2.2.

a) Derive the strong form for the specific problem.
b) Solve and obtain the displacement field, $u(x)$, and the axial stress field, $\sigma(x)$.
c) Derive the weak form for this problem.
d) Obtain an approximate solution to the problem, assuming:

$$\tilde{u}_1(x) = a_1 + a_2x$$
$$\tilde{w}_1(x) = b_1 + b_2x$$

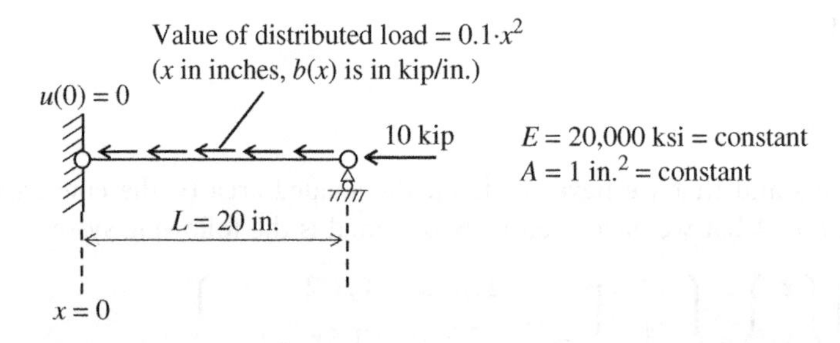

Figure P2.2

e) Obtain an approximate solution to the problem, assuming:

$$\tilde{u}_2(x) = a_0 + a_1 x + a_2 x^2$$
$$\tilde{w}_2(x) = b_0 + b_1 x + b_2 x^2$$

f) What observation can you make for the coefficient matrix in part e?
g) Do the approximate solutions $\tilde{u}_1(x)$ and $\tilde{u}_2(x)$ satisfy the natural boundary condition? Why or why not?
h) Draw (in the same plot) the variation of $u(x)$, $\tilde{u}_1(x)$ and $\tilde{u}_2(x)$. What happens as you "refine" your approximation (use more parameters a_i)?
i) Draw (in the same plot) the variation of $\sigma(x)$ (the exact stress field), $\tilde{\sigma}_1(x)$ (the approximate stress field corresponding to $\tilde{u}_1(x)$) and $\tilde{\sigma}_2(x)$ (the approximate stress field corresponding to $\tilde{u}_2(x)$). What happens as you "refine" your approximation (= use more parameters a_i)? Also, for a given approximation, for which quantity do you obtain a closer match with the exact solution, the displacements or the stresses? Why?

2.3 We usually stipulate the same kind of approximation for $\tilde{u}(x)$ and $\tilde{w}(x)$ in the weak form. That is, if we assume $\tilde{u}(x)$ to be a linear function of x, we assume the same for $\tilde{w}(x)$. Such approximation method is termed **the Bubnov-Galerkin method** or simply the **Galerkin method**. We can also use different approximations for $\tilde{u}(x)$ and $\tilde{w}(x)$ if we wish, at which case we have the **Petrov-Galerkin method**.

Mathematicians—who are notorious for their ability to state the simplest of things in a fashion that ordinary people cannot understand—express this use of different approximations by stating that "$\tilde{u}(x)$ is contained in a collection of functions other than the collection of functions containing $\tilde{w}(x)$." In this problem, we will use a Petrov-Galerkin approach to obtain approximate solutions.

a) Obtain the approximate solution of the structure at Problem 2.2 if:

$$\tilde{u}_3(x) = a_0 + a_1 x + a_2 x^2$$
$$\tilde{w}_3(x) = b_0 + b_1 x + b_2 x^3$$

What observation do you make regarding the coefficient matrix in your linear system?

b) What system of equations is obtained for the structure of Problem 2.2 if we set:

$$\tilde{u}_4(x) = a_0 + a_1 x + a_2 x^2$$
$$\tilde{w}_4(x) = b_0 + b_1 x + b_2 x^2 + b_3 x^3$$

Can you invert the coefficient matrix to obtain the approximate solution for this case?

2.4 We are given the system in Figure P2.3. There is no heat source.

a) Derive the strong form for the specific problem
b) Obtain the temperature field, $\Theta(x)$, and the heat flux, $q_x(x)$.
c) Derive the weak form for this problem.
d) Obtain an approximate solution to the problem, assuming:

$$\tilde{\Theta}(x) = a_0 + a_1 x + a_2 x^2$$
$$\tilde{w}(x) = b_0 + b_1 x + b_2 x^2$$

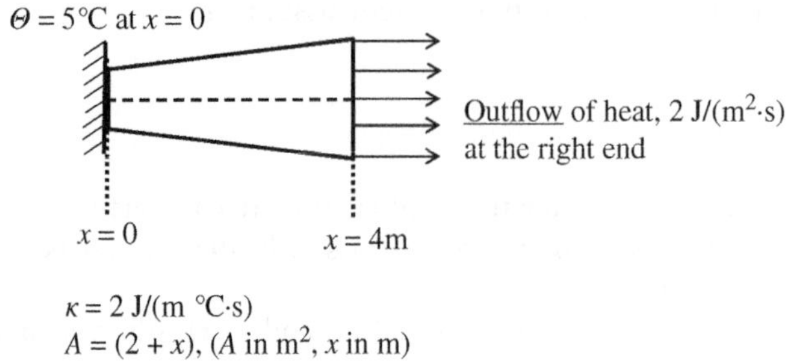

$\Theta = 5°C$ at $x = 0$

Outflow of heat, 2 J/(m^2·s) at the right end

$x = 0$ $x = 4$m

$\kappa = 2$ J/(m °C·s)
$A = (2 + x)$, (A in m^2, x in m)

Figure P2.3

2.5 a) We are given the following strong form: $\frac{d^3u}{dx^3} + s(x) = 0$, where $s(x)$ is a given function and $u(x)$ is the function that we want to solve for. If we introduced a cubic approximation for $\tilde{u}(x)$ and $\widetilde{w}(x)$ in the weak form, would the coefficient matrix be symmetric? Why or why not?

b) Same as in problem a, for the following strong form: $\frac{d^4u}{dx^4} + s(x) = 0$.

c) Based on the findings in a and b, and assuming that we use the same type of approximation for $\tilde{u}(x)$ and $\widetilde{w}(x)$ in the weak form, establish a general criterion that depends on the order of derivatives in the strong form to decide if the coefficient matrix will be symmetric.

References

Boyce, W., and DiPrima, R. (2012). *Elementary Differential Equations and Boundary Value Problems*, 10th ed. Hoboken, NJ: John Wiley & Sons.

Stewart, J. (2015). *Calculus*, 8th ed. Boston: Brooks Col7e.

3

Finite Element Formulation for One-Dimensional Problems

3.1 Introduction—Piecewise Approximation

In the previous chapter, we saw the procedure to obtain the weak form for one-dimensional problems. We also saw (in Example 2.4) how to obtain an approximate solution once we make an approximation for $u(x)$ and $w(x)$ in the weak form. This chapter will provide a more systematic approach for obtaining approximate solutions by means of the finite element method. Before proceeding further in this chapter, the reader may wish to review Appendix A, providing a brief overview of matrix algebra, and Appendix B (especially Sections B.1, B.2, and B.3), which provides a review of matrix analysis of discrete systems.

In the finite element method, the domain of interest is divided into subdomains called elements, as shown in Figure 3.1, and we approximate the function in each element separately. The subdivision of the domain into elements is termed *discretization*. The set of all elements representing a domain constitutes the *finite element mesh*. As also shown in Figure 3.1, if we want to obtain improved accuracy from a finite element analysis, we can establish a different discretization for the domain, using a smaller element size (and thus having a larger number of finite elements in the analysis). The process of re-discretizing using smaller elements is termed *mesh refinement*. If several conditions are met by our finite element approximations, then a refinement of the mesh will correspond to more accurate results. We will be saying that *as we refine the mesh, the approximate, finite element solution will be converging to the exact solution*.

Each finite element has a number of points called *nodes* (e.g., the end points of a truss element), and it is convenient to establish the *nodal values of the field function* as the constant parameters of the approximate solution. For elasticity, we will have the nodal displacements, while for heat conduction we will have the nodal temperatures.

Remark 3.1.1: It is deemed desirable to assign a local number to the nodal points of each element. In our descriptions for one-dimensional finite element analysis, we will always number the left end of an element as node 1 and the right end as node 2. ▮

The following section will show how to establish function approximations in one-dimensional finite elements. It is worth mentioning that in the approximations of the

Fundamentals of Finite Element Analysis: Linear Finite Element Analysis, First Edition.
Ioannis Koutromanos, James McClure, and Christopher Roy.

Companion website: www.wiley.com/go/koutromanos/linear

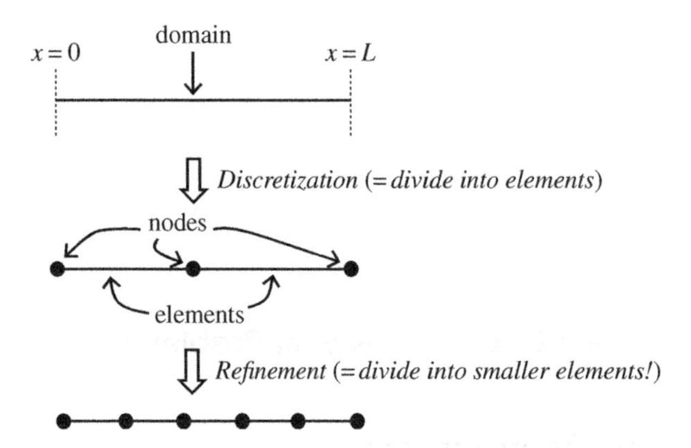

Figure 3.1 Finite element discretization procedure.

weak form considered in Example 2.4, we were stipulating an approximate expression for trial and test functions over the entire domain of the problem. These types of approximations are termed *global approximations.* Thus, a global approximation for a field $\upsilon(x)$ (where $\upsilon(x) = u(x)$ for one-dimensional elasticity, $\upsilon(x) = \Theta(x)$ for one-dimensional heat conduction) is given by the following equation.

$$\upsilon(x) \approx \tilde{\upsilon}(x) = a_0 + a_1 \cdot x + a_2 \cdot x^2 + \ldots, \quad 0 \le x \le L \tag{3.1.1}$$

On the other hand, in finite element analysis, we will be establishing a *piecewise approximation* for the field $\upsilon(x)$, in each element e of the domain:

$$\upsilon(x) \approx \upsilon^e(x) = a_0^{(e)} + a_1^{(e)} \cdot x + a_2^{(e)} \cdot x^2 + \ldots, \quad x_1^{(e)} \le x \le x_2^{(e)} \tag{3.1.2}$$

where $x_1^{(e)}$ is the coordinate of the left endpoint of element e and $x_2^{(e)}$ is the coordinate of the right endpoint of element e.

For piecewise approximations, we must satisfy continuity requirements both in the interior of the elements and across the inter-element boundaries. For example, if we have a two-element mesh, as shown in Figure 3.2, then node 2 of element 1 and node 1 of element 2 will correspond to the same intermediate point of the mesh, as shown in the same figure. Obviously, for the finite element solution to be continuous, the value of υ at the intermediate nodal point shared by the two elements must be identical

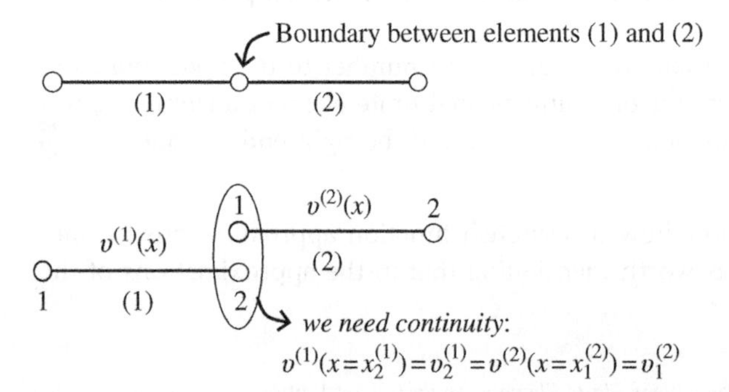

Figure 3.2 Continuity requirements for one-dimensional finite element mesh.

for the approximate field of each element, as shown in Figure 3.2. This requirement is automatically satisfied for the approximations presented in this text, because they ensure that the value of the piecewise approximations of the two elements at the location of the common nodal point is equal to the nodal value of that point.

The selection of the approximation functions in each element will be made to satisfy the necessary conditions guaranteeing the *convergence* of the method—that the approximate solution will eventually get closer and closer to the exact solution as we refine the mesh. *The two necessary conditions for a finite element method to converge are the following:*

1) *Continuity*, that is the trial solutions and weight functions are sufficiently continuous. A function $\upsilon(x)$ is continuous if it does not have any discontinuities over its domain (i.e., for all values that x can obtain). If a function $\upsilon(x)$ is not continuous at some point $x = x_o$, then the plot of the function will exhibit a discrete *jump* (i.e., a discontinuity) at $x = x_o$, as schematically shown in Figure 3.3a. We may also be interested in the continuity of the derivative $\frac{d\upsilon}{dx}$ of a function. For each location x, the value of the derivative of $\upsilon(x)$ is equal to the corresponding slope of the plot of $\upsilon(x)$. A discontinuity in the derivative of $\upsilon(x)$ at some location is graphically manifested through a *kink* (i.e., a sudden slope change, called slope discontinuity) in the plot of the function at the same location, as shown in Figure 3.3b. For any location where the slope of $\upsilon(x)$ is discontinuous, the plot of $\frac{d\upsilon}{dx}$ will have a discontinuity (jump). Based on the above, a function that is continuous and also exhibits continuity of the first derivative will correspond to a plot, which has no jumps and is smooth (without kinks), as shown in Figure 3.3c. We can also extend the discussion to continuity of higher-order derivatives of a function, if necessary.

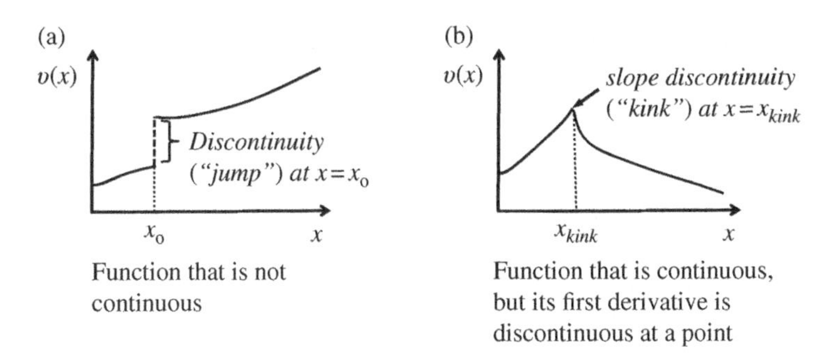

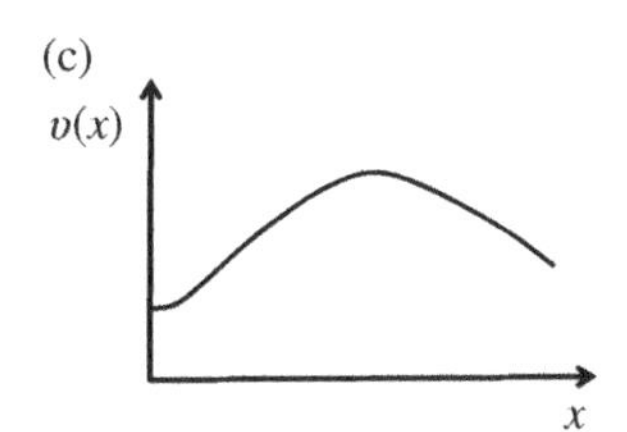

Figure 3.3 Schematic presentation of continuity considerations for a function $\upsilon(x)$.

Now, the continuity requirements for a finite element approximation are essentially stipulated by the highest order of derivative that appears on the weak form. If the highest-order of derivatives in the weak form is k, then we require C^{k-1} continuity. This means that the function and all of its derivatives, up to and including $(k-1)$-order derivative, must be continuous in the interior of the elements and across element boundaries. For the problems that we have examined in Chapter 2, the highest order of derivative in the weak form is 1 (we have first derivatives in the weak form), which implies that we simply need C^0 continuity: the approximate solution itself must be continuous, and no continuity requirements need to be enforced for the derivatives. If we return to Figure 3.3, we can tell that the function in Figure 3.3a does not have C^0 continuity. The function in Figure 3.3b has C^0 continuity, since its derivative is not continuous, but the function itself is (the plot of the function does not have any jumps). The function in Figure 3.3c is continuous and has a continuous first derivative; thus, it must be at least C^1 continuous (we would need to examine the continuity of derivatives of order higher than 1 to determine the exact order of continuity!).

2) *Completeness*, that is the trial solution and the weight functions must be able to perfectly describe a given function in the interior of each element. More specifically, a function is said to be k-th degree- complete if it can represent *any arbitrary* k-th degree polynomial. For example, the function $\tilde{v}(x) = a_0 + a_1 x$ is first-degree complete (or has a degree of completeness equal to 1 or has linear completeness), because it can represent any linear polynomial function (i.e., any polynomial of first degree).

   Similarly, $\tilde{v}(x) = a_0 + a_1 x + a_2 x^2$ is second-degree complete (or has a degree of completeness equal to 2), because it can represent any second-degree (quadratic) polynomial. In general, the degree of completeness is what decides whether and "how fast" an approximation converges (i.e., how much the error in the approximate solution decreases every time we repeat the analysis with a finer mesh!). As an additional example, the approximation $\tilde{v}(x) = a_0 + a_1 x + a_2 x^4$ is first-degree complete, because it cannot represent all the quadratic, cubic, and quartic polynomials. Thus, the approximation $\tilde{v}(x) = a_0 + a_1 x + a_2 x^4$ will generally converge as fast as the linear approximation $\tilde{v}(x) = a_0 + a_1 x$. Having defined the notion of completeness, we can proceed to discuss what is the required degree of completeness.

   For convergence of the finite element method, the trial solutions and weight functions and their derivatives up to and including the highest-order derivative appearing in the weak form must be capable of assuming constant values. So, for elasticity (or heat conduction), this requires that the displacement (or temperature) field and its derivative can take constant values, so the elements can represent rigid-body motion (or constant temperature field) and constant-strain deformation (or constant derivative of temperature). This is equivalent to requiring that the approximate solution has a degree of completeness equal to at least 1. Thus, the approximation $\tilde{v}(x) = a_0 + a_1 x^3 + a_2 x^4$ cannot converge, because it cannot represent a function with constant derivative!

The following section will provide a framework for establishing the piecewise approximation for finite elements, based on the concept of shape functions.

3.2 Shape (Interpolation) Functions and Finite Elements

To facilitate the introduction of the procedure to establish the piecewise finite element approximation, let us examine the simplest case of a finite element—namely, the two-node element. Although this is the simplest case, it is also the most commonly employed one, especially for problems involving very large deformations, inelastic materials, and so on, because of the efficiency and robustness of the resulting discrete equations (robustness of equations means that the analysis does not "blow up" due to numerical issues such as matrices becoming *near-singular*).

For the two-node element, we employ the simplest acceptable type of piecewise approximation of the field function $\upsilon(x)$ that can converge. This is a first-order polynomial (i.e., a linear polynomial):

$$\upsilon^{(e)}(x) = a_0^{(e)} + a_1^{(e)} \cdot x \tag{3.2.1}$$

We can qualitatively draw the variation of $\upsilon^{(e)}(x)$ with x as shown in Figure 3.4. The figure also shows the two nodal values, $\upsilon_1^{(e)}$ and $\upsilon_2^{(e)}$, corresponding to the nodal points, which are located at $x = x_1^{(e)}$ and $x = x_2^{(e)}$.

Now, let us try to cast the finite element approximation in a form that will include the nodal values as the constant parameters. To this end, we know that if we plug $x = x_1^{(e)}$ in Equation (3.2.1), the field function must be equal to the value $\upsilon_1^{(e)}$ of the first nodal point:

$$\upsilon^{(e)}\left(x = x_1^{(e)}\right) = a_0^{(e)} + a_1^{(e)} \cdot x_1^{(e)} = \upsilon_1^{(e)} \tag{3.2.2a}$$

Similarly, if we plug $x = x_2^{(e)}$ in Equation (3.2.1), the field function must equal the value $\upsilon_2^{(e)}$ of the second nodal point:

$$\upsilon^{(e)}\left(x = x_2^{(e)}\right) = a_0^{(e)} + a_1^{(e)} \cdot x_2^{(e)} = \upsilon_2^{(e)} \tag{3.2.2b}$$

We can solve Equations (3.2.2a) and (3.2.2b) for the constants $a_0^{(e)}$ and $a_1^{(e)}$ to obtain:

$$a_0^{(e)} = \frac{\theta_1^{(e)} - \theta_2^{(e)}}{x_2^{(e)} - x_1^{(e)}} x_1^{(e)} - \theta_1^{(e)} \tag{3.2.3a}$$

and

$$a_1^{(e)} = \frac{\theta_2^{(e)} - \theta_1^{(e)}}{x_2^{(e)} - x_1^{(e)}} \tag{3.2.3b}$$

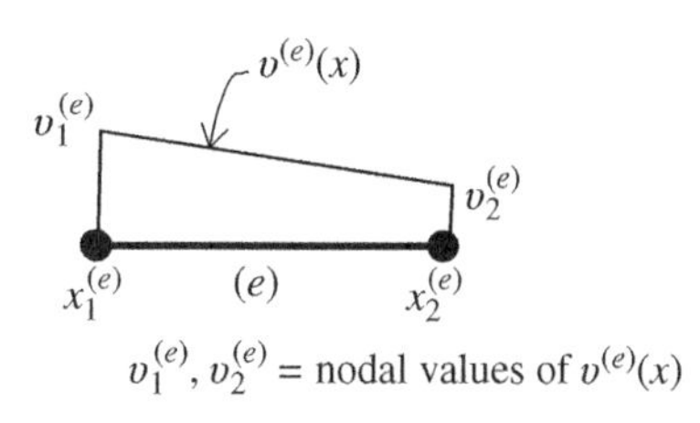

Figure 3.4 Linear finite element approximation for a two-node element.

We can now plug Equations (3.2.3a,b) into Equation (3.2.1) to obtain:

$$\upsilon^{(e)}(x) = \frac{\upsilon_2^{(e)} - \upsilon_1^{(e)}}{x_2^{(e)} - x_1^{(e)}}\left(x - x_1^{(e)}\right) - \upsilon_1^{(e)} \tag{3.2.4}$$

We want to rewrite the equation for $\upsilon^{(e)}(x)$ in a form:

$$\upsilon^{(e)}(x) = N_1^{(e)}(x) \cdot \upsilon_1^{(e)} + N_2^{(e)}(x) \cdot \upsilon_2^{(e)} \tag{3.2.5a}$$

or, equivalently:

$$\upsilon^{(e)}(x) = \left[N^{(e)}\right]\begin{Bmatrix} \upsilon_1^{(e)} \\ \upsilon_2^{(e)} \end{Bmatrix} \tag{3.2.5b}$$

where

$$\left[N^{(e)}\right] = \left[N_1^{(e)}(x) \quad N_2^{(e)}(x)\right].$$

To this end, we separate Equation (3.2.4) into terms multiplying $\upsilon_1^{(e)}$ and terms multiplying $\upsilon_2^{(e)}$:

$$\begin{aligned} \upsilon^{(e)}(x) &= \upsilon_2^{(e)} \frac{x - x_1^{(e)}}{x_2^{(e)} - x_1^{(e)}} + \upsilon_1^{(e)}\left(- \frac{x - x_1^{(e)}}{x_2^{(e)} - x_1^{(e)}} + 1 \right) = \upsilon_1^{(e)} \frac{x - x_2^{(e)}}{x_1^{(e)} - x_2^{(e)}} + \upsilon_2^{(e)} \frac{x - x_1^{(e)}}{x_2^{(e)} - x_1^{(e)}} \\ &= N_1^{(e)}(x) \cdot \upsilon_1^{(e)} + N_2^{(e)}(x) \cdot \upsilon_2^{(e)} \end{aligned} \tag{3.2.6}$$

where the functions $N_1^{(e)}(x)$, $N_2^{(e)}(x)$, called the *shape functions* of the element, are given by:

$$N_1^{(e)}(x) = \frac{x - x_2^{(e)}}{x_1^{(e)} - x_2^{(e)}} \tag{3.2.7a}$$

$$N_2^{(e)}(x) = \frac{x - x_1^{(e)}}{x_2^{(e)} - x_1^{(e)}} \tag{3.2.7b}$$

Remark 3.2.1: Equation (3.2.5a) can also be termed the *finite element interpolation*, because it allows us to interpolate—that is, to directly find the value of the approximate value $\upsilon^{(e)}(x)$ at any location between the element nodal points, provided that we know the nodal values, $\upsilon_1^{(e)}$ and $\upsilon_2^{(e)}$. Accordingly, the shape functions of a finite element are also called *interpolation functions*. ■

The plots of the two shape functions of the two-node linear element are presented in Figure 3.5. Notice that, in the same figure, we show that the first shape function is equal to 1 at node 1, and equal to 0 at node 2. Similarly, the second shape function is equal to 0 at node 1, and equal to 1 at node 2. We can collectively write:

$$N_a^{(e)}\left(x = x_b^{(e)}\right) = \delta_{ab} = \begin{cases} 1, & \text{if } a = b \\ 0, & \text{if } a \neq b \end{cases} \tag{3.2.8}$$

where δ_{ab} (equal to 1 if $a = b$, 0 otherwise) is called the *Kronecker delta*.

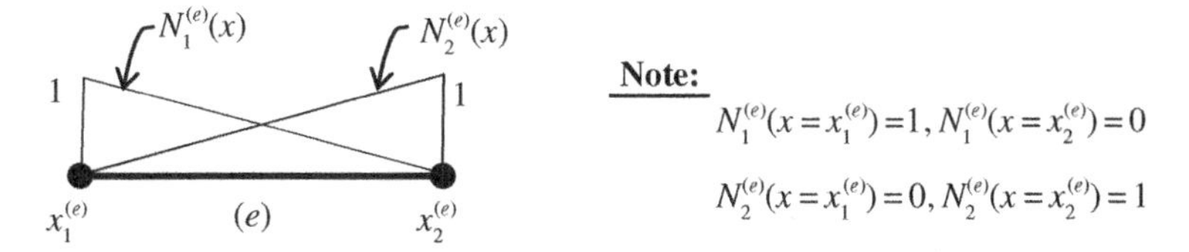

Figure 3.5 Shape functions for a two-node element.

Remark 3.2.2: All the shape functions that we will consider in this text satisfy Equation (3.2.8), which, for obvious reasons, is called the *Kronecker delta* property. This property ensures that the continuity requirements for intermediate nodes in the finite element mesh, schematically summarized in Figure 3.2, will be automatically satisfied. ■

Now, let us establish another important property of the shape functions. If we take the sum of all the shape functions of our element—in the specific case of a two-node element, the sum of both shape functions—for any location x, we obtain:

$$\sum_{i=1}^{2}\left(N_i^{(e)}(x)\right)=N_1^{(e)}(x)+N_2^{(e)}(x)=\frac{x-x_2^{(e)}}{x_1^{(e)}-x_2^{(e)}}+\frac{x-x_1^{(e)}}{x_2^{(e)}-x_1^{(e)}}=\frac{x-x_2^{(e)}-x+x_1^{(e)}}{x_1^{(e)}-x_2^{(e)}}=1$$

This is another basic property of the element shape functions that we will consider. Specifically, the sum of the shape functions for any location in the interior of the element is equal to 1. This property will be called *the partition of unity property* of the shape functions, and it is mathematically expressed as follows.

$$\sum_{i=1}^{n}\left(N_i^{(e)}(x)\right)=1,\ \text{for any value } x \text{ in the interior of element } e \tag{3.2.9}$$

where n is the number of shape functions in the element.

Once the shape functions have been established, we can also obtain the approximate expression for the derivative of the field function υ in an element, by differentiating Equation (3.2.5a):

$$\frac{d\upsilon^{(e)}}{dx}=\frac{d}{dx}\left(N_1^{(e)}(x)\cdot\upsilon_1^{(e)}+N_2^{(e)}(x)\cdot\upsilon_2^{(e)}\right)=\frac{dN_1^{(e)}}{dx}\upsilon_1^{(e)}+\frac{dN_2^{(e)}}{dx}\cdot\upsilon_2^{(e)} \tag{3.2.10}$$

where we have accounted for the fact that the nodal values $\upsilon_1^{(e)}$ and $\upsilon_2^{(e)}$ are constant.

We can also cast Equation (3.2.10) in matrix form to obtain:

$$\frac{d\upsilon^{(e)}}{dx}=\left[\frac{dN_1^{(e)}}{dx}\ \ \frac{dN_2^{(e)}}{dx}\right]\begin{Bmatrix}\upsilon_1^{(e)}\\ \upsilon_2^{(e)}\end{Bmatrix}=\left[B^{(e)}\right]\begin{Bmatrix}\upsilon_1^{(e)}\\ \upsilon_2^{(e)}\end{Bmatrix} \tag{3.2.11}$$

where the array $[B^{(e)}]$ gives the first derivative of the approximate field, by premultiplying the vector with the nodal values of the field. For elasticity problems, the $[B^{(e)}]$ array is termed the *strain-displacement array*. We can take the derivatives of the shape functions and write:

$$[B^e] = \left[\frac{dN_1^{(e)}}{dx} \quad \frac{dN_2^{(e)}}{dx} \right] = \left[\frac{1}{x_1^{(e)} - x_2^{(e)}} \quad \frac{1}{x_2^{(e)} - x_1^{(e)}} \right] = \left[-\frac{1}{\ell^{(e)}} \quad \frac{1}{\ell^{(e)}} \right] \tag{3.2.12}$$

where $\ell^{(e)} = x_2^{(e)} - x_1^{(e)}$ is the length of the element.

Remark 3.2.3: The $[B^{(e)}]$ array of a two-node element is *constant*. ■

We will now move on and establish the same expressions describing the piecewise approximation for a *three-node* element, schematically shown in Figure 3.6. The third node of that element does not have to lie at the middle, as also shown in Figure 3.6.

For such an element, the function $v^{(e)}(x)$ will be a *quadratic polynomial* function:

$$v^{(e)}(x) = a_0^{(e)} + a_1^{(e)} \cdot x + a_2^{(e)} \cdot x^2 \tag{3.2.13}$$

We can once again express $v^{(e)}(x)$ such that the constant parameters of the approximation are the nodal values $v_1^{(e)}$, $v_2^{(e)}$, and $v_3^{(e)}$:

$$\left.\begin{array}{l} v^{(e)}\left(x = x_1^{(e)}\right) = a_0^{(e)} + a_1^{(e)} \cdot x_1^{(e)} + a_2^{(e)} \cdot \left(x_1^{(e)}\right)^2 = v_1^{(e)} \\ v^{(e)}\left(x = x_2^{(e)}\right) = a_0^{(e)} + a_1^{(e)} \cdot x_2^{(e)} + a_2^{(e)} \cdot \left(x_2^{(e)}\right)^2 = v_2^{(e)} \\ v^{(e)}\left(x = x_3^{(e)}\right) = a_0^{(e)} + a_1^{(e)} \cdot x_3^{(e)} + a_2^{(e)} \cdot \left(x_3^{(e)}\right)^2 = v_3^{(e)} \end{array}\right\} \rightarrow$$

$$\rightarrow \begin{bmatrix} 1 & x_1^{(e)} & \left(x_1^{(e)}\right)^2 \\ 1 & x_2^{(e)} & \left(x_2^{(e)}\right)^2 \\ 1 & x_3^{(e)} & \left(x_3^{(e)}\right)^2 \end{bmatrix} \begin{Bmatrix} a_0^{(e)} \\ a_1^{(e)} \\ a_2^{(e)} \end{Bmatrix} = \begin{Bmatrix} v_1^{(e)} \\ v_2^{(e)} \\ v_3^{(e)} \end{Bmatrix} \rightarrow [M]\left\{a^{(e)}\right\} = \left\{v^{(e)}\right\} \tag{3.2.14}$$

where

$$[M] = \begin{bmatrix} 1 & x_1^{(e)} & \left(x_1^{(e)}\right)^2 \\ 1 & x_2^{(e)} & \left(x_2^{(e)}\right)^2 \\ 1 & x_3^{(e)} & \left(x_3^{(e)}\right)^2 \end{bmatrix} \tag{3.2.15a}$$

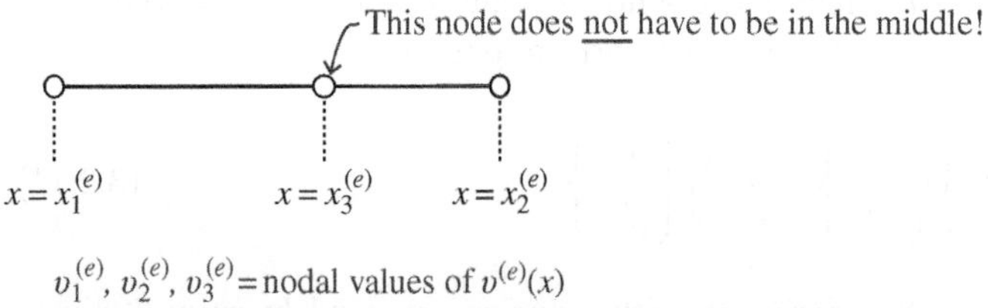

This node does not have to be in the middle!

$x = x_1^{(e)}$ $x = x_3^{(e)}$ $x = x_2^{(e)}$

$v_1^{(e)}, v_2^{(e)}, v_3^{(e)}$ = nodal values of $v^{(e)}(x)$

Figure 3.6 Three-node finite element.

$$\left\{a^{(e)}\right\} = \left[a_0^{(e)} \ a_1^{(e)} \ a_2^{(e)}\right]^T \tag{3.2.15b}$$

and

$$\left\{\upsilon^{(e)}\right\} = \left[\upsilon_1^{(e)} \ \upsilon_2^{(e)} \ \upsilon_3^{(e)}\right]^T \tag{3.2.15c}$$

We eventually obtain:

$$\left\{a^{(e)}\right\} = [M]^{-1}\left\{\upsilon^{(e)}\right\} \tag{3.2.16}$$

Now, we can write Equation (3.2.13) in the following form.

$$\upsilon^{(e)}(x) = a_0^{(e)} + a_1^{(e)} \cdot x + a_2^{(e)} \cdot x^2 = \left[1 \ x \ x^2\right] \begin{Bmatrix} a_0^{(e)} \\ a_1^{(e)} \\ a_2^{(e)} \end{Bmatrix} = [p(x)] \left\{a^{(e)}\right\} \tag{3.2.17}$$

where $[p(x)]$ is a row vector containing the various monomial terms appearing in Equation (3.2.13). Plugging Equation (3.2.16) into (3.2.17) yields:

$$\begin{aligned} \upsilon^{(e)}(x) &= [p(x)]\,[M]^{-1}\left\{\upsilon^{(e)}\right\} = \left[N^{(e)}(x)\right]\left\{\upsilon^{(e)}\right\} \\ &= N_1^{(e)}(x) \cdot \upsilon_1^{(e)} + N_2^{(e)}(x) \cdot \upsilon_2^{(e)} + N_3^{(e)}(x) \cdot \upsilon_3^{(e)} \end{aligned} \tag{3.2.18}$$

where $[N^{(e)}(x)]$ is the *shape function array* for the three-node element, given by the following expression.

$$\left[N^{(e)}(x)\right] = [p(x)]\,[M]^{-1} = \left[N_1^{(e)}(x) \ \ N_2^{(e)}(x) \ \ N_3^{(e)}(x)\right] \tag{3.2.19}$$

If we actually do the math, we obtain the following expressions for the three shape functions of the element.

$$N_1^{(e)}(x) = \frac{\left(x - x_2^{(e)}\right)\left(x - x_3^{(e)}\right)}{\left(x_1^{(e)} - x_2^{(e)}\right)\left(x_1^{(e)} - x_3^{(e)}\right)} \tag{3.2.20a}$$

$$N_2^{(e)}(x) = \frac{\left(x - x_1^{(e)}\right)\left(x - x_3^{(e)}\right)}{\left(x_2^{(e)} - x_1^{(e)}\right)\left(x_2^{(e)} - x_3^{(e)}\right)} \tag{3.2.20b}$$

$$N_3^{(e)}(x) = \frac{\left(x - x_1^{(e)}\right)\left(x - x_2^{(e)}\right)}{\left(x_3^{(e)} - x_1^{(e)}\right)\left(x_3^{(e)} - x_2^{(e)}\right)} \tag{3.2.20c}$$

The plots for the three shape functions, which are parabolic functions (quadratic polynomials), are schematically presented in Figure 3.7.

We can easily verify that the shape functions given by Equations (3.2.20a–c) satisfy the Kronecker delta property. Also, it can be verified that the partition of unity property is satisfied.

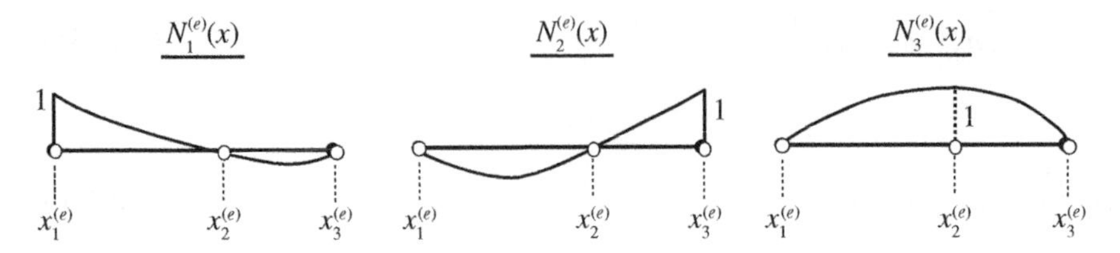

Figure 3.7 Shape functions for three-node finite element.

Remark 3.2.4: Notice something peculiar about the quadratic shape functions of the three-node element: for some of them, *the minima are not obtained at the nodes*! For actual analysis, the same may apply for the maxima of the piecewise approximation. This is one of the reasons that, for several classes of complicated problems, high-order polynomial approximations are not as popular as linear approximations (for linear shape functions, the minimum and maximum values can only be obtained at the nodes). ■

We will now establish the shape functions, for the general case that we have an element with n nodal points. An efficient way to obtain the *polynomial* shape functions for any order of approximation is to use the *Lagrange polynomial interpolation formulas*. These formulas give polynomial shape functions that satisfy the Kroneker delta property and the partition of unity property for a set of n nodal points. The Lagrangian interpolation is given by the following expression.

$$\upsilon^{(e)}(x) = \sum_{i=1}^{n} \left(N_i^{(e)}(x) \cdot \upsilon_i^{(e)} \right) \tag{3.2.21}$$

where

$$N_i^{(e)}(x) = \prod_{\substack{j=1 \\ j \neq i}}^{n} \frac{\left(x - x_j^{(e)}\right)}{\left(x_i^{(e)} - x_j^{(e)}\right)}, \tag{3.2.22}$$

and

$$\prod_{\substack{j=1 \\ j \neq i}}^{n} \frac{\left(x - x_j^{(e)}\right)}{\left(x_i^{(e)} - x_j^{(e)}\right)} = \frac{\left(x - x_1^{(e)}\right)\left(x - x_2^{(e)}\right)\ldots\left(x - x_{i-1}^{(e)}\right)\left(x - x_{i+1}^{(e)}\right)\ldots\left(x - x_{n-1}^{(e)}\right)\left(x - x_n^{(e)}\right)}{\left(x_i^{(e)} - x_1^{(e)}\right)\left(x_i^{(e)} - x_2^{(e)}\right)\ldots\left(x_i^{(e)} - x_{i-1}^{(e)}\right)\left(x_i^{(e)} - x_{i+1}^{(e)}\right)\ldots\left(x_i^{(e)} - x_{n-1}^{(e)}\right)\left(x_i^{(e)} - x_n^{(e)}\right)} \tag{3.2.23}$$

For example, if we set $n = 2$, we obtain:

$$\upsilon^{(e)}(x) = N_1^{(e)}(x) \cdot \upsilon_1^{(e)} + N_2^{(e)}(x) \cdot \upsilon_2^{(e)}, \quad N_1^{(e)}(x) = \frac{\left(x - x_2^{(e)}\right)}{\left(x_1^{(e)} - x_2^{(e)}\right)}, \quad N_2^{(e)}(x) = \frac{\left(x - x_1^{(e)}\right)}{\left(x_2^{(e)} - x_1^{(e)}\right)}$$

One can verify that these are the same expressions as those presented in Equations (3.2.5a), (3.2.7a), and (3.2.7b), for a two-node element.

For the case of general Lagrangian interpolation, we can also establish an expression for the derivative of the approximate field:

$$\frac{dv^{(e)}}{dx} = \frac{d}{dx}\sum_{i=1}^{n}\left[N_i^{(e)}(x)\cdot v_i^{(e)}\right] = \sum_{i=1}^{n}\left(\frac{dN_i^{(e)}}{dx}\cdot v_i^{(e)}\right) = \sum_{i=1}^{n}\left[B_i^{(e)}(x)\cdot v_i^{(e)}\right] \tag{3.2.24}$$

where

$$B_i^{(e)}(x) = \frac{dN_i^{(e)}}{dx} \tag{3.2.25}$$

If we establish the vector with the element nodal values, $\left\{v^{(e)}\right\} = \left[v_1^{(e)} \ v_2^{(e)} \ \ldots \ v_n^{(e)}\right]^T$, we can cast Equations (3.2.21) and (3.2.24) in matrix form:

$$v^{(e)}(x) = \left[N^{(e)}\right]\left\{v^{(e)}\right\} \tag{3.2.26}$$

where

$$\left[N^{(e)}\right] = \left[N_1^{(e)}(x) \ N_2^{(e)}(x) \ \ldots \ N_n^{(e)}(x)\right] \tag{3.2.27}$$

is the shape function array of the element, and

$$\frac{dv^{(e)}}{dx} = \left[B^{(e)}\right]\left\{v^{(e)}\right\} \tag{3.2.28}$$

where

$$\left[B^{(e)}\right] = \left[\frac{dN_1^{(e)}}{dx} \ \frac{dN_2^{(e)}}{dx} \ \ldots \ \frac{dN_n^{(e)}}{dx}\right] \tag{3.2.29}$$

One can verify that, for the special case where $n = 2$, Equations (3.2.26) and (3.2.28) yield Equations (3.2.5a) and (3.2.11), respectively.

Example 3.1: Lagrange polynomial shape functions
In this example, we will establish the shape functions of the four-node element shown in Figure 3.8a. The case at hand corresponds to $n = 4$ (n is the number of nodes in

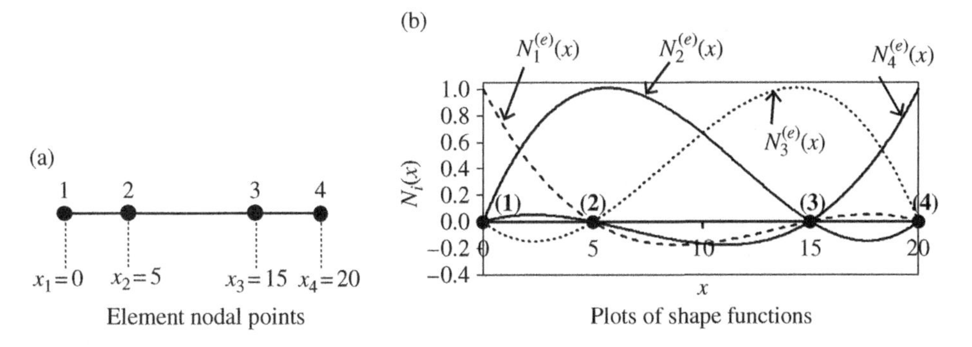

Figure 3.8 Shape functions for example four-node element.

the element). We can use the formula of Equation (3.2.22) for each nodal point i, where $i = 1, 2, 3, 4$:

$$N_i^{(e)}(x) = \prod_{\substack{j=1 \\ j \neq i}}^{n} \frac{(x - x_j)}{(x_i - x_j)} = \prod_{\substack{j=1 \\ j \neq i}}^{4} \frac{(x - x_j)}{(x_i - x_j)}$$

For $i = 1$, we can find the shape function of node 1:

$$N_1^{(e)}(x) = \prod_{\substack{j=1 \\ j \neq 1}}^{4} \frac{(x - x_j)}{(x_1 - x_j)} = \frac{(x-x_2)(x-x_3)(x-x_4)}{(x_1-x_2)(x_1-x_3)(x_1-x_4)} = \frac{(x-5)(x-15)(x-20)}{(0-5)(0-15)(0-20)}$$

$$= -\frac{(x-5)(x-15)(x-20)}{1500}$$

For $i = 2$, we have the shape function of node 2:

$$N_2^{(e)}(x) = \prod_{\substack{j=1 \\ j \neq 2}}^{4} \frac{(x - x_j)}{(x_2 - x_j)} = \frac{(x-x_1)(x-x_3)(x-x_4)}{(x_2-x_1)(x_2-x_3)(x_2-x_4)} = \frac{(x-0)(x-15)(x-20)}{(5-0)(5-15)(5-20)}$$

$$= \frac{x(x-15)(x-20)}{750}$$

The shape function for node 3 is obtained for $i = 3$:

$$N_3^{(e)}(x) = \prod_{\substack{j=1 \\ j \neq 3}}^{4} \frac{(x - x_j)}{(x_3 - x_j)} = \frac{(x-x_1)(x-x_2)(x-x_4)}{(x_3-x_1)(x_3-x_2)(x_3-x_4)} = \frac{(x-0)(x-5)(x-20)}{(15-0)(15-5)(15-20)}$$

$$= -\frac{x(x-5)(x-20)}{750}$$

The shape function of the fourth and final node is obtained for $i = 4$:

$$N_4^{(e)}(x) = \prod_{\substack{j=1 \\ j \neq 4}}^{4} \frac{(x - x_j)}{(x_4 - x_j)} = \frac{(x-x_1)(x-x_2)(x-x_3)}{(x_4-x_1)(x_4-x_2)(x_4-x_3)} = \frac{(x-0)(x-5)(x-15)}{(20-0)(20-5)(20-15)}$$

$$= \frac{x(x-5)(x-15)}{1500}$$

Thus, the shape function array for the four-node element will be:

$$\left[N^{(e)}\right] = \left[N_1^{(e)}(x) \;\; N_2^{(e)}(x) \;\; N_3^{(e)}(x) \;\; N_4^{(e)}(x)\right] \rightarrow$$

$$\rightarrow \left[N^{(e)}\right] = \left[-\frac{(x-5)(x-15)(x-20)}{1500} \;\; \frac{x(x-15)(x-20)}{750} \;\; -\frac{x(x-5)(x-20)}{750} \;\; \frac{x(x-5)(x-15)}{1500}\right]$$

We can plot the variation of our shape functions along the element, as shown in Figure 3.8b.

We can easily verify that our shape functions satisfy:

- The Kronecker delta property
- The partition of unity property

We can also readily obtain the components of the $[B^{(e)}]$-array of the element:

$$\left[B^{(e)}\right] = \left[\frac{dN_1^{(e)}}{dx} \quad \frac{dN_2^{(e)}}{dx} \quad \frac{dN_3^{(e)}}{dx} \quad \frac{dN_4^{(e)}}{dx}\right] \rightarrow$$

$$\rightarrow \left[B^{(e)}\right] = \left[-\frac{3x^2-80x+475}{1500} \quad \frac{3x^2-70x+300}{750} \quad -\frac{3x^2-50x+100}{750} \quad \frac{3x^2-40x+75}{1500}\right]$$ ∎

Remark 3.2.5: The polynomial shape functions for the four-node element in Example 3.1 are third-degree polynomials. For the two-node element, the shape functions are first-degree (linear) polynomials, while the shape functions of a three-node element are second-degree (quadratic) polynomials. Per Equation (3.2.23), if we have a one-dimensional element with n nodes using Lagrangian polynomial shape functions, then each shape function will be a polynomial of degree $n-1$. ∎

3.3 Discrete Equations for Piecewise Finite Element Approximation

This section will show how the introduction of the piecewise finite element approximation to the weak form of the problem transforms the integral equation of the weak form to a system of linear equations. The reader is strongly advised to become familiar with the material presented in Appendix B (especially Sections B.1, B.2, and B.3), before proceeding further.

Let us consider the case that we have discretized a domain under consideration into N_e elements, as shown in Figure 3.9.

Combining all the element subdomains yields the entire domain. This fact allows us to decompose each domain integral into the corresponding element contributions. Specifically, for any integral in the domain, we can write:

$$\int_0^L \ldots dx = \int_{x_1^{(1)}=0}^{x_2^{(1)}} \ldots dx + \int_{x_1^{(2)}}^{x_2^{(2)}} \ldots dx + \ldots + \int_{x_1^{(Ne)}}^{x_2^{(Ne)}=L} \ldots dx \tag{3.3.1}$$

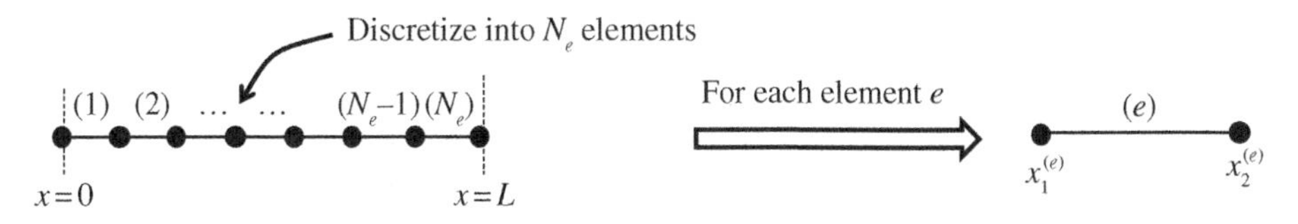

Figure 3.9 Finite element discretization for model with N_e elements.

Remark 3.3.1: As we will see later on, there is a way to numerically calculate the values of the element contributions to the integral of Equation (3.3.1). ▮

Let us assume that we have a one-dimensional elasticity problem to solve. The weak form for one-dimensional elasticity, which was presented in Box 2.3.2, includes the following integral expression.

$$\int_0^L \frac{dw}{dx} EA \frac{du}{dx} dx = \int_0^L w \cdot b dx + \left[w \cdot A \cdot \bar{t}\right]_{\Gamma_t} \tag{3.3.2}$$

We can use Equation (3.3.1) for the left-hand side of the weak form, to obtain:

$$\int_0^L \frac{dw}{dx} EA \frac{du}{dx} dx = \int_{x_1^{(1)}=0}^{x_2^{(1)}} \frac{dw}{dx} EA \frac{du}{dx} dx + \ldots + \int_{x_1^{(Ne)}}^{x_2^{(Ne)}=L} \frac{dw}{dx} EA \frac{du}{dx} dx = \sum_{e=1}^{N_e} \left(\int_{x_1^{(e)}}^{x_2^{(e)}} \frac{dw}{dx} EA \frac{du}{dx} dx \right) \tag{3.3.3}$$

Notice that each of the integral terms in Equation (3.3.3) corresponds to a contribution of an element to the left-hand side of the weak form. We will now express the effect of our piecewise finite element approximation. Specifically, $u(x)$ will be obtained from a global nodal displacement vector, $\{U\}$, in the finite element mesh, as shown in Figure 3.10. Thus, the vector $\{U\}$ has the values of axial displacement at all the nodal points of the mesh. Along the same lines, as shown in the same figure, $w(x)$ will be obtained from a global vector of nodal values, $\{W\}$, in the finite element mesh. As also shown in Figure 3.10 (and in accordance with Section B.2), we can obtain the nodal displacement vector, $\{U^{(e)}\}$, and weighting function vector, $\{W^{(e)}\}$, of each element e, by means of a gather operation, that is, by premultiplying the corresponding global vectors by the gather array, $[L^{(e)}]$, of the element.

To obtain the finite element solution, we only need to know the value of $\{U\}$. Thus, $\{U\}$ contains a set of unknown (constant) parameters of our solution. One of the advantages of using the nodal values $\{U\}$ as our unknown parameters is that we can easily satisfy the essential boundary conditions by setting the displacements of the nodal points that lie on the essential boundary equal to the corresponding prescribed values, in accordance with the rule provided in Box 3.3.1.

Additionally, since $w(x)$ is completely described (as we will explain) from the nodal values $\{W\}$ in our mesh, the statement "$\forall\, w(x)$, which vanishes on the essential boundary"

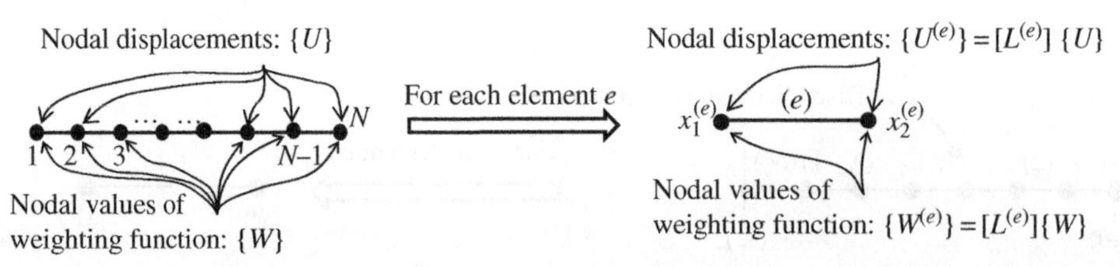

Figure 3.10 Global and element nodal values for finite element analysis.

Box 3.3.1 Treatment of Essential Boundary Conditions in One-Dimensional Finite Element Analysis

If nodal point m lies at the essential boundary Γ_u, for which we have $u = \bar{u}$, then we can directly set the nodal displacement U_m equal to the prescribed value $\bar{u}$.

in the weak form of Box 2.3.1, will be replaced by "$\forall$ $\{W\}$, such that it vanishes at *all the nodes that lie on the essential boundary*." Thus, we can establish the considerations provided in Box 3.3.2.

Box 3.3.2 Values of $\{W\}$ for Nodal Points on the Essential Boundary

If nodal point m lies at the essential boundary Γ_u, for which we have $w = 0$, then we can directly set the nodal weighting function value W_m to zero.

We will now examine each integral element contribution to Equation (3.3.3)—that is, the term $\int_{x_1^{(e)}=0}^{x_2^{(e)}} \frac{dw}{dx} EA \frac{du}{dx} dx$. Since this term corresponds to integration in the interior of the element e, we can plug the piecewise finite element approximation for the displacement field:

$$u(x) \approx u^{(e)}(x) = \sum_{j=1}^{n} \left(N_j^{(e)}(x) \cdot u_j^{(e)} \right) = \left[N^{(e)}\right]\left\{U^{(e)}\right\} \tag{3.3.4}$$

and for the strain field:

$$\frac{du}{dx} \approx \frac{du^{(e)}}{dx} = \sum_{j=1}^{n} \left(\frac{dN_j^{(e)}}{dx} \cdot u_j^{(e)} \right) = \sum_{j=1}^{n} \left(B_j^{(e)} \cdot u_j^{(e)} \right) = \left[B^{(e)}\right]\left\{U^{(e)}\right\} \tag{3.3.5}$$

Along the same lines, the approximate variation of $w(x)$ in element e is given by the same piecewise approximation—the same exact shape functions—as for the displacement field:

$$w(x) \approx w^{(e)}(x) = \sum_{i=1}^{n} \left(N_i^{(e)}(x) \cdot w_i^{(e)} \right) = \left[N^{(e)}\right]\left\{W^{(e)}\right\} \tag{3.3.6a}$$

$$\frac{dw}{dx} \approx \frac{dw^{(e)}}{dx} = \sum_{i=1}^{n} \left(\frac{dN_i^{(e)}}{dx} \cdot w_i^{(e)} \right) = \sum_{i=1}^{n} \left(B_i^{(e)} \cdot w_i^{(e)} \right) = \left[B^{(e)}\right]\left\{W^{(e)}\right\} \tag{3.3.6b}$$

For one-dimensional problems, $[N^{(e)}]$ is a row vector and $\{W^{(e)}\}$ is a column vector. This means that $\{W^{(e)}\}^T$ is a row vector and $[N^{(e)}]^T$ is a column vector, and the approximation for the field $w(x)$ can also be written as:

$$w^{(e)}(x) = \left\{W^{(e)}\right\}^T \left[N^{(e)}\right]^T \tag{3.3.7a}$$

Similarly, we can write:

$$\frac{dw^{(e)}}{dx} = \left\{W^{(e)}\right\}^T \left[B^{(e)}\right]^T \tag{3.3.7b}$$

We will prefer using Equations (3.3.7a,b) instead of Equations (3.3.6a,b). The reason will become apparent after we derive the finite element equations, as explained in Remark 3.3.2.

Now, we can operate on the integral of each element e in Equation (3.3.3). We can use Equations (3.3.4), (3.3.5) and (3.3.7a,b) to obtain:

$$\int_{x_1^{(e)}}^{x_2^{(e)}} \frac{dw}{dx} EA \frac{du}{dx} dx \approx \int_{x_1^{(e)}}^{x_2^{(e)}} \frac{dw^{(e)}}{dx} E^{(e)} A^{(e)} \frac{du^{(e)}}{dx} dx = \int_{x_1^{(e)}}^{x_2^{(e)}} \left\{W^{(e)}\right\}^T \left[B^{(e)}\right]^T E^{(e)} A^{(e)} \left[B^{(e)}\right] \left\{U^{(e)}\right\} dx \tag{3.3.8}$$

Now, for Equation (3.3.8), we can take the vectors with the nodal values, $\{U^{(e)}\}$ and $\{W^{(e)}\}^T$, outside of the integral, because they contain constant quantities! Thus, we have:

$$\int_{x_1^{(e)}}^{x_2^{(e)}} \frac{dw}{dx} E \cdot A \frac{du}{dx} dx \approx \left\{W^{(e)}\right\}^T \left(\int_{x_1^{(e)}}^{x_2^{(e)}} \left[B^{(e)}\right]^T E^{(e)} A^{(e)} \left[B^{(e)}\right] dx \right) \left\{U^{(e)}\right\} = \left\{W^{(e)}\right\}^T \left[k^{(e)}\right] \left\{U^{(e)}\right\} \tag{3.3.9}$$

where $[k^{(e)}]$ is called the *coefficient matrix of element e*. For the specific case of elasticity problems, $[k^{(e)}]$ is called the *stiffness matrix of element e* and is given by the following equation.

$$\left[k^{(e)}\right] = \int_{x_1^{(e)}}^{x_2^{(e)}} \left[B^{(e)}\right]^T E^{(e)} A^{(e)} \left[B^{(e)}\right] dx \tag{3.3.10}$$

Along the same lines, we can separate the first-term of the right-hand side in Equation (3.3.2) in the element contributions, and then plug the finite element approximation of $w(x)$ (given by Equation 3.3.7a) in each term to obtain:

$$\int_0^L w \cdot b\, dx = \sum_{e=1}^{N_e} \left(\int_{x_1^{(e)}}^{x_2^{(e)}} w\, b\, dx \right) = \sum_{e=1}^{N_e} \left(\int_{x_1^{(e)}}^{x_2^{(e)}} w^{(e)}\, b\, dx \right) = \sum_{e=1}^{N_e} \left(\int_{x_1^{(e)}}^{x_2^{(e)}} \left\{W^{(e)}\right\}^T \left[N^{(e)}\right]^T b\, dx \right) \rightarrow$$

$$\rightarrow \int_0^L w \cdot b\, dx = \sum_{e=1}^{N_e} \left(\left\{W^{(e)}\right\}^T \int_{x_1^{(e)}}^{x_2^{(e)}} \left[N^{(e)}\right]^T b\, dx \right) \tag{3.3.11a}$$

Notice that, to obtain Equation (3.3.11a), we have accounted for the fact that $\{W^{(e)}\}^T$ does not vary with x.

Finally, we can also separate the natural boundary term of Equation (3.3.2) into element contributions:

$$[w \cdot A \cdot \bar{t}]_{\Gamma_t} = \sum_{e=1}^{N_e} \left([w \cdot A \cdot \bar{t}]_{\Gamma_t^{(e)}} \right) \tag{3.3.11b}$$

The contribution of each element e to the boundary term in Equation (3.3.11b) can be obtained based on the following rationale: If element e contains natural boundaries (i.e., one of the end points of element e belongs on the natural boundary), then $[w \cdot A \cdot \bar{t}]_{\Gamma_t^{(e)}}$ for element e will be nonzero; otherwise, if element e does not contain natural boundary locations, then $[w \cdot A \cdot \bar{t}]_{\Gamma_t^{(e)}}$ will be zero for that element.

Now, we can plug the approximate $w(x)$ given by Equation (3.3.7a) into Equation (3.3.11b) to obtain:

$$\begin{aligned} &[w \cdot A \cdot \bar{t}]_{\Gamma_t} = \sum_{e=1}^{N_e} \left([w \cdot A \cdot \bar{t}]_{\Gamma_t^{(e)}} \right) = \sum_{e=1}^{N_e} \left(\left[w^{(e)}(x) \cdot A \cdot \bar{t} \right]_{\Gamma_t^{(e)}} \right) \rightarrow \\ &\rightarrow [w \cdot A \cdot \bar{t}]_{\Gamma_t} = \sum_{e=1}^{N_e} \left(\left\{ W^{(e)} \right\}^T \left[\left[N^{(e)} \right]^T \cdot A \cdot \bar{t} \right]_{\Gamma_t^{(e)}} \right) \end{aligned} \tag{3.3.12}$$

where the vector $\{W^{(e)}\}^T$ has been taken outside of the boundary evaluation, since the nodal values $\{W^{(e)}\}$ do not vary with x.

Based on the above, the contribution of each element e to the right-hand side of Equation (3.3.2) can be found by summing the right-hand sides of Equations (3.3.11a) and (3.3.12):

$$\begin{aligned} &\int_{x_1^{(e)}}^{x_2^{(e)}} w^{(e)}\, b\, dx + [w \cdot A \cdot \bar{t}]_{\Gamma_t^{(e)}} = \left\{ W^{(e)} \right\}^T \int_{x_1^{(e)}}^{x_2^{(e)}} \left[N^{(e)} \right]^T b dx + \left\{ W^{(e)} \right\}^T \left[\left[N^{(e)} \right]^T \cdot A \cdot \bar{t} \right]_{\Gamma_t} \rightarrow \\ &\rightarrow \int_{x_1^{(e)}}^{x_2^{(e)}} w^{(e)}\, b\, dx + [w \cdot A \cdot \bar{t}]_{\Gamma_t^{(e)}} = \left\{ W^{(e)} \right\}^T \left[\int_{x_1^{(e)}}^{x_2^{(e)}} \left[N^{(e)} \right]^T b dx + \left[\left[N^{(e)} \right]^T \cdot A \cdot \bar{t} \right]_{\Gamma_t} \right] \rightarrow \\ &\rightarrow \int_{x_1^{(e)}}^{x_2^{(e)}} w^{(e)}\, b\, dx + [w \cdot A \cdot \bar{t}]_{\Gamma_t^{(e)}} = \left\{ W^{(e)} \right\}^T \{f^{(e)}\} \end{aligned} \tag{3.3.13}$$

where $\{f^{(e)}\}$ is called *the equivalent right-hand-side vector of element e. For elasticity problems, $\{f^{(e)}\}$ is called the equivalent nodal force vector of element e.* We have:

$$\left\{ f^{(e)} \right\} = \int_{x_1^{(e)}}^{x_2^{(e)}} \left[N^{(e)} \right]^T b dx + \left[\left[N^{(e)} \right]^T \cdot A \cdot \bar{t} \right]_{\Gamma_t} \tag{3.3.14}$$

Plugging Equations (3.3.3), (3.3.9), (3.3.11a,b), and (3.3.13) in Equation (3.3.2), we eventually obtain:

$$\sum_{e=1}^{N_e}\left[\left\{W^{(e)}\right\}^T\left[k^{(e)}\right]\left\{U^{(e)}\right\}\right]=\sum_{e=1}^{N_e}\left(\left\{W^{(e)}\right\}^T\left\{f^{(e)}\right\}\right)$$

$$\rightarrow\sum_{e=1}^{N_e}\left[\left\{W^{(e)}\right\}^T\left[k^{(e)}\right]\left\{U^{(e)}\right\}\right]-\sum_{e=1}^{N_e}\left(\left\{W^{(e)}\right\}^T\left\{f^{(e)}\right\}\right)=0 \tag{3.3.15}$$

In accordance with Figure 3.10 and our discussion of matrix analysis provided in Section B.2, the nodal values of displacement and weighting function for element e can be obtained from a gather operation, using the global nodal values of the structure and the gather array $[L^{(e)}]$ of the element:

$$\left\{U^{(e)}\right\}=\left[L^{(e)}\right]\{U\} \tag{3.3.16}$$

$$\left\{W^{(e)}\right\}=\left[L^{(e)}\right]\{W\} \tag{3.3.17}$$

Now, we notice that Equation (3.3.15) contains the transpose of $\{W^{(e)}\}$. We can then use Equation (A.2.11) in Appendix A to obtain

$$\left\{W^{(e)}\right\}^T=\{W\}^T\left[L^{(e)}\right]^T \tag{3.3.18}$$

If we finally plug Equations (3.3.16) and (3.3.18) into Equation (3.3.15), we have the following expression:

$$\sum_{e=1}^{N_e}\left(\{W\}^T\left[L^{(e)}\right]^T\left[k^{(e)}\right]\left[L^{(e)}\right]\{U\}\right)-\sum_{e=1}^{N_e}\left(\{W\}^T\left[L^{(e)}\right]^T\left\{f^{(e)}\right\}\right)=0 \tag{3.3.19}$$

The global vectors $\{W\}^T$ and $\{U\}$ can be taken outside of the summations (since they are common terms for all elements, we take them outside the summation as *common factors!*):

$$\{W\}^T\sum_{e=1}^{N_e}\left(\left[L^{(e)}\right]^T\left[k^{(e)}\right]\left[L^{(e)}\right]\right)\{U\}-\{W\}^T\sum_{e=1}^{N_e}\left(\left[L^{(e)}\right]^T\left\{f^{(e)}\right\}\right)=0$$

$$\rightarrow\{W\}^T\left[\sum_{e=1}^{N_e}\left(\left[L^{(e)}\right]^T\left[k^{(e)}\right]\left[L^{(e)}\right]\right)\{U\}-\sum_{e=1}^{N_e}\left(\left[L^{(e)}\right]^T\left\{f^{(e)}\right\}\right)\right]=0$$

Finally, we obtain:

$$\{W\}^T[[K]\{U\}-\{f\}]=0\ \forall\ \{W\} \tag{3.3.20}$$

where $[K]$ is called the *global (or structural) stiffness matrix* and $\{f\}$ is called the *equivalent nodal force vector.* In general, $[K]$ and $\{f\}$ are called the *global coefficient matrix* and *the global right-hand-side vector*, respectively, for the finite element equations, and they can be obtained from the element coefficient arrays by means of an assembly operation (as also explained in Section B.2):

$$[K]=\sum_{e=1}^{N_e}\left(\left[L^{(e)}\right]^T\left[k^{(e)}\right]\left[L^{(e)}\right]\right) \tag{3.3.21a}$$

$$\{f\} = \sum_{e=1}^{N_e}\left(\left[L^{(e)}\right]^T\left\{f^{(e)}\right\}\right) \tag{3.3.21b}$$

Now, let us express the obtained matrix expression (3.3.20) as a summation, assuming that we have a total of N nodal displacements in our structure:

$$\sum_{i=1}^{N} W_i\left[\sum_{j=1}^{N}(K_{ij}U_j) - f_i\right] = 0 \ \forall W_i,\ i = 1,2,\ldots,\ N \tag{3.3.22}$$

Equation (3.3.22) can be expanded as follows:

$$W_1\left(\sum_{j=1}^{n}K_{1j}U_j - f_1\right) + W_2\left(\sum_{j=1}^{n}K_{2j}U_j - f_2\right) + \ldots + W_n\left(\sum_{j=1}^{n}K_{nj}U_j - f_n\right) = 0$$
$$\forall W_i,\ i = 1,2,\ldots,N \tag{3.3.23}$$

Now, since W_1, W_2, ..., W_N in Equation (3.3.23) are *arbitrary*, the only way to *always* obtain a zero result in Equation (3.3.23) for all the possible values of the arbitrary parameters is to have each one of the terms in the brackets being equal to zero. In other words, we must satisfy:

$$\left.\begin{array}{l} \sum_{j=1}^{N}(K_{1j}U_j) - f_1 = 0 \\ \sum_{j=1}^{N}(K_{2j}U_j) - f_2 = 0 \\ \vdots \\ \sum_{j=1}^{N}(K_{Nj}U_j) - f_N = 0 \end{array}\right\} \rightarrow \begin{bmatrix} K_{11} & K_{12} & \ldots & K_{1N} \\ K_{21} & K_{22} & \ldots & K_{2N} \\ \vdots & \vdots & \ddots & \vdots \\ K_{N1} & K_{N2} & \ldots & K_{NN} \end{bmatrix}\begin{Bmatrix} U_1 \\ U_2 \\ \vdots \\ U_N \end{Bmatrix} = \begin{Bmatrix} f_1 \\ f_2 \\ \vdots \\ f_N \end{Bmatrix} \tag{3.3.24}$$

Thus, in all cases of finite element analysis, we will eventually obtain the global (structural) equations for the nodal values of the solution:

$$[K]\{U\} = \{f\} \tag{3.3.25}$$

right-hand side vector

Nodal values of finite element mesh

Coefficient matrix

We can then solve the system of Equations (3.3.25), to obtain the nodal values of the solutions (i.e., the nodal displacement vector $\{U\}$).

To sum up, in a finite element analysis, we can find the contribution of each element to the weak form, expressed through the element coefficient matrix $[k^{(e)}]$ and the right-hand-side vector $\{f^{(e)}\}$, then assemble the element contributions to obtain the global (structural) discrete equations, which we can solve for the nodal values of the solution (e.g., nodal displacements). For elasticity problems, the units of $\{f\}$ in Equation (3.3.25) are units of force, the units of $\{U\}$ are units of displacement, and the units of $[K]$ are units of *stiffness* (i.e., force per unit displacement).

Remark 3.3.2: The reason why we used Equations (3.3.7a,b) instead of Equations (3.3.6a,b) may now be apparent. Using the specific equations enabled us to take $\{W^{(e)}\}^T$ outside of each element integral and natural boundary term. We were then able to take the global vector $\{W\}^T$ outside of the summation in Equation (3.3.19), as a common factor of all the element terms. We thus obtained Equation (3.3.20), which allowed us to account for the fact that the components of $\{W\}$ are arbitrary. ▮

Remark 3.3.3: The element stiffness matrix obtained from Equation (3.3.10) is symmetric. This can be easily verified (by showing that the transpose of $[k^{(e)}]$ is equal to $[k^{(e)}]$ itself). ▮

Remark 3.3.4: The element stiffness matrix is *singular*. This is due to the fact that there are rigid-body modes, to which the element cannot develop strains, and consequently, it cannot develop any stiffness. For one-dimensional problems, we have a single rigid-body mode, that is, translation along the x-axis. ▮

Remark 3.3.5: The possibility for rigid-body modes means that the global stiffness matrix, $[K]$, is also singular. To obtain a stiffness matrix that is nonsingular (and thus invertible), we need to have adequate support conditions (restrained nodal displacements), so that the $[K_{ff}]$ submatrix that corresponds to the unrestrained nodal displacements (as explained in Section B.3 and Remark B.3.3) is nonsingular (invertible). ▮

Remark 3.3.6: If we treat each finite element as the domain of a problem, then we obtain the following, very useful equation, which is satisfied for the coefficient matrix of each element:

$$\left[k^{(e)}\right]\left\{U^{(e)}\right\} = \left\{f^{(e)}\right\} \tag{3.3.26}$$ ▮

3.4 Finite Element Equations for Heat Conduction

The procedure of the previous section can also be applied for establishing the finite element equations for one-dimensional, steady-state heat conduction. The only difference is that the unknown field is the temperature field; thus, we have to solve for a nodal temperature vector, $\{\Theta\}$. The global equations for heat conduction can be written as follows:

$$[K]\{\Theta\} = \{f\} \tag{3.4.1}$$

where $[K]$ is called the *global conductance matrix* and $\{f\}$ is called the *global equivalent nodal flux vector* of the system. $[K]$ and $\{f\}$ can be obtained by means of an assembly operation, from element conductance matrices and equivalent nodal flux vectors. That is, Equations (3.3.21a,b) are still valid for heat conduction, and we only need to provide definitions for $[k^{(e)}]$ and $\{f^{(e)}\}$:

$$\left[k^{(e)}\right] = \int_{x_1^{(e)}}^{x_2^{(e)}} \left[B^{(e)}\right]^T \kappa^{(e)} A^{(e)} \left[B^{(e)}\right] dx \tag{3.4.2a}$$

$$\left\{f^{(e)}\right\} = \int_{x_1^{(e)}}^{x_2^{(e)}} \left[N^{(e)}\right]^T s^{(e)}\, dx - \left[\left[N^{(e)}\right]^T \cdot A^{(e)} \cdot \bar{q}^{(e)}\right]_{\Gamma_q^{(e)}} \tag{3.4.2b}$$

Equations (3.4.2a,b) include material conductivity coefficient, κ, heat source term $s(x)$, and prescribed heat outflow $\bar{q}$, which are all defined in Section 2.5.

The proof of Equations (3.4.1) and (3.4.2a,b), obtained after plugging the piecewise finite element approximation to the weak form of Box 2.6.1, is left as an exercise.

Remark 3.4.1: Similar to considerations for one-dimensional elasticity, the element and global conductance matrices are symmetric and singular. The singularity is due to the fact that there are *zero-flux* modes, for which the element has a zero temperature derivative, and consequently, it cannot develop any heat flux or conductance. For heat conduction, we only have a single zero-flux mode (i.e., constant temperature). To ensure that we can obtain a nonsingular array $[K_{ff}]$ for the global discrete equations, we need to ensure that we have an adequate number of temperature restraints. ∎

3.5 Accounting for Nodes with Prescribed Solution Value ("Fixed" Nodes)

One last issue is how to treat nodes on which we have enforced an essential boundary condition (B.C.). For example, if a node m lies on the essential boundary for a one-dimensional elasticity problem, we need to enforce that $U_m = \bar{u}_m$ at node m, where $\bar{u}_m$ is the prescribed displacement at the essential boundary. Since the m-th global displacement, U_m, has a prescribed value (determined from an essential B.C.), then the approximate weighting function must vanish at that same nodal point, thus $W_m = 0$, and then we can *remove* the m-th equation from (3.3.24). This is equivalent to deleting the m-th row of the global equations and considering only the equations of the unrestrained (free) degrees of freedom, as we saw in Section B.3 (note that if U_m has a nonzero prescribed value, then it will still have an effect on our equations). If we do not want to delete the m-th equation corresponding to U_m, then we need to see how to treat the term:

$$W_m \left[\sum_{j=1}^{N} \left(K_{mj} U_j\right) - f_m\right], \text{ with } W_m = 0 \tag{3.5.1}$$

Since $W_m = 0$, the product term in Equation (3.5.1) *will always be zero* and the term in the square bracket *no longer needs to be zero* (remember how we obtained the stipulation that the terms in the brackets of Equation 3.3.23 must be zero!). Thus, we now write:

$$\sum_{j=1}^{N} \left(K_{mj} U_j\right) - f_m = r_m \tag{3.5.2}$$

where r_m is some unknown value. We can realize that r_m will simply be the *reaction* corresponding to U_m, that is, the force quantity allowing to enforce the prescribed fixed value of U_m. For the case of heat conduction—where the nodal field values are nodal temperatures—the quantity r_m will be the nodal restraint flux, enforcing the fixed value of temperature at the essential boundary node.

3.6 Examples on One-Dimensional Finite Element Analysis

We will now proceed to examine several examples, obtaining approximate solutions of one-dimensional elasticity and heat conduction problems with the finite element method.

Example 3.2: Finite element analysis for one-dimensional (1D) elasticity
Let us examine an example of a finite element solution for 1D elasticity. More specifically, we are interested in solving the bar shown in Figure 3.11, using two elements, with each element having two nodes (and linear shape functions).

The weak form of the specific problem is the following:

$$\int_0^{20} \frac{dw}{dx} EA \frac{du}{dx} dx = [wA\bar{t}]_{x=20} + \int_0^{20} w \cdot b(x) dx$$

where $b(x) = 2$, $E = 5000$, $A(x) = 1 + \frac{x^2}{400}$, $\bar{t} = -\frac{10}{[A]_{x=20}} = -\frac{10}{2} = -5\ ksi$, and $[A\bar{t} \cdot n]_{x=20} = 2 \cdot (-5) \cdot 1 = -10$

Thus, the weak form becomes:

$$\int_0^{20} \frac{dw}{dx} EA \frac{du}{dx} dx = [w]_{x=20} \cdot (-10) + \int_0^{20} w \cdot 2dx$$

Now, we can establish the piecewise finite element approximation in each element e:

$$u^{(e)}(x) = N_1^{(e)}(x) \cdot u_1^{(e)} + N_2^{(e)}(x) \cdot u_2^{(e)}, \quad \frac{du^{(e)}}{dx} = \frac{dN_1^{(e)}(x)}{dx} \cdot u_1^{(e)} + \frac{dN_2^{(e)}(x)}{dx} \cdot u_2^{(e)}$$

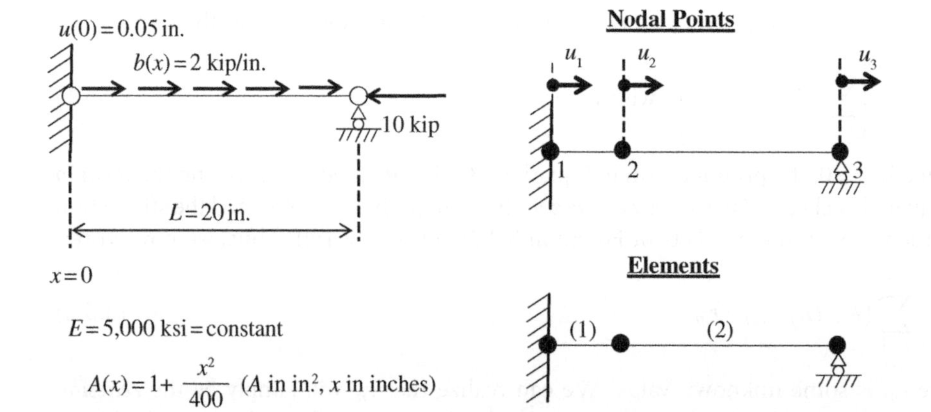

Figure 3.11 Example structure to be solved with finite elements.

We can also write in matrix form:

$$u^{(e)}(x) = \left[N^{(e)}\right]\begin{Bmatrix} u_1^{(e)} \\ u_2^{(e)} \end{Bmatrix} = \left[N^{(e)}\right]\left\{U^{(e)}\right\}, \quad \frac{du^{(e)}(x)}{dx} = \left[B^{(e)}\right]\left\{U^{(e)}\right\}$$

where

$$\left[N^{(e)}\right] = \left[N_1^{(e)}(x) \;\; N_2^{(e)}(x)\right], \quad \left[B^{(e)}\right] = \left[\frac{dN_1^{(e)}(x)}{dx} \;\; \frac{dN_2^{(e)}(x)}{dx}\right], \quad \left\{U^{(e)}\right\} = \begin{Bmatrix} u_1^{(e)} \\ u_2^{(e)} \end{Bmatrix}$$

$$N_1^{(e)}(x) = \frac{x - x_2^{(e)}}{x_1^{(e)} - x_2^{(e)}}, \quad \frac{dN_1^{(e)}(x)}{dx} = \frac{1}{x_1^{(e)} - x_2^{(e)}} = -\frac{1}{x_2^{(e)} - x_1^{(e)}} = -\frac{1}{\ell^{(e)}}$$

$$N_2^{(e)}(x) = \frac{x - x_1^{(e)}}{x_2^{(e)} - x_1^{(e)}}, \quad \frac{dN_2^{(e)}(x)}{dx} = \frac{1}{x_2^{(e)} - x_1^{(e)}} = \frac{1}{\ell^{(e)}}$$

where $\ell^{(e)}$ is the length of element e.

The stiffness matrix, $[k^{(e)}]$, of each element e is obtained from the following expression.

$$\left[k^{(e)}\right] = \int_{x_1^{(e)}}^{x_2^{(e)}} \left[B^{(e)}\right]^T EA \left[B^{(e)}\right] dx$$

$$= \int_{x_1^{(e)}}^{x_2^{(e)}} \left[\frac{dN_1^{(e)}(x)}{dx} \;\; \frac{dN_2^{(e)}(x)}{dx}\right]^T EA \left[\frac{dN_1^{(e)}(x)}{dx} \;\; \frac{dN_2^{(e)}(x)}{dx}\right] dx$$

We can write expressions providing each of the components of the stiffness matrix:

$$k_{11}^{(e)} = \int_{x_1^{(e)}}^{x_2^{(e)}} \frac{dN_1^{(e)}(x)}{dx} EA \frac{dN_1^{(e)}(x)}{dx} dx$$

$$k_{12}^{(e)} = \int_{x_1^{(e)}}^{x_2^{(e)}} \frac{dN_1^{(e)}(x)}{dx} EA \frac{dN_2^{(e)}(x)}{dx} dx$$

$$k_{21}^{(e)} = \int_{x_1^{(e)}}^{x_2^{(e)}} \frac{dN_2^{(e)}(x)}{dx} EA \frac{dN_1^{(e)}(x)}{dx} dx$$

$$k_{22}^{(e)} = \int_{x_1^{(e)}}^{x_2^{(e)}} \frac{dN_2^{(e)}(x)}{dx} EA \frac{dN_2^{(e)}(x)}{dx} dx$$

For element 1, we have:

$$k_{11}^{(1)} = \int_{x_1^{(1)}}^{x_2^{(1)}} \frac{dN_1^{(1)}(x)}{dx} EA \frac{dN_1^{(1)}(x)}{dx} dx = \int_0^5 \frac{dN_1^{(1)}(x)}{dx} EA \frac{dN_1^{(1)}(x)}{dx} dx$$

$$= \int_0^5 \left(-\frac{1}{5}\right) 5000 \cdot \left(1 + \frac{x^2}{400}\right) \left(-\frac{1}{5}\right) dx = 1020.83$$

$$k_{12}^{(1)} = \int_{x_1^{(1)}}^{x_2^{(1)}} \frac{dN_1^{(1)}(x)}{dx} EA \frac{dN_2^{(1)}(x)}{dx} dx = \int_0^5 \frac{dN_1^{(1)}(x)}{dx} EA \frac{dN_2^{(1)}(x)}{dx} dx$$

$$= \int_0^5 \left(-\frac{1}{5}\right) 5000 \cdot \left(1 + \frac{x^2}{400}\right) \frac{1}{5} dx = -1020.83$$

$$k_{21}^{(1)} = \int_{x_1^{(1)}}^{x_2^{(1)}} \frac{dN_2^{(1)}(x)}{dx} EA \frac{dN_1^{(1)}(x)}{dx} dx = k_{12}^{(1)} = -1020.83$$

$$k_{22}^{(1)} = \int_{x_1^{(1)}}^{x_2^{(1)}} \frac{dN_2^{(1)}(x)}{dx} EA \frac{dN_2^{(1)}(x)}{dx} dx = \int_0^5 \frac{dN_2^{(1)}(x)}{dx} EA \frac{dN_2^{(1)}(x)}{dx} dx$$

$$= \int_0^5 \frac{1}{5} 5000 \cdot \left(1 + \frac{x^2}{400}\right) \frac{1}{5} dx = 1020.83$$

$$\left[k^{(1)}\right] = \begin{bmatrix} k_{11}^{(1)} & k_{12}^{(1)} \\ k_{21}^{(1)} & k_{22}^{(1)} \end{bmatrix} = \begin{bmatrix} 1020.83 & -1020.83 \\ -1020.83 & 1020.83 \end{bmatrix}$$

In the same fashion, we can obtain the stiffness matrix of the second element:

$$\left[k^{(2)}\right] = \begin{bmatrix} 479.17 & -479.17 \\ -479.17 & 479.17 \end{bmatrix}$$

We also need to establish the gather/scatter arrays of our elements, in accordance with the considerations of Section B.2:

Element 1:

$$\left.\begin{matrix} u_1^{(1)} = u_1 \\ u_2^{(1)} = u_2 \end{matrix}\right\} \rightarrow \begin{Bmatrix} u_1^{(1)} \\ u_2^{(1)} \end{Bmatrix} = \begin{bmatrix} 1 & 0 & 0 \\ 0 & 1 & 0 \end{bmatrix} \begin{Bmatrix} u_1 \\ u_2 \\ u_3 \end{Bmatrix} \rightarrow \left[L^{(1)}\right] = \begin{bmatrix} 1 & 0 & 0 \\ 0 & 1 & 0 \end{bmatrix}$$

Element 2:

$$\left.\begin{matrix} u_1^{(2)} = u_2 \\ u_2^{(3)} = u_3 \end{matrix}\right\} \rightarrow \begin{Bmatrix} u_1^{(1)} \\ u_2^{(1)} \end{Bmatrix} = \begin{bmatrix} 0 & 1 & 0 \\ 0 & 0 & 1 \end{bmatrix} \begin{Bmatrix} u_1 \\ u_2 \\ u_3 \end{Bmatrix} \rightarrow \left[L^{(2)}\right] = \begin{bmatrix} 0 & 1 & 0 \\ 0 & 0 & 1 \end{bmatrix}$$

We can now calculate the global stiffness matrix, $[K]$:

$$[K] = \left[L^{(1)}\right]^T \left[k^{(1)}\right] \left[L^{(1)}\right] + \left[L^{(2)}\right]^T \left[k^{(2)}\right] \left[L^{(2)}\right]$$

$$= \begin{bmatrix} 1 & 0 \\ 0 & 1 \\ 0 & 0 \end{bmatrix} \begin{bmatrix} k_{11}^{(1)} & k_{12}^{(1)} \\ k_{21}^{(1)} & k_{22}^{(1)} \end{bmatrix} \begin{bmatrix} 1 & 0 & 0 \\ 0 & 1 & 0 \end{bmatrix} + \begin{bmatrix} 0 & 0 \\ 1 & 0 \\ 0 & 1 \end{bmatrix} \begin{bmatrix} k_{11}^{(2)} & k_{12}^{(2)} \\ k_{21}^{(2)} & k_{22}^{(2)} \end{bmatrix} \begin{bmatrix} 0 & 1 & 0 \\ 0 & 0 & 1 \end{bmatrix}$$

$$= \begin{bmatrix} k_{11}^{(1)} & k_{12}^{(1)} & 0 \\ k_{21}^{(1)} & k_{22}^{(1)} & 0 \\ 0 & 0 & 0 \end{bmatrix} + \begin{bmatrix} 0 & 0 & 0 \\ 0 & k_{11}^{(2)} & k_{12}^{(2)} \\ 0 & k_{21}^{(2)} & k_{22}^{(2)} \end{bmatrix} = \begin{bmatrix} k_{11}^{(1)} & k_{12}^{(1)} & 0 \\ k_{21}^{(1)} & k_{22}^{(1)} + k_{11}^{(2)} & k_{12}^{(2)} \\ 0 & k_{21}^{(2)} & k_{22}^{(2)} \end{bmatrix}$$

And if we plug the actual values of the element stiffness matrix components, we have:

$$[K] = \begin{bmatrix} k_{11}^{(1)} & k_{12}^{(1)} & 0 \\ k_{21}^{(1)} & k_{22}^{(1)} + k_{11}^{(2)} & k_{12}^{(2)} \\ 0 & k_{21}^{(2)} & k_{22}^{(2)} \end{bmatrix} = \begin{bmatrix} 1020.83 & -1020.83 & 0 \\ -1020.83 & 1500 & -479.17 \\ 0 & -479.17 & 479.17 \end{bmatrix}$$

We can also calculate the contribution of each element, $\left\{f^{(e)}\right\} = \begin{Bmatrix} f_1^{(e)} \\ f_2^{(e)} \end{Bmatrix}$, to the right-hand-side vector:

$$\left\{f^{(e)}\right\} = \int_{x_1^{(e)}}^{x_2^{(e)}} \left[N^{(e)}\right]^T \cdot b(x)dx + \left[\left[N^{(e)}\right]^T\right]_{\Gamma_t} (A \cdot \bar{t})_{\Gamma_t}$$

$$= \int_{x_1^{(e)}}^{x_2^{(e)}} \left[N_1^{(e)}(x) \;\; N_2^{(e)}(x)\right]^T \cdot b(x)dx + \left[\left[N_1^{(e)}(x) \;\; N_2^{(e)}(x)\right]^T\right]_{\Gamma_t} (A \cdot \bar{t})_{\Gamma_t}$$

$$= \int_{x_1^{(e)}}^{x_2^{(e)}} \begin{Bmatrix} N_1^{(e)}(x) \\ N_2^{(e)}(x) \end{Bmatrix} 2dx + \left[\begin{Bmatrix} N_1^{(e)}(x) \\ N_2^{(e)}(x) \end{Bmatrix}\right]_{x=20} (-10)$$

Element 1:

We have $\left(N_1^{(1)}(x)\right)_{x=20} = \left(N_2^{(1)}(x)\right)_{x=20} = 0$, because the location $x = 20$ lies outside of element 1!

$$f_1^{(1)} = \int_{x_1^{(1)}=0}^{x_2^{(1)}=5} N_1^{(1)}(x)\cdot 2dx + \left(N_1^{(1)}(x)\right)_{x=20}(-10) = \int_0^5 \frac{x-x_2^{(1)}}{x_1^{(1)}-x_2^{(1)}}\cdot 2dx + \left(N_1^{(1)}(x)\right)_{x=20}(-10)$$

$$= \int_0^5 \frac{x-x_2^{(1)}}{x_1^{(1)}-x_2^{(1)}}\cdot 2dx + 0\cdot(-10)$$

Thus:

$$f_1^{(1)} = \int_0^5 \frac{x-5}{0-5}\cdot 2dx = 5\,\text{kip}$$

$$f_2^{(1)} = \int_{x_1^{(1)}=0}^{x_2^{(1)}=5} N_2^{(1)}(x)\cdot 2dx + \left(N_2^{(1)}(x)\right)_{x=20}(-10) = \int_0^5 \frac{x-x_1^{(1)}}{x_2^{(1)}-x_1^{(1)}}\cdot 2dx + \left(N_2^{(1)}(x)\right)_{x=20}(-10)$$

$$= \int_0^5 \frac{x-x_1^{(1)}}{x_2^{(1)}-x_1^{(1)}}\cdot 2dx + 0\cdot(-10) = \int_0^5 \frac{x-0}{5-0}\cdot 2dx = 5\,\text{kip}$$

Thus, we have:

$$\left\{f^{(1)}\right\} = \begin{Bmatrix} f_1^{(1)} \\ f_2^{(1)} \end{Bmatrix} = \begin{Bmatrix} 5 \\ 5 \end{Bmatrix}$$

Element 2:

$$f_1^{(2)} = \int_{x_1^{(2)}=5}^{x_2^{(2)}=20} N_1^{(2)}(x)\cdot 2dx + \left(N_1^{(2)}(x)\right)_{x=20}(-10) = \int_5^{20} \frac{x-x_2^{(2)}}{x_1^{(2)}-x_2^{(2)}}\cdot 2dx + \left(N_1^{(2)}(x)\right)_{x=20}(-10)$$

$$= \int_5^{20} \frac{x-x_2^{(1)}}{x_1^{(1)}-x_2^{(1)}}\cdot 2dx + 0\cdot(-10) = 15\,\text{kip}$$

$$f_2^{(2)} = \int_{x_1^{(2)}=0}^{x_2^{(2)}=5} N_2^{(2)}(x)\cdot 2dx + \left(N_2^{(2)}(x)\right)_{x=20}(-10) = \int_5^{20} \frac{x-x_1^{(2)}}{x_2^{(2)}-x_1^{(2)}}\cdot 2dx + \left(N_2^{(2)}(x)\right)_{x=20}(-10)$$

$$= \int_5^{20} \frac{x-x_1^{(2)}}{x_2^{(2)}-x_1^{(2)}}\cdot 2dx + 1\cdot(-10) = 5\,\text{kip}$$

$$\left\{f^{(2)}\right\} = \begin{Bmatrix} f_1^{(2)} \\ f_2^{(2)} \end{Bmatrix} = \begin{Bmatrix} 15 \\ 5 \end{Bmatrix}$$

We can now assemble the global right-hand-side vector:

$$\{F\} = \left[L^{(1)}\right]^T \begin{Bmatrix} 5 \\ 5 \end{Bmatrix} + \left[L^{(2)}\right]^T \begin{Bmatrix} 15 \\ 5 \end{Bmatrix} = \begin{bmatrix} 1 & 0 \\ 0 & 1 \\ 0 & 0 \end{bmatrix} \begin{Bmatrix} 5 \\ 5 \end{Bmatrix} + \begin{bmatrix} 0 & 0 \\ 1 & 0 \\ 0 & 1 \end{bmatrix} \begin{Bmatrix} 15 \\ 5 \end{Bmatrix}$$

$$= \begin{Bmatrix} 5 \\ 5 \\ 0 \end{Bmatrix} + \begin{Bmatrix} 0 \\ 15 \\ 5 \end{Bmatrix} = \begin{Bmatrix} 5 \\ 20 \\ 5 \end{Bmatrix}$$

The global discrete equations can be written as follows (if we account for the fact that the degree of freedom u_1 is restrained and thus has a nonzero reaction, r_1):

$$\begin{bmatrix} 1020.83 & -1020.83 & 0 \\ -1020.83 & 1500 & -479.17 \\ 0 & -479.17 & 479.17 \end{bmatrix} \begin{Bmatrix} u_1 \\ u_2 \\ u_3 \end{Bmatrix} - \begin{Bmatrix} 5 \\ 20 \\ 5 \end{Bmatrix} = \begin{Bmatrix} r_1 \\ 0 \\ 0 \end{Bmatrix}$$

Now, we can see what requirements are established on the nodal quantities from the *essential boundary condition*. Since we know that $u(x = 0) = 0.05$ in., and since node 1 is at $x = 0$, we can set $u_1 = 0.05$ in., so we have two unknown nodal displacements remaining to be determined:

$$\begin{bmatrix} 1020.83 & -1020.83 & 0 \\ -1020.83 & 1500 & -479.17 \\ 0 & -479.17 & 479.17 \end{bmatrix} \begin{Bmatrix} 0.05 \\ u_2 \\ u_3 \end{Bmatrix} - \begin{Bmatrix} 5 \\ 20 \\ 5 \end{Bmatrix} = \begin{Bmatrix} r_1 \\ 0 \\ 0 \end{Bmatrix}$$

The first equation of the above system is:

$$0 \cdot (2000 \cdot 0.05 - 2000 \cdot u_2 + 0 \cdot u_3 - 5) = r_1$$

We do not need to worry about it (for the time being!). We cast the other two equations in the following form.

$$\begin{bmatrix} 1500 & -479.17 \\ -479.17 & 479.17 \end{bmatrix} \begin{Bmatrix} u_2 \\ u_3 \end{Bmatrix} + 0.05 \begin{Bmatrix} -1020.83 \\ 0 \end{Bmatrix} - \begin{Bmatrix} 20 \\ 5 \end{Bmatrix} = \begin{Bmatrix} 0 \\ 0 \end{Bmatrix}$$

$$\begin{bmatrix} 1500 & -479.17 \\ -479.17 & 479.17 \end{bmatrix} \begin{Bmatrix} u_2 \\ u_3 \end{Bmatrix} = -0.05 \begin{Bmatrix} -1020.83 \\ 0 \end{Bmatrix} + \begin{Bmatrix} 20 \\ 5 \end{Bmatrix}$$

We can solve this system of two equations for the unknown, unrestrained nodal displacements, u_2 and u_3.

The solution of the system of equations gives us:

$$\begin{Bmatrix} u_2 \\ u_3 \end{Bmatrix} = \begin{Bmatrix} 0.0745 \text{ in.} \\ 0.0849 \text{ in.} \end{Bmatrix}$$

Thus, the global nodal displacement vector is:

$$\{U_f\} = \begin{Bmatrix} u_1 \\ u_2 \\ u_3 \end{Bmatrix} = \begin{Bmatrix} 0.05\text{ in.} \\ 0.0745\text{ in.} \\ 0.0849\text{ in.} \end{Bmatrix}$$

We can then determine the approximate displacement field and the approximate stress field inside each element; this procedure corresponds to postprocessing of the solution.

Element 1:

$$u^{(1)}(x) = [N^{(1)}]\{U^{(1)}\} = [N^{(1)}][L^{(1)}]\{U\}$$

$$\sigma^{(1)}(x) = E\frac{du^{(1)}(x)}{dx} = E\left[B^{(1)}\right]\left\{U^{(1)}\right\} = E\left[B^{(1)}\right]\left[L^{(1)}\right]\{U\}$$

Element 2:

$$u^{(2)}(x) = [N^{(2)}]\{U^{(2)}\} = [N^{(2)}][L^{(2)}]\{U\}$$

$$\sigma^{(2)}(x) = E\frac{du^{(2)}(x)}{dx} = E\left[B^{(2)}\right]\left\{U^{(2)}\right\} = E\left[B^{(2)}\right]\left[L^{(2)}\right]\{U\}$$

We can plot the approximate displacement and stress fields, and compare them with the corresponding fields of the exact solution, as shown in Figure 3.12. The exact displacement (the solution to the strong form) and stress fields are given by:

$$u(x) = -0.08\ln\left(1+\frac{x^2}{400}\right) + 0.12\cdot atan\left(\frac{x}{20}\right) + 0.05$$

$$\sigma(x) = E\frac{du(x)}{dx} = \frac{30-2x}{1+\dfrac{x^2}{400}}$$

Notice that, for the specific problem, the nodal displacements obtained by the finite element solution are the *exact* ones! We should keep in mind that this is not always the case with a finite element solution! Also, notice that the stress field in a finite element with linear polynomial shape functions and constant value of E is constant; this gives

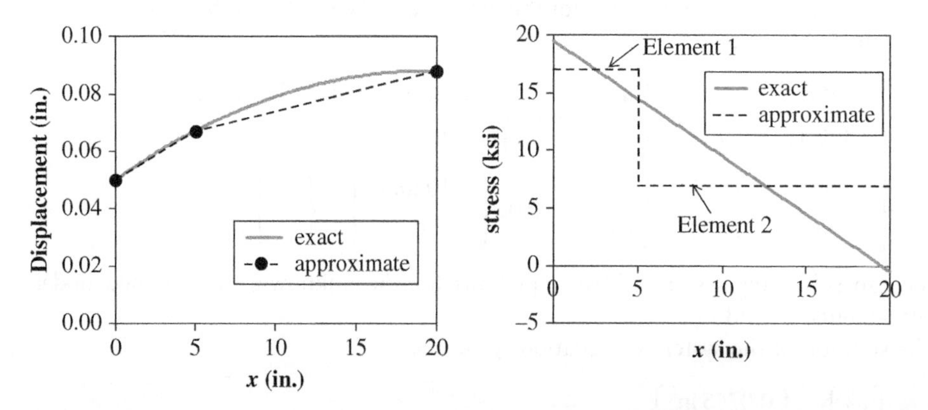

Figure 3.12 Comparison of approximate (finite element) solution and exact solution for the structure of Example 3.2.

rather poor estimates of the exact stress variation. Although we know that our approximation is continuous and complete, and thus it is bound to be convergent, a *coarse mesh* (i.e., a mesh consisting of few elements) may not give satisfactory estimates of the displacement or stress variation.

Remark 3.6.1: In accordance with the discussion in Sections B.5 and B.6, actual finite element programs do not assemble the total coefficient matrix and then get rid of equations corresponding to restrained displacements. Instead, they use the "*ID* and *LM* array" concepts to directly establish the coefficient matrix $[K_{ff}]$ and right-hand-side vector $\{f_f\}$, which correspond to the unrestrained (free) displacements only, as presented in Box B.6.1. ▮

Example 3.3: Analysis for three-node (quadratic) element

We are now going to repeat the analysis for the structure of Example 3.2; we will be using the same three nodal points as in Figure 3.11, but instead of having two-node elements, we will use a single, three-node element.

The approximation for the three-node element is given by the following equations:

$$u^{(1)}(x) = N_1^{(1)}(x)\cdot u_1^{(1)} + N_2^{(1)}(x)\cdot u_2^{(1)} + N_3^{(1)}(x)\cdot u_3^{(1)}, \quad \frac{du^{(1)}}{dx} = \frac{dN_1^{(1)}(x)}{dx}\cdot u_1^{(e)} + \frac{dN_2^{(1)}(x)}{dx}\cdot u_2^{(e)} + \frac{dN_3^{(1)}(x)}{dx}\cdot u_3^{(e)}$$

We can also write:

$$u^{(e)}(x) = \left[N^{(e)}\right]\begin{Bmatrix} u_1^{(e)} \\ u_2^{(e)} \\ u_3^{(e)} \end{Bmatrix} = \left[N^{(e)}\right]\left\{U^{(e)}\right\}, \quad \frac{du^{(e)}(x)}{dx} = \left[B^{(e)}\right]\left\{U^{(e)}\right\}$$

where

$$\left[N^{(1)}\right] = \left[N_1^{(1)}(x) \;\; N_2^{(1)}(x) \;\; N_3^{(1)}(x)\right], \quad \left[B^{(1)}\right] = \left[\frac{dN_1^{(1)}(x)}{dx} \;\; \frac{dN_2^{(1)}(x)}{dx} \;\; \frac{dN_3^{(1)}(x)}{dx}\right], \quad \left\{U^{(1)}\right\} = \begin{Bmatrix} u_1^{(1)} \\ u_2^{(1)} \\ u_3^{(1)} \end{Bmatrix}$$

The three shape functions and their derivatives are given by the following expressions:

$$N_1^{(1)}(x) = \frac{\left(x-x_2^{(1)}\right)\left(x-x_3^{(1)}\right)}{\left(x_1^{(1)}-x_2^{(1)}\right)\left(x_1^{(1)}-x_3^{(1)}\right)} = \frac{x^2-\left(x_2^{(1)}+x_3^{(1)}\right)x + x_2^{(1)}\cdot x_3^{(1)}}{\left(x_1^{(1)}-x_2^{(1)}\right)\left(x_1^{(1)}-x_3^{(1)}\right)} = \frac{x^2-25x+100}{100}, \quad \frac{dN_1^{(1)}(x)}{dx} = \frac{2x-25}{100}$$

$$N_2^{(1)}(x) = \frac{\left(x-x_1^{(1)}\right)\left(x-x_3^{(1)}\right)}{\left(x_2^{(1)}-x_1^{(1)}\right)\left(x_2^{(1)}-x_3^{(1)}\right)} = \frac{x^2-\left(x_1^{(1)}+x_3^{(1)}\right)x + x_1^{(1)}\cdot x_3^{(1)}}{\left(x_2^{(1)}-x_1^{(1)}\right)\left(x_2^{(1)}-x_3^{(1)}\right)} = \frac{-x^2+20x}{75}, \quad \frac{dN_2^{(1)}(x)}{dx} = \frac{-2x+20}{75}$$

$$N_3^{(1)}(x) = \frac{\left(x-x_1^{(1)}\right)\left(x-x_2^{(1)}\right)}{\left(x_3^{(1)}-x_1^{(1)}\right)\left(x_3^{(1)}-x_2^{(1)}\right)} = \frac{x^2-\left(x_1^{(1)}+x_2^{(1)}\right)x + x_1^{(1)}\cdot x_2^{(1)}}{\left(x_3^{(1)}-x_1^{(1)}\right)\left(x_3^{(1)}-x_2^{(1)}\right)} = \frac{x^2-5x}{300}, \quad \frac{dN_3^{(1)}(x)}{dx} = \frac{2x-5}{300}$$

Thus, we can now calculate the stiffness matrix of the element:

$$\left[k^{(1)}\right] = \int_{x_1^{(1)}}^{x_3^{(1)}} \left[B^{(1)}\right]^T EA\left[B^{(1)}\right]dx = \int_{x_1^{(1)}}^{x_3^{(1)}} \left[\frac{dN_1^{(1)}(x)}{dx} \quad \frac{dN_2^{(1)}(x)}{dx} \quad \frac{dN_3^{(1)}(x)}{dx}\right]^T EA\left[\frac{dN_1^{(1)}(x)}{dx} \quad \frac{dN_2^{(1)}(x)}{dx} \quad \frac{dN_3^{(1)}(x)}{dx}\right]dx$$

We can separately calculate each component of $[k^{(1)}]$:

$$k_{11}^{(1)} = \int_{x_1^{(1)}}^{x_3^{(1)}} \frac{dN_1^{(1)}(x)}{dx} EA \frac{dN_1^{(1)}(x)}{dx} dx = \int_0^{20} \frac{dN_1^{(1)}(x)}{dx} EA \frac{dN_1^{(1)}(x)}{dx} dx$$

$$= \int_0^{20} \frac{2x-25}{100} 5000 \cdot \left(1 + \frac{x^2}{400}\right) \frac{2x-25}{100} dx = 1867$$

$$k_{12}^{(1)} = \int_{x_1^{(1)}}^{x_3^{(1)}} \frac{dN_1^{(1)}(x)}{dx} EA \frac{dN_2^{(1)}(x)}{dx} dx = \int_0^{20} \frac{dN_1^{(1)}(x)}{dx} EA \frac{dN_2^{(1)}(x)}{dx} dx$$

$$= \int_0^{20} \frac{2x-25}{100} 5000 \cdot \left(1 + \frac{x^2}{400}\right) \frac{-2x+20}{75} dx = -2267$$

$$k_{13}^{(1)} = \int_{x_1^{(1)}}^{x_3^{(1)}} \frac{dN_1^{(1)}(x)}{dx} EA \frac{dN_3^{(1)}(x)}{dx} dx = \int_0^{20} \frac{dN_1^{(1)}(x)}{dx} EA \frac{dN_3^{(1)}(x)}{dx} dx$$

$$= \int_0^{20} \frac{2x-25}{100} 5000 \cdot \left(1 + \frac{x^2}{400}\right) \frac{2x-5}{300} dx = 400$$

$$k_{21}^{(1)} = \int_{x_1^{(1)}}^{x_3^{(1)}} \frac{dN_2^{(1)}(x)}{dx} EA \frac{dN_1^{(1)}(x)}{dx} dx = k_{12}^{(1)} = -2267$$

$$k_{22}^{(1)} = \int_{x_1^{(1)}}^{x_3^{(1)}} \frac{dN_2^{(1)}(x)}{dx} EA \frac{dN_2^{(1)}(x)}{dx} dx = \int_0^{20} \frac{dN_2^{(1)}(x)}{dx} EA \frac{dN_2^{(1)}(x)}{dx} dx$$

$$= \int_0^{20} \frac{-2x+20}{75} 5000 \cdot \left(1 + \frac{x^2}{400}\right) \frac{-2x+20}{75} dx = 3319$$

$$k_{23}^{(1)} = \int_{x_1^{(1)}}^{x_3^{(1)}} \frac{dN_2^{(1)}(x)}{dx} EA \frac{dN_3^{(1)}(x)}{dx} dx = \int_0^{20} \frac{dN_2^{(1)}(x)}{dx} EA \frac{dN_3^{(1)}(x)}{dx} dx$$

$$= \int_0^{20} \frac{-2x+20}{75} 5000 \cdot \left(1 + \frac{x^2}{400}\right) \frac{2x-5}{300} dx = -1052$$

$$k_{31}^{(1)} = \int_{x_1^{(1)}}^{x_3^{(1)}} \frac{dN_3^{(1)}(x)}{dx} EA \frac{dN_1^{(1)}(x)}{dx} dx = k_{13}^{(1)} = 400$$

$$k_{32}^{(1)} = \int_{x_1^{(1)}}^{x_3^{(1)}} \frac{dN_3^{(1)}(x)}{dx} EA \frac{dN_2^{(1)}(x)}{dx} dx = k_{23}^{(1)} = -1052$$

$$k_{33}^{(1)} = \int_{x_1^{(1)}}^{x_3^{(1)}} \frac{dN_3^{(1)}(x)}{dx} EA \frac{dN_3^{(1)}(x)}{dx} dx = \int_0^{20} \frac{dN_3^{(1)}(x)}{dx} EA \frac{dN_3^{(1)}(x)}{dx} dx$$

$$= \int_0^{20} \frac{2x-5}{300} 5000 \cdot \left(1 + \frac{x^2}{400}\right) \frac{2x-5}{300} dx = 652$$

We will now proceed to obtain the values of the three components of the element's equivalent nodal force vector, $\{f^{(1)}\}$:

$$f_1^{(1)} = \int_{x_1^{(1)}=0}^{x_3^{(1)}=20} N_1^{(1)}(x) \cdot 2dx + \left(N_1^{(1)}(x)\right)_{x=20}(-10) = \int_0^{20} \frac{x^2-25x+100}{100} \cdot 2dx + \left(N_1^{(1)}(x)\right)_{x=20}(-10)$$

$$= \int_0^{20} \frac{x^2-25x+100}{100} \cdot 2dx + 0 \cdot (-10) = \int_0^{20} \frac{x^2-25x+100}{100} \cdot 2dx = -6.67 \text{ kip}$$

$$f_2^{(1)} = \int_{x_1^{(1)}=0}^{x_3^{(1)}=20} N_2^{(1)}(x) \cdot 2dx + \left(N_2^{(1)}(x)\right)_{x=20}(-10) = \int_0^{20} \frac{-x^2+20x}{75} \cdot 2dx + \left(N_2^{(1)}(x)\right)_{x=20}(-10)$$

$$= \int_0^{20} \frac{-x^2+20x}{75} \cdot 2dx + 0 \cdot (-10) = \int_0^{20} \frac{-x^2+20x}{75} \cdot 2dx = 35.56 \text{ kip}$$

$$f_3^{(1)} = \int_{x_1^{(1)}=0}^{x_3^{(1)}=20} N_3^{(1)}(x)\cdot 2dx + \left(N_3^{(1)}(x)\right)_{x=20}(-10) = \int_0^{20} \frac{x^2-5x}{300}\cdot 2dx + \left(N_3^{(1)}(x)\right)_{x=20}(-10)$$

$$= \int_0^{20} \frac{x^2-5x}{300}\cdot 2dx + 1\cdot(-10) = \int_0^{20} \frac{x^2-5x}{300}\cdot 2dx - 10 = 1.11\text{ kip}$$

Thus, we have:

$$\left\{f^{(1)}\right\} = \begin{Bmatrix} f_1^{(1)} \\ f_2^{(1)} \\ f_3^{(1)} \end{Bmatrix} = \begin{Bmatrix} -6.67 \\ 35.56 \\ 1.11 \end{Bmatrix}$$

Next, we conduct the assembly of the global finite element equations. To this end, we find the gather matrix of the element.

$$\left.\begin{matrix} u_1^{(1)} = u_1 \\ u_2^{(1)} = u_2 \\ u_3^{(1)} = u_3 \end{matrix}\right\} \to \begin{Bmatrix} u_1^{(1)} \\ u_2^{(1)} \\ u_3^{(1)} \end{Bmatrix} = \begin{bmatrix} 1 & 0 & 0 \\ 0 & 1 & 0 \\ 0 & 0 & 1 \end{bmatrix} \begin{Bmatrix} u_1 \\ u_2 \\ u_3 \end{Bmatrix} \to \left[L^{(1)}\right] = \begin{bmatrix} 1 & 0 & 0 \\ 0 & 1 & 0 \\ 0 & 0 & 1 \end{bmatrix}$$

Thus:

$$[K] = \left[L^{(1)}\right]^T \left[k^{(1)}\right]\left[L^{(1)}\right] = \begin{bmatrix} 1 & 0 & 0 \\ 0 & 1 & 0 \\ 0 & 0 & 1 \end{bmatrix} \begin{bmatrix} k_{11}^{(1)} & k_{12}^{(1)} & k_{13}^{(1)} \\ k_{21}^{(1)} & k_{22}^{(1)} & k_{23}^{(1)} \\ k_{31}^{(1)} & k_{32}^{(1)} & k_{33}^{(1)} \end{bmatrix} \begin{bmatrix} 1 & 0 & 0 \\ 0 & 1 & 0 \\ 0 & 0 & 1 \end{bmatrix}$$

$$= \begin{bmatrix} k_{11}^{(1)} & k_{12}^{(1)} & k_{13}^{(1)} \\ k_{21}^{(1)} & k_{22}^{(1)} & k_{23}^{(1)} \\ k_{31}^{(1)} & k_{32}^{(1)} & k_{33}^{(1)} \end{bmatrix} = \left[k^{(1)}\right]$$

The equality $[K] = \left[k^{(1)}\right]$ should be expected, since the numbering of the degrees of freedom of element 1 is identical to that of the global degrees of freedom!

The global nodal force vector is:

$$\{F\} = \left[L^{(1)}\right]^T \left\{f^{(1)}\right\} = \begin{bmatrix} 1 & 0 & 0 \\ 0 & 1 & 0 \\ 0 & 0 & 1 \end{bmatrix} \begin{Bmatrix} -6.67 \\ 35.56 \\ 1.11 \end{Bmatrix} = \begin{Bmatrix} -6.67 \\ 35.56 \\ 1.11 \end{Bmatrix}$$

We can now write the following system of equations.

$$\begin{bmatrix} 1867 & -2267 & 400 \\ -2267 & 3319 & -1052 \\ 400 & -1052 & 652 \end{bmatrix} \begin{Bmatrix} u_1 \\ u_2 \\ u_3 \end{Bmatrix} - \begin{Bmatrix} -6.67 \\ 35.56 \\ 1.11 \end{Bmatrix} = \begin{Bmatrix} r_1 \\ 0 \\ 0 \end{Bmatrix}$$

Since we know that $u(x = 0) = 0.05$ in., and since node 1 is at $x = 0$, we can set $u_1 = 0.05$ in., so we have two unknown nodal displacements remaining to be determined:

$$\begin{bmatrix} 1867 & -2267 & 400 \\ -2267 & 3319 & -1052 \\ 400 & -1052 & 652 \end{bmatrix} \begin{Bmatrix} 0.05 \\ u_2 \\ u_3 \end{Bmatrix} - \begin{Bmatrix} -6.67 \\ 35.56 \\ 1.11 \end{Bmatrix} = \begin{Bmatrix} r_1 \\ 0 \\ 0 \end{Bmatrix}$$

$$\begin{bmatrix} 1867 & -2267 & 400 \\ -2267 & 3319 & -1052 \\ 400 & -1052 & 652 \end{bmatrix} \begin{Bmatrix} 0.05 \\ u_2 \\ u_3 \end{Bmatrix} = \begin{Bmatrix} -6.67 + r_1 \\ 35.56 \\ 1.11 \end{Bmatrix}$$

If we now take the two equations corresponding to the unrestrained degrees of freedom:

$$\begin{bmatrix} 3319 & -1052 \\ -1052 & 652 \end{bmatrix} \begin{Bmatrix} u_2 \\ u_3 \end{Bmatrix} = -0.05 \begin{Bmatrix} -2267 \\ 400 \end{Bmatrix} + \begin{Bmatrix} 35.56 \\ 1.11 \end{Bmatrix}$$

And of course, we can solve this discrete system of two equations for the unknown, unrestrained nodal displacements.

$$\begin{Bmatrix} u_2 \\ u_3 \end{Bmatrix} = \begin{Bmatrix} 0.0730\, in \\ 0.0889\, in \end{Bmatrix}$$

Thus, the global displacement vector is

$\{U\} = [u_1 \ \ u_2 \ \ u_3]^T = [0.05\, in \ \ 0.0730\, in \ \ 0.0889\, in]^T$

We can now determine the approximate displacement field and the approximate stress field inside our element.

$$u^{(1)}(x) = \left[N^{(1)}\right]\left\{U^{(1)}\right\} = \left[N^{(1)}\right]\left[L^{(1)}\right]\{U\}$$

$$\sigma^{(1)}(x) = E\frac{du^{(1)}(x)}{dx} = E\left[B^{(1)}\right]\left\{U^{(1)}\right\} = E\left[B^{(1)}\right]\left[L^{(1)}\right]\{U\}$$

Finally, we can compare the approximate and the exact solution, as shown in Figure 3.13.

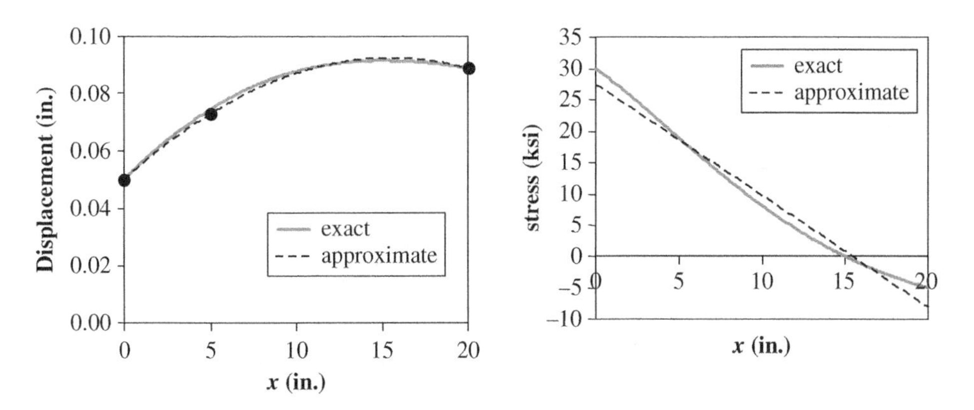

Figure 3.13 Comparison between exact and approximate solutions for structure of Example 3.3.

A comparison of Figures 3.12 and 3.13 indicates that, for the specific problem, the results obtained with a single quadratic (three-node) element are more accurate than those obtained with two linear (two-node) elements.

Example 3.4: Finite element analysis for one-dimensional heat conduction
We consider the system in Figure 3.14, which we want to approximately solve using two two-node elements, as shown in the same figure.

The weak form of the specific problem is:

$$\int_0^2 \frac{dw}{dx}\kappa A\frac{d\Theta}{dx}dx = [-wA\bar{q}]_{x=2} + \int_0^2 w\cdot s(x)dx, \text{ where } s(x) = 2\frac{\text{W}}{\text{m}^2}, \bar{q} = -5\frac{\text{W}}{\text{m}^2}, \text{ and}$$

$$[-A\bar{q}]_{x=2} = -0.4\cdot(-5) = 2\text{W}$$

Thus, the weak form becomes:

$$\int_0^2 \frac{dw}{dx}\kappa A\frac{d\Theta}{dx}dx = [w]_{x=2}\cdot 2 + \int_0^2 w\cdot 2dx$$

The finite element approximation for the temperature field of a two-node element is:

$$\Theta^{(e)}(x) = N_1^{(e)}(x)\cdot\Theta_1^{(e)} + N_2^{(e)}(x)\cdot\Theta_2^{(e)} = \left[N^{(e)}\right]\left\{\Theta^{(e)}\right\}$$

and the corresponding approximation of the derivative of the temperature field is:

$$\frac{d\Theta^{(e)}}{dx} = \frac{dN_1^{(e)}(x)}{dx}\cdot\Theta_1^{(e)} + \frac{dN_2^{(e)}(x)}{dx}\cdot\Theta_2^{(e)} = \left[B^{(e)}\right]\left\{\Theta^{(e)}\right\}$$

where

$$\left[N^{(e)}\right] = \left[N_1^{(e)}(x) \quad N_2^{(e)}(x)\right],\ \left[B^{(e)}\right] = \left[\frac{dN_1^{(e)}(x)}{dx} \quad \frac{dN_2^{(e)}(x)}{dx}\right],\ \left\{\Theta^{(e)}\right\} = \begin{Bmatrix}\Theta_1^{(e)}\\ \Theta_2^{(e)}\end{Bmatrix}$$

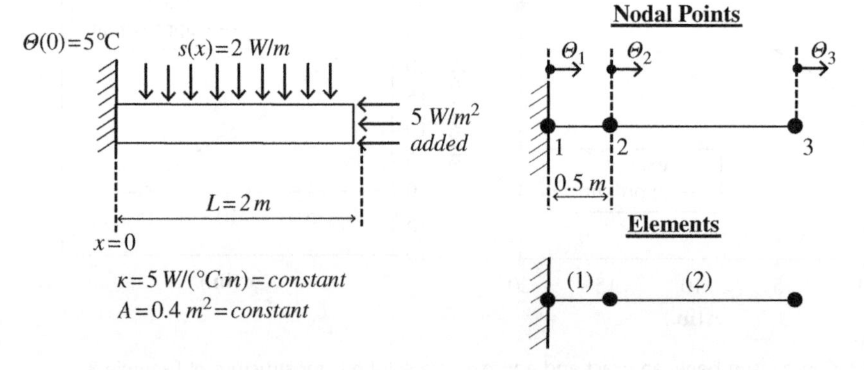

Figure 3.14 Example system to be solved for one-dimensional heat conduction.

$$N_1^{(e)}(x)=\frac{x-x_2^{(e)}}{x_1^{(e)}-x_2^{(e)}},\ \frac{dN_1^{(e)}(x)}{dx}=\frac{1}{x_1^{(e)}-x_2^{(e)}}=-\frac{1}{x_2^{(e)}-x_1^{(e)}}=-\frac{1}{\ell^{(e)}}$$

$$N_2^{(e)}(x)=\frac{x-x_1^{(e)}}{x_2^{(e)}-x_1^{(e)}},\ \frac{dN_2^{(e)}(x)}{dx}=\frac{1}{x_2^{(e)}-x_1^{(e)}}=\frac{1}{\ell^{(e)}}$$

We can now calculate the coefficient (conductance) matrix $[k^{(e)}]$ of each element e:

$$\left[k^{(e)}\right]=\int_{x_1^{(e)}}^{x_2^{(e)}}\left[B^{(e)}\right]^T\kappa^{(e)}A^{(e)}\left[B^{(e)}\right]dx=\int_{x_1^{(e)}}^{x_2^{(e)}}\left[\frac{dN_1^{(e)}(x)}{dx}\ \frac{dN_2^{(e)}(x)}{dx}\right]^T\kappa^{(e)}A^{(e)}\left[\frac{dN_1^{(e)}(x)}{dx}\frac{dN_2^{(e)}(x)}{dx}\right]dx$$

The individual components of $[k^{(e)}]$ are given by:

$$k_{11}^{(e)}=\int_{x_1^{(e)}}^{x_2^{(e)}}\frac{dN_1^{(e)}(x)}{dx}\kappa^{(e)}A^{(e)}\frac{dN_1^{(e)}(x)}{dx}dx$$

$$k_{12}^{(e)}=\int_{x_1^{(e)}}^{x_2^{(e)}}\frac{dN_1^{(e)}(x)}{dx}\kappa^{(e)}A^{(e)}\frac{dN_2^{(e)}(x)}{dx}dx$$

$$k_{21}^{(e)}=\int_{x_1^{(e)}}^{x_2^{(e)}}\frac{dN_2^{(e)}(x)}{dx}\kappa^{(e)}A^{(e)}\frac{dN_1^{(e)}(x)}{dx}dx$$

$$k_{22}^{(e)}=\int_{x_1^{(e)}}^{x_2^{(e)}}\frac{dN_2^{(e)}(x)}{dx}\kappa^{(e)}A^{(e)}\frac{dN_2^{(e)}(x)}{dx}dx$$

For element 1, we have:

$$k_{11}^{(1)}=\int_{x_1^{(1)}}^{x_2^{(1)}}\frac{dN_1^{(1)}(x)}{dx}\kappa^{(1)}A^{(1)}\frac{dN_1^{(1)}(x)}{dx}dx=\int_0^{0.5}\left(-\frac{1}{0.5}\right)5\cdot0.4\left(-\frac{1}{0.5}\right)dx=4$$

$$k_{12}^{(1)}=\int_{x_1^{(1)}}^{x_2^{(1)}}\frac{dN_1^{(1)}(x)}{dx}\kappa^{(1)}A^{(1)}\frac{dN_2^{(1)}(x)}{dx}dx=\int_0^{0.5}\left(-\frac{1}{0.5}\right)5\cdot0.4\frac{1}{0.5}dx=-4$$

$$k_{21}^{(1)}=\int_{x_1^{(1)}}^{x_2^{(1)}}\frac{dN_2^{(1)}(x)}{dx}\kappa^{(1)}A^{(1)}\frac{dN_1^{(1)}(x)}{dx}dx=k_{12}^{(1)}=-4$$

$$k_{22}^{(1)}=\int_{x_1^{(1)}}^{x_2^{(1)}}\frac{dN_2^{(1)}(x)}{dx}\kappa^{(1)}A^{(1)}\frac{dN_2^{(1)}(x)}{dx}dx=\int_0^{0.5}\frac{1}{0.5}5\cdot0.4\frac{1}{0.5}dx=4$$

$$\left[k^{(1)}\right]=\begin{bmatrix}k_{11}^{(1)} & k_{12}^{(1)}\\ k_{21}^{(1)} & k_{22}^{(1)}\end{bmatrix}=\begin{bmatrix}4 & -4\\ -4 & 4\end{bmatrix}$$

In the same fashion, we can obtain the conductance array of the second element:

$$\left[k^{(2)}\right]=\begin{bmatrix}1.33 & -1.33\\ -1.33 & 1.33\end{bmatrix}$$

We can also establish the gather/scatter arrays of our elements:
Element 1:

$$\left.\begin{matrix}\Theta_1^{(1)}=\Theta_1\\ \Theta_2^{(1)}=\Theta_2\end{matrix}\right\}\rightarrow\begin{Bmatrix}\Theta_1^{(1)}\\ \Theta_2^{(1)}\end{Bmatrix}=\begin{bmatrix}1 & 0 & 0\\ 0 & 1 & 0\end{bmatrix}\begin{Bmatrix}\Theta_1\\ \Theta_2\\ \Theta_3\end{Bmatrix}\rightarrow\left[L^{(1)}\right]=\begin{bmatrix}1 & 0 & 0\\ 0 & 1 & 0\end{bmatrix}$$

Element 2:

$$\left.\begin{matrix}\Theta_1^{(2)}=\Theta_2\\ \Theta_2^{(3)}=\Theta_3\end{matrix}\right\}\rightarrow\begin{Bmatrix}\Theta_1^{(1)}\\ \Theta_2^{(1)}\end{Bmatrix}=\begin{bmatrix}0 & 1 & 0\\ 0 & 0 & 1\end{bmatrix}\begin{Bmatrix}\Theta_1\\ \Theta_2\\ \Theta_3\end{Bmatrix}\rightarrow\left[L^{(2)}\right]=\begin{bmatrix}0 & 1 & 0\\ 0 & 0 & 1\end{bmatrix}$$

Now, we can assemble the global conductance matrix, $[K]$:

$$[K]=\left[L^{(1)}\right]^T\left[k^{(1)}\right]\left[L^{(1)}\right]+\left[L^{(2)}\right]^T\left[k^{(2)}\right]\left[L^{(2)}\right]=\begin{bmatrix}1 & 0\\ 0 & 1\\ 0 & 0\end{bmatrix}\begin{bmatrix}k_{11}^{(1)} & k_{12}^{(1)}\\ k_{21}^{(1)} & k_{22}^{(1)}\end{bmatrix}\begin{bmatrix}1 & 0 & 0\\ 0 & 1 & 0\end{bmatrix}$$

$$+\begin{bmatrix}0 & 0\\ 1 & 0\\ 0 & 1\end{bmatrix}\begin{bmatrix}k_{11}^{(2)} & k_{12}^{(2)}\\ k_{21}^{(2)} & k_{22}^{(2)}\end{bmatrix}\begin{bmatrix}0 & 1 & 0\\ 0 & 0 & 1\end{bmatrix}$$

$$=\begin{bmatrix}k_{11}^{(1)} & k_{12}^{(1)} & 0\\ k_{21}^{(1)} & k_{22}^{(1)} & 0\\ 0 & 0 & 0\end{bmatrix}+\begin{bmatrix}0 & 0 & 0\\ 0 & k_{11}^{(2)} & k_{12}^{(2)}\\ 0 & k_{21}^{(2)} & k_{22}^{(2)}\end{bmatrix}=\begin{bmatrix}k_{11}^{(1)} & k_{12}^{(1)} & 0\\ k_{21}^{(1)} & k_{22}^{(1)}+k_{11}^{(2)} & k_{12}^{(2)}\\ 0 & k_{21}^{(2)} & k_{22}^{(2)}\end{bmatrix}=\begin{bmatrix}4 & -4 & 0\\ -4 & 5.33 & -1.33\\ 0 & -1.33 & 1.33\end{bmatrix}$$

We can also calculate the equivalent nodal flux vector of each element, $\{f^{(e)}\} = \begin{Bmatrix} f_1^{(e)} \\ f_2^{(e)} \end{Bmatrix}$:

$$\{f^{(e)}\} = \int_{x_1^{(e)}}^{x_2^{(e)}} \left[N^{(e)}\right]^T \cdot s(x)dx + \left[\left[N^{(e)}\right]^T\right]_{\Gamma_t} (A \cdot \bar{q})_{\Gamma_t}$$

$$= \int_{x_1^{(e)}}^{x_2^{(e)}} \left[N_1^{(e)}(x) \;\; N_2^{(e)}(x)\right]^T \cdot b(x)dx + \left[\left[N_1^{(e)}(x) \;\; N_2^{(e)}(x)\right]^T\right]_{\Gamma_t} (A \cdot \bar{q})_{\Gamma_t}$$

$$= \int_{x_1^{(e)}}^{x_2^{(e)}} \begin{Bmatrix} N_1^{(e)}(x) \\ N_2^{(e)}(x) \end{Bmatrix} 2dx + \left[\begin{Bmatrix} N_1^{(e)}(x) \\ N_2^{(e)}(x) \end{Bmatrix}\right]_{x=2} \cdot 2$$

The two nodal values for element 1 are the following.

$$f_1^{(1)} = \int_{x_1^{(1)}=0}^{x_2^{(1)}=0.5} N_1^{(1)}(x) \cdot 2dx + \left(N_1^{(1)}(x)\right)_{x=2} \cdot 2 = \int_0^{0.5} \frac{x - x_2^{(1)}}{x_1^{(1)} - x_2^{(1)}} \cdot 2dx + \left(N_1^{(1)}(x)\right)_{x=2} \cdot 2$$

$$= \int_0^{0.5} \frac{x - x_2^{(1)}}{x_1^{(1)} - x_2^{(1)}} \cdot 2dx + 0 \cdot 2, \; \textit{because}\, N_1^{(1)}(x) \textit{ is zero outside of element } 1 \; (\textit{and thus at } x = 2!)$$

Thus:

$$f_1^{(1)} = \int_0^{0.5} \frac{x - 0.5}{0 - 0.5} \cdot 2dx = 0.5\,\mathrm{W}$$

Similarly:

$$f_2^{(1)} = \int_{x_1^{(1)}=0}^{x_2^{(1)}=0.5} N_2^{(1)}(x) \cdot 2dx + \left(N_2^{(1)}(x)\right)_{x=2} \cdot 2 = \int_0^{0.5} \frac{x - x_1^{(1)}}{x_2^{(1)} - x_1^{(1)}} \cdot 2dx + \left(N_2^{(1)}(x)\right)_{x=2} \cdot 2$$

$$= \int_0^{0.5} \frac{x - x_1^{(1)}}{x_2^{(1)} - x_1^{(1)}} \cdot 2dx + 0 \cdot 2, \; \textit{because}\, N_2^{(1)}(x = 2) = 0, \textit{ since the location } x = 2 \textit{ lies outside of element } 1!$$

Thus:

$$f_2^{(1)} = \int_0^{0.5} \frac{x - 0}{0.5 - 0} \cdot 2dx = 0.5\,\mathrm{W}$$

$$\{f^{(1)}\} = \begin{Bmatrix} f_1^{(1)} \\ f_2^{(1)} \end{Bmatrix} = \begin{Bmatrix} 0.5 \\ 0.5 \end{Bmatrix}$$

Now, we can obtain the nodal fluxes for element 2:

$$f_1^{(2)} = \int_{x_1^{(2)}=0.5}^{x_2^{(2)}=2} N_1^{(2)}(x)\cdot 2dx + \left(N_1^{(2)}(x)\right)_{x=2}\cdot 2 = \int_{0.5}^{2} \frac{x - x_2^{(2)}}{x_1^{(2)} - x_2^{(2)}}\cdot 2dx + \left(N_1^{(2)}(x)\right)_{x=2}\cdot 2$$

$$= \int_{0.5}^{2} \frac{x - x_2^{(1)}}{x_1^{(1)} - x_2^{(1)}}\cdot 2dx + 0\cdot 2 = \int_{0.5}^{2} \frac{x-2}{0.5-2}\cdot 2dx = 1.5\,\text{W}$$

$$f_2^{(2)} = \int_{x_1^{(2)}=0.5}^{x_2^{(2)}=2} N_2^{(2)}(x)\cdot 2dx + \left(N_2^{(2)}(x)\right)_{x=2}\cdot 2 = \int_{0.5}^{2} \frac{x - x_1^{(2)}}{x_2^{(2)} - x_1^{(2)}}\cdot 2dx + \left(N_2^{(2)}(x)\right)_{x=2}\cdot 2$$

$$= \int_{0.5}^{2} \frac{x - x_1^{(2)}}{x_2^{(2)} - x_1^{(2)}}\cdot 2dx + 1\cdot 2 = \int_{0.5}^{2} \frac{x-0.5}{2-0.5}\cdot 2dx + 2 = 3.5\,\text{W}$$

$$\{f^{(2)}\} = \begin{Bmatrix} f_1^{(2)} \\ f_2^{(2)} \end{Bmatrix} = \begin{Bmatrix} 1.5 \\ 3.5 \end{Bmatrix}$$

We can now assemble the global right-hand-side vector:

$$\{F\} = \left[L^{(1)}\right]^T \begin{Bmatrix} 0.5 \\ 0.5 \end{Bmatrix} + \left[L^{(2)}\right]^T \begin{Bmatrix} 1.5 \\ 3.5 \end{Bmatrix} = \begin{bmatrix} 1 & 0 \\ 0 & 1 \\ 0 & 0 \end{bmatrix} \begin{Bmatrix} 0.5 \\ 0.5 \end{Bmatrix} + \begin{bmatrix} 0 & 0 \\ 1 & 0 \\ 0 & 1 \end{bmatrix} \begin{Bmatrix} 1.5 \\ 3.5 \end{Bmatrix}$$

$$= \begin{Bmatrix} 0.5 \\ 0.5 \\ 0 \end{Bmatrix} + \begin{Bmatrix} 0 \\ 1.5 \\ 3.5 \end{Bmatrix} = \begin{Bmatrix} 0.5 \\ 2 \\ 3.5 \end{Bmatrix}$$

Thus, the global discrete equations (if we account for the fact that the degree of freedom Θ_1 is restrained to a value of 5 and thus has a nonzero restraint flux, r_1) are:

$$\begin{bmatrix} 4 & -4 & 0 \\ -4 & 5.33 & -1.33 \\ 0 & -1.33 & 1.33 \end{bmatrix} \begin{Bmatrix} 5 \\ \Theta_2 \\ \Theta_3 \end{Bmatrix} - \begin{Bmatrix} 0.5 \\ 2 \\ 3.5 \end{Bmatrix} = \begin{Bmatrix} r_1 \\ 0 \\ 0 \end{Bmatrix}$$

We can isolate the first global equation:

$$4\cdot 5 - 4\cdot \Theta_2 + 0\cdot \Theta_3 = r_1 + 0.5 \tag{i}$$

We do not need to worry about this equation; we only need to use it if we want to find the restraint flux r_1! This calculation can be done after we have obtained Θ_2 and Θ_3. If we now take the remaining two global equations:

$$\begin{bmatrix} 5.33 & -1.33 \\ -1.33 & 1.33 \end{bmatrix} \begin{Bmatrix} \Theta_2 \\ \Theta_3 \end{Bmatrix} + 5 \begin{Bmatrix} -4 \\ 0 \end{Bmatrix} - \begin{Bmatrix} 2 \\ 3.5 \end{Bmatrix} = \begin{Bmatrix} 0 \\ 0 \end{Bmatrix}$$

$$\begin{bmatrix} 5.33 & -1.33 \\ -1.33 & 1.33 \end{bmatrix} \begin{Bmatrix} \Theta_2 \\ \Theta_3 \end{Bmatrix} = -5 \begin{Bmatrix} -4 \\ 0 \end{Bmatrix} + \begin{Bmatrix} 2 \\ 3.5 \end{Bmatrix}$$

The solution of the system of the two equations gives us:

$$\begin{Bmatrix} \Theta_2 \\ \Theta_3 \end{Bmatrix} = \begin{Bmatrix} 6.375^{\circ}\mathrm{C} \\ 9.007^{\circ}\mathrm{C} \end{Bmatrix}$$

Since we already know the value of the restrained $\Theta_1 = 5\,^{\circ}\mathrm{C}$, we know the entire global temperature vector for our system.

$$\begin{Bmatrix} \Theta_1 \\ \Theta_2 \\ \Theta_3 \end{Bmatrix} = \begin{Bmatrix} 5^{\circ}\mathrm{C} \\ 6.375^{\circ}\mathrm{C} \\ 9.007^{\circ}\mathrm{C} \end{Bmatrix}$$

We can now return, if we want, to Equation (i) to find the restraint flux of node 1:

$$4 \cdot 5 - 4 \cdot 6.375 + 0 \cdot 9.007 = r_1 + 0.5 \rightarrow r_1 = -6\mathrm{W}$$

Finally, we can continue the postprocessing, by determining and plotting the approximate temperature field and the approximate flux field inside each element:

Element 1:

$$\Theta^{(1)}(x) = \left[N^{(1)}\right]\left\{\Theta^{(1)}\right\} = \left[N^{(1)}\right]\left[L^{(1)}\right]\{\Theta\}$$

$$q^{(1)}(x) = -\kappa \frac{d\Theta^{(1)}(x)}{dx} = -\kappa\left[B^{(1)}\right]\left\{\Theta^{(1)}\right\} = -\kappa\left[B^{(1)}\right]\left[L^{(1)}\right]\{\Theta\}$$

Element 2:

$$\Theta^{(2)}(x) = \left[N^{(2)}\right]\left\{\Theta^{(2)}\right\} = \left[N^{(2)}\right]\left[L^{(2)}\right]\{\Theta\}$$

$$q^{(2)}(x) = -\kappa \frac{d\Theta^{(2)}(x)}{dx} = -\kappa\left[B^{(2)}\right]\left\{\Theta^{(2)}\right\} = -\kappa\left[B^{(2)}\right]\left[L^{(2)}\right]\{\Theta\}$$

The approximate temperature and flux fields are compared with the corresponding fields of the exact solution in Figure 3.15. The nodal temperatures obtained by the finite element solution are the *exact* ones! Also, the flux in a two-node finite element with first-degree (linear) polynomial shape functions and constant value of κ is constant, which gives rather poor estimates of the exact flux variation. Of course, if we refine our mesh consisting of two-node elements, the accuracy of the approximate flux field will also improve. Finally, it is worth emphasizing that the approximate solution does not exactly satisfy the natural boundary condition of the problem.

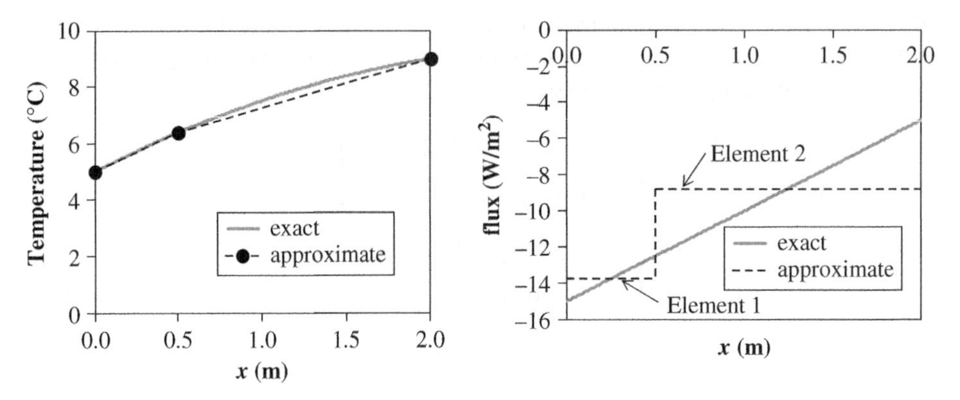

Figure 3.15 Comparison between exact and approximate solutions for system of Example 3.4.

Example 3.5: Effect of mesh refinement for one-dimensional heat conduction
Now, let us repeat the analysis of Example 3.4, for a *refined* mesh (a mesh having more elements than before, and with each element having a smaller size than before), consisting of four two-node elements with linear shape functions. Each element has the same size, as shown in Figure 3.16.

The solution process is the same as before, that is, we calculate the contribution of each element in the global coefficient matrix and the global equivalent nodal flux vector.

For element 1, we have:

$$k_{11}^{(1)} = \int_{x_1^{(1)}}^{x_2^{(1)}} \frac{dN_1^{(1)}(x)}{dx} \kappa^{(1)} A^{(1)} \frac{dN_1^{(1)}(x)}{dx} dx = \int_0^{0.5} \left(-\frac{1}{0.5}\right) 5 \cdot 0.4 \left(-\frac{1}{0.5}\right) dx = 4$$

$$k_{12}^{(1)} = \int_{x_1^{(1)}}^{x_2^{(1)}} \frac{dN_1^{(1)}(x)}{dx} \kappa^{(1)} A^{(1)} \frac{dN_2^{(1)}(x)}{dx} dx = \int_0^{0.5} \left(-\frac{1}{0.5}\right) 5 \cdot 0.4 \frac{1}{0.5} dx = -4$$

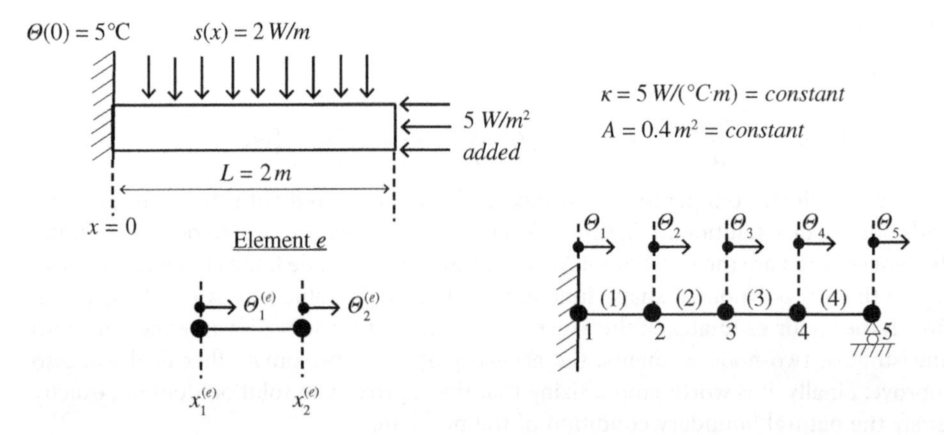

Figure 3.16 Analysis of heat conduction for refined mesh.

$$k_{21}^{(1)} = \int_{x_1^{(1)}}^{x_2^{(1)}} \frac{dN_2^{(1)}(x)}{dx} \kappa^{(1)} A^{(1)} \frac{dN_1^{(1)}(x)}{dx} dx = k_{12}^{(1)} = -4$$

$$k_{22}^{(1)} = \int_{x_1^{(1)}}^{x_2^{(1)}} \frac{dN_2^{(1)}(x)}{dx} \kappa^{(1)} A^{(1)} \frac{dN_2^{(1)}(x)}{dx} dx = \int_0^{0.5} \frac{1}{0.5} 5 \cdot 0.4 \frac{1}{0.5} dx = 4$$

$$\left[k^{(1)}\right] = \begin{bmatrix} k_{11}^{(1)} & k_{12}^{(1)} \\ k_{21}^{(1)} & k_{22}^{(1)} \end{bmatrix} = \begin{bmatrix} 4 & -4 \\ -4 & 4 \end{bmatrix}$$

To calculate $f_1^{(1)}$ and $f_2^{(1)}$, we need to keep in mind that $N_1^{(1)}(x=2) = N_2^{(1)}(x=2) = 0$, since the point $x = 2$ lies outside of element 1. Thus:

$$f_1^{(1)} = \int_{x_1^{(1)}=0}^{x_2^{(1)}=0.5} N_1^{(1)}(x) \cdot 2dx + \left(N_1^{(1)}(x)\right)_{x=2} \cdot 2 = \int_0^{0.5} \frac{x - x_2^{(1)}}{x_1^{(1)} - x_2^{(1)}} \cdot 2dx + \left(N_1^{(1)}(x)\right) \cdot 2$$

$$= \int_0^{0.5} \frac{x - x_2^{(1)}}{x_1^{(1)} - x_2^{(1)}} \cdot 2dx + 0 \cdot 2 = \int_0^{0.5} \frac{x - 0.5}{0 - 0.5} \cdot 2dx = 0.5\,\text{W}$$

$$f_2^{(1)} = \int_{x_1^{(1)}=0}^{x_2^{(1)}=0.5} N_2^{(1)}(x) \cdot 2dx + \left(N_2^{(1)}(x)\right) \cdot 2 = \int_0^{0.5} \frac{x - x_1^{(1)}}{x_2^{(1)} - x_1^{(1)}} \cdot 2dx + \left(N_2^{(1)}(x)\right) \cdot 2$$

$$= \int_0^{0.5} \frac{x - x_1^{(1)}}{x_2^{(1)} - x_1^{(1)}} \cdot 2dx + 0 \cdot 2 = \int_0^{0.5} \frac{x - 0}{0.5 - 0} \cdot 2dx = 0.5\,\text{W}$$

Thus, we have:

$$\left\{f^{(1)}\right\} = \begin{Bmatrix} f_1^{(1)} \\ f_2^{(1)} \end{Bmatrix} = \begin{Bmatrix} 0.5 \\ 0.5 \end{Bmatrix}$$

In the same fashion, we can obtain for the other elements:

$$\left[k^{(2)}\right] = \begin{bmatrix} 4 & -4 \\ -4 & 4 \end{bmatrix}, \begin{Bmatrix} f_1^{(2)} \\ f_2^{(2)} \end{Bmatrix} = \begin{Bmatrix} 0.5 \\ 0.5 \end{Bmatrix}, \left[k^{(3)}\right] = \begin{bmatrix} 4 & -4 \\ -4 & 4 \end{bmatrix},$$

$$\begin{Bmatrix} f_1^{(3)} \\ f_2^{(3)} \end{Bmatrix} = \begin{Bmatrix} 0.5 \\ 0.5 \end{Bmatrix}, \left[k^{(4)}\right] = \begin{bmatrix} 4 & -4 \\ -4 & 4 \end{bmatrix}, \begin{Bmatrix} f_1^{(4)} \\ f_2^{(4)} \end{Bmatrix} = \begin{Bmatrix} 0.5 \\ 2.5 \end{Bmatrix}$$

It is worth emphasizing that we have an effect from the natural boundary only on the nodal flux vector of element 4!

We can now establish the element gather arrays:
Element 1:

$$\left.\begin{matrix}\Theta_1^{(1)} = \Theta_1 \\ \Theta_2^{(1)} = \Theta_2\end{matrix}\right\} \rightarrow \begin{Bmatrix}\Theta_1^{(1)} \\ \Theta_2^{(1)}\end{Bmatrix} = \begin{bmatrix}1 & 0 & 0 & 0 & 0 \\ 0 & 1 & 0 & 0 & 0\end{bmatrix}\begin{Bmatrix}\Theta_1 \\ \Theta_2 \\ \Theta_3 \\ \Theta_4 \\ \Theta_5\end{Bmatrix} \rightarrow \left[L^{(1)}\right] = \begin{bmatrix}1 & 0 & 0 & 0 & 0 \\ 0 & 1 & 0 & 0 & 0\end{bmatrix}$$

For the rest of the elements:

$$\left[L^{(2)}\right] = \begin{bmatrix}0 & 1 & 0 & 0 & 0 \\ 0 & 0 & 1 & 0 & 0\end{bmatrix}, \left[L^{(3)}\right] = \begin{bmatrix}0 & 0 & 1 & 0 & 0 \\ 0 & 0 & 0 & 1 & 0\end{bmatrix}, \left[L^{(4)}\right] = \begin{bmatrix}0 & 0 & 0 & 1 & 0 \\ 0 & 0 & 0 & 0 & 1\end{bmatrix}$$

We can proceed to assemble the global nodal energy equilibrium equations from the corresponding element terms. The global conductance matrix, $[K]$, is:

$$[K] = \left[L^{(1)}\right]^T \left[k^{(1)}\right] \left[L^{(1)}\right] + \left[L^{(2)}\right]^T \left[k^{(2)}\right] \left[L^{(2)}\right] + \left[L^{(3)}\right]^T \left[k^{(3)}\right] \left[L^{(3)}\right] + \left[L^{(4)}\right]^T \left[k^{(4)}\right] \left[L^{(4)}\right]$$

$$= \begin{bmatrix}1 & 0 \\ 0 & 1 \\ 0 & 0 \\ 0 & 0 \\ 0 & 0\end{bmatrix}\begin{bmatrix}4 & -4 \\ -4 & 4\end{bmatrix}\begin{bmatrix}1 & 0 & 0 & 0 & 0 \\ 0 & 1 & 0 & 0 & 0\end{bmatrix} + \begin{bmatrix}0 & 0 \\ 1 & 0 \\ 0 & 1 \\ 0 & 0 \\ 0 & 0\end{bmatrix}\begin{bmatrix}4 & -4 \\ -4 & 4\end{bmatrix}\begin{bmatrix}0 & 1 & 0 & 0 & 0 \\ 0 & 0 & 1 & 0 & 0\end{bmatrix}$$

$$+ \begin{bmatrix}0 & 0 \\ 0 & 0 \\ 1 & 0 \\ 0 & 1 \\ 0 & 0\end{bmatrix}\begin{bmatrix}4 & -4 \\ -4 & 4\end{bmatrix}\begin{bmatrix}0 & 0 & 1 & 0 & 0 \\ 0 & 0 & 0 & 1 & 0\end{bmatrix} + \begin{bmatrix}0 & 0 \\ 0 & 0 \\ 0 & 0 \\ 1 & 0 \\ 0 & 1\end{bmatrix}\begin{bmatrix}4 & -4 \\ -4 & 4\end{bmatrix}\begin{bmatrix}0 & 0 & 0 & 1 & 0 \\ 0 & 0 & 0 & 0 & 1\end{bmatrix}$$

$$[K] = \begin{bmatrix}4 & -4 & 0 & 0 & 0 \\ -4 & 4 & 0 & 0 & 0 \\ 0 & 0 & 0 & 0 & 0 \\ 0 & 0 & 0 & 0 & 0 \\ 0 & 0 & 0 & 0 & 0\end{bmatrix} + \begin{bmatrix}0 & 0 & 0 & 0 & 0 \\ 0 & 4 & -4 & 0 & 0 \\ 0 & -4 & 4 & 0 & 0 \\ 0 & 0 & 0 & 0 & 0 \\ 0 & 0 & 0 & 0 & 0\end{bmatrix} + \begin{bmatrix}0 & 0 & 0 & 0 & 0 \\ 0 & 0 & 0 & 0 & 0 \\ 0 & 0 & 4 & -4 & 0 \\ 0 & 0 & -4 & 4 & 0 \\ 0 & 0 & 0 & 0 & 0\end{bmatrix} + \begin{bmatrix}0 & 0 & 0 & 0 & 0 \\ 0 & 0 & 0 & 0 & 0 \\ 0 & 0 & 0 & 0 & 0 \\ 0 & 0 & 0 & 4 & -4 \\ 0 & 0 & 0 & -4 & 4\end{bmatrix}$$

$$= \begin{bmatrix}4 & -4 & 0 & 0 & 0 \\ -4 & 8 & -4 & 0 & 0 \\ 0 & -4 & 8 & -4 & 0 \\ 0 & 0 & -4 & 8 & -4 \\ 0 & 0 & 0 & -4 & 4\end{bmatrix}$$

Along the same lines, we can assemble the global nodal flux vector $\{f\}$:

$$\{f\} = \left[L^{(1)}\right]^T \begin{Bmatrix} f_1^{(1)} \\ f_2^{(1)} \end{Bmatrix} + \left[L^{(2)}\right]^T \begin{Bmatrix} f_1^{(2)} \\ f_2^{(2)} \end{Bmatrix} + \left[L^{(3)}\right]^T \begin{Bmatrix} f_1^{(3)} \\ f_2^{(3)} \end{Bmatrix} + \left[L^{(4)}\right]^T \begin{Bmatrix} f_1^{(4)} \\ f_2^{(4)} \end{Bmatrix}$$

$$\{f\} = \left[L^{(1)}\right]^T \left\{f^{(1)}\right\} + \left[L^{(2)}\right]^T \left\{f^{(2)}\right\} + \left[L^{(3)}\right]^T \left\{f^{(3)}\right\} + \left[L^{(4)}\right]^T \left\{f^{(4)}\right\}$$

$$= \begin{bmatrix} 1 & 0 \\ 0 & 1 \\ 0 & 0 \\ 0 & 0 \\ 0 & 0 \end{bmatrix} \begin{Bmatrix} 0.5 \\ 0.5 \end{Bmatrix} + \begin{bmatrix} 0 & 0 \\ 1 & 0 \\ 0 & 1 \\ 0 & 0 \\ 0 & 0 \end{bmatrix} \begin{Bmatrix} 0.5 \\ 0.5 \end{Bmatrix} + \begin{bmatrix} 0 & 0 \\ 0 & 0 \\ 1 & 0 \\ 0 & 1 \\ 0 & 0 \end{bmatrix} \begin{Bmatrix} 0.5 \\ 0.5 \end{Bmatrix} + \begin{bmatrix} 0 & 0 \\ 0 & 0 \\ 0 & 0 \\ 1 & 0 \\ 0 & 1 \end{bmatrix} \begin{Bmatrix} 0.5 \\ 2.5 \end{Bmatrix}$$

$$= \begin{Bmatrix} 0.5 \\ 0.5 \\ 0 \\ 0 \\ 0 \end{Bmatrix} + \begin{Bmatrix} 0 \\ 0.5 \\ 0.5 \\ 0 \\ 0 \end{Bmatrix} + \begin{Bmatrix} 0 \\ 0 \\ 0.5 \\ 0.5 \\ 0 \end{Bmatrix} + \begin{Bmatrix} 0 \\ 0 \\ 0 \\ 0.5 \\ 2.5 \end{Bmatrix} = \begin{Bmatrix} 0.5 \\ 1 \\ 1 \\ 1 \\ 2.5 \end{Bmatrix}$$

Thus, the discretized finite element energy balance equations become (if we account for the essential boundary condition):

$$\begin{bmatrix} 4 & -4 & 0 & 0 & 0 \\ -4 & 8 & -4 & 0 & 0 \\ 0 & -4 & 8 & -4 & 0 \\ 0 & 0 & -4 & 8 & -4 \\ 0 & 0 & 0 & -4 & 4 \end{bmatrix} \begin{Bmatrix} 5 \\ \Theta_2 \\ \Theta_3 \\ \Theta_4 \\ \Theta_5 \end{Bmatrix} = \begin{Bmatrix} r_1 + 0.5 \\ 1 \\ 1 \\ 1 \\ 2.5 \end{Bmatrix}$$

Once again we can separate the equations into those corresponding to the unrestrained and the restrained nodal temperatures. The only restrained nodal temperature is that of node 1—thus, we can separate the first global equation of the above system from the other four.

$$4 \cdot 5 - 4 \cdot \Theta_2 + 0 \cdot \Theta_3 + 0 \cdot \Theta_4 + 0 \cdot \Theta_5 = r_1 + 0.5$$

We do not need to worry about this equation, unless we want (during post-processing) to find the restraint flux r_1, just like we did in Example 3.4. If we take the remaining four equations (corresponding to the unrestrained temperatures):

$$\begin{bmatrix} 8 & -4 & 0 & 0 \\ -4 & 8 & -4 & 0 \\ 0 & -4 & 8 & -4 \\ 0 & 0 & -4 & 4 \end{bmatrix} \begin{Bmatrix} \Theta_2 \\ \Theta_3 \\ \Theta_4 \\ \Theta_5 \end{Bmatrix} + 5 \begin{Bmatrix} -4 \\ 0 \\ 0 \\ 0 \end{Bmatrix} = \begin{Bmatrix} 1 \\ 1 \\ 1 \\ 2.5 \end{Bmatrix}$$

$$\rightarrow \begin{bmatrix} 8 & -4 & 0 & 0 \\ -4 & 8 & -4 & 0 \\ 0 & -4 & 8 & -4 \\ 0 & 0 & -4 & 4 \end{bmatrix} \begin{Bmatrix} \Theta_2 \\ \Theta_3 \\ \Theta_4 \\ \Theta_5 \end{Bmatrix} = -5 \begin{Bmatrix} -4 \\ 0 \\ 0 \\ 0 \end{Bmatrix} + \begin{Bmatrix} 1 \\ 1 \\ 1 \\ 2.5 \end{Bmatrix}$$

We can solve this discrete system of four equations for the unknown nodal temperatures:

$$\begin{Bmatrix} \Theta_2 \\ \Theta_3 \\ \Theta_4 \\ \Theta_5 \end{Bmatrix} = \begin{Bmatrix} 6.375 \\ 7.5 \\ 8.375 \\ 9 \end{Bmatrix}, \text{thus } \{\Theta\} = \begin{Bmatrix} \Theta_1 \\ \Theta_2 \\ \Theta_3 \\ \Theta_4 \\ \Theta_5 \end{Bmatrix} = \begin{Bmatrix} 5 \\ 6.375 \\ 7.5 \\ 8.375 \\ 9 \end{Bmatrix}$$

From the global nodal temperature vector, we can determine the approximate temperature field and the approximate flux field inside each element e ($e = 1,2,3,4$):

$$\Theta^{(e)}(x) = \left[N^{(e)}\right]\left\{\Theta^{(e)}\right\} = \left[N^{(e)}\right]\left[L^{(e)}\right]\{\Theta\}$$

$$q^{(e)}(x) = -\kappa \frac{d\Theta^{(e)}(x)}{dx} = -\kappa\left[B^{(e)}\right]\left\{\Theta^{(e)}\right\} = -\kappa\left[B^{(e)}\right]\left[L^{(e)}\right]\{\Theta\}$$

For example, for element 2:

$$\left\{\Theta^{(2)}\right\} = \left[L^{(2)}\right]\{\Theta\} = \begin{bmatrix} 0 & 1 & 0 & 0 & 0 \\ 0 & 0 & 1 & 0 & 0 \end{bmatrix} \begin{Bmatrix} 5 \\ 6.375 \\ 7.5 \\ 8.375 \\ 9 \end{Bmatrix} = \begin{Bmatrix} \Theta_1^{(2)} \\ \Theta_2^{(2)} \end{Bmatrix} = \begin{Bmatrix} 6.375 \\ 7.5 \end{Bmatrix}$$

$$\Theta^{(2)}(x) = \left[N^{(2)}\right]\left\{\Theta^{(2)}\right\} = \left[N_1^{(2)}(x) \quad N_2^{(2)}(x)\right] \begin{Bmatrix} \Theta_1^{(2)} \\ \Theta_2^{(2)} \end{Bmatrix} = N_1^{(2)}(x)\cdot\Theta_1^{(2)} + N_2^{(2)}(x)\cdot\Theta_2^{(2)}$$

$$= \frac{x - x_2^{(2)}}{x_1^{(2)} - x_2^{(2)}}\cdot\Theta_1^{(2)} + \frac{x - x_1^{(2)}}{x_2^{(2)} - x_1^{(2)}}\cdot\Theta_2^{(2)} = \frac{x-1}{0.5-1}\cdot 6.375$$

$$+ \frac{x-0.5}{1-0.5}\cdot 7.5 = 2.25x + 5.25 \; (x \; in \, \text{m}, \; \Theta \; in \, ^\circ\text{C})$$

$$q^{(2)}(x) = -\kappa\frac{d\Theta^{(2)}(x)}{dx} = -\kappa\left[B^{(2)}\right]\left[L^{(2)}\right]\{\Theta\} = -\kappa\left[B^{(2)}\right]\left\{\Theta^{(2)}\right\} = -\kappa\left[-\frac{1}{\ell^{(2)}}\;\frac{1}{\ell^{(2)}}\right]\left\{\begin{matrix}\Theta_1^{(2)}\\ \Theta_2^{(2)}\end{matrix}\right\}$$

$$= -5\left[-\frac{1}{0.5}\;\frac{1}{0.5}\right]\left\{\begin{matrix}6.375\\ 7.5\end{matrix}\right\} = -5\left[\frac{-1}{0.5}\cdot 6.375 + \frac{1}{0.5}\cdot 7.5\right] = -11.25\frac{\text{W}}{\text{m}^2}$$

Using the same approach for the other elements, we can determine the approximate variation of the temperature and flux and compare them to those of the exact solution, as shown in Figure 3.17.

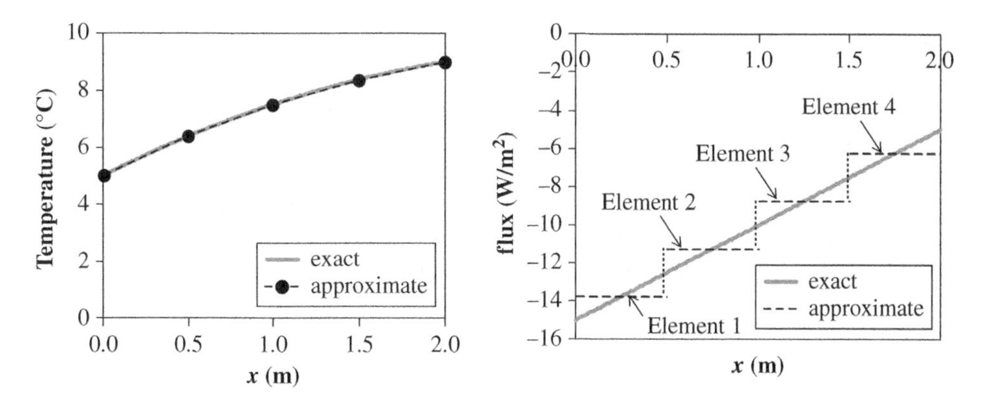

Figure 3.17 Comparison between exact and approximate solutions for system of Example 3.5.

Notice that the approximate temperature field for the refined mesh is almost identical to the exact temperature field. Of course, the approximate flux field still has a piecewise constant variation, and although it shows better match with the exact solution than what we obtained for the two-element mesh in Example 3.4, it is not nearly as close to the exact solution as the temperature field. As we refine more and more the mesh, we will verify that the approximate flux field gets closer and closer to the exact flux field. Eventually, since our approximate solution satisfies the continuity and completeness requirements, we ensure convergence; that is, when our element size becomes "extremely small," both the approximate temperature and flux fields will be practically identical to the exact ones!

3.7 Numerical Integration—Gauss Quadrature

In the previous sections, we saw how to obtain the element coefficient matrix and right-hand-side vector for elasticity and heat conduction. These quantities require the evaluation of integrals, which are frequently cumbersome or even impossible to calculate analytically. Furthermore, analytical evaluation of integrals cannot be easily programmed in general-purpose finite element programs. For this reason, the calculation of the integrals is conducted using *quadrature* (i.e., numerical integration).

Let us suppose that we have to integrate a function of variable ξ, $f(\xi)$, in the interval between ξ_a and ξ_b. Quadrature rules replace the need to calculate the integral with the need to evaluate the function at N_g points $\xi_1, \xi_2, \ldots, \xi_{Ng}$, and then approximate the integral I with a *linear combination* of the evaluated values. That is:

$$I = \int_{\xi_a}^{\xi_b} f(\xi)d\xi \approx f(\xi_1) \cdot W_1 + f(\xi_2) \cdot W_2 + \ldots + f\left(\xi_{N_g}\right) \cdot W_{N_g} = \sum_{i=1}^{N_g} \left(f(\xi_i) \cdot W_i\right) \quad (3.7.1)$$

where the locations $\xi = \xi_1$, $\xi = \xi_2$, ..., $\xi = \xi_{Ng}$, are called the *integration (or quadrature) points*, and W_1, W_2, ..., W_{Ng} are called the *weights* of the integration points.

The most commonly employed method of quadrature for finite element analyses is the *Gauss (more accurately: Gauss-Legendre) quadrature*. A detailed discussion of the Gauss quadrature rule is provided in Stroud and Secrest (1966). For a given number of quadrature points, the specific rule stipulates the location ξ_i of each point, as well as the corresponding weight W_i. For obvious reasons, the quadrature points are also called *Gauss points*. Table 3.1 gives the locations of the Gauss points and the weight values for cases that we have one, two, three, and four Gauss points. It is worth mentioning that quadrature with more than four points is seldom used by finite element programs for the solution of one-dimensional problems. The location and weights of the Gauss points are given by solving a nonlinear optimization problem: "What should be the ξ_i's and W_i's so that we *maximize* the degree of polynomial functions that can be exactly integrated by Gauss quadrature?" Several examples of solving this optimization problem are provided by Fish and Belytschko (2007).

Table 3.1 Gauss Quadrature Rule.

Number of Gauss Points	Locations, ξ_i	Weights, W_i
1	$\xi_1 = 0$	$W_1 = 2$
2	$\xi_1 = \frac{1}{\sqrt{3}} = 0.5773502692$ $\xi_2 = -\frac{1}{\sqrt{3}} = -0.5773502692$	$W_2 = 1$
3	$\xi_1 = -\sqrt{\frac{3}{5}} = -0.7745966692$ $\xi_2 = 0$ $\xi_3 = \sqrt{\frac{3}{5}} = 0.7745966692$	$W_1 = \frac{5}{9} = 0.5555555556$ $W_2 = \frac{8}{9} = 0.8888888889$ $W_3 = \frac{5}{9} = 0.5555555556$
4	$\xi_1 = -\sqrt{\frac{\left(3+2\sqrt{6/5}\right)}{7}} = -0.8611363116$ $\xi_2 = -\sqrt{\frac{\left(3-2\sqrt{6/5}\right)}{7}} = -0.3399810436$ $\xi_3 = +\sqrt{\frac{\left(3-2\sqrt{6/5}\right)}{7}} = 0.3399810436$ $\xi_4 = \sqrt{\frac{\left(3+2\sqrt{6/5}\right)}{7}} = 0.8611363116$	$W_1 = \frac{18-\sqrt{30}}{36} = 0.3478548451$ $W_2 = \frac{18+\sqrt{30}}{36} = 0.6521451549$ $W_3 = \frac{18+\sqrt{30}}{36} = 0.6521451549$ $W_4 = \frac{18-\sqrt{30}}{36} = 0.3478548451$

The accuracy of Gaussian quadrature can be quantified by determining what is the maximum degree of polynomial function that can be exactly integrated for a given number N_q of quadrature points. Box 3.7.1 provides this information.

Box 3.7.1 Accuracy of Gaussian Quadrature
It can be proven that if we use Gauss quadrature with N_q Gauss points, we can exactly integrate a polynomial function of degree up to $N = 2N_q - 1$.

Remark 3.7.1: The sum of the weight coefficients for any type of Gaussian quadrature is equal to 2 (*why?*). ■

Example 3.6: Gauss quadrature for quadratic function
In this example, we will use one- and two-point Gauss quadrature to calculate the integral $I = \int_{-1}^{1} (3\xi^2 - 2)d\xi$.

We have:

$$f(\xi) = 3\xi^2 - 2$$

The exact value of the integral is:

$$I = \int_{-1}^{1} f(\xi)d\xi = \int_{-1}^{1} (3\xi^2 - 2)d\xi = \left[\xi^3 - 2\xi\right]_{-1}^{1} = -2$$

If we use one-point Gauss quadrature, we will be able to exactly calculate polynomials of degree up to $2 \cdot 1 - 1 = 1$. We have $\xi_1 = 0,\ W_1 = 2$

$$I \approx f(\xi_1) \cdot W_1 = \left[3\xi^2 - 2\right]_{\xi_1 = 0} \cdot 2 = \left[3 \cdot 0^2 - 2\right] \cdot 2 = -4$$

As expected, one-point Gauss quadrature did not give the exact value of the integral.

A two-point Gauss quadrature is expected to exactly give the integrals of polynomials of degree up to $2 \cdot 2 - 1 = 3$; thus, it should provide the exact value of the integral. We have:

$$\xi_1 = -\frac{1}{\sqrt{3}},\ W_1 = 1 \qquad \xi_2 = \frac{1}{\sqrt{3}},\ W_2 = 1$$

$$I \approx f(\xi_1) \cdot W_1 + f(\xi_2) \cdot W_2 = \left[3\xi^2 - 2\right]_{\xi_1 = -\frac{1}{\sqrt{3}}} \cdot 1 + \left[3\xi^2 - 2\right]_{\xi_2 = \frac{1}{\sqrt{3}}} \cdot 1$$

$$= \left[3 \cdot \left(-\frac{1}{\sqrt{3}}\right)^2 - 2\right] \cdot 1 + \left[3 \cdot \left(\frac{1}{\sqrt{3}}\right)^2 - 2\right] \cdot 1 = -2$$

■

Example 3.7: Gauss quadrature for fourth-order polynomial function
We are going to use two- and three-point Gauss quadrature to calculate the integral:

$$I = \int_{-1}^{1} (5\xi^4 + 2)d\xi$$

We have:

$$f(\xi) = 5\xi^4 + 2$$

The exact value of the integral can be obtained by analytical evaluation:

$$I = \int_{-1}^{1} f(\xi)d\xi = \int_{-1}^{1} (5\xi^4 + 2)d\xi = \left[\xi^5 + 2\xi\right]_{-1}^{1} = 6$$

If we use two Gauss points:

$$\xi_1 = -\frac{1}{\sqrt{3}},\ W_1 = 1 \qquad \xi_2 = \frac{1}{\sqrt{3}},\ W_2 = 1$$

$$I \approx f(\xi_1)\cdot W_1 + f(\xi_2)\cdot W_2 = \left[5\xi^4 + 2\right]_{\xi_1 = -\frac{1}{\sqrt{3}}}\cdot 1 + \left[5\xi^4 + 2\right]_{\xi_2 = \frac{1}{\sqrt{3}}}\cdot 1$$

$$= \left[5\cdot\left(-\frac{1}{\sqrt{3}}\right)^4 + 2\right]\cdot 1 + \left[5\cdot\left(\frac{1}{\sqrt{3}}\right)^4 + 2\right]\cdot 1 = 5.111$$

If we use three Gauss points:

$$\xi_1 = -\sqrt{\frac{3}{5}},\ W_1 = \frac{5}{9} \qquad \xi_2 = 0,\ W_2 = \frac{8}{9} \qquad \xi_3 = \sqrt{\frac{3}{5}},\ W_3 = \frac{5}{9}$$

$$I \approx f(\xi_1)\cdot W_1 + f(\xi_2)\cdot W_2 + f(\xi_3)\cdot W_3 = \left[5\xi^4 + 2\right]_{\xi_1 = -\sqrt{\frac{3}{5}}}\cdot\frac{5}{9} + \left[5\xi^4 + 2\right]_{\xi_2 = 0}\cdot\frac{8}{9} + \left[5\xi^4 + 2\right]_{\xi_3 = \sqrt{\frac{3}{5}}}\cdot\frac{5}{9}$$

$$= \left[5\cdot\left(-\sqrt{\frac{3}{5}}\right)^4 + 2\right]\cdot\frac{5}{9} + \left[5\cdot(0)^4 + 2\right]\cdot\frac{8}{9} + \left[5\cdot\left(\sqrt{\frac{3}{5}}\right)^4 + 2\right]\cdot\frac{5}{9} = 6$$

As expected, we obtained the exact value of the integral for $N_g = 3$, because the highest order of polynomial that can be exactly integrated with three points is $2\cdot N_g - 1 = 2\cdot 3 - 1 = 5$, and we needed to integrate a fourth-order polynomial. ■

So far, we have seen how Gaussian quadrature can be used for integrals where the interval of integration is between $\xi = -1$ and $\xi = 1$. Of course, the actual integrals that we may encounter can have different limits. For instance, the term $k_{11}^{(1)}$ in Example 3.2 had an integration interval between $x = 0$ and $x = 5$. We can still use Gauss quadrature for such cases, after establishing a *change of coordinates*. We will discuss the general case where we want to calculate the integral $\int_{x_1}^{x_2} f(x)\,dx$. Instead of integrating with respect to x, we integrate with respect to ξ, which obtains values $\xi = -1$ at $x = x_1$ and $\xi = 1$ at $x = x_2$. This change of coordinates will rely on a *mapping*, as shown in Figure 3.18, which will give the *physical* (Cartesian) coordinate x as a function of the coordinate ξ, which we will call the *parent or parametric* coordinate. We can then use the *change of coordinates formula for integration*, which is summarized in Figure 3.18. Per the specific figure, we can also define a parent or parametric domain (corresponding to the values that ξ can obtain) and a physical domain (corresponding to the values that x can obtain).

To establish the mathematical expression for the mapping, $x(\xi)$, we write down two equations, corresponding to the mapping values at the endpoints of the domain. From Figure 3.18, we know that the point $\xi_1 = -1$ is mapped to the point $x = x_1$. Thus, we can write:

$$x(\xi = \xi_1 = -1) = x_1 \tag{3.7.2a}$$

Similarly, we know that the point $\xi_2 = 1$ is mapped to the point $x = x_2$:

$$x(\xi = \xi_2 = 1) = x_2 \tag{3.7.2b}$$

The simplest expression that can satisfy the two mapping conditions of Equation (3.7.2a,b) is a *linear mapping* function:

$$x(\xi) = a \cdot \xi + b \tag{3.7.3}$$

If we use Equation (3.7.3) in Equations (3.7.2a,b), we obtain:

$$a \cdot (-1) + b = x_1 \tag{3.7.4a}$$

$$a \cdot 1 + b = x_2 \tag{3.7.4b}$$

We can then solve and obtain the parameters a and b of the mapping in terms of x_1 and x_2:

$$a = \frac{x_2 - x_1}{2} \tag{3.7.5a}$$

$$b = \frac{x_1 + x_2}{2} \tag{3.7.5b}$$

Remark 3.7.2: We could also use the linear polynomial shape functions of a two-node element for the mapping! That is, we could directly establish the mapping as follows:

$$x(\xi) = N_1(\xi) \cdot x_1 + N_2(\xi) \cdot x_2 \tag{3.7.6}$$

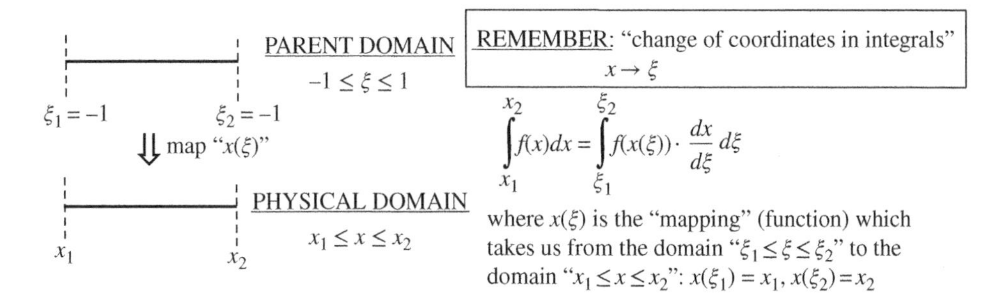

Figure 3.18 Mapping and change of coordinates for integration.

The two shape functions $N_1(\xi)$, $N_2(\xi)$ in Equation (3.7.6) are obtained by treating the parent domain in Figure 3.18 as a two-node finite element. Thus, we can write:

$$N_1(\xi) = \frac{\xi - \xi_2}{\xi_1 - \xi_2} \tag{3.7.7a}$$

and

$$N_2(\xi) = \frac{\xi - \xi_1}{\xi_2 - \xi_1} \tag{3.7.7b}$$

If we now set $\xi_1 = -1$, $\xi_2 = 1$, we obtain:

$$N_1(\xi) = \frac{1}{2}(1 - \xi) \tag{3.7.8a}$$

and

$$N_2(\xi) = \frac{1}{2}(1 + \xi) \tag{3.7.8b}$$

Using the element's shape functions for the coordinate mapping constitutes a core feature of the so-called *isoparametric element* concept, which will be described in detail in Section 6.4. ■

After the mapping is established, we can transform a definite integral of any function $f(x)$ in the physical domain as a corresponding integral in the parametric domain:

$$\int_{x_1}^{x_2} f(x)dx = \int_{-1}^{1} f(x(\xi)) \cdot J d\xi \tag{3.7.9}$$

where J is the *Jacobian* of the mapping, which for the one-dimensional mapping is equal to the derivative of the physical coordinate with the parametric coordinate:

$$J = \frac{dx}{d\xi} \tag{3.7.10}$$

The integral in the right-hand side of Equation (3.7.9) is from -1 to 1. Thus, we can use Gaussian quadrature to evaluate it:

$$\int_{x_1}^{x_2} f(x)dx = \int_{-1}^{1} f(\xi) J d\xi \approx f(\xi_1) \cdot J_1 \cdot W_1 + f(\xi_2) \cdot J_2 \cdot W_2 + \ldots + f\left(\xi_{N_g}\right) \cdot J_n \cdot W_{N_g} = \sum_{i=1}^{N_g} (f(\xi_i) \cdot J_i \cdot W_i) \tag{3.7.11}$$

where

$$J_i = \left.\frac{dx}{d\xi}\right|_{\xi = \xi_i} \tag{3.7.12}$$

Thus, we have reached Equation (3.7.11), by virtue of which we can use Gaussian quadrature to evaluate any definite integral.

Example 3.8: Gauss quadrature for integrals in physical domain
Let us use Gaussian quadrature to calculate the term $k_{11}^{(1)}$ from Example 3.2:

$$k_{11}^{(1)} = \int_0^5 \left(-\frac{1}{5}\right)5000\cdot\left(1+\frac{x^2}{400}\right)\left(-\frac{1}{5}\right)dx = \int_0^5 \frac{1}{5}5000\cdot\left(1+\frac{x^2}{400}\right)\frac{1}{5}dx$$

For the specific integral, we have $x_1 = 0$, $x_2 = 5$. Using Equation (3.7.3) and (3.7.5a,b):

$$x(\xi) = \frac{5-0}{2}\xi + \frac{5+0}{2} = \frac{5}{2}\xi + \frac{5}{2}$$

$$k_{11}^{(1)} = \int_{-1}^{1} 200\cdot\left(1+\frac{\left(\frac{5}{2}\xi+\frac{5}{2}\right)^2}{400}\right)\cdot\frac{5}{2}d\xi = \int_{-1}^{1} f(\xi)\cdot J d\xi, \text{ where } f(\xi) = 200\cdot\left(1+\frac{\left(\frac{5}{2}\xi+\frac{5}{2}\right)^2}{400}\right), J = \frac{dx}{d\xi} = \frac{5}{2}$$

We can now use two-point Gaussian quadrature (*why two points?*) and obtain:

$$k_{11}^{(1)} = f(\xi_1)\cdot J_1\cdot W_1 + f(\xi_2)\cdot J_2\cdot W_2 \rightarrow$$

$$\rightarrow k_{11}^{(1)} = 200\cdot\left(1+\frac{\left(\frac{5}{2}\left(-\frac{1}{\sqrt{3}}\right)+\frac{5}{2}\right)^2}{400}\right)\cdot\frac{5}{2}\cdot 1 + 200\cdot\left(1+\frac{\left(\frac{5}{2}\left(\frac{1}{\sqrt{3}}\right)+\frac{5}{2}\right)^2}{400}\right)\cdot\frac{5}{2}\cdot 1 \rightarrow$$

$$\rightarrow k_{11}^{(1)} = 1020.83,$$

which is the same value that we obtained with exact integration in Example 3.2. ■

We are now ready to establish expressions for Gaussian quadrature to evaluate the element coefficient arrays and right-hand-side vectors in finite element analysis for one-dimensional elasticity. We can simply plug Equation (3.3.10) into Equation (3.7.11) to obtain:

$$\left[k^{(e)}\right] = \int_{x_1^{(e)}}^{x_2^{(e)}} \left[B^{(e)}\right]^T E^{(e)} A^{(e)} \left[B^{(e)}\right] dx = \int_{-1}^{1} \left[B^{(e)}\right]^T E^{(e)} A^{(e)} \left[B^{(e)}\right] J d\xi$$

$$= \sum_{g=1}^{N_g} \left(\left[B^{(e)}\right]_g^T E_g^{(e)} A_g^{(e)} \left[B^{(e)}\right]_g J_g \cdot W_g\right) \tag{3.7.13}$$

where $\left[B^{(e)}\right]_g = \left[B^{(e)}\right]_{x=x_g}$, $E_g^{(e)} = E^{(e)}(x = x_g)$, and $A_g^{(e)} = A^{(e)}(x = x_g)$ are the strain-displacement matrix, Young's modulus and cross-sectional area, respectively, evaluated at the location x_g of the quadrature point g in the physical domain, obtained by plugging $\xi = \xi_g$ in the mapping $x(\xi)$.

Similarly, we can use Gaussian quadrature for the body-force part of the equivalent nodal force vector:

$$\left\{f^{(e)}\right\} = \int_{x_1^{(e)}}^{x_2^{(e)}} \left[N^{(e)}\right]^T \cdot b^{(e)}(x)dx + \left[\left[N^{(e)}\right]^T\right]_{\Gamma_t} \left(A^{(e)} \cdot \bar{t}\right)_{\Gamma_t}$$

$$= \int_{-1}^{1} \left[N^{(e)}\right]^T \cdot b^{(e)} J d\xi + \left[\left[N^{(e)}\right]^T\right]_{\Gamma_t} \left(A^{(e)} \cdot \bar{t}\right)_{\Gamma_t} \rightarrow$$

$$\rightarrow \left\{f^{(e)}\right\} = \sum_{g=1}^{N_g} \left(\left[N^{(e)}\right]_g^T \cdot b_g^{(e)} \cdot J_g \cdot W_g\right) + \left[\left[N^{(e)}\right]^T\right]_{\Gamma_t} \left(A^{(e)} \cdot \bar{t}\right)_{\Gamma_t} \tag{3.7.14}$$

where $\left[N^{(e)}\right]_g = \left[N^{(e)}\right]_{x = x_g}$ is the value of the shape function array at quadrature point g and $b_g^{(e)} = b^{(e)}\left(x = x_g\right)$ is the distributed force value at quadrature point g.

Now, we also need to add that, for each Gauss point g, with parametric coordinate ξ_g, we can find:

- The physical coordinate of the quadrature point:

$$x_g = x\left(\xi_g\right) = \frac{\ell^{(e)}}{2}\xi_g + \frac{x_1^{(e)} + x_2^{(e)}}{2} \tag{3.7.15}$$

where $\ell^{(e)} = x_2^{(e)} - x_1^{(e)}$

- The Jacobian of the mapping:

$$J_g = \frac{\ell^{(e)}}{2} \tag{3.7.16}$$

Parameter $\ell^{(e)}$ in Equations (3.6.15) and (3.6.16) is the length of the element.

We can finally cast an algorithm to obtain the element stiffness matrix and equivalent nodal force vector with Gaussian quadrature, summarized in Box 3.7.2.

Box 3.7.2 Calculation of Element Stiffness Matrix and Equivalent Nodal Force Vector with Gaussian Quadrature

FOR each Gauss point $\xi = \xi_g$, where $g = 1,2,\ldots,N_g$

1) Calculate $x_g = x(\xi_g)$ using Equation (3.7.15)
2) Calculate $\left[B^{(e)}\right]_g = \left[B^{(e)}\right]_{x=x_g}$, $\left[N^{(e)}\right]_g = \left[N^{(e)}\right]_{x=x_g}$, $J_g = \frac{\ell^{(e)}}{2}$, $E_g^{(e)} = E^{(e)}\left(x = x_g\right)$, $A_g^{(e)} = A^{(e)}$ $\left(x = x_g\right)$, $b_g^{(e)} = b^{(e)}\left(x = x_g\right)$
3) Calculate contribution of quadrature point to stiffness matrix:

$$\left[B^{(e)}\right]_g^T E_g^{(e)} A_g^{(e)} \left[B^{(e)}\right]_g J_g \cdot W_g$$

4) Calculate contribution of quadrature point to equivalent nodal force vector:

$$\left[N^{(e)}\right]_g^T \cdot b_g^{(e)} \cdot J_g \cdot W_g$$

END

Calculate:

$$\left[k^{(e)}\right] = \sum_{g=1}^{N_g} \left(\left[B^{(e)}\right]_g^T E_g^{(e)} A_g^{(e)} \left[B^{(e)}\right]_g J_g \cdot W_g \right)$$

and

$$\left\{f^{(e)}\right\} = \sum_{g=1}^{N_g} \left(\left[N^{(e)}\right]_g^T \cdot b_g^{(e)} \cdot J_g \cdot W_g \right) + \left[\left[N^{(e)}\right]^T\right]_{\Gamma_t} \left(A^{(e)} \cdot \bar{t}\right)_{\Gamma_t}$$

We can also use Gaussian quadrature for heat conduction, to find the element conductance matrices and nodal flux vectors. Specifically, we can establish the following two equations.

$$\begin{aligned}\left[k^{(e)}\right] &= \int_{x_1^{(e)}}^{x_2^{(e)}} \left[B^{(e)}\right]^T \kappa^{(e)} A^{(e)} \left[B^{(e)}\right] dx = \int_{-1}^{1} \left[B^{(e)}\right]^T \kappa^{(e)} A^{(e)} \left[B^{(e)}\right] J d\xi \\ &= \sum_{g=1}^{N_g} \left(\left[B^{(e)}\right]_g^T \kappa_g^{(e)} A_g^{(e)} \left[B^{(e)}\right]_g J_g \cdot W_g \right)\end{aligned} \tag{3.7.17}$$

$$\begin{aligned}\left\{f^{(e)}\right\} &= \int_{x_1^{(e)}}^{x_2^{(e)}} \left[N^{(e)}\right]^T \cdot s^{(e)}(x) dx - \left[\left[N^{(e)}\right]^T\right]_{\Gamma_t} \left(A^{(e)} \cdot \bar{q}\right)_{\Gamma_t} \\ &= \int_{-1}^{1} \left[N^{(e)}\right]^T \cdot s^{(e)} J d\xi - \left[\left[N^{(e)}\right]^T\right]_{\Gamma_t} \left(A^{(e)} \cdot \bar{q}\right)_{\Gamma_t}\end{aligned}$$

$$\rightarrow \left\{f^{(e)}\right\} = \sum_{g=1}^{N_g} \left(\left[N^{(e)}\right]_g^T \cdot s_g^{(e)} \cdot J_g \cdot W_g \right) - \left[\left[N^{(e)}\right]^T\right]_{\Gamma_t} \left(A^{(e)} \cdot \bar{q}\right)_{\Gamma_t} \tag{3.7.18}$$

Remark 3.7.3: The required number of Gauss points to exactly integrate the element coefficient array depends on what polynomial functions are to be integrated to obtain each term. For example, if we have a two-node element with linear shape functions and constant A and E (or κ for heat conduction), all the terms in the integral for the coefficient array will be constant, so a single Gauss point will be adequate! For a three-node (quadratic) element with constant A and E (or κ for heat conduction), it can be found that the terms to be integrated are quadratic polynomials; thus, a two-point Gauss quadrature will be required to exactly integrate the terms. Similar considerations can be established for the integrals of the right-hand-side vector of the element. ■

3.8 Convergence of One-Dimensional Finite Element Method

The accuracy of the finite element method has so far been assessed in examples by eye, that is, by comparing the plots of the approximate displacement and stress (or temperature and flux) fields to the corresponding plots of the exact solution. We will now go on to establish quantitative measures of the *error*, that is, the disagreement between the approximate fields and the exact fields. The quantities that we will be using are *norms of functions*.

A norm is a *nonnegative* quantity that provides a measure of the "size" of a mathematical entity. For example, for a vector $\{v\}$ with n components, $v_1, v_2, \ldots, v_n$, we can define a norm as follows:

$$\|\{v\}\| = \sqrt{\sum_{i=1}^{n}\left(v_i^2\right)} = \sqrt{v_1^2 + v_2^2 + \ldots + v_n^2} \tag{3.8.1}$$

Equation (3.8.1) gives the *magnitude* of the vector (there are other types of norms that can be used for vectors and matrices, but the magnitude is the most frequently used one for vectors).

Based on the same rationale, we can define the *norm of a function*, $v(x)$, which is defined over the domain $0 \le x \le L$, as follows:

$$\|v(x)\|_0 = \left(\frac{1}{L}\int_0^L (v(x))^2 dx\right)^{\frac{1}{2}} \tag{3.8.2}$$

The integrand in Equation (3.8.2) is the square of a function, so it is non-negative. The integral of a non-negative function is also a non-negative number. Just like the norm of a vector measures its magnitude, the norm of the function is the measure of the average "magnitude" of the function. We will now proceed to establish some basic error definitions, focusing on elasticity, where the field function is the displacement field.

Let us call $u(x)$ the exact solution of the strong form in one-dimensional elasticity, and $u_{(h)}(x)$ the corresponding approximate, finite element solution. The error in displacements, $e_{(h)}(x)$, is defined as $e_{(h)}(x) = u_{(h)}(x) - u(x)$. We can now establish an *error norm for the displacements* as follows:

$$\|e\|_0 = \left\|e_{(h)}(x)\right\|_0 = \left\|u_{(h)}(x) - u(x)\right\|_0 = \left(\frac{1}{L}\int_0^L \left(u_{(h)}(x) - u(x)\right)^2 dx\right)^{\frac{1}{2}} \tag{3.8.3}$$

The error norm of Equation (3.8.3) is also called the *Lebesque L_2-error norm for the finite element solution* (Fish and Belytschko 2007). In numerical analysis, it is always better to normalize error quantities. To provide a quick, qualitative justification, an error of

0.01 inches may or may not be significant, depending on what is the actual magnitude of the displacements. An error of 0.01 inches for a displacement with a magnitude of 2 inches might be deemed negligible, but the same error for a displacement magnitude of 0.015 inches is another story. In the context of the discussion provided herein, we will use the norm of the exact solution as a measure of that solution's magnitude, and define the *normalized error norm* as follows:

$$\bar{e}_0 = \frac{\left\| u_{(h)}(x) - u(x) \right\|_0}{\left\| u(x) \right\|_0} = \frac{\left(\frac{1}{L} \int_0^L \left(u_{(h)}(x) - u(x) \right)^2 dx \right)^{\frac{1}{2}}}{\left(\frac{1}{L} \int_0^L \left(u(x) \right)^2 dx \right)^{\frac{1}{2}}} \tag{3.8.4}$$

Example 3.9: Calculation of error norms for finite element analysis
We are given the structure shown in Figure 3.19a, for which we have solved the strong form and found the exact displacement field $u(x)$:

$$u(x) = -\frac{x^2}{10000} + 0.0039x + 0.05$$

We have also obtained a finite element solution, using three nodes and two finite elements with linear approximation. We have obtained the nodal displacements, as shown in Figure 3.19b. We will now calculate the error norm and the normalized error norm.

Given the nodal displacements of the finite element solution, we can find the approximate displacement field in each element.

Element 1:

$$x_1^{(1)} = 0,\ x_2^{(1)} = 10,\ \ N_1^{(1)}(x) = \frac{x - x_2^{(1)}}{x_1^{(1)} - x_2^{(1)}} = \frac{10 - x}{10},\ N_2^{(1)}(x) = \frac{x - x_1^{(1)}}{x_2^{(1)} - x_1^{(1)}} = \frac{x}{10}$$

$$u^{(1)}(x) = N_1^{(1)}(x) \cdot u_1^{(1)} + N_2^{(1)}(x) \cdot u_2^{(1)} = \left(1 - \frac{x}{10}\right) \cdot u_1 + \frac{x}{10} \cdot u_2 = \left(1 - \frac{x}{10}\right) \cdot 0.05$$
$$+ \frac{x}{10} \cdot 0.079 = 0.05 + 0.0029x$$

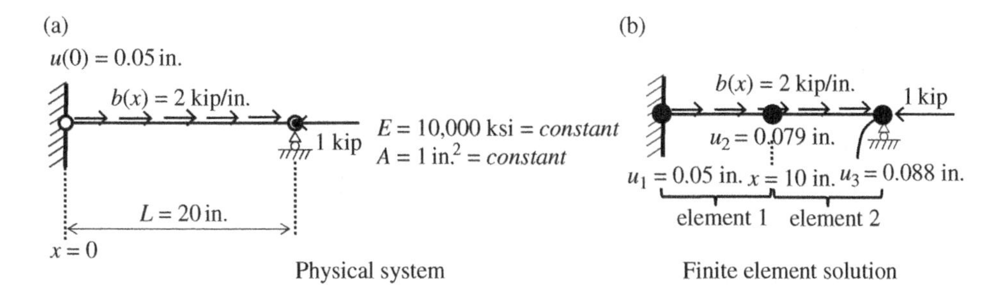

Figure 3.19 Example system for calculation of the error norm.

Element 2:

$$x_1^{(2)} = 10, x_2^{(2)} = 20, \ N_1^{(2)}(x) = \frac{x - x_2^{(2)}}{x_1^{(2)} - x_2^{(2)}} = \frac{20 - x}{10}, \ N_2^{(2)}(x) = \frac{x - x_1^{(2)}}{x_2^{(2)} - x_1^{(2)}} = \frac{x - 10}{10}$$

$$u^{(2)}(x) = N_1^{(2)}(x) \cdot u_1^{(2)} + N_2^{(2)}(x) \cdot u_2^{(2)} = \left(2 - \frac{x}{10}\right) \cdot u_2 + \left(\frac{x}{10} - 1\right) \cdot u_3$$

$$= \left(2 - \frac{x}{10}\right) \cdot 0.079 + \left(\frac{x}{10} - 1\right) \cdot 0.088 \rightarrow u^{(2)}(x) = 0.0009x + 0.07$$

The displacement error norm is then given by use of Equation (3.8.3):

$$\|e\|_0 = \left(\frac{1}{20}\int_0^{20} \left(u_{(h)}(x) - u(x)\right)^2 dx\right)^{\frac{1}{2}}$$

Since we have piecewise (finite element) approximation, we separate the domain integral into the contributions of the two individual elements:

$$\|e\|_0 = \left(\frac{1}{20}\int_0^{20} \left(u_{(h)}(x) - u(x)\right)^2 dx\right)^{\frac{1}{2}} = \left[\frac{1}{20}\int_0^{10} \left(u_{(h)}(x) - u(x)\right)^2 dx + \frac{1}{20}\int_{10}^{20} \left(u_{(h)}(x) - u(x)\right)^2 dx\right]^{\frac{1}{2}}$$

The integral over the interval $0 \le x \le 10$ corresponds to element (1), while that over $10 \le x \le 20$ corresponds to element (2). Now, we have:

$$u_{(h)}(x) = u^{(1)}(x) = 0.05 + 0.0029x, \quad 0 \le x \le 10$$

$$u_{(h)}(x) = u^{(2)}(x) = 0.0009x + 0.07, \quad 10 \le x \le 20$$

Thus, the displacement error norm is:

$$\|e\|_0 = \left[\frac{1}{20}\int_0^{10} \left(u_h(x) - u(x)\right)^2 dx + \frac{1}{20}\int_{10}^{20} \left(u_h(x) - u(x)\right)^2 dx\right]^{\frac{1}{2}}$$

$$= \left[\left(\frac{1}{20}\int_0^{10} \left(u^{(1)}(x) - u(x)\right)^2 dx\right) + \left(\frac{1}{20}\int_{10}^{20} \left(u^{(2)}(x) - u(x)\right)^2 dx\right)\right]^{\frac{1}{2}}$$

$$= \left[\left(\frac{1}{20}\int_0^{10} \left(0.05 + 0.0029x + \frac{x^2}{10000} - 0.0039x - 0.05\right)^2 dx\right)\right.$$

$$\left. + \left(\frac{1}{20}\int_{10}^{20} \left(0.0009x + 0.07 + \frac{x^2}{10000} - 0.0039x - 0.05\right)^2 dx\right)\right]^{\frac{1}{2}}$$

$$= \left(\frac{0.000033 + 0.000033}{20}\right)^{\frac{1}{2}} = 0.00183$$

To find the normalized error norm, we also need to calculate the norm of the exact displacement field:

$$\|u(x)\|_0 = \left(\frac{1}{20}\int_0^{20}(u(x))^2 dx\right)^{\frac{1}{2}} = \left(\frac{1}{20}\int_0^{20}\left(-\frac{x^2}{10000}+0.0039x+0.05\right)^2 dx\right)^{\frac{1}{2}}$$

$$= \left(\frac{1.513}{20}\right)^{\frac{1}{2}} = 0.27504$$

Thus, we have:

$$\bar{e}_0 = \frac{\|u_{(h)}(x)-u(x)\|_0}{\|u(x)\|_0} = \frac{0.00183}{0.27504} = 0.0067$$

■

Let us now continue the discussion of error norms. We can also define the *energy error norm*, $\|e\|_{en}$, which quantifies strain error:

$$\|e\|_{en} = \left(\frac{1}{L}\int_0^L \left(\varepsilon_{(h)}(x)-\varepsilon(x)\right)^2 dx\right)^{\frac{1}{2}} \tag{3.8.5}$$

where $\varepsilon(x)$ and $\varepsilon_{(h)}(x)$ are the exact strain field and approximate strain field, respectively, given by the following expressions.

$$\varepsilon(x) = \frac{du}{dx} \tag{3.8.6a}$$

$$\varepsilon_{(h)}(x) = \frac{du_h}{dx} \tag{3.8.6b}$$

Remark 3.8.1: From Equations (3.8.5) and (3.8.6a,b), we deduce that the energy error norm constitutes a measure of the error in the derivative of the field function. ■

We can also define the *normalized energy error norm*:

$$\bar{e}_{en} = \frac{\|u_{(h)}(x)-u(x)\|_{en}}{\|u(x)\|_{en}} = \frac{\left(\frac{1}{L}\int_0^L \left(\varepsilon_{(h)}(x)-\varepsilon(x)\right)^2 dx\right)^{\frac{1}{2}}}{\left(\frac{1}{L}\int_0^L (\varepsilon(x))^2 dx\right)^{\frac{1}{2}}} \tag{3.8.7}$$

We will now provide a discussion of the *convergence behavior of the finite element method*. In Section 3.1, we qualitatively defined the convergence as the property of the finite element solution to get "closer and closer" to the exact solution, as we refine the mesh. Mesh refinement corresponds to a reduction in the value of a mesh discretization parameter h for the model. The value of h can be taken equal to the dimensionless ratio $\ell^{(e)}/\ell_o$ where $\ell^{(e)}$ is the length of the finite elements in the mesh and ℓ_o is an arbitrarily selected, constant, "reference size value" (e.g., we can set $\ell_o = 1$ inch). Mathematical analysis allows us to obtain closed-form expressions that give the error norms as functions of the mesh size parameter h. We will now provide (without any proof) the statement in Box 3.8.1. For a more detailed discussion of the topic, see Fish and Belytschko (2007) or Strang and Fix (2008).

Box 3.8.1 Convergence of finite element method

It can be proven that if the approximation of a finite element has a degree of completeness equal to p, then the error norm varies as: $\|e\|_0 = C_1 \cdot h^{p+1}$, while the energy norm varies as: $\|e\|_{en} = C_2 \cdot h^p$. The constants C_1 and C_2 depend on the problem and the mesh (but their value is of secondary importance). The exponents of h in the two expressions give us the *rate of convergence* (also called the order) of the finite element approximation.

A direct observation stemming from Box 3.8.1 is that, if we have a finite element approximation which can represent the complete polynomial of order p, then the plots of the logarithms of the norms $\|e\|_0$ and $\|e\|_{en}$ as functions of the logarithm of the element size parameter, h, will be given by the following expressions:

$$\log(\|e\|_0) = \tilde{c}_1 + (p+1)\cdot\log(h) \tag{3.8.8a}$$

$$\log(\|e\|_{en}) = \tilde{c}_2 + p\cdot\log(h) \tag{3.8.8b}$$

where $\tilde{c}_1 = \log(C_1)$ and $\tilde{c}_2 = \log(C_2)$. Thus, based on Equations (3.8.8a,b), if we plot the logarithms of the error norms as functions of the logarithm of h, the plots will be straight lines, as schematically shown in Figure 3.20.

For example, if we have two-node elements with a linear approximation for the displacement, we will obtain that $\|e\|_0 = C_1 \cdot h^2$ and $\|e\|_{en} = C_2 \cdot h$. If we have elements with quadratic approximations for the displacement, we then obtain $\|e\|_0 = C_1 \cdot h^3$ and $\|e\|_{en} = C_2 \cdot h^2$. This shows that, if we have the same number and size of elements in a finite element mesh, then the quadratic elements will have superior convergence properties than linear elements. Furthermore, the constants C_1, C_2 are lower for quadratic elements than for linear elements, so that for a given value of h, the error norms for the quadratic elements are lower than those for linear elements. The reason for preferring quadratic elements for linear analysis is that they have a *higher rate of convergence* (this means that the error norms will exhibit a more rapid reduction as we refine the mesh) than linear elements.

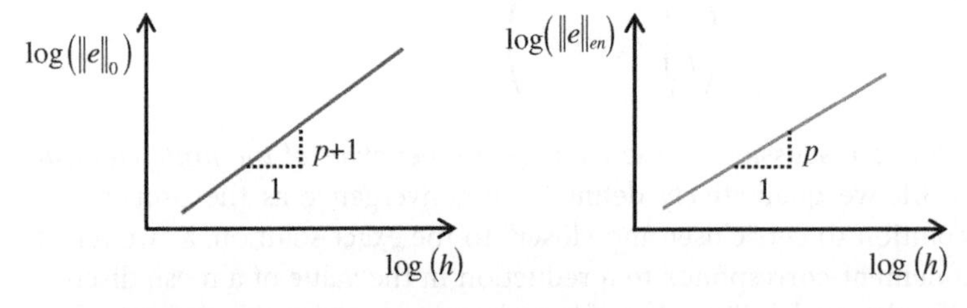

Figure 3.20 Qualitative plots of logarithm of error norms versus logarithm of discretization parameter, h.

Remark 3.8.2: The considerations of Box 3.8.1 also apply for heat conduction problems. In this case, the error norm and energy error norm are defined by replacing $u_{(h)}(x)$ with $\Theta_{(h)}(x)$ and $u(x)$ with $\Theta(x)$ in Equations (3.8.3), (3.8.5) and (3.8.6a,b). ∎

Let us now define a functional[1] $a(w, u)$, which will be called the *inner product form*:

$$a(w,u) = \int_0^L \frac{dw}{dx} EA \frac{du}{dx} dx - [wA\bar{t}]_{\Gamma_t} = \int_0^L \frac{dw}{dx} EA \frac{du}{dx} dx - \left[wA \cdot E \cdot n \frac{du}{dx}\right]_{\Gamma_t} \tag{3.8.9}$$

In this case, the weak form can be written in a version, which is popular in several books:

$$a(w,u) - \int_0^L w \cdot b(x) dx = 0, \forall w(x), \text{ with } w = 0 \text{ at } \Gamma_u \tag{3.8.10}$$

Since the weak form applies for any weighting function, w, which vanishes on the essential boundary, it will also be satisfied for a weighting function, $w_{(h)}$, which is based on a finite element approximation:

$$a\left(w_{(h)},u\right) - \int_0^L w_{(h)} \cdot b(x) dx = 0 \rightarrow \int_0^L \frac{dw_{(h)}}{dx} EA \frac{du}{dx} dx - \left[w_{(h)} A \cdot E \cdot n \frac{du}{dx}\right]_{\Gamma_t} - \int_0^L w_{(h)} \cdot b(x) dx = 0 \tag{3.8.11}$$

We can now also establish the inner product of $w_{(h)}$ with the finite element approximation $u_{(h)}$ to the displacement field. Of course, we need to satisfy an "approximate version of the weak form," which can be mathematically stated as follows:

$$a\left(w_{(h)},u_{(h)}\right) - \int_0^L w_{(h)} \cdot b(x) dx = 0 \rightarrow \int_0^L \frac{dw_{(h)}}{dx} EA \frac{du_{(h)}}{dx} dx - \left[w_{(h)} A \cdot E \cdot n \frac{du_{(h)}}{dx}\right]_{\Gamma_t} - \int_0^L w_{(h)} \cdot b(x) dx = 0 \tag{3.8.12}$$

Now, let us take the error of our finite element approximation, by defining the difference between the exact displacement field and the approximate displacement field:

$$e_{(h)}(x) = u_{(h)}(x) - u(x) \tag{3.8.13}$$

We can also define the error in strain, which, by definition, is equal to the first derivative of the error in displacement:

$$\frac{de_{(h)}}{dx} = \frac{d\left[u_{(h)}(x) - u(x)\right]}{dx} = \frac{du_{(h)}}{dx} - \frac{du}{dx} \tag{3.8.14}$$

1 As explained in Appendix C, a functional is a function of a function of the spatial variables. In the specific case, the form $a(w, u)$ depends on two functions of the spatial variable x, i.e., functions $u(x)$ and $w(x)$.

If we subtract Equation (3.8.11) from Equation (3.8.12), we have:

$$\int_0^L \frac{dw_{(h)}}{dx} EA \frac{du_{(h)}}{dx} dx - \left[w_{(h)} A \cdot E \cdot n \frac{du_{(h)}}{dx} \right]_{\Gamma_t} - \int_0^L w_{(h)} \cdot b(x) dx$$
$$- \left[\int_0^L \frac{dw_{(h)}}{dx} EA \frac{du_{(h)}}{dx} dx - \left[w_{(h)} A \cdot E \cdot n \frac{du_{(h)}}{dx} \right]_{\Gamma_t} - \int_0^L w_{(h)} \cdot b(x) dx \right] = 0$$

We can now group the various terms together, to obtain:

$$\int_0^L \frac{dw_{(h)}}{dx} EA \left(\frac{du_{(h)}}{dx} - \frac{du}{dx} \right) dx - \left[w_{(h)} A \cdot E \cdot n \left(\frac{du_{(h)}}{dx} - \frac{du}{dx} \right) \right]_{\Gamma_t} = 0 \tag{3.8.15}$$

If we account for Equation (3.8.14), then we can write Equation (3.8.15) as follows:

$$\int_0^L \frac{dw_{(h)}}{dx} EA \frac{de_{(h)}}{dx} dx - \left[w_{(h)} A \cdot E \cdot n \frac{de_{(h)}}{dx} \right]_{\Gamma_t} = 0 \tag{3.8.16}$$

But if we remember the definition of the inner product form, provided in Equation (3.8.9), then Equation (3.8.16) becomes:

$$a\left(w_{(h)}, e_{(h)}\right) = 0 \tag{3.8.17}$$

Equation (3.8.17) is a very important finding, sometimes referred to as the *best approximation property* of the finite element solution. It expresses the fact that the error of the solution is *orthogonal—in terms of the inner product form* of Equation (3.8.9)—to the approximation. Mathematically, this is equivalent to stating (the perhaps obvious fact) that the error is the part of the exact solution that cannot be captured (described) by the shape functions of the finite element approximation.

3.9 Effect of Concentrated Forces in One-Dimensional Finite Element Analysis

In elasticity problems, it is very common to idealize applied loads as *concentrated forces*. This means that we treat a nonzero force as being entirely applied at a given point, $x = a$, of the one-dimensional domain, as shown in Figure 3.21. We need to establish a formal procedure to treat such concentrated forces in a finite element analysis. This section describes such a procedure, based on mathematical considerations provided in Fish and Belytschko (2007).

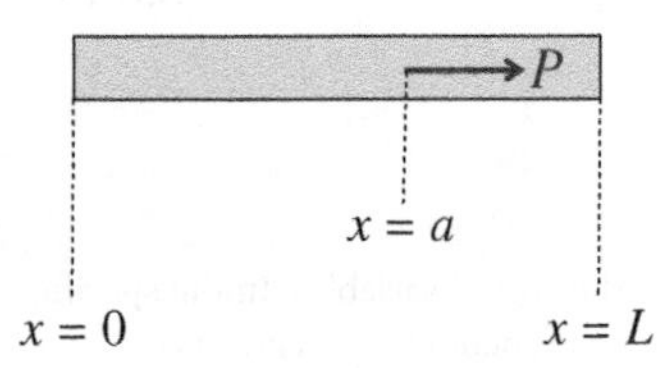

Figure 3.21 Example concentrated force in the interior of a one-dimensional domain.

If we have a concentrated force P at a location $x = a$, then that force can be treated as a distributed (body) force, $b(x)$, given by the following expression.

$$b(x) = P \cdot \delta(x-a) \tag{3.9.1}$$

where $\delta(x-a)$ is the *Dirac delta function* at $x = a$. A qualitative plot of the Dirac delta function inside a domain with $0 \le x \le L$, is shown in Figure 3.22. The specific function represents a "concentrated" force, in the sense that the function has practically zero value at a very small distance, $\frac{\varepsilon}{2}$, from the point of application of the force P. The Dirac delta function takes an infinite value at $x = a$, but its integral over the domain of interest must always satisfy the following equation.

$$\int_0^L \delta(x-a)dx = 1 \tag{3.9.2}$$

A basic property of the Dirac delta function, $\delta(x-a)$ is the following.
for any function $g(x)$:

$$\int_0^L g(x) \cdot \delta(x-a)dx = g(a) \tag{3.9.3}$$

We are now ready to proceed with the treatment of concentrated forces. If the concentrated force P is applied at a location $x = a$ in the interior of an element e with n nodes, we can determine the equivalent nodal force vector $\{f^{(e)}\}$ due to P, simply by plugging Equation (3.9.1) into Equation (3.3.14):

$$\left\{f^{(e)}\right\} = \int_0^L \left[N^{(e)}\right]^T \cdot P \cdot \delta(x-a)dx + \left[\left[N^{(e)}\right]^T \cdot A^{(e)} \cdot \bar{t}\right]_{\Gamma_t} = \left[N^{(e)}\right]^T_{x=a} \cdot P + \left[\left[N^{(e)}\right]^T \cdot A^{(e)} \cdot \bar{t}\right]_{\Gamma_t} \tag{3.9.4}$$

Remark 3.9.1: Notice that, in Equation (3.9.4), we have tacitly set: $\int_{x_1^{(e)}}^{x_2^{(e)}} \left[N^{(e)}\right]^T \cdot b(x)\, dx = \int_0^L \left[N^{(e)}\right]^T \cdot b(x)\, dx$. This is allowed by the fact that the shape function array $[N^{(e)}]$ is zero at any location outside of element e. ▮

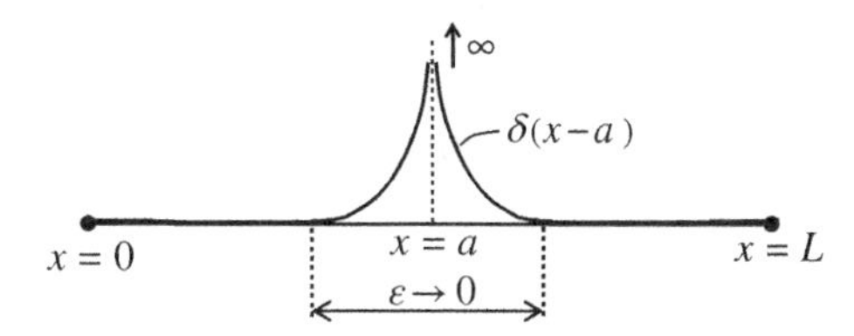

Figure 3.22 Schematic definition of Dirac delta function, $\delta(x-a)$.

The first term in the right-hand side of Equation (3.9.4) is the contribution of the concentrated load P to the equivalent nodal force vector of the element. We can also write the effect of the concentrated load to the equivalent nodal force of each node in the element:

$$\text{For node } 1: N_1^{(e)}\Big|_{x=a} \cdot P \tag{3.9.4-1}$$

$$\text{For node } 2: N_2^{(e)}\Big|_{x=a} \cdot P \tag{3.9.4-2}$$

...

$$\text{For node } n: N_n^{(e)}\Big|_{x=a} \cdot P \tag{3.9.4-n}$$

A concentrated force in the interior of an element only affects the equivalent nodal forces of that specific element. We can now formulate the procedure provided in Box 3.9.1.

Box 3.9.1 Treatment of Concentrated Forces

If we have a concentrated force P applied at $x = a$, then the equivalent nodal force vector due to P for element e is:

$$\left\{f^{(e)}\right\} = \begin{cases} \left[N^{(e)}\right]^T_{x=a} \cdot P, & \text{if } x_1^{(e)} \leq a \leq x_2^{(e)} \\ 0, & \text{otherwise} \end{cases}$$

Remark 3.9.2: As a special case, if $x = a$ corresponds to the location of a nodal point of the structure, we simply obtain that we have to add P to the term corresponding to the nodal point where the concentrated force is applied. ▮

Remark 3.9.3: A procedure to account for concentrated forces in elasticity can also be followed for heat-conduction problems, to treat so-called *concentrated heat sources* in the interior of the domain. ▮

Problems

3.1 We are given the system in Figure P3.1.

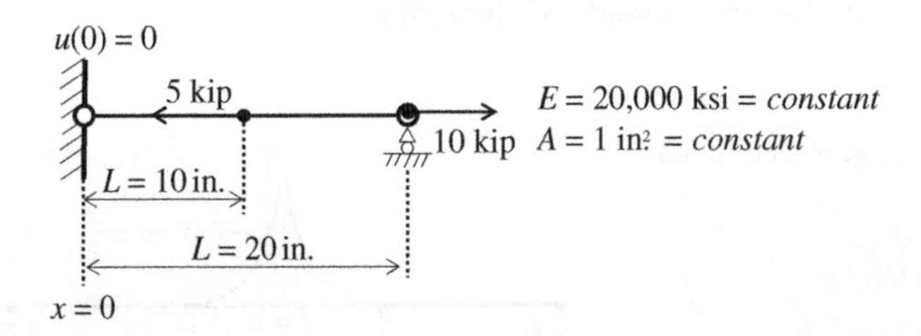

Figure P3.1

a) Derive and solve the strong form for the specific problem. HINT: Define $u_a(x)$, $0 \le x \le 10$ in. and $u_b(x)$, 10 in. $\le x \le 20$ in., then impose appropriate conditions at the point $x = 10$ in. Obtain the displacement field and the stress field.

To obtain an approximate solution to this problem, we establish three nodal points in the domain, as shown in Figure P3.2.

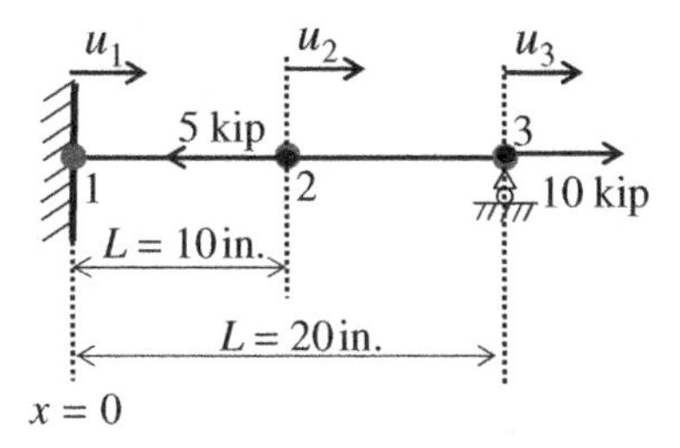

Figure P3.2

b) What information regarding the nodal quantities u_1, u_2, u_3, w_1, w_2, and w_3 can we obtain from the essential boundary conditions?
c) Obtain the approximate solution for the case that we use a single three-node finite element.
d) Obtain the approximate solution for the case that we use two finite elements with linear shape functions.
e) Draw the exact displacement field, and the approximate displacement fields obtained in parts c and d. Also, draw the exact stress field, and the approximate stress fields obtained in parts c and d.
f) Which of the two solutions gives better agreement with the exact solution? Why? Make a general recommendation regarding discretization of regions involving concentrated forces in the interior.

3.2 We are given the structure in Figure P3.3.

We want to solve the structure (i.e., obtain the displacement and strain field) using two-node finite elements. In each case, all the elements in the mesh will have the same size. The element stiffness matrices and right-hand-side vectors must be calculated with Gaussian quadrature.

a) Solve the structure using two finite elements.
b) Solve the structure using four finite elements.

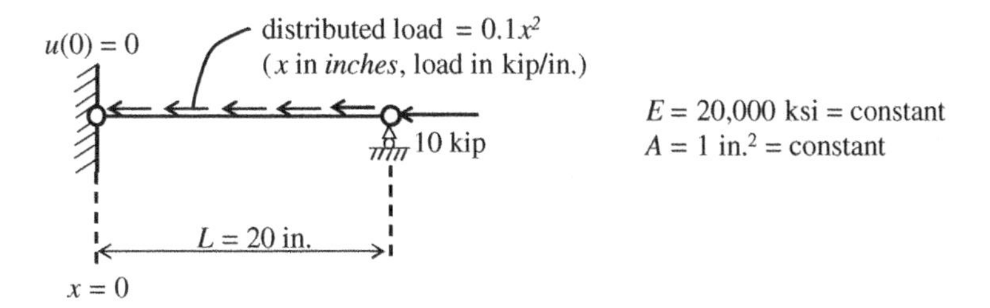

Figure P3.3

c) Solve the structure using eight finite elements.
d) Find the exact solution and calculate (for each finite element mesh) the error norms $\|e\|_0$ and $\|e\|_{en}$ using and summing the element contributions:

$$\|e\|_0 = \left[\frac{1}{20}\int_0^{20}\left(u_{(h)}(x) - u(x)\right)^2 dx\right]^{\frac{1}{2}} = \left[\frac{1}{20}\sum_{e=1}^{N_{el}}\left[\int_{x_1^e}^{x_2^e}\left(u^{(e)}(x) - u(x)\right)^2 dx\right]\right]^{\frac{1}{2}}$$

$$\|e\|_{en} = \left[\frac{1}{20}\int_0^{20}\left(\frac{du_{(h)}}{dx} - \frac{du}{dx}\right)^2 dx\right]^{\frac{1}{2}} = \left[\frac{1}{20}\sum_{e=1}^{N_{el}}\left[\int_{x_1^e}^{x_2^e}\left(\frac{du^{(e)}}{dx} - \frac{du}{dx}\right)^2 dx\right]\right]^{\frac{1}{2}}$$

where N_{el} is the number of elements in the mesh.

NOTE: Use two-point Gaussian quadrature to calculate the element contributions to the error norms.

e) Plot the error norms as functions of the discretization parameter, h.
f) Plot the logarithms of the error norms as functions of the logarithm of the discretization parameter, h.

3.3 We are given the definite integral: $\int_0^5 \frac{1}{x+1} dx$.

a) Calculate the exact value of the integral through analytical integration.
b) Numerically obtain the value of the integral using Gaussian quadrature with one, two, three, and four points.
c) Define an error quantity, describing the accuracy of the Gaussian quadrature for the calculation of the specific integral. Create a plot, providing the error as a function of the number of quadrature points.
d) Explain how many points would be required to obtain the exact value of the given integral with Gaussian quadrature.

3.4 We are given the system in Figure P3.4.

We want to find the approximate temperature field using a single, three-node element (the intermediate node must be located at $x = 2$ m).

a) Calculate the element conductance matrix, using two-point Gaussian quadrature.

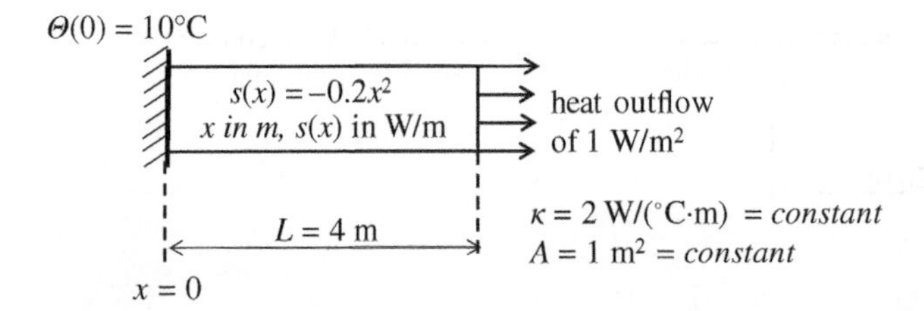

Figure P3.4

b) Do you expect the two-point Gaussian quadrature to have provided the exact conductance matrix?
c) Calculate the element equivalent nodal flux vector, using one-point Gaussian quadrature.
d) Calculate the element equivalent nodal flux vector, using two-point Gaussian quadrature.
e) If we subdivided the mesh into two quadratic (three-node) elements, each element having the same size, how would the L_2-error norm change (by how much would it increase/decrease?)? How would the energy error norm change?

References

Fish, J., and Belytschko, T. (2007). *A First Course in Finite Elements.* Hoboken, NJ: John Wiley & Sons.

Strang, G., and Fix, G. (2008). *An Analysis of the Finite Element Method,* 2nd ed. Cambridge, MA: Wellesley.

Stroud, A., and Secrest, D. (1966). *Gaussian Quadrature Formulas.* Englewood Cliffs, NJ: Prentice-Hall.

4

Multidimensional Problems: Mathematical Preliminaries

4.1 Introduction

This chapter provides the necessary mathematical concepts to allow the description of multidimensional problems. As the name *multidimensional* suggests, we will now have problems where the governing physics occur over domains such as surfaces or solid volumes. This introduces the need to describe our problems with *multiple spatial variables*. In the most general case of three-dimensional space, we will be using a set of three coordinates (x,y,z). We can also use an alternative notation for spatial variables, naming our coordinates as $x_1 = x$, $x_2 = y$, and $x_3 = z$. This will facilitate writing more concise mathematical expressions. Besides the need to use multiple spatial coordinates to describe the position of a point in space, there may be cases where the field functions that we want to obtain are vectors and include multiple components, one along each coordinate direction. The following sections establish mathematical definitions and the extension of basic calculus tools (such as integration by parts) to multidimensional problems. The chapter concludes with an overview of the mathematical procedure to obtain the strong form in multidimensional problems. For obvious reasons, the descriptions provided herein are only meant to provide a basic understanding of the tools necessary for establishing the strong form and weak form in multidimensional problems. For a more rigorous treatment, see textbooks on calculus (e.g., Stewart 2015) or vector calculus (e.g., Marsden and Tromba 2011).

A basic ingredient of the description of multidimensional problems is vector quantities. Basic vector algebra is provided in Appendix A. What needs to be added here is a geometric interpretation of vector quantities, with specific reference to two-dimensional (two-component) and three-dimensional (three-component) vectors. Let us consider a three-component vector, $\vec{F} = \{F\} = [F_1 \;\; F_2 \;\; F_3]^T$. The components F_1, F_2, and F_3 of the vector can be assigned geometrical meaning, if we imagine a system of *Cartesian unit vectors, or Cartesian basis vectors,* $\vec{e}_1 = \vec{i}$, $\vec{e}_2 = \vec{j}$ and $\vec{e}_3 = \vec{k}$, as shown in Figure 4.1a. Each of the basis vectors is a *unit vector*—that is, it has a magnitude equal to 1, and is aligned with the corresponding coordinate axis. Any vector $\vec{F}$ can be thought of as a *linear combination* of the basis vectors, that is:

$$\vec{F} = F_1\vec{e}_1 + F_2\vec{e}_2 + F_3\vec{e}_3 \tag{4.1.1}$$

Fundamentals of Finite Element Analysis: Linear Finite Element Analysis, First Edition.
Ioannis Koutromanos, James McClure, and Christopher Roy.

Companion website: www.wiley.com/go/koutromanos/linear

The components F_1, F, and F_3 can be regarded as the *projections* of $\vec{F}$ along each one of the basis vectors $\vec{e}_1$, $\vec{e}_2$ and $\vec{e}_3$, respectively. Similar considerations apply for two-dimensional vectors. In this case, a vector can be considered as a linear combination of two Cartesian unit vectors, $\vec{e}_1 = \vec{i}$ and $\vec{e}_2 = \vec{j}$, which are shown in Figure 4.1b.

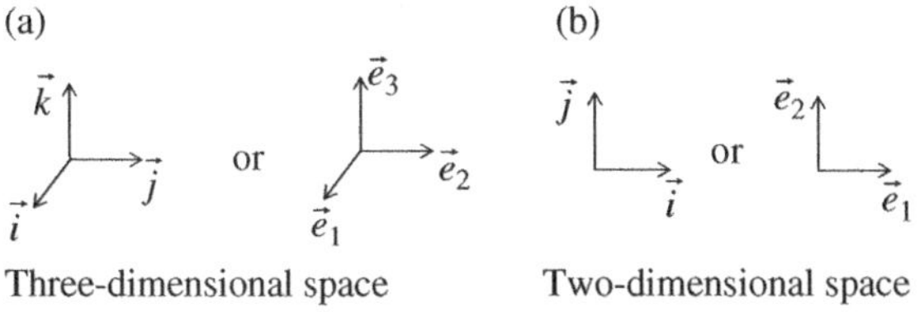

(a) $\vec{k}$, $\vec{j}$, $\vec{i}$ or $\vec{e}_3$, $\vec{e}_2$, $\vec{e}_1$ — Three-dimensional space

(b) $\vec{j}$, $\vec{i}$ or $\vec{e}_2$, $\vec{e}_1$ — Two-dimensional space

Figure 4.1 Cartesian unit (basis) vectors for multidimensional coordinates.

4.2 Basic Definitions

In multidimensional problems, the physical meaning of the governing equations remains the same as for the corresponding one-dimensional ones (i.e., conservation of energy for heat conduction, force equilibrium for elasticity). The only difference is that we now have functions that are multidimensional fields, meaning that instead of having to calculate, for example, $u(x)$ (field functions of one independent variable, x), we will be having to calculate $u(x,y)$ for two-dimensional problems or $u(x,y,z)$ for three-dimensional problems. Additionally, as we will see in subsequent chapters, the field function itself may be scalar-valued or vector-valued. In the latter case, we will be needing to describe (as functions of the spatial independent variables x,y or x,y,z) a set of two or three field functions, expressing the components of a two-dimensional or three-dimensional vector field. Thus, we can distinguish between two major families of problems:

- *Scalar field problems,* where the field function is a scalar quantity, as shown in Figure 4.2.
- *Vector field problems,* where the field function is a vector, and each component of that vector is a scalar field, as shown in Figure 4.3.

Figure 4.2 Field function in scalar field problems.

$u(x, y)$ (2D problems) or $u(x, y, z)$ (3D problems)

$$\underset{\sim}{u}(x,y) = \vec{u}(x,y) = \begin{Bmatrix} u_x(x,y) \\ u_y(x,y) \end{Bmatrix} \quad or \quad \underset{\sim}{u}(x,y,z) = \vec{u}(x,y,z) = \begin{Bmatrix} u_x(x,y,z) \\ u_y(x,y,z) \\ u_z(x,y,z) \end{Bmatrix}$$

2D problems — 3D problems

Figure 4.3 Field function in vector field problems.

Example scalar field problems are multidimensional heat conduction, multidimensional flow in porous media, chemical diffusion, electrostatic problems and potential fluid flow. All of these problems (with the exception of electrostatic problems) will be discussed in Chapter 5. Examples of vector field problems are multidimensional elasticity and electrodynamics. Of these two, only multidimensional elasticity is covered in Chapters 7, 8, and 9.

The description of multidimensional problems entails the need to generalize the notion of the domain of the problem and of the boundaries. In one-dimensional problems, the domain was a line and the boundary consisted of the two endpoints. For three-dimensional problems, the domain Ω is a three-dimensional solid volume, as shown in Figure 4.4a, and the boundary Γ is the surface enclosing the volume. We can also draw, at every location of the boundary surface, the three-component unit normal outward vector $\vec{n}$, as shown in the same figure. Similarly, for two-dimensional problems, the domain Ω is a surface segment, enclosed by a boundary curve Γ, as shown in Figure 4.4b. We also have—at each location of the natural boundary—the two-component unit normal outward vector $\vec{n}$, as also shown in Figure 4.4b.

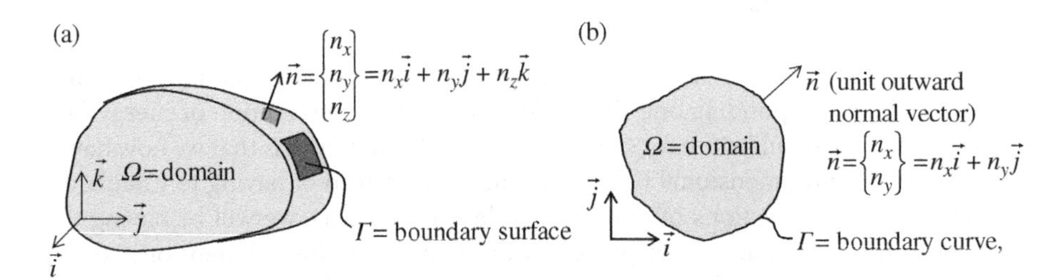

Figure 4.4 Multidimensional domains and boundaries.

Remark 4.2.1: Each of the components of the unit outward normal vector $\vec{n}$ is equal to the cosine of the angle that $\vec{n}$ forms with the corresponding coordinate axis. For example, n_x is equal to the cosine of the angle between $\vec{n}$ and the x-axis. ■

Several basic mathematical operations will now be defined. These operations generalize the concept of differentiation to multidimensional problems.

Definition 4.2.1: Let us imagine that we have a multidimensional field function $f(x, y, z)$, calculated at some point (x, y, z). Let us consider a new location $(x+\Delta x, y, z)$, that is, at a "small distance" Δx (along the x -axis) from the original location. We evaluate the function at that new location, $f(x+\Delta x, y, z)$. We can then calculate the x-change as the difference $\Delta_x f = f(x+\Delta x, y, z) - f(x, y, z)$. The *partial derivative* (or simply *the partial*) *of the field f with respect to x*, $\frac{\partial f}{\partial x}$, is equal to the limit of the ratio of the x-change over the value Δx, as Δx tends to zero:

$$\frac{\partial f}{\partial x} = \lim_{\Delta x \to 0} \frac{f(x+\Delta x, y, z) - f(x, y, z)}{\Delta x} = \lim_{\Delta x \to 0} \frac{\Delta_x f}{\Delta x} \tag{4.2.1a}$$

Along the same lines, we can calculate the *partial derivatives of a field f with respect to y and z*:

$$\frac{\partial f}{\partial y} = \lim_{\Delta y \to 0} \frac{f(x, y+\Delta y, z) - f(x, y, z)}{\Delta y} = \lim_{\Delta y \to 0} \frac{\Delta_y f}{\Delta y} \tag{4.2.1b}$$

$$\frac{\partial f}{\partial z} = \lim_{\Delta z \to 0} \frac{f(x,y,z+\Delta z) - f(x,y,z)}{\Delta z} = \lim_{\Delta z \to 0} \frac{\Delta_z f}{\Delta z} \quad (4.2.1c)$$ ■

Remark 4.2.2: To calculate the partial derivative of a given function $f(x, y, z)$ with respect to one of the coordinates, for example, x, we imagine that the function depends on x alone, and that y and z are constant parameters. Then, we can calculate the partial derivative in the same way that we would calculate the derivative of a one-dimensional function $f(x)$. ■

Example 4.2.1: Partial derivative of multidimensional field
We are given the function $f(x,y,z) = e^{x \cdot y} + 5x^2z^2$. To evaluate the partial derivative with respect to x, we treat y and z as constants, to obtain: $\frac{\partial f}{\partial x} = \frac{\partial(e^{x \cdot y} + 5x^2z^2)}{\partial x} = y \cdot e^{x \cdot y} + 10x \cdot z^2$. Similarly, we can find the partial of the function with respect to z, by treating y and x as constants: $\frac{\partial f}{\partial z} = \frac{\partial(e^{x \cdot y} + 5x^2z^2)}{\partial z} = 0 + 10x^2 \cdot z = 10x^2 \cdot z$ ■

Definition 4.2.2: Based on the rationale of Definition (4.2.1), we can also define *higher-order partial derivatives.* For example, the second-order partial derivative of a function f with respect to x, which is equal to the partial derivative of the function $\frac{\partial f}{\partial x}$ with respect to x, is given by the expression:

$$\frac{\partial^2 f}{\partial x^2} = \frac{\partial}{\partial x}\left(\frac{\partial f}{\partial x}\right) \quad (4.2.2a)$$

Along the same lines, we can define:

$$\frac{\partial^2 f}{\partial x^2} = \frac{\partial}{\partial x}\left(\frac{\partial f}{\partial x}\right) \quad (4.2.2b)$$

$$\frac{\partial^2 f}{\partial x^2} = \frac{\partial}{\partial x}\left(\frac{\partial f}{\partial x}\right) \quad (4.2.2c)$$

Also, we can define *mixed higher-order partial derivatives.* For example, we can define $\frac{\partial^2 f}{\partial x \partial y}$ as the partial derivative of $\frac{\partial f}{\partial x}$ with respect to y, or the partial derivative of $\frac{\partial f}{\partial y}$ with respect to x. Both definitions are equivalent:

$$\frac{\partial^2 f}{\partial x \partial y} = \frac{\partial}{\partial y}\left(\frac{\partial f}{\partial x}\right) = \frac{\partial}{\partial x}\left(\frac{\partial f}{\partial y}\right) \quad (4.2.3)$$

Similarly, we can define other higher-order partial derivatives. ■

As mentioned in Section 1.1, a differential operator is any mathematical object that takes a field function and returns a quantity which includes derivatives of the field. An example differential operator for one-dimensional problems is the derivative operator, $\frac{d}{dx}$, which can be applied to any one-dimensional function $f(x)$ and give us the derivative, $\frac{df}{dx}$. Similarly, we have the *partial derivative operators,* $\frac{\partial}{\partial x}$, $\frac{\partial}{\partial x}$, and $\frac{\partial}{\partial x}$ for multidimensional problems.

A special (and very important) differential operator used in multidimensional problems is the so-called *nabla operator*, denoted by $\vec{\nabla}$ (or simply ∇), given by the following expression:

$$\nabla = \begin{Bmatrix} \partial/\partial x \\ \partial/\partial y \\ \partial/\partial z \end{Bmatrix} = \frac{\partial}{\partial x}\vec{i} + \frac{\partial}{\partial y}\vec{j} + \frac{\partial}{\partial z}\vec{k} \tag{4.2.4}$$

Notice that nabla is a "vector-like" operator, so we can use it in mathematical equations the same way that we would use a vector, only that the components will eventually work as differential operators.

If ∇ operates on a scalar field function (e.g., $T(x,y,z)$), then it yields a vector field, $\vec{\nabla} T = \{\nabla T\}$, called the *gradient of the scalar field T* and given by the following expression.

$$\vec{\nabla} T = \{\nabla T\} = \begin{Bmatrix} \partial T/\partial x \\ \partial T/\partial y \\ \partial T/\partial z \end{Bmatrix} = \frac{\partial T}{\partial x}\vec{i} + \frac{\partial T}{\partial y}\vec{j} + \frac{\partial T}{\partial z}\vec{k} \quad \left(or \ \frac{\partial T}{\partial x_1}\vec{e}_1 + \frac{\partial T}{\partial x_2}\vec{e}_2 + \frac{\partial T}{\partial x_3}\vec{e}_3 \right) \tag{4.2.5}$$

The nabla operator can also operate on a vector field. Specifically, when ∇ operates on a vector field, $\vec{r}(x,y,z)$, with components $r_x(x, y, z)$, $r_y(x, y, z)$ and $r_z(x, y, z)$, then we obtain a scalar field called the *divergence* $\vec{\nabla} \bullet \vec{r}$ of the vector field $\vec{r}(x,y,z)$, given by the following expression:

$$div(\vec{r}) = \vec{\nabla} \bullet \vec{r} = \frac{\partial r_x}{\partial x} + \frac{\partial r_y}{\partial y} + \frac{\partial r_z}{\partial z} \tag{4.2.6}$$

Remark 4.2.3: A convention that is very popular in continuum mechanics textbooks is the *summation convention*. If we have a product, fraction or differentiation term in an expression where one index is repeated once (= *the index appears exactly twice*), then this implies summation with respect to the index.

Let us examine several indicial mathematical expressions involving components of vectors and matrices, assuming that we have adopted the summation convention:

- $\alpha_i \cdot \beta_i$: We have an index, i, that appears exactly twice, thus, summation is implied:

$$\alpha_i \cdot \beta_i = \alpha_1 \cdot \beta_1 + \alpha_2 \cdot \beta_2 + \alpha_3 \cdot \beta_3$$

- $a_i + b_i$: This expression does *not* imply summation, because the term where i is repeated is *not* a product, fraction, or derivative!
- $\alpha_i \cdot \beta_{ii}$: This expression does *not* imply summation, because the index i appears three times, not two.

It is also worth mentioning that, if we have a term corresponding to the component of a matrix and we have an index appearing exactly twice, then we imply summation with respect to the repeated index (e.g., $A_{ii} = A_{11} + A_{22} + A_{33}$).
In the context of summation convention, we can establish the following expression for the divergence of a vector field: $\vec{\nabla} \bullet \vec{r} = \dfrac{\partial r_i}{\partial x_i}$, where index i appears exactly twice, so summation with respect to i is implied. ▮

Other expressions associated with the nabla operator can be established:

- If we have a scalar field $f(x,y,z)$, then we can take the divergence of the gradient of the field:

$$\vec{\nabla} \bullet \left(\vec{\nabla} f\right) = \vec{\nabla}^2 f = \frac{\partial^2 f}{\partial x^2} + \frac{\partial^2 f}{\partial y^2} + \frac{\partial^2 f}{\partial z^2} \tag{4.2.7}$$

  The operator $\vec{\nabla}^2 f = \Delta f$ is called the *Laplacian operator*, and it appears in many scalar field problems.
- If we have a vector field $\vec{r}(x,y,z)$ then we can define the *curl of the field*, $\vec{\nabla} \times \vec{r}$, which is another vector field obtained from the following expression.

$$curl(\vec{r}) = \vec{\nabla} \times \vec{r} = \begin{bmatrix} \dfrac{\partial r_z}{\partial y} - \dfrac{\partial r_y}{\partial z} & \dfrac{\partial r_x}{\partial x} - \dfrac{\partial r_z}{\partial x} & \dfrac{\partial r_y}{\partial x} - \dfrac{\partial r_x}{\partial y} \end{bmatrix}^T \tag{4.2.8}$$

The curl plays a central role in many physical problems, particularly in irrotational fluid flow (Section 5.4.1) and electromagnetism.

Remark 4.2.4: The following rules involving the divergence, the gradient and the curl are very important in multidimensional calculus (and are also useful in real-life situations, as shown for example in Section 5.4.1):

- If $f(x,y,z)$ is a scalar field, then the curl of the gradient of f is zero:

$$curl\left(\vec{\nabla} f\right) = \vec{\nabla} \times \vec{\nabla} f = 0 \tag{4.2.9a}$$

- For every vector field $\vec{r}(x,y,z)$:

$$div\left[curl(\vec{r})\right] = \vec{\nabla} \bullet \left(\vec{\nabla} \times \vec{r}\right) = 0 \tag{4.2.9b}$$

- If we have a vector field $\vec{r}(x,y,z)$, for which $curl(\vec{r}) = \vec{\nabla} \times \vec{r} = 0$, then there is a scalar field $f_r(x,y,z)$, such that:

$$\vec{r} = \vec{\nabla} f_r \tag{4.2.9c}$$ ▮

4.3 Green's Theorem—Divergence Theorem and Green's Formula

A theorem by Green provides the basic tool toward generalizing the concept of integration by parts in multiple dimensions. For any continuous, differentiable function $f(x,y,z)$, defined inside a domain Ω with boundary Γ, Green's theorem states that:

$$\iiint_{\Omega} \nabla f \, dV = \iint_{\Gamma} f \cdot \vec{n} \, dS \tag{4.3.1}$$

Equation (4.3.1) involves vectors, and can be written into three expressions, one for each component of the vectors (i.e., one for each direction):

$$\iiint_{\Omega} \frac{\partial f}{\partial x} dV = \iint_{\Gamma} f \cdot n_x \, dS \tag{4.3.1a}$$

$$\iiint_{\Omega} \frac{\partial f}{\partial y} dV = \iint_{\Gamma} f \cdot n_y \, dS \tag{4.3.1b}$$

$$\iiint_{\Omega} \frac{\partial f}{\partial z} dV = \iint_{\Gamma} f \cdot n_z \, dS \tag{4.3.1c}$$

An alternative way to write Green's theorem is:

$$\iiint_{\Omega} \frac{\partial f}{\partial x_i} dV = \iint_{\Gamma} f \cdot n_i \, dS, \; i = 1,2,3 \tag{4.3.2}$$

Now, imagine that we have a vector field $\vec{r}(x,y,z) = r_x(x,y,z) \cdot \vec{i} + r_y(x,y,z) \cdot \vec{j} + r_z(x,y,z) \cdot \vec{k}$, where all three components are continuous and differentiable functions of x, y, and z. We use Green's theorem as follows.

Write Equation (4.3.1a) for $f(x,y,z) = r_x(x,y,z)$:

$$\iiint_{\Omega} \frac{\partial r_x}{\partial x} dV = \iint_{\Gamma} r_x \cdot n_x \, dS \tag{4.3.3a}$$

Write Equation (4.3.1b) for $f(x,y,z) = r_y(x,y,z)$:

$$\iiint_{\Omega} \frac{\partial r_y}{\partial y} dV = \iint_{\Gamma} r_y \cdot n_y \, dS \tag{4.3.3b}$$

Write Equation (4.3.1c) for $f(x,y,z) = r_z(x,y,z)$:

$$\iiint_{\Omega} \frac{\partial r_z}{\partial z} dV = \iint_{\Gamma} r_z \cdot n_z \, dS \tag{4.3.3c}$$

If we sum equations (4.3.3a, b, c), we obtain:

$$\iiint_{\Omega} \left(\frac{\partial r_x}{\partial x} + \frac{\partial r_y}{\partial y} + \frac{\partial r_z}{\partial z} \right) dV = \iint_{\Gamma} \left(r_x \cdot n_x + r_y \cdot n_y + r_z \cdot n_z \right) dS \tag{4.3.4}$$

The obtained Equation (4.3.4) is called the *Gauss divergence theorem*. We can write Equation (4.3.4) in an alternative form:

$$\iiint_{\Omega} \vec{\nabla} \bullet \vec{r} \, dV = \iint_{\Gamma} (\vec{r} \bullet \vec{n}) \, dS \tag{4.3.5}$$

Finally, we are ready to establish what will prove to be the most useful formula for our discussions of multidimensional problems, namely, *Green's formula*. For any scalar field $w(x, y, z)$ and for any vector field $\vec{q}\,(x,y,z)$, the following equation applies.

$$\iiint_{\Omega} w \cdot \left(\vec{\nabla} \bullet \vec{q}\right) dV = \iint_{\Gamma} w \cdot (\vec{q} \bullet \vec{n}) \, dS - \iiint_{\Omega} \left(\vec{\nabla} w\right) \bullet \vec{q} \, dV \tag{4.3.6}$$

The proof of Green's formula is provided in Box 4.3.1.

Box 4.3.1 Proof of Green's Formula

Set $f = w \cdot q_x$. Then, we have: $\frac{\partial f}{\partial x} = \frac{\partial w}{\partial x} \cdot q_x + w \cdot \frac{\partial q_x}{\partial x}$, and if we plug this expression in Equation (4.3.1a):

$$\iiint_{\Omega} \frac{\partial f}{\partial x} dV = \iint_{\Gamma} f \cdot n_x \, dS \rightarrow \iiint_{\Omega} \left(\frac{\partial w}{\partial x} \cdot q_x + w \cdot \frac{\partial q_x}{\partial x}\right) dV = \iint_{\Gamma} w \cdot q_x \cdot n_x \, dS \tag{4.3.7a}$$

Set $f = w \cdot q_y$. Then, we have: $\frac{\partial f}{\partial y} = \frac{\partial w}{\partial y} \cdot q_y + w \cdot \frac{\partial q_y}{\partial y}$, and if we plug this expression in Equation (4.3.1b):

$$\iiint_{\Omega} \frac{\partial f}{\partial y} dV = \iint_{\Gamma} f \cdot n_y \, dS \rightarrow \iiint_{\Omega} \left(\frac{\partial w}{\partial y} \cdot q_y + w \cdot \frac{\partial q_y}{\partial y}\right) dV = \iint_{\Gamma} w \cdot q_y \cdot n_y \, dS \tag{4.3.7b}$$

Set $f = w \cdot q_z$. Then, we have: $\frac{\partial f}{\partial z} = \frac{\partial w}{\partial z} \cdot q_z + w \cdot \frac{\partial q_z}{\partial z}$, and if we plug this expression in Equation (4.3.1c):

$$\iiint_{\Omega} \frac{\partial f}{\partial z} dV = \iint_{\Gamma} f \cdot n_z \, dS \rightarrow \iiint_{\Omega} \left(\frac{\partial w}{\partial z} \cdot q_z + w \cdot \frac{\partial q_z}{\partial z}\right) dV = \iint_{\Gamma} w \cdot q_z \cdot n_z \, dS \tag{4.3.7c}$$

If we now add the three equations (4.3.7a, b, c):

$$\iiint_{\Omega} \left[\underbrace{\left(\frac{\partial w}{\partial x} \cdot q_x + \frac{\partial w}{\partial y} \cdot q_y + \frac{\partial w}{\partial z} \cdot q_z\right)}_{\left(\vec{\nabla} w\right) \bullet \vec{q}} + w \cdot \underbrace{\left(\frac{\partial q_x}{\partial x} + \frac{\partial q_y}{\partial y} + \frac{\partial q_z}{\partial z}\right)}_{\left(\vec{\nabla} \bullet \vec{q}\right)} \right] dV = \iint_{\Gamma} w \cdot \underbrace{\left(q_x \cdot n_x + q_y \cdot n_y + q_z \cdot n_z\right)}_{\vec{q} \bullet \vec{n}} dS$$

Thus, we eventually obtain:

$\iiint_{\Omega} w \cdot \left(\vec{\nabla} \cdot \vec{q}\right) dV + \iiint_{\Omega} \left(\vec{\nabla} w\right) \bullet \vec{q} \, dV = \iint_{\Gamma} w \cdot (\vec{q} \bullet \vec{n}) dS$, and by taking the second integral of the left-hand side to the right-hand side, we obtain Equation (4.3.6).

Remark 4.3.1: Green's formula is a generalization (to multiple dimensions) of integration by parts. ∎

We have established all the necessary mathematical preliminaries for three-dimensional problems. The various formulas and theorems also apply for two-dimensional problems; in such cases, we simply do not have a third component in the vectors and the fields do not depend on the *z*-coordinate.

Thus, we can establish the gradient of a scalar field $T(x,y)$ and the divergence of a vector field $\vec{r} = \begin{bmatrix} r_x & r_y \end{bmatrix}^T$ for two-dimensional problems:

$$\vec{\nabla} T = \begin{Bmatrix} \partial T/\partial x \\ \partial T/\partial y \end{Bmatrix} = \frac{\partial T}{\partial x}\vec{i} + \frac{\partial T}{\partial y}\vec{j} = \frac{\partial T}{\partial x_1}\vec{e}_1 + \frac{\partial T}{\partial x_2}\vec{e}_2 \tag{4.3.8}$$

$$\vec{\nabla} \bullet \vec{r} = \frac{\partial r_x}{\partial x} + \frac{\partial r_y}{\partial y} \tag{4.3.9}$$

Additionally, for two-dimensional problems, the domain is a two-dimensional surface, while the boundary is a one-dimensional curve. Thus, Green's theorem for two-dimensional problems can be written as follows:

$$\iint_{\Omega} (\nabla f)\, dV = \oint_{\Gamma} f \cdot \vec{n}\, dS \tag{4.3.10}$$

where $\vec{n}$ is the two-component unit-normal outward vector on the boundary curve, as shown in Figure 4.4b. Once again, Equation (4.3.10) can be written in component form:

$$\iint_{\Omega} \frac{\partial f}{\partial x_i} dV = \oint_{\Gamma} f \cdot n_i\, dS,\ i = 1,2 \tag{4.3.11}$$

or:

$$\iint_{\Omega} \frac{\partial f}{\partial x} dV = \oint_{\Gamma} f \cdot n_x\, dS \tag{4.3.11a}$$

and

$$\iint_{\Omega} \frac{\partial f}{\partial y} dV = \oint_{\Gamma} f \cdot n_y\, dS \tag{4.3.11b}$$

Remark 4.3.2: It is important to remember that—as shown in Figure 4.5—when we calculate the integral along the boundary curve, we must sweep the boundary Γ in a counterclockwise fashion. ∎

We can also write the divergence theorem and Green's formula for two-dimensional problems as follows.

Divergence theorem for two-dimensional problems:

$$\iint_{\Omega} \left(\vec{\nabla} \bullet \vec{r}\right) dV = \oint_{\Gamma} \left(\vec{r} \bullet \vec{n}\right) dS \tag{4.3.12}$$

Figure 4.5 Direction of boundary curve integration for two-dimensional problems.

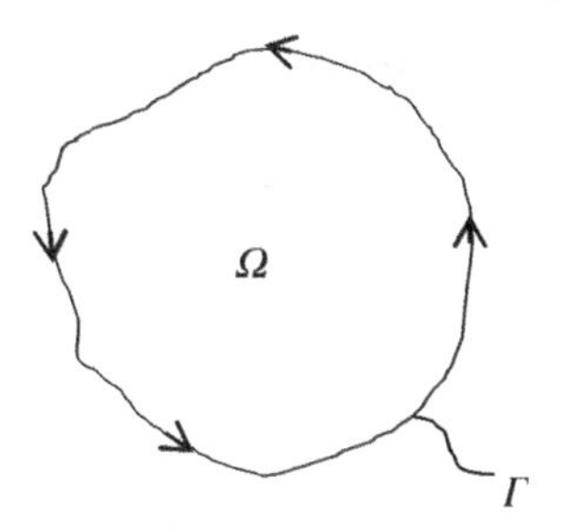

Green's formula for two-dimensional problems:

$$\iint_{\Omega} w \cdot \left(\vec{\nabla} \bullet \vec{q}\right) dV = \oint_{\Gamma} w \cdot \left(\vec{q} \bullet \vec{n}\right) dS - \iint_{\Omega} \left(\vec{\nabla} w\right) \bullet \vec{q}\, dV \tag{4.3.13}$$

4.4 Procedure for Multidimensional Problems

The procedure that we follow for a finite element solution of multidimensional problems is identical to the one that we used for one-dimensional problems—that is, the procedure summarized in Figure 1.8: we formulate the strong form (governing differential equations and boundary conditions) for the problem, then establish the weak form, then we introduce the finite element approximation and obtain a system of linear equations for the nodal values of the field function. To obtain the strong form, we need to examine the governing physics (e.g., force equilibrium for elasticity, energy balance for heat conduction) for a small piece of material cut from the interior of the system under consideration. This piece must be a rectangle for two-dimensional problems or a "box" for three-dimensional problems, as shown in Figure 4.6.

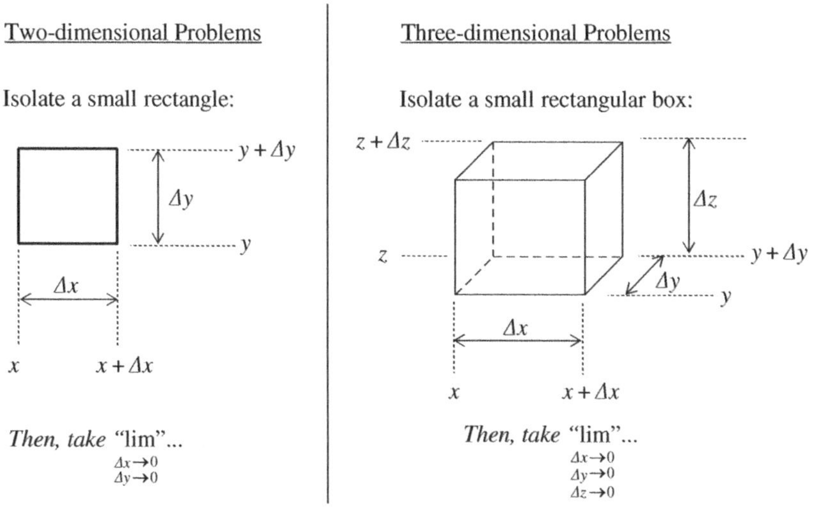

Figure 4.6 Small piece that is isolated for establishing governing physics in multidimensional problems.

Problems

4.1 Use Green's theorem and calculate the following surface integral, for the boundary Γ of any domain Ω:

$$\oint_{\Gamma} \vec{n}\, dS$$

4.2 Obtain an alternative version of Green's theorem, which is often provided in undergraduate analysis textbooks for a two-dimensional domain:

$$\oint_{\Gamma} (Q \cdot dx + R \cdot dy) = \iint_{\Omega} \left(\frac{\partial Q}{\partial x} - \frac{\partial R}{\partial y} \right) d\Omega$$

Hint: Use Remark 4.2.1, and also consider Figure P4.1, where we sweep the boundary Γ in a counterclockwise direction, and consider the tiny piece dS of boundary curve which can be considered as a straight line segment. Then, find a relation between dS and the corresponding lengths dx, dy. Finally, use Equation (4.3.11a) with $f = Q$, then Equation (4.3.11b) with $f = R$, and then sum the two obtained expressions.

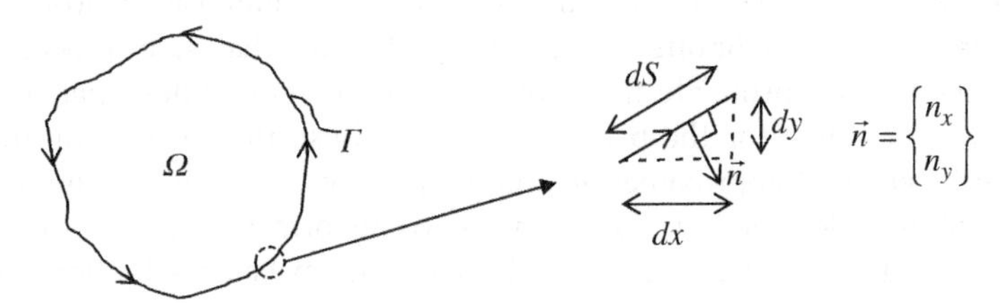

Figure P4.1

References

Marsden, J., and Tromba, A. (2011). *Vector Calculus*, 6th ed. New York: W.H. Freeman.
Stewart, J. (2015). *Calculus*, 8th ed. Boston: Brooks Cole.

5

Two-Dimensional Heat Conduction and Other Scalar Field Problems

Having established the basic mathematical notions for multidimensional problems in the previous chapter, we will proceed in this chapter to obtain the strong and weak form for multidimensional heat conduction, which is formulated in terms of the temperature scalar field, $\Theta(x,y)$. Toward the end of the chapter, we will examine other scalar field problems, for which the mathematical form of the governing equations is the same as that in heat conduction. While our derivation will focus on two-dimensional problems, the equations for the three-dimensional case can be obtained in the same fashion.

Let us consider a two-dimensional domain (i.e., a two dimensional body, such as a thin concrete slab), Ω, which has a boundary Γ and a constant out-of-plane thickness equal to d as shown in Figure 5.1. A heat source, $s(x,y)$, adds heat to the domain; s has units of energy per unit volume and per unit time. At a portion of the boundary, Γ_Θ, we have a prescribed temperature, as shown in Figure 5.1; for the remainder of the boundary, which we call Γ_q, we have a prescribed heat outflow $\bar{q}$, as shown in the same figure, with units of energy per unit surface and per unit time. *The prescribed temperatures and outflows do not have to be spatially constant—they can vary over the boundary.* Additionally, the direction of the boundary outflow at a location of the boundary must be normal to the boundary.

5.1 Strong Form for Two-Dimensional Heat Conduction

Since the problem is two-dimensional, we need to examine the governing physics for a small rectangular segment of the body with dimensions Δx and Δy, as shown in Figure 5.2. We will call this piece of material a *control volume*.

The heat flow in the interior of the domain will be described by a heat flux vector, $\{q\}$. For two-dimensional heat conduction, the heat flux vector will have two components, q_x and q_y. That is, heat can flow along any direction, and this flow is defined by a vector with two components along the two axes, x and y. The components of the heat flux vector give energy flow per unit area and per unit time. It is worth mentioning that, if we have a boundary segment, only the component of the heat flux *normal* to the boundary will correspond to flow across the boundary (i.e., heat entering or leaving the body). We can now draw the control volume, as shown in Figure 5.3, with the corresponding heat flux

Fundamentals of Finite Element Analysis: Linear Finite Element Analysis, First Edition.
Ioannis Koutromanos, James McClure, and Christopher Roy.

Companion website: www.wiley.com/go/koutromanos/linear

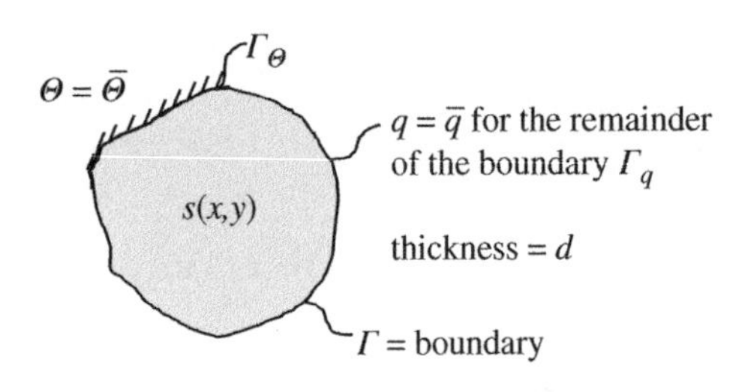

Figure 5.1 Two-dimensional heat conduction problem.

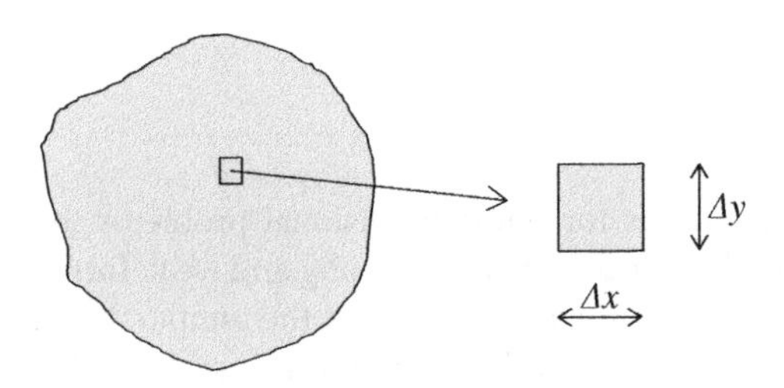

Figure 5.2 Consideration of a control volume from the interior of the two-dimensional domain.

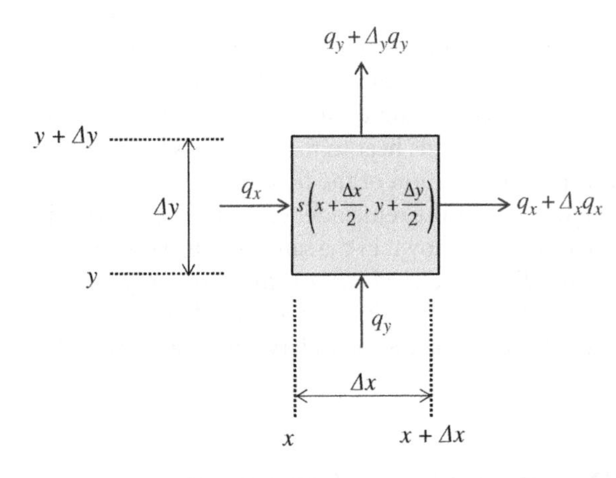

Figure 5.3 Heat inflow and outflow for a small piece of material (control volume).

components on each edge. In the figure, we have also drawn the heat source term, which we can consider as constant and equal to the value of s at the center of the control volume.

We can now set up the equations describing the governing physics. Since we have transfer of heat (energy), we will mathematically state the *conservation of energy principle*[1]—that is, that the amount of energy added to the control volume per unit time minus the amount of energy leaving the volume per unit time must equal zero. Each flux term must be multiplied by the area of the surface through which it flows. For example, the flux q_x in Figure 5.3 flows through a surface with area equal to $(\Delta y) \cdot d$. Similarly, q_y flows through a surface with area equal to $(\Delta x) \cdot d$. The total energy input associated

1 It may be worth mentioning here that the conservation of energy constitutes the *First Law of Thermodynamics*. Although such term will not be used in this text, the reader may encounter it in other references using energy balance expressions.

with the heat source is equal to the product of $s\left(x+\frac{\Delta x}{2},y+\frac{\Delta y}{2}\right)$ times the volume of the piece, $(\Delta x)\cdot(\Delta y)\cdot d$. Thus, we have:

$$\underbrace{q_x\cdot\Delta y\cdot d+q_y\cdot\Delta x\cdot d+s\left(x+\frac{\Delta x}{2},y+\frac{\Delta y}{2}\right)\cdot\Delta x\cdot\Delta y\cdot d}_{\text{Added heat per unit time}}-\underbrace{\left[\left(q_x+\Delta_x q_x\right)\cdot\Delta y\cdot d+\left(q_y+\Delta_y q_y\right)\cdot\Delta x\cdot d\right]}_{\text{Removed heat per unit time}}=0 \tag{5.1.1}$$

We can now conduct mathematical operations in Equation (5.1.1):

$$\Delta_x q_x\cdot\Delta y\cdot d+\Delta_y q_y\cdot\Delta x\cdot d-s\left(x+\frac{\Delta x}{2},y+\frac{\Delta y}{2}\right)\cdot\Delta x\cdot\Delta y\cdot d=0$$

If we divide both sides of the obtained expression by the volume $(\Delta x)\cdot(\Delta y)\cdot d$, we have:

$$\frac{\Delta_x q_x}{\Delta x}+\frac{\Delta_y q_x}{\Delta y}-s\left(x+\frac{\Delta x}{2},y+\frac{\Delta y}{2}\right)=0 \tag{5.1.2}$$

In accordance with Section 4.4, we will now take the limit as the size of the piece—that is, of the two-dimensional control volume tends to zero. Since this size is quantified by the lengths Δx, Δy, we can write:

$$\lim_{\Delta x,\Delta y\to 0}\left[\frac{\Delta_x q_x}{\Delta x}+\frac{\Delta_y q_x}{\Delta y}-s\left(x+\frac{\Delta x}{2},y+\frac{\Delta y}{2}\right)\right]=0 \tag{5.1.3}$$

We now notice that:

$$\lim_{\Delta x,\Delta y\to 0}\left[s\left(x+\frac{\Delta x}{2},y+\frac{\Delta y}{2}\right)\right]=s(x,y) \tag{5.1.4a}$$

If we also account for Equations (4.2.1a, b), we obtain:

$$\lim_{\Delta x,\Delta y\to 0}\left[\frac{\Delta_x q_x}{\Delta x}\right]=\frac{\partial q_x}{\partial x} \tag{5.1.4b}$$

$$\lim_{\Delta x,\Delta y\to 0}\left[\frac{\Delta_y q_y}{\Delta y}\right]=\frac{\partial q_y}{\partial y} \tag{5.1.4c}$$

Plugging Equations (5.1.4a, b, c) into Equation (5.1.3), we finally obtain the *differential equation governing two-dimensional heat conduction*:

$$\frac{\partial q_x}{\partial x}+\frac{\partial q_x}{\partial y}-s(x,y)=0 \tag{5.1.5}$$

From the definition of the divergence in Section 4.2 (Equation 4.2.6), we can also write the governing differential equation in the following form:

$$\vec{\nabla}\bullet\vec{q}-s(x,y)=0 \tag{5.1.6}$$

To complete the strong form for two-dimensional heat conduction, we must supplement the differential equation with boundary conditions. Just like for the one-dimensional problem, we distinguish two types of boundary conditions:

1) Essential boundary conditions, where we prescribe the value of the field function, or the temperature:

$$\Theta = \bar{\Theta} \; on \; \Gamma_{\Theta} \tag{5.1.7}$$

2) Natural boundary conditions, where we have a prescribed heat outflow $\bar{q}$. The value of $\bar{q}$ provides the amount of heat leaving the body, per unit surface of the boundary and per unit time. We will now examine how to set up an equation for natural boundary conditions, with the aid of Figure 5.4. In this figure, we have a prescribed outflow on a part of the boundary of a two-dimensional body. To obtain the pertinent equation, we isolate a small triangular piece, as shown in Figure 5.4, from the body. The piece has two sides aligned with the x and y axes, and a third side lies on the boundary and has a length equal to ΔL. Notice that the heat outflow, $\bar{q}$, points in the direction of the unit normal outward vector, $\vec{n}$, as also shown in Figure 5.4. Because the piece is very small, we consider the inclined boundary segment ΔL as a straight line. The figure also shows that if we define φ as the angle of the $\vec{n}$ vector with the horizontal direction, then we can find the two components, n_x and n_y, of vector $\vec{n}$ in terms of the cosine and sine of φ. Since the two straight sides, with lengths Δx and Δy, are cut from the interior of the body, we need to draw the corresponding heat flux vector components. We also include a term s, expressing the heat source contribution to our triangular piece. Since the piece has very small size, the source term is assumed to be constant. From geometry, we can write the following two expressions:

$$\Delta x = \Delta L \cdot \sin\varphi = \Delta L \cdot n_y \tag{5.1.7a}$$

$$\Delta y = \Delta L \cdot \cos\varphi = \Delta L \cdot n_x \tag{5.1.7b}$$

We can now write down the energy balance equation for the small triangular piece:

$$\underbrace{q_x \cdot \Delta y \cdot d + q_y \cdot \Delta x \cdot d + s \cdot \frac{1}{2} \Delta x \cdot \Delta y \cdot d}_{\text{Added heat per unit time}} - \underbrace{\bar{q} \cdot \Delta L \cdot d}_{\text{Removed heat per unit time}} = 0 \tag{5.1.8}$$

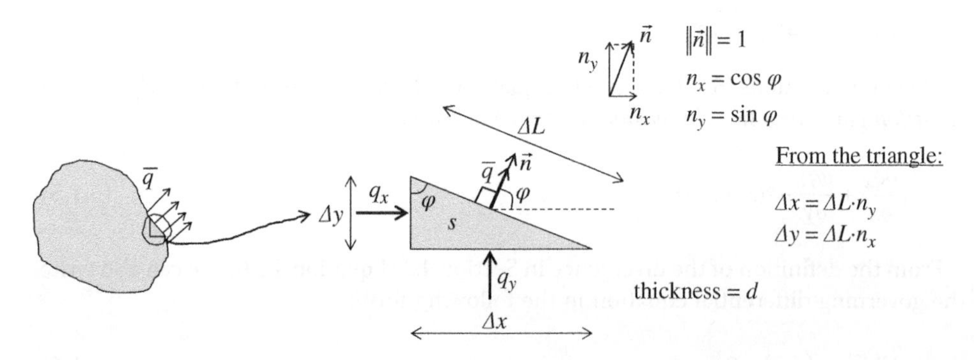

Figure 5.4 Procedure to obtain natural boundary conditions for heat conduction.

We can divide both sides of Equation (5.1.8) by the product $(\Delta L) \cdot d$ and also account for Equations (5.1.7a,b), to obtain:

$$q_x \cdot n_x + q_y \cdot n_y + s \cdot \frac{1}{2}\Delta L \cdot n_x \cdot n_y - \bar{q} = 0$$

Finally, we take the limit as the length ΔL tends to zero, to obtain a mathematical expression of the natural boundary conditions at every point of the natural boundary Γ_q:

$$\lim_{\Delta L \to 0}\left[q_x \cdot n_x + q_y \cdot n_y + s \cdot \frac{1}{2}\Delta L \cdot n_x \cdot n_y - \bar{q}\right] = 0 \to q_x \cdot n_x + q_y \cdot n_y - \bar{q} = 0 \tag{5.1.9}$$

We can now write the following, more concise version of Equation (5.1.9), giving the *natural boundary condition for heat conduction*:

$$\vec{q} \bullet \vec{n} = \bar{q} \text{ on } \Gamma_q \tag{5.1.10}$$

Remark 5.1.1: Whenever we have a natural boundary with $\bar{q} = 0$, i.e., a natural boundary that does not allow any energy flow across it, we will be saying that this is an *insulated boundary.* ▮

All that remains is to formulate the constitutive law, giving a relation between the heat flux vector $\vec{q}$ and the temperature field, Θ. Just as we did for one-dimensional heat conduction in Section 2.5, we can establish a constitutive law—specifically, Fourier's law in two dimensions—which generalizes the statement that heat flows from regions of higher temperature to regions of lower temperature; this constitutive law is described by the following equation.

$$\vec{q} = \{q\} = \begin{Bmatrix} q_x \\ q_y \end{Bmatrix} = -\begin{bmatrix} \kappa_{xx} & \kappa_{xy} \\ \kappa_{yx} & \kappa_{yy} \end{bmatrix}\begin{Bmatrix} \partial\Theta/\partial x \\ \partial\Theta/\partial y \end{Bmatrix} = -[D]\{\nabla\Theta\} \tag{5.1.11}$$

where $[D]$ is the material *thermal conductivity matrix* for two-dimensional heat conduction.

Remark 5.1.2: A special case for the $[D]$-matrix is that of an *isotropic material,* for which $[D]$ is a diagonal matrix:

$$[D] = \kappa\begin{bmatrix} 1 & 0 \\ 0 & 1 \end{bmatrix} \tag{5.1.12}$$

In this case, Fourier's law attains a simpler form than that of Equation (5.1.11):

$$\vec{q} = -\kappa \cdot \vec{\nabla}\,\Theta \tag{5.1.13}$$ ▮

If we want, we can plug the constitutive law (5.1.11) in Equation (5.1.5) and obtain the following form for the governing partial differential equation:

$$\vec{\nabla} \bullet \left(-\begin{bmatrix} \kappa_{xx} & \kappa_{xy} \\ \kappa_{yx} & \kappa_{yy} \end{bmatrix}\begin{Bmatrix} \partial\Theta/\partial x \\ \partial\Theta/\partial y \end{Bmatrix}\right) - s(x,y) = 0 \to \vec{\nabla} \bullet ([D]\{\nabla\Theta\}) + s(x,y) = 0 \tag{5.1.14}$$

Equation (5.1.14) can also be written in the following, expanded form:

$$\frac{\partial}{\partial x}\left(\kappa_{xx}\frac{\partial\Theta}{\partial x}+\kappa_{xy}\frac{\partial\Theta}{\partial y}\right)+\frac{\partial}{\partial y}\left(\kappa_{yx}\frac{\partial\Theta}{\partial x}+\kappa_{yy}\frac{\partial\Theta}{\partial y}\right)+s(x,y)=0 \tag{5.1.15}$$

We can now collectively write the governing differential equations and boundary conditions, to obtain Box 5.1.1, which is the *strong form for two-dimensional heat conduction.*

Box 5.1.1 Strong Form for Two-Dimensional Heat Conduction

$$\vec{\nabla}\bullet\left([D]\vec{\nabla}\Theta\right)+s(x,y)=0$$

$$\Theta=\bar{\Theta} \text{ on } \Gamma_\Theta$$

$$\vec{q}\bullet\vec{n}=-[D]\vec{\nabla}\Theta\bullet\vec{n}=\bar{q} \text{ on } \Gamma_q$$

Remark 5.1.3: For an isotropic material ($\kappa_{xx}=\kappa_{yy}$, $\kappa_{xy}=\kappa_{yx}=0$), which is also *uniform* or *homogeneous* (i.e., κ has the same, constant value everywhere), the governing differential equation for heat conduction becomes:

$$\kappa\frac{\partial^2\Theta}{\partial x^2}+\kappa\frac{\partial^2\Theta}{\partial y^2}+s(x,y)=0 \Leftrightarrow \kappa\cdot\nabla^2\Theta+s(x,y)=0 \quad or \quad \kappa\cdot\Delta\Theta+s(x,y)=0$$

where the Laplacian operator $\Delta T=\nabla^2 T$ was defined in Equation (4.2.7). ■

Example 5.1: Mathematical expression of boundary conditions
We are given the rectangular plate shown in Figure 5.5a. There is a heat source, as shown in the same figure, and the material of the plate is isotropic and uniform with $\kappa = 5$ W/(°C·m). We want to write down the governing differential equation and the expressions for the boundary conditions for the plate.

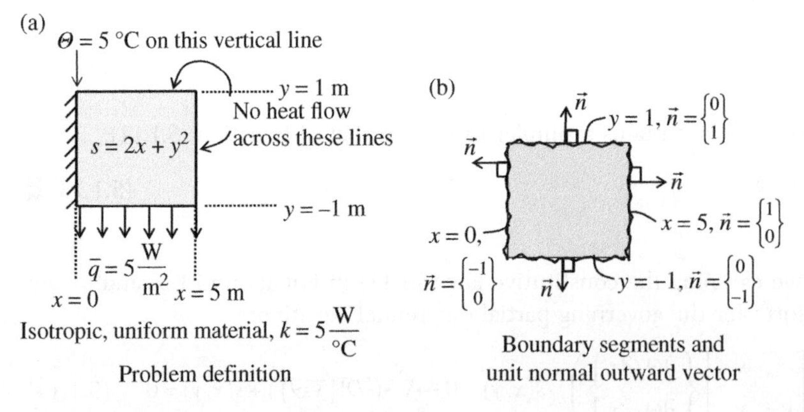

Figure 5.5 Example with two-dimensional heat conduction.

The governing differential equation is: $\kappa \cdot \nabla^2 \Theta + s(x,y) = 0 \Rightarrow$
$5\left(\frac{\partial^2 \Theta}{\partial x^2} + \frac{\partial^2 \Theta}{\partial y^2}\right) + 2x + y^2 = 0$

We can now proceed by writing the mathematical expressions for the four straight-line boundary segments of the rectangular plate. These segments and their respective unit normal outward vectors are presented in Figure 5.5b. The vertical-line boundary segments to the left and right are mathematically identified by the expressions $x = 0$ and $x = 5$ m, respectively (the characteristic of any vertical line is that all of its points have the same, unique value of x-coordinate!). Also, the bottom and top horizontal-line boundary segments are mathematically identified by the expressions $y = -1$ m and $y = 1$ m, respectively (the characteristic of any horizontal line is that all of its points have the same, unique y-coordinate value). We can now notice that one of the boundary segments is an essential one, while the other three are natural boundaries. Thus, we can set up the following expressions.

$$\Theta(x = 0, y) = 5^{\circ}\mathrm{C}$$

$$\vec{q} \bullet \vec{n}\Big|_{(x,y=-1m)} = 5 \Rightarrow -5\left(\frac{\partial \Theta}{\partial x}\Big|_{x,y=-1m} \cdot 0 + \frac{\partial \Theta}{\partial y}\Big|_{x,y=-1m} (-1)\right) = 5 \Rightarrow \frac{\partial \Theta}{\partial y}\Big|_{x,y=-1m} = 1$$

$$\vec{q} \bullet \vec{n}\Big|_{(x=5,y)} = 0 \Rightarrow -5\left(\frac{\partial \Theta}{\partial x}\Big|_{x=5,y} \cdot 1 + \frac{\partial \Theta}{\partial y}\Big|_{x=5,y} \cdot 0\right) = 0 \Rightarrow \frac{\partial \Theta}{\partial x}\Big|_{x=5,y} = 0$$

$$\vec{q} \bullet \vec{n}\Big|_{(x,y=1)} = 0 \Rightarrow -5\left(\frac{\partial \Theta}{\partial x}\Big|_{x,y=1} \cdot 0 + \frac{\partial \Theta}{\partial y}\Big|_{x,y=1} \cdot 1\right) = 0 \Rightarrow \frac{\partial \Theta}{\partial y}\Big|_{x,y=1} = 0$$

■

5.2 Weak Form for Two-Dimensional Heat Conduction

We will now establish the weak form for two-dimensional heat conduction. We consider an *arbitrary* function, $w(x,y)$, which *vanishes on the essential boundary:* $w = 0$ at Γ_Θ. We multiply the governing differential equation in Box (5.1.1) by $w(x,y)$ and then integrate over the two-dimensional domain, Ω:

$$\iint_\Omega w\left(\vec{\nabla} \bullet ([D]\{\nabla \Theta\}) + s(x,y)\right) dV = 0 \rightarrow \iint_\Omega w \vec{\nabla} \bullet ([D]\{\nabla \Theta\}) dV + \iint_\Omega w \cdot s dV = 0 \tag{5.2.1}$$

Now, it is worth noticing the first term in the left-hand side of Equation (5.2.1). We have an "asymmetry" in terms of differentiation of $w(x,y)$ and $\Theta(x,y)$: we have the divergence of a vector which depends on the gradient of $\Theta(x,y)$, while we have no differentiation of $w(x,y)$. To obtain a "symmetry" in terms of the differentiation, we use multidimensional integration by parts—that is, Green's formula (Equation 4.3.6):

$$\iint_\Omega w \cdot \vec{\nabla} \bullet ([D]\{\nabla \Theta\})\, dV = \int_\Gamma w \cdot \left([D]\{\nabla \Theta\} \bullet \vec{n}\right) dS - \iint_\Omega \vec{\nabla} w \bullet [D] \vec{\nabla} \Theta\, dV \tag{5.2.2}$$

The boundary integral in Equation (5.2.2) can be separated into two parts, one over Γ_Θ (where we know that w vanishes) and the other over Γ_q (for which we have the natural boundary condition):

$$\int_{\Gamma} w\cdot([D]\{\nabla\Theta\}\bullet\vec{n})dS = \int_{\Gamma_\Theta} w\cdot([D]\{\nabla\Theta\}\bullet\vec{n})dS + \int_{\Gamma_q} w\cdot([D]\{\nabla\Theta\}\bullet\vec{n})dS \tag{5.2.3}$$

0, because w vanishes on Γ_Θ

If we also account for the natural boundary condition of Box 5.1.1, Equation (5.2.3) becomes:

$$\int_{\Gamma} w\cdot([D]\{\nabla\Theta\}\bullet\vec{n})dS = -\int_{\Gamma_q} w\cdot\bar{q}\,dS \tag{5.2.4}$$

If we plug Equation (5.2.4) into Equation (5.2.2), we have:

$$-\iint_{\Omega}\vec{\nabla}w\bullet[D]\vec{\nabla}\Theta\,dV - \int_{\Gamma_q} w\cdot\bar{q}\,dS + \iint_{\Omega} w\cdot s\,dV = 0 \tag{5.2.5}$$

Finally, rearranging the terms of Equation (5.2.5) and also including the essential boundary condition from Box 5.1.1 yields the *weak form for two-dimensional heat conduction*, provided in Box 5.2.1.

Box 5.2.1 Weak Form for Two-Dimensional Heat Conduction

$$\iint_{\Omega}\left(\vec{\nabla}w\right)\bullet[D]\vec{\nabla}\Theta\,dV = \iint_{\Omega} w\cdot s\,dV - \int_{\Gamma_q} w\cdot\bar{q}\,dS \tag{5.2.6}$$

$\forall w(x)$ with $w = 0$ at Γ_Θ (w vanishes at essential boundary)

$\Theta = \bar{\Theta}$ at essential boundary, Γ_Θ

Remark 5.2.1: From a practical point of view, we can write: $\left(\vec{\nabla}w\right)\bullet[D]\vec{\nabla}\Theta = \{\nabla w\}^T[D]\{\nabla\Theta\}$. This equality can easily be verified by revisiting the definition of the dot product of two (two-dimensional) vectors from Equation (A.2.3) in Appendix A: $\vec{\alpha}\bullet\vec{\beta} = \{\alpha\}^T\{\beta\} = \alpha_1\cdot\beta_1 + \alpha_2\cdot\beta_2$ ■

The weak form thus has two fields (i.e., two functions of x and y)—namely, $\Theta(x,y)$ (which is the temperature field that we want to specify) and $w(x,y)$ (which is the weighting function of the weak form). The finite element solution is readily obtained by introducing an approximation for $\Theta(x,y)$ and $w(x,y)$, as discussed in Chapter 6.

As we shall see in Chapter 6, the obtained weak form leads to symmetric coefficient matrices in the multi-dimensional finite element formulation, if the same type of shape functions is used for the temperature field and the weighting function.

5.3 Equivalence of Strong Form and Weak Form

In the previous section, we obtained the weak form for two-dimensional heat conduction. The next step is to establish the equivalence between the strong and weak form. To do this, we simply need to prove that if we start from the weak form, we can obtain the strong form. So, let us start with the weak form in Box 5.2.1.

If we multiply both sides of Equation (5.2.6) in Box 5.2.1 by –1 and also account for Equation (5.1.11), we have:

$$\iint_{\Omega} \left(\vec{\nabla} w\right) \bullet \vec{q}\, dV = -\iint_{\Omega} w \cdot s\, dV + \int_{\Gamma_q} w \cdot \bar{q}\, dS, \quad \forall\, w(x,y) \;\; with \;\; w = 0 \;\; on \;\; \Gamma_{\Theta} \tag{5.3.1}$$

The left-hand side in Equation (5.3.1) can be transformed using Green's formula:

$$\iint_{\Omega} \left(\vec{\nabla} w\right) \bullet \vec{q}\, dV = \int_{\Gamma} w \cdot (\vec{q} \bullet \vec{n})\, dS - \iint_{\Omega} w \cdot \left(\vec{\nabla} \bullet \vec{q}\right) dV \tag{5.3.2}$$

If we additionally account for the fact that the boundary integral can be separated into two parts, the one over Γ_{Θ} and the other over Γ_q, we obtain:

$$\int_{\Gamma} w \cdot (\vec{q} \bullet \vec{n})\, dS = \underbrace{\int_{\Gamma_{\Theta}} w \cdot (\vec{q} \bullet \vec{n})\, dS}_{0,\ \text{because } w \text{ vanishes on } \Gamma_{\Theta}} + \int_{\Gamma_q} w \cdot (\vec{q} \bullet \vec{n})\, dS = \int_{\Gamma_q} w \cdot (\vec{q} \bullet \vec{n})\, dS \tag{5.3.3}$$

If we plug Equation (5.3.3) into (5.3.2), and then plug the resulting expression into (5.3.1), we have:

$$\int_{\Gamma_q} w \cdot (\vec{q} \bullet \vec{n})\, dS - \iint_{\Omega} w \cdot \left(\vec{\nabla} \bullet \vec{q}\right) dV = \int_{\Gamma_q} w \cdot \bar{q}\, dS - \iint_{\Omega} w \cdot s\, dV$$

$$\Rightarrow \iint_{\Omega} w \cdot \left(\vec{\nabla} \bullet \vec{q} - s\right) dV + \int_{\Gamma_q} w \cdot (\bar{q} - \vec{q} \bullet \vec{n})\, dS = 0 \tag{5.3.4}$$

Since the above equation applies for any $w(x,y)$ that vanishes on the essential boundary, it will also apply for the following special case of $w(x,y)$:

$$w(x,y) = \psi(x,y) \cdot \left(\vec{\nabla} \bullet \vec{q} - s\right), \quad where \;\; \psi(x,y) = 0 \;\; on \;\; \Gamma_q, \quad \psi(x,y) > 0 \;\; in \;\; \Omega \tag{5.3.5}$$

If we plug the special form of $w(x,y)$ from Equation (5.3.5) in Equation (5.3.4), we obtain:

$$\iint_{\Omega} \psi(x,y) \cdot \left(\vec{\nabla} \cdot \vec{q} - s\right)^2 dV + \underbrace{\int_{\Gamma_q} \psi(x,y) \cdot \left(\vec{\nabla} \cdot \vec{q} - s\right)(\bar{q} - \vec{q} \cdot \vec{n})\, dS}_{0,\ \text{because } \psi(x,y) = 0 \text{ on } \Gamma_q} = 0 \tag{5.3.6}$$

Thus, we finally have:

$$\iint_{\Omega} \underbrace{\psi(x,y)}_{>0} \cdot \underbrace{\left(\vec{\nabla} \cdot \vec{q} - s\right)^2}_{\geq 0,\ \text{because it is the square of a function}} dV = 0 \tag{5.3.7}$$

Notice that the integrand in Equation (5.3.7) is non-negative (a positive function multiplied by a non-negative function), so the integral (which is simply found by adding non-negative contributions) must also be non-negative. The only way to have a zero integral in Equation (5.3.7) is for the non-negative term in the integrand to be zero everywhere; that is, we finally obtain:

$$\vec{\nabla}\bullet\vec{q}-s=0 \ in\ \Omega \tag{5.3.8}$$

Also, since the weak form must be satisfied for any arbitrary $w(x,y)$, it must also be satisfied for another special form of $w(x,y)$:

$$w(x,y)=\varphi(x,y)\cdot\left(\bar{q}-\vec{q}\bullet\vec{n}\right),\quad where\quad \varphi(x,y)>0\ \ on\ \ \Gamma_q \tag{5.3.9}$$

If we plug the special form of w(x,y) from Equation (5.3.9) in Equation (5.3.4), we obtain:

$$\iint_{\Omega}\varphi(x,y)\cdot\left(\bar{q}-\vec{q}\cdot\vec{n}\right)\left(\vec{\nabla}\cdot\vec{q}-s\right)dV+\int_{\Gamma_q}\varphi(x,y)\cdot\left(\bar{q}-\vec{q}\cdot\vec{n}\right)^2 dS=0 \tag{5.3.10}$$

0, because we have already found that $\vec{\nabla}\cdot\vec{q}-s=0\ in\ \Omega$!

$$\int_{\Gamma_q}\underbrace{\varphi(x,y)}_{>0}\cdot\underbrace{\left(\bar{q}-\vec{q}\cdot\vec{n}\right)^2}_{\geq 0\text{, because it is the square of a function}} dS=0 \tag{5.3.11}$$

Once again, the integrand in Equation (5.3.11) is non-negative, thus the only way to have a zero integral is for the non-negative term to be zero everywhere!

$$\bar{q}-\vec{q}\bullet\vec{n}=0\ on\ \Gamma_q\rightarrow\vec{q}\bullet\vec{n}=\bar{q}\ \ on\ \ \Gamma_q \tag{5.3.12}$$

Thus, we started with the weak form in Box 5.2.1 and we have obtained the following set of expressions:

$$\vec{\nabla}\bullet\vec{q}-s(x,y)=0$$

$$\Theta=\bar{\Theta}\ on\ \Gamma_\Theta$$

$$\vec{q}\bullet\vec{n}=\bar{q}\ on\ \Gamma_q$$

If we also plug the constitutive equation in the obtained expressions, we finally reach the governing differential equation and boundary conditions that constitute the strong form in Box 5.1.1.

$$\vec{\nabla}\bullet\left[[D]\left(\vec{\nabla}\Theta\right)\right]+s(x,y)=0$$

$$\Theta=\bar{\Theta}\ on\ \Gamma_\Theta$$

$$-[D]\left(\vec{\nabla}\Theta\right)\bullet\vec{n}=\bar{q}\ on\ \Gamma_q$$

This completes the proof of the equivalence between strong form and weak form.

Remark 5.3.1: Just like in one-dimensional problems, the key to establishing the equivalence between the strong form and the weak form is that the latter must be satisfied for any arbitrary $w(x,y)$. ∎

Remark 5.3.2: Similar to what we did for one-dimensional problems, we can define the residual fields r_Ω and r_Γ, associated with the differential equation and natural boundary condition, respectively:

$$r_\Omega = \nabla \cdot \vec{q} - s(x,y) \tag{5.3.13a}$$

$$r_\Gamma = \bar{q} - \vec{q} \cdot \vec{n} \tag{5.3.13b}$$ ∎

Then, the first special form of $w(x,y)$ in Equation (5.3.5) could be written as: $w(x,y) = \psi(x,y) \cdot r_\Omega$ and the second special form of $w(x,y)$ in Equation (5.3.9) could be written as $w(x,y) = \varphi(x,y) \cdot r_\Gamma$. The functions r_Ω and r_Γ provide insight on the finite element formulation, because, just like for the one-dimensional problems in Chapter 2, *the residuals correspond to the conditions of the problem that may not be exactly satisfied by a finite element approximation!* Thus, the governing differential equation and the natural boundary condition may not be exactly satisfied by an approximate finite element solution. In fact, the difference of each residual from zero provides a numerical measure of the error in the satisfaction of the corresponding equation. For example, the absolute value of r_Ω provides a measure of the error of a solution in satisfying the differential equation of energy balance.

5.4 Other Scalar Field Problems

This chapter concludes with a brief presentation of other scalar field problems encountered in practice. These problems involve conservation of quantities other than energy, but they all result in strong forms that are mathematically identical to that obtained in Box 5.1.1 for heat conduction. The only difference is in the physical meaning of various scalar and vector fields and of the boundary conditions.

5.4.1 Two-Dimensional Potential Fluid Flow

We will begin by describing the simplest problem involving fluid flow, that of an *incompressible, inviscid fluid*. An inviscid fluid is one for which we do not have dissipation of energy due to "friction-like" viscous phenomena. In problems involving fluid flow, a central role is played by the *velocity vector field* $\vec{v}(x,y) = \begin{bmatrix} v_x(x,y) & v_y(x,y) \end{bmatrix}^T$, which simply provides the translation of fluid particles along the x- and y-directions per unit time. Each of the components of fluid velocity is positive when it is pointing along the positive direction of the corresponding axis. The term *incompressible* means that the volume of a piece of material with given mass remains practically unchanged during a process. For the current discussion, we will mathematically express this assumption by stipulating that the *density* ρ of the fluid, that is, the amount of mass per unit volume, remains constant. The assumption of incompressibility is satisfactorily accurate for all fluid flow problems, except for cases where the velocities are very large and approach or exceed the *speed of sound*. This latter case, which will not be considered herein, is important for problems involving supersonic or hypersonic aircraft, or for cases of spacecraft during, for example, atmospheric re-entry.

An example case involving inviscid fluid flow is shown in Figure 5.6, wherein we have a solid body that is immersed in a fluid. The fluid domain has four boundary segments, in each one of which we are given one of the two velocity components. Specifically, we are given the velocity component normal to each boundary segment. The two horizontal boundary segments at the top and bottom of this specific problem are *rigid* solid walls, which do not allow any fluid flow across them. For simplicity, we can also assume that the solid body is also *rigid*—that is, it does not experience any deformation.

An additional assumption, which is valid for many classes of fluid problems, is that the flow is *irrotational*. This means that the curl of the velocity vector field is zero:

$$curl(\vec{v}) = \vec{\nabla} \times \vec{v} = 0 \tag{5.4.1}$$

In accordance with Remark (4.2.4), there is some scalar field—let us denote it as $\Phi(x,y)$—such that the following condition applies:

$$\vec{v} = \vec{\nabla}\Phi \tag{5.4.2}$$

As we will see next, we can express the governing equations in a form that involves only the scalar field Φ, which we will name the *flow potential field*. Accordingly, any field flow where the velocity is obtained from Equation (5.4.2) is called a *potential fluid flow*.

We will now proceed to establish the differential equations, which will describe the governing physics—the *conservation of mass principle*. Let us cut a tiny rectangular piece from the interior of the fluid domain, as shown in Figure 5.7a. We also draw the components of fluid velocity on each side of this rectangular piece. We will imagine that this rectangle is a control volume, that is, we remain fixated on this rectangle for a long period of time and notice the fluid flowing into and out of this specific portion of the fluid domain. From Figure 5.7a, we notice that the direction of the fluid velocity is such that, at the left and bottom edges of our rectangle, we have fluid inflow. In fact, we can obtain how much *volume* of fluid is entering from, for example, the left edge. As shown in Figure 5.7b, if we observe the rectangle for a period of time equal to Δt, the volume entering our control region is equal to $(v_x \cdot \Delta t) \cdot \Delta y \cdot d$, where $(v_x \cdot \Delta t)$ is the length along the x-direction of the shaded region in Figure 5.7b and d is the out-of-plane thickness of

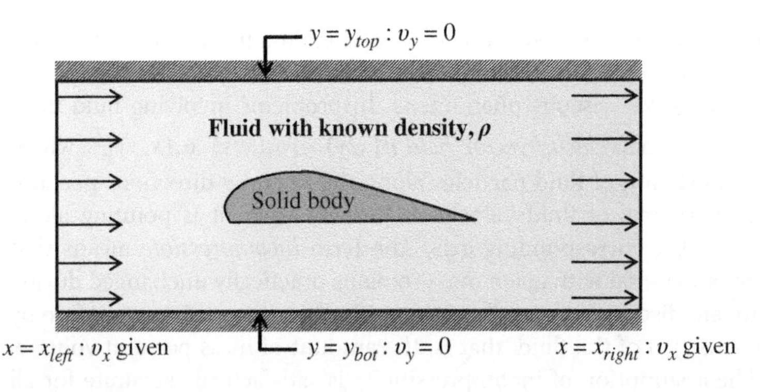

Figure 5.6 Example fluid flow problem.

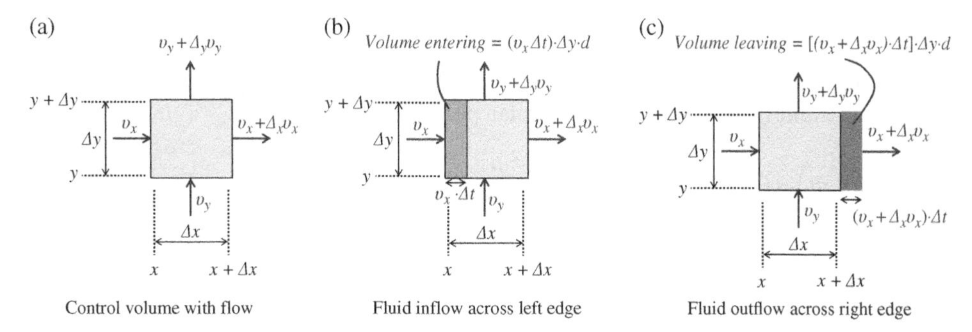

Figure 5.7 Basic considerations for control volume in fluid flow.

the fluid domain. Along the same lines, we have fluid outflow from the right and top edges of our rectangle. As shown in Figure 5.7c, the fluid volume leaving our control region from the right edge is equal to $[(v_x + \Delta_x v_x) \cdot \Delta t] \cdot \Delta y \cdot d$, where $(v_x + \Delta_x v_x) \cdot \Delta t$ is the length along the x-direction of the shaded volume leaving the region in Figure 5.7c.

Now, to convert the volume inflow and outflow into mass flow, we simply need to multiply by the density ρ. Thus, in accordance with the above considerations, we can write:

- Mass "pushed inside" the control volume through left edge = $v_x(x, y) \cdot \Delta t \cdot \Delta y \cdot d \cdot \rho$
- Mass "pushed outside" the region of interest through the right edge = $(v_x + \Delta_x v_x) \cdot \Delta t \cdot \Delta y \cdot d \cdot \rho$

We can establish similar considerations for the velocity components in the y-direction:

- Mass "pushed inside" the control volume through bottom edge = $v_y(x, y) \cdot \Delta t \cdot \Delta y \cdot d \cdot \rho$
- Mass "pushed outside" the region of interest through the top edge = $(v_y + \Delta_y v_y) \cdot \Delta t \cdot \Delta y \cdot d \cdot \rho$

We are now ready to establish the conservation of mass for the control volume region. The total mass entering over the time period Δt must equal the total mass leaving over the same time period:

$$\overbrace{v_x \cdot \Delta t \cdot \Delta y \cdot d \cdot \rho + v_y \cdot \Delta t \cdot \Delta x \cdot d \cdot \rho}^{\text{Mass inflow}} - \overbrace{(v_x + \Delta_x v_x) \cdot \Delta t \cdot \Delta y \cdot d \cdot \rho - (v_y + \Delta_y v_y) \cdot \Delta t \cdot \Delta x \cdot d \cdot \rho}^{\text{Mass outflow}} = 0 \tag{5.4.3}$$

We can now divide both sides of Equation (5.4.3) by $\Delta t \cdot \Delta x \cdot \Delta y \cdot d \cdot \rho$ and then make some mathematical manipulations to obtain:

$$\frac{\Delta_x v_x}{\Delta x} + \frac{\Delta_y v_x}{\Delta y} = 0 \tag{5.4.4}$$

We now take the limit as the size of the control volume tends to zero, and we obtain:

$$\frac{\partial v_x}{\partial x} + \frac{\partial v_x}{\partial y} = 0 \tag{5.4.5}$$

From the definition of the divergence in Section 4.2, we can also write Equation (5.4.5) in the following form.

$$\vec{\nabla} \bullet \vec{\upsilon} = 0 \tag{5.4.6}$$

Finally, if we account for Equation (5.4.2), we obtain the governing partial differential equation for fluid flow in terms of the scalar potential Φ:

$$\vec{\nabla} \bullet \left(\vec{\nabla} \Phi\right) = 0 \rightarrow \nabla^2 \Phi = 0 \rightarrow \frac{\partial^2 \Phi}{\partial x^2} + \frac{\partial^2 \Phi}{\partial y^2} = 0 \tag{5.4.7}$$

Remark 5.4.1: One can easily notice that, from a mathematical point of view, Equation (5.4.7) is equivalent to that obtained in Box 5.1.1, if we set $\Theta = \Phi$, $[D] = [I]$, $s(x,y) = 0$. ∎

Now, the boundary conditions correspond to prescribed velocity in the direction normal to each boundary segment. Thus, we can write the following expression:

$$\vec{\upsilon} \bullet \vec{n} = \left(\vec{\nabla} \Phi\right) \bullet \vec{n} = \bar{\upsilon} \quad \text{at } \Gamma_\upsilon \tag{5.4.8}$$

where $\vec{n}$ is the unit normal outward vector at the boundary Γ_υ and $\bar{\upsilon}$ is the value of the prescribed velocity.

It is worth mentioning that the presence of the rigid solid body in Figure 5.6 is accounting for by treating the perimeter of this body as part of the natural boundary! The reason is that we assume that the fluid cannot penetrate the solid body, and thus we cannot have any fluid flow across the perimeter of the body. This in turn means that, along the perimeter of the body, the velocity component normal to the boundary curve enclosing the rigid solid body must be zero.

Remark 5.4.2: It is worth mentioning that Equation (5.4.6) is a *natural boundary condition*. Thus, strictly speaking, our problem is a very special case, called the *Neumann problem*, wherein we only have natural boundary conditions. To be able to obtain a unique solution, we usually need to also apply an essential boundary condition, at least at one point of the boundary. Interestingly, it can be proven that the value of prescribed Φ that we use for the selected essential boundary location does not affect the obtained velocity components. The reason is that the velocities depend on the derivatives of Φ, and not on Φ itself! For this reason, we will always be saying that we also have an essential boundary, Γ_Φ, where we will be setting $\Phi = 0$. ∎

Based on all the above, we can now establish the strong form for multidimensional potential fluid flow, which is provided in Box 5.4.1.

Box 5.4.1 Strong form for multidimensional inviscid, incompressible, irrotational fluid flow

$$\vec{\nabla} \bullet \left(\vec{\nabla} \Phi\right) = \nabla^2 \Phi = 0 \tag{5.4.9}$$

$$\Phi = \bar{\Phi} \text{ at } \Gamma_\Phi \text{ (essential boundary condition)} \tag{5.4.10}$$

$$\vec{\upsilon} \bullet \vec{n} = \left(\vec{\nabla} \Phi\right) \bullet \vec{n} = \bar{\upsilon} \text{ at } \Gamma_\upsilon \text{ (natural boundary condition)} \tag{5.4.11}$$

Remark 5.4.3: It is also worth mentioning—as a side note—that many problems in science and engineering (other than fluid flow) are described by the partial differential equation (5.4.9), with the entire boundary being an essential one; that is, the value of the scalar field Φ is prescribed and given at each boundary location. These problems are called *Dirichlet problems*. ∎

5.4.2 Fluid Flow in Porous Media

We will now proceed to examine another case, involving fluid flow across a porous medium, such as soil or concrete. In many situations, materials have a porous structure, wherein we have solid particles forming a "skeleton" and in-between the particles we have void (or pore) regions, as shown in Figure 5.8a. The pores contain a fluid (e.g., a liquid), which can flow in the interior of the body, due to the fact that the pores are interconnected, forming a *pore network*. We will focus on steady-state conditions, and we will formulate the governing differential equations in terms of a scalar field $P(x, y)$, which is nothing else than the *pore fluid pressure*. The pore fluid will tend to flow from regions of higher pressure to regions of lower pressure.

We will now derive the governing equations. The key is to establish a flow vector field $\vec{q}_w$, called the *discharge* vector, which gives the flow of volume per unit width and per unit time. We can then consider a small control volume, as shown in Figure 5.8b, to set up equations expressing the conservation of mass. We also need to define a constitutive law, giving the discharge in terms of the pressure scalar field. This is *Darcy's law*, which can be written as follows.

$$\{q_w\} = \vec{q}_w = -\frac{\kappa}{\mu}\{\nabla P\} = -\frac{\kappa}{\mu}\vec{\nabla} P \tag{5.4.12}$$

where κ is the *intrinsic permeability* of the porous medium and μ is the *viscocity* of the fluid. The ratio $\frac{\kappa}{\mu}$ is sometimes referred to as the *permeability*.

Setting up the conservation of mass principle in a small control volume shown in Figure 5.8b, assuming a constant density for the fluid, and also formulating the boundary conditions, yields the strong form of flow in porous media, which is provided in Box 5.4.2.

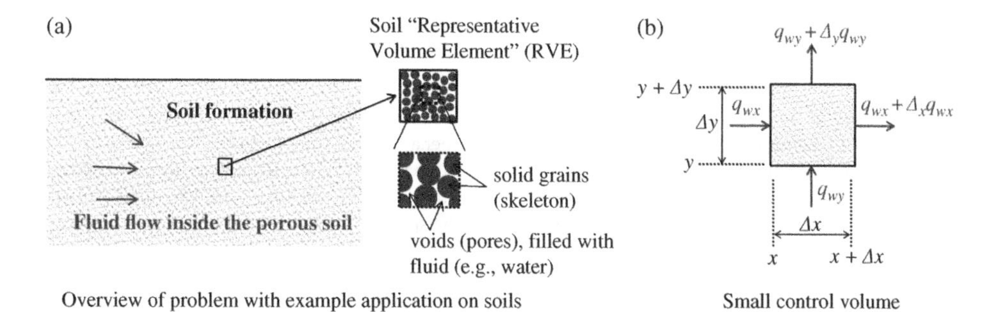

Figure 5.8 Fluid flow in porous solids.

Box 5.4.2 Strong Form for Fluid Flow in a Porous Medium

$$-\vec{\nabla}\cdot\left(\frac{\kappa}{\mu}\vec{\nabla}P\right)=0 \tag{5.4.13}$$

$$P=\bar{P} \text{ at } \Gamma_P \text{(essential boundary condition)} \tag{5.4.14}$$

$$\vec{q}_w\cdot\vec{n}=-\left(\frac{\kappa}{\mu}\vec{\nabla}P\right)\cdot\vec{n}=\bar{q}_w \text{ at } \Gamma_{qw} \text{(natural boundary condition)} \tag{5.4.15}$$

Remark 5.4.4: Equations (5.4.13–15) are mathematically equivalent to those obtained in Box 5.1.1, if we set $\Theta = P$, $[D] = \frac{\kappa}{\mu}[I]$, $s(x, y) = 0$. ∎

5.4.3 Chemical (Molecular) Diffusion-Reaction

A final case that involves conservation of mass and is very important is that of *chemical (or molecular) diffusion*. Let us imagine that we have a domain Ω, which is made of a stationary fluid medium such as water, and we have dissolved a small amount of a substance A. We typically refer to the combined medium and substance as a *solution*, wherein the substance A is called the *solute* and the fluid medium, which is present in much larger quantities than the substance, is called a *solvent*. A central quantity in such situations is the so-called *concentration field*, $C(x,y)$ of substance A, which gives how much mass A per unit volume of medium we have at each location of the domain. Now, we imagine the scenario in Figure 5.9, wherein we have a specific region with a relatively high concentration of substance A, while the remainder of the medium does not contain any amount of A at all. If we were able to have an initial state of this kind, then the molecules of substance A would *spontaneously*[2] want to begin "moving"

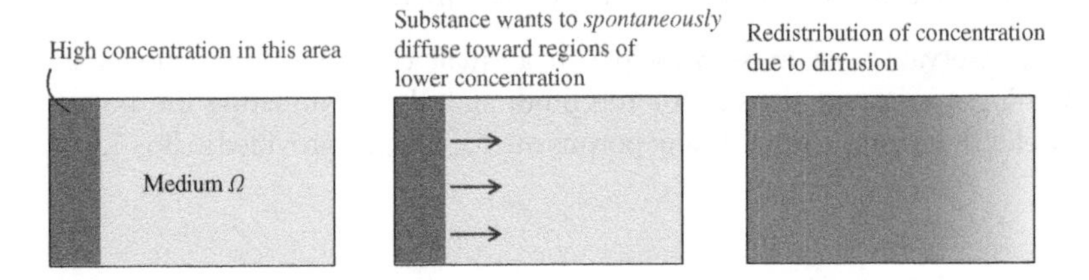

Figure 5.9 Example case involving chemical (molecular) diffusion.

2 The term *spontaneously* can be thought of to imply phenomena that can "automatically" take place in physical systems, without a need of energy addition from an external agent. There is a tool in physics that can help us determine when spontaneous phenomena take place, and that is the *Second Law of Thermodynamics*. Although we will not discuss thermodynamics in detail here, the specific example with spontaneous chemical diffusion can be justified with the Second Law. Along the same lines, we would have, for example, spontaneous heat transfer from regions of higher temperature to regions of lower temperature. Spontaneous processes typically stop when some form of *state balance* (also called *thermodynamic equilibrium*) is obtained. In the context of chemical diffusion, we would not have any spontaneous flow of solute mass if the concentration became identical at every location of our domain.

(this "movement" corresponds to diffusion) toward regions of lower concentration, such that the concentration of these other regions of the medium gradually begins to increase, as also shown in Figure 5.9.

What we establish now is a field quantifying the diffusion (i.e., the movement of molecules, which is movement of mass!). This is the *mass flux vector* $\vec{q}_C$, giving the flow of mass of substance A per unit area and per unit time. Then, we can once again establish a constitutive law, expressing the fact that the diffusion will take place such that molecules "travel" from regions of higher concentration to regions of low concentration. This equation is called *Fick's law*, and it is mathematically stated as follows:

$$\{q_C\} = \vec{q}_C = -D_C\{\nabla C\} = -D_C \vec{\nabla} C \quad (5.4.16)$$

The constant D_C in Equation (5.4.16) is called the *chemical diffusivity coefficient*.

By taking a control volume and establishing the equations expressing conservation of mass of substance A, we obtain the strong form for the chemical diffusion problem, in terms of the scalar concentration field, $C(x, y)$, of the substance, provided in Box 5.4.3. The governing differential equation contains a nonzero source term, $R(x,y)$, which we will call the *reaction term*. It is simply the amount of mass of the substance, per unit volume and per unit time, that is added at each location of the system, due to chemical reactions that may take place in the solution. Of course, for cases where no chemical reaction takes place, $R(x,y)$ is simply zero. Box 5.4.3 also includes the boundary conditions for the chemical diffusion problem. For essential boundaries, we prescribe the value of the concentration, while for natural boundaries, we prescribe the value of the mass outflow, $\bar{q}_C$, across the boundary.

Box 5.4.3 Strong Form for Chemical (Molecular) Diffusion

$$-\vec{\nabla} \bullet \left(D_C \vec{\nabla} C\right) = R(x, y) \quad (5.4.17)$$

$$C = \bar{C} \text{ at } \Gamma_C \text{ (essential boundary condition)} \quad (5.4.18)$$

$$\vec{q}_C \bullet \vec{n} = -\left(D_C \vec{\nabla} C\right) \bullet \vec{n} = \bar{q}_C \text{ at } \Gamma_{qC} \text{ (natural boundary condition)} \quad (5.4.19)$$

Remark 5.4.5: Equations (5.4.17–19) are mathematically equivalent to that obtained in Box 5.1.1, if we set $\Theta = C$, $[D] = D_C[I]$, $s(x,y) = R(x,y)$. ■

Problems

5.1 Consider a steady-state *advection-diffusion-reaction problem*, wherein we have the situation described in Section 5.4.3—that is, a substance dissolved in a fluid, and the concentration field is $C(x,y)$. The only difference now is that the fluid (i.e., the solvent) is no longer stationary; instead, the fluid is in a steady-state (constant in time) motion, and the steady-state velocity field is $\vec{a}(x, y)$. We also assume that the fluid is incompressible, thus, the divergence of the velocity field is zero:

$$\vec{\nabla} \bullet \vec{a} = 0$$

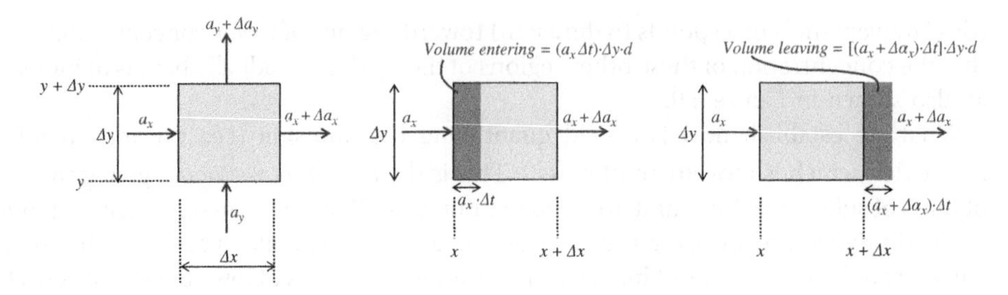

Figure P5.1

a) Obtain the strong form for the problem. You need to rely on the same exact considerations as in Figure 5.8b (mass inflow and outflow for a control volume), but now you also need to account for the mass *transported* by the fluid volume flowing into and out of the region of interest. A potentially helpful illustration (pertaining to transport in the x-direction) is shown in Figure P5.1 (where d is the out-of-plane depth of the considered domain).
b) Obtain the weak form for the problem.
c) What do you observe regarding the symmetry of differentiation between the weighting function, $w(x,y)$, and the actual field, $C(x,y)$ in the weak form? What implications does this observation entail for a finite element analysis of the advection-diffusion-reaction problem?

6

Finite Element Formulation for Two-Dimensional Scalar Field Problems

This chapter will derive the finite element equations for two-dimensional, scalar field problems. The descriptions will be specifically focused to heat conduction, but the same procedures are applicable to the other problems formulated in Section 5.4, as briefly discussed in Section 6.9. The procedure to obtain the finite element equations is identical to that employed for one-dimensional analysis. Thus, we will first introduce a piecewise finite element approximation, then leverage this approximation to convert the integral equations of the weak form to a system of linear equations, where the unknown quantities will be the nodal quantities of the field function (for heat conduction, these are the nodal temperature values).

6.1 Finite Element Discretization and Piecewise Approximation

As shown in Figure 6.1, the salient feature of the finite element solution is that the two-dimensional domain Ω is *discretized* (subdivided) into N_e subdomains, called elements, and each element consists of nodes. We will rely on a piecewise approximation, that is, we will stipulate the approximation of the field for each element separately.

The approximate temperature field, $\Theta^{(e)}(x, y)$, in element e will be given by a polynomial expression involving shape functions and the nodal temperature vector of the element. If the element has n nodal points, then the following equation will apply:

$$\Theta^{(e)}(x,y) = N_1^{(e)}(x,y)\cdot\Theta_1^{(e)} + N_2^{(e)}(x,y)\cdot\Theta_2^{(e)} + \ldots + N_n^{(e)}(x,y)\cdot\Theta_n^{(e)} = \sum_{I=1}^{n}\left(N_I^{(e)}(x,y)\cdot\Theta_I^{(e)}\right) \tag{6.1.1}$$

where the functions $N_I^{(e)}(x,y), I = 1, 2, \ldots, n$ are called the *interpolation functions or shape functions* of element e. We will cast Equation (6.1.1) in a matrix form:

$$\Theta^{(e)}(x,y) = \left[N^{(e)}\right]\left\{\Theta^{(e)}\right\} \tag{6.1.2}$$

Fundamentals of Finite Element Analysis: Linear Finite Element Analysis, First Edition.
Ioannis Koutromanos, James McClure, and Christopher Roy.

Companion website: www.wiley.com/go/koutromanos/linear

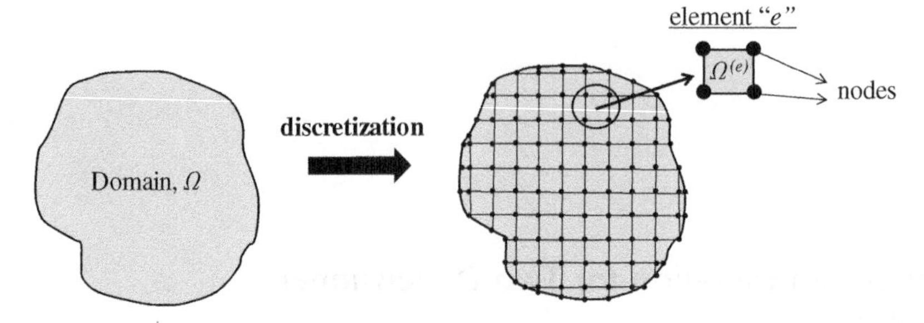

Figure 6.1 Discretization for finite element analysis.

where $[N^{(e)}]$ is the *shape function array of the element,* defined from the following expression:

$$\left[N^{(e)}\right] = \left[N_1^{(e)}(x,y)\;\; N_2^{(e)}(x,y)\;\; \ldots\;\; N_n^{(e)}(x,y)\right] \tag{6.1.3}$$

and $\{\Theta^{(e)}\}$ is the nodal temperature vector of the element:

$$\left\{\Theta^{(e)}\right\} = \left[\Theta_1^{(e)}\;\; \Theta_2^{(e)}\;\; \ldots\;\; \Theta_n^{(e)}\right]^T \tag{6.1.4}$$

Now that we have established the mathematical expressions giving the approximate temperature field in the element, we can conduct partial differentiation of Equation (6.1.1) with x and y to establish the expressions for the components of the approximate gradient of the temperature. For example, for the partial derivative with respect to x, we have:

$$\frac{\partial\Theta^{(e)}}{\partial x} = \frac{\partial}{\partial x}\left[N_1^{(e)}(x,y)\cdot\Theta_1^{(e)} + N_2^{(e)}(x,y)\cdot\Theta_2^{(e)} + \ldots + N_n^{(e)}(x,y)\cdot\Theta_n^{(e)}\right] \tag{6.1.5}$$

Now, we can notice that, in the terms of Equation (6.1.5), only the shape functions depend on x, while the nodal values of temperature are constant numbers. Thus, we have:

$$\frac{\partial\Theta^{(e)}}{\partial x} = \frac{\partial N_1^{(e)}}{\partial x}\cdot\Theta_1^{(e)} + \frac{\partial N_2^{(e)}}{\partial x}\cdot\Theta_2^{(e)} + \ldots + \frac{\partial N_n^{(e)}}{\partial x}\cdot\Theta_n^{(e)} \tag{6.1.6a}$$

Similarly, we obtain:

$$\frac{\partial\Theta^{(e)}}{\partial y} = \frac{\partial N_1^{(e)}}{\partial y}\cdot\Theta_1^{(e)} + \frac{\partial N_2^{(e)}}{\partial y}\cdot\Theta_2^{(e)} + \ldots + \frac{\partial N_n^{(e)}}{\partial y}\cdot\Theta_n^{(e)} \tag{6.1.6b}$$

The left-hand-side terms in Equations (6.1.6a,b) are the components of the gradient of the approximate temperature field. We can combine the two equations into a single matrix expression:

$$\left\{\nabla\Theta^{(e)}\right\} = \begin{Bmatrix} \partial\Theta^{(e)}/\partial x \\ \partial\Theta^{(e)}/\partial x \end{Bmatrix} = \begin{bmatrix} \partial N_1^{(e)}/\partial x & \partial N_2^{(e)}/\partial x & \ldots & \partial N_n^{(e)}/\partial x \\ \partial N_1^{(e)}/\partial y & \partial N_2^{(e)}/\partial y & \ldots & \partial N_n^{(e)}/\partial y \end{bmatrix} \begin{Bmatrix} \Theta_1^{(e)} \\ \Theta_2^{(e)} \\ \vdots \\ \Theta_n^{(e)} \end{Bmatrix} = \left[B^{(e)}\right]\left\{\Theta^{(e)}\right\} \tag{6.1.7}$$

where $[B^{(e)}]$ is the element array that premultiplies the nodal values to give the gradient of the approximate field:

$$\left[B^{(e)}\right] = \begin{bmatrix} \partial N_1^{(e)}/\partial x & \partial N_2^{(e)}/\partial x & \dots & \partial N_n^{(e)}/\partial x \\ \partial N_1^{(e)}/\partial y & \partial N_2^{(e)}/\partial y & \dots & \partial N_n^{(e)}/\partial y \end{bmatrix} \tag{6.1.8}$$

In accordance with Section B.2, the nodal values, $\{\Theta^{(e)}\}$, of each element e, are given from the global nodal temperature vector, $\{\Theta\}$ (i.e., the vector that contains the temperature values of all the nodal points in the finite element mesh), by a gather operation:

$$\left\{\Theta^{(e)}\right\} = \left[L^{(e)}\right]\{\Theta\} \tag{6.1.9}$$

where $[L^{(e)}]$ is the gather array of element e.

The piecewise approximation for the actual field (i.e., the temperature) is also used for the weighting function field, $w(x, y)$. Thus, we can write:

$$w^{(e)}(x,y) = \sum_{i=1}^{n}\left(N_i^{(e)}(x,y)\cdot w_i^{(e)}\right) = \left[N^{(e)}\right]\left\{W^{(e)}\right\} = \left\{W^{(e)}\right\}^T\left[N^{(e)}\right]^T \tag{6.1.10}$$

where $\{W^{(e)}\}$ is the vector with the nodal values of weighting function for element e.

It is noted that the expression involving the transpose of the shape function array in Equation (6.1.10) will be employed in our derivation (because this will later allow us to take an arbitrary vector as a common factor for all our terms).

Similarly, we have the following expression for the gradient of the weighting field function:

$$\left\{\nabla w^{(e)}\right\} = \left[B^{(e)}\right]\left\{W^{(e)}\right\} \tag{6.1.11}$$

We now take the transpose of Equation (6.1.11), because the weak form involves the transpose of the gradient of the weighting function:

$$\left\{\nabla w^{(e)}\right\}^T = \left\{W^{(e)}\right\}^T\left[B^{(e)}\right]^T \tag{6.1.12}$$

Finally, the nodal values $\{W^{(e)}\}$ for each element will be obtained by a gather operation from a global nodal weighting value vector, $\{W\}$:

$$\left\{W^{(e)}\right\} = \left[L^{(e)}\right]\{W\} \tag{6.1.13}$$

We are now ready to introduce our finite element approximation into the weak form. Let us rewrite the integral expression of the weak form for two-dimensional heat conduction, which was obtained in Box 5.2.1, after also accounting for Remark 5.2.1.

$$\iint_{\Omega}\{\nabla w\}^T[D]\{\nabla\Theta\}\,dV = \iint_{\Omega} w\cdot s\,dV - \int_{\Gamma_q} w\cdot\bar{q}\,dS \tag{6.1.14}$$

Just like we did for the one-dimensional cases in Chapter 3, we will separate the domain and natural boundary integrals into the contributions from the various element subdomains, $\Omega^{(e)}$, and natural boundary segments, $\Gamma_q^{(e)}$:

$$\iint_{\Omega}\{\nabla w\}^T[D]\{\nabla\Theta\}\,dV = \sum_{e=1}^{N_e}\left(\iint_{\Omega^{(e)}}\{\nabla w\}^T\left[D^{(e)}\right]\{\nabla\Theta\}\,dV\right) \tag{6.1.15a}$$

$$\iint_{\Omega} w \cdot s dV = \sum_{e=1}^{N_e} \left(\iint_{\Omega^{(e)}} w \cdot s^{(e)} dV \right) \tag{6.1.15b}$$

$$\int_{\Gamma_q} w \cdot \bar{q} dS = \sum_{e=1}^{N_e} \left(\int_{\Gamma_q^{(e)}} w \cdot \bar{q}^{(e)} dS \right) \tag{6.1.15c}$$

Now, for each integral term of element *e*, we can plug in the piecewise approximations for the trial solution (i.e., the actual field Θ) and weight function inside each element *e*. Let us begin with the term given by Equation (6.1.15a). We can plug Equations (6.1.7) and (6.1.12) into the specific expression, at which case we have:

$$\iint_{\Omega^{(e)}} \{\nabla w\}^T \left[D^{(e)}\right] \{\nabla \Theta\} dV = \iint_{\Omega^{(e)}} \left\{W^{(e)}\right\}^T \left[B^{(e)}\right]^T \left[D^{(e)}\right] \left[B^{(e)}\right] \left\{\Theta^{(e)}\right\} dV \tag{6.1.16}$$

The nodal vectors in Equation (6.1.16) are constant, and they can be taken outside of the integral to yield:

$$\iint_{\Omega^{(e)}} \{\nabla w\}^T \left[D^{(e)}\right] \{\nabla \Theta\} dV = \left\{W^{(e)}\right\}^T \iint_{\Omega^{(e)}} \left[B^{(e)}\right]^T \left[D^{(e)}\right] \left[B^{(e)}\right] dV \left\{\Theta^{(e)}\right\}$$

$$= \left\{W^{(e)}\right\}^T \left[k^{(e)}\right] \left\{\Theta^{(e)}\right\} \tag{6.1.17}$$

where $[k^{(e)}]$ is the *element coefficient array*, given by the expression:

$$\left[k^{(e)}\right] = \iint_{\Omega^{(e)}} \left[B^{(e)}\right]^T \left[D^{(e)}\right] \left[B^{(e)}\right] dV \tag{6.1.18}$$

For the special case of heat conduction, $[k^{(e)}]$ is also called the *element conductance array*.

Finally, we can account for Equations (6.1.9) and (6.1.13), to write:

$$\iint_{\Omega^{(e)}} \{\nabla w\}^T \left[D^{(e)}\right] \{\nabla \Theta\} dV = \{W\}^T \left[L^{(e)}\right]^T \left[k^{(e)}\right] \left[L^{(e)}\right] \{\Theta\} \tag{6.1.19}$$

We now add the two contributions of each element to the right-hand side of the weak form, given from Equations (6.1.15b) and (6.1.15c), and then plug Equation (6.1.10) in the resulting expression to obtain:

$$\iint_{\Omega^{(e)}} w \cdot s^{(e)} dV - \int_{\Gamma_q^{(e)}} w \cdot \bar{q}^{(e)} dS = \iint_{\Omega^{(e)}} \left\{W^{(e)}\right\}^T \left[N^{(e)}\right]^T s^{(e)} dV - \int_{\Gamma_q^{(e)}} \left\{W^{(e)}\right\}^T \left[N^{(e)}\right]^T \bar{q}^{(e)} dS$$

$$\rightarrow \iint_{\Omega^{(e)}} w \cdot s^{(e)}\, dV - \int_{\Gamma_q^{(e)}} w \cdot \bar{q}^{(e)}\, dS = \left\{W^{(e)}\right\}^T \left(\iint_{\Omega^{(e)}} \left[N^{(e)}\right]^T s^{(e)}\, dV - \int_{\Gamma_q^{(e)}} \left[N^{(e)}\right]^T \bar{q}^{(e)}\, dS \right) \rightarrow$$

$$\rightarrow \iint_{\Omega^{(e)}} w \cdot s^{(e)}\, dV - \int_{\Gamma_q^{(e)}} w \cdot \bar{q}^{(e)}\, dS = \left\{W^{(e)}\right\}^T \left\{f^{(e)}\right\} \tag{6.1.20}$$

where $\{f^{(e)}\}$ is the *equivalent nodal flux (more generally: the equivalent right-hand-side) vector* of element e, given by the following expression.

$$\left\{f^{(e)}\right\} = \iint_{\Omega^{(e)}} \left[N^{(e)}\right]^T s^{(e)}\, dV - \int_{\Gamma_q^{(e)}} \left[N^{(e)}\right]^T \bar{q}^{(e)}\, dS \tag{6.1.21a}$$

For computation reasons, it is usually more convenient to express the equivalent nodal flux vector as the sum of two terms, one stemming from the heat source and the other from the boundary heat outflow:

$$\left\{f^{(e)}\right\} = \left\{f_{\Omega}^{(e)}\right\} + \left\{f_{\Gamma_q}^{(e)}\right\} \tag{6.1.21b}$$

where

$$\left\{f_{\Omega}^{(e)}\right\} = \iint_{\Omega^{(e)}} \left[N^{(e)}\right]^T s^{(e)} \cdot dV \tag{6.1.21c}$$

and

$$\left\{f_{\Gamma_q}^{(e)}\right\} = -\int_{\Gamma_q} \left[N^{(e)}\right]^T \bar{q}^{(e)} dS \tag{6.1.21d}$$

If we now account for Equation (6.1.13), Equation (6.1.20) gives:

$$\iint_{\Omega^{(e)}} w \cdot s^{(e)}\, dV - \int_{\Gamma_q^{(e)}} w \cdot \bar{q}^{(e)}\, dS = \{W\}^T \left[L^{(e)}\right]^T \left\{f^{(e)}\right\} \tag{6.1.22}$$

Finally, we can Plug Equation (6.1.19) into Equation (6.1.15a):

$$\iint_{\Omega} \{\nabla w\}^T [D] \{\nabla \Theta\}\, dV = \sum_{e=1}^{N_e} \left(\{W\}^T \left[L^{(e)}\right]^T \left[k^{(e)}\right] \left[L^{(e)}\right] \{\Theta\} \right) \tag{6.1.23a}$$

Summing Equations (6.1.15b) and (6.1.15c) and plugging Equation (6.1.22) into the resulting expression yields:

$$\iint_{\Omega} w \cdot s dV - \int_{\Gamma_q} w \cdot \bar{q}\, dS = \sum_{e=1}^{N_e} \left(\{W\}^T \left[L^{(e)}\right]^T \left\{f^{(e)}\right\} \right) \tag{6.1.23b}$$

Finally, we can plug Equations (6.1.23a,b) into the weak form (Equation 6.1.14) to obtain:

$$\sum_{e=1}^{N_e}\left(\{W\}^T\left[L^{(e)}\right]^T\left[k^{(e)}\right]\left[L^{(e)}\right]\{\Theta\}\right)=\sum_{e=1}^{N_e}\left(\{W\}^T\left[L^{(e)}\right]^T\left\{f^{(e)}\right\}\right),\ \forall\ \{W\}^T \quad (6.1.24)$$

where the statement pertaining to an arbitrary weighting field function has been replaced by the statement "for any arbitrary vector $\{W\}^T$". Finally, we take all the terms of Equation (6.1.24) to the left-hand side, and then take $\{W\}^T$ and $\{\Theta\}$ outside of the summation, as common factors (since all the corresponding element contributions are multiplied by these two vectors):

$$\sum_{e=1}^{N_e}\left(\{W\}^T\left[L^{(e)}\right]^T\left[k^{(e)}\right]\left[L^{(e)}\right]\{\Theta\}\right)-\sum_{e=1}^{N_e}\left(\{W\}^T\left[L^{(e)}\right]^T\left\{f^{(e)}\right\}\right)=0$$

$$\rightarrow\{W\}^T\left[\left[\sum_{e=1}^{N_e}\left(\left[L^{(e)}\right]^T\left[k^{(e)}\right]\left[L^{(e)}\right]\right)\right]\{\Theta\}-\sum_{e=1}^{N_e}\left(\left[L^{(e)}\right]^T\left\{f^{(e)}\right\}\right)\right]=0\ \ \forall\{W\}^T$$

(6.1.25)

We thus obtain the equation:

$$\{W\}^T([K]\{\Theta\}-\{f\})=0\ \forall\ \{W\}^T \quad (6.1.26)$$

where $[K]$ is the global conductance array (more generally: the global coefficient array) and $\{f\}$ is the global equivalent nodal flux (or equivalent right-hand-side) vector.

$$[K]=\sum_{e=1}^{N_e}\left(\left[L^{(e)}\right]^T\left[k^{(e)}\right]\left[L^{(e)}\right]\right) \quad (6.1.27)$$

$$\{f\}=\sum_{e=1}^{N_e}\left(\left[L^{(e)}\right]^T\left\{f^{(e)}\right\}\right) \quad (6.1.28)$$

Remark 6.1.1: The arrays $[k^{(e)}]$ and $[K]$ are symmetric. ▮

Since Equation (6.1.26) must be satisfied for any arbitrary vector $\{W\}^T$, the only way to always satisfy it is to have:

$$[K]\{\Theta\}-\{f\}=\{0\}\rightarrow[K]\{\Theta\}=\{f\} \quad (6.1.29)$$

Thus, to obtain a finite element solution for two-dimensional heat conduction, we separate the domain into finite elements, calculate the coefficient (conductance) matrix $[k^{(e)}]$ and the right-hand side (equivalent nodal flux) vector, $\{f^{(e)}\}$, for each element. Then the global system of equations is assembled using the element gather/scatter arrays, $[L^{(e)}]$, and the assembly Equations (6.1.27), (6.1.28). We can then solve Equation (6.1.29) to obtain the nodal solution (i.e., nodal temperature) vector, $\{\Theta\}$

Remark 6.1.2: If we wanted to be perfectly rigorous, we should state that the finite element equations obtained in this section are the equations *per unit thickness of the body*. The actual thickness of the body does not affect the obtained solution. Still, if we are given the constant thickness, d, of the body, then we can account for it by multiplying both sides of Equation (6.1.29) by d. ▮

In subsequent sections, we will see how to establish the shape functions for different types of multidimensional elements. It should be mentioned that, besides having different types of polynomial approximations, we now also have different element shapes. Two commonly employed element shapes that will be discussed in this chapter are the three-node triangular element and the four-node quadrilateral element, as shown in Figure 6.2.

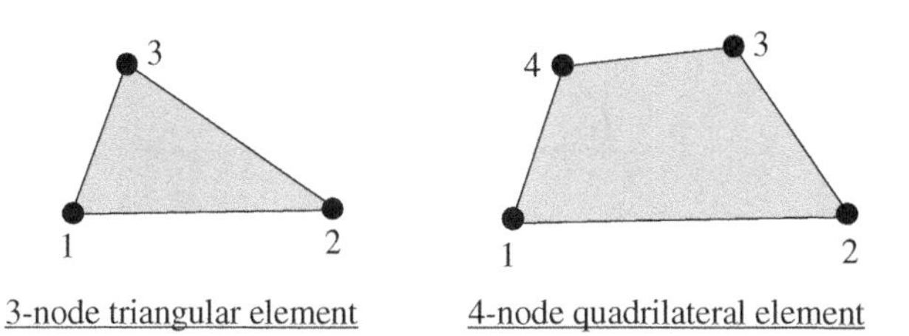

Figure 6.2 Example types of elements employed in two-dimensional finite element analysis.

Before examining specific element formulations, it is important to discuss the requirements for convergence of multidimensional finite element analysis. The same requirements that we had for one-dimensional problems, that is, continuity and completeness, must be satisfied by multidimensional approximate solutions. The required continuity is C^{k-1}, where k is the maximum order of differentiation that appears in the weak form. The weak form for multidimensional heat conduction includes first-order partial derivatives, so we have $k = 1$. Thus, the approximate solution must be C^0, meaning that the approximate temperature field must be continuous inside each element and across inter-element boundaries, as shown in Figure 6.3. Additionally, the piecewise approximation in each finite element must be able to describe constant values of these partial derivatives. This is accomplished if the approximate solution in each element has a completeness of first degree; that is, the piecewise finite element approximation must be capable of exactly representing any polynomial field that is linear with respect to x and linear with respect to y.

A very useful device to help with the determination of the degree of completeness that is provided by a two-dimensional finite element approximation is the so-called *Pascal's triangle*, which is shown in Figure 6.4a. The triangle provides various monomial terms in various rows, and the degree of completeness ensured by these monomial terms. To ensure a specific degree of completeness, all the terms of the triangle up to and including

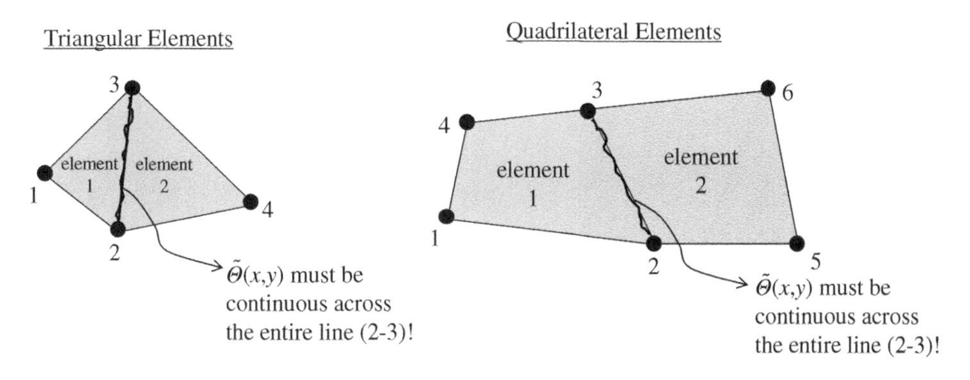

Figure 6.3 Continuity across inter-element boundaries for two-dimensional analysis.

the terms in the row that corresponds to the desired completeness must exist in a polynomial approximation. For example, if we want to have second degree of completeness, all the circled terms shown in Figure 6.4b must be present in the polynomial approximation.

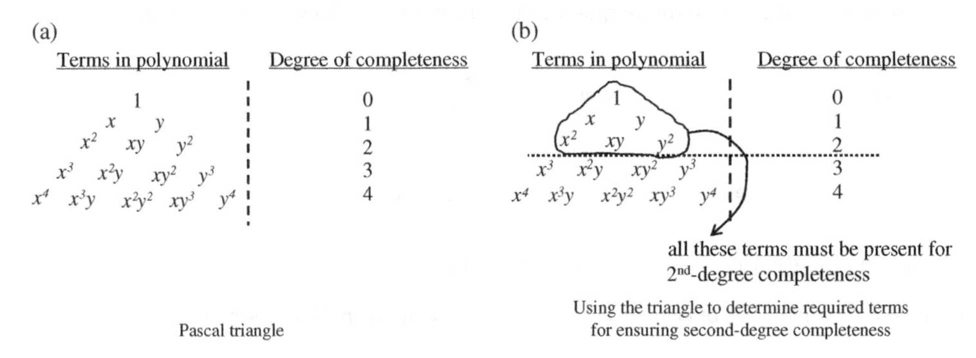

Figure 6.4 Pascal triangle.

Finally, it is important to emphasize that the shape functions of the finite elements for multidimensional analysis considered in the following sections satisfy the same properties as the shape functions in one-dimensional elements, that is: (i) the Kronecker delta property, (ii) the partition of unity property.

We will now discuss three basic types of elements for two-dimensional analysis, namely, the three-node triangle, the four-node rectangle, and the four-node quadrilateral. The latter type of element will allow us to introduce the concept of *isoparametric* finite elements.

6.2 Three-Node Triangular Finite Element

The three-node triangular finite element is the first element to have been used in analysis. Since we have three nodes and want to introduce a polynomial interpolation, we can establish the shape functions in the same way as we did for one-dimensional elements. We will assume that we are given the coordinates of the three nodal points, as shown in Figure 6.5. Specifically, we will start with a polynomial approximation that gives the maximum possible order of completeness. Since we have three nodal points (i.e., three values of nodal temperatures in the element), we need to have three terms in the polynomial approximation. From Pascal's triangle in Figure 6.4a, we understand that we need the

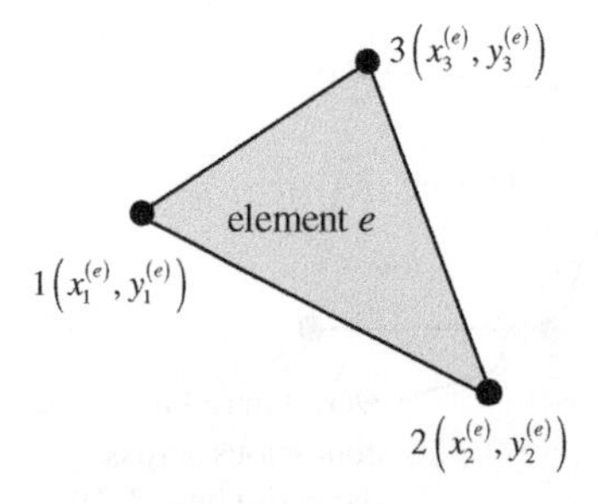

Figure 6.5 Three-node triangular element.

terms of the first two rows to ensure the minimum requirement for completeness, that is, first-degree (linear) completeness. Thus, we have:

$$\Theta^{(e)}(x,y) = a_0^{(e)} + a_1^{(e)}x + a_2^{(e)}y \to \Theta^{(e)}(x,y) = [1 \;\; x \;\; y]\begin{Bmatrix} a_0^{(e)} \\ a_1^{(e)} \\ a_2^{(e)} \end{Bmatrix} = [p(x,y)]\{a\} \tag{6.2.1}$$

where $[p(x,y)] = [1 \;\; x \;\; y]$ and $\{a\} = \left[a_0^{(e)} \;\; a_1^{(e)} \;\; a_2^{(e)}\right]^T$.

We now can write that the value of the approximate temperature field given by Equation (6.2.1) at each one of the nodal points will equal the corresponding nodal temperature. Thus, we can start with node 1:

$$\text{Node 1}: \textit{for } x = x_1^{(e)}, y = y_1^{(e)} \to \Theta^{(e)}\left(x = x_1^{(e)}, y = y_1^{(e)}\right) = \Theta_1^{(e)}$$

$$\to a_0^{(e)} + a_1^{(e)}x_1^{(e)} + a_2^{(e)}y_1^{(e)} = \Theta_1^{(e)} \tag{6.2.2a}$$

Then, we continue with the other two nodal points:

$$\text{Node 2}: \textit{for } x = x_2^{(e)}, y = y_2^{(e)} \to \Theta^{(e)}\left(x = x_2^{(e)}, y = y_2^{(e)}\right) = \Theta_2^{(e)}$$

$$\to a_0^{(e)} + a_1^{(e)}x_2^{(e)} + a_2^{(e)}y_2^{(e)} = \Theta_2^{(e)} \tag{6.2.2b}$$

$$\text{Node 3}: \textit{for } x = x_3^{(e)}, y = y_3^{(e)} \to \Theta^{(e)}\left(x = x_3^{(e)}, y = y_3^{(e)}\right) = \Theta_3^{(e)}$$

$$\to a_0^{(e)} + a_1^{(e)}x_3^{(e)} + a_2^{(e)}y_3^{(e)} = \Theta_3^{(e)} \tag{6.2.2c}$$

We can now collectively cast Equations (6.2.2a,b,c) in the following matrix expression:

$$\left[M^{(e)}\right]\left\{a^{(e)}\right\} = \left\{\Theta^{(e)}\right\} \tag{6.2.3}$$

where

$$\left[M^{(e)}\right] = \begin{bmatrix} 1 & x_1^{(e)} & y_1^{(e)} \\ 1 & x_2^{(e)} & y_2^{(e)} \\ 1 & x_3^{(e)} & y_3^{(e)} \end{bmatrix} \tag{6.2.4}$$

$$\left\{a^{(e)}\right\} = \left[a_0^{(e)} \;\; a_1^{(e)} \;\; a_2^{(e)}\right]^T \tag{6.2.5}$$

and

$$\left\{\Theta^{(e)}\right\} = \left[\Theta_1^{(e)} \;\; \Theta_2^{(e)} \;\; \Theta_3^{(e)}\right]^T \tag{6.2.6}$$

We can solve Equation (6.2.3) for the vector $\{a^{(e)}\}$:

$$\left[M^{(e)}\right]\left\{a^{(e)}\right\} = \left\{\Theta^{(e)}\right\} \to \left\{a^{(e)}\right\} = \left[M^{(e)}\right]^{-1}\left\{\Theta^{(e)}\right\} \tag{6.2.7}$$

If we now plug Equation (6.2.7) into Equation (6.2.1), we obtain:

$$\Theta^{(e)}(x,y) = [p(x,y)]\left[M^{(e)}\right]^{-1}\left\{\Theta^{(e)}\right\} = \left[N^{(e)}\right]\left\{\Theta^{(e)}\right\}$$

$$= \left[N_1^{(e)}(x,y)\ \ N_2^{(e)}(x,y)\ \ N_3^{(e)}(x,y)\right]\begin{Bmatrix}\Theta_1^{(e)}\\ \Theta_2^{(e)}\\ \Theta_3^{(e)}\end{Bmatrix} \tag{6.2.8}$$

where $[N^{(e)}]$ is the shape function array for the three-node triangular element e, given by the following matrix expression.

$$\left[N^{(e)}\right] = [p(x,y)]\left[M^{(e)}\right]^{-1} \tag{6.2.9}$$

If one does the math, the following expressions are obtained for the three shape functions.

$$N_1^{(e)}(x,y) = \frac{1}{2A^{(e)}}\left[x_2^{(e)}y_3^{(e)} - x_3^{(e)}y_2^{(e)} + \left(y_2^{(e)} - y_3^{(e)}\right)x + \left(x_3^{(e)} - x_2^{(e)}\right)y\right] \tag{6.2.10a}$$

$$N_2^{(e)}(x,y) = \frac{1}{2A^{(e)}}\left[x_3^{(e)}y_1^{(e)} - x_1^{(e)}y_3^{(e)} + \left(y_3^{(e)} - y_1^{(e)}\right)x + \left(x_1^{(e)} - x_3^{(e)}\right)y\right] \tag{6.2.10b}$$

$$N_3^{(e)}(x,y) = \frac{1}{2A^{(e)}}\left[x_1^{(e)}y_2^{(e)} - x_2^{(e)}y_1^{(e)} + \left(y_1^{(e)} - y_2^{(e)}\right)x + \left(x_2^{(e)} - x_1^{(e)}\right)y\right] \tag{6.2.10c}$$

where $A^{(e)}$ is the area of the triangular element, obtained from the nodal coordinates as follows.

$$A^{(e)} = \frac{1}{2}\det\left(\left[M^{(e)}\right]\right) = \frac{1}{2}\left[\left(x_2^{(e)}y_3^{(e)} - x_3^{(e)}y_2^{(e)}\right) - \left(x_1^{(e)}y_3^{(e)} - x_3^{(e)}y_1^{(e)}\right) + \left(x_1^{(e)}y_2^{(e)} - x_2^{(e)}y_1^{(e)}\right)\right] \tag{6.2.11}$$

Now that we have the three shape functions, we can also obtain their partial derivatives with respect to x and y:

$$\frac{\partial N_1^{(e)}}{\partial x} = \frac{y_2^{(e)} - y_3^{(e)}}{2A^{(e)}},\quad \frac{\partial N_2^{(e)}}{\partial x} = \frac{y_3^{(e)} - y_1^{(e)}}{2A^{(e)}},\quad \frac{\partial N_3^{(e)}}{\partial x} = \frac{y_1^{(e)} - y_2^{(e)}}{2A^{(e)}} \tag{6.2.12a}$$

$$\frac{\partial N_1^{(e)}}{\partial y} = \frac{x_3^{(e)} - x_2^{(e)}}{2A^{(e)}},\quad \frac{\partial N_2^{(e)}}{\partial y} = \frac{x_1^{(e)} - x_3^{(e)}}{2A^{(e)}},\quad \frac{\partial N_3^{(e)}}{\partial y} = \frac{x_2^{(e)} - x_1^{(e)}}{2A^{(e)}} \tag{6.2.12b}$$

Given the derivatives of the shape functions, we can also write the $[B^{(e)}]$ array for the three-node triangular element:

$$\left[B^{(e)}\right] = \frac{1}{2A^{(e)}}\begin{bmatrix} y_2^{(e)} - y_3^{(e)} & y_3^{(e)} - y_1^{(e)} & y_1^{(e)} - y_2^{(e)} \\ x_3^{(e)} - x_2^{(e)} & x_1^{(e)} - x_3^{(e)} & x_2^{(e)} - x_1^{(e)} \end{bmatrix} \tag{6.2.13}$$

Remark 6.2.1: It is worth mentioning that $[B^{(e)}]$ for the three-node triangular element is *constant.* ∎

The element conductance array, $[k^{(e)}]$, and the part $\left\{f_\Omega^{(e)}\right\}$ of the nodal flux vector due to a heat source (given by Equation 6.1.21c) can be obtained by integration on the triangular domain. If the material conductivity array, $\left[D^{(e)}\right] = \begin{bmatrix} \kappa_{xx} & \kappa_{xy} \\ \kappa_{yx} & \kappa_{xx} \end{bmatrix}^{(e)}$, is constant, then the element's conductance matrix can be easily obtained, since $[B^{(e)}]$, $[D^{(e)}]$ and $[B^{(e)}]^T$ are all constant and can be taken outside of the integral in Equation (6.1.18)! Then, the integration over the triangular domain simply yields:

$$\left[k^{(e)}\right] = \iint_{\Omega^{(e)}} \left[B^{(e)}\right]^T \left[D^{(e)}\right] \left[B^{(e)}\right] dV = \left[B^{(e)}\right]^T \left[D^{(e)}\right] \left[B^{(e)}\right] \iint_{\Omega^{(e)}} dV$$

$$= \left[B^{(e)}\right]^T \left[D^{(e)}\right] \left[B^{(e)}\right] \cdot A^{(e)} \tag{6.2.14}$$

We will now see how to calculate $\left\{f_{\Gamma_q}^{(e)}\right\}$, for the case where a side of a triangular element constitutes a segment of the natural boundary. To do so, we *parameterize* the boundary segment. As an example, we will consider an element with a prescribed boundary outflow at the side (1-3), as shown in Figure 6.6. Since the natural boundary of the element is a line segment, we can describe the position of a point along this segment by means of a single parameter, ξ, with $-1 \le \xi \le 1$, and express the values of x and y along the boundary segment as functions of ξ. The bounds of the parameter ξ are such that we will be able to use Gaussian quadrature. We need to establish a mapping that gives, for any point along the segment (1-3), the x and y coordinates as functions of the ξ-coordinate.

From the parameterization in Figure 6.6, we know that the value $\xi = -1$ corresponds to nodal point (1), while the value $\xi = 1$ corresponds to node 3. Thus, we can establish the following pair of mapping equations.

$$x(\xi) = \frac{x_1^{(e)} + x_3^{(e)}}{2} + \frac{x_3^{(e)} - x_1^{(e)}}{2}\xi, \ -1 \le \xi \le 1 \tag{6.2.15a}$$

$$y(\xi) = \frac{y_1^{(e)} + y_3^{(e)}}{2} + \frac{y_3^{(e)} - y_1^{(e)}}{2}\xi, \ -1 \le \xi \le 1 \tag{6.2.15b}$$

The reader can verify that these mapping expressions take us to the locations of nodes (1) and (3) for $\xi = -1$ and $\xi = 1$, respectively.

We can also express the values of the three shape functions along the natural boundary with respect to the single parameter ξ, by plugging Equations (6.2.15a,b) in Equations (6.2.10a,b,c):

$$N_1^{(e)}(x(\xi), y(\xi)) = \frac{1}{2A^{(e)}} \left[x_2^{(e)} y_3^{(e)} - x_3^{(e)} y_2^{(e)} + \left(y_2^{(e)} - y_3^{(e)}\right) x(\xi) + \left(x_3^{(e)} - x_2^{(e)}\right) y(\xi)\right] \tag{6.2.16a}$$

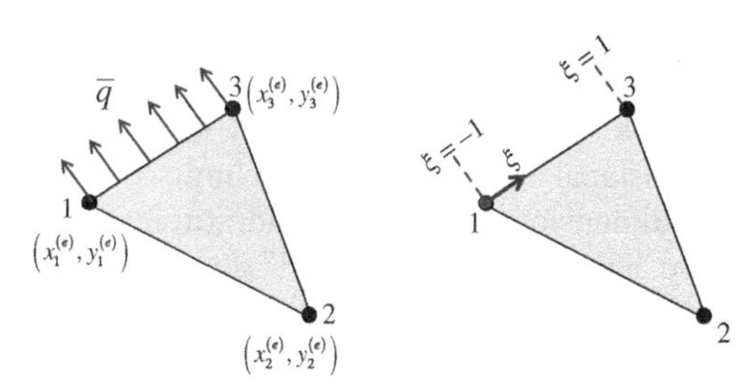

Figure 6.6 Parameterization of natural boundary segment in three-node triangular element.

$$N_2^{(e)}(x(\xi),y(\xi)) = \frac{1}{2A^{(e)}}\left[x_3^{(e)}y_1^{(e)} - x_1^{(e)}y_3^{(e)} + \left(y_3^{(e)} - y_1^{(e)}\right)x(\xi) + \left(x_1^{(e)} - x_3^{(e)}\right)y(\xi)\right] \quad (6.2.16b)$$

$$N_3^{(e)}(x(\xi),y(\xi)) = \frac{1}{2A^{(e)}}\left[x_1^{(e)}y_2^{(e)} - x_2^{(e)}y_1^{(e)} + \left(y_1^{(e)} - y_2^{(e)}\right)x(\xi) + \left(x_2^{(e)} - x_1^{(e)}\right)y(\xi)\right] \quad (6.2.16c)$$

Now, after having established a parameterization for the boundary segment, we use a fundamental theorem of calculus, which states that the differential length dS in the integral of (6.1.21d), which is taken on a boundary segment parameterized in terms of ξ, is given by:

$$dS = \sqrt{\left(\frac{\partial x}{\partial \xi}\right)^2 + \left(\frac{\partial y}{\partial \xi}\right)^2}\, d\xi = \frac{\ell_{31}}{2}d\xi \quad (6.2.17)$$

where ℓ_{31} is the length of the boundary segment (1-3):

$$\ell_{31} = \sqrt{\left(x_3^{(e)} - x_1^{(e)}\right)^2 + \left(y_3^{(e)} - y_1^{(e)}\right)^2} \quad (6.2.18)$$

We can now go ahead and evaluate the integral:

$$\left\{f_{\Gamma_q}^{(e)}\right\} = -\int_{\Gamma_q}\left[N^{(e)}\right]^T \bar{q}\, dS = -\int_{-1}^{1}\begin{Bmatrix} N_1^{(e)}(x(\xi),y(\xi)) \\ N_2^{(e)}(x(\xi),y(\xi)) \\ N_3^{(e)}(x(\xi),y(\xi)) \end{Bmatrix}\bar{q}\frac{L_{31}}{2}d\xi$$

$$= -\begin{Bmatrix} \int_{-1}^{1} N_1^{(e)}(x(\xi),y(\xi))\bar{q}\frac{\ell_{31}}{2}d\xi \\ \int_{-1}^{1} N_2^{(e)}(x(\xi),y(\xi))\bar{q}\frac{\ell_{31}}{2}d\xi \\ \int_{-1}^{1} N_3^{(e)}(x(\xi),y(\xi))\bar{q}\frac{\ell_{31}}{2}d\xi \end{Bmatrix} \quad (6.2.19)$$

Remark 6.2.2: Since Equation (6.2.19) requires the calculation of three one-dimensional integrals with the limits of integration being from –1 to 1, we can use one-dimensional Gaussian quadrature, as explained in Section 3.7. The procedure provided here can also be used if the natural boundary corresponds to another element side (e.g., side (1-2)). ∎

6.3 Four-Node Rectangular Element

Let us now examine another relatively simple case of two-dimensional finite element, namely, the four-node rectangular element, as shown in Figure 6.7. This element was originally formulated in Argyris (1960). We will focus on the special case that the sides of the element are aligned with the two coordinate axes, x and y. Obviously, the polynomial element approximation will be having four terms. We will obtain the shape functions of the element using two alternative approaches. The first approach is that employed for the three-node triangular element in the previous section. Figure 6.7 provides the polynomial approximation for the temperature field for the element, which includes four monomial terms.

We can now write:

$$\Theta^{(e)}(x,y) = \lfloor p(x,y) \rfloor \left\{ a^{(e)} \right\} \tag{6.3.1}$$

where

$$\lfloor p(x,y) \rfloor = \begin{bmatrix} 1 & x & y & x \cdot y \end{bmatrix} \tag{6.3.2a}$$

and

$$\left\{ a^{(e)} \right\} = \begin{bmatrix} a_0^{(e)} & a_1^{(e)} & a_2^{(e)} & a_3^{(e)} \end{bmatrix}^T \tag{6.3.2b}$$

Remark 6.3.1: Note that, for the fourth term in the polynomial, we selected the monomial xy. The reason is that this monomial is "symmetric" in terms of x and y. As shown in Figure 6.8, the specific polynomial approximation includes all the terms ensuring

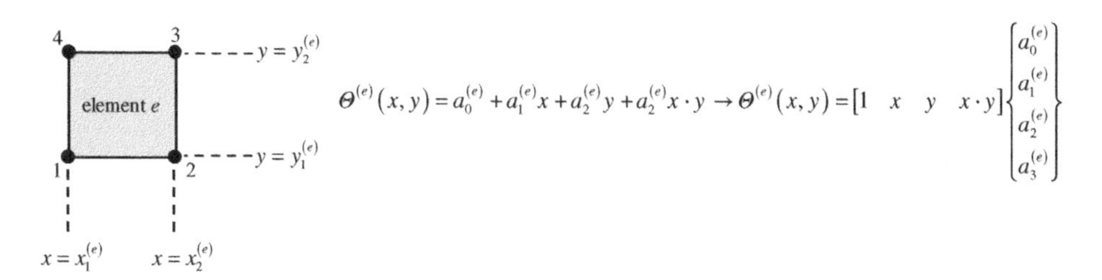

Figure 6.7 Four-node rectangular element aligned with coordinate axes x and y.

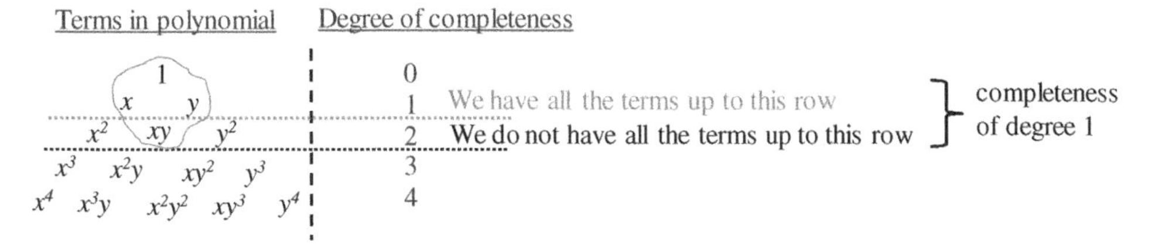

Figure 6.8 Completeness of four-node rectangular element.

first-degree of completeness, but only one of the second-order terms which are also required for second-degree completeness. Hence, the four-node rectangular element approximation is complete to the first degree, which is adequate for the analysis of linear two-dimensional heat conduction. ∎

We will now formulate the approximation in terms of the nodal values of temperature and the corresponding shape functions. We set up four equations, corresponding to the values of temperature at the locations of the four nodes:

Node 1:

$$\Theta^{(e)}\left(x=x_1^{(e)}, y=y_1^{(e)}\right)=\Theta_1^{(e)} \rightarrow a_0^{(e)}+a_1^{(e)} x_1^{(e)}+a_2^{(e)} y_1^{(e)}+a_3^{(e)} x_1^{(e)} \cdot y_1^{(e)}=\Theta_1^{(e)} \tag{6.3.3a}$$

Node 2:

$$\Theta^{(e)}\left(x=x_2^{(e)}, y=y_2^{(e)}\right)=\Theta_2^{(e)} \rightarrow a_0^{(e)}+a_1^{(e)} x_2^{(e)}+a_2^{(e)} y_2^{(e)}+a_3^{(e)} x_2^{(e)} \cdot y_2^{(e)}=\Theta_2^{(e)} \tag{6.3.3b}$$

Node 3:

$$\Theta^{(e)}\left(x=x_3^{(e)}, y=y_3^{(e)}\right)=\Theta_3^{(e)} \rightarrow a_0^{(e)}+a_1^{(e)} x_3^{(e)}+a_2^{(e)} y_3^{(e)}+a_3^{(e)} x_3^{(e)} \cdot y_3^{(e)}=\Theta_3^{(e)} \tag{6.3.3c}$$

Node 4:

$$\Theta^{(e)}\left(x=x_4^{(e)}, y=y_4^{(e)}\right)=\Theta_4^{(e)} \rightarrow a_0^{(e)}+a_1^{(e)} x_4^{(e)}+a_2^{(e)} y_4^{(e)}+a_3^{(e)} x_4^{(e)} \cdot y_4^{(e)}=\Theta_4^{(e)} \tag{6.3.3d}$$

We can collectively write:

$$[M]\left\{a^{(e)}\right\}=\left\{\Theta^{(e)}\right\} \tag{6.3.4}$$

where

$$[M]=\begin{bmatrix} 1 & x_1^{(e)} & y_1^{(e)} & x_1^{(e)} \cdot y_1^{(e)} \\ 1 & x_2^{(e)} & y_2^{(e)} & x_2^{(e)} \cdot y_2^{(e)} \\ 1 & x_3^{(e)} & y_3^{(e)} & x_3^{(e)} \cdot y_3^{(e)} \\ 1 & x_4^{(e)} & y_4^{(e)} & x_4^{(e)} \cdot y_4^{(e)} \end{bmatrix} \tag{6.3.5}$$

and

$$\left\{\Theta^{(e)}\right\}=\left[\Theta_1^{(e)} \quad \Theta_2^{(e)} \quad \Theta_3^{(e)} \quad \Theta_4^{(e)}\right]^T \tag{6.3.6}$$

We can finally obtain:

$$\left\{a^{(e)}\right\}=[M]^{-1}\left\{\Theta^{(e)}\right\} \tag{6.3.7}$$

If we plug (6.3.7) into (6.3.1), we have:

$$\Theta^{(e)}(x,y) = \lfloor p(x,y) \rfloor \left\{a^{(e)}\right\} = \lfloor p(x,y) \rfloor [M]^{-1} \left\{\Theta^{(e)}\right\} = \left[N^{(e)}\right] \left\{\Theta^{(e)}\right\} \tag{6.3.8}$$

where $\left[N^{(e)}\right] = \lfloor p(x,y) \rfloor [M]^{-1}$ is the shape function array for the element:

$$\left[N^{(e)}\right] = \lfloor p(x,y) \rfloor [M]^{-1} = \left[N_1^{(e)}(x,y) \;\; N_2^{(e)}(x,y) \;\; N_3^{(e)}(x,y) \;\; N_4^{(e)}(x,y)\right] \tag{6.3.9}$$

The four shape functions are given by the following expressions:

$$N_1^{(e)}(x,y) = \frac{x - x_2^{(e)}}{x_1^{(e)} - x_2^{(e)}} \cdot \frac{y - y_2^{(e)}}{y_1^{(e)} - y_2^{(e)}} \tag{6.3.10a}$$

$$N_2^{(e)}(x,y) = \frac{x - x_1^{(e)}}{x_2^{(e)} - x_1^{(e)}} \cdot \frac{y - y_2^{(e)}}{y_1^{(e)} - y_2^{(e)}} \tag{6.3.10b}$$

$$N_3^{(e)}(x,y) = \frac{x - x_1^{(e)}}{x_2^{(e)} - x_1^{(e)}} \cdot \frac{y - y_1^{(e)}}{y_2^{(e)} - y_1^{(e)}} \tag{6.3.10c}$$

$$N_4^{(e)}(x,y) = \frac{x - x_2^{(e)}}{x_1^{(e)} - x_2^{(e)}} \cdot \frac{y - y_1^{(e)}}{y_2^{(e)} - y_1^{(e)}} \tag{6.3.10d}$$

If we now notice that the area of the rectangular element, $A^{(e)}$, is given by:

$$A^{(e)} = \left(x_2^{(e)} - x_1^{(e)}\right)\left(y_2^{(e)} - y_1^{(e)}\right) = \left(x_1^{(e)} - x_2^{(e)}\right)\left(y_1^{(e)} - y_2^{(e)}\right) \tag{6.3.11}$$

We can write the following expressions for the shape functions:

$$N_1^{(e)}(x,y) = \frac{\left(x - x_2^{(e)}\right)\left(y - y_2^{(e)}\right)}{A^{(e)}} \tag{6.3.12a}$$

$$N_2^{(e)}(x,y) = -\frac{\left(x - x_1^{(e)}\right)\left(y - y_2^{(e)}\right)}{A^{(e)}} \tag{6.3.12b}$$

$$N_3^{(e)}(x,y) = \frac{\left(x - x_1^{(e)}\right)\left(y - y_1^{(e)}\right)}{A^{(e)}} \tag{6.3.12c}$$

$$N_4^{(e)}(x,y) = -\frac{\left(x - x_2^{(e)}\right)\left(y - y_2^{(e)}\right)}{A^{(e)}} \tag{6.3.12d}$$

We will now see an alternative procedure for obtaining the four shape functions, presented in Fish and Belytschko (2007). Specifically, the shape functions for a rectangular element aligned with axes x and y can be obtained as *tensor products of the one-dimensional two-node element shape functions*. This process is schematically established in Figure 6.9. We can imagine that the four-node rectangle is the "product" of two, two-node, one-dimensional (line) elements. One element is aligned with the x-axis and has shape functions $N_1^{(e-1D)}(x)$ and $N_2^{(e-1D)}(x)$, the other is aligned with the y-axis

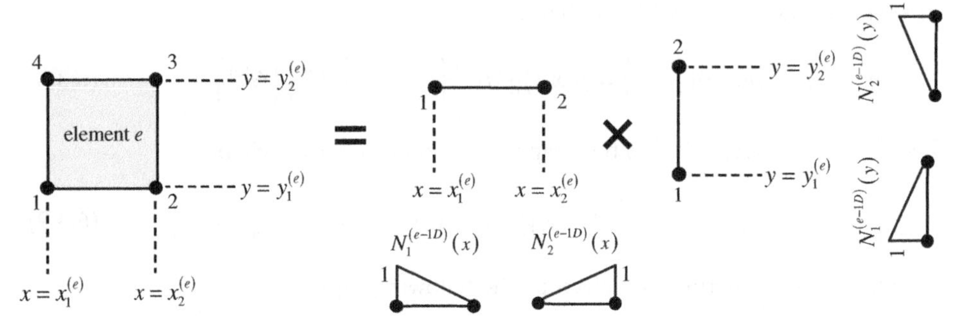

Figure 6.9 Rectangular element considered as "product" of one-dimensional elements aligned with the x- and y-axes.

and has shape functions $N_1^{(e-1D)}(y)$, $N_2^{(e-1D)}(y)$. We can then obtain the shape functions of Equations (6.3.10a–d) by taking appropriate products of the shape functions of the one-dimensional elements, as shown in Figure 6.10. For example, one can verify from the figure that node 1 of the rectangular element corresponds to node 1 ($x = x_1^{(e)}$) of the one-dimensional element along the x-axis and node 1 ($y = y_1^{(e)}$) of the one-dimensional element along the y-axis.

Remark 6.3.2: Note that the "tensor product" approach can only be used for rectangular elements, the sides of which are aligned with the two Cartesian coordinate axes. This peculiarity will be combined with an appropriate mathematical procedure in Section 6.4, for obtaining the shape functions of a general four-node quadrilateral element. ∎

We can finally also calculate the partial derivatives of the shape functions with x and y to obtain the components of the $[B^{(e)}]$-array:

$$\frac{\partial N_1^{(e)}}{\partial x} = \frac{y - y_2^{(e)}}{A^{(e)}}, \quad \frac{\partial N_2^{(e)}}{\partial x} = \frac{y_2^{(e)} - y}{A^{(e)}}, \quad \frac{\partial N_3^{(e)}}{\partial x} = \frac{y - y_1^{(e)}}{A^{(e)}}, \quad \frac{\partial N_4^{(e)}}{\partial x} = \frac{y_1^{(e)} - y}{A^{(e)}} \tag{6.3.13a}$$

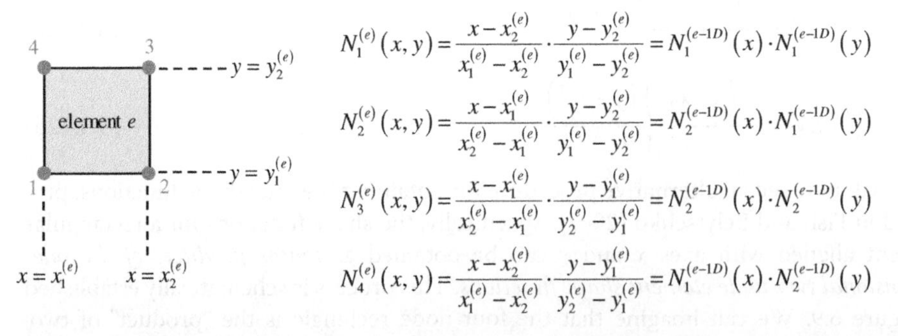

Figure 6.10 Obtaining the shape functions of the four-node rectangular element as a tensor product of shape functions of one-dimensional, two-node elements.

$$\frac{\partial N_1^{(e)}}{\partial y} = \frac{x - x_2^{(e)}}{A^{(e)}}, \quad \frac{\partial N_2^{(e)}}{\partial y} = \frac{x_1^{(e)} - x}{A^{(e)}}, \quad \frac{\partial N_3^{(e)}}{\partial y} = \frac{x - x_1^{(e)}}{A^{(e)}}, \quad \frac{\partial N_4^{(e)}}{\partial y} = \frac{x_2^{(e)} - x}{A^{(e)}} \tag{6.3.13b}$$

$$\left[B^{(e)}\right] = \frac{1}{A^{(e)}} \begin{bmatrix} y - y_2^{(e)} & y_2^{(e)} - y & y - y_1^{(e)} & y_1^{(e)} - y \\ x - x_2^{(e)} & x_1^{(e)} - x & x - x_1^{(e)} & x_2^{(e)} - x \end{bmatrix} \tag{6.3.14}$$

Remark 6.3.3: It can be easily verified that the shape functions for the three-node triangular element (presented in the previous section) and the four-node rectangular element satisfy the Kronecker delta and the partition of unity properties. ▮

Having established all the necessary expressions, we can briefly discuss the procedure to obtain the element conductance array (Equation 6.1.18) and equivalent nodal flux vectors (Equations 6.1.21a–d). We will begin with the array $[k^{(e)}]$ and the vector $\left\{f_\Omega^{(e)}\right\}$, both of which require integration over the element domain. Since our element is a nicely shaped rectangular domain with the boundaries aligned with the coordinate axes, the domain integrals can be obtained as standard double integrals:

$$\left[k^{(e)}\right] = \iint_{\Omega^{(e)}} \left[B^{(e)}\right]^T \left[D^{(e)}\right]\left[B^{(e)}\right] dV = \int_{y_1^{(e)}}^{y_2^{(e)}} \int_{x_1^{(e)}}^{x_2^{(e)}} \left[B^{(e)}\right]^T \left[D^{(e)}\right]\left[B^{(e)}\right] dxdy \tag{6.3.15}$$

$$\left\{f_\Omega^{(e)}\right\} = \iint_{\Omega^{(e)}} \left[N^{(e)}\right]^T s \cdot dV = \int_{y_1^{(e)}}^{y_2^{(e)}} \int_{x_1^{(e)}}^{x_2^{(e)}} \left[N^{(e)}\right]^T s \cdot dxdy \tag{6.3.16}$$

We will now continue with a discussion regarding the calculation of $\left\{f_{\Gamma_q}^{(e)}\right\}$, for cases where sides of a rectangular element belong to a natural boundary segment. For rectangular elements aligned with the x- and y-axes, the boundary segments can be parameterized using either the x or y coordinate. For example, if side (2-3), which is mathematically described by $x = x_2^{(e)}$, constitutes a natural boundary segment as shown in Figure 6.11, then we can simply set $x = x_2^{(e)}$ in our shape functions, as shown in the same figure.

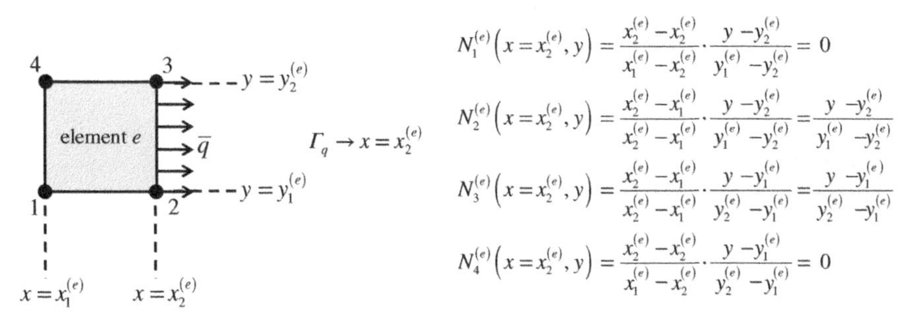

Figure 6.11 Example natural boundary segment for four-node rectangular element.

One can easily verify that, after setting $x = x_2^{(e)}$ for the segment (2-3), the shape functions for the specific boundary segment depend on a single variable, namely, y. Also, the boundary segment coincides with the y-axis, so the integral over the boundary can be set equal to an one-dimensional integral with respect to y. Eventually, the following expression is obtained for $\left\{f_{\Gamma_q}^{(e)}\right\}$.

$$\left\{f_{\Gamma_q}^{(e)}\right\} = -\int_{\Gamma_q} \left[N^{(e)}\right]^T \bar{q} ds = -\int_{y_1^{(e)}}^{y_2^{(e)}} \left\{ \begin{matrix} 0 \\ \dfrac{y - y_2^{(e)}}{y_1^{(e)} - y_2^{(e)}} \\ \dfrac{y - y_1^{(e)}}{y_2^{(e)} - y_1^{(e)}} \\ 0 \end{matrix} \right\} \bar{q} dy = -\left\{ \begin{matrix} 0 \\ \displaystyle\int_{y_1^{(e)}}^{y_2^{(e)}} \dfrac{y - y_2^{(e)}}{y_1^{(e)} - y_2^{(e)}} \bar{q} dy \\ \displaystyle\int_{y_1^{(e)}}^{y_2^{(e)}} \dfrac{y - y_1^{(e)}}{y_2^{(e)} - y_1^{(e)}} \bar{q} dy \\ 0 \end{matrix} \right\} \qquad (6.3.17)$$

Similar considerations apply for cases where other sides of the rectangular element constitute natural boundary segments.

Remark 6.3.4: Per Equation (6.3.17), the values of $\left\{f_{\Gamma_q}^{(e)}\right\}$ corresponding to nodes 1 and 4 are zero for the case that the natural boundary segment corresponds to side (2-3). This should come as no surprise, since a prescribed boundary outflow should have no effect on nodes that do not lie on the natural boundary! ▮

6.4 Isoparametric Finite Elements and the Four-Node Quadrilateral (4Q) Element

So far, we have seen how to obtain the shape functions for the two simplest cases of two-dimensional elements, namely, the three-node triangular element and the four-node rectangular element. We will now proceed to the case of a general, four-node quadrilateral (4Q) element with straight sides, as shown in Figure 6.12.

The shape functions that we obtained for a rectangular element aligned with the x- and y-axes are no longer valid for this element. One approach to obtain the shape functions

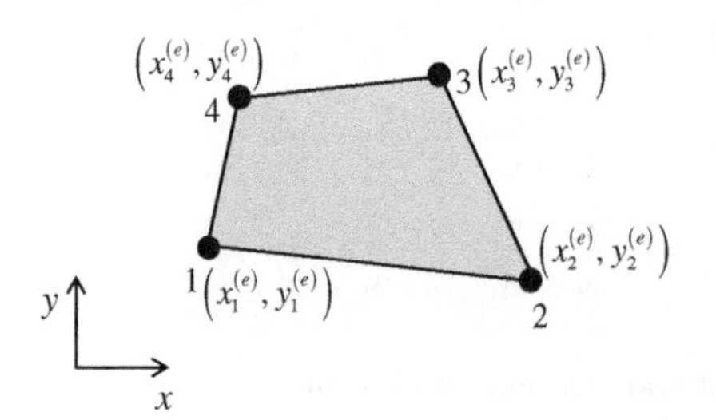

Figure 6.12 Four-node quadrilateral element.

would be to follow the same procedure as the one we presented for the three-node triangle (establish the $\{a^{(e)}\}$ vector, etc.). However, there is a much more efficient approach that we will focus on instead: the *isoparametric element concept,* attributed to Taig (1960) and Irons (1966). What we come up with is *a change of coordinates,* also called *mapping,* from the physical coordinate system (x,y) to a system (ξ,η), which will be called the *parent* or *parametric coordinate space.* One may wonder why make such a change. The answer is that the mapping can be so chosen that the shape of the element in the parametric coordinate space $(\xi\eta)$ becomes a nice square whose sides are aligned with the coordinate ξ- and η-axes, as shown in Figure 6.13. As shown in the same figure, the values of the parametric coordinates are between -1 and 1.

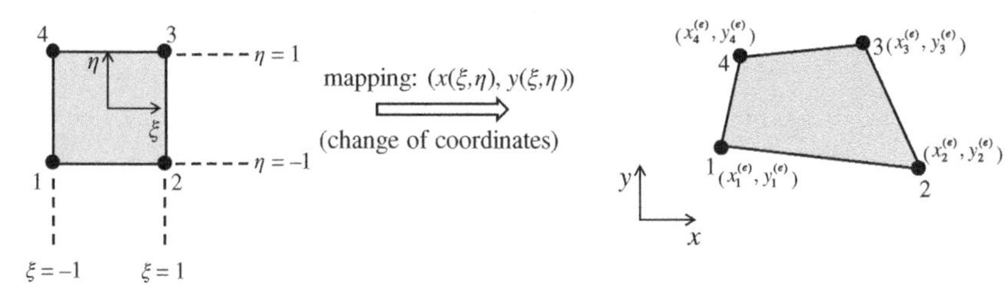

Figure 6.13 Isoparametric element in parent and physical space.

Using an isoparametric element has four advantages:

1) It allows us to easily construct the shape functions in the $(\xi\text{-}\eta)$ space as tensor products of one-dimensional shape functions.
2) It allows us to verify continuity across inter-element boundaries.
3) It allows us to use Gaussian quadrature in multiple dimensions.
4) It can allow us to efficiently model elements with curved sides.

To ensure "well-behaved" elements, *each point in the parametric space must be mapped to a unique point in the physical space and vice versa.* Also, *each point in the parametric space must have a corresponding "image" in the physical space and vice versa.* In mathematic terms, these two conditions are equivalent to stating that the mapping from the physical space to the parametric space is *one-to-one and onto.*

Another important aspect that characterizes isoparametric elements is that the approximate field (for heat conduction this is the approximate temperature field) inside the element is given in terms of the ξ and η coordinates. Thus, for the four-node quadrilateral element, we can write the following expression:

$$\Theta^{(e)}(\xi,\eta) = N_1^{(4Q)}(\xi,\eta)\cdot\Theta_1^{(e)} + N_2^{(4Q)}(\xi,\eta)\cdot\Theta_2^{(e)} + N_3^{(4Q)}(\xi,\eta)\cdot\Theta_3^{(e)} + N_4^{(4Q)}(\xi,\eta)\cdot\Theta_4^{(e)}$$

$$= \sum_{i=1}^{4}\left(N_i^{(4Q)}(\xi,\eta)\cdot\Theta_i^{(e)}\right) \tag{6.4.1}$$

Per Equation (6.4.1), the temperature at a point inside element e can be found once we find the position of that point in the parametric space.

Remark 6.4.1: Notice that from now on, for the shape functions (and every other quantity that depends on the shape functions) of quadrilateral isoparametric elements we will be using the superscript "4Q" instead of e (we will follow a similar approach for other isoparametric elements). ∎

The shape functions in the parametric space can be obtained using the tensor product approach for one-dimensional shape functions, as shown in Figure 6.14. Specifically, the element will be formulated as the product of two one-dimensional elements, each aligned with one of the parametric coordinate axes.

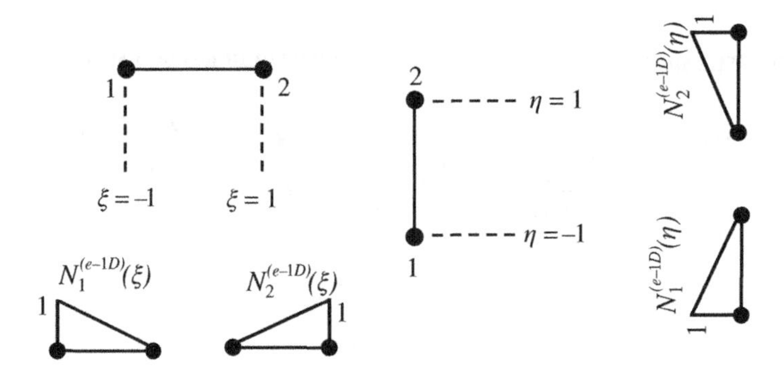

Figure 6.14 One-dimensional elements and one-dimensional shape functions used to obtain the shape functions of the four-node quadrilateral isoparametric element as tensor products.

Using the tensor product approach in the same fashion as in Section (6.3) yields the following expressions for the four shape functions.

$$N_1^{(4Q)}(\xi,\eta) = N_1^{(e-1D)}(\xi)\cdot N_1^{(e-1D)}(\eta) \rightarrow N_1^{(4Q)}(\xi,\eta) = \frac{1-\xi}{2}\cdot\frac{1-\eta}{2} = \frac{1}{4}(1-\xi)(1-\eta) \quad (6.4.2a)$$

$$N_2^{(4Q)}(\xi,\eta) = N_2^{(e-1D)}(\xi)\cdot N_1^{(e-1D)}(\eta) \rightarrow N_2^{(4Q)}(\xi,\eta) = \frac{1+\xi}{2}\cdot\frac{1-\eta}{2} = \frac{1}{4}(1+\xi)(1-\eta) \quad (6.4.2b)$$

$$N_3^{(4Q)}(\xi,\eta) = N_2^{(e-1D)}(\xi)\cdot N_2^{(e-1D)}(\eta) \rightarrow N_3^{(4Q)}(\xi,\eta) = \frac{1+\xi}{2}\cdot\frac{1+\eta}{2} = \frac{1}{4}(1+\xi)(1+\eta) \quad (6.4.2c)$$

$$N_4^{(4Q)}(\xi,\eta) = N_1^{(e-1D)}(\xi)\cdot N_2^{(e-1D)}(\eta) \rightarrow N_4^{(4Q)}(\xi,\eta) = \frac{1-\xi}{2}\cdot\frac{1+\eta}{2} = \frac{1}{4}(1-\xi)(1+\eta) \quad (6.4.2d)$$

Remark 6.4.2: We can easily verify that the shape functions of the isoparametric quadrilateral element satisfy the Kronecker delta property and the partition of unity property. ∎

The question now is how to establish the mapping—that is, the rules for the change of coordinates from the parametric space, ($\xi\eta$), to the physical space, (xy). We want the mapping expressions to yield the nodal physical coordinates of the element at the parametric locations of the element. For example, if we have $\xi = -1$, $\eta = -1$, which correspond to the location of node 1 in the parametric space as shown in Figure 6.13, the mapping must yield $x(-1,-1) = x_1^{(e)}$, and $y(-1,-1) = y_1^{(e)}$. To obtain such a mapping in isoparametric elements, we can directly use the shape functions of the parametric space that we established for the approximation of the field function! More specifically, we can write the following two expressions for the mapping:

$$x(\xi,\eta) = N_1^{(4Q)}(\xi,\eta)\cdot x_1^{(e)} + N_2^{(4Q)}(\xi,\eta)\cdot x_2^{(e)} + N_3^{(4Q)}(\xi,\eta)\cdot x_3^{(e)} + N_4^{(4Q)}(\xi,\eta)\cdot x_4^{(e)} \quad (6.4.3a)$$

$$y(\xi,\eta) = N_1^{(4Q)}(\xi,\eta)\cdot y_1^{(e)} + N_2^{(4Q)}(\xi,\eta)\cdot y_2^{(e)} + N_3^{(4Q)}(\xi,\eta)\cdot y_3^{(e)} + N_4^{(4Q)}(\xi,\eta)\cdot y_4^{(e)} \quad (6.4.3b)$$

Equations (6.4.3a,b) can be written in the more concise form:

$$x(\xi,\eta) = \sum_{i=1}^{4}\left(N_i^{(4Q)}(\xi,\eta)\cdot x_i^{(e)}\right) \quad (6.4.4a)$$

$$y(\xi,\eta) = \sum_{i=1}^{4}\left(N_i^{(4Q)}(\xi,\eta)\cdot y_i^{(e)}\right) \quad (6.4.4b)$$

Since the parametric shape functions satisfy the Kronecker delta property, it is obvious that Equations (6.4.4a,b) will give, for each node "I": $x(\xi_I,\eta_I) = x_I^{(e)}$, $y(\xi_I,\eta_I) = y_I^{(e)}$.

To summarize the above, establishing the shape functions in the parametric space allows us to approximate both the field function (in the parametric space) and the coordinate mapping.

An important quantity associated with the coordinate mapping of the 4Q element is the *Jacobian matrix* of the mapping, given by the following expression.

$$[J] = \begin{bmatrix} J_{11} & J_{12} \\ J_{21} & J_{22} \end{bmatrix} = \begin{bmatrix} \frac{\partial x}{\partial \xi} & \frac{\partial y}{\partial \xi} \\ \frac{\partial x}{\partial \eta} & \frac{\partial y}{\partial \eta} \end{bmatrix} = \begin{bmatrix} x_{,\xi} & y_{,\xi} \\ x_{,\eta} & y_{,\eta} \end{bmatrix} \quad (6.4.5)$$

The four components of the Jacobian matrix can be obtained using the following expressions.

$$\frac{\partial x}{\partial \xi} = \frac{\partial N_1^{(4Q)}}{\partial \xi}\cdot x_1^{(e)} + \frac{\partial N_2^{(4Q)}}{\partial \xi}\cdot x_2^{(e)} + \frac{\partial N_3^{(4Q)}}{\partial \xi}\cdot x_3^{(e)} + \frac{\partial N_4^{(4Q)}}{\partial \xi}\cdot x_4^{(e)}$$

$$\rightarrow \frac{\partial x}{\partial \xi} = \sum_{i=1}^{4}\left(\frac{\partial N_I^{(4Q)}}{\partial \xi}\cdot x_I^{(e)}\right) \quad (6.4.6a)$$

$$\frac{\partial x}{\partial \eta} = \frac{\partial N_1^{(4Q)}}{\partial \eta}\cdot x_1^{(e)} + \frac{\partial N_2^{(4Q)}}{\partial \eta}\cdot x_2^{(e)} + \frac{\partial N_3^{(4Q)}}{\partial \eta}\cdot x_3^{(e)} + \frac{\partial N_4^{(4Q)}}{\partial \eta}\cdot x_4^{(e)}$$

$$\rightarrow \frac{\partial x}{\partial \eta} = \sum_{i=1}^{4}\left(\frac{\partial N_I^{(4Q)}}{\partial \eta}\cdot x_I^{(e)}\right) \quad (6.4.6b)$$

$$\frac{\partial y}{\partial \xi} = \frac{\partial N_1^{(4Q)}}{\partial \xi} \cdot y_1^{(e)} + \frac{\partial N_2^{(4Q)}}{\partial \xi} \cdot y_2^{(e)} + \frac{\partial N_3^{(4Q)}}{\partial \xi} \cdot y_3^{(e)} + \frac{\partial N_4^{(4Q)}}{\partial \xi} \cdot y_4^{(e)}$$

$$\rightarrow \frac{\partial y}{\partial \xi} = \sum_{i=1}^{4} \left(\frac{\partial N_I^{(4Q)}}{\partial \xi} \cdot y_I^{(e)} \right) \tag{6.4.6c}$$

$$\frac{\partial y}{\partial \eta} = \frac{\partial N_1^{(4Q)}}{\partial \eta} \cdot y_1^{(e)} + \frac{\partial N_2^{(4Q)}}{\partial \eta} \cdot y_2^{(e)} + \frac{\partial N_3^{(4Q)}}{\partial \eta} \cdot y_3^{(e)} + \frac{\partial N_4^{(4Q)}}{\partial \eta} \cdot y_4^{(e)}$$

$$\rightarrow \frac{\partial y}{\partial \eta} = \sum_{i=1}^{4} \left(\frac{\partial N_I^{(4Q)}}{\partial \eta} \cdot y_I^{(e)} \right) \tag{6.4.6d}$$

Equations (6.4.6a–d) can be collectively written in matrix form:

$$[J] = \begin{bmatrix} \frac{\partial x}{\partial \xi} & \frac{\partial y}{\partial \xi} \\ \frac{\partial x}{\partial \eta} & \frac{\partial y}{\partial \eta} \end{bmatrix} = \left[N_{,\xi}^{(4Q)} \right] \left[\{x^{(e)}\} \ \{y^{(e)}\} \right] \tag{6.4.7}$$

where

$$\left[N_{,\xi}^{(4Q)} \right] = \begin{bmatrix} \frac{\partial N_1^{(4Q)}}{\partial \xi} & \frac{\partial N_2^{(4Q)}}{\partial \xi} & \frac{\partial N_3^{(4Q)}}{\partial \xi} & \frac{\partial N_4^{(4Q)}}{\partial \xi} \\ \frac{\partial N_1^{(4Q)}}{\partial \eta} & \frac{\partial N_2^{(4Q)}}{\partial \eta} & \frac{\partial N_3^{(4Q)}}{\partial \eta} & \frac{\partial N_4^{(4Q)}}{\partial \eta} \end{bmatrix} \tag{6.4.8a}$$

and

$$\left\{ x^{(e)} \right\} = \left[x_1^{(e)} \ x_2^{(e)} \ x_3^{(e)} \ x_4^{(e)} \right]^T \tag{6.4.8b}$$

$$\left\{ y^{(e)} \right\} = \left[y_1^{(e)} \ y_2^{(e)} \ y_3^{(e)} \ y_4^{(e)} \right]^T \tag{6.4.8c}$$

If we are given the Jacobian array, $[J]$, at one point, we can also calculate its inverse, which will be denoted as $\left[\tilde{J}\right]$. By definition, the inverse of the Jacobian matrix of the coordinate mapping includes the derivatives of the parametric coordinates with respect to the physical coordinates:

$$[J]^{-1} = \left[\tilde{J}\right] = \begin{bmatrix} \tilde{J}_{11} & \tilde{J}_{12} \\ \tilde{J}_{21} & \tilde{J}_{22} \end{bmatrix} = \begin{bmatrix} \xi_{,x} & \eta_{,x} \\ \xi_{,y} & \eta_{,y} \end{bmatrix} = \begin{bmatrix} \frac{\partial \xi}{\partial x} & \frac{\partial \eta}{\partial x} \\ \frac{\partial \xi}{\partial y} & \frac{\partial \eta}{\partial y} \end{bmatrix} \tag{6.4.9}$$

Now that the mapping and the approximate field have been established, we also need to create the expressions giving the components of the gradient of the approximate temperature field, so that we can then calculate the element coefficient matrix (for heat conduction, this is the conductance array). To this end, we can simply use

Equations (6.1.7) and (6.1.8) for the special case of $n = 4$ and for the shape functions of the 4Q element:

$$\left\{\nabla\Theta^{(e)}\right\} = \left[B^{(4Q)}\right]\left\{\Theta^{(e)}\right\} \tag{6.4.10}$$

where

$$\left[B^{(4Q)}\right] = \begin{bmatrix} \partial N_1^{(4Q)}/\partial x & \partial N_2^{(4Q)}/\partial x & \partial N_3^{(4Q)}/\partial x & \partial N_4^{(4Q)}/\partial x \\ \partial N_1^{(4Q)}/\partial y & \partial N_2^{(4Q)}/\partial y & \partial N_3^{(4Q)}/\partial y & \partial N_4^{(4Q)}/\partial y \end{bmatrix} \tag{6.4.11}$$

and

$$\left\{\Theta^{(e)}\right\} = \left[\Theta_1^{(e)} \;\; \Theta_2^{(e)} \;\; \Theta_3^{(e)} \;\; \Theta_4^{(e)}\right]^T \tag{6.4.12}$$

At this stage, we may realize that there is a peculiarity regarding the isoparametric element: while the shape functions have been established as functions of the parametric coordinates, the $[B^{(e)}]$ array (which is used for the calculation of the gradient) contains partial derivatives with respect to the physical coordinates, x and y! This is no real problem, because we can obtain the partial derivatives with physical coordinates by means of the *chain rule of differentiation*. For any shape function $N_I^{(4Q)}$, we can write:

$$\frac{\partial N_I^{(4Q)}}{\partial x} = \frac{\partial N_I^{(4Q)}}{\partial \xi}\cdot\frac{\partial \xi}{\partial x} + \frac{\partial N_I^{(4Q)}}{\partial \eta}\cdot\frac{\partial \eta}{\partial x} \tag{6.4.13a}$$

$$\frac{\partial N_I^{(4Q)}}{\partial y} = \frac{\partial N_I^{(4Q)}}{\partial \xi}\cdot\frac{\partial \xi}{\partial y} + \frac{\partial N_I^{(4Q)}}{\partial \eta}\cdot\frac{\partial \eta}{\partial y} \tag{6.4.13b}$$

If we account for Equation (6.4.9), Equations (6.4.13a,b) become:

$$\frac{\partial N_I^{(4Q)}}{\partial x} = \frac{\partial N_I^{(4Q)}}{\partial \xi}\cdot\tilde{J}_{11} + \frac{\partial N_I^{(4Q)}}{\partial \eta}\cdot\tilde{J}_{12} \tag{6.4.14a}$$

$$\frac{\partial N_I^{(4Q)}}{\partial y} = \frac{\partial N_I^{(4Q)}}{\partial \xi}\cdot\tilde{J}_{21} + \frac{\partial N_I^{(4Q)}}{\partial \eta}\cdot\tilde{J}_{22} \tag{6.4.14b}$$

The terms in Equations (6.4.14a,b) include the derivatives of the shape functions with the parametric coordinates. These can be readily obtained, through partial differentiation of Equations (6.4.2a–d), and are provided in Table 6.1.

Table 6.1 Derivatives of Shape Functions of 4Q Element with Respect to Parametric Coordinates.

I	1	2	3	4
$\frac{\partial N_I^{(4Q)}}{\partial \xi}$	$-\frac{1}{4}(1-\eta)$	$\frac{1}{4}(1-\eta)$	$\frac{1}{4}(1+\eta)$	$-\frac{1}{4}(1+\eta)$
$\frac{\partial N_I^{(4Q)}}{\partial \eta}$	$-\frac{1}{4}(1-\xi)$	$-\frac{1}{4}(1+\xi)$	$\frac{1}{4}(1+\xi)$	$\frac{1}{4}(1-\xi)$

In light of Equations (6.4.14a,b), we can write Equation (6.4.11) in the following form.

$$\left[B^{(4Q)}\right] = \begin{bmatrix} \frac{\partial N_1^{(4Q)}}{\partial \xi}\cdot\tilde{J}_{11} + \frac{\partial N_1^{(4Q)}}{\partial \eta}\cdot\tilde{J}_{12} & \frac{\partial N_2^{(4Q)}}{\partial \xi}\cdot\tilde{J}_{11} + \frac{\partial N_2^{(4Q)}}{\partial \eta}\cdot\tilde{J}_{12} & \frac{\partial N_3^{(4Q)}}{\partial \xi}\cdot\tilde{J}_{11} + \frac{\partial N_3^{(4Q)}}{\partial \eta}\cdot\tilde{J}_{12} & \frac{\partial N_4^{(4Q)}}{\partial \xi}\cdot\tilde{J}_{11} + \frac{\partial N_4^{(4Q)}}{\partial \eta}\cdot\tilde{J}_{12} \\ \frac{\partial N_1^{(4Q)}}{\partial \xi}\cdot\tilde{J}_{21} + \frac{\partial N_1^{(4Q)}}{\partial \eta}\cdot\tilde{J}_{22} & \frac{\partial N_2^{(4Q)}}{\partial \xi}\cdot\tilde{J}_{21} + \frac{\partial N_2^{(4Q)}}{\partial \eta}\cdot\tilde{J}_{22} & \frac{\partial N_3^{(4Q)}}{\partial \xi}\cdot\tilde{J}_{21} + \frac{\partial N_3^{(4Q)}}{\partial \eta}\cdot\tilde{J}_{22} & \partial\frac{N_4^{(4Q)}}{\partial \xi}\cdot\tilde{J}_{21} + \frac{\partial N_4^{(4Q)}}{\partial \eta}\cdot\tilde{J}_{22} \end{bmatrix} \tag{6.4.15}$$

By inspection, we can write the right-hand side of Equation (6.4.15) as the product of two arrays:

$$\left[B^{(4Q)}\right] = \begin{bmatrix} \tilde{J}_{11} & \tilde{J}_{12} \\ \tilde{J}_{21} & \tilde{J}_{22} \end{bmatrix} \begin{bmatrix} \frac{\partial N_1^{(4Q)}}{\partial \xi} & \frac{\partial N_2^{(4Q)}}{\partial \xi} & \frac{\partial N_3^{(4Q)}}{\partial \xi} & \frac{\partial N_4^{(4Q)}}{\partial \xi} \\ \frac{\partial N_1^{(4Q)}}{\partial \eta} & \frac{\partial N_2^{(4Q)}}{\partial \eta} & \frac{\partial N_3^{(4Q)}}{\partial \eta} & \frac{\partial N_4^{(4Q)}}{\partial \eta} \end{bmatrix} = \left[\tilde{J}\right]\left[N_{,\xi}^{(4Q)}\right] \tag{6.4.16}$$

where $\left[N_{,\xi}^{(4Q)}\right]$ contains the partial derivatives of the shape functions with respect to the parametric coordinates:

$$\left[N_{,\xi}^{(4Q)}\right] = \begin{bmatrix} \frac{\partial N_1^{(4Q)}}{\partial \xi} & \frac{\partial N_2^{(4Q)}}{\partial \xi} & \frac{\partial N_3^{(4Q)}}{\partial \xi} & \frac{\partial N_4^{(4Q)}}{\partial \xi} \\ \frac{\partial N_1^{(4Q)}}{\partial \eta} & \frac{\partial N_2^{(4Q)}}{\partial \eta} & \frac{\partial N_3^{(4Q)}}{\partial \eta} & \frac{\partial N_4^{(4Q)}}{\partial \eta} \end{bmatrix} \tag{6.4.17}$$

Remark 6.4.3: It is important to remember that the components of $[J]$ are functions of ξ and η, so the inverse of $[J]$, that is, $\left[\tilde{J}\right]$ will also be a function of ξ and η! Consequently, the $[B^{(4Q)}]$ array obtained from Equation (6.4.16) will be a function of ξ and η. ∎

Now, we are ready to establish expressions for the coefficient array and for the equivalent right-hand-side vector, which are obtained with the integrals of Equations (6.1.18) and (6.1.21a–d). For the general case that the material conductivity array, $[D^{(e)}]$, is a function of x and y, we can use the mapping equations $x(\xi,\eta)$, $y(\xi,\eta)$ —Equations (6.4.4a,b)— to obtain:

$$\left[D^{(e)}(x,y)\right] = \left[D^{(e)}(x(\xi,\eta),y(\xi,\eta))\right] = \left[D^{(e)}(\xi,\eta)\right] \tag{6.4.18}$$

We can do the same for the source term (i.e. express it as a function of the parametric coordinates):

$$s^{(e)}(x,y) = s^{(e)}(x(\xi,\eta),y(\xi,\eta)) = s^{(e)}(\xi,\eta) \tag{6.4.19}$$

The calculation of domain integrals in the physical coordinate space, xy, is hard, since the sides of a quadrilateral element are not in general aligned with either of the coordinate axes. However, there is something that we can exploit: our element also "resides" in the parametric space $\xi\eta$, where it attains a square shape aligned with the parametric coordinate axis. Since nothing prevents us from calculating the domain integrals in whichever coordinate space we please, we will evaluate the integral of $[k^{(e)}]$ in the parametric space! From multidimensional calculus, the integral in the parametric space is obtained as follows:

$$\left[k^{(e)}\right] = \int_{-1}^{1}\int_{-1}^{1} \left[B^{(e)}(\xi,\eta)\right]^T \left[D^{(e)}(\xi,\eta)\right] \left[B^{(e)}(\xi,\eta)\right] J(\xi,\eta)\cdot d\xi d\eta \tag{6.4.20}$$

where $J(\xi, \eta)$ is the *determinant of the Jacobian matrix of the mapping*, $[J]$.

What we accomplish by using the parametric domain for the integration, is that we now need to calculate a standard double integral. Also, since the integration for both ξ and η is conducted from -1 to 1, we can (and we *will*!) use Gaussian quadrature for the calculation of the integrals, as explained in the next section.

Along the same lines, the equivalent nodal flux vector, $\{f^{(e)}\}$, can be obtained from the following expression:

$$\left\{f^{(e)}\right\} = \left\{f_\Omega^{(e)}\right\} + \left\{f_{\Gamma_q}^{(e)}\right\} \tag{6.4.21}$$

where

$$\left\{f_\Omega^{(e)}\right\} = \int_{-1}^{1}\int_{-1}^{1} \left[N^{(e)}(\xi,\eta)\right]^T s(\xi,\eta)\cdot J(\xi,\eta)\cdot d\xi d\eta \tag{6.4.22}$$

and $\left\{f_{\Gamma_q}^{(e)}\right\}$ is a one-dimensional integral with respect to ξ or η, depending on which of the two parameters varies over each boundary segment. This happens because, per Figure 6.13, each element side in the parametric domain corresponds to a constant value of ξ or η. The procedure to obtain boundary integrals will be described in more detail in the following section.

6.5 Numerical Integration for Isoparametric Quadrilateral Elements

The Gaussian quadrature procedure, which we used in Section 3.7 for the calculation of one-dimensional integrals, can be extended to multiple dimensions. Since isoparametric elements involve integration with respect to ξ and η, each of which takes values between -1 and 1, and also since we can separately integrate with respect to ξ and η (due to the fact that the integration domain has a nice square shape aligned with the parametric coordinate axes), we can use one-dimensional Gaussian quadrature for each of the two directions. Specifically, we can write the definite integral of a function $f(\xi,\eta)$ in the parametric element domain in the following form.

$$\int_{-1}^{1}\int_{-1}^{1} f(\xi,\eta)d\xi d\eta = \int_{-1}^{1}\left(\int_{-1}^{1} f(\xi,\eta)d\xi\right)d\eta = \int_{-1}^{1} I(\eta)d\eta \tag{6.5.1}$$

where

$$I(\eta) = \int_{-1}^{1} f(\xi,\eta)d\xi \tag{6.5.2}$$

Since $I(\eta)$ is obtained from one-dimensional integration with respect to ξ, we can use Gaussian quadrature with $N_{g\xi}$ points to numerically obtain it:

$$I(\eta) \approx \sum_{i=1}^{Ng_\xi} (f(\xi_i,\eta) W_i) \tag{6.5.3}$$

Similarly, we can use one-dimensional Gaussian quadrature with $N_{g\eta}$ points with respect to η, to evaluate the right-hand-side of Equation (6.5.1):

$$\int_{-1}^{1}\int_{-1}^{1} f(\xi,\eta)d\xi d\eta = \int_{-1}^{1} I(\eta)d\eta \approx \sum_{j=1}^{Ng_\eta} \left(I\left(\eta_j\right)\cdot W_j\right) \tag{6.5.4}$$

Plugging Equation (6.5.3) into (6.5.4) gives:

$$\int_{-1}^{1}\int_{-1}^{1} f(\xi,\eta)d\xi d\eta \approx \sum_{j=1}^{Ng_\eta} \left(\sum_{i=1}^{Ng_\xi} \left(f\left(\xi_i,\eta_j\right) W_i\right) \cdot W_j \right) \tag{6.5.5}$$

Equation (6.5.5) constitutes a *two-dimensional Gaussian quadrature formula*, which allows the numerical integration of any function in the two-dimensional parametric domain, using $N_{g\xi}$ points in the ξ-direction and $N_{g\eta}$ points in the η-direction.

For a quadrilateral element, we commonly use two quadrature points along the ξ-direction and two quadrature points along the η-direction,[1] eventually obtaining a total of four Gauss quadrature points, the location of which in the parametric space is shown in Figure 6.15a. A general rule of thumb is that for an isoparametric element, we need to use as many Gauss points as the nodal points in the element. For example, for the nine-node quadrilateral isoparametric element, we need a total of nine Gauss points as shown in Figure 6.15b. Using as many quadrature points as the number of nodal points is referred to as *full integration*.

If we now say that we have a total of N_g quadrature points, then for each quadrature point g, we will have $\xi_g = \xi_i$, $\eta_g = \eta_j$, and the weight coefficient will be the product of the weight coefficients in the ξ- and η-directions:

$$W_g = W_i \cdot W_j. \tag{6.5.6}$$

The two-dimensional Gaussian quadrature formula can be written in the following, more concise form.

$$\int_{-1}^{1}\int_{-1}^{1} f(\xi,\eta)d\xi d\eta = \sum_{g=1}^{Ng} \left(f\left(\xi_g,\eta_g\right) W_g\right) \tag{6.5.7}$$

1 In general, the quadrature points are arranged *symmetrically*, so we have a (1 × 1), (2 × 2), (3 × 3), etc. quadrature point grid. The reason is practical: We want to obtain *identical* results no matter how the nodes are numbered (if we do not have symmetric quadrature point grid, the stiffness terms that correspond to a given node of the element will change a "tiny bit" if we change the nodal numbering).

(a)

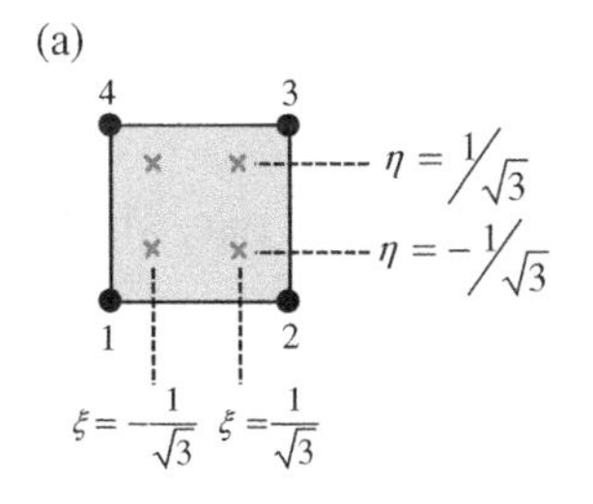

Four-node element (2×2)-point quadrature

(b)

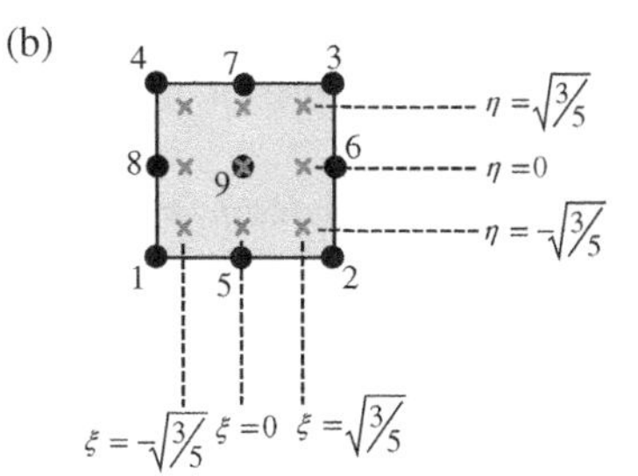

Nine-node element (3×3)-point quadrature

Figure 6.15 Quadrature points for quadrilateral isoparametric elements.

The locations and weight coefficients of the four quadrature points for a case where we use a (2 × 2) Gaussian quadrature (i.e., with two quadrature points along each one of the two parametric directions), are provided in Table 6.2.

Remark 6.5.1: The coordinates and weight coefficients are those obtained from one-dimensional Gaussian quadrature along the ξ and η direction. For example, the reader can verify that the coordinates provided in Table 6.2 correspond to the locations of one-dimensional quadrature with two points, provided in Table 3.1. The weight coefficients of Table 6.2 are also products of the corresponding coefficients of two-point, one-dimensional quadrature (it so happens that all the specific coefficients are equal to 1). ■

Remark 6.5.2: The sum of the weight coefficients of two-dimensional quadrature is always equal to 4 (*why?*). ■

Remark 6.5.3: As discussed in Section 3.7, Gaussian quadrature for one-dimensional integration is optimal: for a given number of quadrature points, we have the maximum possible order of polynomial function that can be exactly integrated. This "optimality property" of Gaussian quadrature does *not* generally apply for multidimensional quadrature! ■

Remark 6.5.4: Strictly speaking, we may be allowed to resort to an integration rule of lower order than that of full integration. This topic is discussed in Section 11.5. ■

Table 6.2 Locations and Weight Coefficients for (2 × 2) Gaussian Quadrature Rule.

g	ξ_g	η_g	W_g
1	$-1/\sqrt{3}$	$-1/\sqrt{3}$	1
2	$1/\sqrt{3}$	$-1/\sqrt{3}$	1
3	$1/\sqrt{3}$	$1/\sqrt{3}$	1
4	$-1/\sqrt{3}$	$1/\sqrt{3}$	1

Example 6.1: Two-dimensional Gaussian quadrature
We are going to demonstrate the use of two-dimensional Gaussian quadrature by calculating the integral $\int_{-1}^{1}\int_{-1}^{1}(\xi^2\eta+3)\,d\xi d\eta$. We will specifically use the 2 × 2 rule, which corresponds to $N_g = 4$; thus, per Equation (6.5.7):

$$\int_{-1}^{1}\int_{-1}^{1} f(\xi,\eta)d\xi d\eta = \sum_{g=1}^{4}\left(f\left(\xi_g,\eta_g\right)W_g\right)$$

$$= f(\xi_1,\eta_1)W_1 + f(\xi_2,\eta_2)W_2 + f(\xi_3,\eta_3)W_3 + f(\xi_4,\eta_4)W_4$$

We have the coordinates and weight coefficients of each of the quadrature points provided in Table 6.2. Thus, we can obtain the values of the function at the location of each quadrature point:

$$f(\xi_1,\eta_1) = \left[\xi_1^{\,2}\eta_1+3\right]_{\substack{\xi_1=-1/\sqrt{3}\\ \eta_1=-1/\sqrt{3}}} = -\frac{1}{3\sqrt{3}}+3, \qquad W_1 = 1$$

$$f(\xi_2,\eta_2) = \left[\xi_2^{\,2}\eta_2+3\right]_{\substack{\xi_2=1/\sqrt{3}\\ \eta_2=-1/\sqrt{3}}} = -\frac{1}{3\sqrt{3}}+3, \qquad W_2 = 1$$

$$f(\xi_3,\eta_3) = \left[\xi_3^{\,2}\eta_3+3\right]_{\substack{\xi_3=1/\sqrt{3}\\ \eta_3=1/\sqrt{3}}} = \frac{1}{3\sqrt{3}}+3, \qquad W_3 = 1$$

$$f(\xi_4,\eta_4) = \left[\xi_4^{\,2}\eta_4+3\right]_{\substack{\xi_4=-1/\sqrt{3}\\ \eta_4=1/\sqrt{3}}} = \frac{1}{3\sqrt{3}}+3, \qquad W_4 = 1$$

We can finally obtain:

$$\sum_{g=1}^{4}\left(f\left(\xi_g,\eta_g\right)W_g\right) = f(\xi_1,\eta_1)W_1 + f(\xi_2,\eta_2)W_2 + f(\xi_3,\eta_3)W_3 + f(\xi_4,\eta_4)W_4$$

$$= \left(-\frac{1}{3\sqrt{3}}+3\right)\cdot 1 + \left(-\frac{1}{3\sqrt{3}}+3\right)\cdot 1 + \left(\frac{1}{3\sqrt{3}}+3\right)\cdot 1 + \left(\frac{1}{3\sqrt{3}}+3\right)\cdot 1 = 12$$

In the specific case, we can also analytically evaluate the integral:

$$\int_{-1}^{1}\int_{-1}^{1}(\xi^2\eta+3)\,d\xi d\eta = \int_{-1}^{1}\left[\int_{-1}^{1}(\xi^2\eta+3)\,d\xi\right]d\eta = \int_{-1}^{1}\left[\frac{\xi^3\eta}{3}+3\xi\right]_{\xi=-1}^{\xi=1} d\eta$$

$$= \int_{-1}^{1}\left(\frac{2\eta}{3}+6\right)d\eta = \left[\frac{\eta^2}{3}+6\eta\right]_{\eta=-1}^{\eta=1} = 12$$

For the specific example, a 2 × 2 quadrature rule provided the exact value of the integral. ∎

Now that two-dimensional Gaussian quadrature has been established, it can be used to find the domain integrals for the element coefficient array (Equation 6.4.20) and the element equivalent right-hand-side vector $\left\{f_\Omega^{(e)}\right\}$, which includes a domain integral (Equation 6.4.22). Thus, we can establish the following expressions:

$$\left[k^{(e)}\right] \approx \sum_{g=1}^{Ng} \left(\left[B^{(e)}\left(\xi_g, \eta_g\right)\right]^T \left[D^{(e)}\left(\xi_g, \eta_g\right)\right] \left[B^{(e)}\left(\xi_g, \eta_g\right)\right] \cdot J\left(\xi_g, \eta_g\right) \cdot W_g \right) \tag{6.5.8}$$

$$\left\{f_\Omega^{(e)}\right\} \approx \sum_{g=1}^{Ng} \left(\left[N^{(e)}\left(\xi_g, \eta_g\right)\right]^T \cdot s\left(\xi_g, \eta_g\right) \cdot J\left(\xi_g, \eta_g\right) \cdot W_g \right) + \left\{f_{\Gamma_q}^{(e)}\right\} \tag{6.5.9}$$

It is now deemed necessary to establish an algorithm with all the steps for the calculation of $[k^{(e)}]$ and $\left\{f_\Omega^{(e)}\right\}$. This algorithm is provided in Box 6.5.1.

Box 6.5.1 Calculation of $[k^{(e)}]$, $\left\{f_\Omega^{(e)}\right\}$ for 4Q Element

FOR EACH QUADRATURE POINT g ($g = 1, 2, \ldots, N_g$):

1) Establish parametric coordinates, ξ_g, η_g.
2) Calculate the values of the shape functions at the specific point:

$$N_{1g}^{(4Q)} = N_1^{(4Q)}\left(\xi_g, \eta_g\right) = \frac{1}{4}\left(1 - \xi_g\right)\left(1 - \eta_g\right)$$

$$N_{2g}^{(4Q)} = N_2^{(4Q)}\left(\xi_g, \eta_g\right) = \frac{1}{4}\left(1 + \xi_g\right)\left(1 - \eta_g\right)$$

$$N_{3g}^{(4Q)} = N_3^{(4Q)}\left(\xi_g, \eta_g\right) = \frac{1}{4}\left(1 + \xi_g\right)\left(1 + \eta_g\right)$$

$$N_{4g}^{(4Q)} = N_4^{(4Q)}\left(\xi_g, \eta_g\right) = \frac{1}{4}\left(1 - \xi_g\right)\left(1 + \eta_g\right)$$

Also, set:

$$\left[N_g\right] = \left[N_{1g}^{(4Q)} \quad N_{2g}^{(4Q)} \quad N_{3g}^{(4Q)} \quad N_{4g}^{(4Q)}\right]$$

3) Calculate values of coordinates in physical space:

$$x_g = N_1^{(4Q)}\left(\xi_g, \eta_g\right)x_1^{(e)} + N_2^{(4Q)}\left(\xi_g, \eta_g\right)x_2^{(e)} + N_3^{(4Q)}\left(\xi_g, \eta_g\right)x_3^{(e)} + N_4^{(4Q)}\left(\xi_g, \eta_g\right)x_4^{(e)}$$

$$y_g = N_1^{(4Q)}\left(\xi_g, \eta_g\right)y_1^{(e)} + N_2^{(4Q)}\left(\xi_g, \eta_g\right)y_2^{(e)} + N_3^{(4Q)}\left(\xi_g, \eta_g\right)y_3^{(e)} + N_4^{(4Q)}\left(\xi_g, \eta_g\right)y_4^{(e)}$$

4) Calculate the Jacobian matrix of the mapping at that point, $\left[J_g\right] = \left[N_{,\xi}^{(4Q)}\right]_{\xi_g, \eta_g} \left[\{x^{(e)}\} \quad \{y^{(e)}\}\right]$, as well as the Jacobian determinant, $J_g = \det([J_g])$, and the inverse of the Jacobian matrix, $\left[\tilde{J}_g\right] = \left[J_g\right]^{-1}$

5) Calculate the derivatives of the shape functions with respect to the *physical* coordinates, x and y, at $\xi = \xi_g$, $\eta = \eta_g$:

$$\frac{\partial N_i^{(4Q)}}{\partial x} = \frac{\partial N_i^{(4Q)}}{\partial \xi}\frac{\partial \xi}{\partial x} + \frac{\partial N_i^{(4Q)}}{\partial \eta}\frac{\partial \eta}{\partial x} = \frac{\partial N_i^{(4Q)}}{\partial \xi}\tilde{J}_{g,11} + \frac{\partial N_i^{(4Q)}}{\partial \eta}\tilde{J}_{g,12}$$

$$\frac{\partial N_i^{(4Q)}}{\partial y} = \frac{\partial N_i^{(4Q)}}{\partial \xi}\frac{\partial \xi}{\partial y} + \frac{\partial N_i^{(4Q)}}{\partial \eta}\frac{\partial \eta}{\partial y} = \frac{\partial N_i^{(4Q)}}{\partial \xi}\tilde{J}_{g,21} + \frac{\partial N_i^{(4Q)}}{\partial \eta}\tilde{J}_{g,22}$$

Or collectively: $\left[N_{,x}^{(4Q)}\right]_{\xi_g,\eta_g} = \left[\tilde{J}_g\right]\left[N_{,\xi}^{(4Q)}\right]_{\xi_g,\eta_g}$, where

$$\left[N_{,x}^{(4Q)}\right]_{\xi_g,\eta_g} = \begin{bmatrix} \frac{\partial N_1^{(4Q)}}{\partial x} & \frac{\partial N_2^{(4Q)}}{\partial x} & \frac{\partial N_3^{(4Q)}}{\partial x} & \frac{\partial N_4^{(4Q)}}{\partial x} \\ \frac{\partial N_1^{(4Q)}}{\partial y} & \frac{\partial N_2^{(4Q)}}{\partial y} & \frac{\partial N_3^{(4Q)}}{\partial y} & \frac{\partial N_4^{(4Q)}}{\partial y} \end{bmatrix}_{\xi_g,\eta_g}$$

6) Set $\left[B_g\right] = \left[N_{,x}^{(4Q)}\right]_{\xi_g,\eta_g}$

7) Calculate the value of heat source s_g and of material conductivity matrix, $[D_g]$ at point g:

$$s_g = s\left(x = x_g, y = y_g\right)\ ,\ \left[D_g\right] = \left[D\left(x = x_g, y = y_g\right)\right]$$

END

Finally, combine the contributions of all the Gauss points to calculate the element conductance matrix, $[k^{(e)}]$, and heat-source contribution to the equivalent nodal force vector, $\{f^{(e)}\}$:

$$\left[k^{(e)}\right] = \int_{-1}^{1}\int_{-1}^{1}\left(\left[B^{(e)}\right]^T\left[D^{(e)}\right]\left[B^{(e)}\right]Jd\xi d\eta\right) \approx \sum_{g=1}^{Ng}\left(\left[B_g\right]^T\left[D_g\right]\left[B_g\right]J_gW_g\right)$$

$$\left\{f_\Omega^{(e)}\right\} = \int_{-1}^{1}\int_{-1}^{1}\left(\left[N^{(e)}\right]^T s\,Jd\xi d\eta\right) \approx \sum_{g=1}^{Ng}\left(\left[N_g\right]^T s_gJ_gW_g\right)$$

After the use of the algorithm in Box 6.5.1, the only remaining task for each element is to calculate $\left\{f_{\Gamma_q}^{(e)}\right\}$, that is, the part of the equivalent nodal flux vector due to prescribed outflow $\bar{q}$ at the natural boundary. Just as we did in Section 6.2 for triangular elements, we establish a parameterization of the boundary with respect to ξ or η (depending on which one of the two varies over the boundary segment), and then use one-dimensional Gaussian quadrature to find the contribution to $\left\{f_{\Gamma_q}^{(e)}\right\}$. The procedure for the case that we have N_g quadrature points for the boundary integration is summarized in Box 6.5.2.

Box 6.5.2 Calculation of $\left\{f_{\Gamma_q}^{(e)}\right\}$ for 4Q Element

- If the natural boundary segment corresponds to a constant value of η, $\eta=\bar{\eta}$ (where $\bar{\eta}=-1$ or 1), we need to evaluate a one-dimensional integral with respect to ξ:

$$\left\{f_{\Gamma_q}^{(e)}\right\}=\int_{\Gamma_q}\left[N^{(4Q)}\right]^T\bar{q}^{(e)}\,ds$$

$$=\int_{-1}^{1}\left(\left[N^{(4Q)}(\xi,\bar{\eta})\right]^T\bar{q}^{(e)}\frac{\ell}{2}d\xi\right)\approx\sum_{g=1}^{Ng}\left(\left[N^{(4Q)}(\xi_g,\bar{\eta})\right]^T\bar{q}^{(e)}\frac{\ell}{2}W_g\right)$$

where ℓ is the length of the natural boundary segment.

- If the natural boundary segment corresponds to a constant value of ξ, $\xi=\bar{\xi}$ (where $\bar{\xi}=-1$ or 1), we need to evaluate a one-dimensional integral with respect to η:

$$\left\{f_{\Gamma_q}^{(e)}\right\}=\int_{\Gamma_q}\left[N^{(4Q)}\right]^T\bar{q}^{(e)}\,ds$$

$$=\int_{-1}^{1}\left(\left[N^{(4Q)}(\bar{\xi},\eta)\right]^T\bar{q}^{(e)}\frac{\ell}{2}d\eta\right)\approx\sum_{g=1}^{Ng}\left(\left[N^{(4Q)}(\bar{\xi},\eta_g)\right]^T\bar{q}^{(e)}\frac{\ell}{2}W_g\right)$$

Remark 6.5.5: The prescribed outflow $\bar{q}$ on the natural boundary can vary as a function of x and y. In this case, we can use the mapping expressions of Equations (6.4.4a,b) to express the outflow as a function of the parametric coordinate that varies over the boundary. Of course, we will need to set $\xi=\bar{\xi}$ or $\eta=\bar{\eta}$ in the mapping expressions (depending on which parametric coordinate has a constant value over the boundary segment). ∎

After we conduct all the computations from Boxes 6.5.1 and 6.5.2, we can calculate the vector $\{f^{(e)}\}=\left\{f_{\Omega}^{(e)}\right\}+\left\{f_{\Gamma_q}^{(e)}\right\}$, and then use the element gather-scatter array $[L^{(e)}]$ to find the contributions of the element to the global conductance matrix, $[K]$, and the global equivalent nodal flux vector, $\{f\}$.

Example 6.2: Computations for quadrilateral isoparametric element

We will now apply the considerations for isoparametric elements summarized in Boxes 6.5.1 and 6.5.2 to solve the single-element mesh shown in Figure 6.16. We are given that the material is isotropic, with $\kappa=2(1+x\cdot y)$, where k is in $\frac{\mathrm{W}}{\mathrm{m}\cdot{}^{\circ}\mathrm{C}}$ and x, y are in meters. We will use 2×2 quadrature (four Gauss points) for domain integrals and two-point quadrature (two Gauss points) for the boundary integrals.

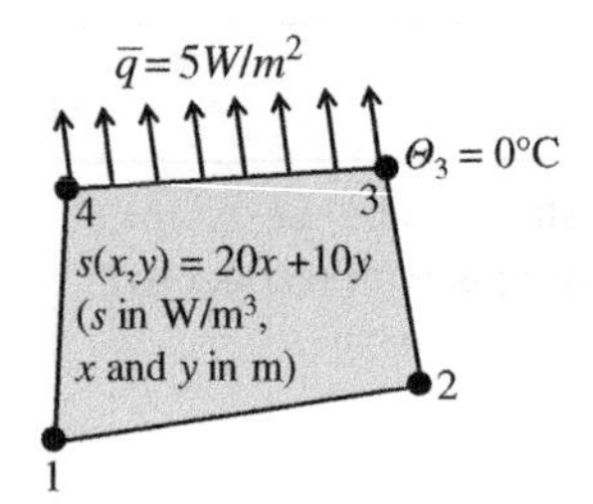

Node	x_i (m)	y_i (m)
1	0	0
2	1	0.2
3	0.8	1
4	0.2	0.9

Figure 6.16 Example mesh with a single, quadrilateral, isoparametric element.

Since the material is isotropic, we have:

$$[D] = \kappa \begin{bmatrix} 1 & 0 \\ 0 & 1 \end{bmatrix} = 2(1 + x \cdot y) \begin{bmatrix} 1 & 0 \\ 0 & 1 \end{bmatrix} \left(\text{values in } \frac{\text{W}}{\text{m} \cdot {}^\circ\text{C}} \right)$$

We establish the matrix with the nodal coordinates, wherein the first column contains the x-coordinate values of the nodes and the second column contains the y-coordinate values:

$$\left[\{x^{(e)}\} \ \{y^{(e)}\} \right] = \begin{bmatrix} x_1^{(e)} & y_1^{(e)} \\ x_2^{(e)} & y_2^{(e)} \\ x_3^{(e)} & y_3^{(e)} \\ x_4^{(e)} & y_4^{(e)} \end{bmatrix} = \begin{bmatrix} 0 & 0 \\ 1 & 0.2 \\ 0.8 & 1 \\ 0.2 & 0.9 \end{bmatrix}$$

We will begin with the computations of each Gauss point. We have 2 × 2 Gaussian quadrature, so we can use Table 6.2 for the location and the weight coefficient values of the quadrature points.

Gauss point 1: $\xi_1 = -\frac{1}{\sqrt{3}}$, $\eta_1 = -\frac{1}{\sqrt{3}}$, $W_1 = 1$

Find physical coordinates of the specific Gauss point:

$$x_1 = x(\xi_1, \eta_1) = \frac{1}{4}\left[1 - \left(-\frac{1}{\sqrt{3}}\right)\right]\left[1 - \left(-\frac{1}{\sqrt{3}}\right)\right] \cdot 0 + \frac{1}{4}\left[1 + \left(-\frac{1}{\sqrt{3}}\right)\right]\left[1 - \left(-\frac{1}{\sqrt{3}}\right)\right] \cdot 1$$
$$+ \frac{1}{4}\left[1 + \left(-\frac{1}{\sqrt{3}}\right)\right]\left[1 + \left(-\frac{1}{\sqrt{3}}\right)\right] \cdot 0.8 + \frac{1}{4}\left[1 - \left(-\frac{1}{\sqrt{3}}\right)\right]\left[1 + \left(-\frac{1}{\sqrt{3}}\right)\right] \cdot 0.2$$
$$= 0.2357 \text{ m}$$

$$y_1 = y(\xi_1, \eta_1) = \frac{1}{4}\left[1 - \left(-\frac{1}{\sqrt{3}}\right)\right]\left[1 - \left(-\frac{1}{\sqrt{3}}\right)\right] \cdot 0 + \frac{1}{4}\left[1 + \left(-\frac{1}{\sqrt{3}}\right)\right]\left[1 - \left(-\frac{1}{\sqrt{3}}\right)\right] \cdot 0.2$$
$$+ \frac{1}{4}\left[1 + \left(-\frac{1}{\sqrt{3}}\right)\right]\left[1 + \left(-\frac{1}{\sqrt{3}}\right)\right] \cdot 1 + \frac{1}{4}\left[1 - \left(-\frac{1}{\sqrt{3}}\right)\right]\left[1 + \left(-\frac{1}{\sqrt{3}}\right)\right] \cdot 0.9$$
$$= 0.2280 \text{ m}$$

Find material conductivity array: $[D_1] = 2(1 + x_1 \cdot y_1) \begin{bmatrix} 1 & 0 \\ 0 & 1 \end{bmatrix} = \begin{bmatrix} 2.1075 & 0 \\ 0 & 2.1075 \end{bmatrix}$

Calculate the shape function array values at Gauss point 1:
$[N_1] = [0.622 \;\; 0.167 \;\; 0.045 \;\; 0.166]$

Also, calculate the array $\left[N_{,\xi}^{(4Q)}\right]_1$, containing the derivatives of the shape functions with respect to the parametric coordinates at the location of Gauss point 1, that is, for $\xi_1 = -\frac{1}{\sqrt{3}}$, $\eta_1 = -\frac{1}{\sqrt{3}}$:

$$\left[N_{,\xi}^{(4Q)}\right]_1 = \begin{bmatrix} -0.3943 & 0.3943 & 0.1057 & -0.1057 \\ -0.3943 & -0.1057 & 0.1057 & 0.3943 \end{bmatrix}$$

We can also obtain the Jacobian array $[J_1]$ and its determinant J_1 of the mapping at the location of Gauss point 1:

$$[J_1] = \left[N_{,\xi}^{(4Q)}\right]_1 \left[\{x^{(e)}\} \;\; \{y^{(e)}\}\right] = \begin{bmatrix} 0.458 & 0.089 \\ 0.058 & 0.439 \end{bmatrix}$$

$$J_1 = 0.196, \quad \left[\tilde{J}_1\right] = [J_1]^{-1} = \begin{bmatrix} 2.242 & -0.456 \\ -0.295 & 2.336 \end{bmatrix}$$

$$[B_1] = \left[\tilde{J}_1\right][N_{,\xi}]_1 = \begin{bmatrix} -0.7042 & 0.9324 & 0.1887 & -0.4169 \\ -0.8048 & -0.3630 & 0.2157 & 0.9521 \end{bmatrix}$$

Also, evaluate the source term at Gauss point 1:

$$s_1 = s(x_1, y_1) = 20 \cdot x_1 + 10 \cdot y_1 = 20 \cdot 0.2357 + 10 \cdot 0.2280 = 6.994$$

We continue with the same computations for the other three Gauss points of the element.

<u>Gauss point 2</u>: $\xi_2 = \frac{1}{\sqrt{3}}$, $\eta_2 = -\frac{1}{\sqrt{3}}$, $W_2 = 1$

$$x_2 = x(\xi_2, \eta_2) = 0.7643\,\text{m}$$

$$y_2 = y(\xi_2, \eta_2) = 0.3313\,\text{m}$$

$$[D_2] = 2(1 + x_2 y_2)\begin{bmatrix} 1 & 0 \\ 0 & 1 \end{bmatrix} = \begin{bmatrix} 2.5063 & 0 \\ 0 & 2.5063 \end{bmatrix}$$

$$[N_2] = [0.167 \;\; 0.622 \;\; 0.167 \;\; 0.045]$$

$$[N_{,\xi}]_2 = \begin{bmatrix} -0.3943 & 0.3943 & 0.1057 & -0.1057 \\ -0.1057 & -0.3943 & 0.3943 & 0.1057 \end{bmatrix}$$

$$[J_2] = \begin{bmatrix} 0.458 & 0.089 \\ -0.058 & 0.411 \end{bmatrix}, \quad J_2 = 0.193, \quad \left[\tilde{J}_2\right] = \begin{bmatrix} 2.126 & -0.463 \\ 0.299 & 2.371 \end{bmatrix}$$

$$[B_2] = \left[\tilde{J}_2\right][N_{,\xi}]_2 = \begin{bmatrix} -0.7895 & 1.0211 & 0.042 & -0.2736 \\ -0.3684 & -0.8169 & 0.9664 & 0.2189 \end{bmatrix}$$

$$s_2 = 20 \cdot x_2 + 10 \cdot y_2 = 20 \cdot 0.7643 + 10 \cdot 0.3313 = 18.598$$

Gauss point 3: $\xi_3 = \frac{1}{\sqrt{3}}, \; \eta_3 = \frac{1}{\sqrt{3}}, \; W_3 = 1$

$$x_3 = x(\xi_3,\eta_3) = 0.6976\,\text{m}$$

$$y_3 = y(\xi_3,\eta_3) = 0.8053\,\text{m}$$

$$[D_3] = 2(1 + x_3 y_3)\begin{bmatrix} 1 & 0 \\ 0 & 1 \end{bmatrix} = \begin{bmatrix} 3.1236 & 0 \\ 0 & 3.1236 \end{bmatrix}$$

$$[N_3] = [0.045 \;\; 0.167 \;\; 0.622 \;\; 0.167]$$

$$[N_{,\xi}]_3 = \begin{bmatrix} -0.1057 & 0.1057 & 0.3943 & -0.3943 \\ -0.1057 & -0.3943 & 0.3943 & 0.1057 \end{bmatrix}$$

$$[J_3] = \begin{bmatrix} 0.342 & 0.061 \\ -0.058 & 0.411 \end{bmatrix}, \; J_3 = 0.144, \; \left[\tilde{J}_3\right] = \begin{bmatrix} 2.851 & -0.421 \\ 0.401 & 2.377 \end{bmatrix}$$

$$[B_3] = \left[\tilde{J}_3\right][N_{,\xi}]_3 = \begin{bmatrix} -0.2568 & 0.4671 & 0.9583 & -1.1686 \\ -0.2935 & -0.8948 & 1.0952 & 0.0930 \end{bmatrix}$$

$$s_3 = 20 \cdot x_3 + 10 \cdot y_3 = 20 \cdot 0.6976 + 10 \cdot 0.8053 = 22.006$$

Gauss point 4: $\xi_4 = -\frac{1}{\sqrt{3}}, \; \eta_4 = \frac{1}{\sqrt{3}}, \; W_4 = 1$

$$x_4 = x(\xi_4,\eta_4) = 0.3024\,\text{m}$$

$$y_4 = y(\xi_4,\eta_4) = 0.7354\,\text{m}$$

$$[D_4] = 2(1 + x_4 y_4)\begin{bmatrix} 1 & 0 \\ 0 & 1 \end{bmatrix} = \begin{bmatrix} 2.4448 & 0 \\ 0 & 2.4448 \end{bmatrix}$$

$$[N_4] = [0.167 \;\; 0.045 \;\; 0.167 \;\; 0.622]$$

$$[N_{,\xi}]_4 = \begin{bmatrix} -0.1057 & 0.1057 & 0.3943 & -0.3943 \\ -0.3943 & -0.1057 & 0.1057 & 0.3943 \end{bmatrix}$$

$$[J_4] = \begin{bmatrix} 0.342 & 0.061 \\ 0.058 & 0.439 \end{bmatrix}, \; J_4 = 0.147, \; \left[\tilde{J}_4\right] = \begin{bmatrix} 2.9913 & -0.412 \\ -0.393 & 2.330 \end{bmatrix}$$

$$[B_4] = \left[\tilde{J}_4\right][N_{,\xi}]_4 = \begin{bmatrix} -0.1535 & 0.3596 & 1.1360 & -1.3421 \\ -0.8772 & -0.2877 & 0.0912 & 1.0737 \end{bmatrix}$$

$$s_4 = s(x_4,y_4) = 20 \cdot x_4 + 10 \cdot y_4 = 20 \cdot 0.3024 + 10 \cdot 0.7354 = 13.402$$

Now, we can calculate the element conductance matrix per unit thickness and the contribution of the heat source to the equivalent nodal flux vector per unit thickness:

$$\left[k^{(e)}\right] = \int_{-1}^{1}\int_{-1}^{1}\left(\left[B^{(e)}\right]^T\left[D^{(e)}\right]\left[B^{(e)}\right]J d\xi d\eta\right) \approx \sum_{g=1}^{Ng}\left([B_g]^T[D_g][B_g]J_g W_g\right)$$

$$= [B_1]^T[D_1][B_1]J_1 W_1 + [B_2]^T[D_2][B_2]J_2 W_2 + [B_3]^T[D_3][B_3]J_3 W_3$$

$$+ [B_4]^T[D_4][B_4]J_4 W_4$$

$$\rightarrow \left[k^{(e)}\right] = \begin{bmatrix} 1.1930 & -0.2601 & -0.6616 & -0.2713 \\ -0.2601 & 1.7755 & -0.4232 & -1.0923 \\ -0.6616 & -0.4232 & 1.9060 & -0.8213 \\ -0.2713 & -1.0923 & -0.8213 & 2.1849 \end{bmatrix} \left(\text{values in } \frac{\text{W}}{\text{m} \cdot {}^{o}\text{C}}\right)$$

$$\left\{f_{\Omega}^{(e)}\right\} = \int_{-1}^{1}\int_{-1}^{1}\left(\left[N^{(4Q)}\right]^{T} s \cdot J d\xi d\eta\right) \approx \sum_{g=1}^{Ng}\left(\left[N_g\right]^{T} s_g \cdot J_g W_g\right)$$

$$= \left[N_1\right]^{T} s_1 \cdot J_1 W_1 + \left[N_2\right]^{T} s_2 \cdot J_2 W_2 + \left[N_3\right]^{T} s_3 \cdot J_3 W_3 + \left[N_4\right]^{T} s_4 \cdot J_4 W_4$$

$$\left\{f_{\Omega}^{(e)}\right\} = \begin{Bmatrix} 1.921 \\ 3.078 \\ 2.959 \\ 2.142 \end{Bmatrix} \text{(values in W/m)}$$

Finally, we can move on to calculate the contribution of the boundary outflow across segment (3-4) to the equivalent nodal flux vector. In accordance with the element shape in the parametric space shown in Figure 6.13, the side (3-4) corresponds to $\eta = 1 = \text{constant}$, so we must use ξ as the variable of the parameterization.

The length of segment (3-4) is

$$\ell_{34} = \sqrt{\left(x_4^{(e)} - x_3^{(e)}\right)^2 + \left(y_4^{(e)} - y_3^{(e)}\right)^2} = \sqrt{(0.2-0.8)^2 + (0.9-1)^2} = 0.6083\,m$$

We can now use Box 6.5.2 to obtain:

$$\left\{f_{\Gamma_{t,34}}^{(e)}\right\} = \left\{f_{\Gamma_{ty,34}}^{(e)}\right\} = -\int_{\Gamma_{t,34}}\left[N^{(4Q)}\right]^{T}\bar{q}\,ds = -\int_{-1}^{1}\left(\left[N^{(4Q)}\right]^{T}\bar{q}\frac{0.6083\,m}{2}d\xi\right)$$

$$\approx -\sum_{g=1}^{2}\left(\left[N^{(4Q)}\left(\xi_g, \eta = 1\right)\right]^{T}\bar{q}\frac{0.6083\,m}{2}W_g\right)$$

$$= -\sum_{g=1}^{2}\left(\left[N^{(4Q)}\left(\xi_g, \eta = 1\right)\right]^{T} 5 \cdot \frac{0.6083\,m}{2}W_g\right)$$

$$= -\left[N^{(4Q)}\left(\xi_1 = -\frac{1}{\sqrt{3}}, \eta = 1\right)\right]^{T} 5 \cdot \frac{0.6083\,m}{2} \cdot 1$$

$$-\left[N^{(4Q)}\left(\xi_2 = \frac{1}{\sqrt{3}}, \eta = 1\right)\right]^{T} 5 \cdot \frac{0.6083\,m}{2} \cdot 1$$

$$\rightarrow \left\{f_{\Gamma,34}^{(e)}\right\} = \begin{Bmatrix} 0 \\ 0 \\ -1.521 \\ -1.521 \end{Bmatrix} \text{(values in W/m)}$$

We could be expecting that the values of $\left\{f_{\Gamma,34}^{(e)}\right\}$ corresponding to nodes 1 and 2 are zero, since these nodes do not lie on the natural boundary segment (3-4)!

Thus, the total equivalent nodal flux vector of the element (per unit thickness) is:

$$\left\{f^{(e)}\right\} = \left\{f_{\Omega}^{(e)}\right\} + \left\{f_{\Gamma_{34}}^{(e)}\right\} = \begin{Bmatrix} 1.921 \\ 3.078 \\ 2.959 \\ 2.142 \end{Bmatrix} + \begin{Bmatrix} 0 \\ 0 \\ -1.521 \\ -1.521 \end{Bmatrix} = \begin{Bmatrix} 1.921 \\ 3.078 \\ 1.438 \\ 0.621 \end{Bmatrix} \text{(values in W/m)}$$

Now, we have the total conductance matrix and equivalent nodal flux vector. The temperature at node 3 is restrained (see Figure 6.16), so the only degrees of freedom that are not restrained are Θ_1, Θ_2, and Θ_4. We can establish the equations for the free degrees of freedom in accordance with Section B.3. Since the restrained temperature Θ_3 is zero, we only need to keep rows and columns 1, 2, and 4 from the conductance matrix and rows 1, 2, and 4 from the equivalent nodal flux vector:

$$\begin{bmatrix} 1.1930 & -0.2601 & -0.2713 \\ -0.2601 & 1.7755 & -1.0923 \\ -0.2713 & -1.0923 & 2.1849 \end{bmatrix} \begin{Bmatrix} \Theta_1 \\ \Theta_2 \\ \Theta_4 \end{Bmatrix} = \begin{Bmatrix} 1.921 \\ 3.078 \\ 0.621 \end{Bmatrix}$$

and we can solve for the temperatures of the unrestrained degrees-of-freedom to obtain:

$$\begin{Bmatrix} \Theta_1 \\ \Theta_2 \\ \Theta_4 \end{Bmatrix} = \begin{Bmatrix} 2.99 \\ 3.72 \\ 2.52 \end{Bmatrix} {}^{\text{o}}\text{C}$$

■

6.6 Higher-Order Isoparametric Quadrilateral Elements

The four-node quadrilateral is not the only type of element based on the isoparametric concept. Higher-order elements, which have more nodal points and also higher-order polynomial shape functions, can be established and are used in research and practice. The shape functions for higher-order elements can still be obtained as tensor products of higher-order one-dimensional shape functions, and these elements are called *Lagrangian isoparametric elements* (because the higher-order one-dimensional shape functions are obtained as Lagrangian polynomials). One example is the nine-node quadrilateral element (9Q), shown in Figure 6.17, for which the nine shape functions are quadratic Lagrangian polynomials in terms of ξ and η. As shown in Figure 6.17, for higher-order isoparametric elements, the sides of the element in the physical space no longer need to be straight lines (of course, they can be if we want). The fact that the sides of elements such as the 9Q can be curves allows to efficiently model geometries with curved boundaries.

One practical disadvantage of higher-order Lagrangian elements like the 9Q is the existence of *interior* nodes (nodes not lying on the perimeter of the element), such as node 9 in Figure 6.17. For this reason, analysts prefer to resort to a family of higher-order isoparametric elements called *serendipity* elements, which only include nodes along the

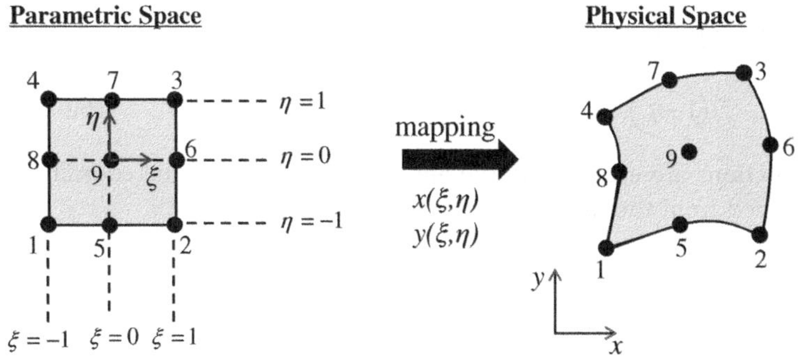

Figure 6.17 Nine-node quadrilateral (9Q) isoparametric element.

perimeter. One such example is the quadratic, eight-node (8Q) serendipity element, shown in Figure 6.18, which is the higher-order quadrilateral element most commonly used in practice.

The shape functions for the 8Q serendipity element are the following (they were originally obtained by inspection, but there is a systematic approach to obtain them, as explained in Hughes 2000).

$$N_1^{(8Q)}(\xi,\eta) = -\frac{1}{4}(1-\xi)(1-\eta)(1+\xi+\eta) \tag{6.6.1a}$$

$$N_2^{(8Q)}(\xi,\eta) = -\frac{1}{4}(1+\xi)(1-\eta)(1-\xi+\eta) \tag{6.6.1b}$$

$$N_3^{(8Q)}(\xi,\eta) = \frac{1}{4}(1+\xi)(1+\eta)(\xi+\eta-1) \tag{6.6.1c}$$

$$N_4^{(8Q)}(\xi,\eta) = -\frac{1}{4}(1-\xi)(1+\eta)(1+\xi-\eta) \tag{6.6.1d}$$

$$N_5^{(8Q)}(\xi,\eta) = \frac{1}{2}(1-\xi^2)(1-\eta) \tag{6.6.1e}$$

$$N_6^{(8Q)}(\xi,\eta) = \frac{1}{2}(1+\xi)(1-\eta^2) \tag{6.6.1f}$$

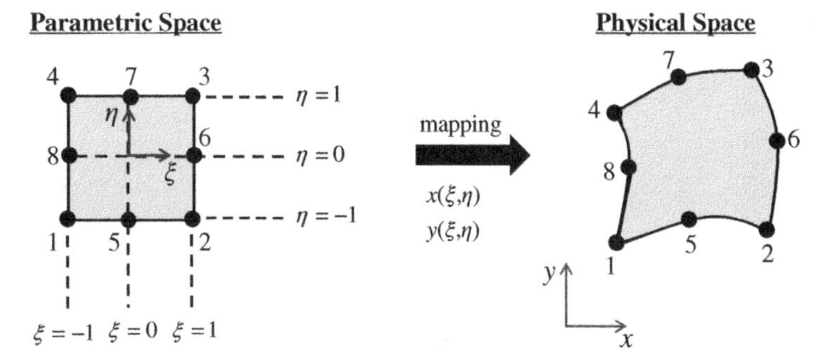

Figure 6.18 Eight-node quadrilateral (8Q) serendipity isoparametric element.

$$N_7^{(8Q)}(\xi,\eta) = \frac{1}{2}(1-\xi^2)(1+\eta) \tag{6.6.1g}$$

$$N_8^{(8Q)}(\xi,\eta) = \frac{1}{2}(1-\xi)(1-\eta^2) \tag{6.6.1h}$$

Once the shape functions are established, the approximate field and the coordinate mapping can be obtained from the following expressions.

$$\Theta^{(e)}(\xi,\eta) = \sum_{I=1}^{8}\left(N_I^{(8Q)}(\xi,\eta)\cdot\Theta_I^{(e)}\right) \tag{6.6.2}$$

$$x(\xi,\eta) = \sum_{I=1}^{8}\left(N_I^{(8Q)}(\xi,\eta)\cdot x_I^{(e)}\right) \tag{6.6.3a}$$

$$y(\xi,\eta) = \sum_{I=1}^{8}\left(N_I^{(8Q)}(\xi,\eta)\cdot y_I^{(e)}\right) \tag{6.6.3b}$$

The process of obtaining the element coefficient array and equivalent right-hand-side vector of higher-order elements is identical to that presented for the 4Q element in Section 6.5. The only difference is that we have more shape functions and nodal points; thus, the number of columns in the shape function array and in the $[B^{(e)}]$-array will be equal to 9 for a 9Q element and 8 for an 8Q element.

6.7 Isoparametric Triangular Elements

The isoparametric element concept can also be used for triangular elements. In this case, the parametric coordinates will be defined in a different fashion than for quadrilateral elements. Specifically, we can rely on the so-called triangular coordinate concept, also called the *natural coordinates* approach. Let us imagine that we have a point P, in the interior of a three-node triangular element, as shown in Figure 6.19a. As shown in the same figure, joining P with each of the nodal points defines three sub-triangles, with areas A_1, A_2, and A_3. Obviously, the sum of the three areas will give the total area, $A^{(e)}$, enclosed by the element:

$$A_1 + A_2 + A_3 = A^{(e)} \tag{6.7.1}$$

We now define the following, dimensionless area ratios:

$$\xi_1 = \frac{A_1}{A^{(e)}} \tag{6.7.2a}$$

$$\xi_2 = \frac{A_2}{A^{(e)}} \tag{6.7.2b}$$

$$\xi_3 = \frac{A_3}{A^{(e)}} = \frac{A^{(e)} - A_1 - A_2}{A^{(e)}} = 1 - \xi_1 - \xi_2 \tag{6.7.2c}$$

Thus, all three of the dimensionless area ratios are expressed in terms of two *natural coordinates*, ξ_1 and ξ_2, which can be used to establish an isoparametric element. The following applies for the range of values of the two natural coordinates.

$$
\begin{aligned}
&0 \le \xi_1 \le 1 \\
&0 \le \xi_2 \le 1
\end{aligned}
\tag{6.7.3}
$$

The *coordinate lines of a triangular isoparametric element*, the lines corresponding to a constant value of one of the parametric coordinates, are conceptually presented in Figure 6.19b.

We can now establish an isoparametric element and the associated mapping, as shown in Figure 6.20. Just like we did for the quadrilateral elements, we will establish our shape functions in the parametric (natural) coordinate space. Specifically, the following three shape functions can be established for the three-node triangular (3 T) element in the parametric space, ξ_1–ξ_2:

$$N_1^{(3T)}(\xi_1, \xi_2) = \xi_1 \tag{6.7.4a}$$

$$N_2^{(3T)}(\xi_1, \xi_2) = \xi_2 \tag{6.7.4b}$$

$$N_3^{(3T)}(\xi_1, \xi_2) = 1 - \xi_1 - \xi_2 \tag{6.7.4c}$$

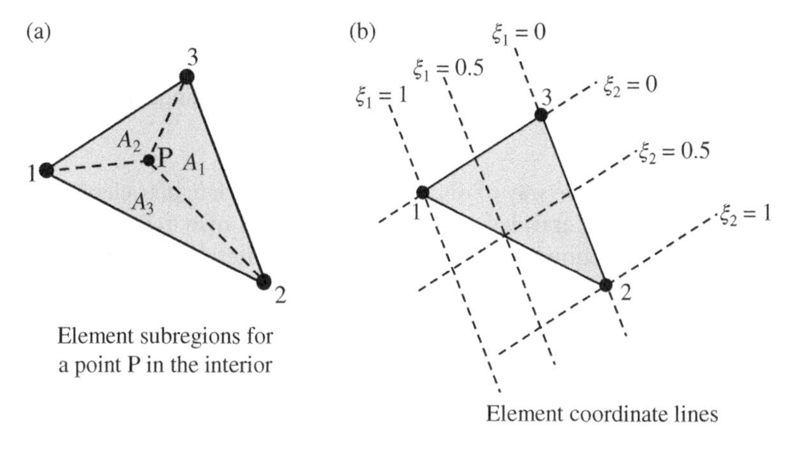

Figure 6.19 Three-node triangular isoparametric element.

Element in parent (parametric) space

ξ_2

2 (ξ_1=0, ξ_2=1)

1 (ξ_1=1, ξ_2=0)

3 (ξ_1=0, ξ_2=0)

ξ_1

mapping
(change of coordinates)

$x(\xi_1, \xi_2)$
$y(\xi_1, \xi_2)$

Element in physical space

3 $(x_3^{(e)}, y_3^{(e)})$

1 $(x_1^{(e)}, y_1^{(e)})$

2 $(x_2^{(e)}, y_2^{(e)})$

y x

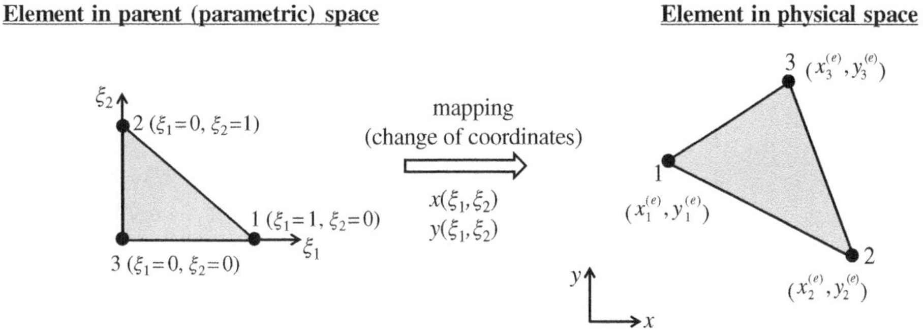

Figure 6.20 Three-node triangular (3 T) isoparametric element.

All that remains is a discussion of how integration can be conducted in the natural coordinate space of the (3T) element. For any function $f(x,y)$, we can use the change-of-coordinates formula (and account for the shape of the element domain in the parametric space) to obtain:

$$\iint_{\Omega^{(e)}} f(x,y)dV = \int_0^1 \int_0^{1-\xi_1} f(x(\xi_1,\xi_2),y(\xi_1,\xi_2)) \cdot J d\xi_1 d\xi_2 \tag{6.7.5}$$

where J is the Jacobian determinant of the coordinate mapping:

$$J = \det([J]),\ [J] = \begin{bmatrix} \partial x/\partial\xi_1 & \partial x/\partial\xi_2 \\ \partial y/\partial\xi_1 & \partial y/\partial\xi_2 \end{bmatrix} \tag{6.7.6}$$

Numerical quadrature can then be used to evaluate domain integrals like the one in Equation (6.7.5). Of course, since the range of possible values for ξ_1 and ξ_2 is between 0 and 1, and also since the form of the integral in Equation (6.7.5) does not allow us to fully separate the variables in the integration, different quadrature rules are established for triangular coordinates (Cowper 1973) than for quadrilateral parametric domains. These rules, and the order of polynomial accuracy that they ensure, are provided in Table 6.3.

Remark 6.7.1: The sum of the weight coefficients for quadrature in a triangular domain is 1 (*why?*). ■

Remark 6.7.2: There is an alternative way to obtain isoparametric triangular elements, by "collapsing" the shape of quadrilateral elements. For example, as shown in Figure 6.21, the three-node triangle can be obtained from a four-node quadrilateral element by stipulating that two of the nodes (e.g., nodes 3 and 4) coincide, i.e. they have the same physical coordinates and the same nodal value of field function. After enforcing these conditions, the algorithms for a 4Q element in Boxes 6.5.1 and 6.5.2 can be used for obtaining the coefficient array and right-hand-side vector of a triangular element. ■

Table 6.3 Two-Dimensional Quadrature Rules for Triangular Parametric Space.

Rule	Order of Polynomial Exactly Integrated	Point	Locations	Weight Coefficients
1-point	1	1	$\xi_1 = 1/3 = 0.33333333$, $\xi_2 = 1/3 = 0.33333333$	$W_1 = 1$
3-point	2	1	$\xi_1 = 2/3 = 0.66666667$, $\xi_2 = 1/6 = 0.16666667$	$W_1 = 1/3 = 0.3333333$
		2	$\xi_1 = 1/6 = 0.16666667$, $\xi_2 = 2/3 = 0.66666667$	$W_2 = 1/3 = 0.3333333$
		3	$\xi_1 = 1/6 = 0.16666667$, $\xi_2 = 1/6 = 0.16666667$	$W_3 = 1/3 = 0.3333333$
4-point	3	1	$\xi_1 = 1/3 = 0.33333333$, $\xi_2 = 1/3 = 0.33333333$	$W_1 = -0.56250000$
		2	$\xi_1 = 0.60000000$, $\xi_2 = 0.20000000$	$W_2 = 0.52083333$
		3	$\xi_1 = 0.20000000$, $\xi_2 = 0.60000000$	$W_2 = 0.52083333$
		4	$\xi_1 = 0.20000000$, $\xi_2 = 0.20000000$	$W_3 = 0.52083333$

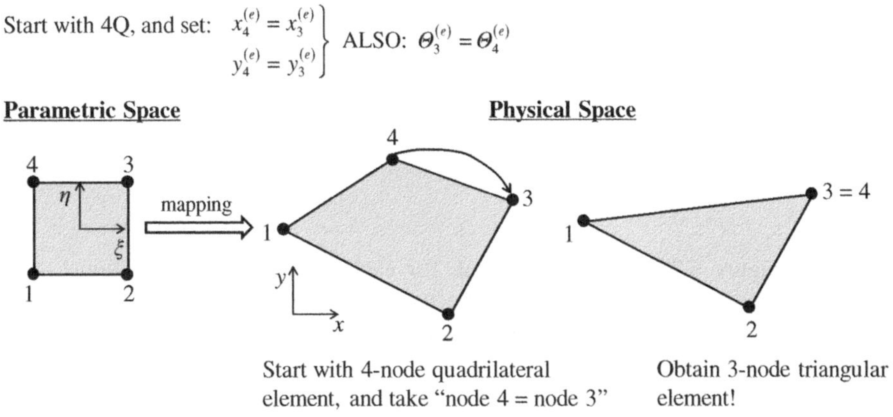

Start with 4Q, and set: $\left.\begin{array}{l} x_4^{(e)} = x_3^{(e)} \\ y_4^{(e)} = y_3^{(e)} \end{array}\right\}$ ALSO: $\Theta_3^{(e)} = \Theta_4^{(e)}$

Parametric Space

Physical Space

mapping

Start with 4-node quadrilateral element, and take "node 4 = node 3"

Obtain 3-node triangular element!

Figure 6.21 Three-node triangular element as a "collapsed" quadrilateral element with two coincident nodes.

6.8 Continuity and Completeness of Isoparametric Elements

We will now provide a brief discussion of the continuity and completeness ensured through isoparametric elements. The discussion will specifically focus on quadrilateral elements, but the same exact considerations apply for triangular isoparametric elements like the (3T) element in Section 6.7. The highest order of partial derivative in the weak form for heat conduction is 1, which means that we need C^0 continuity and first-degree completeness from the isoparametric elements.

Let us begin by considering continuity in the interior of an element. The mapping $x(\xi,\eta)$, $y(\xi,\eta)$ in isoparametric elements is continuous (since the shape functions in the parametric space are continuous functions of ξ and η). It must also be one-to-one and onto ("onto" means that *every* point in the physical space xy is the map of a point in the parametric space $\xi\eta$) and we need to have $J > 0$ (i.e., the Jacobian determinant of the coordinate transformation is positive everywhere). In this case, the inverse mapping $\xi(x, y)$, $\eta(x, y)$ exists and is also continuous with respect to x and y. This means that the shape functions in an isoparametric element are continuous functions of x and y in the interior of the element, since they are continuous functions of ξ and η, which, in turn, are continuous functions of x and y. Thus, the approximate field is continuous in the interior of each element.

In addition to continuity in the interior of the elements, we must satisfy continuity across inter-element boundaries. Isoparametric elements automatically ensure such continuity of the approximate field, for example, the temperature field Θ. For example, let us assume that we have two quadrilateral elements that share a common edge, (AB), as shown in Figure 6.22. We will prove the continuity by showing that, for each point P lying along the edge (AB), as shown in Figure 6.22, the approximate value of temperature at P obtained from each of the two elements is the same.

Since the shape functions are expressed in the parametric space, we examine each element in that space, as shown in Figure 6.23. From the orientation and nodal numbering of the two elements, it can be found that point P corresponds to the edge with $\xi = 1$ for element 1, and to the edge with $\xi = -1$ for element 2. We can then obtain the expressions

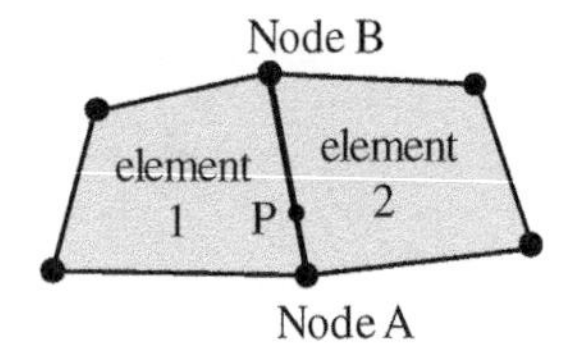

Figure 6.22 Example of point P lying on the inter-element boundary between two elements.

(*i*) and (*ii*) in Figure 6.23, which show that, for each one of the two elements, the temperature of point P only depends on the nodal temperatures at A and B. Using Equations (*i*) and (*ii*) of Figure (6.23), and for any point P along the side AB (i.e., for any value of η_P), we obtain:

$$\Theta_P^{(1)} = \Theta_P^{(2)} \tag{6.8.1}$$

where $\Theta_P^{(1)}$ and $\Theta_P^{(2)}$ are the temperature values at point P, obtained for elements (1) and (2), respectively. Thus, we have proven the necessary condition for having continuity across inter-element boundaries.

Remark 6.8.1: The same procedure can be used to prove continuity across the element boundaries for higher-order isoparametric elements. ∎

(a)

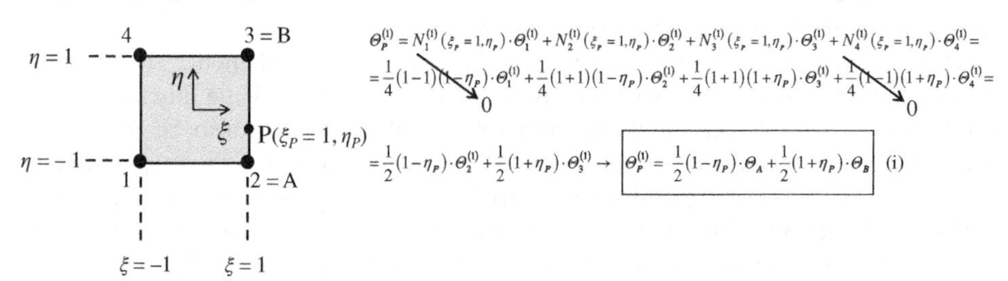

(b)

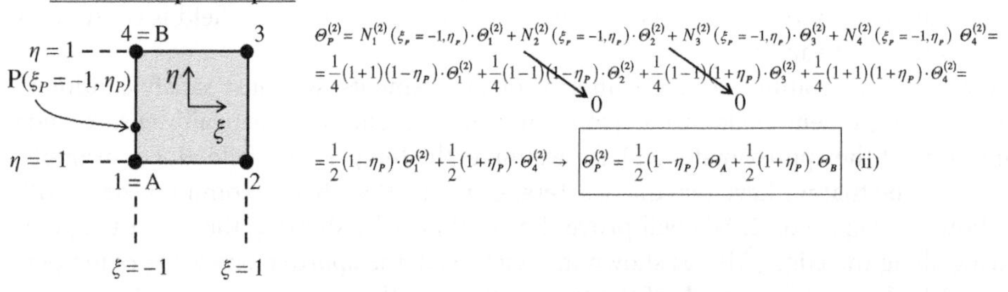

Figure 6.23 Value of approximate solution at intermediate point P, obtained for each of the two elements sharing the boundary.

It can also be proven that all isoparametric element approximations are *at least* first-order complete. The mathematical procedure to prove this statement is a bit more complicated, and will not be pursued here. The interested reader is referred to Strang and Fix (2008).

Based on the above, isoparametric finite elements satisfy both the continuity and completeness requirements for multidimensional heat conduction; thus, they are bound to ensure convergence.

6.9 Concluding Remarks: Finite Element Analysis for Other Scalar Field Problems

This chapter has provided the finite element equations for scalar field problems. The case of two-dimensional heat conduction was selected as the specific physical problem of interest, to provide context for the various quantities. The same exact equations and element formulations presented in the previous Sections can be used for any other type of scalar field problem described in Section 5.4. Although the detailed description of the finite element equations for these other physical problems will not be pursued here, a discussion of how they are related to heat conduction is deemed necessary. Table 6.4 provides a summary of the correspondence between the finite element analysis of heat conduction and that of other scalar field problems. This correspondence facilitates the use of the finite element equations presented in previous sections for the analysis of potential fluid flow, flow in porous media, and chemical diffusion.

For example, if the finite element equations for flow in porous media are to be obtained, one needs to replace nodal temperature vectors $\{\Theta\}$ with nodal pressure vectors $\{P\}$, also replace the material conductance array $[D]$ with the array $\frac{\kappa}{\mu}[I]$, and set the source term to zero, since such term is not present for flow in porous media.

Table 6.4 Correspondence between Various Multidimensional Scalar-Field Problems.

Physical Problem	Nodal Solution Vector	Material Array $[D]$	Source Term
Heat conduction	$\{\Theta\}$	$[D]$	$s(x,y)$
Potential fluid flow	$\{\Phi\}$	$[I]$	0
Flow in porous media	$\{P\}$	$\frac{\kappa}{\mu}[I]$	0
Chemical diffusion	$\{C\}$	$D_C[I]$	$R(x,y)$

Problems

6.1 Determine the degree (= order) of completeness of the following polynomial approximations:

a) $z(x,y) = a_o + a_1x + a_2x \cdot y + a_3x^2 + a_4y^2$.
b) $z(x,y) = a_o + a_1x + a_2x \cdot y + a_3x^2 + a_4y^2 + a_5y + a_6y^3 + a_7x^3 + a_8y^4 + a_9x^4$.

6.2 We are given a rectangular, four-node, two-dimensional element as shown in Figure P6.1. The origin of the coordinates coincides with node 1. We are also given the four two-dimensional shape functions for this element.

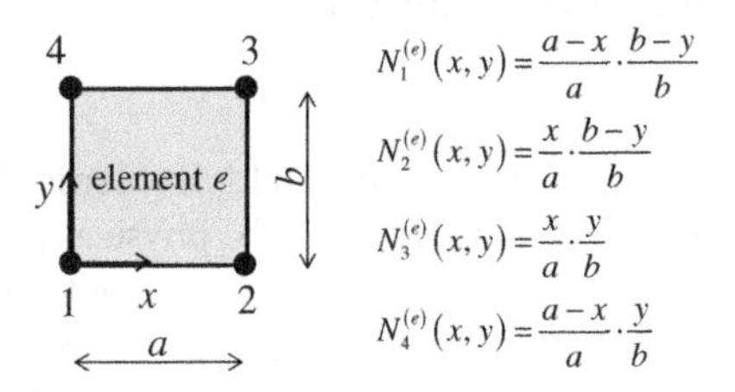

Figure P6.1

a) Show that the shape functions satisfy the partition of unity property.
b) Show that the shape functions satisfy the Kronecker delta property.
c) Find the expressions for the components of the $[B]$ array.

6.3 We are given the steady-state heat conduction system shown in Figure P6.2. The temperature along the line $x = 0$ has a prescribed zero value. We want to analyze the behavior of this system using two square finite elements, as shown in the figure.

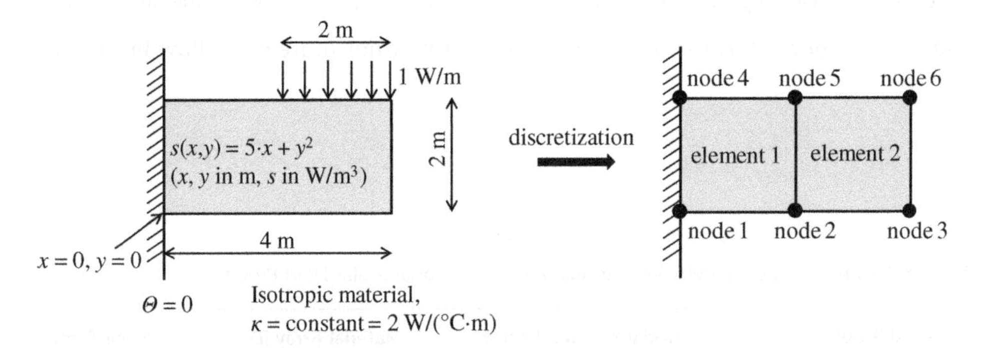

Figure P6.2

a) Obtain the coefficient (conductance) matrix and right-hand-side (equivalent nodal flux) vector for each one of the two elements.
b) Establish the gather/scatter arrays, $[L^{(e)}]$, for each one of the two elements, 1 and 2, and assemble the global conductance matrix and equivalent nodal flux vector.
c) Calculate the nodal temperature vector, $\{\Theta\}$.
d) Create a contour plot for the approximate temperature field.

e) Create a vector arrow plot for the approximate flux field.
f) What part of the strong form is expected to be satisfied exactly and what part is expected not to be satisfied exactly by our finite element solution?

6.4 We are given the three shape functions of the three-node triangular element shown in Figure P6.3, which is used for two-dimensional heat conduction analysis. The figure also provides the nodal coordinates, as well as the *nodal temperature values* (so we have somehow managed to solve the problem, and we already know the nodal temperatures!).

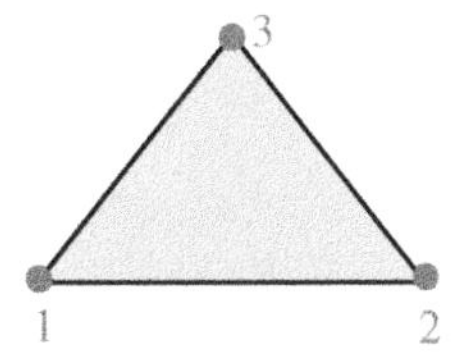

Node	x_i (m)	y_i (m)	Θ_i (°C)
1	0	0	5
2	2	0	8
3	1	2	10

Figure P6.3

$$N_1^{(e)}(x,y) = \frac{1}{4}(4-2x-y)$$

$$N_2^{(e)}(x,y) = \frac{1}{4}(2x-y)$$

$$N_3^{(e)}(x,y) = \frac{1}{2}y$$

The material in the element is isotropic and uniform, and the conductivity is $\kappa = 2$ W/(m·°C).

a) Verify that the shape functions satisfy the Kronecker delta property.
b) Verify that the shape functions satisfy the partition of unity property.
c) Determine the $[B^{(e)}]$ array for the element.
d) Determine the temperature and heat flux at the location $x = 0.5$, $y = 0.5$.

6.5 We are given the two-element mesh for the heat conduction problem shown in Figure P6.4. The material in both elements is isotropic and uniform (the conductivity values κ_1 and κ_2 are given in W/(m·°C)). The rectangular body shown in Figure P6.4 is also transmitting heat to its surroundings, at a rate of 1 W/m^3.

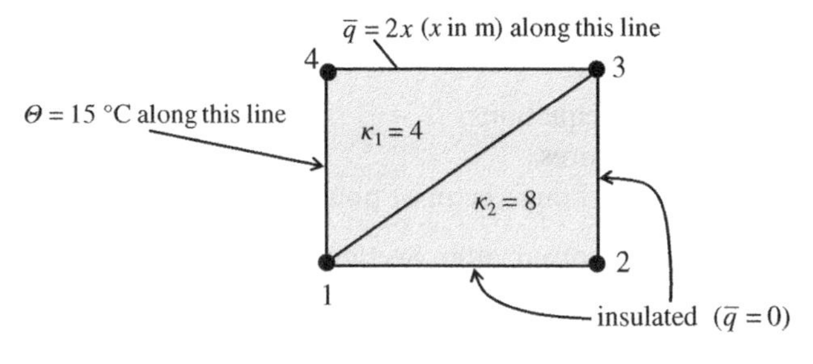

Node	x_i (m)	y_i (m)
1	0	0
2	5	0
3	5	5
4	0	5

Figure P6.4

a) Determine the global discrete equations.
b) Determine the nodal temperatures.
c) Prove that the approximate temperature field is continuous across the interelement boundary.

6.6 We are given a rectangular, nine-node, two-dimensional element as shown in Figure P6.5. The origin of the coordinates coincides with node 9. The coordinate values are in meters.

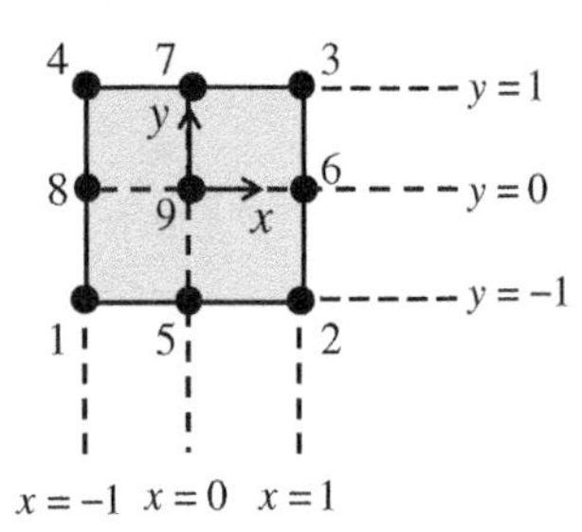

Figure P6.5

a) Obtain the shape functions of the element (in the physical domain) using the "tensor product approach."
b) Find what order of completeness corresponds to the specific finite element approximation. To do so, take the shape functions corresponding to nodes 1, 5, 6, and 9, find which *nine* different monomial terms are present, and then establish the order of completeness from Pascal's triangle.
c) Find the expressions for the components of the $[B^{(e)}]$ array.
d) Calculate the components (1,1), (4,3), and (5,6) of the element conductance array, $[k^{(e)}]$, if the material is homogeneous and isotropic with $\kappa = 2$ W/(m·°C).

6.7 We are given the two-element mesh shown in Figure P6.6. The material in both elements is isotropic and uniform. The conductivity values κ_1 and κ_2 are in W/(m·°C).

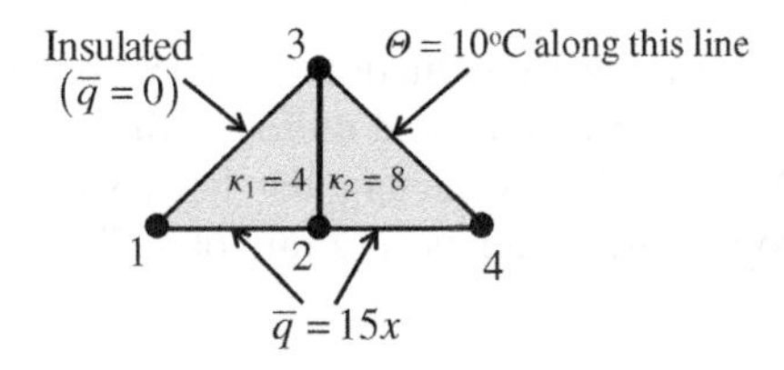

Node	x_i (m)	y_i (m)
1	0	0
2	2	0
3	2	2
4	4	0

Figure P6.6

a) Determine the global discrete equations.
b) Determine the nodal temperatures.
c) Determine the temperature and flux vector at point P($x = 1$ m, $y = 0.5$ m).

6.8 (*Note:* This problem should be solved after Sections 5.4.2 and 6.9 have been studied).

We have a solid porous body with $\kappa = 0.002$ in.^{2}. The body is saturated in water ($\mu = 2.73$ psi·sec), for which we know the pressure along the line segments AB and

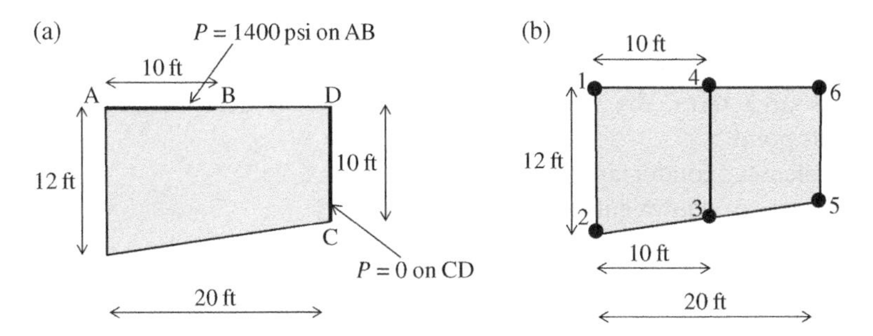

Figure P6.7

CD, as shown in Figure P6.7a. The remaining perimeter of the body (besides segments AB and CD) is covered with an impervious membrane (the membrane does not allow any fluid to flow through it).

a) Establish the global (structural) equations for a finite element analysis of the problem, using two 4-node isoparametric quadrilateral elements for the discretization shown in Figure P6.7b.
b) Calculate the nodal water pressure values.
c) What is the volume of water that flows through the segment CD in one hour?

6.9 Establish a program routine (in MATLAB, Mathcad, Excel spreadsheet, Fortran, C++, Visual Basic etc.—the choice is yours) to conduct the calculations for isoparametric, four-node quadrilateral elements in heat conduction, for *isotropic materials.*

The following input information must be provided in our program:

1) The numbers of the four nodal points.
2) The physical coordinates, x and y, of the four nodal points.
3) The variation of the scalar conductivity coefficient, κ, of the element. We want to allow any κ that has the following variation: $\kappa = a_o + a_1x + a_2y + a_3xy$, where a_o, a_1, a_2, and a_3 are constants. Thus, we need to provide as input the values of the four a parameters.
4) The variation of the source term, s, inside the element. We want to allow any source term that has the following variation: $s(x,y) = b_o + b_1x + b_2y + b_3xy + b_4x^2 + b_5y^2$, where b_o, b_1, b_2, b_3, b_4, and b_5 are constants. Thus, we need to provide as input the values of the six b parameters.
5) Information on whether each one of the sides of the element is a natural boundary segment. Obviously, if a side of the element is part of the natural boundary, we need to also give as input the value of the prescribed flux. We only allow constant prescribed fluxes on natural boundaries.
6) The number of integration points in the element.

Our routines must be capable of computing and providing as *output* the following information:

General information:

a) The coordinates of the Gauss points in the physical space, xy.

Preprocessing information:

b) The values of the shape functions, the [*B*] array, the conductivity value, the source term, the [*J*] array, and the Jacobian determinant, *J*, at each Gauss point.
c) The element conductance array, $[k^{(e)}]$
d) The element equivalent nodal flux vector, $\{f^{(e)}\}$

Postprocessing information (where we also need to pass as input the nodal temperatures of the element):

e) The values of temperature at the integration points.
f) The components of the flux vector, {*q*}, at the integration points.

6.10 Use the program you developed in Problem 6.9 to solve (find the nodal temperatures) the problem shown in Figure P6.8. We are given that $\kappa = 8$ W/(m·°C).

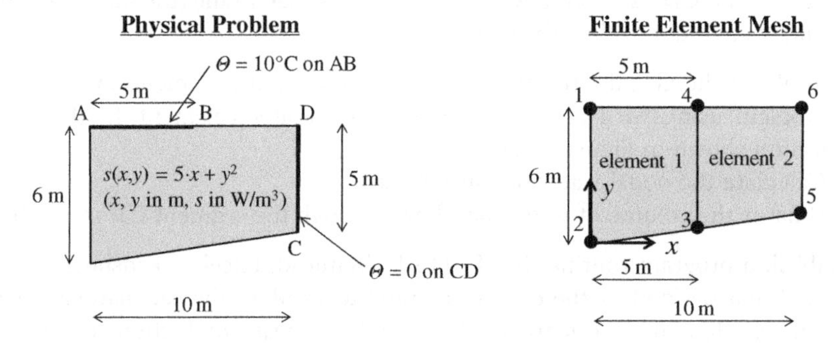

Figure P6.8

References

Argyris, J. (1960). *Energy Theorems and Structural Analysis*. London: Butterworth Scientific Publications.

Cowper, G. (1973). "Gaussian Quadrature Formulas for Triangles," *International Journal for Numerical Methods in Engineering* 7, 405–408.

Fish, J., and Belytschko, T. (2007). *A First Course in Finite Elements*. Hoboken, NJ: John Wiley & Sons.

Irons, B. (1966). "Engineering Application of Numerical Integration in Stiffness Method," *Journal of the American Institute of Aeronautics and Astronautics* 14, 2035–2037.

Strang, G., and Fix, G. (2008), *An Analysis of the Finite Element Method*, 2nd ed. Cambridge, MA: Wellesley.

Taig, I. (1961). "Structural Analysis by the Matrix Displacement Method," *Report No. S017*, English Electric Aviation.

7

Multidimensional Elasticity

7.1 Introduction

This chapter will provide a framework for the mathematical description of problems in continuum solid mechanics. In solid mechanics, we want to specify a *displacement vector field*, $\{u(x,y,z)\} = \begin{Bmatrix} u_x(x,y,z) \\ u_y(x,y,z) \\ u_z(x,y,z) \end{Bmatrix}$, which is assumed to be a continuous function of x, y, and z (continuity ensures that we will not have the formation of "holes" or "material overlapping" in the deformed configuration of the solid body). We will focus on *small-displacement, small-strain problems* (i.e., problems where the deformations are much smaller than the dimensions of the undeformed body and the strains are also very small). We also assume that we have *static loading*, that is, all the loadings and deformations change very slowly with time, so inertial effects can be neglected. Finally, we will focus on *linear elasticity*, which implies that the behavior of a solid body is linear (e.g., if we double the loads the displacements, strains and stresses will also become twice as large, etc.), and that after removal of the applied loads the solid will return to its original, undeformed configuration.

This chapter is meant to provide a basic, practical understanding of fundamental aspects of elasticity theory. A more complete and mathematically rigorous treatment can be found in special textbooks dedicated to the subject, such as Love (1927), Sokolnikoff (1956), Malvern (1977), Fung and Tong (2001).

7.2 Definition of Strain Tensor

The behavior of a solid body is quantified by two *second-order tensors*[1], namely the *strain tensor* (which is a measure of the magnitude of deformations), and the *stress tensor* (which is a measure of the intensity of forces developing in the interior of a body as a

---

1 From a practical point of view, a second-order tensor is a square matrix that follows a specific transformation rule from one coordinate system to another. This rule is discussed in Section 7.6. The number of indices (subscripts) that define the individual components of a tensor is called the *order* of the tensor. Since each stress and strain component includes two indices, the stress and strain tensors are second-order ones. Vectors are first-order tensors, since each of their components is defined by a single index.

Fundamentals of Finite Element Analysis: Linear Finite Element Analysis, First Edition.
Ioannis Koutromanos, James McClure, and Christopher Roy.

Companion website: www.wiley.com/go/koutromanos/linear

result of the deformations). For the general, three-dimensional problem, the strain tensor $[\varepsilon]$ is a (3×3) matrix that includes a total of nine components.

$$[\varepsilon] = \begin{bmatrix} \varepsilon_{xx} & \varepsilon_{xy} & \varepsilon_{xz} \\ \varepsilon_{yx} & \varepsilon_{yy} & \varepsilon_{yz} \\ \varepsilon_{zx} & \varepsilon_{zy} & \varepsilon_{zz} \end{bmatrix} \tag{7.2.1}$$

We can use tensor notation ($x_1 = x$, $x_2 = y$, $x_3 = z$), and write the following generic definition for each component of the strain tensor.

$$\varepsilon_{ij} = \frac{1}{2}\left(\frac{\partial u_i}{\partial x_j} + \frac{\partial u_j}{\partial x_i}\right) \tag{7.2.2}$$

For example, we have: $\varepsilon_{xx} = \frac{\partial u_x}{\partial x}$, $\varepsilon_{xz} = \frac{1}{2}\left(\frac{\partial u_x}{\partial z} + \frac{\partial u_z}{\partial x}\right)$.

The *strain-displacement relation* of Equation (7.2.2) is valid for infinitesimally small strains.

The geometric meaning of several strain components pertaining to the x- and y-direction is presented in Figure 7.1, by means of a small piece of material with dimensions Δx and Δy. One can understand that the *normal strains* (i.e., the strains ε_{ij}, where $i = j$) measure the *stretching* of a piece of material along each one of the coordinate axes, while the shear strains (i.e., the strains $\gamma_{ij} = 2\varepsilon_{ij}$, where $i \neq j$) measure the *distortion*, that is, how the angles of an initially rectangular piece of material differ from a right angle in the deformed configuration.

We know from Equation (7.2.2) that the strain components are obtained from partial differentiation of the components of the displacement vector field, with respect to the

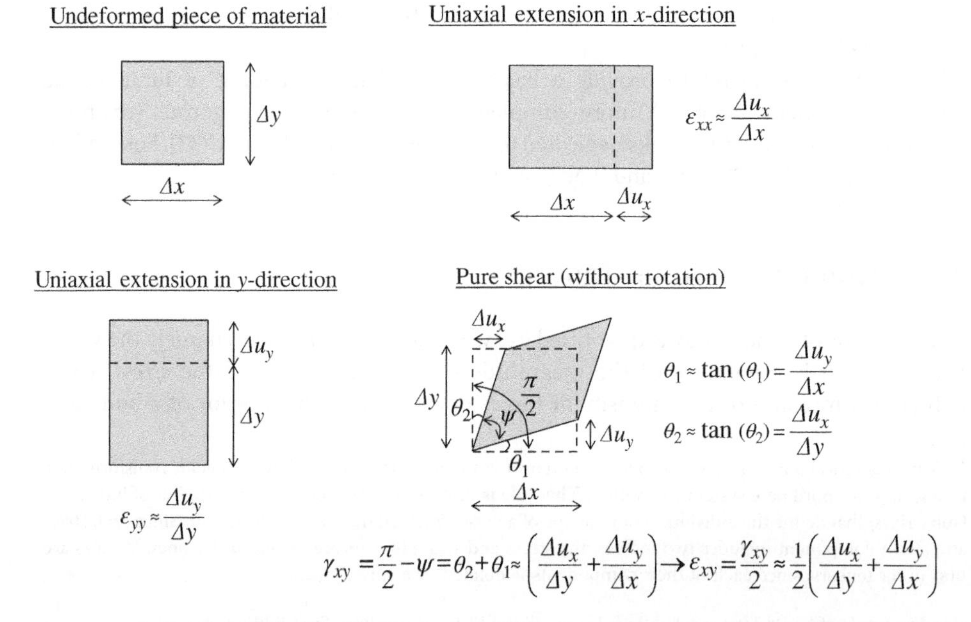

Figure 7.1 Definition of several of the strain components.

spatial coordinates. This leads to several mathematical conditions that must be satisfied by the strain components. These conditions are commonly referred to as *compatibility conditions*, because they express the fact that the strain components are obtained from a continuous displacement field, by means of the strain-displacement relation. For a general three-dimensional problem, we can establish the following six compatibility conditions.

$$\frac{\partial^2 \varepsilon_{xx}}{\partial y \partial z} = \frac{\partial}{\partial x}\left(-\frac{\partial \varepsilon_{yz}}{\partial x} + \frac{\partial \varepsilon_{zx}}{\partial y} + \frac{\partial \varepsilon_{xy}}{\partial z}\right) \tag{7.2.3a}$$

$$\frac{\partial^2 \varepsilon_{yy}}{\partial z \partial x} = \frac{\partial}{\partial y}\left(-\frac{\partial \varepsilon_{zx}}{\partial y} + \frac{\partial \varepsilon_{xy}}{\partial z} + \frac{\partial \varepsilon_{yz}}{\partial x}\right) \tag{7.2.3b}$$

$$\frac{\partial^2 \varepsilon_{zz}}{\partial x \partial y} = \frac{\partial}{\partial z}\left(-\frac{\partial \varepsilon_{xy}}{\partial z} + \frac{\partial \varepsilon_{yz}}{\partial x} + \frac{\partial \varepsilon_{zx}}{\partial y}\right) \tag{7.2.3c}$$

$$2\frac{\partial^2 \varepsilon_{xy}}{\partial x \partial y} = \frac{\partial^2 \varepsilon_{xx}}{\partial y^2} + \frac{\partial^2 \varepsilon_{yy}}{\partial x^2} \tag{7.2.3d}$$

$$2\frac{\partial^2 \varepsilon_{yz}}{\partial y \partial z} = \frac{\partial^2 \varepsilon_{yy}}{\partial z^2} + \frac{\partial^2 \varepsilon_{zz}}{\partial y^2} \tag{7.2.3e}$$

$$2\frac{\partial^2 \varepsilon_{zx}}{\partial z \partial x} = \frac{\partial^2 \varepsilon_{zz}}{\partial x^2} + \frac{\partial^2 \varepsilon_{xx}}{\partial z^2} \tag{7.2.3f}$$

7.3 Definition of Stress Tensor

Another notion that we examined in one-dimensional elasticity in Chapter 2 and must be generalized in multidimensional solid mechanics is that of stress. Just like we did for strain, we need to define (for the general three-dimensional problem) a (3 × 3) stress tensor, which is a measure of the intensity of forces developing in the interior of a body as a result of the deformations. The stress tensor, $[\sigma]$, contains the following nine components.

$$[\sigma] = \begin{bmatrix} \sigma_{xx} & \sigma_{xy} & \sigma_{xz} \\ \sigma_{yx} & \sigma_{yy} & \sigma_{yz} \\ \sigma_{zx} & \sigma_{zy} & \sigma_{zz} \end{bmatrix} \tag{7.3.1}$$

Each stress component, σ_{ij}, provides the limit value of force in the i-direction divided by the area of a face whose normal vector is pointing in the j-direction, as the value of the area tends to zero. The various components (with their sign convention) are shown for a box-shaped piece of material from the interior of a three-dimensional solid body in Figure 7.2. Notice from the figure that the first subscript in a stress component refers to the direction of the normalized force, while the second subscript refers to the direction of the unit normal to the plane on which the stress is acting.

The stress tensor in solid mechanics can be defined indirectly, based on a procedure established by Cauchy. Let us imagine that we cut a slice through a continuous, three-dimensional body, as shown in Figure 7.3a. Then, the plane of the slice, which is formally called the *section*, will have distributed forces (which are simply the forces exerted by the

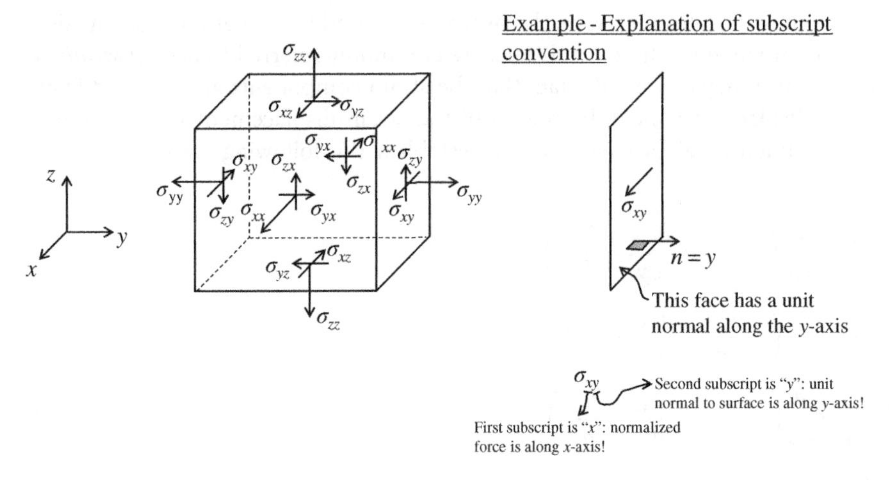

Figure 7.2 Components of stress tensor.

piece of the body on one side of the section to the piece on the other side). Each tiny region on the section with area ds, will have a unit normal outward vector $\vec{n}$ and a small amount of sectional force $d\vec{f}$, as shown in the same figure. Then, the following equation applies for the force $d\vec{f}$ and the stress components at the location of the tiny area:

$$d\vec{f} = (ds) \cdot \vec{n} \bullet [\sigma] \quad \text{or} \quad \vec{\Sigma}_n \equiv \frac{d\vec{f}}{ds} = \vec{n} \bullet [\sigma] \tag{7.3.2}$$

To obtain Equation (7.3.2), let us imagine that we cut a small tetrahedron from the interior of the sliced body, as shown in Figure 7.3b, such that one the tetrahedron's faces lies on the plane of the section. Each of the three faces of the tetrahedron that correspond to the "interior" of the body is normal to one of the three coordinate axes x, y and z. The fourth face has the unit normal outward vector $\vec{n}$ of the section. We can draw the stress components on each of the three "interior faces", as shown in Figure 7.3b, and we also

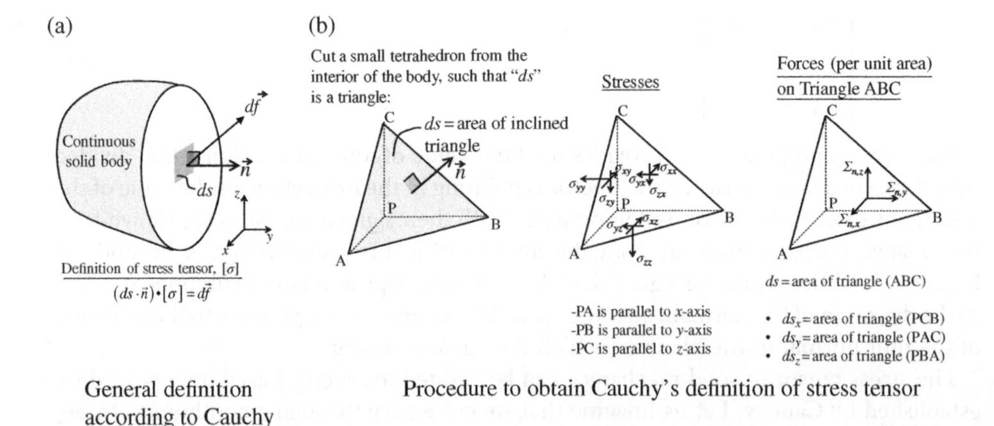

Figure 7.3 Definition of stress tensor.

draw the force components Σ_{nx}, Σ_{ny} and Σ_{nz} per unit area of the inclined face, shown in the same figure.

Based on geometric considerations (or from calculus), the projected areas ds_x, ds_y, and ds_z in Figure 7.3b are given by:

$$ds_x = ds\cdot\cos(\vec{n},x) = ds\cdot n_x \quad (7.3.3a)$$

$$ds_y = ds\cdot\cos(\vec{n},y) = ds\cdot n_y \quad (7.3.3b)$$

$$ds_z = ds\cdot\cos(\vec{n},z) = ds\cdot n_z \quad (7.3.3c)$$

Now, if we take equilibrium along the three directions for the free body of the tetrahedron, we have:

$$\sum F_x = 0 \rightarrow -\sigma_{xx}\cdot ds_x - \sigma_{xy}\cdot ds_y - \sigma_{xz}\cdot ds_z + \Sigma_{nx}\cdot ds = 0$$

$$\rightarrow \sigma_{xx}\cdot ds_x + \sigma_{xy}\cdot ds_y + \sigma_{xz}\cdot ds_z = \Sigma_{n,x}\cdot ds \quad (7.3.4a)$$

$$\sum F_y = 0 \rightarrow -\sigma_{yx}\cdot ds_x - \sigma_{yy}\cdot ds_y - \sigma_{yz}\cdot ds_z + \Sigma_{n,y}\cdot ds = 0$$

$$\rightarrow \sigma_{yx}\cdot ds_x + \sigma_{yy}\cdot ds_y + \sigma_{yz}\cdot ds_z = \Sigma_{n,y}\cdot ds \quad (7.3.4b)$$

$$\sum F_z = 0 \rightarrow -\sigma_{zx}\cdot ds_x - \sigma_{zy}\cdot ds_y - \sigma_{zz}\cdot ds_z + \Sigma_{n,z}\cdot ds = 0$$

$$\rightarrow \sigma_{zx}\cdot ds_x + \sigma_{zy}\cdot ds_y + \sigma_{zz}\cdot ds_z = \Sigma_{n,z}\cdot ds \quad (7.3.4c)$$

If we finally divide both sides of Equations (7.3.4a–c) by ds, account for Equations (7.3.3a–c), and then combine the three resulting equations into a single matrix expression, we obtain Equation (7.3.2).

Thus, the stress tensor is a quantity that allows us to completely define the vector $\vec{\Sigma}_n$, for any infinitesimal surface area ds of any section in the interior of the body (or in the boundary, as we will later see!).

7.4 Representing Stress and Strain as Column Vectors—The Voigt Notation

In finite element analysis, it is often deemed preferable to represent the stress and strain with column vectors. This allows us to obtain expressions which turn out to be "reminiscent" of the equations obtained for one-dimensional elasticity, without a need to resort to operations of second-order (or even higher-order) tensors. To express the strain and stress as column vectors, we resort to the so-called *Voigt notation*. The vector representation of the strain is as follows.

$$\underline{\underline{\varepsilon}} = [\varepsilon] = \begin{bmatrix} \varepsilon_{xx} & \varepsilon_{xy} & \varepsilon_{xz} \\ \varepsilon_{yx} & \varepsilon_{yy} & \varepsilon_{yz} \\ \varepsilon_{zx} & \varepsilon_{zy} & \varepsilon_{zz} \end{bmatrix} \rightarrow \{\varepsilon\} = \begin{Bmatrix} \varepsilon_{xx} \\ \varepsilon_{yy} \\ \varepsilon_{zz} \\ \gamma_{xy} \\ \gamma_{yz} \\ \gamma_{zx} \end{Bmatrix} \quad \begin{pmatrix} \gamma_{xy} = 2\varepsilon_{xy} \\ \gamma_{yz} = 2\varepsilon_{yz} \\ \gamma_{zx} = 2\varepsilon_{zx} \end{pmatrix} \quad (7.4.1)$$

Similarly, the following vector form can be established for the stress.

$$\underline{\underline{\sigma}} = [\sigma] = \begin{bmatrix} \sigma_{xx} & \sigma_{xy} & \sigma_{xz} \\ \sigma_{yx} & \sigma_{yy} & \sigma_{yz} \\ \sigma_{zx} & \sigma_{zy} & \sigma_{zz} \end{bmatrix} \rightarrow \{\sigma\} = \begin{Bmatrix} \sigma_{xx} \\ \sigma_{yy} \\ \sigma_{zz} \\ \sigma_{xy} \\ \sigma_{yz} \\ \sigma_{zx} \end{Bmatrix} \quad (7.4.2)$$

Remark 7.4.1: Note that the Voigt notation for strain requires that we use the values of $\gamma_{ij} = 2\varepsilon_{ij}$ for the shear strain components. ■

7.5 Constitutive Law (Stress-Strain Relation) for Multidimensional Linear Elasticity

Any mathematical relation between the stress and the strain provides information on the nature of the material and is called a *constitutive model or constitutive law*. For a linearly elastic material, the stress vector is "proportional" to the strain vector, and the proportionality is expressed by means of a *material stiffness matrix*, $[D]$. Thus, we can write:

$$\{\sigma\} = [D]\{\varepsilon\} \quad (7.5.1)$$

The simplest case of multiaxial elasticity is that corresponding to an *isotropic linearly elastic material*, for which all the components of the material stiffness matrix, $[D]$, can be determined in terms of two constants (e.g., of Young's modulus or modulus of elasticity E, and Poisson's ratio, ν). We need to remember the conceptual meaning of E and ν. Both of these constants refer to a uniaxial-stress loading scenario (e.g., uniaxial tension on a bar). Then, E (which has dimensions of stress) is equal to the ratio of the applied axial stress over the corresponding axial strain, while ν (which is dimensionless) is equal to the opposite of the ratio of the transverse strain over the axial strain.

In general, a linearly elastic material will not be isotropic. A general elastic material that is not isotropic is termed *anisotropic or aelotropic* (this second term primarily occurs in older textbooks). The only requirement placed on the $[D]$ matrix of an anisotropic material is that it is symmetric.[2] Thus, we can write:

$$[D] = \begin{bmatrix} D_{11} & D_{12} & D_{13} & D_{14} & D_{15} & D_{16} \\ D_{21} = D_{12} & D_{22} & D_{23} & D_{24} & D_{25} & D_{26} \\ D_{31} = D_{13} & D_{32} = D_{23} & D_{33} & D_{34} & D_{35} & D_{36} \\ D_{41} = D_{14} & D_{42} = D_{24} & D_{43} = D_{34} & D_{44} & D_{45} & D_{46} \\ D_{51} = D_{15} & D_{52} = D_{25} & D_{53} = D_{35} & D_{54} = D_{45} & D_{55} & D_{56} \\ D_{61} = D_{16} & D_{62} = D_{26} & D_{63} = D_{36} & D_{64} = D_{46} & D_{65} = D_{56} & D_{66} \end{bmatrix} \quad (7.5.2)$$

2 This requirement stems from strain-energy principles which apply for a linearly elastic material.

Given the symmetry conditions, one ends up with the existence of 21 different constants in the stiffness matrix of a general anisotropic material. In real life, even if a material is not isotropic, it will usually be characterized by some form of *symmetry* with respect to specific planes. This means that, if we establish a new coordinate system that results from a reflection about a specific plane, then the stiffness matrix in the new coordinate system will be the same as in our original coordinate system. For example, if we have a symmetry for a single reflection about the plane normal to the y-axis, then it can be proven that the stiffness matrix will be given by the following equation:

$$[D] = \begin{bmatrix} D_{11} & D_{12} & D_{13} & 0 & 0 & D_{16} \\ D_{21}=D_{12} & D_{22} & D_{23} & 0 & 0 & D_{26} \\ D_{31}=D_{13} & D_{32}=D_{23} & D_{33} & 0 & 0 & D_{36} \\ 0 & 0 & 0 & D_{44} & D_{45} & 0 \\ 0 & 0 & 0 & D_{54}=D_{45} & D_{55} & 0 \\ D_{61}=D_{16} & D_{62}=D_{26} & D_{63}=D_{36} & 0 & 0 & D_{66} \end{bmatrix} \quad (7.5.3a)$$

Along the same lines, if the material is symmetric with respect to reflection about the plane normal to the z-axis, then we obtain:

$$[D] = \begin{bmatrix} D_{11} & D_{12} & D_{13} & D_{14} & 0 & 0 \\ D_{21}=D_{12} & D_{22} & D_{23} & D_{24} & 0 & 0 \\ D_{31}=D_{13} & D_{32}=D_{23} & D_{33} & D_{34} & 0 & 0 \\ D_{41}=D_{14} & D_{42}=D_{24} & D_{43}=D_{34} & D_{44} & 0 & 0 \\ 0 & 0 & 0 & 0 & D_{55} & D_{56} \\ 0 & 0 & 0 & 0 & D_{65}=D_{56} & D_{66} \end{bmatrix} \quad (7.5.3b)$$

Equations (7.5.3a and b) (symmetry about a single plane of reflection) characterize a so-called *monoclinic material.*

If we now count the number of different elastic constants for the case of a monoclinic material, we will find out that we have a total of 13. Further simplification is obtained if we have symmetry for reflection about three different, mutually perpendicular planes. In this case, we have an *orthotropic material,* and if the three axes of orthotropy (i.e., the axes normal to the planes of reflection) are the x, y, and z axes, then the stiffness matrix can be written as follows:

$$[D] = \begin{bmatrix} D_{11} & D_{12} & D_{13} & 0 & 0 & 0 \\ D_{21}=D_{12} & D_{22} & D_{23} & 0 & 0 & 0 \\ D_{31}=D_{13} & D_{32}=D_{23} & D_{33} & 0 & 0 & 0 \\ 0 & 0 & 0 & D_{44} & 0 & 0 \\ 0 & 0 & 0 & 0 & D_{55} & 0 \\ 0 & 0 & 0 & 0 & 0 & D_{66} \end{bmatrix} \quad \text{(orthotropic material)} \quad (7.5.4)$$

It can be verified by inspection that, for an orthotropic material, the stress-strain relation for each of the three shear components is independent of that of the other components. For example, the relation between σ_{xy} and γ_{xy} is linear (the constant of linear proportionality being D_{44}) and independent of the other stress or strain components. Additionally, the normal stresses do not depend on the shear strains, and the normal strains are not related to the shear stresses. Orthotropic elastic laws are typically used for fiber-reinforced polymer (FRP) materials or for wood.

From inspection, it can be verified that an orthotropic material includes nine different material constants. Usually, a special notation is employed for the material constants of orthotropic materials. Thus, we typically define three values of moduli of elasticity, E_1, E_2, and E_3, corresponding to uniaxial stress along each of the three axes of orthotropy, and also six "Poisson ratio" values, $\nu_{12} = \nu_{xy}$, $\nu_{13} = \nu_{xz}$, $\nu_{21} = \nu_{yx}$, $\nu_{23} = \nu_{yx}$, and $\nu_{31} = \nu_{zx}$, $\nu_{32} = \nu_{zy}$. The Poisson ratio ν_{ij} is defined as "the ratio of strain in the j-direction over the strain in the i-direction, when we apply a uniaxial stress in the i-direction." Furthermore, we define three *shear moduli*, $G_{12} = G_{xy}$, $G_{23} = G_{yz}$, and $G_{31} = G_{zx}$, which define the shear stress-strain laws for the three components of shear strain in the orthotropic coordinate system.

The following relations apply for the various constants of an orthotropic material.

$$\frac{\nu_{21}}{E_2} = \frac{\nu_{12}}{E_1}, \quad \frac{\nu_{31}}{E_3} = \frac{\nu_{13}}{E_1}, \quad \frac{\nu_{23}}{E_2} = \frac{\nu_{32}}{E_3} \tag{7.5.5}$$

This means that, after we define the three moduli of elasticity and three out of six "Poisson ratios," we can calculate the values of the remaining three Poisson's ratios using Equation (7.5.5). The values of the material constants can be obtained from experimental material tests. At least six different tests are required: three tests with uniaxial stress applied along each of the three axes of orthotropy, and three simple-shear tests (the latter will give the values of the three shear moduli).

The material stiffness matrix of an orthotropic material can now be written as:

$$[D] = \begin{bmatrix} \frac{1-\nu_{23}\nu_{32}}{E_2E_3D_1} & \frac{\nu_{21}+\nu_{31}\nu_{23}}{E_2E_3D_1} & \frac{\nu_{31}+\nu_{21}\nu_{32}}{E_2E_3D_1} & 0 & 0 & 0 \\ \frac{\nu_{21}+\nu_{31}\nu_{23}}{E_2E_3D_1} & \frac{1-\nu_{13}\nu_{31}}{E_1E_3D_1} & \frac{\nu_{32}+\nu_{12}\nu_{31}}{E_1E_3D_1} & 0 & 0 & 0 \\ \frac{\nu_{31}+\nu_{21}\nu_{32}}{E_2E_3D_1} & \frac{\nu_{32}+\nu_{12}\nu_{31}}{E_1E_3D_1} & \frac{1-\nu_{13}\nu_{31}}{E_1E_3D_1} & 0 & 0 & 0 \\ 0 & 0 & 0 & G_{12} & 0 & 0 \\ 0 & 0 & 0 & 0 & G_{23} & 0 \\ 0 & 0 & 0 & 0 & 0 & G_{31} \end{bmatrix} \text{(orthotropic material)} \tag{7.5.6}$$

where

$$D_1 = \frac{1-\nu_{12}\nu_{21}-\nu_{23}\nu_{32}-\nu_{13}\nu_{31}-2\nu_{21}\nu_{32}\nu_{13}}{E_1E_2E_3} \tag{7.5.7}$$

Another type of material symmetry is *transverse isotropy*, which means that we have symmetry for reflection with respect to any plane, which is perpendicular to another, given plane. This latter plane is called the *plane of isotropy*. For example, if we have a transversely isotropic material with the transverse isotropy corresponding to the xy-plane, we have:

$$[D]=\begin{bmatrix} D_{11} & D_{12} & D_{13} & 0 & 0 & 0 \\ D_{21}=D_{12} & D_{22}=D_{11} & D_{23}=D_{13} & 0 & 0 & 0 \\ D_{31}=D_{13} & D_{32}=D_{23} & D_{33} & 0 & 0 & 0 \\ 0 & 0 & 0 & D_{44}=\dfrac{D_{11}-D_{12}}{2} & 0 & 0 \\ 0 & 0 & 0 & 0 & D_{55} & 0 \\ 0 & 0 & 0 & 0 & 0 & D_{66}=D_{55} \end{bmatrix} \tag{7.5.8a}$$

The stiffness matrix of Equation (7.5.8a) can also be written in the following, alternative form, which is obtained from Equation (7.5.6) after setting $E_2 = E_1$, $\nu_{12} = \nu_{21}$, $\nu_{23} = \nu_{13}$, $\nu_{31} = \nu_{32}$, $G_{23} = G_{31}$:

$$[D]=\begin{bmatrix} \dfrac{1-\nu_{13}\cdot\nu_{31}}{E_1E_3D_1} & \dfrac{\nu_{12}+\nu_{13}\cdot\nu_{31}}{E_1E_3D_1} & \dfrac{\nu_{31}+\nu_{12}\nu_{13}}{E_1E_3D_1} & 0 & 0 & 0 \\ \dfrac{\nu_{12}+\nu_{13}\cdot\nu_{31}}{E_1E_3D_1} & \dfrac{1-\nu_{13}\cdot\nu_{31}}{E_1E_3D_1} & \dfrac{\nu_{31}+\nu_{12}\nu_{31}}{E_1E_3D_1} & 0 & 0 & 0 \\ \dfrac{\nu_{31}+\nu_{12}\nu_{13}}{E_1E_3D_1} & \dfrac{\nu_{31}+\nu_{12}\nu_{31}}{E_1E_3D_1} & \dfrac{1-\nu_{13}\cdot\nu_{31}}{E_1E_3D_1} & 0 & 0 & 0 \\ 0 & 0 & 0 & G_{12}=\dfrac{E_1}{2(1+\nu_{12})} & 0 & 0 \\ 0 & 0 & 0 & 0 & G_{31} & 0 \\ 0 & 0 & 0 & 0 & 0 & G_{31} \end{bmatrix} \tag{7.5.8b}$$

where

$$D_1=\frac{1-(\nu_{12})^2-2\cdot\nu_{13}\cdot\nu_{31}-2\nu_{12}\cdot\nu_{13}\cdot\nu_{31}}{(E_1)^2E_3} \tag{7.5.9a}$$

and

$$\frac{\nu_{31}}{E_3}=\frac{\nu_{13}}{E_1} \tag{7.5.9b}$$

One can verify that, for a transversely isotropic material, we have a total of 5 independent material parameters to specify, namely, E_1, E_3, ν_{12}, G_{31} and ν_{31} (we can then find ν_{13} from Equation 7.5.9b).

Materials that can be considered as transversely isotropic are the so-called unidirectional FRP composites, where a set of parallel fibers made of a strong material (e.g., glass, Kevlar or carbon) is embedded inside a matrix material (e.g., epoxy resin). During fabrication, the fibers are laid in the wet matrix. After a curing period, the matrix sets and the result is a composite material with very high resistance along the direction of the fibers, and not-so-high resistance (but not necessarily low) along the directions perpendicular to the fibers. The isotropy plane for such materials is the one perpendicular to the axis of the fibers. Of course, FRP composites may have fibers aligned with more than a single direction, at which case an orthotropic material law may be more appropriate.

Finally, we can have the aforementioned case of *isotropic material*, for which we assume to have symmetry about all possible planes! In such case, the material stiffness matrix is completely defined in terms of two parameters, and can be written as follows:

$$[D] = \begin{bmatrix} \lambda+2\mu & \lambda & \lambda & 0 & 0 & 0 \\ \lambda & \lambda+2\mu & \lambda & 0 & 0 & 0 \\ \lambda & \lambda & \lambda+2\mu & 0 & 0 & 0 \\ 0 & 0 & 0 & \mu & 0 & 0 \\ 0 & 0 & 0 & 0 & \mu & 0 \\ 0 & 0 & 0 & 0 & 0 & \mu \end{bmatrix} \tag{7.5.10a}$$

where λ, μ are material constants called *Lamé's constants*. In fact, the constant μ is the well-known *shear modulus* of a linear elastic, isotropic material: $\mu = G$.

Of course, as mentioned above, the stiffness matrix of isotropic elastic materials can be written in terms of the more familiar parameters E (Young's modulus) and ν (Poisson ratio):

$$[D] = \frac{E}{(1+\nu)(1-2\nu)} \begin{bmatrix} 1-\nu & \nu & \nu & 0 & 0 & 0 \\ \nu & 1-\nu & \nu & 0 & 0 & 0 \\ \nu & \nu & 1-\nu & 0 & 0 & 0 \\ 0 & 0 & 0 & \frac{1-2\nu}{2} & 0 & 0 \\ 0 & 0 & 0 & 0 & \frac{1-2\nu}{2} & 0 \\ 0 & 0 & 0 & 0 & 0 & \frac{1-2\nu}{2} \end{bmatrix} \tag{7.5.10b}$$

Equations (7.5.10a, b) allow us to deduce that the shear modulus of an isotropic, linearly elastic material, is $G = \frac{E}{2(1+\nu)}$.

Remark 7.5.1: At this stage, we need to mention that the stiffness matrix is in fact a *fourth-order tensor*. Since the stress is a second-order tensor and the strain is also a second-order tensor, then the correct stress-strain law must be written as:

$$[\sigma] = [D] : [\varepsilon]$$

or, in component form, we can write:

$$\sigma_{ij} = D_{ijkl} \cdot \varepsilon_{kl}$$

Where, in accordance with the summation convention described in Remark 4.2.3, we imply summation for the repeated indices k, l.

Of course, we can use the Voigt notation to establish a correspondence between the matrix form of $[D]$ and the fourth-order tensor form of the same entity. Specifically, we can write the following rule:

If the a component of $\{\sigma\}$ corresponds to the stress component σ_{ij}, and the b component of $\{\varepsilon\}$ corresponds to the kl component of the strain tensor, then $D_{ijkl} = D_{ab}$. Along the same lines, if we know the components of the fourth order tensor, we can establish the corresponding components of the material stiffness matrix with Voigt notation. ∎

Remark 7.5.2: The finite element equations that we will encounter for multiaxial linear elasticity will apply for all the types of material symmetry and for anisotropic materials. This will be the case because we will be relying on the generic expression $\{\sigma\} = [D]\{\varepsilon\}$, which, of course, is valid for all types of linearly elastic materials described in this section. ∎

7.6 Coordinate Transformation Rules for Stress, Strain, and Material Stiffness Matrix

We saw in Section 4.1 that the components of a vector refer to a given coordinate system that we use. The same applies for the components of the stress and strain tensors. If we decide to change the coordinate system that we use, then the values of stress and strain tensor components must also change, in accordance with specific *coordinate transformation rules* for second-order tensors. Let us imagine that we use a Cartesian coordinate system with unit basis vectors $\vec{e}_1$, $\vec{e}_2$, and $\vec{e}_3$. We know the value of all the components of the stress tensor, $[\sigma]$, and the strain tensor, $[\varepsilon]$, in this coordinate system. Now, we want to switch to a new Cartesian coordinate system with unit basis vectors $\vec{e}'_1$, $\vec{e}'_2$, and $\vec{e}'_3$. If we have a vector $\{x\}$ expressed in the original coordinate system, the components of the same vector $\{x'\}$ in the new coordinate system are obtained using the following transformation equation.

$$\{x'\} = [R]\{x\} \tag{7.6.1}$$

where $[R]$ is a coordinate transformation tensor, expressing the modification when we switch from the system with the original basis vectors to the system with the new basis vectors. For three-dimensional problems, $[R]$ is a (3×3) array, and its components are given from the following equation.

$$[R] = \begin{bmatrix} R_{11} & R_{12} & R_{13} \\ R_{21} & R_{22} & R_{23} \\ R_{31} & R_{32} & R_{33} \end{bmatrix} = \begin{bmatrix} \vec{e}'_1 \bullet \vec{e}_1 & \vec{e}'_1 \bullet \vec{e}_2 & \vec{e}'_1 \bullet \vec{e}_3 \\ \vec{e}'_2 \bullet \vec{e}_1 & \vec{e}'_2 \bullet \vec{e}_2 & \vec{e}'_2 \bullet \vec{e}_3 \\ \vec{e}'_3 \bullet \vec{e}_1 & \vec{e}'_3 \bullet \vec{e}_2 & \vec{e}'_3 \bullet \vec{e}_3 \end{bmatrix} \tag{7.6.2}$$

We can concisely write the following expression:

$$R_{ij} = \vec{e}'_i \bullet \vec{e}_j \tag{7.6.3}$$

The components of the inverse of the transformation matrix, $[R]^{-1}$, are given by:

$$R_{ij}^{-1} = \vec{e}_i \bullet \vec{e}'_j \tag{7.6.4}$$

Remark 7.6.1: It shouldn't take long to realize that $R_{ij}^{-1} = \vec{e}_i \bullet \vec{e}'_j = \vec{e}'_j \bullet \vec{e}_i = R_{ji}$, thus:

$$[R]^{-1} = [R]^T \tag{7.6.5}$$

In other words, the coordinate transformation array $[R]$ is *orthogonal.* ∎

The coordinate transformation expression for a vector can also be written in component form:

$$v'_i = \sum_{j=1}^{3} \left(R_{ij} \cdot v_j\right) = R_{i1} \cdot v_1 + R_{i2} \cdot v_2 + R_{i3} \cdot v_3 \tag{7.6.6}$$

We will now present the transformation rules for second-order tensors. The components of the stress tensor expressed in the new coordinate system are obtained from the following expression.

$$[\sigma'] = [R][\sigma][R]^T \tag{7.6.7}$$

We can also write a component-based expression:

$$\sigma'_{ij} = \sum_{k=1}^{3} \sum_{l=1}^{3} \left(R_{ik} \cdot \sigma_{kl} \cdot R_{jl}\right) \tag{7.6.8}$$

If we rely on the summation convention defined in Remark 4.2.3 (indices that appear exactly twice imply summation), we can write a more concise version of Equation (7.6.8):

$$\sigma'_{ij} = R_{ik} \cdot \sigma_{kl} \cdot R_{jl}$$

A similar coordinate transformation rule can be established for the strain tensor; that is:

$$[\varepsilon'] = [R][\varepsilon][R]^T \tag{7.6.9}$$

Remark 7.6.2: For a second-order tensor, there is a coordinate system in which the shear components (i.e., the off-diagonal components) of the tensor vanish, and thus the tensor becomes a diagonal matrix. The directions of the unit vectors defining the coordinate system in which the off-diagonal terms vanish are called the *principal directions* of the tensor. The principal direction vectors define the so-called *principal system* of the tensor. The three diagonal components of the tensor in the principal coordinate system are called the *principal values of the tensor*. It can be proven that the three principal values of the tensor are the eigenvalues of the array representing the tensor (a discussion of eigenvalues and eigenvectors for matrix algebra is provided in Section A.3). Each of the three eigenvectors of the matrix representing the tensor will be parallel to the corresponding principal direction. The 3×3 stress tensor $[\sigma]$ can be written as:

$$[\sigma] = [R][\hat{\sigma}][R]^T \tag{7.6.10}$$

where $[\hat{\sigma}] = \begin{bmatrix} \hat{\sigma}_1 & 0 & 0 \\ 0 & \hat{\sigma}_2 & 0 \\ 0 & 0 & \hat{\sigma}_3 \end{bmatrix}$ is called the *principal stress tensor,* which is a diagonal array containing the three principal values and $[R] = [\{\varphi_1\} \ \{\varphi_2\} \ \{\varphi_3\}]$ is an orthogonal matrix, each column of which is the corresponding eigenvector of the array $[\sigma]$, normalized so that its magnitude is equal to 1. The principal stresses $\hat{\sigma}_i$ and vectors $\{\varphi_i\}$, $i = 1, 2, 3$ can be found by solving the following eigenvalue problem.

$$([\sigma] - \hat{\sigma}_i[I])\{\varphi_i\} = \{0\} \ , i = 1,2,3 \tag{7.6.11}$$

The following equation is also satisfied:

$$[\hat{\sigma}] = [R]^T[\sigma][R] \tag{7.6.12}$$

Along the same lines, we can also define the principal strain values and the corresponding principal directions of the strain tensor. ∎

As mentioned earlier, in finite element analysis, we prefer to adopt the Voigt (vector-based) notation for the stresses and strains. In this case, we can establish the following coordinate transformation rule for the stress vector, $\{\sigma\}$:

$$\{\sigma'\} = [R_\sigma]\{\sigma\} \tag{7.6.13}$$

where

$$[R_\sigma] = \begin{bmatrix} R_{11}^2 & R_{12}^2 & R_{13}^2 & 2R_{11}\cdot R_{12} & 2R_{12}\cdot R_{13} & 2R_{11}\cdot R_{13} \\ R_{21}^2 & R_{22}^2 & R_{23}^2 & 2R_{21}\cdot R_{22} & 2R_{22}\cdot R_{23} & 2R_{21}\cdot R_{23} \\ R_{31}^2 & R_{32}^2 & R_{33}^2 & 2R_{31}\cdot R_{32} & 2R_{32}\cdot R_{33} & 2R_{31}\cdot R_{33} \\ R_{11}\cdot R_{21} & R_{12}\cdot R_{22} & R_{13}\cdot R_{23} & R_{11}\cdot R_{22}+R_{12}\cdot R_{21} & R_{12}\cdot R_{23}+R_{13}\cdot R_{22} & R_{11}\cdot R_{23}+R_{13}\cdot R_{21} \\ R_{21}\cdot R_{31} & R_{22}\cdot R_{32} & R_{23}\cdot R_{33} & R_{21}\cdot R_{32}+R_{22}\cdot R_{31} & R_{22}\cdot R_{33}+R_{23}\cdot R_{32} & R_{21}\cdot R_{33}+R_{23}\cdot R_{31} \\ R_{11}\cdot R_{31} & R_{12}\cdot R_{32} & R_{13}\cdot R_{33} & R_{11}\cdot R_{32}+R_{12}\cdot R_{31} & R_{12}\cdot R_{33}+R_{13}\cdot R_{32} & R_{11}\cdot R_{33}+R_{13}\cdot R_{31} \end{bmatrix} \tag{7.6.14}$$

Along the same lines, we can establish a coordinate transformation rule for the strain vector:

$$\{\varepsilon'\} = [R_\varepsilon]\{\varepsilon\} \tag{7.6.15}$$

where

$$[R_\varepsilon] = \begin{bmatrix} R_{11}^2 & R_{12}^2 & R_{13}^2 & R_{11}\cdot R_{12} & R_{12}\cdot R_{13} & R_{11}\cdot R_{13} \\ R_{21}^2 & R_{22}^2 & R_{23}^2 & R_{21}\cdot R_{22} & R_{22}\cdot R_{23} & R_{21}\cdot R_{23} \\ R_{31}^2 & R_{32}^2 & R_{33}^2 & R_{31}\cdot R_{32} & R_{32}\cdot R_{33} & R_{31}\cdot R_{33} \\ 2R_{11}\cdot R_{21} & 2R_{12}\cdot R_{22} & 2R_{13}\cdot R_{23} & R_{11}\cdot R_{22}+R_{12}\cdot R_{21} & R_{12}\cdot R_{23}+R_{13}\cdot R_{22} & R_{11}\cdot R_{23}+R_{13}\cdot R_{21} \\ 2R_{21}\cdot R_{31} & 2R_{22}\cdot R_{32} & 2R_{23}\cdot R_{33} & R_{21}\cdot R_{32}+R_{22}\cdot R_{31} & R_{22}\cdot R_{33}+R_{23}\cdot R_{32} & R_{21}\cdot R_{33}+R_{23}\cdot R_{31} \\ 2R_{11}\cdot R_{31} & 2R_{12}\cdot R_{32} & 2R_{13}\cdot R_{33} & R_{11}\cdot R_{32}+R_{12}\cdot R_{31} & R_{12}\cdot R_{33}+R_{13}\cdot R_{32} & R_{11}\cdot R_{33}+R_{13}\cdot R_{31} \end{bmatrix} \tag{7.6.16}$$

It can be found that the following two equations apply:

$$[R_\sigma]^{-1} = [R_\varepsilon]^T \tag{7.6.17a}$$

$$[R_\varepsilon]^{-1} = [R_\sigma]^T \tag{7.6.17b}$$

Remark 7.6.3: The matrix expressing the coordinate transformation rules is not the same for stresses and strains; this discrepancy is caused by the fact that the shear strains are multiplied by 2 in the vector representation of the strains. ∎

Since the stress and strain components depend on the coordinate system that we use, the same will generally apply for the components of the material stiffness matrix, $[D]$. Specifically, the following equation can be used for the transformation of $[D]$:

$$[D'] = [R_\sigma][D][R_\varepsilon]^{-1} = [R_\sigma][D][R_\sigma]^T \tag{7.6.18}$$

Remark 7.6.4: The proof of Equation (7.6.18) is straightforward. We have: $\{\sigma\} = [D]\{\varepsilon\}$. By inverting the stress and strain transformation relations of Equations (7.6.13) and (7.6.15), we have: $\{\sigma\} = [R_\sigma]^{-1}\{\sigma'\}$ and $\{\varepsilon\} = [R_\varepsilon]^{-1}\{\varepsilon'\}$. Plugging these equations into the stress-strain law, we obtain:

$[R_\sigma]^{-1}\{\sigma'\} = [D][R_\varepsilon]^{-1}\{\varepsilon'\}$. If we premultiply both sides of the equation by $[R_\sigma]$, we will obtain the final relation: $\{\sigma'\} = [R_\sigma][D][R_\varepsilon]^{-1}\{\varepsilon'\}$. Thus, we can set $\{\sigma'\} = [D']\{\varepsilon'\}$, where $[D'] = [R_\sigma][D][R_\varepsilon]^{-1}$. ∎

7.7 Stress, Strain, and Constitutive Models for Two-Dimensional (Planar) Elasticity

A significant number of problems in solid mechanics and elasticity allow us to only consider two dimensions, x and y. In such cases, the displacement vector field only has two components, $u_x(x, y)$ and $u_y(x, y)$, and it suffices to focus on the corresponding *planar strain* and *stress tensors*. The symmetric planar stress and strain tensors contain the following components:

$$[\sigma] = \begin{bmatrix} \sigma_{xx} & \sigma_{xy} \\ \sigma_{yx} & \sigma_{yy} \end{bmatrix} \tag{7.7.1}$$

$$[\varepsilon] = \begin{bmatrix} \varepsilon_{xx} & \varepsilon_{xy} \\ \varepsilon_{yx} & \varepsilon_{yy} \end{bmatrix} \tag{7.7.2}$$

We can also establish a Voigt notation for planar problems, and write the stress and strain as column vectors:

$$\{\sigma\} = \begin{Bmatrix} \sigma_{xx} \\ \sigma_{yy} \\ \sigma_{xy} \end{Bmatrix}, \quad \{\varepsilon\} = \begin{Bmatrix} \varepsilon_{xx} \\ \varepsilon_{yy} \\ \gamma_{xy} \end{Bmatrix}$$

We can distinguish two cases in two-dimensional solid mechanics/elasticity problems:

1) *Plane-stress problems:* In these problems, the thickness of the solid body (dimension along the z-direction) is much smaller than the in-plane dimensions. In such situations, it is reasonable to assume (based on results obtained from the three-dimensional solution) that all stress components corresponding to the z-direction—that is, σ_{zz}, σ_{zx}, and σ_{yz}—are equal to zero! Thus, the stress tensor only has three nonzero components, and we only need focus to the corresponding in-plane components of the strain tensor. Textbooks in the literature refer to thin, two-dimensional bodies as "membranes" or "disks." Example cases where we can consider plane-stress conditions are shown in Figure 7.4.
2) *Plane-strain problems:* In these problems, the thickness of the solid body (dimension along the z-direction) is much larger than the in-plane dimensions. In such situations, it is reasonable to assume (based on the solution of the three-dimensional problem) that all strain components corresponding to the z-direction—that is, ε_{zz}, γ_{zx}, and γ_{yz}—are equal to zero! Thus, the strain tensor only has three nonzero components, and we only need focus to the corresponding in-plane components of the stress tensor. An example case where we can consider plane-strain conditions is shown in Figure 7.5.

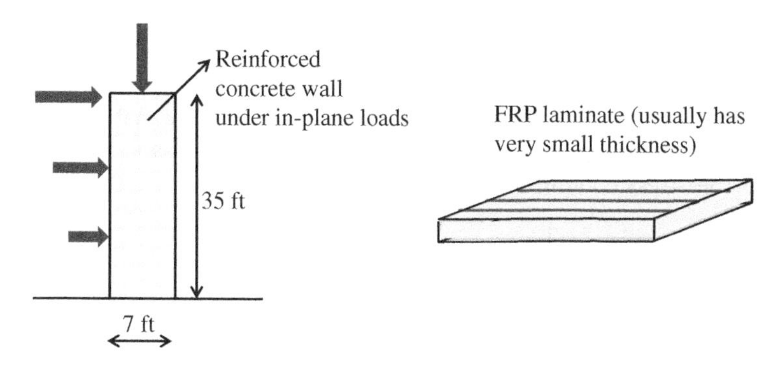

Figure 7.4 Example cases where plane-stress conditions can be assumed.

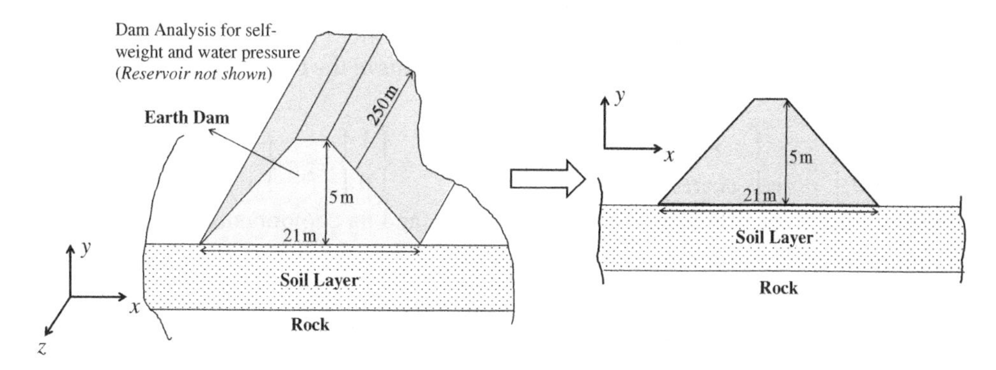

Figure 7.5 Example case where plane-strain conditions can be assumed.

The constitutive equations of isotropic linear elasticity for *plane stress* conditions can be obtained by stipulating that $\sigma_{zz} = \sigma_{zx} = \sigma_{yz} = 0$:

$$\{\sigma\} = [D]\{\varepsilon\}, \text{ where } [D] = \frac{E}{1-\nu^2}\begin{bmatrix} 1 & \nu & 0 \\ \nu & 1 & 0 \\ 0 & 0 & \frac{1-\nu}{2} \end{bmatrix} \tag{7.7.3}$$

Remark 7.7.1: For plane-stress, we generally have $\varepsilon_{zz} \neq 0$. ∎

The constitutive equations of isotropic elasticity for *plane strain* conditions can be obtained by stipulating that $\varepsilon_{zz} = \gamma_{zx} = \gamma_{yz} = 0$:

$$\{\sigma\} = [D]\{\varepsilon\}, \textit{ where } [D] = \frac{E}{(1+\nu)(1-2\nu)}\begin{bmatrix} 1-\nu & \nu & 0 \\ \nu & 1-\nu & 0 \\ 0 & 0 & \frac{1-2\nu}{2} \end{bmatrix} \tag{7.7.4}$$

Remark 7.7.2: For plane-strain, we generally have $\sigma_{zz} \neq 0$. ∎

We can also take the plane-stress and plane-strain version of the stiffness matrix, for cases other than isotropy. For example, the stiffness matrix for linearly elastic, orthotropic materials, and plane-stress conditions attains the following form.

$$[D] = \begin{bmatrix} \frac{E_x}{1-\nu_{xy}\cdot\nu_{yx}} & \frac{\nu_{yx}\cdot E_x}{1-\nu_{xy}\cdot\nu_{yx}} & 0 \\ \frac{\nu_{xy}\cdot E_y}{1-\nu_{xy}\cdot\nu_{yx}} & \frac{E_y}{1-\nu_{xy}\cdot\nu_{yx}} & 0 \\ 0 & 0 & G_{xy} \end{bmatrix} \tag{7.7.5}$$

We can also specialize the coordinate transformation equations for two-dimensional problems. In this case, the only modification in the coordinate system can stem from a rotation of the two basis vectors by some angle θ, as shown in Figure 7.6. From the same figure, we can obtain the actual values of the components of the two new coordinate basis vectors, $\vec{e}'_1$ and $\vec{e}'_2$:

$\vec{e}'_1 = \begin{Bmatrix} \cos\theta \\ \sin\theta \end{Bmatrix}, \vec{e}'_2 = \begin{Bmatrix} -\sin\theta \\ \cos\theta \end{Bmatrix}$. We also know that $\vec{e}_1 = \begin{Bmatrix} 1 \\ 0 \end{Bmatrix}, \vec{e}_2 = \begin{Bmatrix} 0 \\ 1 \end{Bmatrix}$. In this case, the coordinate transformation array is a (2×2) matrix, and its components are given from the following equation.

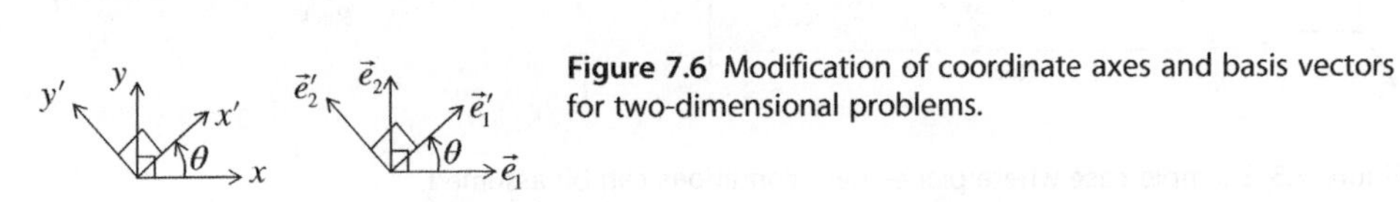

Figure 7.6 Modification of coordinate axes and basis vectors for two-dimensional problems.

$$[R]=\begin{bmatrix}R_{11} & R_{12}\\ R_{21} & R_{22}\end{bmatrix}=\begin{bmatrix}\vec{e}'_1\bullet\vec{e}_1 & \vec{e}'_1\bullet\vec{e}_2\\ \vec{e}'_2\bullet\vec{e}_1 & \vec{e}'_2\bullet\vec{e}_2\end{bmatrix}=\begin{bmatrix}\cos\theta & \sin\theta\\ -\sin\theta & \cos\theta\end{bmatrix} \tag{7.7.6}$$

We can establish transformation equations for two-dimensional, second-order tensors, too. For example, to find the components of the two-dimensional stress tensor in the new (rotated) two-dimensional coordinate system, we have the following expressions.

$[\sigma']=\begin{bmatrix}\sigma'_{xx} & \sigma'_{xy}\\ \sigma'_{yx} & \sigma'_{yy}\end{bmatrix}$, where

$$\sigma'_{xx}=\sigma_{xx}\cdot\cos^2\theta+\sigma_{yy}\cdot\sin^2\theta+\sigma_{xy}\cdot 2\cos\theta\cdot\sin\theta \tag{7.7.7a}$$

$$\sigma'_{yy}=\sigma_{xx}\cdot\sin^2\theta+\sigma_{yy}\cdot\cos^2\theta-\sigma_{xy}\cdot 2\cos\theta\cdot\sin\theta \tag{7.7.7b}$$

$$\sigma'_{xy}=-\sigma_{xx}\cdot\cos\theta\cdot\sin\theta+\sigma_{yy}\cdot\cos\theta\cdot\sin\theta+\sigma_{xy}\cdot\left(\cos^2\theta-\sin^2\theta\right) \tag{7.7.7c}$$

If we also use the trigonometric identities:

$\cos^2\theta=\dfrac{1+\cos(2\theta)}{2}$, $\sin^2\theta=\dfrac{1-\cos(2\theta)}{2}$, $2\cos\theta\cdot\sin\theta=\sin(2\theta)$, we finally obtain:

$$\sigma'_{xx}=\frac{\sigma_{xx}+\sigma_{yy}}{2}+\left(\frac{\sigma_{xx}-\sigma_{yy}}{2}\right)\cdot\cos(2\theta)+\sigma_{xy}\cdot\sin(2\theta) \tag{7.7.8a}$$

$$\sigma'_{yy}=\frac{\sigma_{xx}+\sigma_{yy}}{2}-\left(\frac{\sigma_{xx}-\sigma_{yy}}{2}\right)\cdot\cos(2\theta)+\sigma_{xy}\cdot\sin(2\theta) \tag{7.7.8b}$$

$$\sigma'_{xy}=\left(\frac{\sigma_{yy}-\sigma_{xx}}{2}\right)\cdot\sin(2\theta)+\sigma_{xy}\cdot\cos(2\theta) \tag{7.7.8c}$$

The finally obtained expressions are the well-known coordinate transformation equations for the two-dimensional stress-tensor, commonly presented in introductory engineering mechanics textbooks.

Remark 7.7.3: We can additionally establish coordinate transformation rules in Voigt notation for planar problems. All we need to do is to keep the first, second and fourth rows and columns of the (6×6) transformation arrays that we established in Equations (7.6.14) and (7.6.16):

$$[R_\sigma]=\begin{bmatrix}R_{11}^2 & R_{12}^2 & 2R_{11}\cdot R_{12}\\ R_{21}^2 & R_{22}^2 & 2R_{21}\cdot R_{22}\\ R_{11}\cdot R_{21} & R_{12}\cdot R_{22} & R_{11}\cdot R_{22}+R_{12}\cdot R_{21}\end{bmatrix}=\begin{bmatrix}\cos^2\theta & \sin^2\theta & 2\cos\theta\cdot\sin\theta\\ \sin^2\theta & \cos^2\theta & -2\cos\theta\cdot\sin\theta\\ -\cos\theta\cdot\sin\theta & \cos\theta\cdot\sin\theta & \cos^2\theta-\sin^2\theta\end{bmatrix} \tag{7.7.9a}$$

$$[R_\varepsilon]=\begin{bmatrix}R_{11}^2 & R_{12}^2 & R_{11}\cdot R_{12}\\ R_{21}^2 & R_{22}^2 & R_{21}\cdot R_{22}\\ 2R_{11}\cdot R_{21} & 2R_{12}\cdot R_{22} & R_{11}\cdot R_{22}+R_{12}\cdot R_{21}\end{bmatrix}=\begin{bmatrix}\cos^2\theta & \sin^2\theta & \cos\theta\cdot\sin\theta\\ \sin^2\theta & \cos^2\theta & -\cos\theta\cdot\sin\theta\\ -2\cos\theta\cdot\sin\theta & 2\cos\theta\cdot\sin\theta & \cos^2\theta-\sin^2\theta\end{bmatrix} \tag{7.7.9b}$$

■

Remark 7.7.4: We can also define the principal stress and strain values for planar problems, wherein the stress and strain tensors are (2×2) matrices; this means that we will have two principal values for each tensor. In fact, there are explicit formulas, which can provide the principal values and also the transformation matrix $[R]$ defining the principal system. Specifically, the principal values of the stress tensor can be found by:

$$\hat{\sigma}_1 = \frac{\sigma_{xx} + \sigma_{yy}}{2} + \sqrt{\left(\frac{\sigma_{yy} - \sigma_{yy}}{2}\right)^2 + \left(\sigma_{xy}\right)^2} \tag{7.7.10a}$$

$$\hat{\sigma}_2 = \frac{\sigma_{xx} + \sigma_{yy}}{2} - \sqrt{\left(\frac{\sigma_{yy} - \sigma_{yy}}{2}\right)^2 + \left(\sigma_{xy}\right)^2} \tag{7.7.10b}$$

Equations (7.6.10) and (7.6.12) also apply for planar problems, if we define:

$$[\hat{\sigma}] = \begin{bmatrix} \hat{\sigma}_1 & 0 \\ 0 & \hat{\sigma}_2 \end{bmatrix} \tag{7.7.11}$$

$$[R] = \begin{bmatrix} \cos\theta_p & -\sin\theta_p \\ \sin\theta_p & \cos\theta_p \end{bmatrix} \tag{7.7.12}$$

and

$$\theta_p = \frac{1}{2} atan\left(\frac{2\sigma_{xy}}{\sigma_{xx} - \sigma_{yy}}\right) \tag{7.7.13}$$

The angle θ_p in Equation (7.7.13) gives the orientation of the planes on which the principal stresses act. Specifically, θ_p is the angle between the x-axis and the axis perpendicular to the plane on which $\hat{\sigma}_1$ acts; it is also the angle between the y-axis and the axis perpendicular to the plane on which $\hat{\sigma}_2$ acts. Along the same lines, we can establish the mathematical expressions providing the principal values of the strain tensor for planar problems. ∎

Example 7.1: Stiffness matrix for composite material
We want to find the stiffness matrix (in the x-y system) for a thin composite laminate, which consists of an isotropic epoxy resin matrix with $E_m = 500$ ksi, $\nu_m = 0.2$, and contains E-glass fibers aligned with the axes x' and y' as shown in Figure 7.7. The fiber content (i.e., the volume of fibers in each direction over the volume of the matrix) is $\rho_f = 20\%$. We are also given that the modulus of elasticity of the fibers is $E_f = 10500$ ksi and that the fibers develop a uniaxial state of stress.

We will make the assumption that there is *perfect bond* between the matrix and the fibers. This means that the strain in the matrix and in the fibers is identical, and it is equal to the strain of the composite material. This allows us to write the stiffness matrix of the composite material as the sum:

$$[D] = [D_{matrix}] + \rho_f [D_{fibers}]$$

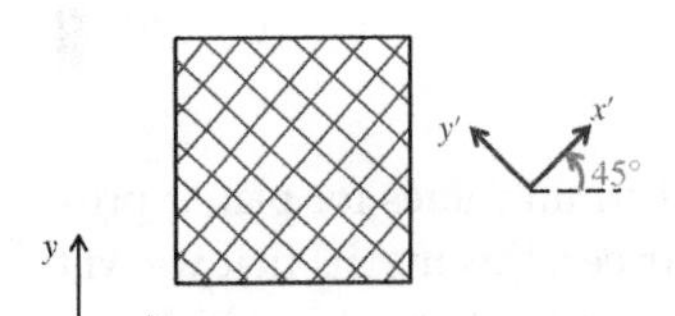

Figure 7.7 Composite material with fibers aligned at an angle with axes x and y.

where $[D_{matrix}]$ is the stiffness of the matrix alone, and $[D_{fibers}]$ is the stiffness of the fibers alone. Plane-stress conditions will be assumed, since the thickness of composite laminates is usually small, compared to the in-plane dimensions.

The stiffness matrix $[D_{matrix}]$ in the x-y system is simply given from Equation (7.7.3):

$$[D_{matrix}] = \frac{E_m}{1-\nu_m^2}\begin{bmatrix} 1 & \nu_m & 0 \\ \nu_m & 1 & 0 \\ 0 & 0 & \frac{1-\nu_m}{2} \end{bmatrix} = \frac{500}{1-0.2^2}\begin{bmatrix} 1 & 0.2 & 0 \\ 0.2 & 1 & 0 \\ 0 & 0 & \frac{1-0.2}{2} \end{bmatrix}$$

$$= \begin{bmatrix} 520.833 & 104.167 & 0 \\ 104.167 & 520.833 & 0 \\ 0 & 0 & 208.333 \end{bmatrix} ksi$$

Since the fibers are aligned with the axes x' and y', it is more convenient to first establish their stiffness matrix $[D_{fiber,x'y'}]$ in the $x'y'$ system, then transform it to the xy system. Since we have fibers in a uniaxial state of stress aligned with the x' axis, the stiffness that these fibers develop will simply be equal to the modulus of elasticity E_f. The same applies for the fibers aligned with the y' axis.

Thus, we can write:

$$\left[D_{fiber,x'y'}\right] = \begin{bmatrix} E_f & 0 & 0 \\ 0 & E_f & 0 \\ 0 & 0 & 0 \end{bmatrix} = \begin{bmatrix} 10500 & 0 & 0 \\ 0 & 10500 & 0 \\ 0 & 0 & 0 \end{bmatrix} ksi.$$

Now, we need to establish the coordinate transformation array, that takes us from the $x'y'$ system (for which we know the stiffness matrix) to the xy system (for which we want to find the stiffness matrix). We can observe from Figure 7.7 that the xy system is obtained by rotating the $x'y'$ system by -45°; thus, we need to set $\theta = -45°$ in Equation (7.7.9a):

$$[R_\sigma] = \begin{bmatrix} \cos^2(-45^o) & \sin^2(-45^o) & 2\cos(-45^o)\cdot\sin(-45^o) \\ \sin^2(-45^o) & \cos^2(-45^o) & -2\cos(-45^o)\cdot\sin(-45^o) \\ -\cos(-45^o)\cdot\sin(-45^o) & \cos(-45^o)\cdot\sin(-45^o) & \cos^2(-45^o)-\sin^2(-45^o) \end{bmatrix}$$

$$= \begin{bmatrix} 0.5 & 0.5 & -1 \\ 0.5 & 0.5 & 1 \\ 0.5 & -0.5 & 0 \end{bmatrix}$$

We can finally transform the fiber stiffness matrix to the xy system:

$$\left[D_{fiber}\right] = [R_\sigma]\left[D_{fiber,x'y'}\right][R_\sigma]^T = \begin{bmatrix} 5250 & 5250 & 0 \\ 5250 & 5250 & 0 \\ 0 & 0 & 5250 \end{bmatrix} ksi$$

It is worth noticing that, for the xy system, the fibers have a nonzero (3,3)-term in their material stiffness, which means that the fibers oriented at an angle with the x and y directions have a contribution to the shear stiffness of the composite material.

Finally, we can sum the contributions of the epoxy resin and fibers to obtain the requested composite material stiffness matrix:

$$[D] = [D_{matrix}] + \rho_f [D_{fiber}]$$

$$= \begin{bmatrix} 520.833 & 104.167 & 0 \\ 104.167 & 520.833 & 0 \\ 0 & 0 & 208.333 \end{bmatrix} ksi + 0.2 \cdot \begin{bmatrix} 5250 & 5250 & 0 \\ 5250 & 5250 & 0 \\ 0 & 0 & 5250 \end{bmatrix} ksi$$

$$= \begin{bmatrix} 1570.833 & 1154.167 & 0 \\ 1154.167 & 1570.833 & 0 \\ 0 & 0 & 1258.333 \end{bmatrix} ksi$$

■

7.8 Strong Form for Two-Dimensional Elasticity

Having established the definitions of the previous sections, let us examine the problem of Figure 7.8, where a planar body is subjected to in-plane body forces, $\{b\} = [b_x, b_y]^T$, and also to in-plane boundary tractions $\{t\} = [t_x, t_y]^T$ along a part of the boundary, called Γ_t (the traction boundary or natural boundary). The displacement vector, $\{u\} = [u_x, u_y]^T$, is assumed to be restrained to a given value along a part of the boundary, called Γ_u (geometric or essential boundary).

Each part of the boundary, Γ, belongs to either Γ_u or Γ_t (in mathematical set theory, this statement is made by stipulating that the union of Γ_u and Γ_t is equal to Γ, $\Gamma_u \cup \Gamma_t = \Gamma$). Also, at each portion of the boundary, we cannot simultaneously have a prescribed displacement in a direction and a traction in the same direction (in mathematical set theory, this statement is made by stipulating that the intersection of Γ_u and Γ_t is a null or empty set, $\Gamma_u \cap \Gamma_t = \emptyset$).

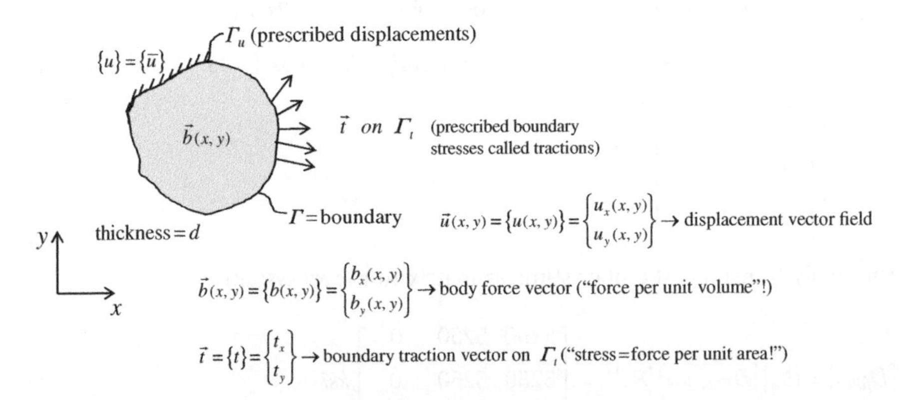

Figure 7.8 Summary of two-dimensional elasticity problem.

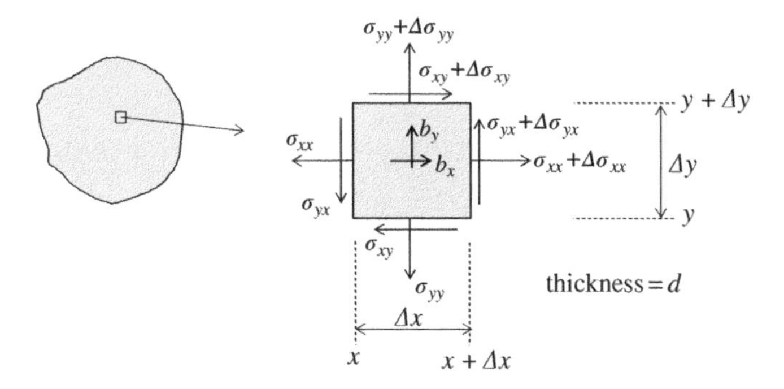

Figure 7.9 Equilibrium of a small piece of material cut from the interior of a two-dimensional body.

Let us also assume that the body has a thickness equal to d. For plane-stress problems, d is the actual thickness of the two-dimensional solid. For plane-strain problems, we can imagine that we isolate a "slice" with thickness equal to d and we examine its behavior. The governing differential equations can be obtained if we isolate a small portion of the body, with dimensions Δx, Δy, as shown in Figure 7.9, and formulate the pertinent equilibrium equations.

Specifically, we need to consider force equilibrium along each one of the two directions, x and y. We begin with direction x:

$$\sum F_x = 0 \rightarrow (\sigma_{xx} + \Delta\sigma_{xx}) \cdot (\Delta y \cdot d) + (\sigma_{xy} + \Delta\sigma_{xy}) \cdot (\Delta x \cdot d) - \sigma_{xx} \cdot (\Delta y \cdot d)$$

$$- \sigma_{xy} \cdot (\Delta x \cdot d) + b_x \cdot (\Delta x \cdot \Delta y \cdot d) = 0$$

$$\rightarrow \Delta\sigma_{xx} \cdot (\Delta y \cdot d) + \Delta\sigma_{xy} \cdot (\Delta x \cdot d) + b_x \cdot (\Delta x \cdot \Delta y \cdot d) = 0$$

$$\rightarrow \frac{\Delta\sigma_{xx}}{\Delta x} + \frac{\Delta\sigma_{xy}}{\Delta y} + b_x = 0$$

If we take the limit as the size of the piece of material tends to zero, that is, $\lim_{\substack{\Delta x \rightarrow 0, \\ \Delta y \rightarrow 0}}$ we obtain the *differential equation of equilibrium in the x-direction*:

$$\frac{\partial\sigma_{xx}}{\partial x} + \frac{\partial\sigma_{xy}}{\partial y} + b_x = 0 \tag{7.8.1a}$$

Along the same lines, we can consider the equilibrium along the y-direction:

$$\sum F_y = 0 \rightarrow (\sigma_{yx} + \Delta\sigma_{yx}) \cdot (\Delta y \cdot d) + (\sigma_{yy} + \Delta\sigma_{yy}) \cdot (\Delta x \cdot d)$$

$$- \sigma_{yx} \cdot (\Delta y \cdot d) - \sigma_{yy} \cdot (\Delta x \cdot d) + b_y \cdot (\Delta x \cdot \Delta y \cdot d) = 0$$

$$\rightarrow \Delta\sigma_{yx} \cdot (\Delta y \cdot d) + \Delta\sigma_{yy} \cdot (\Delta x \cdot d) + b_y \cdot (\Delta x \cdot \Delta y \cdot d) = 0$$

$$\rightarrow \frac{\Delta\sigma_{yx}}{\Delta x} + \frac{\Delta\sigma_{yy}}{\Delta y} + b_y = 0$$

If we take the limit as the size of the piece of material tends to zero, that is, $\lim_{\substack{\Delta x \to 0 \\ \Delta y \to 0}}$, we obtain the *differential equation of equilibrium in the y-direction*:

$$\frac{\partial \sigma_{yx}}{\partial x} + \frac{\partial \sigma_{yy}}{\partial y} + b_y = 0 \tag{7.8.1b}$$

Equations (7.8.1a) and (7.8.1b) give the system of partial differential equations of equilibrium, which can be collectively written as follows.

$$\sum_{j=1}^{2} \left(\frac{\partial \sigma_{ij}}{\partial x_j} \right) + b_i = 0, \ i = 1,2 \tag{7.8.2}$$

If we use the summation convention of Remark 4.2.3, we can also write:

$$\frac{\partial \sigma_{ij}}{\partial x_j} + b_i = 0, \ (\text{"}j\text{" is repeated once in the first term, so we imply summation with } j) \tag{7.8.3}$$

To see how to cast the natural (traction) boundary conditions in mathematical form, we isolate a *small* inclined segment of the natural boundary, with length equal to ΔL, and a unit normal outward vector equal to $\{n\}$, and establish the equilibrium of a small piece of material in the vicinity of the boundary segment, as shown in Figure 7.10. Since the volume of the segment is small (we will set $\Delta L \to 0$ in the end), the contribution of the body forces to the equilibrium equations is negligible and vanishes!

We have the following equilibrium equations.

$$\sum F_x = 0 \to -\sigma_{xx} \cdot d \cdot \Delta y - \sigma_{xy} \cdot d \cdot \Delta x + t_x \cdot d \cdot \Delta L = 0$$
$$\to -\sigma_{xx} \cdot d \cdot \Delta L \cdot n_x - \sigma_{xy} \cdot d \cdot \Delta L \cdot n_y + t_x \cdot d \cdot \Delta L = 0 \to$$

$$\boxed{\sigma_{xx} \cdot n_x + \sigma_{xy} \cdot n_y = t_x} \tag{7.8.4a}$$

$$\sum F_y = 0 \to -\sigma_{yy} \cdot d \cdot \Delta x - \sigma_{yx} \cdot d \cdot \Delta y + t_y \cdot d \cdot \Delta L = 0$$
$$\to \sigma_{yx} \cdot d \cdot \Delta L \cdot n_x + \sigma_{yy} \cdot d \cdot \Delta L \cdot n_y = t_y \cdot d \cdot \Delta L \to$$

$$\boxed{\sigma_{yx} \cdot n_x + \sigma_{yy} \cdot n_y = t_y} \tag{7.8.4b}$$

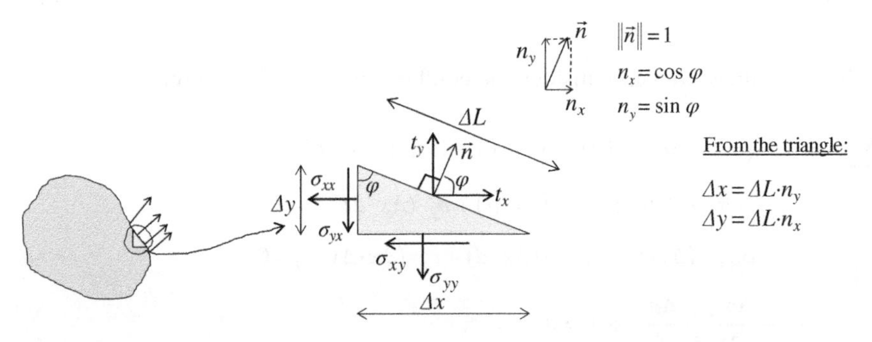

Figure 7.10 Equilibrium of a small piece of material taken from the natural boundary.

If we combine Equations (7.8.4a,b), we obtain:

$$[\sigma]\vec{n} = \vec{t} \tag{7.8.5}$$

Equation (7.8.5) is independent of ΔL, and thus it is valid for an infinitesimal boundary length, too, that is, if we set $\Delta L \to 0$.

We can now combine the expressions for the governing differential equations and boundary conditions to obtain the strong form for two-dimensional solid mechanics:

$$\frac{\partial \sigma_{xx}}{\partial x} + \frac{\partial \sigma_{xy}}{\partial y} + b_x = 0$$

$$\frac{\partial \sigma_{yx}}{\partial x} + \frac{\partial \sigma_{yy}}{\partial y} + b_y = 0$$

$\vec{u} = \{\bar{u}\}$ $on\, \Gamma_u$ (essential boundary conditions)

$[\sigma]\vec{n} = \vec{t}$ $on\, \Gamma_t$ (natural boundary conditions)

In general, the natural boundary and essential boundary can be different for the x- and y-directions. For example, at a given location, of the boundary, we may have a prescribed displacement u_x and a prescribed traction t_y. Thus, we can establish an essential and natural boundary for the x-direction, Γ_{ux} and Γ_{tx}, respectively, and also an essential and natural boundary for the y-direction, Γ_{uy} and Γ_{ty}, respectively. In this case, the strong form for two-dimensional solid mechanics problems is described by the expressions in Box 7.8.1.

Remark 7.8.1: The only condition in Box 7.8.1 that corresponds to the special case of linearly elastic material is at the very last line of the Box. If we remove this last line, then the strong form in Box 7.8.1 is valid for any type of material behavior. ∎

Box 7.8.1 Strong Form for Two-Dimensional Solid Mechanics

$$\frac{\partial \sigma_{xx}}{\partial x} + \frac{\partial \sigma_{xy}}{\partial y} + b_x = 0 \tag{7.8.1a}$$

$$\frac{\partial \sigma_{yx}}{\partial x} + \frac{\partial \sigma_{yy}}{\partial y} + b_y = 0 \tag{7.8.1b}$$

$u_x = \bar{u}_x$ $on\, \Gamma_{ux}$, $u_y = \bar{u}_y$ $on\, \Gamma_{uy}$ (essential boundary conditions)

$\vec{\sigma}_x \bullet \vec{n} = t_x$ $on\, \Gamma_{tx}$, $\vec{\sigma}_y \bullet \vec{n} = t_y$ $on\, \Gamma_{ty}$ (natural boundary conditions)

where $\vec{\sigma}_x = \begin{Bmatrix} \sigma_{xx} \\ \sigma_{xy} \end{Bmatrix}$, $\vec{\sigma}_y = \begin{Bmatrix} \sigma_{yx} \\ \sigma_{yy} \end{Bmatrix}$

$$\Gamma_{ux} \cup \Gamma_{tx} = \Gamma_{uy} \cup \Gamma_{ty} = \Gamma,\ \Gamma_{ux} \cap \Gamma_{tx} = \Gamma_{uy} \cap \Gamma_{ty} = \emptyset$$

for the special case of linear elasticity: $\{\sigma\} = [D]\{\varepsilon\}$ (constitutive equation)

Remark 7.8.2: We can actually use the constitutive equations and obtain a strong form where the differential equation only includes the components of the displacement vector field (and not stress components). For the case of an isotropic linearly elastic material, the differential equations in terms of the displacement field are referred to as *Navier's equations* of linear elasticity. ∎

7.9 Weak Form for Two-Dimensional Elasticity

Having established the strong form for two-dimensional elasticity in Box 7.8.1, we will now proceed to obtain the corresponding weak form. We consider an arbitrary *vector* field, $\vec{w}(x,y) = \{w(x,y)\} = \begin{Bmatrix} w_x(x,y) \\ w_y(x,y) \end{Bmatrix}$, whose components vanish at the corresponding essential boundary segments:

$$w_x = 0 \text{ on } \Gamma_{ux} \tag{7.9.1a}$$

$$w_x = 0 \text{ on } \Gamma_{uy} \tag{7.9.1b}$$

We multiply Equation (7.8.1a) in Box 7.8.1 by w_x and integrate the resulting expression over Ω:

$$\iint_\Omega w_x\left(\frac{\partial\sigma_{xx}}{\partial x} + \frac{\partial\sigma_{xy}}{\partial y} + b_x\right)dV = 0 \rightarrow \iint_\Omega w_x\left(\frac{\partial\sigma_{xx}}{\partial x} + \frac{\partial\sigma_{xy}}{\partial y}\right)dV + \iint_\Omega w_x b_x dV = 0 \rightarrow$$

$$\rightarrow \iint_\Omega w_x\left(\nabla\bullet\vec{\sigma}_x\right)dV + \iint_\Omega w_x b_x dV = 0 \tag{7.9.2}$$

The first domain integral in Equation (7.9.2) includes an expression where w_x is not differentiated, while we have derivatives of stresses, which stresses are obtained from the strains that result after differentiation of $\{u\}$. To obtain symmetric coefficient arrays after the introduction of a finite element approximation, we must have "symmetry" in terms of differentiation of the displacement field and w_x. To this end, we use Green's formula to the first integral term of Equation (7.9.2):

$$\iint_\Omega w_x\left(\nabla\bullet\vec{\sigma}_x\right)dV = \int_\Gamma w_x\vec{\sigma}_x\bullet\vec{n}\,dS - \iint_\Omega\left(\nabla w_x\right)\bullet\vec{\sigma}_x dV \tag{7.9.3}$$

The boundary integral in Equation (7.9.3) can be separated into the individual contributions of the essential boundary Γ_{ux} and the natural boundary Γ_{tx}. We have:

$$\int_\Gamma w_x\vec{\sigma}_x\bullet\vec{n}\,dS = \int_{\Gamma_{ux}} w_x\vec{\sigma}_x\bullet\vec{n}\,dS + \int_{\Gamma_{tx}} w_x\vec{\sigma}_x\bullet\vec{n}\,dS = \int_{\Gamma_{tx}} w_x\vec{\sigma}_x\bullet\vec{n}\,dS = \int_{\Gamma_{tx}} w_x\cdot t_x\,dS,$$

since $w_x = 0$ on Γ_{ux} and $\vec{\sigma}_x\bullet\vec{n} = t_x$ on Γ_{tx}!

Thus, equation (7.9.2) eventually gives:

$$\int_{\Gamma_{tx}} w_x t_x \, dS - \iint_{\Omega} (\nabla w_x) \bullet \vec{\sigma}_x dV + \iint_{\Omega} w_x b_x dV = 0 \rightarrow$$

$$\iint_{\Omega} (\nabla w_x) \bullet \vec{\sigma}_x dV = \iint_{\Omega} w_x b_x dV + \int_{\Gamma_{tx}} w_x t_x \, dS \rightarrow$$

$$\iint_{\Omega} \left(\frac{\partial w_x}{\partial x} \sigma_{xx} + \frac{\partial w_x}{\partial y} \sigma_{xy} \right) dV = \iint_{\Omega} w_x b_x dV + \int_{\Gamma_{tx}} w_x t_x \, dS \tag{7.9.4}$$

We then multiply Equation (7.8.1b) in Box 7.8.1 by w_y and integrate the resulting expression over Ω:

$$\iint_{\Omega} w_y \left(\frac{\partial \sigma_{yx}}{\partial x} + \frac{\partial \sigma_{yy}}{\partial y} + b_y \right) dV = 0 \rightarrow \iint_{\Omega} w_y \left(\frac{\partial \sigma_{yx}}{\partial x} + \frac{\partial \sigma_{yy}}{\partial y} \right) dV + \iint_{\Omega} w_y b_y dV = 0$$

$$\rightarrow \iint_{\Omega} w_y (\nabla \bullet \vec{\sigma}_y) dV + \iint_{\Omega} w_y b_y dV = 0 \tag{7.9.5}$$

If we now use Green's formula to the first integral term in Equation (7.9.5):

$$\iint_{\Omega} w_y (\nabla \bullet \vec{\sigma}_y) dV = \int_{\Gamma} w_y \vec{\sigma}_y \bullet \vec{n} \, dS - \iint_{\Omega} (\nabla w_y) \bullet \vec{\sigma}_y dV \tag{7.9.6}$$

The boundary integral in (7.9.6) becomes:

$$\int_{\Gamma} w_y \vec{\sigma}_y \bullet \vec{n} \, dS = \int_{\Gamma_{uy}} w_y \vec{\sigma}_y \bullet \vec{n} \, dS + \int_{\Gamma_{ty}} w_y \vec{\sigma}_y \bullet \vec{n} \, dS = \int_{\Gamma_{ty}} w_y \vec{\sigma}_y \bullet \vec{n} \, dS = \int_{\Gamma_{ty}} w_y \cdot t_y \, dS,$$

since $w_y = 0$ on Γ_{uy} and $\vec{\sigma}_y \bullet \vec{n} = t_y$ on Γ_{ty}!

Thus, equation (7.9.5) gives:

$$\int_{\Gamma_{ty}} w_y t_y \, dS - \iint_{\Omega} (\nabla w_y) \bullet \vec{\sigma}_y dV + \iint_{\Omega} w_y b_y dV = 0$$

$$\rightarrow \iint_{\Omega} (\nabla w_y) \bullet \vec{\sigma}_y dV = \iint_{\Omega} w_y b_y dV + \int_{\Gamma_{ty}} w_y t_y \, dS$$

$$\rightarrow \iint_{\Omega} \left(\frac{\partial w_y}{\partial x} \sigma_{yx} + \frac{\partial w_y}{\partial y} \sigma_{yy} \right) dV = \iint_{\Omega} w_y b_y dV + \int_{\Gamma_{ty}} w_y t_y \, dS \tag{7.9.7}$$

If we now sum Equations (7.9.4) and (7.9.7) we obtain:

$$\iint_{\Omega} \left(\frac{\partial w_x}{\partial x} \sigma_{xx} + \frac{\partial w_x}{\partial y} \sigma_{xy} + \frac{\partial w_y}{\partial x} \sigma_{yx} + \frac{\partial w_y}{\partial y} \sigma_{yy} \right) dV$$

$$= \iint_{\Omega} (w_x b_x + w_y b_y) dV + \int_{\Gamma_{tx}} w_x t_x \, dS + \int_{\Gamma_{ty}} w_y t_y \, dS$$

If we also account for the fact that $\sigma_{xy} = \sigma_{yx}$, we have:

$$\iint_{\Omega} \left[\frac{\partial w_x}{\partial x}\sigma_{xx} + \left(\frac{\partial w_x}{\partial y} + \frac{\partial w_y}{\partial x} \right)\sigma_{xy} + \frac{\partial w_y}{\partial y}\sigma_{yy} \right] dV$$

$$= \iint_{\Omega} \left(w_x b_x + w_y b_y \right) dV + \int_{\Gamma_{tx}} w_x t_x \, dS + \int_{\Gamma_{ty}} w_y t_y \, dS$$

Finally, if we define: $\bar{\varepsilon}_{xx} = \frac{\partial w_x}{\partial x}$, $\bar{\varepsilon}_{yy} = \frac{\partial w_y}{\partial y}$, $\bar{\gamma}_{xy} = \frac{\partial w_x}{\partial y} + \frac{\partial w_y}{\partial x}$, we have the following expression.

$$\iint_{\Omega} \{\bar{\varepsilon}\}^T \{\sigma\} dV = \iint_{\Omega} \{w\}^T \{b\} dV + \int_{\Gamma_{tx}} w_x t_x \, dS + \int_{\Gamma_{ty}} w_y t_y \, dS, \quad \text{where} \quad \{\bar{\varepsilon}\} = \begin{Bmatrix} \bar{\varepsilon}_{xx} \\ \bar{\varepsilon}_{yy} \\ \bar{\gamma}_{xy} \end{Bmatrix},$$

$$\{\sigma\} = \begin{Bmatrix} \sigma_{xx} \\ \sigma_{yy} \\ \sigma_{xy} \end{Bmatrix}$$

We also need to carry the constitutive law and the essential boundary conditions (because the strong form with which we started also includes them!), and obtain:

$$\iint_{\Omega} \{\bar{\varepsilon}\}^T \{\sigma\} dV = \iint_{\Omega} \{w\}^T \{b\} dV + \int_{\Gamma_{tx}} w_x t_x \, dS + \int_{\Gamma_{ty}} w_y t_y \, dS \tag{7.9.8}$$

$$\forall \{w(x,y)\} = \begin{Bmatrix} w_x(x,y) \\ w_y(x,y) \end{Bmatrix} \text{ with } w_x = 0 \text{ on } \Gamma_{ux},\ w_y = 0 \text{ on } \Gamma_{uy}$$

$$\{\sigma\} = [D]\{\varepsilon\}, \text{ where } \{\sigma\} = \begin{Bmatrix} \sigma_{xx} \\ \sigma_{yy} \\ \sigma_{xy} \end{Bmatrix}, \ \{\varepsilon\} = \begin{Bmatrix} \varepsilon_{xx} \\ \varepsilon_{yy} \\ \gamma_{xy} \end{Bmatrix}, \ \begin{Bmatrix} \varepsilon_{xx} = \partial u_x / \partial x \\ \varepsilon_{yy} = \partial u_y / \partial y \\ \gamma_{xy} = \partial u_x / \partial y + \partial u_y / \partial x \end{Bmatrix}$$

$$u_x = \bar{u}_x \quad on\ \Gamma_{ux}$$

$$u_y = \bar{u}_y \quad on\ \Gamma_{uy}$$

Remark 7.9.1: In Equation (7.9.8), we can identify $\{w(x, y)\}$ with an *arbitrary, kinematically admissible displacement field*; we then directly realize that Equation (7.9.8) is the *principle of virtual work* (more accurately: *the principle of virtual displacements*), where the left-hand side is the internal virtual work and the right-hand side is the external virtual work. ■

If we plug the constitutive law in Equation (7.9.8), we eventually obtain the *weak form for two-dimensional elasticity*, provided in Box (7.9.1).

Box 7.9.1 Weak Form for Two-Dimensional Elasticity

$$\iint_{\Omega} \{\bar{\varepsilon}\}^T [D] \{\varepsilon\} dV = \iint_{\Omega} \{w\}^T \{b\} dV + \int_{\Gamma_{tx}} w_x t_x \, dS + \int_{\Gamma_{ty}} w_y t_y \, dS$$

$$\forall \{w\} = \begin{Bmatrix} w_x \\ w_y \end{Bmatrix} \text{ with } w_x = 0 \text{ on } \Gamma_{ux},\ w_y = 0 \text{ on } \Gamma_{uy}$$

$$u_x = \bar{u}_x \quad on\ \Gamma_{ux}$$

$$u_y = \bar{u}_y \quad on\ \Gamma_{uy}$$

Remark 7.9.2: The weak form of Box 7.9.1 gives the virtual work principle per unit thickness of the two-dimensional solid body. If we are given the actual thickness, *d*, then the total internal and external work quantities are obtained by multiplying the integral equation in Box 7.9.1 by *d*:

$$d \cdot \iint_{\Omega} \{\bar{\varepsilon}\}^T [D] \{\varepsilon\} dV = d \cdot \iint_{\Omega} \{w\}^T \{b\} dV + d \cdot \int_{\Gamma_{tx}} w_x t_x \, dS + d \cdot \int_{\Gamma_{ty}} w_y t_y \, dS$$

■

7.10 Equivalence between the Strong Form and the Weak Form

In the previous section, we started with the strong form for two-dimensional elasticity and obtained the corresponding weak form. The proof of the equivalence between weak and strong form can be established if we start from the weak form and obtain the strong form. To do so, we once again rely on the fact that $\{w(x, y)\}$ is *arbitrary*, so the weak form must also apply for some "convenient" special cases of $\{w(x, y)\}$. Let us establish the following definitions:

$$r_x = \frac{\partial \sigma_{xx}}{\partial x} + \frac{\partial \sigma_{xy}}{\partial y} + b_x = \nabla \bullet \vec{\sigma}_x + b_x \ , \ r_{\Gamma x} = t_x - \vec{\sigma}_x \bullet \vec{n} \tag{7.10.1a}$$

$$r_y = \frac{\partial \sigma_{yx}}{\partial x} + \frac{\partial \sigma_{yy}}{\partial y} + b_y = \nabla \bullet \vec{\sigma}_y + b_y \ , \ r_{\Gamma y} = t_y - \vec{\sigma}_y \bullet \vec{n} \tag{7.10.1b}$$

where $\vec{\sigma}_x$, $\vec{\sigma}_y$ have been defined in Box 7.8.1.

We now start with the weak form, provided in Box 7.9.1, for the two-dimensional elasticity problem.

We account for the stress-strain law, $[D]\{\varepsilon\} = \{\sigma\}$, then manipulate the integral term of the left-hand side:

$$\iint_{\Omega} \{\bar{\varepsilon}\}^T \{\sigma\} dV = \iint_{\Omega} \frac{\partial w_x}{\partial x} \sigma_{xx} + \left(\frac{\partial w_x}{\partial y} + \frac{\partial w_y}{\partial x} \right) \sigma_{xy} + \frac{\partial w_y}{\partial y} \sigma_{yy} dV =$$

$$\iint_{\Omega}\left(\frac{\partial w_x}{\partial x}\sigma_{xx}+\frac{\partial w_x}{\partial y}\sigma_{xy}\right)dV+\iint_{\Omega}\left(\frac{\partial w_y}{\partial x}\sigma_{yx}+\frac{\partial w_y}{\partial y}\sigma_{yy}\right)dV$$
$$=\iint_{\Omega}(\nabla w_x)\bullet\vec{\sigma}_x dV+\iint_{\Omega}(\nabla w_y)\bullet\vec{\sigma}_y dV \qquad (7.10.2)$$

Using Green's formula for each one of the two terms in the right-hand side of Equation (7.10.2) and accounting for the fact that w_x vanishes on Γ_{ux}, w_y vanishes on Γ_{uy}, we obtain:

$$\iint_{\Omega}(\nabla w_x)\bullet\vec{\sigma}_x dV+\iint_{\Omega}(\nabla w_y)\bullet\vec{\sigma}_y dV$$
$$=\int_{\Gamma} w_x\vec{\sigma}_x\bullet\vec{n}\,dS-\iint_{\Omega} w_x\cdot(\nabla\bullet\vec{\sigma}_x)dV+\int_{\Gamma} w_y\vec{\sigma}_y\bullet\vec{n}\,dS-\iint_{\Omega} w_y\cdot(\nabla\bullet\vec{\sigma}_y)dV$$
$$=\int_{\Gamma_{ux}} w_x\vec{\sigma}_x\bullet\vec{n}\,dS+\int_{\Gamma_{tx}} w_x\vec{\sigma}_x\bullet\vec{n}\,dS+\int_{\Gamma_{uy}} w_y\vec{\sigma}_y\bullet\vec{n}\,dS+\int_{\Gamma_{ty}} w_y\vec{\sigma}_y\bullet\vec{n}\,dS$$
$$-\iint_{\Omega} w_x\cdot(\nabla\bullet\vec{\sigma}_x)dV-\iint_{\Omega} w_y\cdot(\nabla\bullet\vec{\sigma}_y)dV$$
$$=\int_{\Gamma_{tx}} w_x\vec{\sigma}_x\bullet\vec{n}\,dS+\int_{\Gamma_{ty}} w_y\vec{\sigma}_y\bullet\vec{n}\,dS-\iint_{\Omega} w_x\cdot(\nabla\bullet\vec{\sigma}_x)dV-\iint_{\Omega} w_y\cdot(\nabla\bullet\vec{\sigma}_y)dV$$

If we substitute the obtained expression in the left-hand side of Equation (7.10.2), then plug the resulting expression into Equation (7.9.8), we have:

$$\int_{\Gamma_{tx}} w_x\vec{\sigma}_x\bullet\vec{n}\,dS+\int_{\Gamma_{ty}} w_y\vec{\sigma}_y\bullet\vec{n}\,dS-\iint_{\Omega} w_x\cdot(\nabla\bullet\vec{\sigma}_x)dV-\iint_{\Omega} w_y\cdot(\nabla\bullet\vec{\sigma}_y)dV$$
$$=\iint_{\Omega}\{w\}^T\{b\}dV+\int_{\Gamma_{tx}} w_x t_x\,dS+\int_{\Gamma_{ty}} w_y t_y\,dS$$

We now group the integral terms, transfer everything to the right-hand side, and reverse the two sides of the obtained expression to obtain:

$$\iint_{\Omega} w_x\cdot[(\nabla\bullet\vec{\sigma}_x)+b_x]dV+\iint_{\Omega} w_y\cdot[(\nabla\bullet\vec{\sigma}_y+b_y)]dV+\iint_{\Omega}\{w\}^T\{b\}dV$$
$$+\int_{\Gamma_{tx}} w_x\left(t_x-\vec{\sigma}_y\bullet\vec{n}\right)dS+\int_{\Gamma_{ty}} w_y\left(t_y-\vec{\sigma}_y\bullet\vec{n}\right)dS=0$$
$$\rightarrow\iint_{\Omega} w_x\cdot r_x dV+\iint_{\Omega} w_y\cdot r_y dV+\iint_{\Omega}\{w\}^T\{b\}dV+\int_{\Gamma_{tx}} w_x\cdot r_{\Gamma x}\,dS+\int_{\Gamma_{ty}} w_y\cdot r_{\Gamma y}\,dS=0$$
$$(7.10.3)$$

Now, since Equation (7.10.3) is satisfied for any arbitrary $\{w(x, y)\}$, it must also apply for the following special form of $\{w(x, y)\}$:

$$\begin{Bmatrix} w_x(x,y) \\ w_y(x,y) \end{Bmatrix} = \begin{Bmatrix} \varphi_x \cdot r_x \\ \varphi_y \cdot r_y \end{Bmatrix}, \text{ with } \begin{matrix} \varphi_x(x,y) > 0 \; in \; \Omega, \; \varphi_x(x,y) = 0 \; on \; \Gamma \\ \varphi_y(x,y) > 0 \; in \; \Omega, \; \varphi_y(x,y) = 0 \; on \; \Gamma \end{matrix} \tag{7.10.4}$$

If we plug this special expression for $\{w(x, y)\}$ in Equation (7.10.3), all the boundary integrals vanish (because φ_x and φ_y are zero on Γ_{tx} and Γ_{ty}, respectively) and we obtain:

$$\iint_\Omega \varphi_x \cdot (r_x)^2 dV + \iint_\Omega \varphi_y \cdot (r_y)^2 dV = 0 \tag{7.10.5}$$

The integrant in each of the two integrals in Equation (7.10.5) is non-negative; thus, both integrals are non-negative, and the only way for them to give a zero sum is to satisfy:

$$\left.\begin{matrix} \varphi_x \cdot (r_x)^2 = 0 \\ \varphi_y \cdot (r_y)^2 = 0 \end{matrix}\right\} \; everywhere \; in \; \Omega \tag{7.10.6}$$

Since φ_x and φ_y cannot become zero (by definition, they are positive!), Equation (7.10.6) gives:

$$r_x = 0 \rightarrow \frac{\partial \sigma_{xx}}{\partial x} + \frac{\partial \sigma_{xy}}{\partial y} + b_x = 0 \; in \; \Omega \tag{7.10.7a}$$

$$r_y = 0 \rightarrow \frac{\partial \sigma_{yx}}{\partial x} + \frac{\partial \sigma_{yy}}{\partial y} + b_y = 0 \; in \; \Omega \tag{7.10.7b}$$

If we did not satisfy the weak form for the special case of $\{w(x, y)\}$ in Equation (7.10.4), this would mean that there would be at least one special case of $\{w(x, y)\}$ not satisfying the weak form. This, in turn, would contradict our starting assumption that the weak form is satisfied for any arbitrary $\{w(x, y)\}$! Thus, the only way to satisfy the weak form for any arbitrary $\{w(x, y)\}$ is to satisfy Equations (7.10.7a, b), which are the governing partial differential equations of equilibrium for the problem!

Now, since the integral equation in Box 7.9.1 is satisfied for any arbitrary $\{w(x, y)\}$, it must also apply for another special form of $\{w(x, y)\}$:

$$\begin{Bmatrix} w_x \\ w_y \end{Bmatrix} = \begin{Bmatrix} \psi_x \cdot r_{\Gamma x} \\ \psi_y \cdot r_{\Gamma y} \end{Bmatrix}, \text{ with } \begin{matrix} \psi_x(x,y) = 0 \; in \; \Omega, \; \psi_x(x,y) > 0 \; on \; \Gamma_{tx} \\ \psi_y(x,y) = 0 \; in \; \Omega, \; \psi_y(x,y) > 0 \; on \; \Gamma_{ty} \end{matrix} \tag{7.10.8}$$

If we now plug this special case of $\{w(x, y)\}$ in Equation (7.10.3), all the domain integrals vanish (because ψ_x and ψ_y are zero in Ω) and we obtain:

$$\int_{\Gamma_{tx}} \psi_x \cdot (r_{\Gamma x})^2 dS + \int_{\Gamma_{ty}} \psi_y \cdot (r_{\Gamma y})^2 dS = 0 \tag{7.10.9}$$

The integrand in each of the two integrals of Equation (7.10.9) is non-negative; thus, both integrals are non-negative, and the only way for them to give a zero sum is to have:

$$\psi_x \cdot (r_x)^2 = 0 \; on \; \Gamma_{tx} \tag{7.10.10a}$$

$$\psi_y \cdot (r_y)^2 = 0 \; on \; \Gamma_{ty} \tag{7.10.10b}$$

and because ψ_x and ψ_y cannot become zero (by definition, they are positive on the corresponding natural boundary segments!), we obtain:

$$r_{\Gamma x} = 0 \ on \ \Gamma_{tx} \rightarrow t_x - \vec{\sigma}_x \bullet \vec{n} = 0 \rightarrow \vec{\sigma}_x \bullet \vec{n} = t_x \ on \ \Gamma_{tx}, \ where \ \vec{\sigma}_x = \begin{Bmatrix} \sigma_{xx} \\ \sigma_{xy} \end{Bmatrix} \qquad (7.10.11a)$$

$$r_{\Gamma y} = 0 \ on \ \Gamma_{ty} \rightarrow t_y - \vec{\sigma}_y \bullet \vec{n} = 0 \rightarrow \vec{\sigma}_y \bullet \vec{n} = t_y \ on \ \Gamma_{ty}, \ where \ \vec{\sigma}_y = \begin{Bmatrix} \sigma_{yx} \\ \sigma_{yy} \end{Bmatrix} \qquad (7.10.11b)$$

If we did not satisfy the weak form for the special case of $\{w(x, y)\}$ in Equation (7.10.8), this would mean that there would be at least one special case of $\{w\}$ not satisfying the weak form, contradicting our starting assumption that the weak form is satisfied for any arbitrary $\{w(x, y)\}$! Thus, the only way to satisfy the weak form for any arbitrary $\{w(x, y)\}$ is to satisfy Equations (7.10.11a, b), which correspond to the natural boundary conditions of the problem.

Combining the sets of Equations (7.10.7a,b) and (7.10.11a,b), and also accounting for the fact that the weak form also includes the essential boundary conditions, we eventually reach the strong form of the problem, that is, the expressions in Box 7.8.1! Thus, beginning from the weak form, we obtained the strong form; this means that we have established the equivalence between the strong form and the weak form.

Remark 7.10.1: The four terms r_x, $r_{\Gamma x}$, r_y, and $r_{\Gamma y}$ are the *residuals of the two-dimensional solid mechanics problem* and—just like for the problems considered in previous chapters—they correspond to the conditions of the problem that *are not generally satisfied by a finite element formulation!* Thus, the governing differential equations and the natural boundary conditions are generally not exactly satisfied by a finite element formulation. In fact, the difference of each residual from zero provides a numerical measure of the *error* in the satisfaction of the corresponding equation. For example, the absolute value of r_x provides a measure of the error of a solution in satisfying the differential equation of equilibrium along the x-direction. ■

7.11 Strong Form for Three-Dimensional Elasticity

The same exact procedure that we followed in Section 7.8 can yield the strong form for three-dimensional elasticity. The problem setting is schematically established in Figure 7.11. We have a three-dimensional solid body, made of linearly elastic material. The body is subjected to body forces $\{b\} = \begin{bmatrix} b_x & b_y & b_z \end{bmatrix}^T$ in its interior, and to prescribed displacements on the essential boundary, Γ_u, and prescribed tractions $\{t\} = \begin{bmatrix} t_x & t_y & t_z \end{bmatrix}^T$ on the natural boundary, Γ_t.

To obtain the governing differential equations of equilibrium, we cut a small, box-shaped piece of material, with dimensions Δx, Δy and Δz, as shown in Figure 7.12. We can then write down the force equilibrium equations for each one of the three directions, and take the limit as the size of the piece of material tends to zero. To take the equations for the natural boundary conditions, we cut a wedge-shaped piece from the natural boundary surface, as shown in Figure 7.13. We then write down the three

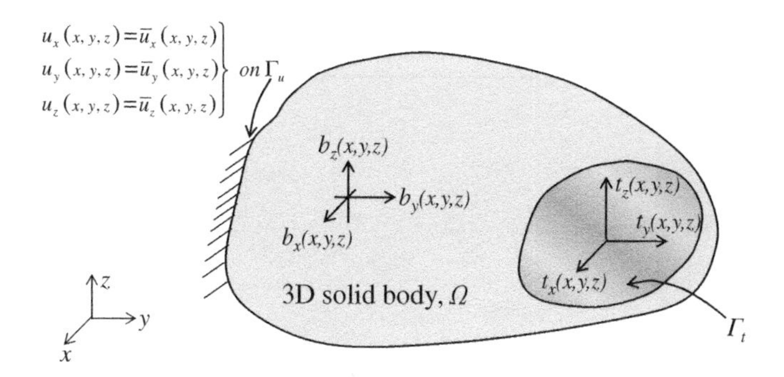

Figure 7.11 Three-dimensional solid body subjected to loadings and with prescribed displacements/tractions at the boundary.

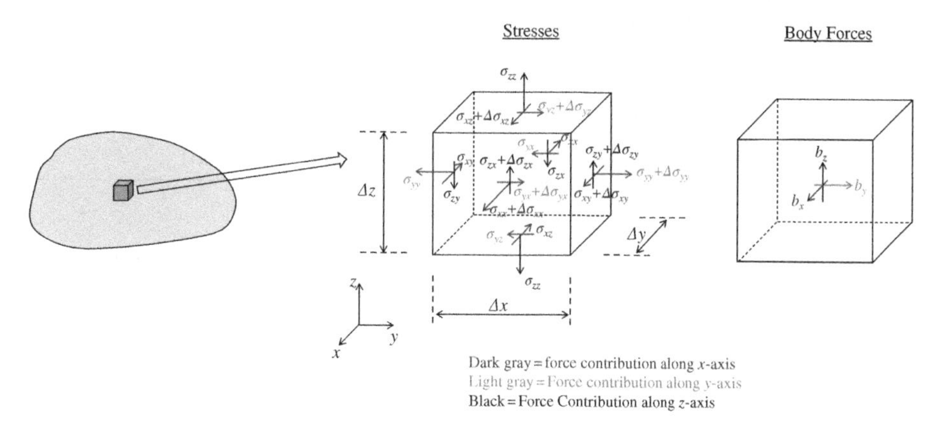

Figure 7.12 Loadings on small, box-shaped piece of three-dimensional solid body.

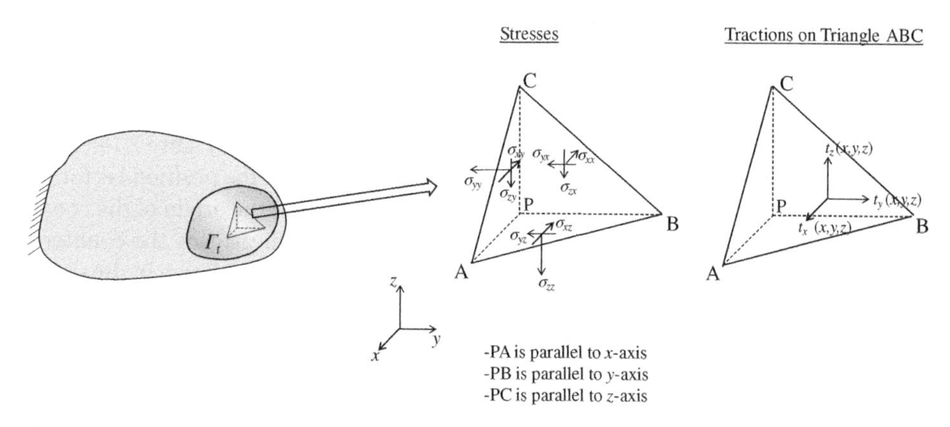

Figure 7.13 Loadings on small tetrahedral segment isolated from the natural boundary.

Box 7.11.1 Strong Form for Three-Dimensional Solid Mechanics

$$\frac{\partial \sigma_{xx}}{\partial x}+\frac{\partial \sigma_{xy}}{\partial y}+\frac{\partial \sigma_{xz}}{\partial z}+b_x=0 \tag{7.11.1}$$

$$\frac{\partial \sigma_{yx}}{\partial x}+\frac{\partial \sigma_{yy}}{\partial y}+\frac{\partial \sigma_{yz}}{\partial z}+b_y=0 \tag{7.11.2}$$

$$\frac{\partial \sigma_{zx}}{\partial x}+\frac{\partial \sigma_{zy}}{\partial y}+\frac{\partial \sigma_{zz}}{\partial z}+b_z \tag{7.11.3}$$

$u_x=\bar{u}_x$ on Γ_{ux}, $u_y=\bar{u}_y$ on Γ_{uy}, $u_z=\bar{u}_z$ on Γ_{uz} (essential boundary conditions)

$\vec{\sigma}_x \bullet \vec{n}=t_x$ on Γ_{tx}, $\vec{\sigma}_y \bullet \vec{n}=t_y$ on Γ_{ty}, $\vec{\sigma}_z \bullet \vec{n}=t_z$ on Γ_{tz} (natural boundary conditions)

where $\vec{\sigma}_x=\begin{Bmatrix}\sigma_{xx}\\ \sigma_{xy}\\ \sigma_{xz}\end{Bmatrix}$, $\vec{\sigma}_y=\begin{Bmatrix}\sigma_{yx}\\ \sigma_{yy}\\ \sigma_{yz}\end{Bmatrix}$, $\vec{\sigma}_y=\begin{Bmatrix}\sigma_{zx}\\ \sigma_{zy}\\ \sigma_{zz}\end{Bmatrix}$

$\Gamma_{ux}\cup\Gamma_{tx}=\Gamma_{uy}\cup\Gamma_{ty}=\Gamma_{uz}\cup\Gamma_{tz}=\Gamma$, $\Gamma_{ux}\cap\Gamma_{tx}=\Gamma_{uy}\cap\Gamma_{ty}=\Gamma_{uz}\cap\Gamma_{tz}=\emptyset$

for the special case of linear elasticity: $\{\sigma\}=[D]\{\varepsilon\}$ (constitutive equation)

equilibrium equations and take the limit as the surface area of the wedge belonging to the natural boundary, dS, tends to zero. We eventually obtain the strong form provided in Box 7.11.1.

7.12 Using Polar (Cylindrical) Coordinates

In several problems of solid mechanics, it is convenient to resort to a different set of orthogonal coordinate axes than the (xyz) system considered in previous sections. The most typical situations of this kind are curved structures with, for example, a cylindrical or spherical geometry. We will only discuss the former case here, which is applicable to the commonly encountered situation of *axisymmetry*, discussed in Section 10.9. To facilitate the mathematical description of curved structures, we replace the two planar coordinates, x and y, with a pair of coordinates r and θ. As shown in Figure 7.14a for a two-dimensional space, the coordinate r is measuring the length of the position vector of a point (the position vector of a point is simply a vector starting at the origin of the coordinate space and ending at the point), while the coordinate θ is measuring the counterclockwise rotation of the position vector with respect to the x-axis. As shown in the same figure, we can establish a pair of basis vectors, $\vec{e}_r$ and $\vec{e}_\theta$, for our polar coordinate system. The vector $\vec{e}_r$ is pointing in the *radial direction*, that is, in the direction of the position vector of a point. The direction of $\vec{e}_\theta$ is called the *tangential direction*, that is, the direction perpendicular to the radial one. The pair of polar coordinates r and θ can be combined with the z-coordinate to allow the description of three-dimensional problems, as shown in Figure 7.14b. Having the polar coordinates, we can cut and consider the equilibrium of curvilinear pieces of material from the interior of the body, with the corresponding body force and stress components, as shown in Figure 7.14c.

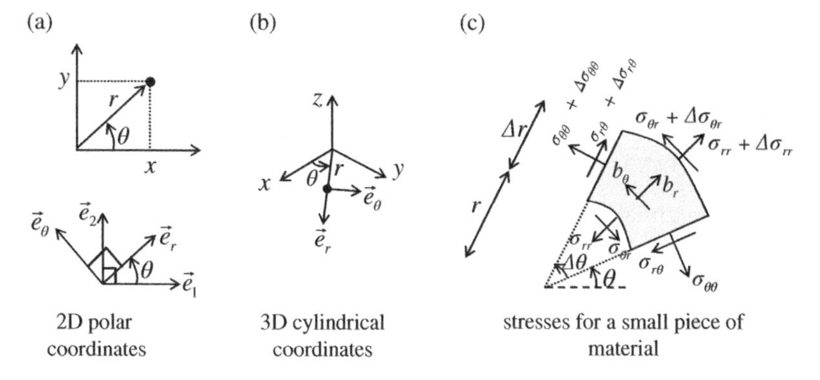

Figure 7.14 Polar coordinates and stress components.

To formulate two-dimensional elasticity problems in polar coordinates, we must first establish coordinate transformation rules from the (xy) system to the (r-θ) system. To this end, we begin by expressing the Cartesian coordinates x and y in terms of the polar coordinates. Using Figure 7.14a, we have:

$$x = r \cdot \cos\theta \tag{7.12.1a}$$

$$y = r \cdot \sin\theta \tag{7.12.1b}$$

We can also obtain the Jacobian matrix for the above "change-of-coordinates formula" (mapping):

$$\begin{bmatrix} \frac{\partial x}{\partial r} & \frac{\partial y}{\partial r} \\ \frac{\partial x}{\partial \theta} & \frac{\partial y}{\partial \theta} \end{bmatrix} = \begin{bmatrix} \cos\theta & \sin\theta \\ -r \cdot \sin\theta & r \cdot \cos\theta \end{bmatrix} \tag{7.12.2}$$

We can invert the matrix in Equation (7.12.2), to obtain the derivatives of the polar coordinates with respect to coordinates x, y:

$$\begin{bmatrix} \frac{\partial x}{\partial r} & \frac{\partial y}{\partial r} \\ \frac{\partial x}{\partial \theta} & \frac{\partial y}{\partial \theta} \end{bmatrix}^{-1} = \begin{bmatrix} \frac{\partial r}{\partial x} & \frac{\partial \theta}{\partial x} \\ \frac{\partial r}{\partial y} & \frac{\partial \theta}{\partial y} \end{bmatrix} = \begin{bmatrix} \cos\theta & -\frac{\sin\theta}{r} \\ \sin\theta & \frac{\cos\theta}{r} \end{bmatrix} \tag{7.12.3}$$

Now, let us establish a coordinate transformation matrix from the Cartesian coordinates to the polar ones. The unit basis vectors for the polar coordinate system are: $\vec{e}_r = \begin{Bmatrix} \cos\theta \\ \sin\theta \end{Bmatrix}, \vec{e}_\theta = \begin{Bmatrix} -\sin\theta \\ \cos\theta \end{Bmatrix}$. We also know the unit vectors for the Cartesian coordinate system, that is, $\vec{e}_1 = \begin{Bmatrix} 1 \\ 0 \end{Bmatrix}, \vec{e}_2 = \begin{Bmatrix} 0 \\ 1 \end{Bmatrix}$. The coordinate transformation array is a 2×2

matrix; its components are given from the following expressions, in accordance with Equation (7.7.6):

$$[R] = \begin{bmatrix} R_{11} & R_{12} \\ R_{21} & R_{22} \end{bmatrix} = \begin{bmatrix} \vec{e}_r \cdot \vec{e}_1 & \vec{e}_r \cdot \vec{e}_2 \\ \vec{e}_\theta \cdot \vec{e}_1 & \vec{e}_\theta \cdot \vec{e}_2 \end{bmatrix} = \begin{bmatrix} \cos\theta & \sin\theta \\ -\sin\theta & \cos\theta \end{bmatrix} \tag{7.12.4}$$

We also have the differential equations of equilibrium in the x- and y- directions, obtained in Section 7.8.

$$\frac{\partial \sigma_{xx}}{\partial x} + \frac{\partial \sigma_{xy}}{\partial y} + b_x = 0 \tag{7.12.5a}$$

$$\frac{\partial \sigma_{xy}}{\partial x} + \frac{\partial \sigma_{yy}}{\partial y} + b_y = 0 \tag{7.12.5b}$$

We will now obtain the stresses in the x-y system, in terms of the stresses in polar coordinates: to this end, we have to notice that, per Figure 7.14a, the xy coordinate system is obtained if we rotate the polar coordinate system by an angle equal to $-\theta$. Thus, if we use Equations (7.7.7a–c), we have the following three expressions:

$$\sigma_{xx} = \sigma_{rr} \cdot \cos^2\theta + \sigma_{\theta\theta} \cdot \sin^2\theta - \sigma_{r\theta} \cdot 2\cos\theta \cdot \sin\theta \tag{7.12.6a}$$

$$\sigma_{yy} = \sigma_{rr} \cdot \sin^2\theta + \sigma_{\theta\theta} \cdot \cos^2\theta + \sigma_{r\theta} \cdot 2\cos\theta \cdot \sin\theta \tag{7.12.6b}$$

$$\sigma_{xy} = \sigma_{rr} \cdot \cos\theta \cdot \sin\theta - \sigma_{\theta\theta} \cdot \cos\theta \cdot \sin\theta + \sigma_{r\theta} \cdot \left(\cos^2\theta - \sin^2\theta\right) \tag{7.12.6c}$$

We will now obtain the equations of equilibrium along the r- and θ-directions, that is, along the radial and the circumferential direction, respectively. We define a vector whose components are Equations (7.12.5a,b). The components of this vector are simply "equilibrium equations along the x- and y-directions." If we premultiply this vector by $[R]$, we will obtain the corresponding equilibrium equations along the radial and tangential directions:

$$[R]\begin{Bmatrix} \frac{\partial \sigma_{xx}}{\partial x} + \frac{\partial \sigma_{xy}}{\partial y} + b_x \\ \frac{\partial \sigma_{xy}}{\partial x} + \frac{\partial \sigma_{yy}}{\partial y} + b_y \end{Bmatrix} = [R]\begin{Bmatrix} 0 \\ 0 \end{Bmatrix} \to \begin{bmatrix} \cos\theta & \sin\theta \\ -\sin\theta & \cos\theta \end{bmatrix} \begin{Bmatrix} \frac{\partial \sigma_{xx}}{\partial x} + \frac{\partial \sigma_{xy}}{\partial y} + b_x \\ \frac{\partial \sigma_{xy}}{\partial x} + \frac{\partial \sigma_{yy}}{\partial y} + b_y \end{Bmatrix} = \begin{Bmatrix} 0 \\ 0 \end{Bmatrix}$$

This yields the following two equations:

$$\frac{\partial \sigma_{xx}}{\partial x} \cdot \cos\theta + \frac{\partial \sigma_{xy}}{\partial y} \cdot \cos\theta + \frac{\partial \sigma_{xy}}{\partial x} \cdot \sin\theta + \frac{\partial \sigma_{yy}}{\partial y} \cdot \sin\theta + b_r = 0 \tag{7.12.7a}$$

$$-\frac{\partial \sigma_{xx}}{\partial x} \cdot \sin\theta - \frac{\partial \sigma_{xy}}{\partial y} \cdot \sin\theta + \frac{\partial \sigma_{xy}}{\partial x} \cdot \cos\theta + \frac{\partial \sigma_{yy}}{\partial y} \cdot \cos\theta + b_\theta = 0 \tag{7.12.7b}$$

Now, we can use the chain rule of differentiation to express the partial derivatives of the stresses σ_{xx}, σ_{yy}, and σ_{xy} with respect to x and y, in terms of derivatives with respect to r and θ:

$$\frac{\partial \sigma_{xx}}{\partial x} = \frac{\partial \sigma_{xx}}{\partial r} \cdot \frac{\partial r}{\partial x} + \frac{\partial \sigma_{xx}}{\partial \theta} \cdot \frac{\partial \theta}{\partial x} = \frac{\partial \sigma_{xx}}{\partial r} \cdot \cos\theta - \frac{\partial \sigma_{xx}}{\partial \theta} \cdot \frac{\sin\theta}{r} \tag{7.12.8a}$$

$$\frac{\partial \sigma_{xy}}{\partial x} = \frac{\partial \sigma_{xy}}{\partial r} \cdot \frac{\partial r}{\partial x} + \frac{\partial \sigma_{xy}}{\partial \theta} \cdot \frac{\partial \theta}{\partial x} = \frac{\partial \sigma_{xy}}{\partial r} \cdot \cos\theta - \frac{\partial \sigma_{xy}}{\partial \theta} \cdot \frac{\sin\theta}{r} \tag{7.12.8b}$$

$$\frac{\partial \sigma_{xy}}{\partial y} = \frac{\partial \sigma_{xy}}{\partial r} \cdot \frac{\partial r}{\partial y} + \frac{\partial \sigma_{xy}}{\partial \theta} \cdot \frac{\partial \theta}{\partial y} = \frac{\partial \sigma_{xy}}{\partial r} \cdot \sin\theta + \frac{\partial \sigma_{xy}}{\partial \theta} \cdot \frac{\cos\theta}{r} \tag{7.12.8c}$$

$$\frac{\partial \sigma_{yy}}{\partial y} = \frac{\partial \sigma_{yy}}{\partial r} \cdot \frac{\partial r}{\partial y} + \frac{\partial \sigma_{yy}}{\partial \theta} \cdot \frac{\partial \theta}{\partial y} = \frac{\partial \sigma_{yy}}{\partial r} \cdot \sin\theta + \frac{\partial \sigma_{yy}}{\partial \theta} \cdot \frac{\cos\theta}{r} \tag{7.12.8d}$$

If we now plug Equations (7.12.8a–d) into Equation (7.12.7a), we have:

$$\left(\frac{\partial \sigma_{xx}}{\partial r} \cdot \cos\theta - \frac{\partial \sigma_{xx}}{\partial \theta} \cdot \frac{\sin\theta}{r}\right) \cdot \cos\theta + \left(\frac{\partial \sigma_{xy}}{\partial r} \cdot \sin\theta + \frac{\partial \sigma_{xy}}{\partial \theta} \cdot \frac{\cos\theta}{r}\right) \cdot \cos\theta$$
$$+ \left(\frac{\partial \sigma_{xy}}{\partial r} \cdot \cos\theta - \frac{\partial \sigma_{xy}}{\partial \theta} \cdot \frac{\sin\theta}{r}\right) \cdot \sin\theta + \left(\frac{\partial \sigma_{yy}}{\partial r} \cdot \sin\theta + \frac{\partial \sigma_{yy}}{\partial \theta} \cdot \frac{\cos\theta}{r}\right) \cdot \sin\theta + b_r = 0$$
$$\rightarrow \frac{\partial \sigma_{xx}}{\partial r} \cdot \cos^2\theta - \frac{1}{r}\frac{\partial \sigma_{xx}}{\partial \theta} \cdot \sin\theta \cdot \cos\theta + \frac{\partial \sigma_{xy}}{\partial r} \cdot \sin\theta \cdot \cos\theta + \frac{1}{r}\frac{\partial \sigma_{xx}}{\partial \theta} \cdot \cos^2\theta$$
$$+ \frac{\partial \sigma_{xy}}{\partial r} \cdot \sin\theta \cdot \cos\theta - \frac{\partial \sigma_{xy}}{\partial \theta} \cdot \frac{\sin^2\theta}{r} + \frac{\partial \sigma_{yy}}{\partial r} \cdot \sin^2\theta + \frac{1}{r} \cdot \frac{\partial \sigma_{yy}}{\partial \theta} \cdot \sin\theta \cdot \cos\theta + b_r = 0 \tag{7.12.9a}$$

If we follow the same approach for Equation (7.12.7b):

$$-\left(\frac{\partial \sigma_{xx}}{\partial r} \cdot \cos\theta - \frac{\partial \sigma_{xx}}{\partial \theta} \cdot \frac{\sin\theta}{r}\right) \cdot \sin\theta - \left(\frac{\partial \sigma_{xy}}{\partial r} \cdot \sin\theta + \frac{\partial \sigma_{xy}}{\partial \theta} \cdot \frac{\cos\theta}{r}\right) \cdot \sin\theta$$
$$+ \left(\frac{\partial \sigma_{xy}}{\partial r} \cdot \cos\theta - \frac{\partial \sigma_{xy}}{\partial \theta} \cdot \frac{\sin\theta}{r}\right) \cdot \cos\theta + \left(\frac{\partial \sigma_{yy}}{\partial r} \cdot \sin\theta + \frac{\partial \sigma_{yy}}{\partial \theta} \cdot \frac{\cos\theta}{r}\right) \cdot \cos\theta + b_\theta = 0$$
$$\rightarrow -\frac{\partial \sigma_{xx}}{\partial r} \cdot \sin\theta \cdot \cos\theta + \frac{1}{r}\frac{\partial \sigma_{xx}}{\partial \theta} \cdot \sin^2\theta - \frac{\partial \sigma_{xy}}{\partial r} \cdot \sin^2\theta - \frac{1}{r}\frac{\partial \sigma_{xy}}{\partial \theta} \cdot \sin\theta \cdot \cos\theta$$
$$+ \frac{\partial \sigma_{xy}}{\partial r} \cdot \cos^2\theta - \frac{1}{r}\frac{\partial \sigma_{xy}}{\partial \theta} \cdot \sin\theta \cdot \cos\theta + \frac{\partial \sigma_{yy}}{\partial r} \cdot \sin\theta \cdot \cos\theta + \frac{1}{r}\frac{\partial \sigma_{yy}}{\partial \theta} \cdot \cos^2\theta + b_\theta = 0 \tag{7.12.9b}$$

We can now obtain alternative expressions for the partial derivatives of the stresses σ_{xx}, σ_{yy}, and σ_{xy} with respect to r and θ, by simply differentiating Equations (7.12.6a–c):

$$\frac{\partial \sigma_{xx}}{\partial r} = \frac{\partial \sigma_{rr}}{\partial r} \cdot \cos^2\theta + \frac{\partial \sigma_{\theta\theta}}{\partial r} \cdot \sin^2\theta - \frac{\partial \sigma_{r\theta}}{\partial r} \cdot 2\cos\theta \cdot \sin\theta \tag{7.12.10a}$$

$$\frac{\partial \sigma_{yy}}{\partial r} = \frac{\partial \sigma_{rr}}{\partial r} \cdot \sin^2\theta + \frac{\partial \sigma_{\theta\theta}}{\partial r} \cdot \cos^2\theta + \frac{\partial \sigma_{r\theta}}{\partial r} \cdot 2\cos\theta \cdot \sin\theta \tag{7.12.10b}$$

$$\frac{\partial \sigma_{xy}}{\partial r} = \frac{\partial \sigma_{rr}}{\partial r} \cdot \cos\theta \cdot \sin\theta - \frac{\partial \sigma_{\theta\theta}}{\partial r} \cdot \cos\theta \cdot \sin\theta + \frac{\partial \sigma_{r\theta}}{\partial r} \cdot \left(\cos^2\theta - \sin^2\theta\right) \tag{7.12.10c}$$

$$\frac{\partial \sigma_{xx}}{\partial \theta} = \frac{\partial \sigma_{rr}}{\partial \theta} \cdot \cos^2\theta + \frac{\partial \sigma_{\theta\theta}}{\partial \theta} \cdot \sin^2\theta - \frac{\partial \sigma_{r\theta}}{\partial \theta} \cdot 2\cos\theta \cdot \sin\theta - \sigma_{rr} \cdot 2\cos\theta \cdot \sin\theta$$
$$+ \sigma_{\theta\theta} \cdot 2\cos\theta \cdot \sin\theta - \sigma_{r\theta} \cdot 2\cos(2\theta) \tag{7.12.10d}$$

$$\frac{\partial\sigma_{yy}}{\partial\theta}=\frac{\partial\sigma_{rr}}{\partial\theta}\cdot\sin^2\theta+\frac{\partial\sigma_{\theta\theta}}{\partial\theta}\cdot\cos^2\theta+\frac{\partial\sigma_{r\theta}}{\partial\theta}\cdot 2\cos\theta\cdot\sin\theta+\sigma_{rr}\cdot 2\cos\theta\cdot\sin\theta$$
$$-\sigma_{\theta\theta}\cdot 2\cos\theta\cdot\sin\theta+\sigma_{r\theta}\cdot 2\cos(2\theta) \tag{7.12.10e}$$

$$\frac{\partial\sigma_{xy}}{\partial\theta}=\frac{\partial\sigma_{rr}}{\partial\theta}\cdot\cos\theta\cdot\sin\theta-\frac{\partial\sigma_{\theta\theta}}{\partial\theta}\cdot\cos\theta\cdot\sin\theta+\frac{\partial\sigma_{r\theta}}{\partial\theta}\cdot\left(\cos^2\theta-\sin^2\theta\right)+\sigma_{rr}\cdot\cos(2\theta)$$
$$-\sigma_{\theta\theta}\cdot\cos(2\theta)-\sigma_{r\theta}\cdot 2\cdot\sin(2\theta) \tag{7.12.10f}$$

Finally, substituting Equations (7.12.10a–f) into the two Equations (7.12.9a, b) and after additional mathematical manipulations, we reach the two *differential equations of equilibrium in polar coordinates*:

$$\frac{\partial\sigma_{rr}}{\partial r}+\frac{\sigma_{rr}-\sigma_{\theta\theta}}{r}+\frac{1}{r}\frac{\partial\sigma_{r\theta}}{\partial\theta}+b_r=0 \tag{7.12.11a}$$

$$\frac{\partial\sigma_{r\theta}}{\partial r}+\frac{2\sigma_{r\theta}}{r}+\frac{1}{r}\frac{\partial\sigma_{\theta\theta}}{\partial\theta}+b_\theta=0 \tag{7.12.11b}$$

We can also describe a three-dimensional problem with polar coordinates. In this case, we use the coordinate axes r, θ, and z, as shown in Figure 7.14b, which are termed the *cylindrical coordinates*. Notice that the third coordinate axis, z, is identical to that used for the three-dimensional Cartesian coordinates! The following differential equations of equilibrium can be obtained for cylindrical coordinates.

$$\frac{\partial\sigma_{rr}}{\partial r}+\frac{\sigma_{rr}-\sigma_{\theta\theta}}{r}+\frac{1}{r}\frac{\partial\sigma_{r\theta}}{\partial\theta}+\frac{\partial\sigma_{rz}}{\partial z}+b_r=0 \tag{7.12.12a}$$

$$\frac{\partial\sigma_{r\theta}}{\partial r}+\frac{2\sigma_{r\theta}}{r}+\frac{1}{r}\frac{\partial\sigma_{\theta\theta}}{\partial\theta}+\frac{\partial\sigma_{\theta z}}{\partial z}+b_\theta=0 \tag{7.12.12b}$$

$$\frac{\partial\sigma_{zr}}{\partial r}+\frac{\sigma_{zr}+\sigma_{z\theta}}{r}+\frac{1}{r}\frac{\partial\sigma_{z\theta}}{\partial\theta}+\frac{\partial\sigma_{zz}}{\partial z}+b_z=0 \tag{7.12.12c}$$

We can also transform the displacement components in the cylindrical coordinate system, and consider the corresponding components of the displacement vector field, $u_r(r,\theta,z)$, $u_\theta(r,\theta,z)$, and $u_z(r,\theta,z)$. Leveraging the chain rule of differentiation and the coordinate transformation equations, we can then obtain the strain tensor components in cylindrical coordinates in terms of the corresponding displacement components:

$$\left[\varepsilon_{polar}\right]=\begin{bmatrix}\varepsilon_{rr} & \varepsilon_{r\theta} & \varepsilon_{rz}\\ \varepsilon_{\theta r}=\varepsilon_{r\theta} & \varepsilon_{\theta\theta} & \varepsilon_{\theta z}\\ \varepsilon_{zr} & \varepsilon_{z\theta}=\varepsilon_{\theta z} & \varepsilon_{zz}\end{bmatrix}$$
$$=\begin{bmatrix}\dfrac{\partial u_r}{\partial r} & \dfrac{1}{2}\left(\dfrac{1}{r}\dfrac{\partial u_r}{\partial\theta}+\dfrac{\partial u_\theta}{\partial r}-\dfrac{u_\theta}{r}\right) & \dfrac{1}{2}\left(\dfrac{\partial u_r}{\partial z}+\dfrac{\partial u_z}{\partial r}\right)\\ \dfrac{1}{2}\left(\dfrac{1}{r}\dfrac{\partial u_r}{\partial\theta}+\dfrac{\partial u_\theta}{\partial r}-\dfrac{u_\theta}{r}\right) & \dfrac{1}{r}\dfrac{\partial u_\theta}{\partial\theta}+\dfrac{u_r}{r} & \dfrac{1}{2}\left(\dfrac{\partial u_\theta}{\partial z}+\dfrac{1}{r}\dfrac{\partial u_z}{\partial\theta}\right)\\ \dfrac{1}{2}\left(\dfrac{\partial u_r}{\partial z}+\dfrac{\partial u_z}{\partial r}\right) & \dfrac{1}{2}\left(\dfrac{\partial u_\theta}{\partial z}+\dfrac{1}{r}\dfrac{\partial u_z}{\partial\theta}\right) & \dfrac{\partial u_z}{\partial z}\end{bmatrix} \tag{7.12.13}$$

We will return to the use of cylindrical coordinates in Section 10.9, where we examine the finite element formulation of axisymmetric problems.

References

Fung, Y., and Tong, P. (2001). *Classical and Computational Solid Mechanics.* Singapore: World Scientific Publishing Company.

Love, A. (1927). *A Treatise on the Mathematical Theory of Elasticity,* 4th edition, Cambridge: Cambridge University Press.

Malvern, L. (1977). *Introduction to the Mechanics of a Continuous Medium.* New York: Pearson.

Sokolnikoff, I. (1956). *Mathematical Theory of Elasticity,* 2nd edition. New York: McGraw-Hill.

8

Finite Element Formulation for Two-Dimensional Elasticity

In the previous chapter, we obtained the weak form for two-dimensional elasticity, which we repeat in Box 8.1 for convenience.

Box 8.1 Weak Form for Two-Dimensional Elasticity

$$\iint_{\Omega} \{\bar{\varepsilon}\}^T [D] \{\varepsilon\} dV = \iint_{\Omega} \{w\}^T \{b\} dV + \int_{\Gamma_{tx}} w_x t_x \, dS + \int_{\Gamma_{ty}} w_y t_y \, dS$$

$$\forall \{w(x,y)\} = \begin{Bmatrix} w_x(x,y) \\ w_y(x,y) \end{Bmatrix} \text{ with } w_x = 0 \text{ on } \Gamma_{ux}, \; w_y = 0 \text{ on } \Gamma_{uy}$$

$$u_x = \bar{u}_x \quad on \; \Gamma_{ux}$$

$$u_y = \bar{u}_y \quad on \; \Gamma_{uy}$$

Using this weak form as our starting point, we will now proceed to the description of the finite element formulation and analysis for two-dimensional elasticity problems.

8.1 Piecewise Finite Element Approximation—Assembly Equations

We already saw in Figure 6.1 that in a finite element mesh, the two-dimensional domain is discretized into N_e subdomains, which are the elements. Each element has n nodes. We then establish a *piecewise approximation* for the displacement field, $\{u(x,y)\}$ and for the virtual displacement[1] field, $\{w(x,y)\}$, inside each element e. Since each of these fields is a

1 As explained in Remark 7.9.1, the weak form for two-dimensional elasticity constitutes the principle of virtual displacements. Accordingly, the weighting function vector field $\{w(x,y)\}$ is the virtual displacement field.

Fundamentals of Finite Element Analysis: Linear Finite Element Analysis, First Edition.
Ioannis Koutromanos, James McClure, and Christopher Roy.

Companion website: www.wiley.com/go/koutromanos/linear

vector field, we will simply establish a piecewise approximation for each component of the displacement and virtual displacement.

The displacement field in the interior of each element will be approximated by means of polynomial shape (interpolation) functions, just like we did for two-dimensional heat conduction in Chapter 6. Thus, the approximate displacement vector field $\{u^{(e)}(x,y)\}$ in the interior of an element e with n nodal points is given by the following equation.

$$\begin{Bmatrix} u_x^{(e)}(x,y) \\ u_y^{(e)}(x,y) \end{Bmatrix} \approx \begin{bmatrix} N_1^{(e)}(x,y) & 0 & N_2^{(e)}(x,y) & 0 & \ldots & N_n^{(e)}(x,y) & 0 \\ 0 & N_1^{(e)}(x,y) & 0 & N_2^{(e)}(x,y) & \ldots & 0 & N_n^{(e)}(x,y) \end{bmatrix} \begin{Bmatrix} u_{x1}^{(e)} \\ u_{y1}^{(e)} \\ u_{x2}^{(e)} \\ u_{y2}^{(e)} \\ \vdots \\ u_{xn}^{(e)} \\ u_{yn}^{(e)} \end{Bmatrix} \tag{8.1.1}$$

We can also write Equation (8.1.1) in a more concise form:

$$\left\{u^{(e)}\right\} \approx \left[N^{(e)}\right]\left\{U^{(e)}\right\} \tag{8.1.2}$$

where $\left\{U^{(e)}\right\} = \left[u_{x1}^{(e)} \quad u_{y1}^{(e)} \quad u_{x2}^{(e)} \quad u_{y2}^{(e)} \quad \ldots \quad u_{xn}^{(e)} \quad u_{yn}^{(e)}\right]^T$ is the nodal displacement vector of the element and $[N^{(e)}]$ is the shape function array:

$$\left[N^{(e)}\right] = \begin{bmatrix} N_1^{(e)}(x,y) & 0 & N_2^{(e)}(x,y) & 0 & \ldots & N_n^{(e)}(x,y) & 0 \\ 0 & N_1^{(e)}(x,y) & 0 & N_2^{(e)}(x,y) & \ldots & 0 & N_n^{(e)}(x,y) \end{bmatrix} \tag{8.1.3}$$

Remark 8.1.1: Equation (8.1.1) is equivalent to a pair of expressions, each providing the finite element approximation of the corresponding component of the displacement field:

$$u_x^{(e)}(x,y) = N_1^{(e)}(x,y)\cdot u_{x1}^{(e)} + N_2^{(e)}(x,y)\cdot u_{x2}^{(e)} + \ldots + N_n^{(e)}(x,y)\cdot u_{xn}^{(e)} = \sum_{i=1}^{n}\left(N_i^{(e)}(x,y)\cdot u_{xi}^{(e)}\right) \tag{8.1.4a}$$

$$u_y^{(e)}(x,y) = N_1^{(e)}(x,y)\cdot u_{y1}^{(e)} + N_2^{(e)}(x,y)\cdot u_{y2}^{(e)} + \ldots + N_n^{(e)}(x,y)\cdot u_{yn}^{(e)} = \sum_{i=1}^{n}\left(N_i^{(e)}(x,y)\cdot u_{yi}^{(e)}\right) \tag{8.1.4b}$$

■

After establishing the displacement approximation, we can readily formulate the approximate strain field in each element. Specifically, the components of the strain vector, $\left\{\varepsilon^{(e)}\right\} = \left[\varepsilon_{xx}^{(e)} \quad \varepsilon_{yy}^{(e)} \quad \gamma_{xy}^{(e)}\right]^T$, are given by the following equations.

$$\varepsilon_{xx}^{(e)} = \frac{\partial u_x^{(e)}}{\partial x} = \frac{\partial}{\partial x}\left[\sum_{i=1}^{n}\left(N_i^{(e)}(x,y)\cdot u_{xi}^{(e)}\right)\right] = \sum_{i=1}^{n}\left(\frac{\partial N_i^{(e)}}{\partial x}\cdot u_{xi}^{(e)}\right) \tag{8.1.5a}$$

$$\varepsilon_{yy}^{(e)} = \frac{\partial u_y^{(e)}}{\partial y} = \frac{\partial}{\partial y}\left[\sum_{i=1}^{n}\left(N_i^{(e)}(x,y)\cdot u_{yi}^{(e)}\right)\right] = \sum_{i=1}^{n}\left(\frac{\partial N_i^{(e)}}{\partial y}\cdot u_{yi}\right)u_{yi}^{(e)} \tag{8.1.5b}$$

$$\gamma_{xy}^{(e)} = \frac{\partial u_x^{(e)}}{\partial y} + \frac{\partial u_y^{(e)}}{\partial x} = \frac{\partial}{\partial y}\left[\sum_{i=1}^{n}\left(N_i(x,y)\cdot u_{xi}^{(e)}\right)\right] + \frac{\partial}{\partial x}\left[\sum_{i=1}^{n}\left(N_i(x,y)\cdot u_{yi}^{(e)}\right)\right]$$
$$= \sum_{i=1}^{n}\left(\frac{\partial N_i}{\partial y}\cdot u_{xi}^{(e)}\right) + \sum_{i=1}^{n}\left(\frac{\partial N_i}{\partial x}\cdot u_{yi}^{(e)}\right) \tag{8.1.5c}$$

We can cast Equations (8.1.5a–c) in matrix form to obtain a single, concise expression:

$$\left\{\varepsilon^{(e)}\right\} \approx \left[B^{(e)}\right]\left\{U^{(e)}\right\} \tag{8.1.6}$$

where $[B^{(e)}]$ is the strain-displacement array of the element, defined from the following equation.

$$\left[B^{(e)}\right] = \begin{bmatrix} \frac{\partial N_1^{(e)}}{\partial x} & 0 & \frac{\partial N_2^{(e)}}{\partial x} & 0 & \cdots & \frac{\partial N_n^{(e)}}{\partial x} & 0 \\ 0 & \frac{\partial N_1^{(e)}}{\partial y} & 0 & \frac{\partial N_2^{(e)}}{\partial y} & \cdots & 0 & \frac{\partial N_n^{(e)}}{\partial y} \\ \frac{\partial N_1^{(e)}}{\partial y} & \frac{\partial N_1^{(e)}}{\partial x} & \frac{\partial N_2^{(e)}}{\partial y} & \frac{\partial N_2^{(e)}}{\partial x} & \cdots & \frac{\partial N_n^{(e)}}{\partial y} & \frac{\partial N_n^{(e)}}{\partial x} \end{bmatrix} \tag{8.1.7}$$

Now, if we stipulate that we have the same type of approximation for $\{w\}$ as for $\{u\}$, we obtain:

$$\left\{w^{(e)}\right\} = \begin{Bmatrix} w_x^{(e)}(x,y) \\ w_y^{(e)}(x,y) \end{Bmatrix} \approx \left[N^{(e)}\right]\left\{W^{(e)}\right\} \tag{8.1.8}$$

where, per Remark 7.9.1, $\{W^{(e)}\}$ is the nodal virtual displacement vector of element e. Along the same lines, the virtual strain components in the element are approximated by the equation:

$$\left\{\bar{\varepsilon}^{(e)}\right\} = \left[\bar{\varepsilon}_{xx}^{(e)} \;\; \bar{\varepsilon}_{yy}^{(e)} \;\; \bar{\gamma}_{xy}^{(e)}\right]^T = \left[\frac{\partial w_x^{(e)}}{\partial x} \;\; \frac{\partial w_y^{(e)}}{\partial y} \;\; \frac{\partial w_x^{(e)}}{\partial y} + \frac{\partial w_y^{(e)}}{\partial x}\right]^T \approx \left[B^{(e)}\right]\left\{W^{(e)}\right\} \tag{8.1.9}$$

Just like the other finite element formulations that we have previously examined in this text, we define a Boolean gather array, $[L^{(e)}]$, for each element e, such that (per Section B.2) the element nodal values $\{U^{(e)}\}$ and $\{W^{(e)}\}$ are given as:

$$\left\{U^{(e)}\right\} = \left[L^{(e)}\right]\{U\} \tag{8.1.10a}$$

$$\left\{W^{(e)}\right\} = \left[L^{(e)}\right]\{W\} \rightarrow \left\{W^{(e)}\right\}^T = \{W\}^T\left[L^{(e)}\right]^T \tag{8.1.10b}$$

where $\{U\}$ and $\{W\}$ are the corresponding global (structural) nodal displacement and virtual displacement vectors (i.e., the vectors containing the displacements and virtual displacements of nodal points of the finite element mesh).

We can now break the integrals of Box 8.1 into the contributions of the N_e elements in the mesh, and we express the total domain integral as the sum of the integral contributions of the elements comprising the mesh:

$$\sum_{e=1}^{Ne}\left(\iint_{\Omega^{(e)}}\{\bar{\varepsilon}\}^T[D]\{\varepsilon\}dV\right)=\sum_{e=1}^{Ne}\left(\iint_{\Omega^{(e)}}\{w\}^T\{b\}dV\right)+\sum_{e=1}^{Ne}\left(\int_{\Gamma_{tx}^{(e)}}\{w\}^T\{t_x\}\,dS\right)$$
$$+\sum_{e=1}^{Ne}\left(\int_{\Gamma_{ty}^{(e)}}\{w\}^T\{t_y\}\,dS\right) \tag{8.1.11}$$

where we have set $\{t_x\}=\begin{Bmatrix}t_x\\0\end{Bmatrix}$ and $\{t_y\}=\begin{Bmatrix}0\\t_y\end{Bmatrix}$.

We can account for the fact that, inside element e, we have $\{u\}\approx\{u^{(e)}\}=\left[N^{(e)}\right]\{U^{(e)}\}$, $\{\varepsilon\}\approx\{\varepsilon^{(e)}\}=\left[B^{(e)}\right]\{U^{(e)}\}$, $\{w\}\approx\{w^{(e)}\}=\left[N^{(e)}\right]\{W^{(e)}\}$, $\{\bar{\varepsilon}\}\approx\{\bar{\varepsilon}^{(e)}\}=\left[B^{(e)}\right]\{W^{(e)}\}$, to obtain:

$$\sum_{e=1}^{Ne}\left(\iint_{\Omega^{(e)}}\left(\left[B^{(e)}\right]\{W^{(e)}\}\right)^T\left[D^{(e)}\right]\left[B^{(e)}\right]\{U^{(e)}\}dV\right)$$
$$=\sum_{e=1}^{Ne}\left(\iint_{\Omega^{(e)}}\left(\left[N^{(e)}\right]\{W^{(e)}\}\right)^T\{b\}dV\right)$$
$$+\sum_{e=1}^{Ne}\left(\int_{\Gamma_{tx}^{(e)}}\left(\left[N^{(e)}\right]\{W^{(e)}\}\right)^T\{t_x\}\,dS\right)+\sum_{e=1}^{Ne}\left(\int_{\Gamma_{ty}^{(e)}}\left(\left[N^{(e)}\right]\{W^{(e)}\}\right)^T\{t_y\}\,dS\right)$$

If we also plug Equations (8.1.10a, b) into the obtained expression, we have:

$$\rightarrow\sum_{e=1}^{Ne}\left(\iint_{\Omega^{(e)}}\{W\}^T\left[L^{(e)}\right]^T\left[B^{(e)}\right]^T\left[D^{(e)}\right]\left[B^{(e)}\right]\left[L^{(e)}\right]\{U\}dV\right)$$
$$-\sum_{e=1}^{Ne}\left(\iint_{\Omega^{(e)}}\{W\}^T\left[L^{(e)}\right]^T\left[N^{(e)}\right]^T\{b\}dV\right)$$
$$-\sum_{e=1}^{Ne}\left(\int_{\Gamma_{tx}^{(e)}}\{W\}^T\left[L^{(e)}\right]^T\left[N^{(e)}\right]^T\{t_x\}\,dS\right)-\sum_{e=1}^{Ne}\left(\int_{\Gamma_{ty}^{(e)}}\{W\}^T\left[L^{(e)}\right]^T\left[N^{(e)}\right]^T\{t_y\}\,dS\right)=0$$
$$\tag{8.1.12}$$

Finally, since $\{W\}^T$, $\{U\}$ and $[L^{(e)}]$ are constant, they can be taken outside of the integrals in Equation (8.1.12). Furthermore, the vectors $\{W\}^T$ and $\{U\}$ are common in all the corresponding terms, so they can be taken as common factors outside of the summation. Thus, we have:

$$\rightarrow \{W\}^T \sum_{e=1}^{Ne} \left[\left[L^{(e)}\right]^T \left(\iint_{\Omega^{(e)}} \left[B^{(e)}\right]^T \left[D^{(e)}\right] \left[B^{(e)}\right] dV \right) \left[L^{(e)}\right] \right] \{U\}$$

$$-\{W\}^T \sum_{e=1}^{Ne} \left[\left[L^{(e)}\right]^T \left(\iint_{\Omega^{(e)}} \left[N^{(e)}\right]^T \{b\} dV + \int_{\Gamma_{tx}^{(e)}} \left[N^{(e)}\right]^T \{t_x\} dS + \int_{\Gamma_{ty}^{(e)}} \left[N^{(e)}\right]^T \{t_y\} dS \right) \right] = 0$$

$$\rightarrow \{W\}^T \left(\sum_{e=1}^{Ne} \left[\left[L^{(e)}\right]^T \left[k^{(e)}\right] \left[L^{(e)}\right] \right] \{U\} - \left[L^{(e)}\right]^T \left\{f^{(e)}\right\} \right) = 0 \qquad (8.1.13)$$

where $[k^{(e)}]$ the element stiffness matrix and $\{f^{(e)}\}$ is the equivalent nodal force vector. The following equations apply.

$$\left[k^{(e)}\right] = \iint_{\Omega^{(e)}} \left[B^{(e)}\right]^T \left[D^{(e)}\right] \left[B^{(e)}\right] dV \qquad (8.1.14)$$

$$\left\{f^{(e)}\right\} = \iint_{\Omega^{(e)}} \left[N^{(e)}\right]^T \{b\} dV + \int_{\Gamma_{tx}^{(e)}} \left[N^{(e)}\right]^T \{t_x\} dS + \int_{\Gamma_{ty}^{(e)}} \left[N^{(e)}\right]^T \{t_y\} dS \qquad (8.1.15)$$

We can now write Equation (8.1.13) in the following form:

$$\{W\}^T [[K]\{U\} - \{f\}] = 0 \qquad (8.1.16)$$

where $[K]$ is the global stiffness matrix:

$$[K] = \sum_{e=1}^{Ne} \left(\left[L^{(e)}\right]^T \left[k^{(e)}\right] \left[L^{(e)}\right] \right) \qquad (8.1.17)$$

and $\{f\}$ is the global equivalent nodal force vector:

$$\{f\} = \sum_{e=1}^{Ne} \left(\left[L^{(e)}\right]^T \left\{f^{(e)}\right\} \right) \qquad (8.1.18)$$

Thus, once again, we obtained the global stiffness (coefficient) matrix and equivalent nodal force (right-hand-side) vector as assemblies of the corresponding element contributions.

Equation (8.1.16) involves vectors and arrays with dimensions equal to N, which is the total number of degrees of freedom in the system. This equation can be expanded in the following form.

$$\sum_{i=1}^{N} W_i \left(\sum_{j=1}^{N} (K_{ij} \cdot U_j) - f_i \right) = 0 \rightarrow W_1 \cdot (K_{11} \cdot U_1 + K_{12} \cdot U_2 + \ldots + K_{1N} \cdot U_N - f_1)$$
$$+ W_2 \cdot (K_{21} \cdot U_1 + K_{22} \cdot U_2 + \ldots + K_{2N} \cdot U_N - f_2) + \ldots$$
$$\ldots + W_N \cdot (K_{N1} \cdot U_1 + K_{N2} \cdot U_2 + \ldots + K_{NN} \cdot U_N - f_N) = 0$$
$$(8.1.19)$$

Since each parameter W_1, W_2, ..., W_N in Equation (8.1.19) is arbitrary, the only way to obtain a zero result in the above expression *for any arbitrary* vector $\{W\}$ is to simultaneously satisfy the following equations.

$$K_{11} \cdot U_1 + K_{12} \cdot U_2 + \ldots + K_{1N} \cdot U_N - f_1 = 0 \tag{8.1.20a$_1$}$$

$$K_{21} \cdot U_1 + K_{22} \cdot U_2 + \ldots + K_{2N} \cdot U_N - f_2 = 0 \tag{8.1.20a$_2$}$$

$$\ldots$$

$$K_{N1} \cdot U_1 + K_{N2} \cdot U_2 + \ldots + K_{NN} \cdot U_N - f_N = 0 \tag{8.1.20a$_N$}$$

Equations (8.1.20a$_1$–a$_N$) can be cast into the following global system of equations for the nodal displacements.

$$[K]\{U\} - \{f\} = 0 \tag{8.1.21}$$

Equation (8.1.21) provides the discrete finite element equations for two-dimensional linear elasticity, also called *the finite element equilibrium or stiffness equations for linear elasticity.*

Remark 8.1.2: The above equations were obtained from the weak form *per unit thickness.* Thus, the nodal forces are actually forces per unit thickness. If we know the thickness of the body, d, then the stiffness matrix and right-hand-side vector (which will now have units of force) *must be multiplied by* d:

$$\left[k^{(e)}\right] = d\left[\iint_{\Omega^{(e)}} \left[B^{(e)}\right]^T \left[D^{(e)}\right] \left[B^{(e)}\right] dV\right] \tag{8.1.22}$$

$$\left\{f^{(e)}\right\} = d\left[\iint_{\Omega^{(e)}} \left[N^{(e)}\right]^T \{b\}\, dV + \int_{\Gamma_{tx}^{(e)}} \left[N^{(e)}\right]^T \{t_x\}\, dS + \int_{\Gamma_{ty}^{(e)}} \left[N^{(e)}\right]^T \{t_y\}\, dS\right] \tag{8.1.23}$$

■

8.2 Accounting for Restrained (Fixed) Displacements

At this point, it is deemed necessary to discuss how to handle restrained displacements, that is, the displacements at nodes that lie on the essential boundary. If a specific component of $\{W\}$, say W_k, corresponds to a restrained degree of freedom (essential boundary condition), then we have $W_k = 0$. Thus, the term $W_k \cdot (K_{k1} \cdot U_1 + K_{k2} \cdot U_2 + \ldots + K_{kN} \cdot U_N - f_k)$ in Equation (8.1.19) is *always* zero, no matter what the value of the sum in the bracket is! In this case, we cannot stipulate what the value of the sum in the bracket is, and we can write:

$$K_{k1} \cdot U_1 + K_{k2} \cdot U_2 + \ldots + K_{kN} \cdot U_N - f_k = r_k \tag{8.2.1}$$

where r_k is the *reaction force* corresponding to the restrained global degree of freedom k! In accordance with Equation (B.3.10), we can partition the global displacement vector, $\{U\}$, into two parts, namely, the part $\{U_f\}$ corresponding to the unrestrained (free) degrees of freedom, and the part $\{U_s\}$ corresponding to the restrained (fixed) degrees of freedom. The values of $\{U_s\}$ are known. The solution procedure after the partition is described in Section B.3.

8.3 Postprocessing

After we have solved the global equations and obtained the nodal displacement vector $\{U\}$ of the structure, we can proceed with a *post-processing* of the solution (i.e., obtain important information that requires knowledge of $\{U\}$). Specifically, for each element e, we can determine:

- The element nodal displacements:

$$\left\{U^{(e)}\right\} = \left[L^{(e)}\right]\{U\} \tag{8.3.1}$$

- The element displacement field:

$$\left\{u^{(e)}\right\} = \left[N^{(e)}\right]\left\{U^{(e)}\right\} = \left[N^{(e)}\right]\left[L^{(e)}\right]\{U\} \tag{8.3.2}$$

- The element strain field:

$$\left\{\varepsilon^{(e)}\right\} = \left[B^{(e)}\right]\left\{U^{(e)}\right\} = \left[B^{(e)}\right]\left[L^{(e)}\right]\{U\} \tag{8.3.3}$$

- The element stress field:

$$\left\{\sigma^{(e)}\right\} = \left[D^{(e)}\right]\left\{\varepsilon^{(e)}\right\} = \left[D^{(e)}\right]\left[B^{(e)}\right]\left\{U^{(e)}\right\} = \left[D^{(e)}\right]\left[B^{(e)}\right]\left[L^{(e)}\right]\{U\} \tag{8.3.4}$$

8.4 Continuity—Completeness Requirements

For linear elasticity, the highest order of partial derivatives in the weak form is 1. For this reason, the continuity and completeness requirements for the approximations are the same as those for two-dimensional heat flow. In other words, *C^0 continuity and first-degree completeness are adequate to ensure that a finite element approximation for two-dimensional (2D) linear elasticity is convergent.* Based on the above, the same types of finite elements that we established for 2D heat conduction can also be used for 2D elasticity!

8.5 Finite Elements for Two-Dimensional Elasticity

We will now proceed to discuss how to use the various types of elements of Chapter 6 for two-dimensional elasticity problems. We will examine how to calculate the stiffness matrix per unit thickness, given by Equation (8.1.14), and the equivalent nodal force vector per unit thickness, given by Equation (8.1.15). It is convenient to separate the latter into two terms, one corresponding to the contribution from the body forces and the other to the contribution of the boundary tractions:

$$\left\{f^{(e)}\right\} = \left\{f_{\Omega}^{(e)}\right\} + \left\{f_{\Gamma_t}^{(e)}\right\} \tag{8.5.1}$$

where

$$\left\{f_{\Omega}^{(e)}\right\} = \iint\limits_{\Omega^{(e)}} \left[N^{(e)}\right]^T \{b\} \cdot dV \tag{8.5.2}$$

and

$$\left\{f_{\Gamma_t}^{(e)}\right\} = \int\limits_{\Gamma_{tx}^{(e)}} \left[N^{(e)}\right]^T \{t_x\} dS + \int\limits_{\Gamma_{ty}^{(e)}} \left[N^{(e)}\right]^T \{t_y\} dS \tag{8.5.3}$$

If we are given the actual thickness, *d*, we simply multiply the obtained stiffness matrix and nodal force vector by *d*. Otherwise, we complete our calculations for a unit thickness, knowing that the deformations and stresses remain constant over the thickness.

The remainder of this section will specifically discuss the use of the three-node triangular and four-node quadrilateral elements for elasticity. The procedures used for these types of elements can also be applied, for example, for the higher-order isoparametric elements of Section 6.6.

8.5.1 Three-Node Triangular Element (Constant Strain Triangle)

In Section 6.2, we obtained the shape functions of a three-node triangular element for two-dimensional heat conduction. We can use the same type of element and the same exact shape functions for two-dimensional elasticity. In fact, the three-node triangular element was the one originally used in analyses of solid mechanics, and it still enjoys popularity today. The element and the displacements for the three nodal points are schematically shown in Figure 8.1.

The element nodal displacement vector, $\{U^{(e)}\}$, has the following six components:

$$\left\{U^{(e)}\right\} = \left[U_1^{(e)}\ U_2^{(e)}\ U_3^{(e)}\ U_4^{(e)}\ U_5^{(e)}\ U_6^{(e)}\right]^T = \left[u_{x1}^{(e)}\ u_{y1}^{(e)}\ u_{x2}^{(e)}\ u_{y2}^{(e)}\ u_{x3}^{(e)}\ u_{y3}^{(e)}\right]^T \tag{8.5.4}$$

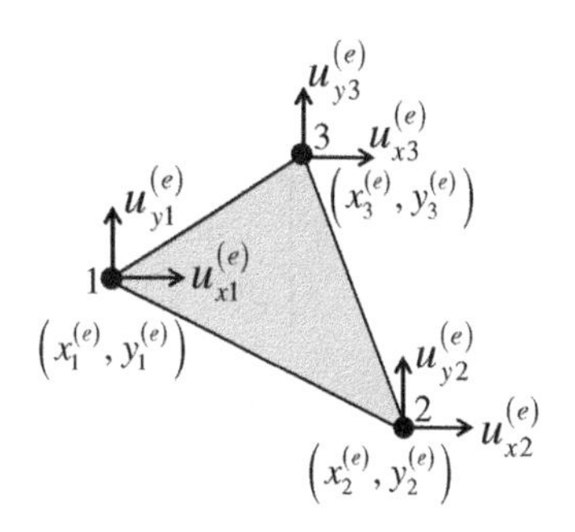

Figure 8.1 Nodal coordinates and displacements for three-node triangular element.

The following three shape functions are used for the three-node triangular element:

$$N_1^{(e)}(x,y) = \frac{1}{2A^{(e)}}\left[x_2^{(e)}y_3^{(e)} - x_3^{(e)}y_2^{(e)} + \left(y_2^{(e)} - y_3^{(e)}\right)x + \left(x_3^{(e)} - x_2^{(e)}\right)y\right] \quad (8.5.5a)$$

$$N_2^{(e)}(x,y) = \frac{1}{2A^{(e)}}\left[x_3^{(e)}y_1^{(e)} - x_1^{(e)}y_3^{(e)} + \left(y_3^{(e)} - y_1^{(e)}\right)x + \left(x_1^{(e)} - x_3^{(e)}\right)y\right] \quad (8.5.5b)$$

$$N_3^{(e)}(x,y) = \frac{1}{2A^{(e)}}\left[x_1^{(e)}y_2^{(e)} - x_2^{(e)}y_1^{(e)} + \left(y_1^{(e)} - y_2^{(e)}\right)x + \left(x_2^{(e)} - x_1^{(e)}\right)y\right] \quad (8.5.5c)$$

where

$$A^{(e)} = \frac{1}{2}\det\left(\left[M^{(e)}\right]\right) = \frac{1}{2}\left[\left(x_2^{(e)}y_3^{(e)} - x_3^{(e)}y_2^{(e)}\right) - \left(x_1^{(e)}y_3^{(e)} - x_3^{(e)}y_1^{(e)}\right) + \left(x_1^{(e)}y_2^{(e)} - x_2^{(e)}y_1^{(e)}\right)\right] \quad (8.5.6)$$

We can now proceed to write down the strain-displacement relation for the element, using a special case (for $n = 3$) of Equations (8.1.6) and (8.1.7):

$$\begin{Bmatrix} \varepsilon_{xx}^{(e)} \\ \varepsilon_{yy}^{(e)} \\ \gamma_{xy}^{(e)} \end{Bmatrix} = \begin{bmatrix} \frac{\partial N_1^{(e)}}{\partial x} & 0 & \frac{\partial N_2^{(e)}}{\partial x} & 0 & \frac{\partial N_3^{(e)}}{\partial x} & 0 \\ 0 & \frac{\partial N_1^{(e)}}{\partial y} & 0 & \frac{\partial N_2^{(e)}}{\partial y} & 0 & \frac{\partial N_3^{(e)}}{\partial y} \\ \frac{\partial N_1^{(e)}}{\partial y} & \frac{\partial N_1^{(e)}}{\partial x} & \frac{\partial N_2^{(e)}}{\partial y} & \frac{\partial N_2^{(e)}}{\partial x} & \frac{\partial N_3^{(e)}}{\partial y} & \frac{\partial N_3^{(e)}}{\partial x} \end{bmatrix} \begin{Bmatrix} u_{x1}^{(e)} \\ u_{y1}^{(e)} \\ u_{x2}^{(e)} \\ u_{y2}^{(e)} \\ u_{x3}^{(e)} \\ u_{y3}^{(e)} \end{Bmatrix} \quad (8.5.7)$$

The derivatives of the shape functions with respect to x and y can be readily obtained:

$$\frac{\partial N_1^{(e)}}{\partial x} = \frac{y_2^{(e)} - y_3^{(e)}}{2A^{(e)}}, \quad \frac{\partial N_2^{(e)}}{\partial x} = \frac{y_3^{(e)} - y_1^{(e)}}{2A^{(e)}}, \quad \frac{\partial N_3^{(e)}}{\partial x} = \frac{y_1^{(e)} - y_2^{(e)}}{2A^{(e)}}$$

$$\frac{\partial N_1^{(e)}}{\partial y} = \frac{x_3^{(e)} - x_2^{(e)}}{2A^{(e)}}, \quad \frac{\partial N_2^{(e)}}{\partial y} = \frac{x_1^{(e)} - x_3^{(e)}}{2A^{(e)}}, \quad \frac{\partial N_3^{(e)}}{\partial y} = \frac{x_2^{(e)} - x_1^{(e)}}{2A^{(e)}}$$

Thus, we can finally write:

$$\left\{\varepsilon^{(e)}\right\} = \left[B^{(e)}\right]\left\{U^{(e)}\right\},$$

where

$$\left[B^{(e)}\right] = \frac{1}{2A^{(e)}}\begin{bmatrix} y_2^{(e)} - y_3^{(e)} & 0 & y_3^{(e)} - y_1^{(e)} & 0 & y_1^{(e)} - y_2^{(e)} & 0 \\ 0 & x_3^{(e)} - x_2^{(e)} & 0 & x_1^{(e)} - x_3^{(e)} & 0 & x_2^{(e)} - x_1^{(e)} \\ x_3^{(e)} - x_2^{(e)} & y_2^{(e)} - y_3^{(e)} & x_1^{(e)} - x_3^{(e)} & y_3^{(e)} - y_1^{(e)} & x_2^{(e)} - x_1^{(e)} & y_1^{(e)} - y_2^{(e)} \end{bmatrix} \quad (8.5.8)$$

Since $[B^{(e)}]$ is constant, the strain vector, $\{\varepsilon^{(e)}\}$, is also constant in the interior of the element. For this reason, the three-node triangular element is frequently referred to as the *constant strain triangle* (CST).

As explained in Section 6.7, it is also possible to establish an isoparametric formulation using triangular coordinates ξ_1, ξ_2. We will have a mapping from a parent (parametric) domain, ($\xi_1\xi_2$), to the physical coordinate space (xy), as shown in Figure 8.2.

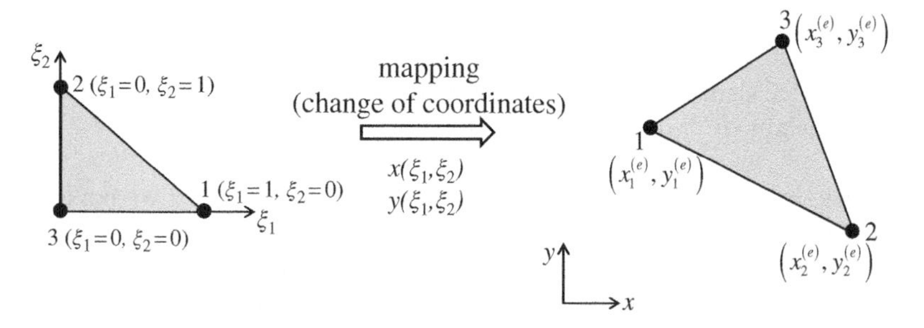

Figure 8.2 Isoparametric three-node triangular (3T) element.

For an isoparametric three-node triangular (3T) element, the displacement field is established in the parametric domain. Per the discussion in Section 6.7, the three shape functions in this formulation are expressed in terms of ξ_1 and ξ_2:

$$N_1^{(3T)}(\xi_1,\xi_2) = \xi_1 \tag{8.5.9a}$$

$$N_2^{(3T)}(\xi_1,\xi_2) = \xi_2 \tag{8.5.9b}$$

$$N_3^{(3T)}(\xi_1,\xi_2) = 1-\xi_1-\xi_2 \tag{8.5.9c}$$

We use the same shape functions to establish the coordinate mapping for the element:

$$x(\xi_1,\xi_2) = N_1^{(3T)}(\xi_1,\xi_2)\cdot x_1^{(e)} + N_2^{(3T)}(\xi_1,\xi_2)\cdot x_2^{(e)} + N_3^{(3T)}(\xi_1,\xi_2)\cdot x_3^{(e)} \tag{8.5.10a}$$

$$y(\xi_1,\xi_2) = N_1^{(3T)}(\xi_1,\xi_2)\cdot y_1^{(e)} + N_2^{(3T)}(\xi_1,\xi_2)\cdot y_2^{(e)} + N_3^{(3T)}(\xi_1,\xi_2)\cdot y_3^{(e)} \tag{8.5.10b}$$

We can also readily calculate the components of the Jacobian array, $[J]$, of the coordinate mapping. If we account for Equations (8.5.9a–c) and (8.5.10a,b), the Jacobian array is:

$$[J] = \begin{bmatrix} \partial x/\partial\xi_1 & \partial y/\partial\xi_1 \\ \partial x/\partial\xi_2 & \partial y/\partial\xi_2 \end{bmatrix} = \begin{bmatrix} x_1^{(e)}-x_3^{(e)} & y_1^{(e)}-y_3^{(e)} \\ x_2^{(e)}-x_3^{(e)} & y_2^{(e)}-y_3^{(e)} \end{bmatrix} = \begin{bmatrix} x_{13}^{(e)} & y_{13}^{(e)} \\ x_{23}^{(e)} & y_{23}^{(e)} \end{bmatrix} \tag{8.5.11}$$

where $x_{ij}^{(e)} = x_i^{(e)} - x_j^{(e)}$, $y_{ij}^{(e)} = y_i^{(e)} - y_j^{(e)}$.

We can invert the Jacobian array to obtain the following expression.

$$[J]^{-1} = \left[\tilde{J}\right] = \begin{bmatrix} \partial\xi_1/\partial x & \partial\xi_2/\partial x \\ \partial\xi_1/\partial y & \partial\xi_2/\partial y \end{bmatrix} = \frac{1}{x_{13}^{(e)}\cdot y_{23}^{(e)} - x_{23}^{(e)}\cdot y_{13}^{(e)}} \begin{bmatrix} y_{23}^{(e)} & -y_{13}^{(e)} \\ -x_{23}^{(e)} & x_{13}^{(e)} \end{bmatrix} \tag{8.5.12}$$

The strain-displacement array $[B^{(e)}]$ of the element contains the derivatives of the shape functions with respect to the physical coordinates, x and y. For each shape function, $N_i^{(3T)}$, we can obtain these partial derivatives using the chain rule of differentiation:

$$\frac{\partial N_i^{(3T)}}{\partial x} = \frac{\partial N_i^{(3T)}}{\partial \xi_1}\frac{\partial \xi_1}{\partial x} + \frac{\partial N_i^{(3T)}}{\partial \xi_2}\frac{\partial \xi_2}{\partial x} = \frac{\partial N_i^{(3T)}}{\partial \xi_1}\tilde{J}_{11} + \frac{\partial N_i^{(3T)}}{\partial \xi_2}\tilde{J}_{12} \quad (8.5.13a)$$

$$\frac{\partial N_i^{(3T)}}{\partial y} = \frac{\partial N_i^{(3T)}}{\partial \xi_1}\frac{\partial \xi_1}{\partial y} + \frac{\partial N_i^{(3T)}}{\partial \xi_2}\frac{\partial \xi_2}{\partial y} = \frac{\partial N_i^{(3T)}}{\partial \xi_1}\tilde{J}_{21} + \frac{\partial N_i^{(3T)}}{\partial \xi_2}\tilde{J}_{22} \quad (8.5.13b)$$

Remark 8.5.1: Calculating the derivatives with Equations (8.5.13a, b) eventually yields the same constant $[B^{(e)}]$ of Equation (8.5.8). ■

Parametric coordinates for triangular elements are useful when we have a material stiffness matrix, $[D^{(e)}]$, which is not constant inside the element or we have distributed body forces. The reason is that these coordinates allow us to calculate domain integrals for any triangular element shape. Specifically, the domain integrals for the 3 T element are given from the following equations:

$$\left[k^{(e)}\right] = \iint_{\Omega^{(e)}} \left[B^{(e)}\right]^T \left[D^{(e)}\right] \left[B^{(e)}\right] dV = \int_0^1 \int_0^{1-\xi_1} \left(\left[B^{(e)}\right]^T \left[D^{(e)}\right] \left[B^{(e)}\right] J \, d\xi_2 d\xi_1 \right) \quad (8.5.14)$$

$$\left\{f_\Omega^{(e)}\right\} = \iint_{\Omega^{(e)}} \left[N^{(e)}\right]^T \{b\} \cdot dV = \int_0^1 \int_0^{1-\xi_1} \left(\left[N^{(e)}\right]^T \{b\} J \, d\xi_2 d\xi_1 \right) \quad (8.5.15)$$

where J is the determinant of the Jacobian matrix, $[J]$, of the coordinate mapping from parametric space to the physical space.

If a side of a triangular element constitutes a segment of the natural boundary, we need to also calculate the contribution of the natural boundary tractions to the nodal force vector of the element. To do so, we *parameterize* the boundary segment, just like we did for the case of two-dimensional heat conduction in Section 6.2.

For example, if we have prescribed tractions t_x and t_y on the side (1-3) of an element, as shown in Figure 8.3, we can parameterize the specific side where we have the applied tractions, as shown in the same figure. The value of the shape functions of Equations (8.5.5a–c) on a boundary segment is expressed in terms of a single parameter ξ, as follows.

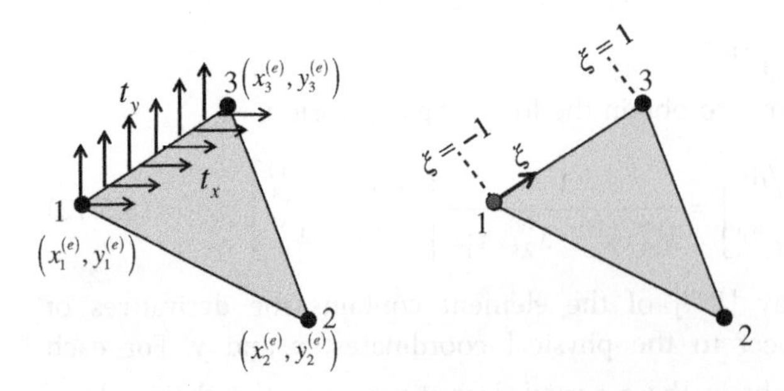

Figure 8.3 Example parameterization of natural boundary segment for three-node triangular element.

$$N_1^{(e)}(x(\xi),y(\xi)) = \frac{1}{2A^{(e)}}\left[x_2^{(e)}y_3^{(e)} - x_3^{(e)}y_2^{(e)} + \left(y_2^{(e)} - y_3^{(e)}\right)x(\xi) + \left(x_3^{(e)} - x_2^{(e)}\right)y(\xi)\right] \quad (8.5.16a)$$

$$N_2^{(e)}(x(\xi),y(\xi)) = \frac{1}{2A^{(e)}}\left[x_3^{(e)}y_1^{(e)} - x_1^{(e)}y_3^{(e)} + \left(y_3^{(e)} - y_1^{(e)}\right)x(\xi) + \left(x_1^{(e)} - x_3^{(e)}\right)y(\xi)\right] \quad (8.5.16b)$$

$$N_3^{(e)}(x(\xi),y(\xi)) = \frac{1}{2A^{(e)}}\left[x_1^{(e)}y_2^{(e)} - x_2^{(e)}y_1^{(e)} + \left(y_1^{(e)} - y_2^{(e)}\right)x(\xi) + \left(x_2^{(e)} - x_1^{(e)}\right)y(\xi)\right] \quad (8.5.16c)$$

where

$$\left.\begin{aligned} x(\xi) &= \frac{x_3^{(e)} + x_1^{(e)}}{2} + \frac{x_3^{(e)} - x_1^{(e)}}{2}\xi \\ y(\xi) &= \frac{y_3^{(e)} + y_1^{(e)}}{2} + \frac{y_3^{(e)} - y_1^{(e)}}{2}\xi \end{aligned}\right\}, \; -1 \le \xi \le 1 \quad (8.5.16d)$$

We can always verify that if a node is not part of the boundary segment under consideration, then the shape function of that node is zero over the segment! For the specific example in Figure 8.3, the shape function of node 2 vanishes on the segment (1-3). In general, the nodal forces due to boundary tractions can be obtained from the following expression.

$$\left\{f_{\Gamma_t}^{(e)}\right\} = \int_{\Gamma_t^{(e)}} \left[N^{(e)}\right]^T \begin{Bmatrix} t_x \\ t_y \end{Bmatrix} ds = \int_{-1}^{1} \left(\begin{bmatrix} N_1^{(e)}(x(\xi),y(\xi)) & 0 \\ 0 & N_1^{(e)}(x(\xi),y(\xi)) \\ N_2^{(e)}(x(\xi),y(\xi)) & 0 \\ 0 & N_2^{(e)}(x(\xi),y(\xi)) \\ N_3^{(e)}(x(\xi),y(\xi)) & 0 \\ 0 & N_3^{(e)}(x(\xi),y(\xi)) \end{bmatrix} \begin{Bmatrix} t_x \\ t_y \end{Bmatrix} \frac{\ell}{2} d\xi \right)$$

(8.5.17)

where ℓ is the length of the boundary segment with the applied tractions. The integral of Equation (8.5.17) can be calculated with one-dimensional Gaussian quadrature.

8.5.2 Quadrilateral Isoparametric Element

We can also use the four-node quadrilateral (4Q) isoparametric element, described in Section 6.4, for two-dimensional elasticity. The 4Q element is available in the vast majority of commercial programs. As shown in Figure 8.4, the element relies on the mapping from a parent (parametric) space, ($\xi\eta$), to the physical (Cartesian) space (xy). We establish the displacement approximation and the coordinate mapping using shape functions in the parametric space.

The approximate displacement field inside the element is given by the following expressions.

$$u_x^{(e)}(\xi,\eta) = N_1^{(4Q)}(\xi,\eta)\cdot u_{x1}^{(e)} + N_2^{(4Q)}(\xi,\eta)\cdot u_{x2}^{(e)} + N_3^{(4Q)}(\xi,\eta)\cdot u_{x3}^{(e)} + N_4^{(4Q)}(\xi,\eta)\cdot u_{x4}^{(e)}$$

(8.5.18a)

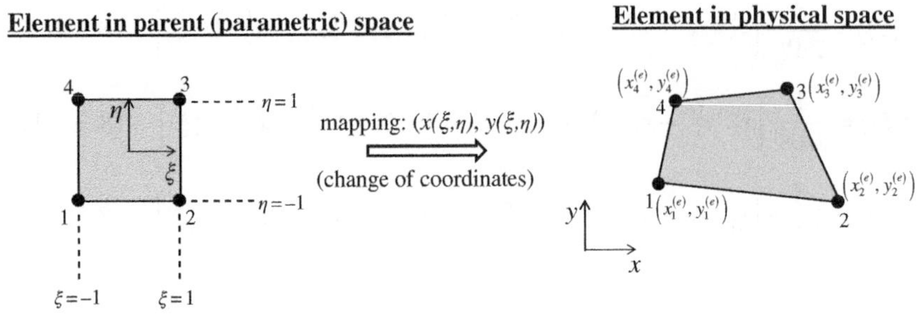

Figure 8.4 Nodal coordinates and displacements for four-node quadrilateral element.

$$u_y^{(e)}(\xi,\eta) = N_1^{(4Q)}(\xi,\eta)\cdot u_{y1}^{(e)} + N_2^{(4Q)}(\xi,\eta)\cdot u_{y2}^{(e)} + N_3^{(4Q)}(\xi,\eta)\cdot u_{y3}^{(e)} + N_4^{(4Q)}(\xi,\eta)\cdot u_{y4}^{(e)} \tag{8.5.18b}$$

Equations (8.5.18a, b) can also be written in the following, more concise form.

$$u_x^{(e)}(\xi,\eta) = \sum_{i=1}^{4}\left(N_i^{(4Q)}(\xi,\eta)\cdot u_{xi}^{(e)}\right)$$

$$u_y^{(e)}(\xi,\eta) = \sum_{i=1}^{4}\left(N_i^{(4Q)}(\xi,\eta)\cdot u_{yi}^{(e)}\right)$$

We can now establish a matrix expression for the approximate displacement field in the element:

$$\left\{u^{(e)}\right\} = \begin{bmatrix} N_1^{(4Q)}(\xi,\eta) & 0 & N_2^{(4Q)}(\xi,\eta) & 0 & N_3^{(4Q)}(\xi,\eta) & 0 & N_4^{(4Q)}(\xi,\eta) & 0 \\ 0 & N_1^{(4Q)}(\xi,\eta) & 0 & N_2^{(4Q)}(\xi,\eta) & 0 & N_3^{(4Q)}(\xi,\eta) & 0 & N_4^{(4Q)}(\xi,\eta) \end{bmatrix} \begin{Bmatrix} u_{x1}^{(e)} \\ u_{y1}^{(e)} \\ u_{x2}^{(e)} \\ u_{y2}^{(e)} \\ u_{x3}^{(e)} \\ u_{y3}^{(e)} \\ u_{x4}^{(e)} \\ u_{y4}^{(e)} \end{Bmatrix}$$

or more concisely:

$$\left\{u^{(e)}\right\} = \left[N^{(4Q)}(\xi,\eta)\right]\left\{U^{(e)}\right\} \tag{8.5.19}$$

Per Section 6.4, the four shape functions of the element are given from the following equations.

$$N_1^{(4Q)}(\xi,\eta) = \frac{1}{4}(1-\xi)(1-\eta) \tag{8.5.20a}$$

$$N_2^{(4Q)}(\xi,\eta) = \frac{1}{4}(1+\xi)(1-\eta) \tag{8.5.20b}$$

$$N_3^{(4Q)}(\xi,\eta) = \frac{1}{4}(1+\xi)(1+\eta) \tag{8.5.20c}$$

$$N_4^{(4Q)}(\xi,\eta) = \frac{1}{4}(1-\xi)(1+\eta) \tag{8.5.20d}$$

As described in Section 6.4, the same shape functions can be used for the coordinate mapping. Thus, we have:

$$x(\xi,\eta) = N_1^{(4Q)}(\xi,\eta)\cdot x_1^{(e)} + N_2^{(4Q)}(\xi,\eta)\cdot x_2^{(e)} + N_3^{(4Q)}(\xi,\eta)\cdot x_3^{(e)} + N_4^{(4Q)}(\xi,\eta)\cdot x_4^{(e)} \tag{8.5.21a}$$

$$y(\xi,\eta) = N_1^{(4Q)}(\xi,\eta)\cdot y_1^{(e)} + N_2^{(4Q)}(\xi,\eta)\cdot y_2^{(e)} + N_3^{(4Q)}(\xi,\eta)\cdot y_3^{(e)} + N_4^{(4Q)}(\xi,\eta)\cdot y_4^{(e)} \tag{8.5.21b}$$

We can use the shape function array of the element to write a concise version of the coordinate mapping equations:

$$\begin{Bmatrix} x(\xi,\eta) \\ y(\xi,\eta) \end{Bmatrix} = \left[N^{(4Q)}(\xi,\eta)\right]\left\{x^{(e)}\right\} \tag{8.5.22}$$

where

$$\left\{x^{(e)}\right\} = \left[x_1^{(e)} \ y_1^{(e)} \ x_2^{(e)} \ y_2^{(e)} \ x_3^{(e)} \ y_3^{(e)} \ x_4^{(e)} \ y_4^{(e)}\right]^T$$

The Jacobian matrix of the coordinate mapping is given by the following expression.

$$[J] = \begin{bmatrix} J_{11} & J_{12} \\ J_{21} & J_{22} \end{bmatrix} = \begin{bmatrix} \frac{\partial x}{\partial \xi} & \frac{\partial y}{\partial \xi} \\ \frac{\partial x}{\partial \eta} & \frac{\partial y}{\partial \eta} \end{bmatrix} = \begin{bmatrix} x,_\xi & y,_\xi \\ x,_\eta & y,_\eta \end{bmatrix} = \begin{bmatrix} \sum_{i=1}^{4}\left(\frac{\partial N_i^{(4Q)}}{\partial \xi}\cdot x_i^{(e)}\right) & \sum_{i=1}^{4}\left(\frac{\partial N_i^{(4Q)}}{\partial \xi}\cdot y_i^{(e)}\right) \\ \sum_{i=1}^{4}\left(\frac{\partial N_i^{(4Q)}}{\partial \eta}\cdot x_i^{(e)}\right) & \sum_{i=1}^{4}\left(\frac{\partial N_i^{(4Q)}}{\partial \eta}\cdot y_i^{(e)}\right) \end{bmatrix} \tag{8.5.23}$$

We can expand the equations in the right-hand side of Equation (8.5.23) to obtain:

$$\frac{\partial x}{\partial \xi} = \frac{\partial N_1^{(4Q)}}{\partial \xi}\cdot x_1^{(e)} + \frac{\partial N_2^{(4Q)}}{\partial \xi}\cdot x_2^{(e)} + \frac{\partial N_3^{(4Q)}}{\partial \xi}\cdot x_3^{(e)} + \frac{\partial N_4^{(4Q)}}{\partial \xi}\cdot x_4^{(e)} \tag{8.5.24a}$$

$$\frac{\partial x}{\partial \eta} = \frac{\partial N_1^{(4Q)}}{\partial \eta}\cdot x_1^{(e)} + \frac{\partial N_2^{(4Q)}}{\partial \eta}\cdot x_2^{(e)} + \frac{\partial N_3^{(4Q)}}{\partial \eta}\cdot x_3^{(e)} + \frac{\partial N_4^{(4Q)}}{\partial \eta}\cdot x_4^{(e)} \tag{8.5.24b}$$

$$\frac{\partial y}{\partial \xi} = \frac{\partial N_1^{(4Q)}}{\partial \xi}\cdot y_1^{(e)} + \frac{\partial N_2^{(4Q)}}{\partial \xi}\cdot y_2^{(e)} + \frac{\partial N_3^{(4Q)}}{\partial \xi}\cdot y_3^{(e)} + \frac{\partial N_4^{(4Q)}}{\partial \xi}\cdot y_4^{(e)} \tag{8.5.24c}$$

$$\frac{\partial y}{\partial \eta} = \frac{\partial N_1^{(4Q)}}{\partial \eta} \cdot y_1^{(e)} + \frac{\partial N_2^{(4Q)}}{\partial \eta} \cdot y_2^{(e)} + \frac{\partial N_3^{(4Q)}}{\partial \eta} \cdot y_3^{(e)} + \frac{\partial N_4^{(4Q)}}{\partial \eta} \cdot y_4^{(e)} \tag{8.5.24d}$$

We can cast Equations (8.5.24a–d) in a more concise, matrix form:

$$[J] = \begin{bmatrix} \frac{\partial x}{\partial \xi} & \frac{\partial y}{\partial \xi} \\ \frac{\partial x}{\partial \eta} & \frac{\partial y}{\partial \eta} \end{bmatrix} = \left[N_{,\xi}^{(4Q)}\right]\left[\{x^{(e)}\} \quad \{y^{(e)}\}\right] \tag{8.5.25}$$

where

$$\left[N_{,\xi}^{(4Q)}\right] = \begin{bmatrix} \frac{\partial N_1^{(4Q)}}{\partial \xi} & \frac{\partial N_2^{(4Q)}}{\partial \xi} & \frac{\partial N_3^{(4Q)}}{\partial \xi} & \frac{\partial N_4^{(4Q)}}{\partial \xi} \\ \frac{\partial N_1^{(4Q)}}{\partial \eta} & \frac{\partial N_2^{(4Q)}}{\partial \eta} & \frac{\partial N_3^{(4Q)}}{\partial \eta} & \frac{\partial N_4^{(4Q)}}{\partial \eta} \end{bmatrix} \tag{8.5.26}$$

and

$$\left\{x^{(e)}\right\} = \left[x_1^{(e)} \; x_2^{(e)} \; x_3^{(e)} \; x_4^{(e)}\right]^T, \; \left\{y^{(e)}\right\} = \left[y_1^{(e)} \; y_2^{(e)} \; y_3^{(e)} \; y_4^{(e)}\right]^T \tag{8.5.27}$$

Given the Jacobian array, $[J]$, we can also establish its inverse:

$$[J]^{-1} = [\tilde{J}] = \begin{bmatrix} \tilde{J}_{11} & \tilde{J}_{12} \\ \tilde{J}_{21} & \tilde{J}_{22} \end{bmatrix} = \begin{bmatrix} \xi_{,x} & \eta_{,x} \\ \xi_{,y} & \eta_{,y} \end{bmatrix} = \begin{bmatrix} \frac{\partial \xi}{\partial x} & \frac{\partial \eta}{\partial x} \\ \frac{\partial \xi}{\partial y} & \frac{\partial \eta}{\partial y} \end{bmatrix} \tag{8.5.28}$$

The strain-displacement relation in the element can be written as follows.

$$\begin{Bmatrix} \varepsilon_{xx}^{(e)} \\ \varepsilon_{yy}^{(e)} \\ \gamma_{xy}^{(e)} \end{Bmatrix} = \begin{bmatrix} \frac{\partial N_1^{(4Q)}}{\partial x} & 0 & \frac{\partial N_2^{(4Q)}}{\partial x} & 0 & \frac{\partial N_3^{(4Q)}}{\partial x} & 0 & \frac{\partial N_4^{(4Q)}}{\partial x} & 0 \\ 0 & \frac{\partial N_1^{(4Q)}}{\partial y} & 0 & \frac{\partial N_2^{(4Q)}}{\partial y} & 0 & \frac{\partial N_3^{(4Q)}}{\partial y} & 0 & \frac{\partial N_4^{(4Q)}}{\partial y} \\ \frac{\partial N_1^{(4Q)}}{\partial y} & \frac{\partial N_1^{(4Q)}}{\partial x} & \frac{\partial N_2^{(4Q)}}{\partial y} & \frac{\partial N_2^{(4Q)}}{\partial x} & \frac{\partial N_3^{(4Q)}}{\partial y} & \frac{\partial N_3^{(4Q)}}{\partial x} & \frac{\partial N_4^{(4Q)}}{\partial y} & \frac{\partial N_4^{(4Q)}}{\partial x} \end{bmatrix} \begin{Bmatrix} u_{x1}^{(e)} \\ u_{y1}^{(e)} \\ u_{x2}^{(e)} \\ u_{y2}^{(e)} \\ u_{x3}^{(e)} \\ u_{y3}^{(e)} \\ u_{x4}^{(e)} \\ u_{y4}^{(e)} \end{Bmatrix} \tag{8.5.29}$$

We can collectively write $\{\varepsilon^{(e)}\} = [B^{(e)}]\{U^{(e)}\}$, where the strain-displacement array $[B^{(e)}]$ is given by:

$$[B^{(e)}] = \begin{bmatrix} \frac{\partial N_1^{(4Q)}}{\partial x} & 0 & \frac{\partial N_2^{(4Q)}}{\partial x} & 0 & \frac{\partial N_3^{(4Q)}}{\partial x} & 0 & \frac{\partial N_4^{(4Q)}}{\partial x} & 0 \\ 0 & \frac{\partial N_1^{(4Q)}}{\partial y} & 0 & \frac{\partial N_2^{(4Q)}}{\partial y} & 0 & \frac{\partial N_3^{(4Q)}}{\partial y} & 0 & \frac{\partial N_4^{(4Q)}}{\partial y} \\ \frac{\partial N_1^{(4Q)}}{\partial y} & \frac{\partial N_1^{(4Q)}}{\partial x} & \frac{\partial N_2^{(4Q)}}{\partial y} & \frac{\partial N_2^{(4Q)}}{\partial x} & \frac{\partial N_3^{(4Q)}}{\partial y} & \frac{\partial N_3^{(4Q)}}{\partial x} & \frac{\partial N_4^{(4Q)}}{\partial y} & \frac{\partial N_4^{(4Q)}}{\partial x} \end{bmatrix} \tag{8.5.30}$$

The derivatives of the shape functions with the physical coordinates are obtained using the chain rule of differentiation:

$$\frac{\partial N_i^{(4Q)}}{\partial x} = \frac{\partial N_i^{(4Q)}}{\partial \xi_1}\frac{\partial \xi}{\partial x} + \frac{\partial N_i^{(4Q)}}{\partial \eta}\frac{\partial \eta}{\partial x} \tag{8.5.31a}$$

$$\frac{\partial N_i^{(4Q)}}{\partial y} = \frac{\partial N_i^{(4Q)}}{\partial \xi}\frac{\partial \xi}{\partial y} + \frac{\partial N_i^{(4Q)}}{\partial \eta}\frac{\partial \eta}{\partial y} \tag{8.5.31b}$$

Since we know that the partial derivatives of the parametric coordinates with the physical coordinates are the components of the $[\tilde{J}]$ array, we can write Equations (8.5.31a, b) in the following form.

$$\frac{\partial N_i^{(4Q)}}{\partial x} = \frac{\partial N_i^{(4Q)}}{\partial \xi}\tilde{J}_{11} + \frac{\partial N_i^{(4Q)}}{\partial \eta}\tilde{J}_{12} \tag{8.5.32a}$$

$$\frac{\partial N_i^{(4Q)}}{\partial y} = \frac{\partial N_i^{(4Q)}}{\partial \xi}\tilde{J}_{21} + \frac{\partial N_i^{(4Q)}}{\partial \eta}\tilde{J}_{22} \tag{8.5.32b}$$

Similar to what we did for heat conduction in Section 6.4, we can evaluate the matrix containing all the partial derivatives of the shape functions with the physical coordinates using the expression:

$$\left[N_{,x}^{(4Q)}\right] = \begin{bmatrix} \tilde{J}_{11} & \tilde{J}_{12} \\ \tilde{J}_{21} & \tilde{J}_{22} \end{bmatrix} \begin{bmatrix} \frac{\partial N_1^{(4Q)}}{\partial \xi} & \frac{\partial N_2^{(4Q)}}{\partial \xi} & \frac{\partial N_3^{(4Q)}}{\partial \xi} & \frac{\partial N_4^{(4Q)}}{\partial \xi} \\ \frac{\partial N_1^{(4Q)}}{\partial \eta} & \frac{\partial N_2^{(4Q)}}{\partial \eta} & \frac{\partial N_3^{(4Q)}}{\partial \eta} & \frac{\partial N_4^{(4Q)}}{\partial \eta} \end{bmatrix} = [\tilde{J}]\left[N_{,\xi}^{(4Q)}\right] \tag{8.5.33}$$

where

$$\left[N_{,x}^{(4Q)}\right] = \begin{bmatrix} \frac{\partial N_1^{(4Q)}}{\partial x} & \frac{\partial N_2^{(4Q)}}{\partial x} & \frac{\partial N_3^{(4Q)}}{\partial x} & \frac{\partial N_4^{(4Q)}}{\partial x} \\ \frac{\partial N_1^{(4Q)}}{\partial y} & \frac{\partial N_2^{(4Q)}}{\partial y} & \frac{\partial N_3^{(4Q)}}{\partial y} & \frac{\partial N_4^{(4Q)}}{\partial y} \end{bmatrix} \tag{8.5.34}$$

Once we obtain the $\left[N_{,x}^{(4Q)}\right]$ array, we can establish the $[B^{(e)}]$ array of Equation (8.5.30). The element stiffness matrix is given by:

$$\left[k^{(e)}\right] = \iint_{\Omega^{(e)}} \left[B^{(e)}\right]^T \left[D^{(e)}\right] \left[B^{(e)}\right] dV = \int_{-1}^{1}\int_{-1}^{1} \left(\left[B^{(e)}\right]^T \left[D^{(e)}\right] \left[B^{(e)}\right] J d\xi d\eta\right) \tag{8.5.35}$$

and the element equivalent nodal force vector is given by:

$$\left\{f^{(e)}\right\} = \iint_{\Omega^{(e)}} \left[N^{(e)}\right]^T \{b\} dV + \int_{\Gamma_{tx}} \left[N^{(e)}\right]^T \{t_x\} dS + \int_{\Gamma_{ty}} \left[N^{(e)}\right]^T \{t_y\} dS \tag{8.5.36}$$

The equivalent nodal force vector can also be obtained as the sum of a domain term and a natural boundary term, in accordance with Equations (8.5.1–3):

$$\left\{f^{(e)}\right\} = \left\{f_{\Omega}^{(e)}\right\} + \left\{f_{\Gamma_t}^{(e)}\right\}$$

where

$$\left\{f_{\Omega}^{(e)}\right\} = \iint_{\Omega^{(e)}} \left[N^{(4Q)}\right]^T \{b\} \cdot dV \text{ and} \left\{f_{\Gamma_t}^{(e)}\right\} = \int_{\Gamma_{tx}^{(e)}} \left[N^{(4Q)}\right]^T \{t_x\} dS + \int_{\Gamma_{ty}^{(e)}} \left[N^{(4Q)}\right]^T \{t_y\} dS$$

We can transform the element domain integral of $\left\{f_{\Omega}^{(e)}\right\}$ to an integral over the parametric domain:

$$\left\{f_{\Omega}^{(e)}\right\} = \iint_{\Omega^{(e)}} \left[N^{(4Q)}\right]^T \{b\} \cdot dV = \int_{-1}^{1}\int_{-1}^{1} \left(\left[N^{(4Q)}\right]^T \{b\} J \, d\xi d\eta\right)$$

Additionally, the contribution of a natural boundary segment (with length equal to ℓ) to the equivalent nodal forces is given by:

$$\left\{f_{\Gamma_t}^{(e)}\right\} = \int_{\Gamma_t^{(e)}} \left[N^{(4Q)}\right]^T \begin{Bmatrix} t_x \\ t_y \end{Bmatrix} ds = \int_{-1}^{1} \left(\left[N^{(4Q)}\right]^T \begin{Bmatrix} t_x \\ t_y \end{Bmatrix} \frac{\ell}{2} d\xi\right)$$ if the natural boundary segment corresponds to a constant value of η (at which case ξ can be used for the parameterization of the boundary), and

$$\left\{f_{\Gamma_t}^{(e)}\right\} = \int_{-1}^{1} \left(\left[N^{(4Q)}\right]^T \begin{Bmatrix} t_x \\ t_y \end{Bmatrix} \frac{\ell}{2} d\eta\right)$$ if the natural boundary segment corresponds to a constant value of ξ.

Remark 8.5.2: For an element natural boundary edge corresponding to a constant value of ξ or η, only two of the shape functions will be nonzero, namely, the shape functions of the nodes comprising $\Gamma_t^{(e)}$! ∎

Remark 8.5.3: It is important to reiterate that $[k^{(e)}]$ and $\{f^{(e)}\}$ are values per unit thickness of the element. ∎

The domain and the natural boundary contributions to $\{f^{(e)}\}$ are calculated using two-dimensional and one-dimensional Gaussian quadrature, respectively. The necessary calculations for $[k^{(e)}]$ and $\left\{f_{\Omega}^{(e)}\right\}$ in a 4Q element are provided in Box 8.5.1. The same algorithmic steps can be used for the three-node triangular (3 T) isoparametric element described in the previous section; the only difference for the 3 T element is that we have three shape functions and ξ_1, ξ_2 as the parametric coordinates.

Box 8.5.1 Calculation of $[k^{(e)}]$, $\left\{f_{\Omega}^{(e)}\right\}$ for 4Q Element

FOR EACH QUADRATURE POINT g ($g = 1, 2, \ldots, N_g$):

1) Establish parametric coordinates, ξ_g, η_g.
2) Calculate the values of the shape functions at the specific point:

$$N_{1g}^{(4Q)} = N_1^{(4Q)}\left(\xi_g,\eta_g\right) = \frac{1}{4}\left(1-\xi_g\right)\left(1-\eta_g\right)$$

$$N_{2g}^{(4Q)} = N_2^{(4Q)}\left(\xi_g,\eta_g\right) = \frac{1}{4}\left(1+\xi_g\right)\left(1-\eta_g\right)$$

$$N_{3g}^{(4Q)} = N_3^{(4Q)}\left(\xi_g,\eta_g\right) = \frac{1}{4}\left(1+\xi_g\right)\left(1+\eta_g\right)$$

$$N_{4g}^{(4Q)} = N_4^{(4Q)}\left(\xi_g,\eta_g\right) = \frac{1}{4}\left(1-\xi_g\right)\left(1+\eta_g\right)$$

Also, set:

$$\left[N_g\right] = \begin{bmatrix} N_{1g}^{(4Q)} & 0 & N_{2g}^{(4Q)} & 0 & N_{3g}^{(4Q)} & 0 & N_{4g}^{(4Q)} & 0 \\ 0 & N_{1g}^{(4Q)} & 0 & N_{2g}^{(4Q)} & 0 & N_{3g}^{(4Q)} & 0 & N_{4g}^{(4Q)} \end{bmatrix}$$

3) Calculate values of coordinates in physical space:

$$x_g = N_1^{(4Q)}\left(\xi_g,\eta_g\right)x_1^{(e)} + N_2^{(4Q)}\left(\xi_g,\eta_g\right)x_2^{(e)} + N_3^{(4Q)}\left(\xi_g,\eta_g\right)x_3^{(e)} + N_4^{(4Q)}\left(\xi_g,\eta_g\right)x_4^{(e)}$$

$$y_g = N_1^{(4Q)}\left(\xi_g,\eta_g\right)y_1^{(e)} + N_2^{(4Q)}\left(\xi_g,\eta_g\right)y_2^{(e)} + N_3^{(4Q)}\left(\xi_g,\eta_g\right)y_3^{(e)} + N_4^{(4Q)}\left(\xi_g,\eta_g\right)y_4^{(e)}$$

4) Calculate the Jacobian matrix of the mapping at that point, as well as the Jacobian determinant, $J_g = det([J_g])$, and the inverse of the Jacobian matrix, $\left[\tilde{J}_g\right] = \left[J_g\right]^{-1}$
5) Calculate the derivatives of the shape functions with respect to the parametric coordinates, ξ and η, at the location $\xi = \xi_g$, $\eta = \eta_g$:

$$\left[N_{,\xi}^{(4Q)}\right]_{\xi_g,\eta_g} = \begin{bmatrix} \dfrac{\partial N_1^{(4Q)}}{\partial \xi} & \dfrac{\partial N_2^{(4Q)}}{\partial \xi} & \dfrac{\partial N_3^{(4Q)}}{\partial \xi} & \dfrac{\partial N_4^{(4Q)}}{\partial \xi} \\ \dfrac{\partial N_1^{(4Q)}}{\partial \eta} & \dfrac{\partial N_2^{(4Q)}}{\partial \eta} & \dfrac{\partial N_3^{(4Q)}}{\partial \eta} & \dfrac{\partial N_4^{(4Q)}}{\partial \eta} \end{bmatrix}_{\xi_g,\eta_g}$$

6) Calculate the derivatives of the shape functions with respect to the physical coordinates, x and y, at $\xi = \xi_g$, $\eta = \eta_g$: $\left[N_{,x}^{(4Q)}\right]_{\xi_g,\eta_g} = \left[\tilde{J}_g\right]\left[N_{,\xi}^{(4Q)}\right]_{\xi_g,\eta_g}$, with

$$\left[N_{,x}^{(4Q)}\right]_{\xi_g,\eta_g} = \begin{bmatrix} \frac{\partial N_1^{(4Q)}}{\partial x} & \frac{\partial N_2^{(4Q)}}{\partial x} & \frac{\partial N_3^{(4Q)}}{\partial x} & \frac{\partial N_4^{(4Q)}}{\partial x} \\ \frac{\partial N_1^{(4Q)}}{\partial y} & \frac{\partial N_2^{(4Q)}}{\partial y} & \frac{\partial N_3^{(4Q)}}{\partial y} & \frac{\partial N_4^{(4Q)}}{\partial y} \end{bmatrix}_{\xi_g,\eta_g}$$

7) Establish the $[B^{(e)}]$ array at Gauss point g:

$$[B_g] = \begin{bmatrix} \frac{\partial N_1^{(4Q)}}{\partial x} & 0 & \frac{\partial N_2^{(4Q)}}{\partial x} & 0 & \frac{\partial N_3^{(4Q)}}{\partial x} & 0 & \frac{\partial N_4^{(4Q)}}{\partial x} & 0 \\ 0 & \frac{\partial N_1^{(4Q)}}{\partial y} & 0 & \frac{\partial N_2^{(4Q)}}{\partial y} & 0 & \frac{\partial N_3^{(4Q)}}{\partial y} & 0 & \frac{\partial N_4^{(4Q)}}{\partial y} \\ \frac{\partial N_1^{(4Q)}}{\partial y} & \frac{\partial N_1^{(4Q)}}{\partial x} & \frac{\partial N_2^{(4Q)}}{\partial y} & \frac{\partial N_2^{(4Q)}}{\partial x} & \frac{\partial N_3^{(4Q)}}{\partial y} & \frac{\partial N_3^{(4Q)}}{\partial x} & \frac{\partial N_4^{(4Q)}}{\partial y} & \frac{\partial N_4^{(4Q)}}{\partial x} \end{bmatrix}_{\xi_g,\eta_g}$$

8) Calculate the value of body force and of material stiffness matrix, $[D^{(e)}]$ at point g:

$$\{b_g\} = \begin{Bmatrix} b_x \\ b_y \end{Bmatrix}_g = \begin{Bmatrix} b_x(x=x_g, y=y_g) \\ b_y(x=x_g, y=y_g) \end{Bmatrix}, \quad [D_g] = [D(x=x_g, y=y_g)]$$

END

Finally, combine the contributions of all the Gauss points to calculate the element stiffness matrix, $[k^{(e)}]$, and body-force contribution to the equivalent nodal force vector, $\left\{f_\Omega^{(e)}\right\}$:

$$[k^{(e)}] = \int_{-1}^{1}\int_{-1}^{1}\left(\left[B^{(e)}\right]^T\left[D^{(e)}\right]\left[B^{(e)}\right]Jd\xi d\eta\right) \approx \sum_{g=1}^{Ng}\left([B_g]^T[D_g][B_g]J_g W_g\right)$$

$$\left\{f_\Omega^{(e)}\right\} = \int_{-1}^{1}\int_{-1}^{1}\left(\left[N^{(e)}\right]^T\{b\}Jd\xi d\eta\right) \approx \sum_{g=1}^{Ng}\left([N_g]^T\{b_g\}J_g W_g\right)$$

We also need to use the algorithm in Box 8.5.2 to find the nodal forces $\left\{f_{\Gamma_t}^{(e)}\right\}$ due to prescribed tractions at edges of the element. The procedure corresponds to the case that we use N_g quadrature points for the one-dimensional integration on the boundary. Finally, we can calculate the equivalent nodal force vector for element e, $\{f^{(e)}\} = \left\{f_\Omega^{(e)}\right\} + \left\{f_{\Gamma_t}^{(e)}\right\}$, and then use the element gather-scatter array $[L^{(e)}]$ to find the contributions of the element to the global stiffness matrix, $[K]$, and the global equivalent nodal force vector, $\{f\}$. In actual analysis programs, instead of using the gather-scatter array, we may rely on the *ID* and *LM* arrays, as explained in Sections B.4 and B.5.

To facilitate understanding of the considerations for isoparametric elements in two-dimensional elasticity, the following subsection will present a quantitative example.

Box 8.5.2 Calculation of $\{f^{(e)}_{\Gamma t}\}$ for 4Q Element

- If the natural boundary segment corresponds to a constant value of η, $\eta = \bar{\eta}$ (where $\bar{\eta} = -1$ or 1), we need to evaluate a one-dimensional integral with respect to ξ:

$$\left\{f^{(e)}_{\Gamma_t}\right\} = \int_{\Gamma_t^{(e)}} \left[N^{(4Q)}\right]^T \begin{Bmatrix} t_x \\ t_y \end{Bmatrix} ds = \int_{-1}^{1} \left(\left[N^{(4Q)}(\xi, \bar{\eta})\right]^T \begin{Bmatrix} t_x \\ t_y \end{Bmatrix} \frac{\ell}{2} d\xi\right) \approx \sum_{g=1}^{Ng} \left(\left[N^{(4Q)}(\xi_g, \bar{\eta})\right]^T \begin{Bmatrix} t_x \\ t_y \end{Bmatrix} \frac{\ell}{2} W_g\right)$$

where ℓ is the length of the boundary segment.

- If the natural boundary segment corresponds to a constant value of ξ, $\xi = \bar{\xi}$ (where $\bar{\xi} = -1$ or 1), we need to evaluate a one-dimensional integral with respect to η:

$$\left\{f^{(e)}_{\Gamma_t}\right\} = \int_{\Gamma_t^{(e)}} \left[N^{(4Q)}\right]^T \begin{Bmatrix} t_x \\ t_y \end{Bmatrix} ds = \int_{-1}^{1} \left(\left[N^{(4Q)}(\bar{\xi}, \eta)\right]^T \begin{Bmatrix} t_x \\ t_y \end{Bmatrix} \frac{\ell}{2} d\eta\right) \approx \sum_{g=1}^{Ng} \left(\left[N^{(4Q)}(\bar{\xi}, \eta_g)\right]^T \begin{Bmatrix} t_x \\ t_y \end{Bmatrix} \frac{\ell}{2} W_g\right)$$

8.5.3 Example: Calculation of Stiffness Matrix and Equivalent Nodal Forces for Four-Node Quadrilateral Isoparametric Element

We want to find the nodal displacements for the single-element mesh shown in Figure 8.5. We are given that the material is linearly elastic, isotropic, and uniform (homogeneous), with $E = 29{,}000$ ksi and $\nu = 0.3$. We assume a plane-stress condition (*why?*). For the quadrilateral element, we will use 2 × 2 quadrature (four Gauss points) for domain integrals and two-point quadrature for the boundary integrals.

The material stiffness matrix is *constant* (uniform material) and is given by (plane-stress condition):

$$[D] = \frac{E}{1-\nu^2}\begin{bmatrix} 1 & \nu & 0 \\ \nu & 1 & 0 \\ 0 & 0 & \frac{1-\nu}{2} \end{bmatrix} = \begin{bmatrix} 31868 & 9560 & 0 \\ 9560 & 31868 & 0 \\ 0 & 0 & 11154 \end{bmatrix} \text{(values in ksi)}$$

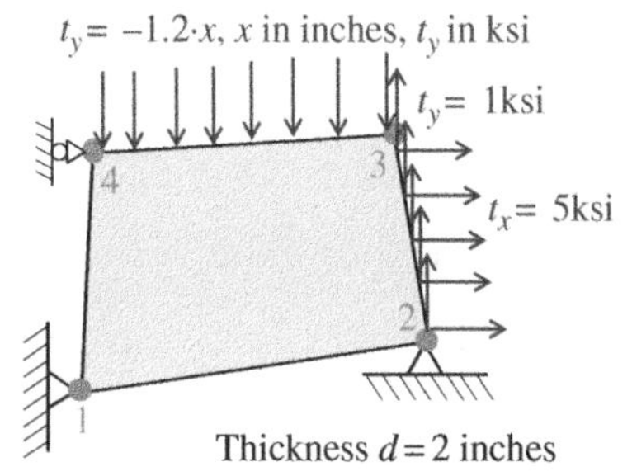

Node	x_i(in.)	y_i(in.)
1	0	0
2	25	5
3	20	25
4	5	20

Figure 8.5 Example single-element mesh for two-dimensional elasticity problem.

We can write the equations for the coordinate mapping:

$$x(\xi,\eta) = \frac{1}{4}(1-\xi)(1-\eta)\cdot x_1 + \frac{1}{4}(1+\xi)(1-\eta)\cdot x_2 + \frac{1}{4}(1+\xi)(1+\eta)\cdot x_3 + \frac{1}{4}(1-\xi)(1+\eta)\cdot x_4$$
$$= \frac{1}{4}(1-\xi)(1-\eta)\cdot 0 + \frac{1}{4}(1+\xi)(1-\eta)\cdot 25 + \frac{1}{4}(1+\xi)(1+\eta)\cdot 20 + \frac{1}{4}(1-\xi)(1+\eta)\cdot 5$$

$$y(\xi,\eta) = \frac{1}{4}(1-\xi)(1-\eta)\cdot y_1 + \frac{1}{4}(1+\xi)(1-\eta)\cdot y_2 + \frac{1}{4}(1+\xi)(1+\eta)\cdot y_3 + \frac{1}{4}(1-\xi)(1+\eta)\cdot y_4$$
$$= \frac{1}{4}(1-\xi)(1-\eta)\cdot 0 + \frac{1}{4}(1+\xi)(1-\eta)\cdot 5 + \frac{1}{4}(1+\xi)(1+\eta)\cdot 25 + \frac{1}{4}(1-\xi)(1+\eta)\cdot 20$$

We can now also find the components of the [*J*]-array:

$$\frac{\partial x}{\partial \xi} = -\frac{1}{4}(1-\eta)\cdot x_1 + \frac{1}{4}(1-\eta)\cdot x_2 + \frac{1}{4}(1+\eta)\cdot x_3 - \frac{1}{4}(1+\eta)\cdot x_4$$
$$= -\frac{1}{4}(1-\eta)\cdot 0 + \frac{1}{4}(1-\eta)\cdot 25 + \frac{1}{4}(1+\eta)\cdot 20 - \frac{1}{4}(1+\eta)\cdot 5$$
$$\frac{\partial y}{\partial \xi} = -\frac{1}{4}(1-\eta)\cdot y_1 + \frac{1}{4}(1-\eta)\cdot y_2 + \frac{1}{4}(1+\eta)\cdot y_3 - \frac{1}{4}(1+\eta)\cdot y_4$$
$$= -\frac{1}{4}(1-\eta)\cdot 0 + \frac{1}{4}(1-\eta)\cdot 5 + \frac{1}{4}(1+\eta)\cdot 25 - \frac{1}{4}(1+\eta)\cdot 20$$
$$\frac{\partial x}{\partial \eta} = -\frac{1}{4}(1-\xi)\cdot x_1 - \frac{1}{4}(1+\xi)\cdot x_2 + \frac{1}{4}(1+\xi)\cdot x_3 + \frac{1}{4}(1-\xi)\cdot x_4$$
$$= -\frac{1}{4}(1-\xi)\cdot 0 - \frac{1}{4}(1+\xi)\cdot 25 + \frac{1}{4}(1+\xi)\cdot 20 + \frac{1}{4}(1-\xi)\cdot 5$$
$$\frac{\partial y}{\partial \eta} = -\frac{1}{4}(1-\xi)\cdot y_1 - \frac{1}{4}(1+\xi)\cdot y_2 + \frac{1}{4}(1+\xi)\cdot y_3 + \frac{1}{4}(1-\xi)\cdot y_4$$
$$= -\frac{1}{4}(1-\xi)\cdot 0 - \frac{1}{4}(1+\xi)\cdot 5 + \frac{1}{4}(1+\xi)\cdot 25 + \frac{1}{4}(1-\xi)\cdot 20$$

We now go ahead with the computations of each Gauss point.

Gauss point 1: $\xi_1 = -\frac{1}{\sqrt{3}}$, $\eta_1 = -\frac{1}{\sqrt{3}}$, $W_1 = 1$

$$[D_1] = \frac{E}{1-\nu^2}\begin{bmatrix} 1 & \nu & 0 \\ \nu & 1 & 0 \\ 0 & 0 & \frac{1-\nu}{2} \end{bmatrix} = \begin{bmatrix} 31868 & 9560 & 0 \\ 9560 & 31868 & 0 \\ 0 & 0 & 11154 \end{bmatrix}$$

$$[N_1] = \begin{bmatrix} 0.62 & 0.00 & 0.17 & 0.00 & 0.04 & 0.00 & 0.17 & 0.00 \\ 0.00 & 0.62 & 0.00 & 0.17 & 0.00 & 0.04 & 0.00 & 0.17 \end{bmatrix}$$

$$[J_1] = \begin{bmatrix} 11.44 & 2.50 \\ 1.44 & 10.00 \end{bmatrix}, \ J_1 = 110.83, \ \left[\tilde{J}_1\right] = \begin{bmatrix} 0.09 & -0.02 \\ -0.01 & 0.10 \end{bmatrix}$$

$$[B_1] = \begin{bmatrix} -0.027 & 0.000 & 0.038 & 0.000 & 0.007 & 0.000 & -0.018 & 0.000 \\ 0.000 & -0.036 & 0.000 & -0.016 & 0.000 & 0.010 & 0.000 & 0.042 \\ -0.036 & -0.027 & -0.016 & 0.038 & 0.010 & 0.007 & 0.042 & -0.018 \end{bmatrix}$$

$x_1 = 5.89$ in., $y_1 = 5.28$ in.
$b_{x1} = 17.07$ kip/in.^{3}, $b_{y1} = 0$
Gauss point 2: $\xi_2 = \dfrac{1}{\sqrt{3}}$, $\eta_2 = -\dfrac{1}{\sqrt{3}}$, $W_2 = 1$

$$[D_2] = \frac{E}{1-\nu^2}\begin{bmatrix} 1 & \nu & 0 \\ \nu & 1 & 0 \\ 0 & 0 & \dfrac{1-\nu}{2} \end{bmatrix} = \begin{bmatrix} 31868 & 9560 & 0 \\ 9560 & 31868 & 0 \\ 0 & 0 & 11154 \end{bmatrix}$$

$$[N_2] = \begin{bmatrix} 0.17 & 0.00 & 0.62 & 0.00 & 0.17 & 0.00 & 0.04 & 0.00 \\ 0.00 & 0.17 & 0.00 & 0.62 & 0.00 & 0.17 & 0.00 & 0.04 \end{bmatrix}$$

$$[J_2] = \begin{bmatrix} 11.44 & 2.50 \\ -1.44 & 10.00 \end{bmatrix}, \quad J_2 = 118.04, \quad \left[\tilde{J}_2\right] = \begin{bmatrix} 0.08 & -0.02 \\ 0.01 & 0.10 \end{bmatrix}$$

$$[B_2] = \begin{bmatrix} -0.031 & 0.000 & 0.042 & 0.000 & 0.001 & 0.000 & -0.011 & 0.000 \\ 0.000 & -0.015 & 0.000 & -0.033 & 0.000 & 0.040 & 0.000 & 0.009 \\ -0.015 & -0.031 & -0.033 & 0.042 & 0.040 & 0.001 & 0.009 & -0.011 \end{bmatrix}$$

$x_2 = 19.11$ in., $y_2 = 8.17$ in.
$b_{x2} = 46.38$ kip/in.^{3}, $b_{y2} = 0$
Gauss point 3: $\xi_3 = \dfrac{1}{\sqrt{3}}$, $\eta_3 = \dfrac{1}{\sqrt{3}}$, $W_3 = 1$

$$[D_3] = \frac{E}{1-\nu^2}\begin{bmatrix} 1 & \nu & 0 \\ \nu & 1 & 0 \\ 0 & 0 & \dfrac{1-\nu}{2} \end{bmatrix} = \begin{bmatrix} 31868 & 9560 & 0 \\ 9560 & 31868 & 0 \\ 0 & 0 & 11154 \end{bmatrix}$$

$$[N_3] = \begin{bmatrix} 0.04 & 0.00 & 0.17 & 0.00 & 0.62 & 0.00 & 0.17 & 0.00 \\ 0.00 & 0.04 & 0.00 & 0.17 & 0.00 & 0.62 & 0.00 & 0.17 \end{bmatrix}$$

$$[J_3] = \begin{bmatrix} 8.56 & 2.50 \\ -1.44 & 10.00 \end{bmatrix}, \quad J_3 = 89.17, \quad \left[\tilde{J}_3\right] = \begin{bmatrix} 0.11 & -0.03 \\ 0.02 & 0.10 \end{bmatrix}$$

$$[B_3] = \begin{bmatrix} -0.009 & 0.000 & 0.023 & 0.000 & 0.033 & 0.000 & -0.047 & 0.000 \\ 0.000 & -0.012 & 0.000 & -0.036 & 0.000 & 0.044 & 0.000 & 0.004 \\ -0.012 & -0.009 & -0.036 & 0.023 & 0.044 & 0.033 & 0.004 & -0.047 \end{bmatrix}$$

$x_3 = 17.44$ in., $y_3 = 19.72$ in.
$b_{x3} = 54.60$ kip/in.^{3}, $b_{y3} = 0$
Gauss point 4: $\xi_4 = -\dfrac{1}{\sqrt{3}}$, $\eta_4 = \dfrac{1}{\sqrt{3}}$, $W_4 = 1$

$$[D_4] = \frac{E}{1-\nu^2}\begin{bmatrix} 1 & \nu & 0 \\ \nu & 1 & 0 \\ 0 & 0 & \dfrac{1-\nu}{2} \end{bmatrix} = \begin{bmatrix} 31868 & 9560 & 0 \\ 9560 & 31868 & 0 \\ 0 & 0 & 11154 \end{bmatrix}$$

$$[N_4]=\begin{bmatrix}0.17 & 0.00 & 0.04 & 0.00 & 0.17 & 0.00 & 0.62 & 0.00\\ 0.00 & 0.17 & 0.00 & 0.04 & 0.00 & 0.17 & 0.00 & 0.62\end{bmatrix}$$

$$[J_4]=\begin{bmatrix}8.56 & 2.50\\ 1.44 & 10.00\end{bmatrix},\quad J_4=81.96,\quad \left[\tilde{J}_4\right]=\begin{bmatrix}0.12 & -0.03\\ -0.02 & 0.10\end{bmatrix}$$

$$[B_4]=\begin{bmatrix}-0.001 & 0.000 & 0.016 & 0.000 & 0.045 & 0.000 & -0.060 & 0.000\\ 0.000 & -0.039 & 0.000 & -0.013 & 0.000 & 0.004 & 0.000 & 0.048\\ -0.039 & -0.001 & -0.013 & 0.016 & 0.004 & 0.045 & 0.048 & -0.060\end{bmatrix}$$

$x_4 = 7.56$ in., $y_4 = 16.83$ in.

$b_{x4} = 31.95$ kip/in.^{3}, $b_{y4} = 0$

Now, we can calculate the element stiffness matrix and the contribution of the body forces to the equivalent nodal force vector. Since we are also given the thickness, $d = 2$ in., we can multiply the integrals obtained from Gaussian quadrature by 2 in.:

$$\left[k^{(e)}\right]=d\int_{-1}^{1}\int_{-1}^{1}\left(\left[B^{(e)}\right]^T\left[D^{(e)}\right]\left[B^{(e)}\right]Jd\xi d\eta\right)\approx d\sum_{g=1}^{Ng}\left([B_g]^T[D_g][B_g]J_gW_g\right)$$

$$=2\text{ in.}\cdot\left([B_1]^T[D_1][B_1]J_1W_1+[B_2]^T[D_2][B_2]J_2W_2+[B_3]^T[D_3][B_3]J_3W_3+[B_4]^T[D_4][B_4]J_4W_4\right)$$

$$\rightarrow\left[k^{(e)}\right]=\begin{bmatrix}19624 & 7160 & -13664 & -2872 & -7109 & -8656 & 1148 & 4367\\ 7160 & 23997 & -1278 & 6537 & -8656 & -11871 & 2774 & -18664\\ -13664 & -1278 & 34114 & -13381 & 3073 & 2544 & -23524 & 12116\\ -2872 & 6537 & -13381 & 28175 & 4137 & -16798 & 12116 & -17914\\ -7109 & -8656 & 3073 & 4137 & 25401 & 6470 & -21365 & -1952\\ -8656 & -11871 & 2544 & -16798 & 6470 & 29594 & -358 & -925\\ 1148 & 2774 & -23524 & 12116 & -21365 & -358 & 43742 & -14532\\ 4367 & -18664 & 12116 & -17914 & -1952 & -925 & -14532 & 37503\end{bmatrix}\text{ (values in kip/in.)}$$

$$\left\{f_\Omega^{(e)}\right\}=d\int_{-1}^{1}\int_{-1}^{1}\left(\left[N^{(e)}\right]^T\{b\}Jd\xi d\eta\right)\approx d\sum_{g=1}^{Ng}\left([N_g]^T\{b_g\}J_gW_g\right)$$

$$=2\text{ in.}\cdot\left([N_1]^T\begin{Bmatrix}b_{x1}\\0\end{Bmatrix}J_1W_1+[N_2]^T\begin{Bmatrix}b_{x2}\\0\end{Bmatrix}J_2W_2+[N_3]^T\begin{Bmatrix}b_{x3}\\0\end{Bmatrix}J_3W_3+[N_4]^T\begin{Bmatrix}b_{x4}\\0\end{Bmatrix}J_4W_4\right)$$

$$\left\{f_\Omega^{(e)}\right\}=\begin{Bmatrix}5486.1\\0.0\\9298.6\\0.0\\8923.6\\0.0\\6000.0\\0.0\end{Bmatrix}\text{ (values in kip)}$$

Note that, since only b_x is nonzero, the body force vector only gives nodal forces in the x-direction!

Finally, we can move on to calculate *the contribution of the boundary tractions to the equivalent nodal force vector:*

Segment 2-3

We have $\xi = 1 =$ constant on that segment (see element shape in parametric space in Figure 6.13), so we must use η as the variable of the parameterization. The expressions $x(\eta)$, $y(\eta)$ are obtained if we simply set $\xi = 1$ in our shape functions:

$$x(\eta) = \frac{1}{4}(1-1)(1-\eta)\cdot x_1 + \frac{1}{4}(1+1)(1-\eta)\cdot x_2 + \frac{1}{4}(1+1)(1+\eta)\cdot x_3 + \frac{1}{4}(1-1)(1+\eta)\cdot x_4$$

$$= \frac{1}{2}(1-\eta)\cdot 25 + \frac{1}{2}(1+\eta)\cdot 20$$

$$y(\eta) = \frac{1}{2}(1-\eta)\cdot 5 + \frac{1}{2}(1+\eta)\cdot 25$$

The length of segment 2-3 is

$$\ell_{23} = \sqrt{(x_3-x_2)^2 + (y_3-y_2)^2} = \sqrt{(20-25)^2 + (25-5)^2} = 20.6\,\text{in.}$$

In this segment, we have $t_x = 5$ ksi = constant, $t_y = 1$ ksi = constant; thus, this is a natural boundary segment where both t_x and t_y have been prescribed. We can therefore proceed using the second formula from Box 8.5.2:

$$\left\{f^{(e)}_{\Gamma_{t,23}}\right\} = d\int_{\Gamma_{t,23}} \left[N^{(4Q)}\right]^T \begin{Bmatrix} t_x \\ t_y \end{Bmatrix} ds = d\int_{-1}^{1} \left(\left[N^{(4Q)}\right]^T \begin{Bmatrix} 5 \\ 1 \end{Bmatrix} \frac{20.6}{2} d\eta\right)$$

$$\approx d\sum_{g=1}^{2}\left(\left[N^{(4Q)}\left(\xi = 1, \eta_g\right)\right]^T \begin{Bmatrix} 5 \\ 1 \end{Bmatrix} \frac{20.6}{2} W_g\right)$$

$$\left\{f^{(e)}_{\Gamma_{t,23}}\right\} = 2\,\text{in.}\cdot\sum_{g=1}^{2}\left(\left[N^{(4Q)}\left(\xi = 1, \eta_g\right)\right]^T \begin{Bmatrix} 5 \\ 1 \end{Bmatrix} \frac{20.6}{2} W_g\right)$$

$$= 2\,\text{in.}\cdot\left[N^{(4Q)}\left(\xi = 1, \eta_1 = -\frac{1}{\sqrt{3}}\right)\right]^T \begin{Bmatrix} 5 \\ 1 \end{Bmatrix} \frac{20.6}{2}\cdot 1$$

$$+\ 2\,\text{in.}\cdot\left[N^{(4Q)}\left(\xi = 1, \eta_2 = \frac{1}{\sqrt{3}}\right)\right]^T \begin{Bmatrix} 5 \\ 1 \end{Bmatrix} \frac{20.6}{2}\cdot 1$$

$$\rightarrow \left\{f^{(e)}_{\Gamma_{t,23}}\right\} = \begin{Bmatrix} 0 \\ 0 \\ 103 \\ 20.6 \\ 103 \\ 20.6 \\ 0 \\ 0 \end{Bmatrix} \text{(values in kip)}$$

Notice that there is no force at nodes 1 and 4, which do not belong on the boundary segment 2-3!

Segment 3-4

We have $\eta = 1 =$ constant on that segment (see element shape in parametric space in Figure 6.13), so we must use ξ for the parameterization. The expressions $x(\xi)$, $y(\xi)$ are obtained if we simply set $\eta = 1$ in our shape functions:

$$x(\xi) = \frac{1}{4}(1-\xi)(1-1)\cdot x_1 + \frac{1}{4}(1+\xi)(1-1)\cdot x_2 + \frac{1}{4}(1+\xi)(1+1)\cdot x_3 + \frac{1}{4}(1-\xi)(1+1)\cdot x_4$$

$$= \frac{1}{2}(1+\xi)\cdot 20 + \frac{1}{2}(1-\xi)\cdot 5 = 12.5 + 7.5\xi$$

$$y(\xi) = \frac{1}{2}(1+\xi)\cdot 25 + \frac{1}{2}(1-\xi)\cdot 20$$

The length of segment 3-4 is

$$\ell_{34} = \sqrt{(x_4-x_3)^2 + (y_4-y_3)^2} = \sqrt{(5-20)^2 + (20-25)^2} = 15.8\,\text{in.}$$

We now use the first formula of Box 8.5.2 to obtain:

$$\left\{f_{\Gamma_{t,34}}^{(e)}\right\} = \left\{f_{\Gamma_{ty,34}}^{(e)}\right\} = d\int_{\Gamma_{t,34}} \left[N^{(4Q)}\right]^T \begin{Bmatrix} t_x \\ t_y \end{Bmatrix} ds = d\int_{-1}^{1}\left(\left[N^{(4Q)}\right]^T \begin{Bmatrix} 0 \\ -1.2\cdot x \end{Bmatrix} \frac{15.8}{2} d\xi\right)$$

$$\approx d\sum_{g=1}^{2}\left(\left[N^{(4Q)}(\xi_g, \bar{\eta}=1)\right]^T \begin{Bmatrix} 0 \\ -1.2\cdot x(\xi_g) \end{Bmatrix} \frac{15.8}{2} W_g\right)$$

$$= d\sum_{g=1}^{2}\left(\left[N^{(4Q)}(\xi_g, \bar{\eta}=1)\right]^T \begin{Bmatrix} 0 \\ -1.2(12.5+7.5\xi_g) \end{Bmatrix} \frac{15.8}{2} W_g\right)$$

$$= 2\,\text{in.}\cdot\left[N^{(4Q)}\left(\xi_1 = -\frac{1}{\sqrt{3}}, \bar{\eta}=1\right)\right]^T \begin{Bmatrix} 0 \\ -1.2\left(12.5+7.5\left(-\frac{1}{\sqrt{3}}\right)\right) \end{Bmatrix} \frac{15.8}{2}\cdot 1$$

$$+ 2\,\text{in.}\cdot\left[N^{(4Q)}\left(\xi_2 = \frac{1}{\sqrt{3}}, \bar{\eta}=1\right)\right]^T \begin{Bmatrix} 0 \\ -1.2\left(12.5+7.5\left(\frac{1}{\sqrt{3}}\right)\right) \end{Bmatrix} \frac{15.8}{2}\cdot 1$$

$$\rightarrow \left\{f_{\Gamma_{t,34}}^{(e)}\right\} = \begin{Bmatrix} 0 \\ 0 \\ 0 \\ 0 \\ 0 \\ -284.4 \\ 0 \\ -189.6 \end{Bmatrix} \text{(values in kip)}$$

Notice that there is no force at nodes 1 and 2, which do not belong on boundary segment 3-4!

Thus, the total equivalent nodal force vector of the element is:

$$\left\{f^{(e)}\right\} = \left\{f_{\Omega}^{(e)}\right\} + \left\{f_{\Gamma_{t,23}}^{(e)}\right\} + \left\{f_{\Gamma_{t,34}}^{(e)}\right\} = \begin{Bmatrix} 5486.1 \\ 0.0 \\ 9298.6 \\ 0.0 \\ 8923.6 \\ 0.0 \\ 6000.0 \\ 0.0 \end{Bmatrix} + \begin{Bmatrix} 0 \\ 0 \\ 103 \\ 20.6 \\ 103 \\ 20.6 \\ 0 \\ 0 \end{Bmatrix} + \begin{Bmatrix} 0 \\ 0 \\ 0 \\ 0 \\ 0 \\ -284.4 \\ 0 \\ -189.6 \end{Bmatrix} = \begin{Bmatrix} 5486.1 \\ 0.0 \\ 9401.6 \\ 20.6 \\ 9026.6 \\ -263.8 \\ 6000.0 \\ -189.6 \end{Bmatrix} \text{(values in kip)}$$

Now, we have the stiffness matrix and equivalent nodal force vector. Since we have a single-element mesh, we have $\{U\} = \left\{U^{(e)}\right\} = \left[u_{x1}\ u_{y1}\ u_{x2}\ u_{y2}\ u_{x3}\ u_{y3}\ u_{x4}\ u_{y4}\right]^T$. Both displacements at nodes 1 and 2 and the displacement of node 4 along the x-direction are restrained, so the only degrees of freedom which are not restrained are $U_5 = u_{x3}$, $U_6 = u_{y3}$, and $U_8 = u_{y4}$. Since all the restrained displacements are zero, we can establish the equations for the free degrees of freedom using rows and columns 5, 6, and 8 from the stiffness matrix and rows 5, 6, and 8 from the equivalent nodal force vector:

$$\begin{bmatrix} 25401 & 6470 & -1952 \\ 6470 & 29594 & -925 \\ -1952 & -925 & 37503 \end{bmatrix} \begin{Bmatrix} U_5 \\ U_6 \\ U_8 \end{Bmatrix} = \begin{Bmatrix} 9026.6 \\ -263.8 \\ -189.6 \end{Bmatrix}$$

We can solve for the displacements of the unrestrained degrees of freedom to obtain:

$$\begin{Bmatrix} U_5 \\ U_6 \\ U_8 \end{Bmatrix} = \begin{Bmatrix} u_{x3} \\ u_{y3} \\ u_{y4} \end{Bmatrix} = \begin{Bmatrix} 0.380 \\ -0.092 \\ 0.013 \end{Bmatrix} \text{in.}$$

Problems

8.1 We are given the structure in Figure P8.1. The nodes and elements are numbered as shown in the figure. The only forces applied on the structure are body forces in the interior of element 1, as described in the figure. All elements are formulated using the isoparametric concept.

a) Establish a local node numbering for each element and also the gather-scatter array, $[L^{(e)}]$, for each element e, $e = 1, 2, 3, 4$.
b) Calculate the global right-hand-side (force) vector, using 2 × 2 Gaussian quadrature (four integration points) where necessary.

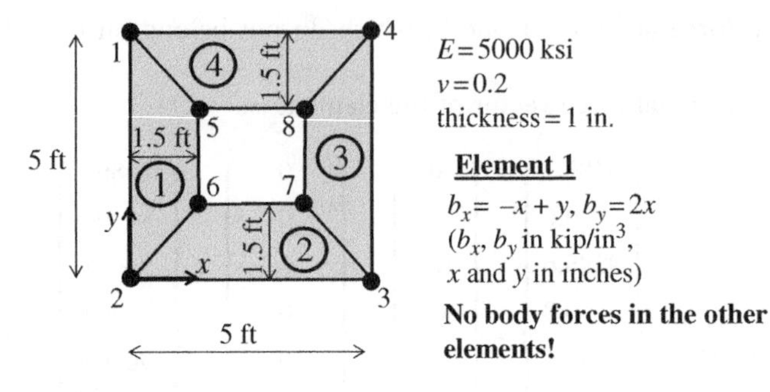

Figure P8.1

c) Which degrees of freedom are restrained and which ones are unrestrained? Is the stiffness subarray, $[K_{ff}]$, corresponding to the unrestrained degrees-of-freedom, invertible or not? Justify your answer *without any calculation.*
d) Which are the unrestrained degrees of freedom in Figure P8.2? Will $[K_{ff}]$ for the structure in Figure P8.2 be invertible or not? Justify your answer *without any calculation.*

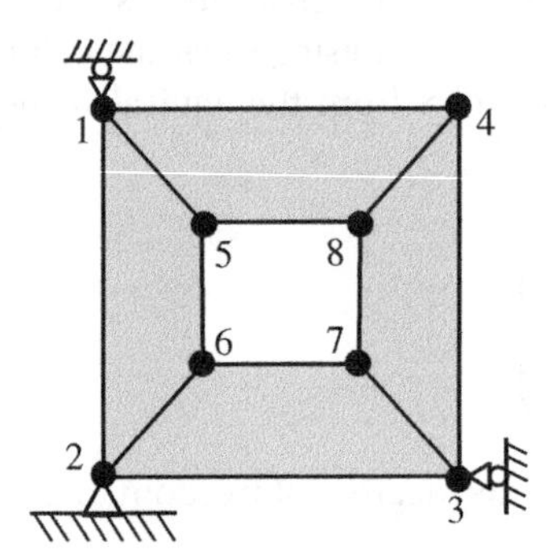

Figure P8.2

8.2 We are given the structure shown in Figure P8.3 (the coordinate values are given in feet). The structure is subjected to concentrated nodal forces on the three

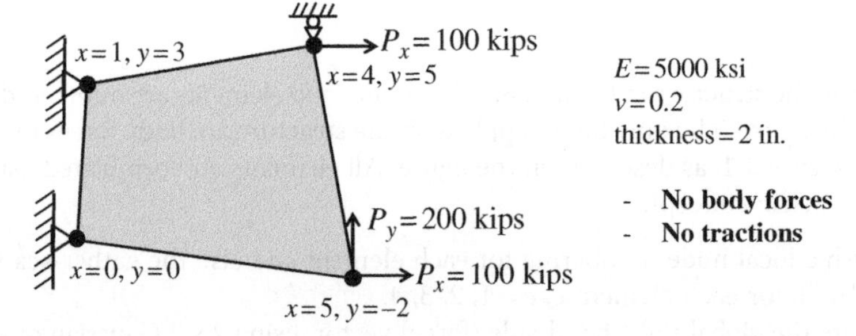

Figure P8.3

unrestrained degrees of freedom. The effect of a nodal force is accounted for by simply adding this force to the right-hand-side vector, at the corresponding degree of freedom (so, if for example we have a concentrated force P in the x-direction at node 2, we must add P to the term of the $\{f^{(e)}\}$ vector corresponding to degree of freedom u_{x2}).

a) Find the nodal displacements if we are given the isoparametric mapping in Figure P8.4.

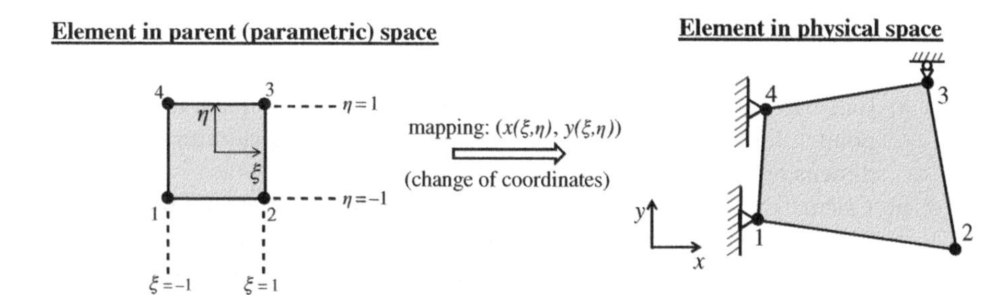

Figure P8.4

b) Is the stiffness matrix of the element symmetric? If yes, did you expect it to be symmetric, and why?
c) All finite element programs use the parametric space nodal numbering in a counter-clockwise fashion shown in Figure P8.4. Now, let us examine what will happen if the nodal numbering in the physical space is done in a clockwise fashion, as shown in Figure P8.5. Calculate the nodal displacements for this case. Compare the nodal displacements obtained for this case to those obtained in part a).

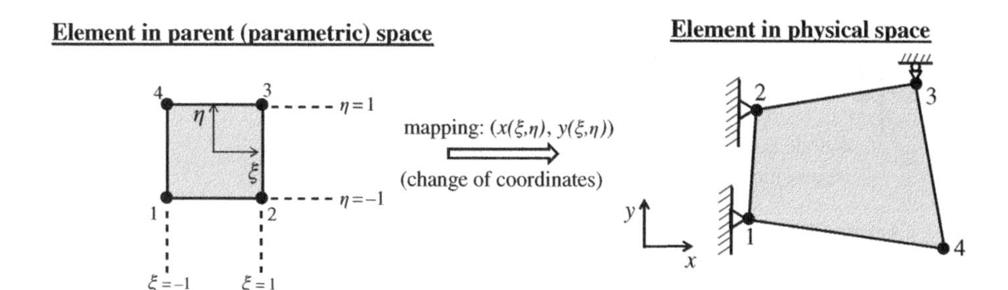

Figure P8.5

8.3 We are given the single-element mesh shown in Figure P8.6. The element uses an isoparametric formulation.

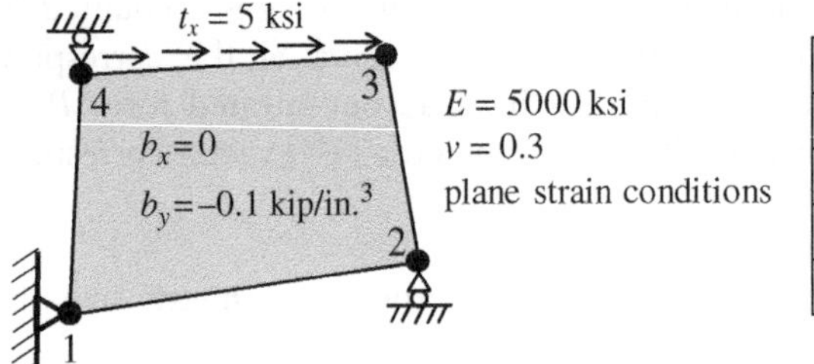

Node	x_i (in.)	y_i (in.)
1	−5	0
2	25	5
3	17	22
4	0	18

Figure P8.6

a) Calculate the nodal displacements. Use 2 × 2 Gaussian quadrature (four Gauss points) for the domain integrals and two-point quadrature for the one-dimensional boundary integral.
b) Calculate the physical coordinates (x,y), the strain vector, and the stress vector corresponding to the point with parametric coordinates $(\xi = 0, \eta = 0)$.
c) Maintain the right-hand-side vector that you calculated in part a, and calculate the stiffness matrix again, using 1 × 1 Gaussian quadrature (one Gauss point). *This integration corresponds to the reduced integration procedure, which is common in many commercial programs.* Obtain the nodal displacement vector. What do you observe for the case that you use a single quadrature point for the calculation of the stiffness matrix?

8.4 We are given the finite element mesh in Figure P8.7. We intend to repeat our calculation many times, and each time we subdivide each element into four elements of the same type. You are asked to determine (*without any calculation*) whether we are bound to have convergence.

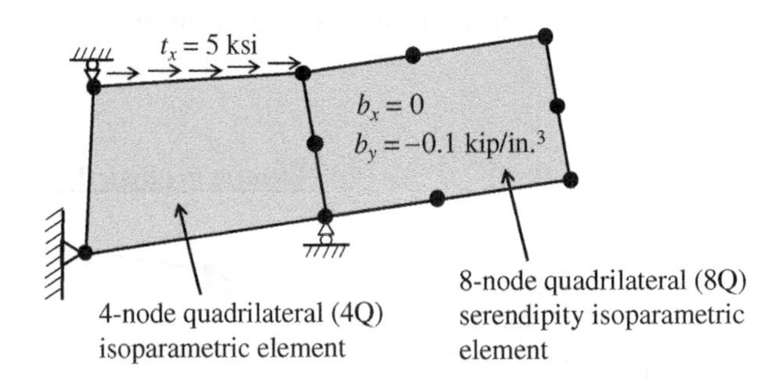

Figure P8.7

8.5 We are given an eight-node quadrilateral element, shown in Figure P8.8, using an isoparametric formulation (see Section 6.6). The element has been used for the analysis of a solid for which plane-stress conditions apply. We are given that the material of the solid is uniform and orthotropic (see Sections 7.5 and 7.7), with $E_x = 5000$ ksi, $\nu_{xy} = 0.2$, $E_y = 4000$ ksi, $G_{xy} = 2000$ ksi. The nodal coordinates and also the element nodal values of the solution are provided in the table of Figure P8.8.

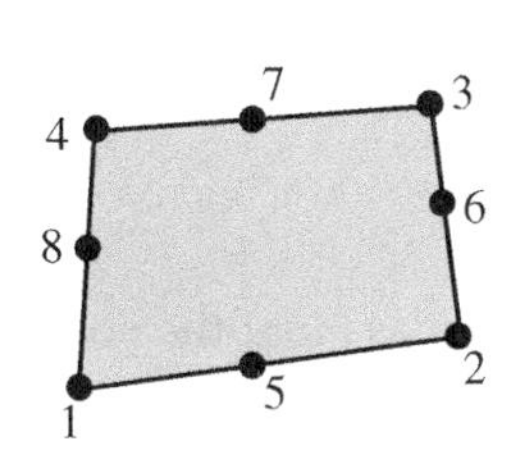

Node	x_i (in.)	y_i (in.)
1	0	0
2	25	5
3	20	25
4	5	20
5	12	2.4
6	22	17
7	12.5	22.5
8	2.5	10

Node	u_{xi} (in.)	u_{yi} (in.)
1	0	0
2	0	0
3	0.05	–0.01
4	0.04	0.01
5	0	0.01
6	0.03	–0.008
7	0.04	0
8	0	0.005

Figure P8.8

a) Calculate the *strains* at the element centroid (i.e., the point with parametric coordinates $\xi = 0$, $\eta = 0$).
b) Calculate the *stresses* at the element centroid (i.e., the point with parametric coordinates $\xi = 0$, $\eta = 0$).
c) Calculate the *principal strains* at the element centroid, and the principal strain directions.
d) Calculate the *principal stresses* at the element centroid, and the principal stress directions.

8.6 Create a programming routine conducting the finite element calculations for two-dimensional elasticity.
The following input information must be provided in our program:

1) The numbers of the four nodal points.
2) The physical coordinates, x and y, of the four nodal points.
3) The Young's modulus, E, the Poisson ratio, ν, and the thickness d of the material. We only allow *constant* E, ν, and d.
4) Information on whether we have a plane-stress or a plane-strain element.
5) The variation of the body force vector, $\{b\}$, inside the element. We want to allow any body force vector whose components have the following variation: $b_x(x,y) = b_o + b_1x + b_2y + b_3xy + b_4x^2 + b_5y^2$, where b_o, b_1, b_2, b_3, b_4, and b_5 are constants and $b_y(x,y) = c_o + c_1x + c_2y + c_3xy + c_4x^2 + c_5y^2$, where c_o, c_1, c_2, c_3, c_4, and c_5 are constants. Thus, we need to provide as input the values of the six b parameters and the six c parameters.
6) Information on whether each one of the sides of the element is a natural boundary segment. Obviously, if a side of the element is part of the natural boundary, we need to also give as input the value of the prescribed traction vector. We only allow *constant* prescribed tractions on natural boundaries.
7) The number of integration points in the element.

Our routines must be capable of computing and providing as OUTPUT the following information:

General information:

a) The coordinates of the Gauss points in the PHYSICAL SPACE, xy.

Pre-processing information:

b) The values of the shape functions, the [B]-array, the body-force vector, the [J] array and the Jacobian determinant, J, at each Gauss point.
c) The element stiffness matrix, $[k^{(e)}]$
d) The element equivalent nodal force vector, $\{f^{(e)}\}$

Post-processing information (where we also need to pass as input the nodal displacements of the element):

e) The displacement vector at the integration points.
f) The strain vector, $\{\varepsilon\}$, and the stress vector, $\{\sigma\}$, at the integration points.

8.7 We have the three-node triangular element and the nodal coordinates shown in Figure P8.9. We know that the displacements at nodes 1 and 3 are zero. Based on the provided information, determine whether it is possible to have the following strain vector at a point P, in the interior of the element (we do not know the exact location of P):

$$\begin{Bmatrix} \varepsilon_{xx} \\ \varepsilon_{yy} \\ \gamma_{xy} \end{Bmatrix}_P = \begin{Bmatrix} 0.001 \\ 0.0015 \\ -0.0012 \end{Bmatrix}$$

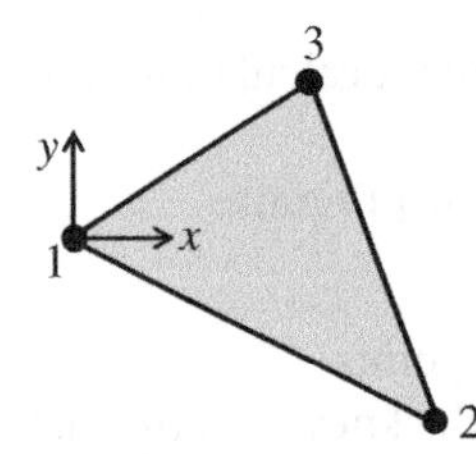

Node	x_i (in.)	y_i (in.)
1	0	0
2	10	–5
3	5	5

Figure P8.9

8.8 Repeat the solution of Problem 8.2, using the analysis program provided in the companion web page.

9

Finite Element Formulation for Three-Dimensional Elasticity

In this chapter, we will examine the finite element analysis of three-dimensional, linearly elastic, solid bodies. In this case, we want to determine the three-component *displacement vector field*, $\{u(x,y,z)\} = \begin{Bmatrix} u_x(x,y,z) \\ u_y(x,y,z) \\ u_z(x,y,z) \end{Bmatrix}$, which is assumed to be a continuous function of x, y, and z. Before proceeding with the finite element equations, the reader is advised to read the material in Chapter 7, particularly the introductory definitions of stress and strain (Sections 7.2, 7.3), the constitutive law of multiaxial linear elasticity (Section 7.5), and the strong form for three-dimensional elasticity (Section 7.11).

We will first examine the weak form for the problem. Then, we will proceed with the finite element equations and a discussion of different types of finite elements commonly encountered in practice. Many of the derivations (e.g., the equations corresponding to restrained degrees of freedom, DOF) are identical to those presented for two-dimensional elasticity in the previous chapter, but they will also be provided here for completeness.

9.1 Weak Form for Three-Dimensional Elasticity

Given the strong form for three-dimensional elasticity, provided in Box 7.11.1, we can establish the weak form. One can verify that the derivation is merely an extension of that for two-dimensional elasticity. Let us consider an arbitrary vector field, $\vec{w}(x,y,z) = \{w(x,y,z)\} = \left[w_x(x,y,z) \;\; w_y(x,y,z) \;\; w_z(x,y,z)\right]^T$, whose components vanish at the corresponding essential boundary segments: $w_x = 0$ on Γ_{ux}, $w_y = 0$ on Γ_{uy}, $w_z = 0$ on Γ_{uz}. The arbitrary field is merely a *virtual displacement field*. Then, we can obtain the following weak form for three-dimensional elasticity, which is summarized in Box 9.1.1.

The equivalence of the weak form and the strong form can be established in exactly the same way as for two-dimensional elasticity in Section 7.10 (the only difference is that

Fundamentals of Finite Element Analysis: Linear Finite Element Analysis, First Edition.
Ioannis Koutromanos, James McClure, and Christopher Roy.

Companion website: www.wiley.com/go/koutromanos/linear

Box 9.1.1 Weak Form for Three-Dimensional Elasticity

$$\iiint_{\Omega} \{\bar{\varepsilon}\}^T [D] \{\varepsilon\} dV = \iiint_{\Omega} \{w\}^T \{b\} dV + \iint_{\Gamma_{tx}} w_x t_x \, dS + \iint_{\Gamma_{ty}} w_y t_y \, dS + \iint_{\Gamma_{tz}} w_z t_z \, dS$$

$$\forall \{w(x,y,z)\} = [w_x(x,y,z) \;\; w_y(x,y,z) \;\; w_z(x,y,z)]^T$$

with $w_x = 0$ on Γ_{ux}, $w_y = 0$ on Γ_{uy}, $w_z = 0$ on Γ_{uz}

$$u_x = \bar{u}_x \quad \text{on } \Gamma_{ux}$$

$$u_y = \bar{u}_y \quad \text{on } \Gamma_{uy}$$

$$u_z = \bar{u}_z \quad \text{on } \Gamma_{uz}$$

each of the two "special" forms of $\{w\}$ that we must use will now have three components instead of two).

9.2 Piecewise Finite Element Approximation—Assembly Equations

In a finite element mesh, the three-dimensional domain is discretized into N_e element subdomains, as shown in Figure 9.1. We then establish a *piecewise approximation* for $\{u(x, y, z)\}$ and for $\{w(x, y, z)\}$ inside each element e. We will discuss the general situation where each element has n nodes. Of course, since the displacement vector has three components, each nodal point will also have three degrees of freedom, namely, the three nodal displacement components.

The piecewise approximation of the displacement field in element e is described by the following expressions.

$$u_x(x,y,z) \approx u_x^{(e)}(x,y,z) = \sum_{i=1}^{n} \left(N_i^{(e)}(x,y,z) \cdot u_{xi}^{(e)} \right) \tag{9.2.1a}$$

$$u_y(x,y,z) \approx u_y^{(e)}(x,y,z) = \sum_{i=1}^{n} \left(N_i^{(e)}(x,y,z) \cdot u_{yi}^{(e)} \right) \tag{9.2.1b}$$

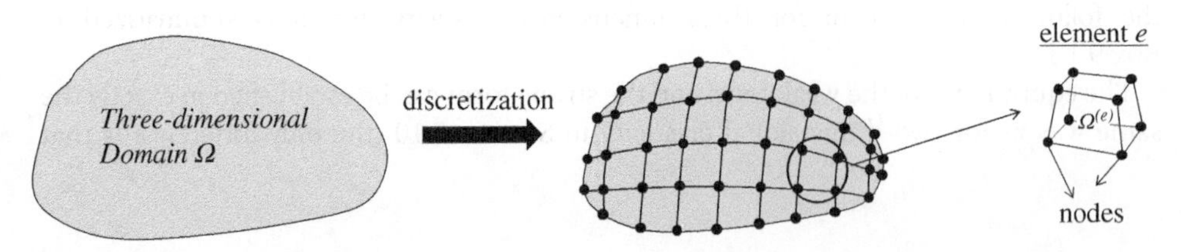

Figure 9.1 Finite element discretization.

$$u_z(x,y,z) \approx u_z^{(e)}(x,y,z) = \sum_{i=1}^{n}\left(N_i^{(e)}(x,y,z)\cdot u_{zi}^{(e)}\right) \tag{9.2.1c}$$

We can write Equations (9.2.1a–c) in a combined matrix form:

$$\begin{Bmatrix} u_x^{(e)}(x,y,z) \\ u_y^{(e)}(x,y,z) \\ u_z^{(e)}(x,y,z) \end{Bmatrix} \approx \begin{bmatrix} N_1^{(e)}(x,y,z) & 0 & 0 & N_2^{(e)}(x,y,z) & 0 & 0 & \dots & N_n^{(e)}(x,y,z) & 0 & 0 \\ 0 & N_1^{(e)}(x,y,z) & 0 & 0 & N_2^{(e)}(x,y,z) & 0 & \dots & 0 & N_n^{(e)}(x,y,z) & 0 \\ 0 & 0 & N_1^{(e)}(x,y,z) & 0 & 0 & N_2^{(e)}(x,y,z) & \dots & 0 & 0 & N_n^{(e)}(x,y,z) \end{bmatrix} \begin{Bmatrix} u_{x1}^{(e)} \\ u_{y1}^{(e)} \\ u_{z1}^{(e)} \\ u_{x2}^{(e)} \\ u_{y2}^{(e)} \\ u_{z2}^{(e)} \\ \vdots \\ u_{xn}^{(e)} \\ u_{yn}^{(e)} \\ u_{zn}^{(e)} \end{Bmatrix}$$

or

$$\left\{u^{(e)}\right\} \approx \left[N^{(e)}\right]\left\{U^{(e)}\right\} \tag{9.2.2}$$

where $\left\{u^{(e)}\right\} = \left[u_x^{(e)}(x,y,z) \;\; u_y^{(e)}(x,y,z) \;\; u_z^{(e)}(x,y,z)\right]^T$ is the approximate displacement field, $\left\{U^{(e)}\right\} = \left[u_{x1}^{(e)} \; u_{y1}^{(e)} \; u_{z1}^{(e)} \; u_{x2}^{(e)} \; u_{y2}^{(e)} \; u_{z2}^{(e)} \; \dots \; u_{xn}^{(e)} \; u_{yn}^{(e)} \; u_{zn}^{(e)}\right]^T$ is the nodal displacement vector, and $[N^{(e)}]$ is the shape function array of element e:

$$\left[N^{(e)}\right] = \begin{bmatrix} N_1^{(e)}(x,y,z) & 0 & 0 & N_2^{(e)}(x,y,z) & 0 & 0 & \dots & N_n^{(e)}(x,y,z) & 0 & 0 \\ 0 & N_1^{(e)}(x,y,z) & 0 & 0 & N_2^{(e)}(x,y,z) & 0 & \dots & 0 & N_n^{(e)}(x,y,z) & 0 \\ 0 & 0 & N_1^{(e)}(x,y,z) & 0 & 0 & N_2^{(e)}(x,y,z) & \dots & 0 & 0 & N_n^{(e)}(x,y,z) \end{bmatrix}$$

The components of the approximate strain field, $\{\varepsilon^{(e)}\}$, in each element e, are given by the following equations.

$$\varepsilon_{xx}^{(e)} = \frac{\partial u_x^{(e)}}{\partial x} \approx \frac{\partial}{\partial x}\left[\sum_{i=1}^{n}\left(N_i^{(e)}(x,y,z)\cdot u_{xi}^{(e)}\right)\right] = \sum_{i=1}^{n}\left(\frac{\partial N_i^{(e)}}{\partial x}\cdot u_{xi}^{(e)}\right) \tag{9.2.3a}$$

$$\varepsilon_{yy}^{(e)} = \frac{\partial u_y^{(e)}}{\partial y} \approx \frac{\partial}{\partial y}\left[\sum_{i=1}^{n}\left(N_i^{(e)}(x,y,z)\cdot u_{yi}^{(e)}\right)\right] = \sum_{i=1}^{n}\left(\frac{\partial N_i^{(e)}}{\partial y}\cdot u_{yi}^{(e)}\right) \tag{9.2.3b}$$

$$\varepsilon_{zz}^{(e)} = \frac{\partial u_z^{(e)}}{\partial z} \approx \frac{\partial}{\partial z}\left[\sum_{i=1}^{n}\left(N_i^{(e)}(x,y,z)\cdot u_{zi}^{(e)}\right)\right] = \sum_{i=1}^{n}\left(\frac{\partial N_i^{(e)}}{\partial z}\cdot u_{zi}^{(e)}\right) \tag{9.2.3c}$$

$$\gamma_{xy}^{(e)} = \frac{\partial u_x^{(e)}}{\partial y} + \frac{\partial u_y^{(e)}}{\partial x} \approx \frac{\partial}{\partial y}\left[\sum_{i=1}^{n}\left(N_i^{(e)}(x,y,z)\cdot u_{xi}^{(e)}\right)\right] + \frac{\partial}{\partial x}\left[\sum_{i=1}^{n}\left(N_i^{(e)}(x,y,z)\cdot u_{yi}^{(e)}\right)\right]$$

$$= \sum_{i=1}^{n}\left(\frac{\partial N_i^{(e)}}{\partial y}\cdot u_{xi}^{(e)}\right) + \sum_{i=1}^{n}\left(\frac{\partial N_i^{(e)}}{\partial x}\cdot u_{yi}^{(e)}\right) \tag{9.2.3d}$$

$$\gamma_{yz}^{(e)} = \frac{\partial u_y^{(e)}}{\partial z} + \frac{\partial u_z^{(e)}}{\partial y} \approx \frac{\partial}{\partial z}\left[\sum_{i=1}^{n}\left(N_i^{(e)}(x,y,z)\cdot u_{yi}^{(e)}\right)\right] + \frac{\partial}{\partial y}\left[\sum_{i=1}^{n}\left(N_i^{(e)}(x,y,z)\cdot u_{zi}^{(e)}\right)\right]$$

$$= \sum_{i=1}^{n}\left(\frac{\partial N_i^{(e)}}{\partial z}\cdot u_{yi}^{(e)}\right) + \sum_{i=1}^{n}\left(\frac{\partial N_i^{(e)}}{\partial y}\cdot u_{zi}^{(e)}\right) \tag{9.2.3e}$$

$$\gamma_{zx}^{(e)} = \frac{\partial u_z^{(e)}}{\partial x} + \frac{\partial u_x^{(e)}}{\partial z} \approx \frac{\partial}{\partial x}\left[\sum_{i=1}^{n}\left(N_i^{(e)}(x,y,z)\cdot u_{zi}^{(e)}\right)\right] + \frac{\partial}{\partial z}\left[\sum_{i=1}^{n}\left(N_i^{(e)}(x,y,z)\cdot u_{xi}^{(e)}\right)\right]$$

$$= \sum_{i=1}^{n}\left(\frac{\partial N_i^{(e)}}{\partial x}\cdot u_{zi}^{(e)}\right) + \sum_{i=1}^{n}\left(\frac{\partial N_i^{(e)}}{\partial z}\cdot u_{xi}^{(e)}\right) \tag{9.2.3f}$$

We can cast Equations (9.2.3a–f) in a combined, matrix expression:

$$\left\{\varepsilon^{(e)}\right\} \approx \left[B^{(e)}\right]\left\{U^{(e)}\right\} \tag{9.2.4}$$

where the strain-displacement matrix $[B^{(e)}]$ for an element with n nodes is given by the following equation.

$$\left[B^{(e)}\right] = \begin{bmatrix}
\frac{\partial N_1^{(e)}}{\partial x} & 0 & 0 & \frac{\partial N_2^{(e)}}{\partial x} & 0 & 0 & \cdots & \frac{\partial N_n^{(e)}}{\partial x} & 0 & 0 \\
0 & \frac{\partial N_1^{(e)}}{\partial y} & 0 & 0 & \frac{\partial N_2^{(e)}}{\partial y} & 0 & \cdots & 0 & \frac{\partial N_n^{(e)}}{\partial y} & 0 \\
0 & 0 & \frac{\partial N_1^{(e)}}{\partial z} & 0 & 0 & \frac{\partial N_2^{(e)}}{\partial z} & \cdots & 0 & 0 & \frac{\partial N_n^{(e)}}{\partial z} \\
\frac{\partial N_1^{(e)}}{\partial y} & \frac{\partial N_1^{(e)}}{\partial x} & 0 & \frac{\partial N_2^{(e)}}{\partial y} & \frac{\partial N_2^{(e)}}{\partial x} & 0 & \cdots & \frac{\partial N_n^{(e)}}{\partial y} & \frac{\partial N_n^{(e)}}{\partial x} & 0 \\
0 & \frac{\partial N_1^{(e)}}{\partial z} & \frac{\partial N_1^{(e)}}{\partial y} & 0 & \frac{\partial N_2^{(e)}}{\partial z} & \frac{\partial N_2^{(e)}}{\partial y} & \cdots & 0 & \frac{\partial N_n^{(e)}}{\partial z} & \frac{\partial N_n^{(e)}}{\partial y} \\
\frac{\partial N_1^{(e)}}{\partial z} & 0 & \frac{\partial N_1^{(e)}}{\partial x} & \frac{\partial N_2^{(e)}}{\partial z} & 0 & \frac{\partial N_2^{(e)}}{\partial x} & \cdots & \frac{\partial N_n^{(e)}}{\partial z} & 0 & \frac{\partial N_n^{(e)}}{\partial x}
\end{bmatrix} \tag{9.2.5}$$

Now, if we stipulate that the virtual displacement vector field, $\{w(x, y, z)\}$, uses the same approximation as that for the displacement field $\{u(x, y, z)\}$, we obtain the piecewise approximation for the virtual displacements $\{w^{(e)}\}$ and strains $\left\{\bar{\varepsilon}^{(e)}\right\}$:

$$\left\{w^{(e)}\right\} \approx \left[N^{(e)}\right]\left\{W^{(e)}\right\} \tag{9.2.6a}$$

$$\left\{\bar{\varepsilon}^{(e)}\right\} \approx \left[B^{(e)}\right]\left\{W^{(e)}\right\} \tag{9.2.6b}$$

where $\{W^{(e)}\}$ is the nodal virtual displacement vector of the element.

We now define the Boolean gather array, $[L^{(e)}]$, for each element e, which allows us to associate the nodal vectors of element e to the corresponding global nodal vectors of the entire finite element mesh:

$$\left\{U^{(e)}\right\} = \left[L^{(e)}\right]\{U\} \tag{9.2.7a}$$

$$\left\{W^{(e)}\right\} = \left[L^{(e)}\right]\{W\} \tag{9.2.7b}$$

We can now proceed to obtaining the finite element equations, by introducing the approximations into the integral expression of the weak form in Box 9.1.1. We begin by separating the various integrals into the contributions of the N_e elements in the mesh:

$$\sum_{e=1}^{Ne}\left(\iiint_{\Omega^{(e)}}\{\bar{\varepsilon}\}^T[D]\{\varepsilon\}dV\right) = \sum_{e=1}^{Ne}\left(\iiint_{\Omega^{(e)}}\{w\}^T\{b\}dV\right) + \sum_{e=1}^{Ne}\left(\iint_{\Gamma_{tx}^{(e)}}\{w\}^T\{t_x\}\,dS\right)$$
$$+\sum_{e=1}^{Ne}\left(\iint_{\Gamma_{ty}^{(e)}}\{w\}^T\{t_y\}\,dS\right) + \sum_{e=1}^{Ne}\left(\iint_{\Gamma_{tz}^{(e)}}\{w\}^T\{t_z\}\,dS\right) \tag{9.2.8}$$

Each quantity in Equation (9.2.8) corresponds to a summation of N_e terms, each term involving an integral over an element subdomain.

If we are inside element e, we have:

$$\{u(x,y,z)\} \approx \left\{u^{(e)}(x,y,z)\right\} = \left[N^{(e)}\right]\left\{U^{(e)}\right\} \tag{9.2.9a}$$

$$\{\varepsilon(x,y,z)\} \approx \left\{\varepsilon^{(e)}(x,y,z)\right\} = \left[B^{(e)}\right]\left\{U^{(e)}\right\} \tag{9.2.9b}$$

$$\{w(x,y,z)\} \approx \left\{w^{(e)}(x,y,z)\right\} = \left[N^{(e)}\right]\left\{W^{(e)}\right\} \tag{9.2.9c}$$

$$\{\bar{\varepsilon}(x,y,z)\} \approx \left\{\bar{\varepsilon}^{(e)}(x,y,z)\right\} = \left[B^{(e)}\right]\left\{W^{(e)}\right\} \tag{9.2.9d}$$

In light of Equations (9.2.9a–d), Equation (9.2.8) becomes:

$$\sum_{e=1}^{Ne}\left(\iiint_{\Omega^{(e)}}\left(\left[B^{(e)}\right]\left\{W^{(e)}\right\}\right)^T\left[D^{(e)}\right]\left[B^{(e)}\right]\left\{U^{(e)}\right\}dV\right)$$
$$= \sum_{e=1}^{Ne}\left(\iiint_{\Omega^{(e)}}\left(\left[N^{(e)}\right]\left\{W^{(e)}\right\}\right)^T\{b\}dV\right) + \sum_{e=1}^{Ne}\left(\iint_{\Gamma_{tx}^{(e)}}\left(\left[N^{(e)}\right]\left\{W^{(e)}\right\}\right)^T\{t_x\}\,dS\right)$$
$$+\sum_{e=1}^{Ne}\left(\iint_{\Gamma_{ty}^{(e)}}\left(\left[N^{(e)}\right]\left\{W^{(e)}\right\}\right)^T\{t_y\}\,dS\right) + \sum_{e=1}^{Ne}\left(\iint_{\Gamma_{tz}^{(e)}}\left(\left[N^{(e)}\right]\left\{W^{(e)}\right\}\right)^T\{t_z\}\,dS\right)$$

$$\rightarrow \sum_{e=1}^{Ne}\left(\iiint_{\Omega^{(e)}} \{W\}^T \left[L^{(e)}\right]^T \left[B^{(e)}\right]^T \left[D^{(e)}\right]\left[B^{(e)}\right]\left[L^{(e)}\right]\{U\} dV\right)$$

$$= \sum_{e=1}^{Ne}\left(\iiint_{\Omega^{(e)}} \left(\left[N^{(e)}\right]\left\{w^{(e)}\right\}\right)^T \{b\} dV\right) + \sum_{e=1}^{Ne}\left(\iint_{\Gamma_{tx}^{(e)}} \{W\}^T \left[L^{(e)}\right]^T \left[N^{(e)}\right]^T \{t_x\} dS\right)$$

$$+ \sum_{e=1}^{Ne}\left(\iint_{\Gamma_{ty}^{(e)}} \{W\}^T \left[L^{(e)}\right]^T \left[N^{(e)}\right]^T \{t_y\} dS\right) + \sum_{e=1}^{Ne}\left(\iint_{\Gamma_{tz}^{(e)}} \{W\}^T \left[L^{(e)}\right]^T \left[N^{(e)}\right]^T \{t_z\} dS\right)$$

Finally, since $\{W\}^T$, $\{U\}$, and $[L^{(e)}]$ are constant, they can be taken outside of the integrals. Also, $\{W\}^T$ and $\{U\}$ are common in all the corresponding terms, so they can be taken as common factors outside of the summation operation:

$$\{W\}^T \sum_{e=1}^{Ne}\left[\left[L^{(e)}\right]^T \left(\iiint_{\Omega^{(e)}} \left[B^{(e)}\right]^T \left[D^{(e)}\right]\left[B^{(e)}\right] dV\right)\left[L^{(e)}\right]\right]\{U\}$$

$$-\{W\}^T \sum_{e=1}^{Ne}\left[\left[L^{(e)}\right]^T \left(\iiint_{\Omega^{(e)}} \left[N^{(e)}\right]^T \{b\} dV + \iint_{\Gamma_{tx}^{(e)}} \left[N^{(e)}\right]^T \{t_x\} dS\right.\right.$$

$$\left.\left. + \iint_{\Gamma_{ty}^{(e)}} \left[N^{(e)}\right]^T \{t_y\} dS + \iint_{\Gamma_{tz}^{(e)}} \left[N^{(e)}\right]^T \{t_z\} dS\right)\right] = 0$$

$$\rightarrow \{W\}^T \left(\sum_{e=1}^{Ne}\left[\left[L^{(e)}\right]^T \left[k^{(e)}\right]\left[L^{(e)}\right]\right]\{U\} - \left[L^{(e)}\right]^T \left\{f^{(e)}\right\}\right) = 0 \quad (9.2.10)$$

where we have defined the *element stiffness matrix*, $[k^{(e)}]$:

$$\left[k^{(e)}\right] = \iiint_{\Omega^{(e)}} \left[B^{(e)}\right]^T \left[D^{(e)}\right]\left[B^{(e)}\right] dV \quad (9.2.11)$$

and the element *equivalent nodal force vector*, $\{f^{(e)}\}$:

$$\left\{f^{(e)}\right\} = \iiint_{\Omega^{(e)}} \left[N^{(e)}\right]^T \{b\} dV + \iint_{\Gamma_{tx}^{(e)}} \left[N^{(e)}\right]^T \{t_x\} dS + \iint_{\Gamma_{ty}^{(e)}} \left[N^{(e)}\right]^T \{t_y\} dS + \iint_{\Gamma_{tz}^{(e)}} \left[N^{(e)}\right]^T \{t_z\} dS \quad (9.2.12)$$

Remark 9.2.1: Just like for two-dimensional elasticity, it is preferable to separate the equivalent nodal force vector of each element e into two parts:

$$\left\{f^{(e)}\right\} = \left\{f_{\Omega}^{(e)}\right\} + \left\{f_{\Gamma t}^{(e)}\right\} \tag{9.2.13}$$

where

$$\left\{f_{\Omega}^{(e)}\right\} = \iiint\limits_{\Omega^{(e)}} \left[N^{(e)}\right]^T \{b\}\, dV \tag{9.2.14a}$$

$$\left\{f_{\Gamma t}^{(e)}\right\} = \iint\limits_{\Gamma_{tx}^{(e)}} \left[N^{(e)}\right]^T \{t_x\}\, dS + \iint\limits_{\Gamma_{ty}^{(e)}} \left[N^{(e)}\right]^T \{t_y\}\, dS + \iint\limits_{\Gamma_{tz}^{(e)}} \left[N^{(e)}\right]^T \{t_z\}\, dS \tag{9.2.14b}$$

■

We can further write Equation (9.2.10) in the following form:

$$\{W\}^T[[K]\{U\} - \{f\}] = 0 \tag{9.2.15}$$

where we have established the *global stiffness matrix,* $[K]$:

$$[K] = \sum_{e=1}^{Ne} \left(\left[L^{(e)}\right]^T \left[k^{(e)}\right]\left[L^{(e)}\right]\right) \tag{9.2.16}$$

and the *global equivalent nodal force vector,* $\{f\}$:

$$\{f\} = \sum_{e=1}^{Ne} \left(\left[L^{(e)}\right]^T \left\{f^{(e)}\right\}\right) \tag{9.2.17}$$

Finally, accounting for the fact that the global vector $\{W\}^T$ in Equation (9.2.15) is arbitrary yields the matrix form of the global finite element equations for three-dimensional (3D) elasticity, i.e. *the finite element equilibrium or stiffness equations for 3D elasticity*:

$$[K]\{U\} - \{f\} = 0 \tag{9.2.18}$$

Remark 9.2.2: In accordance with the pertinent discussion in Section 8.2, if a specific component of $\{W\}$, say W_k, corresponds to a restrained degree of freedom (essential boundary condition), then the corresponding global stiffness equation attains the following form:

$$K_{k1} \cdot U_1 + K_{k2} \cdot U_2 + \ldots + K_{kN} \cdot U_N - f_k = r_k \tag{9.2.19}$$

where r_k is the reaction force corresponding to the restrained global degree of freedom k.

■

Thus, once again, plugging the piecewise finite element approximation to the weak form transformed the integral expression into a system of linear equations for the nodal displacements (Equation 9.2.18).

After solving the global equations for the unrestrained degrees of freedom and knowing the nodal displacement vector $\{U\}$ of the structure, we can conduct the post-processing computations of each element e and obtain:

- The element nodal displacements:

$$\left\{U^{(e)}\right\} = \left[L^{(e)}\right]\{U\} \tag{9.2.19a}$$

- The element displacement field:

$$\left\{u^{(e)}\right\} = \left[N^{(e)}\right]\left\{U^{(e)}\right\} = \left[N^{(e)}\right]\left[L^{(e)}\right]\{U\} \tag{9.2.19b}$$

- The element strain field:

$$\left\{\varepsilon^{(e)}\right\} = \left[B^{(e)}\right]\left\{U^{(e)}\right\} = \left[B^{(e)}\right]\left[L^{(e)}\right]\{U\} \tag{9.2.19c}$$

- The element stress field:

$$\left\{\sigma^{(e)}\right\} = \left[D^{(e)}\right]\left\{\varepsilon^{(e)}\right\} = \left[D^{(e)}\right]\left[B^{(e)}\right]\left\{U^{(e)}\right\} = \left[D^{(e)}\right]\left[B^{(e)}\right]\left[L^{(e)}\right]\{U\} \tag{9.2.19d}$$

Remark 9.2.3: We can verify that the highest order of partial derivative appearing in the weak form is 1. Thus, just like for two-dimensional linear elasticity, *C^0 continuity and first-order completeness are adequate to ensure that a finite element approximation is convergent.* ■

The following section will focus on several types of finite elements for 3D elasticity. All elements considered rely on the isoparametric concept.

9.3 Isoparametric Finite Elements for Three-Dimensional Elasticity

9.3.1 Eight-Node Hexahedral Element

The first type of three-dimensional finite element that we will examine is the hexahedral isoparametric element, also termed *brick element*. The simplest formulation of this type is the eight-node hexahedral (8H) isoparametric element. Just as we did for two-dimensional isoparametric elements, we establish a parent (parametric) space, $\xi\eta\zeta$, wherein the element attains a nice cubic shape, as shown in Figure 9.2. The key to formulate the element is to have a coordinate mapping from the parametric space to the physical (Cartesian) space, as shown in the same figure. We will establish the shape functions (and the finite element approximations for the displacement field) in the parametric space.

Note that, just like we did for the 4Q element in Sections 6.4 and 8.5.2, the orientation of the node numbering in the element must be the same for the parametric and for the physical space. The element has eight nodes, so it should also have eight shape functions. Since the element has a "nice shape" in the parametric space, we can once again obtain

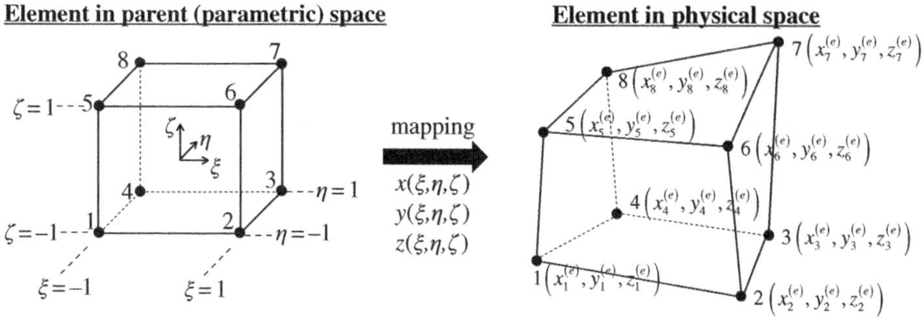

Element in parent (parametric) space

Element in physical space

mapping

$x(\xi,\eta,\zeta)$

$y(\xi,\eta,\zeta)$

$z(\xi,\eta,\zeta)$

Figure 9.2 Eight-node hexahedral (8H) isoparametric element.

the shape functions as *tensor products* of one-dimensional shape functions, as shown in Figure 9.3.

The mathematical expressions for the shape functions are the following:

$$N_1^{(8H)}(\xi,\eta,\zeta) = N_1^{(e-1D)}(\xi)\cdot N_1^{(e-1D)}(\eta)\cdot N_1^{(e-1D)}(\zeta) \tag{9.3.1a}$$

$$N_2^{(8H)}(\xi,\eta,\zeta) = N_2^{(e-1D)}(\xi)\cdot N_1^{(e-1D)}(\eta)\cdot N_1^{(e-1D)}(\zeta) \tag{9.3.1b}$$

$$N_3^{(8H)}(\xi,\eta,\zeta) = N_2^{(e-1D)}(\xi)\cdot N_2^{(e-1D)}(\eta)\cdot N_1^{(e-1D)}(\zeta) \tag{9.3.1c}$$

$$N_4^{(8H)}(\xi,\eta,\zeta) = N_1^{(e-1D)}(\xi)\cdot N_2^{(e-1D)}(\eta)\cdot N_1^{(e-1D)}(\zeta) \tag{9.3.1d}$$

$$N_5^{(8H)}(\xi,\eta,\zeta) = N_1^{(e-1D)}(\xi)\cdot N_1^{(e-1D)}(\eta)\cdot N_2^{(e-1D)}(\zeta) \tag{9.3.1e}$$

$$N_6^{(8H)}(\xi,\eta,\zeta) = N_2^{(e-1D)}(\xi)\cdot N_1^{(e-1D)}(\eta)\cdot N_2^{(e-1D)}(\zeta) \tag{9.3.1f}$$

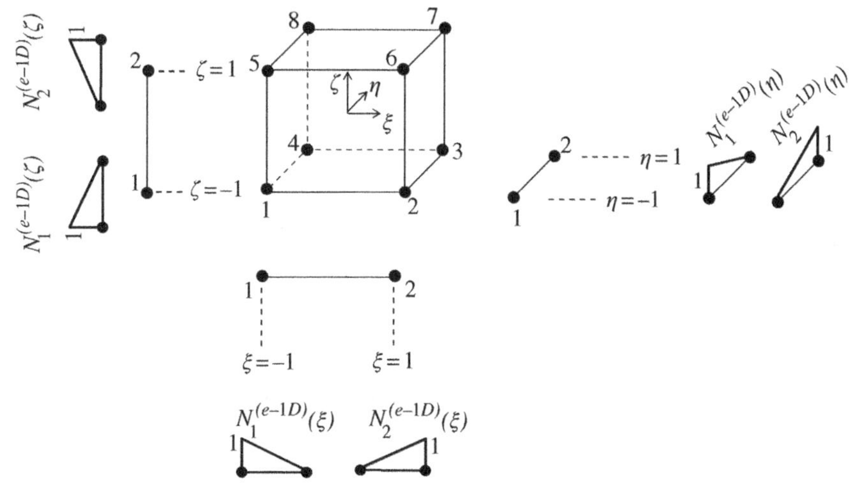

Figure 9.3 One-dimensional shape functions used as the basis for the eight shape functions of the 8H element.

$$N_7^{(8H)}(\xi,\eta,\zeta) = N_2^{(e-1D)}(\xi)\cdot N_2^{(e-1D)}(\eta)\cdot N_2^{(e-1D)}(\zeta) \tag{9.3.1g}$$

$$N_8^{(8H)}(\xi,\eta,\zeta) = N_1^{(e-1D)}(\xi)\cdot N_2^{(e-1D)}(\eta)\cdot N_2^{(e-1D)}(\zeta) \tag{9.3.1h}$$

We can replace the expressions for the one-dimensional shape functions in the tensor products to finally obtain:

$$N_1^{(8H)}(\xi,\eta,\zeta) = \frac{1-\xi}{2}\cdot\frac{1-\eta}{2}\cdot\frac{1-\zeta}{2} = \frac{1}{8}(1-\xi)(1-\eta)(1-\zeta) \tag{9.3.2a}$$

$$N_2^{(8H)}(\xi,\eta,\zeta) = \frac{1+\xi}{2}\cdot\frac{1-\eta}{2}\cdot\frac{1-\zeta}{2} = \frac{1}{8}(1+\xi)(1-\eta)(1-\zeta) \tag{9.3.2b}$$

$$N_3^{(8H)}(\xi,\eta,\zeta) = \frac{1+\xi}{2}\cdot\frac{1+\eta}{2}\cdot\frac{1-\zeta}{2} = \frac{1}{8}(1+\xi)(1+\eta)(1-\zeta) \tag{9.3.2c}$$

$$N_4^{(8H)}(\xi,\eta,\zeta) = \frac{1-\xi}{2}\cdot\frac{1+\eta}{2}\cdot\frac{1-\zeta}{2} = \frac{1}{8}(1-\xi)(1+\eta)(1-\zeta) \tag{9.3.2d}$$

$$N_5^{(8H)}(\xi,\eta,\zeta) = \frac{1-\xi}{2}\cdot\frac{1-\eta}{2}\cdot\frac{1+\zeta}{2} = \frac{1}{8}(1-\xi)(1-\eta)(1+\zeta) \tag{9.3.2e}$$

$$N_6^{(8H)}(\xi,\eta,\zeta) = \frac{1+\xi}{2}\cdot\frac{1-\eta}{2}\cdot\frac{1+\zeta}{2} = \frac{1}{8}(1+\xi)(1-\eta)(1+\zeta) \tag{9.3.2f}$$

$$N_7^{(8H)}(\xi,\eta,\zeta) = \frac{1+\xi}{2}\cdot\frac{1+\eta}{2}\cdot\frac{1+\zeta}{2} = \frac{1}{8}(1+\xi)(1+\eta)(1+\zeta) \tag{9.3.2g}$$

$$N_8^{(8H)}(\xi,\eta,\zeta) = \frac{1-\xi}{2}\cdot\frac{1+\eta}{2}\cdot\frac{1+\zeta}{2} = \frac{1}{8}(1-\xi)(1+\eta)(1+\zeta) \tag{9.3.2h}$$

Having established the eight shape functions, we can now formulate the approximation for the displacements in the parametric space. The approximate displacement field is given by the following equation.

$$\begin{Bmatrix} u_x^{(e)}(\xi,\eta,\zeta) \\ u_y^{(e)}(\xi,\eta,\zeta) \\ u_z^{(e)}(\xi,\eta,\zeta) \end{Bmatrix} = \begin{bmatrix} N_1^{(8H)}(\xi,\eta,\zeta) & 0 & 0 & \ldots\ldots & N_8^{(8H)}(\xi,\eta,\zeta) & 0 & 0 \\ 0 & N_1^{(8H)}(\xi,\eta,\zeta) & 0 & \ldots\ldots & 0 & N_8^{(8H)}(\xi,\eta,\zeta) & 0 \\ 0 & 0 & N_1^{(8H)}(\xi,\eta,\zeta) & \ldots\ldots & 0 & 0 & N_8^{(8H)}(\xi,\eta,\zeta) \end{bmatrix} \begin{Bmatrix} u_{x1}^{(e)} \\ u_{y1}^{(e)} \\ u_{z1}^{(e)} \\ \vdots \\ u_{x8}^{(e)} \\ u_{y8}^{(e)} \\ u_{z8}^{(e)} \end{Bmatrix} \tag{9.3.3a}$$

We can group the quantities corresponding to each node and write more concisely a *block matrix equation* for the approximate displacement field:

$$\begin{Bmatrix} u_x^{(e)}(\xi,\eta,\zeta) \\ u_y^{(e)}(\xi,\eta,\zeta) \\ u_z^{(e)}(\xi,\eta,\zeta) \end{Bmatrix} = \left[\left[N_1^{(8H)}(\xi,\eta,\zeta)\right] \; \left[N_2^{(8H)}(\xi,\eta,\zeta)\right] \; \cdots \; \left[N_8^{(8H)}(\xi,\eta,\zeta)\right] \right] \begin{Bmatrix} \left\{U_1^{(e)}\right\} \\ \left\{U_2^{(e)}\right\} \\ \vdots \\ \left\{U_8^{(e)}\right\} \end{Bmatrix} \tag{9.3.3b}$$

or

$$\begin{Bmatrix} u_x^{(e)}(\xi,\eta,\zeta) \\ u_y^{(e)}(\xi,\eta,\zeta) \\ u_z^{(e)}(\xi,\eta,\zeta) \end{Bmatrix} = \left[N_1^{(8H)}(\xi,\eta,\zeta)\right]\left\{U_1^{(e)}\right\} + \ldots + \left[N_8^{(8H)}(\xi,\eta,\zeta)\right]\left\{U_8^{(e)}\right\} \tag{9.3.3c}$$

where each (3 × 3) *block matrix*, $\left[N_i^{(8H)}(\xi,\eta,\zeta)\right]$, is given from the following expression:

$$\left[N_i^{(8H)}(\xi,\eta,\zeta)\right] = \begin{bmatrix} N_i^{(8H)}(\xi,\eta,\zeta) & 0 & 0 \\ 0 & N_i^{(8H)}(\xi,\eta,\zeta) & 0 \\ 0 & 0 & N_i^{(8H)}(\xi,\eta,\zeta) \end{bmatrix} \tag{9.3.4}$$

and each (3 × 1) *block vector*, $\left\{U_i^{(e)}\right\}$, contains the three nodal displacement components of node i:

$$\left\{U_i^{(e)}\right\} = \begin{Bmatrix} u_{xi}^{(e)} \\ u_{yi}^{(e)} \\ u_{zi}^{(e)} \end{Bmatrix} \tag{9.3.5}$$

Note that we can also write a set of three equations for the components of the displacement field in the interior of the element:

$$u_x^{(e)}(\xi,\eta,\zeta) = \sum_{I=1}^{8}\left[N_I^{(8H)}(\xi,\eta,\zeta)\cdot u_{xI}^{(e)}\right] = N_1^{(8H)}(\xi,\eta,\zeta)\cdot u_{x1}^{(e)} + \ldots + N_8^{(8H)}(\xi,\eta,\zeta)\cdot u_{x8}^{(e)} \tag{9.3.6a}$$

$$u_y^{(e)}(\xi,\eta,\zeta) = \sum_{I=1}^{8}\left[N_I^{(8H)}(\xi,\eta,\zeta)\cdot u_{yI}^{(e)}\right] = N_1^{(8H)}(\xi,\eta,\zeta)\cdot u_{y1}^{(e)} + \ldots + N_8^{(8H)}(\xi,\eta,\zeta)\cdot u_{y8}^{(e)} \tag{9.3.6b}$$

$$u_z^{(e)}(\xi,\eta,\zeta) = \sum_{I=1}^{8}\left[N_I^{(8H)}(\xi,\eta,\zeta)\cdot u_{zI}^{(e)}\right] = N_1^{(8H)}(\xi,\eta,\zeta)\cdot u_{z1}^{(e)} + \ldots + N_8^{(8H)}(\xi,\eta,\zeta)\cdot u_{z8}^{(e)} \tag{9.3.6c}$$

We will now proceed with establishing the coordinate mapping, $x = x(\xi,\eta,\zeta)$, $y = y(\xi,\eta,\zeta)$ and $z = z(\xi,\eta,\zeta)$. Since we have an isoparametric element, the coordinate mapping uses the same shape functions as those for the approximate displacements. We can establish a block matrix equation for the mapping:

$$\begin{Bmatrix} x(\xi,\eta,\zeta) \\ y(\xi,\eta,\zeta) \\ z(\xi,\eta,\zeta) \end{Bmatrix} = \left[\left[N_1^{(8H)}(\xi,\eta,\zeta)\right] \quad \left[N_2^{(8H)}(\xi,\eta,\zeta)\right] \quad \ldots \quad \left[N_8^{(8H)}(\xi,\eta,\zeta)\right]\right] \begin{Bmatrix} \left\{x_1^{(e)}\right\} \\ \left\{x_2^{(e)}\right\} \\ \vdots \\ \left\{x_8^{(e)}\right\} \end{Bmatrix} \tag{9.3.7}$$

where

$$\left\{x_i^{(e)}\right\} = \begin{Bmatrix} x_i^{(e)} \\ y_i^{(e)} \\ z_i^{(e)} \end{Bmatrix} \tag{9.3.8}$$

Equation (9.3.7) can be written as a set of three expressions:

$$x(\xi,\eta,\zeta) = \sum_{I=1}^{8}\left[N_I^{(8H)}(\xi,\eta,\zeta)\cdot x_I^{(e)}\right] = N_1^{(8H)}(\xi,\eta,\zeta)\cdot x_1^{(e)} + \ldots + N_8^{(8H)}(\xi,\eta,\zeta)\cdot x_8^{(e)} \tag{9.3.7a}$$

$$y(\xi,\eta,\zeta) = \sum_{I=1}^{8}\left[N_I^{(8H)}(\xi,\eta,\zeta)\cdot y_I^{(e)}\right] = N_1^{(8H)}(\xi,\eta,\zeta)\cdot y_1^{(e)} + \ldots + N_8^{(8H)}(\xi,\eta,\zeta)\cdot y_8^{(e)} \tag{9.3.7b}$$

$$z(\xi,\eta,\zeta) = \sum_{I=1}^{8}\left[N_I^{(8H)}(\xi,\eta,\zeta)\cdot z_I^{(e)}\right] = N_1^{(8H)}(\xi,\eta,\zeta)\cdot z_1^{(e)} + \ldots + N_8^{(8H)}(\xi,\eta,\zeta)\cdot z_8^{(e)} \tag{9.3.7c}$$

Another set of necessary quantities for the element formulation is the partial derivatives of the mapping expressions with respect to the parametric coordinates. Given Equations (9.3.7a–c), we can obtain these derivatives as follows.

$$\frac{\partial x}{\partial \xi} = \frac{\partial}{\partial \xi}\sum_{i=1}^{8}\left(N_i^{(8H)}(\xi,\eta,\zeta)\cdot x_i^{(e)}\right) \rightarrow \frac{\partial x}{\partial \xi} = \sum_{i=1}^{8}\left(\frac{\partial N_i^{(8H)}}{\partial \xi}\cdot x_i^{(e)}\right) \tag{9.3.9a}$$

$$\frac{\partial y}{\partial \xi} = \frac{\partial}{\partial \xi}\sum_{i=1}^{8}\left(N_i^{(8H)}(\xi,\eta,\zeta)\cdot y_i^{(e)}\right) \rightarrow \frac{\partial y}{\partial \xi} = \sum_{i=1}^{8}\left(\frac{\partial N_i^{(8H)}}{\partial \xi}\cdot y_i^{(e)}\right) \tag{9.3.9b}$$

$$\frac{\partial z}{\partial \xi} = \frac{\partial}{\partial \xi}\sum_{i=1}^{8}\left(N_i^{(8H)}(\xi,\eta,\zeta)\cdot z_i^{(e)}\right) \rightarrow \frac{\partial z}{\partial \xi} = \sum_{i=1}^{8}\left(\frac{\partial N_i^{(8H)}}{\partial \xi}\cdot z_i^{(e)}\right) \tag{9.3.9c}$$

$$\frac{\partial x}{\partial \eta} = \frac{\partial}{\partial \eta}\sum_{i=1}^{8}\left(N_i^{(8H)}(\xi,\eta,\zeta)\cdot x_i^{(e)}\right) \rightarrow \frac{\partial x}{\partial \eta} = \sum_{i=1}^{8}\left(\frac{\partial N_i^{(8H)}}{\partial \eta}\cdot x_i^{(e)}\right) \tag{9.3.9d}$$

$$\frac{\partial y}{\partial \eta} = \frac{\partial}{\partial \eta}\sum_{i=1}^{8}\left(N_i^{(8H)}(\xi,\eta,\zeta)\cdot y_i^{(e)}\right) \rightarrow \frac{\partial y}{\partial \eta} = \sum_{i=1}^{8}\left(\frac{\partial N_i^{(8H)}}{\partial \eta}\cdot y_i^{(e)}\right) \tag{9.3.9e}$$

$$\frac{\partial z}{\partial \eta} = \frac{\partial}{\partial \eta}\sum_{i=1}^{8}\left(N_i^{(8H)}(\xi,\eta,\zeta)\cdot z_i^{(e)}\right) \rightarrow \frac{\partial z}{\partial \eta} = \sum_{i=1}^{8}\left(\frac{\partial N_i^{(8H)}}{\partial \eta}\cdot z_i^{(e)}\right) \tag{9.3.9f}$$

$$\frac{\partial x}{\partial \zeta} = \frac{\partial}{\partial \zeta}\sum_{i=1}^{8}\left(N_i^{(8H)}(\xi,\eta,\zeta)\cdot x_i^{(e)}\right) \rightarrow \frac{\partial x}{\partial \zeta} = \sum_{i=1}^{8}\left(\frac{\partial N_i^{(8H)}}{\partial \zeta}\cdot x_i^{(e)}\right) \tag{9.3.9g}$$

$$\frac{\partial y}{\partial \zeta} = \frac{\partial}{\partial \zeta}\sum_{i=1}^{8}\left(N_i^{(8H)}(\xi,\eta,\zeta)\cdot y_i^{(e)}\right) \rightarrow \frac{\partial y}{\partial \zeta} = \sum_{i=1}^{8}\left(\frac{\partial N_i^{(8H)}}{\partial \zeta}\cdot y_i^{(e)}\right) \tag{9.3.9h}$$

$$\frac{\partial z}{\partial \zeta} = \frac{\partial}{\partial \zeta}\sum_{i=1}^{8}\left(N_i^{(8H)}(\xi,\eta,\zeta)\cdot z_i^{(e)}\right) \rightarrow \frac{\partial z}{\partial \zeta} = \sum_{i=1}^{8}\left(\frac{\partial N_i^{(8H)}}{\partial \zeta}\cdot z_i^{(e)}\right) \tag{9.3.9i}$$

Since we are given the expressions for the shape functions in Equations (9.3.2a–h), we can readily find their derivatives with respect to the parametric coordinates:

$$\frac{\partial N_1^{(8H)}}{\partial \xi} = -\frac{1}{8}(1-\eta)(1-\zeta),\ \frac{\partial N_1^{(8H)}}{\partial \eta} = -\frac{1}{8}(1-\xi)(1-\zeta),\ \frac{\partial N_1^{(8H)}}{\partial \zeta} = -\frac{1}{8}(1-\xi)(1-\eta)$$

$$\frac{\partial N_2^{(8H)}}{\partial \xi} = \frac{1}{8}(1-\eta)(1-\zeta),\ \frac{\partial N_2^{(8H)}}{\partial \eta} = -\frac{1}{8}(1+\xi)(1-\zeta),\ \frac{\partial N_2^{(8H)}}{\partial \zeta} = -\frac{1}{8}(1+\xi)(1-\eta)$$

$$\frac{\partial N_3^{(8H)}}{\partial \xi} = \frac{1}{8}(1+\eta)(1-\zeta),\ \frac{\partial N_3^{(8H)}}{\partial \eta} = \frac{1}{8}(1+\xi)(1-\zeta),\ \frac{\partial N_3^{(8H)}}{\partial \zeta} = -\frac{1}{8}(1+\xi)(1+\eta)$$

$$\frac{\partial N_4^{(8H)}}{\partial \xi} = -\frac{1}{8}(1+\eta)(1-\zeta),\ \frac{\partial N_4^{(8H)}}{\partial \eta} = \frac{1}{8}(1-\xi)(1-\zeta),\ \frac{\partial N_4^{(8H)}}{\partial \zeta} = -\frac{1}{8}(1-\xi)(1+\eta)$$

$$\frac{\partial N_5^{(8H)}}{\partial \xi} = -\frac{1}{8}(1-\eta)(1+\zeta),\ \frac{\partial N_5^{(8H)}}{\partial \eta} = -\frac{1}{8}(1-\xi)(1+\zeta),\ \frac{\partial N_5^{(8H)}}{\partial \zeta} = \frac{1}{8}(1-\xi)(1-\eta)$$

$$\frac{\partial N_6^{(8H)}}{\partial \xi} = \frac{1}{8}(1-\eta)(1+\zeta),\ \frac{\partial N_6^{(8H)}}{\partial \eta} = -\frac{1}{8}(1+\xi)(1+\zeta),\ \frac{\partial N_6^{(8H)}}{\partial \zeta} = \frac{1}{8}(1+\xi)(1-\eta)$$

$$\frac{\partial N_7^{(8H)}}{\partial \xi} = \frac{1}{8}(1+\eta)(1+\zeta),\ \frac{\partial N_7^{(8H)}}{\partial \eta} = \frac{1}{8}(1+\xi)(1+\zeta),\ \frac{\partial N_7^{(8H)}}{\partial \zeta} = \frac{1}{8}(1+\xi)(1+\eta)$$

$$\frac{\partial N_8^{(8H)}}{\partial \xi} = -\frac{1}{8}(1+\eta)(1+\zeta),\ \frac{\partial N_8^{(8H)}}{\partial \eta} = \frac{1}{8}(1-\xi)(1+\zeta),\ \frac{\partial N_8^{(8H)}}{\partial \zeta} = \frac{1}{8}(1-\xi)(1+\eta)$$

Once we have the expressions in Equations (9.3.9a–i), we can establish the *Jacobian array* $[J]$ of the coordinate mapping:

$$[J] = \begin{bmatrix} \partial x/\partial \xi & \partial y/\partial \xi & \partial z/\partial \xi \\ \partial x/\partial \eta & \partial y/\partial \eta & \partial z/\partial \eta \\ \partial x/\partial \zeta & \partial y/\partial \zeta & \partial z/\partial \zeta \end{bmatrix} \tag{9.3.10}$$

We can also define the inverse of the Jacobian array:

$$[\tilde{J}] = [J]^{-1} = \begin{bmatrix} \partial \xi/\partial x & \partial \eta/\partial x & \partial \zeta/\partial x \\ \partial \xi/\partial y & \partial \eta/\partial y & \partial \zeta/\partial y \\ \partial \xi/\partial z & \partial \eta/\partial z & \partial \zeta/\partial z \end{bmatrix} = \begin{bmatrix} \tilde{J}_{11} & \tilde{J}_{12} & \tilde{J}_{13} \\ \tilde{J}_{21} & \tilde{J}_{22} & \tilde{J}_{23} \\ \tilde{J}_{31} & \tilde{J}_{32} & \tilde{J}_{33} \end{bmatrix} \tag{9.3.11}$$

Given the approximate field from Equations (9.3.6a–c), we can also obtain the strain-displacement matrix equation, as a special case ($n = 8$) of Equation (9.2.5):

$$\left\{\varepsilon^{(e)}\right\} \approx \left[B^{(e)}\right]\left\{U^{(e)}\right\},$$

where

$$\left[B^{(e)}\right]=\begin{bmatrix}\frac{\partial N_1^{(8H)}}{\partial x} & 0 & 0 & \frac{\partial N_2^{(8H)}}{\partial x} & 0 & 0 & \dots & \frac{\partial N_8^{(8H)}}{\partial x} & 0 & 0\\ 0 & \frac{\partial N_1^{(8H)}}{\partial y} & 0 & 0 & \frac{\partial N_2^{(8H)}}{\partial y} & 0 & \dots & 0 & \frac{\partial N_8^{(8H)}}{\partial y} & 0\\ 0 & 0 & \frac{\partial N_1^{(8H)}}{\partial z} & 0 & 0 & \frac{\partial N_2^{(8H)}}{\partial z} & \dots & 0 & 0 & \frac{\partial N_8^{(8H)}}{\partial z}\\ \frac{\partial N_1^{(8H)}}{\partial y} & \frac{\partial N_1^{(8H)}}{\partial x} & 0 & \frac{\partial N_2^{(8H)}}{\partial y} & \frac{\partial N_2^{(8H)}}{\partial x} & 0 & \dots & \frac{\partial N_8^{(8H)}}{\partial y} & \frac{\partial N_8^{(8H)}}{\partial x} & 0\\ 0 & \frac{\partial N_1^{(8H)}}{\partial z} & \frac{\partial N_1^{(8H)}}{\partial y} & 0 & \frac{\partial N_2^{(8H)}}{\partial z} & \frac{\partial N_2^{(8H)}}{\partial y} & \dots & 0 & \frac{\partial N_8^{(8H)}}{\partial z} & \frac{\partial N_8^{(8H)}}{\partial y}\\ \frac{\partial N_1^{(8H)}}{\partial z} & 0 & \frac{\partial N_1^{(8H)}}{\partial x} & \frac{\partial N_2^{(8H)}}{\partial z} & 0 & \frac{\partial N_2^{(8H)}}{\partial x} & \dots & \frac{\partial N_8^{(8H)}}{\partial z} & 0 & \frac{\partial N_8^{(8H)}}{\partial x}\end{bmatrix} \tag{9.3.12}$$

The $[B^{(e)}]$ array includes derivatives of the shape functions with respect to the *physical* coordinates, while the shape functions have been established in terms of the *parametric* coordinates. So, just as we did for two-dimensional isoparametric elements in Section 8.5.2, we use the chain rule of differentiation, so the derivatives of the shape function of node i with respect to the physical coordinates are as follows:

$$\frac{\partial N_i^{(8H)}}{\partial x}=\frac{\partial N_i^{(8H)}}{\partial \xi}\frac{\partial \xi}{\partial x}+\frac{\partial N_i^{(8H)}}{\partial \eta}\frac{\partial \eta}{\partial x}+\frac{\partial N_i^{(8H)}}{\partial \zeta}\frac{\partial \zeta}{\partial x}=\frac{\partial N_i^{(8H)}}{\partial \xi}\tilde{J}_{11}+\frac{\partial N_i^{(8H)}}{\partial \eta}\tilde{J}_{12}+\frac{\partial N_i^{(8H)}}{\partial \zeta}\tilde{J}_{13} \tag{9.3.13a}$$

$$\frac{\partial N_i^{(8H)}}{\partial y}=\frac{\partial N_i^{(8H)}}{\partial \xi}\frac{\partial \xi}{\partial y}+\frac{\partial N_i^{(8H)}}{\partial \eta}\frac{\partial \eta}{\partial y}+\frac{\partial N_i^{(8H)}}{\partial \zeta}\frac{\partial \zeta}{\partial y}=\frac{\partial N_i^{(8H)}}{\partial \xi}\tilde{J}_{21}+\frac{\partial N_i^{(8H)}}{\partial \eta}\tilde{J}_{22}+\frac{\partial N_i^{(8H)}}{\partial \zeta}\tilde{J}_{23} \tag{9.3.13b}$$

$$\frac{\partial N_i^{(8H)}}{\partial z}=\frac{\partial N_i^{(8H)}}{\partial \xi}\frac{\partial \xi}{\partial z}+\frac{\partial N_i^{(8H)}}{\partial \eta}\frac{\partial \eta}{\partial z}+\frac{\partial N_i^{(8H)}}{\partial \zeta}\frac{\partial \zeta}{\partial z}=\frac{\partial N_i^{(8H)}}{\partial \xi}\tilde{J}_{31}+\frac{\partial N_i^{(8H)}}{\partial \eta}\tilde{J}_{32}+\frac{\partial N_i^{(8H)}}{\partial \zeta}\tilde{J}_{33} \tag{9.3.13c}$$

We can establish the following matrix expression to obtain the derivatives of all the shape functions with respect to the physical coordinates.

$$\left[N_{,x}^{(8H)}\right]=\begin{bmatrix}\tilde{J}_{11} & \tilde{J}_{12} & \tilde{J}_{13}\\ \tilde{J}_{21} & \tilde{J}_{22} & \tilde{J}_{23}\\ \tilde{J}_{31} & \tilde{J}_{32} & \tilde{J}_{33}\end{bmatrix}\begin{bmatrix}\frac{\partial N_1^{(8H)}}{\partial \xi} & \frac{\partial N_2^{(8H)}}{\partial \xi} & \frac{\partial N_3^{(8H)}}{\partial \xi} & \frac{\partial N_4^{(8H)}}{\partial \xi} & \frac{\partial N_5^{(8H)}}{\partial \xi} & \frac{\partial N_6^{(8H)}}{\partial \xi} & \frac{\partial N_7^{(8H)}}{\partial \xi} & \frac{\partial N_8^{(8H)}}{\partial \xi}\\ \frac{\partial N_1^{(8H)}}{\partial \eta} & \frac{\partial N_2^{(8H)}}{\partial \eta} & \frac{\partial N_3^{(8H)}}{\partial \eta} & \frac{\partial N_4^{(8H)}}{\partial \eta} & \frac{\partial N_5^{(8H)}}{\partial \eta} & \frac{\partial N_6^{(8H)}}{\partial \eta} & \frac{\partial N_7^{(8H)}}{\partial \eta} & \frac{\partial N_8^{(8H)}}{\partial \eta}\\ \frac{\partial N_1^{(8H)}}{\partial \zeta} & \frac{\partial N_2^{(8H)}}{\partial \zeta} & \frac{\partial N_3^{(8H)}}{\partial \zeta} & \frac{\partial N_4^{(8H)}}{\partial \zeta} & \frac{\partial N_5^{(8H)}}{\partial \zeta} & \frac{\partial N_6^{(8H)}}{\partial \zeta} & \frac{\partial N_7^{(8H)}}{\partial \zeta} & \frac{\partial N_8^{(8H)}}{\partial \zeta}\end{bmatrix}$$

$$=\left[\tilde{J}\right]\left[N_{,\xi}^{(8H)}\right] \tag{9.3.14}$$

where $\left[N_{,\xi}^{(8H)}\right]$ is the (3 × 8) matrix containing the derivatives of the shape functions with respect to the parametric coordinates:

$$\left[N_{,\xi}^{(8H)}\right] = \begin{bmatrix} \frac{\partial N_1^{(8H)}}{\partial \xi} & \frac{\partial N_2^{(8H)}}{\partial \xi} & \frac{\partial N_3^{(8H)}}{\partial \xi} & \frac{\partial N_4^{(8H)}}{\partial \xi} & \frac{\partial N_5^{(8H)}}{\partial \xi} & \frac{\partial N_6^{(8H)}}{\partial \xi} & \frac{\partial N_7^{(8H)}}{\partial \xi} & \frac{\partial N_8^{(8H)}}{\partial \xi} \\ \frac{\partial N_1^{(8H)}}{\partial \eta} & \frac{\partial N_2^{(8H)}}{\partial \eta} & \frac{\partial N_3^{(8H)}}{\partial \eta} & \frac{\partial N_4^{(8H)}}{\partial \eta} & \frac{\partial N_5^{(8H)}}{\partial \eta} & \frac{\partial N_6^{(8H)}}{\partial \eta} & \frac{\partial N_7^{(8H)}}{\partial \eta} & \frac{\partial N_8^{(8H)}}{\partial \eta} \\ \frac{\partial N_1^{(8H)}}{\partial \zeta} & \frac{\partial N_2^{(8H)}}{\partial \zeta} & \frac{\partial N_3^{(8H)}}{\partial \zeta} & \frac{\partial N_4^{(8H)}}{\partial \zeta} & \frac{\partial N_5^{(8H)}}{\partial \zeta} & \frac{\partial N_6^{(8H)}}{\partial \zeta} & \frac{\partial N_7^{(8H)}}{\partial \zeta} & \frac{\partial N_8^{(8H)}}{\partial \zeta} \end{bmatrix} \quad (9.3.15)$$

and $\left[N_{,x}^{(8H)}\right]$ contains the corresponding derivatives with respect to the physical coordinates:

$$\left[N_{,x}^{(8H)}\right] = \begin{bmatrix} \frac{\partial N_1^{(8H)}}{\partial x} & \frac{\partial N_2^{(8H)}}{\partial x} & \frac{\partial N_3^{(8H)}}{\partial x} & \frac{\partial N_4^{(8H)}}{\partial x} & \frac{\partial N_5^{(8H)}}{\partial x} & \frac{\partial N_6^{(8H)}}{\partial x} & \frac{\partial N_7^{(8H)}}{\partial x} & \frac{\partial N_8^{(8H)}}{\partial x} \\ \frac{\partial N_1^{(8H)}}{\partial y} & \frac{\partial N_2^{(8H)}}{\partial y} & \frac{\partial N_3^{(8H)}}{\partial y} & \frac{\partial N_4^{(8H)}}{\partial y} & \frac{\partial N_5^{(8H)}}{\partial y} & \frac{\partial N_6^{(8H)}}{\partial y} & \frac{\partial N_7^{(8H)}}{\partial y} & \frac{\partial N_8^{(8H)}}{\partial y} \\ \frac{\partial N_1^{(8H)}}{\partial z} & \frac{\partial N_2^{(8H)}}{\partial z} & \frac{\partial N_3^{(8H)}}{\partial z} & \frac{\partial N_4^{(8H)}}{\partial z} & \frac{\partial N_5^{(8H)}}{\partial z} & \frac{\partial N_6^{(8H)}}{\partial z} & \frac{\partial N_7^{(8H)}}{\partial z} & \frac{\partial N_8^{(8H)}}{\partial z} \end{bmatrix} \quad (9.3.16)$$

We will now proceed to the computation of the element stiffness matrix and equivalent nodal force vector, which are given by the following expressions.

$$\left[k^{(e)}\right] = \iiint\limits_{\Omega^{(e)}} \left[B^{(e)}\right]^T \left[D^{(e)}\right] \left[B^{(e)}\right] dV$$

$$\left\{f^{(e)}\right\} = \left\{f_{\Omega}^{(e)}\right\} + \left\{f_{\Gamma t}^{(e)}\right\} = \iiint\limits_{\Omega^{(e)}} \left[N^{(e)}\right]^T \{b\} dV + \iint\limits_{\Gamma_t^{(e)}} \left[N^{(e)}\right]^T \{t\}\, dS$$

We will first go through the procedure to calculate the three-dimensional integrals over $\Omega^{(e)}$, and then see how to obtain the boundary surface integrals. For the integrals over $\Omega^{(e)}$, we must express the material stiffness matrix $[D^{(e)}]$ and the body force vector $\{b\}$ as functions of the parametric coordinates, so that we can change coordinates and conduct the three-dimensional element domain integration in the parametric space. Thus, we need to use the coordinate mapping and obtain:

$$\left[D^{(e)}(\xi,\eta,\zeta)\right] = \left[D^{(e)}(x(\xi,\eta,\zeta), y(\xi,\eta,\zeta), z(\xi,\eta,\zeta))\right] \quad (9.3.17)$$

$$\{b(\xi,\eta,\zeta)\} = \{b(x(\xi,\eta,\zeta), y(\xi,\eta,\zeta), z(\xi,\eta,\zeta))\} \quad (9.3.18)$$

We can express the element domain (volume) integrals for the stiffness matrix $[k^{(e)}]$ and equivalent nodal force vector $\left\{f_{\Omega}^{(e)}\right\}$ as triple integrals in the parametric coordinate space.

$$\left[k^{(e)}\right] = \int_{-1}^{1}\int_{-1}^{1}\int_{-1}^{1} \left[B^{(e)}(\xi,\eta,\zeta)\right]^T \left[D^{(e)}(\xi,\eta,\zeta)\right] \left[B^{(e)}(\xi,\eta,\zeta)\right] J(\xi,\eta,\zeta) d\xi\, d\eta\, d\zeta \qquad (9.3.19)$$

$$\left\{f_\Omega^{(e)}\right\} = \iiint_{\Omega^{(e)}} \left[N^{(e)}\right]^T \{b\} dV = \int_{-1}^{1}\int_{-1}^{1}\int_{-1}^{1} \left[N^{(8H)}(\xi,\eta,\zeta)\right]^T \{b(\xi,\eta,\zeta)\} \cdot J(\xi,\eta,\zeta) d\xi\, d\eta\, d\zeta$$

(9.3.20)

As explained in the following subsection, these integrals are calculated numerically, using Gaussian quadrature.

9.3.2 Numerical (Gaussian) Quadrature for Hexahedral Isoparametric Elements

A major advantage of the isoparametric concept is that we can use Gaussian (numerical) quadrature to calculate the element contributions to the stiffness matrix and right-hand-side vector. We will first discuss how to integrate a general function, and then specialize our discussion to the evaluation of the element integrals.

Let us imagine that we want to integrate a function $f(\xi, \eta, \zeta)$ in the parametric space. Since the parametric domain is a nice cube whose sides are aligned with the coordinate axes, we can conduct the integration with respect to each one of the parametric variables separately. Since the range of values of each parametric coordinate is from −1 to 1, we can use one-dimensional quadrature for each one-dimensional integration. Let us imagine that we have a one-dimensional quadrature using $N_{g\xi}$ quadrature points in the ξ-direction, $N_{g\eta}$ quadrature points in the η-direction and $N_{g\zeta}$ quadrature points in the ζ-direction. We will imagine that we first conduct the integration in the ξ-direction:

$$\int_{-1}^{1}\int_{-1}^{1}\int_{-1}^{1} f(\xi,\eta,\zeta) d\xi d\eta d\zeta = \int_{-1}^{1}\int_{-1}^{1} \left(\int_{-1}^{1} f(\xi,\eta,\zeta) d\xi\right) d\eta d\zeta = \int_{-1}^{1}\int_{-1}^{1} I_1(\eta,\zeta) d\eta d\zeta \qquad (9.3.21)$$

where we have set:

$$I_1(\eta,\zeta) = \int_{-1}^{1} f(\xi,\eta,\zeta) d\xi \approx \sum_{i=1}^{Ng_\xi} \left(f(\xi_i,\eta,\zeta) W_i\right) \qquad (9.3.22)$$

We can then proceed to integrate I_1 with respect to η:

$$\int_{-1}^{1}\int_{-1}^{1} I_1(\eta,\zeta) d\eta d\zeta = \int_{-1}^{1} \left(\int_{-1}^{1} I_1(\eta,\zeta) d\eta\right) d\zeta = \int_{-1}^{1} I_2(\zeta) d\zeta \qquad (9.3.23)$$

where we have set:

$$I_2(\zeta) = \int_{-1}^{1} I_1(\eta,\zeta) d\eta \approx \sum_{j=1}^{Ng_\eta} \left(I_1\left(\eta_j,\zeta\right) W_j\right) \qquad (9.3.24)$$

If we set $\eta = \eta_j$ in Equation (9.3.22), we obtain:

$$I_1\left(\eta_j,\zeta\right) = \sum_{i=1}^{Ng_\xi}\left(f\left(\xi_i,\eta_j,\zeta\right)W_i\right) \tag{9.3.25}$$

Plugging Equation (9.3.25) into Equation (9.3.24) yields:

$$I_2(\zeta) \approx \sum_{j=1}^{Ng_\eta}\left(I_1\left(\eta_j,\zeta\right)W_j\right) = \sum_{j=1}^{Ng_\eta}\left[\sum_{i=1}^{Ng_\xi}\left(f\left(\xi_i,\eta_j,\zeta\right)W_i\right)W_j\right] \tag{9.3.26}$$

The final step is to use quadrature to integrate I_2 with respect to ζ:

$$\int_{-1}^{1} I_2(\zeta)d\zeta \approx \sum_{k=1}^{N_{g\zeta}}\left(I_2(\zeta_k)\cdot W_k\right) \tag{9.3.27}$$

From Equation (9.3.26), if we set $\zeta = \zeta_k$, we have:

$$I_2(\zeta_k) \approx \sum_{j=1}^{Ng_\eta}\left(I_1\left(\eta_j,\zeta_k\right)W_j\right) = \sum_{j=1}^{Ng_\eta}\left[\sum_{i=1}^{Ng_\xi}\left(f\left(\xi_i,\eta_j,\zeta_k\right)W_i\right)W_j\right] \tag{9.3.28}$$

Now, remembering that the left-hand side of Equation (9.3.27) equals the integral $\int_{-1}^{1}\int_{-1}^{1}\int_{-1}^{1} f(\xi,\eta,\zeta)d\xi d\eta d\zeta$ that we were trying to calculate, we finally obtain:

$$\int_{-1}^{1}\int_{-1}^{1}\int_{-1}^{1} f(\xi,\eta,\zeta)d\xi d\eta d\zeta \approx \sum_{k=1}^{Ng_\zeta}\sum_{j=1}^{Ng_\eta}\sum_{i=1}^{Ng_\xi}\left(f\left(\xi_i,\eta_j,\zeta_k\right)W_i W_j W_k\right) \tag{9.3.29}$$

If we have a total of N_g quadrature points, with $N_g = N_{g\xi}\cdot N_{g\eta}\cdot N_{g\zeta}$, then for each quadrature point g: $\xi_g = \xi_i, \eta_g = \eta_j, \zeta_g = \zeta_k, W_g = W_i\cdot W_j\cdot W_k$

Thus, the integral now becomes:

$$\int_{-1}^{1}\int_{-1}^{1}\int_{-1}^{1} f(\xi,\eta,\zeta)d\xi\, d\eta\, d\zeta \approx \sum_{i=1}^{Ng}\left(f\left(\xi_g,\eta_g,\zeta_g\right)W_g\right) \tag{9.3.30}$$

Since the Gaussian quadrature procedure applies for the domain integral of any quantity, we will use it for the calculation of the stiffness matrix $[k^{(e)}]$ and of the equivalent nodal forces due to body forces, $\left\{f_\Omega^{(e)}\right\}$:

$$\left[k^{(e)}\right] = \int_{-1}^{1}\int_{-1}^{1}\int_{-1}^{1}\left[B^{(e)}(\xi,\eta,\zeta)\right]^T\left[D^{(e)}(\xi,\eta,\zeta)\right]\left[B^{(e)}(\xi,\eta,\zeta)\right]J(\xi,\eta,\zeta)d\xi d\eta d\zeta$$

$$\approx \sum_{g=1}^{Ng}\left(\left[B_g\right]^T\left[D_g\right]\left[B_g\right]J_g W_g\right) \tag{9.3.31}$$

$$\left\{f_{\Omega}^{(e)}\right\} = \iiint_{\Omega^{(e)}} \left[N^{(e)}\right]^T \{b\} dV = \int_{-1}^{1}\int_{-1}^{1}\int_{-1}^{1} \left[N^{(8H)}(\xi,\eta,\zeta)\right]^T \{b(\xi,\eta,\zeta)\} \cdot J(\xi,\eta,\zeta) d\xi d\eta d\zeta$$

$$\approx \sum_{g=1}^{Ng} \left(\left[N_g\right]^T \{b_g\} J_g W_g \right) \tag{9.3.32}$$

Just like for two-dimensional isoparametric elements, a nice practical rule of thumb is to use a symmetric grid of quadrature points and have the number of quadrature points be equal (or almost equal) to the number of nodal points. So, for an eight-node hexahedral element, we have the eight-point (2 × 2 × 2) Gaussian quadrature rule summarized in Figure 9.4. The algorithm to obtain $[k^{(e)}]$ and $\left\{f_{\Omega}^{(e)}\right\}$ is summarized in Box 9.3.1.

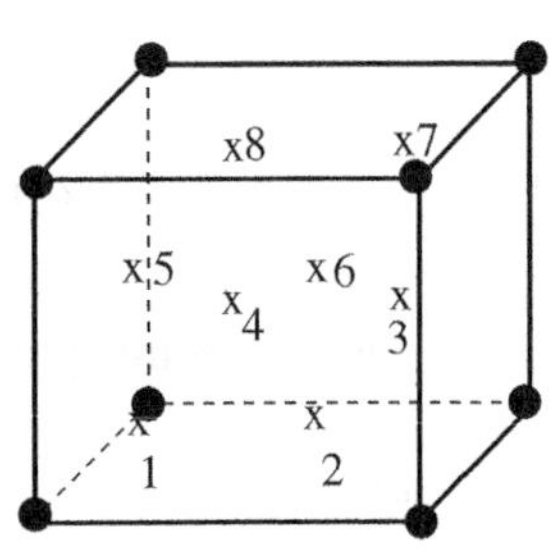

Locations of quadrature points (in parametric space)

g	ξ_g	η_g	ζ_g	W_g
1	$-1/\sqrt{3}$	$-1/\sqrt{3}$	$-1/\sqrt{3}$	1
2	$1/\sqrt{3}$	$-1/\sqrt{3}$	$-1/\sqrt{3}$	1
3	$1/\sqrt{3}$	$1/\sqrt{3}$	$-1/\sqrt{3}$	1
4	$-1/\sqrt{3}$	$1/\sqrt{3}$	$-1/\sqrt{3}$	1
5	$-1/\sqrt{3}$	$-1/\sqrt{3}$	$1/\sqrt{3}$	1
6	$1/\sqrt{3}$	$-1/\sqrt{3}$	$1/\sqrt{3}$	1
7	$1/\sqrt{3}$	$1/\sqrt{3}$	$1/\sqrt{3}$	1
8	$-1/\sqrt{3}$	$1/\sqrt{3}$	$1/\sqrt{3}$	1

Figure 9.4 Eight-point (2 × 2 × 2) Gaussian quadrature rule for 8H element.

Box 9.3.1 Calculation of $[k^{(e)}]$, $\left\{f_{\Omega}^{(e)}\right\}$ for 8H Element

FOR EACH QUADRATURE POINT g ($g = 1, 2, \ldots, N_g$):

1) Establish parametric coordinates, ξ_g, η_g, ζ_g
2) Calculate the values of the shape functions at the specific point:

$$\left[N^{(8H)}\right]_g = \begin{bmatrix} N_1^{(8H)}(\xi,\eta,\zeta) & 0 & 0 & \ldots\ldots & N_8^{(8H)}(\xi,\eta,\zeta) & 0 & 0 \\ 0 & N_1^{(8H)}(\xi,\eta,\zeta) & 0 & \ldots\ldots & 0 & N_8^{(8H)}(\xi,\eta,\zeta) & 0 \\ 0 & 0 & N_1^{(8H)}(\xi,\eta,\zeta) & \ldots\ldots & 0 & 0 & N_8^{(8H)}(\xi,\eta,\zeta) \end{bmatrix}\Bigg|_{\substack{\xi=\xi_g \\ \eta=\eta_g \\ \zeta=\zeta_g}}$$

Calculate values of coordinates in physical space:

$$\begin{Bmatrix} x_g \\ y_g \\ z_g \end{Bmatrix} = \left[N^{(8H)}\right]_g \begin{Bmatrix} \left\{x_1^{(e)}\right\} \\ \vdots \\ \left\{x_8^{(e)}\right\} \end{Bmatrix}$$

3) Find the matrix containing the derivatives of shape functions with the parametric coordinates at the location of the Gauss point:

$$\left[N_{,\xi}^{(8H)}\right]_g = \left.\begin{bmatrix} \frac{\partial N_1^{(8H)}}{\partial \xi} & \frac{\partial N_2^{(8H)}}{\partial \xi} & \frac{\partial N_3^{(8H)}}{\partial \xi} & \frac{\partial N_4^{(8H)}}{\partial \xi} & \frac{\partial N_5^{(8H)}}{\partial \xi} & \frac{\partial N_6^{(8H)}}{\partial \xi} & \frac{\partial N_7^{(8H)}}{\partial \xi} & \frac{\partial N_8^{(8H)}}{\partial \xi} \\ \frac{\partial N_1^{(8H)}}{\partial \eta} & \frac{\partial N_2^{(8H)}}{\partial \eta} & \frac{\partial N_3^{(8H)}}{\partial \eta} & \frac{\partial N_4^{(8H)}}{\partial \eta} & \frac{\partial N_5^{(8H)}}{\partial \eta} & \frac{\partial N_6^{(8H)}}{\partial \eta} & \frac{\partial N_7^{(8H)}}{\partial \eta} & \frac{\partial N_8^{(8H)}}{\partial \eta} \\ \frac{\partial N_1^{(8H)}}{\partial \zeta} & \frac{\partial N_2^{(8H)}}{\partial \zeta} & \frac{\partial N_3^{(8H)}}{\partial \zeta} & \frac{\partial N_4^{(8H)}}{\partial \zeta} & \frac{\partial N_5^{(8H)}}{\partial \zeta} & \frac{\partial N_6^{(8H)}}{\partial \zeta} & \frac{\partial N_7^{(8H)}}{\partial \zeta} & \frac{\partial N_8^{(8H)}}{\partial \zeta} \end{bmatrix}\right|_{\substack{\xi=\xi_g \\ \eta=\eta_g \\ \zeta=\zeta_g}}$$

4) Calculate the Jacobian matrix $[J_g]$ of the mapping at the Gauss point, as well as the Jacobian determinant, $J_g = det([J_g])$, and the inverse of the Jacobian matrix, $\left[\tilde{J}_g\right] = \left[J_g\right]^{-1}$

5) Calculate the matrix containing the derivatives of shape functions with the physical coordinates at the location of the Gauss point.

$$\left[N_{,x}^{(8H)}\right]_g = \left[\tilde{J}_g\right]\left[N_{,\xi}^{(8H)}\right]_g$$

6) Having obtained the derivatives with the physical coordinates, establish the $[B^{(e)}]$ array at point g:

$$\left[B^{(e)}\right]_g = \begin{bmatrix} \frac{\partial N_1^{(8H)}}{\partial x} & 0 & 0 & \frac{\partial N_2^{(8H)}}{\partial x} & 0 & 0 & \cdots & \frac{\partial N_8^{(8H)}}{\partial x} & 0 & 0 \\ 0 & \frac{\partial N_1^{(8H)}}{\partial y} & 0 & 0 & \frac{\partial N_2^{(8H)}}{\partial y} & 0 & \cdots & 0 & \frac{\partial N_8^{(8H)}}{\partial y} & 0 \\ 0 & 0 & \frac{\partial N_1^{(8H)}}{\partial z} & 0 & 0 & \frac{\partial N_2^{(8H)}}{\partial z} & \cdots & 0 & 0 & \frac{\partial N_8^{(8H)}}{\partial z} \\ \frac{\partial N_1^{(8H)}}{\partial y} & \frac{\partial N_1^{(8H)}}{\partial x} & 0 & \frac{\partial N_2^{(8H)}}{\partial y} & \frac{\partial N_2^{(8H)}}{\partial x} & 0 & \cdots & \frac{\partial N_8^{(8H)}}{\partial y} & \frac{\partial N_8^{(8H)}}{\partial x} & 0 \\ 0 & \frac{\partial N_1^{(8H)}}{\partial z} & \frac{\partial N_1^{(8H)}}{\partial y} & 0 & \frac{\partial N_2^{(8H)}}{\partial z} & \frac{\partial N_2^{(8H)}}{\partial y} & \cdots & 0 & \frac{\partial N_8^{(8H)}}{\partial z} & \frac{\partial N_8^{(8H)}}{\partial y} \\ \frac{\partial N_1^{(8H)}}{\partial z} & 0 & \frac{\partial N_1^{(8H)}}{\partial x} & \frac{\partial N_2^{(8H)}}{\partial z} & 0 & \frac{\partial N_2^{(8H)}}{\partial x} & \cdots & \frac{\partial N_8^{(8H)}}{\partial z} & 0 & \frac{\partial N_8^{(8H)}}{\partial x} \end{bmatrix}_g$$

7) Calculate the value of body force and of material stiffness matrix, $[D^{(e)}]$ at point g:

$$\{b\}_g = \begin{Bmatrix} b_x \\ b_y \\ b_z \end{Bmatrix}_g = \begin{Bmatrix} b_x(x=x_g, y=y_g, z=z_g) \\ b_y(x=x_g, y=y_g, z=z_g) \\ b_z(x=x_g, y=y_g, z=z_g) \end{Bmatrix}, \quad [D_g] = [D(x=x_g, y=y_g, z=z_g)]$$

END

Finally, combine the contributions of all the Gauss points to calculate the element stiffness matrix, $[k^{(e)}]$, and body-force contribution to the equivalent nodal force vector, $\left\{f_\Omega^{(e)}\right\}$:

$$\left[k^{(e)}\right] = \int_{-1}^{1}\int_{-1}^{1}\int_{-1}^{1}\left(\left[B^{(e)}\right]^T\left[D^{(e)}\right]\left[B^{(e)}\right]J d\xi d\eta d\zeta\right) \approx \sum_{g=1}^{Ng}\left(\left(\left[B^{(e)}\right]_g\right)^T\left[D_g\right]\left[B^{(e)}\right]_g J_g W_g\right)$$

$$\left\{f_\Omega^{(e)}\right\} = \int_{-1}^{1}\int_{-1}^{1}\int_{-1}^{1}\left(\left[N^{(8H)}\right]^T\{b\} J d\xi d\eta d\zeta\right) \approx \sum_{g=1}^{Ng}\left(\left(\left[N^{(8H)}\right]_g\right)^T\{b_g\} J_g W_g\right)$$

9.3.3 Calculation of Boundary Integral Contributions to Nodal Forces

The last task remaining for the eight-node hexahedral (8H) element is the calculation of the equivalent nodal forces, $\left\{f_{\Gamma_t}^{(e)}\right\}$, due to tractions at natural boundary surfaces. Since we now have to calculate surface integrals, the parameterization of the boundary must use two variables (a surface is a two-dimensional geometric entity and thus requires two coordinates for its parameterization). A convenient aspect of the isoparametric formulation is that in each one of the element faces we have one of the three parametric coordinates fixed to a constant value, so the other two parametric coordinates can be readily used for the parameterization of the boundary! We will now rely on calculus theory for the integration over an element surface, which is parameterized by two variables. We are trying to calculate the integral $\iint_{\Gamma_t^{(e)}} f(x,y,z)dS$, for a function $f(x, y, z)$. The parametric coordinate space of the 8H element can be readily used to parameterize the boundary surfaces. We can distinguish the following three cases:

- If the boundary surface segment in an element corresponds to a *fixed value* of ξ, then η and ζ are used for the parameterization. The integral can be calculated from the expression:

$$\iint_{\Gamma_t^{(e)}} f(x,y,z)dS = \int_{-1}^{1}\int_{-1}^{1} f(x(\eta,\zeta),y(\eta,\zeta),z(\eta,\zeta))\left\|\frac{\partial\vec{x}}{\partial\eta}\times\frac{\partial\vec{x}}{\partial\zeta}\right\| d\eta\, d\zeta \tag{9.3.33}$$

where $\frac{\partial\vec{x}}{\partial\eta} = \begin{bmatrix}\frac{\partial x}{\partial\eta} & \frac{\partial y}{\partial\eta} & \frac{\partial z}{\partial\eta}\end{bmatrix}^T$, $\frac{\partial\vec{x}}{\partial\zeta} = \begin{bmatrix}\frac{\partial x}{\partial\zeta} & \frac{\partial y}{\partial\zeta} & \frac{\partial z}{\partial\zeta}\end{bmatrix}^T$ and

$$\left\|\frac{\partial\vec{x}}{\partial\eta}\times\frac{\partial\vec{x}}{\partial\zeta}\right\| = \sqrt{\left(\frac{\partial x}{\partial\eta}\cdot\frac{\partial y}{\partial\zeta}-\frac{\partial y}{\partial\eta}\cdot\frac{\partial x}{\partial\zeta}\right)^2+\left(\frac{\partial z}{\partial\eta}\cdot\frac{\partial x}{\partial\zeta}-\frac{\partial x}{\partial\eta}\cdot\frac{\partial z}{\partial\zeta}\right)^2+\left(\frac{\partial y}{\partial\eta}\cdot\frac{\partial z}{\partial\zeta}-\frac{\partial z}{\partial\eta}\cdot\frac{\partial y}{\partial\zeta}\right)^2} \tag{9.3.34}$$

- If the boundary segment corresponds to a *fixed value* of η, then ξ and ζ are used for the parameterization. Then:

$$\iint_{\Gamma_t^{(e)}} f(x,y,z)dS = \int_{-1}^{1}\int_{-1}^{1} f(x(\xi,\zeta),y(\xi,\zeta),z(\xi,\zeta))\left\|\frac{\partial\vec{x}}{\partial\zeta}\times\frac{\partial\vec{x}}{\partial\xi}\right\| d\xi d\zeta \tag{9.3.35}$$

where $\frac{\partial\vec{x}}{\partial\xi} = \begin{bmatrix}\frac{\partial x}{\partial\xi} & \frac{\partial y}{\partial\xi} & \frac{\partial z}{\partial\xi}\end{bmatrix}^T$ and

$$\left\|\frac{\partial\vec{x}}{\partial\zeta}\times\frac{\partial\vec{x}}{\partial\xi}\right\| = \sqrt{\left(\frac{\partial x}{\partial\zeta}\cdot\frac{\partial y}{\partial\xi}-\frac{\partial y}{\partial\zeta}\cdot\frac{\partial x}{\partial\xi}\right)^2+\left(\frac{\partial z}{\partial\zeta}\cdot\frac{\partial x}{\partial\xi}-\frac{\partial x}{\partial\zeta}\cdot\frac{\partial z}{\partial\xi}\right)^2+\left(\frac{\partial y}{\partial\zeta}\cdot\frac{\partial z}{\partial\xi}-\frac{\partial z}{\partial\zeta}\cdot\frac{\partial y}{\partial\xi}\right)^2} \tag{9.3.36}$$

- Finally, if the boundary corresponds to a *fixed value* of ζ, then ξ and η are used for the parameterization:

$$\iint_{\Gamma_t^{(e)}} f(x,y,z)dS = \int_{-1}^{1}\int_{-1}^{1} f(x(\xi,\eta),y(\xi,\eta),z(\xi,\eta)) \left\| \frac{\partial \vec{x}}{\partial \xi} \times \frac{\partial \vec{x}}{\partial \eta} \right\| d\xi d\eta \tag{9.3.37}$$

where

$$\left\| \frac{\partial \vec{x}}{\partial \xi} \times \frac{\partial \vec{x}}{\partial \eta} \right\| = \sqrt{\left(\frac{\partial x}{\partial \xi}\cdot\frac{\partial y}{\partial \eta} - \frac{\partial y}{\partial \xi}\cdot\frac{\partial x}{\partial \eta}\right)^2 + \left(\frac{\partial z}{\partial \xi}\cdot\frac{\partial x}{\partial \eta} - \frac{\partial x}{\partial \xi}\cdot\frac{\partial z}{\partial \eta}\right)^2 + \left(\frac{\partial y}{\partial \xi}\cdot\frac{\partial z}{\partial \eta} - \frac{\partial z}{\partial \xi}\cdot\frac{\partial y}{\partial \eta}\right)^2} \tag{9.3.38}$$

If, for example, we have a prescribed traction vector $\{t(x, y, z)\}$ (notice that, in general, $\{t\}$ can be a function of the physical coordinates) and the natural boundary surface segment Γ_t corresponds to $\zeta = \bar{\zeta}$ = constant, we have:

$$\left\{f_{\Gamma_t}^{(e)}\right\} = \iint_{\Gamma_t^{(e)}} \left[N^{(e)}\right]^T \{t\}\, dS$$

$$\approx \int_{-1}^{1}\int_{-1}^{1} \left[N^{(8H)}(\xi,\eta,\bar{\zeta})\right]^T \{t(x(\xi,\eta,\bar{\zeta}),y(\xi,\eta,\bar{\zeta}),z(\xi,\eta,\bar{\zeta}))\} \left\| \frac{\partial \vec{x}}{\partial \xi} \times \frac{\partial \vec{x}}{\partial \eta} \right\| d\xi d\eta \tag{9.3.39}$$

which, of course, becomes an integral of quantities that are functions of ξ and η only, and the integration is conducted with respect to ξ and η!

The surface integrals can be numerically calculated using the well-known two-dimensional Gaussian quadrature presented in Section 6.5 (e.g., 2 × 2 points).

9.3.4 Higher-Order Hexahedral Isoparametric Elements

Higher-order hexahedral isoparametric elements can also be used in three-dimensional analysis, in case we want increased order of convergence or we want to describe curved element sides and faces. The higher-order hexahedral elements resulting from tensor products of higher-order one-dimensional polynomial approximations are called *Lagrangian elements*, and they include nodes in the interior of the element. As discussed in Section 6.6, it may be preferable to use the more efficient, three-dimensional *serendipity* elements, which do not include nodes in their interior. The quadratic hexahedral Lagrangian element has 27 nodes, while the quadratic serendipity hexahedral element, shown in Figure 9.5, has 20 nodes.

9.3.5 Tetrahedral Isoparametric Elements

Another popular type of elements for three-dimensional analyses is tetrahedral elements. In fact, generating a mesh consisting of tetrahedral elements alone is easier than generating a mesh including hexahedral elements. Just as the eight-node hexahedral element can be thought of as an extension (to three dimensions) of the two-dimensional,

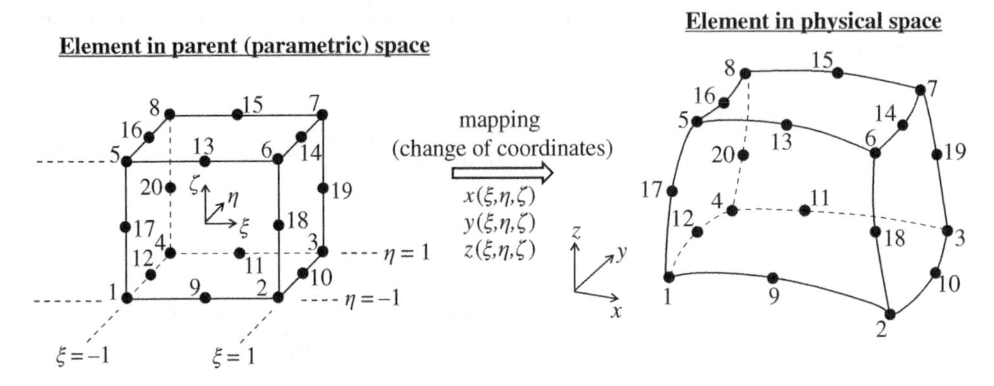

Figure 9.5 Three-dimensional hexahedral serendipity element.

four-node quadrilateral element, the tetrahedral element can be thought of as an extension of the two-dimensional triangular element. Thus, similarly to the process followed for an isoparametric 3 T element in Section 6.7, we will establish a four-node isoparametric tetrahedron by defining an appropriate parametric space with coordinates ξ_1, ξ_2, and ξ_3, as shown in Figure 9.6.

The parametric coordinates ξ_1, ξ_2, and ξ_3 for a point P in the interior of the element, shown in Figure 9.7, give ratios of volumes of the subtetrahedra defined by P and the corner nodes over the total volume of the tetrahedron:

$$\xi_1 = \frac{\text{Volume of (342P)}}{\text{Volume of (1234)}} = \frac{V_1}{V^{(e)}} \tag{9.3.40a}$$

$$\xi_2 = \frac{\text{Volume of (134P)}}{\text{Volume of (1234)}} = \frac{V_2}{V^{(e)}} \tag{9.3.40b}$$

$$\xi_3 = \frac{\text{Volume of (124P)}}{\text{Volume of (1234)}} = \frac{V_3}{V^{(e)}} \tag{9.3.40c}$$

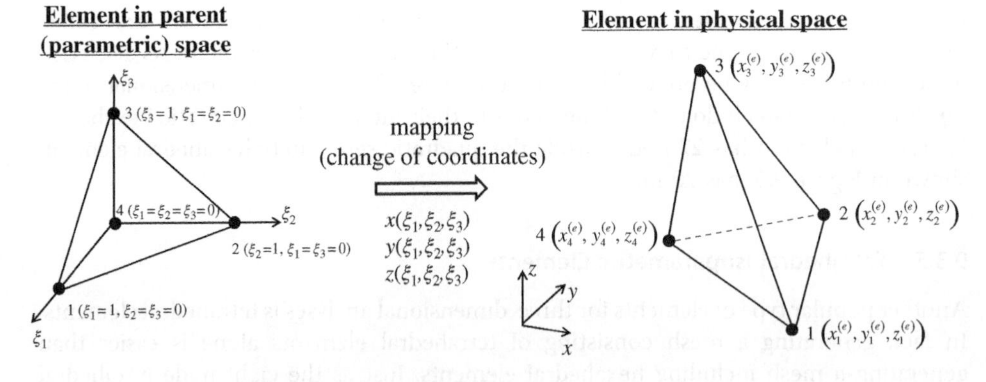

Figure 9.6 Four-node, tetrahedral isoparametric element.

Figure 9.7 Interior point for a four-node, tetrahedral isoparametric element.

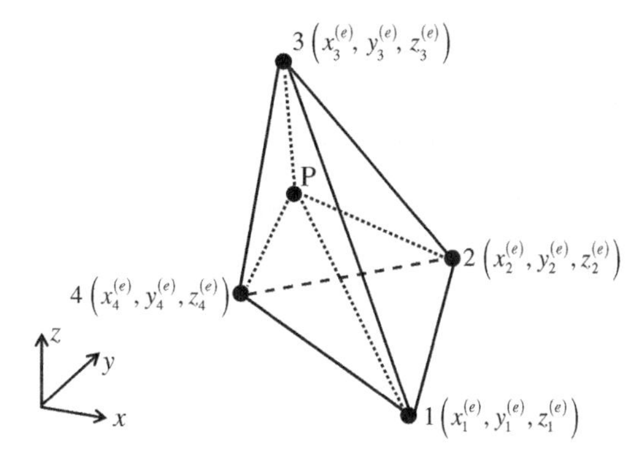

The volume of the remaining sub-tetrahedron, V_4 = Volume of (123P), satisfies:

$$\frac{V_4}{V^{(e)}} = 1 - \frac{V_1}{V^{(e)}} - \frac{V_2}{V^{(e)}} - \frac{V_3}{V^{(e)}} = 1 - \xi_1 - \xi_2 - \xi_3 \tag{9.3.40d}$$

We now define the following four shape functions (since the element has four nodes) in the parametric space:

$$N_1^{(4T)}(\xi_1,\xi_2,\xi_3) = \xi_1 \tag{9.3.41a}$$

$$N_2^{(4T)}(\xi_1,\xi_2,\xi_3) = \xi_2 \tag{9.3.41b}$$

$$N_3^{(4T)}(\xi_1,\xi_2,\xi_3) = \xi_3 \tag{9.3.41c}$$

$$N_4^{(4T)}(\xi_1,\xi_2,\xi_3) = 1 - \xi_1 - \xi_2 - \xi_3 \tag{9.3.41d}$$

The above shape functions satisfy the Kronecker Delta property and the partition of unity property. By virtue of the isoparametric concept, we use these shape functions both for the displacement approximation and for the coordinate mapping. The procedure to obtain the element arrays and vectors is conceptually the same as that employed for hexahedral elements.

The integral of a function $f(x, y, z)$ over a tetrahedral physical domain can be calculated by evaluation in the parametric space which has a much more convenient shape. Thus, we can write:

$$\iiint_{\Omega^{(e)}} f(x,y,z)dV = \int_0^1 \int_0^{1-\xi_1} \int_0^{1-\xi_1-\xi_2} f(x(\xi_1,\xi_2,\xi_3), y(\xi_1,\xi_2,\xi_3), z(\xi_1,\xi_2,\xi_3)) J d\xi_3 d\xi_2 d\xi_1$$

$$\rightarrow \iiint_{\Omega^{(e)}} f(x,y,z)dV = \int_0^1 \int_0^{1-\xi_1} \int_0^{1-\xi_1-\xi_2} f(\xi_1,\xi_2,\xi_3) J \, d\xi_3 d\xi_2 d\xi_1 \tag{9.3.42}$$

Quadrature rules for tetrahedral domains (Hammer et al. 1956) allow us to numerically evaluate the triple integrals like that in Equation (9.3.42). If we use N_g quadrature points, we have the following rule:

$$\int_0^1 \int_0^{1-\xi_1} \int_0^{1-\xi_1-\xi_2} f(\xi_1,\xi_2,\xi_3) J d\xi_3 d\xi_2 d\xi_1 \approx \sum_{g=1}^{Ng} \left(f\left(\xi_{1g},\xi_{2g},\xi_{3g}\right) J_g W_g \right) \tag{9.3.43}$$

Several tetrahedral quadrature rules, together with the corresponding order of polynomials that can be exactly integrated, are provided in Table 9.1.

Higher-order tetrahedral elements can also be defined, by introducing additional nodal points and increasing the polynomial order of the shape functions.

Remark 9.3.1: If we have tractions on a boundary surface segment of a tetrahedral element, we employ a two-variable parameterization using the two appropriate *triangular* coordinates. We can then employ the rules for quadrature in triangular domains, presented in Section 6.7. For example, if we have tractions on the surface segment (134), we can verify from Figure 9.6 that this segment corresponds to $\xi_2 = 0 =$ constant, thus we can employ ξ_1 and ξ_3 for our parameterization. Similarly, for surface (124), we have $\xi_3 = 0 =$ constant, and for surface (234), we have $\xi_1 = 0 =$ constant. An issue arises when we have a boundary integral over the triangular segment (123), because all three parametric coordinates vary over that segment. For that case, we can resolve the issue, using either of the following two procedures:

1) Use a new set of triangular coordinates, η_1 and η_2, for the triangular surface (123), and then establish and use a mapping to calculate the integral in the new triangular parametric space, or
2) Renumber the nodes of the element so that the surface with the prescribed tractions is one of the other three surfaces in the parametric space. ∎

9.3.6 Three-Dimensional Elements from Collapsed (Degenerated) Hexahedral Elements

Other types of isoparametric three-dimensional elements can also be obtained by "collapsing" (or degenerating) an eight-node hexahedral element, so that multiple nodes in the parametric space correspond to the same node in the physical space. The same

Table 9.1 Quadrature Rules for Tetrahedral Parametric Space.

Rule	Order of Polynomial Exactly Integrated	Point	Locations	Weight Coefficients
1-point	1	1	$\xi_1 = 1/4 = 0.25$, $\xi_2 = 1/4 = 0.25$, $\xi_3 = 1/4 = 0.25$	$W_1 = 1$
4-point	2	1	$\xi_1 = 0.5854102$, $\xi_2 = 0.1381966$, $\xi_3 = 0.1381966$	$W_1 = 1/4 = 0.25$
		2	$\xi_1 = 0.1381966$, $\xi_2 = 0.5854102$, $\xi_3 = 0.1381966$	$W_2 = 1/4 = 0.25$
		3	$\xi_1 = 0.1381966$, $\xi_2 = 0.1381966$, $\xi_3 = 0.5854102$	$W_3 = 1/4 = 0.25$
		4	$\xi_1 = 0.1381966$, $\xi_2 = 0.1381966$, $\xi_3 = 0.1381966$	$W_4 = 1/4 = 0.25$
5-point	3	1	$\xi_1 = 1/4 = 0.25$, $\xi_2 = 1/4 = 0.25$, $\xi_3 = 1/4 = 0.25$	$W_1 = -4/5 = -0.80$
		2	$\xi_1 = 1/2 = 0.50$, $\xi_2 = \xi_3 = 1/6 = 0.1666667$	$W_2 = 9/20 = 0.45$
		3	$\xi_1 = \xi_3 = 1/6 = 0.1666667$, $\xi_2 = 1/2 = 0.50$	$W_3 = 9/20 = 0.45$
		4	$\xi_1 = \xi_2 = 1/6 = 0.1666667$, $\xi_3 = 1/2 = 0.50$	$W_4 = 9/20 = 0.45$
		5	$\xi_1 = \xi_2 = \xi_3 = 1/6 = 0.1666667$	$W_5 = 9/20 = 0.45$

approach was discussed in Remark 6.7.2 for two-dimensional analysis, where triangular elements were obtained from collapsing quadrilateral isoparametric elements. A commonly employed three-dimensional element of this type is the six-node wedge element shown in Figure 9.8. This element is formulated as a standard eight-node hexahedron, where the Cartesian (physical) coordinates for two nodes (7 and 8) at the top surface are identical, and the same applies for two nodes (3 and 4) at the bottom surface. Of course, we also need to enforce that the nodal displacements are identical for each pair of coincident nodes. Along the same lines, the four-node tetrahedron can be obtained as a collapsed hexahedron, as shown in Figure 9.9. In this case, two of the nodes in the bottom surface (nodes 3 and 4 in the figure) coincide in the Cartesian coordinate space, and the same applies for all four nodes of the top surface.

9.3.7 Concluding Remark: Continuity and Completeness Ensured by Three-Dimensional Isoparametric Elements and Use for Other Problems

The requirements of C^0 continuity and degree-1 completeness for convergence are satisfied by the types of three-dimensional isoparametric elements that we examined in this chapter; thus, these types of elements are bound to be convergent for elasticity problems.

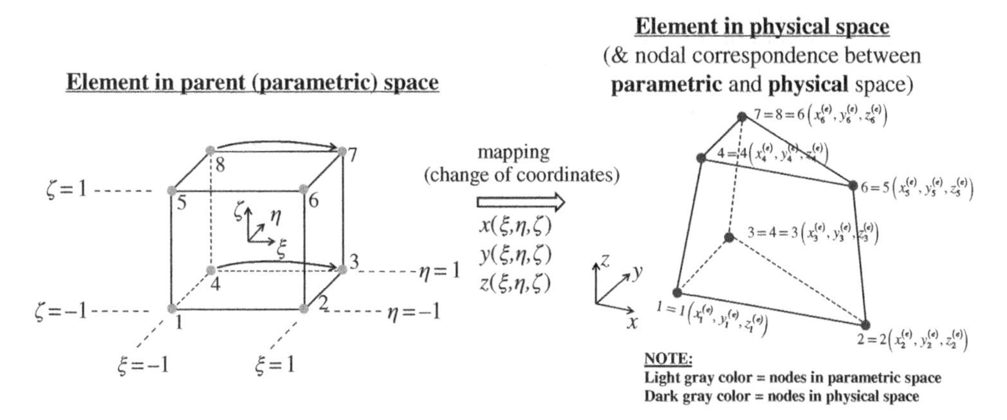

Figure 9.8 Wedge-shaped, six-node solid element obtained from a collapsed 8H isoparametric element, in which two nodes coincide at the top surface and another two nodes coincide at the bottom surface.

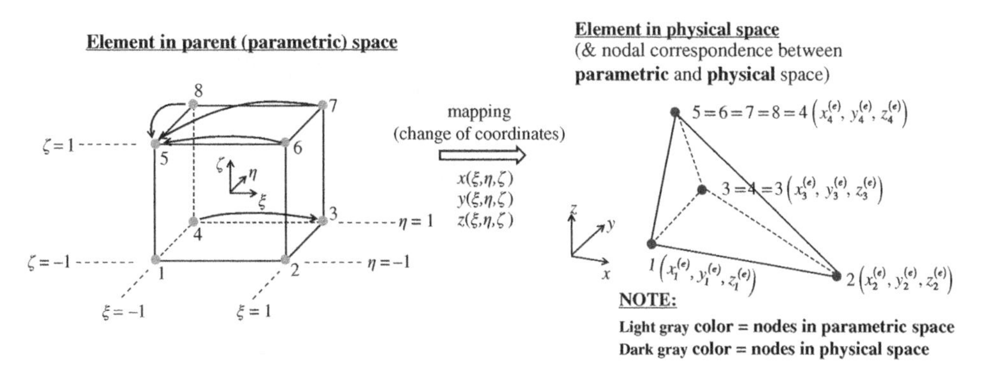

Figure 9.9 Four-node tetrahedral element obtained as a collapsed 8H isoparametric element.

The same continuity and completeness requirements correspond to three-dimensional versions of the scalar-field problems discussed in Chapter 5, so the elements presented herein can be used for these problems as well.

Example 9.1: Determination of stiffness matrix and equivalent nodal force vector for 8H element

We want to find the stiffness matrix and equivalent nodal force vector of the hexahedral element shown in Figure 9.10. The figure also provides the coordinates (in the physical space) of the eight nodal points. The material is isotropic, linearly elastic, with $E(x,y,z) = 5000\left(1+\frac{xyz}{100}\right)$ (E in ksi, and x, y, z, in inches) and $\nu = 0.2$. The element is subjected to body force $b_x(x,y,z) = 0.1(1+0.2x+0.4z)$ (where b_x is in kip/in.^{3} and x, y are in inches). All other body force components are equal to zero.

We will calculate $[k^{(e)}]$ and $\{f^{(e)}\} = \{f_\Omega^{(e)}\}$ (there are no boundary tractions), using an eight-point Gauss quadrature and the algorithm of Box 9.3.1. We will present in detail the computations for the first Gauss point, with parametric coordinates $\xi_1 = -\frac{1}{\sqrt{3}}$, $\eta_1 = -\frac{1}{\sqrt{3}}$, $\zeta_1 = -\frac{1}{\sqrt{3}}$.

We find the values of the shape functions at the location of the first Gauss point.

$$N_1^{(8H)}(\xi=\xi_1,\eta=\eta_1,\zeta=\zeta_1) = 0.4906 \quad N_2^{(8H)}(\xi=\xi_1,\eta=\eta_1,\zeta=\zeta_1) = 0.1314$$
$$N_3^{(8H)}(\xi=\xi_1,\eta=\eta_1,\zeta=\zeta_1) = 0.0352 \quad N_4^{(8H)}(\xi=\xi_1,\eta=\eta_1,\zeta=\zeta_1) = 0.1314$$
$$N_5^{(8H)}(\xi=\xi_1,\eta=\eta_1,\zeta=\zeta_1) = 0.1314 \quad N_6^{(8H)}(\xi=\xi_1,\eta=\eta_1,\zeta=\zeta_1) = 0.0352$$
$$N_7^{(8H)}(\xi=\xi_1,\eta=\eta_1,\zeta=\zeta_1) = 0.0094 \quad N_8^{(8H)}(\xi=\xi_1,\eta=\eta_1,\zeta=\zeta_1) = 0.0352$$

Next, we establish the shape function array at the location of Gauss point 1, whose components are the shape functions evaluated at the same location:

$$\left[N^{(8H)}\right]_1 = \left.\begin{bmatrix} N_1^{(8H)}(\xi,\eta,\zeta) & 0 & 0 & \ldots\ldots & N_8^{(8H)}(\xi,\eta,\zeta) & 0 & 0 \\ 0 & N_1^{(8H)}(\xi,\eta,\zeta) & 0 & \ldots\ldots & 0 & N_8^{(8H)}(\xi,\eta,\zeta) & 0 \\ 0 & 0 & N_1^{(8H)}(\xi,\eta,\zeta) & \ldots\ldots & 0 & 0 & N_8^{(8H)}(\xi,\eta,\zeta) \end{bmatrix}\right|_{\substack{\xi=\xi_1\\ \eta=\eta_1\\ \zeta=\zeta_1}}$$

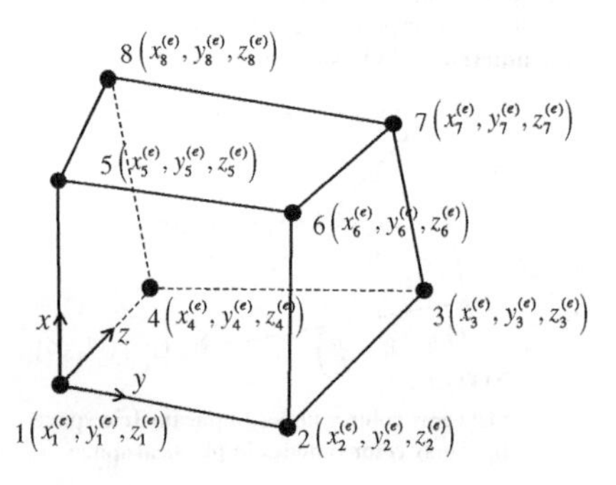

Node	x (in)	y (in)	z (in)
1	0	0	0
2	0	4	0
3	0	5	6
4	0	0	5
5	4	0	0
6	4	4	0
7	4	4	5
8	5	–1	5

Figure 9.10 Example eight-node hexahedral element in physical space.

We can also find the physical coordinates of the Gauss point:

$$\begin{Bmatrix} x_1 \\ y_1 \\ z_1 \end{Bmatrix} = \left[N^{(8H)}\right]_1 \begin{Bmatrix} \left\{x_1^{(e)}\right\} \\ \vdots \\ \left\{x_8^{(e)}\right\} \end{Bmatrix} = \begin{Bmatrix} 0.8805 \\ 0.8453 \\ 1.0918 \end{Bmatrix}$$

We also find the array containing the derivatives of the shape functions at the location of Gauss point 1:

$$\left[N_{,\xi}^{(8H)}\right]_1 = \left.\begin{bmatrix} \frac{\partial N_1^{(8H)}}{\partial \xi} & \frac{\partial N_2^{(8H)}}{\partial \xi} & \frac{\partial N_3^{(8H)}}{\partial \xi} & \frac{\partial N_4^{(8H)}}{\partial \xi} & \frac{\partial N_5^{(8H)}}{\partial \xi} & \frac{\partial N_6^{(8H)}}{\partial \xi} & \frac{\partial N_7^{(8H)}}{\partial \xi} & \frac{\partial N_8^{(8H)}}{\partial \xi} \\ \frac{\partial N_1^{(8H)}}{\partial \eta} & \frac{\partial N_2^{(8H)}}{\partial \eta} & \frac{\partial N_3^{(8H)}}{\partial \eta} & \frac{\partial N_4^{(8H)}}{\partial \eta} & \frac{\partial N_5^{(8H)}}{\partial \eta} & \frac{\partial N_6^{(8H)}}{\partial \eta} & \frac{\partial N_7^{(8H)}}{\partial \eta} & \frac{\partial N_8^{(8H)}}{\partial \eta} \\ \frac{\partial N_1^{(8H)}}{\partial \zeta} & \frac{\partial N_2^{(8H)}}{\partial \zeta} & \frac{\partial N_3^{(8H)}}{\partial \zeta} & \frac{\partial N_4^{(8H)}}{\partial \zeta} & \frac{\partial N_5^{(8H)}}{\partial \zeta} & \frac{\partial N_6^{(8H)}}{\partial \zeta} & \frac{\partial N_7^{(8H)}}{\partial \zeta} & \frac{\partial N_8^{(8H)}}{\partial \zeta} \end{bmatrix}\right|_{\substack{\xi = \xi_1 \\ \eta = \eta_1 \\ \zeta = \zeta_1}}$$

$$\left[N_{,\xi}^{(8H)}\right]_1 = \begin{bmatrix} -0.3110 & 0.3110 & 0.0833 & -0.0833 & -0.0833 & 0.0833 & 0.0223 & -0.0223 \\ -0.3110 & -0.0833 & 0.0833 & 0.3110 & -0.0833 & -0.0223 & 0.0223 & 0.0833 \\ -0.3110 & -0.0833 & -0.0223 & -0.0833 & 0.3110 & 0.0833 & 0.0223 & 0.0833 \end{bmatrix}$$

Next, we find the Jacobian array at Gauss point 1:

$$[J_1] = \begin{bmatrix} -0.0223 & 2.1057 & 0.0833 \\ 0.0833 & 0.0000 & 2.5833 \\ 2.0833 & -0.1057 & -0.0223 \end{bmatrix}$$, and the corresponding Jacobian determinant:

$J_1 = 11.3296$, as well as the inverse of the Jacobian,

$$\left[\tilde{J}_1\right] = \begin{bmatrix} 0.0241 & 0.0034 & 0.4801 \\ 0.4752 & -0.0153 & 0.0057 \\ -0.0008 & 0.3870 & -0.0155 \end{bmatrix}$$. We can now find the array

$\left[N_{,x}^{(8H)}\right]_1 = \left[\tilde{J}_1\right]\left[N_{,\xi}^{(8H)}\right]_1$, containing the derivatives of the shape functions with respect to the physical coordinates at Gauss point 1:

$$\left[N_{,x}^{(8H)}\right]_1 = \begin{bmatrix} -0.1579 & -0.0328 & -0.0084 & -0.0410 & 0.1470 & 0.0419 & 0.0113 & 0.0398 \\ -0.1448 & 0.1486 & 0.0382 & -0.0448 & -0.0366 & 0.0404 & 0.0104 & -0.0114 \\ -0.1153 & -0.0312 & 0.0325 & 0.1217 & -0.0370 & -0.0100 & 0.0083 & 0.0310 \end{bmatrix}$$

We can now establish the strain-displacement matrix at Gauss point 1,

$$\left[B^{(e)}\right]_1 = \begin{bmatrix} -0.158 & 0.000 & 0.000 & -0.033 & 0.000 & 0.000 & -0.008 & 0.000 & 0.000 & -0.041 & 0.000 & 0.000 & 0.147 & 0.000 & 0.000 & 0.042 & 0.000 & 0.000 & 0.011 & 0.000 & 0.000 & 0.040 & 0.000 & 0.000 \\ 0.000 & -0.145 & 0.000 & 0.000 & 0.149 & 0.000 & 0.000 & 0.038 & 0.000 & 0.000 & -0.045 & 0.000 & 0.000 & -0.037 & 0.000 & 0.000 & 0.040 & 0.000 & 0.000 & 0.010 & 0.000 & 0.000 & -0.011 & 0.000 \\ 0.000 & 0.000 & -0.115 & 0.000 & 0.000 & -0.031 & 0.000 & 0.000 & 0.033 & 0.000 & 0.000 & 0.122 & 0.000 & 0.000 & -0.037 & 0.000 & 0.000 & -0.010 & 0.000 & 0.000 & 0.008 & 0.000 & 0.000 & 0.031 \\ -0.145 & -0.158 & 0.000 & 0.149 & -0.033 & 0.000 & 0.038 & -0.008 & 0.000 & -0.045 & -0.041 & 0.000 & -0.037 & 0.147 & 0.000 & 0.040 & 0.042 & 0.000 & 0.010 & 0.011 & 0.000 & -0.011 & 0.040 & 0.000 \\ 0.000 & -0.115 & -0.145 & 0.000 & -0.031 & 0.149 & 0.000 & 0.033 & 0.038 & 0.000 & 0.122 & -0.045 & 0.000 & -0.037 & -0.037 & 0.000 & -0.010 & 0.040 & 0.000 & 0.008 & 0.010 & 0.000 & 0.031 & -0.011 \\ -0.115 & 0.000 & -0.158 & -0.031 & 0.000 & -0.033 & 0.033 & 0.000 & -0.008 & 0.122 & 0.000 & -0.041 & -0.037 & 0.000 & 0.147 & -0.010 & 0.000 & 0.042 & 0.008 & 0.000 & 0.011 & 0.031 & 0.000 & 0.040 \end{bmatrix}$$

Finally, we can also determine the material stiffness matrix and the body force vector at the location of Gauss point 1:

$E_1 = 5000\left(1 + \frac{x_1 y_1 z_1}{100}\right) = 5040.6\,\text{ksi}$. Since the material is isotropic, the stiffness matrix is given by:

$$[D_1] = \frac{E_1}{(1+\nu)(1-2\nu)}\begin{bmatrix} 1-\nu & \nu & \nu & 0 & 0 & 0 \\ \nu & 1-\nu & \nu & 0 & 0 & 0 \\ \nu & \nu & 1-\nu & 0 & 0 & 0 \\ 0 & 0 & 0 & \frac{1-2\nu}{2} & 0 & 0 \\ 0 & 0 & 0 & 0 & \frac{1-2\nu}{2} & 0 \\ 0 & 0 & 0 & 0 & 0 & \frac{1-2\nu}{2} \end{bmatrix}$$

$$= \begin{bmatrix} 5600.7 & 1400.2 & 1400.2 & 0.0 & 0.0 & 0.0 \\ 1400.2 & 5600.7 & 1400.2 & 0.0 & 0.0 & 0.0 \\ 1400.2 & 1400.2 & 5600.7 & 0.0 & 0.0 & 0.0 \\ 0.0 & 0.0 & 0.0 & 2100.3 & 0.0 & 0.0 \\ 0.0 & 0.0 & 0.0 & 0.0 & 2100.3 & 0.0 \\ 0.0 & 0.0 & 0.0 & 0.0 & 0.0 & 2100.3 \end{bmatrix}$$

$$\{b\}_1 = \begin{Bmatrix} b_x \\ b_y \\ b_z \end{Bmatrix}_1 = \begin{Bmatrix} b_x(x=x_1, y=y_1, z=z_1) \\ b_y(x=x_1, y=y_1, z=z_1) \\ b_z(x=x_1, y=y_1, z=z_1) \end{Bmatrix}$$

$$= \begin{Bmatrix} 0.1(1+0.2\cdot 0.8805+0.4\cdot 1.0918) \\ 0 \\ 0 \end{Bmatrix} = \begin{Bmatrix} 0.1613 \\ 0 \\ 0 \end{Bmatrix}$$

We can conduct the same type of computations for the remaining seven Gauss points of the element. Eventually, we can obtain the stiffness matrix and equivalent nodal force vector:

$$\left[k^{(e)}\right] = \sum_{g=1}^{8}\left(\left(\left[B^{(e)}\right]_g\right)^T [D_g]\left[B^{(e)}\right]_g J_g W_g\right) =$$

$$
= \begin{bmatrix}
5695 & 1686 & 1244 & 973 & 280 & 568 & 91 & 62 & -60 & 1766 & 921 & 299 & -2218 & -271 & 0 & -2541 & -1683 & -29 & -1961 & -934 & -776 & -1805 & -61 & -1247 \\
1686 & 5413 & 997 & -381 & -2250 & -167 & -249 & -1346 & -1184 & 982 & 1509 & 381 & 327 & 845 & 555 & -1682 & -2386 & -41 & -930 & -1626 & -649 & 247 & -159 & 108 \\
1244 & 997 & 4762 & 630 & 314 & 430 & -310 & -1186 & -1127 & -299 & -160 & -1021 & 536 & 615 & 550 & 221 & 208 & -924 & -775 & -651 & -1383 & -1246 & -138 & -1286 \\
973 & -381 & 630 & 5954 & -1596 & 1421 & 1704 & -724 & 215 & -251 & -182 & 264 & -1817 & 1519 & -47 & -2619 & 313 & -221 & -2688 & 222 & -1627 & -1257 & 830 & -634 \\
280 & -2250 & 314 & -1596 & 5800 & -1407 & -724 & 1540 & -201 & 193 & -2151 & 1456 & 1540 & -1853 & 61 & -369 & 762 & -657 & -177 & -565 & -211 & 853 & -1283 & 645 \\
568 & -167 & 430 & 1421 & -1407 & 5211 & -305 & 297 & -910 & -51 & 1434 & -1750 & 201 & -186 & -604 & 358 & -654 & 514 & -1555 & 55 & -1776 & -636 & 628 & -1116 \\
91 & -249 & -310 & 1704 & -724 & -305 & 5158 & -1046 & -1218 & 1127 & -435 & -823 & -1021 & 699 & 633 & -1898 & 66 & 1424 & -3779 & 226 & 443 & -1382 & 1464 & 156 \\
62 & -1346 & -1186 & -724 & 1540 & 297 & -1046 & 4717 & 1110 & 210 & -2115 & -284 & 719 & -1064 & -637 & -266 & -343 & 243 & -480 & 64 & 704 & 1526 & -1453 & -249 \\
-60 & -1184 & -1127 & 215 & -201 & -910 & -1218 & 1110 & 4327 & -761 & 278 & 814 & 651 & -632 & -1151 & 1478 & -17 & -1696 & -198 & 630 & -38 & -107 & 15 & -220 \\
1766 & 982 & -299 & -251 & 193 & -51 & 1127 & 210 & -761 & 6969 & 2284 & -1766 & -1843 & -201 & 1423 & -1821 & -932 & 811 & -3669 & -2162 & 281 & -2279 & -374 & 361 \\
921 & 1509 & -160 & -182 & -2151 & 1434 & -435 & -2115 & 278 & 2284 & 6867 & -1749 & 105 & -333 & -2 & -931 & -1822 & 790 & -2082 & -3122 & 198 & 319 & 1168 & -789 \\
299 & 381 & -1021 & 264 & 1456 & -1750 & -823 & -284 & 814 & -1766 & -1749 & 5947 & 1495 & 305 & -1821 & 831 & 811 & -1576 & -114 & -130 & -1282 & -188 & -789 & 688 \\
-2218 & 327 & 536 & -1817 & 1540 & 201 & -1021 & 719 & 651 & -1843 & 105 & 1495 & 5322 & -1301 & -1372 & 895 & -423 & -668 & -377 & -374 & -446 & 1059 & -594 & -398 \\
-271 & 845 & 615 & 1519 & -1853 & -186 & 699 & -1064 & -632 & -201 & -333 & 305 & -1301 & 5389 & 1387 & 193 & -2209 & -429 & -46 & -1901 & -1474 & -591 & 1125 & 413 \\
0 & 555 & 550 & -47 & 61 & -604 & 633 & -637 & -1151 & 1423 & -2 & -1821 & -1372 & 1387 & 5520 & -604 & 142 & 665 & -120 & -1417 & -1802 & 86 & -90 & -1358 \\
-2541 & -1682 & 221 & -2619 & -369 & 358 & -1898 & -266 & 1478 & -1821 & -931 & 831 & 895 & 193 & -604 & 6209 & 1805 & -1444 & 2003 & 1132 & -777 & -228 & 118 & -63 \\
-1683 & -2386 & 208 & 313 & 762 & -654 & 66 & -343 & -17 & -932 & -1822 & 811 & -423 & -2209 & 142 & 1805 & 6088 & -1447 & 1057 & 1837 & -435 & -204 & -1926 & 1392 \\
-29 & -41 & -924 & -221 & -657 & 514 & 1424 & 243 & -1696 & 811 & 790 & -1576 & -668 & -429 & 665 & -1444 & -1447 & 5601 & -66 & 199 & -1092 & 193 & 1341 & -1493 \\
-1961 & -930 & -775 & -2688 & -177 & -1555 & -3779 & -480 & -198 & -3669 & -2082 & -114 & -377 & -46 & -120 & 2003 & 1057 & -66 & 9330 & 2783 & 2060 & 1140 & -126 & 767 \\
-934 & -1626 & -651 & 222 & -565 & 55 & 226 & 64 & 630 & -2162 & -3122 & -130 & -374 & -1901 & -1417 & 1132 & 1837 & 199 & 2783 & 7740 & 1696 & -893 & -2427 & -383 \\
-776 & -649 & -1383 & -1627 & -211 & -1776 & 443 & 704 & -38 & 281 & 198 & -1282 & -446 & -1474 & -1802 & -777 & -435 & -1092 & 2060 & 1696 & 6716 & 841 & 170 & 656 \\
-1805 & 247 & -1246 & -1257 & 853 & -636 & -1382 & 1526 & -107 & -2279 & 319 & -188 & 1059 & -591 & 86 & -228 & -204 & 193 & 1140 & -893 & 841 & 4751 & -1257 & 1057 \\
-61 & -159 & -138 & 830 & -1283 & 628 & 1464 & -1453 & 15 & -374 & 1168 & -789 & -594 & 1125 & -90 & 118 & -1926 & 1341 & -126 & -2427 & 170 & -1257 & 4956 & -1137 \\
-1247 & 108 & -1286 & -634 & 645 & -1116 & 156 & -249 & -220 & 361 & -789 & 688 & -398 & 413 & -1358 & -63 & 1392 & -1493 & 767 & -383 & 656 & 1057 & -1137 & 4128
\end{bmatrix}
$$

$$\left\{f^{(e)}\right\} = \left\{f_{\Omega}^{(e)}\right\} = \sum_{g=1}^{8}\left(\left(\left[N^{(8H)}\right]_g\right)^T \{b_g\} J_g W_g\right) =$$

$$\begin{bmatrix} 2.41 & 0.00 & 0.00 & 2.45 & 0.00 & 0.00 & 3.65 & 0.00 & 0.00 & 3.66 & 0.00 & 0.00 & 2.67 & 0.00 & 0.00 & 2.63 & 0.00 & 0.00 & 3.78 & 0.00 & 0.00 & 3.92 & 0.00 & 0.00 \end{bmatrix}^T$$

Since we only have body force along the x-direction, the equivalent nodal forces corresponding to the y- and z- directions are all equal to zero. ∎

Example 9.2: Equivalent nodal force vector for natural boundary conditions in a 8H element

We are given that, for the element shown in Figure 9.10, there is a traction $t_z = 2(1 + 0.05y \cdot z)$ (t_z in ksi, y and z in inches) on the face (4378) of the element, while all other traction components are zero.

The face (4378) in the parametric space corresponds to $\eta = \bar{\eta} = 1 = \text{constant}$. We will calculate $\left\{f_{\Gamma}^{(e)}\right\}$ using a four-point, two-dimensional Gauss quadrature and the procedure outlined in Section 9.3.3. Since the boundary corresponds to a constant η, the integration must be conducted on the $\xi\zeta$-plane. We will present in detail the computations for the first Gauss point, with parametric coordinates $\xi_1 = -\frac{1}{\sqrt{3}}$, $\zeta_1 = -\frac{1}{\sqrt{3}}$, $\eta_1 = \bar{\eta} = 1$.

We find the values of the shape functions at the location of the first Gauss point.

$$N_1^{(8H)}(\xi = \xi_1, \eta = \bar{\eta}, \zeta = \zeta_1) = 0 \qquad N_2^{(8H)}(\xi = \xi_1, \eta = \bar{\eta}, \zeta = \zeta_1) = 0$$

$$N_2^{(8H)}(\xi = \xi_1, \eta = \bar{\eta}, \zeta = \zeta_1) = 0.1667 \qquad N_4^{(8H)}(\xi = \xi_1, \eta = \bar{\eta}, \zeta = \zeta_1) = 0.6220$$

$$N_4^{(8H)}(\xi = \xi_1, \eta = \bar{\eta}, \zeta = \zeta_1) = 0 \qquad N_6^{(8H)}(\xi = \xi_1, \eta = \bar{\eta}, \zeta = \zeta_1) = 0$$

$$N_7^{(8H)}(\xi = \xi_1, \eta = \bar{\eta}, \zeta = \zeta_1) = 0.0447 \qquad N_8^{(8H)}(\xi = \xi_1, \eta = \bar{\eta}, \zeta = \zeta_1) = 0.1667$$

One may notice that the shape functions of the nodes that do not lie on the natural boundary face vanish. Next, we establish the shape function array at the location of Gauss point 1, whose components are the shape functions evaluated at the same location:

$$\left[N^{(8H)}\right]_1 = \begin{bmatrix} N_1^{(8H)}(\xi,\eta,\zeta) & 0 & 0 & \ldots\ldots & N_8^{(8H)}(\xi,\eta,\zeta) & 0 & 0 \\ 0 & N_1^{(8H)}(\xi,\eta,\zeta) & 0 & \ldots\ldots & 0 & N_8^{(8H)}(\xi,\eta,\zeta) & 0 \\ 0 & 0 & N_1^{(8H)}(\xi,\eta,\zeta) & \ldots\ldots & 0 & 0 & N_8^{(8H)}(\xi,\eta,\zeta) \end{bmatrix}\Bigg|_{\substack{\xi=\xi_1\\ \eta=\bar{\eta}\\ \zeta=\zeta_1}}$$

We can also find the physical coordinates of the Gauss point:

$$\begin{Bmatrix} x_1 \\ y_1 \\ z_1 \end{Bmatrix} = \left[N^{(8H)}\right]_1 \begin{Bmatrix} \left\{x_1^{(e)}\right\} \\ \vdots \\ \left\{x_8^{(e)}\right\} \end{Bmatrix} = \begin{Bmatrix} 1.0120 \\ 0.8453 \\ 5.1667 \end{Bmatrix}$$

We also find the array containing the derivatives of the shape functions at the location of Gauss point 1 for the natural boundary surface:

$$\left[N_{,\xi}^{(8H)}\right]_1 = \begin{bmatrix} \frac{\partial N_1^{(8H)}}{\partial \xi} & \frac{\partial N_2^{(8H)}}{\partial \xi} & \frac{\partial N_3^{(8H)}}{\partial \xi} & \frac{\partial N_4^{(8H)}}{\partial \xi} & \frac{\partial N_5^{(8H)}}{\partial \xi} & \frac{\partial N_6^{(8H)}}{\partial \xi} & \frac{\partial N_7^{(8H)}}{\partial \xi} & \frac{\partial N_8^{(8H)}}{\partial \xi} \\ \frac{\partial N_1^{(8H)}}{\partial \eta} & \frac{\partial N_2^{(8H)}}{\partial \eta} & \frac{\partial N_3^{(8H)}}{\partial \eta} & \frac{\partial N_4^{(8H)}}{\partial \eta} & \frac{\partial N_5^{(8H)}}{\partial \eta} & \frac{\partial N_6^{(8H)}}{\partial \eta} & \frac{\partial N_7^{(8H)}}{\partial \eta} & \frac{\partial N_8^{(8H)}}{\partial \eta} \\ \frac{\partial N_1^{(8H)}}{\partial \zeta} & \frac{\partial N_2^{(8H)}}{\partial \zeta} & \frac{\partial N_3^{(8H)}}{\partial \zeta} & \frac{\partial N_4^{(8H)}}{\partial \zeta} & \frac{\partial N_5^{(8H)}}{\partial \zeta} & \frac{\partial N_6^{(8H)}}{\partial \zeta} & \frac{\partial N_7^{(8H)}}{\partial \zeta} & \frac{\partial N_8^{(8H)}}{\partial \zeta} \end{bmatrix}\Bigg|_{\substack{\xi=\xi_1\\ \eta=\bar{\eta}\\ \zeta=\zeta_1}}$$

$$\left[N_{,\xi}^{(8H)}\right]_1 = \begin{bmatrix} 0.0000 & 0.0000 & 0.3943 & -0.3943 & 0.0000 & 0.0000 & 0.1057 & -0.1057 \\ -0.3110 & -0.0833 & 0.0833 & 0.3110 & -0.0833 & -0.0223 & 0.0223 & 0.0833 \\ 0.0000 & 0.0000 & -0.1057 & -0.3943 & 0.0000 & 0.0000 & 0.1057 & 0.3943 \end{bmatrix}$$

Next, we find the Jacobian array at Gauss point 1 of the natural boundary:

$$[J_1] = \begin{bmatrix} -0.1057 & 2.5000 & 0.3943 \\ 0.0833 & 0.0000 & 2.5833 \\ 2.3943 & -0.5000 & -0.1057 \end{bmatrix}$$

Since the natural boundary corresponds to a constant value of η, we need to find the vectors $\left.\frac{\partial \vec{x}}{\partial \xi}\right|_1 = \left.\begin{bmatrix} \frac{\partial x}{\partial \xi} & \frac{\partial y}{\partial \xi} & \frac{\partial z}{\partial \xi} \end{bmatrix}^T\right|_1$ and $\left.\frac{\partial \vec{x}}{\partial \zeta}\right|_1 = \left.\begin{bmatrix} \frac{\partial x}{\partial \zeta} & \frac{\partial y}{\partial \zeta} & \frac{\partial z}{\partial \zeta} \end{bmatrix}^T\right|_1$. From the definition of the Jacobian matrix of the mapping, one can deduce that the first row of $[J_1]$ is the $\left.\frac{\partial \vec{x}}{\partial \xi}\right|_1$ vector, and the third row of $[J_1]$ is the $\left.\frac{\partial \vec{x}}{\partial \zeta}\right|_1$ vector. Thus, we have: $\left.\frac{\partial \vec{x}}{\partial \xi}\right|_1 = \begin{Bmatrix} -0.1057 \\ 2.5000 \\ 0.3943 \end{Bmatrix}$ and $\left.\frac{\partial \vec{x}}{\partial \zeta}\right|_1 = \begin{Bmatrix} 2.3943 \\ -0.5000 \\ -0.1057 \end{Bmatrix}$. We also calculate the quantity $\left\|\frac{\partial \vec{x}}{\partial \zeta} \times \frac{\partial \vec{x}}{\partial \xi}\right\|_1 = 6.0063$

Finally, we can also determine the traction vector at the location of Gauss point 1 of the natural boundary:

$$\{t\}_1 = \begin{Bmatrix} t_x \\ t_y \\ t_z \end{Bmatrix}_1 = \begin{Bmatrix} t_x(x=x_1, y=y_1, z=z_1) \\ t_y(x=x_1, y=y_1, z=z_1) \\ t_z(x=x_1, y=y_1, z=z_1) \end{Bmatrix} = \begin{Bmatrix} 0 \\ 0 \\ 2(1+0.05\cdot 0.8453\cdot 5.1667) \end{Bmatrix}$$

$$= \begin{Bmatrix} 0 \\ 0 \\ 2.4367 \end{Bmatrix}$$

We can conduct the same type of computations for the remaining three Gauss points of the natural boundary surface of the element. Gauss points 2, 3, and 4 have parametric coordinates $(\xi_2, \eta_2 = \bar{\eta}, \zeta_2) = \left(\frac{1}{\sqrt{3}}, 1, -\frac{1}{\sqrt{3}}\right)$, $(\xi_3, \eta_3 = \bar{\eta}, \zeta_3) = \left(\frac{1}{\sqrt{3}}, 1, \frac{1}{\sqrt{3}}\right)$, and $(\xi_4, \eta_4 = \bar{\eta}, \zeta_4) = \left(-\frac{1}{\sqrt{3}}, 1, \frac{1}{\sqrt{3}}\right)$, respectively. Eventually, we can obtain the equivalent nodal force vector $\left\{f_\Gamma^{(e)}\right\}$ due to the natural boundary conditions:

$$\left\{f_\Gamma^{(e)}\right\} = \sum_{g=1}^{4} \left(\left(\left[N^{(8H)}\right]_g \right)^T \{t_g\} \left\| \frac{\partial \vec{x}}{\partial \zeta} \times \frac{\partial \vec{x}}{\partial \xi} \right\|_g W_g \right) =$$

$$\begin{bmatrix} 0.00 & 0.00 & 0.00 & 0.00 & 0.00 & 0.00 & 0.00 & 0.00 & 19.69 & 0.00 & 0.00 & 15.64 & 0.00 & 0.00 & 0.00 & 0.00 & 0.00 & 0.00 & 0.00 & 0.00 & 17.99 & 0.00 & 0.00 & 14.23 \end{bmatrix}^T$$

Since we only have traction along the z-direction, the equivalent nodal forces corresponding to the x- and y-directions are all equal to zero. Additionally, the only nodal points that have a nonzero force are 3, 4, 7, and 8, that is, the points that comprise the natural boundary face of the element. ∎

Problems

9.1 We are given the element in Figure P9.1, for which we know that nodes 1, 2, 3, and 4 are *fully restrained*.

a) Calculate the stiffness submatrix, $[K_{ff}]$, corresponding to the unrestrained degrees of freedom.
b) Calculate the nodal force subvector $\{f_f\}$, corresponding to the unrestrained degrees of freedom.
c) Obtain the nodal displacements.
d) Calculate the strain vector at the point with parametric coordinates $\xi = 0$, $\eta = 0$, $\zeta = 0$.

Element in physical space

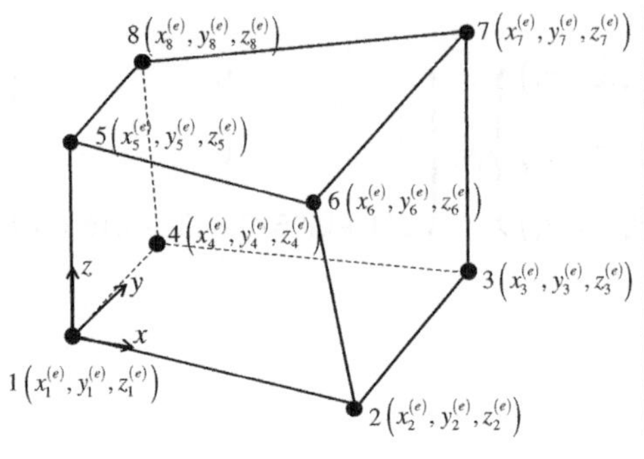

Node	x (ft)	y (ft)	z (ft)
1	0	0	0
2	5	0	–1
3	6	5	–0.5
4	1	4	0
5	0	0	4
6	4	–1	3
7	6	5	5
8	–1	3.5	4.5

$E = 29000$ ksi, $v = 0.3$

Body Forces

$b_x = 0.05$ (kip/in.^{3})

$b_y = 0.02x - 0.01z$ (b_y in kip/in.^{3}, x and z in inches)

$b_z = 0.01$ (kip/in.^{3})

Tractions on face (2376):

$t_x = 0.5$ (ksi)

$t_y = t_z = 0$

Figure P9.1

9.2 Calculate the nodal displacements for the mesh consisting of the nodes and loads shown in Figure P9.1, using a mesh consisting of four-node tetrahedral (4 T) elements. You need to decide on how exactly to discretize the domain.

9.3 Calculate the nodal displacements for the mesh consisting of the nodes and loads shown in Figure P9.1, using a mesh consisting of two six-node, wedge-shaped elements, like the one shown in Figure 9.8. The first element must be using nodes 1, 2, 3, 5, 6, and 7, while the second element must be using nodes 1, 3, 4, 5, 7, and 8.

Reference

Hammer, P., Marlowe, O., and Stroud, A. (1956). "Numerical Integration over Simplexes and Cones." *Mathematical Tables and Aids to Computation,* 10, 130–137.

10

Topics in Applied Finite Element Analysis

So far, we have discussed several families of problems, such as scalar field problems and vector field problems (for the latter, we focused on multidimensional elasticity), and their respective finite element approximations and solutions. This chapter will provide information on several useful topics, which often arise when using the finite element method in engineering practice. Each of the sections provided below can be read independently. The only exception is Section 10.7, which partially relies on Section 10.6.

10.1 Concentrated Loads in Multidimensional Analysis

A topic that was not discussed in Chapters 8 and 9 was how to treat concentrated loads in multidimensional finite element analysis. A concentrated load is described by a force vector $\{P_C\}$. Similar to the procedure in Section 3.9 for one-dimensional analysis, we will treat $\{P_C\}$ as a body force. The mathematical instrument that will allow this consideration is the generalization of the Dirac delta function to multi-dimensional domains, as a scalar function which depends on the spatial coordinates (x, y, z). It is more efficient to group the three spatial variables into a single vector variable $\vec{r} = [x \;\; y \;\; z]^T$, which we will call the *position vector*. Let us now imagine that the concentrated load $\{P_C\}$ is applied at a point with position vector $\vec{r}_P = [x_P \;\; y_P \;\; z_P]^T$. Then, we can define the multidimensional Dirac delta function, $\delta(\vec{r}-\vec{r}_P)$, for a concentrated source at point $\vec{r}_P$, which must satisfy the following two conditions.

$$\iiint_{\Omega} \delta(\vec{r}-\vec{r}_P)\,dV = 1 \tag{10.1.1}$$

for any function $g(\vec{r}) = g(x,y,z)$:

$$\int_0^L g(\vec{r}) \cdot \delta(\vec{r}-\vec{r}_P)\,dx = g(\vec{r}_P) = g(x_P,y_P,z_P) \tag{10.1.2}$$

Now, the concentrated force at location $\vec{r}_P$ will be considered as a body-force vector $\{b_P\}$, given by the following expression.

$$\{b_P\} = \{P_C\} \cdot \delta(\vec{r}-\vec{r}_P) \tag{10.1.3}$$

Fundamentals of Finite Element Analysis: Linear Finite Element Analysis, First Edition.
Ioannis Koutromanos, James McClure, and Christopher Roy.

Companion website: www.wiley.com/go/koutromanos/linear

In this case, if the location $\vec{r}_P$ corresponds to the interior of some element e, then the equivalent nodal force vector of that element due to the concentrated force $\{P_C\}$ can be obtained by plugging Equation (10.1.2) into Equation (9.2.14a):

$$\left\{f_\Omega^{(e)}\right\} = \iiint_{\Omega^{(e)}} \left[N^{(e)}(x,y,z)\right]^T \cdot \delta(\vec{r}-\vec{r}_P) \cdot \{P_C\} dV = \left[N^{(e)}(x_P,y_P,z_P)\right]^T \{P_C\} \quad (10.1.4)$$

Remark 10.1.1: Notice that Equation (10.1.4) indicates that we need to evaluate the shape functions at given physical coordinates. For isoparametric elements, we do *not* need to find these expressions; instead, we can find the values of the shape functions after we know the parametric coordinates ξ_P, η_P, and ζ_P of the point where the force is applied. If we have the physical coordinates of $\vec{r}_P$ where $\{P_C\}$ is applied, then we can calculate ξ_P, η_P, and ζ_P. To this end, we can solve a system of three nonlinear equations to find the corresponding parametric coordinate values. The procedure to set up the specific system of equations for an eight-node hexahedral (8H) element is presented in Box 10.1.1. Similar equations can be set up for other types of isoparametric elements, such as tetrahedral or serendipity hexahedral elements. The solution process will not be discussed here.

Box 10.1.1

Inverse mapping problem for finding parametric coordinates of a point with physical (Cartesian) coordinates (x_P, y_P, z_P) in the interior of an 8H element

Find the values of ξ_P, η_P and ζ_P, such that the following three equations are satisfied:

$$x(\xi_P,\eta_P,\zeta_P) = \sum_{I=1}^{8}\left[N_I^{(8H)}(\xi_P,\eta_P,\zeta_P)\cdot x_I^{(e)}\right] = x_P$$

$$y(\xi_P,\eta_P,\zeta_P) = \sum_{I=1}^{8}\left[N_I^{(8H)}(\xi_P,\eta_P,\zeta_P)\cdot y_I^{(e)}\right] = y_P$$

$$z(\xi_P,\eta_P,\zeta_P) = \sum_{I=1}^{8}\left[N_I^{(8H)}(\xi_P,\eta_P,\zeta_P)\cdot z_I^{(e)}\right] = z_P$$

After we find the coordinates ξ_P, η_P, and ζ_P, we can simply obtain the equivalent nodal force vector due to the concentrated force:

$$\left\{f_\Omega^{(e)}\right\} = \left[N^{(e)}(x_P,y_P,z_P)\right]^T \{P_C\} = \left[N^{(e)}(\xi_P,\eta_P,\zeta_P)\right]^T \{P_C\} \quad (10.1.5)$$

An alternative approach would be to directly calculate the eight shape functions of the $8H$ element in the physical space, using the procedure described in Sections 6.2 and 6.3 for two-dimensional elements. ■

Remark 10.1.2: For the most usual case, in which the concentrated force is applied at one of the nodal points, say at node k, then the Kronecker delta properties of the shape functions mean that the entirety of the concentrated loads will be carried by node k, and thus, we can simply add the concentrated forces to the corresponding components of the global $\{f\}$-vector corresponding to node k! ▮

Remark 10.1.3: The multidimensional Dirac delta function can also be used in scalar field problems, to describe a *concentrated source.* For example, for heat conduction problems, we can use the expression $s_P(\vec{r}) = S \cdot \delta(\vec{r} - \vec{r}_P)$ to define a concentrated heat source located at $\vec{r} = \vec{r}_P$ and having an intensity equal to S. ▮

10.2 Effect of Autogenous (Self-Induced) Strains—The Special Case of Thermal Strains

In Chapters 7, 8, and 9, we examined the behavior of elastic solid bodies for which the entirety of the strains is caused by mechanical loads. There are several types of strain that occur from processes other than mechanical loading and are thus independent of the applied stresses. These strains are called *autogenous or self-induced strains,* due to the fact that they can develop "by themselves," without the occurrence of stresses in the interior of a solid. To account for the presence of autogenous strains, we need to generalize the stress-strain relation of linearly elastic materials as follows.

$$\{\sigma\} = [D](\{\varepsilon\} - \{\varepsilon_o\}) \tag{10.2.1}$$

where $\{\varepsilon_o\}$ is the autogenous part of the strains in the material. The self-induced strains are usually given from an appropriate law. For the simplest cases that we will examine here, $\{\varepsilon_o\}$ is independent of the displacement field and is known *a priori* in an analysis. The most common case of autogenous strains is that of *thermal strains* due to temperature changes. If the material at a location of a *three-dimensional body* had an initial temperature (during construction) equal to Θ_o and now has a different temperature equal to Θ, then the autogenous thermal (or temperature-induced) strains for three-dimensional problems are given by the following vector:

$$\{\varepsilon_o\} = \left[a_T(\Theta - \Theta_o) \;\; a_T(\Theta - \Theta_o) \;\; a_T(\Theta - \Theta_o) \;\; 0 \;\; 0 \;\; 0\right]^T \tag{10.2.2}$$

where a_T is a constant called the *thermal coefficient* of the material, which must be given as a material parameter.

For two-dimensional problems (plane stress and plane strain), the autogenous thermal strain vector is given as a three-component vector:

$$\{\varepsilon_o\} = \left[a_T(\Theta - \Theta_o) \;\; a_T(\Theta - \Theta_o) \;\; 0\right]^T \tag{10.2.3}$$

Remark 10.2.1: Although Equation (10.2.3) indicates that we have two nonzero components for the autogenous strain of two-dimensional elasticity, it turns out that—strictly speaking—this is not correct for plane-strain conditions. Specifically, we need to remember than plane-strain stems from the stipulation that $\varepsilon_{zz} = \gamma_{xz} = \gamma_{yz} = 0$.

This would imply that we would need to consider a nonzero autogenous strain in the z-direction as well, and write a (4×4) stress-strain matrix relation. This relation must include the third row (corresponding to σ_{zz}) of the three-dimensional stress-strain law to give:

$$\begin{Bmatrix} \sigma_{xx} \\ \sigma_{yy} \\ \sigma_{zz} \\ \sigma_{xy} \end{Bmatrix} = \begin{bmatrix} D_{11} & D_{12} & D_{13} & D_{14} \\ D_{21} & D_{22} & D_{23} & D_{24} \\ D_{31} & D_{32} & D_{33} & D_{34} \\ D_{41} & D_{42} & D_{43} & D_{44} \end{bmatrix} \left(\begin{Bmatrix} \varepsilon_{xx} \\ \varepsilon_{yy} \\ 0 \\ \gamma_{xy} \end{Bmatrix} - \begin{Bmatrix} a_T(\Theta - \Theta_o) \\ a_T(\Theta - \Theta_o) \\ a_T(\Theta - \Theta_o) \\ 0 \end{Bmatrix} \right) \quad (10.2.4)$$

Despite the fact that the rigorous treatment of plane-strain would require use of Equation (10.2.4), we often resort to a much simpler approach, using the 3×3 stress-strain law that we established in Section 7.7 and the three-component vector of Equation (10.2.3). ■

We need to keep in mind that the strain-displacement equation $\{\varepsilon\} = [B^{(e)}]\{U^{(e)}\}$ still applies for each element e, for the *total* strains (because the total strains represent the actual change in shape of a solid body).

Now, let us return to the finite element equations for multidimensional elasticity, described in Chapters 8 and 9. Specifically, let us revisit an expression that led to Equation (9.2.18) in Chapter 9:

$$\sum_{e=1}^{Ne} \left[\int_{\Omega^{(e)}} \{W\}^T \left[L^{(e)}\right]^T \left[B^{(e)}\right]^T \left[D^{(e)}\right] \left[B^{(e)}\right] \left[L^{(e)}\right] \{U\}\, dV \right] = \{W\}^T \sum_{e=1}^{Ne} \left(\left[L^{(e)}\right]^T \left\{f^{(e)}\right\} \right) \quad (10.2.5)$$

where "$\int_{\Omega^{(e)}} \ldots dV$" means integral over the domain (triple integral if we have three-dimensional elasticity, double integral if we have two-dimensional elasticity).

If we focus on the left-hand side of Equation (10.2.5), we may be able to discern that a specific term gives:

$$\left[D^{(e)}\right] \left[B^{(e)}\right] \left[L^{(e)}\right] \{U\} = \left[D^{(e)}\right] \left[B^{(e)}\right] \left\{U^{(e)}\right\} = \left[D^{(e)}\right] \left\{\varepsilon^{(e)}\right\} = \left\{\sigma^{(e)}\right\} \quad (10.2.6)$$

where we have accounted for the stress-strain relation, $\left\{\sigma^{(e)}\right\} = \left[D^{(e)}\right] \left\{\varepsilon^{(e)}\right\}$, in the absence of autogenous strains. To incorporate the autogenous strains, all we need to do is plug Equation (10.2.6) into Equation (10.2.5):

$$\sum_{e=1}^{Ne} \left[\int_{\Omega^{(e)}} \{W\}^T \left[L^{(e)}\right]^T \left[B^{(e)}\right]^T \left\{\sigma^{(e)}\right\} dV \right] = \{W\}^T \sum_{e=1}^{Ne} \left(\left[L^{(e)}\right]^T \left\{f^{(e)}\right\} \right) \quad (10.2.7)$$

and then introduce the "new" version of the stress-strain relation, that is, Equation (10.2.1), which accounts for autogenous strains:

$$\left\{\sigma^{(e)}\right\} = \left[D^{(e)}\right]\left(\left\{\varepsilon^{(e)}\right\} - \left\{\varepsilon_o^{(e)}\right\}\right) = \left[D^{(e)}\right]\left(\left[B^{(e)}\right]\left\{U^{(e)}\right\} - \left\{\varepsilon_o^{(e)}\right\}\right)$$
$$= \left[D^{(e)}\right]\left(\left[B^{(e)}\right]\left[L^{(e)}\right]\{U\} - \left\{\varepsilon_o^{(e)}\right\}\right) \quad (10.2.8)$$

Plugging Equation (10.2.8) into (10.2.7), we obtain:

$$\sum_{e=1}^{Ne}\left[\int_{\Omega^{(e)}} \{W\}^T\left[L^{(e)}\right]^T\left[B^{(e)}\right]^T\left[D^{(e)}\right]\left(\left[B^{(e)}\right]\left[L^{(e)}\right]\left\{\varepsilon^{(e)}\right\} - \left\{\varepsilon_o^{(e)}\right\}\right) dV\right]$$
$$= \{W\}^T\sum_{e=1}^{Ne}\left(\left[L^{(e)}\right]^T\left\{f^{(e)}\right\}\right) \quad (10.2.9)$$

Now, we rearrange Equation (10.2.9), and we bring all the terms in the left-hand side, while also separating the terms that depend on the nodal displacements, $\{U\}$ from all other terms:

$$\{W\}^T\sum_{e=1}^{Ne}\left[\left[L^{(e)}\right]^T\left(\int_{\Omega^{(e)}}\left[B^{(e)}\right]^T\left[D^{(e)}\right]\left[B^{(e)}\right] dV\right)\left[L^{(e)}\right]\{U\} - \left[L^{(e)}\right]^T\int_{\Omega^{(e)}}\left[B^{(e)}\right]^T\left[D^{(e)}\right]\left\{\varepsilon_o^{(e)}\right\} dV\right]$$
$$= \{W\}^T\sum_{e=1}^{Ne}\left(\left[L^{(e)}\right]^T\left\{f^{(e)}\right\}\right) \rightarrow$$

$$\rightarrow \{W\}^T\sum_{e=1}^{Ne}\left[\left[L^{(e)}\right]^T\left[k^{(e)}\right]\left[L^{(e)}\right]\{U\} - \left[L^{(e)}\right]^T\left(\int_{\Omega^{(e)}}\left[B^{(e)}\right]^T\left[D^{(e)}\right]\left\{\varepsilon_o^{(e)}\right\} dV + \left\{f^{(e)}\right\}\right)\right] = 0$$

and since the vector $\{W\}$ in the obtained expression is arbitrary, the only way to always get a zero result is to have

$$[K]\{U\} - \{f\} = 0 \rightarrow [K]\{U\} = \{f\}$$

The stiffness matrix $[K]$ in the obtained expression is given by Equations (9.2.11) and (9.2.16), while the equivalent nodal force vector $\{f\}$ now includes an additional term:

$$\{f\} = \sum_{e=1}^{N_e}\left[\left[L^{(e)}\right]^T\left(\underline{\int_{\Omega^{(e)}}\left[B^{(e)}\right]^T\left[D^{(e)}\right]\left\{\varepsilon_o^{(e)}\right\} dV} + \left\{f_\Omega^{(e)}\right\} + \left\{f_{\Gamma t}^{(e)}\right\}\right)\right] \quad (10.2.10)$$

Notice that the underlined term in Equation (10.2.10) is due to the autogenous strains. Thus, nonzero autogenous strains can be incorporated in an analysis simply by adding the term $\int_{\Omega^{(e)}}\left[B^{(e)}\right]^T\left[D^{(e)}\right]\{\varepsilon_o\} dV$ to the equivalent nodal force vector of each element! This is reflected in Box 10.2.1, which summarizes how to account for autogenous strains in a finite element analysis.

Box 10.2.1 Accounting for Autogenous Strains in Analysis

Whenever we have autogenous strains $\left\{\varepsilon_o^{(e)}\right\}$ in our finite elements, then the only modification in our equations is that the equivalent nodal force vector, $\{f^{(e)}\}$, of each element, will include an extra term due to autogenous strains and will be given by:

$$\left\{f^{(e)}\right\} = \int_{\Omega^{(e)}} \left[B^{(e)}\right]^T \left[D^{(e)}\right] \left\{\varepsilon_o^{(e)}\right\} dV + \left\{f_\Omega^{(e)}\right\} + \left\{f_{\Gamma t}^{(e)}\right\} \qquad (10.2.11)$$

where $\left\{f_\Omega^{(e)}\right\}$ is the contribution of body forces and $\left\{f_{\Gamma t}^{(e)}\right\}$ is the contribution of boundary tractions, as explained in Chapters 8 (for two-dimensional elasticity) and 9 (for three-dimensional elasticity).

10.3 The Patch Test for Verification of Finite Element Analysis Software

Finite element analysis requires the design, implementation, and use of computer software. It is important to ensure that a finite element program works as planned. The process by which we ensure that an analysis software works as intended is called *verification*. Each finite element program developer conducts a set of verification analyses to establish the accuracy of the program. One of the most common methods to verify a finite element formulation is the *patch test*, originally presented by Bazeley et al. (1965) and formulated in a more rigorous mathematical setting by Strang and Fix (2008). As shown in Figure 10.1, we create a patch of finite elements and apply a loading condition that will lead to constant strains (and consequently constant stresses) in the entire mesh. We then check the output of the analysis, to verify that we do obtain constant strains and stresses everywhere. As mentioned in Fish and Belytschko (2007), the differences in stress values in the interior of the mesh should be within the precision of the computer used in the analysis ($\sim 10^{-8} - 10^{-10}$), and a difference in stress values in the order of even 10^{-3} should raise concerns about the existence of bugs.

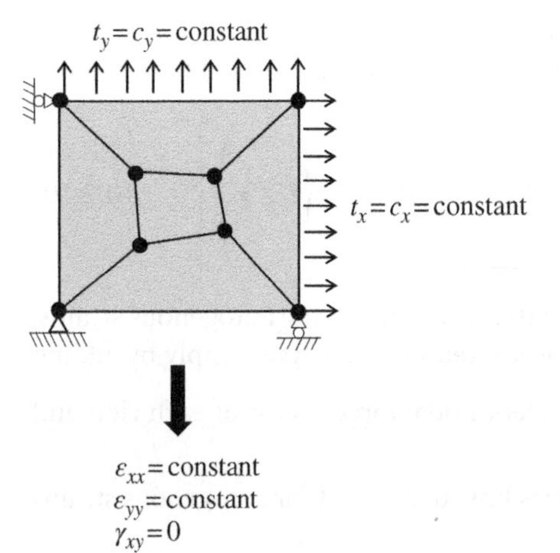

Figure 10.1 Example of mesh for patch test of plane stress elements.

Remark 10.3.1: When a type of elements is used in a patch test and captures the state of constant stress and constant strain in the interior of the entire mesh, we say that *the specific element type passes the patch test.* ∎

Remark 10.3.2: The shape of the elements must not be rectangular or box-shaped, because some element types may pass the patch test when the elements have a "nice" rectangular or box shape and not pass it for a general, nonrectangular shape. ∎

Remark 10.3.3: Any new element type that is created must pass the patch test. Otherwise, the element formulation may not ensure convergence, and thus may be inappropriate for use in analysis. ∎

Figure 10.2 illustrates an example patch test in the commercial analysis program *ABAQUS* for a four-element mesh subjected to tractions that generate constant normal stresses and strains, together with the printed stresses from the analysis. Per Figure 10.2, our analysis can accurately reproduce the state of constant stress in the element. This should come as no surprise, since commercial software such as *ABAQUS* has undergone extensive verification by its manufacturer; still, it is always better for users to make sure that everything is in order in an analysis program!

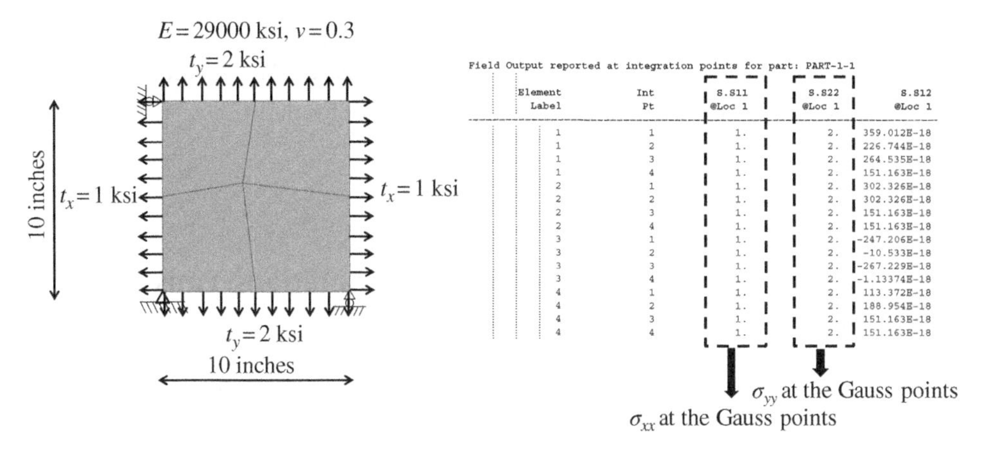

Field Output reported at integration points for part: PART-1-1

Element Label	Int Pt	S.S11 @Loc 1	S.S22 @Loc 1	S.S12 @Loc 1
1	1	1.	2.	359.012E-18
1	2	1.	2.	226.744E-18
1	3	1.	2.	264.535E-18
1	4	1.	2.	151.163E-18
2	1	1.	2.	302.326E-18
2	2	1.	2.	302.326E-18
2	3	1.	2.	151.163E-18
2	4	1.	2.	151.163E-18
3	1	1.	2.	-247.206E-18
3	2	1.	2.	-10.533E-18
3	3	1.	2.	-267.229E-18
3	4	1.	2.	-1.13374E-18
4	1	1.	2.	113.372E-18
4	2	1.	2.	188.954E-18
4	3	1.	2.	151.163E-18
4	4	1.	2.	151.163E-18

Figure 10.2 Patch test of plane-stress elements in the commercial program *ABAQUS*.

The patch test merely constitutes a rudimentary form of verification. A more detailed discussion of verification for finite element analysis and the related concept of *validation* is provided in Chapter 18.

10.4 Subparametric and Superparametric Elements

In Chapters 6 and 9, we examined the isoparametric element concept, which relies on the use of identical shape functions for the field (temperature, displacement etc.) approximations and for the geometry mapping. This explains the prefix "iso-" (which means "equal") in *isoparametric*. Strictly speaking, having the same functions for the approximate field and for the geometry mapping is not necessary. For example, we can have a higher order of polynomials (with respect to the parametric coordinates) for the field function than for the coordinate mapping, at which case we have a *subparametric finite element*. Alternatively, we may try to use lower-order polynomials for the field than for the coordinate mapping, at which case we have *superparametric finite elements*.

Superparametric elements do *not* satisfy the requirements for first-order completeness, which is a prerequisite for convergence. Thus, such elements do not guarantee convergence and are not used in practice. Subparametric elements *do converge*, and are used in practice. For some applications and for a given number of nodes used for the geometry mapping, subparametric elements may exhibit superior accuracy than the corresponding isoparametric elements. For example, the quadrilateral subparametric element in Figure 10.3, which uses four nodes for the geometry mapping and eight nodes for the approximate field, will have identical or superior accuracy to that of the isoparametric 4Q element described in Section 6.4.

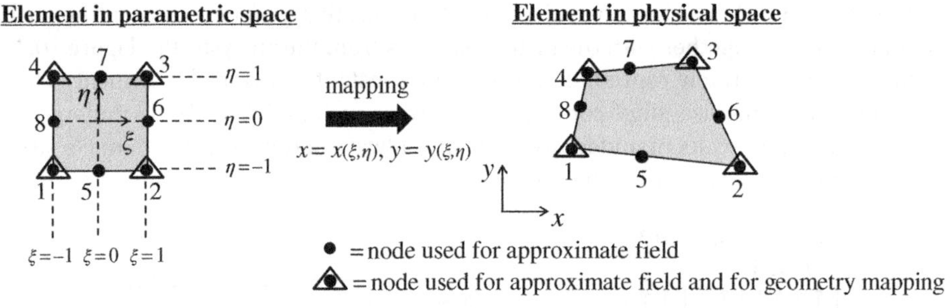

Figure 10.3 Example of a quadrilateral subparametric element.

10.5 Field-Dependent Natural Boundary Conditions: Emission Conditions and Compliant Supports

In many applications of engineering, it may not be physically correct to prescribe a value for the natural boundary conditions which does not depend on the field value at the natural boundary. For example, in heat conduction, if we have a warm body that is immersed in a region with lower temperature, then heat will be *emitted* across the boundary. We may all have seen science fiction movies, where an unlucky hero or villain is ejected out of a spacecraft without a suit, and then quickly freezes. This happens because all the heat in the body of the hero (or villain) is quickly emitted to the space void, which has a temperature of almost absolute zero (zero degrees Kelvin, which corresponds to –273 °C, the lowest possible temperature that can occur in nature). Along the same lines, if we swim into the sea (which typically has lower temperature than that of the human body), we feel a pleasant "cooling" sensation. This sensation also stems from loss of heat from our body, due to the fact that we are in contact with and surrounded by water that has a lower temperature. Similarly, if we have a warm bath, we feel a more "comfortably warm" sensation, which stems from the fact that the temperature of the water is similar to that of our body, and we do not have significant heat outflow from our body to the water. The above considerations are meant to demonstrate a general rule: whenever a body with a temperature field Θ is immersed in an environment (a surrounding medium) with different temperature Θ_{env}, then we will have heat exchange with the environment, in the form of heat flow across the boundary surface of the body, as shown in Figure 10.4a. If the temperature at the boundary is higher than Θ_{env}, then we will have heat outflow from the body toward the environment, as shown in the same figure. The magnitude of the heat outflow $\bar{q}$ will depend on the difference $(\Theta - \Theta_{env})$. We can establish a condition that mathematically expresses this phenomenon. We will call the above condition an *emissivity* condition, and we will focus on the simplest case, where the heat outflow is linearly proportional to the difference $(\Theta - \Theta_{env})$. Thus, we will write that $\bar{q}$ is equal to the product of a constant k_q times the temperature difference:

$$\bar{q} = \vec{q} \bullet \vec{n} = k_q(\Theta - \Theta_{env}) \text{ at } \Gamma_q \tag{10.5.1}$$

Note that Equation (10.5.1) constitutes a *natural boundary condition*, in which the boundary outflow depends on the field function, Θ. The same type of *field-dependent natural boundary condition* can be established for problems involving mass flow, for example, moisture (water) flow in porous media, where the field function is the liquid water pressure field, P, of the water. If we have a porous medium such as a concrete beam, as shown in Figure 10.4b, exposed to the atmospheric air, the moisture flow across the boundary can be coupled to the difference of the *relative humidity* h_R on the surface of the beam and the relative humidity $h_{R(env)}$ of the surrounding air, by means of the following expression.

$$\bar{q} = \vec{q} \bullet \vec{n} = k_q\left(h_R - h_{R(env)}\right) \text{ at } \Gamma_q \tag{10.5.2}$$

Through a series of assumptions, the relative humidity h_R can be expressed in terms of the liquid water pressure, P, by means of the *Kelvin equation*:

$$P = \frac{RT}{\bar{V}_L} \cdot \ln(h_R) \tag{10.5.3}$$

The specific equation can also be written in the following form.

$$h_R = \exp\left[\frac{\bar{V}_L}{RT} \cdot P\right] \tag{10.5.4}$$

Plugging Equation (10.5.4) into Equation (10.5.2) yields a boundary condition that is a nonlinear function of the water pressure field P. Still, we can *linearize* the relation between h_R and P. For example, if we know that the average value of relative humidity is equal to 60%, we can take a Taylor series expansion about that value and only keep the linear (first-order) terms. An example involving such a linearization approach has been presented in Section 1.8.

We can also establish field-dependent natural boundary conditions for elasticity problems. Specifically, we can stipulate that the traction components on the natural boundary depend on the corresponding components of the displacement field. This corresponds to what we call *compliant* (deformable) supports. For the simplest case, we can take the tractions to be linearly proportional to the displacements, at which case we have *elastic* (or spring) supports, as shown for example in Figure 10.4c. For the case of spring supports, the natural boundary conditions can attain the general form:

$$\{t\} = [\sigma] \bullet \vec{n} = -[k_s]\{u\} \text{ at } \Gamma_t \tag{10.5.5}$$

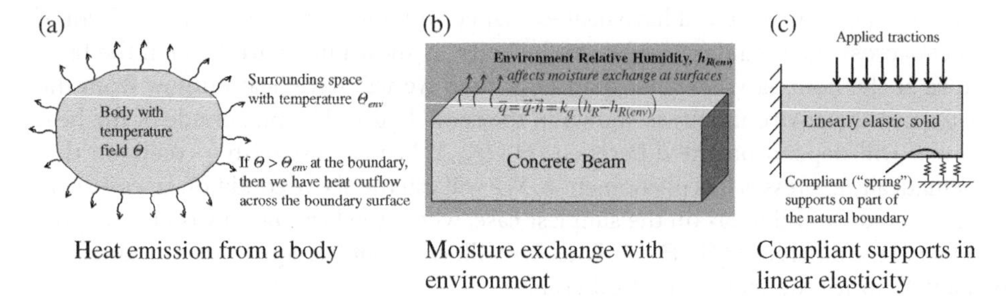

Figure 10.4 Examples of generalized natural boundary conditions in analyses.

where $\{t\}$ is the traction vector, $\{u\}$ is the displacement vector, and $[k_s]$ is a symmetric *foundation stiffness matrix* (2×2 for two-dimensional problems; 3×3 for three-dimensional ones). In the simplest case, we can consider a diagonal $[k_s]$; that is, for three-dimensional problems, we have:

$$[k_s] = \begin{bmatrix} k_{sx} & 0 & 0 \\ 0 & k_{sy} & 0 \\ 0 & 0 & k_{sz} \end{bmatrix} \tag{10.5.6}$$

If the foundation stiffness matrix is given by Equation (10.5.6), Equation (10.5.5) can be written as a set of the following three expressions.

$$t_x = -k_{sx} \cdot u_x \text{ at } \Gamma_{tx} \tag{10.5.7a}$$

$$t_y = -k_{sy} \cdot u_y \text{ at } \Gamma_{ty} \tag{10.5.7b}$$

$$t_z = -k_{sz} \cdot u_z \text{ at } \Gamma_{tz} \tag{10.5.7c}$$

We need to remember that, in general, we can have different natural boundary segments for each one of the three directions x, y and z.

Field-dependent natural boundary conditions like those described in Figure 10.4 are also called *generalized natural boundary conditions.*

Remark 10.5.1: Notice the negative (–) sign in Equations (10.5.5) and (10.5.7a–c). This should be expected because, at each point, the elastic support will apply tractions that *oppose* the displacements. For example, in Figure 10.4c, if a point of the elastic support with the vertical springs moved "downward," then the tractions that the support would exert on the solid would be upward. ∎

Now, we will rewrite the finite element equations for elasticity, for the case of field-dependent natural boundary conditions, that is, elastic supports. Specifically, we will write the equations for the general three-dimensional problem, where a part of the natural boundary, Γ_{t1}, has prescribed tractions, and another part of the natural boundary, Γ_{t2}, has elastic supports. Notice that the entire natural boundary, Γ_t, must either include prescribed tractions or elastic supports. For this reason, the combination of the two natural boundary parts must give us Γ_t:

$$\Gamma_{t1} \cup \Gamma_{t2} = \Gamma_t \tag{10.5.8}$$

We revisit Equation (9.2.10) which gave the finite element equations for three-dimensional elasticity:

$$\{W\}^T\left(\sum_{e=1}^{Ne}\left[\left[L^{(e)}\right]^T\left[k^{(e)}\right]\left[L^{(e)}\right]\right]\{U\}-\left[L^{(e)}\right]^T\left\{f^{(e)}\right\}\right)=0 \tag{9.2.10}$$

where, from Equation (9.2.13), the element equivalent nodal force vector is given by:

$$\left\{f^{(e)}\right\}=\left\{f_{\Omega}^{(e)}\right\}+\left\{f_{\Gamma_t}^{(e)}\right\} \tag{9.2.13}$$

and the natural boundary contribution to the nodal forces of each element, $\left\{f_{\Gamma_t}^{(e)}\right\}$, is obtained from Equation (9.2.14b). For brevity, we will assume that the natural boundary for the three directions x, y, and z is identical, at which case we can write:

$$\left\{f_{\Gamma_t}^{(e)}\right\}=\iint_{\Gamma_t^{(e)}}\left[N^{(e)}\right]^T\{t\}\,dS \tag{9.2.10b}$$

Now, all we need to do is separate the natural boundary into two parts, one corresponding to a prescribed traction and the other one to the elastic supports:

$$\left\{f_{\Gamma_t}^{(e)}\right\}=\iint_{\Gamma_{t1}^{(e)}}\left[N^{(e)}\right]^T\{t\}\,dS+\iint_{\Gamma_{t2}^{(e)}}\left[N^{(e)}\right]^T\{t\}\,dS \tag{10.5.9}$$

The tractions in $\Gamma_{t2}^{(e)}$ are given by the following expression.

$$\{t\}=-[k_s]\{u\} \text{ at } \Gamma_{t2}^{(e)} \tag{10.5.10}$$

Plugging Equation (10.5.10) into Equation (10.5.9), we obtain:

$$\left\{f_{\Gamma t}^{(e)}\right\}=\iint_{\Gamma_{t1}^{(e)}}\left[N^{(e)}\right]^T\{t\}\,dS-\iint_{\Gamma_{t2}^{(e)}}\left[N^{(e)}\right]^T[k_s]\{u\}\,dS \tag{10.5.11}$$

We will now separate $\left\{f_{\Gamma_t}^{(e)}\right\}$ into the part due to prescribed tractions:

$$\left\{f_{\Gamma_{t1}}^{(e)}\right\}=\iint_{\Gamma_{t1}^{(e)}}\left[N^{(e)}\right]^T\{t\}\,dS \tag{10.5.12a}$$

and the part due to compliant supports:

$$\left\{f_{\Gamma_{t2}}^{(e)}\right\}=\iint_{\Gamma_{t2}^{(e)}}\left[N^{(e)}\right]^T[k_s]\{u\}\,dS \tag{10.5.12b}$$

Since the integral in Equation (10.5.12b) corresponds to element e, we know that the displacement vector in this integral will be given by the finite element approximation of element e, given by Equation (9.2.2):

$$\left\{u^{(e)}\right\}\approx\left[N^{(e)}\right]\left\{U^{(e)}\right\}=\left[N^{(e)}\right]\left[L^{(e)}\right]\{U\} \tag{10.5.13}$$

where $[N^{(e)}]$ and $[L^{(e)}]$ are the shape function array and the gather array, respectively, of element e.

In light of Equations (10.5.11), (10.5.12), and (9.2.13), Equation (9.2.10) becomes:

$$\{W\}^T \sum_{e=1}^{Ne} \left[\left(\left[L^{(e)}\right]^T \left[k^{(e)}\right] \left[L^{(e)}\right] \right) \{U\} - \left[L^{(e)}\right]^T \left(\left\{f_\Omega^{(e)}\right\} + \left\{f_{\Gamma t1}^{(e)}\right\} - \left\{f_{\Gamma t2}^{(e)}\right\} \right) \right] = 0 \tag{10.5.14}$$

Now, we will plug Equation (10.5.12b) into Equation (10.5.14):

$$\{W\}^T \left[\sum_{e=1}^{Ne} \left(\left[L^{(e)}\right]^T \left[B^{(e)}\right]^T \left[D^{(e)}\right] \left[B^{(e)}\right] \left[L^{(e)}\right] \right) \right] \{U\}$$
$$- \{W\}^T \left[\sum_{e=1}^{Ne} \left[L^{(e)}\right]^T \left(\left\{f_\Omega^{(e)}\right\} + \left\{f_{\Gamma t1}^{(e)}\right\} - \iint_{\Gamma_{t2}^{e}} \left[N^{(e)}\right]^T [k_s]\{u\}\, dS \right) \right] = 0$$

If we also account for Equation (10.5.13), we have:

$$\{W\}^T \sum_{e=1}^{Ne} \left[\left(\left[L^{(e)}\right]^T \left[B^{(e)}\right]^T \left[D^{(e)}\right] \left[B^{(e)}\right] \left[L^{(e)}\right] \right) \right] \{U\}$$
$$- \sum_{e=1}^{Ne} \left[\left[L^{(e)}\right]^T \left(\left\{f_\Omega^{(e)}\right\} + \left\{f_{\Gamma t1}^{(e)}\right\} - \underline{\iint_{\Gamma_{t2}^{(e)}} \left[N^{(e)}\right]^T [k_s] \left[N^{(e)}\right] \left[L^{(e)}\right] \{U\}\, dS} \right) \right] = 0$$

Now, we notice that the underlined term (due to the compliant support) in the obtained expression depends on the nodal displacements, $\{U\}$! Thus, we can group our terms as follows:

$$\{W\}^T \sum_{e=1}^{Ne} \left[\left(\left[L^{(e)}\right]^T \left[k^{(e)}\right] \left[L^{(e)}\right] \right) \{U\} - \left[L^{(e)}\right]^T \left\{f^{(e)}\right\} \right] = 0$$

where the element stiffness matrix, $[k^{(e)}]$, is now given by:

$$\left[k^{(e)}\right] = \int_{\Omega^{(e)}} \left[B^{(e)}\right]^T \left[D^{(e)}\right] \left[B^{(e)}\right] dV + \int_{\Gamma_{t2}^{(e)}} \left[N^{(e)}\right]^T [k_s] \left[N^{(e)}\right] dV \tag{10.5.15}$$

and the element equivalent nodal force vector, $\{f^{(e)}\}$, is given by:

$$\left\{f^{(e)}\right\} = \left\{f_\Omega^{(e)}\right\} + \left\{f_{\Gamma_{tx1}}^{(e)}\right\} + \left\{f_{\Gamma_{ty1}}^{(e)}\right\} + \left\{f_{\Gamma_{tz1}}^{(e)}\right\} \tag{10.5.16}$$

In other words, the part of the natural boundary that contains compliant (spring) supports contributes to the stiffness matrix! This should come as no surprise, since the tractions of the compliant supports do depend on the values of the displacements.

Now, we can provide the considerations in Box 10.5.1, for the most general case that we have elastic supports for the three directions x, y, and z at different parts of the natural boundary.

Box 10.5.1 General Treatment of Compliant (Spring) Supports

Let us imagine that an element e contains a part of the natural boundary with elastic supports in directions x, y, and z.

$$\{t_x\} = [t_x \;\; 0 \;\; 0]^T = -[k_{sx}]\{u\} \text{ at } \Gamma_{tx2}^{(e)} \tag{10.5.17a}$$

$$\{t_y\} = [0 \;\; t_y \;\; 0]^T = -[k_{sy}]\{u\} \text{ at } \Gamma_{ty2}^{(e)} \tag{10.5.17b}$$

$$\{t_z\} = [0 \;\; 0 \;\; t_z]^T = -[k_{sz}]\{u\} \text{ at } \Gamma_{tz2}^{(e)} \tag{10.5.17c}$$

We also have prescribed tractions at the remainder of the natural boundary:

$$[\sigma]\bullet\vec{n} = \{t_x\} \text{ at } \Gamma_{tx1}^{(e)} \tag{10.5.18a}$$

$$[\sigma]\bullet\vec{n} = \{t_y\} \text{ at } \Gamma_{ty1}^{(e)} \tag{10.5.18b}$$

$$[\sigma]\bullet\vec{n} = \{t_z\} \text{ at } \Gamma_{tz1}^{(e)} \tag{10.5.18c}$$

Then, the stiffness matrix and the equivalent nodal force vector of the element are given by:

$$\begin{aligned}\left[k^{(e)}\right] = &\int_{\Omega^{(e)}} \left[B^{(e)}\right]^T \left[D^{(e)}\right]\left[B^{(e)}\right] dV + \int_{\Gamma_{tx2}^{(e)}} \left[N^{(e)}\right]^T [k_{sx}]\left[N^{(e)}\right] dV \\ &+ \int_{\Gamma_{ty2}^{(e)}} \left[N^{(e)}\right]^T [k_{sy}]\left[N^{(e)}\right] dV + \int_{\Gamma_{tz2}^{(e)}} \left[N^{(e)}\right]^T [k_{sz}]\left[N^{(e)}\right] dV\end{aligned} \tag{10.5.19}$$

and

$$\left\{f^{(e)}\right\} = \left\{f_{\Omega}^{(e)}\right\} + \left\{f_{\Gamma_{tx1}}^{(e)}\right\} + \left\{f_{\Gamma_{ty1}}^{(e)}\right\} + \left\{f_{\Gamma_{tz1}}^{(e)}\right\} \tag{10.5.20}$$

Remark 10.5.2: It can be seen that the presence of compliant (spring) supports introduces terms to the coefficient matrix, not to the right-hand-side vector. It is also important to remember that the additional terms in Equation (10.5.19) will only exist for the nodal points which lie on the generalized natural boundary! Thus, the additional matrix terms will not have any effect on the stiffness values corresponding to the interior nodes of the mesh (i.e., the nodes which do not lie on the natural boundary). ▮

Remark 10.5.3: If we only have a *concentrated* spring support, where we have a single spring at a node, and the concentrated forces exerted from the spring on the node are linearly proportional to the nodal displacements, then we can directly add the spring coefficients to the corresponding global stiffness entries of the nodal point where the spring is located. The proof of this stipulation is left as an exercise in Problem 10.3. ▮

Remark 10.5.4: Similar considerations apply for field-dependent natural boundary conditions of scalar field problems. For example, the presence of boundary conditions (10.5.1) in a heat conduction problem would lead to additional terms in the coefficient (conductance) array. If $\Theta_{\text{env}} \neq 0$, then these boundary conditions will also introduce additional terms in the equivalent nodal flux vector. Specific expressions are provided in Box 10.5.2. ▮

Box 10.5.2 General Treatment of Boundary Conditions (10.5.1) in Heat Conduction

Let us imagine that an element e contains a part of the natural boundary with a boundary condition of the form:

$$\vec{q} \bullet \vec{n} = k_q(\Theta - \Theta_{env}) \text{ at } \Gamma_{q2}^{(e)} \tag{10.5.21}$$

We also have prescribed outflow at the remainder of the natural boundary:

$$\vec{q} \bullet \vec{n} = \bar{q} \text{ at } \Gamma_{q1}^{(e)} \tag{10.5.22}$$

Then, the conductance matrix and the equivalent nodal flux vector of the element are given by:

$$\left[k^{(e)}\right] = \int_{\Omega^{(e)}} \left[B^{(e)}\right]^T \left[D^{(e)}\right] \left[B^{(e)}\right] dV + \int_{\Gamma_{q2}^{(e)}} \left[N^{(e)}\right]^T k_q \left[N^{(e)}\right] dV \tag{10.5.23}$$

$$\left\{f^{(e)}\right\} = \left\{f_{\Omega}^{(e)}\right\} + \left\{f_{\Gamma q1}^{(e)}\right\} \tag{10.5.24}$$

where

$$\left\{f_{\Gamma}^{(e)}\right\} = -\int_{\Gamma_{q1}^{(e)}} \left[N^{(e)}\right]^T \bar{q}\, dV + \int_{\Gamma_{q2}^{(e)}} \left[N^{(e)}\right]^T k_q \cdot T_{env}\, dV \tag{10.5.25}$$

10.6 Treatment of Nodal Constraints

There are cases in elasticity problems where the displacements of various nodes are *constrained* to satisfy a specific mathematical equation, called the *constraint condition*. A constraint condition involves more than one nodal displacements, and it is not the same as a support condition (the latter is also called a *restraint*). In our discussion, we will describe a constraint condition with the following mathematical equation:

$$\{U_{slave}\} = [T]\{U_{master}\} \tag{10.6.1}$$

where $[T]$ is a known matrix (we will see how we can obtain it for several special cases), called the *constraint transformation matrix*, $\{U_{master}\}$ is the displacement vector of a set of nodes, and $\{U_{slave}\}$ is the displacement vector for another set of nodes. Equation (10.6.1) means that, if we know the values of the displacements $\{U_{master}\}$, we can directly calculate the values of the displacements $\{U_{slave}\}$. In other words, the degrees of freedom (DOFs) contained in $\{U_{slave}\}$ are *slaved* to the values of $\{U_{master}\}$, so that for given $\{U_{master}\}$, the displacements $\{U_{slave}\}$ must attain the specific values that are required to satisfy (10.6.1). This is why the DOFs contained in $\{U_{master}\}$ are called the *master degrees of freedom*, while those in $\{U_{slave}\}$ are the *slave degrees of freedom*. Accordingly, we can also speak of *master nodes* and *slave nodes*.

A case that frequently arises is that where two nodes are connected with a *rigid link*, which is aligned with a specific direction. This situation is schematically presented in Figure 10.5.

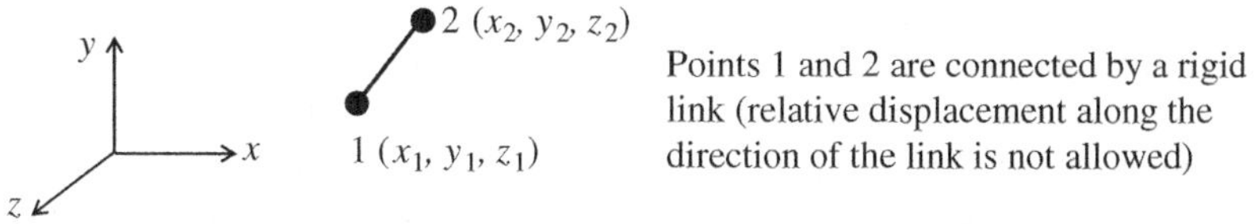

y x z

2 (x_2, y_2, z_2)

1 (x_1, y_1, z_1)

Points 1 and 2 are connected by a rigid link (relative displacement along the direction of the link is not allowed)

Figure 10.5 Rigid link connecting two nodal points.

A rigid link can be thought of as a truss bar that does not allow any deformation along its axis—that is, a truss bar with infinite axial stiffness. Such infinite stiffness will prevent the *relative* motion of the two nodes along the direction (axis) of the link. This leads to one of the nodal DOFs being slaved to the rest. Let us examine the case of a three-dimensional link, connecting two nodal points with coordinates $[x_1\ y_1\ z_1]^T$ and $[x_2\ y_2\ z_2]^T$. The constraint equation, which expresses the fact that the projection of the relative displacement of the two nodes on the direction of the link is zero, is written as follows:

$$\left(\vec{u}_2 - \vec{u}_1\right) \bullet \vec{\boldsymbol{\ell}} = 0 \tag{10.6.2}$$

where $\vec{u}_1$ is the displacement vector of node 1, $\vec{u}_2$ is the displacement vector of node 2, and $\vec{\boldsymbol{\ell}}$ is a unit vector in the direction of the link. Its three components ℓ_1, ℓ_2, and ℓ_3 can be obtained by taking the unit vector connecting the two nodal points. Using the coordinates of the two nodal points, we can write:

$$\vec{\boldsymbol{\ell}} = \frac{\vec{r}_{12}}{\left\|\vec{r}_{12}\right\|} \tag{10.6.3}$$

where

$$\vec{r}_{12} = \left[x_2 - x_1 \quad y_2 - y_1 \quad z_2 - z_1\right]^T \tag{10.6.4}$$

and $\left\|\vec{r}_{12}\right\|$ is the magnitude of vector $\vec{r}_{12}$.

We can rewrite Equation (10.6.2) in terms of the displacement components at the two nodal points:

$$u_{2x} \cdot \ell_1 - u_{1x} \cdot \ell_1 + u_{2y} \cdot \ell_2 - u_{1y} \cdot \ell_2 + u_{2z} \cdot \ell_3 - u_{1z} \cdot \ell_3 = 0 \tag{10.6.5}$$

The constraint equation generally includes all three nodal displacements of both nodal points. The single constraint equation will give one DOF slaved to the others. We will set whichever displacement component we wish as the slave one, provided that this component is multiplied by a nonzero quantity in Equation (10.6.5). For example, if ℓ_1 is nonzero, we can solve (10.6.5) for u_{2x} to obtain:

$$u_{2x} = u_{1x} - u_{2y} \cdot \frac{\ell_2}{\ell_1} + u_{1y} \cdot \frac{\ell_2}{\ell_1} - u_{2z} \cdot \frac{\ell_3}{\ell_1} + u_{1z} \cdot \frac{\ell_3}{\ell_1} = 0 \tag{10.6.6}$$

Thus, the displacement u_{2x} will be the single slave DOF. The other five displacement components will be master DOFs. So, for the specific case of a rigid link, we can set $\{U_{slave}\} = u_{2x}$, $\{U_{master}\} = \left[u_{1x}\ u_{1y}\ u_{1z}\ u_{2y}\ u_{2z}\right]^T$ and $[T] = \left[1\ \ \frac{\ell_2}{\ell_1}\ \frac{\ell_3}{\ell_1}\ \ -\frac{\ell_2}{\ell_1}\ \ -\frac{\ell_3}{\ell_1}\right]^T$ in Equation (10.6.1).

If $\ell_1 = 0$, then we need to solve Equation (10.6.5) for another displacement component, which is multiplied by a nonzero number. Then, the displacement component that we solved for will be the slave DOF, and we can also define $\{U_{master}\}$ and $[T]$ accordingly.

Another common situation occurs when two (or more) nodal points are connected with a *rigid bar* (beam) or a rigid body in general. In this case, we typically treat one of the nodal points as the master one (and its DOFs as the master ones as well), and all the other nodal points in the rigid body (and the DOFs of these points) as slave ones. Again, we can establish a relation with the same form as that of Equation (10.6.1). This case is discussed in detail in Appendix D.

The question that now arises is how nodal constraints affect our finite element equations. The first important part that we must remember is the following: The finite element equations that we solve must only include the master DOFs. After we solve and obtain the values of the master DOFs, the slave DOFs are obtained from Equation (10.6.1). Thus, the question that we need to answer is: "If we are given Equation (10.6.1) and we have the stiffness matrix corresponding to all the DOFs of the structure, how do we formulate the stiffness equations in a form that includes only the master DOFs?" Of course, this form will indirectly account for the fact that the stiffness of the slave DOFs will somehow "affect" the master DOFs. *We will start with the simplest case, where we only have nonzero stiffness terms for the slave DOFs.* Let us imagine that we have assembled the global stiffness matrix, $[K_{slave}]$, and the global equivalent nodal force vector, $\{F_{slave}\}$, which correspond to $\{U_{slave}\}$.

In that case, as discussed in Appendix D, the stiffness matrix corresponding to the master DOFs is given by

$$[K_{master}] = [T]^T [K_{slave}][T] \tag{10.6.7}$$

and the equivalent nodal force vector corresponding to $\{U_{master}\}$ is obtained from the following equation.

$$\{F_{master}\} = [T]^T \{F_{slave}\} \tag{10.6.8}$$

For the more general possible case, where we have stiffness terms for both the master and slave DOFs, we can transform our global stiffness equations in a form that only includes the master DOFs. What we will do is *partition* the global displacement vector, $\{U\}$, into two parts:

$$\{U\} = \begin{Bmatrix} \{U_{master}\} \\ \{U_{slave}\} \end{Bmatrix} \tag{10.6.9}$$

It is important to emphasize that, in the specific case, $\{U_{master}\}$ in Equation (10.6.9) also includes the DOFs that do not participate in any constraint equation!

We can now write both displacement vectors in terms of $\{U_{master}\}$ as follows:

$$\{U\} = \begin{Bmatrix} \{U_{master}\} \\ \{U_{slave}\} \end{Bmatrix} = \begin{Bmatrix} [I]\{U_{master}\} \\ [T]\{U_{master}\} \end{Bmatrix} \tag{10.6.10}$$

where we have accounted for Equation (10.6.1).

Equation (10.6.10) can be recast in the following, partitioned matrix form:

$$\{U\} = \begin{bmatrix} [I] \\ [T] \end{bmatrix} \{U_{master}\} = [E]\{U_{master}\} \tag{10.6.11}$$

where $[E] = \begin{bmatrix} [I] \\ [T] \end{bmatrix}$.

What we can do now is imagine that we assemble the global stiffness matrix, $[K]$, and the global equivalent nodal force vector, $\{F\}$. In that case, the stiffness matrix and nodal force vector can also be partitioned as follows.

$[K] = \begin{bmatrix} [K_{mm}] & [K_{ms}] \\ [K_{sm}] & [K_{ss}] \end{bmatrix}$, $\{F\} = \begin{Bmatrix} \{F_{master}\} \\ \{F_{slave}\} \end{Bmatrix}$, where the subscript m corresponds to the master DOF and subscript s stands for the slave DOF.

The final global stiffness equation, written in terms of the *master* DOFs and accounting for the constraints, is the following:

$$\left[\tilde{K}_{master}\right]\{U_{master}\} = \left\{\tilde{F}_{master}\right\} \tag{10.6.12}$$

where

$$\begin{aligned}\left[\tilde{K}_{master}\right] &= [E]^T[K][E] = \begin{bmatrix} [I] & [T]^T \end{bmatrix} \begin{bmatrix} [K_{mm}] & [K_{ms}] \\ [K_{sm}] & [K_{ss}] \end{bmatrix} \begin{bmatrix} [I] \\ [T] \end{bmatrix} \\ &= [K_{mm}] + [K_{ms}][T] + [T]^T[K_{sm}] + [T]^T[K_{ss}][T]\end{aligned} \tag{10.6.13}$$

and

$$\left\{\tilde{F}_{master}\right\} = [E]^T\{F\} = \begin{bmatrix} [I] & [T]^T \end{bmatrix} \begin{Bmatrix} \{F_{master}\} \\ \{F_{slave}\} \end{Bmatrix} = \{F_{master}\} + [T]^T\{F_{slave}\} \tag{10.6.14}$$

Equations (10.6.13) and (10.6.14) state that each row of $[K]$ and $\{F\}$ corresponding to the slave DOFs, $\{U_{slave}\}$, is premultiplied by the $[T]^T$ array and then added to the rows of the master DOFs, and each column of $[K]$ corresponding to $\{U_{slave}\}$ is post-multiplied by the $[T]$ array and then added to the columns of the master stiffness terms. In practice, the constraints are actually accounted for during the assembly operation, which is conducted by means of the *ID* and *LM* arrays, as described in Sections B.4 and B.5. In such case, the algorithm in Box 10.6.1 allows us to obtain $\left[\tilde{K}_{master}\right]$ and $\left\{\tilde{F}_{master}\right\}$ during the assembly operation.

Box 10.6.1 Algorithm to assembling global stiffness matrix in the presence of constraint equations

(We obtain $[K] = \left[\tilde{K}_{master}\right]$ and $\{F\} = \left\{\tilde{F}_{master}\right\}$, which correspond to the master DOFs)

If we have computed the stiffness matrix $[k^{(e)}]$ and the equivalent nodal force vector $\{f^{(e)}\}$ for each element e, then:

FOR each degree of freedom i, $i = 1, 2, \ldots, N_{de}$ of the element:

IF $I = ID(LM(e,i))$ is slave to several global degrees of freedom $m = 1, 2, \ldots$, then:

FOR each degree of freedom j, $j = 1, 2, \ldots, N_{de}$ of the element:

IF $J = ID(LM(e,j))$ is slave to several global degrees of freedom 'n', then:

$$K_{mn} = K_{mn} + T_{Im} \cdot k_{ij}^{(e)} T_{Jn}, \text{ for each } m \text{ and each } n$$

ELSE

$$K_{mJ} = K_{mJ} + T_{im} \cdot k_{ij}^{(e)} \text{ for each } m$$

ENDIF

$$F_m = F_m + T_{im} \cdot f_i^{(e)} \text{ for each } m$$

END

ELSE

FOR each degree of freedom j, $j = 1, 2, \ldots, N_{de}$ of the element:

IF $J = ID(LM(e,j))$ is slave to several global degrees of freedom n, then:

$$K_{in} = K_{in} + k_{ij}^{(e)} T_{Jn}, \text{ for each } n$$

ENDIF

END

ENDIF

END

Remark 10.6.1: The considerations in Box 10.6.1 apply for the case that we assemble the global stiffness matrix, $[K]$. As explained in Section B.5, in reality we only need assemble the global $[K_{ff}]$ array, corresponding to the unrestrained (free) DOFs. Thus, the algorithm of Box 10.6.1 must be appropriately combined with considerations based on the *ID* array of the master DOFs, to ensure that we only assemble the global equations of the unrestrained DOFs. ▌

Remark 10.6.2: It is worth mentioning that in an analysis, we are not allowed to have an essential boundary condition on a slave DOF of a constraint condition. Most analysis programs abort the solution with an error message when such situation is detected in a model. ▌

We will now finalize our discussion on constraints, focusing on two cases that appear very commonly in finite element analyses. The first case is that of *embedded elements*, wherein one or more elements and their nodal points are assumed to be embedded in the interior of a mesh comprised by another set of elements. As a specific example for a two-dimensional analysis, we present the case in Figure 10.6a, where we have a mesh of five elements. Elements (1) and (2) are four-node quadrilateral (4Q) continuum elements, while elements (3), (4), and (5) are two-node truss elements. Notice that there is no connectivity between nodes 7, 8, 9, and 10 that belong to the truss elements, and nodes 1, 2, 3, 4, 5, and 6 that belong to the quadrilateral elements. We want to enforce a set of *embedment constraint equations*, which simply stipulate that the displacements of the truss elements (the embedded-slave part) will be identical to the corresponding displacements of the host part (the master, continuum region). As an example, we will specifically focus on the equations for node 8. This node lies in the interior of the 4Q element (1). If the physical coordinates of node 8 are (x_8, y_8), then the constraint equation must mathematically express the following:

> "The nodal displacement vector $\{U_8\}$ of node 8 must be equal to the value of the displacement field of the host part, that is, of element (1), at the location (x_8, y_8)."

Thus, we can write the following equation:

$$\{U_8\} = \left[N^{(1)}\right]_{(x_8,y_8)} \left\{U^{(1)}\right\} \tag{10.6.15}$$

where $\left[N^{(1)}\right]_{(x_8,y_8)}$ is the (2 × 8) shape function array of element (1) evaluated at the location (x_8, y_8), and $\{U^{(1)}\}$ is the nodal displacement vector of element (1), that is, the eight-component vector containing the nodal displacements of nodes 1, 2, 3, and 4. Equation (10.6.15) is a special case of the constraint Equation (10.6.1), with $\{U_{slave}\} = \{U_8\}$, $\{U_{master}\} = \left\{U^{(1)}\right\}$, and $[T] = \left[N^{(1)}\right]_{(x_8,y_8)}$. We can similarly proceed with establishing constraint equations for the remaining three embedded-slave nodes 7, 9, and 10.

Remark 10.6.3: Just like in Section 10.1, we have a situation where the shape functions of the 4Q element (1) are typically established in the parametric space, but we need to find the values of the shape functions for a point with given *physical* coordinates (x_8, y_8). We can once again use the considerations of Remark 10.1.1 for this situation. ■

Remark 10.6.4: It must be emphasized here that an additional *detection algorithm* is necessary for the case of embedded elements, because a finite element program must be capable of determining the solid element that serves as a host for each node of the embedded elements. Of course, for simple models like the one in Figure 10.6a, we can use visual inspection and "manually" establish this information. ■

A second case that finds great applicability in practice is that of *tie constraints*. Such constraints allow different, adjacent regions of a model to be tied to each other, so that we do not have displacement discontinuities at the interface of the two regions. This is very useful for simulations where one region must be modeled with a relatively fine mesh, while the other region can be modeled with a coarser mesh. A schematic example for a tie constraint between two regions of a two-dimensional model is shown in Figure 10.6b. One of the two regions, having a larger element size (i.e., the coarser mesh), is treated as a master region, while the other region, with the smaller element size (i.e., the finer mesh), is treated as the slave region. What we do in this case is similar to the treatment of embedded elements. Specifically, we treat the displacements of each node of the slave region that lies on the interface as slaved to the displacements of the *corresponding segment* of the master region. By corresponding segment, we mean a straight line between two neighboring points of the master region. For example, node 8 in Figure 10.6b lies between nodes 4 and 5. Thus, we say that node 8 is slaved to the interface segment defined by nodes 4 and 5. Now, the constraint equations will be established, with the aid of Figure 10.6c. Let us imagine that we have a node s that must be constrained to the segment between two master nodes, A and B. What we will do is treat the segment AB as a one-dimensional isoparametric element, with the coordinate ξ shown in Figure 10.6.c. We are given the spatial (Cartesian) coordinates of nodal points A, B, and s. We need to find the value of the parametric coordinate ξ_s, which can be obtained from the following expression:

$$\xi_s = 2\frac{\|\vec{r}_{sA}\|}{\|\vec{r}_{AB}\|} - 1 \tag{10.6.16}$$

where $\|\vec{r}_{sA}\|$ is the distance between points s and A (see Equations 10.6.3 and 10.6.4) and $\|\vec{r}_{AB}\|$ is the distance between points A and B. The latter is, by definition, equal to the length of the segment AB. Now, we can use the shape functions for a one-dimensional, two-node, isoparametric element, to obtain:

$$\{U_s\} = \begin{Bmatrix} u_{xs} \\ u_{ys} \end{Bmatrix} = \frac{1-\xi_s}{2}\begin{Bmatrix} u_{xA} \\ u_{yA} \end{Bmatrix} + \frac{1+\xi_s}{2}\begin{Bmatrix} u_{xB} \\ u_{yB} \end{Bmatrix} = [T_{tie}]\{U_{master}\} \tag{10.6.17}$$

where

$$\{U_{master}\} = \begin{bmatrix} u_{xA} & u_{yA} & u_{xB} & u_{yB} \end{bmatrix}^T \tag{10.6.18a}$$

and

$$[T_{tie}] = \frac{1}{2}\begin{bmatrix} 1-\xi_s & 0 & 1+\xi_s & 0 \\ 0 & 1-\xi_s & 0 & 1+\xi_s \end{bmatrix} \tag{10.6.18b}$$

Once again, we have obtained a special case of Equation (10.6.1), with $\{U_{slave}\} = \{U_s\}$ and $[T] = [T_{tie}]$.

Remark 10.6.5: It is important to remember that, in tie constraints, the slaved surface must be the one with the *finer mesh*. Otherwise, problems will occur, and we may not be able to enforce the tie constraint. One schematic demonstration of what may go wrong for the case that the finer mesh is treated as the master region is presented in Figure 10.6d. One can verify that the slave region in the figure has a coarser mesh and also three slaved nodes, namely, nodes 4 (slaved to segment between nodes 7 and 8), 5 (slaved to segment between nodes 9 and 10), and 6 (slaved to segment between nodes 11 and 12). Notice from the figure that, since there is no constraint enforced for the segment between nodes 8 and 9 and the segment between nodes 10 and 11, these two segments cannot prevent the separation or interpenetration of the two regions. This specific example demonstrates the importance of having the finer mesh in the slave region. It is also worth mentioning that commercial software may include "smart algorithms," such as "two-way tied constraint," wherein the constraint equations for any two regions, 1 and 2, are established into two passes. In the first pass, region 1 is treated as master and region 2 is treated as slave. In the second pass, region 2 is the master and region 1 is the slave. This "two-pass" algorithm ensures that problems such as that in Figure 10.6d are precluded. ■

Remark 10.6.6: The considerations for the two-dimensional case can be extended to three-dimensions. The difference, then, is that the interface consists of quadrilateral or triangular surface segments. In this case, each node of the slaved region will be tied to a surface segment of the master region, and its displacements will be slaved to those of the nodes that comprise the master surface segment. ■

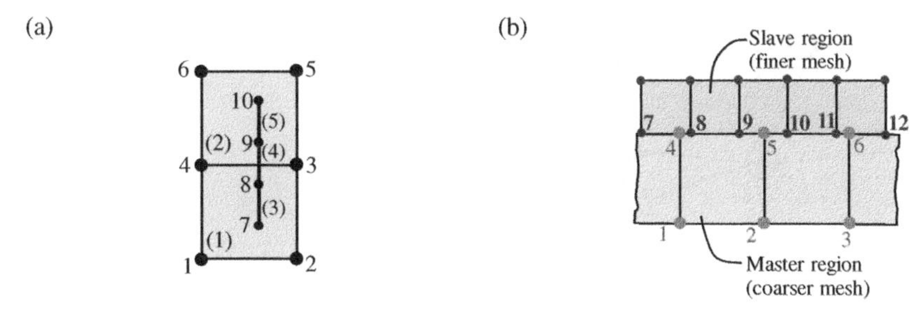

Truss elements (3), (4), and (5) embedded in solid 4Q elements (1) and (2)

Example tie constraint between regions with different element size

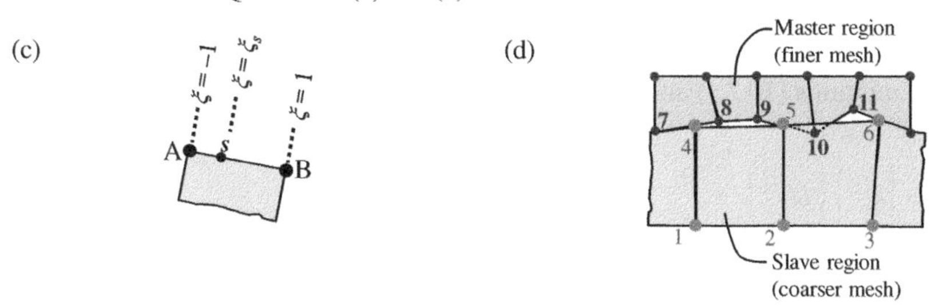

Treating a node s as slaved to a segment between two other nodes, A and B.

Problem in tie constraint where the mesh of master region is finer than that of slave region

Figure 10.6 Examples with embedment and tie constraints.

10.7 Treatment of Compliant (Spring) Connections Between Nodal Points

Another common occurrence in structural analysis is the presence of compliant connections between nodes. These connections allow a relative motion between the nodes, but they develop a connection force that resists this motion. One such situation is shown in Figure 10.7, where two nodes, 2 and 2′, coincide (i.e., they have the same exact coordinates) and they are connected with a spring along the x-axis. Although the nodes have the same exact location, we draw a close-up view with them spaced slightly apart, so that this allows us to illustrate the horizontal spring connecting them. We also assume that the displacements along the y-direction are identical for the two nodes.

The *spring axial force, f_s*, is equal to the product of the spring constant, k_s, times the spring elongation. The latter is simply equal to the difference in x-displacement between the two nodal points:

$$f_s = k_s(u_{x2} - u_{x2'}) = [k_s \quad -k_s]\begin{Bmatrix} u_{x2} \\ u_{x2'} \end{Bmatrix} \tag{10.7.1}$$

The force f_s is positive when it is tensile. Notice that this force is proportional to the *relative displacement of the two nodal points that the spring connects*. Given the spring axial force, we can find the forces that develop at the two nodal points of the spring element:

$$\left\{f_s^{(e)}\right\} = \begin{Bmatrix} f_{s2} \\ f_{s2'} \end{Bmatrix} = \begin{Bmatrix} -f_s \\ f_s \end{Bmatrix} \tag{10.7.2}$$

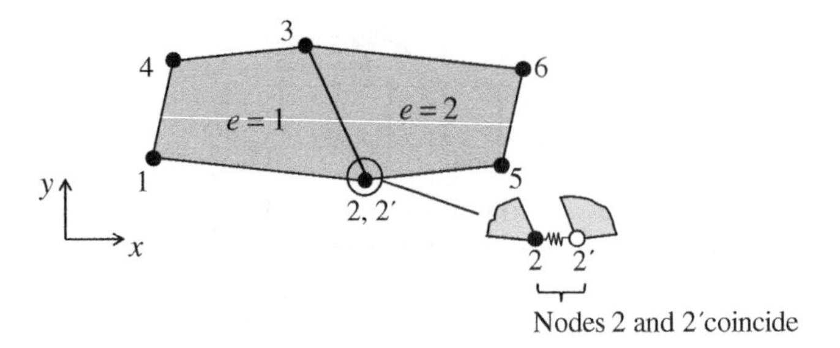

Figure 10.7 Example of Compliant Connection: Nodes 2 and 2′ are connected with a spring along the horizontal direction, while the displacement components in the *y*-direction are identical for the two nodes.

Combining Equations (10.7.1) and (10.7.2), we eventually obtain the stiffness equation for the spring element:

$$\left\{f_s^{(e)}\right\} = \left[k_s^{(e)}\right]\left\{U_s^{(e)}\right\} \tag{10.7.3}$$

where $\left[k_s^{(e)}\right] = \begin{bmatrix} k_s & -k_s \\ -k_s & k_s \end{bmatrix}$ and $\left\{U_s^{(e)}\right\} = \begin{Bmatrix} u_{x2} \\ u_{x2'} \end{Bmatrix}$

The nodal displacements of the spring element can be obtained from a gather operation, just like we did for the element displacements in previous chapters:

$$\left\{U_s^{(e)}\right\} = \left[L^{(e)}\right]\{U\} \tag{10.7.4}$$

We also need to enforce the condition that the two nodes have the same displacement along the *y*-direction. This is accomplished by writing:

$$u_{2'y} = u_{2y} \tag{10.7.5}$$

which is a constraint equation between nodal DOFs and can be treated in accordance with the descriptions in Section 10.6.

The procedure that we followed for the example case of a spring connection can be employed for any general compliant connection between nodal points. Of course, it is not necessary to have only one spring in the connection. More complicated compliant connections can be considered, where the behavior in the various directions is coupled. For example, we may have a connection wherein relative displacement in the *x*-direction causes forces not only along the *x*-axis but also along the other axes. This general situation can be described by a connection stiffness equation, defined by Equation (10.7.3). Of course, for general connections, the size of the vectors and arrays in Equation (10.7.3) must be modified to match the total number of DOFs affected by the connection. In the example that we considered, we could be having a coupled connection which affects both the *x*- and *y*-displacements of the connected nodal points. In such case, the "connection" element would have four DOFs (the two nodal displacements of each nodal point, 2 and 2′), and of course we would *not* enforce the constraint (10.7.5).

Remark 10.7.1: For obvious reasons, the spring connection between two nodal points that coincide (as is schematically shown in Figure 10.7) is referred to as a *zero-length* element. For the case of a single-spring connection along a specific direction (e.g., along the *x*-direction), the locations of the two connected nodes do not generally need to coincide. ■

Remark 10.7.2: Sometimes, we may hear the term *spring connection between elements.* This term implies exactly the same situation as that discussed here, for two nodal points of the connected elements. ∎

Remark 10.7.3: We must always ensure that, for compliant connections, we have equations that describe what happens for all DOFs in the connected nodal points. That is, for the example shown in Figure 10.7, we had the spring element connecting the x-displacements, and we added the constraint for equal y-displacements. If we had not added the latter constraint, this would mean that the two nodes, 2 and 2′, would be completely detached along the y-axis. Neglecting the constraint in the y-direction could lead to physically incorrect results, or it could even lead to a singular $[K_{ff}]$ array, rendering the global equations unsolvable! Still, it may sometimes be physically correct to allow completely free relative motion in several directions, as long as we are aware of this situation in our model and we think that it is a reasonable assumption for the problem at hand. ∎

Remark 10.7.4: There is some conceptual connection between supports and constraints or compliant connections. The major difference is that supports pertain to the displacements of a specific nodal point, while constraints pertain to the relative displacements of multiple nodal points. Along the same lines, the compliant (spring) supports give forces (tractions) that are proportional to the displacements at the nodes of the natural boundary, while the compliant (spring) connections give connection forces that are proportional to the relative displacements of the connected nodes. ∎

10.8 Symmetry in Analysis

An advantageous situation in analysis arises when the problem at hand is *symmetric.* This section is dedicated to the procedures by which we can take advantage of symmetry in an analysis. First of all, we need to provide a descriptive explanation of what we mean by symmetry. Let us consider the solid body shown in Figure 10.8. We imagine that we fold the figure about a plane, as shown in Figure 10.8. The body will be called *symmetric* about the specific plane, if, after folding, the part of the body that was on one side of the plane perfectly coincides with the part that was on the other side. The specific plane is termed the *plane of symmetry.* We can also extend the symmetry considerations in

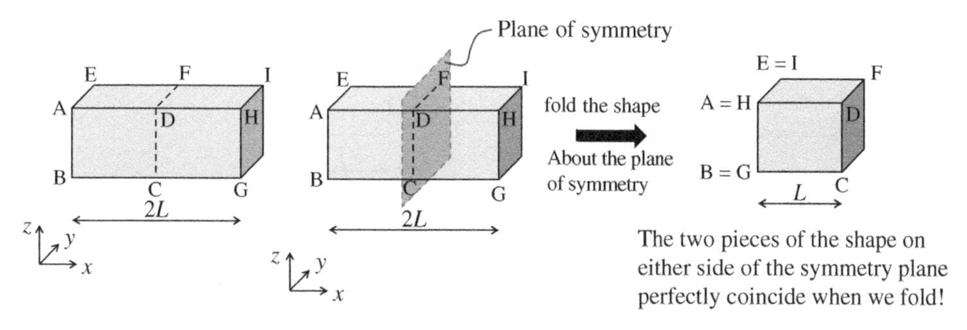

Figure 10.8 Schematic explanation of symmetry.

two-dimensional analyses, wherein we may speak of an *axis of symmetry*, as shown in Figure 10.9.

Now, besides the symmetry of the geometry of a system, we also need to discuss the symmetry of the loadings (or source terms) and boundary conditions. We will first lay emphasis on solid and structural mechanics, then we will generalize to other field problems such as heat conduction. We can distinguish two particular cases of load arrangement, allowing us to exploit symmetry. The first case corresponds to a symmetric structure with *symmetric loads*, as shown in Figure 10.10a. The second case corresponds to a symmetric structure with *antisymmetric loads*, as shown in Figure 10.10b. The term *antisymmetric loads* expresses the fact that, after folding the figure about the plane

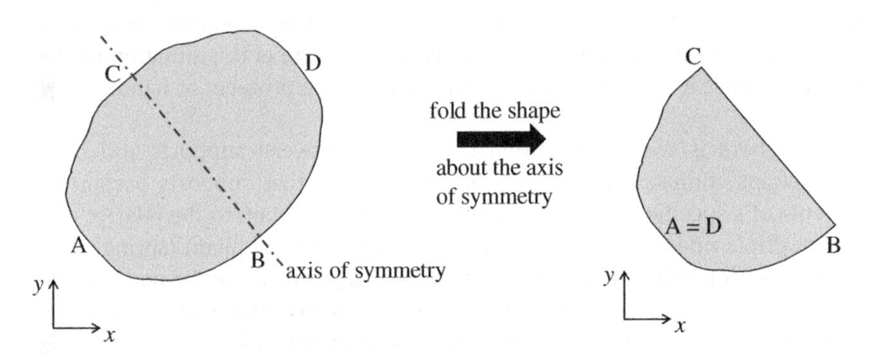

Figure 10.9 Symmetry for a two-dimensional solid.

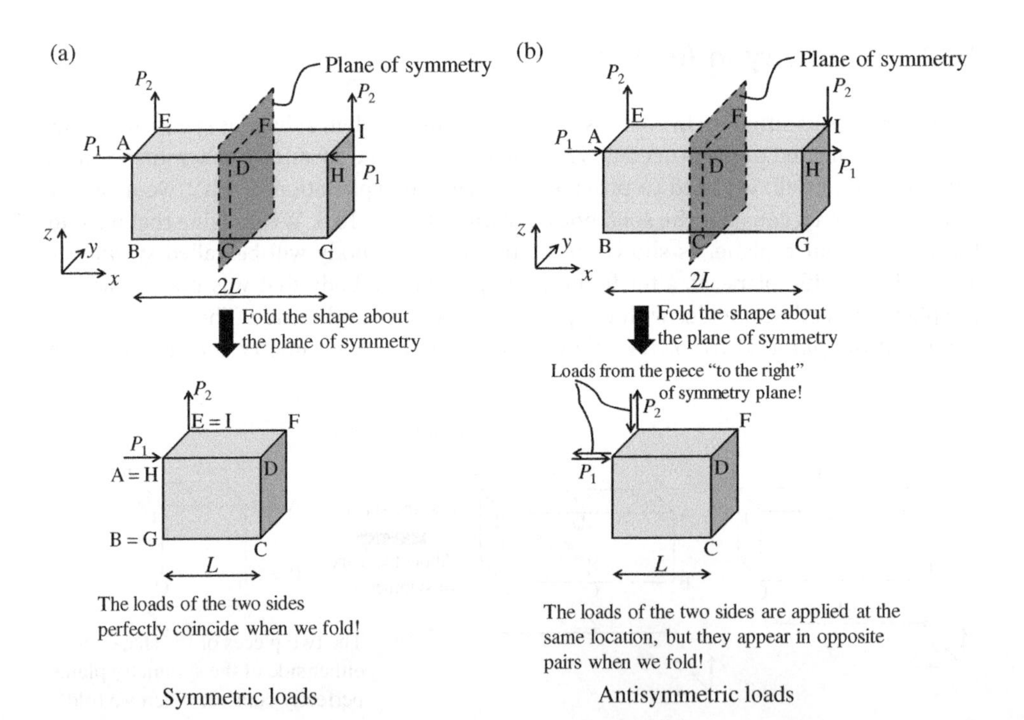

Figure 10.10 Cases of load symmetry.

(or axis) of symmetry, the loads on one side of the symmetry plane become equal and opposite to those on the other side. It is also worth mentioning that, to have "symmetric" or "antisymmetric" loads, the restraint (support) conditions such as prescribed displacements and the applied boundary tractions must also be symmetric or antisymmetric, respectively.

Whenever we have a symmetric solid with symmetric or antisymmetric loads, we are allowed to solve *only half of the system*—that is, the piece on either side of the symmetry plane or axis. The symmetry is accounted for with appropriate conditions at the intersection of the domain with the symmetry plane. The considerations that we need to use when solving half of the structure are summarized in Box 10.8.1, and they will not be proven here (they stem from mathematical analysis on symmetric domains).

As an example, let us consider the two-dimensional solid in Figure 10.11a. We can verify that the solid, the loadings, and the boundary conditions are symmetric about the y-axis. Thus, as shown in Figure 10.11a, we can conduct the analysis only for half of the solid (i.e., the piece to the right of the axis of symmetry). The rules in Box 10.8.1 lead to the realization that, along the symmetry axis, the displacements in the direction perpendicular to the axis—that is, the displacements in the x-direction— must be zero.

It is also worth mentioning that we may have two axes of symmetry, as shown in Figure 10.11b, where we have a thin tension coupon subjected to a stress of 30 ksi at the top and bottom edges. We can verify that the structure and loads exhibit double symmetry, with respect to both the x- and y-axis! This allows us to only analyze a quarter of the solid, as shown in Figure 10.11b. Symmetry conditions for the specific model lead to the requirement that, for the points lying on each symmetry axis, the displacements in the direction perpendicular to that symmetry axis vanish.

Box 10.8.1 Symmetric Problems in Elasticity

Whenever we have a *symmetric structure with symmetric loading*, then the following apply:

- The displacement fields are symmetric with respect to the plane (or axis) of symmetry.
- At every location on the plane (or axis) of symmetry, the displacements perpendicular to that plane (or axis) are equal to zero.
- If the plane (or axis) of symmetry is perpendicular to one of the coordinate axes (x, y, or z), then the distribution of normal strains and stresses is symmetric, while that of shear stresses and strains is antisymmetric.

Whenever we have a *symmetric structure with antisymmetric loading*, then the following apply:

- The displacement fields are antisymmetric with respect to the plane (or axis) of symmetry.
- At every location on the plane (or axis) of symmetry, the displacements parallel to that plane (or axis) are equal to zero.
- If the plane (or axis) of symmetry is perpendicular to one of the coordinate axes (x, y, or z), then the distribution of normal strains and stresses is antisymmetric, while that of shear stresses and strains is symmetric.

A question that may arise now is how we could exploit the symmetry for cases where the structure is symmetric but the loadings are not symmetric or antisymmetric. In such case, we can always consider the general (nonsymmetric) loading as the sum of a symmetric and an antisymmetric part, as shown in Figure 10.12. Doing so allows us to exploit symmetry and solve half of the structure. The only additional complexity stems from the need to conduct the analysis twice (first for the symmetric part of the loads, then for the antisymmetric part). Note that, for each of the two solutions, we will need to consider different restraints along the symmetry plane (or axis). In one case, the restraints will be those for symmetric loading (no displacements perpendicular to the symmetry plane); in the other case, the restraints will be those for antisymmetric loading (no displacements parallel to the symmetry plane). Solving half the structure can be advantageous for cases of very large structures, since we need to keep in computer memory (and invert) a much smaller stiffness matrix $[K_{ff}]$ than what we would have for solving the entire structure.

The notion of symmetry can be generalized to other physical problems. In such cases, instead of speaking of "symmetry of loads," we will be speaking of "symmetry of source terms and boundary conditions." One such example is presented in Figure 10.13, where we show a symmetric heat conduction problem with symmetric boundary conditions, wherein the axis y is the axis of symmetry. We can also have a source term, $s(x,y)$, but this term should also be symmetric. The descriptions of Box 10.8.2 apply for analysis of general mathematical problems with symmetric domain geometry, symmetric (or antisymmetric) *source terms*, and symmetric (or antisymmetric) boundary conditions.

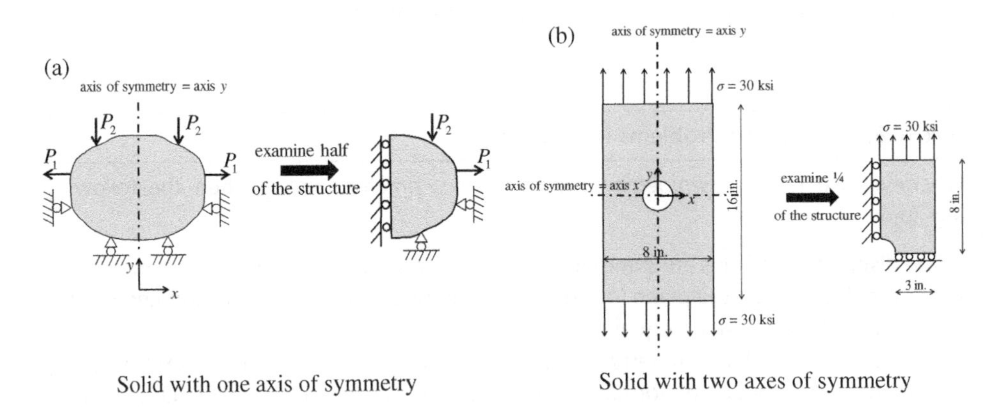

Figure 10.11 Example cases of load symmetry.

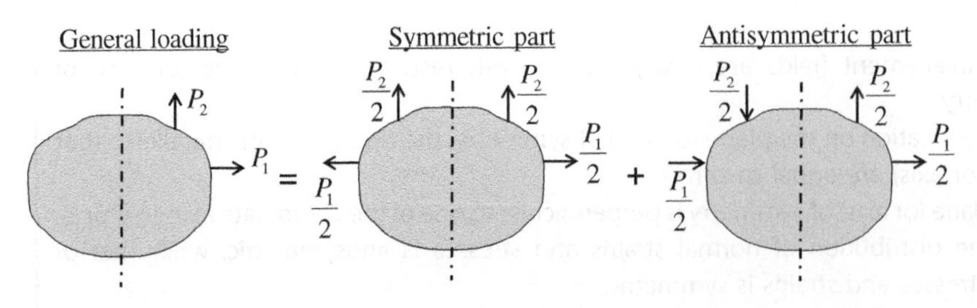

Figure 10.12 Consideration of general loading as the sum of symmetric and antisymmetric loads.

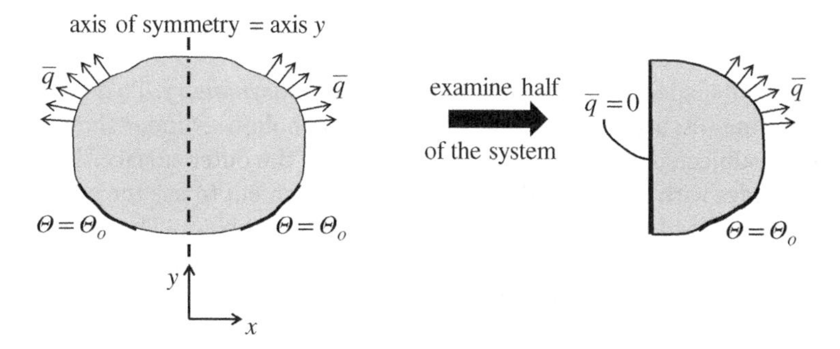

Figure 10.13 Example heat conduction problem with symmetry of geometry and boundary conditions about the *y*-axis.

Box 10.8.2 Generalization of Symmetric Problems for a Field Problem

Whenever we have a *symmetric domain with symmetric source terms and symmetric boundary conditions*, the following apply:

- The field function and its gradient vector field are symmetric with respect to the plane (or axis) of symmetry.
- At every location on the plane (or axis) of symmetry, the component of the gradient of the field function in the direction perpendicular to the plane (or axis) of symmetry is zero. In other words, the partial derivative of the field with respect to a coordinate aligned with the direction perpendicular to the plane (or axis) of symmetry is zero. A direct consequence of this condition for isotropic materials is that the axis of symmetry will be a natural boundary with prescribed zero outflow, $\bar{q} = 0$.

On the other hand, whenever we have a *symmetric domain with antisymmetric source terms and antisymmetric boundary conditions*, then the following apply:

- The field function and its gradient vector field are antisymmetric with respect to the plane (or axis) of symmetry.
- At every location on the plane (or axis) of symmetry, the component of the gradient of the field function in directions parallel to the plane (or axis) of symmetry must be zero. In other words, the partial derivatives of the field with respect to coordinates in the plane (or in the direction of the axis) of symmetry will be zero.

Remark 10.8.1: An observation that may have been made by this step is that all the figures and considerations provided in this section corresponded to the case where the axis of symmetry (for two-dimensional problems) or the axis perpendicular to the plane of symmetry (for three-dimensional problems) is parallel to one of the coordinate axes, x, y, or z. The question that may arise is how we could treat a problem where this is not the case. Although this problem can be treated computationally, the most efficient approach is still to take the coordinate axes such that one of them is perpendicular to the plane (or axis) of symmetry. This facilitates the finite element analysis exploiting the symmetry conditions in accordance with Boxes 10.8.1 and 10.8.2. ∎

10.9 Axisymmetric Problems and Finite Element Analysis

We now proceed with another special case of symmetry, namely, *axisymmetry*. To define the problem, let us examine—as a schematic example case—the hollow cylinder shown in Figure 10.14, which is subjected to a radial inward pressure on the outer surface. The axis of the cylinder coincides with the z-coordinate axis. It is convenient to use the polar coordinates r, θ, and z, explained in Section 7.12, for the specific problem. The cross-section of the cylinder lies in the r-θ plane. The cylinder of Figure 10.14 is *axisymmetric*, in the sense that the shape and dimensions of the section of the cylinder with *any* plane defined by the z-axis and the radial direction, r, is always the same. Such a section is shown in Figure 10.14. If the loading is also the same for each of these sections—as is the case for the cylinder in Figure 10.14—we can actually exploit this axisymmetry and only solve a two-dimensional slice (section) of the structure with a r-z plane.

In axisymmetric problems, the various quantities do *not* depend on the θ-direction. A direct consequence of this condition is that the following strain and stress components are zero.

$$\varepsilon_{r\theta} = \varepsilon_{z\theta} = 0 \tag{10.9.1a}$$

$$\sigma_{r\theta} = \sigma_{z\theta} = 0 \tag{10.9.1b}$$

Thus, for axisymmetric cases, the strain and stress vectors contain only four components:

$$\{\varepsilon\} = \begin{bmatrix} \varepsilon_{rr} & \varepsilon_{zz} & 2\varepsilon_{rz} & \varepsilon_{\theta\theta} \end{bmatrix}^T = \begin{bmatrix} \dfrac{\partial u_r}{\partial r} & \dfrac{\partial u_z}{\partial z} & \dfrac{\partial u_r}{\partial z} + \dfrac{\partial u_z}{\partial r} & \dfrac{u_r}{r} \end{bmatrix}^T \tag{10.9.2}$$

$$\{\sigma\} = \begin{bmatrix} \sigma_{rr} & \sigma_{zz} & \sigma_{rz} & \sigma_{\theta\theta} \end{bmatrix}^T \tag{10.9.3}$$

In Section (7.5), we obtained the (6 × 6) material stiffness matrix $[D]$, which can premultiply the strain vector to give the stress vector. The stress-strain law for linearly elastic material and axisymmetric behavior can be obtained, by eliminating the columns of $[D]$ corresponding to the zero strains and the rows of the same matrix corresponding to the zero stresses. Thus, we can write the following (4 × 4) material law for an axisymmetric elastic solid:

$$\begin{Bmatrix} \sigma_{rr} \\ \sigma_{zz} \\ \sigma_{rz} \\ \sigma_{\theta\theta} \end{Bmatrix} = \begin{bmatrix} D_{11} & D_{12} & D_{14} & D_{13} \\ D_{21} & D_{22} & D_{24} & D_{23} \\ D_{41} & D_{42} & D_{44} & D_{43} \\ D_{31} & D_{32} & D_{34} & D_{33} \end{bmatrix} \begin{Bmatrix} \varepsilon_{rr} \\ \varepsilon_{zz} \\ 2\varepsilon_{rz} \\ \varepsilon_{\theta\theta} \end{Bmatrix} \tag{10.9.4}$$

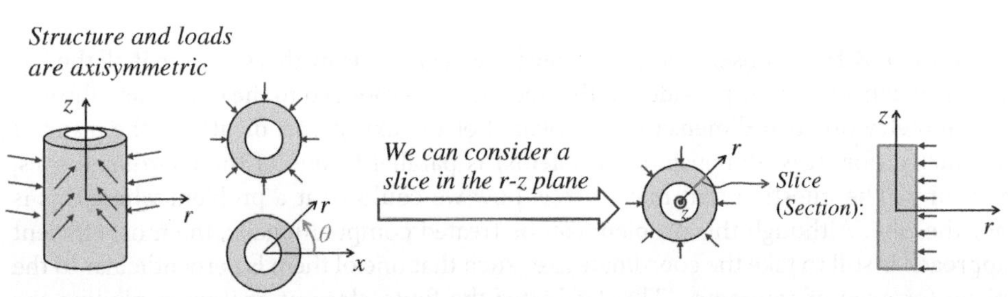

Figure 10.14 Example case of axisymmetric structure and loading.

Exploiting axisymmetry, we express any three-dimensional volume integral of any function $f(r,z)$ for the axisymmetric structure in the polar coordinates, and also account for the fact that the function does not depend on θ.

$$\iiint_{\Omega} f(r,z)dV = \iiint_{\Omega} f(r,z)drdz \cdot rd\theta = \int_0^{2\pi}\left(\iint_{\Omega_{2D}} f(r,z)\cdot rdrdz\right)d\theta = 2\pi\iint_{\Omega_{2D}} f(r,z)\cdot r\,drdz \tag{10.9.5}$$

where Ω_{2D} is the domain of the two-dimensional slice with the r-z plane that we consider to solve the axisymmetric problem.

Along the same lines, the boundary surface integral of any axisymmetric function $f(r,z)$ yields the following expression.

$$\iint_{\Gamma} f(r,z)dS = \int_0^{2\pi}\left(\int_{\Gamma_c} f(r,z)\cdot r\,d\ell\right)d\theta = 2\pi\int_{\Gamma_c} f(r,z)\cdot r\,d\ell \tag{10.9.6}$$

where Γ_c is the boundary curve of the two-dimensional slice with the r-z plane that we use for our solution.

If we now revisit the weak form, that is, the principle of virtual work, for a three-dimensional elastic solid (Box 9.1.1), and account for the fact that the structure is axisymmetric, we obtain the following *principle of virtual work equation for axisymmetric problems*:

$$2\pi\iint_{\Omega_{2D}}\{\bar{\varepsilon}\}^T[D]\{\varepsilon\}\cdot r\,drdz - 2\pi\iint_{\Omega_{2D}}\{w\}^T\{b\}\cdot r\,drdz - 2\pi\int_{\Gamma_C}\{w\}^T\{t\}\cdot r\,d\ell = 0 \tag{10.9.7}$$

Finally, dividing Equation (10.9.7) by (2π) yields the following expression.

$$\iint_{\Omega_{2D}}\{\bar{\varepsilon}\}^T[D]\{\varepsilon\}\cdot r\,drdz - \iint_{\Omega_{2D}}\{w\}^T\{b\}\cdot r\,drdz - \int_{\Gamma_C}\{w\}^T\{t\}\cdot r\,d\ell = 0 \tag{10.9.8}$$

Since Equation (10.9.8) corresponds to a two-dimensional r-z Cartesian space, the coordinates of all points (e.g., the nodal points) for axisymmetric problems are expressed through a pair of (r,z) values.

The conceptual steps used for the finite element analysis of three-dimensional elasticity are also valid for axisymmetric problems. Thus, we can establish a finite element approximation and discretization and obtain the following global stiffness equation:

$$[K]\{U\} = \{f\} \tag{10.9.9}$$

where the global displacement vector, $\{U\}$, contains two displacement components for each node, one along the radial direction r and another along the z-direction. That is, for a mesh with N nodes, we have the following $2N$ components in $\{U\}$:

$$\{U\} = \left[u_{1r}\ u_{1z}\ u_{2r}\ u_{2z}\ \ldots\ u_{Nr}\ u_{Nz}\right]^T \tag{10.9.10}$$

For an axisymmetric element e, the element stiffness matrix is given by the following expression:

$$\left[k^{(e)}\right] = \iint_{\Omega_{2D}^{(e)}} \left[B^{(e)}\right]^T \left[D^{(e)}\right] \left[B^{(e)}\right] \cdot r\, drdz \tag{10.9.11}$$

where the $[D^{(e)}]$ array is the (4×4) axisymmetric material stiffness matrix, and the strain-displacement array, $[B^{(e)}]$, is given by the following expression (for an element with n nodal points).

$$\left[B^{(e)}\right] = \begin{bmatrix} \frac{\partial N_1^{(e)}}{\partial r} & 0 & \frac{\partial N_2^{(e)}}{\partial r} & 0 & \cdots & \frac{\partial N_n^{(e)}}{\partial r} & 0 \\ 0 & \frac{\partial N_1^{(e)}}{\partial z} & 0 & \frac{\partial N_2^{(e)}}{\partial z} & \cdots & 0 & \frac{\partial N_n^{(e)}}{\partial z} \\ \frac{\partial N_1^{(e)}}{\partial z} & \frac{\partial N_1^{(e)}}{\partial r} & \frac{\partial N_2^{(e)}}{\partial z} & \frac{\partial N_2^{(e)}}{\partial r} & \cdots & \frac{\partial N_n^{(e)}}{\partial z} & \frac{\partial N_n^{(e)}}{\partial r} \\ \frac{N_1^{(e)}}{r} & 0 & \frac{N_2^{(e)}}{r} & 0 & \cdots & \frac{N_n^{(e)}}{r} & 0 \end{bmatrix} \tag{10.9.12}$$

Finally, the element equivalent nodal force vector is given by the following equation.

$$\left\{f^{(e)}\right\} = \iint_{\Omega_{2D}} \left[N^{(e)}\right]^T \left\{b^{(e)}\right\} \cdot r\, drdz + \int_{\Gamma_C} \left[N^{(e)}\right]^T \left\{t^{(e)}\right\} \cdot r\, d\ell = 0 \tag{10.9.13}$$

Remark 10.9.1: Equation (10.9.13) gives the equivalent nodal forces for *area-weighted axisymmetric element formulations.* The body forces and tractions have the same meaning and units as those for three-dimensional problems. That is, the body forces are forces per unit volume, while the boundary tractions are forces per unit surface area of the boundary. An alternative formulation that is available in commercial programs is that of a *volume-weighted axisymmetric element.* The only difference for volume-weighted formulations is that, instead of defining the body forces $\{b\}$ and boundary tractions $\{t\}$, we give the forces $\left\{\tilde{b}\right\} = \{b\} \cdot r$ and $\{\tilde{t}\} = \{t\} \cdot r$, respectively. These can be thought of as the loadings per radian of angle θ! For volume-weighted formulations, the equivalent nodal force vector of each element e is given by the following expression.

$$\left\{f^{(e)}\right\} = \iint_{\Omega_{2D}} \left[N^{(e)}\right]^T \left\{\tilde{b}^{(e)}\right\} drdz + \int_{\Gamma_C} \left[N^{(e)}\right]^T \left\{\tilde{t}^{(e)}\right\} d\ell = 0 \tag{10.9.14}$$

A simplification of Equation (10.9.14) compared to Equation (10.9.13) is that we no longer need to multiply by r inside the integrals that give the equivalent nodal force vector. ∎

Remark 10.9.2: Based on the above, an axisymmetric problem requires a two-dimensional mesh, corresponding to the domain of the slice. We can use any of the two-dimensional finite element formulations presented in Chapter 8 for axisymmetric problems, provided that we use the definitions of $[B^{(e)}]$ and $[D^{(e)}]$ presented in the current section (Equations 10.9.4 and 10.9.12). An additional difference is in the name of the axes. Instead of calling our Cartesian axes x and y, we call them r and z! Most finite element programs do not even make this distinction, and simply retain the names x and y for the axes of the axisymmetric mesh. The user may even have the capability to decide which one of the two axes is the *axis of symmetry*, that is, the z-axis of our axisymmetric case in Figure 10.14. ■

Remark 10.9.3: Although we will not discuss it here, the notion of axisymmetry is also applicable and can be exploited in scalar field problems, such as heat conduction. ■

10.10 A Brief Discussion on Efficient Mesh Refinement

Refining a finite element mesh (i.e., using more elements with smaller size in our analysis) generally leads to increased accuracy. This kind of refinement is called *h-refinement,* and it is distinguished from *p-refinement,* for which we retain the exact same mesh, but use higher-order polynomial shape functions in our elements. The p-refinement was popular in the earlier days of finite element analysis, when automated mesh generators did not have the same capabilities that they do today. In fact, clever and efficient *hierarchical approaches* had been established to increase the order of the approximation in each element without a need to re-mesh (and introduce new nodes).

Nowadays, especially in the case of elastic analyses, re-meshing and repetition of an analysis is relatively easy to do. It is still worth keeping some practical rules in mind to ensure an efficient mesh refinement, that is, to make sure that we use a fine mesh only at locations where it is necessary. A useful guideline is to pursue increased number of elements in regions where the field functions change abruptly as we move in the interior of the domain. In the context of elasticity, an "abrupt" change corresponds to *high gradients of strain and stress.* For regions where we expect the strain and stress values to vary abruptly (and their corresponding gradients are large), we will need a greatly refined mesh. On the other hand, for regions away from abrupt changes, a much coarser mesh will do the job.

A potential source of inaccuracies that commonly appears in finite element analyses is the use of elements with straight sides to represent curved boundaries. One would expect that the more straight-sided elements we use to represent a curve, the closer our mesh geometry will be to the correct one. Of course, a crude representation of the geometry is an additional source of inaccuracy (besides the discretization error).

This section will attempt a very brief explanation of the above considerations, by means of an example case presented in Figure 10.11b. We have a steel coupon subjected to a constant tensile stress (traction) at its two ends. The coupon has a circular crack (i.e., a circular hole) at its center. This crack is expected to lead to a *stress concentration*—that is, to an increase in the stress values in the vicinity of the crack compared to the corresponding values "away" from the crack. Specifically, for the case in Figure 10.15a, it may be obvious that away from the crack, we will have a stress $\sigma_{yy} = \sigma_o$ =30 ksi. However, the stress at a crack tip is given by the expression $\sigma_{yy} = K \cdot \sigma_o$, where K is called the *stress concentration factor.* For circular cracks, we have $K = 3$. The symmetry of the problem allows us to examine one quarter of the entire geometry, as shown in Figure 10.15a. The solution using a relatively coarse mesh, shown in Figure 10.15b, gives rather poor estimates of the stress concentration at point 1, which is at the tip of the crack. If we repeat the analysis with a mesh appropriately refined such that we have smaller and smaller elements as we get closer to the hole, we obtain the results of Figure 10.15c, from which we can deduce that we have very good estimate of the stress concentration at point 1. It is important to emphasize that the refinement of the mesh in Figure 10.15c is more pronounced in the vicinity of the circular crack.

The rather poor stress estimate (about half the correct value) obtained in Figure 10.15b for the σ_{yy} stress at point 1 is due to both the coarse mesh at a region where the stresses and strains are expected to change abruptly and to the crude representation of the geometry of the circular hole with two straight-line segments. The local refinement of the mesh in the vicinity of the hole tackles both these issues (we now have much smaller elements in the region of abrupt stress changes and also have a much more accurate representation of the circular geometry of the hole).

Now, let us check the stress at point 3 in Figure 10.15b, which is far from the circular crack and very close to the boundary where uniform traction is applied. We can immediately observe that even the coarse mesh can give very accurate estimates of the stress at that point. The reason is because point 3 is at a region with very "smooth" change of the solution (i.e., small gradients of strains and stresses), and thus a finer mesh is not necessary.

Finally, another point that is worth our attention is point 2 in Figure 10.15c, which is relatively close to the circular crack (hole). This point has a σ_{yy} – stress, which is very

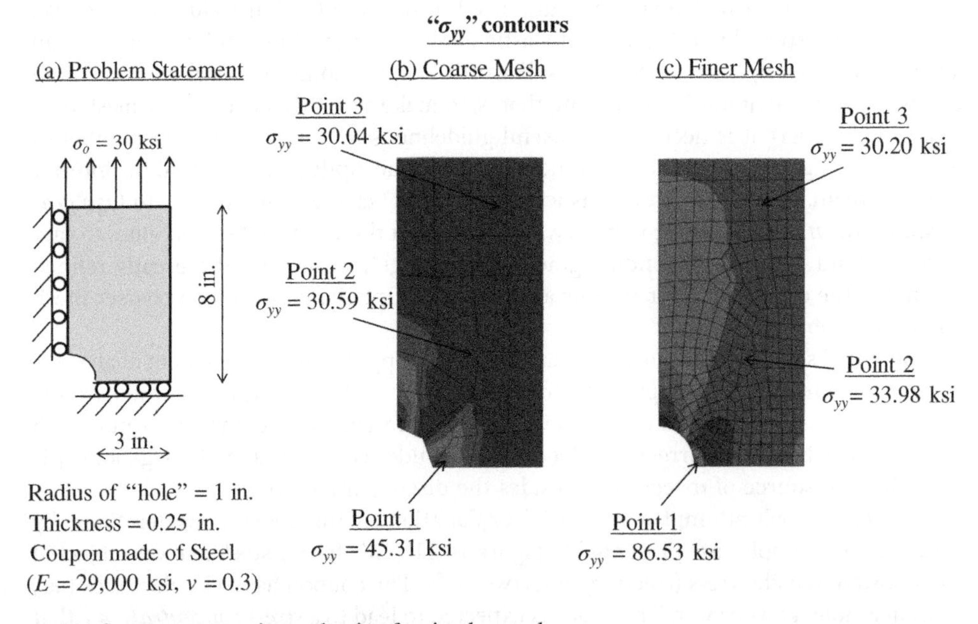

Figure 10.15 Analysis of a stress concentration problem using plane-stress finite elements and effect of mesh refinement.

close to the uniformly applied traction of 30 ksi at the top (for the refined mesh, the stress at point 2 is only 13% higher than the stress at point 3). The reason is that we can imagine the existence of the hole as a "disturbance" in a smooth stress distribution. This means that, if we did not have the circular crack, we would expect the stress σ_{yy} to be constant everywhere and equal to 30 ksi! The disturbance only has a significant effect on the stresses and strains in the vicinity of the crack's location. This means that σ_{yy} changes abruptly only in a small region, which is approximately equal to the diameter of the circular crack, that is, 2 inches.

This "local" nature of the effect of disturbances on the stresses and strains is a direct consequence of *Saint-Venant's principle* for linear elasticity. Saint-Venant's principle cannot be proven formally, but it can be observed in any problem in linear elasticity. It states that, if we have a disturbance (e.g., application of concentrated forces or change in geometry such as corners, cracks and holes) over a region with a size equal to D (for our example, D can be thought of as the diameter of the circular crack), then the effects of the disturbance become negligible at a distance (from the disturbance), which has the same order of magnitude as D (e.g., a value between D and $3D$). The principle of Saint-Venant is valid for any problem whose governing partial differential equations are *elliptic*. This is the case for all the problems examined in this text up to this point, including elasticity.

Considerations based on Saint-Venant's principle allow us to approximately determine the regions at which we expect to have effects from "disturbances" and thus may need a locally refined mesh to capture the possibly abrupt changes of the strain and stress components. Disturbances that might require a locally fine mesh include not only cracks, holes, and corners but also abrupt changes in material properties (e.g., the value of modulus of elasticity for an isotropic, linearly elastic material). Large gradients of stress and strain components are also associated with large values of body forces; a fine mesh is also necessary for such cases.

Problems

10.1 We have the beam shown in Figure P10.1, which has a thickness of 1 m, and for which we assume that plane-stress conditions apply. The beam is made of an isotropic, linearly elastic material for which $E = 30$ GPa, $\nu = 0.2$, $a_T = 10^{-5}/°C$,

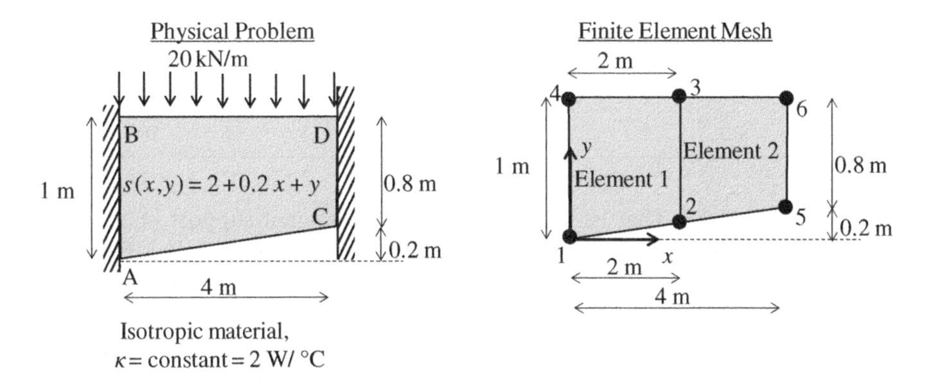

Figure P10.1

and $\Theta_o = 15\,°\text{C}$. **We do not know the actual temperature variation of the beam.** All we know is that the material has a constant conductivity $\kappa = 2\,\text{W/(°C·m)}$, that **the temperature at points A and B is kept equal to 15 °C**, that all the sides are insulated (meaning that the flux across these sides is zero!), and that we have the addition of heat due to an external source, given by

$$s(x,y) = 2 + 0.2x + y\left(x \text{ and } y \text{ are in meters, } s \text{ in W/m}^3\right).$$

Calculate the nodal displacements, and the stresses and strains at the Gauss points of the finite element mesh shown in Figure P10.1.

Hint: This is a *coupled* heat-elasticity problem. The nature of the coupling, though, is *one-way*: The temperatures affect the elasticity solution (in accordance with Section 10.2), but the displacements/stresses/strains *do not* affect the thermal behavior. Thus, we should be able to solve the two separate physical problems *sequentially*—that is, solve the first problem, then plug the solution from the first problem into the second problem as a known field.

10.2 We have the structure shown in Figure P10.2, which has a thickness of 0.4 m., and for which we can assume that plane-stress conditions apply. The structure is made of an isotropic, linearly elastic material for which $E = 30$ GPa, $\nu = 0.2$. We are also given the nodal coordinates of the finite element model, as provided in Table P10.1. There is also a distributed load (force per unit length) at the top of the structure. Solve the structure (i.e., find the nodal displacements), after taking advantage of the symmetry.

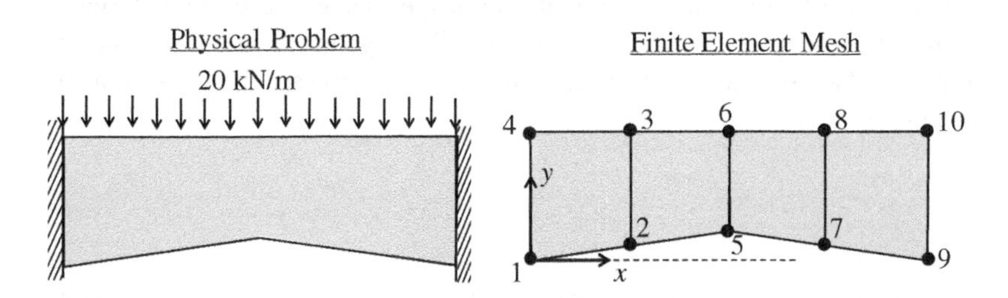

Figure P10.2

Table P10.1 Nodal Coordinates of Model Shown in Figure P10.2

Node "*i*"	1	2	3	4	5	6	7	8	9	10
x_i **(m)**	0	2	2	0	4	4	6	6	8	8
y_i **(m)**	0	0.1	1	1	0.2	1	0.1	1	0	1

10.3 Prove Remark 10.5.3. That is, if we have a point compliant support (a concentrated spring support), with a compliant support stiffness matrix $[k_s]$, then we need to add the spring stiffness coefficients to the corresponding global stiffness entries of the nodal point that is supported by the spring.

10.4 ***Note:*** Before solving Problem 10.4, the reader is advised to work on Problem 10.3.

We are given the finite element mesh shown in Figure P10.3, which includes two elements, 1 and 2. The thickness is equal to 6 in., and we assume plane-stress

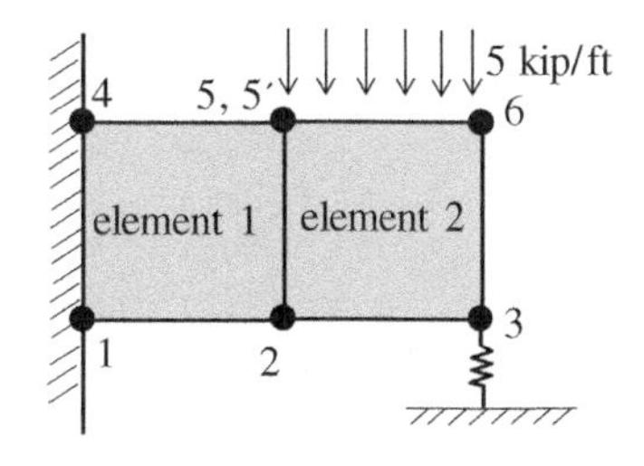

Node "i"	x_i (ft)	y_i (ft)
1	0	0
2	5	0
3	10	0
4	0	5
5	5	5
5′	5	5
6	4	1

Figure P10.3

conditions. The material is isotropic, linearly elastic, with $E = 6000$ ksi, $\nu = 0.2$. Element 1 contains nodes 1, 2, 5, and 4, and Element 2 contains nodes 2, 3, 6, and 5′. Nodes 5 and 5′ are at the same exact location, their displacements in the y-direction are the same, while they are connected with a spring connection along the x-direction, with stiffness equal to 2000 kip/in. Node 3 has a compliant (spring) support in the y-direction, with $k_{sy} = 1000$ kip/in.

a) Calculate the nodal displacements of the structure.
b) Calculate the support reactions (including the reaction of the elastic support at node 3).
c) Calculate the stresses at the Gauss points of element 1.

References

Bazeley, G., Cheung, Y., Irons, B., and Zienkiewicz, O. (1965). "Triangular Elements in Plate Bending—Conforming and Nonconforming Solutions." *Proceedings of the First Conference on Matrix Methods in Structural Mechanics*, Wright-Patterson Air Force Base, Dayton, Ohio.

Fish, J., and Belytschko, T. (2007). *A First Course in Finite Elements.* Hoboken, NJ: John Wiley & Sons.

Strang, G., and Fix, G. (2008). *An Analysis of the Finite Element Method*, 2nd ed. Cambridge, MA: Wellesley.

11

Convergence of Multidimensional Finite Element Analysis, Locking Phenomena in Multidimensional Solids and Reduced Integration

This chapter will consider several advanced topics on the use of finite elements in analysis of multidimensional elasticity. We will first discuss the convergence of multidimensional finite element analysis. We will then proceed to examine situations where the finite element method may give *overstiff* results for beam bending analysis and for *incompressible elasticity*. Commonly encountered remedies such as inclusion of *incompatible modes* in quadrilateral elements and *reduced integration* procedures will be discussed. Reduced integration may lead to an undesired side effect, namely, *spurious zero-energy modes or hourglass modes*, which may render the stiffness matrix singular, even if we have adequate support conditions in our structure. To eliminate potential problems associated with spurious zero-energy modes, we need to include *hourglass stiffness* in our elements. The chapter will be concluded with the description of methods that use reduced integration only for the part of the element stiffness that locks, for example, the stiffness related to volumetric strains for incompressible elasticity. Specifically, we will examine the selective-reduced integration procedures and the B-bar method, both of which are popular in analysis software.

11.1 Convergence of Multidimensional Finite Elements

In Section 3.8, we discussed the convergence of one-dimensional finite element formulations. We will now establish a more general rule for the convergence of the finite element method for multidimensional domains. We will specifically focus on three-dimensional elasticity, wherein the displacement field has three components, and we also have three spatial coordinates x, y, z for a domain Ω with volume equal to V_Ω. The one-dimensional and two-dimensional cases can be obtained by reducing the number of displacement components and spatial coordinates. Just as we did in Section 3.8, we need to establish appropriate norms of field functions for our discussion. Specifically, for any vector field $\{v(x,y,z)\} = \left[v_x(x,y,z) \;\; v_y(x,y,z) \;\; v_z(x,y,z)\right]^T$ we define the following three norms.

$$\|\{v\}\|_0 = \sqrt{\frac{1}{V_\Omega}\iiint\limits_{\Omega}\left[\left(v_x(x,y,z)\right)^2 + \left(v_y(x,y,z)\right)^2 + \left(v_z(x,y,z)\right)^2\right]dV} = \left[\frac{1}{V_\Omega}\iiint\limits_{\Omega}\sum_{i=1}^{3}\left(v_i^2\right)dV\right]^{0.5} \tag{11.1.1}$$

Fundamentals of Finite Element Analysis: Linear Finite Element Analysis, First Edition.
Ioannis Koutromanos, James McClure, and Christopher Roy.

Companion website: www.wiley.com/go/koutromanos/linear

$$\|\{v\}\|_1 = \left[\frac{1}{V_\Omega}\iiint_\Omega \left[\sum_{i=1}^{3}(v_i^2) + \sum_{k=1}^{3}\sum_{m=1}^{3}\left(\frac{\partial v_k}{\partial x_m}\right)^2\right]dV\right]^{0.5} \tag{11.1.2}$$

and

$$\|\{v\}\|_2 = \left[\frac{1}{V_\Omega}\iiint_\Omega \left[\sum_{i=1}^{3}(v_i^2) + \sum_{k=1}^{3}\sum_{m=1}^{3}\left(\frac{\partial v_k}{\partial x_m}\right)^2 + \sum_{n=1}^{3}\sum_{k=1}^{3}\sum_{m=1}^{3}\left(\frac{\partial^2 v_k}{\partial x_m \partial x_n}\right)^2\right]dV\right]^{0.5} \tag{11.1.3}$$

Now, let us imagine that we use finite element analysis for a three-dimensional problem, in which the exact displacement field is $\{u(x,y,z)\} = \left[u_x(x,y,z) \;\; u_y(x,y,z) \;\; u_z(x,y,z)\right]^T$. Let us also assume that the approximate displacement field, for a dimensionless mesh size parameter h, is $\{u_{(h)}(x,y,z)\} = \left[u_{(h)x}(x,y,z) \;\; u_{(h)y}(x,y,z) \;\; u_{(h)z}(x,y,z)\right]^T$. Parameter h can be thought of as the dimensionless ratio $\ell^{(e)}/\ell_o$ where $\ell^{(e)}$ is the element size[1] and ℓ_o is an arbitrarily selected "reference size value," as also described in Section 3.8. We can now define the solution (displacement) error vector field, $\{e_{(h)}(x, y, z)\}$, from the following expression.

$$\{e_{(h)}(x,y,z)\} = \{u(x,y,z)\} - \{u_{(h)}(x,y,z)\} \tag{11.1.4}$$

Equation (11.1.4) can also be written in component form:

$$e_{(h)x}(x,y,z) = u_{(h)x}(x,y,z) - u_x(x,y,z) \tag{11.1.5a}$$

$$e_{(h)y}(x,y,z) = u_{(h)y}(x,y,z) - u_y(x,y,z) \tag{11.1.5b}$$

$$e_{(h)z}(x,y,z) = u_{(h)z}(x,y,z) - u_z(x,y,z) \tag{11.1.5c}$$

The following rule applies for the error $\{e_{(h)}(x, y, z)\}$: For a "sufficiently smooth" exact displacement field (i.e., a displacement field with appropriate order of continuity), $\{u(x,y,z)\}$, we satisfy:

$$\left\|\{e\}_{(h)}\right\|_m \le C_{(m)} \cdot h^{k+1-m} \|\{u\}\|_{k+1} \tag{11.1.6}$$

where k is the maximum degree of completeness in the element shape functions, and $C_{(m)}$ is a constant (its exact value is not of primary importance).

Remark 11.1.1: Section 4.1 from the textbook by Hughes (2000) provides detailed explanations on the role of "smoothness" for satisfaction of inequality (11.1.6). ∎

Now, we are particularly interested in the norms with $m = 0$ and $m = 1$, that is, the *displacement error norm*, $\|\{e\}_{(h)}\|_0$, and the norm $\|\{e\}_{(h)}\|_1$, which includes first derivatives of the displacement field, and thus provides a measure of the *strain error* and *stress error* (since the stresses are merely the product of a given material stiffness matrix times

---

1 For a three-dimensional mesh, we can take $\ell^{(e)}$ as the maximum value of side length in the elements of a mesh, or even as the cubic root of the volume of the largest element in the mesh.

the strains, the errors of these two quantities are related). Setting $m = 0$ and $m = 1$ in Equation (11.1.6), we obtain:

$$\left\|\{e\}_{(h)}\right\|_0 \leq C_{(0)} \cdot h^{k+1} \|\{u\}\|_{k+1} \tag{11.1.7a}$$

$$\left\|\{e\}_{(h)}\right\|_1 \leq C_{(1)} \cdot h^{k} \|\{u\}\|_{k+1} \tag{11.1.7b}$$

For a given problem, the norm $\|\{u\}\|_{k+1}$ in Equations (11.1.5a, b) has a constant value (because the exact solution of a problem, $\{u(x,y,z)\}$, is unique). For the case that we use the lowest-possible order of shape functions (two-node elements for one-dimensional problems, four-node quadrilaterals or three-node triangles for two-dimensional problems, and eight-node hexahedra or four-node tetrahedral for three-dimensional problems), we have $k = 1$ (first-degree completeness). Thus, we can verify that for $k = 1$, we have a *quadratic rate of convergence* for the displacements, in the sense that, if the element size parameter h reduced by a factor of 2, then the norm $\|\{e\}_{(h)}\|_0$ would become $2^2 = 4$ times smaller. Similarly, we have a linear rate of convergence for the strains (and stresses).

Remark 11.1.2: The above considerations are based on the assumption that all the element integrals are exactly evaluated! In actual finite element computations, the *integrals are evaluated with numerical quadrature.* It can be proven that the minimum required order of quadrature to ensure the validity of inequality (11.1.6) is $\bar{k} + k - 2m$, where k is the degree of completeness of the finite element approximation, $m = 1$ (for solid mechanics) is the highest order of derivatives appearing in the weak form, and $\bar{k}$ is the maximum degree of monomials appearing in the isoparametric shape functions. The value of $\bar{k}$ is equal to 2 for 4Q elements, 3 for 8Q serendipity elements, and 3 for 8H hexahedral elements). ∎

Remark 11.1.3: We can extend the discussion of this section to scalar field problems. We can take the convergence rules of these problems by considering the special case of a vector field problem with a single component. ∎

Remark 11.1.4: The problems presented in the previous chapters all constitute *elliptic boundary value problems.* If we have an elliptic problem involving a field function $\upsilon(x,y,z)$, and a corresponding finite element solution $\upsilon_{(h)}(x,y,z)$, then the following equation applies:

$$\left\|e_{en(h)}\right\| \leq C \cdot h^k \tag{11.1.8}$$

where $\|e_{en(h)}\|$ is called the *energy error norm* and is defined from the following expression.

$$\left\|e_{en(h)}\right\| = \left[\frac{1}{V_\Omega}\int_\Omega \left\|\vec{\nabla}\upsilon_{(h)} - \vec{\nabla}\upsilon\right\|^2 d\Omega\right]^{1/2} \tag{11.1.9}$$

The energy error norm that was defined in Equation (3.8.5) can be regarded as a one-dimensional version of Equation (11.1.9). ∎

Further discussion on the error and convergence of finite element analysis is provided in Chapter 18.

11.2 Effect of Element Shape in Multidimensional Analysis

The behavior of an isoparametric element in an analysis strongly depends on the element shape. For example, in two-dimensional analysis, the closer the shape of a quadrilateral element gets to a rectangle, the better the behavior (i.e., accuracy) of that element. The same applies as the shape of a hexahedral element for three-dimensional analysis becomes more and more similar to a box. It is interesting to mention that sometimes, we may observe that the reduction in error norms as we refine the mesh is greater than that suggested by Equations (11.1.7a, b). This phenomenon is caused by the fact that a refinement of the mesh may make the shape of each element to get closer to the "optimal" one (rectangle for two-dimensions, box for three dimensions), and thus we may have a combined beneficial effect due to refined mesh and due to better element shape. Such an "element shape" effect does not exist, e.g., in one-dimensional analysis.

Very distorted element shapes, like the one shown in the physical space in Figure 11.1, should be avoided in analyses. In the specific example case, the Jacobian determinant of the mapping will be attaining negative (and also zero) values in the vicinity of node 4, so the $[B^{(e)}]$ array may not behave well: Some of its terms may approach infinity (*why?*). In general, quadrilaterals with angles exceeding 180°, as in the case for the angle at node 4 of Figure 11.1, should be avoided. Similarly, the shape of three-dimensional elements must not be overly distorted. Commercial software may include automated mesh generators with the capability to avoid the existence of severely distorted elements. However, for complicated geometries, we may still have situations where some of the elements have a severely distorted shape (at which case we may get a warning from the mesh generator).

An additional concern for higher-order isoparametric elements (e.g., eight-node serendipity quadrilaterals) is the position of the intermediate nodes at the element sides. These nodes should not be very far away from the middle of the line or curve joining the corresponding corner nodes; otherwise, the Jacobian of the isoparametric mapping (and, consequently, the $[B^{(e)}]$ array) may not have the desired behavior. In other words, the Jacobian may still become zero at locations of an element with intermediate nodes far from the midpoints of the corresponding sides.

Remark 11.2.1: The fact that placing the intermediate nodes of higher-order elements at "inappropriate locations" causes some components of the $[B^{(e)}]$ array to tend to infinity may be desirable for several types of analysis. For example, if we take an 8Q serendipity element and establish the nodal correspondence with the triangular six-node element shown in Figure 11.2, it can be proven that the strains and stresses at node 1 are linearly proportional to the term $1/\sqrt{r}$, where r is the distance (in the physical space) from

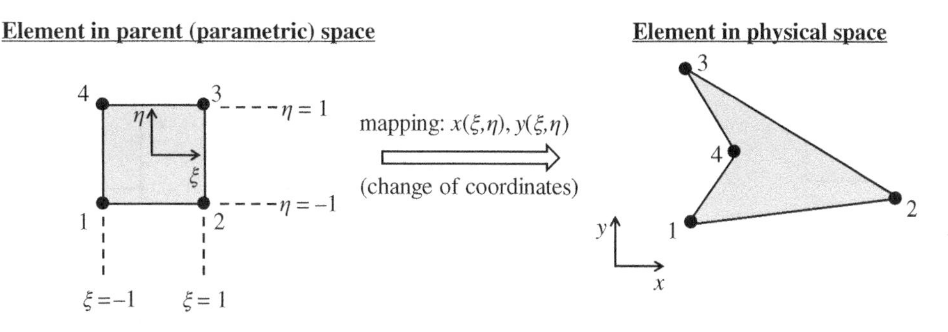

Figure 11.1 Severely distorted 4Q element shape.

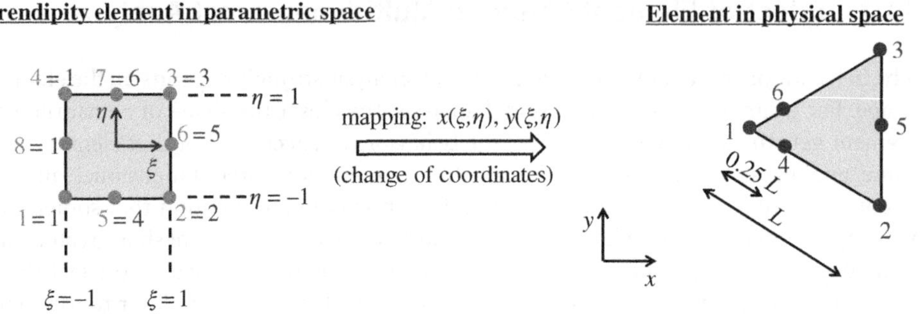

Figure 11.2 Triangular element with "stress singularity" at node 1, suitable for linearly elastic fracture mechanics (LEFM) simulations.

point 1. So, as we approach node 1, the strains and stresses have a *singularity*, that is, they tend to infinity. This kind of strain and stress singularities is found at the tip of line cracks in linearly elastic fracture mechanics (LEFM). For this reason, elements like the one in Figure 11.2 are often available in commercial finite element programs for LEFM simulations. ∎

11.3 Incompatible Modes for Quadrilateral Finite Elements

Quadrilateral finite elements are often used for simulations of slender beams (slender means that the cross-sectional depth is much smaller than the length of the member). Slender beams are deformed in such a way that the shear strains are very small and can be neglected, at which case the *Euler-Bernoulli beam theory* is valid, as explained in more detail in Section 13.3. According to this theory, any segment of a beam that is subjected to bending moments is deformed in such a fashion that plane sections of the beam remain plane after the deformation, and they also remain perpendicular to the deformed axis of the beam. A beam segment with this type of deformation is shown in Figure 11.3a. It is worth emphasizing that the shear strains for the specific deformed configurations are equal to zero. On the other hand, if we used a single 4Q element to capture the behavior of the specific slender beam segment, then our element would attain the deformed shapes shown in Figure 11.3b.

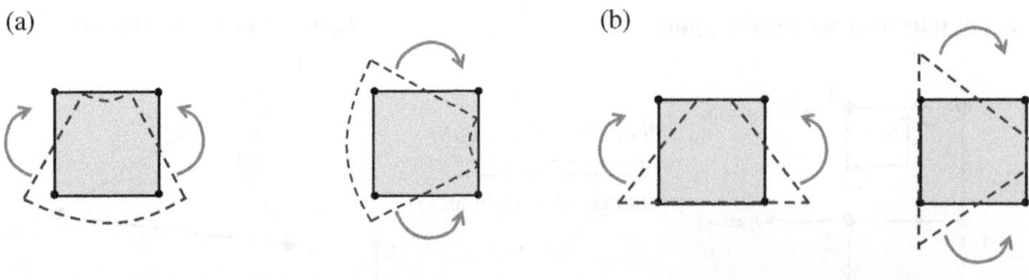

Figure 11.3 Deformed shapes for an actual beam segment and for a 4Q element undergoing bending.

The deformed shapes of the 4Q element in Figure 11.3b correspond to nonzero shear strains. The sections in Figure 11.3b remain planar, but are no longer perpendicular to the deformed axis of the beam. Since a portion of the deformations is bound to cause shear strains, the shear stiffness (which, for slender beams, is much higher than the flexural stiffness) is triggered, leading to overstiff behavior. This overstiff behavior is frequently referred to as *shear locking,* although a more appropriate term may be *parasitic shear stiffness*. It mainly stems from the inability of a single quadrilateral element to represent the deformed shape of a slender beam segment. Of course, the problem can be circumvented by refining the finite element mesh, so that we have at least four to six elements over the sectional depth of the beam.

To represent pure bending (zero-shear-strain) deformation modes with relatively few quadrilateral elements, professor Ed Wilson and his coworkers at the University of California–Berkeley, modified the formulation of a 4Q isoparametric element, establishing the following approximation for the displacement components:

$$u_x(\xi,\eta)=\sum_{i=1}^{4}\left(N_i^{(4Q)}(\xi,\eta)\cdot u_{xi}^{(e)}\right)+\underline{N_5(\xi,\eta)\cdot a_{x5}^{(e)}+N_6(\xi,\eta)\cdot a_{x6}^{(e)}} \tag{11.3.1a}$$

$$u_y(\xi,\eta)=\sum_{i=1}^{4}\left(N_i^{(4Q)}(\xi,\eta)\cdot u_{xi}^{(e)}\right)+\underline{N_5(\xi,\eta)\cdot a_{y5}^{(e)}+N_6(\xi,\eta)\cdot a_{y6}^{(e)}} \tag{11.3.1b}$$

The underlined quantities in Equations (11.3.1a, b) correspond to extra terms added in the 4Q element's approximation. The two shape functions $N_5(\xi,\eta)$, $N_6(\xi,\eta)$ in these terms are given by:

$$N_5(\xi,\eta)=1-\xi^2 \tag{11.3.2a}$$

$$N_6(\xi,\eta)=1-\eta^2 \tag{11.3.2b}$$

Remark 11.3.1: The shape functions in Equations (11.3.2a, b) are meant to reproduce an element geometry with "curved" edges, shown in Figure 11.3a. For the case where axes ξ and η coincide with axes x and y, the terms of Equations (11.3.1a, b) multiplied by $N_5(\xi,\eta)$ reproduce the left deformed shape of Figure 11.3a, while the terms multiplied by $N_6(\xi,\eta)$ reproduce the right deformed shape of the same figure. ∎

Remark 11.3.2: It is important to emphasize that *only the four shape functions of the isoparametric 4Q element are used for the mapping of the geometry.* In other words, the two additional shape functions of Equations (11.3.2a, b) are only used for the displacement approximation. ∎

We can now cast Equations (11.3.2a, b) in the following, partitioned matrix form.

$$\left\{u^{(e)}(\xi,\eta)\right\}=\left[\left[N^{(4Q)}(\xi,\eta)\right]\ \left[N_a^{(e)}(\xi,\eta)\right]\right]\begin{Bmatrix}\{U^{(e)}\}\\ \{a^{(e)}\}\end{Bmatrix} \tag{11.3.3}$$

where

$$\left[N^{(4Q)}(\xi,\eta)\right]=\begin{bmatrix}N_1^{(4Q)} & 0 & N_2^{(4Q)} & 0 & N_3^{(4Q)} & 0 & N_4^{(4Q)} & 0\\ 0 & N_1^{(4Q)} & 0 & N_2^{(4Q)} & 0 & N_3^{(4Q)} & 0 & N_4^{(4Q)}\end{bmatrix} \tag{11.3.4a}$$

$$\left[N_a^{(e)}(\xi,\eta)\right] = \begin{bmatrix} N_5^{(e)} & 0 & N_6^{(e)} & 0 \\ 0 & N_5^{(e)} & 0 & N_6^{(e)} \end{bmatrix} \tag{11.3.4b}$$

$$\left\{U^{(e)}\right\} = \left[u_{x1}^{(e)} \ u_{y1}^{(e)} \ u_{x2}^{(e)} \ u_{y2}^{(e)} \ u_{x3}^{(e)} \ u_{y3}^{(e)} \ u_{x4}^{(e)} \ u_{y4}^{(e)}\right]^T \tag{11.3.4c}$$

and

$$\left\{a^{(e)}\right\} = \left[a_{x5}^{(e)} \ a_{y5}^{(e)} \ a_{x6}^{(e)} \ a_{y6}^{(e)}\right]^T \tag{11.3.4d}$$

In addition to the nodal displacements, $\{U^{(e)}\}$, we have four parameters (degrees of freedom, or DOF) $a_{x5}^{(e)}, a_{y5}^{(e)}, a_{x6}^{(e)}, a_{y6}^{(e)}$ to specify for each element. These parameters do not correspond to nodal values of displacement; instead, they have the same conceptual meaning as the parameters a in the approximate solution that we established in Example 2.4! In other words, the four parameters can be thought of as "constants multiplying monomial terms in an approximate solution."

The four additional parameters $a_{x5}^{(e)}, a_{y5}^{(e)}, a_{x6}^{(e)}, a_{y6}^{(e)}$ of each element e do *not* affect the approximate displacement fields of the other elements in the mesh. *For this reason, they can be eliminated* (we can express their effect indirectly by modifying our equations accordingly) at the element level. By establishing the $[B^{(e)}]$ array in the same fashion as we did for the 4Q element in Section 8.5.2 (only now we have six shape functions), we can then calculate the element stiffness matrix $[k^{(e)}]$ and the element equivalent nodal force vector, $\{f^{(e)}\}$. The element's stiffness matrix can be partitioned into parts corresponding to $\{U^{(e)}\}$ and parts corresponding to the additional parameters $\{a^{(e)}\}$:

$$\left[k^{(e)}\right] = \begin{bmatrix} \left[k_{uu}^{(e)}\right] & \left[k_{ua}^{(e)}\right] \\ \left[k_{au}^{(e)}\right] & \left[k_{aa}^{(e)}\right] \end{bmatrix} \tag{11.3.5}$$

If we have body forces and boundary tractions in the element, the element's equivalent force vector $\{f^{(e)}\}$ is calculated by setting $N_5 = N_6 = 0$ in the array $[N^{(e)}]^T$ (the reason for doing this is left as an exercise). We can then exploit the fact that the additional displacement terms due to $\{a^{(e)}\}$ are only inside the element e (they do not affect the other elements of the mesh) and the right-hand-side vector corresponding to these degrees of freedom is zero. Specifically, Equation (3.3.26) for a 4Q element with incompatible modes obtains the following special form.

$$\begin{bmatrix} \left[k_{uu}^{(e)}\right] & \left[k_{ua}^{(e)}\right] \\ \left[k_{au}^{(e)}\right] & \left[k_{aa}^{(e)}\right] \end{bmatrix} \begin{Bmatrix} \{U^{(e)}\} \\ \{a^{(e)}\} \end{Bmatrix} = \begin{Bmatrix} \{f^{(e)}\} \\ \{0\} \end{Bmatrix} \tag{11.3.6}$$

which can be written as two separate expressions:

$$\left[k_{uu}^{(e)}\right]\left\{U^{(e)}\right\} + \left[k_{ua}^{(e)}\right]\left\{a^{(e)}\right\} = \left\{f^{(e)}\right\} \tag{11.3.6a}$$

and

$$\left[k_{au}^{(e)}\right]\left\{U^{(e)}\right\} + \left[k_{aa}^{(e)}\right]\left\{a^{(e)}\right\} = \{0\} \tag{11.3.6b}$$

We can solve Equation (11.3.6b) for $\{a^{(e)}\}$ to obtain:

$$\left\{a^{(e)}\right\} = -\left[k_{aa}^{(e)}\right]^{-1}\left[k_{au}^{(e)}\right]\left\{U^{(e)}\right\} \tag{11.3.7}$$

We can now obtain the following transformation equation for the element's degrees of freedom:

$$\left\{\begin{matrix}\{U^{(e)}\}\\ \{a^{(e)}\}\end{matrix}\right\} = \left\{\begin{matrix}\{U^{(e)}\}\\ -\left[k_{aa}^{(e)}\right]^{-1}\left[k_{au}^{(e)}\right]\{U^{(e)}\}\end{matrix}\right\} = \begin{bmatrix}[I]\\ -\left[k_{aa}^{(e)}\right]^{-1}\left[k_{au}^{(e)}\right]\end{bmatrix}\left\{U^{(e)}\right\}$$

$$\rightarrow \left\{V^{(e)}\right\} = \left[T^{(e)}\right]\left\{U^{(e)}\right\} \tag{11.3.8}$$

where $\{V^{(e)}\}$ is an expanded element DOF vector, given by the following expression.

$$\left\{V^{(e)}\right\} = \left\{\begin{matrix}\{U^{(e)}\}\\ \{a^{(e)}\}\end{matrix}\right\} \tag{11.3.9a}$$

and

$$\left[T^{(e)}\right] = \begin{bmatrix}[I]\\ -\left[k_{aa}^{(e)}\right]^{-1}\left[k_{au}^{(e)}\right]\end{bmatrix} \tag{11.3.9b}$$

Equation (11.3.8) can be thought of as a special case of *constraint equation*, as discussed in Section 10.6, and in particular Equation (10.6.11), with $\{U\} = \left\{V^{(e)}\right\}$, $\{U_{master}\} = \left\{U^{(e)}\right\}$, and $[E] = \left[T^{(e)}\right]$. Thus, we can write a transformation equation that will give us a stiffness matrix corresponding to the "master" DOFs alone, that is, an (8×8) stiffness matrix $\left[\tilde{K}_{master}\right] = \left[\tilde{k}_u^{(e)}\right]$, corresponding to $\{U^{(e)}\}$. If we use Equation (10.6.13), we can obtain:

$$\left[\tilde{k}_u^{(e)}\right] = \left[T^{(e)}\right]^T\left[k^{(e)}\right]\left[T^{(e)}\right] = \left[k_{uu}^{(e)}\right] - \left[k_{ua}^{(e)}\right]\left[k_{aa}^{(e)}\right]^{-1}\left[k_{au}^{(e)}\right] \tag{11.3.9}$$

We have thus obtained a *reduced* element stiffness matrix, $\left[\tilde{k}_u^{(e)}\right]$, which only includes terms corresponding to the nodal displacement vector, $\{U^{(e)}\}$, while indirectly accounting for the existence of the additional DOFs, $\{a^{(e)}\}$. This "reduction" of a stiffness matrix to include only a set of desired DOFs (in our case, the nodal displacements of the element) for the case that the "slave" DOF are given from an equation like (11.3.7), is called *static condensation,* and it is a popular procedure in matrix structural mechanics and dynamics. Accordingly, the matrix $\left[\tilde{k}_u^{(e)}\right]$ is referred to as the *condensed stiffness matrix of the element.* Static condensation for the specific quadrilateral element allows us to have a displacement vector $\{U\}$ in our global stiffness equations which only includes the nodal values. When we assemble the global stiffness matrix $[K]$, we use $\left[\tilde{k}_u^{(e)}\right]$ for each element. Once we solve the structure and find $\{U\}$, we can find the displacements of each element e using the gather operation $\{U^{(e)}\} = [L^{(e)}]\{U\}$, and then use equation (11.3.7) to find the values of $\{a^{(e)}\}$. We can then continue (if we wish) and determine the displacement, strain, and stress distributions inside our element.

Remark 11.3.3: The existence of the additional terms (with shape functions $N_5(\xi, \eta)$, $N_6(\xi, \eta)$) in the element displacement field leads to a "dangerous" situation: This type of element *does not satisfy continuity* of displacements across interelement boundaries. This is why these elements are referred to as *incompatible elements*, and the four extra deformation modes that we added in Equations (11.3.1a, b) are called *incompatible modes*. The reason why we use the term *incompatibility* is that the deformed mesh will generally be showing "overlapping" of material or "holes," which violate one of our basic assumptions for solid mechanics—namely, the compatibility condition. The formulation of the incompatible quadrilateral element belongs to the category of *variational crimes*, a term used by Strang and Fix (2008) to imply that one of the requirements for convergence of an approximate method based on a variational (i.e., weak) form is not met! Particularly, the specific type of elements does not satisfy C^0 continuity.

It turns out, however, that as we refine a mesh, the incompatible modes have smaller and smaller participation, and in the limit they vanish. To ensure that the incompatible elements pass the patch test, professor R. Taylor and his coworkers at the University of California–Berkeley established some practical modifications in the calculation of the $[B^{(e)}]$ array and the integration rules. The fact that the practical modifications allow the incompatible elements to pass the patch test led Strang and Fix (2008) to make a famous[2] statement: "Two wrongs do make a right in California," meaning that the combination of the (theoretically incorrect) incompatible modes with the (theoretically unjustified) practical modifications led to a well-behaved element. ■

As an example of the efficiency of quadrilateral elements with incompatible modes, we analyze a horizontal cantilever beam, shown in Figure 11.4a, using a relatively coarse mesh of 4Q, plane-stress elements, where we have only one element over the cross-sectional depth of the beam. We can see in Figure 11.4b that the parasitic shear stiffness of the 4Q elements leads to overstiff response—that is, to underestimation of the tip vertical displacements (by 32%). A much finer mesh using six 4Q elements over the cross-sectional depth leads to very good estimate of the tip displacement (10% error), as shown in Figure 11.4c. On the other hand, using the coarse mesh with quadrilateral elements including incompatible modes leads to a practically perfect agreement with the beam theory solution, as shown in Figure 11.4d.

11.4 Volumetric Locking in Continuum Elements

The stress and strain tensor in solid mechanics can be separated into two parts, namely, a *volumetric part* and a *deviatoric part*. The volumetric strain (for small strains) is given by:

$$\varepsilon_v = \varepsilon_{xx} + \varepsilon_{yy} + \varepsilon_{zz} = tr([\varepsilon]) \tag{11.4.1}$$

where $tr([\varepsilon])$ is called the *trace* of the strain tensor (the trace of a tensor is simply the sum of all its diagonal components).

The 3×3 strain tensor $[\varepsilon]$ can be written as follows:

$$[\varepsilon] = \frac{\varepsilon_v}{3}[I] + \begin{bmatrix} e_{xx} & e_{xy} & e_{xz} \\ e_{yx} & e_{yy} & e_{yz} \\ e_{zx} & e_{zy} & e_{zz} \end{bmatrix} = \frac{\varepsilon_v}{3}[I] + [e] \tag{11.4.2}$$

2 A clarification is needed: famous within the finite element community!

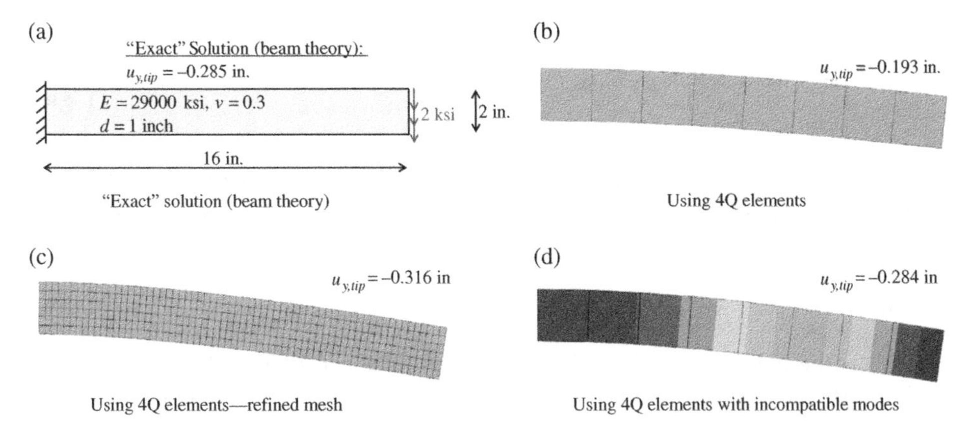

Figure 11.4 Solution of a cantilever beam using quadrilateral elements.

where $\frac{\varepsilon_V}{3}[I]$ is the *volumetric part* of the strain tensor ($[I]$ is a 3 × 3 identity array) and $[e]$ is the *deviatoric part* of the strain tensor. Each component of the deviatoric strain tensor is given by:

$$e_{ij} = \varepsilon_{ij} - \frac{\varepsilon_V}{3} \cdot \delta_{ij} \tag{11.4.3}$$

where δ_{ij} is the Kronecker delta. Equation (11.4.3) can also be cast in the following form.

$$e_{ij} = \begin{cases} \varepsilon_{ij} - \frac{\varepsilon_V}{3}, \ if\ i = j \\ \varepsilon_{ij}, \ if\ i \neq j \end{cases} \tag{11.4.4}$$

The volumetric stress is given in terms of the trace of the stress tensor, $tr[\sigma]$, as follows.

$$\sigma_V = \frac{\sigma_{xx} + \sigma_{yy} + \sigma_{zz}}{3} = \frac{tr([\sigma])}{3} = -p \tag{11.4.5}$$

The value p in Equation (11.4.5) is the *pressure* (note that it is defined so that it is positive in compression!).

Remark 11.4.1: Notice that we divide the trace of $[\sigma]$ by 3 to obtain the volumetric stress, while we do not do so when we define the volumetric strain. ∎

We can now write the stress tensor as the summation of a volumetric part and a deviatoric part:

$$[\sigma] = \sigma_V[I] + \begin{bmatrix} s_{xx} & s_{xy} & s_{xz} \\ s_{yx} & s_{yy} & s_{yz} \\ s_{zx} & s_{zy} & s_{zz} \end{bmatrix} \tag{11.4.6}$$

Each component of the deviatoric stress tensor, $[s]$, is defined by the following expression.

$$s_{ij} = \sigma_{ij} - \sigma_V \cdot \delta_{ij} \tag{11.4.7}$$

or

$$s_{ij} = \begin{cases} \sigma_{ij} - \sigma_v \cdot \delta_{ij} = \sigma_{ij} + p \cdot \delta_{ij}, \; if \; i = j \\ \sigma_{ij}, \; if \; i \neq j \end{cases} \quad (11.4.8)$$

The elastic constitutive equations *for isotropic linear elasticity* (described in Section 7.5) can be separated into two independent parts. The first part corresponds to the volumetric constitutive law:

$$\sigma_v = \frac{3\lambda + 2\mu}{3} \varepsilon_v = K_V \cdot \varepsilon_v \quad (11.4.9)$$

where K_V is the material's *bulk modulus*. This modulus can be expressed in terms of Lamé's constants, λ and μ, or in terms of modulus of elasticity E and Poisson's ratio ν:

$$K_V = \frac{3\lambda + 2\mu}{3} = \frac{E}{3(1-2\nu)} \quad (11.4.10)$$

The second part of the constitutive equation corresponds to the deviatoric stress-strain law:

$$s_{ij} = 2\mu \cdot e_{ij} = 2G \cdot e_{ij} \quad (11.4.11)$$

A material is called *incompressible* when it deforms in such a fashion that $\varepsilon_v = 0$. Thus, *for an incompressible material, the strain tensor is purely deviatoric!* In reality, there are no materials that are completely incompressible, but there are materials that are practically incompressible; that is, they satisfy $\varepsilon_v \to 0$. This is the case for materials in which the deviatoric material stiffness, quantified by $\mu = G$, is much lower than the volumetric material stiffness, quantified by K_V. One example is water and fluids under "slow loading" conditions, for which the shear stiffness is negligible compared to the volumetric material stiffness, and thus the deformation process is practically equivolumetric and the material can be treated as incompressible. There are also cases in solid mechanics where incompressibility is relevant: specifically, most metals (such as steel) used in building construction, when deformed in the inelastic regime, undergo straining in an almost equivolumetric fashion. To describe these materials, we separate the strain tensor into two parts, an elastic part $[\varepsilon^{(el)}]$ and a plastic part $[\varepsilon^{(pl)}]$:

$$[\varepsilon] = \left[\varepsilon^{(el)}\right] + \left[\varepsilon^{(pl)}\right] \quad (11.4.12a)$$

Then, by stipulating that the plastic part of the strain tensor is equivolumetric, we reach the conclusion that the plastic strain tensor is a deviatoric tensor (i.e., the volumetric part of that tensor is zero):

$$\varepsilon_v^{(pl)} = 0, \text{ thus } [\varepsilon^{(pl)}] = [e^{(pl)}] \quad (11.4.12b)$$

Now, due to the fact that the plastic strain magnitudes are typically much greater than the elastic ones (e.g., 10 times greater or even more), we reach the conclusion that the response of elastoplastic metals will be near-incompressible. Based on the above, incompressibility (or rather near-incompressibility) is an important aspect of finite element analysis that should concern practitioners and researchers working on solid and structural mechanics. We can also establish incompressibility or near-incompressibility for a linearly elastic isotropic material, by setting $\nu \to \frac{1}{2}$. In this case, K_V tends to infinity and the volumetric stress can no longer be calculated from Equation (11.4.9).

A question that may arise now is how our finite element equations should be modified to account for incompressibility. It turns out that incompressibility introduces *constraint equations,* involving the nodal displacements of the mesh. For example, let us consider the example finite element mesh shown in Figure 11.5. The mesh includes a total of eight constant-strain triangular (CST) elements. Each element has a single quadrature point, shown in the figure. The nodal displacements of nodes 1, 2, 3, 4, and 7 are restrained (fixed) to a zero value. Thus, the vector with the unrestrained degrees-of-freedom, $\{U_f\}$, would have a total of eight components (two displacements for each of the four unrestrained nodal points). Now, the requirement for incompressibility leads to equations demanding that we have zero volumetric strain for each location where we calculate the strains. These locations are simply the quadrature points of the elements. For example, for element (1), we can write that:

$$\varepsilon_v^{(1)} = 0 \rightarrow \left[b_{vol}^{(1)}\right]\left\{U^{(1)}\right\} = 0 \tag{11.4.13}$$

where $\left[b_{vol}^{(e)}\right]$ is a 6×1 row vector[3], giving the volumetric strain of the quadrature point of each element e, in terms of the nodal displacements of the same element. We can similarly write down another seven constraint equations, for the quadrature point of each one of the seven other elements in the mesh.

Remark 11.4.2: Similar to the above considerations, when we have 4Q elements, and we use four-point Gaussian quadrature, then we must satisfy the incompressibility (or near-compressibility) condition at four distinct locations. This would mean that, in addition to our global stiffness equations, we would have an additional four constraint equations per element, for the unrestrained nodal displacements. ∎

Now, a very interesting situation arises for the example in Figure 11.5: due to the fact that we have a total of eight constraint equations for the nodal displacements, and we have a total of eight unrestrained nodal displacements, the only way to satisfy all eight constraint equations is to have $\{U_f\} = \{0\}$! That is, even though we have applied forces on our mesh, we will have zero displacements everywhere, due to the fact that there is an

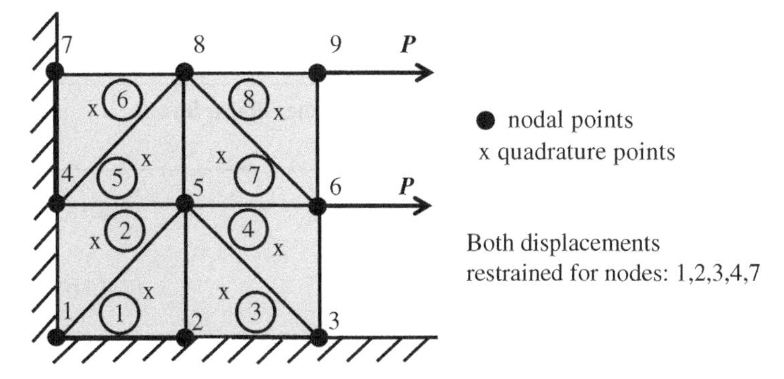

Figure 11.5 Example case where an incompressible material would lead to complete locking (zero displacements for all the unrestrained nodes) due to incompressibility constraint equations.

3 See Remark 11.8.1 on how to obtain the $\left[b_{vol}^{(e)}\right]$ vector.

excessive number of incompressibility constraint equations! This is an extreme case of the situation where the additional constraints due to incompressibility may lead to overly stiff solutions (i.e., solutions where the displacements are severely underestimated or they are even equal to zero everywhere). This situation is termed *volumetric locking*. In general, the term *locking* in finite elements refers to overly stiff solutions for cases that we need to satisfy kinematic constraints (such as zero volumetric strain) in our solution. Remember that we also encountered "shear locking" in Section 11.3.

One quick, practical way to determine whether an element will have a propensity to lock is the *constraint counting* method (Hughes 2000), based on a *constraint ratio R*, that is, the number of nodal equilibrium equations (which is equal to the number of unrestrained nodal displacements) divided by the number of incompressibility constraints. The optimal value of R in a finite element solution can be obtained by establishing the governing differential equations of incompressible problems. For example, in a two-dimensional body with incompressible material and plane-strain conditions, we have two differential equations of equilibrium and one incompressibility constraint for the displacement field, as shown in Box 11.4.1. The ideal (optimal) value of R is equal to the ratio of the number of differential equations over the number of constraint equations in the strong form of the problem. In Box 11.4.1, we have two differential equations and one constraint equation; thus, the optimal value of R is 2. For three-dimensional incompressible solids, we have three differential equations of equilibrium and one incompressibility constraint; thus, the optimal value of R is equal to 3.

If we have a finite element mesh, we need to count the number of unrestrained nodal displacements, N_f, and the number of incompressibility constraints, N_c. The latter is simply equal to the total number of Gauss (quadrature) points in the elements of the mesh. Then, we can calculate:

$$R = \frac{N_f}{N_c} \tag{11.4.14}$$

For plane-strain analysis, experience suggests that:

- If $R = 2$, we have an optimal condition (we will not have locking in general).
- If $1 < R \leq 2$, we have an *overconstrained mesh* (more constraints than those corresponding to optimal behavior, and we will have *overstiff* response).

Box 11.4.1 Governing Differential Equations for Incompressible, Plane-Strain Solid Mechanics

$$\frac{\partial \sigma_{xx}}{\partial x} + \frac{\partial \sigma_{xy}}{\partial y} + b_x = 0 \tag{7.8.1a}$$

$$\frac{\partial \sigma_{yx}}{\partial x} + \frac{\partial \sigma_{yy}}{\partial y} + b_y = 0 \tag{7.8.1b}$$

Incompressibility constraint: $\varepsilon_V = \varepsilon_{xx} + \varepsilon_{yy} = \frac{\partial u_x}{\partial x} + \frac{\partial u_y}{\partial y} = 0$

Thus, we have two (2) differential equations and one (1) constraint equation, therefore, $R = 2$.

- If $R \leq 1$, we have a locking mesh (much more constraints than those corresponding to optimal behavior; these excessive constraints lead to severe locking, which may even give a solution with zero displacements everywhere, despite the fact that we apply loads on the unrestrained degrees of freedom).
- If $R > 2$, this means that we have an underconstrained mesh (meaning that the incompressibility constraint will not be satisfied everywhere exactly!). This case is more of a theoretical consideration, since it is unlikely to occur in practice.

11.5 Uniform Reduced Integration and Spurious Zero-Energy (Hourglass) Modes

One obvious approach to avoid the overstiff, locking behavior for incompressible or near-incompressible materials, is to use *reduced integration*. More specifically, since we saw in the previous Section that the number of incompressibility constraints in our analysis is equal to the number of quadrature points in the mesh, we can obviously use less quadrature points to reduce the number of constraints. In the *uniform reduced integration (URI)* procedure, the number of quadrature points in each direction is one-order lower than that required for full integration (discussed in Section 6.5). For example, for the 4Q isoparametric element, a one-point (1×1) quadrature is used instead of the standard, four-point (2×2), full integration. So, if we use one Gauss point in two-dimensional analysis, we have $\xi_1 = 0$, $\eta_1 = 0$, $W_1 = 4$. Similarly, for a 9Q Lagrangian or 8Q serendipity quadrilateral element, four quadrature points (2×2) are used instead of the nine (3×3) required for full integration. The use of URI was originally proposed by Zienkiewicz et al. (1971) to alleviate locking problems in analysis of shells (shell theory and finite element analysis are discussed in Chapter 14).

The use of URI leads to an undesired side effect: *rank deficiency* of the stiffness matrix $[K_{ff}]$, associated with the existence of *spurious zero-energy modes*. Remember that a two-dimensional 4Q solid element (unrestrained) has eight degrees of freedom. So, the element can exhibit eight independent displaced configurations. Of these configurations, three correspond to rigid-body motion (two translations and one rotation). Thus, a quadrilateral element must have five distinct modes of deformation (i.e., five independent displacement configurations must lead to the development of strains and stiffness in the element). In the context of stiffness matrices, the number of independent displaced configurations to which the element can develop stiffness is called the *rank* of the stiffness matrix (the more generic definition of rank in a square matrix is equal to the number of nonzero eigenvalues that the matrix has). To ensure a good understanding of the rank and the considerations presented below, the reader is strongly encouraged to read Section A.4.

Now, let us examine the case of uniform reduced integration in a (4Q) element. The stiffness matrix is given by:

$$\left[k^{(e)}\right] = \int_{-1}^{1}\int_{-1}^{1}\left(\left[B^{(e)}\right]^T\left[D^{(e)}\right]\left[B^{(e)}\right]Jd\xi d\eta\right) \approx [B_1]^T[D_1][B_1]J_1 \cdot 4 \tag{11.5.1}$$

For a single quadrature point at $\xi = 0$, $\eta = 0$, we can establish an expression for the strain-displacement matrix $[B_1]$ at that point:

$$[B_1] = \frac{1}{2A^{(e)}} \begin{bmatrix} y_{24}^{(e)} & 0 & y_{31}^{(e)} & 0 & -y_{24}^{(e)} & 0 & -y_{31}^{(e)} & 0 \\ 0 & x_{42}^{(e)} & 0 & x_{13}^{(e)} & 0 & -x_{42}^{(e)} & 0 & -x_{13}^{(e)} \\ x_{42}^{(e)} & y_{24}^{(e)} & x_{13}^{(e)} & y_{31}^{(e)} & -x_{42}^{(e)} & -y_{24}^{(e)} & -x_{13}^{(e)} & -y_{31}^{(e)} \end{bmatrix} \quad (11.5.2)$$

where

$$x_{ij}^{(e)} = x_i^{(e)} - x_j^{(e)}, \quad y_{ij}^{(e)} = y_i^{(e)} - y_j^{(e)} \quad (11.5.3)$$

and $A^{(e)}$ is the area enclosed by the quadrilateral element, given by the following expression:

$$A^{(e)} = \frac{1}{2}\left(x_{42}^{(e)} \cdot y_{13}^{(e)} + x_{13}^{(e)} \cdot y_{24}^{(e)}\right) \quad (11.5.4)$$

Given $A^{(e)}$, the Jacobian J_1 at the single quadrature point can be obtained from the following equation.

$$J_1 = \frac{A^{(e)}}{4} \quad (11.5.5)$$

The rank of a matrix is equal to the number of linearly independent rows or columns that the matrix has (whichever of the two is smaller). $[B_1]$ has three rows, which are linearly independent, thus it has a rank equal to 3, and $[D_1]$ also has a rank equal to 3. When a matrix is the product of other matrices, its rank cannot be greater than the minimum rank of the matrices in the product. It turns out that the rank of $[k^{(e)}]$ evaluated with one-point quadrature is exactly equal to 3. Thus, for the case that we use URI for a 4Q element, we have a problem of *rank deficiency*: the rank of the stiffness matrix is lower than the theoretically required one, which is equal to 5! Thus, there are two spurious zero energy modes, for which the element cannot develop strains or stiffness. The term *zero-energy* refers to the zero *strain energy* of the modes. These two modes are shown in Figure 11.6a for a 4Q element, and for obvious reasons they are frequently referred to as *hourglass modes*. For an eight-node serendipity (8Q) element, reduced quadrature leads to a single spurious zero-energy mode, shown in Figure 11.6b.

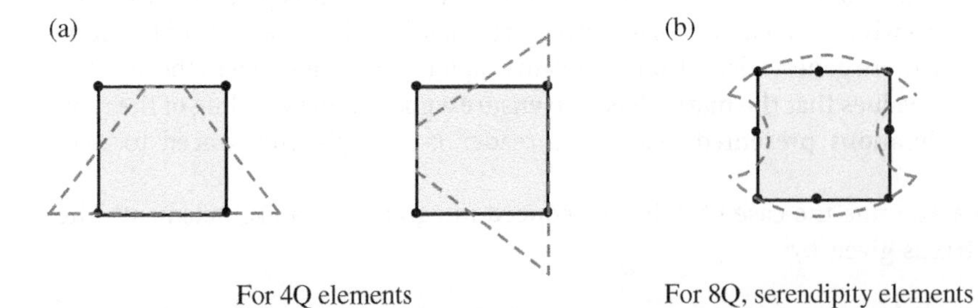

Figure 11.6 Spurious zero-energy modes for quadrilateral isoparametric elements using uniform reduced integration.

Uniform reduced integration is also used in three-dimensional analysis, where for the eight-node hexahedral (8H) element, we have a single quadrature point. In this case, we also have rank deficiency and spurious zero-energy modes (12 in total), four of which are shown in Figure 11.7.

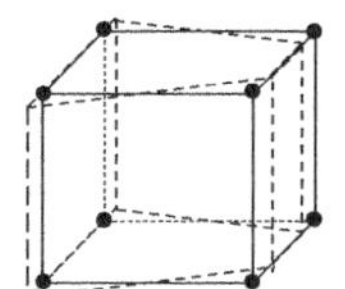 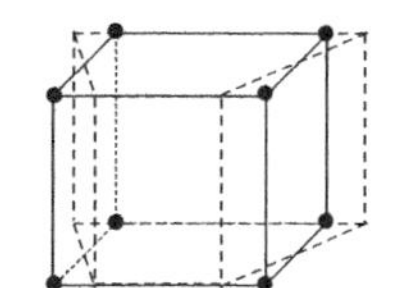 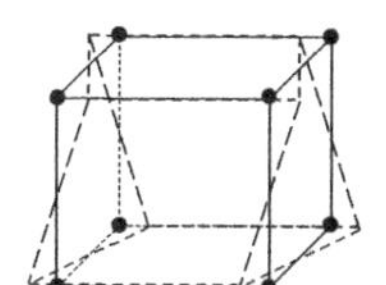 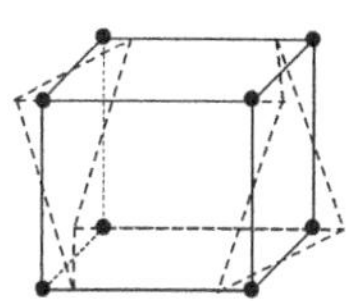

Figure 11.7 Examples of spurious zero-energy modes for hexahedral isoparametric elements using uniform reduced integration. *Source:* Flanagan and Belytschko 1981. Reproduced with permission from John Wiley & Sons.

11.6 Resolving the Problem of Hourglass Modes: Hourglass Stiffness

The rank deficiency issue in elements with URI (4Q for two-dimensional problems, 8H for three-dimensional problems) is tackled with the introduction of *hourglass stiffness* in the element stiffness matrix. More specifically, a procedure exists to establish the vectors corresponding to the spurious zero-energy modes. These are simply linearly independent deformation shapes for which the unique quadrature point does not "see" any strain. Then, an artificial hourglass stiffness is introduced to resist the spurious zero-energy modes and restore the correct rank for the element stiffness matrix. The addition of hourglass stiffness belongs to a family of procedures called *hourglass control* and is common in many commercial finite element programs.

Remark 11.6.1: The use of URI with hourglass control in 4Q elements leads to behavior for beam bending similar to that of incompatible-mode elements, at a much lower computational cost. ∎

Remark 11.6.2: Large-scale nonlinear finite element computations using the so-called *explicit finite element method* (e.g., Belytschko et al. 2014) typically rely on URI, because this increases the efficiency of the analysis (we have fewer quadrature points, and thus much less computations to conduct for each element). In such cases, hourglass control is necessary to suppress spurious modes. ∎

In the remainder of this section, we will describe a procedure providing the hourglass stiffness. We will specifically focus to the case of a 4Q element with one-point quadrature. The first step for hourglass control is to determine the hourglass mode vectors, for an element with arbitrary shape. Let us begin by focusing on the nodal displacements and forces in the x-direction. We will express the nodal displacements in the x-direction, $\left\{U_x^{(e)}\right\} = \left[u_{x1}^{(e)} \; u_{x2}^{(e)} \; u_{x3}^{(e)} \; u_{x4}^{(e)}\right]^T$, as a linear combination of four *basis vectors*. These four vectors, $\vec{e}_1, \vec{e}_2, \vec{e}_3$ and $\vec{e}_4$ are linearly independent, but *nonorthogonal* basis vectors, called *covariant basis vectors*, given by the following expressions.

$$\vec{e}_1 = \frac{1}{4}[1 \ \ 1 \ \ 1 \ \ 1]^T \tag{11.6.1a}$$

$$\vec{e}_2 = \left[x_1^{(e)} - \bar{x} \quad x_2^{(e)} - \bar{x} \quad x_3^{(e)} - \bar{x} \quad x_4^{(e)} - \bar{x}\right]^T, \text{ where } \bar{x} = \frac{x_1^{(e)} + x_2^{(e)} + x_3^{(e)} + x_4^{(e)}}{4} \tag{11.6.1b}$$

$$\vec{e}_3 = \left[y_1^{(e)} - \bar{y} \quad y_2^{(e)} - \bar{y} \quad y_3^{(e)} - \bar{y} \quad y_4^{(e)} - \bar{y}\right]^T, \text{ where } \bar{y} = \frac{y_1^{(e)} + y_2^{(e)} + y_3^{(e)} + y_4^{(e)}}{4} \tag{11.6.1c}$$

$$\vec{e}_4 = \vec{h}, \text{where } \vec{h} = [1 \ \ -1 \ \ 1 \ \ -1]^T \tag{11.6.1d}$$

We have:

$$\left\{U_x^{(e)}\right\} = \sum_{i=1}^{4} \left(a^i \cdot \vec{e}_i\right) = a^1 \cdot \vec{e}_1 + a^2 \cdot \vec{e}_2 + a^3 \cdot \vec{e}_3 + a^4 \cdot \vec{e}_4 \tag{11.6.2}$$

Remark 11.6.3: Notice that the first three terms in Equation (11.6.2) would collectively express the part of $\left\{U_x^{(e)}\right\}$ corresponding to a complete linear two-dimensional polynomial field $u_x(x,y)$. Specifically, the term $a^1 \cdot \vec{e}_1$ corresponds to constant nodal values of x-displacement (i.e., rigid-body translation along the x-direction), the term $a^2 \cdot \vec{e}_2$ corresponds to nodal x-displacements that vary linearly with x, while the term $a^3 \cdot \vec{e}_3$ corresponds to nodal x-displacements that vary linearly with y. The part of the field $u_x(x,y)$ that corresponds to constant strain is *linear* with respect to x and y. Constant-strain modes of deformation are captured by the single quadrature point in the element. Similarly, the part of the field $u_x(x,y)$, which is associated with rigid-body translation or rotation, is also linear with respect to x and y! Eventually, the final term $a^4 \cdot \vec{e}_4$ in Equation (11.6.5) is the part of $\left\{U_x^{(e)}\right\}$ *not* associated with a linear field $u_x(x,y)$. By implication, since the term $a^4 \cdot \vec{e}_4$ is the part of the nodal displacement vector that is not associated with constant-strain or rigid-body modes, it must be the part of the nodal displacement vector associated with hourglass modes. ∎

Given the four covariant basis vectors, we can establish an associated set of *contravariant basis vectors* $\vec{e}^1, \vec{e}^2, \vec{e}^3$, and $\vec{e}^4$, which will allow us to find the coefficients a^1, a^2, a^3, and a^4 in Equation (11.6.2). The first three contravariant basis vectors are given by:

$$\vec{e}^1 = [1 \ \ 1 \ \ 1 \ \ 1]^T \tag{11.6.3a}$$

$$\vec{e}^2 = \frac{1}{2A^{(e)}}\left[y_{24}^{(e)} \quad y_{31}^{(e)} \quad -y_{24}^{(e)} \quad -y_{31}^{(e)}\right]^T \tag{11.6.3b}$$

$$\vec{e}^3 = \frac{1}{2A^{(e)}}\left[x_{42}^{(e)} \quad x_{13}^{(e)} \quad -x_{42}^{(e)} \quad -x_{13}^{(e)}\right]^T \tag{11.6.3c}$$

where $x_{ij}^{(e)}, y_{ij}^{(e)}$ have been defined in Equation (11.5.3) and $A^{(e)}$ is the area enclosed by the quadrilateral element, given by Equation (11.5.4).

The vector $\vec{e}^4$ will correspond to the *hourglass contravariant basis vector*. The hourglass nodal force vector (i.e., the element force vector resisting the hourglass mode) will be parallel to $\vec{e}^4$. The hourglass contravariant basis vector can be obtained with various alternative methods, summarized in Box 11.6.1.

Box 11.6.1 Alternative Methods to Obtain the Hourglass Contravariant Basis Vector, $\vec{e}^4$

- Method 1: Solve a system of equations (Kosloff and Frazier 1978)

$$\vec{e}^4 \bullet \vec{e}_1 = 0 \rightarrow e_1^4 \cdot 1 + e_2^4 \cdot 1 + e_3^4 \cdot 1 + e_4^4 \cdot 1 = 0 \rightarrow e_1^4 + e_2^4 + e_3^4 + e_4^4 = 0 \quad (11.6.4a)$$

$$\vec{e}^4 \bullet \vec{e}_2 = 0 \rightarrow e_1^4 \cdot \left(x_1^{(e)} - \bar{x}\right) + e_2^4 \cdot \left(x_2^{(e)} - \bar{x}\right) + e_3^4 \cdot \left(x_3^{(e)} - \bar{x}\right) + e_4^4 \cdot \left(x_4^{(e)} - \bar{x}\right) = 0 \quad (11.6.4b)$$

$$\vec{e}^4 \bullet \vec{e}_3 = 0 \rightarrow e_1^4 \cdot \left(y_1^{(e)} - \bar{y}\right) + e_2^4 \cdot \left(y_2^{(e)} - \bar{y}\right) + e_3^4 \cdot \left(y_3^{(e)} - \bar{y}\right) + e_4^4 \cdot \left(y_4^{(e)} - \bar{y}\right) = 0 \quad (11.6.4c)$$

The fourth equation, can be written based on *one of two options:*
Option a: set $e_1^4 = 1$, then normalize the vector such that

$$\sum_{i=1}^{4} \left[\left(e_i^4\right)^2 \right] = 4 \quad (11.6.4d-i)$$

Option b: use the equation:

$$\vec{e}^4 \bullet \vec{e}_4 = 1 \rightarrow e_1^4 - e_2^4 + e_3^4 - e_4^4 = 1 \quad (11.6.4d-ii)$$

- Method 2: Use *Gram-Schmidt orthogonalization* (Flanagan and Belytschko 1981)

$$\vec{e}^4 = \vec{h} - \sum_{i=1}^{3} \left[\left(\vec{h} \bullet \vec{e}_i\right) \cdot \vec{e}^i \right] = \vec{h} - \left(\vec{h} \bullet \vec{e}_1\right) \cdot \vec{e}^1 - \left(\vec{h} \bullet \vec{e}_2\right) \cdot \vec{e}^2 - \left(\vec{h} \bullet \vec{e}_3\right) \cdot \vec{e}^3 \quad (11.6.5)$$

The following condition must be satisfied by the sets of covariant and contravariant basis vectors.

$$\vec{e}^i \bullet \vec{e}_j = \delta^i{}_j = \begin{cases} 1, \text{ if } i = j \\ 0, \text{ if } i \neq j \end{cases} \quad (11.6.6)$$

Now, we return to the linear combination of Equation (11.6.2). Each coefficient a^k in this expression can be found by taking the dot product of $\left\{U_x^{(e)}\right\}$ and $\vec{e}^k$, and exploiting Equation (11.6.6):

$$\vec{U}_x^{(e)} \bullet \vec{e}^k = \left[\sum_{i=1}^{4} \left(a^i \cdot \vec{e}_i\right) \right] \bullet \vec{e}^k = \left[\sum_{i=1}^{4} \left[a^i \left(\vec{e}_i \bullet \vec{e}^k\right) \right] \right] = \sum_{i=1}^{4} \left(a^i \cdot \delta_i^k\right) = a^k \quad (11.6.7)$$

The projection of the x-displacements on the hourglass mode in the x-direction is:

$$a^4 = \vec{U}_x^{(e)} \bullet \vec{e}^4 \quad (11.6.8)$$

Now, the vector containing the nodal hourglass-resisting forces in the x-direction can be written as:

$$\vec{f}_{x,HG}^{(e)} = q_{HG} \cdot \vec{e}^4 \quad (11.6.9)$$

where the scalar coefficient, q_{HG}, can be found by:

$$q_{HG} = K_{HG} \cdot a^4 \quad (11.6.10)$$

and the term K_{HG} is called the *hourglass stiffness* term; it is a user-defined parameter in an analysis.

We can write the vector with the x-hourglass-resisting nodal forces as follows:

$$\vec{f}^{(e)}_{x,HG} = q_{HG} \cdot \vec{e}^{4} = K_{HG} \cdot a^4 \cdot \vec{e}^{4} = K_{HG} \cdot \left(\vec{U}^{(e)}_x \bullet \vec{e}^{4}\right) \cdot \vec{e}^{4} = \left[K^{(e)}_{x,HG}\right]\left\{\vec{U}^{(e)}_x\right\} \quad (11.6.11)$$

where the (IJ) component of the x-hourglass stiffness matrix, $\left[K^{(e)}_{x,HG}\right]$, is given by the following expression.

$$K^{(e)}_{x,HG\ IJ} = K_{HG} \cdot e^4_I \cdot e^4_J \quad (11.6.12)$$

Subscripts I and J in Equation (11.6.12) take values between 1 and 4.

For an isotropic material, the hourglass stiffness term, K_{HG}, can be taken equal to:

$$K_{HG} = C_{HG} \cdot \frac{1}{2} E \cdot A \cdot \left(\vec{e}^{2} \bullet \vec{e}^{2} + \vec{e}^{3} \bullet \vec{e}^{3}\right) = \frac{C_{HG} \cdot E}{4A} \cdot \left[\left(x^{(e)}_{42}\right)^2 + \left(x^{(e)}_{13}\right)^2 + \left(y^{(e)}_{24}\right)^2 + \left(y^{(e)}_{31}\right)^2\right] \quad (11.6.13)$$

where E is the modulus of elasticity of the material and C_{HG} is a dimensionless *hourglass coefficient*. This coefficient is typically assigned values between 0.05 and 0.15.

Remark 11.6.4: The x-components of hourglass stiffness are only related to the x-nodal displacements. This means that the (IJ) component given by Equation (11.6.12) corresponds to the x-nodal displacement of node I and the x-nodal displacement of node J. In other words, when we want to add the x-hourglass stiffness to the (8 × 8) stiffness matrix of the 4Q element, then each term (IJ) of Equation (11.6.12) must be added to the term $(2I\text{-}1,\ 2J\text{-}1)$ of $[k^{(e)}]$. ■

The same procedure is employed for hourglass control in the y-direction. The only difference is that we need to define the four-dimensional vector with the nodal displacements in the y-direction, $\left\{U^{(e)}_y\right\} = \left[u^{(e)}_{y1}\ u^{(e)}_{y2}\ u^{(e)}_{y3}\ u^{(e)}_{y4}\right]^T$. Eventually, we can establish the (8 × 8) hourglass stiffness matrix, $\left[k^{(e)}_{HG}\right]$, of the element, using the procedure outlined in Box 11.6.2.

Box 11.6.2 Determination of Hourglass Stiffness Matrix, $\left[k^{(e)}_{HG}\right]$ for 4Q Element

1) Find vectors $\vec{e}_1, \vec{e}_2, \vec{e}_3$ and $\vec{e}_4$ using Equations (11.6.1).
2) Find vectors $\vec{e}^{1}, \vec{e}^{2}$ and $\vec{e}^{3}$ using Equations (11.6.3).
3) Find vector $\vec{e}^{4}$ using Box 11.6.1.
4) Find term a^4 using Equation (11.6.8).
5) Calculate the terms of $\left[k^{(e)}_{HG}\right]$: For each pair of element nodes I, J:

$$k^{(e)}_{HG\ 2I-1,\ 2J-1} = k^{(e)}_{HG\ 2I,\ 2J} = K_{x,HG\ IJ} = K_{HG} \cdot e^4_I \cdot e^4_J$$

Remark 11.6.5: The hourglass stiffness matrix $\left[k_{HG}^{(e)}\right]$ in Box 11.6.2 is *per unit thickness*. If we are given the thickness value, d, for the two-dimensional solid at hand, then we need to multiply $\left[k_{HG}^{(e)}\right]$ of Box 11.6.2 by d to obtain the actual hourglass stiffness matrix. ∎

Remark 11.6.6: An alternative definition for the second and third covariant basis vectors, $\vec{e}_2$ and $\vec{e}_3$, is commonly employed in practice:

$$\vec{e}_2 = \left[x_1^{(e)} \quad x_2^{(e)} \quad x_3^{(e)} \quad x_4^{(e)}\right]^T \tag{11.6.14a}$$

$$\vec{e}_3 = \left[y_1^{(e)} \quad y_2^{(e)} \quad y_3^{(e)} \quad y_4^{(e)}\right]^T \tag{11.6.14b}$$

This alternative definition for the two vectors leads to violation of Equation (11.6.6), because, in general, we will have $\vec{e}^1 \bullet \vec{e}_2 \neq 0$ and $\vec{e}^1 \bullet \vec{e}_3 \neq 0$. Still, it can be verified (and formally proven) that the alternative definitions for $\vec{e}_2$ and $\vec{e}_3$ lead to the same exact values of components of $\vec{e}^4$ and $\left[k_{HG}^{(e)}\right]$ as those obtained with Equations (11.6.1b, c). ∎

As an illustrative example, we analyze a horizontal beam with a downward tip load and plot the deformed configurations for various cases, as shown in Figure 11.8. In our analyses, we only use 4Q elements. We can see that with full integration (Figure 11.8a), we have a "reasonable" deformed configuration. If we use uniform reduced integration (Figure 11.8b), we can observe that we have a "strange" deformed shape, where the hourglass modes dominate the behavior. Finally, if we use reduced integration with hourglass stiffness (Figure 11.8c), we can suppress the hourglass modes and prevent them from seriously affecting our solution.

Another interesting phenomenon is worth mentioning: If we solve this example problem using eight-node serendipity (8Q) elements with URI, we obtain a reasonable deformed mesh (Figure 11.8d), not dominated by the spurious zero-energy modes. This happens because the zero-energy mode for the 8Q element shown in Figure 11.6b, is *noncommunicable,* which means it cannot occur into a multielement mesh! This is the reason why analysis programs may not require hourglass stiffness for eight-node

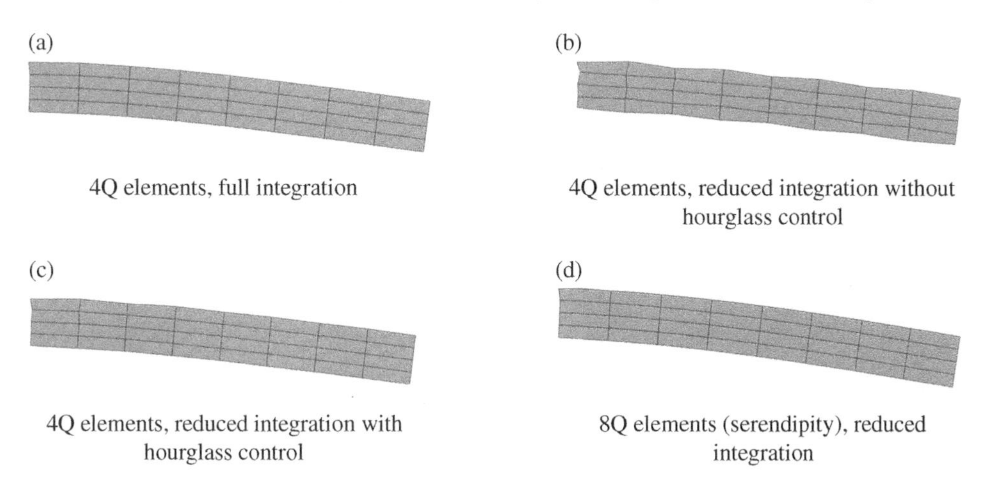

Figure 11.8 Analysis of a beam bending problem using different types of quadrilateral elements.

serendipity elements! Still, caution is required, because we may still sometimes obtain very low stiffness mobilized for the mode in Figure 11.6b, as described in Zienkiewicz and Taylor (2005).

Example 11.1: Calculation of hourglass stiffness
We want to calculate the hourglass stiffness matrix for the quadrilateral element shown in Figure 11.9. The material is elastic isotropic, with $E = 5000$ ksi, $\nu = 0.3$, and plane-strain conditions can be assumed. We are also given $C_{HG} = 0.1$, and that the procedure by Flanagan and Belytschko (Box 11.6.1) to find $\vec{e}^4$.

We first find the area A, enclosed by the element, which can be calculated from the expression:

$$A^{(e)} = \frac{1}{2}\left(x_{42}^{(e)} \cdot y_{13}^{(e)} + x_{13}^{(e)} \cdot y_{24}^{(e)}\right) = \frac{1}{2}[(0-25)\cdot(0-22) + (-5-17)\cdot(5-18)] = 418\, in^2$$

We can now establish the three contravariant basis vectors:

$$\vec{e}^1 = [1 \;\; 1 \;\; 1 \;\; 1]^T$$

$$\vec{e}^2 = \frac{1}{2A^{(e)}}\left[y_{24}^{(e)} \;\; y_{31}^{(e)} \;\; -y_{24}^{(e)} \;\; -y_{31}^{(e)}\right]^T = \frac{1}{2\cdot 418}[5-18 \;\; 22-0 \;\; -(5-18) \;\; -(22-0)]^T$$

$$= \frac{1}{836}[-13 \;\; 22 \;\; 13 \;\; -22]^T$$

$$\vec{e}^3 = \frac{1}{2A^{(e)}}\left[x_{42}^{(e)} \;\; x_{13}^{(e)} \;\; -x_{42}^{(e)} \;\; -x_{13}^{(e)}\right]^T = \frac{1}{2\cdot 418}[0-25 \;\; -5-17 \;\; -(0-25) \;\; -(-5-17)]^T$$

$$= \frac{1}{836}[-25 \;\; -22 \;\; 25 \;\; 22]^T$$

Next, we will obtain the hourglass basis vector $\vec{e}^4$. We begin by setting $\vec{h} = [1 \;\; -1 \;\; 1 \;\; -1]^T$. Then, we find the covariant basis vectors $\vec{e}_1$, $\vec{e}_2$, $\vec{e}_3$, and the dot products of these vectors with $\vec{h}$ (these dot products are required in the method of Flanagan and Belytshcko):

$$\vec{e}_1 = \frac{1}{4}[1 \;\; 1 \;\; 1 \;\; 1]^T$$

$$\bar{x} = \frac{x_1^{(e)} + x_2^{(e)} + x_3^{(e)} + x_4^{(e)}}{4} = \frac{-5+25+17+0}{4} = 9.25\, in,$$

$$\bar{y} = \frac{y_1^{(e)} + y_2^{(e)} + y_3^{(e)} + y_4^{(e)}}{4} = \frac{0+5+22+18}{4} = 11.25\, in$$

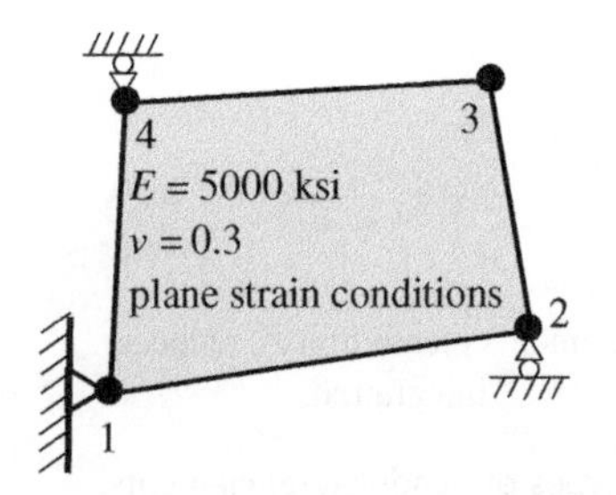

Node	x_i (in)	y_i (in)
1	–5	0
2	25	5
3	17	22
4	0	18

Figure 11.9 Example case for calculation of hourglass stiffness.

$$\vec{e}_2 = \left[x_1^{(e)} - \bar{x} \;\; x_2^{(e)} - \bar{x} \;\; x_3^{(e)} - \bar{x} \;\; x_4^{(e)} - \bar{x}\right]^T = [-14.25 \;\; 15.75 \;\; 7.75 \;\; -9.25]^T$$

$$\vec{e}_3 = \left[y_1^{(e)} - \bar{y} \;\; y_2^{(e)} - \bar{y} \;\; y_3^{(e)} - \bar{y} \;\; y_4^{(e)} - \bar{y}\right]^T = [-11.25 \;\; -6.25 \;\; 10.75 \;\; 6.75]^T$$

$$\vec{h} \bullet \vec{e}_1 = 0, \vec{h} \bullet \vec{e}_2 = -13, \vec{h} \bullet \vec{e}_3 = -1$$

Thus, we can find $\vec{e}^4$ using Equation (11.6.5):

$$\vec{e}^4 = \vec{h} - \left(\vec{h} \bullet \vec{e}_1\right) \cdot \vec{e}^1 - \left(\vec{h} \bullet \vec{e}_2\right) \cdot \vec{e}^2 - \left(\vec{h} \bullet \vec{e}_3\right) \cdot \vec{e}^3 = [0.767943 \;\; -0.68421 \;\; 1.232057 \;\; -1.31579]^T$$

We also calculate K_{HG}:

$$K_{HG} = \frac{C_{HG} \cdot E}{4A} \cdot \left[\left(x_{42}^{(e)}\right)^2 + \left(x_{13}^{(e)}\right)^2 + \left(y_{24}^{(e)}\right)^2 + \left(y_{31}^{(e)}\right)^2\right]$$

$$= \frac{0.1 \cdot 5000}{4 \cdot 418} \cdot \left[(-25)^2 + (-22)^2 + (-13)^2 + (22)^2\right] = 526.91\,kip$$

and we are ready to obtain the hourglass stiffness matrix for the x- and for the y-directions with the aid of Box 11.6.2:

$$\left[k_{x,HG}^{(e)}\right] = \begin{bmatrix} 310.7 & 0 & -276.9 & 0 & 498.5 & 0 & -532.4 & 0 \\ 0 & 0 & 0 & 0 & 0 & 0 & 0 & 0 \\ -276.9 & 0 & 246.7 & 0 & -444.2 & 0 & 474.4 & 0 \\ 0 & 0 & 0 & 0 & 0 & 0 & 0 & 0 \\ 498.5 & 0 & -444.2 & 0 & 799.8 & 0 & -854.2 & 0 \\ 0 & 0 & 0 & 0 & 0 & 0 & 0 & 0 \\ -532.4 & 0 & 474.4 & 0 & -854.2 & 0 & 912.2 & 0 \\ 0 & 0 & 0 & 0 & 0 & 0 & 0 & 0 \end{bmatrix},$$

$$\left[k_{y,HG}^{(e)}\right] = \begin{bmatrix} 0 & 0 & 0 & 0 & 0 & 0 & 0 & 0 \\ 0 & 310.7 & 0 & -276.9 & 0 & 498.5 & 0 & -532.4 \\ 0 & 0 & 0 & 0 & 0 & 0 & 0 & 0 \\ 0 & -276.9 & 0 & 246.7 & 0 & -444.2 & 0 & 474.4 \\ 0 & 0 & 0 & 0 & 0 & 0 & 0 & 0 \\ 0 & 498.5 & 0 & -444.2 & 0 & 799.8 & 0 & -854.2 \\ 0 & 0 & 0 & 0 & 0 & 0 & 0 & 0 \\ 0 & -532.4 & 0 & 474.4 & 0 & -854.2 & 0 & 912.2 \end{bmatrix}$$

It is worth noticing that, as expected, the hourglass stiffness matrix for the x-direction only includes stiffness terms in the components corresponding to the x-displacements. Similarly, the hourglass stiffness matrix for the y-direction only includes terms for the y-displacements. The hourglass stiffness matrix $\left[k_{HG}^{(e)}\right]$ of the element is given as:

$$\left[k_{HG}^{(e)}\right] = \left[k_{x,HG}^{(e)}\right] + \left[k_{y,HG}^{(e)}\right]$$

■

11.7 Selective-Reduced Integration

The previous two sections were focused on the case where we use a single quadrature point in elements, to resolve issues associated with locking. We saw that the use of uniform reduced integration entails the need for hourglass control, to prevent problems associated with spurious zero-energy modes. An intriguing question that could arise at this point is: Why not try to use reduced integration *only for the part of the stiffness that locks*? This is the basic premise behind *selective-reduced integration* (SRI) procedures. Our discussion will focus on the context of volumetric locking, where SRI uses a reduced integration rule only for the volumetric part of the stiffness matrix—the part that is associated with volumetric strains.

The constitutive law for three-dimensional linear elasticity and isotropic material can be written in the following form, based on Equation (7.5.10a):

$$\begin{Bmatrix} \sigma_{xx} \\ \sigma_{yy} \\ \sigma_{zz} \\ \sigma_{xy} \\ \sigma_{yz} \\ \sigma_{zx} \end{Bmatrix} = \begin{bmatrix} \lambda+2\mu & \lambda & \lambda & 0 & 0 & 0 \\ \lambda & \lambda+2\mu & \lambda & 0 & 0 & 0 \\ \lambda & \lambda & \lambda+2\mu & 0 & 0 & 0 \\ 0 & 0 & 0 & \mu & 0 & 0 \\ 0 & 0 & 0 & 0 & \mu & 0 \\ 0 & 0 & 0 & 0 & 0 & \mu \end{bmatrix} \begin{Bmatrix} \varepsilon_{xx} \\ \varepsilon_{yy} \\ \varepsilon_{zz} \\ \gamma_{xy} \\ \gamma_{yz} \\ \gamma_{zx} \end{Bmatrix} \tag{11.7.1}$$

We can separate the material stiffness matrix in two parts:

$$\{\sigma\} = [D]\{\varepsilon\} = \left([\bar{D}] + \left[\bar{\bar{D}}\right]\right)\{\varepsilon\} \tag{11.7.2}$$

where

$$[\bar{D}] = \begin{bmatrix} 2\mu & 0 & 0 & 0 & 0 & 0 \\ 0 & 2\mu & 0 & 0 & 0 & 0 \\ 0 & 0 & 2\mu & 0 & 0 & 0 \\ 0 & 0 & 0 & \mu & 0 & 0 \\ 0 & 0 & 0 & 0 & \mu & 0 \\ 0 & 0 & 0 & 0 & 0 & \mu \end{bmatrix} \tag{11.7.3a}$$

and

$$\left[\bar{\bar{D}}\right] = \begin{bmatrix} \lambda & \lambda & \lambda & 0 & 0 & 0 \\ \lambda & \lambda & \lambda & 0 & 0 & 0 \\ \lambda & \lambda & \lambda & 0 & 0 & 0 \\ 0 & 0 & 0 & 0 & 0 & 0 \\ 0 & 0 & 0 & 0 & 0 & 0 \\ 0 & 0 & 0 & 0 & 0 & 0 \end{bmatrix} \tag{11.7.3b}$$

Remark 11.7.1: One can verify that, just like the bulk modulus K_V, Lamé's constant λ also tends to infinity for incompressible materials (i.e., when $\nu \rightarrow \frac{1}{2}$), while the constant $\mu = G$ remains finite (and retains a "reasonable" value) even for incompressible materials. ▮

We can establish a similar separation of the material stiffness for the plane-strain case. The constitutive law is:

$$\begin{Bmatrix} \sigma_{xx} \\ \sigma_{yy} \\ \sigma_{xy} \end{Bmatrix} = \begin{bmatrix} \lambda + 2\mu & \lambda & 0 \\ \lambda & \lambda + 2\mu & 0 \\ 0 & 0 & \mu \end{bmatrix} \begin{Bmatrix} \varepsilon_{xx} \\ \varepsilon_{yy} \\ \gamma_{xy} \end{Bmatrix} \tag{11.7.4a}$$

and we can separate the law into two parts:

$$\{\sigma\} = [D]\{\varepsilon\} = \left([\bar{D}] + \left[\bar{\bar{D}}\right]\right)\{\varepsilon\} \tag{11.7.5}$$

where

$$[\bar{D}] = \begin{bmatrix} 2\mu & 0 & 0 \\ 0 & 2\mu & 0 \\ 0 & 0 & \mu \end{bmatrix} \tag{11.7.6a}$$

and

$$\left[\bar{\bar{D}}\right] = \begin{bmatrix} \lambda & \lambda & 0 \\ \lambda & \lambda & 0 \\ 0 & 0 & 0 \end{bmatrix} \tag{11.7.6b}$$

Given Equation (11.7.5), the stiffness matrix (per unit thickness) of a 4Q element with plane-strain conditions can also be separated into two contributions:

$$\left[k^{(e)}\right] = \iint_{\Omega^{(e)}} \left[B^{(e)}\right]^T \left[D^{(e)}\right] \left[B^{(e)}\right] dV = \iint_{\Omega^{(e)}} \left[B^{(e)}\right]^T \left(\left[\bar{\bar{D}}^{(e)}\right] + \left[\bar{D}^{(e)}\right]\right) \left[B^{(e)}\right] dV = \left[k_\lambda^{(e)}\right] + \left[k_\mu^{(e)}\right] \tag{11.7.7}$$

where

$$\left[k_\lambda^{(e)}\right] = \iint_{\Omega^{(e)}} \left[B^{(e)}\right]^T \left[\bar{\bar{D}}^{(e)}\right] \left[B^{(e)}\right] dV \tag{11.7.8a}$$

$$\left[k_\mu^{(e)}\right] = \iint_{\Omega^{(e)}} \left[B^{(e)}\right]^T \left[\bar{D}^{(e)}\right] \left[B^{(e)}\right] dV \tag{11.7.8b}$$

The matrix $\left[k_\lambda^{(e)}\right]$ corresponds to the part of the stiffness matrix for volumetric strains—the part that can "lock" (i.e., become overstiff) for incompressible or near-incompressible conditions—while the stiffness matrix $\left[k_\mu^{(e)}\right]$ corresponds to deviatoric strains and does not "lock."

In the SRI procedure, the $\left[k_{\lambda}^{(e)}\right]$-part of the stiffness matrix is integrated using reduced integration (for 4Q elements, using one-point quadrature) and the $\left[k_{\mu}^{(e)}\right]$-part of the stiffness matrix is integrated using full integration (for 4Q elements, this is the standard four-point quadrature that we saw in Section 6.5). We then avoid locking problems, and at the same time retain the correct rank for the element stiffness matrix. The SRI procedure for 4Q and 8Q elements is schematically summarized in Figure 11.10.

Selective-reduced integration can also be used for the eight-node hexahedral (8H) element in three-dimensional analysis, as well as the analysis of beams (see Section 13.14) and shells. The only problem of the method is that it cannot be easily extended to analyses involving materials for which the constitutive law cannot be decomposed into a volumetric and a deviatoric part. This is the case for anisotropic materials or for several types of elastoplastic materials.

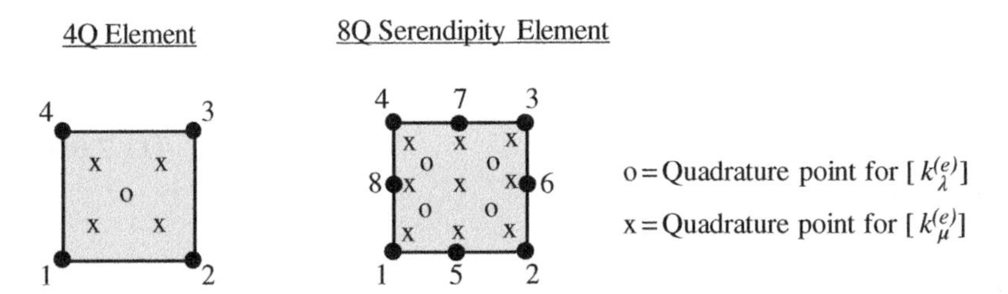

Figure 11.10 Selective-reduced integration for quadrilateral elements.

11.8 The B-bar Method for Resolving Locking

The selective-reduced integration procedure can help us avoid volumetric locking for incompressible materials, without the deleterious effect of the spurious zero-energy modes and rank deficiency. While the method is very efficient for isotropic elastic materials, it may not be possible to extend the SRI to the general case where, for example, the volumetric and deviatoric parts of the constitutive model are coupled. For this reason, another method, called the *B-bar method* (Hughes 1980), is commonly employed for four-node quadrilateral (4Q) elements in lieu of SRI. The B-bar method is equivalent to SRI for isotropic elastic materials, but it can also be used unaltered for general linearly elastic or inelastic materials. The method will be presented here for two-dimensional, plane-strain elasticity. We use a version of the plane-strain constitutive equations that includes four stress components and the corresponding strain components: specifically, we take Equation (11.7.1) and eliminate the fifth and sixth rows and columns, that is, the rows and columns corresponding to γ_{yz} and γ_{zx}, respectively. We thus write:

$$\{\sigma\} = [D]\{\varepsilon\} \tag{11.8.1}$$

where

$$\{\sigma\} = \left[\sigma_{xx} \; \sigma_{yy} \; \sigma_{zz} \; \sigma_{xy}\right]^T \tag{11.8.2a}$$

$$\{\varepsilon\} = \begin{bmatrix} \varepsilon_{xx} & \varepsilon_{yy} & \varepsilon_{zz} & \gamma_{xy} \end{bmatrix}^T \tag{11.8.2b}$$

and

$$[D] = \begin{bmatrix} \lambda + 2\mu & \lambda & \lambda & 0 \\ \lambda & \lambda + 2\mu & \lambda & 0 \\ \lambda & \lambda & \lambda + 2\mu & 0 \\ 0 & 0 & 0 & \mu \end{bmatrix} \tag{11.8.2c}$$

We now separate the strain vector of Equation (11.8.2b) into two parts, one corresponding to the volumetric part of the strains and the other corresponding to the deviatoric part of the strains.

$$\{\varepsilon\} = \{\varepsilon_v\} + \{e\} \tag{11.8.3}$$

where

$$\{\varepsilon_v\} = \frac{1}{3}[\varepsilon_v \ \ \varepsilon_v \ \ \varepsilon_v \ \ 0]^T \tag{11.8.4a}$$

and

$$\{e\} = \begin{bmatrix} e_{xx} & e_{yy} & e_{zz} & \gamma_{xy} \end{bmatrix}^T \tag{11.8.4b}$$

We know that, for plane strain problems, the total normal strain in the z-direction, ε_{zz}, is identically equal to zero. The only way to generally satisfy this requirement is to have $e_{zz} = -\varepsilon_v = -\frac{\varepsilon_{xx} + \varepsilon_{yy}}{3}$; thus, given that $\varepsilon_{zz} = 0$, we have:

$$\varepsilon_v = \varepsilon_{xx} + \varepsilon_{yy} \tag{11.8.5}$$

$$e_{xx} = \varepsilon_{xx} - \frac{\varepsilon_v}{3} = \frac{2}{3}\frac{\partial u_x}{\partial x} - \frac{1}{3}\frac{\partial u_y}{\partial y} \tag{11.8.6a}$$

$$e_{yy} = \varepsilon_{yy} - \frac{\varepsilon_v}{3} = -\frac{1}{3}\frac{\partial u_x}{\partial x} + \frac{2}{3}\frac{\partial u_y}{\partial y} \tag{11.8.6b}$$

$$e_{zz} = \varepsilon_{zz} - \frac{\varepsilon_v}{3} = 0 - \frac{\varepsilon_v}{3} = -\frac{1}{3}\left(\frac{\partial u_x}{\partial x} + \frac{\partial u_y}{\partial y}\right) \tag{11.8.6c}$$

$$\gamma_{xy} = \frac{\partial u_x}{\partial y} + \frac{\partial u_y}{\partial x} \tag{11.8.6d}$$

Per Equation (11.8.3), the strain vector is separated into two parts. We can now do the same for the strain-displacement law of an element; that is, we will write the strain-displacement law as the sum of a term giving the volumetric part of the strain and a term that gives the deviatoric part of the strain:

$$\{\varepsilon\} = \{\varepsilon_v\} + \{e\} = \left[\bar{\bar{B}}^{(e)}\right]\left\{U^{(e)}\right\} + \left[\bar{B}^{(e)}\right]\left\{U^{(e)}\right\} \tag{11.8.7}$$

where $\left[\bar{\bar{B}}^{(e)}\right]$ is the part of the strain-displacement array that gives the volumetric strain, while $\left[\bar{B}^{(e)}\right]$ is the part of the strain-displacement array that gives the deviatoric strain. The following expressions apply.

$$\left[\bar{\bar{B}}^{(e)}\right] = \begin{bmatrix} \frac{1}{3}\frac{\partial N_1^{(e)}}{\partial x} & \frac{1}{3}\frac{\partial N_1^{(e)}}{\partial y} & \frac{1}{3}\frac{\partial N_2^{(e)}}{\partial x} & \frac{1}{3}\frac{\partial N_2^{(e)}}{\partial y} & \frac{1}{3}\frac{\partial N_3^{(e)}}{\partial x} & \frac{1}{3}\frac{\partial N_3^{(e)}}{\partial y} & \frac{1}{3}\frac{\partial N_4^{(e)}}{\partial x} & \frac{1}{3}\frac{\partial N_4^{(e)}}{\partial y} \\ \frac{1}{3}\frac{\partial N_1^{(e)}}{\partial x} & \frac{1}{3}\frac{\partial N_1^{(e)}}{\partial y} & \frac{1}{3}\frac{\partial N_2^{(e)}}{\partial x} & \frac{1}{3}\frac{\partial N_2^{(e)}}{\partial y} & \frac{1}{3}\frac{\partial N_3^{(e)}}{\partial x} & \frac{1}{3}\frac{\partial N_3^{(e)}}{\partial y} & \frac{1}{3}\frac{\partial N_4^{(e)}}{\partial x} & \frac{1}{3}\frac{\partial N_4^{(e)}}{\partial y} \\ \frac{1}{3}\frac{\partial N_1^{(e)}}{\partial x} & \frac{1}{3}\frac{\partial N_1^{(e)}}{\partial y} & \frac{1}{3}\frac{\partial N_2^{(e)}}{\partial x} & \frac{1}{3}\frac{\partial N_2^{(e)}}{\partial y} & \frac{1}{3}\frac{\partial N_3^{(e)}}{\partial x} & \frac{1}{3}\frac{\partial N_3^{(e)}}{\partial y} & \frac{1}{3}\frac{\partial N_4^{(e)}}{\partial x} & \frac{1}{3}\frac{\partial N_4^{(e)}}{\partial y} \\ 0 & 0 & 0 & 0 & 0 & 0 & 0 & 0 \end{bmatrix} \quad (11.8.8)$$

$$\left[\bar{B}^{(e)}\right] = \begin{bmatrix} \frac{2}{3}\frac{\partial N_1^{(e)}}{\partial x} & -\frac{1}{3}\frac{\partial N_1^{(e)}}{\partial y} & \frac{2}{3}\frac{\partial N_2^{(e)}}{\partial x} & -\frac{1}{3}\frac{\partial N_2^{(e)}}{\partial y} & \frac{2}{3}\frac{\partial N_3^{(e)}}{\partial x} & -\frac{1}{3}\frac{\partial N_3^{(e)}}{\partial y} & \frac{2}{3}\frac{\partial N_4^{(e)}}{\partial x} & -\frac{1}{3}\frac{\partial N_4^{(e)}}{\partial y} \\ -\frac{1}{3}\frac{\partial N_1^{(e)}}{\partial x} & \frac{2}{3}\frac{\partial N_1^{(e)}}{\partial y} & -\frac{1}{3}\frac{\partial N_2^{(e)}}{\partial x} & \frac{2}{3}\frac{\partial N_2^{(e)}}{\partial y} & -\frac{1}{3}\frac{\partial N_3^{(e)}}{\partial x} & \frac{2}{3}\frac{\partial N_3^{(e)}}{\partial y} & -\frac{1}{3}\frac{\partial N_4^{(e)}}{\partial x} & \frac{2}{3}\frac{\partial N_4^{(e)}}{\partial y} \\ -\frac{1}{3}\frac{\partial N_1^{(e)}}{\partial x} & -\frac{1}{3}\frac{\partial N_1^{(e)}}{\partial y} & -\frac{1}{3}\frac{\partial N_2^{(e)}}{\partial x} & -\frac{1}{3}\frac{\partial N_2^{(e)}}{\partial y} & -\frac{1}{3}\frac{\partial N_3^{(e)}}{\partial x} & -\frac{1}{3}\frac{\partial N_3^{(e)}}{\partial y} & -\frac{1}{3}\frac{\partial N_4^{(e)}}{\partial x} & -\frac{1}{3}\frac{\partial N_4^{(e)}}{\partial y} \\ \frac{\partial N_1^{(e)}}{\partial y} & \frac{\partial N_1^{(e)}}{\partial x} & \frac{\partial N_2^{(e)}}{\partial y} & \frac{\partial N_2^{(e)}}{\partial x} & \frac{\partial N_3^{(e)}}{\partial y} & \frac{\partial N_3^{(e)}}{\partial x} & \frac{\partial N_4^{(e)}}{\partial y} & \frac{\partial N_4^{(e)}}{\partial x} \end{bmatrix} \quad (11.8.9)$$

Remark 11.8.1: Notice that the first three rows of $\left[\bar{\bar{B}}^{(e)}\right]$ in Equation (11.8.8) are identical. In fact, we can define a row vector $\left[b_{vol}^{(e)}\right]$, giving the volumetric strain ε_v from the nodal displacements of the element:

$$\varepsilon_v^{(e)} = \left[b_{vol}^{(e)}\right]\left\{U^{(e)}\right\} \quad (11.8.10)$$

where

$$\left[b_{vol}^{(e)}\right] = \begin{bmatrix} \frac{\partial N_1^{(e)}}{\partial x} & \frac{\partial N_1^{(e)}}{\partial y} & \frac{\partial N_2^{(e)}}{\partial x} & \frac{\partial N_2^{(e)}}{\partial y} & \frac{\partial N_3^{(e)}}{\partial x} & \frac{\partial N_3^{(e)}}{\partial y} & \frac{\partial N_4^{(e)}}{\partial x} & \frac{\partial N_4^{(e)}}{\partial y} \end{bmatrix} \quad (11.8.11)$$

one can then verify that each of the three nonzero rows of array $\left[\bar{\bar{B}}^{(e)}\right]$ in Equation (11.8.8) is simply equal to one third of $\left[b_{vol}^{(e)}\right]$. Along the same lines, we can also establish the $\left[b_{vol}^{(e)}\right]$ row vector for any other type of finite element. ■

In the B-bar method, the value of $\left[\bar{\bar{B}}^{(e)}\right]$ at any location, (ξ, η) in the interior of the element, is set equal to the value of the specific array at the origin of the element, $\xi = 0, \eta = 0$. In other words, the strain-displacement array of the element is defined as follows.

$$\left[B^{(e)}(\xi,\eta)\right] = \left[\bar{\bar{B}}^{(e)}(\xi,\eta)\right] + \left[\bar{B}^{(e)}(\xi,\eta)\right] \approx \left[\bar{\bar{B}}_{(0,0)}^{(e)}\right] + \left[\bar{B}^{(e)}(\xi,\eta)\right] \quad (11.8.12)$$

where

$$\left[\bar{\bar{B}}^{(e)}_{(0,0)}\right] = \left[\bar{\bar{B}}^{(e)}(\xi = 0, \eta = 0)\right] \tag{11.8.13}$$

Equation (11.8.12) implies that the strain at each point is found by using the actual deviatoric strain at that same point and the volumetric strain at the centroid of the element in the parametric space—that is, the point with $\xi = 0$, $\eta = 0$. After this modification, we can calculate the element stiffness matrix using full integration, without encountering volumetric locking. The vast majority of analysis programs use the B-bar method for 4Q and 8H elements (obviously, the method is applicable to 8H elements as well).

Problems

11.1 The rank of an eight-node, hexahedral (8*H*) isoparametric solid element with uniform reduced integration ($1 \times 1 \times 1 = 1$ quadrature point) is equal to 6 (it can capture a constant strain state, so the rank is simply equal to the number of strain components, which is 6).

a) Prove that the number of spurious zero-energy modes is 12. *Hint:* The element has a total of 24 degrees of freedom. The correct rank of the element would be equal to the difference of the number of possible rigid-body-modes from 24.

b) For each of the three directions, x, y, and z, we have four spurious zero-energy modes. The corresponding covariant basis vectors are $\vec{e}_5 = \vec{h}_1 = [1\ 1\ -1\ -1\ -1\ -1\ 1\ 1]^T$, $\vec{e}_6 = \vec{h}_2 = [1\ -1\ -1\ 1 -1\ 1\ 1 -1]^T$, $\vec{e}_7 = \vec{h}_3 = [1\ -1\ 1\ -1\ 1\ -1\ 1\ -1]^T$, and $\vec{e}_8 = \vec{h}_4 = [1\ -1\ 1\ -1\ -1\ 1\ -1\ 1]^T$. Given the covariant and contravariant basis vectors in Table P11.1, find the remaining contravariant basis vectors $\vec{e}^{\,5}, \vec{e}^{\,6}, \vec{e}^{\,7}$, and $\vec{e}^{\,8}$.

Use these vectors and an hourglass stiffness parameter $K_{HG} = 500$ to obtain the hourglass stiffness matrix of the element in Problem 9.1, if we use one-point quadrature for that element.

Table P11.1 (note: the derivatives of the shape functions are evaluated at $\xi = \eta = \zeta = 0$).

Covariant Basis Vectors	Contrariant Basis Vectors
$\vec{e}_1 = \frac{1}{8}[1\ 1\ 1\ 1\ 1\ 1\ 1\ 1]^T$	$\vec{e}^{\,1} = [1\ 1\ 1\ 1\ 1\ 1\ 1\ 1]^T$
$\vec{e}_2 = \left[x_1^{(e)}\ x_2^{(e)}\ x_3^{(e)}\ x_4^{(e)}\ x_5^{(e)}\ x_6^{(e)}\ x_7^{(e)}\ x_8^{(e)}\right]^T$	$\vec{e}^{\,2} = \left[\frac{\partial N_1^{(8H)}}{\partial x}\ \frac{\partial N_2^{(8H)}}{\partial x}\ \cdots\ \frac{\partial N_7^{(8H)}}{\partial x}\ \frac{\partial N_8^{(8H)}}{\partial x}\right]^T$
$\vec{e}_3 = \left[y_1^{(e)}\ y_2^{(e)}\ y_3^{(e)}\ y_4^{(e)}\ y_5^{(e)}\ y_6^{(e)}\ y_7^{(e)}\ y_8^{(e)}\right]^T$	$\vec{e}^{\,3} = \left[\frac{\partial N_1^{(8H)}}{\partial y}\ \frac{\partial N_2^{(8H)}}{\partial y}\ \cdots\ \frac{\partial N_7^{(8H)}}{\partial y}\ \frac{\partial N_8^{(8H)}}{\partial y}\right]^T$
$\vec{e}_4 = \left[z_1^{(e)}\ z_2^{(e)}\ z_3^{(e)}\ z_4^{(e)}\ z_5^{(e)}\ z_6^{(e)}\ z_7^{(e)}\ z_8^{(e)}\right]^T$	$\vec{e}^{\,4} = \left[\frac{\partial N_1^{(8H)}}{\partial z}\ \frac{\partial N_2^{(8H)}}{\partial z}\ \cdots\ \frac{\partial N_7^{(8H)}}{\partial z}\ \frac{\partial N_8^{(8H)}}{\partial z}\right]^T$

c) Obtain the stiffness matrix (including the hourglass stiffness) of the element in Problem 9.1, for the case that we use one-point quadrature.

References

Belytschko, T., Liu, W., Moran, B., and Elkhodary, K. (2014). *Nonlinear Finite Elements for Continua and Structures*, 2nd ed. Hoboken, NJ: John Wiley & Sons.

Flanagan, D., and Belytschko, T. (1981). "A Uniform Strain Hexahedron and Quadrilateral with Orthogonal Hourglass Control." *International Journal for Numerical Methods in Engineering*, 17, 679–706.

Hughes, T. J. R. (1980). "Generalization of Selective Integration Procedures to Anisotropic and Nonlinear Media." *International Journal for Numerical Methods in Engineering*, 15, 1413–1418.

Hughes, T. J. R. (2000). *The Finite Element Method—Linear Static and Dynamic Finite Element Analysis*, 2nd ed. Mineola, NY: Dover.

Kosloff, D., and Frazier, G. (1978). "Treatment of Hourglass Patterns in Low Order Finite Element Codes." *International Journal for Numerical and Analytical Methods in Geomechanics*, 2, 57–72.

Strang, G., and Fix, G. (2008). *An Analysis of the Finite Element Method*, 2nd ed. Cambridge, MA: Wellesley.

Zienkiewicz, O., Taylor, R., and Too, J. (1971). "Reduced Integration Technique in General Analysis of Plates and Shells." *International Journal for Numerical Methods in Engineering*, 3, 275–290.

Zienkiewicz, O., and Taylor, R. (2005). *The Finite Element Method for Solid and Structural Mechanics*, 6th ed. Oxford: Butterworth-Heinemann.

12

Multifield (Mixed) Finite Elements

So far, we have examined elements for which we approximate a single field, for example, the temperature field for heat conduction and the displacement field for elasticity. In many situations, for a given problem (e.g., elasticity) we may need to establish a finite element approximation (by means of shape functions) for more than one field. For example, we may want to establish separate finite element approximations for the displacement, strain and/or stress fields (instead of following the approach we have seen so far, where the strains are obtained by applying the strain-displacement equations to the approximate displacement field, and then the stresses are obtained by applying the material constitutive law to the approximate strain field). Of course, as we will see in the following sections, using independent approximations for different fields that are in fact related (e.g., by means of strain-displacement equations or constitutive laws) introduces the need to weakly satisfy the mathematical relation between these fields. Elements with independent (separate) approximations for fields such as stresses, strains, and displacements are called *multifield elements or mixed elements*. The corresponding weak forms that allow multifield element approximations are called *multifield weak forms*. The reason why mixed elements are employed in practice is to resolve potential difficulties that may arise due to, for example, parasitic shear stiffness in analyses of flexure in beams or volumetric locking in incompressible elasticity, without a need to employ the approaches of the previous chapter (e.g., incompatible modes or uniform reduced integration with hourglass control) and without a need to resort to an overly fine mesh.

The present chapter will focus on elasticity problems. We will begin by establishing a multifield weak form for the general case where we want to introduce independent approximations for the displacements, stresses and strains. We will then identify this weak form with the celebrated *Hu-Washizu variational principle*. We will also discuss a two-field (displacement-stress) weak form, corresponding to the *Hellinger-Reissner variational principle*. The general multifield finite element equations (where we introduce separate approximations for the displacements, strains and stresses) will then be formulated. Subsequently, the chapter will discuss several popular, mixed finite element formulations.

Fundamentals of Finite Element Analysis: Linear Finite Element Analysis, First Edition.
Ioannis Koutromanos, James McClure, and Christopher Roy.

Companion website: www.wiley.com/go/koutromanos/linear

12.1 Multifield Weak Forms for Elasticity

Before proceeding, let us revisit the strong form—that is, the conditions that must be satisfied by the exact solution—of a three-dimensional elasticity problem. These conditions are presented in Box 12.1.1.

Box 12.1.1 Strong Form for Three-Dimensional Elasticity

- Differential equations of equilibrium:

$$\sum_{j=1}^{3}\left(\frac{\partial \sigma_{ij}}{\partial x_j}\right) + b_i = 0, \text{for } i = 1, 2, 3$$

- Essential (prescribed displacement) and natural (prescribed traction) boundary conditions:

$$u_i = \bar{u}_i \text{ on } \Gamma_{ui} \text{(essential)}, \sum_{j=1}^{3}\left(\sigma_{ij} \cdot n_j\right) = t_i \text{ on } \Gamma_{ti} \text{(natural)}$$

- Strain-displacement relation:

$$\varepsilon_{ij} = \frac{1}{2}\left(\frac{\partial u_i}{\partial x_j} + \frac{\partial u_j}{\partial x_i}\right)$$

- Stress-strain relation (i.e., the constitutive law of the material):

$$\{\sigma\} = [D]\{\varepsilon\}$$

The equations in Box 12.1.1 involve three vector fields:

1) The displacement field, $\{u\}$
2) The stress field, $\{\sigma\}$
3) The strain field, $\{\varepsilon\}$

Remark 12.1.1: If we wanted to be perfectly rigorous, we would state that the stress and strain fields are tensor fields, not vector fields. ▮

So far, we have been establishing the strong and weak forms in terms of one field alone, the displacement field. Once we have the displacement field, we can use the strain-displacement relation to find the strain field. Then, using the stress-strain law, we can calculate the stress field. Such approaches are commonly referred to as *single-field* or *irreducible formulations*.

Before proceeding, we cast the strain-displacement relation in vector form, in accordance with the following expression.

$$\{\varepsilon\} = \{\nabla_\varepsilon u\} \tag{12.1.1}$$

where

$$\{\varepsilon\} = \begin{bmatrix} \varepsilon_{xx} & \varepsilon_{yy} & \varepsilon_{zz} & \gamma_{xy} & \gamma_{yz} & \gamma_{zx} \end{bmatrix}^T \tag{12.1.2}$$

and

$$\{\nabla_\varepsilon u\} = \begin{bmatrix} \frac{\partial u_x}{\partial x} & \frac{\partial u_y}{\partial y} & \frac{\partial u_z}{\partial z} & \frac{\partial u_x}{\partial y} + \frac{\partial u_y}{\partial x} & \frac{\partial u_y}{\partial z} + \frac{\partial u_z}{\partial y} & \frac{\partial u_z}{\partial x} + \frac{\partial u_x}{\partial z} \end{bmatrix}^T \tag{12.1.3}$$

We will now establish a different version of weak form, called a *multifield* weak form or *reducible* weak form. A reducible weak form will include the displacements, the stresses and the strains as separate fields. While this may appear strange at first, this weak form allows us to introduce separate finite element approximations for the displacements, stresses and strains, which in turn proves to be "convenient" for several situations. The term *reducible weak form* means that, at any stage of our derivations, we have the capability to reduce the number of fields, by plugging in the strain-displacement law and/or the stress-strain law.

As a first step to obtain a multifield weak form, we consider an arbitrary virtual displacement field, $\{w\}$, which vanishes on the essential boundary. Using this virtual displacement field, we then establish the weak form of the governing equations, just like we did in Section 7.9:

$$\iiint_\Omega \{\nabla_\varepsilon w\}^T \{\sigma\} dV - \left(\iiint_\Omega \{w\}^T \{b\} dV + \iint_{\Gamma_{tx}} w_x t_x \, dS + \iint_{\Gamma_{ty}} w_y t_y \, dS + \iint_{\Gamma_{tz}} w_z t_z \, dS \right) = 0 \tag{12.1.4}$$

where

$$\{\nabla_\varepsilon w\} = \begin{bmatrix} \frac{\partial w_x}{\partial x} & \frac{\partial w_y}{\partial y} & \frac{\partial w_z}{\partial z} & \frac{\partial w_x}{\partial y} + \frac{\partial w_y}{\partial x} & \frac{\partial w_y}{\partial z} + \frac{\partial w_z}{\partial y} & \frac{\partial w_z}{\partial x} + \frac{\partial w_x}{\partial z} \end{bmatrix}^T \tag{12.1.5}$$

$$\{\sigma\} = \begin{bmatrix} \sigma_{xx} & \sigma_{yy} & \sigma_{zz} & \sigma_{xy} & \sigma_{yz} & \sigma_{zx} \end{bmatrix}^T \tag{12.1.6}$$

Next, we take all the terms of Equation (12.1.1) to the right-hand side and premultiply by the transpose of an arbitrary, six-component, vector field $\{\bar{\sigma}\}$ (the selection of the letter σ is not by chance—it may be obvious what is the physical meaning of $\{\bar{\sigma}\}$!):

$$\{\bar{\sigma}\}^T (\{\nabla_\varepsilon u\} - \{\varepsilon\}) = 0 \rightarrow \{\bar{\sigma}\}^T \{\nabla_\varepsilon u\} - \{\bar{\sigma}\}^T \{\varepsilon\} = 0 \tag{12.1.7}$$

We can now integrate Equation (12.1.7) over the domain of the problem to obtain the following expression:

$$\iiint_\Omega \left(\{\bar{\sigma}\}^T \{\nabla_\varepsilon u\} - \{\bar{\sigma}\}^T \{\varepsilon\} \right) dV = 0 \tag{12.1.8}$$

Finally, we cast the stress-strain law from Box 12.1.1 in the following form:

$$[D]\{\varepsilon\} - \{\sigma\} = \{0\} \tag{12.1.9}$$

and then premultiply it by the transpose of an arbitrary six-component vector $\{\bar{\varepsilon}\}$ (once again, the physical meaning of $\{\bar{\varepsilon}\}$ may be obvious) to obtain:

$$\{\bar{\varepsilon}\}^T([D]\{\varepsilon\}-\{\sigma\})=0 \rightarrow \{\bar{\varepsilon}\}^T[D]\{\varepsilon\}-\{\bar{\varepsilon}\}^T\{\sigma\}=0 \tag{12.1.10}$$

If we then integrate over the entire domain, we have the following equation.

$$\iiint_{\Omega}\left(\{\bar{\varepsilon}\}^T[D]\{\varepsilon\}-\{\bar{\varepsilon}\}^T\{\sigma\}\right)dV=0 \tag{12.1.11}$$

Finally, we add Equations (12.1.4), (12.1.8) and (12.1.11) to obtain the following *multi-field* (more precisely: three-field) *weak form*:

$$\iiint_{\Omega}\left(\{\nabla_\varepsilon w\}^T\{\sigma\}+\{\bar{\sigma}\}^T\{\nabla_\varepsilon u\}-\{\bar{\sigma}\}^T\{\varepsilon\}+\{\bar{\varepsilon}\}^T[D]\{\varepsilon\}-\{\bar{\varepsilon}\}^T\{\sigma\}\right)dV$$

$$-\left(\iiint_{\Omega}\{w\}^T\{b\}dV+\iint_{\Gamma_{tx}}w_x t_x\,dS+\iint_{\Gamma_{ty}}w_y t_y\,dS+\iint_{\Gamma_{tz}}w_z t_z\,dS\right)=0\rightarrow$$

$$\rightarrow \boxed{\begin{aligned}&\iiint_{\Omega}\{\bar{\varepsilon}\}^T[D]\{\varepsilon\}dV+\iiint_{\Omega}\left(\{\nabla_\varepsilon w\}^T-\{\bar{\varepsilon}\}^T\right)\{\sigma\}dV+\iiint_{\Omega}\{\bar{\sigma}\}^T(\{\nabla_\varepsilon u\}-\{\varepsilon\})dV\\&=\iiint_{\Omega}\{w\}^T\{b\}dV+\iint_{\Gamma_{tx}}w_x t_x\,dS+\iint_{\Gamma_{ty}}w_y t_y\,dS+\iint_{\Gamma_{tz}}w_z t_z\,dS\end{aligned}}$$

$$\tag{12.1.12}$$

Remark 12.1.2: Equation (12.1.12) is a *generalized version of the principle of virtual work.* In the specific equation, we have additional virtual work terms, stemming from violation of the strain-displacement law and of the stress-strain law! Specifically, our multifield finite element equations will not exactly (strongly) satisfy the strain-displacement relation and the stress-strain law. We will only weakly satisfy these conditions (by virtue of the integral Equations 12.1.8 and 12.1.11). The field $\{\bar{\varepsilon}\}$ is a *virtual strain field*, while $\{\bar{\sigma}\}$ is a *virtual stress field.* ■

We now want to try and establish the multifield weak form of Equation (12.1.12) as a variational one, in a fashion similar to that presented in Appendix C for one-dimensional elasticity problems. In the specific appendix, we obtain the weak form from the minimization of a strain energy functional. Along the same lines, we consider the following potential energy functional, W, which depends on $\{u\}$, $\{\sigma\}$, and $\{\varepsilon\}$.

$$W(\{u\},\{\varepsilon\},\{\sigma\})=\iiint_{\Omega}\frac{1}{2}\{\varepsilon\}^T[D]\{\varepsilon\}dV+\iiint_{\Omega}\frac{1}{2}\{\sigma\}^T(\{\varepsilon\}-\{\nabla_\varepsilon u\})\,dV$$

$$-\iiint_{\Omega}\{u\}^T\{b\}dV-\iint_{\Gamma_{tx}}u_x t_x\,dS-\iint_{\Gamma_{ty}}u_y t_y\,dS-\iint_{\Gamma_{tz}}u_z t_z\,dS \tag{12.1.13}$$

Now, if we consider arbitrary small variations $\{\delta u\}$, $\{\delta\sigma\}$, and $\{\delta\varepsilon\}$ to the three fields, we can find the corresponding modified value of the potential energy functional:

$$
\begin{aligned}
&W(\{u\}+\{\delta u\},\{\varepsilon\}+\{\delta\varepsilon\},\{\sigma\}+\{\delta\sigma\})\\
&=\iiint_{\Omega}\frac{1}{2}(\{\varepsilon\}+\{\delta\varepsilon\})^T[D](\{\varepsilon\}+\{\delta\varepsilon\})dV\\
&+\iiint_{\Omega}\frac{1}{2}(\{\sigma\}+\{\delta\sigma\})^T[(\{\varepsilon\}+\{\delta\varepsilon\})-(\{\nabla_\varepsilon u\}+\{\nabla_\varepsilon\delta u\})]\,dV\\
&-\iiint_{\Omega}(\{u\}+\{\delta u\})^T\{b\}dV-\iint_{\Gamma_{tx}}(u_x+\delta u_x)t_x\,dS-\iint_{\Gamma_{ty}}(u_y+\delta u_y)t_y\,dS\\
&-\iint_{\Gamma_{tz}}(u_z+\delta u_z)t_z\,dS
\end{aligned}
\tag{12.1.14}
$$

We now define the variation δW of the functional $W(\{u\},\{\varepsilon\},\{\sigma\})$:

$$\delta W=W(\{u\}+\{\delta u\},\{\varepsilon\}+\{\delta\varepsilon\},\{\sigma\}+\{\delta\sigma\})-W(\{u\},\{\varepsilon\},\{\sigma\}) \tag{12.1.15}$$

In accordance with Appendix C, we can once again obtain an integral expression, by stipulating that the actual solution fields $\{u\}$, $\{\varepsilon\}$ and $\{\sigma\}$ must give a zero first variation to the functional, for any arbitrary set of small variations $\{\delta u\}$, $\{\delta\varepsilon\}$, $\{\delta\sigma\}$:

$$
\begin{aligned}
\delta W=0\rightarrow&\iiint_{\Omega}\{\delta\varepsilon\}^T[D]\{\varepsilon\}dV+\underline{\frac{1}{2}\iiint_{\Omega}\{\delta\varepsilon\}^T[D]\{\delta\varepsilon\}dV}\\
&+\iiint_{\Omega}\left(\{\nabla_\varepsilon\delta u\}^T-\{\delta\varepsilon\}^T\right)\{\sigma\}dV+\iiint_{\Omega}\{\delta\sigma\}^T(\{\nabla_\varepsilon u\}-\{\varepsilon\})dV\\
&-\left(\iiint_{\Omega}\{\delta u\}^T\{b\}dV+\iint_{\Gamma_{tx}}\delta u_x t_x\,dS+\iint_{\Gamma_{ty}}\delta u_y t_y\,dS+\iint_{\Gamma_{tz}}\delta u_z t_z\,dS\right)=0
\end{aligned}
$$

The underlined term in the above expression is much smaller than the other terms (because the variations are very small, and this term is the "product of two variations!"), thus, we can neglect it to eventually obtain the expression in Box 12.1.2, which corresponds to the *three-field, Hu-Washizu variational principle*. The Hu-Washizu principle is commonly referred to in textbooks, when it comes to mixed element formulations.

Box 12.1.2 Hu-Washizu Three-Field Variational Principle

$$\iiint_{\Omega}\{\delta\varepsilon\}^T[D]\{\varepsilon\}dV+\iiint_{\Omega}\left(\{\nabla_\varepsilon\delta u\}^T-\{\delta\varepsilon\}^T\right)\{\sigma\}dV+\iiint_{\Omega}\{\delta\sigma\}^T(\{\nabla_\varepsilon u\}-\{\varepsilon\})dV$$

$$-\left(\iiint_{\Omega}\{\delta u\}^T\{b\}dV+\iint_{\Gamma_{tx}}\delta u_x t_x\,dS+\iint_{\Gamma_{ty}}\delta u_y t_y\,dS+\iint_{\Gamma_{tz}}\delta u_z t_z\,dS\right)=0 \tag{12.1.16}$$

Remark 12.1.3: One can easily verify that Equations (12.1.12) and (12.1.16) are identical, if we set $\{\delta u\}=c\cdot\{w\}$, $\{\delta\varepsilon\}=c\cdot\{\bar{\varepsilon}\}$, $\{\delta\sigma\}=c\cdot\{\bar{\sigma}\}$, where c is a very small number. ∎

Now, it is worth mentioning that there is a "reduced" version of the general Hu-Washizu form, called the *Hellinger-Reissner variational principle*. The specific principle is established for two fields, that is, the stresses and the displacements. The strains are assumed to be obtained from the exact satisfaction of the strain-displacement law. That is, Equation (12.1.1) is *exactly* (strongly) satisfied at all times. What will be weakly satisfied is the equilibrium conditions and the stress-strain (constitutive) law. Notice that, when we established a "weak" term for the constitutive law in Equation (12.1.11), we premultiplied by the transpose of a virtual strain field, $\{\bar{\varepsilon}\}$. Now, we want to premultiply by the transpose of the virtual stress field $\{\bar{\sigma}\}$, but there is a need for modification: Since $\{\bar{\sigma}\}$ has the physical meaning of "stress," we must use it to premultiply a strain quantity and then integrate, otherwise the meaning of the corresponding integral "weak" term will no longer correspond to a virtual work from violation of the constitutive law. Thus, we need to somehow modify the left-hand side of Equation (12.1.9), $[D]\{\varepsilon\}-\{\sigma\}$, so that it becomes a strain quantity. To this end, we premultiply Equation (12.1.9) by the inverse of the material stiffness matrix, $[D]$:

$$[D]^{-1}([D]\{\varepsilon\}-\{\sigma\})=\{0\}\rightarrow\{\varepsilon\}-[D]^{-1}\{\sigma\}=\{0\} \tag{12.1.17}$$

The quantities at the left-hand side of Equation (12.1.17) are indeed strain quantities (if we premultiply the stress vector by the inverse of the material stiffness matrix, we obtain a strain vector). We now pre-multiply this equation by the transpose of the virtual stress field, $\{\bar{\sigma}\}$, and then integrate over the domain of the problem:

$$\iiint_{\Omega}\left(\{\bar{\sigma}\}^T\{\varepsilon\}-\{\bar{\sigma}\}^T[D]^{-1}\{\sigma\}\right)dV=0 \tag{12.1.18}$$

The displacement-stress mixed weak form is then obtained by summing Equations (12.1.18) and (12.1.4), after setting $\{\bar{\varepsilon}\}-\{\nabla_\varepsilon w\}=\{0\}$ (since the strain-displacement Equations 12.1.1 are now exactly satisfied for both the actual and the virtual displacement field). We eventually establish the following *two-field (stress-displacement) weak form*:

$$\begin{aligned}&\iiint_{\Omega}\{\bar{\varepsilon}\}^T\{\sigma\}dV+\iiint_{\Omega}\{\bar{\sigma}\}^T\left(\{\varepsilon\}-[D]^{-1}\{\sigma\}\right)dV\\&=\iiint_{\Omega}\{w\}^T\{b\}dV+\iint_{\Gamma_{tx}}w_xt_x\,dS+\iint_{\Gamma_{ty}}w_yt_y\,dS+\iint_{\Gamma_{tz}}w_zt_z\,dS\end{aligned}\tag{12.1.19}$$

The weak form (12.1.19) can also be obtained from variational considerations to give the celebrated Hellinger-Reissner variational principle, provided in Box 12.1.3.

Box 12.1.3 Hellinger-Reissner Two-Field Variational Principle

$$\begin{aligned}&\iiint_{\Omega}\{\delta\varepsilon\}^T\{\sigma\}dV+\iiint_{\Omega}\{\delta\sigma\}^T\left(\{\varepsilon\}-[D]^{-1}\{\sigma\}\right)dV\\&-\left(\iiint_{\Omega}\{\delta u\}^T\{b\}dV+\iint_{\Gamma_{tx}}\delta u_xt_x\,dS+\iint_{\Gamma_{ty}}\delta u_yt_y\,dS+\iint_{\Gamma_{tz}}\delta u_zt_z\,dS\right)=0\end{aligned}\tag{12.1.20}$$

Remark 12.1.4: If we want to be perfectly rigorous, we need to accompany the integral expression of a multifield weak form with the stipulation that it applies for any arbitrary $\{w\}$ whose components vanish at the corresponding essential boundary segments, for any arbitrary virtual stress $\{\bar{\sigma}\}$ and (if we have the three-field weak form) for any arbitrary virtual strain field $\{\bar{\varepsilon}\}$. ∎

12.2 Mixed (Multifield) Finite Element Formulations

The next step after having obtained the multifield weak form is to introduce the finite element approximations for the various fields. We will present the general case where all three fields (displacements, stresses, and strains) are approximated. We consider a finite element discretization into N_e elements. Each element has n nodal points. We will introduce approximations for the displacements, stresses and strains using shape functions. In general, the shape functions that we use for the displacements, stresses and strains are not the same! Thus, we can write the following expressions for the approximate displacement field.

$$u_x(x,y,z)\approx u_x^{(e)}(x,y,z)=\sum_{i=1}^{n}\left(N_i^{(e)}(x,y,z)\cdot u_{xi}^{(e)}\right)\tag{12.2.1a}$$

$$u_y(x,y,z) \approx u_y^{(e)}(x,y,z) = \sum_{i=1}^{n} \left(N_i^{(e)}(x,y,z) \cdot u_{yi}^{(e)} \right) \tag{12.2.1b}$$

$$u_z(x,y,z) \approx u_z^{(e)}(x,y,z) = \sum_{i=1}^{n} \left(N_i^{(e)}(x,y,z) \cdot u_{zi}^{(e)} \right) \tag{12.2.1c}$$

We can collectively write:

$$\left\{u^{(e)}\right\} = \left[u_x^{(e)}(x,y,z) \;\; u_y^{(e)}(x,y,z) \;\; u_z^{(e)}(x,y,z)\right]^T \approx \left[N^{(e)}\right]\left\{U^{(e)}\right\} \tag{12.2.2}$$

where $\{U^{(e)}\}$ is the nodal displacement vector of the element:

$$\left\{U^{(e)}\right\} = \left[u_{x1}^{(e)} \;\; u_{y1}^{(e)} \;\; u_{z1}^{(e)} \;\; u_{x2}^{(e)} \;\; u_{y2}^{(e)} \;\; u_{z2}^{(e)} \;\; \ldots \;\; u_{xn}^{(e)} \;\; u_{yn}^{(e)} \;\; u_{zn}^{(e)}\right]^T$$

and $[N^{(e)}]$ is the shape function array that has been defined in Section 9.2:

$$\left[N^{(e)}\right] = \begin{bmatrix} N_1^{(e)} & 0 & 0 & N_2^{(e)} & 0 & 0 & \ldots & N_n^{(e)} & 0 & 0 \\ 0 & N_1^{(e)} & 0 & 0 & N_2^{(e)} & 0 & \ldots & 0 & N_n^{(e)} & 0 \\ 0 & 0 & N_1^{(e)} & 0 & 0 & N_2^{(e)} & \ldots & 0 & 0 & N_n^{(e)} \end{bmatrix} \tag{12.2.3}$$

It can be verified that, given Equation (12.2.2), the vector field $\{\nabla_\varepsilon u\}$ (defined in Equation 12.1.1), can be obtained from the following expression.

$$\left\{\nabla_\varepsilon u^{(e)}\right\} \approx \left[B^{(e)}\right]\left\{U^{(e)}\right\} \tag{12.2.4}$$

where $[B^{(e)}]$ is the "strain-displacement matrix" that we defined in Section 9.2.

Since the three-field weak form includes a total of three fields, we now need to establish finite element approximations for the strains and stresses:

$$\left\{\varepsilon^{(e)}\right\} = \left[\varepsilon_{xx}^{(e)} \;\; \varepsilon_{yy}^{(e)} \;\; \varepsilon_{zz}^{(e)} \;\; \gamma_{xy}^{(e)} \;\; \gamma_{yz}^{(e)} \;\; \gamma_{zx}^{(e)}\right]^T \approx \left[N_\varepsilon^{(e)}\right]\left\{E^{(e)}\right\} \tag{12.2.5}$$

where $\{E^{(e)}\}$ is a vector containing the strain degrees of freedom (DOFs) of element e and

$$\left\{\sigma^{(e)}\right\} = \left[\sigma_{xx}^{(e)} \;\; \sigma_{yy}^{(e)} \;\; \sigma_{zz}^{(e)} \;\; \sigma_{xy}^{(e)} \;\; \sigma_{yz}^{(e)} \;\; \sigma_{zx}^{(e)}\right]^T \approx \left[N_\sigma^{(e)}\right]\left\{\Sigma^{(e)}\right\} \tag{12.2.6}$$

where $\{\Sigma^{(e)}\}$ is a vector containing the stress DOFs of element e.

Remark 12.2.1: The strain and stress DOFs, $\{E^{(e)}\}$ and $\{\Sigma^{(e)}\}$, respectively, are not necessarily nodal values of stresses and strains. That is, we cannot generally say that $\{E^{(e)}\}$ is a nodal strain vector and $\{\Sigma^{(e)}\}$ is a nodal stress vector. Instead, these two vectors contain constant parameters of an approximate polynomial solution, similar to the parameters a in the approximate solution that we established in Example 2.4 and the parameters $\{a^{(e)}\}$ of the incompatible-mode quadrilateral element in Section 11.3. ■

Next, we use the same finite element approximations for virtual displacements, stresses, and strains inside each element e:

$$\left\{w^{(e)}\right\} \approx \left[N^{(e)}\right]\left\{W^{(e)}\right\} \rightarrow \left\{w^{(e)}\right\}^T \approx \left\{W^{(e)}\right\}^T \left[N^{(e)}\right]^T \tag{12.2.7a}$$

also:

$$\left\{\nabla_\varepsilon w^{(e)}\right\}^T \approx \left\{W^{(e)}\right\}^T \left[B^{(e)}\right]^T \tag{12.2.7b}$$

$$\left\{\bar{\varepsilon}^{(e)}\right\} \approx \left[N_\varepsilon^{(e)}\right]\left\{\bar{E}^{(e)}\right\} \rightarrow \left\{\bar{\varepsilon}^{(e)}\right\}^T \approx \left\{\bar{E}^{(e)}\right\}^T \left[N_\varepsilon^{(e)}\right]^T \tag{12.2.7c}$$

$$\left\{\bar{\sigma}^{(e)}\right\} \approx \left[N_\sigma^{(e)}\right]\left\{\bar{\Sigma}^{(e)}\right\} \rightarrow \left\{\bar{\sigma}^{(e)}\right\}^T \approx \left\{\bar{\Sigma}^{(e)}\right\}^T \left[N_\sigma^{(e)}\right]^T \tag{12.2.7d}$$

The vector $\{W^{(e)}\}$ contains the nodal virtual displacements of the element, while $\{\bar{E}^{(e)}\}$ and $\left\{\bar{\Sigma}^{(e)}\right\}$ contain constant virtual strain and stress parameters, respectively. Now, we can plug the finite element approximation into the three-field weak form, that is, Equation (12.1.12). We can separate the integrals into the individual element contributions and write an equation of the following form.

$$\sum_{e=1}^{Ne}\left[\iiint_{\Omega^{(e)}}\left[\{\bar{\varepsilon}\}^T[D]\{\varepsilon\}+\left(\{\nabla_\varepsilon w\}^T-\{\bar{\varepsilon}\}^T\right)\{\sigma\}+\{\bar{\sigma}\}^T(\{\nabla_\varepsilon u\}-\{\varepsilon\})\right]dV\right]$$

$$=\sum_{e=1}^{Ne}\left(\iiint_{\Omega^{(e)}}\{w\}^T\{b\}dV\right)+\sum_{e=1}^{Ne}\left(\iint_{\Gamma_{tx}^{(e)}}\{w\}^T\{t_x\}\,dS\right)+\sum_{e=1}^{Ne}\left(\iint_{\Gamma_{ty}^{(e)}}\{w\}^T\{t_y\}\,dS\right)$$

$$+\sum_{e=1}^{Ne}\left(\iint_{\Gamma_{tz}^{(e)}}\{w\}^T\{t_z\}\,dS\right)$$

$$\sum_{e=1}^{Ne}\left(\bar{W}_{int}^{(e)}\right)=\sum_{e=1}^{Ne}\left(\bar{W}_{ext}^{(e)}\right) \tag{12.2.8}$$

where $\bar{W}_{int}^{(e)}$ is the contribution of element e to the left-hand side of the weak form, that is, the element contribution to the internal virtual work, and $\bar{W}_{ext}^{(e)}$ is the contribution of element e to the right-hand side of the weak form, that is, the element contribution to the external virtual work.

For each element e, we have:

$$\bar{W}_{int}^{(e)}=\iiint_{\Omega^{(e)}}\left[\{\bar{\varepsilon}\}^T[D]\{\varepsilon\}+\left(\{\nabla_\varepsilon w\}^T-\{\bar{\varepsilon}\}^T\right)\{\sigma\}+\{\bar{\sigma}\}^T(\{\nabla_\varepsilon u\}-\{\varepsilon\})\right]dV$$

$$=\iiint_{\Omega^{(e)}}\{\bar{\varepsilon}\}^T[D]\{\varepsilon\}dV+\iiint_{\Omega^{(e)}}\{\nabla_\varepsilon w\}^T\{\sigma\}dV-\iiint_{\Omega^{(e)}}\{\bar{\varepsilon}\}^T\{\sigma\}dV$$

$$+\iiint_{\Omega^{(e)}}\{\bar{\sigma}\}^T\{\varepsilon\}dV-\iiint_{\Omega^{(e)}}\{\bar{\sigma}\}^T\{\nabla_\varepsilon u\}dV$$

If we now plug the finite element approximations for the actual and virtual displacement, stress and strain fields, we obtain:

$$\bar{W}_{int}^{(e)} = \iiint\limits_{\Omega^{(e)}} \left\{\bar{E}^{(e)}\right\}^T \left[N_\varepsilon^{(e)}\right]^T \left[D^{(e)}\right] \left[N_\varepsilon^{(e)}\right] \left\{E^{(e)}\right\} dV$$

$$+ \iiint\limits_{\Omega^{(e)}} \left\{W^{(e)}\right\}^T \left[B^{(e)}\right]^T \left[N_\sigma^{(e)}\right] \left\{\Sigma^{(e)}\right\} dV - \iiint\limits_{\Omega^{(e)}} \left\{\bar{E}^{(e)}\right\}^T \left[N_\varepsilon^{(e)}\right]^T \left[N_\sigma^{(e)}\right] \left\{\Sigma^{(e)}\right\} dV$$

$$+ \iiint\limits_{\Omega^{(e)}} \left\{\bar{\Sigma}^{(e)}\right\}^T \left[N_\sigma^{(e)}\right]^T \left[B^{(e)}\right] \left\{U^{(e)}\right\} dV - \iiint\limits_{\Omega^{(e)}} \left\{\bar{\Sigma}^{(e)}\right\}^T \left[N_\sigma^{(e)}\right]^T \left[N_\varepsilon^{(e)}\right] \left\{E^{(e)}\right\} dV$$

Now, we can take the constant vectors $\{U^{(e)}\}$, $\{E^{(e)}\}$, $\{\Sigma^{(e)}\}$, $\{W^{(e)}\}^T$, $\{\bar{E}^{(e)}\}^T$ and $\left\{\bar{\Sigma}^{(e)}\right\}^T$ outside of the integrals to obtain the following equation.

$$\bar{W}_{int}^{(e)} = \left\{\bar{E}^{(e)}\right\}^T \left[\left(\iiint\limits_{\Omega^{(e)}} \left[N_\varepsilon^{(e)}\right]^T \left[D^{(e)}\right] \left[N_\varepsilon^{(e)}\right] dV \right) \left\{E^{(e)}\right\} - \left(\iiint\limits_{\Omega^{(e)}} \left[N_\varepsilon^{(e)}\right]^T \left[N_\sigma^{(e)}\right] dV \right) \left\{\Sigma^{(e)}\right\} \right] + \left\{W^{(e)}\right\}^T \left(\iiint\limits_{\Omega^{(e)}} \left[B^{(e)}\right]^T \left[N_\sigma^{(e)}\right] dV \right) \left\{\Sigma^{(e)}\right\}$$

$$+ \left\{\bar{\Sigma}^{(e)}\right\}^T \left[\left(\iiint\limits_{\Omega^{(e)}} \left[N_\sigma^{(e)}\right]^T \left[B^{(e)}\right] dV \right) \left\{U^{(e)}\right\} - \left(\iiint\limits_{\Omega^{(e)}} \left[N_\sigma^{(e)}\right]^T \left[N_\varepsilon^{(e)}\right] dV \right) \left\{E^{(e)}\right\} \right] \tag{12.2.9}$$

We now define the following *multifield element arrays*:

$$\left[k_{\varepsilon\varepsilon}^{(e)}\right] = \iiint\limits_{\Omega^{(e)}} \left[N_\varepsilon^{(e)}\right]^T \left[D^{(e)}\right] \left[N_\varepsilon^{(e)}\right] dV \tag{12.2.10a}$$

$$\left[k_{\varepsilon\sigma}^{(e)}\right] = -\iiint\limits_{\Omega^{(e)}} \left[N_\varepsilon^{(e)}\right]^T \left[N_\sigma^{(e)}\right] dV \tag{12.2.10b}$$

$$\left[k_{\sigma\varepsilon}^{(e)}\right] = -\iiint\limits_{\Omega^{(e)}} \left[N_\sigma^{(e)}\right]^T \left[N_\varepsilon^{(e)}\right] dV \tag{12.2.10c}$$

$$\left[k_{u\sigma}^{(e)}\right] = \iiint\limits_{\Omega^{(e)}} \left[B^{(e)}\right]^T \left[N_\sigma^{(e)}\right] dV \tag{12.2.10d}$$

$$\left[k_{\sigma u}^{(e)}\right] = \iiint\limits_{\Omega^{(e)}} \left[N_\sigma^{(e)}\right]^T \left[B^{(e)}\right] dV \tag{12.2.10e}$$

If we account for Equations (12.2.10a–e), Equation (12.2.9) can be written as follows:

$$\bar{W}_{int}^{(e)} = \left\{\bar{E}^{(e)}\right\}^T \left(\left[k_{\varepsilon\varepsilon}^{(e)}\right]\left\{E^{(e)}\right\} + \left[k_{\varepsilon\sigma}^{(e)}\right]\left\{\Sigma^{(e)}\right\}\right) + \left\{W^{(e)}\right\}^T \left[k_{u\sigma}^{(e)}\right]\left\{\Sigma^{(e)}\right\}$$
$$+ \left\{\bar{\Sigma}^{(e)}\right\}^T \left(\left[k_{\sigma u}^{(e)}\right]\left\{U^{(e)}\right\} + \left[k_{\sigma\varepsilon}^{(e)}\right]\left\{E^{(e)}\right\}\right)$$

and we can cast the obtained expression in a partitioned-matrix equation form:

$$\bar{W}_{int}^{(e)} = \begin{Bmatrix} \left\{\bar{E}^{(e)}\right\} \\ \left\{\bar{\Sigma}^{(e)}\right\} \\ \left\{W^{(e)}\right\} \end{Bmatrix}^T \begin{bmatrix} \left[k_{\varepsilon\varepsilon}^{(e)}\right] & \left[k_{\varepsilon\sigma}^{(e)}\right] & [0] \\ \left[k_{\sigma\varepsilon}^{(e)}\right] & [0] & \left[k_{\sigma u}^{(e)}\right] \\ [0] & \left[k_{u\sigma}^{(e)}\right] & [0] \end{bmatrix} \begin{Bmatrix} \left\{E^{(e)}\right\} \\ \left\{\Sigma^{(e)}\right\} \\ \left\{U^{(e)}\right\} \end{Bmatrix} = \left\{\bar{U}_{\varepsilon\sigma u}^{(e)}\right\}^T \left[k^{(e)}\right]\left\{U_{\varepsilon\sigma u}^{(e)}\right\} \tag{12.2.11}$$

where the *coefficient array, $[k^{(e)}]$, of the three-field mixed element* is given by:

$$\left[k^{(e)}\right] = \begin{bmatrix} \left[k_{\varepsilon\varepsilon}^{(e)}\right] & \left[k_{\varepsilon\sigma}^{(e)}\right] & [0] \\ \left[k_{\sigma\varepsilon}^{(e)}\right] & [0] & \left[k_{\sigma u}^{(e)}\right] \\ [0] & \left[k_{u\sigma}^{(e)}\right] & [0] \end{bmatrix} \tag{12.2.12}$$

and the element DOF vector, $\left\{U_{\varepsilon\sigma u}^{(e)}\right\}$, includes all the element solution quantities

$$\left\{U_{\varepsilon\sigma u}^{(e)}\right\} = \begin{Bmatrix} \left\{E^{(e)}\right\} \\ \left\{\Sigma^{(e)}\right\} \\ \left\{U^{(e)}\right\} \end{Bmatrix} \tag{12.2.13}$$

Similarly, $\left\{\bar{U}_{\varepsilon\sigma u}^{(e)}\right\}$ in Equation (12.2.11) is the virtual element DOF vector, including all the element virtual parameters (i.e., the virtual strain DOFs, virtual stress DOFs, and virtual nodal displacements).

Remark 12.2.2: We can easily verify that the array $[k^{(e)}]$ in Equation (12.2.12) is symmetric! This is verified as follows. First, we can establish the following expressions from the definitions of the various subarrays:

$$\left[k_{\varepsilon\varepsilon}^{(e)}\right]^T = \left[k_{\varepsilon\varepsilon}^{(e)}\right] \left(\left[k_{\varepsilon\varepsilon}^{(e)}\right] \text{ is symmetric}\right) \tag{12.2.14a}$$

$$\left[k_{\sigma\varepsilon}^{(e)}\right] = \left[k_{\varepsilon\sigma}^{(e)}\right]^T \tag{12.2.14b}$$

$$\left[k_{u\sigma}^{(e)}\right] = \left[k_{\sigma u}^{(e)}\right]^T \tag{12.2.14c}$$

Then, taking the transpose of $[k^{(e)}]$, in accordance with matrix linear algebra for block matrices:

$$\left[k^{(e)}\right]^T = \begin{bmatrix} \left[k_{\varepsilon\varepsilon}^{(e)}\right] & \left[k_{\varepsilon\sigma}^{(e)}\right] & [0] \\ \left[k_{\sigma\varepsilon}^{(e)}\right] & [0] & \left[k_{\sigma u}^{(e)}\right] \\ [0] & \left[k_{u\sigma}^{(e)}\right] & [0] \end{bmatrix}^T = \begin{bmatrix} \left[k_{\varepsilon\varepsilon}^{(e)}\right]^T & \left[k_{\sigma\varepsilon}^{(e)}\right]^T & [0] \\ \left[k_{\varepsilon\sigma}^{(e)}\right]^T & [0] & \left[k_{u\sigma}^{(e)}\right]^T \\ [0] & \left[k_{\sigma u}^{(e)}\right]^T & [0] \end{bmatrix}$$

$$= \begin{bmatrix} \left[k_{\varepsilon\varepsilon}^{(e)}\right] & \left[k_{\varepsilon\sigma}^{(e)}\right] & [0] \\ \left[k_{\sigma\varepsilon}^{(e)}\right] & [0] & \left[k_{\sigma u}^{(e)}\right] \\ [0] & \left[k_{u\sigma}^{(e)}\right] & [0] \end{bmatrix} = \left[k^{(e)}\right]$$

■

Now, we can write the global, multifield, finite element equations, by establishing a gather operation for each element:

$$\left\{U_{\varepsilon\sigma u}^{(e)}\right\} = \left[L^{(e)}\right]\{U_{\varepsilon\sigma u}\} \tag{12.2.15}$$

where $\{U_{\varepsilon\sigma u}\}$ it the *global degree-of-freedom vector,* containing all the DOFs (displacements, strains, and stresses) of the entire structure. Equation (12.2.15) can be separated into three distinct equations, one for displacement DOF, another for stress DOF, and a third one for strain DOF.

$$\left\{U^{(e)}\right\} = \left[L_u^{(e)}\right]\{U\} \tag{12.2.16a}$$

$$\left\{\Sigma^{(e)}\right\} = \left[L_\sigma^{(e)}\right]\{\Sigma\} \tag{12.2.16b}$$

$$\left\{E^{(e)}\right\} = \left[L_\varepsilon^{(e)}\right]\{E\} \tag{12.2.16c}$$

where $\{\Sigma\}$ and $\{E\}$ are global vectors containing all the stress and strain DOFs, respectively.

Finally, we are ready to return to Equation (12.2.8). One can verify that the terms $\bar{W}_{ext}^{(e)}$ are identical to those for the right-hand-side of the elasticity equations described in Section 9.2. This means that the right-hand-side terms corresponding to $\{U^{(e)}\}$ will be the standard equivalent nodal force vectors $\{f^{(e)}\}$ that we obtained in Section 9.2.

After defining a global virtual DOF vector $\{W_{\varepsilon\sigma u}\}^T$, and also establish the gather operations for the virtual DOF, we can eventually obtain:

$$\{W_{\varepsilon\sigma u}\}^T[[K]\{U_{\varepsilon\sigma u}\} - \{F_{\varepsilon\sigma u}\}] = 0 \tag{12.2.17}$$

where the *global coefficient matrix,* $[K]$, is given by:

$$[K] = \sum_{e=1}^{Ne}\left(\left[L^{(e)}\right]^T\left[k^{(e)}\right]\left[L^{(e)}\right]\right) \tag{12.2.18}$$

and the *global equivalent right-hand-side vector,* $\{F_{\varepsilon\sigma u}\}$, is given by the following expression.

$$\{F_{\varepsilon\sigma u}\} = \sum_{e=1}^{Ne}\left(\left[L^{(e)}\right]^T\left\{f_{\varepsilon\sigma u}^{(e)}\right\}\right) \tag{12.2.19}$$

The element right-hand-side vector, $\left\{f_{\varepsilon\sigma u}^{(e)}\right\}$, can be written in block-form:

$$\left\{f_{\varepsilon\sigma u}^{(e)}\right\} = \left\{\begin{array}{c}\{0\}\\ \{0\}\\ \left\{f^{(e)}\right\}\end{array}\right\} \tag{12.2.20}$$

where $\{f^{(e)}\}$ is the equivalent nodal force vector obtained from Equation (9.2.12).

Finally, accounting for the fact that the global vector $\{W_{\varepsilon\sigma u}\}^T$ is arbitrary yields the matrix form of the global finite element equations for a three-field mixed formulation:

$$[K]\{U_{\varepsilon\sigma u}\} - \{F_{\varepsilon\sigma u}\} = 0 \tag{12.2.21}$$

Remark 12.2.3: We can order the terms in the global vector $\{U_{\varepsilon\sigma u}\}$ in such fashion that the displacements are first, the stress DOFs are second, and the strain DOFs are third. In this case, we can also cast the global finite element equations in the following form.

$$\begin{bmatrix}[K_{uu}] & [K_{u\sigma}] & [0]\\ [K_{\sigma u}] & [0] & [K_{\sigma\varepsilon}]\\ [0] & [K_{\varepsilon\sigma}] & [0]\end{bmatrix}\left\{\begin{array}{c}\{U\}\\ \{\Sigma\}\\ \{E\}\end{array}\right\} = \left\{\begin{array}{c}\{f\}\\ \{0\}\\ \{0\}\end{array}\right\} \tag{12.2.22}$$

where the various global subarrays $[K_{uu}]$, $[K_{u\sigma}]$, $[K_{\sigma u}]$, $[K_{\varepsilon\sigma}]$, $[K_{\sigma\varepsilon}]$ are given by:

$$[K_{uu}] = \sum_{e=1}^{Ne}\left(\left[L_u^{(e)}\right]^T\left[k_{uu}^{(e)}\right]\left[L_u^{(e)}\right]\right) \tag{12.2.23a}$$

$$[K_{u\sigma}] = \sum_{e=1}^{Ne}\left(\left[L_u^{(e)}\right]^T\left[k_{u\sigma}^{(e)}\right]\left[L_\sigma^{(e)}\right]\right) \tag{12.2.23b}$$

$$[K_{\sigma u}] = \sum_{e=1}^{Ne}\left(\left[L_\sigma^{(e)}\right]^T\left[k_{\sigma u}^{(e)}\right]\left[L_u^{(e)}\right]\right) = [K_{u\sigma}]^T \tag{12.2.23c}$$

$$[K_{\sigma\varepsilon}] = \sum_{e=1}^{Ne}\left(\left[L_\sigma^{(e)}\right]^T\left[k_{\sigma\varepsilon}^{(e)}\right]\left[L_\varepsilon^{(e)}\right]\right) \tag{12.2.23d}$$

$$[K_{\varepsilon\sigma}] = \sum_{e=1}^{Ne}\left(\left[L_\varepsilon^{(e)}\right]^T\left[k_{\varepsilon\sigma}^{(e)}\right]\left[L_\sigma^{(e)}\right]\right) = [K_{\sigma\varepsilon}]^T \tag{12.2.23e}$$

$$\{f\} = \sum_{e=1}^{Ne}\left(\left[L_u^{(e)}\right]^T\left\{f^{(e)}\right\}\right) \tag{12.2.23f}$$

▮

Remark 12.2.4: It is worth examining the continuity requirements for our three-field finite element approximation. These are decided based on the maximum order of derivatives (for the various approximate fields) in the weak form. If we check

Equation (12.1.12), we will see that the maximum orders of partial derivatives for the displacement, strain, and stress fields are equal to 1, 0, and 0, respectively. Thus, the approximate displacement field must be C^0-continuous, and the types of finite element approximations described in previous chapters can be used for the displacements. Now, for strains and stresses, we need C^{-1} continuity! This type of continuity means that we only need to have stress and strain fields for which the various integral terms in the weak form are bounded. Thus, we do not need to have continuous stress or strain fields across the elements. This is also meaningful from a physical point of view, because the stress and strain field does not have to satisfy continuity even in the exact solution. A simple example case where stress and strain discontinuities exist in the exact solution for one-dimensional elasticity is the structure in Problem 3.1. ∎

Remark 12.2.5: A direct consequence of the fact that we do not need to have continuity of the approximate stress and strain fields is that the stress and strain approximation can be established independently for each element. This simply means that the stress and strain DOFs, $\{\Sigma^{(e)}\}$ and $\{E^{(e)}\}$, respectively, of an element will be *local* to that element (and will not be coupled to the DOFs of the other elements in the mesh). This is a similar situation to that involving the additional DOFs $\{a^{(e)}\}$ for incompatible modes, discussed in Section 11.3. Just like in that section, we can establish a *condensed stiffness matrix* that only includes stiffness terms corresponding to $\{U^{(e)}\}$, while also accounting (indirectly) for the stress and strain DOFs of the element:

$$\left[k^{(e)}\right] = \left[\tilde{k}_u^{(e)}\right] = \begin{bmatrix} [I] \\ -\begin{bmatrix} \left[k_{\varepsilon\varepsilon}^{(e)}\right] & \left[k_{\varepsilon\sigma}^{(e)}\right] \\ \left[k_{\sigma\varepsilon}^{(e)}\right] & [0] \end{bmatrix}^{-1} \begin{bmatrix} [0] \\ \left[k_{\sigma u}^{(e)}\right] \end{bmatrix} \end{bmatrix}^T \begin{bmatrix} \left[k_{\varepsilon\varepsilon}^{(e)}\right] & \left[k_{\varepsilon\sigma}^{(e)}\right] & [0] \\ \left[k_{\sigma\varepsilon}^{(e)}\right] & [0] & \left[k_{\sigma u}^{(e)}\right] \\ [0] & \left[k_{u\sigma}^{(e)}\right] & [0] \end{bmatrix} \begin{bmatrix} [I] \\ -\begin{bmatrix} \left[k_{\varepsilon\varepsilon}^{(e)}\right] & \left[k_{\varepsilon\sigma}^{(e)}\right] \\ \left[k_{\sigma\varepsilon}^{(e)}\right] & [0] \end{bmatrix}^{-1} \begin{bmatrix} [0] \\ \left[k_{\sigma u}^{(e)}\right] \end{bmatrix} \end{bmatrix} \tag{12.2.24}$$

Equation (12.2.24) is obtained after establishing the following relation:

$$\begin{Bmatrix} \{E^{(e)}\} \\ \{\Sigma^{(e)}\} \end{Bmatrix} = -\begin{bmatrix} \left[k_{\varepsilon\varepsilon}^{(e)}\right] & \left[k_{\varepsilon\sigma}^{(e)}\right] \\ \left[k_{\sigma\varepsilon}^{(e)}\right] & [0] \end{bmatrix}^{-1} \begin{bmatrix} [0] \\ \left[k_{\sigma u}^{(e)}\right] \end{bmatrix} \left\{U^{(e)}\right\} \tag{12.2.25}$$

Equation (12.2.25) is a direct consequence of the second and third rows of the partitioned matrix Equation (12.2.22), written at the element level. If we use the condensed element stiffness matrices from Equation (12.2.24), then our global stiffness equations will attain the same mathematical form as those in Section 9.2, which applied for the case where we only approximated the displacement field. ∎

The finite element equations that we established in this section are valid for any three-field approximation. In the following sections, we will focus on three of the most commonly encountered multifield finite element approaches. Specifically, we will discuss the stress-displacement mixed quadrilateral element by Pian and Sumihara, the *u-p* (displacement-pressure) approximation, and the assumed (enhanced) strain formulation.

12.3 Two-Field (Stress-Displacement) Formulations and the Pian-Sumihara Quadrilateral Element

Before proceeding further, we will write *the finite element equations for two-field (stress-displacement) mixed fornulations.* These are obtained by establishing two piecewise finite element approximations, one for the displacement field (Equation 12.2.2) and another for the stress field (Equation 12.2.6), and then plug these approximations in the two-field weak form (12.1.19). The finite element Equations for a two-field weak form can be derived (the derivation follows the same exact steps as for the three-field case that was presented in the previous Section) if we plug the finite element approximations in Equation (12.1.19). We can eventually obtain:

$$\begin{bmatrix} [0] & [K_{u\sigma}] \\ [K_{\sigma u}] & -[K_{\sigma\sigma}] \end{bmatrix} \begin{Bmatrix} \{U\} \\ \{\Sigma\} \end{Bmatrix} = \begin{Bmatrix} \{f\} \\ \{0\} \end{Bmatrix} \quad (12.3.1)$$

The expressions giving the various arrays in Equation (12.3.1) are provided below.

$$[K_{u\sigma}] = \sum_{e=1}^{Ne} \left(\left[L_u^{(e)}\right]^T \left[k_{u\sigma}^{(e)}\right] \left[L_\sigma^{(e)}\right] \right) \quad (12.3.2a)$$

$$[K_{\sigma u}] = \sum_{e=1}^{Ne} \left(\left[L_\sigma^{(e)}\right]^T \left[k_{\sigma u}^{(e)}\right] \left[L_u^{(e)}\right] \right) = [K_{u\sigma}]^T \quad (12.3.2b)$$

$$[K_{\sigma\sigma}] = \sum_{e=1}^{Ne} \left(\left[L_\sigma^{(e)}\right]^T \left[k_{\sigma\sigma}^{(e)}\right] \left[L_\sigma^{(e)}\right] \right) \quad (12.3.2c)$$

$$\{f\} = \sum_{e=1}^{Ne} \left(\left[L_u^{(e)}\right]^T \left\{f^{(e)}\right\} \right) \quad (12.3.2d)$$

The element arrays $\left[k_{u\sigma}^{(e)}\right]$, $\left[k_{\sigma u}^{(e)}\right]$, and $\left[k_{\sigma\sigma}^{(e)}\right]$ in Equations (12.3.2a–c) are obtained from the following expressions.

$$\left[k_{\sigma\sigma}^{(e)}\right] = \iiint_{\Omega^{(e)}} \left[N_\sigma^{(e)}\right]^T \left[D^{(e)}\right]^{-1} \left[N_\sigma^{(e)}\right] dV \quad (12.3.3a)$$

$$\left[k_{u\sigma}^{(e)}\right] = \iiint_{\Omega^{(e)}} \left[B^{(e)}\right]^T \left[N_\sigma^{(e)}\right] dV \quad (12.3.3b)$$

$$\left[k_{\sigma u}^{(e)}\right] = \iiint_{\Omega^{(e)}} \left[N_\sigma^{(e)}\right]^T \left[B^{(e)}\right] dV \quad (12.3.3c)$$

Remark 12.3.1: Similar to Remark 12.2.5, if we have a discontinuous approximate stress field, we can establish a condensed element coefficient array, corresponding to the nodal displacements alone:

$$\left[k^{(e)}\right]=\left[\tilde{k}_u^{(e)}\right]=\begin{bmatrix}[I]\\ \left[k_{\sigma\sigma}^{(e)}\right]^{-1}\left[k_{\sigma u}^{(e)}\right]\end{bmatrix}^T\begin{bmatrix}[0] & \left[k_{u\sigma}^{(e)}\right]\\ \left[k_{\sigma u}^{(e)}\right] & \left[k_{\sigma\sigma}^{(e)}\right]\end{bmatrix}\begin{bmatrix}[I]\\ \left[k_{\sigma\sigma}^{(e)}\right]^{-1}\left[k_{\sigma u}^{(e)}\right]\end{bmatrix} \tag{12.3.4}$$

which is obtained after establishing the relation:

$$\left\{\Sigma^{(e)}\right\}=\left[k_{\sigma\sigma}^{(e)}\right]^{-1}\left[k_{\sigma u}^{(e)}\right]\left\{U^{(e)}\right\} \tag{12.3.5}$$

Equation (12.3.5) is a direct consequence of the second row of the partitioned matrix Equation (12.3.1), written at the element level. ■

We will devote the remainder of this subsection to a very popular two-field finite element, the *Pian-Sumihara* (1985) element. We begin with a four-node quadrilateral (4Q) element, which uses the usual isoparametric shape functions for the displacements—that is, we obtain the arrays $[N^{(e)}]$ and $[B^{(e)}]$ provided in Section 8.5.2. To complete the two-field element approximation, we introduce the following approximate stress field in the parametric space, $\xi\eta$:

$$\left\{\tilde{\sigma}^{(e)}\right\}=\begin{Bmatrix}\sigma_{\xi\xi}^{(e)}\\ \sigma_{\eta\eta}^{(e)}\\ \sigma_{\xi\eta}^{(e)}\end{Bmatrix}=\begin{Bmatrix}\tilde{\Sigma}_1^{(e)}+\tilde{\Sigma}_4^{(e)}\cdot\eta\\ \tilde{\Sigma}_2^{(e)}+\tilde{\Sigma}_5^{(e)}\cdot\xi\\ \tilde{\Sigma}_3^{(e)}\end{Bmatrix}=\begin{bmatrix}\sigma_{\xi\xi}^{(e)} & \sigma_{\xi\eta}^{(e)}\\ \sigma_{\eta\xi}^{(e)} & \sigma_{\eta\eta}^{(e)}\end{bmatrix}=\begin{bmatrix}\tilde{\Sigma}_1^{(e)}+\tilde{\Sigma}_4^{(e)}\cdot\eta & \tilde{\Sigma}_3^{(e)}\\ \tilde{\Sigma}_3^{(e)} & \tilde{\Sigma}_2^{(e)}+\tilde{\Sigma}_5^{(e)}\cdot\xi\end{bmatrix} \tag{12.3.6}$$

Notice that the components of the stress vector (or stress tensor) of Equation (12.3.6) are expressed in the directions of the parametric coordinate system! To obtain the components of the stress field in the Cartesian coordinate axes, xy, we will need to use a two-dimensional stress coordinate transformation, using an appropriate (2×2) array, $[R^{(e)}]$, in accordance with Equation (7.6.7).

$$\left[\sigma^{(e)}\right]=\begin{bmatrix}\sigma_{xx}^{(e)} & \sigma_{xy}^{(e)}\\ \sigma_{yx}^{(e)} & \sigma_{yy}^{(e)}\end{bmatrix}=\left[R^{(e)}\right]\left[\tilde{\sigma}^{(e)}\right]\left[R^{(e)}\right]^T \tag{12.3.7}$$

As explained in Zienkiewicz et al. (2005), the transformation matrix $[R^{(e)}]$ must be such that the following two conditions are satisfied:

- The transformed stress tensor is independent of the element orientation and nodal numbering.
- The element must be able to exactly reproduce a constant stress condition.

The approach by Pian and Sumihara (1985) uses $[R^{(e)}] = [J_o]$, where $[J_o]$ is the Jacobian matrix of the coordinate mapping evaluated at the centroid ($\xi = 0, \eta = 0$) of the element.

Using the derivatives of the 4Q shape functions evaluated at the specific location, the following expression is obtained.

$$\left[R^{(e)}\right] = \left[J_o^{(e)}\right] = \begin{bmatrix} J_{o,11}^{(e)} & J_{o,12}^{(e)} \\ J_{o,21}^{(e)} & J_{o,22}^{(e)} \end{bmatrix} = \frac{1}{4}\begin{bmatrix} -x_1^{(e)} + x_2^{(e)} + x_3^{(e)} - x_4^{(e)} & -y_1^{(e)} - y_2^{(e)} + y_3^{(e)} + y_4^{(e)} \\ -y_1^{(e)} + y_2^{(e)} + y_3^{(e)} - y_4^{(e)} & -x_1^{(e)} - x_2^{(e)} + x_3^{(e)} + x_4^{(e)} \end{bmatrix} \tag{12.3.8}$$

We then plug Equations (12.3.6) and (12.3.8) into Equation (12.3.7), to establish the following column vector (in Voigt notation) with the stress components of the element:

$$\left\{\sigma^{(e)}\right\} = \begin{Bmatrix} \sigma_{xx}^{(e)} \\ \sigma_{yy}^{(e)} \\ \sigma_{xy}^{(e)} \end{Bmatrix} = \begin{Bmatrix} \left(J_{o,11}^{(e)}\right)^2 \cdot \tilde{\sigma}_{\xi\xi}^{(e)} + 2J_{o,11}^{(e)} \cdot J_{o,12}^{(e)} \cdot \tilde{\sigma}_{\xi\eta}^{(e)} + \left(J_{o,12}^{(e)}\right)^2 \cdot \tilde{\sigma}_{\eta\eta}^{(e)} \\ \left(J_{o,21}^{(e)}\right)^2 \cdot \tilde{\sigma}_{\xi\xi}^{(e)} + 2J_{o,21}^{(e)} \cdot J_{o,22}^{(e)} \cdot \tilde{\sigma}_{\xi\eta}^{(e)} + \left(J_{o,22}^{(e)}\right)^2 \cdot \tilde{\sigma}_{\eta\eta}^{(e)} \\ J_{o,11}^{(e)} \cdot J_{o,21}^{(e)} \cdot \tilde{\sigma}_{\xi\xi}^{(e)} + \left(J_{o,21}^{(e)} \cdot J_{o,12}^{(e)} + J_{o,11}^{(e)} \cdot J_{o,22}^{(e)}\right) \cdot \tilde{\sigma}_{\xi\eta}^{(e)} + J_{o,12}^{(e)} \cdot J_{o,22}^{(e)} \cdot \tilde{\sigma}_{\eta\eta}^{(e)} \end{Bmatrix} \tag{12.3.9}$$

We also define the following modified constant parameters of the stress approximation:

$$\Sigma_1^{(e)} = \left(J_{o,11}^{(e)}\right)^2 \cdot \tilde{\Sigma}_1^{(e)} + \left(J_{o,12}^{(e)}\right)^2 \cdot \tilde{\Sigma}_2^{(e)} + 2J_{o,11}^{(e)} \cdot J_{o,12}^{(e)} \cdot \tilde{\Sigma}_3^{(e)} \tag{12.3.10a}$$

$$\Sigma_2^{(e)} = \left(J_{o,21}^{(e)}\right)^2 \cdot \tilde{\Sigma}_1^{(e)} + \left(J_{o,22}^{(e)}\right)^2 \cdot \tilde{\Sigma}_2^{(e)} + 2J_{o,21}^{(e)} \cdot J_{o,22}^{(e)} \cdot \tilde{\Sigma}_3^{(e)} \tag{12.3.10b}$$

$$\Sigma_3^{(e)} = J_{o,21}^{(e)} \cdot J_{o,11}^{(e)} \cdot \tilde{\Sigma}_1^{(e)} + \left(J_{o,21}^{(e)} \cdot J_{o,12}^{(e)} + J_{o,11}^{(e)} \cdot J_{o,22}^{(e)}\right) \cdot \tilde{\Sigma}_2^{(e)} + J_{o,12}^{(e)} \cdot J_{o,22}^{(e)} \cdot \tilde{\Sigma}_3^{(e)} \tag{12.3.10c}$$

$$\Sigma_4^{(e)} = \tilde{\Sigma}_4^{(e)} \tag{12.3.10d}$$

$$\Sigma_5^{(e)} = \tilde{\Sigma}_5^{(e)} \tag{12.3.10e}$$

We eventually obtain the following expression for the three stress components along the x and y axes of the Cartesian space, expressed as functions of the parametric element coordinates ξ and η.

$$\left\{\sigma^{(e)}\right\} = \begin{Bmatrix} \sigma_{xx}^{(e)} \\ \sigma_{yy}^{(e)} \\ \sigma_{xy}^{(e)} \end{Bmatrix} = \begin{bmatrix} 1 & 0 & 0 & \left(J_{o,11}^{(e)}\right)^2 \cdot \eta & \left(J_{o,12}^{(e)}\right)^2 \cdot \xi \\ 0 & 1 & 0 & \left(J_{o,21}^{(e)}\right)^2 \cdot \eta & \left(J_{o,22}^{(e)}\right)^2 \cdot \xi \\ 0 & 0 & 1 & J_{o,11}^{(e)} \cdot J_{o,21}^{(e)} \cdot \eta & J_{o,12}^{(e)} \cdot J_{o,22}^{(e)} \cdot \xi \end{bmatrix} \begin{Bmatrix} \Sigma_1^{(e)} \\ \Sigma_2^{(e)} \\ \Sigma_3^{(e)} \\ \Sigma_4^{(e)} \\ \Sigma_5^{(e)} \end{Bmatrix} = \left[N_\sigma^{(e)}(\xi,\eta)\right]\left\{\Sigma^{(e)}\right\} \tag{12.3.11}$$

where $\left[N_\sigma^{(e)}(\xi,\eta)\right]$ is the matrix with the polynomial interpolation functions of the stress approximation:

$$\left[N_\sigma^{(e)}(\xi,\eta)\right] = \begin{bmatrix} 1 & 0 & 0 & \left(J_{o,11}^{(e)}\right)^2\cdot\eta & \left(J_{o,12}^{(e)}\right)^2\cdot\xi \\ 0 & 1 & 0 & \left(J_{o,21}^{(e)}\right)^2\cdot\eta & \left(J_{o,22}^{(e)}\right)^2\cdot\xi \\ 0 & 0 & 1 & J_{o,11}^{(e)}\cdot J_{o,21}^{(e)}\cdot\eta & J_{o,12}^{(e)}\cdot J_{o,22}^{(e)}\cdot\xi \end{bmatrix} \tag{12.3.12}$$

and $\left\{\Sigma^{(e)}\right\} = \left[\Sigma_1^{(e)}\ \Sigma_2^{(e)}\ \Sigma_3^{(e)}\ \Sigma_4^{(e)}\ \Sigma_5^{(e)}\right]^T$ is the vector with the stress degrees of freedom (DOFs) for the element.

Eventually, we can condense the stress DOFs at the element level, per Remark 12.3.1.

To demonstrate the accuracy of the Pian-Sumihara two-field element, the analysis of the cantilever beam of Figure 11.4 has been repeated, using the two-field formulation and an irregular mesh (i.e., with nonrectangular elements). As shown in Figure 12.1, the Pian-Sumihara element gives very good estimates of the "exact" vertical displacement at the tip (which is –0.285 in.), and also gives very good estimates of the stress distribution.

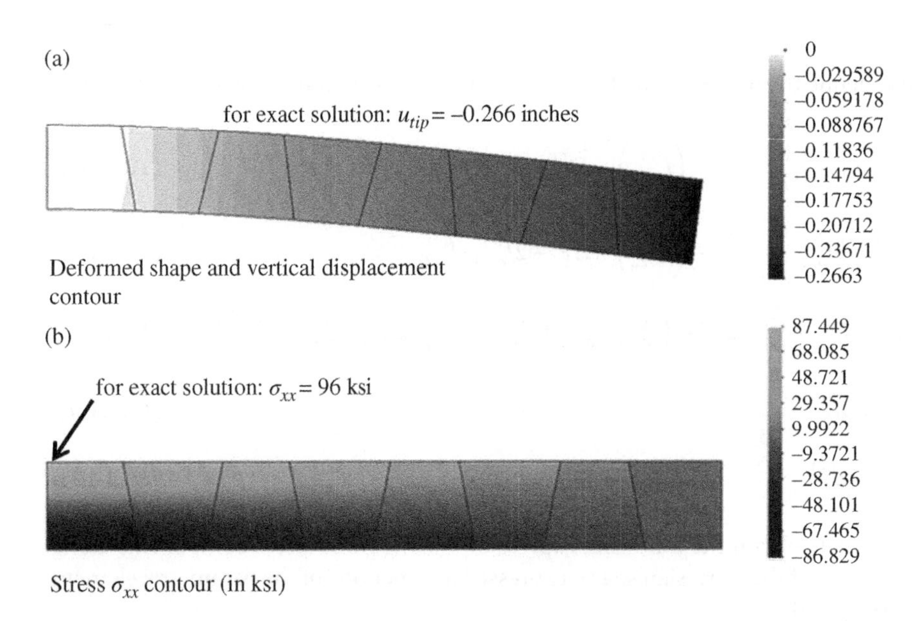

Figure 12.1 Solution of the cantilever beam of Figure 11.4 using a mesh of nonrectangular elements and the Pian-Sumihara stress-displacement formulation.

12.4 Displacement-Pressure (*u-p*) Formulations and Finite Element Approximations

Another popular mixed formulation (perhaps the most popular for this type of elements) uses approximations for the displacement and pressure fields. Such elements are typically employed to address the issue of volumetric locking, which was also discussed in

Section 11.4. To facilitate our discussion, we will rely on definitions of Section 11.4 and write the stress tensor of the material as the sum of two components:

$$[\sigma] = [s] - p[I] \tag{12.4.1a}$$

where $[s]$ is the deviatoric stress tensor and p is the pressure. We can also cast this equation into vector form as follows.

$$\{\sigma\} = \{s\} - p\{I\} \tag{12.4.1b}$$

where $\{I\} = [1\ 1\ 0]^T$ for two-dimensional, plane-strain problems, and $\{I\} = [1\ 1\ 1\ 0\ 0\ 0]^T$ for three-dimensional problems.

We will limit our discussion to an isotropic linearly elastic solid. In accordance with Section 11.4, the pressure and the volumetric strain are related through the equation $p = -K_V \cdot \varepsilon_V$. However, this equation cannot be used in a finite element computation for incompressible materials, because the pressure will then be nonzero, whereas the volumetric strain will be zero. This would imply that $K_V \rightarrow \infty$, which is not acceptable in a numerical computation. *This difficulty can be easily circumvented with the two-field, displacement-pressure (u-p) formulation.* In this two-field formulation, we do not treat the entire stress vector as an independent approximate field. Instead, we only treat the pressure (associated to the volumetric part of the stress) as a second (besides the displacement), independently approximated, scalar field quantity.

Let us write the two sets of equations that describe the problem. We will start with the weak form for the differential equations of equilibrium, that is, Equation (12.1.4). If we plug Equation (12.4.1b) into (12.1.4):

$$\iiint_\Omega \{\nabla_\varepsilon w\}^T(\{s\} - p\{I\})dV - \left(\iiint_\Omega \{w\}^T\{b\}dV + \iint_{\Gamma_{tx}} w_x t_x\, dS + \iint_{\Gamma_{ty}} w_y t_y\, dS + \iint_{\Gamma_{tz}} w_z t_z\, dS\right) = 0 \tag{12.4.2}$$

It can be proven that

$$\{\nabla_\varepsilon w\}^T\{s\} = \{\bar{e}\}^T\{s\} \tag{12.4.3}$$

The vector $\{\bar{e}\}$ in Equation (12.4.3) is the deviatoric part of the field $\{\nabla_\varepsilon w\}$, given by the expression:

$$\{\bar{e}\} = \{\nabla_\varepsilon w\} - \frac{\bar{\varepsilon}_V}{3}\{I\} \tag{12.4.4}$$

where

$$\bar{\varepsilon}_V = \frac{\partial w_x}{\partial x} + \frac{\partial w_y}{\partial x} + \frac{\partial w_z}{\partial x} \tag{12.4.5}$$

The fields $\bar{\varepsilon}_V$ and $\{\bar{e}\}$ can be thought of as virtual volumetric and deviatoric strain fields, respectively. We can also prove that

$$\{\nabla_\varepsilon w\}^T p\{I\} = \bar{\varepsilon}_V \cdot p \tag{12.4.6}$$

We now write—in accordance with the considerations of Section 11.4—the *incompressibility constraint*, that is, the equation stipulating that the volumetric strain must be zero at every location:

$$\varepsilon_V = \varepsilon_{xx} + \varepsilon_{yy} + \varepsilon_{zz} = \frac{\partial u_x}{\partial x} + \frac{\partial u_y}{\partial x} + \frac{\partial u_z}{\partial x} = 0 \tag{12.4.7}$$

We multiply Equation (12.4.7) with an arbitrary function w_p (the physical meaning may be obvious from the subscript p) and integrate over the domain Ω to obtain:

$$\iiint_{\Omega} w_p \left(\frac{\partial u_x}{\partial x} + \frac{\partial u_y}{\partial x} + \frac{\partial u_z}{\partial x} \right) dV = 0 \tag{12.4.8}$$

Equation (12.4.8) is a *weak form of the incompressibility condition*, wherein the scalar field w_p is an arbitrary *virtual pressure field*. If we plug Equations (12.4.3) and (12.4.6) into Equation (12.4.2) and add the resulting expression to Equation (12.4.8), we obtain *a two-field weak form of the incompressible elasticity problem*:

$$\begin{aligned} &\iiint_{\Omega} \{\bar{e}\}^T \{s\} dV - \iiint_{\Omega} \bar{\varepsilon}_V \cdot p dV + \iiint_{\Omega} w_p \left(\frac{\partial u_x}{\partial x} + \frac{\partial u_y}{\partial x} + \frac{\partial u_z}{\partial x} \right) dV \\ &= \iiint_{\Omega} \{w\}^T \{b\} dV + \iint_{\Gamma_{tx}} w_x t_x \, dS + \iint_{\Gamma_{ty}} w_y t_y \, dS + \iint_{\Gamma_{tz}} w_z t_z \, dS \end{aligned} \tag{12.4.9}$$

Equation (12.4.9) is the principle of virtual work for the constrained, incompressible solid mechanics problem. For linear isotropic elasticity, we can also express the deviatoric stress as a function of the deviatoric part of the strains, in accordance with Section 11.7:

$$\{s\} = [\bar{D}]\{e\} \tag{12.4.10}$$

We can finally write the following weak form for incompressible linear elasticity:

$$\begin{aligned} &\iiint_{\Omega} \{\bar{e}\}^T [\bar{D}]\{e\} dV - \iiint_{\Omega} \bar{\varepsilon}_V \cdot p dV + \iiint_{\Omega} w_p \left(\frac{\partial u_x}{\partial x} + \frac{\partial u_y}{\partial x} + \frac{\partial u_z}{\partial x} \right) dV \\ &= \iiint_{\Omega} \{w\}^T \{b\} dV + \iint_{\Gamma_{tx}} w_x t_x \, dS + \iint_{\Gamma_{ty}} w_y t_y \, dS + \iint_{\Gamma_{tz}} w_z t_z \, dS \end{aligned} \tag{12.4.11}$$

Now, if we introduce a finite element approximation of the weak form given by Equation (12.4.11), we can establish a mixed finite element formulation involving the displacements and pressures, hence the term *u-p* formulation. Specifically, in each element, we will establish a piecewise approximation for the displacement and pressure fields:

$$\left\{u^{(e)}\right\} \approx \left[N^{(e)}\right]\left\{U^{(e)}\right\} \tag{12.4.12a}$$

$$p^{(e)}(x,y,z) \approx \left[N_p^{(e)}\right]\left\{P^{(e)}\right\} \tag{12.4.12b}$$

It is worth mentioning that $\left[N_p^{(e)}\right]$ in Equation (12.4.12b) is a row vector!

The element nodal displacements $\{U^{(e)}\}$ and pressure degrees-of-freedom (DOFs), $\{P^{(e)}\}$, are obtained through a gather operation from corresponding global DOFs:

$$\left\{U^{(e)}\right\} = \left[L_u^{(e)}\right]\{U\} \tag{12.4.13a}$$

$$\left\{P^{(e)}\right\} = \left[L_p^{(e)}\right]\{P\} \tag{12.4.13b}$$

Using the procedure in Section 11.7, we will separate the strain-displacement relation of the finite element into two parts, one part giving the volumetric strain:

$$\varepsilon_v^{(e)} = \left[b_{vol}^{(e)}\right]\left\{U^{(e)}\right\} \tag{12.4.14a}$$

and the other part giving the deviatoric strain vector:

$$\{e\} = \left[\bar{B}^{(e)}\right]\left\{U^{(e)}\right\} \tag{12.4.14b}$$

Remark 12.4.1: The array $\left[b_{vol}^{(e)}\right]$ is a row vector with the same number of components as $\{U^{(e)}\}$. The values of $\left[b_{vol}^{(e)}\right]$ for the special case of a four-node quadrilateral element were provided in Remark 11.8.1. ■

Eventually, we obtain the following global, two-field finite element equations:

$$\begin{bmatrix} [K_{uu}] & [K_{up}] \\ [K_{pu}] & [0] \end{bmatrix} \begin{Bmatrix} \{U\} \\ \{P\} \end{Bmatrix} = \begin{Bmatrix} \{f\} \\ \{0\} \end{Bmatrix} \tag{12.4.15}$$

where $\{f\}$ is the global equivalent nodal force vector (given by the same expressions as in Section 9.2), and the various global subarrays can be obtained through an assembly operation of element contributions:

$$[K_{uu}] = \sum_{e=1}^{Ne}\left(\left[L_u^{(e)}\right]^T\left[k_{uu}^{(e)}\right]\left[L_u^{(e)}\right]\right) \tag{12.4.16a}$$

$$[K_{up}] = \sum_{e=1}^{Ne}\left(\left[L_u^{(e)}\right]^T\left[k_{up}^{(e)}\right]\left[L_p^{(e)}\right]\right) \tag{12.4.16b}$$

$$[K_{pu}] = \sum_{e=1}^{Ne}\left(\left[L_p^{(e)}\right]^T\left[k_{pu}^{(e)}\right]\left[L_u^{(e)}\right]\right) = [K_{up}]^T \tag{12.4.16c}$$

$$\{f\} = \sum_{e=1}^{Ne}\left(\left[L_u^{(e)}\right]^T\left\{f^{(e)}\right\}\right) \tag{12.4.16d}$$

The element arrays in Equations (12.4.16a–c) are obtained as:

$$\left[k_{uu}^{(e)}\right] = \iiint_{\Omega^{(e)}}\left[\bar{B}^{(e)}\right]^T\left[\bar{D}^{(e)}\right]\left[\bar{B}^{(e)}\right]dV \tag{12.4.17a}$$

$$\left[k_{up}^{(e)}\right] = \iiint_{\Omega^{(e)}} \left[b_{vol}^{(e)}\right]^T \left[N_p^{(e)}\right] dV \tag{12.4.17b}$$

$$\left[k_{pu}^{(e)}\right] = \iiint_{\Omega^{(e)}} \left[N_p^{(e)}\right]^T \left[b_{vol}^{(e)}\right] dV = \left[k_{up}^{(e)}\right]^T \tag{12.4.17c}$$

12.5 Stability of Mixed *u-p* Formulations—the inf-sup Condition

Having established the finite element equations for the two-field, mixed *u-p* formulation, we will now provide a brief discussion pertaining to convergence of *u-p* approximations. Our discussion will rely on considerations in Strang and Fix (2008). We have the global system of Equations (12.4.15) for the problem. We imagine that these equations correspond to the unrestrained displacement degrees of freedom (in other words, we have already accounted for the effect of the support conditions). Then, to ensure convergence of the displacement-pressure approximation, the condition in Box 12.5.1 must be satisfied.

Box 12.5.1 Necessary Condition to Have Convergence of Mixed *u-p* Finite Element Approximations

For every pressure vector $\{P\}$, there must be a nonzero displacement vector $\{U\}$, such that:

$$\{U\}^T [K_{up}] \{P\} \geq \beta \sqrt{\{U\}^T [K_{uu}] \{U\}} \cdot \sqrt{\{P\}^T [K_{pp}] \{P\}} \tag{12.5.1}$$

where $\beta > 0$ is a constant and $[K_{pp}]$ is a "pressure coefficient matrix" given by:

$$[K_{pp}] = \sum_{e=1}^{Ne} \left(\left[L_p^{(e)}\right]^T \left[k_{pp}^{(e)}\right] \left[L_p^{(e)}\right] \right) \tag{12.5.2}$$

with

$$\left[k_{pp}^{(e)}\right] = \iiint_{\Omega^{(e)}} \left[N_p^{(e)}\right]^T \left[N_p^{(e)}\right] dV \tag{12.5.3}$$

It is worth noticing that the condition (12.5.1) will immediately fail, if there is a nonzero pressure vector $\{P\}$ with $[K_{up}]\{P\} = \{0\}$.

If we set $\{Y\} = [K_{uu}]^{1/2}\{U\}$, then[1] we can cast Equation (12.5.1) in the following equivalent form (after also accounting for the symmetry of $[K_{uu}]^{1/2}$).

1 $[K_{uu}]^{1/2}$ is the symmetric matrix that satisfies: $[K_{uu}]^{1/2}[K_{uu}]^{1/2} = [K_{uu}]$. In the present discussion, we will assume that given $[K_{uu}]$, we are somehow able to obtain $[K_{uu}]^{1/2}$!

$$\{Y\}^T[K_{uu}]^{-1/2}[K_{up}]\{P\} \geq \beta\|\{Y\}\| \cdot \sqrt{\{P\}^T[K_{pp}]\{P\}} \tag{12.5.4}$$

We then divide both sides of (12.5.4) by the magnitude of $\{Y\}$ to obtain:

$$\frac{\{Y\}^T[K_{uu}]^{-1/2}[K_{up}]\{P\}}{\|\{Y\}\|} \geq \beta\sqrt{\{P\}^T[K_{pp}]\{P\}} \tag{12.5.5}$$

Next, we establish a useful lemma:

Lemma 12.5.1: The maximum value (the *supremum*) obtained for the left-hand side of Equation (12.5.5) among all possible vectors $\{Y\}$ is equal to $\left\|[K_{uu}]^{-1/2}[K_{up}]\{P\}\right\|$:

$$\sup_{\{Y\}}\frac{\{Y\}^T[K_{uu}]^{-1/2}[K_{up}]\{P\}}{\|\{Y\}\|} = \left\|[K_{uu}]^{-1/2}[K_{up}]\{P\}\right\| \tag{12.5.6}$$

Proof: let us set $[K_{uu}]^{-1/2}[K_{up}]\{P\} = \{V_P\}$. Then: $\{Y\}^T[K_{uu}]^{-1/2}[K_{up}]\{P\} = \{Y\}^T\{V_P\}$, which by definition is equal to the inner (dot) product of vectors $\{Y\}$ and $\{V_P\}$. In accordance with Equation (A.2.4) in Appendix A, the dot product can be written in the form: $\{Y\}^T\{V_P\} = \|\{Y\}\|\|\{V_P\}\|\cos\left(\vec{Y},\vec{V}_P\right)$, where $\cos\left(\vec{Y},\vec{V}_P\right)$ is the cosine of the angle formed by the two vectors, $\vec{Y}$ and $\vec{V}_P$. Obviously, the maximum possible value of this cosine is equal to 1. Accordingly, we can write that the maximum possible value of the ratio $\frac{\{Y\}^T\{V_P\}}{\|\{Y\}\|} = \frac{\|\{Y\}\|\|\{V_P\}\|\cos\left(\vec{Y},\vec{V}_P\right)}{\|\{Y\}\|}$ is obtained for the cosine being equal to 1 (i.e., when the two vectors are parallel!). Thus, we have obtained:

$$\sup_{\{Y\}}\frac{\{Y\}^T\{V_P\}}{\|\{Y\}\|} = \frac{\|\{Y\}\|\|\{V_P\}\|\cdot 1}{\|\{Y\}\|} = \|\{V_P\}\|, \text{ Q.E.D.}$$

■

Now, it suffices to ensure that the supremum of the left-hand side of Inequality (12.5.5) satisfies the specific inequality; thus, we can modify the convergence condition as follows. For every pressure solution vector, $\{P\}$, we must satisfy:

$$\frac{\left\|[K_{uu}]^{-1/2}[K_{up}]\{P\}\right\|}{\sqrt{\{P\}^T[K_{pp}]\{P\}}} \geq \beta \tag{12.5.7}$$

If we square both sides of Inequality (12.5.7), we have:

$$\frac{\left\|[K_{uu}]^{-1/2}[K_{up}]\{P\}\right\|^2}{\left(\sqrt{\{P\}^T[K_{pp}]\{P\}}\right)^2} = \frac{\{P\}^T[K_{up}]^T[K_{uu}]^{-1}[K_{up}]\{P\}}{\{P\}^T[K_{pp}]\{P\}} \geq \beta^2 \tag{12.5.8}$$

Of course, since inequality (12.5.8) must be satisfied for every vector $\{P\}$, it suffices to satisfy it for the $\{P\}$ vector that gives the minimum possible value (*the infimum*) for the left-hand side of the inequality. Since our derivation included obtaining a supremum (among possible vectors $\{Y\}$) and an infimum (among all the pressure solution vectors $\{P\}$), the condition we have for convergence is called the *inf-sup condition.* It is also mentioned as the *Ladyzenskaja-Babuska-Brezzi (LBB)* condition. Usually, it is hard to prove whether an element satisfies or not the inf-sup condition. For this reason, simple rules have been devised to provide an idea of whether an element is stable or not.

A summary of several basic types of elements for *u-p* approximations, together with a description of whether or not they satisfy the inf-sup condition, is provided in Figure 12.2. When an element fails to satisfy the inf-sup condition, this means that we will have several possible pressure solution vectors, $\{P_{null}\}$, for which

$$[K_{up}]\{P_{null}\} = \{0\} \tag{12.5.9}$$

The vectors $\{P_{null}\}$ are called *pressure modes,* and can be thought of as distributions of the pressure parameters, $\{P\}$ that are "not perceived" by the element elastic stiffness and do not mobilize strains and stresses. This is somehow similar to the spurious zero-energy (hourglass) modes that we saw in Section 11.5. The existence of pressure modes is also referred to as *numerical instability for u-p elements.* To provide a schematic demonstration of such pressure modes, Figure 12.3a shows the results obtained for an analysis of a rectangular domain, using the mixed *u-p* triangular elements of case "3/3c" in Figure 12.2. The specific element formulation is also termed a P1-P1 one, because both the displacement field and the pressure field rely on an approximation that has a first-degree of completeness. As shown in the pressure contour of Figure 12.3a, the specific formulation leads to spurious fluctuations of the pressure field, visually manifested through the alternation of lighter-colored and darker-colored bands in the contour plot. These fluctuations are the pressure modes, caused by the fact that the P1-P1 formulation does not satisfy the inf-sup condition, thus entailing numerical instability. The analysis of the same problem with a P2-P1 formulation (second-degree completeness for the displacement and first-degree completeness for the pressure), which corresponds to case "6/3c" of Figure 12.2 and does satisfy the inf-sup condition, gives a nice, smooth distribution of the contour values, without the presence of any pressure modes, as shown in Figure 12.3b.

For elements that do not satisfy the inf-sup condition, there is a remedy for pressure modes, which is mathematically similar to that employed for hourglass control. That is, we establish a *stabilization* part in our element coefficient arrays. This allows even elements that do not satisfy the inf-sup condition in Figure 12.2 to ultimately give convergent solutions and avoid numerical instabilities.

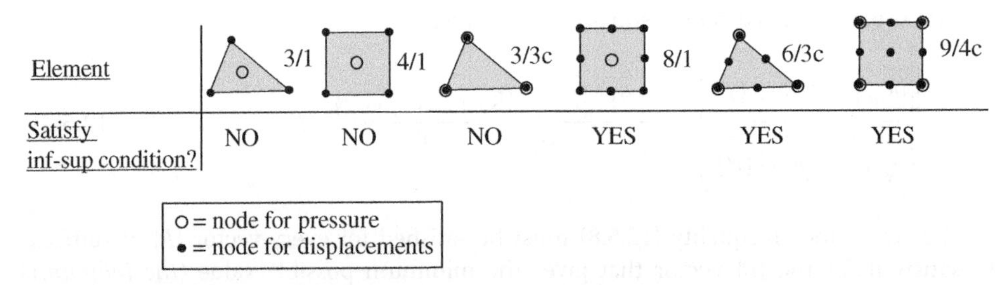

Figure 12.2 Satisfaction of inf-sup condition for some simple types of *u-p* elements (sufficiently adapted from Bathe 1996).

(a)

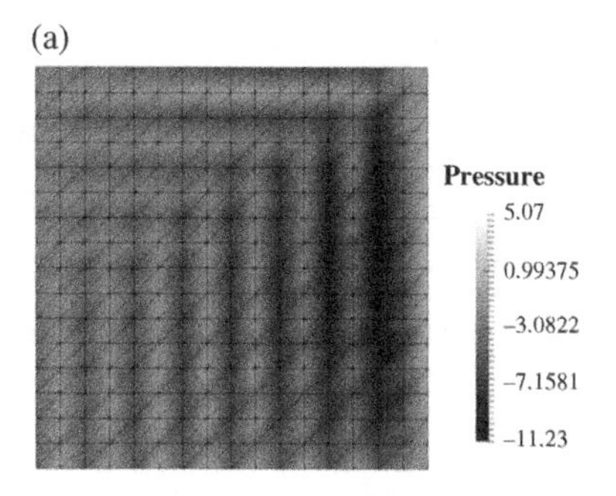

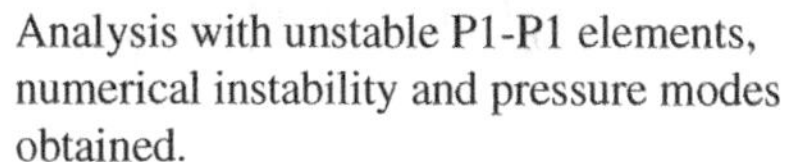
Analysis with unstable P1-P1 elements, numerical instability and pressure modes obtained.

(b)

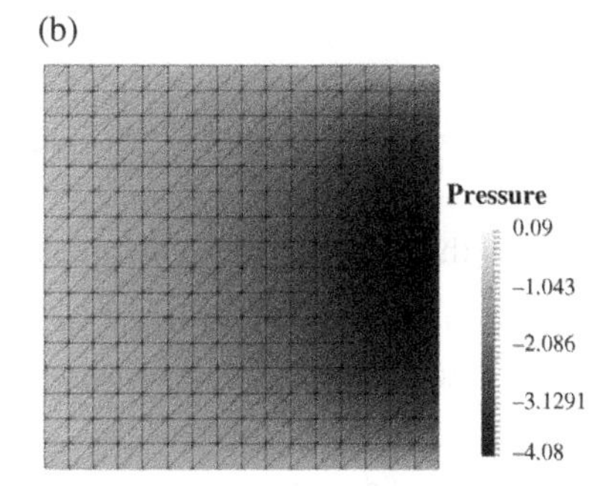

Analysis with stable P2-P1 elements, no numerical instability.

Figure 12.3 Example demonstration of unstable and stable *u-p* mixed formulations and manifestation of pressure instability for the former (analysis results courtesy of Mr. G. Deskos, Imperial College, London, UK).

12.6 Assumed (Enhanced)-Strain Methods and the B-bar Method as a Special Case

Another very popular type of finite element approximation, based on a three-field weak form, is the *assumed (or enhanced)-strain formulation* (Simo and Rifai 1990). In this formulation, the strain field $\{\varepsilon^{(e)}\}$ of each element is assumed to be given as the sum of two contributions:

$$\left\{\varepsilon^{(e)}\right\} = \left[B^{(e)}\right]\left\{U^{(e)}\right\} + \left\{\varepsilon^{(e)}_{enh}\right\} \tag{12.6.1}$$

where the field $\left\{\varepsilon^{(e)}_{enh}\right\}$ is called the *strain enhancement* field. Notice that the first term in the right-hand side of Equation (12.6.1) is the standard approximate strains that we would obtain in a finite element where we only approximate the displacements (i.e., like the ones described in Sections 8.1 and 9.2). In other words, we *assume* (hence the name "assumed strain" methods) that the approximate strain in each element is the sum of the strain field of a "standard" finite element plus the strain enhancement field. Now, we need to establish an approximation for the strain enhancement, not for the strain! In other words, we will have the element approximations for the displacements (Equation 12.2.2), the stresses (Equation 12.2.6), and we will also provide an approximation for the strain enhancement:

$$\left\{\varepsilon^{(e)}_{enh}\right\} = \left[N^{(e)}_{enh}\right]\left\{E^{(e)}_{enh}\right\} \tag{12.6.2}$$

where $\left\{E^{(e)}_{enh}\right\}$ contains the strain-enhancement degrees-of-freedom (DOFs) for the element. In light of Equation (12.6.2), Equation (12.6.1) becomes:
and

$$\left\{\varepsilon^{(e)}\right\} = \left[B^{(e)}\right]\left\{U^{(e)}\right\} + \left[N^{(e)}_{enh}\right]\left\{E^{(e)}_{enh}\right\} \tag{12.6.3}$$

Similarly, the virtual strain field in each element is given by the following expression.

$$\left\{\bar{\varepsilon}^{(e)}\right\}=\left[B^{(e)}\right]\left\{W^{(e)}\right\}+\left[N_{enh}^{(e)}\right]\left\{\bar{E}_{enh}^{(e)}\right\} \tag{12.6.4}$$

where $\left\{\bar{E}_{enh}^{(e)}\right\}$ is the virtual strain-enhancement DOF vector.

Now, we can write the contribution $\bar{W}_{int}^{(e)}$ of element e to the left-hand side of the three-field weak form (Equation 12.2.8) as follows.

$$\begin{aligned}\bar{W}_{int}^{e}=&\left(\iiint_{\Omega^{(e)}}\left(\left\{\bar{E}_{enh}^{(e)}\right\}^T\left[N_{enh}^{(e)}\right]^T+\left\{W^{(e)}\right\}^T\left[B^{(e)}\right]^T\right)\left[D^{(e)}\right]\left[N_{\varepsilon}^{(e)}\right]\left\{E_{enh}^{(e)}\right\}+\left[B^{(e)}\right]\left\{U^{(e)}\right\}dV\right)\\&-\left(\iiint_{\Omega^{(e)}}\left(\left\{\bar{E}_{enh}^{(e)}\right\}^T\left[N_{enh}^{(e)}\right]^T\right)\left[N_{\sigma}^{(e)}\right]\left\{\Sigma^{(e)}\right\}dV\right)\\&+\left(\iiint_{\Omega^{(e)}}\left\{\bar{\Sigma}^{(e)}\right\}^T\left[N_{\sigma}^{(e)}\right]^T\left[N_{enh}^{(e)}\right]\left\{E_{enh}^{(e)}\right\}dV\right)\end{aligned} \tag{12.6.5}$$

Next, we stipulate that, for our assumed (enhanced)-strain formulation, the approximations for the stress and for the strain enhancement are *orthogonal*. This means that the following condition is satisfied for the virtual approximate stress field and the approximate strain enhancement field:

$$\iiint_{\Omega^{(e)}}\left(\left\{\bar{\sigma}^{(e)}\right\}^T\left\{\varepsilon_{enh}^{(e)}\right\}\right)dV=0\rightarrow\iiint_{\Omega^{(e)}}\left(\left\{\bar{\Sigma}^{(e)}\right\}^T\left[N_{\sigma}^{(e)}\right]^T\left[N_{enh}^{(e)}\right]\left\{E_{enh}^{(e)}\right\}\right)dV=0 \tag{12.6.6}$$

Now, since the vectors $\{\Sigma^{(e)}\}^T$ and $\left\{E_{enh}^{(e)}\right\}$ are constant, they can be taken outside the integral of Equation (12.6.6):

$$\left\{\Sigma^{(e)}\right\}^T\left[\iiint_{\Omega^{(e)}}\left(\left[N_{\sigma}^{(e)}\right]^T\left[N_{enh}^{(e)}\right]\right)dV\right]\left\{E_{enh}^{(e)}\right\}=0 \tag{12.6.7}$$

The only way to always satisfy Equation (12.6.7), is to have:

$$\iiint_{\Omega^{(e)}}\left(\left[N_{\sigma}^{(e)}\right]^T\left[N_{enh}^{(e)}\right]\right)dV=0 \tag{12.6.8a}$$

Similarly, if we take the transpose of Equation (12.6.8a), we have:

$$\iiint_{\Omega^{(e)}}\left(\left[N_{enh}^{(e)}\right]^T\left[N_{\sigma}^{(e)}\right]\right)dV=0 \tag{12.6.8b}$$

Equation (12.6.8b) leads to the conclusion that the *approximate virtual strain enhancement is also orthogonal to the actual approximate stress field:*

$$\iiint_{\Omega^{(e)}} \left(\left\{ \bar{E}_{enh}^{(e)} \right\}^T \left[N_{enh}^{(e)} \right]^T \right) \left[N_{\sigma}^{(e)} \right] \left\{ \Sigma^{(e)} \right\} dV = \iiint_{\Omega^{(e)}} \left(\left\{ \bar{\varepsilon}_{enh}^{(e)} \right\}^T \left\{ \sigma^{(e)} \right\} \right) dV = 0 \qquad (12.6.9)$$

Equations (12.6.6) and (12.6.9) lead to the conclusion that the two last terms of the right-hand side in Equation (12.6.5) vanish, so we obtain:

$$\bar{W}_{int}^{(e)} = \iiint_{\Omega^{(e)}} \left(\left\{ \bar{E}_{enh}^{(e)} \right\}^T \left[N_{enh}^{(e)} \right]^T + \left\{ W^{(e)} \right\}^T \left[B^{(e)} \right]^T \right) \left[D^{(e)} \right] \left(\left[N_{\varepsilon}^{(e)} \right] \left\{ E_{enh}^{(e)} \right\} + \left[B^{(e)} \right] \left\{ U^{(e)} \right\} \right) dV$$

$$\begin{aligned}
\bar{W}_{int}^{(e)} = & \iiint_{\Omega^{(e)}} \left\{ \bar{E}_{enh}^{(e)} \right\}^T \left[N_{enh}^{(e)} \right]^T \left[D^{(e)} \right] \left[N_{\varepsilon}^{(e)} \right] \left\{ E_{enh}^{(e)} \right\} dV \\
& + \iiint_{\Omega^{(e)}} \left\{ \bar{E}_{enh}^{(e)} \right\}^T \left[N_{enh}^{(e)} \right]^T \left[D^{(e)} \right] \left[B^{(e)} \right] \left\{ U^{(e)} \right\} dV \\
& + \iiint_{\Omega^{(e)}} \left\{ W^{(e)} \right\}^T \left[B^{(e)} \right]^T \left[D^{(e)} \right] \left[N_{\varepsilon}^{(e)} \right] \left\{ E_{enh}^{(e)} \right\} dV \\
& + \iiint_{\Omega^{(e)}} \left\{ W^{(e)} \right\}^T \left[B^{(e)} \right]^T \left[D^{(e)} \right] \left[B^{(e)} \right] \left\{ U^{(e)} \right\} dV
\end{aligned}$$

$$\begin{aligned}
\rightarrow \bar{W}_{int}^{(e)} = & \left\{ W^{(e)} \right\}^T \left(\left[k_{u,u}^{(e)} \right] \left\{ U^{(e)} \right\} + \left[k_{u,enh}^{(e)} \right] \left\{ E_{enh}^{(e)} \right\} \right) \\
& + \left\{ \bar{E}_{enh}^{(e)} \right\}^T \left(\left[k_{enh,u}^{(e)} \right] \left\{ U^{(e)} \right\} + \left[k_{enh,enh}^{(e)} \right] \left\{ E_{enh}^{(e)} \right\} \right)
\end{aligned} \qquad (12.6.10)$$

In Equation (12.6.10), we have defined the following element coefficient arrays.

$$\left[k_{enh,u}^{(e)} \right] = \iiint_{\Omega^{(e)}} \left[N_{enh}^{(e)} \right]^T \left[D^{(e)} \right] \left[B^{(e)} \right] dV \qquad (12.6.11a)$$

$$\left[k_{enh,enh}^{(e)} \right] = \iiint_{\Omega^{(e)}} \left[N_{enh}^{(e)} \right]^T \left[D^{(e)} \right] \left[N_{enh}^{(e)} \right] dV \qquad (12.6.11b)$$

$$\left[k_{u,u}^{(e)} \right] = \iiint_{\Omega^{(e)}} \left[B^{(e)} \right]^T \left[D^{(e)} \right] \left[B^{(e)} \right] dV \qquad (12.6.11c)$$

$$\left[k_{u,enh}^{(e)} \right] = \iiint_{\Omega^{(e)}} \left[B^{(e)} \right]^T \left[D^{(e)} \right] \left[N_{enh}^{(e)} \right] dV = \left[k_{enh,u}^{(e)} \right]^T \qquad (12.6.11d)$$

We can rewrite Equation (12.6.10) as a block-matrix expression:

$$\bar{W}_{int}^{(e)} = \begin{Bmatrix} \{W^{(e)}\} \\ \{\bar{E}_{enh}^{(e)}\} \end{Bmatrix}^T \begin{bmatrix} [k_{u,u}^{(e)}] & [k_{u,enh}^{(e)}] \\ [k_{enh,u}^{(e)}] & [k_{enh,enh}^{(e)}] \end{bmatrix} \begin{Bmatrix} \{U^{(e)}\} \\ \{E_{enh}^{(e)}\} \end{Bmatrix} = \{\bar{U}_{enh-u}^{(e)}\}^T [k^{(e)}] \{U_{enh-u}^{(e)}\} \tag{12.6.12}$$

where $\{U_{enh-u}^{(e)}\} = \begin{Bmatrix} \{U^{(e)}\} \\ \{E_{enh}^{(e)}\} \end{Bmatrix}$, $\{\bar{U}_{enh-u}^{(e)}\} = \begin{Bmatrix} \{W^{(e)}\} \\ \{\bar{E}_{enh}^{(e)}\} \end{Bmatrix}$.

Finally, stipulating that the element DOFs are obtained from a gather operation of global DOF vectors, we can write:

$$\{U^{(e)}\} = [L_u^{(e)}]\{U\} \tag{12.6.13a}$$

$$\{E_{enh}^{(e)}\} = [L_{enh}^{(e)}]\{E_{enh}\} \tag{12.6.13b}$$

Establishing gather operations for the virtual displacements and virtual strain enhancement DOFs, and exploiting the fact that the weak form must be satisfied for any arbitrary value of virtual DOFs, we can finally obtain the global stiffness equations for a mesh consisting of assumed-strain elements:

$$\begin{bmatrix} [K_{u,u}] & [K_{u,enh}] \\ [K_{enh,u}] & [K_{enh,enh}] \end{bmatrix} \begin{Bmatrix} \{U\} \\ \{E_{enh}\} \end{Bmatrix} = \begin{Bmatrix} \{f\} \\ \{0\} \end{Bmatrix} \tag{12.6.14}$$

where $\{f\}$ is the global equivalent nodal force vector and we also have the following assembly equations.

$$[K_{u,u}] = \sum_{e=1}^{Ne} \left([L_u^{(e)}]^T [k_{u,u}^{(e)}] [L_u^{(e)}] \right) \tag{12.6.15a}$$

$$[K_{u,enh}] = \sum_{e=1}^{Ne} \left([L_u^{(e)}]^T [k_{u,enh}^{(e)}] [L_{enh}^{(e)}] \right) \tag{12.6.15b}$$

$$[K_{enh,u}] = \sum_{e=1}^{Ne} \left([L_{enh}^{(e)}]^T [k_{enh,u}^{(e)}] [L_u^{(e)}] \right) = [K_{u,enh}]^T \tag{12.6.15c}$$

$$[K_{enh,enh}] = \sum_{e=1}^{Ne} \left([L_{enh}^{(e)}]^T [k_{enh,enh}^{(e)}] [L_{enh}^{(e)}] \right) \tag{12.6.15d}$$

$$\{f\} = \sum_{e=1}^{Ne} \left([L_u^{(e)}]^T \{f^{(e)}\} \right) \tag{12.6.15e}$$

Remark 12.6.1: There is an intriguing observation for assumed (enhanced)-strain methods: Although we begin with a three-field mixed element formulation, *we do not* *need to actually establish the approximation for the stresses!* It suffices to stipulate the assumption of orthogonality between approximate stresses and approximate strain enhancement. If we wish, we can always imagine that the stress approximation will be tacitly established to enforce the satisfaction of Equations (12.6.6) and (12.6.9). Since the approximate stress field does not participate in the element expression for $\bar{W}^{(e)}_{\text{int}}$ (Equation 12.6.10) or in the global equations, we never actually need to establish the specific approximation. ■

Remark 12.6.2: The B-bar method, which we discussed in Section 11.8, can result as a *special case* of an enhanced-strain formulation, in which we set:

$$\{E_{enh}\} = \left\{U^{(e)}\right\} \tag{12.6.16}$$

$$\left[N^{(e)}_{enh}\right] = \left[\bar{B}^{(e)}\right] + \left[\bar{\bar{B}}^{(e)}_{o}\right] - \left[B^{(e)}\right] \tag{12.6.17}$$

where $\left[\bar{B}^{(e)}\right]$ is the strain-displacement matrix that corresponds to the part of the strains that doesn't lock (e.g., the deviatoric strains if we have incompressible behavior) and $\left[\bar{\bar{B}}^{(e)}_{o}\right]$ is the part of the strain-displacement matrix corresponding to the strains that lock (e.g., the volumetric strains). The latter, locking part, is evaluated at the single quadrature point at the center of a 4Q or 8H element in the parametric space!

Notice that this special case of enhanced-strain approach has the additional advantage of not including additional "strain enhancement" DOFs. In other words, the only vector with element constants for the solution is the nodal displacement vector, $\{U^{(e)}\}$. The relation between the B-bar method and mixed finite element methods has been established in Simo and Hughes (1986). ■

Remark 12.6.3: As mentioned in Section 11.8, the B-bar method is equivalent to the selective-reduced integration (SRI) procedure for isotropic, linearly elastic behavior. Combined with Remark 12.6.2, this fact leads to the conclusion that SRI can be thought of as a special case of a multifield (mixed) element method. ■

12.7 A Concluding Remark for Multifield Elements

The previous sections have discussed different multifield formulations, and how these formulations can provide improved solutions for several cases such as incompressible elasticity. This may create the impression that multifield elements are bound to have superior accuracy compared to that of conventional elements (i.e., formulations based on a displacement approximation). This is not the case. In fact, a statement referred to as the *principle of limitation* (De Veubeke 1965) stipulates that the convergence of multifield formulations is not superior to that of conventional elements. This explains why—given the availability of multifield elements—analysts still greatly rely on conventional elements, wherein we approximate a single field (the displacement). The power (and usefulness) of multifield elements lies to the fact that they can provide improved accuracy for "challenging" situations, without a need to resort to an overly fine mesh.

References

Bathe, K. J. (1996). *Finite Element Procedures.* Englewood Cliffs, NJ: Prentice-Hall.

De Veubeke, F. (1965). “Displacement and Equilibrium Models in the Finite Element Method.” In: Zienkiewicz, O. *Stress Analysis: Recent Developments in Numerical and Experimental Methods.* London: Wiley.

Pian, T., and Sumihara, K. (1985). “Rational Approach for Assumed Stress Elements.” *International Journal for Numerical Methods in Engineering,* 20, 1685–1695.

Simo, J. C., and Hughes, T. J. R. (1986). “On the Variational Foundations of Assumed Strain Methods.” *ASME Journal of Applied Mechanics,* 53(1), 51–54.

Simo, J. C., and Rifai, M. (1990). “A Class of Mixed Assumed Strain Methods and the Method of Incompatible Modes.” *International Journal for Numerical Methods in Engineering,* 29, 1595–1638.

Strang, G., and Fix, G. (2008). *An Analysis of the Finite Element Method,* 2nd ed. Cambridge, MA: Wellesley.

Zienkiewicz, O., Taylor, R., and Zhu, J. (2005). *The Finite Element Method: Its Basis and Fundamentals,* 6th ed. London: Butterworth-Heinemann.

13

Finite Element Analysis of Beams

In this (and the next) chapter, we will examine *structural elements*, that is, elements for which the nodal degrees of freedom include *rotations* besides displacements. The two types of structural elements that we will consider are beam and shell elements. The former type is examined in this chapter, while the latter will be discussed in Chapter 14.

13.1 Basic Definitions for Beams

Beams are line members (members for which one of the dimensions, the length, is significantly greater than the other two). We will examine beams with endpoints at $x = 0$ and $x = L$, as shown in Figure 13.1. The deformation of a beam can be described with respect to a *reference line*, as shown in the same figure. Beams are generally subjected to both *axial and transverse loads*. The axial loading (i.e., forces parallel to the reference line) causes axial internal forces just like in truss structures, while the transverse loading gives rise to shear forces and bending (flexural) moments.

Our discussion will be limited to two-dimensional (2D) beams (i.e., beams that can be described as members on a plane). In our derivations, we will assume that a beam structure is aligned with a coordinate system (xyz), such that the reference line is parallel to the x-axis, as shown in Figure 13.2. The coordinate system that is aligned with the beam in this fashion is the *local coordinate system* of the beam. In particular, the local x-axis is also termed the beam axis. At each location x along the reference line, we can consider a planar slice perpendicular to the axis of the beam. We will call such a slice the *cross-section* of the beam. The cross-sectional plane can be described by the local y- and z-axes. We also have the cross-sectional dimensions; for example, Figure 13.2 shows the cross-sectional depth, d, of a beam, as the sectional dimension along the local y-axis. The local y- and z- axes at each section of the beam are assumed to pass through the centroid of the section, so they are *centroidal axes*! If the local coordinate axes y and z are centroidal for a cross-section, A, it can easily be proven *(how?)* that the following equations are satisfied, involving surface integrals in the cross-sectional plane.

$$\iint_A y dS = 0 \tag{13.1.1a}$$

Fundamentals of Finite Element Analysis: Linear Finite Element Analysis, First Edition.
Ioannis Koutromanos, James McClure, and Christopher Roy.

Companion website: www.wiley.com/go/koutromanos/linear

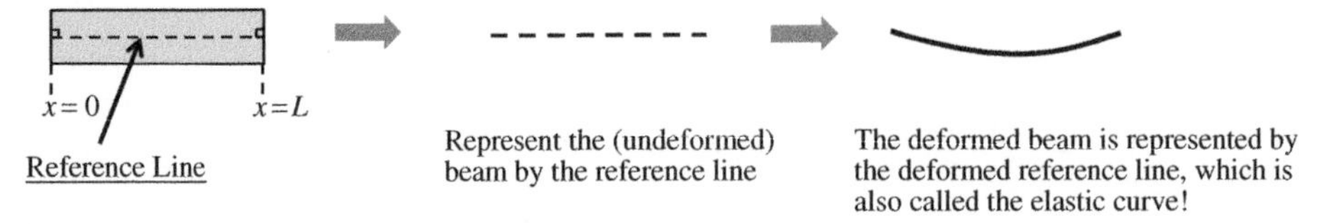

Figure 13.1 Beam geometry and reference line.

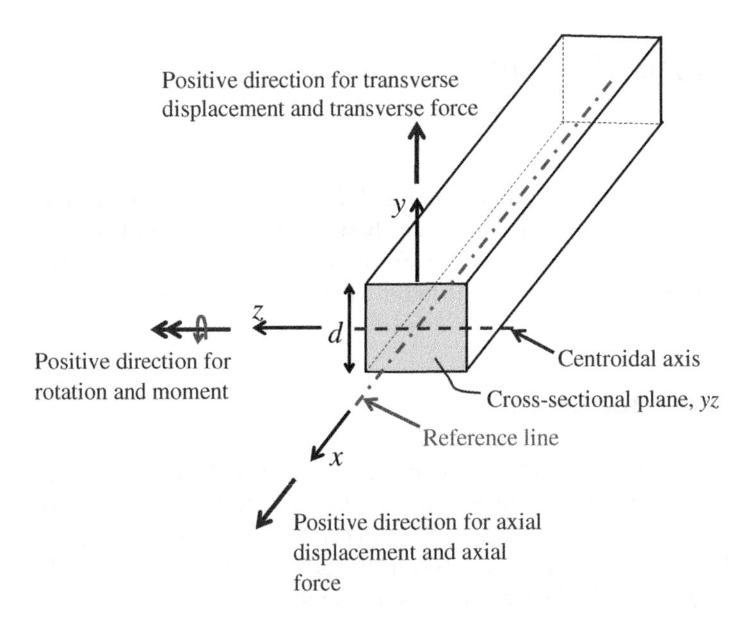

Figure 13.2 Three-dimensional beam geometry, reference line, and local axes.

$$\iint_A z dS = 0 \tag{13.1.1b}$$

Figure 13.2 also shows the sign convention for the displacements and forces and moments in a 2D beam structure. It is worth mentioning that a moment or rotation is represented by a vector pointing toward the positive z-axis. This is in accordance with the *right-hand rule*, which states that if we keep the thumb of our right hand straight, pointing toward the positive z-axis and curve the other fingers of our hand, then these other fingers will be showing the corresponding direction of the moment. Per Figure 13.2, the right-hand rule also applies for rotations.

Remark 13.1.1: A beam with constant sectional dimensions is called a *prismatic beam*. In this chapter, we will generally assume that the sectional dimensions vary along the beam length. ■

A 2D beam can be deformed in two basic modes, shown in Figure 13.3, namely, *axial deformation* and *deflection*. For a linearly elastic beam, these two modes of deformation can be examined independently from one another. This is why many textbooks only examine the flexural mode of deformation for beam structures. The axial mode of

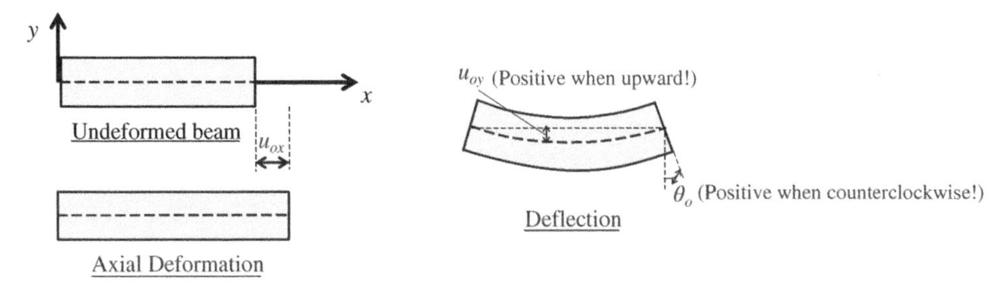

Figure 13.3 Modes of deformation for a two-dimensional beam.

deformation gives the same equations as those obtained for one-dimensional elasticity! For 2D beam problems, we only have displacements along the x- and y-directions. Since the deformation is described in terms of the reference line, at any location, x, along the length of the beam, we will be having an axial displacement (elongation) of the reference line, $u_{ox}(x)$, and a transverse displacement of the reference line, $u_{oy}(x)$. Each of the sections in a beam also undergoes a rotation, $\theta_o(x)$. These three kinematic quantities fully describe the deformed configuration of a 2D beam.

Before we move on to formulate the governing equations, it is imperative to establish a *positive sign convention for the internal loadings* (axial force, bending moment, and shear force) that develop in the interior of a beam as a result of applied loadings. The positive directions of the internal axial force N, shear force V, and bending moment M for a beam segment are presented in Figure 13.4. One can verify that the directions of N, V, and M are toward the positive local axes when we have a section at the right end of a beam segment, while they have the opposite directions when we have a section at the left end of a beam segment.

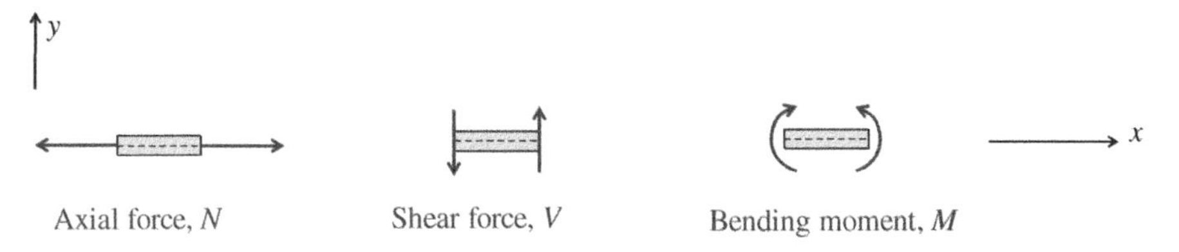

Figure 13.4 Sign convention for internal loadings in a beam.

13.2 Differential Equations and Boundary Conditions for 2D Beams

The governing differential equations of equilibrium for a beam can be obtained if we isolate a small segment of the beam, with length equal to Δx, as shown in Figure 13.5, and examine its equilibrium. Notice that the beam is subjected to distributed loads (forces per unit length), $p_x(x)$ and $p_y(x)$. The equations that we will obtain are generic, in the sense that, they apply for both elastic and inelastic materials, and they are valid for all beam theories (each beam theory introduces several additional assumptions regarding the way in which a beam can deform).

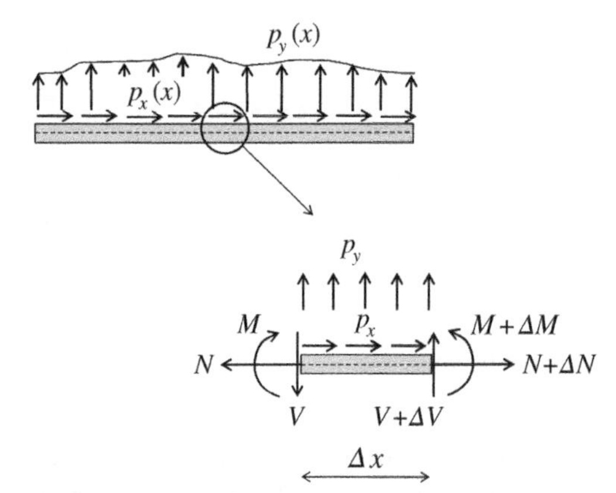

Figure 13.5 Beam segment with applied loadings.

Since the length of the segment is small, we will assume that the value of the distributed loads is constant and equal to the value at the midpoint of the segment. Let us first establish the force equilibrium equation in the axial direction, x, for the segment with length Δx in Figure 13.5:

$$N + \Delta N - N + p_x \cdot \Delta x = 0 \rightarrow \frac{\Delta N}{\Delta x} + p_x = 0 \tag{13.2.1}$$

If we now formulate the equilibrium equation for forces in the y-direction (taking upward forces as positive), we have:

$$V + \Delta V - V + p_y \cdot \Delta x = 0 \rightarrow \frac{\Delta V}{\Delta x} + p_y = 0 \tag{13.2.2}$$

Finally, if we establish the moment equilibrium equations about the right end of the segment (taking the counterclockwise moments as positive) we have:

$$M + \Delta M - M - p_y \cdot \frac{(\Delta x)^2}{2} + V \cdot \Delta x = 0 \rightarrow \frac{\Delta M}{\Delta x} + V - p_y \cdot \frac{\Delta x}{2} = 0 \tag{13.2.3}$$

If we take the limit as Δx tends to zero in Equations (13.2.1), (13.2.2), and (13.2.3), we establish the differential equations of equilibrium for a two-dimensional beam. Equation (13.2.1) becomes:

$$\frac{dN}{dx} + p_x = 0 \tag{13.2.4a}$$

which is the differential equation of equilibrium for the axial force. Similarly, Equations (13.2.2) and (13.2.3) yield:

$$\frac{dV}{dx} + p_y = 0 \tag{13.2.4b}$$

$$\frac{dM}{dx} + V = 0 \tag{13.2.4c}$$

Equations (13.2.4a–c) are *the differential equations of equilibrium* for a two-dimensional beam.

Now let us formulate the mathematical expressions for the boundary conditions. Just like for one-dimensional problems in Chapter 2, the boundary consists of the two endpoints, $x = 0$ and $x = L$. The essential boundary conditions can correspond to three prescribed kinematic quantities.

$$\text{Prescribed axial displacement}: u_{ox} = \bar{u}_{ox} \text{ at } \Gamma_{ux} \tag{13.2.5a}$$

$$\text{Prescribed transverse displacement}: u_{oy} = \bar{u}_{oy} \text{ at } \Gamma_{uy} \tag{13.2.5b}$$

$$\text{Prescribed sectional rotation}: \theta_o = \bar{\theta}_o \text{ at } \Gamma_{u\theta} \tag{13.2.5c}$$

Remark 13.2.1: For a beam structure, we must have at least one point with prescribed u_{ox} and at least one point with prescribed u_{oy}. Otherwise, the beam cannot preclude rigid-body translation along the x- and y-directions. Additionally, to preclude rigid-body rotation of the beam, we must at least have either prescribed θ_o at a point or prescribed y-displacement at a second point. ■

To derive the mathematical expressions of the natural boundary conditions for a beam, we distinguish two cases, as shown in Figure 13.6. We may have prescribed forces F_x and F_y and prescribed moments, M_z, at either of the two beam ends. Now, we work exactly as we did for one-dimensional elasticity: we take a "cut" (section) at a very small distance, $dx \to 0$, from the point where we apply the forces/moments, we draw the free-body diagram of the beam segment with length dx (which will include the internal loadings drawn in accordance with the sign convention) and establish the equilibrium equations:

If we have applied "tractions" at the left end, $x = 0$:
$N(0) + F_x + p_x(0) \cdot dx = 0$, and since $dx \to 0$:

$$N(0) + F_x = 0 \quad \to \quad N(0) = -F_x \tag{13.2.6a}$$

$V(0) + F_y + p_y(0) \cdot dx = 0$, and since $dx \to 0$:

$$V(0) + F_y = 0 \quad \to \quad V(0) = -F_y \tag{13.2.6b}$$

$M(0) + M_z + V(0) \cdot dx + p_y(0) \cdot \dfrac{(dx)^2}{2} = 0$, and since $dx \to 0$:

$$M(0) + M_z = 0 \quad \to \quad M(0) = -M_z \tag{13.2.6c}$$

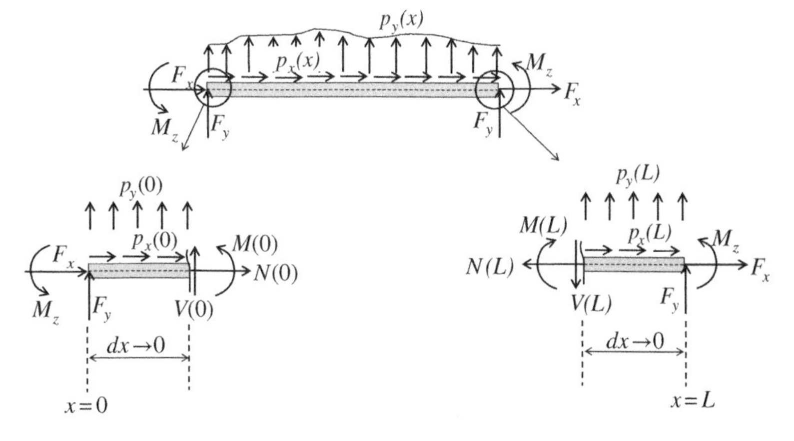

Figure 13.6 Natural boundary conditions for two-dimensional beam.

If we have applied "tractions" at the right end, $x = L$:
$-N(L) + F_x + p_x(L) \cdot dx = 0$, and since $dx \to 0$:

$$-N(L) + F_x = 0 \;\rightarrow\; N(L) = F_x \tag{13.2.6d}$$

$-V(L) + F_y + p_y(L) \cdot dx = 0$, and since $dx \to 0$:

$$-V(L) + F_y = 0 \;\rightarrow\; V(L) = F_y \tag{13.2.6e}$$

$$-M(L) + M_z + V(L) \cdot dx - p_y(L) \cdot \frac{(dx)^2}{2} = 0, \text{ and since } dx \to 0:$$

$$-M(L) + M_z = 0 \;\rightarrow\; M(L) = M_z \tag{13.2.6f}$$

We can collectively write:

$$\begin{Bmatrix} N \cdot n = F_x \;\; at \; \Gamma_{tx} \\ V \cdot n = F_y \;\; at \; \Gamma_{ty} \\ M \cdot n = M_z \;\; at \; \Gamma_{t\theta} \end{Bmatrix}, \text{ where } n = -1 \;\; at\, x = 0 \;\; and \;\; n = 1 \;\; at \;\; x = L \tag{13.2.7}$$

The combined essential and natural boundary of a beam give the total boundary, Γ. Thus, we can write:

$$\Gamma_{ux} \cup \Gamma_{tx} = \Gamma_{uy} \cup \Gamma_{ty} = \Gamma_{u\theta} \cup \Gamma_{t\theta} = \Gamma$$

In accordance with considerations of previous chapters, we cannot simultaneously prescribe the displacement and traction in the same direction for a given point. We can mathematically express this fact by writing:

$$\Gamma_{ux} \cap \Gamma_{tx} = \Gamma_{uy} \cap \Gamma_{ty} = \Gamma_{u\theta} \cap \Gamma_{t\theta} = \emptyset$$

The boundary, Γ, for a beam can be thought of as the combination of the two end points, $x = 0$ and $x = L$.

What we have seen so far *applies for any beam theory* (i.e., it is valid for all the possible assumptions that we can additionally make, regarding the kinematics of a beam). Next, we will begin our discussion for the two main beam theories, namely, the Euler-Bernoulli theory and the Timoshenko beam theory. We will initially establish the strong and weak form and the finite element formulation for the former theory; then, we will move on to the latter theory.

13.3 Euler-Bernoulli Beam Theory

The first beam theory that we will examine is the Euler-Bernoulli (E-B) theory, which is the oldest theory and is valid for slender beams (beams for which the cross-sectional depth, *d*, is much smaller than the length of the beam, at least 3 to 5 times smaller).

The basic kinematic assumptions (assumptions regarding the displacement field) of E-B theory are the following:

i) Plane sections remain plane and perpendicular to the deformed reference line of the beam, as shown for two sections A_1A_2 and B_1B_2 in Figure 13.7.
ii) The normal stress parallel to the cross-sectional plane is equal to zero, $\sigma_{yy} = 0$.

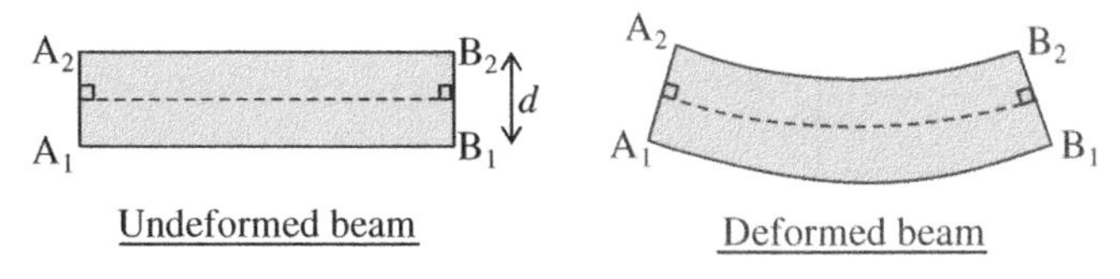

Figure 13.7 Euler-Bernoulli beam segment.

iii) When we determine the axial displacement at a point in a cross-section, we neglect the change of the depth of the cross-section, that is, we stipulate that the y-displacement at every point of the cross section is equal to the y-displacement of the reference line at the same cross-section, $u_y(x,y) = u_{oy}(x)$.

A consequence of the first kinematic assumption is that the shear strain, γ_{xy}, is zero everywhere. One may think that this would also lead (if we use the constitutive laws for elasticity) to zero shear stress, $\sigma_{xy} = 0$. However, this is not the case. It turns out that we can obtain a nonzero shear stress distribution to satisfy the equilibrium equations. The way to allow the existence of nonzero shear stresses while the shear strain is zero is to assume that the material's shear modulus, G, tends to infinity! In other words, an E-B beam is infinitely rigid against shear strains. This allows a nonzero shear stress, even if the corresponding shear strain is zero. A second consequence of the first assumption is that the rotation, θ_o, of any section is equal to the corresponding slope of the deformed reference curve:

$$\theta_o = \frac{du_{oy}}{dx} \tag{13.3.1}$$

Another interesting point pertains to the third kinematic assumption. A direct consequence of stating that $u_y(x,y) = u_{oy}(x)$ is that $\varepsilon_{yy} = 0$. However, this leads to a conceptual inconsistency: We have both $\sigma_{yy} = 0$ and $\varepsilon_{yy} = 0$! This cannot happen, because it is equivalent to stating that we restrain the displacement at a point, and we simultaneously have a prescribed force at that same point. It turns out, as also explained in Hughes (2000), that we can establish the displacement field of a beam without worrying about ε_{yy}, and then determine the distribution of the nonzero ε_{yy}! Neglecting the effect of the change of the sectional thickness (i.e., the effect of ε_{yy}) when we establish the beam equations simplifies the descriptions and gives results that are in satisfactory agreement with experimental tests.

Now let us imagine that we want to establish the expression that gives the axial displacement, $u_x(x, y)$, for a point P that is located at a vertical distance y from the reference line, as shown in Figure 13.8. The figure geometrically demonstrates that the axial displacement u_{Px} of P stems from the combined effect of axial deformation, u_{ox}, and of sectional rotation, θ_o.

Using geometric considerations, we can obtain from Figure 13.8:
$u_{Px} = u_{ox} - (PP') = u_{ox} - (CP)\cdot\tan(\theta_o)$, and for small rotations ($tan(\theta_o) \approx \theta_o$):

$$u_{Px} = u_{ox} - y\cdot\theta_o \tag{13.3.2}$$

Thus, for any point with a coordinate x and at a distance y along the depth of the section from the reference line, we have:

$$u_x(x,y) = u_{ox}(x) - y\cdot\theta_o(x) = u_{ox}(x) - y\cdot\frac{du_{oy}}{dx} \tag{13.3.3}$$

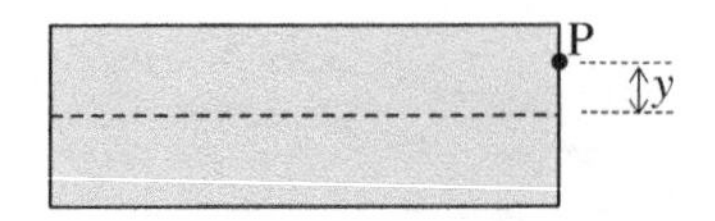

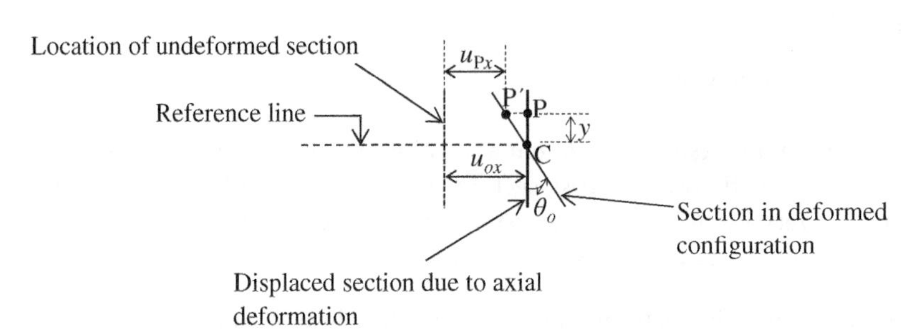

Figure 13.8 Determination of axial displacement at a point P.

We can now determine the axial strain $\varepsilon_{xx}(x, y)$, which is nothing more than the derivative of u_x with respect to x:

$$\varepsilon_{xx}(x,y) = \frac{du_x}{dx} = \frac{d}{dx}(u_{ox}(x) - y\cdot\theta_o(x)) = \frac{du_{ox}}{dx} - y\cdot\frac{d\theta_o}{dx} \rightarrow \varepsilon_{xx}(x,y) = \varepsilon_o - y\cdot\varphi \tag{13.3.4}$$

where $\frac{du_{ox}}{dx} = \varepsilon_o$ is the *axial strain* of the reference line and $\varphi = \frac{d\theta_o}{dx} = \frac{d^2u_{oy}}{dx^2}$ is the *curvature* of the beam. The curvature locally quantifies how "curved" the geometry of the deformed beam becomes. The reciprocal of the curvature is called the *radius of curvature*, ρ, of the deformed beam, because it gives the radius of the arc segment whose curvature is equal to φ. The fact that the curvature, $\varphi = 1/\rho$, is equal to the derivative of the slope θ_o with respect to x, can be proven based on geometric considerations, as shown for a beam segment in Figure 13.9.

Remark 13.3.1: The equality between φ and the second derivative of u_{oy} is only valid for small deformations! ∎

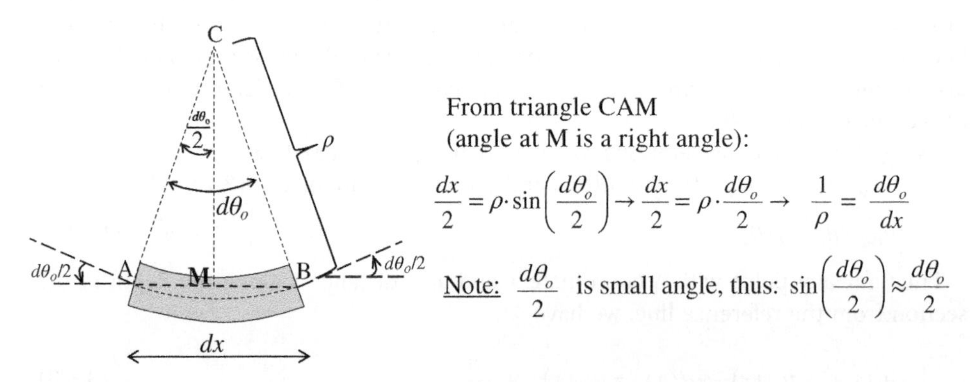

Figure 13.9 Geometric relation between curvature, $\varphi = 1/\rho$, and slope field $\theta_o(x)$.

If we have $\sigma_{yy} = \sigma_{zz} = 0$, we can verify (*how?*) that the constitutive equation for the normal stresses and for linear elasticity reduces to:

$$\sigma_{xx} = E \cdot \varepsilon_{xx} \tag{13.3.5}$$

Thus, the axial stress, $\sigma_{xx}(x, y)$ at a location of a beam is given by:

$$\sigma_{xx}(x,y) = E \cdot \varepsilon_{xx}(x,y) = E \cdot (\varepsilon_o - y \cdot \varphi) \tag{13.3.6}$$

For E-B beams, the axial force N and bending moment M can be obtained as *stress resultants*. If we imagine the cross-section of the beam, A, to consist of the combination of many points (x, y, z), each point having a corresponding infinitesimal sectional area $dS = (dy)\cdot(dz)$, we can "sum" the axial force contributions of each small area dS, as shown in Figure 13.10 to obtain the axial force, N, at a location x: $N = \iint_A \sigma_{xx} dS$, and if we plug Equation (13.3.5) in the integral, we obtain:

$$N = \iint_A \sigma_{xx} dS = \iint_A E \cdot \varepsilon_{xx} dS = \iint_A E \cdot (\varepsilon_{ox} - y \cdot \varphi) dS \rightarrow N = E \cdot \varepsilon_{ox} \iint_A dS - E \cdot \varphi \iint_A y dS \tag{13.3.7}$$

Now, we know that the sum of all the infinitesimal areas dS is equal to the cross-sectional area, A:

$$\iint_A dS = A \tag{13.3.8}$$

Plugging Equations (13.1.1a) and (13.3.8) into (13.3.7), we obtain:

$$N = EA \cdot \varepsilon_{ox} = EA \cdot \frac{du_{ox}}{dx} \tag{13.3.9}$$

We also have a bending moment about the z-axis. The contribution of every point with stress $\sigma_{xx}(x, y)$ to the bending moment is found by taking the force of that point and multiplying it by the corresponding lever arm in the y-direction, as shown in Figure 13.10. If we then sum the contributions of all the points, we have:

$$M_z = M = \iint_A -y \cdot \sigma_{xx} dS = \iint_A -y \cdot E \cdot \varepsilon_{xx} dS = \iint_A -y \cdot E \cdot (\varepsilon_{ox} - y \cdot \varphi) dS$$

$$M = -E \cdot \varepsilon_{ox} \iint_A y dS + E \cdot \varphi \iint_A y^2 dS \tag{13.3.10}$$

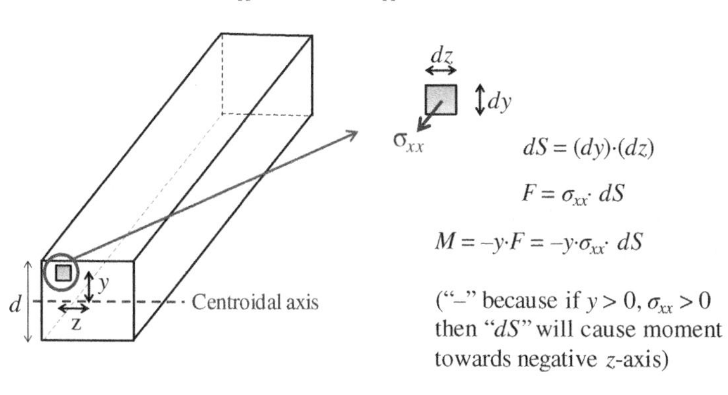

Figure 13.10 Contribution of "small piece" of cross-section to axial force and bending moment.

Now, we define the *moment of inertia about the z-axis, I*:

$$\iint_A y^2 dS = I \tag{13.3.11}$$

If we plug Equations (13.1.1a) and (13.3.11) into (13.3.10):

$$M = EI \cdot \varphi = EI \cdot \frac{d^2 u_{oy}}{dx^2} \tag{13.3.12}$$

Remark 13.3.2: For a beam section, we can define a moment about the y-axis, too! That is, $M_y = \iint_A z \cdot \sigma_{xx} dS$. For the case that we are examining, we will assume that the value of M_y is zero everywhere. ■

We can write a matrix equation, combining Equations (13.3.9) and (13.3.12):

$$\begin{Bmatrix} N \\ M \end{Bmatrix} = \begin{bmatrix} EA & 0 \\ 0 & EI \end{bmatrix} \begin{Bmatrix} \varepsilon_o \\ \varphi \end{Bmatrix} \rightarrow \{\hat{\sigma}\} = [\hat{D}]\{\hat{\varepsilon}\} \tag{13.3.13}$$

which is the sectional constitutive equation for a two-dimensional E-B beam, where we have established a *generalized strain* vector, $\{\hat{\varepsilon}\}$:

$$\{\hat{\varepsilon}\} = \begin{Bmatrix} \varepsilon_o \\ \varphi \end{Bmatrix} = \begin{bmatrix} \frac{du_{ox}}{dx} & \frac{d^2 u_{oy}}{dx^2} \end{bmatrix}^T \tag{13.3.14}$$

and a *generalized stress* vector, $\{\hat{\sigma}\}$:

$$\{\hat{\sigma}\} = \begin{Bmatrix} N \\ M \end{Bmatrix} \tag{13.3.15}$$

Remark 13.3.3: To obtain Equations (13.3.9) and (13.3.12), we tacitly assumed that we have a constant modulus of elasticity, E. The specific equations also apply for a variable E, provided that E only varies with x (and not with y or z). ■

13.4 Strong Form for Two-Dimensional Euler-Bernoulli Beams

The governing equations of beam theory, that is, Equations (13.2.4), (13.2.5) and (13.2.6), can be specialized to the case of Euler-Bernoulli (E-B) beam theory. Since for E-B beams we only have two kinematic fields, $u_{ox}(x)$ and $u_{oy}(x)$, it is desirable (and possible) to have a system of two differential equations (with two unknown field functions!). To this end, we differentiate equation (13.2.4c) with x and obtain:

$$\frac{d^2 M}{dx^2} + \frac{dV}{dx} = 0 \tag{13.4.1a}$$

and if we plug Equation (13.2.4b) into Equation (13.4.1a), we eventually obtain:

$$\frac{d^2M}{dx^2} - p_y = 0 \tag{13.4.1b}$$

Equations (13.2.4a) and (13.4.1b) constitute the two governing differential equations of equilibrium for an E-B beam. These equations are valid for both elastic and inelastic materials.

For an E-B beam, all types of essential boundary conditions at the left or right end can be written in terms of u_{ox}, u_{oy}:

$$u_{ox} = \bar{u}_{ox} \ at \ \Gamma_{ux} \tag{13.4.2a}$$

$$u_{oy} = \bar{u}_{oy} \ at \ \Gamma_{uy} \tag{13.4.2b}$$

In light of Equation (13.3.1), we can transform the rotational essential boundary conditions, i.e., Equation (13.2.5c), into the following form.

$$\frac{du_{oy}}{dx} = \bar{\theta}_o \quad at \ \Gamma_{u\theta} \tag{13.4.2c}$$

It is also desirable to have all the natural boundary conditions expressed with respect to N and M:

$$N \cdot n = F_x \quad at \ \Gamma_{tx} \tag{13.4.3a}$$

$$V \cdot n = -\frac{dM}{dx} \cdot n = F_y \quad at \ \Gamma_{ty} \tag{13.4.3b}$$

$$M \cdot n = M_z \ at \ \Gamma_{t\theta} \tag{13.4.3c}$$

Thus, we can now write the strong form for a two-dimensional E-B beam theory, provided in Box 13.4.1, if we also account for the sectional constitutive laws of Equation (13.3.13).

Box 13.4.1 Strong Form for Two-Dimensional, Euler-Bernoulli Beam Theory

$$\frac{d}{dx}\left(EA\frac{du_{ox}}{dx}\right) + p_x = 0 \tag{13.4.4a}$$

$$\frac{d^2}{dx^2}\left(EI\frac{d^2u_{oy}}{dx^2}\right) - p_y = 0 \tag{13.4.4b}$$

$$u_{ox} = \bar{u}_{ox} \quad on \ \Gamma_{ux}$$

$$u_{oy} = \bar{u}_{oy} \quad on \ \Gamma_{uy}$$

$$\frac{du_{oy}}{dx} = \bar{\theta} \quad on \ \Gamma_{u\theta}$$

$$N \cdot n = F_x \ at \ \Gamma_{tx}$$

$$V \cdot n = -\frac{dM}{dx} \cdot n = F_y \ at \ \Gamma_{ty}$$

$$M \cdot n = M_z \ at \ \Gamma_{t\theta}$$

Remark 13.4.1: The differential equation and boundary conditions corresponding to the axial direction are identical to those obtained for one-dimensional elasticity. ∎

13.5 Weak Form for Two-Dimensional Euler-Bernoulli Beams

We will now establish the weak form for two-dimensional Euler-Bernoulli (E-B) theory. We will consider an arbitrary displacement vector field $\{w(x)\}$, with components w_{ox} and w_{oy}, for which we have:

$$w_{ox} = 0 \text{ on } \Gamma_{ux}, w_{oy} = 0 \text{ on } \Gamma_{uy}, \frac{dw_{oy}}{dx} = 0 \text{ on } \Gamma_{u\theta} \tag{13.5.1}$$

First, we multiply the differential equation of axial equilibrium, (13.4.4a), by w_{ox} and integrate over the domain:

$$\int_0^L w_{ox}\left(\frac{dN}{dx} + p_x\right)dx = 0 \rightarrow \int_0^L w_{ox}\frac{dN}{dx}dx + \int_0^L w_{ox}\cdot p_x dx = 0 \tag{13.5.2}$$

The first integral term in Equation (13.5.2) is "unsymmetric" in terms of differentiation of u_{ox} and w_{ox}, and also includes second-order derivatives for u_{ox} (and we would prefer to have lower-order derivatives, since they would reduce the continuity requirements for the finite element approximation). Thus, we conduct integration by parts for this term:

$$\int_0^L w_{ox}\frac{dN}{dx}dx = [w_{ox}\cdot N]_0^L - \int_0^L \frac{dw_{ox}}{dx}\cdot N\,dx \tag{13.5.3}$$

We know that the two endpoints of a beam, $x = 0$ and $x = L$, generally comprise the essential and natural boundaries, Γ_{ux} and Γ_{tx}, and we know that we have $w_{ox} = 0$ on Γ_{ux}. Thus, the first term in the right-hand side of Equation (13.5.3) becomes:

$$[w_{ox}\cdot N]_0^L = [w_{ox}\cdot N\cdot n]_{\Gamma_{ux}} + [w_{ox}\cdot N\cdot n]_{\Gamma_{tx}} = [w_{ox}\cdot N\cdot n]_{\Gamma_{tx}} = [w_{ox}\cdot F_x]_{\Gamma_{tx}} \tag{13.5.4}$$

Thus, if we plug Equation (13.5.4) into Equation (13.5.3), then plug the resulting expression in Equation (13.5.2), we obtain:

$$[w_{ox}\cdot F_x]_{\Gamma_{tx}} - \int_0^L \frac{dw_{ox}}{dx}\cdot N\,dx + \int_0^L w_{ox}\cdot p_x dx = 0 \rightarrow \int_0^L \frac{dw_{ox}}{dx}\cdot N\,dx = \int_0^L w_{ox}\cdot p_x dx + [w_{ox}\cdot F_x]_{\Gamma_{tx}} \tag{13.5.5}$$

We now multiply the differential equation of equilibrium for flexure, (13.4.4b), by w_{oy} and integrate over the domain:

$$\int_0^L w_{oy}\left(\frac{d^2M}{dx^2} - p_y\right)dx = 0 \rightarrow \int_0^L w_{oy}\frac{d^2M}{dx^2}dx - \int_0^L w_{oy}\cdot p_y dx = 0 \tag{13.5.6}$$

The first term in Equation (13.5.6) is unsymmetric in terms of differentiation of u_{oy} and w_{oy}, and also includes fourth-order derivatives for u_{oy} (remember how the moment M is related to the transverse displacement field, by virtue of Equation 13.3.12). We would prefer the order of the derivatives to be as low as possible. Thus, we conduct integration by parts in Equation (13.5.6):

$$\int_0^L w_{oy}\frac{d^2M}{dx^2}dx = \left[w_{oy}\cdot\frac{dM}{dx}\right]_0^L - \int_0^L \frac{dw_{oy}}{dx}\cdot\frac{dM}{dx}dx$$

$$= \left[w_{oy}\cdot\frac{dM}{dx}\cdot n\right]_{\Gamma_{ty}} + \left[w_{oy}\cdot\frac{dM}{dx}\cdot n\right]_{\Gamma_{uy}} - \int_0^L \frac{dw_{oy}}{dx}\cdot\frac{dM}{dx}dx \tag{13.5.7}$$

We know that $w_{oy}=0$ on Γ_{uy}; thus, the first term in the right-hand side of Equation (13.5.7) becomes:

$$\left[w_{oy}\cdot\frac{dM}{dx}\cdot n\right]_{\Gamma_{ty}} + \left[w_{oy}\cdot\frac{dM}{dx}\cdot n\right]_{\Gamma_{uy}} = \left[w_{oy}\cdot\frac{dM}{dx}\cdot n\right]_{\Gamma_{ty}} + 0 = -\left[w_{oy}\cdot F_y\right]_{\Gamma_{ty}} \tag{13.5.8}$$

where we accounted for Equation (13.4.3b).

Thus, what we have so far is:

$$-\int_0^L \frac{dw_{oy}}{dx}\cdot\frac{dM}{dx}dx - \left[w_{oy}\cdot F_y\right]_{\Gamma_{ty}} - \int_0^L w_{oy}\cdot p_y dx = 0 \tag{13.5.9}$$

Equation (13.5.9) is still unsymmetric in terms of differentiation of u_{oy} and w_{oy}; thus, we once again use integration by parts for the first term:

$$-\int_0^L \frac{dw_{oy}}{dx}\cdot\frac{dM}{dx}dx = -\left[\frac{dw_{oy}}{dx}\cdot M\right]_0^L + \int_0^L \frac{d^2w_{oy}}{dx^2}\cdot M dx = -\left[\frac{dw_{oy}}{dx}\cdot M\cdot n\right]_{\Gamma_{t\theta}}$$

$$-\left[\frac{dw_{oy}}{dx}\cdot M\cdot n\right]_{\Gamma_{u\theta}} + \int_0^L \frac{d^2w_{oy}}{dx^2}\cdot M dx \rightarrow -\int_0^L \frac{dw_{oy}}{dx}\cdot\frac{dM}{dx}dx = -\left[\frac{dw_{oy}}{dx}\cdot M\cdot n\right]_{\Gamma_{t\theta}} + \int_0^L \frac{d^2w_{oy}}{dx^2}\cdot M dx \tag{13.5.10}$$

where we have accounted for the fact that $\frac{dw_{oy}}{dx}=0$ on $\Gamma_{u\theta}$.

If we plug Equation (13.5.10) into Equation (13.5.9), we obtain:

$$-\left[\frac{dw_{oy}}{dx}\cdot M_z\right]_{\Gamma_{t\theta}} - \left[w_{oy}\cdot F_y\right]_{\Gamma_{ty}} + \int_0^L \frac{d^2w_{oy}}{dx^2}\cdot M dx = \int_0^L w_{oy}\cdot p_y dx$$

$$\rightarrow \int_0^L \frac{d^2w_{oy}}{dx^2}\cdot M dx = \int_0^L w_{oy}\cdot p_y dx + \left[\frac{dw_{oy}}{dx}\cdot M_z\right]_{\Gamma_{t\theta}} + \left[w_{oy}\cdot F_y\right]_{\Gamma_{ty}} \tag{13.5.11}$$

Adding equations (13.5.5) and (13.5.11), we obtain:

$$\int_0^L \left(\frac{dw_{ox}}{dx}\cdot N + \frac{d^2w_{oy}}{dx^2}\cdot M\right)dx = \int_0^L \left(w_{ox}\cdot p_x + w_{oy}\cdot p_y\right)dx + \left[w_{ox}\cdot F_x\right]_{\Gamma_{tx}} + \left[w_{oy}\cdot F_y\right]_{\Gamma_{ty}} + \left[\frac{dw_{oy}}{dx}\cdot M_z\right]_{\Gamma_{t\theta}}$$

(13.5.12)

Now, we take Equation (13.5.12), account for the constitutive equations for an Euler-Bernoulli beam (Equation 13.3.13), account for the definition of the generalized strain vector (Equation 13.3.14), include the essential boundary conditions, and also define the *virtual generalized strain field* $\{\bar{\varepsilon}\}$:

$$\{\bar{\varepsilon}\} = \begin{Bmatrix} \bar{\varepsilon}_o \\ \bar{\varphi} \end{Bmatrix} = \begin{bmatrix} \dfrac{dw_{ox}}{dx} & \dfrac{d^2w_{oy}}{dx^2} \end{bmatrix}^T \tag{13.5.13}$$

With all the above, we eventually obtain the *weak form for Euler-Bernoulli beam theory*, which is provided in Box 13.5.1. Notice that the weak form of the governing equations for an Euler-Bernoulli beam are the same as those of multidimensional elasticity, with appropriate definitions for the generalized strains and stresses. This should be expected, because beam problems are a special case of solid mechanics, so their weak form should also be a "virtual work" expression, with sectional force resultants (forces and/or moments) playing the role of generalized stresses and tractions.

The integral expression in the left-hand side of the weak form in Box 13.5.1 can be separated into two contributions, one due to "axial virtual work" and the other due to "flexural virtual work," to yield the following version of the weak form.

$$\int_0^L \frac{dw_{ox}}{dx}EA\frac{du_{ox}}{dx}dx + \int_0^L \frac{d^2w_{oy}}{dx^2}EI\frac{d^2u_{oy}}{dx^2}dx = \int_0^L \{w\}^T\{p\}dx + \left[w_{ox}F_x\right]_{\Gamma_{tx}} + \left[w_{oy}F_y\right]_{\Gamma_{ty}} + \left[\frac{dw_{oy}}{dx}M_z\right]_{\Gamma_{t\theta}}$$

(13.5.14)

Remark 13.5.1: For the axial displacement u_{ox} and weight function w_{ox}, the highest order of derivative appearing in the weak form is 1, while for the transverse displacement and weight function components, u_{oy} and w_{oy}, respectively, the highest order of derivatives appearing in the weak form is 2. This means that the finite element approximation for the axial components must be first-degree complete and be characterized by C^0 continuity, just as for elasticity problems. On the other hand, the element approximation for the transverse components must have second-degree completeness and be C^1 continuous. In the following section, we will see how to obtain beam element shape functions that are C^1 and satisfy the completeness requirements. ∎

Remark 13.5.2: The equivalence between the weak form and the strong form can once again be established by starting with the weak form and then obtaining the strong form.

The key is to exploit the arbitrariness of $\{w\}$ in the weak form, then state that the weak form must also be satisfied for convenient special forms of $\{w\}$, which, in turn, will give us the strong form. ■

Box 13.5.1 Weak Form for Euler-Bernoulli Beam Theory

$$\int_0^L \{\bar{\hat{\varepsilon}}\}^T [\hat{D}]\,\{\hat{\varepsilon}\}dx = \int_0^L \{w\}^T\{p\}dx + [w_{ox}F_x]_{\Gamma_{tx}} + [w_{oy}F_y]_{\Gamma_{ty}} + \left[\frac{dw_{oy}}{dx}M_z\right]_{\Gamma_{t\theta}}$$

where $[\hat{D}] = \begin{bmatrix} EA & 0 \\ 0 & EI \end{bmatrix}$, $\{\hat{\varepsilon}\} = \begin{Bmatrix} \varepsilon_o \\ \varphi \end{Bmatrix} = \begin{Bmatrix} \frac{du_{ox}}{dx} \\ \frac{d^2u_{oy}}{dx^2} \end{Bmatrix}$, $\{\bar{\hat{\varepsilon}}\} = \begin{Bmatrix} \bar{\varepsilon}_o \\ \bar{\varphi} \end{Bmatrix} = \begin{Bmatrix} \frac{dw_{ox}}{dx} \\ \frac{d^2w_{oy}}{dx^2} \end{Bmatrix}$, $\{p\} = \begin{Bmatrix} p_x(x) \\ p_y(x) \end{Bmatrix}$

$\forall\{w\} = \begin{Bmatrix} w_{ox} \\ w_{oy} \end{Bmatrix}$ with $w_{ox} = 0$ on Γ_{ux}, $w_{oy} = 0$ on Γ_{uy}, $\frac{dw_{oy}}{dx} = 0$ on $\Gamma_{u\theta}$

$u_{ox} = \bar{u}_{ox}$ on Γ_{ux}

$u_{oy} = \bar{u}_{oy}$ on Γ_{uy}

$\frac{du_{oy}}{dx} = \bar{\theta}$ on $\Gamma_{u\theta}$

13.6 Finite Element Formulation: Two-Node Euler-Bernoulli Beam Element

It is straightforward to prove that a piecewise finite element approximation for a beam structure leads to a global stiffness equation of the form:

$$[K]\{U\} = \{f\} \tag{13.6.1}$$

where the global stiffness matrix, $[K]$, is given as an assembly operation of the element stiffness matrices, $[k^{(e)}]$

$$[K] = \sum_{e=1}^{Ne}\left(\left[L^{(e)}\right]^T\left[k^{(e)}\right]\left[L^{(e)}\right]\right) \tag{13.6.2}$$

and the equivalent nodal force vector $\{f\}$ is given through assembly of the corresponding element vectors, $\{f^{(e)}\}$.

$$\{f\} = \sum_{e=1}^{Ne}\left(\left[L^{(e)}\right]^T\left\{f^{(e)}\right\}\right) \tag{13.6.3}$$

The element displacement vector field $\{u^{(e)}(x)\}$, containing the axial and transverse displacement component fields, will be expressed as the product of a shape function array and a vector $\{U^{(e)}\}$ containing the nodal displacements and rotations for the element.

We will still use the term *nodal displacements* for $\{U^{(e)}\}$, even though the vector also contains nodal rotations! Thus, we write:

$$\left\{u^{(e)}(x)\right\} \approx \left[N^{(e)}(x)\right]\left\{U^{(e)}\right\} \tag{13.6.4}$$

For an element with n nodal points, the vector $\{U^{(e)}\}$ has $3n$ components (two displacements and one rotation for each node) and the dimension of the shape function array, $[N^{(e)}(x)]$, is $(2 \times 3n)$.

The generalized strain vector, $\left\{\hat{\varepsilon}^{(e)}\right\}$, is given by premultiplying the nodal displacement vector by a generalized strain-displacement array $[B^{(e)}(x)]$:

$$\left\{\hat{\varepsilon}^{(e)}\right\} = \left[B^{(e)}(x)\right]\left\{U^{(e)}\right\} \tag{13.6.5}$$

Given the above, and assuming that the virtual displacement field is given by the same approximation as the actual displacement field, we eventually obtain the following expressions for the element stiffness matrix and the equivalent nodal force vector:

$$\left[k^{(e)}\right] = \int_0^{\ell^{(e)}} \left[B^{(e)}\right]^T \left[\hat{D}^{(e)}\right] \left[B^{(e)}\right] dx \tag{13.6.6}$$

where $\ell^{(e)}$ is the element length, and

$$\left\{f^{(e)}\right\} = \int_0^{\ell^{(e)}} \left[N^{(e)}\right]^T \{p\}\, dx + \left[N^{(e)}\right]^T_{\Gamma^e_{tx}} \begin{Bmatrix} F_x \\ 0 \end{Bmatrix} + \left[N^{(e)}\right]^T_{\Gamma^e_{ty}} \begin{Bmatrix} 0 \\ F_y \end{Bmatrix} + \left[N_\theta^{(e)}\right]^T_{\Gamma^e_{t\theta}} M_z \tag{13.6.7}$$

In Equation (13.6.7), $\left[N_\theta^{(e)}\right]$ is a row vector containing the first derivatives of the shape functions in the second row of $[N^{(e)}]$, which second row is used for the approximation of the transverse displacement, u_{oy}.

The simplest type of E-B finite element is the two-node element, shown in Figure 13.11. The element can be idealized as a line segment having a length equal to $\ell^{(e)}$, and the material and cross-sectional properties can be assumed to vary over the length of the beam, that is, $E = E(x)$, $A = A(x)$, $I = I(x)$. The two-node beam element has two degrees of freedom in the (local) axial direction, x, and four degrees of freedom that affect the (local) transverse response, namely the two nodal transverse displacements and the two nodal rotations (for which we know that they must be equal to

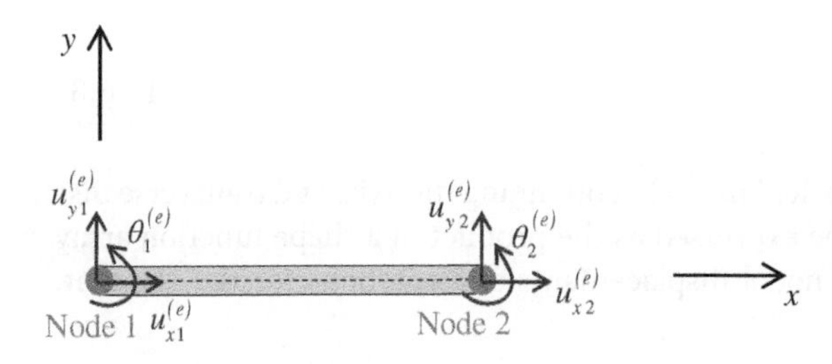

Figure 13.11 Nodal numbering, nodal degrees of freedom, and local axes for two-node Euler-Bernoulli beam element.

the derivative of the transverse displacement at the nodal points). We will need two shape functions for the axial displacement field and four shape functions for the transverse displacement field in the element.

We can now establish the following polynomial expressions for the axial and transverse displacement fields, $u_{ox}(x)$ and $u_{oy}(x)$, respectively:

$$u_{ox}(x) \approx N_{ux1}^{(e)}(x) \cdot u_{x1}^{(e)} + N_{ux2}^{(e)}(x) \cdot u_{x2}^{(e)} \tag{13.6.8a}$$

$$u_{oy}(x) \approx N_{uy1}^{(e)}(x) \cdot u_{y1}^{(e)} + N_{\theta 1}^{(e)}(x) \cdot \theta_1^{(e)} + N_{uy2}^{(e)}(x) \cdot u_{y2}^{(e)} + N_{\theta 2}^{(e)}(x) \cdot \theta_2^{(e)} \tag{13.6.8b}$$

The two shape functions used for the approximation of the axial displacement field in Equation (15.6.8a) are the same as those introduced in two-node, one-dimensional elements in Section 3.2. The four shape functions used for the approximation of the transverse displacement are the C^1–continuous family of *Hermitian polynomials*:

$$N_{uy1}^{(e)}(x) = 1 - \frac{3x^2}{\left(\ell^{(e)}\right)^2} + \frac{2x^3}{\left(\ell^{(e)}\right)^3} \tag{13.6.9a}$$

$$N_{\theta 1}^{(e)}(x) = x - \frac{2x^2}{\ell^{(e)}} + \frac{x^3}{\left(\ell^{(e)}\right)^2} \tag{13.6.9b}$$

$$N_{uy2}^{(e)}(x) = \frac{3x^2}{\left(\ell^{(e)}\right)^2} - \frac{2x^3}{\left(\ell^{(e)}\right)^3} \tag{13.6.9c}$$

$$N_{\theta 2}^{(e)}(x) = -\frac{x^2}{\ell^{(e)}} + \frac{x^3}{\left(\ell^{(e)}\right)^2} \tag{13.6.9d}$$

The subscript in each of the four Hermitian polynomials in Equations (13.6.9a–d) indicates the nodal displacement/rotation to which each shape function corresponds. Since these polynomials are characterized by C^1 continuity, *both the transverse displacement and its first derivative, that is, the slope of the elastic curve, are continuous across interelement boundaries.*

As shown in Figure 13.12, each shape function $N_{uyi}^{(e)}$ represents the deformed geometry of the beam for a unit transverse displacement at node i, while all the other degrees of freedom are zero, and each shape function $N_{\theta i}^{(e)}$ represents the deformed geometry of the beam for a unit rotation at node i, while all the other nodal degrees of freedom are zero.

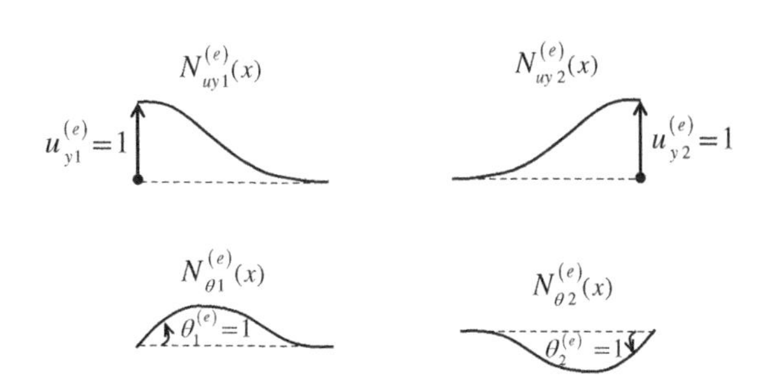

Figure 13.12 Hermitian polynomial shape functions for a two-node E-B beam element.

We now return to Equation (13.6.4) and set:

$$\left\{u^{(e)}(x)\right\} = \begin{Bmatrix} u_{ox}^{(e)}(x) \\ u_{oy}^{(e)}(x) \end{Bmatrix} \tag{13.6.10}$$

$$\left[N^{(e)}\right] = \begin{bmatrix} N_{ux1}^{(e)}(x) & 0 & 0 & N_{ux2}^{(e)}(x) & 0 & 0 \\ 0 & N_{uy1}^{(e)}(x) & N_{\theta 1}^{(e)}(x) & 0 & N_{uy2}^{(e)}(x) & N_{\theta 2}^{(e)}(x) \end{bmatrix} \tag{13.6.11}$$

and $\left\{U^{(e)}\right\} = \left[u_{x1}^{(e)} \; u_{y1}^{(e)} \; \theta_1^{(e)} \; u_{x2}^{(e)} \; u_{y2}^{(e)} \; \theta_2^{(e)}\right]^T$.

Now, the generalized strain vector is given by the following equation.

$$\left\{\hat{\varepsilon}^{(e)}\right\} = \begin{Bmatrix} \varepsilon_o^{(e)} \\ \varphi^{(e)} \end{Bmatrix} = \begin{Bmatrix} \dfrac{du_{ox}^{(e)}}{dx} \\ \dfrac{d^2u_{oy}^{(e)}}{dx^2} \end{Bmatrix} \approx \begin{Bmatrix} \dfrac{dN_{ux1}^{(e)}}{dx} \cdot u_{x1}^{(e)} + \dfrac{dN_{ux2}^{(e)}}{dx} \cdot u_{x2}^{(e)} \\ \dfrac{d^2N_{uy1}^{(e)}}{dx^2} \cdot u_{y1}^{(e)} + \dfrac{d^2N_{\theta 1}^{(e)}}{dx^2} \cdot \theta_1^{(e)} + \dfrac{d^2N_{uy2}^{(e)}}{dx^2} \cdot u_{y2}^{(e)} + \dfrac{d^2N_{\theta 2}^{(e)}}{dx^2} \cdot \theta_2^{(e)} \end{Bmatrix} \tag{13.6.12}$$

We can cast Equation (13.6.12) in the form of Equation (13.6.5), where:

$$\left[B^{(e)}\right] = \begin{bmatrix} \dfrac{dN_{ux1}^{(e)}}{dx} & 0 & 0 & \dfrac{dN_{ux2}^{(e)}}{dx} & 0 & 0 \\ 0 & \dfrac{d^2N_{uy1}^{(e)}}{dx^2} & \dfrac{d^2N_{\theta 1}^{(e)}}{dx^2} & 0 & \dfrac{d^2N_{uy2}^{(e)}}{dx^2} & \dfrac{d^2N_{\theta 2}^{(e)}}{dx^2} \end{bmatrix} \tag{13.6.13}$$

Since we know the polynomial expressions of the six shape functions, we can also find the required derivatives for the determination of the $[B^{(e)}]$ array.

$$\frac{dN_{ux1}^{(e)}}{dx} = -\frac{1}{\ell^{(e)}} \tag{13.6.14a}$$

$$\frac{dN_{ux2}^{(e)}}{dx} = \frac{1}{\ell^{(e)}} \tag{13.6.14b}$$

$$\frac{d^2N_{uy1}^{(e)}}{dx^2} = -\frac{6}{\left(\ell^{(e)}\right)^2} + \frac{12x}{\left(\ell^{(e)}\right)^3} = -\frac{6}{\left(\ell^{(e)}\right)^2} + \frac{6(\xi+1)}{\left(\ell^{(e)}\right)^2} \tag{13.6.14c}$$

$$\frac{d^2N_{\theta 1}^{(e)}}{dx^2} = -\frac{4}{\ell^{(e)}} + \frac{6x}{\left(\ell^{(e)}\right)^2} = -\frac{4}{\ell^{(e)}} + \frac{3(\xi+1)}{\ell^{(e)}} \tag{13.6.14d}$$

$$\frac{d^2N_{uy2}^{(e)}}{dx^2} = \frac{6}{\left(\ell^{(e)}\right)^2} - \frac{12x}{\left(\ell^{(e)}\right)^3} = \frac{6}{\left(\ell^{(e)}\right)^2} - \frac{6(\xi+1)}{\left(\ell^{(e)}\right)^2} \tag{13.6.14e}$$

$$\frac{d^2N_{\theta 2}^{(e)}}{dx^2} = -\frac{2}{\ell^{(e)}} + \frac{6x}{\left(\ell^{(e)}\right)^2} = -\frac{2}{\ell^{(e)}} + \frac{3(\xi+1)}{\ell^{(e)}} \tag{13.6.14f}$$

In Equations (13.6.14a–f), we also express the shape functions in terms of the parametric coordinate, ξ. This allows us to readily program an isoparametric one-dimensional formulation using Gaussian quadrature. The minimum required quadrature rule for E-B beam elements is a two-point one. If we use N_g quadrature points for a beam element subjected to distributed loads $p_x(x)$ and $p_y(x)$, then we can calculate the element stiffness matrix and equivalent nodal force vector with the algorithm in Box 13.6.1.

Box 13.6.1 Calculation of Element Stiffness Matrix and Nodal Force Vector Using Gaussian Quadrature

For each quadrature point g, $g = 1, \ldots, N_g$:

- Determine parametric coordinate, ξ_g, and weight coefficient, W_g.
- If we have a $\left[\hat{D}^{(e)}\right]$-array or a $\{p\}$-vector whose components are functions of x, calculate $x_g = x(\xi_g) = \frac{\ell^{(e)}}{2}(1+\xi_g)$.
- Determine the values of the $[B^{(e)}]$ array and of the $\left[\hat{D}^{(e)}\right]$ array at $\xi = \xi_g$: $[B_g] = [B^{(e)}(\xi_g)]$, $[\hat{D}_g] = \left[\hat{D}^{(e)}(x_g)\right]$
- If we have distributed loads, calculate $[N_g] = [N^{(e)}(\xi_g)]$ and $\{p_g\} = \{p(x_g)\}$.

After we complete the above computations for all Gauss points, we can obtain the element's stiffness matrix, $[k^{(e)}]$, and the element's equivalent nodal force vector, $\{f^{(e)}\}$:

$$\left[k^{(e)}\right] = \int_{-1}^{1} \left[B^{(e)}(\xi)\right]^T \left[\hat{D}^{(e)}(\xi)\right] \left[B^{(e)}(\xi)\right] \frac{\ell^{(e)}}{2} d\xi \approx \sum_{g=1}^{Ng} \left([B_g]^T [\hat{D}_g] [B_g] \frac{\ell^{(e)}}{2} W_g \right)$$

$$\left\{f^{(e)}\right\} = \int_{-1}^{1} \left[N^{(e)}(\xi)\right]^T \{p(\xi)\} \frac{\ell^{(e)}}{2} d\xi + \left[N^{(e)}\right]^T_{\Gamma^e_{tx}} \begin{Bmatrix} F_x \\ 0 \end{Bmatrix} + \left[N^{(e)}\right]^T_{\Gamma^e_{ty}} \begin{Bmatrix} 0 \\ F_y \end{Bmatrix} + \left[N^{(e)}_\theta\right]^T_{\Gamma^e_{t\theta}} M_z$$

$$\approx \sum_{g=1}^{Ng} \left([N_g]^T \{p_g\} \frac{\ell^{(e)}}{2} W_g \right) + \left[N^{(e)}\right]^T_{\Gamma^e_{tx}} \begin{Bmatrix} F_x \\ 0 \end{Bmatrix} + \left[N^{(e)}\right]^T_{\Gamma^e_{ty}} \begin{Bmatrix} 0 \\ F_y \end{Bmatrix} + \left[N^{(e)}_\theta\right]^T_{\Gamma^e_{t\theta}} M_z$$

where $\left[N^{(e)}_\theta\right] = \begin{bmatrix} 0 & \frac{dN^{(e)}_{uy1}}{dx} & \frac{dN^{(e)}_{\theta 1}}{dx} & 0 & \frac{dN^{(e)}_{uy2}}{dx} & \frac{dN^{(e)}_{\theta 2}}{dx} \end{bmatrix}$

Remark 13.6.1: Some additional discussion is necessary for the natural boundary terms in Box 13.6.1. If there is a natural boundary for an element, it will correspond to one of the two nodal points. It can be proven that, at either of these two points of the two-node E-B beam element, we have:

$$\left.\frac{dN^{(e)}_{uy1}}{dx}\right|_{x=0} = \left.\frac{dN^{(e)}_{uy1}}{dx}\right|_{x=\ell^{(e)}} = 0, \quad \left.\frac{dN^{(e)}_{uy2}}{dx}\right|_{x=0} = \left.\frac{dN^{(e)}_{uy2}}{dx}\right|_{x=\ell^{(e)}} = 0$$

$$\left.\frac{dN_{\theta 1}^{(e)}}{dx}\right|_{x=0}=1,\quad \left.\frac{dN_{\theta 1}^{(e)}}{dx}\right|_{x=\ell^{(e)}}=0,\quad \left.\frac{dN_{\theta 2}^{(e)}}{dx}\right|_{x=0}=0,\quad \left.\frac{dN_{\theta 2}^{(e)}}{dx}\right|_{x=\ell^{(e)}}=1$$

If we also notice that $\left.N_{uy1}^{(e)}\right|_{x=0}=1$, $\left.N_{uy1}^{(e)}\right|_{x=\ell^{(e)}}=0$, $\left.N_{uy2}^{(e)}\right|_{x=0}=0$, $\left.N_{uy2}^{(e)}\right|_{x=\ell^{(e)}}=1$, $\left.N_{ux1}^{(e)}\right|_{x=0}=1$, $\left.N_{ux1}^{(e)}\right|_{x=\ell^{(e)}}=0$, $\left.N_{ux2}^{(e)}\right|_{x=0}=0$, $\left.N_{ux2}^{(e)}\right|_{x=\ell^{(e)}}=1$, we can reach the conclusion that *if a nodal point of a beam element is part of the natural boundary, i.e. we have an applied force or moment at that node, we simply need to add the prescribed nodal force/moment to the corresponding component of* $\{f^{(e)}\}$! ■

Remark 13.6.2: For an E-B beam element with constant cross-section and constant Young's modulus (E, A, I are constant), we can obtain a closed-form expression for the stiffness matrix that is commonly provided in matrix structural analysis textbooks:

$$\left[k^{(e)}\right]=\begin{bmatrix} \frac{EA}{\ell^{(e)}} & 0 & 0 & -\frac{EA}{\ell^{(e)}} & 0 & 0 \\ 0 & \frac{12EI}{\left(\ell^{(e)}\right)^3} & \frac{6EI}{\left(\ell^{(e)}\right)^2} & 0 & -\frac{12EI}{\left(\ell^{(e)}\right)^3} & \frac{6EI}{\left(\ell^{(e)}\right)^2} \\ 0 & \frac{6EI}{\left(\ell^{(e)}\right)^2} & \frac{4EI}{\ell^{(e)}} & 0 & -\frac{6EI}{\left(\ell^{(e)}\right)^2} & \frac{2EI}{\ell^{(e)}} \\ -\frac{EA}{\ell^{(e)}} & 0 & 0 & \frac{EA}{\ell^{(e)}} & 0 & 0 \\ 0 & -\frac{12EI}{\left(\ell^{(e)}\right)^3} & -\frac{6EI}{\left(\ell^{(e)}\right)^2} & 0 & \frac{12EI}{\left(\ell^{(e)}\right)^3} & -\frac{6EI}{\left(\ell^{(e)}\right)^2} \\ 0 & \frac{6EI}{\left(\ell^{(e)}\right)^2} & \frac{2EI}{\ell^{(e)}} & 0 & -\frac{6EI}{\left(\ell^{(e)}\right)^2} & \frac{4EI}{\ell^{(e)}} \end{bmatrix} \tag{13.6.15}$$

■

Example 13.1: Stiffness matrix of Euler-Bernoulli Beam Element
Let us use two-point Gaussian quadrature to calculate the stiffness matrix of a two-node Euler-Bernoulli element, with $\ell^{(e)}=120$ in., $A=120$ in.^{2} and $I=1440$ in.^{4}. The material of the beam is linearly elastic, with $E=5000\left[1+\left(\frac{x}{120}\right)^2\right]$, where E is in ksi and x is in inches. We can readily use the algorithm of Box 13.6.1 to conduct the computations of each quadrature point.

Gauss point 1 ($\xi_1=-\frac{1}{\sqrt{3}}$, $W_1=1$)

$$x_1=x(\xi_1)=\frac{\ell^{(e)}}{2}(1+\xi_1)=\frac{120\,\text{in.}}{2}\left(1-\frac{1}{\sqrt{3}}\right)=25.36\text{ in.}$$

$$E_1=5000\left[1+\left(\frac{x_1}{120}\right)^2\right]=5000\left[1+\left(\frac{25.36}{120}\right)^2\right]=5223\text{ ksi}$$

$$\left[\hat{D}_1\right] = \begin{bmatrix} E_1 A & 0 \\ 0 & E_1 I \end{bmatrix} = \begin{bmatrix} 5223\cdot 120 & 0 \\ 0 & 5223\cdot 1440 \end{bmatrix} = \begin{bmatrix} 626795\,\text{kip} & 0 \\ 0 & 7521539\,\text{kip}-\text{in.}^2 \end{bmatrix}$$

$$\left.\frac{dN_{ux1}^{(e)}}{dx}\right|_{x=x_1} = -\frac{1}{\ell^{(e)}} = -\frac{1}{120} = -0.00833$$

$$\left.\frac{dN_{ux2}^{(e)}}{dx}\right|_{x=x_1} = \frac{1}{\ell^{(e)}} = \frac{1}{120} = 0.00833$$

$$\left.\frac{d^2N_{uy1}^{(e)}}{dx^2}\right|_{x=x_1} = -\frac{6}{\left(\ell^{(e)}\right)^2} + \frac{12x_1}{\left(\ell^{(e)}\right)^3} = -\frac{6}{120^2} + \frac{12\cdot 25.36}{120^3} = -0.0002406$$

$$\left.\frac{d^2N_{\theta 1}^{(e)}}{dx^2}\right|_{x=x_1} = -\frac{4}{\ell^{(e)}} + \frac{6x_1}{\left(\ell^{(e)}\right)^2} = -\frac{4}{120} + \frac{6\cdot 25.36}{120^2} = -0.0227271$$

$$\left.\frac{d^2N_{uy2}^{(e)}}{dx^2}\right|_{x=x_1} = \frac{6}{\left(\ell^{(e)}\right)^2} - \frac{12x_1}{\left(\ell^{(e)}\right)^3} = \frac{6}{120^2} - \frac{12\cdot 25.36}{120^3} = -0.0002406$$

$$\left.\frac{d^2N_{\theta 2}^{(e)}}{dx^2}\right|_{x=x_1} = -\frac{2}{\ell^{(e)}} + \frac{6x_1}{\left(\ell^{(e)}\right)^2} = -\frac{2}{120} + \frac{6\cdot 25.36}{120^2} = -0.0061004$$

Thus, we have:

$$[B_1] = \left.\begin{bmatrix} \frac{dN_{ux1}^{(e)}}{dx} & 0 & 0 & \frac{dN_{ux2}^{(e)}}{dx} & 0 & 0 \\ 0 & \frac{d^2N_{uy1}^{(e)}}{dx^2} & \frac{d^2N_{\theta 1}^{(e)}}{dx^2} & 0 & \frac{d^2N_{uy2}^{(e)}}{dx^2} & \frac{d^2N_{\theta 2}^{(e)}}{dx^2} \end{bmatrix}\right|_{x=x_1} \rightarrow$$

$$[B_1] = \begin{bmatrix} -0.00833 & 0 & 0 & 0.00833 & 0 & 0 \\ 0 & -0.0002406 & -0.0227271 & 0 & -0.0002406 & -0.0061004 \end{bmatrix}$$

Similarly, we have for <u>Gauss point 2</u> ($\xi_2 = \frac{1}{\sqrt{3}}$, $W_2 = 1$):

$x_2 = 94.64\,\text{in.}$

$E_2 = 8110\,\text{ksi}$

$$\left[\hat{D}_2\right] = \begin{bmatrix} 973205\,\text{kip} & 0 \\ 0 & 11678461\,\text{kip}-\text{in.}^2 \end{bmatrix}$$

$$[B_2] = \left.\begin{bmatrix} \frac{dN_{ux1}^{(e)}}{dx} & 0 & 0 & \frac{dN_{ux2}^{(e)}}{dx} & 0 & 0 \\ 0 & \frac{d^2N_{uy1}^{(e)}}{dx^2} & \frac{d^2N_{\theta 1}^{(e)}}{dx^2} & 0 & \frac{d^2N_{uy2}^{(e)}}{dx^2} & \frac{d^2N_{\theta 2}^{(e)}}{dx^2} \end{bmatrix}\right|_{x=x_2} \rightarrow$$

$$[B_2] = \begin{bmatrix} -0.00833 & 0 & 0 & 0.00833 & 0 & 0 \\ 0 & 0.0002406 & 0.0061004 & 0 & 0.0002406 & 0.0227671 \end{bmatrix}$$

Finally, we have: $\left[k^{(e)}\right] = \sum_{g=1}^{2}\left([B_g]^T\left[\hat{D}_g\right][B_g]\frac{\ell^{(e)}}{2}W_g\right) \rightarrow$

$$\rightarrow \left[k^{(e)}\right] = [B_1]^T\left[\hat{D}_1\right][B_1]\frac{120}{2}\cdot 1 + [B_2]^T\left[\hat{D}_2\right][B_2]\frac{120}{2}\cdot 1 \rightarrow$$

$$\rightarrow \left[k^{(e)}\right] = \begin{bmatrix} 6667 & 0 & 0 & -6667 & 0 & 0 \\ 0 & 66.7 & 3500 & 0 & -66.7 & 4500 \\ 0 & 3500 & 260000 & 0 & -3500 & 160000 \\ -6667 & 0 & 0 & 6667 & 0 & 0 \\ 0 & -66.7 & -3500 & 0 & 66.7 & -4500 \\ 0 & 4500 & 160000 & 0 & -4500 & 380000 \end{bmatrix} \text{(units in kip, in.).}$$

■

13.7 Coordinate Transformation Rules for Two-Dimensional Beam Elements

The stiffness matrices of beam elements are typically obtained in a local coordinate system, for which the x-axis coincides with the axis of the beam. For a general frame structure (structure consisting of beam members that are subjected to both flexural and axial loads) the local coordinate systems will be different for the various members! One such example is shown for a four-element frame in Figure 13.13. In such cases, per Section B.8, it is necessary to establish a *global* (or structural) coordinate system, XY, as shown in the figure, which is unique and on which we will assemble the global equilibrium (or stiffness) equations. Then, we need to find *coordinate transformation rules*, which give the relation between the element nodal force vector, nodal displacement vector, and stiffness matrix in the local coordinate system and the corresponding entities in the global coordinate system.

We first need to establish the element local coordinate axes, $x^{(e)}$ and $y^{(e)}$, as shown in Figure 13.14. Then, we determine the orientation angle, φ, which is equal to the angle by which we need to rotate the global X-axis so that it coincides with the local $x^{(e)}$ axis of the element, as also shown in Figure 13.14. As explained in the same figure, if we know the nodal coordinates in the global system, we can determine the sine and the cosine of the element's orientation angle, φ.

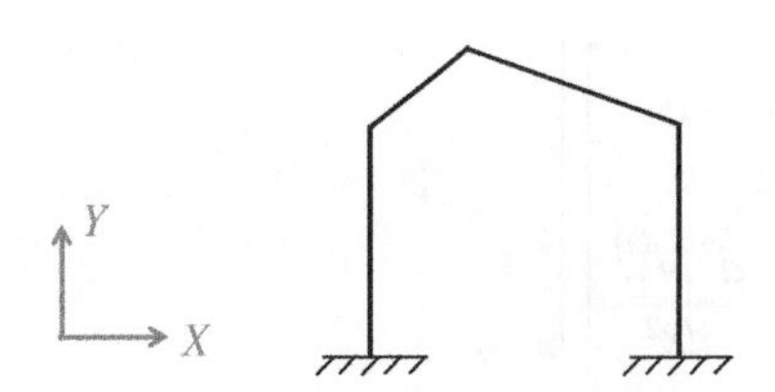

Figure 13.13 Global coordinate system for a frame structure.

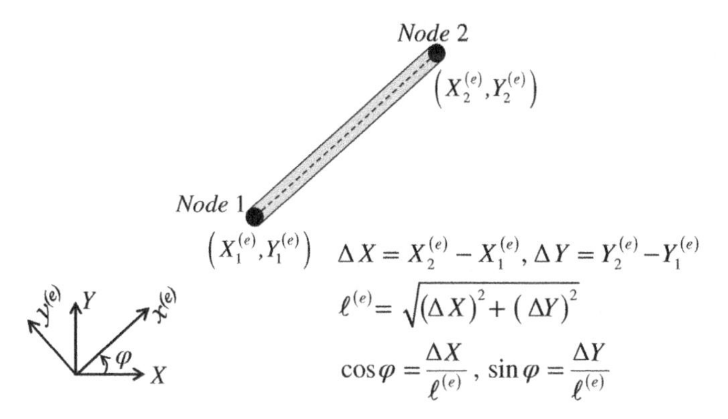

Figure 13.14 Determination of orientation angle φ for beam element.

Now, to establish the *transformation rule for nodal displacements*, we can use compatibility equations. To this end, we isolate node 1 and impose a displacement $u_{X1}^{(e)}$ along the global X-axis, as shown in Figure 13.15. We can then determine the corresponding displacement components $u'^{(e)}_{x1}$ and $u'^{(e)}_{y1}$ along the local axes. As shown in that figure, we then repeat the same process for the case that we have a displacement $u_{Y1}^{(e)}$ along the global Y-axis. A similar procedure yields the coordinate transformation rules for node 2, as shown in Figure 13.16.

The z-axis is common (perpendicular to the xy plane) for the local and global coordinate system, thus we can stipulate that the nodal rotations (which take place about the z-axis) are identical in the local and global coordinate systems:

$$\theta'^{(e)}_1 = \theta_1^{(e)} \tag{13.7.1a}$$

$$\theta'^{(e)}_2 = \theta_2^{(e)} \tag{13.7.1b}$$

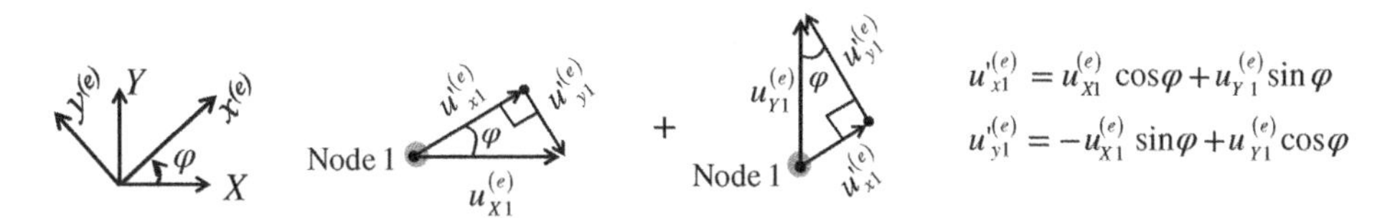

Figure 13.15 Establishing coordinate transformation rules for displacements of node 1.

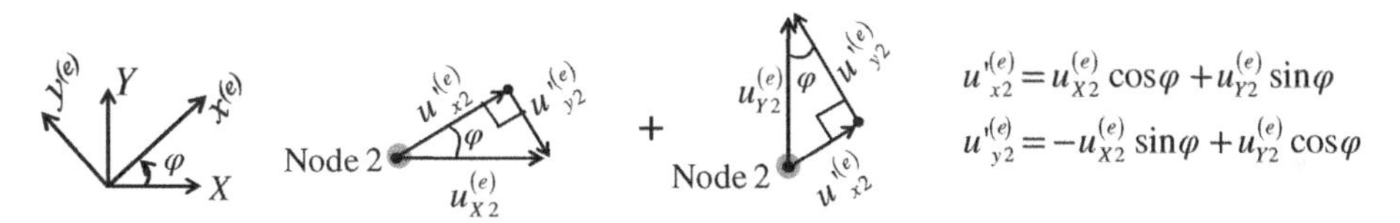

Figure 13.16 Establishing coordinate transformation rules for displacements of node 2.

We can combine Figures 13.15 and 13.16 and Equations (13.7.1a,b) to collectively establish the following coordinate transformation equation:

$$\begin{Bmatrix} u'^{(e)}_{x1} \\ u'^{(e)}_{y1} \\ \theta'^{(e)}_{1} \\ u'^{(e)}_{x2} \\ u'^{(e)}_{y2} \\ \theta'^{(e)}_{2} \end{Bmatrix} = \begin{bmatrix} \cos\varphi & \sin\varphi & 0 & 0 & 0 & 0 \\ -\sin\varphi & \cos\varphi & 0 & 0 & 0 & 0 \\ 0 & 0 & 1 & 0 & 0 & 0 \\ 0 & 0 & 0 & \cos\varphi & \sin\varphi & 0 \\ 0 & 0 & 0 & -\sin\varphi & \cos\varphi & 0 \\ 0 & 0 & 0 & 0 & 0 & 1 \end{bmatrix} \begin{Bmatrix} u^{(e)}_{X1} \\ u^{(e)}_{Y1} \\ \theta^{(e)}_{1} \\ u^{(e)}_{X2} \\ u^{(e)}_{Y2} \\ \theta^{(e)}_{2} \end{Bmatrix} \rightarrow \left\{U'^{(e)}\right\} = \left[R^{(e)}\right]\left\{U^{(e)}\right\} \tag{13.7.2}$$

where $\{U'^{(e)}\}$ is the nodal displacement vector in the local coordinate system, $\{U^{(e)}\}$ is the nodal displacement vector in the global coordinate system, and $[R^{(e)}]$ is the *coordinate transformation array*, given by:

$$\left[R^{(e)}\right] = \begin{bmatrix} \cos\varphi & \sin\varphi & 0 & 0 & 0 & 0 \\ -\sin\varphi & \cos\varphi & 0 & 0 & 0 & 0 \\ 0 & 0 & 1 & 0 & 0 & 0 \\ 0 & 0 & 0 & \cos\varphi & \sin\varphi & 0 \\ 0 & 0 & 0 & -\sin\varphi & \cos\varphi & 0 \\ 0 & 0 & 0 & 0 & 0 & 1 \end{bmatrix} \tag{13.7.3}$$

To establish the transformation equations for the nodal forces and moments, we follow a procedure shown in Figure 13.17. Specifically, we make two cuts at very small (practically zero) distances, dx', from the two nodes, and we draw the internal forces at the

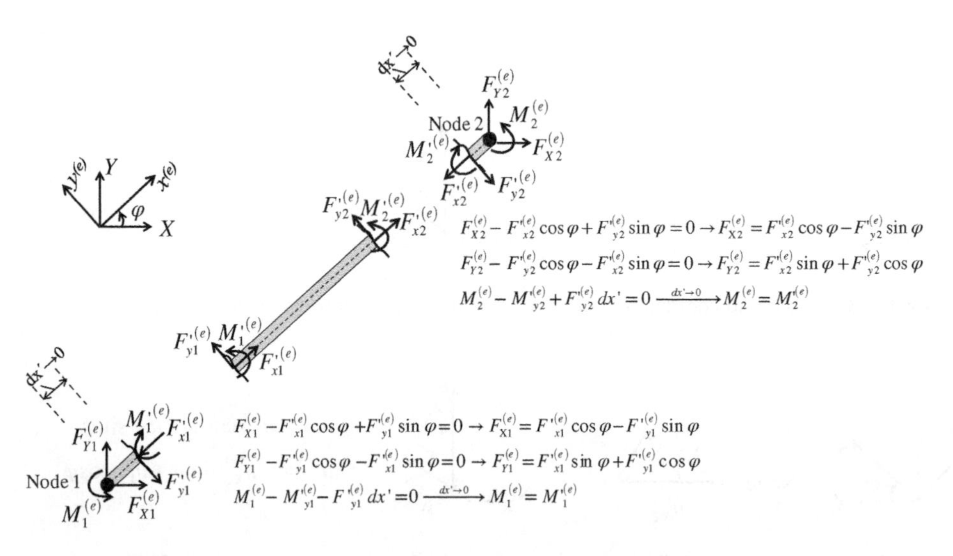

Figure 13.17 Establishing coordinate transformation rules for nodal forces.

locations of the cuts in accordance with the local coordinate system, while the nodal forces are drawn according to the global coordinate axes. Now, we can establish the three equilibrium equations for each one of the two small pieces dx' that we have isolated, as shown in Figure 13.17.

We can collectively write the two sets of transformation equations of Figure (13.17) in a unique matrix equation:

$$\begin{Bmatrix} F_{X1}^{(e)} \\ F_{Y1}^{(e)} \\ M_1^{(e)} \\ F_{X2}^{(e)} \\ F_{Y2}^{(e)} \\ M_2^{(e)} \end{Bmatrix} = \begin{bmatrix} \cos\varphi & -\sin\varphi & 0 & 0 & 0 & 0 \\ \sin\varphi & \cos\varphi & 0 & 0 & 0 & 0 \\ 0 & 0 & 1 & 0 & 0 & 0 \\ 0 & 0 & 0 & \cos\varphi & -\sin\varphi & 0 \\ 0 & 0 & 0 & \sin\varphi & \cos\varphi & 0 \\ 0 & 0 & 0 & 0 & 0 & 1 \end{bmatrix} \begin{Bmatrix} F'^{(e)}_{x1} \\ F'^{(e)}_{y1} \\ M'^{(e)}_1 \\ F'^{(e)}_{x2} \\ F'^{(e)}_{y2} \\ M'^{(e)}_2 \end{Bmatrix} \tag{13.7.4}$$

If we now define the nodal force vector in the local coordinate system as $\{F'^{(e)}\}$ and the corresponding vector in the global coordinate system as $\{F^{(e)}\}$, we can use inspection to understand that Equation (13.7.4) can be written as follows.

$$\left\{F^{(e)}\right\} = \left[R^{(e)}\right]^T \left\{F'^{(e)}\right\} \tag{13.7.5}$$

where

$$\left\{F^{(e)}\right\} = \left[F_{X1}^{(e)} \; F_{Y1}^{(e)} \; M_1^{(e)} \; F_{X2}^{(e)} \; F_{Y2}^{(e)} \; M_2^{(e)}\right]^T \tag{13.7.6a}$$

and

$$\left\{F'^{(e)}\right\} = \left[F'^{(e)}_{x1} \; F'^{(e)}_{y1} \; M'^{(e)}_1 \; F'^{(e)}_{x2} \; F'^{(e)}_{y2} \; M'^{(e)}_2\right]^T \tag{13.7.6b}$$

Finally, we know that the element stiffness matrix has been formulated in the local coordinate system, xy. For this reason, we can write Equation (3.3.26) for the special case of the beam element in the local coordinate system:

$$\left\{F'^{(e)}\right\} = \left[k'^{(e)}\right] \left\{U'^{(e)}\right\} \tag{13.7.7}$$

where $[k'^{(e)}]$ is the array given by Equation (13.6.6).

Plugging Equation (13.7.7) into Equation (13.7.5), we have:

$$\left\{F^{(e)}\right\} = \left[R^{(e)}\right]^T \left\{F'^{(e)}\right\} = \left[R^{(e)}\right]^T \left[k'^{(e)}\right] \left\{U'^{(e)}\right\}$$

Finally, plugging Equation (13.7.2) in the right-hand side of the resulting expression, we obtain:

$$\left\{F^{(e)}\right\} = \left[R^{(e)}\right]^T \left[k'^{(e)}\right] \left[R^{(e)}\right] \left\{U^{(e)}\right\} \rightarrow \left\{F^{(e)}\right\} = \left[k^{(e)}\right] \left\{U^{(e)}\right\}$$

where

$$\left[k^{(e)}\right] = \left[R^{(e)}\right]^T \left[k'^{(e)}\right] \left[R^{(e)}\right] \tag{13.7.8}$$

is the element stiffness matrix in the global coordinate system. Equation (13.7.8) constitutes the coordinate transformation equation for the stiffness matrix. The equivalent nodal force vector of the element must also be transformed to the global coordinate system, using Equation (13.7.5).

Remark 13.7.1: If a beam element has constant E, A, and I, then we can write a general expression for the stiffness matrix in the global coordinate system:

$$\left[k^{(e)}\right] = \begin{bmatrix} c^2\frac{EA}{\ell^{(e)}} + s^2\frac{12EI}{\left(\ell^{(e)}\right)^3} & cs\frac{EA}{\ell^{(e)}} - cs\frac{12EI}{\left(\ell^{(e)}\right)^3} & -s\frac{6EI}{\left(\ell^{(e)}\right)^2} & -c^2\frac{EA}{\ell^{(e)}} - s^2\frac{12EI}{\left(\ell^{(e)}\right)^3} & -cs\frac{EA}{\ell^{(e)}} + cs\frac{12EI}{\left(\ell^{(e)}\right)^3} & -s\frac{6EI}{\left(\ell^{(e)}\right)^2} \\ cs\frac{EA}{\ell^{(e)}} - cs\frac{12EI}{\left(\ell^{(e)}\right)^3} & s^2\frac{EA}{\ell^{(e)}} + c^2\frac{12EI}{\left(\ell^{(e)}\right)^3} & c\frac{6EI}{\left(\ell^{(e)}\right)^2} & -cs\frac{EA}{\ell^{(e)}} + cs\frac{12EI}{\left(\ell^{(e)}\right)^3} & -s^2\frac{EA}{\ell^{(e)}} - c^2\frac{12EI}{\left(\ell^{(e)}\right)^3} & c\frac{6EI}{\left(\ell^{(e)}\right)^2} \\ -s\frac{6EI}{\left(\ell^{(e)}\right)^2} & c\frac{6EI}{\left(\ell^{(e)}\right)^2} & \frac{4EI}{\ell^{(e)}} & s\frac{6EI}{\left(\ell^{(e)}\right)^2} & -c\frac{6EI}{\left(\ell^{(e)}\right)^2} & \frac{2EI}{\ell^{(e)}} \\ -c^2\frac{EA}{\ell^{(e)}} - s^2\frac{12EI}{\left(\ell^{(e)}\right)^3} & -cs\frac{EA}{\ell^{(e)}} + cs\frac{12EI}{\left(\ell^{(e)}\right)^3} & s\frac{6EI}{\left(\ell^{(e)}\right)^2} & c^2\frac{EA}{\ell^{(e)}} + s^2\frac{12EI}{\left(\ell^{(e)}\right)^3} & cs\frac{EA}{\ell^{(e)}} - cs\frac{12EI}{\left(\ell^{(e)}\right)^3} & s\frac{6EI}{\left(\ell^{(e)}\right)^2} \\ -cs\frac{EA}{\ell^{(e)}} + cs\frac{12EI}{\left(\ell^{(e)}\right)^3} & -s^2\frac{EA}{\ell^{(e)}} - c^2\frac{12EI}{\left(\ell^{(e)}\right)^3} & -c\frac{6EI}{\left(\ell^{(e)}\right)^2} & cs\frac{EA}{\ell^{(e)}} - cs\frac{12EI}{\left(\ell^{(e)}\right)^3} & s^2\frac{EA}{\ell^{(e)}} + c^2\frac{12EI}{\left(\ell^{(e)}\right)^3} & -c\frac{6EI}{\left(\ell^{(e)}\right)^2} \\ -s\frac{6EI}{\left(\ell^{(e)}\right)^2} & c\frac{6EI}{\left(\ell^{(e)}\right)^2} & \frac{2EI}{\ell^{(e)}} & s\frac{6EI}{\left(\ell^{(e)}\right)^2} & -c\frac{6EI}{\left(\ell^{(e)}\right)^2} & \frac{4EI}{\ell^{(e)}} \end{bmatrix} \tag{13.7.8}$$

where $c^2 = (cos\varphi)^2, s^2 = (sin\varphi)^2, cs = cos\varphi \cdot sin\varphi$. ∎

Remark 13.7.2: To calculate the equivalent nodal force vector in the local coordinate system, the distributed load vector $\{p\}$ and the applied forces/moments at the natural boundary locations must also be expressed in the local coordinate system, before using Equation (13.6.7). ∎

After we calculate the element stiffness and equivalent nodal force values in the global coordinate system, we can then use the element's gather-scatter array, $[L^{(e)}]$, to add the values to the correct locations of the global stiffness matrix and equivalent nodal force vector.

13.8 Timoshenko Beam Theory

The second beam theory that we will examine is called Timoshenko beam theory. Its major difference from the Euler-Bernoulli (E-B) theory is that the shear deformations are nonzero. For relatively squat beam members (beams with length-to-sectional-depth ratio less than 3 or 5), significant shear deformations may develop. The E-B theory may significantly overestimate the stiffness for such members. For this reason, Timoshenko beam theory should be used to properly account for the shear deformability.

There are four basic assumptions in Timoshenko beam theory:

1) Plane sections remain plane, but not necessarily perpendicular to the deformed reference line of the beam.
2) The normal stress parallel to the cross-sectional plane is equal to zero, $\sigma_{yy} = 0$.
3) When we determine the axial displacement at a point in a cross-section, we neglect the change of the depth of the cross-section; that is, we stipulate that the y-displacement at every point of the cross section is equal to the y-displacement of the reference line at the same cross-section:

$$u_y(x,y) = u_{oy}(x).$$

4) The shear stress at a specific cross section is constant (it does not change with y or z).

The fact that plane sections do not remain perpendicular to the reference line after deformation means that the slope of the reference line, which is the first derivative of the transverse displacement field with x, differs from the rotation of the cross-section, θ_o. This, in turn, leads to nonzero shear strains, γ_{xy}. In fact, γ_{xy} is the difference of the rotation of the section from the slope of the reference line, as shown in Figure 13.18.

For any point with a coordinate "x" and at a distance "y" (along the y-axis) from the reference line, we have:

$$u_x(x,y) = u_{ox}(x) - y \cdot \theta_o(x) \tag{13.8.1}$$

The strain ε_{xx} for a Timoshenko beam is given by the following equation.

$$\varepsilon_{xx}(x,y) = \frac{du_x}{dx} = \frac{d}{dx}\left(u_{ox}(x) - y \cdot \theta_o(x)\right) = \frac{du_{ox}}{dx} - y \cdot \frac{d\theta_o}{dx} = \varepsilon_o - y \cdot \varphi \tag{13.8.2}$$

where $\frac{du_{ox}}{dx} = \varepsilon_o$ is the *axial strain* of the reference line and $\varphi = \frac{d\theta_o}{dx}$ is the *curvature*.

Remark 13.8.1: We will still use the term *curvature* for φ, although it is no longer equal to the geometric curvature of the elastic curve (i.e., the second derivative of the transverse displacement field)! ∎

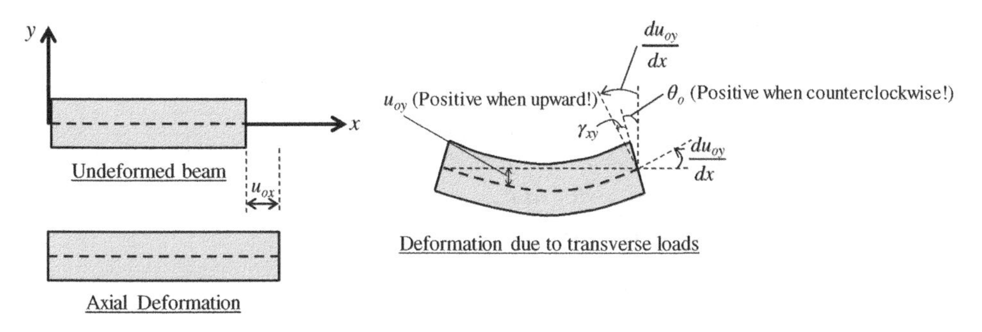

Figure 13.18 Timoshenko beam segment.

The strain γ_{xy} is defined by the equation presented in Section 7.2:

$$\gamma_{xy}(x,y) = \frac{\partial u_x}{\partial y} + \frac{\partial u_y}{\partial x} = \frac{\partial}{\partial y}(u_{ox}(x) - y \cdot \theta_o(x)) + \frac{\partial u_{oy}}{\partial x} = \frac{du_{oy}}{dx} - \theta_o \tag{13.8.3}$$

which is the same value as that shown schematically in Figure 13.18.

The axial stress, σ_{xx}, at a location of a beam cross-section is given by multiplying the axial strain with the modulus of elasticity:

$$\sigma_{xx}(x,y) = E \cdot \varepsilon_{xx}(x,y) = E \cdot (\varepsilon_o - y \cdot \varphi) \tag{13.8.4}$$

This is the same exact expression as that provided in Equation (13.3.6) for E-B theory. For this reason, the axial force N and bending moment M can once be obtained with the same expressions as those presented for E-B theory in Section 13.3:

$$N = EA \cdot \varepsilon_{ox} = EA \cdot \frac{du_{ox}}{dx} \tag{13.8.5a}$$

$$M = EI \cdot \varphi = EI \cdot \frac{d\theta_o}{dx} \tag{13.8.5b}$$

Since the shear stress, σ_{xy}, is assumed to be constant at each cross-section, the shear force V can be found by multiplying the shear stress by the corresponding *shear area of the cross-section, A_s*. The shear stress is equal to the product of the shear modulus, G, times γ_{xy}. Thus, we can write:

$$V = A_s \cdot \sigma_{xy} = A_s \cdot G \cdot \gamma_{xy} = \kappa \cdot A \cdot G \cdot \gamma_{xy} \tag{13.8.6}$$

where κ is an adjustment coefficient, referred to as the *shear factor*. The shear factor is the ratio between the shear area, A_s, and the cross-sectional area, A. This factor is used is to improve the accuracy of the results using Timoshenko beam theory (the assumption for constant shear stress over the cross-sectional depth simplifies the calculations, but is not perfectly accurate). For rectangular cross-sections, it can be proven that we need to set $\kappa = 5/6$. We can now collectively write the three equations giving the internal sectional forces of a Timoshenko beam as functions of appropriate kinematic quantities, to obtain the *generalized stress-strain law (or constitutive law) for Timoshenko beams*:

$$\begin{Bmatrix} N \\ M \\ V \end{Bmatrix} = \begin{bmatrix} EA & 0 & 0 \\ 0 & EI & 0 \\ 0 & 0 & \kappa \cdot GA \end{bmatrix} \begin{Bmatrix} \varepsilon_o \\ \varphi \\ \gamma_{xy} \end{Bmatrix} \rightarrow \{\hat{\sigma}\} = [\hat{D}]\{\hat{\varepsilon}\} \tag{13.8.7}$$

where we have established the generalized strain vector,

$$\{\hat{\varepsilon}\} = \begin{bmatrix} \varepsilon_o & \varphi & \gamma_{xy} \end{bmatrix}^T = \begin{bmatrix} \frac{du_{ox}}{dx} & \frac{d\theta_o}{dx} & \frac{du_{oy}}{dx} - \theta_o \end{bmatrix}^T \tag{13.8.8}$$

and the generalized stress vector,

$$\{\hat{\sigma}\} = [N \ \ M \ \ V]^T \tag{13.8.9}$$

13.9 Strong Form for Two-Dimensional Timoshenko Beam Theory

For a Timoshenko beam, the three differential equations of equilibrium involve three kinematic field functions, $u_{ox}(x)$, $u_{oy}(x)$, and $\theta_o(x)$. If we plug the obtained sectional constitutive law (Equation 13.8.7) into the three differential equations of equilibrium for a beam, (13.2.4a–c), and supplement the obtained expressions with the essential and natural boundary conditions, we obtain the strong form for Timoshenko beam theory, as presented in Box 13.9.1.

Box 13.9.1 Strong Form for Timoshenko Beam Theory

$$\frac{d}{dx}\left(EA\frac{du_{ox}}{dx}\right)+p_x=0 \tag{E1}$$

$$\frac{d}{dx}\left[\kappa AG\left(\frac{du_{oy}}{dx}-\theta_o\right)\right]+p_y=0 \tag{E2}$$

$$\frac{d}{dx}\left(EI\frac{d\theta_o}{dx}\right)+\kappa AG\left(\frac{du_{oy}}{dx}-\theta_o\right)=0 \tag{E3}$$

Essential Boundary Conditions:

$u_{ox}=\bar{u}_{ox}$ on Γ_{ux}

$u_{oy}=\bar{u}_{oy}$ on Γ_{uy}

$\theta_o=\bar{\theta}$ on $\Gamma_{u\theta}$

Natural Boundary Conditions:

$N\cdot n=F_x$ at Γ_{tx}

$V\cdot n=F_y$ at Γ_{ty}

$M\cdot n=M_z$ at $\Gamma_{t\theta}$

Remark 13.9.1: It should come as no surprise that the use of the Timoshenko kinematic assumptions has no effect on the differential equation and boundary conditions in the axial direction, which are identical to those of the E-B theory. ∎

13.10 Weak Form for Two-Dimensional Timoshenko Beam Theory

To obtain the weak form for the Timoshenko theory, we will first work with the general differential Equations of equilibrium for a beam, Equations (13.2.4a–c). This will allow us to obtain a generic form that is valid for any type of material behavior (elastic or inelastic). We will then plug the constitutive law (Equation 13.8.7) in the resulting expression to obtain the weak form for the special case of linearly elastic material.

Once again, we will consider an arbitrary displacement vector field $\{w\}$, which now has three components w_{ox}, w_{oy}, and $w_{o\theta}$, which satisfy:

$$w_{ox}=0 \text{ on } \Gamma_{ux}, w_{oy}=0 \text{ on } \Gamma_{uy},\ w_{o\theta}=0 \text{ on } \Gamma_{u\theta} \tag{13.10.1}$$

Remark 13.10.1: In the context of virtual work, w_{ox} and w_{oy} can be thought of as the components of a virtual displacement field and $w_{o\theta}$ can be thought of as a virtual sectional rotation field. ■

First, we multiply Equation (13.2.4a), that is, the differential equation of axial equilibrium, by w_{ox} and integrate over the domain:

$$\int_0^L w_{ox}\left(\frac{dN}{dx}+p_x\right)dx=0 \rightarrow \int_0^L w_{ox}\frac{dN}{dx}dx+\int_0^L w_{ox}\cdot p_x dx=0 \qquad (13.10.2)$$

The first integral term in Equation (13.10.2) is "unsymmetric" in terms of differentiation of u_{ox} and w_{ox}, and also includes second-order derivatives for u_{ox}. Thus, we conduct integration by parts to reduce the order of differentiation and ensure symmetry:

$$\int_0^L w_{ox}\frac{dN}{dx}dx=[w_{ox}\cdot N]_0^L-\int_0^L \frac{dw_{ox}}{dx}\cdot N\,dx \qquad (13.10.3)$$

Let us now focus on the first term of the right-hand side of Equation (13.10.3). Since we know that the two endpoints of a beam, $x = 0$ and $x = L$, are the boundaries Γ_{ux} and Γ_{tx}, and we know that we have $w_{ox} = 0$ on Γ_{ux}, we can write:

$$[w_{ox}\cdot N]_0^L=[w_{ox}\cdot N\cdot n]_{\Gamma_{ux}}+[w_{ox}\cdot N\cdot n]_{\Gamma_{tx}}=[w_{ox}\cdot N\cdot n]_{\Gamma_{tx}}=[w_{ox}\cdot F_x]_{\Gamma_{tx}} \qquad (13.10.4)$$

where we also accounted for Equation (13.4.3a).

Plugging Equation (13.10.4) into Equation (13.10.3), we obtain:

$$[w_{ox}\cdot F_x]_{\Gamma_{tx}}-\int_0^L \frac{dw_{ox}}{dx}\cdot N\,dx+\int_0^L w_{ox}\cdot p_x dx=0 \rightarrow \int_0^L \frac{dw_{ox}}{dx}\cdot N\,dx=\int_0^L w_{ox}\cdot p_x dx+[w_{ox}\cdot F_x]_{\Gamma_{tx}} \qquad (13.10.5)$$

We now multiply Equation (13.2.4b), that is, the differential equation of equilibrium for shear, by w_{oy} and integrate over the domain:

$$\int_0^L w_{oy}\left(\frac{dV}{dx}+p_y\right)dx=0 \rightarrow \int_0^L w_{oy}\frac{dV}{dx}dx+\int_0^L w_{oy}\cdot p_y dx=0 \qquad (13.10.6)$$

We conduct integration by parts for the first term of Equation (13.10.6):

$$\int_0^L w_{oy}\frac{dV}{dx}dx=[w_{oy}V]_0^L-\int_0^L \frac{dw_{oy}}{dx}V dx \qquad (13.10.7)$$

We know that $x = 0$ and $x = L$ correspond to Γ_{uy} and Γ_{ty}, and we know that we have $w_{oy} = 0$ on Γ_{uy}; thus:

$$\left[\frac{dw_{oy}}{dx}\cdot V\right]_0^L=[w_{oy}\cdot V\cdot n]_{\Gamma_{uy}}+[w_{oy}\cdot V\cdot n]_{\Gamma_{ty}}=[w_{oy}\cdot V\cdot n]_{\Gamma_{ty}}=[w_{oy}\cdot F_y]_{\Gamma_{ty}} \qquad (13.10.8)$$

where we accounted for Equation (13.4.3b).

Plugging Equation (13.10.8) into Equation (13.10.7), then plugging the resulting expression into Equation (13.10.6) yields:

$$\left[w_{oy}\cdot F_y\right]_{\Gamma_{ty}} - \int_0^L \frac{dw_{oy}}{dx}\cdot V\,dx + \int_0^L w_{oy}\cdot p_y dx = 0 \rightarrow \int_0^L \frac{dw_{oy}}{dx}\cdot V\,dx = \int_0^L w_{oy}\cdot p_y dx + \left[w_{oy}\cdot F_y\right]_{\Gamma_{ty}} \tag{13.10.9}$$

Finally, if we multiply equation (13.2.4c) by $w_{o\theta}$ and integrate over the length of the beam, we have:

$$\int_0^L w_{o\theta}\left(\frac{dM}{dx}+V\right)dx = 0 \rightarrow \int_0^L w_{o\theta}\frac{dM}{dx}dx + \int_0^L w_{o\theta}\cdot Vdx = 0 \tag{13.10.10}$$

Conducting integration by parts for the first term of Equation (13.10.10):

$$\int_0^L w_{o\theta}\frac{dM}{dx}dx = \left[w_{o\theta}\cdot M\right]_0^L - \int_0^L \frac{dw_{o\theta}}{dx}\cdot Mdx = \left[w_{o\theta}\cdot M\cdot n\right]_{\Gamma_{t\theta}} + \left[w_{o\theta}\cdot M\cdot n\right]_{\Gamma_{u\theta}} - \int_0^L \frac{dw_{o\theta}}{dx}\cdot Mdx \rightarrow$$

$$\rightarrow \int_0^L w_{o\theta}\frac{dM}{dx}dx = \left[w_{o\theta}\cdot M_z\right]_{\Gamma_{t\theta}} - \int_0^L \frac{dw_{o\theta}}{dx}\cdot Mdx \tag{13.10.11}$$

where we accounted for the natural boundary condition for moment, that is, Equation (13.4.3c). If we plug (13.10.11) into (13.10.10), we finally obtain:

$$\left[w_{o\theta}\cdot M_z\right]_{\Gamma_{t\theta}} - \int_0^L \frac{dw_{o\theta}}{dx}\cdot Mdx + \int_0^L w_{o\theta}\cdot Vdx = 0 \;\rightarrow\; \int_0^L \frac{dw_{o\theta}}{dx}\cdot Mdx - \int_0^L w_{o\theta}\cdot Vdx = \left[w_{o\theta}\cdot M_z\right]_{\Gamma_{t\theta}} \tag{13.10.12}$$

We can now substitute the constitutive equations (13.8.7) for a Timoshenko beam in the three integral equations (13.10.5), (13.10.9), and (13.10.12):

$$\int_0^L \frac{dw_{ox}}{dx}\cdot EA\frac{du_{ox}}{dx}\,dx = \int_0^L w_{ox}\cdot p_x dx + \left[w_{ox}\cdot F_x\right]_{\Gamma_{tx}} \tag{13.10.13a}$$

$$\int_0^L \frac{dw_{oy}}{dx}\cdot \kappa AG\left(\frac{du_{oy}}{dx}-\theta_o\right)dx = \int_0^L w_{oy}\cdot p_y dx + \left[w_{oy}\cdot F_y\right]_{\Gamma_{ty}} \tag{13.10.13b}$$

$$\int_0^L \frac{dw_{o\theta}}{dx}\cdot EI\frac{d\theta_o}{dx}dx - \int_0^L w_{o\theta}\cdot \kappa AG\left(\frac{du_{oy}}{dx}-\theta_o\right)dx = \left[w_{o\theta}\cdot M_z\right]_{\Gamma_{t\theta}} \tag{13.10.13c}$$

If we now sum the three equations (13.10.13a–c) and group all the integral terms multiplied by the axial rigidity, all the integral terms multiplied by the flexural rigidity and all the integral terms multiplied by κAG, we have:

$$\int_0^L \frac{dw_{ox}}{dx}\cdot EA\frac{du_{ox}}{dx}dx+\int_0^L\left(\frac{dw_{oy}}{dx}-w_{o\theta}\right)\cdot\kappa AG\left(\frac{du_{oy}}{dx}-\theta_o\right)dx+\int_0^L\frac{dw_{o\theta}}{dx}\cdot EI\frac{d\theta_o}{dx}dx$$

$$=\int_0^L w_{oy}\cdot p_y dx+\int_0^L w_{ox}\cdot p_x dx+\left[w_{ox}\cdot F_x\right]_{\Gamma_{tx}}+\left[w_{oy}\cdot F_y\right]_{\Gamma_{ty}}+\left[w_{o\theta}\cdot M_z\right]_{\Gamma_{t\theta}} \tag{13.10.14}$$

We can finally write a more concise matrix multiplication form for the left-hand side of Equation (13.10.14), and also provide the essential boundary conditions, to establish the weak form for Timoshenko beam theory, summarized in Box (13.10.1).

Box 13.10.1 Weak Form for Timoshenko Beam Theory

$$\int_0^L \{\bar{\varepsilon}\}^T[\hat{D}]\{\hat{\varepsilon}\}dx=\int_0^L\{w\}^T\{p\}dx+\left[w_{ox}F_x\right]_{\Gamma_{tx}}+\left[w_{oy}F_y\right]_{\Gamma_{ty}}+\left[w_{o\theta}M_z\right]_{\Gamma_{t\theta}}$$

where $[\hat{D}]=\begin{bmatrix}EA & 0 & 0\\ 0 & EI & 0\\ 0 & 0 & \kappa GA\end{bmatrix}$, $\{\hat{\varepsilon}\}=\left[\varepsilon_o\ \ \varphi\ \ \gamma_{xy}\right]^T=\left[\frac{du_{ox}}{dx}\ \ \frac{d\theta_o}{dx}\ \ \frac{du_{oy}}{dx}-\theta_o\right]^T$, $\{p\}=\left[p_x\ \ p_y\ \ 0\right]^T$

and $\{\bar{\varepsilon}\}=\left[\bar{\varepsilon}_o\ \ \bar{\varphi}\ \ \bar{\gamma}_{xy}\right]^T=\left[\frac{dw_{ox}}{dx}\ \ \frac{dw_{o\theta}}{dx}\ \ \frac{dw_{oy}}{dx}-w_{o\theta}\right]^T$

$\forall\{w\}=\left[w_{ox}\ \ w_{oy}\ \ w_{o\theta}\right]^T$ with $w_{ox}=0$ on Γ_{ux}, $w_{oy}=0$ on Γ_{uy}, $w_{o\theta}=0$ on $\Gamma_{u\theta}$

$u_{ox}=\bar{u}_{ox}$ on Γ_{ux}

$u_{oy}=\bar{u}_{oy}$ on Γ_{uy}

$\theta_o=\bar{\theta}$ on $\Gamma_{u\theta}$

Remark 13.10.2: Just like in the case of E-B theory, the weak form for a Timoshenko beam is a "virtual work" expression, with sectional force resultants (forces and/or moments) playing the role of generalized stresses and tractions. ■

Remark 13.10.3: Contrary to the weak form for E-B beams, the highest order of derivatives in the weak form for Timoshenko beams is 1. This means that the finite element approximations for all three kinematic quantities u_{ox}, u_{oy}, and θ_o must have first-degree completeness and be characterized by C^0 continuity. Thus, the standard Lagrangian polynomial shape functions presented in Chapter 3 are adequate to ensure the convergence of a finite element formulation for a Timoshenko beam problem. ■

The equivalence between the weak form and the strong form can once again be established by starting with the weak form and then obtaining the strong form. The detailed derivation of the equivalence will not be pursued here and is left as an exercise.

13.11 Two-Node Timoshenko Beam Finite Element

The generic finite element Equations (13.6.1) to (13.6.6) for Euler-Bernoulli theory are also valid for Timoshenko theory. What changes is the expressions for the shape function array, generalized strain-displacement matrix and sectional stiffness matrix. The present section will discuss the simplest type of Timoshenko beam finite element (i.e., the two-node, six-degree-of-freedom element). The specific beam element uses the same linear shape functions for the three kinematic fields inside each element, *e*. Specifically, we have the following expressions:

$$u_{ox}^{(e)}(x) \approx N_1^{(e)}(x) \cdot u_{x1}^{(e)} + N_2^{(e)}(x) \cdot u_{x2}^{(e)} \tag{13.11.1a}$$

$$u_{oy}^{(e)}(x) \approx N_1^{(e)}(x) \cdot u_{y1}^{(e)} + N_2^{(e)}(x) \cdot u_{y2}^{(e)} \tag{13.11.1b}$$

$$\theta_o^{(e)}(x) \approx N_1^{(e)}(x) \cdot \theta_1^{(e)} + N_2^{(e)}(x) \cdot \theta_2^{(e)} \tag{13.11.1c}$$

where the two shape functions are defined from the following equations.

$$N_1^{(e)}(x) = 1 - \frac{x}{\ell^{(e)}} \tag{13.11.2a}$$

$$N_2^{(e)}(x) = \frac{x}{\ell^{(e)}} \tag{13.11.2b}$$

The two-node Timoshenko beam element with its two shape functions is shown in Figure 13.19.

We can now establish the displacement field approximation for the finite element in matrix form:

$$\left\{u^{(e)}(x)\right\} \approx \left[N^{(e)}(x)\right]\left\{U^{(e)}\right\} \tag{13.11.3}$$

where

$$\left\{u^{(e)}(x)\right\} = \left[u_{ox}^{(e)}(x) \;\; u_{oy}^{(e)}(x) \;\; \theta_o^{(e)}(x)\right]^T \tag{13.11.4}$$

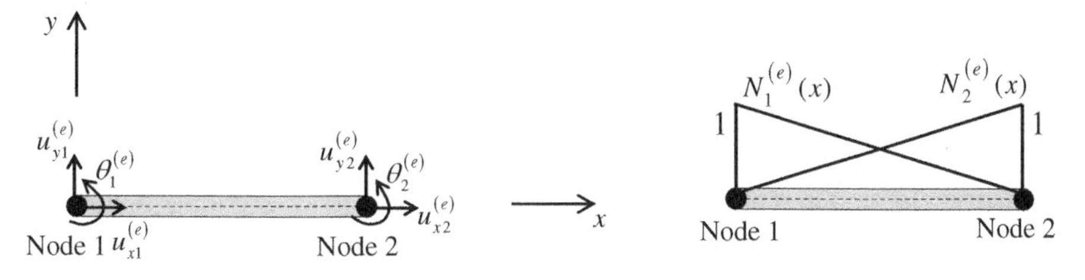

Figure 13.19 Two-node Timoshenko beam element and shape functions.

$$\left\{U^{(e)}\right\} = \left[u_{x1}^{(e)} \quad u_{y1}^{(e)} \quad \theta_1^{(e)} \quad u_{x2}^{(e)} \quad u_{y2}^{(e)} \quad \theta_2^{(e)}\right]^T \tag{13.11.5}$$

and

$$\left[N^{(e)}\right] = \begin{bmatrix} N_1^{(e)}(x) & 0 & 0 & N_2^{(e)}(x) & 0 & 0 \\ 0 & N_1^{(e)}(x) & 0 & 0 & N_2^{(e)}(x) & 0 \\ 0 & 0 & N_1^{(e)}(x) & 0 & 0 & N_2^{(e)}(x) \end{bmatrix} \tag{13.11.6}$$

The generalized strain vector can be given by appropriate products of shape function derivatives and nodal displacements:

$$\left\{\hat{\varepsilon}^{(e)}\right\} = \begin{Bmatrix} \varepsilon_o^{(e)} \\ \varphi^{(e)} \\ \gamma_{xy}^{(e)} \end{Bmatrix} = \begin{Bmatrix} \dfrac{du_{ox}^{(e)}}{dx} \\ \dfrac{d\theta_o^{(e)}}{dx} \\ \dfrac{du_{oy}^{(e)}}{dx} - \theta_o^{(e)} \end{Bmatrix} \approx \begin{Bmatrix} \dfrac{dN_1^{(e)}}{dx} u_{x1}^{(e)} + \dfrac{dN_2^{(e)}}{dx} \cdot u_{x2}^{(e)} \\ \dfrac{dN_1^{(e)}}{dx} \theta_1^{(e)} + \dfrac{dN_2^{(e)}}{dx} \cdot \theta_2^{(e)} \\ \dfrac{dN_1^{(e)}}{dx} \cdot u_{y1}^{(e)} + \dfrac{dN_2^{(e)}}{dx} \cdot u_{y2}^{(e)} - N_1^{(e)} \cdot \theta_1^{(e)} - N_2^{(e)} \cdot \theta_2^{(e)} \end{Bmatrix} \tag{13.11.7}$$

Equation (13.11.7) can be cast in matrix form:

$$\left\{\hat{\varepsilon}^{(e)}\right\} = \left[B^{(e)}\right]\left\{U^{(e)}\right\} \tag{13.11.8}$$

where the generalized strain-displacement matrix, $[B^{(e)}]$, is given by the following expression.

$$\left[B^{(e)}\right] = \begin{bmatrix} \dfrac{dN_1^{(e)}}{dx} & 0 & 0 & \dfrac{dN_2^{(e)}}{dx} & 0 & 0 \\ 0 & 0 & \dfrac{dN_1^{(e)}}{dx} & 0 & 0 & \dfrac{dN_2^{(e)}}{dx} \\ 0 & \dfrac{dN_1^{(e)}}{dx} & -N_1^{(e)} & 0 & \dfrac{dN_2^{(e)}}{dx} & -N_2^{(e)} \end{bmatrix} \tag{13.11.9}$$

The derivatives of the two shape functions in Equation (13.11.9) are constant and are given by the following expressions.

$$\frac{dN_1^{(e)}}{dx} = -\frac{1}{\ell^{(e)}} \tag{13.11.10a}$$

$$\frac{dN_2^{(e)}}{dx} = \frac{1}{\ell^{(e)}} \tag{13.11.10b}$$

The assembly equations for the global stiffness matrix and equivalent nodal force vector are the same as those presented in Section 13.6. Equation (13.6.6) applies for the element stiffness matrix, $[k^{(e)}]$, for Timoshenko beam theory, with $[B^{(e)}]$ given by Equation (13.11.9), and the sectional stiffness matrix $\left[\hat{D}^{(e)}\right]$ defined by Equation (13.8.7).

The expression providing the equivalent element nodal forces and moments of the Timoshenko beam element is the following.

$$\left\{f^{(e)}\right\} = \int_0^{\ell^{(e)}} \left[N^{(e)}\right]^T \{p\}\, dx + \left[N^{(e)}\right]^T_{\Gamma^e_{tx}} \begin{Bmatrix} F_x \\ 0 \\ 0 \end{Bmatrix} + \left[N^{(e)}\right]^T_{\Gamma^e_{ty}} \begin{Bmatrix} 0 \\ F_y \\ 0 \end{Bmatrix} + \left[N^{(e)}\right]^T_{\Gamma^e_{t\theta}} \begin{Bmatrix} 0 \\ 0 \\ M_z \end{Bmatrix} \tag{13.11.11}$$

To evaluate the integrals for $[k^{(e)}]$ and $\{f^{(e)}\}$, we can use a one-dimensional parametric representation for the beam, with a single parametric coordinate, ξ, $-1 \le \xi \le 1$. This parametric coordinate space will allow us to use Gaussian quadrature for the element arrays.

Remark 13.11.1: The two shape functions can be expressed in terms of the parametric coordinate, ξ, as follows.

$$N_1^{(e)}(\xi) = \frac{1}{2}(1 - \xi) \tag{13.11.12a}$$

$$N_2^{(e)}(\xi) = \frac{1}{2}(1 + \xi) \tag{13.11.12b}$$

■

We can now readily program an isoparametric, one-dimensional formulation for the Timoshenko beam element. The minimum required quadrature rule for Timoshenko beam elements is a two-point one. For the general case that we use N_g quadrature points for a beam subjected to distributed loads $p_x(x)$ and $p_y(x)$, then we can calculate the element stiffness matrix and equivalent nodal force vector with the algorithm described in Box 13.11.1.

Remark 13.11.2: If we have beam elements that are not all aligned with the same direction (e.g., the case shown in Figure 13.13), then the considerations of this section give the stiffness matrix $[k'^{(e)}]$ and equivalent nodal force vector $\{f'^{(e)}\}$ in the *local* coordinate system. Before conducting the assembly operations, we need to calculate the stiffness matrix and equivalent nodal force vector in the global coordinate system, using the transformation equations (13.7.8) and (13.7.5), respectively. These equations are valid for any beam theory. ■

13.12 Continuum-Based Beam Elements

A beam element can also be obtained as a *constrained continuum formulation*. Specifically, we can describe the deformation and stiffness of a two-dimensional beam element by enforcing the kinematics of a beam to a conventional, plane-stress, two-dimensional continuum element. Our discussion will be limited to Timoshenko beam theory. The basic rationale for the constrained continuum approach is presented in Figure 13.20, where the side elevation of a two-node beam element is drawn. At each node I, we draw a unit vector $\vec{e}_f^{\,I}$ along the sectional depth, which we will be referring to as the *fiber* vector. The fiber vector simply gives the direction of the cross-sectional depth. Based on the kinematic assumptions that we made for beam theories, no elongation develops in the fiber direction! For this reason, we can imagine that, at each location in the beam, we have an axially rigid (= nondeformable) bar aligned with the fiber unit vector $\vec{e}_f^{\,I}$, as also shown in Figure 13.21. The figure shows the case where the reference line of a beam passes through the middle of the sectional depth, d. This case will be the only one considered

Box 13.11.1 Calculation of Element Stiffness Matrix and Nodal Force Vector for Timoshenko Beam Element Using Gaussian Quadrature

For each quadrature point g, $g = 1, \ldots, N_g$:

- Determine parametric coordinate, ξ_g, and weight coefficient, W_g.
- If we have a $\left[\hat{D}^{(e)}\right]$ -array or a $\{p\}$-vector whose components are functions of x, calculate $x_g = x(\xi_g) = \frac{\ell^{(e)}}{2}(1+\xi_g)$.
- Determine the values of the $[B^{(e)}]$ array and of the $\left[\hat{D}^{(e)}\right]$ array at $\xi = \xi_g$: $[B_g] = [B^{(e)}(\xi_g)]$, $[\hat{D}_g] = \left[\hat{D}^{(e)}(x_g)\right]$
- If we have distributed loads, calculate $[N_g] = [N^{(e)}(\xi_g)]$ and $\{p_g\} = \{p(x_g)\}$.

After we complete the above computations for all Gauss points, we can obtain the element's stiffness matrix, $[k^{(e)}]$, and the element's equivalent nodal force vector, $\{f^{(e)}\}$:

$$\left[k^{(e)}\right] = \int_{-1}^{1} \left[B^{(e)}(\xi)\right]^T \left[\hat{D}^{(e)}(\xi)\right] \left[B^{(e)}(\xi)\right] \frac{\ell^{(e)}}{2} d\xi \approx \sum_{g=1}^{Ng} \left([B_g]^T [\hat{D}_g] [B_g] \frac{\ell^{(e)}}{2} W_g \right)$$

$$\{f^{(e)}\} = \int_{-1}^{1} \left[N^{(e)}(\xi)\right]^T \begin{Bmatrix} p_x(\xi) \\ p_y(\xi) \\ 0 \end{Bmatrix} \frac{\ell^{(e)}}{2} d\xi + \left[N^{(e)}\right]^T_{\Gamma^e_{tx}} \begin{Bmatrix} F_x \\ 0 \\ 0 \end{Bmatrix} + \left[N^{(e)}\right]^T_{\Gamma^e_{ty}} \begin{Bmatrix} 0 \\ F_y \\ 0 \end{Bmatrix} + \left[N^{(e)}\right]^T_{\Gamma^e_{t\theta}} \begin{Bmatrix} 0 \\ 0 \\ M_z \end{Bmatrix}$$

$$\approx \sum_{g=1}^{Ng} \left([N_g]^T \begin{Bmatrix} p_x(x_g) \\ p_y(x_g) \\ 0 \end{Bmatrix} \frac{\ell^{(e)}}{2} W_g \right) + \left[N^{(e)}\right]^T_{\Gamma^e_{tx}} \begin{Bmatrix} F_x \\ 0 \\ 0 \end{Bmatrix} + \left[N^{(e)}\right]^T_{\Gamma^e_{ty}} \begin{Bmatrix} 0 \\ F_y \\ 0 \end{Bmatrix} + \left[N^{(e)}\right]^T_{\Gamma^e_{t\theta}} \begin{Bmatrix} 0 \\ 0 \\ M_z \end{Bmatrix}$$

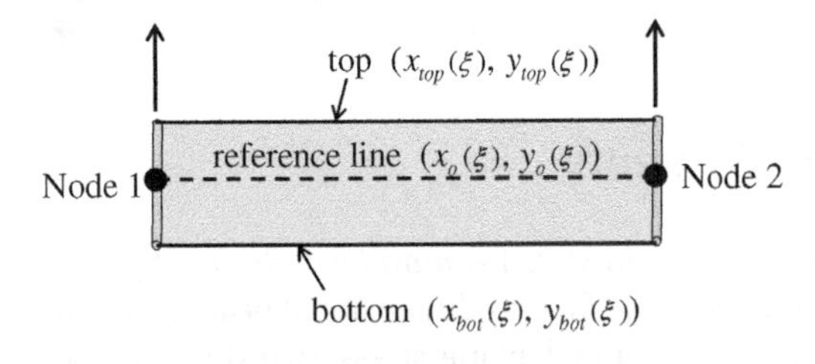

Figure 13.20 Two-node, continuum-based beam element.

herein, although the reference line does not necessarily need to pass through the middle of the sectional depth. Knowledge of the coordinates x_o and y_o for a point on the reference line, of the fiber unit vector $\vec{e}^{\,I}_f$, and of the sectional depth d can provide us with the spatial coordinates of any point at the top and bottom of the beam section, as shown in Figure 13.21. In the specific figure, we have expressed the spatial coordinates of the reference point, the fiber vector, and the sectional depth as functions of the parametric coordinate, ξ. Notice that we define a second parametric coordinate, η for our element. This second parametric coordinate essentially gives the location along the fiber vector, as shown in Figure 13.21.

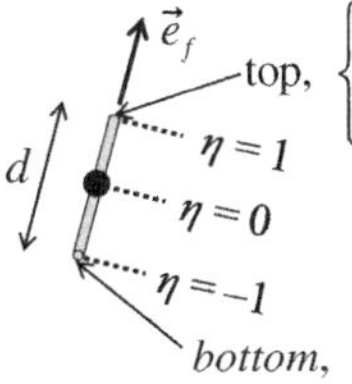

$\vec{e}_f$, d

top, $\begin{Bmatrix} x_{top}(\xi) \\ y_{top}(\xi) \end{Bmatrix} = \begin{Bmatrix} x(\xi,\eta=1) \\ y(\xi,\eta=1) \end{Bmatrix} = \begin{Bmatrix} x_o(\xi) \\ y_o(\xi) \end{Bmatrix} + 1 \cdot \frac{d(\xi)}{2} \begin{Bmatrix} e_{f,x}(\xi) \\ e_{f,y}(\xi) \end{Bmatrix}$

$\eta = 1$

$\eta = 0$

$\eta = -1$

bottom, $\begin{Bmatrix} x_{bot}(\xi) \\ y_{bot}(\xi) \end{Bmatrix} = \begin{Bmatrix} x(\xi,\eta=-1) \\ y(\xi,\eta=-1) \end{Bmatrix} = \begin{Bmatrix} x_o(\xi) \\ y_o(\xi) \end{Bmatrix} + (-1) \cdot \frac{d(\xi)}{2} \begin{Bmatrix} e_{f,x}(\xi) \\ e_{f,y}(\xi) \end{Bmatrix}$

Figure 13.21 Definition of a fiber vector passing through the middle of the sectional depth.

Now, the physical coordinates of any point with parametric coordinates (ξ,η) can be obtained as follows.

$$\{x(\xi,\eta)\} = \{x(\xi)\}_o + \eta\left[\frac{d(\xi)}{2}\vec{e}_f(\xi)\right] \tag{13.12.1a}$$

Equation (13.12.1a) suggests that the fiber orientation can be a function of ξ (i.e., the fiber orientation can vary along an element). This is the case for *curved beam elements*, which we will examine in Section 13.13.

Now, we will set up the specifics of a constrained continuum approach; specifically, we are going to *imagine that the beam element is in fact a four-node quadrilateral (4Q) continuum element*, as shown in Figure 13.22, with the four corner nodes, 1^-, 1^+, 2^+, and 2^- being the end points of the fiber vectors at the two nodal points, 1 and 2, of the beam element! The thickness of the quadrilateral element is equal to the sectional width, b_w, of the beam[1]. The coordinates of nodes 1^- and 1^+ can be obtained in terms of the coordinates $\{x_1\}$ of node 1:

$$\{x_{1^-}\} = \{x_1\} - \frac{d_1}{2}\vec{e}_f^{\,1} \tag{13.12.1b}$$

$$\{x_{1^+}\} = \{x_1\} + \frac{d_1}{2}\vec{e}_f^{\,1} \tag{13.12.1c}$$

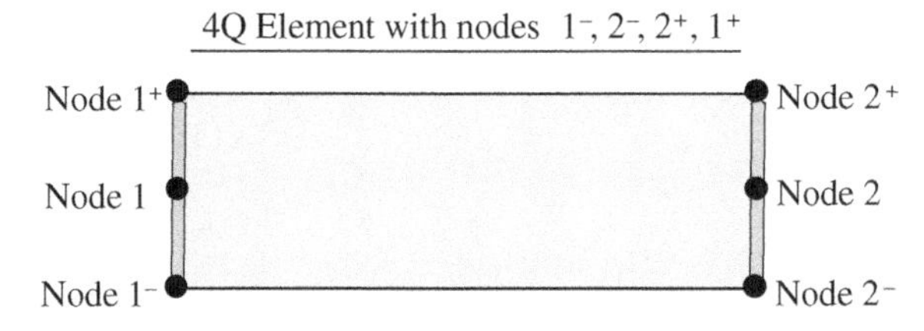

Nodes 1^- and 1^+ are constrained (slaves) to node 1
Nodes 2^- and 2^+ are constrained (slaves) to node 2

Figure 13.22 Two-node beam element treated as a four-node quadrilateral (4Q), continuum element.

---

1 The width b_w can generally vary in the interior of the beam.

The quantity d_1 in Equations (13.12.1b) and (13.12.1c) is the sectional depth at the location of node 1. Similarly, we can find the coordinates of nodes 2^- and 2^+ in terms of the coordinates $\{x_2\}$ of node 2:

$$\{x_{2^-}\} = \{x_2\} - \frac{d_2}{2}\vec{e}_f^{\,2} \tag{13.12.1d}$$

$$\{x_{2^+}\} = \{x_2\} + \frac{d_2}{2}\vec{e}_f^{\,2} \tag{13.12.1e}$$

Nodes 1^- and 1^+ can be thought of as being connected to node 1 with a rigid bar. In accordance with Appendix D, the displacements at nodes 1^- and 1^+ can be obtained from the displacements and rotation of node 1 as follows.

$$\vec{u}_{1^-} = \vec{u}_1 + \vec{\theta}_1 \times \Delta\vec{r}_{1^-} \tag{13.12.2a}$$

$$\vec{u}_{1^+} = \vec{u}_1 + \vec{\theta}_1 \times \Delta\vec{r}_{1^+} \tag{13.12.2b}$$

where

$$\vec{u}_1 = \begin{bmatrix} u_{x1}^{(e)} & u_{y1}^{(e)} & 0 \end{bmatrix}^T \tag{13.12.3a}$$

$$\vec{\theta}_1 = \begin{bmatrix} 0 & 0 & \theta_1^{(e)} \end{bmatrix}^T \tag{13.12.3b}$$

$$\Delta\vec{r}_{1^-} = -\frac{d}{2}\vec{e}_f^{\,1} = -\frac{d_1}{2}\begin{bmatrix} e_{fx}^1 & e_{fy}^1 & 0 \end{bmatrix}^T \tag{13.12.3c}$$

$$\text{and } \Delta\vec{r}_{1^+} = \frac{d_1}{2}\vec{e}_f^{\,1} = \frac{d_1}{2}\begin{bmatrix} e_{fx}^1 & e_{fy}^1 & 0 \end{bmatrix}^T \tag{13.12.3d}$$

Remark 13.12.1: Notice that Equations (13.12.2) apply for three-dimensional space, which is why we expressed the two-dimensional vectors of our problems with equivalent three-dimensional ones! ■

Along the same lines, we can express the displacements at nodes 2^- and 2^+ in terms of the displacements and rotation of node 2:

$$\vec{u}_{2^-} = \vec{u}_2 + \vec{\theta}_2 \times \Delta\vec{r}_{2^-} \tag{13.12.4a}$$

$$\vec{u}_{2^+} = \vec{u}_2 + \vec{\theta}_2 \times \Delta\vec{r}_{2^+} \tag{13.12.4b}$$

where

$$\vec{u}_2 = \begin{bmatrix} u_{x2}^{(e)} & u_{y2}^{(e)} & 0 \end{bmatrix}^T \tag{13.12.5a}$$

$$\vec{\theta}_2 = \begin{bmatrix} 0 & 0 & \theta_2^{(e)} \end{bmatrix}^T \tag{13.12.5b}$$

$$\Delta\vec{r}_{2^-} = -\frac{d_2}{2}\vec{e}_f^{\,2} = -\frac{d_2}{2}\begin{bmatrix} e_{fx}^2 & e_{fy}^2 & 0 \end{bmatrix}^T \tag{13.12.5c}$$

$$\text{and } \Delta\vec{r}_{2^+} = \frac{d_2}{2}\vec{e}_f^{\,2} = \frac{d_2}{2}\begin{bmatrix} e_{fx}^2 & e_{fy}^2 & 0 \end{bmatrix}^T \tag{13.12.5d}$$

We can make some mathematical manipulations and combine Equations (13.12.2a,b) into a single matrix expression, which will provide the displacements of the nodes of the underlying 4Q continuum element from the displacements and rotations of the two nodes of the beam element.

$$\begin{Bmatrix} \{u_{I^-}\} \\ \{u_{I^+}\} \end{Bmatrix} = [T_I]\{u_I\} \tag{13.12.6}$$

where

$$\{u_{I^-}\} = \begin{bmatrix} u_{xI^-} & u_{yI^-} \end{bmatrix}^T \tag{13.12.7a}$$

$$\{u_{I^+}\} = \begin{bmatrix} u_{xI^+} & u_{yI^+} \end{bmatrix}^T \tag{13.12.7b}$$

$$\{u_I\} = \begin{bmatrix} u_{xI} & u_{yI} & \theta_I \end{bmatrix}^T \tag{13.12.7c}$$

$$\text{and } [T_I] = \begin{bmatrix} 1 & 0 & -d_I/2 \cdot e_{fy}^I \\ 0 & 1 & d_I/2 \cdot e_{fx}^I \\ 1 & 0 & -d_I/2 \cdot e_{fy}^I \\ 0 & 1 & d_I/2 \cdot e_{fx}^I \end{bmatrix} \tag{13.12.7d}$$

We can set $I = 1$, then $I = 2$ in Equations (13.12.6), and collectively write the resulting expressions as a single matrix equation:

$$\left\{U^{(4Q)}\right\} = \begin{Bmatrix} \{u_{1^-}\} \\ \{u_{2^-}\} \\ \{u_{2^+}\} \\ \{u_{1^+}\} \end{Bmatrix} = \left[T^{4Q}\right]\left\{U^{(e)}\right\} \tag{13.12.8}$$

where

$$\left[T^{4Q}\right] = \begin{bmatrix} 1 & 0 & d_1/2 \cdot e_{fy}^1 & 0 & 0 & 0 \\ 0 & 1 & -d_1/2 \cdot e_{fx}^1 & 0 & 0 & 0 \\ 0 & 0 & 0 & 1 & 0 & d_2/2 \cdot e_{fy}^2 \\ 0 & 0 & 0 & 0 & 1 & -d_2/2 \cdot e_{fx}^2 \\ 0 & 0 & 0 & 1 & 0 & -d_2/2 \cdot e_{fy}^2 \\ 0 & 0 & 0 & 0 & 1 & d_2/2 \cdot e_{fx}^2 \\ 1 & 0 & -d_1/2 \cdot e_{fy}^1 & 0 & 0 & 0 \\ 0 & 1 & d_1/2 \cdot e_{fx}^1 & 0 & 0 & 0 \end{bmatrix} \tag{13.12.9}$$

and $\{U^{(e)}\}$ contains the nodal displacements and rotations of the beam element (Equation 13.11.5).

Similarly, we can cast an equation connecting the forces and moments of the two nodes of the beam element to the forces of the four nodes of the underlying quadrilateral element:

$$\left\{F^{(e)}\right\} = \left[T^{4Q}\right]^T \left\{F^{(4Q)}\right\} \tag{13.12.10}$$

where $\{F^{(e)}\}$ contains the nodal forces and moments of the beam element in the global coordinate system, given by Equation (13.7.6a), and

$$\left\{F^{(4Q)}\right\} = \begin{Bmatrix} \left\{f_{1^-}^{(4Q)}\right\} \\ \left\{f_{2^-}^{(4Q)}\right\} \\ \left\{f_{2^+}^{(4Q)}\right\} \\ \left\{f_{1^+}^{(4Q)}\right\} \end{Bmatrix} \tag{13.12.11}$$

where

$$\left\{f_I^{(4Q)}\right\} = \begin{Bmatrix} f_{xI}^{(4Q)} \\ f_{yI}^{(4Q)} \end{Bmatrix} \tag{13.12.12}$$

Finally, we can calculate the stiffness matrix of the continuum-based beam element. To this end, we must first obtain the stiffness matrix of the underlying quadrilateral element, $[k^{(4Q)}]$ using the procedures described in Chapter 8. After that, we can once again use the transformation equations described in Appendix D, to express the stiffness in terms of the degrees of freedom of the beam element:

$$\left[k^{(e)}\right] = \left[T^{4Q}\right]^T \left[k^{(4Q)}\right] \left[T^{4Q}\right] \tag{13.12.13}$$

Some discussion is required for the stress-strain law of the material in the element. We will focus on the case of a beam element where the fiber vector at all locations is aligned with the local y-axis, and the beam axis is aligned with the local x-axis. A plane-stress assumption is employed. At the same time, the condition $\sigma_{yy} = 0$ must be enforced. The fact that the fibers are assumed to be inextensible will not give any strain ε_{yy}. Notice the inconsistency mentioned in Section 13.3, since we cannot simultaneously have $\sigma_{yy} = 0$ and $\varepsilon_{yy} = 0$! For this reason, we only strictly enforce $\sigma_{yy} = 0$, and can calculate the required value of ε_{yy} from the stress-strain law after we have obtained a finite element solution for the displacement field. The underlying 4Q element is used for the computation of two strain components, ε_{xx} and γ_{xy}, and the corresponding stress components. For example, if we start with the constitutive law corresponding to plane-stress conditions and consider the case of linear isotropic elasticity (Equation 7.7.3):

$$\begin{Bmatrix} \sigma_{xx}^{(4Q)} \\ \sigma_{yy}^{(4Q)} \\ \sigma_{xy}^{(4Q)} \end{Bmatrix} = \frac{E}{1-\nu^2} \begin{bmatrix} 1 & \nu & 0 \\ \nu & 1 & 0 \\ 0 & 0 & \dfrac{1-\nu}{2} \end{bmatrix} \begin{Bmatrix} \varepsilon_{xx}^{(4Q)} \\ \varepsilon_{yy}^{(4Q)} \\ \gamma_{xy}^{(4Q)} \end{Bmatrix}$$

If we now enforce the condition that $\sigma_{yy}^{(4Q)} = 0$, Equation (7.7.3) yields the following set of equations.

$$\sigma_{xx}^{(4Q)} = \frac{E}{1-\nu^2}\varepsilon_{xx}^{(4Q)} + \frac{\nu E}{1-\nu^2}\varepsilon_{yy}^{(4Q)} \quad (13.12.14a)$$

$$0 = \frac{\nu E}{1-\nu^2}\varepsilon_{xx}^{(4Q)} + \frac{E}{1-\nu^2}\varepsilon_{yy}^{(4Q)} \quad (13.12.14b)$$

$$\sigma_{xy}^{(4Q)} = \frac{E}{1-\nu^2}\frac{1-\nu}{2}\gamma_{xy}^{(4Q)} = G \cdot \gamma_{xy}^{(4Q)} \quad (13.12.14c)$$

Equation (13.12.14b) can give:

$$\varepsilon_{yy}^{(4Q)} = -\nu \cdot \varepsilon_{xx}^{(4Q)} \quad (13.12.15)$$

which is a well-known expression corresponding to uniaxial stress conditions, that is, when $\sigma_{xx} \neq 0$ and $\sigma_{yy} = \sigma_{zz} = 0$. Of course, if we then plug Equation (13.12.15) in equation (13.12.14a), we obtain:

$$\sigma_{xx}^{(4Q)} = E \cdot \varepsilon_{xx}^{(4Q)} \quad (13.12.16)$$

Equations (13.12.16) and (13.12.14c) comprise the stress-strain law that we need to implement for the underlying continuum element. We need two strain components, $\varepsilon_{xx}^{(4Q)}$ and $\gamma_{xy}^{(4Q)}$. The strain $\varepsilon_{yy}^{(4Q)}$ can then be obtained through the stipulation that $\sigma_{yy}^{(4Q)} = 0$, which—for a linearly elastic, isotropic material—leads to Equation (13.12.15).

Since we only need to work with two strain components and the corresponding stress components, we can define the *reduced strain-displacement matrix* $\left[B_{red}^{(4Q)}\right]$, corresponding to the specific two components of strain:

$$\left[B_{red}^{(4Q)}\right] = \begin{bmatrix} \frac{\partial N_1^{(4Q)}}{\partial x} & 0 & \frac{\partial N_2^{(4Q)}}{\partial x} & 0 & \frac{\partial N_3^{(4Q)}}{\partial x} & 0 & \frac{\partial N_4^{(4Q)}}{\partial x} & 0 \\ \frac{\partial N_1^{(4Q)}}{\partial y} & \frac{\partial N_1^{(4Q)}}{\partial x} & \frac{\partial N_2^{(4Q)}}{\partial y} & \frac{\partial N_2^{(4Q)}}{\partial x} & \frac{\partial N_3^{(4Q)}}{\partial y} & \frac{\partial N_3^{(4Q)}}{\partial x} & \frac{\partial N_4^{(4Q)}}{\partial y} & \frac{\partial N_4^{(4Q)}}{\partial x} \end{bmatrix} \quad (13.12.17)$$

such that:

$$\begin{Bmatrix} \varepsilon_{xx}^{(4Q)} \\ \gamma_{xy}^{(4Q)} \end{Bmatrix} = \left[B_{red}^{(4Q)}\right]\left\{U^{(4Q)}\right\} \quad (13.12.18)$$

The shape functions in Equation (13.12.17) are those of a quadrilateral isoparametric element, described in Section 8.5.2. We can also define a *reduced material stiffness matrix*, $\left[D_{red}^{(4Q)}\right] = \begin{bmatrix} E & 0 \\ 0 & G \end{bmatrix}$, such that:

$$\begin{Bmatrix} \sigma_{xx}^{(4Q)} \\ \sigma_{xy}^{(4Q)} \end{Bmatrix} = \left[D_{red}^{(4Q)}\right]\begin{Bmatrix} \varepsilon_{xx}^{(4Q)} \\ \gamma_{xy}^{(4Q)} \end{Bmatrix} \quad (13.12.19)$$

The stiffness matrix of the underlying element can now be calculated using the expression

$$\left[k^{(4Q)}\right]=\iint_{\Omega^{(e)}}\left[B_{red}^{(4Q)}\right]^{T}\left[D_{red}^{(4Q)}\right]\left[B_{red}^{(4Q)}\right]b_{w}dV=\int_{-1}^{1}\int_{-1}^{1}\left(\left[B_{red}^{(4Q)}\right]^{T}\left[D_{red}^{(4Q)}\right]\left[B_{red}^{(4Q)}\right]b_{w}Jd\xi d\eta\right) \tag{13.12.20}$$

and two-dimensional Gaussian quadrature can be used for the evaluation of the stiffness matrix.

13.13 Extension of Continuum-Based Beam Elements to General Curved Beams

The continuum-based beam element formulation described in the previous section can also be employed for the analysis of *curved beams*, that is, beams in which the undeformed geometry is curved. Let us consider the most general case of a curved beam element, which is shown in Figure 13.23. The undeformed geometry is no longer represented by a reference straight line, but by a reference curve. *We examine the most general case of a beam element, in which plane sections in the undeformed configuration are not necessarily normal to the undeformed reference curve.* This case can still be described with beam theory.

The beam element will be defined with an isoparametric formulation, as shown in Figure 13.23. The element includes three nodes: this is the minimum number of nodes required to represent a curved geometry. A larger number of nodal points can also be used, if necessary.

We are given the material properties, E and G and the coordinates in the physical space of the three nodes. We are also given the sectional depth, d, at the three nodal points and the fiber direction, $\vec{e}_f^{\,I}$, for each nodal point $I = 1,2,3$. We are also given the sectional width, b_w, which must be set equal to the thickness of the underlying continuum element. We now define the three one-dimensional shape functions, $N_1^{(1D)}(\xi)$, $N_2^{(1D)}(\xi)$ and $N_3^{(1D)}(\xi)$ in the parametric space:

$$N_1^{(1D)}(\xi)=-\frac{1}{2}\xi(1-\xi) \tag{13.13.1a}$$

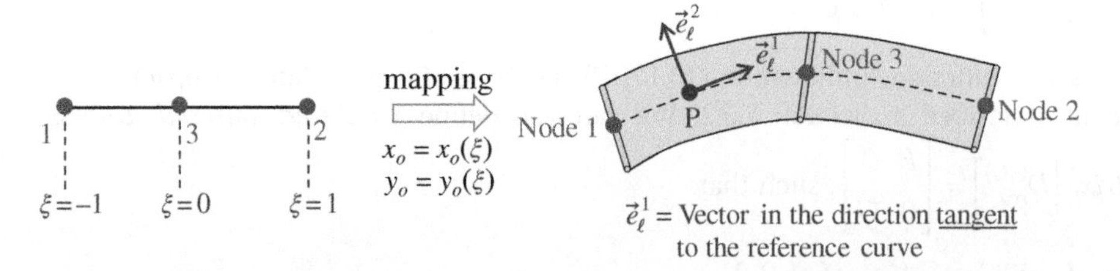

Figure 13.23 Three-node curved beam element and laminar vector for a location P along the reference curve.

$$N_2^{(1D)}(\xi) = \frac{1}{2}\xi(1+\xi) \tag{13.13.1b}$$

$$N_3^{(1D)}(\xi) = 1-\xi^2 \tag{13.13.1c}$$

These one-dimensional shape functions are used to determine the variation of the physical coordinates of the reference curve and of the fiber vector along the length of the beam. The physical coordinates, $x_o(\xi)$, $y_o(\xi)$, of any point on the reference curve are given by the following expression.

$$\begin{Bmatrix} x_o(\xi) \\ y_o(\xi) \end{Bmatrix} = N_1^{(1D)}(\xi)\{x_{o1}\} + N_2^{(1D)}(\xi)\{x_{o2}\} + N_3^{(1D)}(\xi)\{x_{o3}\} \tag{13.13.2}$$

where, for each nodal point $I = 1,2,3$, we have

$$\{x_{oI}\} = \begin{Bmatrix} x_I \\ y_I \end{Bmatrix} \tag{13.13.3}$$

At each location P on the reference curve, we define a pair of mutually perpendicular vectors $\vec{e}_\ell^{\,1}$, $\vec{e}_\ell^{\,2}$, such that $\vec{e}_\ell^{\,1}$ is tangent to the reference curve at P, as shown in Figure 13.23. The coordinate system defined by these vectors will be called the *laminar coordinate system*. We must keep in mind that, in the most general case which we examine here, for a given ξ, the fiber direction, $\vec{e}_f(\xi)$, may not coincide with the direction $\vec{e}_\ell^{\,2}(\xi)$!

According to differential geometry theory (e.g., O'Neill 2006), if the reference curve is described by Equation (13.13.2), then the components of the tangent vector $\vec{e}_\ell^{\,1}$ at any location are given by:

$$\vec{e}_\ell^{\,1} = \begin{Bmatrix} e_{\ell x}^1 \\ e_{\ell y}^1 \end{Bmatrix} = \frac{\vec{x}_{,\xi}}{\left\|\vec{x}_{,\xi}\right\|} \tag{13.13.4a}$$

where

$$\vec{x}_{,\xi} = \frac{\partial}{\partial \xi}\begin{Bmatrix} x_o(\xi) \\ y_o(\xi) \end{Bmatrix} = \frac{\partial N_1^{(1D)}}{\partial \xi}\{x_{o1}\} + \frac{\partial N_2^{(1D)}}{\partial \xi}\{x_{o2}\} + \frac{\partial N_3^{(1D)}}{\partial \xi}(\xi)\{x_{o3}\} \tag{13.13.4b}$$

The vector $\vec{e}_\ell^{\,2}$ can then be found, by stipulating that it is normal to $\vec{e}_\ell^{\,1}$ and that their cross-product is the unit vector toward the positive z-axis (*why?*).

$$\vec{e}_\ell^{\,1} \bullet \vec{e}_\ell^{\,2} = 0 \tag{13.13.5a}$$

$$\vec{e}_\ell^{\,1} \times \vec{e}_\ell^{\,2} = \vec{e}_z \tag{13.13.5b}$$

Remark 13.13.1: For Equation (13.13.5b), we need to treat the vectors $\vec{e}_\ell^{\,1}$ and $\vec{e}_\ell^{\,2}$ as three-dimensional (i.e., provide a third, zero component along the z-axis). ▮

One can verify (from the definition of the cross-product in Equation A.2.6) that, to satisfy Equation (13.13.5b), we must have:

$$\vec{e}_\ell^{\,2} = \begin{Bmatrix} e_{\ell x}^2 \\ e_{\ell y}^2 \end{Bmatrix} = \begin{Bmatrix} -e_{\ell y}^1 \\ e_{\ell x}^1 \end{Bmatrix} \tag{13.13.5c}$$

Now, we establish the formulation of an underlying six-node quadrilateral (6Q) element with two curved sides, as shown in Figure 13.24. The six nodes of the quadrilateral element are simply the endpoints of the fiber vectors at each nodal location of our beam element! The coordinates of the six nodal points of the 6Q element can be obtained from those of nodes 1, 2 and 3, based on Equations (13.12.1b-e) for nodes 1 and 2, and by establishing similar expressions for node 3. We can then set up an isoparametric mapping for the continuum element, as shown in Figure 13.25. The six shape functions of the 6Q element can be found through a generic procedure described in Section 3.7 of the textbook by Hughes (2000).

$$N_1^{(6Q)}(\xi,\eta) = -\frac{1}{4}\xi(1-\xi)(1-\eta) \tag{13.13.6a}$$

$$N_2^{(6Q)}(\xi,\eta) = \frac{1}{4}\xi(1+\xi)(1-\eta) \tag{13.13.6b}$$

$$N_3^{(6Q)}(\xi,\eta) = \frac{1}{4}\xi(1+\xi)(1+\eta) \tag{13.13.6c}$$

$$N_4^{(6Q)}(\xi,\eta) = -\frac{1}{4}\xi(1-\xi)(1+\eta) \tag{13.13.6d}$$

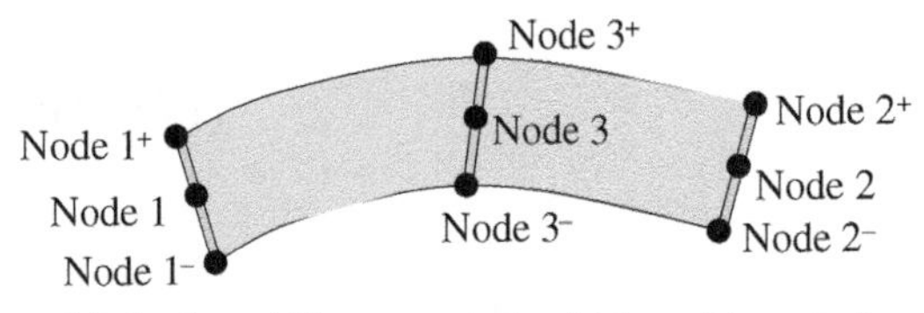

Figure 13.24 Three-node, curved beam element with underlying, six-node quadrilateral (6Q) continuum element.

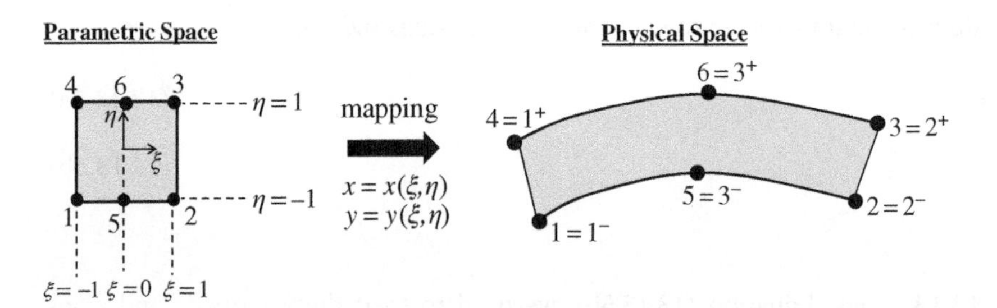

Figure 13.25 Isoparametric mapping for underlying continuum element.

$$N_5^{(6Q)}(\xi,\eta) = \frac{1}{2}\left(1-\xi^2\right)\left(1-\eta\right) \tag{13.13.6e}$$

$$N_6^{(6Q)}(\xi,\eta) = \frac{1}{2}\left(1-\xi^2\right)\left(1+\eta\right) \tag{13.13.6f}$$

Similar to the considerations described in Section 13.12, we establish a *constraint* equation giving the nodal displacements of the underlying 6Q element in terms of the nodal degrees of freedom of the beam element:

$$\left\{U^{(6Q)}\right\} = \begin{Bmatrix} \{u_{1^-}\} \\ \{u_{2^-}\} \\ \{u_{2^+}\} \\ \{u_{1^+}\} \\ \{u_{3^-}\} \\ \{u_{3^+}\} \end{Bmatrix} = \left[T^{6Q}\right]\left\{U^{(e)}\right\} \tag{13.13.7}$$

where

$$\left[T^{6Q}\right] = \begin{bmatrix} 1 & 0 & d_1/2\cdot e_{fy}^1 & 0 & 0 & 0 & 0 & 0 & 0 \\ 0 & 1 & -d_1/2\cdot e_{fx}^1 & 0 & 0 & 0 & 0 & 0 & 0 \\ 0 & 0 & 0 & 1 & 0 & d_2/2\cdot e_{fy}^2 & 0 & 0 & 0 \\ 0 & 0 & 0 & 0 & 1 & -d_2/2\cdot e_{fx}^2 & 0 & 0 & 0 \\ 0 & 0 & 0 & 1 & 0 & -d_2/2\cdot e_{fy}^2 & 0 & 0 & 0 \\ 0 & 0 & 0 & 0 & 1 & d_2/2\cdot e_{fx}^2 & 0 & 0 & 0 \\ 1 & 0 & -d_1/2\cdot e_{fy}^1 & 0 & 0 & 0 & 0 & 0 & 0 \\ 0 & 1 & d_1/2\cdot e_{fx}^1 & 0 & 0 & 0 & 0 & 0 & 0 \\ 0 & 0 & 0 & 0 & 0 & 0 & 1 & 0 & d_3/2\cdot e_{fy}^3 \\ 0 & 0 & 0 & 0 & 0 & 0 & 0 & 1 & -d_3/2\cdot e_{fx}^3 \\ 0 & 0 & 0 & 0 & 0 & 0 & 1 & 0 & -d_3/2\cdot e_{fy}^3 \\ 0 & 0 & 0 & 0 & 0 & 0 & 0 & 1 & d_3/2\cdot e_{fx}^3 \end{bmatrix} \tag{13.13.8}$$

and $\left\{U^{(e)}\right\} = \left[u_{x1}^{(e)} \; u_{y1}^{(e)} \; \theta_1^{(e)} \; u_{x2}^{(e)} \; u_{y2}^{(e)} \; \theta_2^{(e)} \; u_{x3}^{(e)} \; u_{y3}^{(e)} \; \theta_3^{(e)}\right]^T$ (13.13.9)

There is now a tricky aspect, which stems from the fact that *the stress normal to the lamina must be zero*. Since the laminar direction generally varies along the element, *the calculation of the stresses and strains at each location must be conducted at the local laminar coordinate system*. In other words, the strains that we need to calculate should correspond to the coordinate system defined by vectors $\vec{e}_\ell^{\,1}, \vec{e}_\ell^{\,2}$. We need to establish a coordinate transformation rule from the xy coordinate system to the system of laminar coordinates $x_\ell^{(1)}$, $x_\ell^{(2)}$, and vice versa:

$$\begin{Bmatrix} x_\ell^{(1)} \\ x_\ell^{(2)} \end{Bmatrix} = \begin{bmatrix} e_{\ell x}^1 & e_{\ell y}^1 \\ e_{\ell x}^2 & e_{\ell y}^2 \end{bmatrix} \begin{Bmatrix} x \\ y \end{Bmatrix} \leftrightarrow \begin{Bmatrix} x \\ y \end{Bmatrix} = \begin{bmatrix} e_{\ell x}^1 & e_{\ell x}^2 \\ e_{\ell y}^1 & e_{\ell y}^2 \end{bmatrix} \begin{Bmatrix} x_\ell^{(1)} \\ x_\ell^{(2)} \end{Bmatrix} \tag{13.13.10a}$$

The same coordinate transformation law can be used for the displacement components:

$$\begin{Bmatrix} u_\ell^{(1)} \\ u_\ell^{(2)} \end{Bmatrix} = \begin{bmatrix} e_{\ell x}^1 & e_{\ell y}^1 \\ e_{\ell x}^2 & e_{\ell y}^2 \end{bmatrix} \begin{Bmatrix} u_x \\ u_y \end{Bmatrix} \tag{13.13.10b}$$

Now, the laminar strain components can be defined from the following equations:

$$\varepsilon_{11,\ell} = \frac{\partial u_\ell^{(1)}}{\partial x_\ell^{(1)}}, \tag{13.13.11a}$$

$$\varepsilon_{22,\ell} = \frac{\partial u_\ell^{(2)}}{\partial x_\ell^{(2)}} \tag{13.13.11b}$$

$$\gamma_{12,\ell} = \frac{\partial u_\ell^{(1)}}{\partial x_\ell^{(2)}} + \frac{\partial u_\ell^{(2)}}{\partial x_\ell^{(1)}} \tag{13.13.11c}$$

We can use the chain rule of differentiation to calculate the partial derivatives with respect to the laminar coordinates. For any function f, we have:

$$\frac{\partial f}{\partial x_\ell^{(1)}} = \frac{\partial f}{\partial x}\frac{\partial x}{\partial x_\ell^{(1)}} + \frac{\partial f}{\partial y}\frac{\partial y}{\partial x_\ell^{(1)}} \tag{13.13.12a}$$

and
$$\frac{\partial f}{\partial x_\ell^{(2)}} = \frac{\partial f}{\partial x}\frac{\partial x}{\partial x_\ell^{(2)}} + \frac{\partial f}{\partial y}\frac{\partial y}{\partial x_\ell^{(2)}} \tag{13.13.12b}$$

We can use Equation (13.13.10a) to obtain $x = e_{\ell x}^1 x_\ell^{(1)} + e_{\ell x}^2 x_\ell^{(2)}$ and $y = e_{\ell y}^1 x_\ell^{(1)} + e_{\ell y}^2 x_\ell^{(2)}$; thus:

$$\frac{\partial x}{\partial x_\ell^{(1)}} = e_{\ell x}^1 \tag{13.13.13a}$$

$$\frac{\partial x}{\partial x_\ell^{(2)}} = e_{\ell x}^2 \tag{13.13.13b}$$

$$\frac{\partial y}{\partial x_\ell^{(1)}} = e_{\ell y}^1 \tag{13.13.13c}$$

$$\frac{\partial y}{\partial x_\ell^{(2)}} = e_{\ell y}^2 \tag{13.13.13d}$$

Let us now define the arrays with the derivatives of the shape functions with respect to x, y for the underlying 6Q element:

$$\left[N_{,x}^{(6Q)}\right]=\begin{bmatrix}\tilde{J}_{11} & \tilde{J}_{12}\\ \tilde{J}_{21} & \tilde{J}_{22}\end{bmatrix}\begin{bmatrix}\frac{\partial N_1^{(6Q)}}{\partial\xi} & \frac{\partial N_2^{(6Q)}}{\partial\xi} & \frac{\partial N_3^{(6Q)}}{\partial\xi} & \frac{\partial N_4^{(6Q)}}{\partial\xi} & \frac{\partial N_5^{(6Q)}}{\partial\xi} & \frac{\partial N_6^{(6Q)}}{\partial\xi}\\ \frac{\partial N_1^{(6Q)}}{\partial\eta} & \frac{\partial N_2^{(6Q)}}{\partial\eta} & \frac{\partial N_3^{(6Q)}}{\partial\eta} & \frac{\partial N_4^{(6Q)}}{\partial\eta} & \frac{\partial N_5^{(6Q)}}{\partial\eta} & \frac{\partial N_6^{(6Q)}}{\partial\eta}\end{bmatrix}=\left[\tilde{J}\right]\left[N_{,\xi}^{(6Q)}\right] \tag{13.13.14}$$

where

$$\left[N_{,x}^{(6Q)}\right]=\begin{bmatrix}\frac{\partial N_1^{(6Q)}}{\partial x} & \frac{\partial N_2^{(6Q)}}{\partial x} & \frac{\partial N_3^{(6Q)}}{\partial x} & \frac{\partial N_4^{(6Q)}}{\partial x} & \frac{\partial N_5^{(6Q)}}{\partial x} & \frac{\partial N_6^{(6Q)}}{\partial x}\\ \frac{\partial N_1^{(6Q)}}{\partial y} & \frac{\partial N_2^{(6Q)}}{\partial y} & \frac{\partial N_3^{(6Q)}}{\partial y} & \frac{\partial N_4^{(6Q)}}{\partial y} & \frac{\partial N_5^{(6Q)}}{\partial y} & \frac{\partial N_6^{(6Q)}}{\partial y}\end{bmatrix} \tag{13.13.15}$$

and the array $\left[\tilde{J}\right]$ is the inverse of the Jacobian matrix $[J]$ of the isoparametric coordinate mapping; $[J]$ can be obtained from the expression:

$$[J]=\left[N_{,\xi}^{(6Q)}\right]\left[\left\{x^{(6Q)}\right\}\ \left\{y^{(6Q)}\right\}\right]$$

where $\left\{x^{(6Q)}\right\}=\left[x_{1^-}\ x_{2^-}\ x_{2^+}\ x_{1^+}\ x_{3^-}\ x_{3^+}\right]^T$, $\left\{y^{(6Q)}\right\}=\left[y_{1^-}\ y_{2^-}\ y_{2^+}\ y_{1^+}\ y_{3^-}\ y_{3^+}\right]^T$.

We can now leverage Equations (13.13.12) and (13.13.13) to obtain an array $\left[N_{,x\ell}^{(6Q)}\right]$ with the derivatives of the shape functions with respect to the laminar coordinates:

$$\left[N_{,x\ell}^{(6Q)}\right]=\begin{bmatrix}e_{\ell x}^1 & e_{\ell y}^1\\ e_{\ell x}^2 & e_{\ell y}^2\end{bmatrix}\left[N_{,x}^{(6Q)}\right] \tag{13.13.16}$$

where

$$\left[N_{,x\ell}^{(6Q)}\right]=\begin{bmatrix}\frac{\partial N_1^{(6Q)}}{\partial x_\ell^{(1)}} & \frac{\partial N_2^{(6Q)}}{\partial x_\ell^{(1)}} & \frac{\partial N_3^{(6Q)}}{\partial x_\ell^{(1)}} & \frac{\partial N_4^{(6Q)}}{\partial x_\ell^{(1)}} & \frac{\partial N_5^{(6Q)}}{\partial x_\ell^{(1)}} & \frac{\partial N_6^{(6Q)}}{\partial x_\ell^{(1)}}\\ \frac{\partial N_1^{(6Q)}}{\partial x_\ell^{(2)}} & \frac{\partial N_2^{(6Q)}}{\partial x_\ell^{(2)}} & \frac{\partial N_3^{(6Q)}}{\partial x_\ell^{(2)}} & \frac{\partial N_4^{(6Q)}}{\partial x_\ell^{(2)}} & \frac{\partial N_5^{(6Q)}}{\partial x_\ell^{(2)}} & \frac{\partial N_6^{(6Q)}}{\partial x_\ell^{(2)}}\end{bmatrix} \tag{13.13.17}$$

Finally, a *laminar strain – displacement array*, which gives the two laminar strains, $\{\varepsilon_\ell\}=\left[\varepsilon_{11,\ell}\ \gamma_{12,\ell}\right]^T$, from the displacements in the xy coordinate system, can be established. The laminar strain-displacement relation can be written as:

$$\begin{Bmatrix}\varepsilon_{11,\ell}\\ \gamma_{12,\ell}\end{Bmatrix}=\begin{bmatrix}\frac{\partial N_1^{(6Q)}}{\partial x_\ell^{(1)}} & 0 & \frac{\partial N_2^{(6Q)}}{\partial x_\ell^{(1)}} & 0 & \frac{\partial N_3^{(6Q)}}{\partial x_\ell^{(1)}} & 0 & \frac{\partial N_4^{(6Q)}}{\partial x_\ell^{(1)}} & 0 & \frac{\partial N_5^{(6Q)}}{\partial x_\ell^{(1)}} & 0 & \frac{\partial N_6^{(6Q)}}{\partial x_\ell^{(1)}} & 0\\ \frac{\partial N_1^{(6Q)}}{\partial x_\ell^{(2)}} & \frac{\partial N_1^{(6Q)}}{\partial x_\ell^{(1)}} & \frac{\partial N_2^{(6Q)}}{\partial x_\ell^{(2)}} & \frac{\partial N_2^{(6Q)}}{\partial x_\ell^{(1)}} & \frac{\partial N_3^{(6Q)}}{\partial x_\ell^{(2)}} & \frac{\partial N_3^{(6Q)}}{\partial x_\ell^{(1)}} & \frac{\partial N_4^{(6Q)}}{\partial x_\ell^{(2)}} & \frac{\partial N_4^{(6Q)}}{\partial x_\ell^{(1)}} & \frac{\partial N_5^{(6Q)}}{\partial x_\ell^{(2)}} & \frac{\partial N_5^{(6Q)}}{\partial x_\ell^{(1)}} & \frac{\partial N_6^{(6Q)}}{\partial x_\ell^{(2)}} & \frac{\partial N_6^{(6Q)}}{\partial x_\ell^{(1)}}\end{bmatrix}\left\{U_\ell^{(6Q)}\right\} \tag{13.13.18}$$

where $\left\{U_\ell^{(6Q)}\right\}$ is the nodal displacement vector of the 6Q element expressed in the local laminar coordinate system. This vector can be obtained from the displacement vector $\{U^{(6Q)}\}$, using the following expression:

$$\left\{U_\ell^{(6Q)}\right\} = \left[R_\ell^{(6Q)}\right]\left\{U^{(6Q)}\right\} \tag{13.13.19}$$

where the transformation array $\left[R_\ell^{(6Q)}\right]$ is given from the following block-matrix expression:

$$\left[R_\ell^{(6Q)}\right] = \begin{bmatrix} [r_\ell] & [0] & [0] & [0] & [0] & [0] \\ [0] & [r_\ell] & [0] & [0] & [0] & [0] \\ [0] & [0] & [r_\ell] & [0] & [0] & [0] \\ [0] & [0] & [0] & [r_\ell] & [0] & [0] \\ [0] & [0] & [0] & [0] & [r_\ell] & [0] \\ [0] & [0] & [0] & [0] & [0] & [r_\ell] \end{bmatrix} \tag{13.13.20}$$

where all the block submatrices have dimensions (2×2) and $[r_\ell] = \begin{bmatrix} e_{\ell x}^1 & e_{\ell y}^1 \\ e_{\ell x}^2 & e_{\ell y}^2 \end{bmatrix}$.

Plugging Equation (13.13.19) into Equation (13.13.18) yields:

$$\{\varepsilon_\ell\} = \left[B_\ell^{(6Q)}\right]\left\{U^{(6Q)}\right\} \tag{13.13.21}$$

where:

$$\left[B_\ell^{(6Q)}\right] = \begin{bmatrix} \frac{\partial N_1^{(6Q)}}{\partial x_\ell^{(1)}} & 0 & \frac{\partial N_2^{(6Q)}}{\partial x_\ell^{(1)}} & 0 & \frac{\partial N_3^{(6Q)}}{\partial x_\ell^{(1)}} & 0 & \frac{\partial N_4^{(6Q)}}{\partial x_\ell^{(1)}} & 0 & \frac{\partial N_5^{(6Q)}}{\partial x_\ell^{(1)}} & 0 & \frac{\partial N_6^{(6Q)}}{\partial x_\ell^{(1)}} & 0 \\ \frac{\partial N_1^{(6Q)}}{\partial x_\ell^{(2)}} & \frac{\partial N_1^{(6Q)}}{\partial x_\ell^{(1)}} & \frac{\partial N_2^{(6Q)}}{\partial x_\ell^{(2)}} & \frac{\partial N_2^{(6Q)}}{\partial x_\ell^{(1)}} & \frac{\partial N_3^{(6Q)}}{\partial x_\ell^{(2)}} & \frac{\partial N_3^{(6Q)}}{\partial x_\ell^{(1)}} & \frac{\partial N_4^{(6Q)}}{\partial x_\ell^{(2)}} & \frac{\partial N_4^{(6Q)}}{\partial x_\ell^{(1)}} & \frac{\partial N_5^{(6Q)}}{\partial x_\ell^{(2)}} & \frac{\partial N_5^{(6Q)}}{\partial x_\ell^{(1)}} & \frac{\partial N_6^{(6Q)}}{\partial x_\ell^{(2)}} & \frac{\partial N_6^{(6Q)}}{\partial x_\ell^{(1)}} \end{bmatrix}\left[R_\ell^{(6Q)}\right] \tag{13.13.22}$$

Remark 13.13.2: It can be found that the components (1,2I-1), and (1,2I) ($I = 1, 2, \ldots, 6$) of the $\left[B_\ell^{(6Q)}\right]$ array are given by the following expressions.

$$B_{\ell\,1,2I-1}^{(6Q)} = \frac{\partial N_I^{(6Q)}}{\partial x_\ell^{(1)}} e_{\ell x}^1, \tag{13.13.23a}$$

$$B_{\ell\,1,2I}^{(6Q)} = \frac{\partial N_I^{(6Q)}}{\partial x_\ell^{(1)}} e_{\ell y}^1 \tag{13.13.23b}$$

Similarly, the components (2,2I-1), and (2,2I) of the same array are given by:

$$B_{\ell\,2,2I-1}^{(6Q)} = \frac{\partial N_I^{(6Q)}}{\partial x_\ell^{(1)}} e_{\ell x}^2 + \frac{\partial N_I^{(6Q)}}{\partial x_\ell^{(2)}} e_{\ell x}^1 \tag{13.13.23c}$$

$$B_{\ell\,2,2I}^{(6Q)} = \frac{\partial N_I^{(6Q)}}{\partial x_\ell^{(1)}} e_{\ell y}^2 + \frac{\partial N_I^{(6Q)}}{\partial x_\ell^{(2)}} e_{\ell y}^1 \tag{13.13.23d}$$

Thus, we can collectively write:

$$\left[B_\ell^{(6Q)}\right] = \begin{bmatrix} \frac{\partial N_1^{(6Q)}}{\partial x_\ell^{(1)}} e_{\ell x}^1 & \frac{\partial N_1^{(6Q)}}{\partial x_\ell^{(1)}} e_{\ell y}^1 & \cdots\cdots & \frac{\partial N_6^{(6Q)}}{\partial x_\ell^{(1)}} e_{\ell x}^1 & \frac{\partial N_6^{(6Q)}}{\partial x_\ell^{(1)}} e_{\ell y}^1 \\ \frac{\partial N_1^{(6Q)}}{\partial x_\ell^{(1)}} e_{\ell x}^2 + \frac{\partial N_1^{(6Q)}}{\partial x_\ell^{(2)}} e_{\ell x}^1 & \frac{\partial N_1^{(6Q)}}{\partial x_\ell^{(1)}} e_{\ell y}^2 + \frac{\partial N_1^{(6Q)}}{\partial x_\ell^{(2)}} e_{\ell y}^1 & \cdots\cdots & \frac{\partial N_6^{(6Q)}}{\partial x_\ell^{(1)}} e_{\ell x}^2 + \frac{\partial N_6^{(6Q)}}{\partial x_\ell^{(2)}} e_{\ell x}^1 & \frac{\partial N_6^{(6Q)}}{\partial x_\ell^{(1)}} e_{\ell y}^2 + \frac{\partial N_6^{(6Q)}}{\partial x_\ell^{(2)}} e_{\ell y}^1 \end{bmatrix} \tag{13.13.24}$$ ▮

The stiffness matrix in the xy coordinate system can be obtained from Gaussian quadrature, similar to the descriptions in Chapter 8:

$$\begin{aligned}\left[k^{(6Q)}\right] &= \iint_{\Omega^{(e)}} \left[B_\ell^{(6Q)}\right]^T \left[D_{\ell,red}^{(6Q)}\right]\left[B_\ell^{(6Q)}\right] \cdot b_w\, dV \\ &= \int_{-1}^{1}\int_{-1}^{1} \left(\left[B_\ell^{(6Q)}\right]^T \left[D_{\ell,red}^{(6Q)}\right]\left[B_\ell^{(6Q)}\right] \cdot b_w \cdot J d\xi d\eta \right)\end{aligned} \tag{13.13.25}$$

where, as mentioned earlier, b_w is the sectional width.

After this computation, the stiffness matrix of the curved beam element in the global coordinate system is obtained using the following equation.

$$\left[k^{(e)}\right] = \left[T^{6Q}\right]^T \left[k^{(6Q)}\right]\left[T^{6Q}\right] \tag{13.13.26}$$

Example 13.2: Stiffness matrix for a curved beam element
In this example, we will obtain the stiffness matrix for the three-node beam element, shown in Figure 13.26. The element is a circular arc segment, whose center C is also

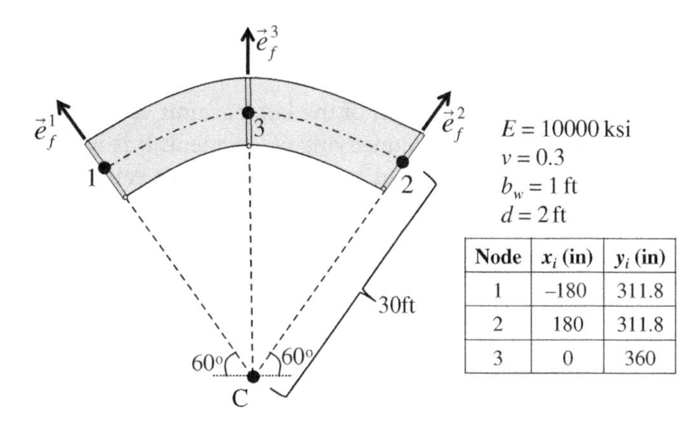

Node	x_i (in)	y_i (in)
1	–180	311.8
2	180	311.8
3	0	360

Figure 13.26 Example geometry of curved beam element.

the origin of the coordinate space. The coordinates for the three nodal points are also provided in Figure 13.26.

We can first establish the fiber vector for each one of the three nodal points. For the specific geometry, where the beam element is a circular arc, the fiber vector at each nodal point is in *the radial direction* (i.e., the direction of a vector from the arc center C to the nodal point of interest).

$$\vec{e}_f^{\,1} = \begin{Bmatrix} \cos 150^o \\ \sin 150^o \end{Bmatrix} = \begin{Bmatrix} -0.5 \\ 0.866 \end{Bmatrix},\ \vec{e}_f^{\,2} = \begin{Bmatrix} \cos 60^o \\ \sin 60^o \end{Bmatrix} = \begin{Bmatrix} 0.5 \\ 0.866 \end{Bmatrix},\ \vec{e}_f^{\,3} = \begin{Bmatrix} \cos 90^o \\ \sin 90^o \end{Bmatrix} = \begin{Bmatrix} 0 \\ 1 \end{Bmatrix}$$

We can now find the coordinates of the six nodes in the underlying continuum element:

$$\{x_{1^-}\} = \{x_1\} - \frac{d_1}{2}\vec{e}_f^{\,1} = \begin{Bmatrix} -180 \\ 311.8 \end{Bmatrix} - 12 \cdot \begin{Bmatrix} -0.5 \\ 0.866 \end{Bmatrix} = \begin{Bmatrix} -174 \\ 301.4 \end{Bmatrix}$$

$$\{x_{1^+}\} = \{x_1\} + \frac{d_1}{2}\vec{e}_f^{\,1} = \begin{Bmatrix} -180 \\ 311.8 \end{Bmatrix} + 12 \cdot \begin{Bmatrix} -0.5 \\ 0.866 \end{Bmatrix} = \begin{Bmatrix} -186 \\ 322.2 \end{Bmatrix}$$

$$\{x_{2^-}\} = \{x_2\} - \frac{d_2}{2}\vec{e}_f^{\,2} = \begin{Bmatrix} 180 \\ 311.8 \end{Bmatrix} - 12 \cdot \begin{Bmatrix} 0.5 \\ 0.866 \end{Bmatrix} = \begin{Bmatrix} 174 \\ 301.4 \end{Bmatrix}$$

$$\{x_{2^+}\} = \{x_2\} + \frac{d_2}{2}\vec{e}_f^{\,2} = \begin{Bmatrix} 180 \\ 311.8 \end{Bmatrix} + 12 \cdot \begin{Bmatrix} 0.5 \\ 0.866 \end{Bmatrix} = \begin{Bmatrix} 186 \\ 322.2 \end{Bmatrix}$$

$$\{x_{3^-}\} = \{x_3\} - \frac{d_3}{2}\vec{e}_f^{\,3} = \begin{Bmatrix} 0 \\ 360 \end{Bmatrix} - 12 \cdot \begin{Bmatrix} 0 \\ 1 \end{Bmatrix} = \begin{Bmatrix} 0 \\ 348 \end{Bmatrix}$$

$$\{x_{3^+}\} = \{x_3\} + \frac{d_3}{2}\vec{e}_f^{\,3} = \begin{Bmatrix} 0 \\ 360 \end{Bmatrix} + 12 \cdot \begin{Bmatrix} 0 \\ 1 \end{Bmatrix} = \begin{Bmatrix} 0 \\ 372 \end{Bmatrix}$$

We can also establish the derivatives of the three shape functions of the beam element—given by Equations (13.13.1a, b, c)—with respect to ξ:

$$\frac{\partial N_1^{(1D)}}{\partial \xi} = -\frac{1}{2} + \xi$$

$$\frac{\partial N_2^{(1D)}}{\partial \xi} = \frac{1}{2} + \xi$$

$$\frac{\partial N_3^{(1D)}}{\partial \xi} = -2\xi$$

These derivatives will allow us to obtain the direction of the laminar unit vectors.

We will use a (2 × 2) Gaussian quadrature for the underlying 6Q element. It is worth remembering that, when it comes to continuum-based beam elements, we can use a different quadrature rule for the ξ and η directions. In fact, the total quadrature points in the ξ direction are the quadrature points along the beam reference curve, while the number of quadrature points in the η direction correspond to integration along the sectional depth.

We can establish the expressions giving the derivatives of the shape functions in the 6Q element with the parametric coordinates:

$$\frac{\partial N_1^{(6Q)}}{\partial \xi} = -\left(\frac{1}{4} - \frac{1}{2}\xi\right)(1-\eta),\quad \frac{\partial N_1^{(6Q)}}{\partial \eta} = -\frac{1}{4}\xi(1-\xi)$$

$$\frac{\partial N_2^{(6Q)}}{\partial \xi} = \left(\frac{1}{4} + \frac{1}{2}\xi\right)(1-\eta), \quad \frac{\partial N_2^{(6Q)}}{\partial \eta} = -\frac{1}{4}\xi(1+\xi)$$

$$\frac{\partial N_3^{(6Q)}}{\partial \xi} = \left(\frac{1}{4} + \frac{1}{2}\xi\right)(1+\eta), \quad \frac{\partial N_3^{(6Q)}}{\partial \eta} = \frac{1}{4}\xi(1+\xi)$$

$$\frac{\partial N_4^{(6Q)}}{\partial \xi} = \left(-\frac{1}{4} + \frac{1}{2}\xi\right)(1+\eta), \quad \frac{\partial N_4^{(6Q)}}{\partial \eta} = -\frac{1}{4}\xi(1-\xi)$$

$$\frac{\partial N_5^{(6Q)}}{\partial \xi} = -\xi(1-\eta), \quad \frac{\partial N_5^{(6Q)}}{\partial \eta} = -\frac{1}{2}(1-\xi^2)$$

$$\frac{\partial N_6^{(6Q)}}{\partial \xi} = -\xi(1+\eta), \quad \frac{\partial N_6^{(6Q)}}{\partial \eta} = \frac{1}{2}(1-\xi^2)$$

We will now conduct the necessary computations for each quadrature point of the 6Q element.

Let us begin with the computations of quadrature point (Gauss point) 1 $\left(\xi = -\frac{1}{\sqrt{3}}, \eta = -\frac{1}{\sqrt{3}}\right)$:

$$\left[N_{,\xi}^{(6Q)}\right]_1 = \left[N_{,\xi}^{(6Q)}\right]_{\xi=-\frac{1}{\sqrt{3}},\eta=-\frac{1}{\sqrt{3}}}$$

$$= \begin{bmatrix} -0.8497 & -0.0610 & -0.0163 & -0.2277 & 0.9107 & 0.2440 \\ -0.2277 & 0.0610 & -0.0610 & 0.2277 & -0.3333 & 0.3333 \end{bmatrix}$$

$$[J]_1 = [J]_{\xi=-\frac{1}{\sqrt{3}},\eta=-\frac{1}{\sqrt{3}}} = \left[N_{,\xi}^{(6Q)}\right]_1 \begin{bmatrix} x_{1^-} & y_{1^-} \\ x_{2^-} & y_{2^-} \\ x_{2^+} & y_{2^+} \\ x_{1^+} & y_{1^+} \\ x_{3^-} & y_{3^-} \\ x_{3^+} & x_{3^+} \end{bmatrix} = \begin{bmatrix} 176.54 & 54.62 \\ -3.46 & 11.46 \end{bmatrix}, \text{ and we also have}$$

$$J_1 = 2213.04$$

We can also obtain: $\left[\tilde{J}\right]_1 = [J]_1^{-1} = \begin{bmatrix} 0.0052 & -0.0247 \\ 0.0016 & 0.0798 \end{bmatrix}$.

We now find the laminar basis vectors for Gauss point 1:

$$\vec{x}_{,\xi} = \left.\frac{\partial N_1^{(1D)}}{\partial \xi}\right|_{\xi=-1/\sqrt{3}} \{x_{o1}\} + \left.\frac{\partial N_2^{(1D)}}{\partial \xi}\right|_{\xi=-1/\sqrt{3}} \{x_{o2}\} + \left.\frac{\partial N_3^{(1D)}}{\partial \xi}\right|_{\xi=-1/\sqrt{3}} \{x_{o3}\} = \begin{Bmatrix} 180 \\ 55.7 \end{Bmatrix}$$

Thus, we have:

$$\vec{e}_\ell^{\,1} = \frac{\vec{x}_{,\xi}}{\left\|\vec{x}_{,\xi}\right\|} = \begin{Bmatrix} 0.955 \\ 0.295 \end{Bmatrix}, \vec{e}_\ell^{\,2} = \begin{Bmatrix} -e_{\ell y}^1 \\ e_{\ell x}^1 \end{Bmatrix} = \begin{Bmatrix} -0.295 \\ 0.955 \end{Bmatrix}$$

$$[r_\ell] = \begin{bmatrix} e^1_{\ell x} & e^1_{\ell y} \\ e^2_{\ell x} & e^2_{\ell y} \end{bmatrix} = \begin{bmatrix} 0.955 & 0.295 \\ -0.295 & 0.955 \end{bmatrix}$$, and we can define:

$$\left[R_\ell^{(6Q)}\right] = \begin{bmatrix} [r_\ell] & [0] & [0] & [0] & [0] & [0] \\ [0] & [r_\ell] & [0] & [0] & [0] & [0] \\ [0] & [0] & [r_\ell] & [0] & [0] & [0] \\ [0] & [0] & [0] & [r_\ell] & [0] & [0] \\ [0] & [0] & [0] & [0] & [r_\ell] & [0] \\ [0] & [0] & [0] & [0] & [0] & [r_\ell] \end{bmatrix}:$$

$$\left[R_\ell^{(6Q)}\right]_1 = \begin{bmatrix} 0.955 & 0.295 & 0 & 0 & 0 & 0 & 0 & 0 & 0 & 0 & 0 & 0 \\ -0.295 & 0.955 & 0 & 0 & 0 & 0 & 0 & 0 & 0 & 0 & 0 & 0 \\ 0 & 0 & 0.955 & 0.295 & 0 & 0 & 0 & 0 & 0 & 0 & 0 & 0 \\ 0 & 0 & -0.295 & 0.955 & 0 & 0 & 0 & 0 & 0 & 0 & 0 & 0 \\ 0 & 0 & 0 & 0 & 0.955 & 0.295 & 0 & 0 & 0 & 0 & 0 & 0 \\ 0 & 0 & 0 & 0 & -0.295 & 0.955 & 0 & 0 & 0 & 0 & 0 & 0 \\ 0 & 0 & 0 & 0 & 0 & 0 & 0.955 & 0.295 & 0 & 0 & 0 & 0 \\ 0 & 0 & 0 & 0 & 0 & 0 & -0.295 & 0.955 & 0 & 0 & 0 & 0 \\ 0 & 0 & 0 & 0 & 0 & 0 & 0 & 0 & 0.955 & 0.295 & 0 & 0 \\ 0 & 0 & 0 & 0 & 0 & 0 & 0 & 0 & -0.295 & 0.955 & 0 & 0 \\ 0 & 0 & 0 & 0 & 0 & 0 & 0 & 0 & 0 & 0 & 0.955 & 0.295 \\ 0 & 0 & 0 & 0 & 0 & 0 & 0 & 0 & 0 & 0 & -0.295 & 0.955 \end{bmatrix}$$

We now establish the array with the derivatives of the shape functions with respect to x, y:

$$\left[N_{,x}^{(6Q)}\right]_1 = \left[\tilde{J}\right]_1 \left[N_{,\xi}^{(6Q)}\right]_1$$

which gives:

$$\left[N_{,x}^{(6Q)}\right]_1 = \begin{bmatrix} 0.001218 & -0.001822 & 0.001421 & -0.006799 & 0.012945 & -0.006963 \\ -0.019492 & 0.004771 & -0.004892 & 0.017805 & -0.025165 & 0.026972 \end{bmatrix}$$

Next, we find the array with the derivatives of shape functions with respect to laminar coordinates:

$$\left[N_{,x\ell}^{(6Q)}\right] = \begin{bmatrix} e^1_{\ell x} & e^1_{\ell y} \\ e^2_{\ell x} & e^2_{\ell y} \end{bmatrix} \left[N_{,x}^{(6Q)}\right]$$

$$\left[N_{,x\ell}^{(6Q)}\right]_1 = \begin{bmatrix} -0.004598 & -0.000330 & -0.000088 & -0.001232 & 0.004928 & 0.001320 \\ -0.018981 & 0.005096 & -0.005093 & 0.019019 & -0.027867 & 0.027825 \end{bmatrix}$$

Finally, we obtain the laminar strain-displacement array for Gauss point 1:

$$\left[B_\ell^{(6Q)}\right]_1 = \begin{bmatrix} \frac{\partial N_1^{(6Q)}}{\partial x_\ell^{(1)}} & 0 & \frac{\partial N_2^{(6Q)}}{\partial x_\ell^{(1)}} & 0 & \frac{\partial N_3^{(6Q)}}{\partial x_\ell^{(1)}} & 0 & \frac{\partial N_4^{(6Q)}}{\partial x_\ell^{(1)}} & 0 & \frac{\partial N_5^{(6Q)}}{\partial x_\ell^{(1)}} & 0 & \frac{\partial N_6^{(6Q)}}{\partial x_\ell^{(1)}} & 0 \\ \frac{\partial N_1^{(6Q)}}{\partial x_\ell^{(2)}} & \frac{\partial N_1^{(6Q)}}{\partial x_\ell^{(1)}} & \frac{\partial N_2^{(6Q)}}{\partial x_\ell^{(2)}} & \frac{\partial N_2^{(6Q)}}{\partial x_\ell^{(1)}} & \frac{\partial N_3^{(6Q)}}{\partial x_\ell^{(2)}} & \frac{\partial N_3^{(6Q)}}{\partial x_\ell^{(1)}} & \frac{\partial N_4^{(6Q)}}{\partial x_\ell^{(2)}} & \frac{\partial N_4^{(6Q)}}{\partial x_\ell^{(1)}} & \frac{\partial N_5^{(6Q)}}{\partial x_\ell^{(2)}} & \frac{\partial N_5^{(6Q)}}{\partial x_\ell^{(1)}} & \frac{\partial N_6^{(6Q)}}{\partial x_\ell^{(2)}} & \frac{\partial N_6^{(6Q)}}{\partial x_\ell^{(1)}} \end{bmatrix} \left[R_\ell^{(6Q)}\right]_1$$

$$\left[B_\ell^{(6Q)}\right]_1 = \begin{bmatrix} -0.00439 & -0.00136 & -0.00032 & -0.00010 & -0.00008 & -0.00003 & -0.00118 & -0.00036 & 0.00471 & 0.00146 & 0.00126 & 0.00039 \\ -0.01677 & -0.01000 & 0.00497 & 0.00119 & -0.00484 & -0.00159 & 0.01853 & 0.00444 & -0.02808 & -0.00353 & 0.02619 & 0.00949 \end{bmatrix}$$

We must also find the reduced material stiffness matrix, $\left[D_{\ell,red}^{(6Q)}\right]_1$:

$$\left[D_{\ell,red}^{(6Q)}\right]_1 = \begin{bmatrix} E & 0 \\ 0 & G \end{bmatrix}_1 = \begin{bmatrix} E & 0 \\ 0 & \dfrac{E}{2(1+\nu)} \end{bmatrix}_1 = \begin{bmatrix} E & 0 \\ 0 & \dfrac{E}{2(1+\nu)} \end{bmatrix}_1 = \begin{bmatrix} 10000 & 0 \\ 0 & 3846 \end{bmatrix}$$

We now repeat the same computations for the other quadrature points of the underlying continuum element:

Quadrature point 2 $\left(\xi = \dfrac{1}{\sqrt{3}},\ \eta = -\dfrac{1}{\sqrt{3}}\right)$:

$$\left[N_{,\xi}^{(6Q)}\right]_2 = \left[N_{,\xi}^{(6Q)}\right]_{\xi = \frac{1}{\sqrt{3}}, \eta = -\frac{1}{\sqrt{3}}}$$

$$= \begin{bmatrix} 0.0610 & 0.8497 & 0.2277 & 0.0163 & -0.9107 & -0.2440 \\ 0.0610 & -0.2277 & 0.2277 & -0.0610 & -0.3333 & 0.3333 \end{bmatrix}$$

$$[J]_2 = [J]_{\xi = \frac{1}{\sqrt{3}}, \eta = -\frac{1}{\sqrt{3}}} = \left[N_{,\xi}^{(6Q)}\right]_2 \begin{bmatrix} x_{1^-} & y_{1^-} \\ x_{2^-} & y_{2^-} \\ x_{2^+} & y_{2^+} \\ x_{1^+} & y_{1^+} \\ x_{3^-} & y_{3^-} \\ x_{3^+} & x_{3^+} \end{bmatrix} = \begin{bmatrix} 176.54 & -54.62 \\ 3.46 & 11.46 \end{bmatrix}, J_2 = 2213.04,$$

$$\left[\widetilde{J}\right]_2 = [J]_2^{-1} = \begin{bmatrix} 0.0052 & 0.0247 \\ -0.0016 & 0.0798 \end{bmatrix}.$$

$$\vec{e}_\ell^{\,1} = \begin{Bmatrix} 0.955 \\ -0.295 \end{Bmatrix},\ \vec{e}_\ell^{\,2} = \begin{Bmatrix} 0.295 \\ 0.955 \end{Bmatrix},\ [r_\ell] = \begin{bmatrix} e_{\ell x}^1 & e_{\ell y}^1 \\ e_{\ell x}^2 & e_{\ell y}^2 \end{bmatrix} = \begin{bmatrix} 0.955 & -0.295 \\ 0.295 & 0.955 \end{bmatrix}$$

$$\left[R_\ell^{(6Q)}\right]_2 = \begin{bmatrix} 0.955 & -0.295 & 0 & 0 & 0 & 0 & 0 & 0 & 0 & 0 & 0 & 0 \\ 0.295 & 0.955 & 0 & 0 & 0 & 0 & 0 & 0 & 0 & 0 & 0 & 0 \\ 0 & 0 & 0.955 & -0.295 & 0 & 0 & 0 & 0 & 0 & 0 & 0 & 0 \\ 0 & 0 & 0.295 & 0.955 & 0 & 0 & 0 & 0 & 0 & 0 & 0 & 0 \\ 0 & 0 & 0 & 0 & 0.955 & -0.295 & 0 & 0 & 0 & 0 & 0 & 0 \\ 0 & 0 & 0 & 0 & 0.295 & 0.955 & 0 & 0 & 0 & 0 & 0 & 0 \\ 0 & 0 & 0 & 0 & 0 & 0 & 0.955 & -0.295 & 0 & 0 & 0 & 0 \\ 0 & 0 & 0 & 0 & 0 & 0 & 0.295 & 0.955 & 0 & 0 & 0 & 0 \\ 0 & 0 & 0 & 0 & 0 & 0 & 0 & 0 & 0.955 & -0.295 & 0 & 0 \\ 0 & 0 & 0 & 0 & 0 & 0 & 0 & 0 & 0.295 & 0.955 & 0 & 0 \\ 0 & 0 & 0 & 0 & 0 & 0 & 0 & 0 & 0 & 0 & 0.955 & -0.295 \\ 0 & 0 & 0 & 0 & 0 & 0 & 0 & 0 & 0 & 0 & 0.295 & 0.955 \end{bmatrix}$$

$$\left[N_{,x}^{(6Q)}\right]_2 = \begin{bmatrix} 0.001822 & -0.001218 & 0.006799 & -0.001421 & -0.012945 & 0.006963 \\ 0.004771 & -0.019492 & 0.017805 & -0.004892 & -0.025165 & 0.026972 \end{bmatrix}$$

$$\left[N_{,x\ell}^{(6Q)}\right]_2 = \begin{bmatrix} 0.000330 & 0.004598 & 0.001232 & 0.000088 & -0.004928 & -0.001320 \\ 0.005096 & -0.018981 & 0.019019 & -0.005093 & -0.027867 & 0.027825 \end{bmatrix}$$

$$\left[B_{\ell}^{(6Q)}\right]_2 = \begin{bmatrix} 0.00032 & -0.00010 & 0.00439 & -0.00136 & 0.00118 & -0.00036 & 0.00008 & -0.00003 & -0.00471 & 0.00146 & -0.00126 & 0.00039 \\ 0.00497 & -0.00119 & -0.01677 & 0.01000 & 0.01853 & -0.00444 & -0.00484 & 0.00159 & -0.02808 & 0.00353 & 0.02619 & -0.00949 \end{bmatrix}$$

$$\left[D_{\ell,red}^{(6Q)}\right]_2 = \begin{bmatrix} E & 0 \\ 0 & G \end{bmatrix}_2 = \begin{bmatrix} 10000 & 0 \\ 0 & 3846 \end{bmatrix}$$

Quadrature point 3 $\left(\xi = \frac{1}{\sqrt{3}},\ \eta = \frac{1}{\sqrt{3}}\right)$:

$$\left[N_{,\xi}^{(6Q)}\right]_3 = \left[N_{,\xi}^{(6Q)}\right]_{\xi=\frac{1}{\sqrt{3}},\eta=\frac{1}{\sqrt{3}}}$$

$$= \begin{bmatrix} 0.0163 & 0.2277 & 0.8497 & 0.0610 & -0.2440 & -0.9107 \\ 0.0610 & -0.2277 & 0.2277 & -0.0610 & -0.3333 & 0.3333 \end{bmatrix}$$

$$[J]_3 = [J]_{\xi=\frac{1}{\sqrt{3}},\eta=\frac{1}{\sqrt{3}}} = \left[N_{,\xi}^{(6Q)}\right]_3 \begin{bmatrix} x_{1^-} & y_{1^-} \\ x_{2^-} & y_{2^-} \\ x_{2^+} & y_{2^+} \\ x_{1^+} & y_{1^+} \\ x_{3^-} & y_{3^-} \\ x_{3^+} & x_{3^+} \end{bmatrix} = \begin{bmatrix} 183.46 & -56.76 \\ 3.46 & 11.46 \end{bmatrix},\ J_3 = 2299.89,$$

$$\left[\tilde{J}\right]_3 = [J]_3^{-1} = \begin{bmatrix} 0.0050 & 0.0247 \\ -0.0015 & 0.0798 \end{bmatrix}.$$

$$\vec{e}_{\ell}^{\,1} = \begin{Bmatrix} 0.955 \\ -0.295 \end{Bmatrix},\ \vec{e}_{\ell}^{\,2} = \begin{Bmatrix} 0.295 \\ 0.955 \end{Bmatrix},\ [r_{\ell}] = \begin{bmatrix} e_{\ell x}^1 & e_{\ell y}^1 \\ e_{\ell x}^2 & e_{\ell y}^2 \end{bmatrix} = \begin{bmatrix} 0.955 & -0.295 \\ 0.295 & 0.955 \end{bmatrix}$$

$$\left[R_{\ell}^{(6Q)}\right]_3 = \begin{bmatrix} 0.955 & -0.295 & 0 & 0 & 0 & 0 & 0 & 0 & 0 & 0 & 0 & 0 \\ 0.295 & 0.955 & 0 & 0 & 0 & 0 & 0 & 0 & 0 & 0 & 0 & 0 \\ 0 & 0 & 0.955 & -0.295 & 0 & 0 & 0 & 0 & 0 & 0 & 0 & 0 \\ 0 & 0 & 0.295 & 0.955 & 0 & 0 & 0 & 0 & 0 & 0 & 0 & 0 \\ 0 & 0 & 0 & 0 & 0.955 & -0.295 & 0 & 0 & 0 & 0 & 0 & 0 \\ 0 & 0 & 0 & 0 & 0.295 & 0.955 & 0 & 0 & 0 & 0 & 0 & 0 \\ 0 & 0 & 0 & 0 & 0 & 0 & 0.955 & -0.295 & 0 & 0 & 0 & 0 \\ 0 & 0 & 0 & 0 & 0 & 0 & 0.295 & 0.955 & 0 & 0 & 0 & 0 \\ 0 & 0 & 0 & 0 & 0 & 0 & 0 & 0 & 0.955 & -0.295 & 0 & 0 \\ 0 & 0 & 0 & 0 & 0 & 0 & 0 & 0 & 0.295 & 0.955 & 0 & 0 \\ 0 & 0 & 0 & 0 & 0 & 0 & 0 & 0 & 0 & 0 & 0.955 & -0.295 \\ 0 & 0 & 0 & 0 & 0 & 0 & 0 & 0 & 0 & 0 & 0.295 & 0.955 \end{bmatrix}$$

$$\left[N_{,x}^{(6Q)}\right]_3 = \begin{bmatrix} 0.001587 & -0.004484 & 0.009855 & -0.001202 & -0.009443 & 0.003688 \\ 0.004842 & -0.018504 & 0.016882 & -0.004958 & -0.026223 & 0.027962 \end{bmatrix}$$

$$\left[N_{,x\ell}^{(6Q)}\right]_3 = \begin{bmatrix} 0.000085 & 0.001186 & 0.004424 & 0.000318 & -0.001271 & -0.004742 \\ 0.005095 & -0.019003 & 0.019040 & -0.005092 & -0.027842 & 0.027803 \end{bmatrix}$$

$$\left[B_\ell^{(6Q)}\right]_3 = \begin{bmatrix} 0.00008 & -0.00003 & 0.00113 & -0.00035 & 0.00423 & -0.00131 & 0.00030 & -0.00009 & -0.00121 & 0.00038 & -0.00453 & 0.00140 \\ 0.00489 & -0.00142 & -0.01780 & 0.00675 & 0.01950 & -0.00140 & -0.00477 & 0.00181 & -0.02697 & 0.00702 & 0.02516 & -0.01275 \end{bmatrix}$$

$$\left[D_{\ell,red}^{(6Q)}\right]_3 = \begin{bmatrix} E & 0 \\ 0 & G \end{bmatrix}_3 = \begin{bmatrix} 10000 & 0 \\ 0 & 3846 \end{bmatrix}$$

Quadrature point 4 $\left(\xi = -\frac{1}{\sqrt{3}},\ \eta = \frac{1}{\sqrt{3}}\right)$:

$$\left[N_{,\xi}^{(6Q)}\right]_4 = \left[N_{,\xi}^{(6Q)}\right]_{\xi=-\frac{1}{\sqrt{3}},\eta=\frac{1}{\sqrt{3}}}$$

$$= \begin{bmatrix} -0.2277 & -0.0163 & -0.0610 & -0.8497 & 0.2440 & 0.9107 \\ -0.2277 & 0.0610 & -0.0610 & 0.2277 & -0.3333 & 0.3333 \end{bmatrix}$$

$$[J]_4 = [J]_{\xi=-\frac{1}{\sqrt{3}},\eta=\frac{1}{\sqrt{3}}} = \left[N_{,\xi}^{(6Q)}\right]_4 \begin{bmatrix} x_{1^-} & y_{1^-} \\ x_{2^-} & y_{2^-} \\ x_{2^+} & y_{2^+} \\ x_{1^+} & y_{1^+} \\ x_{3^-} & y_{3^-} \\ x_{3^+} & x_{3^+} \end{bmatrix} = \begin{bmatrix} 183.46 & 56.76 \\ -3.46 & 11.46 \end{bmatrix},\ J_4 = 2299.89$$

$$\left[\tilde{J}\right]_4 = [J]_4^{-1} = \begin{bmatrix} 0.0050 & -0.0247 \\ 0.0015 & 0.0798 \end{bmatrix}.$$

$$\vec{e}_\ell^{\,1} = \begin{Bmatrix} 0.955 \\ 0.295 \end{Bmatrix},\ \vec{e}_\ell^{\,2} = \begin{Bmatrix} -0.295 \\ 0.955 \end{Bmatrix},\ [r_\ell] = \begin{bmatrix} e_{\ell x}^1 & e_{\ell y}^1 \\ e_{\ell x}^2 & e_{\ell y}^2 \end{bmatrix} = \begin{bmatrix} 0.955 & 0.295 \\ -0.295 & 0.955 \end{bmatrix}$$

$$\left[R_\ell^{(6Q)}\right]_4 = \begin{bmatrix} 0.955 & 0.295 & 0 & 0 & 0 & 0 & 0 & 0 & 0 & 0 & 0 & 0 \\ -0.295 & 0.955 & 0 & 0 & 0 & 0 & 0 & 0 & 0 & 0 & 0 & 0 \\ 0 & 0 & 0.955 & 0.295 & 0 & 0 & 0 & 0 & 0 & 0 & 0 & 0 \\ 0 & 0 & -0.295 & 0.955 & 0 & 0 & 0 & 0 & 0 & 0 & 0 & 0 \\ 0 & 0 & 0 & 0 & 0.955 & 0.295 & 0 & 0 & 0 & 0 & 0 & 0 \\ 0 & 0 & 0 & 0 & -0.295 & 0.955 & 0 & 0 & 0 & 0 & 0 & 0 \\ 0 & 0 & 0 & 0 & 0 & 0 & 0.955 & 0.295 & 0 & 0 & 0 & 0 \\ 0 & 0 & 0 & 0 & 0 & 0 & -0.295 & 0.955 & 0 & 0 & 0 & 0 \\ 0 & 0 & 0 & 0 & 0 & 0 & 0 & 0 & 0.955 & 0.295 & 0 & 0 \\ 0 & 0 & 0 & 0 & 0 & 0 & 0 & 0 & -0.295 & 0.955 & 0 & 0 \\ 0 & 0 & 0 & 0 & 0 & 0 & 0 & 0 & 0 & 0 & 0.955 & 0.295 \\ 0 & 0 & 0 & 0 & 0 & 0 & 0 & 0 & 0 & 0 & -0.295 & 0.955 \end{bmatrix}$$

$$\left[N_{,x}^{(6Q)}\right]_4 = \begin{bmatrix} 0.004484 & -0.001587 & 0.001202 & -0.009855 & 0.009443 & -0.003688 \\ -0.018504 & 0.004842 & -0.004958 & 0.016882 & -0.026223 & 0.027962 \end{bmatrix}$$

$$\left[N_{,x\ell}^{(6Q)}\right]_4 = \begin{bmatrix} -0.001186 & -0.000085 & -0.000318 & -0.004424 & 0.001271 & 0.004742 \\ -0.019003 & 0.005095 & -0.005092 & 0.019040 & -0.027842 & 0.027803 \end{bmatrix}$$

$$\left[B_{\ell}^{(6Q)}\right]_4 = \begin{bmatrix} -0.00113 & -0.00035 & -0.00008 & -0.00003 & -0.00030 & -0.00009 & -0.00423 & -0.00131 & 0.00121 & 0.00038 & 0.00453 & 0.00140 \\ -0.01780 & -0.00675 & 0.00489 & 0.00142 & -0.00477 & -0.00181 & 0.01950 & 0.00140 & -0.02697 & -0.00702 & 0.02516 & 0.01275 \end{bmatrix}$$

$$\left[D_{\ell,red}^{(6Q)}\right]_4 = \begin{bmatrix} E & 0 \\ 0 & G \end{bmatrix}_4 = \begin{bmatrix} 10000 & 0 \\ 0 & 3846 \end{bmatrix}$$

We can finally obtain the stiffness matrix $[k^{(6Q)}]$ of the underlying continuum element:

$$\left[k^{(6Q)}\right] = \int_{-1}^{1}\int_{-1}^{1}\left(\left[B_{\ell}^{(6Q)}\right]^T\left[D_{\ell,red}^{(6Q)}\right]\left[B_{\ell}^{(6Q)}\right]b_w \cdot J d\xi d\eta\right)$$

$$= \sum_{g=1}^{4}\left(\left[B_{\ell}^{(6Q)}\right]_g^T\left[D_{\ell,red}^{(6Q)}\right]_g\left[B_{\ell}^{(6Q)}\right]_g b_{wg} J_g W_g\right)$$

$$= \left[B_{\ell}^{(6Q)}\right]_1^T\left[D_{\ell,red}^{(6Q)}\right]_1\left[B_{\ell}^{(6Q)}\right]_1 b_{w1} \cdot J_1 \cdot W_1$$

$$+ \left[B_{\ell}^{(6Q)}\right]_2^T\left[D_{\ell,red}^{(6Q)}\right]_2\left[B_{\ell}^{(6Q)}\right]_2 b_{w2} \cdot J_2 \cdot W_2$$

$$+ \left[B_{\ell}^{(6Q)}\right]_3^T\left[D_{\ell,red}^{(6Q)}\right]_3\left[B_{\ell}^{(6Q)}\right]_3 b_{w3} \cdot J_3 \cdot W_3$$

$$+ \left[B_{\ell}^{(6Q)}\right]_4^T\left[D_{\ell,red}^{(6Q)}\right]_4\left[B_{\ell}^{(6Q)}\right]_4 b_{w4} \cdot J_4 \cdot W_4$$

$$\left[k^{(6Q)}\right] = \begin{bmatrix} 72948 & 30235 & -34720 & 3846 & 37219 & 3160 & -70823 & -7688 & 64538 & 23051 & -69163 & -52604 \\ 30235 & 15943 & -3846 & -4399 & 3160 & 3710 & -30765 & -5752 & 53820 & 6539 & -52604 & -16040 \\ -34720 & -3846 & 72948 & -30235 & -70823 & 7688 & 37219 & -3160 & 64538 & -23051 & -69163 & 52604 \\ 3846 & -4399 & -30235 & 15943 & 30765 & -5752 & -3160 & 3710 & -53820 & 6539 & 52604 & -16040 \\ 37219 & 3160 & -70823 & 30765 & 85569 & -11243 & -37308 & 3846 & -84534 & 27326 & 69876 & -53854 \\ 3160 & 3710 & 7688 & -5752 & -11243 & 3341 & -3846 & -1909 & 27326 & -1022 & -23085 & 1632 \\ -70823 & -30765 & 37219 & -3160 & -37308 & -3846 & 85569 & 11243 & -84534 & -27326 & 69876 & 53854 \\ -7688 & -5752 & -3160 & 3710 & 3846 & -1909 & 11243 & 3341 & -27326 & -1022 & 23085 & 1632 \\ 64538 & 53820 & 64538 & -53820 & -84534 & 27326 & -84534 & -27326 & 328101 & 0 & -288111 & 0 \\ 23051 & 6539 & -23051 & 6539 & 27326 & -1022 & -27326 & -1022 & 0 & 14198 & 0 & -25232 \\ -69163 & -52604 & -69163 & 52604 & 69876 & -23085 & 69876 & 23085 & -288111 & 0 & 286686 & 0 \\ -52604 & -16040 & 52604 & -16040 & -53854 & 1632 & 53854 & 1632 & 0 & -25232 & 0 & 54048 \end{bmatrix}$$

We establish the constraint transformation matrix $[T^{6Q}]$, given by Equation (13.13.8):

$$
\left[T^{6Q}\right]=\begin{bmatrix}
1 & 0 & 12\cdot 0.866 & 0 & 0 & 0 & 0 & 0 & 0\\
0 & 1 & -12\cdot(-0.5) & 0 & 0 & 0 & 0 & 0 & 0\\
0 & 0 & 0 & 1 & 0 & 12\cdot 0.866 & 0 & 0 & 0\\
0 & 0 & 0 & 0 & 1 & -12\cdot 0.5 & 0 & 0 & 0\\
0 & 0 & 0 & 1 & 0 & -12\cdot 0.866 & 0 & 0 & 0\\
0 & 0 & 0 & 0 & 1 & 12\cdot 0.5 & 0 & 0 & 0\\
1 & 0 & -12\cdot 0.866 & 0 & 0 & 0 & 0 & 0 & 0\\
0 & 1 & 12\cdot(-0.5) & 0 & 0 & 0 & 0 & 0 & 0\\
0 & 0 & 0 & 0 & 0 & 0 & 1 & 0 & 12\cdot 1\\
0 & 0 & 0 & 0 & 0 & 0 & 0 & 1 & -12\cdot 0\\
0 & 0 & 0 & 0 & 0 & 0 & 1 & 0 & -12\cdot 1\\
0 & 0 & 0 & 0 & 0 & 0 & 0 & 1 & 12\cdot 0
\end{bmatrix}
=\begin{bmatrix}
1 & 0 & 10.4 & 0 & 0 & 0 & 0 & 0 & 0\\
0 & 1 & 6 & 0 & 0 & 0 & 0 & 0 & 0\\
0 & 0 & 0 & 1 & 0 & 10.4 & 0 & 0 & 0\\
0 & 0 & 0 & 0 & 1 & -6 & 0 & 0 & 0\\
0 & 0 & 0 & 1 & 0 & -10.4 & 0 & 0 & 0\\
0 & 0 & 0 & 0 & 1 & 6 & 0 & 0 & 0\\
1 & 0 & -10.4 & 0 & 0 & 0 & 0 & 0 & 0\\
0 & 1 & -6 & 0 & 0 & 0 & 0 & 0 & 0\\
0 & 0 & 0 & 0 & 0 & 0 & 1 & 0 & 12\\
0 & 0 & 0 & 0 & 0 & 0 & 0 & 1 & 0\\
0 & 0 & 0 & 0 & 0 & 0 & 1 & 0 & -12\\
0 & 0 & 0 & 0 & 0 & 0 & 0 & 1 & 0
\end{bmatrix}
$$

After this computation, the stiffness matrix of the beam element is obtained using Equation (13.13.26).

$$\left[k^{(e)}\right] = \left[T^{6Q}\right]^T \left[k^{(6Q)}\right] \left[T^{6Q}\right]$$

$$\left[k^{(e)}\right] = \begin{bmatrix} 16872 & 3025 & -155671 & 2410 & 0 & 18663 & -19282 & -3025 & -248495 \\ 3025 & 7779 & 512802 & 0 & 1111 & -130671 & -3025 & -8890 & 672163 \\ -155671 & 512802 & 43493824 & 18663 & 130671 & -15324131 & 137008 & -643473 & 47221776 \\ 2410 & 0 & 18663 & 16872 & -3025 & -155671 & -19282 & 3025 & -248495 \\ 0 & 1111 & 130671 & -3025 & 7779 & -512802 & 3025 & -8890 & -672163 \\ 18663 & -130671 & -15324131 & -155671 & -512802 & 43493824 & 137008 & 643473 & 47221776 \\ -19282 & -3025 & 137008 & -19282 & 3025 & 137008 & 38564 & 0 & 496990 \\ -3025 & -8890 & -643473 & 3025 & -8890 & 643473 & 0 & 17781 & 0 \\ -248495 & 672163 & 47221776 & -248495 & -672163 & 47221776 & 496990 & 0 & 171505403 \end{bmatrix}$$

■

13.14 Shear Locking and Selective-Reduced Integration for Thin Timoshenko Beam Elements

In this chapter, we examined the Euler-Bernoulli (E-B) theory, which is accurate for "thin" (slender) beams, and the Timoshenko beam theory, which is better suited for squat beams (where shear strains are nonzero and must be accounted for). An intriguing question that may arise is what might happen if we use Timoshenko beam elements for analyses of slender beams. Will we be able to capture the thin beam behavior with Timoshenko beam elements? To investigate this issue, we analyze the behavior of a slender beam, shown in Figure 13.27, when subjected to a single downward tip load. We are interested in obtaining the nodal deflection at the tip of the beam with E-B and Timoshenko beam elements. Our finite element mesh consists of two elements (with equal size). Figure 13.27 shows that using two E-B beam elements yields nodal displacements that are equal to the corresponding values of the exact solution. This is not the case when we use two Timoshenko beam elements; instead, the use of two elements of this type leads to significant underestimation of the nodal deflections. One could say that we have an *overstiff* solution (because we underestimate the deformations for

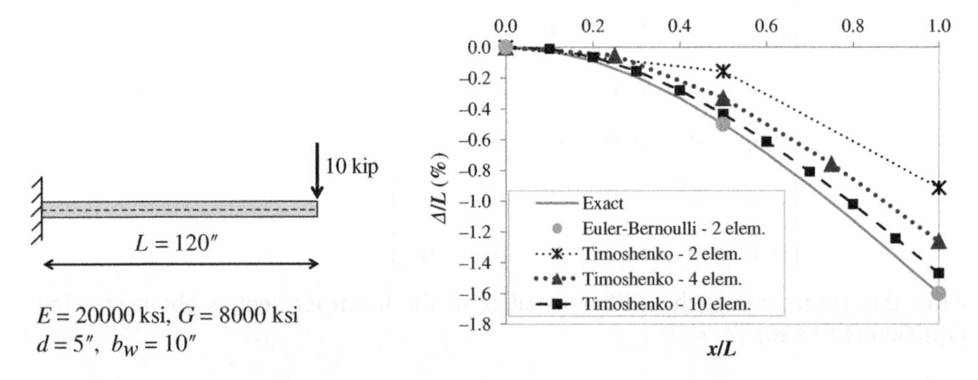

Figure 13.27 Analysis of slender cantilever beam.

given loadings, so we must somehow overestimate the stiffness!). Increasing the number of Timoshenko beam elements improves the accuracy, as shown in Figure 13.27, but even if we use 10 Timoshenko beam elements, we cannot obtain a solution that is as accurate as the one for two E-B elements.

To understand the nature of the problem, we will try to separate the effect of shear strains from the effect of axial-flexural strains in a Timoshenko beam element. To this end, we isolate the row of the generalized strain-displacement relation corresponding to the shear strains. This is merely the third row of Equation (13.11.7):

$$\gamma^{(e)} = \begin{bmatrix} 0 & \frac{dN_1^{(e)}}{dx} & -N_1^{(e)} & 0 & \frac{dN_2^{(e)}}{dx} & -N_2^{(e)} \end{bmatrix} \left\{U^{(e)}\right\} = \left[B_s^{(e)}\right]\left\{U^{(e)}\right\} \tag{13.14.1}$$

We also combine the other two equations of the strain-displacement law—that is, rows 1 and 2 of Equation (13.11.7)—which will correspond to the strains associated with axial and flexural deformation:

$$\begin{Bmatrix} \varepsilon_o^{(e)} \\ \varphi^{(e)} \end{Bmatrix} = \begin{bmatrix} \frac{dN_1^{(e)}}{dx} & 0 & 0 & \frac{dN_2^{(e)}}{dx} & 0 & 0 \\ 0 & 0 & \frac{dN_1^{(e)}}{dx} & 0 & 0 & \frac{dN_2^{(e)}}{dx} \end{bmatrix} \left\{U^{(e)}\right\} = \left[B_{af}^{(e)}\right]\left\{U^{(e)}\right\} \tag{13.14.2}$$

The separation of shear strains from the axial-flexural ones further allows us to write the stiffness matrix as the summation of two terms:

$$\left[k^{(e)}\right] = \left[k_{af}^{(e)}\right] + \left[k_s^{(e)}\right] \tag{13.14.3}$$

where $\left[k_{af}^{(e)}\right]$ is the part of $[k^{(e)}]$ expressing the axial-flexural stiffness:

$$\left[k_{af}^{(e)}\right] = \int_0^{\ell^{(e)}} \left[B_{af}^{(e)}\right]^T \left[\hat{D}_{af}^{(e)}\right] \left[B_{af}^{(e)}\right] dx \tag{13.14.4}$$

and $\left[k_s^{(e)}\right]$ is the part of $[k^{(e)}]$ expressing the shear stiffness:

$$\left[k_s^{(e)}\right] = \int_0^{\ell^{(e)}} \left[B_s^{(e)}\right]^T \kappa \cdot G \cdot A \left[B_s^{(e)}\right] dx \tag{13.14.5}$$

Remark 13.14.1: For an element with constant cross-sectional and material properties, it can be proven that the flexural (bending) stiffness terms are proportional to the quantity $E\left(\frac{d}{L}\right)^3$, while the shear stiffness terms are proportional to the quantity $\kappa \cdot G \cdot \frac{d}{L}$. ∎

Given the above, let us try to provide an explanation of what is causing the overstiff response shown in Figure 13.27 for Timoshenko beam elements. When we have such elements, the kinematic field is bound to "always see" a nonzero shear strain. In fact, the coarser the mesh is (i.e., the less elements we use to model our beam), the larger the magnitude of this spurious, or *"parasitic," shear strain* will be! Now, the presence of nonzero shear strains means that the shear stiffness will be mobilized when we use

Timoshenko beam elements. For the specific values provided in Figure 13.27, and in accordance with Remark 13.14.1, the flexural stiffness is proportional to $E\left(\frac{d}{L}\right)^3 = 20000\left(\frac{5}{120}\right)^3 = 1.447$, while the shear stiffness is proportional to $\kappa \cdot G \cdot \frac{d}{L} = \frac{5}{6} \cdot 8000 \cdot \frac{5}{120} = 277.8$. This would mean that the shear stiffness in slender beams is orders of magnitude greater than the flexural stiffness. In summary, the parasitic shear strain in the analysis of "thin" (slender) beams leads to the development of *parasitic shear stiffness*, which, in turn, leads to the overstiff response of Timoshenko beam elements! Although the term parasitic shear stiffness is common, several textbooks and papers may use the term *shear locking* to describe the overstiff response discussed herein. The use of large number of Timoshenko beam elements improves the accuracy of the approximate deflection field, reducing the value of the parasitic shear strain and partially alleviating the overstiff response. Still, we are always bound to "slightly" mobilize the parasitic shear stiffness.

The problem of parasitic shear stiffness is conceptually similar to the volumetric locking problem for solid elements, described in Section 11.4. This indicates that the issue of parasitic shear stiffness can be addressed through similar methods as those employed to resolve volumetric locking issues. For example, we can *use selective-reduced integration (SRI) for the analysis of slender beams with Timoshenko beam elements.* Specifically, we will use full integration (two-point quadrature) for the axial-flexural stiffness part (the part that is "not locking") and reduced (one-point) integration for the shear stiffness part (the part that is "locking"). This SRI approach is schematically described in Figure 13.28a. The quadrature rules will give the following expressions for the flexural part and the shear part of the element stiffness matrix:

$$\left[k_{af}^{(e)}\right] = \int_{-1}^{1} \left[B_{af}^{(e)}(\xi)\right]^T \left[\hat{D}_{af}^{(e)}(\xi)\right] \left[B_{af}^{(e)}(\xi)\right] \frac{\ell^{(e)}}{2} d\xi \approx \sum_{g=1}^{2} \left(\left[B_{af}^{(e)}\right]_g^T \left[\hat{D}_{af}^{(e)}\right]_g \left[B_{af}^{(e)}\right]_g \frac{\ell^{(e)}}{2} W_g \right) \tag{13.14.6}$$

$$\left[k_s^{(e)}\right] = \int_{-1}^{1} \left[B_s^{(e)}(\xi)\right]^T \kappa \cdot G(\xi) \cdot A(\xi) \left[B_s^{(e)}(\xi)\right] \frac{\ell^{(e)}}{2} d\xi$$

$$\approx \left[B_s^{(e)}\right]_{\xi=0}^T \kappa \cdot G_{\xi=0} \cdot A_{\xi=0} \left[B_s^{(e)}\right]_{\xi=0} \frac{\ell^{(e)}}{2} \cdot 2 \rightarrow$$

$$\rightarrow \left[k_s^{(e)}\right] = \left[B_s^{(e)}\right]_{\xi=0}^T \kappa \cdot G_{\xi=0} \cdot A_{\xi=0} \left[B_s^{(e)}\right]_{\xi=0} \ell^{(e)} \tag{13.14.7}$$

Using Timoshenko beam elements with SRI for the analysis of the beam in Figure 13.27 gives very accurate results, even if we use two elements, as shown in Figure 13.28b. The reason for this very good accuracy stems from the fact that the use of a single Gauss point (which is located at the middle of the element, as shown in Figure 13.28a) for the shear part of the stiffness leads to a situation where that single Gauss point does not "see" parasitic shear strains and does not mobilize the parasitic shear stiffness. Thus, problems associated with overstiff response are avoided.

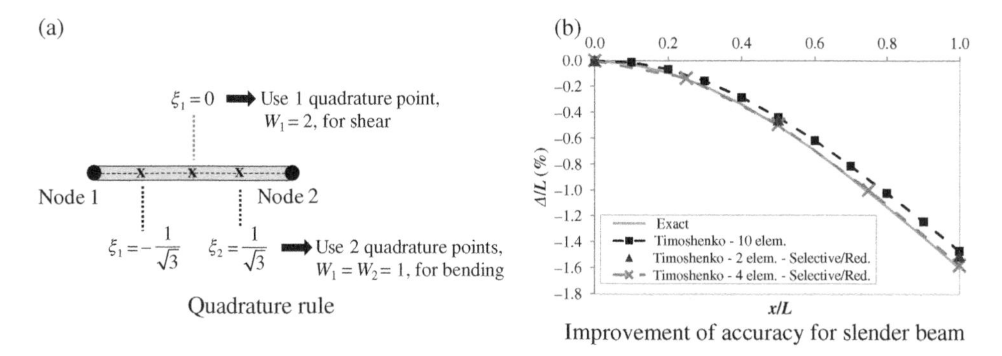

Figure 13.28 Selective-reduced integration for Timoshenko beam elements and improvement in accuracy for analysis of slender beams.

Remark 13.14.2: The selective-reduced integration procedure can be extended to the case of continuum-based beam elements. Specifically, we establish the stiffness matrix as the sum of two parts, one with $G = 0$ (which will give the axial-flexure part of the stiffness and for which we can use full integration in the underlying continuum element) and one with $E = 0$ (which will give the shear part of the stiffness and for which we can use one-point quadrature in the ξ-direction for the underlying continuum element). ∎

Remark 13.14.3: Other remedies, such as mixed elements or even *force-based beam elements* (which will not be described in detail here), can also be used to resolve the issue of parasitic shear stiffness. ∎

Problems

13.1 We are given the beam shown in Figure P13.1. The beam's local coordinate system, xy, is also shown. The beam has a rectangular cross-section with depth equal to d and width equal to b_w. We use Euler-Bernoulli beam theory for the beam.

Figure P13.1

a) Determine the variation of axial displacement, $u_{ox}(x)$, transverse displacement, $u_{oy}(x)$ (assume that E, d, b_w, and L are known), and determine the end reaction forces and moments when the beam is subjected to a constant transverse load, $p_y(x) = p_o$ = constant, as shown in Figure P13.2.

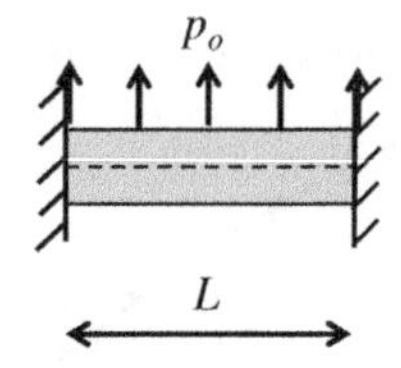

Figure P13.2

b) We are also given the constitutive law for a linearly elastic material under uniaxial stress and in the presence of autogenous thermal strains:

$$\sigma_{xx} = E(\varepsilon_{xx} - a_T \cdot \Delta\Theta)$$

where $\Delta\Theta = \Theta - \Theta_o$ is the temperature change from the reference temperature of the material (we can imagine Θ_o to be the temperature of the material "during construction") and a_T is the thermal coefficient (a given material constant). Modify the section constitutive law (the law which gives generalized stresses as functions of generalized strains) when we have a temperature change $\Delta\Theta(x, y) = \Delta\Theta_C$ = constant. Then, establish the beam differential equations in the presence of this uniform temperature change, solve and obtain the variation of axial and transverse displacements (assume E, d, b_w, L, and a_T are known) and also the end reaction forces and moments.

c) Repeat part b, for $\Delta\Theta(x, y) = \delta\Theta \cdot y$, where $\delta\Theta$ is a constant value.

13.2 We are given the member shown in Figure P13.3. The member has a rectangular cross-section with a constant width, b_w, and a linearly varying depth, d, as shown in the figure. The beam will be analyzed using a single Euler-Bernoulli element.

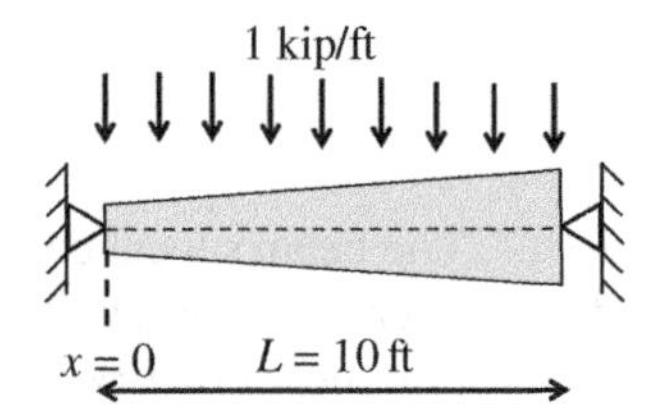

$E = 5000$ ksi
$d(x) = 10 + x/8$, d in inches, x in inches
$b_w = 10$ inches = constant

Figure P13.3

a) Determine the displacement variation along the beam.
b) Determine the curvature at the two quadrature points.
c) Determine the bending moment at the two quadrature points using the section constitutive equation.
d) Repeat the calculations using a single Timoshenko beam element. What do you observe?

13.3 We are given the frame shown in Figure P13.4. The frame consists of members with constant rectangular cross-section (identical E, sectional depth d and sectional width b_w for all members) and is subjected to a horizontal force as shown in the same figure. Note that all members are relatively slender.

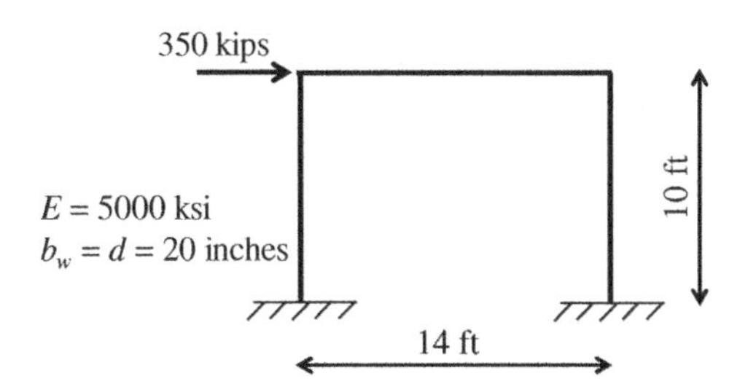

Figure P13.4

a) Determine the nodal displacements and rotations.
b) Determine the nodal reactions.
c) Determine the curvature field in the horizontal member.

13.4 We are given the frame shown in Figure P13.5. The frame consists of members with constant rectangular cross-section (identical E, G, sectional depth d and width b_w for all members) and is subjected to a horizontal force as shown in the same figure. We need to use one Euler-Bernoulli element for the more slender member, and one Timoshenko beam element for the more "squat" member.

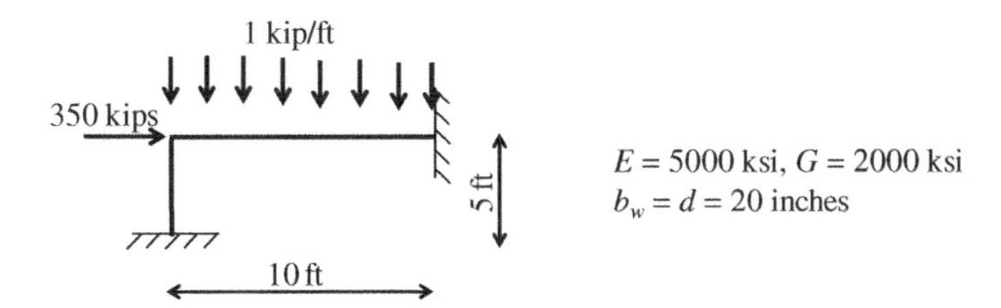

Figure P13.5

a) Determine the nodal displacements and rotations.
b) Determine the nodal reactions.
c) Determine the curvature field in the horizontal member.

13.5 We have the structure given in Figure P13.6, which consists of a horizontal Euler-Bernoulli beam and an inclined truss member. We want to use a single finite element for each one of the two members. The beam member has a rectangular cross-section, and the variation of sectional depth d and width b_w along the length of the member is provided in Figure P13.6.

a) Obtain the stiffness matrices of the two elements in the global coordinate system. Use a two-point quadrature for the beam element.

b) How many quadrature points would be required if you wanted to exactly integrate the stiffness terms in the beam element?
c) Calculate the nodal displacements/rotations of the structure.

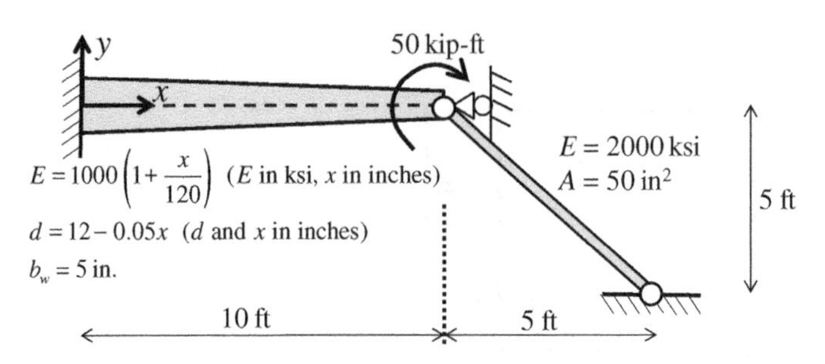

Figure P13.6

13.6 We have the finite element mesh shown in Figure P13.7, which consists of an Euler-Bernoulli beam element, connected to a set of quadrilateral elements at node 2. The loads applied on the mesh are also given in Figure P13.7. Explain (without calculations) whether the specific finite element model will be able to enforce the connection of the beam element to the quadrilateral elements, yielding meaningful results. If not, explain how we would be able to enforce the connection between the two types of elements (*Hint:* Useful information may be found in Section 10.6).

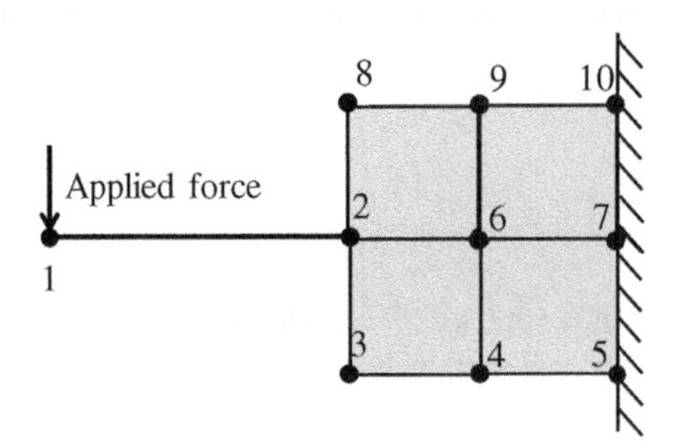

Figure P13.7

References

Hughes, T. (2000). *The Finite Element Method—Linear Static and Dynamic Finite Element Analysis*, 2nd ed. New York: Dover.

O'Neill, B. (2006). *Differential Geometry*, 2nd ed. Burlington, MA: Academic Press.

14

Finite Element Analysis of Shells

14.1 Introduction

Shells are surface structures, that is, structures for which one dimension, the thickness, is much smaller than the other two dimensions. In the previous Chapter, we saw that the deformation of a beam is quantified based on the configuration of the reference line. In the same fashion, the deformation of a shell is quantified from the configuration of a *reference surface*, as shown in Figure 14.1. The figure also shows the local coordinate system of a flat shell, as well as various forces that can possibly act on the shell. The reference surface lies in the local xy-plane, while the direction of the shell's thickness is along the local z-axis.

As shown in Figure 14.1, a shell resists forces that lie in the plane of the reference surface by means of its *membrane resistance*, and forces perpendicular to the reference surface by means of its *flexural (plate) resistance*. A shell that is only subjected to perpendicular loading and only develops flexural resistance is also called a *plate*.

Many textbooks are dedicated *to plate theory and analysis, in which the membrane resistance is always neglected*, but this text will not separately examine plates. The major reason is that the concept of a plate is merely a mathematical convenience (if a surface structure has flexural stiffness and resistance, it must also have membrane resistance!). We can always analyze plate problems as a special case of shell problems where there is no membrane loading. We will begin our discussion with the simplest case of a *planar shell*, also called a *flat* shell, for which the shell is a flat planar body.

There are two theories regarding the kinematics of a shell's flexural deformation. The first one is the *Reissner-Mindlin* shell theory, which can be thought of as the extension of Timoshenko beam theory to shells. The basic assumptions of the Reissner-Mindlin theory are the following:

1) Plane sections normal to the undeformed *mid-surface* remain plane, but not normal to the mid-surface, in the deformed configuration.
2) The normal stress along the direction of the thickness is zero, $\sigma_{zz} = 0$.
3) The effect of the change of thickness to the displacements of the shell is neglected, which simply means that when we establish the kinematics of the shell, we will be assuming that the displacement along the local z-axis will not be varying along the thickness: $u_z(x, y, z) \approx u_{oz}(x, y)$.

Fundamentals of Finite Element Analysis: Linear Finite Element Analysis, First Edition.
Ioannis Koutromanos, James McClure, and Christopher Roy.

Companion website: www.wiley.com/go/koutromanos/linear

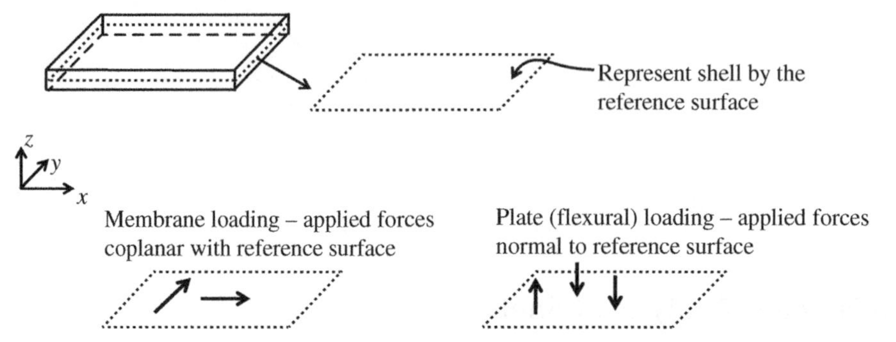

Figure 14.1 Shell reference surface and types of loading.

4) The through-thickness shear stresses of the shell, σ_{yz} and σ_{zx}, are assumed to be constant over the thickness (similar to the assumption for Timoshenko beams in Section 13.8).

The stress components (with example distributions along the thickness d) of a Reissner-Mindlin flat shell are shown in Figure 14.2. The example stress distributions are those corresponding to linear elasticity and homogeneous material throughout the entire thickness of the shell.

The second popular shell theory is the *Kirchhoff-Love* (K-L) theory, which is an extension of the Euler-Bernoulli beam theory (i.e., we do not have through-thickness shear deformations). This theory is *valid for thin shells,* that is, shells wherein the thickness is much smaller than the in-plan dimensions. The finite element analysis of K-L shells requires multidimensional shape functions that satisfy C^1 continuity. It turns out that it is very hard to obtain such shape functions in multiple dimensions. A recently developed method, called *Isogeometric Analysis* (Cottrell et al. 2007) relies on the use of B-splines for shape functions and can resolve this difficulty. If we are to use the standard finite element method for K-L shells, we need to resort to specifically designed *hybrid*

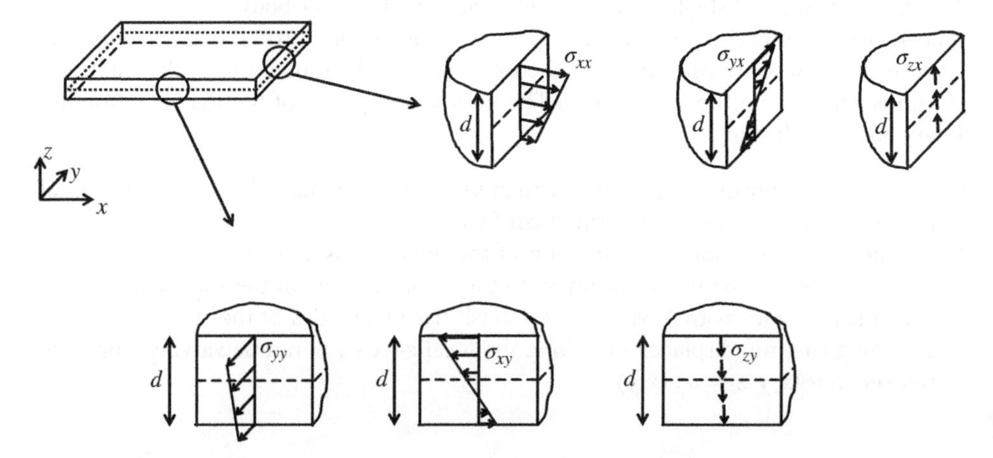

Figure 14.2 Planar Reissner-Mindlin Shell and stress components acting on two of the shell's faces.

finite elements,[1] using the so-called *discrete Kirchhoff quadrilateral (DKQ) or discrete Kirchhoff triangular (DKT)* element formulation. In a discrete Kirchhoff element, the condition for zero through-thickness shear strain is satisfied at a set of discrete points (not over the entire domain). It turns out that this approach leads to convergent finite element solutions for thin shells. The present Chapter is primarily focused on Reissner-Mindlin shell theory. A description of the DKQ element formulation is provided in Section 14.9.

Let us now continue with some further kinematic considerations for shells. The kinematic quantities of the mid-surface are shown in Figure 14.3. We have five quantities, $u_{ox}(x, y)$, $u_{oy}(x, y)$, $u_{oz}(x, y)$, $\theta_1(x, y)$, and $\theta_2(x, y)$, drawn in Figure 14.3 with the directions that will be assumed as positive. The rotation field θ_1 is drawn in the negative y-direction, because we want the deformed shape of the shell for each of the two elevation views, 1-1 and 2-2 in Figure 14.3, to have exactly the same form of deformations as that of two-dimensional Timoshenko beams aligned with the x-axis (elevation 1-1) or y-axis (elevation 2-2).

In the remainder of this chapter, we will be assuming that the reference surface coincides with the *mid-surface* of the shell—that is, a surface that passes through the middle of the shell's thickness at every point. This assumption means that the following mathematical condition will apply.

$$\int_{-d/2}^{d/2} z dz = \left[\frac{z^2}{2}\right]_{-d/2}^{d/2} = \frac{1}{2}\left(\frac{d}{2}\right)^2 - \frac{1}{2}\left(-\frac{d}{2}\right)^2 = 0 \tag{14.1.1}$$

where d is the thickness of the shell.

The three components of the displacement field, $u_x(x, y, z)$, $u_y(x, y, z)$, and $u_z(x, y, z)$, at any location (x, y, z) in a shell, can be determined from the five kinematic quantities of the mid-surface as follows.

$$u_x(x,y,z) = u_{ox}(x,y) - z \cdot \theta_1(x,y) \tag{14.1.2a}$$

$$u_y(x,y,z) = u_{oy}(x,y) - z \cdot \theta_2(x,y) \tag{14.1.2b}$$

$$u_z(x,y,z) \approx u_{oz}(x,y) \tag{14.1.2c}$$

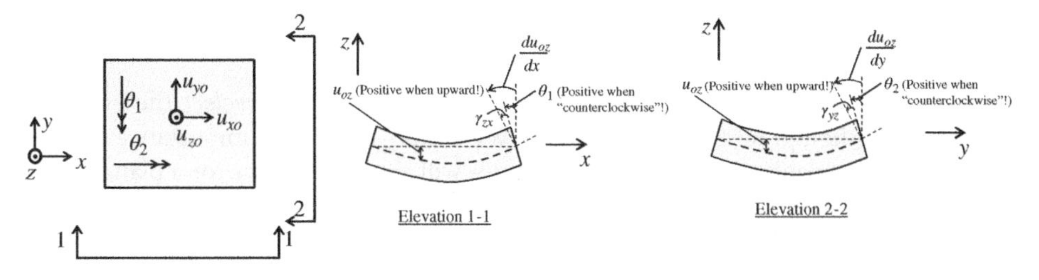

Figure 14.3 Plan view of shell mid-surface, elevation views of deformed shell and definition of kinematic quantities of the mid-surface.

---

1 In a hybrid finite element, we approximate a specific field in the interior of the element and a different field along the boundary of the element (for example, we may have approximation for the stresses in the interior of an element and for the displacements along the boundary of the element).

We can also define the strain components, through appropriate partial differentiation (as explained in Section 7.2) of the displacement field defined by Equations (14.1.2a–c):

$$\varepsilon_{xx} = \frac{\partial u_x}{\partial x} = \frac{\partial u_{ox}}{\partial x} - z \cdot \frac{\partial \theta_1}{\partial x} = \varepsilon_{ox} - z \cdot \varphi_{11} \tag{14.1.3a}$$

$$\varepsilon_{yy} = \frac{\partial u_y}{\partial y} = \frac{\partial u_{oy}}{\partial y} - z \cdot \frac{\partial \theta_2}{\partial y} = \varepsilon_{oy} - z \cdot \varphi_{22} \tag{14.1.3b}$$

$$\gamma_{xy} = \frac{\partial u_x}{\partial y} + \frac{\partial u_y}{\partial x} = \frac{\partial u_{ox}}{\partial y} + \frac{\partial u_{oy}}{\partial x} - z \cdot \frac{\partial \theta_1}{\partial y} - z \cdot \frac{\partial \theta_2}{\partial x} = \gamma_{oxy} - z \cdot 2\varphi_{12} \tag{14.1.3c}$$

$$\varepsilon_{zz} = \frac{\partial u_z}{\partial z} \approx 0 \tag{14.1.3d}$$

$$\gamma_{yz} = \frac{\partial u_z}{\partial y} + \frac{\partial u_y}{\partial z} = \frac{\partial u_{oz}}{\partial y} - \theta_2 \tag{14.1.3e}$$

$$\gamma_{zx} = \frac{\partial u_z}{\partial y} + \frac{\partial u_x}{\partial z} = \frac{\partial u_{oz}}{\partial x} - \theta_1 \tag{14.1.3f}$$

In Equations (14.13a–f), we have defined:

$$\varepsilon_{ox} = \frac{\partial u_{ox}}{\partial x} \tag{14.1.4a}$$

$$\varepsilon_{oy} = \frac{\partial u_{oy}}{\partial y} \tag{14.1.4b}$$

$$\gamma_{oxy} = \frac{\partial u_{ox}}{\partial y} + \frac{\partial u_{oy}}{\partial x} \tag{14.1.4c}$$

$$\varphi_{11} = \frac{\partial \theta_1}{\partial x} \tag{14.1.4d}$$

$$\varphi_{22} = \frac{\partial \theta_2}{\partial y} \tag{14.1.4e}$$

$$\varphi_{12} = \frac{1}{2}\left(\frac{\partial \theta_2}{\partial x} + \frac{\partial \theta_1}{\partial y}\right) \tag{14.1.4f}$$

The quantities defined by Equations (14.1.4a–f) are collectively termed the *generalized strains of a shell.*

Remark 14.1.1: Equations (14.1.2c) and (14.1.3d) imply that we *neglect* the effect of the strain ε_{zz}, although it is not exactly zero. We simply neglect its importance for the kinematics of the problem. The condition that we will actually enforce for a planar (flat) shell is that the stress σ_{zz} will be zero, where z is the axis perpendicular to the planar mid-surface of the shell. ∎

Remark 14.1.2: The quantities ε_{ox}, ε_{oy}, and γ_{oxy} are termed the *membrane strain fields* of the shell, and they do quantify how much "in-plane stretching and distortion" an initially flat shell incurs due to the deformation. We can also define the symmetric, second-order *membrane strain tensor,* $[\varepsilon_0] = \begin{bmatrix} \varepsilon_{ox} & \varepsilon_{oxy} \\ \varepsilon_{oyx} & \varepsilon_{oy} \end{bmatrix}$, where $\varepsilon_{oij} = \frac{1}{2}\left(\frac{\partial u_{oi}}{\partial x_j} + \frac{\partial u_{oj}}{\partial x_i}\right)$,

$i, j = 1, 2$. This tensor follows the same exact transformation equations as the two-dimensional strain tensor in Section 7.7. ∎

Remark 14.1.3: The quantities φ_{11}, φ_{22}, and φ_{12} are termed the *curvature fields* of the shell, and they do quantify how much curved an initially flat shell becomes after the deformation. We can define a symmetric, second-order *curvature tensor*, $[\varphi] = \begin{bmatrix} \varphi_{11} & \varphi_{12} \\ \varphi_{21} & \varphi_{22} \end{bmatrix}$, where $\varphi_{ij} = \frac{1}{2}\left(\frac{\partial \theta_i}{\partial x_j} + \frac{\partial \theta_j}{\partial x_i}\right)$. This tensor also follows the same exact transformation equations as the two-dimensional strain tensor. ∎

14.2 Stress Resultants for Shells

For shells, it is convenient to express the effect of stresses in terms of *stress resultants*, which can be seen as a generalization of the internal loadings in a beam. Resultants are obtained after appropriate integration of stress terms over the thickness of the shell. Specifically, we can establish three components of *shell membrane forces*:

$$\hat{n}_{xx} = \int_{-d/2}^{d/2} \sigma_{xx} dz \tag{14.2.1a}$$

$$\hat{n}_{xy} = \int_{-d/2}^{d/2} \sigma_{xy} dz \tag{14.2.1b}$$

$$\hat{n}_{yy} = \int_{-d/2}^{d/2} \sigma_{yy} dz \tag{14.2.1c}$$

We can also define three components of *shell bending moments*:

$$\hat{m}_{xx} = -\int_{-d/2}^{d/2} \sigma_{xx} \cdot z dz \tag{14.2.2a}$$

$$\hat{m}_{yy} = -\int_{-d/2}^{d/2} \sigma_{yy} \cdot z dz \tag{14.2.2b}$$

$$\hat{m}_{xy} = -\int_{-d/2}^{d/2} \sigma_{xy} \cdot z dz \tag{14.2.2c}$$

Remark 14.2.1: The membrane forces have units of "force per unit width" of a shell section. Similarly, the shell moments have units of "moment per unit width" of a shell section. ▮

Remark 14.2.2: We can define a symmetric, second-order *membrane force tensor*, $[\hat{n}] = \begin{bmatrix} \hat{n}_{xx} & \hat{n}_{xy} \\ \hat{n}_{yx} = \hat{n}_{xy} & \hat{n}_{yy} \end{bmatrix}$ and a symmetric, second-order *moment tensor*, $[\hat{m}] = \begin{bmatrix} \hat{m}_{xx} & \hat{m}_{xy} \\ \hat{m}_{yx} = \hat{m}_{xy} & \hat{m}_{yy} \end{bmatrix}$. ▮

We can also establish the *shell shear forces per unit width*, $\hat{q}_x$ and $\hat{q}_y$, given by through-thickness integration of the contributions of the shear stresses σ_{xz} and σ_{yz}:

$$\hat{q}_x = \int_{-d/2}^{d/2} \sigma_{xy} dz = \sigma_{xz} \cdot d \tag{14.2.3a}$$

$$\hat{q}_y = \int_{-d/2}^{d/2} \sigma_{yz} dz = \sigma_{yz} \cdot d \tag{14.2.3b}$$

The representation of a segment of the mid-surface of the shell with the corresponding membrane force and bending moment/shear force stress resultants is shown in Figure 14.4. In accordance with the introductory descriptions of Section 14.1, the resistance of a shell to membrane forces is termed *membrane action*, while the bending resistance is also termed *plate action*.

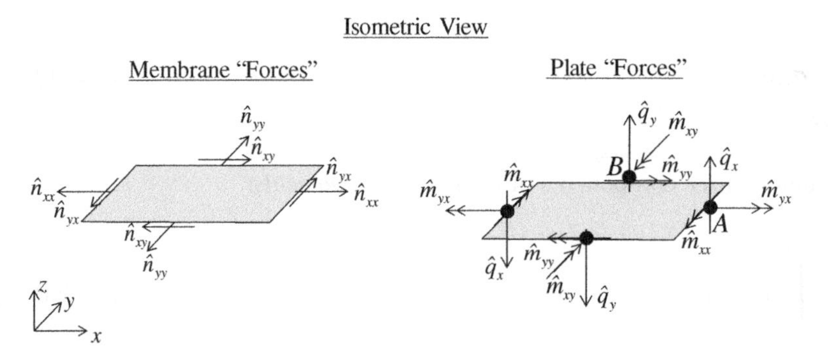

Figure 14.4 Stress resultants on a flat shell segment.

14.3 Differential Equations of Equilibrium and Boundary Conditions for Flat Shells

We will now obtain the strong form, that is, the differential equations and boundary conditions for a planar (flat) shell. *For planar shells, the governing equations for membrane action are independent of the corresponding expressions for plate action.* Let us first examine the membrane behavior of a shell. To this end, we isolate a small segment of a shell with planar dimensions (Δx) and (Δy), as shown in Figure 14.5, and examine its equilibrium. The figure includes membrane loads (forces per unit surface area) $p_x(x, y)$ and $p_y(x, y)$. Since the segment is small, the value of p_x and p_y is assumed to be constant.

We can write two equilibrium equations for the in-plane forces of the shell. One equation corresponds to equilibrium along the x-direction:

$$\sum F_x = 0 \rightarrow (n_{xx} + \Delta n_{xx}) \cdot \Delta y + \left(n_{xy} + \Delta n_{xy}\right) \cdot \Delta x - n_{xx} \cdot \Delta y - n_{xy} \cdot \Delta x + p_x \cdot \Delta x \cdot \Delta y = 0$$

$$\rightarrow \Delta n_{xx} \cdot \Delta y + \Delta n_{xy} \cdot \Delta x + p_x \cdot \Delta x \cdot \Delta y = 0 \rightarrow \frac{\Delta n_{xx}}{\Delta x} + \frac{\Delta n_{xy}}{\Delta y} + p_x = 0$$

If we take the limit as the size of the segment tends to zero, $\lim_{\substack{\Delta x \rightarrow 0 \\ \Delta x \rightarrow 0}}$, we obtain:

$$\frac{\partial n_{xx}}{\partial x} + \frac{\partial n_{xy}}{\partial y} + p_x = 0 \qquad (14.3.1a)$$

We now write a second equilibrium equation, corresponding to forces along the y-direction:

$$\sum F_y = 0 \rightarrow \left(n_{yx} + \Delta n_{yx}\right) \cdot \Delta y + \left(n_{yy} + \Delta n_{yy}\right) \cdot \Delta x - n_{yx} \cdot \Delta y - n_{yy} \cdot \Delta x + p_y \cdot \Delta x \cdot \Delta y = 0$$

$$\rightarrow \Delta n_{yx} \cdot \Delta y + \Delta n_{yy} \cdot \Delta x + p_y \cdot \Delta x \cdot \Delta y = 0 \rightarrow \frac{\Delta n_{yx}}{\Delta x} + \frac{\Delta n_{yy}}{\Delta y} + p_y = 0$$

If we take the limit as the size of the piece tends to zero, $\lim_{\substack{\Delta x \rightarrow 0 \\ \Delta x \rightarrow 0}}$:

$$\frac{\partial n_{yx}}{\partial x} + \frac{\partial n_{yy}}{\partial y} + p_y = 0 \qquad (14.3.1b)$$

Equations (14.3.1a,b) collectively comprise the system of partial differential equations of equilibrium for membrane (in-plane) loading of shells. These equations must be supplemented with essential and natural boundary conditions. The former correspond to prescribed values of u_{ox}, u_{oy} and are given by the expressions:

$$u_{ox} = \bar{u}_{ox} \; on \; \Gamma_{ux} \qquad (14.3.2a)$$

$$u_{oy} = \bar{u}_{oy} \; on \; \Gamma_{uy} \qquad (14.3.2b)$$

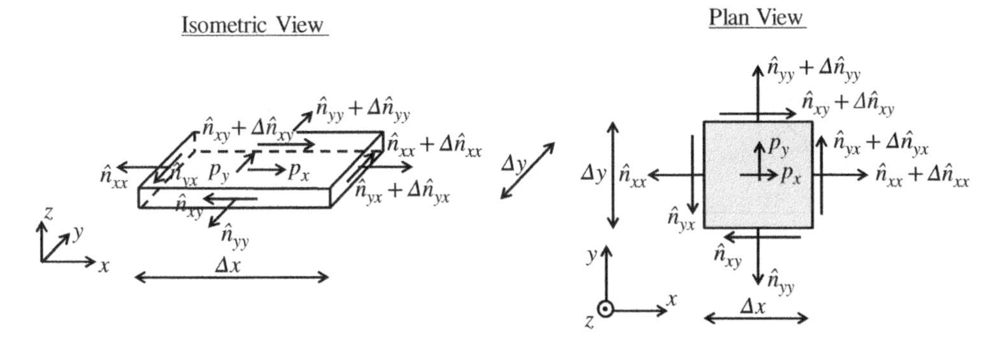

Figure 14.5 Membrane forces on shell segment.

The corresponding natural boundary conditions are written as follows.

$$\hat{n}_{xx} \cdot n_x + \hat{n}_{xy} \cdot n_y = \hat{f}_x \quad on \ \Gamma_{tx} \tag{14.3.3a}$$

$$\hat{n}_{yx} \cdot n_x + \hat{n}_{yy} \cdot n_y = \hat{f}_y \quad on \ \Gamma_{ty} \tag{14.3.3b}$$

where $\hat{f}_x$ and $\hat{f}_y$ are prescribed forces per unit length of the natural boundary and n_x, n_y are the components of the unit normal outward vector on the boundary of the shell.

Now let us continue with the differential equations and boundary conditions for plate (flexural) behavior of flat shells. To this end, we isolate a small segment of the shell with dimensions Δx, Δy, as shown in Figure 14.6, and examine the equilibrium for flexure. Figure 14.6 includes applied loads p_z (i.e., forces per unit surface area in the z-direction). We can establish three equilibrium equations, namely, force equilibrium along the z-axis and moment equilibrium about the x- and y- axes. Let us begin with the force equilibrium equation along the z-axis:

$$\sum F_z = 0 \rightarrow (\hat{q}_x + \Delta\hat{q}_x)\cdot\Delta y + \left(\hat{q}_y + \Delta\hat{q}_y\right)\cdot\Delta x - \hat{q}_x\cdot\Delta y - \hat{q}_y\cdot\Delta x + p_z\cdot\Delta x\cdot\Delta y = 0$$

$$\rightarrow \Delta\hat{q}_x\cdot\Delta y + \Delta\hat{q}_y\cdot\Delta x + p_z\cdot\Delta x\cdot\Delta y = 0 \rightarrow \frac{\Delta\hat{q}_x}{\Delta x} + \frac{\Delta\hat{q}_y}{\Delta y} + p_z = 0$$

If we take the limit as the size of the segment tends to zero, $\lim_{\substack{\Delta x \to 0 \\ \Delta y \to 0}}$, we obtain:

$$\frac{\partial\hat{q}_x}{\partial x} + \frac{\partial\hat{q}_y}{\partial y} + p_z = 0 \tag{14.3.4a}$$

We continue with the equilibrium of moments about the y-axis, using point A in Figure 14.6 as the reference point:

$$\left(\sum M_y\right)^A = 0 \rightarrow (\hat{q}_x\cdot\Delta y)\cdot\Delta x + \hat{q}_y\cdot\frac{(\Delta x)^2}{2} - \left(\hat{q}_y + \Delta\hat{q}_y\right)\cdot\frac{(\Delta x)^2}{2} - p_z\cdot\Delta x\cdot\Delta y\cdot\frac{\Delta x}{2}$$

$$+ (\hat{m}_{xx} + \Delta\hat{m}_{xx})\cdot\Delta y - \hat{m}_{xx}\cdot\Delta y + \left(\hat{m}_{xy} + \Delta\hat{m}_{xy}\right)\cdot\Delta x - \hat{m}_{xy}\cdot\Delta x = 0$$

$$\rightarrow \hat{q}_x\cdot\Delta x\cdot\Delta y + \Delta\hat{q}_y\frac{(\Delta x)^2}{2} - p_z\cdot\Delta x\cdot\Delta y\cdot\frac{\Delta x}{2} + \hat{m}_{xx}\cdot\Delta y + \Delta\hat{m}_{xy}\cdot\Delta x = 0$$

$$\rightarrow \hat{q}_x + \frac{\Delta\hat{q}_y\Delta x}{2\Delta y} - p_z\cdot\frac{\Delta x}{2} + \frac{\Delta\hat{m}_{xx}}{\Delta x} + \frac{\Delta\hat{m}_{xy}}{\Delta y} = 0$$

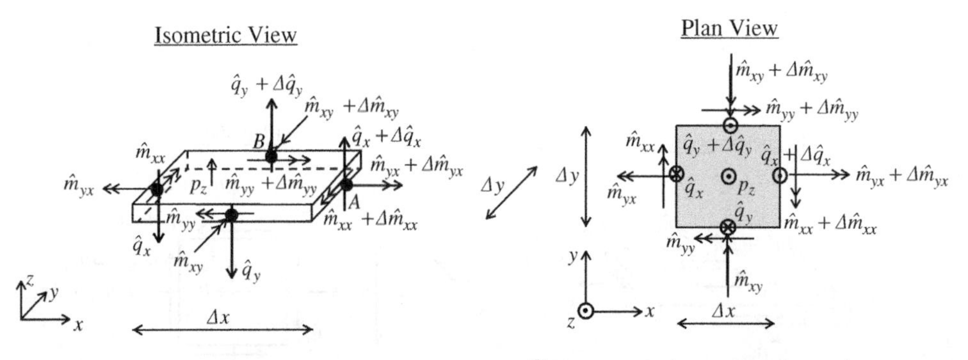

Figure 14.6 Plate bending forces on a shell segment.

If we take the limit as the size of the piece tends to zero, $\lim_{\substack{\Delta x \to 0 \\ \Delta y \to 0}}$, we have:

$$\frac{\partial \hat{m}_{xx}}{\partial x} + \frac{\partial \hat{m}_{xy}}{\partial y} + \hat{q}_x = 0$$

where we have accounted for the fact that $\lim_{\substack{\Delta x \to 0 \\ \Delta y \to 0}} \left(\frac{\Delta \hat{q}_y \Delta x}{2\Delta y} \right) = \lim_{\Delta x \to 0} \left(\frac{\partial \hat{q}_y}{\partial y} \frac{\Delta x}{2} \right) = 0$,

$\lim_{\substack{\Delta x \to 0 \\ \Delta y \to 0}} \left(-p_z \cdot \frac{\Delta x}{2} \right) = 0$

In several cases, we also have an applied loading in the form of distributed moment per unit area, $\hat{m}_1$, which is applied in the direction shown in Figure 14.7. We thus can modify the obtained expression and write the general differential equation of moment equilibrium about the x-axis, for the case where we also have a distributed moment $\hat{m}_1$:

$$\frac{\partial \hat{m}_{xx}}{\partial x} + \frac{\partial \hat{m}_{xy}}{\partial y} + \hat{q}_x + \hat{m}_1 = 0 \tag{14.3.4b}$$

We now write the equation of moment equilibrium about the x-axis, using point B in Figure 14.6 as the reference:

$$\left(\sum M_x\right)^B = 0 \to \left(\hat{q}_y \cdot \Delta x\right) \cdot \Delta y + \hat{q}_x \cdot \frac{(\Delta y)^2}{2} - (\hat{q}_x + \Delta \hat{q}_x) \cdot \frac{(\Delta y)^2}{2} - p_z \cdot \Delta x \cdot \Delta y \cdot \frac{\Delta y}{2}$$

$$+ \left(\hat{m}_{yy} + \Delta \hat{m}_{yy}\right) \cdot \Delta x - \hat{m}_{yy} \cdot \Delta x + \left(\hat{m}_{yx} + \Delta \hat{m}_{yx}\right) \cdot \Delta y - \hat{m}_{yy} \cdot \Delta y = 0$$

$$\to \hat{q}_y \cdot \Delta x \cdot \Delta y + \Delta \hat{q}_x \frac{(\Delta y)^2}{2} - p_z \cdot \Delta x \cdot \Delta y \cdot \frac{\Delta y}{2} + \hat{m}_{yy} \cdot \Delta x + \Delta \hat{m}_{yx} \cdot \Delta y = 0$$

$$\to \hat{q}_y + \frac{\Delta \hat{q}_x \Delta y}{2\Delta x_x} - p_z \cdot \frac{\Delta y}{2} + \frac{\Delta \hat{m}_{yy}}{\Delta y} + \frac{\Delta \hat{m}_{yx}}{\Delta x} = 0$$

If we take the limit as the size of the piece tends to zero, $\lim_{\substack{\Delta y \to 0 \\ \Delta x \to 0}}$,

we obtain: $\frac{\partial \hat{m}_{yx}}{\partial x} + \frac{\partial \hat{m}_{yy}}{\partial y} + \hat{q}_y = 0$

where we have accounted for the fact that

$\lim_{\substack{\Delta y \to 0 \\ \Delta x \to 0}} \left(\frac{\Delta \hat{q}_x \Delta y}{2\Delta x} \right) = \lim_{\Delta y \to 0} \left(\frac{\partial \hat{q}_x}{\partial x} \frac{\Delta y}{2} \right) = 0$, $\lim_{\substack{\Delta y \to 0 \\ \Delta x \to 0}} \left(-p_z \cdot \frac{\Delta y}{2} \right) = 0$.

We will once again write the differential equation for the most general case, where we also have a distributed moment per unit area, $\hat{m}_2$, as shown in Figure 14.7:

$$\frac{\partial \hat{m}_{yx}}{\partial x} + \frac{\partial \hat{m}_{yy}}{\partial y} + \hat{q}_y + \hat{m}_2 = 0 \tag{14.3.4c}$$

In summary, the flexural (bending or plate) response of a flat shell is described by the system of differential equations (14.3.4a–c).

Plan View

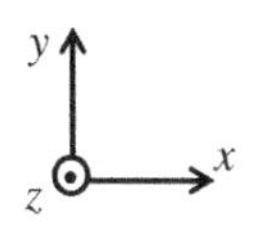

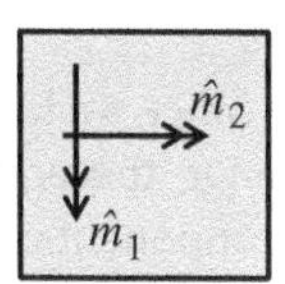

Figure 14.7 Applied distributed moments (per unit area) in a shell.

The essential boundary conditions for flexural (plate) action can be written as follows.

$$u_{oz} = \bar{u}_{oz} \quad on \quad \Gamma_{uz} \tag{14.3.5a}$$

$$\theta_1 = \bar{\theta}_1 \quad on \quad \Gamma_{u\theta 1} \tag{14.3.5b}$$

$$\theta_2 = \bar{\theta}_2 \quad on \quad \Gamma_{u\theta 2} \tag{14.3.5c}$$

The natural boundary conditions for plate behavior of a planar shell are:

$$\hat{q}_x \cdot n_x + \hat{q}_y \cdot n_y = \bar{\hat{q}} \; on \; \Gamma_{tz} \tag{14.3.6a}$$

$$\hat{m}_{xx} \cdot n_x + \hat{m}_{xy} \cdot n_y = \bar{\hat{m}}_x \quad on \quad \Gamma_{t\theta 1} \tag{14.3.6b}$$

$$\hat{m}_{yx} \cdot n_x + \hat{m}_{yy} \cdot n_y = \bar{\hat{m}}_y \quad on \quad \Gamma_{t\theta 2} \tag{14.3.6c}$$

Remark 14.3.1: The differential equations and boundary conditions that we obtained in this Section are not based on an assumption regarding the material behavior, thus they are valid for all types of material! ■

Remark 14.3.2: A special note is necessary for the natural boundary conditions. In real life, we most frequently have prescribed tractions normal and tangential to a boundary segment, which leads to corresponding directions of the membrane forces and moments in the natural boundary. ■

14.4 Constitutive Law for Linear Elasticity in Terms of Stress Resultants and Generalized Strains

As mentioned in Section 14.1, a basic assumption made for the theory of flat (planar) shells is that $\sigma_{zz} = 0$. This condition is referred to as the *quasi-plane-stress assumption* and allows us to establish a reduced constitutive (stress-strain) law for linearly elastic materials in shells. Let us revisit the generic stress-strain law for linear elasticity, provided in Section 7.5.

$$\{\sigma\} = [D]\{\varepsilon\} \tag{14.4.1}$$

where $\{\sigma\} = \begin{bmatrix}\sigma_{xx} & \sigma_{yy} & \sigma_{zz} & \sigma_{xy} & \sigma_{yz} & \sigma_{zx}\end{bmatrix}^T$, $\{\varepsilon\} = \begin{bmatrix}\varepsilon_{xx} & \varepsilon_{yy} & \varepsilon_{zz} & 2\varepsilon_{xy} & 2\varepsilon_{yz} & 2\varepsilon_{zx}\end{bmatrix}^T$ and $[D]$ is the (6×6) material stiffness matrix.

To account for the quasi-plane stress assumption, we will first expand the third row of Equation (14.4.1), which yields the *zz*-stress component:

$$\sigma_{zz} = D_{31}\varepsilon_{xx} + D_{32}\varepsilon_{yy} + D_{33}\varepsilon_{zz} + D_{34} \cdot 2\varepsilon_{xy} + D_{35} \cdot 2\varepsilon_{yz} + D_{36} \cdot 2\varepsilon_{zx} \tag{14.4.2}$$

Since we know that the left-hand side of Equation (14.4.2) is zero, $\sigma_{zz} = 0$, we can rearrange the equation to express the strain component ε_{zz} in terms of the other strain components:

$$D_{31}\varepsilon_{xx} + D_{32}\varepsilon_{yy} + D_{33}\varepsilon_{zz} + D_{34} \cdot 2\varepsilon_{xy} + D_{35} \cdot 2\varepsilon_{yz} + D_{36} \cdot 2\varepsilon_{zx} = 0$$

$$\rightarrow \varepsilon_{zz} = -\frac{1}{D_{33}}\left(D_{31}\varepsilon_{xx} + D_{32}\varepsilon_{yy} + D_{34} \cdot 2\varepsilon_{xy} + D_{35} \cdot 2\varepsilon_{yz} + D_{36} \cdot 2\varepsilon_{zx}\right) \tag{14.4.3}$$

We will now define a *reduced stress vector*, $\{\tilde{\sigma}\}$:

$$\{\tilde{\sigma}\} = \left[\sigma_{xx} \;\; \sigma_{yy} \;\; \sigma_{zz} \;\; \sigma_{xy} \;\; \sigma_{yz} \;\; \sigma_{zx}\right]^T \tag{14.4.4a}$$

which includes the stress components except for $\sigma_{zz} = 0$. We also define the corresponding *reduced strain vector*,

$$\{\tilde{\varepsilon}\} = \left[\varepsilon_{xx} \;\; \varepsilon_{yy} \;\; 2\varepsilon_{xy} \;\; 2\varepsilon_{yz} \;\; 2\varepsilon_{zx}\right]^T \tag{14.4.4b}$$

We will establish the constitutive law for a linearly elastic material in a shell, in terms of the reduced vectors $\{\tilde{\sigma}\}$ and $\{\tilde{\varepsilon}\}$.

First of all, we note that we can isolate the rows of the matrix Equation (14.4.1) corresponding to the components of $\{\tilde{\sigma}\}$:

$$\{\sigma\} = \begin{Bmatrix} \sigma_{xx} \\ \sigma_{yy} \\ \sigma_{xy} \\ \sigma_{yz} \\ \sigma_{zx} \end{Bmatrix} = \begin{bmatrix} D_{11} & D_{12} & D_{13} & D_{14} & D_{15} & D_{16} \\ D_{21} & D_{22} & D_{23} & D_{24} & D_{25} & D_{26} \\ D_{41} & D_{42} & D_{43} & D_{44} & D_{45} & D_{46} \\ D_{51} & D_{52} & D_{53} & D_{54} & D_{55} & D_{56} \\ D_{61} & D_{62} & D_{63} & D_{64} & D_{65} & D_{66} \end{bmatrix} \begin{Bmatrix} \varepsilon_{xx} \\ \varepsilon_{yy} \\ \varepsilon_{zz} \\ 2\varepsilon_{xy} \\ 2\varepsilon_{yz} \\ 2\varepsilon_{zx} \end{Bmatrix} = [D_{red}]\{\varepsilon\} \tag{14.4.5}$$

where $[D_{red}]$ is a *reduced* version of the material stiffness matrix $[D]$, obtained after removing the third row of that matrix (the specific row corresponds to Equation 14.4.2):

$$[D_{red}] = \begin{bmatrix} D_{11} & D_{12} & D_{13} & D_{14} & D_{15} & D_{16} \\ D_{21} & D_{22} & D_{23} & D_{24} & D_{25} & D_{26} \\ D_{41} & D_{42} & D_{43} & D_{44} & D_{45} & D_{46} \\ D_{51} & D_{52} & D_{53} & D_{54} & D_{55} & D_{56} \\ D_{61} & D_{62} & D_{63} & D_{64} & D_{65} & D_{66} \end{bmatrix} \tag{14.4.6}$$

Given Equation (14.4.3), we can write a *matrix transformation relation between* $\{\varepsilon\}$ *and* $\{\tilde{\varepsilon}\}$:

$$\{\varepsilon\} = \begin{Bmatrix} \varepsilon_{xx} \\ \varepsilon_{yy} \\ \varepsilon_{zz} \\ 2\varepsilon_{xy} \\ 2\varepsilon_{yz} \\ 2\varepsilon_{zx} \end{Bmatrix} = \begin{bmatrix} 1 & 0 & 0 & 0 & 0 \\ 0 & 1 & 0 & 0 & 0 \\ -\frac{D_{31}}{D_{33}} & -\frac{D_{32}}{D_{33}} & -\frac{D_{34}}{D_{33}} & -\frac{D_{35}}{D_{33}} & -\frac{D_{36}}{D_{33}} \\ 0 & 0 & 1 & 0 & 0 \\ 0 & 0 & 0 & 1 & 0 \\ 0 & 0 & 0 & 0 & 1 \end{bmatrix} \begin{Bmatrix} \varepsilon_{xx} \\ \varepsilon_{yy} \\ 2\varepsilon_{xy} \\ 2\varepsilon_{yz} \\ 2\varepsilon_{zx} \end{Bmatrix} = \left[T_{pl-s}\right]\{\tilde{\varepsilon}\} \tag{14.4.7}$$

where

$$\left[T_{pl-s}\right] = \begin{bmatrix} 1 & 0 & 0 & 0 & 0 \\ 0 & 1 & 0 & 0 & 0 \\ -\frac{D_{31}}{D_{33}} & -\frac{D_{32}}{D_{33}} & -\frac{D_{34}}{D_{33}} & -\frac{D_{35}}{D_{33}} & -\frac{D_{36}}{D_{33}} \\ 0 & 0 & 1 & 0 & 0 \\ 0 & 0 & 0 & 1 & 0 \\ 0 & 0 & 0 & 0 & 1 \end{bmatrix} \quad (14.4.8)$$

is a (6 × 5) transformation matrix giving the full strain vector $\{\varepsilon\}$ in terms of the reduced strain vector $\{\tilde{\varepsilon}\}$.

We can now plug Equation (14.4.7) into Equation (14.4.5) to finally obtain:

$$\{\tilde{\sigma}\} = [D_{red}]\left[T_{pl-s}\right]\{\tilde{\varepsilon}\} \rightarrow \{\tilde{\sigma}\} = \left[\tilde{D}\right]\{\tilde{\varepsilon}\} \quad (14.4.9)$$

where $\left[\tilde{D}\right] = [D_{red}]\left[T_{pl-s}\right]$ is a reduced (*condensed*) material stiffness matrix, expressing the relation between the stress components of $\{\tilde{\sigma}\}$ (i.e., the stress components that can attain nonzero values) and the corresponding strain components of $\{\tilde{\varepsilon}\}$. We can directly write an equation providing each of the components of $\left[\tilde{D}\right]$ in terms of the corresponding components of $[D]$:

$$\tilde{D}_{ij} = D_{ij} - D_{i3}\frac{D_{3j}}{D_{33}} \quad (14.4.10)$$

The remainder of this section will consider the simplest possible case of an isotropic, linearly elastic material, in which the reduced constitutive law only includes two elastic constants, namely, the modulus of elasticity E and Poisson's ratio ν. In this case, $[D]$ is given by Equation (7.5.10b), and one can verify that Equation (14.4.9) attains the form:

$$\begin{Bmatrix} \sigma_{xx} \\ \sigma_{yy} \\ \sigma_{xy} \\ \sigma_{yz} \\ \sigma_{zx} \end{Bmatrix} = \frac{E}{1-\nu^2}\begin{bmatrix} 1 & \nu & 0 & 0 & 0 \\ \nu & 1 & 0 & 0 & 0 \\ 0 & 0 & \frac{1-\nu}{2} & 0 & 0 \\ 0 & 0 & 0 & \frac{1-\nu}{2} & 0 \\ 0 & 0 & 0 & 0 & \frac{1-\nu}{2} \end{bmatrix}\begin{Bmatrix} \varepsilon_{xx} \\ \varepsilon_{yy} \\ \gamma_{xy} \\ \gamma_{yz} \\ \gamma_{zx} \end{Bmatrix} \quad (14.4.11)$$

Now, since the stress resultants can be obtained by through-thickness integration of appropriate stress components, and since the latter can be obtained from the strain components (through Equation 14.4.11), we can establish a relation between the stress resultants and the generalized strains defined in Equations (14.1.4a–f). In our derivation, we will assume that the elastic constants (E and ν) can vary over the shell surface, that is, they are functions of x and y, but they do not change over the thickness at a given location of the shell; thus, E and ν are independent of z!

We will first work with the bending moment $\hat{m}_{xx}$. From Equation (14.2.2a), we had:

$$\hat{m}_{xx} = -\int_{-d/2}^{d/2} \sigma_{xx} \cdot z dz.$$

We use Equation (14.4.11), to express the stress σ_{xx} in terms of the strains:

$$\sigma_{xx} = \frac{E}{1-\nu^2}\varepsilon_{xx} + \frac{E\cdot\nu}{1-\nu^2}\varepsilon_{yy} \tag{14.4.12}$$

We can then plug Equations (14.1.3a–b), giving the strains ε_{xx} and ε_{yy} in terms of the generalized strains, into Equation (14.4.12):

$$\sigma_{xx} = \frac{E}{1-\nu^2}(\varepsilon_{ox} - \varphi_{11}\cdot z) + \frac{E\cdot\nu}{1-\nu^2}(\varepsilon_{oy} - \varphi_{22}\cdot z) \tag{14.4.13}$$

If we plug Equation (14.4.13) into Equation (14.2.2a), we eventually obtain:

$$\hat{m}_{xx} = -\frac{E}{1-\nu^2}\int_{-d/2}^{d/2}\varepsilon_{ox}\cdot zdz - \frac{E}{1-\nu^2}\int_{-d/2}^{d/2}(-\varphi_{11}\cdot z)\cdot zdz - \frac{E\cdot\nu}{1-\nu^2}\int_{-d/2}^{d/2}\varepsilon_{oy}\cdot zdz - \frac{E\cdot\nu}{1-\nu^2}\int_{-d/2}^{d/2}(-\varphi_{22}\cdot z)\cdot zdz$$

$$= -\frac{E}{1-\nu^2}\cdot\varepsilon_{ox}\int_{-d/2}^{d/2} zdz + \frac{E}{1-\nu^2}\cdot\varphi_{11}\int_{-d/2}^{d/2} z^2dz - \frac{E\cdot\nu}{1-\nu^2}\cdot\varepsilon_{oy}\int_{-d/2}^{d/2} zdz + \frac{E\cdot\nu}{1-\nu^2}\cdot\varphi_{22}\int_{-d/2}^{d/2} z^2dz$$

Finally, if we account for Equation (14.1.1), that is, $\int_{-d/2}^{d/2} zdz = 0$, we have:

$$\hat{m}_{xx} = \frac{E}{1-\nu^2}\cdot\frac{d^3}{12}\varphi_{11} + \frac{E\cdot\nu}{1-\nu^2}\cdot\frac{d^3}{12}\varphi_{22} \tag{14.4.14}$$

Along the same lines, the moment $\hat{m}_{yy}$ is obtained from Equation (14.2.2b), $\hat{m}_{yy} = -\int_{-d/2}^{d/2}\sigma_{yy}\cdot zdz$, where the stress σ_{yy} can be expressed in terms of the generalized strains, by means of Equations (14.4.11) and (14.1.3a–b):

$$\sigma_{yy} = \frac{E\cdot\nu}{1-\nu^2}\varepsilon_{xx} + \frac{E}{1-\nu^2}\varepsilon_{yy} = \frac{E\cdot\nu}{1-\nu^2}(\varepsilon_{ox} - \varphi_{11}\cdot z) + \frac{E}{1-\nu^2}(\varepsilon_{oy} - \varphi_{22}\cdot z) \tag{14.4.15}$$

Plugging Equation (14.4.15) into Equation (14.2.2b) yields:

$$\hat{m}_{yy} = -\frac{E\cdot\nu}{1-\nu^2}\int_{-d/2}^{d/2}\varepsilon_{ox}\cdot zdz - \frac{E\cdot\nu}{1-\nu^2}\int_{-d/2}^{d/2}(-\varphi_{11}\cdot z)\cdot zdz$$

$$-\frac{E}{1-\nu^2}\int_{-d/2}^{d/2}\varepsilon_{oy}\cdot zdz - \frac{E}{1-\nu^2}\int_{-d/2}^{d/2}(-\varphi_{22}\cdot z)\cdot zdz$$

$$= -\frac{E\cdot\nu}{1-\nu^2}\cdot\varepsilon_{ox}\int_{-d/2}^{d/2} zdz + \frac{E\cdot\nu}{1-\nu^2}\cdot\varphi_{11}\int_{-d/2}^{d/2} z^2dz - \frac{E}{1-\nu^2}\cdot\varepsilon_{oy}\int_{-d/2}^{d/2} zdz + \frac{E}{1-\nu^2}\cdot\varphi_{22}\int_{-d/2}^{d/2} z^2dz$$

and if we account for Equation (14.1.1), we have:

$$\hat{m}_{yy} = \frac{E \cdot \nu}{1-\nu^2} \cdot \frac{d^3}{12} \varphi_{11} + \frac{E}{1-\nu^2} \cdot \frac{d^3}{12} \varphi_{22} \tag{14.4.16}$$

We continue with the expression for the moment $\hat{m}_{xy}$; we use Equations (14.2.2c), (14.4.11), and (14.1.3c), and also set $G = \frac{E}{2(1+\nu)}$, to obtain:

$$\hat{m}_{xy} = -\int_{-d/2}^{d/2} \sigma_{xy} \cdot z dz = -\int_{-d/2}^{d/2} G \cdot \gamma_{xy} \cdot z dz = -\int_{-d/2}^{d/2} G\left(\gamma_{oxy} - 2\varphi_{12} \cdot z\right) \cdot z dz$$

$$\hat{m}_{xy} = -G \int_{-d/2}^{d/2} \gamma_{oxy} \cdot z dz - G \int_{-d/2}^{d/2} (-2\varphi_{12} \cdot z) \cdot z dz = -G \cdot \gamma_{oxy} \int_{-d/2}^{d/2} z dz + G \cdot 2\varphi_{12} \int_{-d/2}^{d/2} z^2 dz$$

Accounting for Equation (14.1.1) yields:

$$\hat{m}_{xy} = 2G \cdot \frac{d^3}{12} \varphi_{12} \tag{14.4.17}$$

For the through-thickness shear force $\hat{q}_x$ we use Equations (14.2.3a), (14.4.11), and (14.1.3f):

$$\hat{q}_x = \int_{-d/2}^{d/2} \sigma_{zx} dz = \int_{-d/2}^{d/2} G \cdot \gamma_{zx} dz$$

We will finally write:

$$\hat{q}_x = \kappa \cdot G \cdot d \cdot \gamma_{zx} \tag{14.4.18}$$

where, just like for Timoshenko beam theory in Section 13.8, we have used a shear correction factor, κ, (equal to 5/6 for linear elasticity) to improve the accuracy of the results using the Reissner-Mindlin theory. The shear correction factor is necessary, because the simplifying assumption that the through-thickness shear stresses are constant over the thickness is not perfectly accurate.

Finally, for the shear force $\hat{q}_y$, we use Equations (14.2.3b), (14.4.11), and (14.1.3e):

$$\hat{q}_y = \int_{-d/2}^{d/2} \sigma_{yz} dz = \int_{-d/2}^{d/2} G \cdot \gamma_{yz} dz = G \int_{-d/2}^{d/2} \gamma_{yz} dz \rightarrow \hat{q}_y = \kappa \cdot G \cdot d \cdot \gamma_{yz} \tag{14.4.19}$$

Equations (14.4.14), (14.4.16–19), collectively define a *generalized stress-strain relation for bending (plate) behavior*:

$$\{\hat{\sigma}_b\} = \left[\hat{D}_b\right]\{\hat{\varepsilon}_b\} \tag{14.4.20}$$

where $\{\hat{\sigma}_b\}$ is the *generalized stress vector for bending (plate) behavior*:

$$\{\hat{\sigma}_b\} = \left[\hat{m}_{xx} \ \ \hat{m}_{yy} \ \ \hat{m}_{xy} \ \ \hat{q}_x \ \ \hat{q}_y\right]^T \tag{14.4.21}$$

$\{\hat{\varepsilon}_b\}$ is the corresponding *generalized strain vector for bending behavior*:

$$\{\hat{\varepsilon}_b\} = \left[\varphi_{11} \quad \varphi_{22} \quad 2\varphi_{12} \quad \gamma_{zx} \quad \gamma_{zy}\right]^T \tag{14.4.22}$$

and $\left[\hat{D}_b\right]$ is the *generalized stiffness matrix for bending behavior and for linearly elastic isotropic material*:

$$\left[\hat{D}_b\right] = \begin{bmatrix} \frac{E \cdot d^3}{12(1-\nu^2)} & \frac{\nu \cdot E \cdot d^3}{12(1-\nu^2)} & 0 & 0 & 0 \\ 0 & 0 & 0 & 0 & 0 \\ 0 & 0 & G \cdot \frac{d^3}{12} & 0 & 0 \\ 0 & 0 & 0 & \kappa \cdot G \cdot d & 0 \\ 0 & 0 & 0 & 0 & \kappa \cdot G \cdot d \end{bmatrix} \tag{14.4.23}$$

It is also possible to write Equation (14.4.20) in a *partitioned form*, separating the moment (flexural) and shear-force terms:

$$\begin{Bmatrix} \{\hat{\sigma}_f\} \\ \{\hat{\sigma}_s\} \end{Bmatrix} = \begin{Bmatrix} \left[\hat{D}_f\right]\{\varepsilon_f\} \\ \left[\hat{D}_s\right]\{\varepsilon_s\} \end{Bmatrix} = \begin{bmatrix} \left[\hat{D}_f\right] & [0] \\ [0] & \left[\hat{D}_s\right] \end{bmatrix} \begin{Bmatrix} \{\hat{\varepsilon}_f\} \\ \{\hat{\varepsilon}_s\} \end{Bmatrix} \tag{14.4.24}$$

where

$$\{\hat{\sigma}_f\} = \left[\hat{m}_{xx} \quad \hat{m}_{yy} \quad \hat{m}_{xy}\right]^T \tag{14.4.25}$$

$$\{\hat{\sigma}_s\} = \left[\hat{q}_x \quad \hat{q}_y\right]^T \tag{14.4.26}$$

$$\{\hat{\varepsilon}_f\} = \left[\varphi_{11} \quad \varphi_{22} \quad 2\varphi_{12}\right]^T \tag{14.4.27}$$

$$\{\hat{\varepsilon}_s\} = \left[\gamma_{zx} \quad \gamma_{zy}\right]^T \tag{14.4.28}$$

$$\left[\hat{D}_f\right] = \begin{bmatrix} \frac{E \cdot d^3}{12(1-\nu^2)} & \frac{\nu \cdot E \cdot d^3}{12(1-\nu^2)} & 0 \\ \frac{\nu \cdot E \cdot d^3}{12(1-\nu^2)} & \frac{E \cdot d^3}{12(1-\nu^2)} & 0 \\ 0 & 0 & G \cdot \frac{d^3}{12} \end{bmatrix} = \frac{E \cdot d^3}{12(1-\nu^2)} \begin{bmatrix} 1 & \nu & 0 \\ \nu & 1 & 0 \\ 0 & 0 & \frac{1-\nu}{2} \end{bmatrix} \tag{14.4.29}$$

and

$$\left[\hat{D}_s\right] = \begin{bmatrix} \kappa \cdot G \cdot d & 0 \\ 0 & \kappa \cdot G \cdot d \end{bmatrix} = \kappa \cdot G \cdot d \begin{bmatrix} 1 & 0 \\ 0 & 1 \end{bmatrix} \tag{14.4.30}$$

We will now establish a generalized stress-strain law for membrane (in-plane) behavior and linearly elastic, isotropic material. We begin with the membrane force component $\hat{n}_{xx}$. We use Equations (14.2.1a), (14.1.4a–b), and (14.4.11), to obtain:

$$\hat{n}_{xx} = \int_{-d/2}^{d/2} \sigma_{xx} dz = \int_{-d/2}^{d/2} \left[\frac{E}{1-\nu^2} \varepsilon_{xx} + \frac{E \cdot \nu}{1-\nu^2} \varepsilon_{yy} \right] dz$$

$$= \int_{-d/2}^{d/2} \left[\frac{E}{1-\nu^2} (\varepsilon_{ox} - \varphi_{11} \cdot z) + \frac{E \cdot \nu}{1-\nu^2} (\varepsilon_{oy} - \varphi_{22} \cdot z) \right] dz$$

$$= \frac{E}{1-\nu^2} \int_{-d/2}^{d/2} \varepsilon_{ox} dz + \frac{E}{1-\nu^2} \int_{-d/2}^{d/2} (-\varphi_{11}) \cdot z dz + \frac{E \cdot \nu}{1-\nu^2} \int_{-d/2}^{d/2} \varepsilon_{oy} dz + \frac{E \cdot \nu}{1-\nu^2} \int_{-d/2}^{d/2} (-\varphi_{22}) \cdot z dz$$

$$= \frac{E}{1-\nu^2} \cdot \varepsilon_{ox} \int_{-d/2}^{d/2} dz - \frac{E}{1-\nu^2} \cdot \varphi_{11} \int_{-d/2}^{d/2} z dz + \frac{E \cdot \nu}{1-\nu^2} \cdot \varepsilon_{oy} \int_{-d/2}^{d/2} dz - \frac{E \cdot \nu}{1-\nu^2} \cdot \varphi_{22} \int_{-d/2}^{d/2} z dz$$

If we account for Equation (14.1.1), we have:

$$\hat{n}_{xx} = \frac{E}{1-\nu^2} \cdot d \cdot \varepsilon_{ox} + \frac{E \cdot \nu}{1-\nu^2} \cdot d \cdot \varepsilon_{oy} \tag{14.4.31}$$

If we use Equations (14.2.1b), (14.1.4a–b), and (14.4.11), and also account for Equation (14.1.1), we obtain a similar equation for the membrane force $\hat{n}_{yy}$:

$$\hat{n}_{yy} = \frac{E \cdot \nu}{1-\nu^2} \cdot d \cdot \varepsilon_{ox} + \frac{E}{1-\nu^2} \cdot d \cdot \varepsilon_{oy} \tag{14.4.32}$$

Finally, we use Equations (14.2.1c), (14.1.4c), and (14.4.11), to establish an expression for the force $\hat{n}_{xy}$:

$$\hat{n}_{xy} = \int_{-d/2}^{d/2} \sigma_{xy} dz = \int_{-d/2}^{d/2} G \cdot \gamma_{xy} dy = \int_{-d/2}^{d/2} G\left(\gamma_{oxy} - 2\varphi_{12} \cdot z\right) dz$$

$$= G \int_{-d/2}^{d/2} \gamma_{oxy} dz + G \int_{-d/2}^{d/2} (-2\varphi_{12}) \cdot z dz = G \cdot \gamma_{oxy} \int_{-d/2}^{d/2} dz - G \cdot 2\varphi_{12} \int_{-d/2}^{d/2} z dz$$

and, if we account for Equation (14.1.1):

$$\hat{n}_{xy} = G \cdot d \cdot \gamma_{oxy} \tag{14.4.33}$$

We can now combine Equations (14.4.31), (14.4.32), and (14.4.33) into a unique matrix expression:

$$\{\hat{\sigma}_m\} = \left[\hat{D}_m\right] \{\hat{\varepsilon}_m\} \tag{14.4.34}$$

where

$$\{\hat{\varepsilon}_m\} = \begin{bmatrix}\varepsilon_{ox} & \varepsilon_{oy} & \varepsilon_{oxy}\end{bmatrix}^T \tag{14.4.35}$$

is the generalized membrane strain vector,

$$\{\hat{\sigma}_m\} = \begin{bmatrix}\hat{n}_{xx} & \hat{n}_{yy} & \hat{n}_{xy}\end{bmatrix}^T \tag{14.4.36}$$

is the generalized stress vector, and

$$\left[\hat{D}_m\right] = \frac{E \cdot d}{1-\nu^2}\begin{bmatrix}1 & \nu & 0 \\ \nu & 1 & 0 \\ 0 & 0 & \dfrac{1-\nu}{2}\end{bmatrix} \tag{14.4.37}$$

Remark 14.4.1: The constitutive law for membrane behavior is the same as the one we had obtained for two-dimensional plane-stress elasticity, except that the stiffness terms are multiplied by the thickness d! ∎

Remark 14.4.2: It is important to keep in mind that, if the material elastic constants vary over the thickness of the shell, that is, E and ν are functions of z, then there is *coupling* between the bending and membrane behavior. For example, the generalized membrane strains will affect the value of the generalized bending stresses, and vice versa. In that case, we generally cannot write separate constitutive laws for the membrane stress resultants and for the bending stress resultants. ∎

Now, we will briefly discuss the most general situation, where we have linearly elastic (but not necessarily isotropic) material. In this case, we can always write a constitutive relation of the form:

$$\{\hat{\sigma}\} = \left[\hat{D}\right]\{\hat{\varepsilon}\} \tag{14.4.38}$$

where $\{\hat{\varepsilon}\}$ is the generalized strain vector of the shell (containing all the generalized strains):

$$\{\hat{\varepsilon}\} = \begin{bmatrix}\varepsilon_{ox} & \varepsilon_{oy} & \gamma_{oxy} & \varphi_{11} & \varphi_{22} & 2\varphi_{12} & \gamma_{zx} & \gamma_{zy}\end{bmatrix}^T \tag{14.4.39}$$

and $\{\hat{\sigma}\}$ is the generalized stress vector of the shell (containing all the generalized stresses):

$$\{\hat{\sigma}\} = \begin{bmatrix}\hat{n}_{xx} & \hat{n}_{yy} & \hat{n}_{xy} & \hat{m}_{xx} & \hat{m}_{yy} & \hat{m}_{xy} & \hat{q}_x & \hat{q}_y\end{bmatrix}^T \tag{14.4.40}$$

The special case of isotropic linear elasticity (and for material constants E and ν, which do not vary with z) leads to generalized stress-strain relations for membrane terms, flexural terms and shear terms that are uncoupled from each other; this in turn allows us to cast Equation (14.4.38) in the following diagonal, block-matrix form:

$$\{\hat{\sigma}\} = \begin{Bmatrix} \{\hat{\sigma}_m\} \\ \{\hat{\sigma}_f\} \\ \{\hat{\sigma}_s\} \end{Bmatrix} = \begin{bmatrix} [\hat{D}_m] & [0] & [0] \\ [0] & [\hat{D}_f] & [0] \\ [0] & [0] & [\hat{D}_s] \end{bmatrix} \begin{Bmatrix} \{\hat{\varepsilon}_m\} \\ \{\hat{\varepsilon}_f\} \\ \{\hat{\varepsilon}_s\} \end{Bmatrix} \tag{14.4.41}$$

Thus, the case of isotropic linear elasticity can be considered as a special case of Equation (14.4.38), with:

$$[\hat{D}] = \begin{bmatrix} [\hat{D}_m] & [0] & [0] \\ [0] & [\hat{D}_f] & [0] \\ [0] & [0] & [\hat{D}_s] \end{bmatrix} \tag{14.4.42}$$

14.5 Weak Form of Shell Equations

We will now proceed to obtain the weak form for flat shells, using the same conceptual procedure as that employed for elasticity and beam problems. We first establish an arbitrary vector field $\{w\} = \begin{bmatrix} w_{ox} & w_{oy} & w_{oz} & w_{\theta 1} & w_{\theta 2} \end{bmatrix}^T$, with each of the components vanishing at the corresponding essential boundary segment:

$$w_{ox} = 0 \text{ at } \Gamma_{ux} \tag{14.5.1a}$$

$$w_{oy} = 0 \text{ at } \Gamma_{uy} \tag{14.5.1b}$$

$$w_{oz} = 0 \text{ at } \Gamma_{uz} \tag{14.5.1c}$$

$$w_{\theta 1} = 0 \text{ at } \Gamma_{u\theta 1} \tag{14.5.1d}$$

$$w_{\theta 2} = 0 \text{ at } \Gamma_{u\theta 2} \tag{14.5.1e}$$

We then multiply each of the five differential equations of equilibrium—that is, Equations (14.3.1a,b) for membrane action and Equations (14.3.4a,b,c) for plate behavior—by the corresponding component of the $\{w\}$-vector. Subsequently, we integrate each expression over the two-dimensional domain defined by the mid-surface of the shell structure and obtain the weak form. There is a more straightforward way to obtain the weak form, namely, the *continuum-based formulation*, in which we use the (known) weak form for three-dimensional elasticity (which was presented in Section 9.1 and must apply for any three-dimensional solid elastic body), then impose the kinematic constraints that we have for shell theory, as well as the condition $\sigma_{zz} = 0$. Since $\{w\}$ can be thought of as an arbitrary, virtual vector field with mid-surface deformations, we can define the corresponding three-dimensional virtual displacement components, such that they satisfy the kinematic assumptions of shell theory:

$$w_x = (x,y,z) = w_{ox}(x,y) - z \cdot w_{\theta 1} \tag{14.5.2a}$$

$$w_y = (x,y,z) = w_{oy}(x,y) - z \cdot w_{\theta 2} \tag{14.5.2b}$$

$$w_z(x,y,z) \approx w_{oz}(x,y) \tag{14.5.2c}$$

We can also establish the virtual strain components, obtained from partial differentiation of the virtual displacement field:

$$\bar{\varepsilon}_{xx}=\frac{\partial w_x}{\partial x}=\frac{\partial w_{ox}}{\partial x}-z\cdot\frac{\partial w_{\theta 1}}{\partial x}=\bar{\varepsilon}_{ox}-z\cdot\bar{\varphi}_{11} \tag{14.5.3a}$$

$$\bar{\varepsilon}_{yy}=\frac{\partial w_y}{\partial y}=\frac{\partial w_{oy}}{\partial y}-z\cdot\frac{\partial w_{\theta 2}}{\partial y}=\bar{\varepsilon}_{oy}-z\cdot\bar{\varphi}_{22} \tag{14.5.3b}$$

$$\bar{\gamma}_{xy}=\frac{\partial w_x}{\partial y}+\frac{\partial w_y}{\partial x}=\frac{\partial w_{ox}}{\partial y}+\frac{\partial w_{oy}}{\partial x}-z\cdot\frac{\partial w_{\theta 1}}{\partial y}-z\cdot\frac{\partial w_{\theta 2}}{\partial x}=\bar{\gamma}_{oxy}-z\cdot 2\bar{\varphi}_{12} \tag{14.5.3c}$$

$$\bar{\varepsilon}_{zz}=\frac{\partial w_z}{\partial z}\approx 0 \tag{14.5.3d}$$

$$\bar{\gamma}_{yz}=\frac{\partial w_z}{\partial y}+\frac{\partial w_y}{\partial z}=\frac{\partial w_{oz}}{\partial y}-w_{\theta 2} \tag{14.5.3e}$$

$$\bar{\gamma}_{zx}=\frac{\partial w_z}{\partial x}+\frac{\partial w_x}{\partial z}=\frac{\partial w_{oz}}{\partial x}-w_{\theta 1} \tag{14.5.3f}$$

In Equations (14.5.3a–f), we have used the following *virtual generalized strain components:*

$$\bar{\varepsilon}_{ox}=\frac{\partial w_{ox}}{\partial x} \tag{14.5.4a}$$

$$\bar{\varepsilon}_{oy}=\frac{\partial w_{oy}}{\partial y} \tag{14.5.4b}$$

$$\bar{\gamma}_{oxy}=\frac{\partial w_{ox}}{\partial y}+\frac{\partial w_{oy}}{\partial x} \tag{14.5.4c}$$

$$\bar{\varphi}_{11}=\frac{\partial w_{\theta 1}}{\partial x} \tag{14.5.4d}$$

$$\bar{\varphi}_{22}=\frac{\partial w_{\theta 2}}{\partial y} \tag{14.5.4e}$$

$$\bar{\varphi}_{12}=\frac{1}{2}\left(\frac{\partial w_{\theta 2}}{\partial x}+\frac{\partial w_{\theta 1}}{\partial y}\right) \tag{14.5.4f}$$

We will now examine the *internal virtual work* $\bar{W}_{int}$, that is, the left-hand side in the weak form for three-dimensional elasticity in Box 9.1.1. If the three-dimensional domain defined by the volume of the shell is denoted by V and the domain defined by the shell mid-surface is denoted by Ω, as shown in Figure 14.8, then we can conduct the volume integration into two stages, one stage corresponding to integration over the thickness (integration with z) and the other to integration over Ω, then the internal virtual work is given by:

$$\bar{W}_{int}=\iiint\limits_V\left(\bar{\varepsilon}_{xx}\cdot\sigma_{xx}+\bar{\varepsilon}_{yy}\cdot\sigma_{yy}+\bar{\gamma}_{xy}\cdot\sigma_{xy}+\bar{\gamma}_{yz}\cdot\sigma_{yz}+\bar{\gamma}_{zx}\cdot\sigma_{zx}\right)dV\rightarrow$$

$$\rightarrow=\iint\limits_\Omega\left[\int_{-d/2}^{d/2}\left(\bar{\varepsilon}_{xx}\cdot\sigma_{xx}+\bar{\varepsilon}_{yy}\cdot\sigma_{yy}+\bar{\gamma}_{xy}\cdot\sigma_{xy}+\bar{\gamma}_{yz}\cdot\sigma_{yz}+\bar{\gamma}_{zx}\cdot\sigma_{zx}\right)dz\right]d\Omega \tag{14.5.5}$$

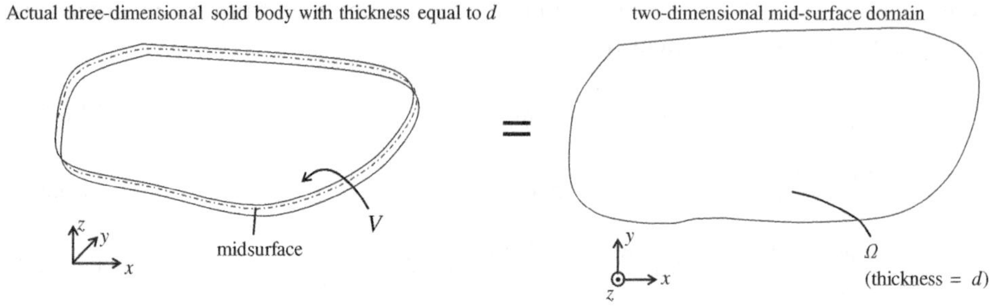

Actual three-dimensional solid body with thickness equal to d

two-dimensional mid-surface domain

midsurface

V

Ω (thickness = d)

Figure 14.8 Conversion of a three-dimensional volume integral to a surface integral of a shell mid-surface.

Now, let us examine each one of the five internal virtual work terms in the integrand of the left-hand side of Equation (14.5.5) and see how it can be expressed using the loading and deformation quantities of the mid-surface. The first term in Equation (14.5.5) is:

$$\int_{-d/2}^{d/2} \bar{\varepsilon}_{xx}\cdot\sigma_{xx}dz = \int_{-d/2}^{d/2} (\bar{\varepsilon}_{ox}-\bar{\varphi}_{11}\cdot z)\cdot\sigma_{xx}dz = \int_{-d/2}^{d/2} \bar{\varepsilon}_{ox}\cdot\sigma_{xx}dz + \int_{-d/2}^{d/2} (-\bar{\varphi}_{11}\cdot z\cdot\sigma_{xx})dz$$

Since $\bar{\varepsilon}_{ox}$ and $\bar{\varphi}_{11}$ are independent of z, we can take them outside of the through-thickness integrals:

$$\int_{-d/2}^{d/2} \bar{\varepsilon}_{xx}\cdot\sigma_{xx}dz = \bar{\varepsilon}_{ox}\cdot\int_{-d/2}^{d/2} \sigma_{xx}dz + \bar{\varphi}_{11}\cdot\int_{-d/2}^{d/2} (-z\cdot\sigma_{xx})dz = \bar{\varepsilon}_{ox}\cdot\hat{n}_{xx} + \bar{\varphi}_{11}\cdot\hat{m}_{xx} \qquad (14.5.6a)$$

where we have also accounted for Equations (14.2.1a) and (14.2.2a).

In the same fashion, we can express the remaining through-thickness integrals in the left-hand side of Equation (14.5.5) in terms of stress resultants and virtual generalized strains:

$$\int_{-d/2}^{d/2} \bar{\varepsilon}_{yy}\cdot\sigma_{yy}dz = \int_{-d/2}^{d/2} (\bar{\varepsilon}_{oy}-\bar{\varphi}_{22}\cdot z)\cdot\sigma_{yy}dz = \bar{\varepsilon}_{oy}\cdot\int_{-d/2}^{d/2} \sigma_{yy}dz + \bar{\varphi}_{22}\cdot\int_{-d/2}^{d/2} (-z\cdot\sigma_{yy})dz \rightarrow$$

$$\rightarrow \int_{-d/2}^{d/2} \bar{\varepsilon}_{yy}\cdot\sigma_{yy}dz = \bar{\varepsilon}_{oy}\cdot\hat{n}_{yy} + \bar{\varphi}_{22}\cdot\hat{m}_{yy} \qquad (14.5.6b)$$

$$\int_{-d/2}^{d/2} \bar{\gamma}_{xy}\cdot\sigma_{xy}dz = \int_{-d/2}^{d/2} \left(\bar{\gamma}_{oxy}-2\bar{\varphi}_{12}\cdot z\right)\cdot\sigma_{xy}dz = \bar{\gamma}_{oxy}\cdot\int_{-d/2}^{d/2} \sigma_{xy}dz + 2\bar{\varphi}_{12}\cdot\int_{-d/2}^{d/2} (-z\cdot\sigma_{xy})dz \rightarrow$$

$$\rightarrow \int_{-d/2}^{d/2} \bar{\gamma}_{xy}\cdot\sigma_{xy}dz = \bar{\gamma}_{oxy}\cdot\hat{n}_{xy} + 2\bar{\varphi}_{12}\cdot\hat{m}_{xy} \qquad (14.5.6c)$$

$$\int_{-d/2}^{d/2} \bar{\gamma}_{yz} \cdot \sigma_{yz} dz = \bar{\gamma}_{yz} \cdot \int_{-d/2}^{d/2} \sigma_{yz} dz = \bar{\gamma}_{yz} \cdot \hat{q}_y \tag{14.5.6d}$$

$$\int_{-d/2}^{d/2} \bar{\gamma}_{zx} \cdot \sigma_{zx} dz = \bar{\gamma}_{zx} \cdot \int_{-d/2}^{d/2} \sigma_{zx} dz = \bar{\gamma}_{zx} \cdot \hat{q}_x \tag{14.5.6e}$$

Thus, if we account for Equations (14.5.6a–e), the right-hand side of Equation (14.5.5) becomes:

$$\begin{aligned}
\bar{W}_{int} &= \iiint_V \left(\bar{\varepsilon}_{xx} \cdot \sigma_{xx} + \bar{\varepsilon}_{yy} \cdot \sigma_{yy} + \bar{\gamma}_{xy} \cdot \sigma_{xy} + \bar{\gamma}_{yz} \cdot \sigma_{yz} + \bar{\gamma}_{zx} \cdot \sigma_{zx} \right) dV = \\
&= \iint_\Omega \left[\int_{-d/2}^{d/2} \left(\bar{\varepsilon}_{xx} \cdot \sigma_{xx} + \bar{\varepsilon}_{yy} \cdot \sigma_{yy} + \bar{\gamma}_{xy} \cdot \sigma_{xy} + \bar{\gamma}_{yz} \cdot \sigma_{yz} + \bar{\gamma}_{zx} \cdot \sigma_{zx} \right) dz \right] d\Omega = \\
&= \iint_\Omega \left(\bar{\varepsilon}_{ox} \cdot \hat{n}_{xx} + \bar{\varepsilon}_{oy} \cdot \hat{n}_{yy} + \bar{\gamma}_{oxy} \cdot \hat{n}_{xy} \right) d\Omega \\
&+ \iint_\Omega \left(\bar{\varphi}_{11} \cdot \hat{m}_{xx} + \bar{\varphi}_{22} \cdot \hat{m}_{yy} + 2\bar{\varphi}_{12} \cdot \hat{m}_{xy} + \bar{\gamma}_{zx} \cdot \hat{q}_x + \bar{\gamma}_{yz} \cdot \hat{q}_y \right) d\Omega
\end{aligned}$$

Thus, we obtain:

$$\begin{aligned}
\bar{W}_{int} &= \iiint_V \left(\bar{\varepsilon}_{xx} \cdot \sigma_{xx} + \bar{\varepsilon}_{yy} \cdot \sigma_{yy} + \bar{\gamma}_{xy} \cdot \sigma_{xy} + \bar{\gamma}_{yz} \cdot \sigma_{yz} + \bar{\gamma}_{zx} \cdot \sigma_{zx} \right) dV \\
&= \iint_\Omega \{\bar{\hat{\varepsilon}}_m\}^T \{\hat{\sigma}_m\} d\Omega + \iint_\Omega \{\bar{\hat{\varepsilon}}_b\}^T \{\hat{\sigma}_b\} d\Omega
\end{aligned} \tag{14.5.7}$$

where we have defined the virtual *generalized membrane strain vector*, $\{\bar{\hat{\varepsilon}}_m\}$:

$$\{\bar{\hat{\varepsilon}}_m\} = \begin{bmatrix} \bar{\varepsilon}_{ox} & \bar{\varepsilon}_{oy} & \bar{\gamma}_{oxy} \end{bmatrix}^T \tag{14.5.8a}$$

and the *virtual generalized bending (plate) strain vector* $\{\bar{\hat{\varepsilon}}_b\}$:

$$\{\bar{\hat{\varepsilon}}_b\} = \begin{bmatrix} \bar{\varphi}_{11} & \bar{\varphi}_{22} & 2\bar{\varphi}_{12} & \bar{\gamma}_{zx} & \bar{\gamma}_{zy} \end{bmatrix}^T \tag{14.5.8b}$$

The first integral term in Equation (14.5.7) represents the internal virtual work due to membrane behavior and the second integral term represents the internal virtual work due to bending (plate) behavior. We can use the definitions of Equations (14.4.39), (14.4.40), to write Equation (14.5.7) in the following form:

$$\bar{W}_{int} = \iint_\Omega \{\bar{\hat{\varepsilon}}\}^T \{\hat{\sigma}\} d\Omega \tag{14.5.9a}$$

If we also account for Equation (14.4.38), which applies for shells made of linearly elastic material, we have:

$$\bar{W}_{int} = \iint_{\Omega} \{\hat{\bar{\varepsilon}}\}^T [\hat{D}] \{\hat{\varepsilon}\} d\Omega \tag{14.5.9b}$$

Remark 14.5.1: For the special case where Equation (14.4.41) applies, we can separate the virtual work terms associated with membrane action, flexure, and shear, and write:

$$\bar{W}_{int} = \iint_{\Omega} \{\hat{\bar{\varepsilon}}_m\}^T [\hat{D}_m] \{\hat{\varepsilon}_m\} d\Omega + \iint_{\Omega} \{\hat{\bar{\varepsilon}}_f\}^T [\hat{D}_f] \{\hat{\varepsilon}_f\} d\Omega + \iint_{\Omega} \{\hat{\bar{\varepsilon}}_s\}^T [\hat{D}_s] \{\hat{\varepsilon}_s\} d\Omega \tag{14.5.9c}$$

■

Let us now examine the "external work" terms, that is, the right-hand side of the weak form for three-dimensional elasticity, given in Box 9.1.1:

$$\bar{W}_{ext} = \iiint_V (w_x b_x + w_y b_y + w_z b_z) dV + \iint_{\Gamma_{tx}} w_x t_x d\Gamma + \iint_{\Gamma_{ty}} w_y t_y d\Gamma + \iint_{\Gamma_{tz}} w_z t_z d\Gamma \tag{14.5.10}$$

We will once again apply the considerations based on Figure 14.8 for the three-dimensional volume integrals of the body forces:

$$\iiint_V (w_x b_x + w_y b_y + w_z + b_z) dV = \iint_{\Omega} \left[\int_{-d/2}^{d/2} (w_x b_x + w_y b_y + w_z b_z) dz \right] d\Omega$$
$$= \iint_{\Omega} \left[\int_{-d/2}^{d/2} (w_x b_x) dz \right] d\Omega + \iint_{\Omega} \left[\int_{-d/2}^{d/2} (w_y b_y) dz \right] d\Omega + \iint_{\Omega} \left[\int_{-d/2}^{d/2} (w_z b_z) dz \right] d\Omega \tag{14.5.11}$$

The right-hand side of Equation (14.5.11) has three terms. For the first term, we have:

$$\iint_{\Omega} \left[\int_{-d/2}^{d/2} (w_x b_x) dz \right] d\Omega = \iint_{\Omega} \left[\int_{-d/2}^{d/2} (w_{ox} - z \cdot w_{\theta 1}) b_x dz \right] d\Omega$$
$$= \iint_{\Omega} w_{ox} \left[\int_{-d/2}^{d/2} b_x dz \right] d\Omega + \iint_{\Omega} w_{\theta 1} \left[\int_{-d/2}^{d/2} (-z \cdot b_x) dz \right] d\Omega \rightarrow$$
$$\rightarrow \iint_{\Omega} \left[\int_{-d/2}^{d/2} (w_x b_x) dz \right] d\Omega = \iint_{\Omega} w_{ox} p_x d\Omega + \iint_{\Omega} w_{\theta 1} \hat{m}_1 d\Omega \tag{14.5.12}$$

where we have accounted for Equation (14.5.2a) and have tacitly relied on the stipulation that the *distributed membrane force p_x per unit shell surface area* is given by integrating the body force over the shell thickness:

$$p_x(x,y) = \int_{-d/2}^{d/2} b_x dz \tag{14.5.13a}$$

and that the *distributed moment $\hat{m}_1$ per unit shell surface area* is defined by the following equation.

$$\hat{m}_1(x,y) = \int_{-d/2}^{d/2} (-z \cdot b_x) dz \tag{14.5.13b}$$

Similarly, for the second term of the right-hand side in Equation (14.5.11), we obtain:

$$\iint_\Omega \left[\int_{-d/2}^{d/2} \left(w_y b_y\right) dz \right] d\Omega = \iint_\Omega w_{oy} p_y d\Omega + \iint_\Omega w_{\theta 2} \hat{m}_2 d\Omega \tag{14.5.14}$$

where we used the definitions:

$$p_y(x,y) = \int_{-d/2}^{d/2} b_x dz \tag{14.5.15a}$$

and

$$\hat{m}_2(x,y) = \int_{-d/2}^{d/2} \left(-z \cdot b_y\right) dz \tag{14.5.15b}$$

Finally, the third term in the right-hand side of Equation (14.5.11) yields:

$$\iint_\Omega \left[\int_{-d/2}^{d/2} (w_z b_z) dz \right] d\Omega = \iint_\Omega w_{oz} \left[\int_{-d/2}^{d/2} b_z dz \right] d\Omega = \iint_\Omega w_{oz} p_z d\Omega \tag{14.5.16}$$

where we tacitly defined the *distributed normal force p_z per unit shell surface area* as follows.

$$p_z(x,y) = \int_{-d/2}^{d/2} b_z dz \tag{14.5.17}$$

Let us now consider the terms of Equation (14.5.10) giving work from prescribed tractions over natural boundary surface segments, Γ_t. The surface Γ_t can be thought of as a line segment Γ with a thickness equal to d, as shown in Figure 14.9. Just like we converted volume integrals to mid-surface ones, we will transform the boundary surface integrals of a three-dimensional domain to one-dimensional line boundary terms over the edges of the shell mid-surface.

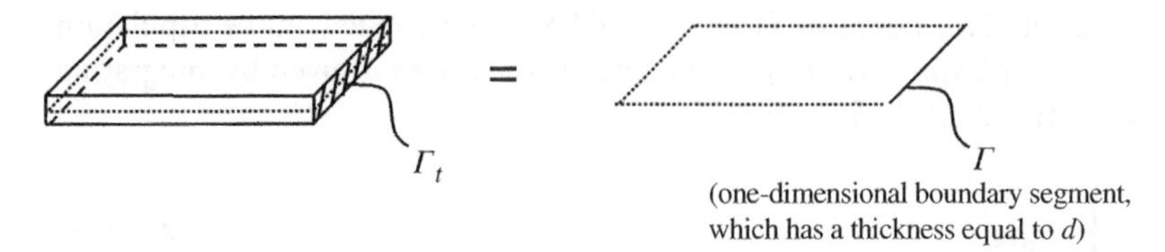

Figure 14.9 Consideration of a three-dimensional boundary surface segment as a one-dimensional boundary segment for the shell mid-surface.

For the first traction boundary term in Equation (14.5.10), we account for Equation (14.5.2a) to obtain:

$$\iint_{\Gamma_t} w_x t_x d\Gamma = \int_{\Gamma}\left(\int_{-d/2}^{d/2} w_x t_x dz\right) dS = \int_{\Gamma}\left[\int_{-d/2}^{d/2}(w_{ox} - z\cdot w_{\theta 1})t_x dz\right] dS$$

$$= \int_{\Gamma}\left[\int_{-d/2}^{d/2} w_{ox} t_x dz\right] dS + \int_{\Gamma}\left[\int_{-d/2}^{d/2}(-z\cdot w_{\theta 1})t_x dz\right] dS$$

$$\rightarrow \iint_{\Gamma_t} w_x t_x d\Gamma = \int_{\Gamma} w_{ox}\left[\int_{-d/2}^{d/2} t_x dz\right] dS + \int_{\Gamma} w_{\theta 1}\left[\int_{-d/2}^{d/2} -z\cdot t_x dz\right] dS$$

$$= \int_{\Gamma} w_{ox}\cdot \hat{f}_x dS + \int_{\Gamma} w_{\theta 1}\cdot \bar{m}_x dS \tag{14.5.18}$$

where $\hat{f}_x$ is the *distributed membrane force in the x-direction per unit width of the shell boundary line, given by the expression*:

$$\hat{f}_x = \int_{-d/2}^{d/2} t_x dz \tag{14.5.19a}$$

and $\bar{m}_x$ is the *prescribed moment in the direction of* θ_1 *per unit width of the shell boundary line*, given by:

$$\bar{m}_x = \int_{-d/2}^{d/2} -z\cdot t_x dz \tag{14.5.19b}$$

The second traction boundary term in the right-hand side of Equation (14.5.10) gives:

$$\iint_{\Gamma_t} w_y t_y d\Gamma = \int_{\Gamma}\left(\int_{-d/2}^{d/2} w_y t_y dz\right) dS = \int_{\Gamma}\left[\int_{-d/2}^{d/2} \left(w_{oy} - z \cdot w_{\theta 2}\right) t_y dz\right] dS$$

$$= \int_{\Gamma}\left[\int_{-d/2}^{d/2} w_{oy} t_y dz\right] dS + \int_{\Gamma}\left[\int_{-d/2}^{d/2} (-z \cdot w_{\theta 2}) t_y dz\right] dS$$

$$\rightarrow \iint_{\Gamma_t} w_y t_y d\Gamma = \int_{\Gamma} w_{oy}\left[\int_{-d/2}^{d/2} t_y dz\right] dS + \int_{\Gamma} w_{\theta 2}\left[\int_{-d/2}^{d/2} -z \cdot t_y dz\right] dS$$

$$= \int_{\Gamma} w_{oy} \cdot \hat{f}_y dS + \int_{\Gamma} w_{\theta 2} \cdot \bar{\hat{m}}_y dS \qquad (14.5.20)$$

where $\hat{f}_y$ and $\bar{\hat{m}}_y$ are the *distributed prescribed membrane force in the y-direction and moment in the direction of θ_2 per unit width of the shell boundary*, given by the expressions:

$$\hat{f}_y = \int_{-d/2}^{d/2} t_y dz \qquad (14.5.21a)$$

and

$$\bar{\hat{m}}_y = \int_{-d/2}^{d/2} -z \cdot t_y dz \qquad (14.5.21b)$$

Finally, the third traction boundary term in the right-hand side of Equation (14.5.10) yields:

$$\iint_{\Gamma_t} w_z t_z d\Gamma = \int_{\Gamma}\left[\int_{-d/2}^{d/2} w_{oz} t_z dz\right] dS = \int_{\Gamma} w_{oz}\left[\int_{-d/2}^{d/2} t_z dz\right] dS = \int_{\Gamma} w_{oz} \cdot \hat{f}_z dS \qquad (14.5.22)$$

where the *distributed prescribed shear force $\hat{f}_z$ in the z-direction, per unit width of the shell boundary edge, is given by the expression*:

$$\hat{f}_z = \int_{-d/2}^{d/2} t_z dz \qquad (14.5.23)$$

If we plug Equations (14.5.9b), (14.5.12), (14.5.14), (14.5.16), and (14.5.18), (14.5.20), and (14.5.22) into the weak form for three-dimensional elasticity (Box 9.1.1), and account for the shell essential boundary conditions, we eventually obtain the *weak form for shell problems*, which is the principle of virtual work for such structures and is provided in Box 14.5.1.

Box 14.5.1 Weak Form for Shell Problems

$$\iint_{\Omega} \{\bar{\hat{\varepsilon}}\}^T [\hat{D}] \{\hat{\varepsilon}\} d\Omega = \iint_{\Omega} \{w\}^T \{p\} dV + \int_{\Gamma_{tx}} w_{ox}\hat{f}_x dS + \int_{\Gamma_{ty}} w_{oy}\hat{f}_y dS + \int_{\Gamma_{tz}} w_{oz}\hat{f}_z dS$$

$$+ \int_{\Gamma_{t\theta 1}} w_{\theta 1}\bar{\hat{m}}_x dS + \int_{\Gamma_{t\theta 2}} w_{\theta 2}\bar{\hat{m}}_y dS$$

where $\{p\} = \begin{bmatrix} p_x & p_y & p_z & \hat{m}_1 & \hat{m}_2 \end{bmatrix}^T$

$$\forall \ \{w\} = \begin{bmatrix} w_{ox} & w_{ox} & w_{ox} & w_{\theta 1} & w_{\theta 2} \end{bmatrix}^T$$

with $w_x = 0$ on Γ_{ux}, $w_y = 0$ on Γ_{uy}, $w_z = 0$ on Γ_{uz}, $w_{\theta 1} = 0$ on $\Gamma_{u\theta 1}$, $w_{\theta 2} = 0$ at $\Gamma_{u\theta 2}$

$$u_{ox} = \bar{u}_{ox} \quad on \quad \Gamma_{ux}$$

$$u_{oy} = \bar{u}_{oy} \quad on \quad \Gamma_{uy}$$

$$u_{oz} = \bar{u}_{oz} \quad on \quad \Gamma_{uz}$$

$$\theta_1 = \bar{\theta}_1 \quad on \quad \Gamma_{u\theta 1}$$

$$\theta_2 = \bar{\theta}_2 \quad on \quad \Gamma_{u\theta 2}$$

14.6 Finite Element Formulation for Shell Structures

We will now move on to examine the finite element analysis of shell structures. Once again, we assume that we subdivide the structure into N_e subdomains called the elements, and also that we have a piecewise approximation for the kinematic quantities of the reference surface in each element e. We will also assume that each element has n nodes, with six degrees of freedom per node (three translations and three rotations), that the shell surface lies in the xy plane, and that we use an isoparametric formulation, that is, the shape functions are polynomials with respect to parametric coordinates ξ and η (in accordance with Section 6.4). The finite element approximation in each element e is defined by the following expressions.

$$u_{ox}^{(e)}(\xi,\eta) = \sum_{i=1}^{n} \left(N_i^{(e)}(\xi,\eta) \cdot u_{xi}^{(e)} \right) \tag{14.6.1a}$$

$$u_{oy}^{(e)}(\xi,\eta) = \sum_{i=1}^{n} \left(N_i^{(e)}(\xi,\eta) \cdot u_{yi}^{(e)} \right) \tag{14.6.1b}$$

$$u_{oz}^{(e)}(\xi,\eta) = \sum_{i=1}^{n} \left(N_i^{(e)}(\xi,\eta) \cdot u_{zi}^{(e)} \right) \tag{14.6.1c}$$

$$\theta_1^{(e)}(\xi,\eta) = -\sum_{i=1}^{n} \left(N_i^{(e)}(\xi,\eta) \cdot \theta_{yi}^{(e)} \right) \tag{14.6.1d}$$

$$\theta_2^{(e)}(\xi,\eta) = \sum_{i=1}^{n} \left(N_i^{(e)}(\xi,\eta) \cdot \theta_{xi}^{(e)} \right) \tag{14.6.1e}$$

We can cast Equations (14.6.1a–e) in matrix form, as shown in Figure 14.10.

Shape functions for node 1 | Shape functions for node 2 | ...... | Shape functions for node 'n'

$$\begin{Bmatrix} u_{xo}(\xi,\eta) \\ u_{yo}(\xi,\eta) \\ u_{zo}(\xi,\eta) \\ \theta_1(\xi,\eta) \\ \theta_2(\xi,\eta) \end{Bmatrix} = \begin{bmatrix} N_1^{(4Q)}(\xi,\eta) & 0 & 0 & 0 & 0 & N_2^{(4Q)}(\xi,\eta) & 0 & 0 & 0 & 0 & \cdots & N_n^{(4Q)}(\xi,\eta) & 0 & 0 & 0 & 0 \\ 0 & N_1^{(4Q)}(\xi,\eta) & 0 & 0 & 0 & 0 & N_2^{(4Q)}(\xi,\eta) & 0 & 0 & 0 & \cdots & 0 & N_n^{(4Q)}(\xi,\eta) & 0 & 0 & 0 \\ 0 & 0 & N_1^{(4Q)}(\xi,\eta) & 0 & 0 & 0 & 0 & N_2^{(4Q)}(\xi,\eta) & 0 & 0 & \cdots & 0 & 0 & N_n^{(4Q)}(\xi,\eta) & 0 & 0 \\ 0 & 0 & 0 & -N_1^{(4Q)}(\xi,\eta) & 0 & 0 & 0 & 0 & -N_2^{(4Q)}(\xi,\eta) & 0 & \cdots & 0 & 0 & 0 & -N_n^{(4Q)}(\xi,\eta) & 0 \\ 0 & 0 & 0 & 0 & N_1^{(4Q)}(\xi,\eta) & 0 & 0 & 0 & 0 & N_2^{(4Q)}(\xi,\eta) & \cdots & 0 & 0 & 0 & 0 & N_n^{(4Q)}(\xi,\eta) \end{bmatrix} \begin{Bmatrix} u_{x1}^{(e)} \\ u_{y1}^{(e)} \\ u_{z1}^{(e)} \\ \theta_{x1}^{(e)} \\ \theta_{y1}^{(e)} \\ u_{x2}^{(e)} \\ u_{y2}^{(e)} \\ u_{z2}^{(e)} \\ \theta_{x2}^{(e)} \\ \theta_{y2}^{(e)} \\ \vdots \\ u_{xn}^{(e)} \\ u_{yn}^{(e)} \\ u_{zn}^{(e)} \\ \theta_{xn}^{(e)} \\ \theta_{yn}^{(e)} \end{Bmatrix}$$

Figure 14.10 Matrix expression giving the approximate displacement field for a shell element.

We can also write a block matrix form for the finite element approximation. We can write:

$$\begin{Bmatrix} u_{ox}^{(e)}(\xi,\eta) \\ u_{oy}^{(e)}(\xi,\eta) \\ u_{oz}^{(e)}(\xi,\eta) \\ \theta_1^{(e)}(\xi,\eta) \\ \theta_2^{(e)}(\xi,\eta) \end{Bmatrix} = \left[N_1^{(e)}(\xi,\eta)\right]\left\{U_1^{(e)}\right\} + \left[N_2^{(e)}(\xi,\eta)\right]\left\{U_2^{(e)}\right\} + \ldots + \left[N_n^{(e)}(\xi,\eta)\right]\left\{U_n^{(e)}\right\} \rightarrow$$

$$\rightarrow \begin{Bmatrix} u_{ox}^{(e)}(\xi,\eta) \\ u_{oy}^{(e)}(\xi,\eta) \\ u_{oz}^{(e)}(\xi,\eta) \\ \theta_1^{(e)}(\xi,\eta) \\ \theta_2^{(e)}(\xi,\eta) \end{Bmatrix} = \left[\left[N_1^{(e)}(\xi,\eta)\right] \left[N_2^{(e)}(\xi,\eta)\right] \cdots \left[N_n^{(e)}(\xi,\eta)\right]\right] \begin{Bmatrix} \left\{U_1^{(e)}\right\} \\ \left\{U_2^{(e)}\right\} \\ \vdots \\ \left\{U_n^{(e)}\right\} \end{Bmatrix} \tag{14.6.2}$$

where $\left[N_i^{(e)}(\xi,\eta)\right]$ is a (5 × 6) array, with the shape functions expressing the contribution of node i to the approximate fields:

$$\left[N_i^{(e)}(\xi,\eta)\right] = \begin{bmatrix} N_i^{(e)}(\xi,\eta) & 0 & 0 & 0 & 0 & 0 \\ 0 & N_i^{(e)}(\xi,\eta) & 0 & 0 & 0 & 0 \\ 0 & 0 & N_i^{(e)}(\xi,\eta) & 0 & 0 & 0 \\ 0 & 0 & 0 & 0 & -N_i^{(e)}(\xi,\eta) & 0 \\ 0 & 0 & 0 & N_i^{(e)}(\xi,\eta) & 0 & 0 \end{bmatrix} \tag{14.6.3}$$

and $\left\{U_i^{(e)}\right\}$ is a vector with the six degrees of freedom of nodal point i:

$$\left\{U_i^{(e)}\right\} = \left[u_{xi}^{(e)} \quad u_{yi}^{(e)} \quad u_{zi}^{(e)} \quad \theta_{xi}^{(e)} \quad \theta_{yi}^{(e)} \quad \theta_{zi}^{(e)}\right]^T \tag{14.6.4}$$

In our derivations, we will also use the more concise equation:

$$\left\{u^{(e)}\right\} = \left[N^{(e)}\right]\left\{U^{(e)}\right\} \tag{14.6.5}$$

where

$$\left[N^{(e)}\right] = \left[\left[N_1^{(e)}(\xi,\eta)\right] \quad \left[N_2^{(e)}(\xi,\eta)\right] \quad \cdots \quad \left[N_n^{(e)}(\xi,\eta)\right]\right] \tag{14.6.6}$$

$$\left\{u^{(e)}\right\} = \left[u_{ox}^{(e)}(\xi,\eta) \quad u_{oy}^{(e)}(\xi,\eta) \quad u_{oz}^{(e)}(\xi,\eta) \quad \theta_1^{(e)}(\xi,\eta) \quad \theta_2^{(e)}(\xi,\eta)\right]^T \tag{14.6.7}$$

and

$$\left\{U^{(e)}\right\} = \begin{Bmatrix} \left\{U_1^{(e)}\right\} \\ \left\{U_2^{(e)}\right\} \\ \vdots \\ \left\{U_n^{(e)}\right\} \end{Bmatrix} \tag{14.6.8}$$

Remark 14.6.1: In Equations (14.6.2) and (14.6.8), we have included the nodal rotations $\theta_{zi}^{(e)}$ along the z-axis, for each nodal point i. Although these rotations do not affect any of the approximate fields for a flat shell, we will still retain them for consistency (usually, three-dimensional shell models include six degrees of freedom at each node) and also to facilitate transition to discussion of nonplanar shell formulations, where the kinematics may depend on all rotational degrees of freedom of the nodal points! ■

Remark 14.6.2: It is important to notice the negative (–) sign on the shape functions multiplying the nodal rotations in the y-direction. This sign stems from the positive sign convention for $\theta_1^{(e)}(\xi,\eta)$. ■

The generalized strain vectors for a shell element are given by appropriate differentiations of the kinematic fields of the mid-surface. Specifically, we can establish expressions for the approximate membrane generalized strains:

$$\left\{\hat{\varepsilon}_m^{(e)}\right\} = \begin{Bmatrix} \varepsilon_{ox}^{(e)} \\ \varepsilon_{oy}^{(e)} \\ \gamma_{oxy}^{(e)} \end{Bmatrix} = \begin{Bmatrix} \dfrac{\partial u_{xo}^{(e)}}{\partial x} \\ \dfrac{\partial u_{yo}^{(e)}}{\partial y} \\ \dfrac{\partial u_{xo}^{(e)}}{\partial y} + \dfrac{\partial u_{yo}^{(e)}}{\partial x} \end{Bmatrix} = \begin{Bmatrix} \dfrac{\partial}{\partial x}\sum_{i=1}^{n}\left(N_i^{(e)}(\xi,\eta)\cdot u_{xi}^{(e)}\right) \\ \dfrac{\partial}{\partial y}\sum_{i=1}^{n}\left(N_i^{(e)}(\xi,\eta)\cdot u_{yi}^{(e)}\right) \\ \dfrac{\partial}{\partial y}\sum_{i=1}^{n}\left(N_i^{(e)}(\xi,\eta)\cdot u_{xi}^{(e)}\right) + \dfrac{\partial}{\partial x}\sum_{i=1}^{n}\left(N_i^{(e)}(\xi,\eta)\cdot u_{yi}^{(e)}\right) \end{Bmatrix} \tag{14.6.8a}$$

the approximate bending (flexure) generalized strains:

$$\left\{\hat{\varepsilon}_f^{(e)}\right\} = \left\{\begin{array}{c} \varphi_{11}^{(e)} \\ \varphi_{22}^{(e)} \\ 2\varphi_{12}^{(e)} \end{array}\right\} = \left\{\begin{array}{c} \dfrac{\partial\theta_1^{(e)}}{\partial x} \\ \dfrac{\partial\theta_2^{(e)}}{\partial y} \\ \dfrac{\partial\theta_1^{(e)}}{\partial y} + \dfrac{\partial u_2^{(e)}}{\partial x} \end{array}\right\} = \left\{\begin{array}{c} \dfrac{\partial}{\partial x}\sum_{i=1}^{n}\left(N_i^{(e)}(\xi,\eta)\cdot\theta_{yi}^{(e)}\right) \\ \dfrac{\partial}{\partial y}\sum_{i=1}^{n}\left(N_i^{(e)}(\xi,\eta)\cdot\theta_{xi}^{(e)}\right) \\ \dfrac{\partial}{\partial y}\sum_{i=1}^{n}\left(-N_i^{(e)}(\xi,\eta)\cdot\theta_{yi}^{(e)}\right) + \dfrac{\partial}{\partial x}\sum_{i=1}^{n}\left(N_i^{(e)}(\xi,\eta)\cdot\theta_{xi}^{(e)}\right) \end{array}\right\}$$

(14.6.8b)

and the approximate shear generalized strains:

$$\left\{\hat{\varepsilon}_s^{(e)}\right\} = \left\{\begin{array}{c} \gamma_{zx}^{(e)} \\ \gamma_{zy}^{(e)} \end{array}\right\} = \left\{\begin{array}{c} \dfrac{\partial u_{oz}^{(e)}}{\partial x} - \theta_1^{(e)} \\ \dfrac{\partial u_{oz}^{(e)}}{\partial x} - \theta_2^{(e)} \end{array}\right\} = \left\{\begin{array}{c} \dfrac{\partial}{\partial x}\sum_{i=1}^{n}\left(N_i^{(e)}(\xi,\eta)\cdot u_{zi}^{(e)}\right) - \sum_{i=1}^{n}\left(-N_i^{(e)}(\xi,\eta)\cdot\theta_{yi}^{(e)}\right) \\ \dfrac{\partial}{\partial x}\sum_{i=1}^{n}\left(N_i^{(e)}(\xi,\eta)\cdot u_{zi}^{(e)}\right) - \sum_{i=1}^{n}\left(-N_i^{(e)}(\xi,\eta)\cdot\theta_{xi}^{(e)}\right) \end{array}\right\}$$

(14.6.8c)

To obtain finite element equations, we will combine Equations (14.6.8a–c) in a block matrix form, where a matrix premultiplies the vector $\{U^{(e)}\}$ defined by Equation (14.6.7). First we write three distinct expressions for the membrane, bending, and shear strains:

$$\left\{\hat{\varepsilon}_m^{(e)}\right\} = \left[B_m^{(e)}(\xi,\eta)\right]\left\{U^{(e)}\right\} = \left[\left[B_{m1}^{(e)}(\xi,\eta)\right] \quad \left[B_{m2}^{(e)}(\xi,\eta)\right] \quad \cdots \quad \left[B_{mn}^{(e)}(\xi,\eta)\right]\right]\left\{\begin{array}{c} \left\{U_1^{(e)}\right\} \\ \left\{U_2^{(e)}\right\} \\ \vdots \\ \left\{U_n^{(e)}\right\} \end{array}\right\}$$

(14.6.9a)

$$\left\{\hat{\varepsilon}_f^{(e)}\right\} = \left[B_f^{(e)}(\xi,\eta)\right]\left\{U^{(e)}\right\} = \left[\left[B_{f1}^{(e)}(\xi,\eta)\right] \quad \left[B_{f2}^{(e)}(\xi,\eta)\right] \quad \cdots \quad \left[B_{fn}^{(e)}(\xi,\eta)\right]\right]\left\{\begin{array}{c} \left\{U_1^{(e)}\right\} \\ \left\{U_2^{(e)}\right\} \\ \vdots \\ \left\{U_n^{(e)}\right\} \end{array}\right\}$$

(14.6.9b)

and

$$\left\{\hat{\varepsilon}_s^{(e)}\right\} = \left[B_s^{(e)}(\xi,\eta)\right]\left\{U^{(e)}\right\} = \left[\left[B_{s1}^{(e)}(\xi,\eta)\right] \; \left[B_{s2}^{(e)}(\xi,\eta)\right] \; \cdots \; \left[B_{sn}^{(e)}(\xi,\eta)\right]\right]\begin{Bmatrix} \left\{U_1^{(e)}\right\} \\ \left\{U_2^{(e)}\right\} \\ \vdots \\ \left\{U_n^{(e)}\right\} \end{Bmatrix} \tag{14.6.9c}$$

The matrices $\left[B_{mi}^{(e)}(\xi,\eta)\right]$, $\left[B_{fi}^{(e)}(\xi,\eta)\right]$ and $\left[B_{si}^{(e)}(\xi,\eta)\right]$ in Equations (14.6.9a–c) are the block strain-displacement arrays for nodal point i, corresponding to membrane strains, flexural strains, and shear strains, respectively. These arrays are given by the following equations.

$$\left[B_{mi}^{(e)}(\xi,\eta)\right] = \begin{bmatrix} \frac{\partial N_i^{(e)}}{\partial x} & 0 & 0 & 0 & 0 & 0 \\ 0 & \frac{\partial N_i^{(e)}}{\partial y} & 0 & 0 & 0 & 0 \\ \frac{\partial N_i^{(e)}}{\partial y} & \frac{\partial N_i^{(e)}}{\partial x} & 0 & 0 & 0 & 0 \end{bmatrix} \tag{14.6.10a}$$

$$\left[B_{fi}^{(e)}(\xi,\eta)\right] = \begin{bmatrix} 0 & 0 & 0 & 0 & -\frac{\partial N_i^{(e)}}{\partial x} & 0 \\ 0 & 0 & 0 & \frac{\partial N_i^{(e)}}{\partial y} & 0 & 0 \\ 0 & 0 & 0 & \frac{\partial N_i^{(e)}}{\partial x} & -\frac{\partial N_i^{(e)}}{\partial y} & 0 \end{bmatrix} \tag{14.6.10b}$$

$$\left[B_{si}^{(e)}(\xi,\eta)\right] = \begin{bmatrix} 0 & 0 & \frac{\partial N_i^{(e)}}{\partial x} & 0 & N_i^{(e)} & 0 \\ 0 & 0 & \frac{\partial N_i^{(e)}}{\partial y} & -N_i^{(e)} & 0 & 0 \end{bmatrix} \tag{14.6.10c}$$

We can also combine Equations (14.6.9a–c) in a unique, block matrix expression:

$$\left\{\hat{\varepsilon}^{(e)}\right\}=\left\{\begin{matrix}\left\{\hat{\varepsilon}_m^{(e)}\right\}\\ \left\{\hat{\varepsilon}_f^{(e)}\right\}\\ \left\{\hat{\varepsilon}_s^{(e)}\right\}\end{matrix}\right\}=\left[\begin{matrix}\left[\begin{matrix}\left[B_{m1}^{(e)}(\xi,\eta)\right]\\ \left[B_{f1}^{(e)}(\xi,\eta)\right]\\ \left[B_{s1}^{(e)}(\xi,\eta)\right]\end{matrix}\right] & \left[\begin{matrix}\left[B_{m2}^{(e)}(\xi,\eta)\right]\\ \left[B_{f2}^{(e)}(\xi,\eta)\right]\\ \left[B_{s2}^{(e)}(\xi,\eta)\right]\end{matrix}\right] & \cdots & \left[\begin{matrix}\left[B_{mn}^{(e)}(\xi,\eta)\right]\\ \left[B_{fn}^{(e)}(\xi,\eta)\right]\\ \left[B_{sn}^{(e)}(\xi,\eta)\right]\end{matrix}\right]\end{matrix}\right]\left\{\begin{matrix}\left\{U_1^{(e)}\right\}\\ \left\{U_2^{(e)}\right\}\\ \vdots\\ \left\{U_n^{(e)}\right\}\end{matrix}\right\}\rightarrow$$

$$\rightarrow\left\{\hat{\varepsilon}^{(e)}\right\}=\left[B^{(e)}(\xi,\eta)\right]\left\{U^{(e)}\right\} \tag{14.6.10d}$$

Equation (14.6.10d) establishes the generalized strain-displacement matrix $[B^{(e)}]$ for the shell element e.

Now, all the integral terms in the weak form (Box 14.5.1) can be separated into the element contributions. We also assume the same approximation for the virtual displacement/rotation fields as for the actual displacement/rotation fields. Thus, the virtual generalized strain vector can be written as:

$$\left\{\bar{\hat{\varepsilon}}^{(e)}\right\}^T=\left\{W^{(e)}\right\}^T\left[B^{(e)}\right]^T \tag{14.6.11}$$

where $\{W^{(e)}\}$ is a virtual nodal displacement/rotation vector for element e. The left-hand side of the weak form can be written as a sum of the individual element contributions:

$$\iint_\Omega\{\bar{\hat{\varepsilon}}\}^T[\hat{D}]\{\hat{\varepsilon}\}dV=\sum_{e=1}^{N_e}\left(\iint_{\Omega^{(e)}}\{\bar{\hat{\varepsilon}}\}^T[\hat{D}]\{\hat{\varepsilon}\}dV\right)$$

If we also account for the gather operation of each element (per Section B.2), that is, $\{U^{(e)}\}=[L^{(e)}]^T\{U\}$ and $\{W^{(e)}\}^T=\{W\}^T[L^{(e)}]^T$, we have:

$$\iint_\Omega\{\bar{\hat{\varepsilon}}\}^T[\hat{D}]\{\hat{\varepsilon}\}d\Omega=\sum_{e=1}^{N_e}\left(\iint_{\Omega^{(e)}}\left(\{W\}^T\left[L^{(e)}\right]^T\left[B^{(e)}\right]^T\left[\hat{D}^{(e)}\right]\left[B^{(e)}\right]\left[L^{(e)}\right]\{U\}\right)dV\right)$$

$$=\{W\}^T\sum_{e=1}^{N_e}\left(\left[L^{(e)}\right]^T\iint_{\Omega^{(e)}}\left[B^{(e)}\right]^T\left[\hat{D}^{(e)}\right]\left[B^{(e)}\right]dV\left[L^{(e)}\right]\right)\{U\}$$

$$\rightarrow\iint_\Omega\{\bar{\hat{\varepsilon}}\}^T[\hat{D}]\{\hat{\varepsilon}\}d\Omega=\{W\}^T\sum_{e=1}^{N_e}\left(\left[L^{(e)}\right]^T\left[k^{(e)}\right]\left[L^{(e)}\right]\right)\{U\} \tag{14.6.12}$$

where $[k^{(e)}]$ is the *stiffness matrix for element e*, given by:

$$\left[k^{(e)}\right] = \iint_{\Omega^{(e)}} \left[B^{(e)}\right]^T \left[\hat{D}^{(e)}\right] \left[B^{(e)}\right] dV \tag{14.6.13}$$

The external virtual work (the right-hand side of the weak form in Box 14.5.1) becomes:

$$\iint_{\Omega} \{w\}^T \{p\} dV + \int_{\Gamma_{tx}} w_{ox} \hat{f}_x dS + \int_{\Gamma_{ty}} w_{oy} \hat{f}_y dS + \int_{\Gamma_{tz}} w_{oz} \hat{f}_z dS + \int_{\Gamma_{t\theta 1}} w_{\theta 1} \bar{m}_x dS + \int_{\Gamma_{t\theta 2}} w_{\theta 2} \bar{m}_y dS$$

$$= \iint_{\Omega} \{w\}^T \{p\} d\Omega + \int_{\Gamma_{tx}} \{w\}^T \left\{\hat{f}_x\right\} dS + \int_{\Gamma_{ty}} \{w\}^T \left\{\hat{f}_y\right\} dS + \int_{\Gamma_{tz}} \{w\}^T \left\{\hat{f}_z\right\} dS$$

$$+ \int_{\Gamma_{t\theta 1}} \{w\}^T \{\hat{m}_x\} dS + \int_{\Gamma_{t\theta 2}} \{w\}^T \{\hat{m}_y\} dS \tag{14.6.14}$$

where we have defined the following expanded vectors:

$$\left\{\hat{f}_x\right\} = \left[\hat{f}_x \ 0 \ 0 \ 0 \ 0\right]^T \tag{14.6.15a}$$

$$\left\{\hat{f}_y\right\} = \left[0 \ \hat{f}_y \ 0 \ 0 \ 0\right]^T \tag{14.6.15b}$$

$$\left\{\hat{f}_y\right\} = \left[0 \ 0 \ \hat{f}_z \ 0 \ 0\right]^T \tag{14.6.15c}$$

$$\{\hat{m}_x\} = \left[0 \ 0 \ 0 \ \bar{m}_x \ 0\right]^T \tag{14.6.15d}$$

$$\{\hat{m}_y\} = \left[0 \ 0 \ 0 \ \bar{m}_y \ 0\right]^T \tag{14.6.15e}$$

Now, we take the first integral term in the right-hand side of Equation (14.6.14), and we separate it into individual element contributions. We also account for the fact that the virtual displacement field in each element is approximated with the same approximation (the same shape functions) as the actual displacement field:

$$\left\{w^{(e)}\right\}^T = \left[w_{ox}^{(e)}(\xi,\eta) \ w_{oy}^{(e)}(\xi,\eta) \ w_{oz}^{(e)}(\xi,\eta) \ w_{\theta 1}^{(e)}(\xi,\eta) \ w_{\theta 2}^{(e)}(\xi,\eta)\right] = \left\{W^{(e)}\right\}^T \left[N^{(e)}\right]^T \tag{14.6.16}$$

We now have:

$$\iint_{\Omega} \{w\}^T \{p\} dV = \sum_{e=1}^{N_e} \left(\iint_{\Omega^{(e)}} \{w\}^T \{p\} dV \right) = \sum_{e=1}^{N_e} \left(\iint_{\Omega^{(e)}} \{W\}^T \left[L^{(e)}\right]^T \left[N^{(e)}\right]^T \{p\} dV \right)$$

$$= \{W\}^T \sum_{e=1}^{N_e} \left(\left[L^{(e)}\right]^T \iint_{\Omega^{(e)}} \left[N^{(e)}\right]^T \{p\} dV \right) \rightarrow$$

$$\rightarrow \iint_{\Omega} \{w\}^T \{p\} dV = \{W\}^T \sum_{e=1}^{N_e} \left(\left[L^{(e)}\right]^T \left\{f_{\Omega}^{(e)}\right\} \right) \tag{14.6.17}$$

where

$$\left\{f_{\Omega}^{(e)}\right\} = \iint_{\Omega^{(e)}} \left[N^{(e)}\right]^T \{p\} dV \tag{14.6.18}$$

The first natural boundary integral term in the right-hand-side of Equation (14.6.14) becomes:

$$\int_{\Gamma_{tx}} \{w\}^T \left\{\hat{f}_x\right\} dS = \sum_{e=1}^{N_e} \left(\int_{\Gamma_{tx}^{(e)}} \{w\}^T \left\{\hat{f}_x\right\} dS \right) = \sum_{e=1}^{N_e} \left(\int_{\Gamma_{tx}^{(e)}} \{W\}^T \left[L^{(e)}\right]^T \left[N^{(e)}\right] \left\{\hat{f}_x\right\} dS \right) \rightarrow$$

$$\rightarrow \int_{\Gamma_{tx}} \{w\}^T \left\{\hat{f}_x\right\} dS = \{W\}^T \sum_{e=1}^{N_e} \left(\left[L^{(e)}\right] \int_{\Gamma_{tx}^{(e)}} \left[N^{(e)}\right]^T \left\{\hat{f}_x\right\} dS \right) \tag{14.6.19a}$$

In the same fashion:

$$\int_{\Gamma_{ty}} \{w\}^T \left\{\hat{f}_y\right\} dS = \{W\}^T \sum_{e=1}^{N_e} \left(\left[L^{(e)}\right]^T \int_{\Gamma_{ty}^{(e)}} \left[N^{(e)}\right]^T \left\{\hat{f}_y\right\} dS \right) \tag{14.6.19b}$$

$$\int_{\Gamma_{tz}} \{w\}^T \left\{\hat{f}_z\right\} dS = \{W\}^T \sum_{e=1}^{N_e} \left(\left[L^{(e)}\right]^T \int_{\Gamma_{tz}^{(e)}} \left[N^{(e)}\right]^T \left\{\hat{f}_y\right\} dS \right) \tag{14.6.19c}$$

$$\int_{\Gamma_{t\theta 1}} \{w\}^T \{\hat{m}_x\} dS = \{W\}^T \sum_{e=1}^{N_e} \left(\left[L^{(e)}\right]^T \int_{\Gamma_{t\theta 1}^{(e)}} \left[N^{(e)}\right]^T \{\hat{m}_x\} dS \right) \tag{14.6.19d}$$

$$\int_{\Gamma_{t\theta 2}} \{w\}^T \{\hat{m}_y\} dS = \{W\}^T \sum_{e=1}^{N_e} \left(\left[L^{(e)}\right]^T \int_{\Gamma_{t\theta 2}^{(e)}} \left[N^{(e)}\right]^T \{\hat{m}_y\} dS \right) \tag{14.6.19e}$$

Plugging Equations (14.6.18) and (14.6.19) into (14.6.14) yields:

$$\begin{aligned}
&\iint_{\Omega} \{w\}^T \{p\} dV + \int_{\Gamma_{tx}} \{w\}^T \left\{\hat{f}_x\right\} dS + \int_{\Gamma_{ty}} \{w\}^T \left\{\hat{f}_y\right\} dS + \int_{\Gamma_{tz}} \{w\}^T \left\{\hat{f}_z\right\} dS \\
&+ \int_{\Gamma_{t\theta 1}} \{w\}^T \{\hat{m}_x\} dS + \int_{\Gamma_{t\theta 2}} \{w\}^T \{\hat{m}_y\} dS \\
&= \{W\}^T \sum_{e=1}^{N_e} \left(\left[L^{(e)}\right]^T \left\{f^{(e)}\right\} \right)
\end{aligned} \tag{14.6.20}$$

where $\{f^{(e)}\}$ is the element's *equivalent nodal force/moment vector*:

$$\left\{f^{(e)}\right\} = \left\{f_{\Omega}^{(e)}\right\} + \left\{f_{\Gamma}^{(e)}\right\} \tag{14.6.21}$$

and we have also defined a vector $\left\{f_{\Gamma}^{(e)}\right\}$, containing the equivalent nodal forces/moments due to natural boundary conditions:

$$\left\{f_{\Gamma}^{(e)}\right\} = \int_{\Gamma_{tx}^{(e)}} \left[N^{(e)}\right]^T \left\{\hat{f}_x\right\} dS + \int_{\Gamma_{ty}^{(e)}} \left[N^{(e)}\right]^T \left\{\hat{f}_y\right\} dS + \int_{\Gamma_{tz}^{(e)}} \left[N^{(e)}\right]^T \left\{\hat{f}_z\right\} dS$$
$$+ \int_{\Gamma_{t\theta 1}^{(e)}} \left[N^{(e)}\right]^T \{\hat{m}_x\} dS + \int_{\Gamma_{t\theta 2}^{(e)}} \left[N^{(e)}\right]^T \{\hat{m}_y\} dS \tag{14.6.22}$$

Accounting for Equations (14.6.12) and (14.6.20) allows us to write the finite element approximation of the weak form:

$$\{W\}^T[K]\{U\} = \{W\}^T\{f\} \rightarrow \{W\}^T([K]\{U\} - \{f\}) = 0 \tag{14.6.23}$$

where $[K]$ is the *global stiffness matrix*:

$$[K] = \sum_{e=1}^{N_e} \left(\left[L^{(e)}\right]^T \left[k^{(e)}\right] \left[L^{(e)}\right]\right) \tag{14.6.24}$$

and $\{f\}$ is the *global equivalent nodal force vector*:

$$\{f\} = \sum_{e=1}^{N_e} \left(\left[L^{(e)}\right]^T \left\{f^{(e)}\right\}\right) \tag{14.6.25}$$

Since Equation (14.6.23) must apply for all arbitrary global virtual nodal displacement/rotation vectors, $\{W\}$, the only way to always obtain a zero right-hand side in the equation is to have:

$$[K]\{U\} - \{f\} = \{0\} \tag{14.6.26}$$

Equation (14.6.26) constitutes the *global (structural) finite element equations* for a shell structure. Once again, this equation is mathematically identical to that obtained for the finite element solution of other problems in previous chapters. In the following section, we will specifically examine the case of a four-node, quadrilateral shell element.

14.7 Four-Node Planar (Flat) Shell Finite Element

One of the simplest cases of shell finite elements is the four-node planar shell element, shown in Figure 14.11. For this element, we can use an isoparametric formulation and employ the same shape functions that we used for the four-node quadrilateral (4Q) element in Section 6.4. The nodal degrees of freedom for a four-node, three-dimensional, flat shell element are shown in Figure 14.11. It may be worth emphasizing that the

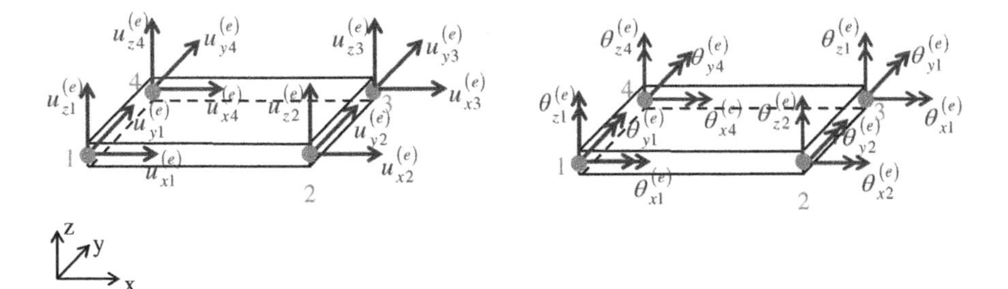

Figure 14.11 Four-node flat shell finite element and nodal degrees of freedom.

formulation that we will present is valid for a general quadrilateral flat shell element (i.e., the element does not need to be rectangular as that of Figure 14.11).

The approximation for the displacement fields is a special case of Equation (14.6.2), if we set $n = 4$:

$$\begin{Bmatrix} u_{ox}^{(e)}(\xi,\eta) \\ u_{oy}^{(e)}(\xi,\eta) \\ u_{oz}^{(e)}(\xi,\eta) \\ \theta_1^{(e)}(\xi,\eta) \\ \theta_2^{(e)}(\xi,\eta) \end{Bmatrix} = \left[\left[N_1^{(4Q)}(\xi,\eta)\right] \; \left[N_2^{(4Q)}(\xi,\eta)\right] \; \left[N_3^{(4Q)}(\xi,\eta)\right] \; \left[N_n^{(4Q)}(\xi,\eta)\right]\right] \begin{Bmatrix} \left\{U_1^{(e)}\right\} \\ \left\{U_2^{(e)}\right\} \\ \left\{U_3^{(e)}\right\} \\ \left\{U_4^{(e)}\right\} \end{Bmatrix} \tag{14.7.1}$$

Since the formulation is isoparametric, we can also establish a parametric representation of the geometry, through a coordinate mapping from the parametric domain to the physical domain:

$$x(\xi,\eta) = N_1^{(4Q)}(\xi,\eta)\cdot x_1^{(e)} + N_2^{(4Q)}(\xi,\eta)\cdot x_2^{(e)} + N_3^{(4Q)}(\xi,\eta)\cdot x_3^{(e)} + N_4^{(4Q)}(\xi,\eta)\cdot x_4^{(e)} \tag{14.7.2a}$$

$$y(\xi,\eta) = N_1^{(4Q)}(\xi,\eta)\cdot y_1^{(e)} + N_2^{(4Q)}(\xi,\eta)\cdot y_2^{(e)} + N_3^{(4Q)}(\xi,\eta)\cdot y_3^{(e)} + N_4^{(4Q)}(\xi,\eta)\cdot y_4^{(e)} \tag{14.7.2b}$$

Equations (14.7.2a,b) can also be written as follows:

$$x(\xi,\eta) = \left[N^{(4Q)}(\xi,\eta)\right]\left\{x^{(e)}\right\} \tag{14.7.3a}$$

$$y(\xi,\eta) = \left[N^{(4Q)}(\xi,\eta)\right]\left\{y^{(e)}\right\} \tag{14.7.3b}$$

where

$$\left\{x^{(e)}\right\} = \left[x_1^{(e)} \; x_2^{(e)} \; x_3^{(e)} \; x_4^{(e)}\right]^T \tag{14.7.4a}$$

$$\left\{y^{(e)}\right\} = \left[y_1^{(e)} \; y_2^{(e)} \; y_3^{(e)} \; y_4^{(e)}\right]^T \tag{14.7.4b}$$

and

$$\left[N^{(4Q)}(\xi,\eta)\right] = \left[N_1^{(4Q)}(\xi,\eta) \; N_2^{(4Q)}(\xi,\eta) \; N_3^{(4Q)}(\xi,\eta) \; N_4^{(4Q)}(\xi,\eta)\right] \tag{14.7.4c}$$

It is reminded that the shape functions of a 4Q element are:

$$N_1^{(4Q)}(\xi,\eta) = \frac{1}{4}(1-\xi)(1-\eta) \tag{14.7.5a}$$

$$N_2^{(4Q)}(\xi,\eta) = \frac{1}{4}(1-\xi)(1-\eta) \tag{14.7.5b}$$

$$N_3^{(4Q)}(\xi,\eta) = \frac{1}{4}(1-\xi)(1-\eta) \tag{14.7.5c}$$

$$N_4^{(4Q)}(\xi,\eta) = \frac{1}{4}(1-\xi)(1-\eta) \tag{14.7.5d}$$

We also need to establish the Jacobian matrix $[J]$ of the coordinate mapping:

$$[J] = \begin{bmatrix} \frac{\partial x}{\partial \xi} & \frac{\partial y}{\partial \xi} \\ \frac{\partial x}{\partial \eta} & \frac{\partial y}{\partial \eta} \end{bmatrix} = \begin{bmatrix} \sum_{i=1}^{4}\left(\frac{\partial N_i^{(4Q)}}{\partial \xi} \cdot x_i^{(e)}\right) & \sum_{i=1}^{4}\left(\frac{\partial N_i^{(4Q)}}{\partial \xi} \cdot y_i^{(e)}\right) \\ \sum_{i=1}^{4}\left(\frac{\partial N_i^{(4Q)}}{\partial \eta} \cdot x_i^{(e)}\right) & \sum_{i=1}^{4}\left(\frac{\partial N_i^{(4Q)}}{\partial \eta} \cdot y_i^{(e)}\right) \end{bmatrix} = \left[N_{,\xi}^{(4Q)}\right]\left[\left\{x^{(e)}\right\} \; \left\{y^{(e)}\right\}\right] \tag{14.7.6}$$

where

$$\left[N_{,\xi}^{(4Q)}\right] = \begin{bmatrix} \frac{\partial N_1^{(4Q)}}{\partial \xi} & \frac{\partial N_2^{(4Q)}}{\partial \xi} & \frac{\partial N_3^{(4Q)}}{\partial \xi} & \frac{\partial N_4^{(4Q)}}{\partial \xi} \\ \frac{\partial N_1^{(4Q)}}{\partial \eta} & \frac{\partial N_2^{(4Q)}}{\partial \eta} & \frac{\partial N_3^{(4Q)}}{\partial \eta} & \frac{\partial N_4^{(4Q)}}{\partial \eta} \end{bmatrix} \tag{14.7.7}$$

Given the above, we can establish the components of the $[B^{(e)}]$ array as a special case of Equation (14.6.10) with $n = 4$. Since this array contains derivatives of shape functions with respect to the physical coordinates, we have to use the chain rule of differentiation, and of course the components of the $[\tilde{J}]$ array (i.e., the inverse of $[J]$). The key is to use the following expression, previously provided as Equation (8.5.33), to find the partial derivatives of the shape functions with respect to the spatial (Cartesian) coordinates.

$$\left[N_{,x}^{(4Q)}\right] = \begin{bmatrix} \frac{\partial N_1^{(4Q)}}{\partial x} & \frac{\partial N_2^{(4Q)}}{\partial x} & \frac{\partial N_3^{(4Q)}}{\partial x} & \frac{\partial N_4^{(4Q)}}{\partial x} \\ \frac{\partial N_1^{(4Q)}}{\partial y} & \frac{\partial N_2^{(4Q)}}{\partial y} & \frac{\partial N_3^{(4Q)}}{\partial y} & \frac{\partial N_4^{(4Q)}}{\partial y} \end{bmatrix} = [\tilde{J}]\left[N_{,\xi}^{(4Q)}\right] \tag{8.5.33}$$

Finally, we can employ Gaussian quadrature for the calculation of the various integrals (2×2 quadrature is required for full integration). The procedure to obtain the element stiffness matrix, $[k^{(e)}]$, and the part of the equivalent nodal force vector, $\left\{f_\Omega^{(e)}\right\}$, due to distributed surface forces, for the case where we use N_g quadrature points, is summarized in Box 14.7.1.

Box 14.7.1 Calculation of $[k^{(e)}]$, $\{f_{\Omega}^{(e)}\}$ for 4Q flat shell element

FOR EACH QUADRATURE POINT g ($g = 1, 2, \ldots, N_g$):

i) Establish parametric coordinates, ξ_g, η_g.
ii) Calculate the values of the shape functions at the specific point:

$$N_{1g}^{(4Q)} = N_1^{(4Q)}(\xi_g, \eta_g) = \frac{1}{4}(1-\xi_g)(1-\eta_g), \quad N_{2g}^{(4Q)} = N_2^{(4Q)}(\xi_g, \eta_g) = \frac{1}{4}(1+\xi_g)(1-\eta_g)$$

$$N_{3g}^{(4Q)} = N_3^{(4Q)}(\xi_g, \eta_g) = \frac{1}{4}(1+\xi_g)(1+\eta_g), \quad N_{4g}^{(4Q)} = N_4^{(4Q)}(\xi_g, \eta_g) = \frac{1}{4}(1-\xi_g)(1+\eta_g)$$

Also, establish the shape function array $\left[N^{(4Q)}\right]_g = \left[N^{(4Q)}(\xi_g, \eta_g)\right]$ per Equation (14.7.4c).

iii) Find the values of the block shape function array $\left[N_I^{(e)}\right]_g$ of each node $I = 1,2,3,4$ using Equation (14.6.3). Then, find the shell element shape function array:

$$\left[N^{(e)}\right]_g = \left[\left[N_1^{(e)}\right]_g \ \left[N_2^{(e)}\right]_g \ \left[N_3^{(e)}\right]_g \ \left[N_4^{(e)}\right]_g\right]$$

iv) Calculate values of coordinates in physical space, $x_g = x(\xi_g, \eta_g), y_g = y(\xi_g, \eta_g)$:

$$x_g = \left[N^{(4Q)}\right]_g \left\{x^{(e)}\right\}, \quad y_g = \left[N^{(4Q)}\right]_g \left\{y^{(e)}\right\}$$

v) Find the matrix $\left[N_{,\xi}^{(4Q)}\right]_g$ using Equation (14.7.7) for $\xi = \xi_g$, $\eta = \eta_g$.
vi) Calculate the Jacobian matrix of the mapping at that point using Equation (14.7.6), as well as the Jacobian determinant, $J_g = det([J_g])$, and the inverse of the Jacobian matrix, $\left[\tilde{J}_g\right] = \left[J_g\right]^{-1}$.
vii) Calculate the derivatives of the shape functions with respect to the physical coordinates, x and y, at $\xi = \xi_g$, $\eta = \eta_g$, using Equation (8.5.33): $\left[N_{,x}^{(4Q)}\right]_g = \left[\tilde{J}_g\right]\left[N_{,\xi}^{(4Q)}\right]_g$.
viii) Find the block strain-displacement arrays, $\left[B_{mI}^{(e)}\right]_g$, $\left[B_{fI}^{(e)}\right]_g$, $\left[B_{sI}^{(e)}\right]_g$, for each node $I = 1,2,3,4$, using Equations (14.6.10a–c); then, define the shell element strain-displacement array, $[B^{(e)}]_g$, at the location of Gauss point g, using Equation (14.6.10d) for $n = 4$.
ix) Calculate $\left[\hat{D}_g\right] = \left[\hat{D}(x_g, y_g)\right]$.
x) Calculate $\{p_g\} = \{p(x_g, y_g)\}$.

Finally, combine the contributions of all the Gauss points to calculate the element stiffness matrix, $[k^{(e)}]$, and distributed-force contribution to the equivalent nodal force vector, $\left\{f_{\Omega}^{(e)}\right\}$:

$$\left[k^{(e)}\right] = \int_{-1}^{1}\int_{-1}^{1}\left(\left[B^{(e)}\right]^T\left[D^{(e)}\right]\left[B^{(e)}\right] J d\xi d\eta\right) \approx \sum_{g=1}^{Ng}\left(\left[B^{(e)}\right]_g^T\left[D_g\right]\left[B^{(e)}\right]_g J_g W_g\right)$$

$$\left\{f_{\Omega}^{(e)}\right\} = \iint_{\Omega^{(e)}}\left[N^{(e)}\right]^T\{p\} dV = \int_{-1}^{1}\int_{-1}^{1}\left[N^{(e)}(\xi,\eta)\right]^T\{p(\xi,\eta)\} J d\xi d\eta \approx \sum_{g=1}^{Ng}\left(\left[N^{(e)}\right]_g^T\{p_g\} J_g W_g\right)$$

After we have obtained $[k^{(e)}]$ and $\left\{f_\Omega^{(e)}\right\}$, we complete the element computation, by calculating the part $\left\{f_\Gamma^{(e)}\right\}$ of the equivalent nodal forces due to natural boundary conditions. To this end, we use a single-variable parameterization of each boundary segment, and calculate (with one-dimensional Gaussian quadrature) the contribution of the natural boundary forces and moments on $\left\{f_\Gamma^{(e)}\right\}$, as summarized in Box 14.7.2.

Box 14.7.2 Calculation of $\{f_{\Gamma t}^{(e)}\}$ for 4Q Flat Shell Element

First, define a vector $\{f_\Gamma\} = \left[f_x \ f_y \ f_z \ m_{1\Gamma} \ m_{2\Gamma}\right]^T$, containing the prescribed boundary forces and/or moments (note that, if only several of the components of $\{f_\Gamma\}$ are prescribed, we can use the values of these components and simply set the value of the other boundary force/moment components equal to zero).

- If the natural boundary segment corresponds to a constant value of η, $\eta = \bar{\eta}$ (where $\bar{\eta} = -1$ or 1), we need to evaluate a one-dimensional integral with respect to ξ:

$$\left\{f_{\Gamma_t}^{(e)}\right\} = \int_{\Gamma_t^{(e)}} \left[N^{(4Q)}\right]^T \{f_\Gamma\} ds = \int_{-1}^{1} \left(\left[N^{(4Q)}(\xi,\bar{\eta})\right]^T \{f_\Gamma\} \frac{\ell_\Gamma}{2} d\xi\right)$$

$$\approx \sum_{g=1}^{Ng} \left(\left[N^{(4Q)}(\xi_g,\bar{\eta})\right]^T \{f_\Gamma\} \frac{\ell_\Gamma}{2} W_g\right)$$

  where ℓ_Γ is the length of the natural boundary segment.
- If the natural boundary segment corresponds to a constant value of ξ, $\xi = \bar{\xi}$ (where $\bar{\xi} = -1$ or 1), we need to evaluate a one-dimensional integral with respect to η:

$$\left\{f_{\Gamma_t}^{(e)}\right\} = \int_{\Gamma_t^{(e)}} \left[N^{(4Q)}\right]^T \{f_\Gamma\} ds = \int_{-1}^{1} \left(\left[N^{(4Q)}(\bar{\xi},\eta)\right]^T \{f_\Gamma\} \frac{\ell_\Gamma}{2} d\eta\right)$$

$$\approx \sum_{g=1}^{Ng} \left(\left[N^{(4Q)}(\bar{\xi},\eta_g)\right]^T \{f_\Gamma\} \frac{\ell_\Gamma}{2} W_g\right)$$

Remark 14.7.1: One can verify that a shell element using the formulation presented in this section does not develop any stiffness for the rotational degrees of freedom along the local z-axis, called the *drilling degrees of freedom*! If these degrees of freedom are not connected to other elements in a three-dimensional mesh, which develop stiffness, then there may be problems in the solution of the global stiffness equations. Several studies have been focused on the definition of mathematically meaningful drilling stiffness terms for elastic shells (e.g., Hughes and Brezzi (1989)). The incorporation of such drilling stiffness is possible in many commercial finite element programs; however, it will not be discussed in this text. It is worth mentioning that if all the shell elements in a mesh are coplanar (i.e., they are all planar and lie on the same plane xy), then one can simply restrain all the drilling rotational nodal degrees-of-freedom to a zero value. ■

14.8 Coordinate Transformations for Shell Elements

The considerations of the previous Sections were based on the assumption that the shell x-and y-axes coincide with the global x- and y- coordinate axes. In general analyses, the various flat shell elements may not all lie on the same plane, as shown, e.g., in Figure 14.12 for the analysis of a wall structure with shell elements. Obviously, the normal direction (local z-axis) is not identical for the various elements in this case.

The expressions established in Section 14.7 for $[k^{(e)}]$, $\{f^{(e)}\}$, etc. are always valid for the element *local coordinate system,* in which axes x and y lie in the plane of the element and axis z is normal to the plane of the element's mid-surface.

For cases involving elements that do not all have the same normal direction, the global equations must be formulated in a unique, *global (or structural) coordinate system, XYZ.* The stiffness matrix and equivalent nodal force/moment vector of each element are initially established in the local coordinate system. Just like for beams in Section 13.7 (and for two-dimensional trusses in Section B.8), it is necessary to establish a *coordinate transformation equation,* from the local to the global coordinate system. We must apply the coordinate transformation equations to the element stiffness matrix and equivalent nodal force vector before we conduct the assembly operation. In other words, the assembly equations are conducted for the stiffness and nodal force expressed in the global coordinate system. This section is dedicated to the formulation of the coordinate transformation for shells. As a starting point, we will assume that we are given the nodal coordinates in the global coordinate system. Specifically, for each node I, we will have a triad of coordinate values X_I, Y_I, Z_I.

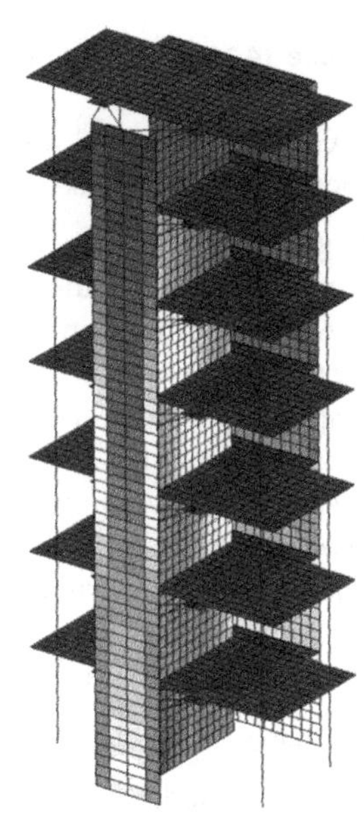

Figure 14.12 Flat shell element model for a reinforced concrete wall.

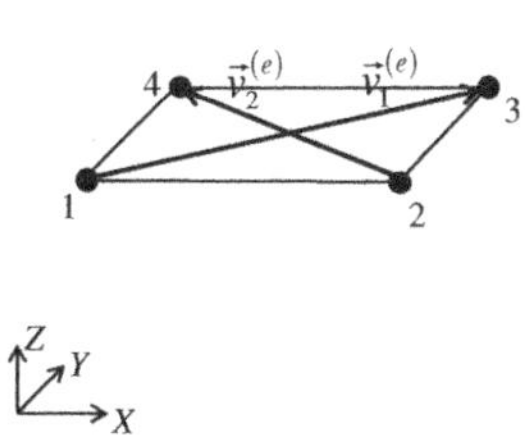

The first step toward establishing the local coordinate system of a quadrilateral shell element is to calculate the unit vectors in the directions of the diagonals, as shown in Figure 14.13. The figure also provides the mathematical

$$\vec{v}_1^{(e)} = \begin{Bmatrix} v_{1x}^{(e)} \\ v_{1y}^{(e)} \\ v_{1z}^{(e)} \end{Bmatrix} = \frac{1}{\ell_{31}^{(e)}} \begin{Bmatrix} X_3^{(e)} - X_1^{(e)} \\ Y_3^{(e)} - Y_1^{(e)} \\ Z_3^{(e)} - Z_1^{(e)} \end{Bmatrix}, \quad \ell_{31}^{(e)} = \sqrt{\left(X_3^{(e)} - X_1^{(e)}\right)^2 + \left(Y_3^{(e)} - Y_1^{(e)}\right)^2 + \left(Z_3^{(e)} - Z_1^{(e)}\right)^2}$$

$$\vec{v}_2^{(e)} = \begin{Bmatrix} v_{2x}^{(e)} \\ v_{2y}^{(e)} \\ v_{2z}^{(e)} \end{Bmatrix} = \frac{1}{\ell_{41}^{(e)}} \begin{Bmatrix} X_4^{(e)} - X_2^{(e)} \\ Y_4^{(e)} - Y_2^{(e)} \\ Z_4^{(e)} - Z_2^{(e)} \end{Bmatrix}, \quad \ell_{41}^{(e)} = \sqrt{\left(X_4^{(e)} - X_2^{(e)}\right)^2 + \left(Y_4^{(e)} - Y_2^{(e)}\right)^2 + \left(Z_4^{(e)} - Z_2^{(e)}\right)^2}$$

Figure 14.13 Establishing unit vectors in the directions of the diagonals of a quadrilateral shell element.

equations giving the unit vector $\vec{v}_1^{(e)}$ in the direction of diagonal (1-3), and the unit vector $\vec{v}_2^{(e)}$ in the direction of diagonal (2-4).

Having obtained the two unit vectors, $\vec{v}_1^{(e)}$ and $\vec{v}_2^{(e)}$, we can use the fact that these vectors lie in the plane of the shell to find the unit vector $\vec{n}^{(e)}$ normal to the plane of the element. This vector is obtained as the cross product (or exterior product) of the two "diagonal" vectors[2]. The normal vector $\vec{n}^{(e)}$ and the mathematical expression yielding this vector are provided in Figure 14.14.

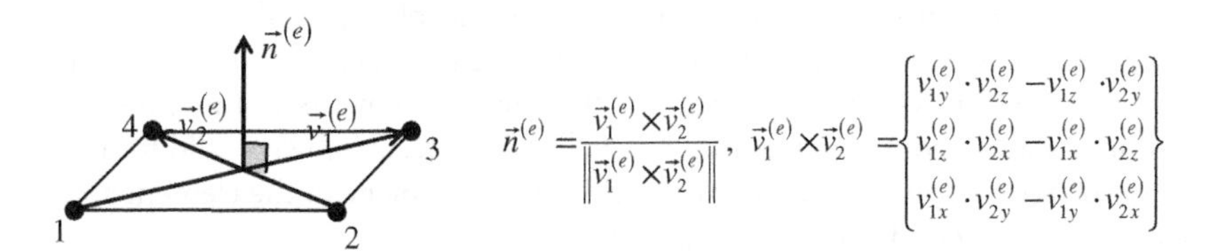

$$\vec{n}^{(e)} = \frac{\vec{v}_1^{(e)} \times \vec{v}_2^{(e)}}{\left\| \vec{v}_1^{(e)} \times \vec{v}_2^{(e)} \right\|}, \quad \vec{v}_1^{(e)} \times \vec{v}_2^{(e)} = \begin{Bmatrix} v_{1y}^{(e)} \cdot v_{2z}^{(e)} - v_{1z}^{(e)} \cdot v_{2y}^{(e)} \\ v_{1z}^{(e)} \cdot v_{2x}^{(e)} - v_{1x}^{(e)} \cdot v_{2z}^{(e)} \\ v_{1x}^{(e)} \cdot v_{2y}^{(e)} - v_{1y}^{(e)} \cdot v_{2x}^{(e)} \end{Bmatrix}$$

Figure 14.14 Establishing the unit normal vector for a quadrilateral shell element.

Given $\vec{v}_1^{(e)}$ and $\vec{v}_2^{(e)}$, we also establish a pair of unit vectors, $\vec{t}_1^{(e)}$ and $\vec{t}_2^{(e)}$, providing the local x- and local y-axes, respectively, of the element. The local z-direction of the element is given by vector $\vec{n}^{(e)}$. The orientation of vectors $\vec{t}_1^{(e)}$ and $\vec{t}_2^{(e)}$ and the mathematical expressions giving these two vectors are shown in Figure 14.15.

After the establishment of the local coordinate axes, we can write the coordinate transformation equations for nodal displacements and rotations. The equations for the nodal displacements are obtained as a special case of Equation (7.6.2). An important consideration pertaining to the nodal rotations will be used here: *for small nodal rotations (which is always an assumption made in linear analysis!), the three components of nodal rotation can be considered as the three components of a vector.* This means that the same coordinate transformation equations can be used for nodal displacements and nodal rotations.

Let us start with the nodal displacements. If, for a nodal point I of an element, we have the nodal displacements $\left[u_{Ix}^{(e)} \; u_{Iy}^{(e)} \; u_{Iz}^{(e)}\right]^T$ in the global coordinate system, then the nodal displacements of the same node in the local coordinate system, $\left[u_{Ix}^{\prime(e)} \; u_{Iy}^{\prime(e)} \; u_{Iz}^{\prime(e)}\right]^T$, are given by the following equation.

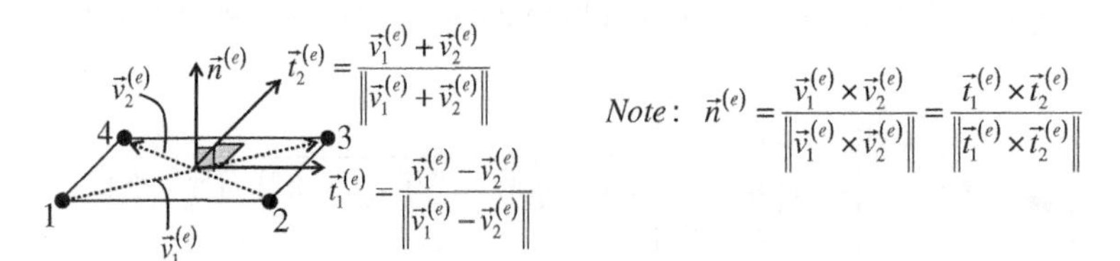

$$\vec{t}_2^{(e)} = \frac{\vec{v}_1^{(e)} + \vec{v}_2^{(e)}}{\left\| \vec{v}_1^{(e)} + \vec{v}_2^{(e)} \right\|}, \quad \vec{t}_1^{(e)} = \frac{\vec{v}_1^{(e)} - \vec{v}_2^{(e)}}{\left\| \vec{v}_1^{(e)} - \vec{v}_2^{(e)} \right\|}$$

$$Note: \; \vec{n}^{(e)} = \frac{\vec{v}_1^{(e)} \times \vec{v}_2^{(e)}}{\left\| \vec{v}_1^{(e)} \times \vec{v}_2^{(e)} \right\|} = \frac{\vec{t}_1^{(e)} \times \vec{t}_2^{(e)}}{\left\| \vec{t}_1^{(e)} \times \vec{t}_2^{(e)} \right\|}$$

Figure 14.15 Establishing the unit vectors giving the local in-plane coordinate axes of a quadrilateral planar shell element.

---

2 It is known from vector calculus that the cross-product of two vectors is a third vector, perpendicular to the plane defined by the two vectors.

$$\begin{Bmatrix} u_{Ix}^{\prime(e)} \\ u_{Iy}^{\prime(e)} \\ u_{Iz}^{\prime(e)} \end{Bmatrix} = \begin{bmatrix} t_{1x}^{(e)} & t_{1y}^{(e)} & t_{1z}^{(e)} \\ t_{2x}^{(e)} & t_{2y}^{(e)} & t_{2z}^{(e)} \\ n_x^{(e)} & n_y^{(e)} & n_z^{(e)} \end{bmatrix} \begin{Bmatrix} u_{Ix}^{(e)} \\ u_{Iy}^{(e)} \\ u_{Iz}^{(e)} \end{Bmatrix} \tag{14.8.1}$$

The first row of the coefficient array in Equation (14.8.1) consists of the three components of the unit vector $\vec{t}_1^{(e)}$, the second row of the array includes the three components of vector $\vec{t}_2^{(e)}$, and the third row includes the components of vector $\vec{n}^{(e)}$.

Similarly, if we have the nodal rotations $\left[\theta_{Ix}^{(e)} \; \theta_{Iy}^{(e)} \; \theta_{Iz}^{(e)}\right]^T$ for a nodal point I in the global coordinate system, then the corresponding rotations in the local coordinate system, $\left[\theta_{Ix}^{\prime(e)} \; \theta_{Iy}^{\prime(e)} \; \theta_{Iz}^{\prime(e)}\right]^T$, are given by the following expression.

$$\begin{Bmatrix} \theta_{Ix}^{\prime(e)} \\ \theta_{Iy}^{\prime(e)} \\ \theta_{Iz}^{\prime(e)} \end{Bmatrix} = \begin{bmatrix} t_{1x}^{(e)} & t_{1y}^{(e)} & t_{1z}^{(e)} \\ t_{2x}^{(e)} & t_{2y}^{(e)} & t_{2z}^{(e)} \\ n_x^{(e)} & n_y^{(e)} & n_z^{(e)} \end{bmatrix} \begin{Bmatrix} \theta_{Ix}^{(e)} \\ \theta_{Iy}^{(e)} \\ \theta_{Iz}^{(e)} \end{Bmatrix} \tag{14.8.2}$$

We can collectively cast Equations (14.8.1) and (14.8.2) into a single matrix expression:

$$\left\{U_I^{\prime(e)}\right\} = \begin{Bmatrix} u_{Ix}^{\prime(e)} \\ u_{Iy}^{\prime(e)} \\ u_{Iz}^{\prime(e)} \\ \theta_{Ix}^{\prime(e)} \\ \theta_{Iy}^{\prime(e)} \\ \theta_{Iz}^{\prime(e)} \end{Bmatrix} = \begin{bmatrix} t_{1x}^{(e)} & t_{1y}^{(e)} & t_{1z}^{(e)} & 0 & 0 & 0 \\ t_{2x}^{(e)} & t_{2y}^{(e)} & t_{2z}^{(e)} & 0 & 0 & 0 \\ n_x^{(e)} & n_y^{(e)} & n_z^{(e)} & 0 & 0 & 0 \\ 0 & 0 & 0 & t_{1x}^{(e)} & t_{1y}^{(e)} & t_{1z}^{(e)} \\ 0 & 0 & 0 & t_{2x}^{(e)} & t_{2y}^{(e)} & t_{2z}^{(e)} \\ 0 & 0 & 0 & n_x^{(e)} & n_y^{(e)} & n_z^{(e)} \end{bmatrix} \begin{Bmatrix} u_{Ix}^{(e)} \\ u_{Iy}^{(e)} \\ u_{Iz}^{(e)} \\ \theta_{Ix}^{(e)} \\ \theta_{Iy}^{(e)} \\ \theta_{Iz}^{(e)} \end{Bmatrix} = \left[r^{(e)}\right]\left\{U_I^{(e)}\right\} \tag{14.8.3}$$

where $[r^{(e)}]$ is the coordinate transformation array for the six degrees of freedom of each node in element e, given by:

$$\left[r^{(e)}\right] = \begin{bmatrix} t_{1x}^{(e)} & t_{1y}^{(e)} & t_{1z}^{(e)} & 0 & 0 & 0 \\ t_{2x}^{(e)} & t_{2y}^{(e)} & t_{2z}^{(e)} & 0 & 0 & 0 \\ n_x^{(e)} & n_y^{(e)} & n_z^{(e)} & 0 & 0 & 0 \\ 0 & 0 & 0 & t_{1x}^{(e)} & t_{1y}^{(e)} & t_{1z}^{(e)} \\ 0 & 0 & 0 & t_{2x}^{(e)} & t_{2y}^{(e)} & t_{2z}^{(e)} \\ 0 & 0 & 0 & n_x^{(e)} & n_y^{(e)} & n_z^{(e)} \end{bmatrix} \tag{14.8.4}$$

We can establish a block-matrix coordinate transformation equation for the full nodal displacement/rotation vector of a four-node flat shell element e (containing the nodal displacements and rotations of all four nodal points):

$$\begin{Bmatrix} \left\{U_1'^{(e)}\right\} \\ \left\{U_2'^{(e)}\right\} \\ \left\{U_3'^{(e)}\right\} \\ \left\{U_4'^{(e)}\right\} \end{Bmatrix} = \begin{bmatrix} [r^{(e)}] & [0] & [0] & [0] \\ [0] & [r^{(e)}] & [0] & [0] \\ [0] & [0] & [r^{(e)}] & [0] \\ [0] & [0] & [0] & [r^{(e)}] \end{bmatrix} \begin{Bmatrix} \left\{U_1^{(e)}\right\} \\ \left\{U_2^{(e)}\right\} \\ \left\{U_3^{(e)}\right\} \\ \left\{U_4^{(e)}\right\} \end{Bmatrix}$$

$$\rightarrow \left\{U'^{(e)}\right\} = \left[R^{(e)}\right]\left\{U^{(e)}\right\} \tag{14.8.5}$$

where $[R^{(e)}]$ is the (24×24) coordinate transformation array for a four-node flat shell element:

$$\left[R^{(e)}\right] = \begin{bmatrix} [r^{(e)}] & [0] & [0] & [0] \\ [0] & [r^{(e)}] & [0] & [0] \\ [0] & [0] & [r^{(e)}] & [0] \\ [0] & [0] & [0] & [r^{(e)}] \end{bmatrix} \tag{14.8.6}$$

It can easily be proven that, for nodal forces and moments, the following transformation equation applies:

$$\left\{F^{(e)}\right\} = \left[R^{(e)}\right]^T \left\{F'^{(e)}\right\} \tag{14.8.7}$$

Finally, we can obtain (with a procedure similar to that employed in Section 13.7) the coordinate transformation equation for the element stiffness matrix:

$$\left[k^{(e)}\right] = \left[R^{(e)}\right]^T \left[k'^{(e)}\right]\left[R^{(e)}\right] \tag{14.8.8}$$

where $[k'^{(e)}]$ is the stiffness matrix in the local coordinate system, and $[k^{(e)}]$ is the same quantity in the global coordinate system. In summary, if we need to conduct coordinate transformation for a shell element, we use Equations (14.6.13) and (14.6.21) to find $[k'^{(e)}]$ and $\{f'^{(e)}\}$, respectively. Then, we find $[k^{(e)}]$ and $\{f^{(e)}\}$ (the stiffness matrix and equivalent nodal force vector in the global coordinate system) using Equations (14.8.8) and (14.8.7), respectively. Finally, we can conduct the assembly operations (Equations 14.6.24 and 14.6.25) with the arrays $[k^{(e)}]$ and vectors $\{f^{(e)}\}$ of all elements.

Remark 14.8.1: It is important to remember that the physical coordinates of the nodal points must be expressed in the local coordinate system, before being used in the calculations described in Section 14.7. If we have the coordinates $\left[X_I^{(e)} \; Y_I^{(e)} \; Z_I^{(e)}\right]^T$ of nodal point I of element e in the global coordinate system, then the corresponding coordinates of the same nodal point in the global coordinate system can be obtained from the following equation.

$$\begin{Bmatrix} x_I'^{(e)} \\ y_I'^{(e)} \\ z_I'^{(e)} \end{Bmatrix} = \begin{bmatrix} t_{1x}^{(e)} & t_{1y}^{(e)} & t_{1z}^{(e)} \\ t_{2x}^{(e)} & t_{2y}^{(e)} & t_{2z}^{(e)} \\ n_x^{(e)} & n_y^{(e)} & n_z^{(e)} \end{bmatrix} \begin{Bmatrix} X_I^{(e)} \\ Y_I^{(e)} \\ Z_I^{(e)} \end{Bmatrix} \tag{14.8.9}$$

▮

Remark 14.8.2: For a flat shell element e, the coordinate $z'^{(e)}$ (corresponding to the coordinate along the local axis $\vec{n}^{(e)}$) has the same value for all nodal points in the element. ▮

Example 14.1: Determination of $[k^{(e)}]$ and $\{f^{(e)}\}$ for a four-node quadrilateral shell element

In this example, we will find the stiffness matrix and equivalent nodal force/moment vector for the four-node element shown in Figure 14.16. The coordinates (in the global coordinate system) of the four nodal points are also provided in the same figure. The material is isotropic, linearly elastic, with $E = 5000$ ksi, $v = 0.2$, and the thickness of the element is $d = 6$ inches. The element is subjected to a distributed force (per unit surface area of the shell) of 1kip/in.^{2}, along the negative global axis Z. No other loadings are applied.

We begin by determining the 3 × 3 coordinate transformation block matrix, $[r^{(e)}]$:

$$\ell_{31}^{(e)} = \sqrt{\left(X_3^{(e)} - X_1^{(e)}\right)^2 + \left(Y_3^{(e)} - Y_1^{(e)}\right)^2 + \left(Z_3^{(e)} - Z_1^{(e)}\right)^2}$$

$$= \sqrt{(5-0)^2 + (4-0)^2 + (2-1)^2} = 6.481\,\text{in.}$$

$$\ell_{41}^{(e)} = \sqrt{\left(X_4^{(e)} - X_2^{(e)}\right)^2 + \left(Y_4^{(e)} - Y_2^{(e)}\right)^2 + \left(Z_4^{(e)} - Z_2^{(e)}\right)^2}$$

$$= \sqrt{(1-5)^2 + (4-0)^2 + (2-1)^2} = 5.745\,\text{in.}$$

We can now find the unit vectors $\vec{v}_1^{(e)}, \vec{v}_2^{(e)}$:

$$\vec{v}_1^{(e)} = \begin{Bmatrix} v_{1x}^{(e)} \\ v_{1y}^{(e)} \\ v_{1z}^{(e)} \end{Bmatrix} = \frac{1}{\ell_{31}^{(e)}} \begin{Bmatrix} X_3^{(e)} - X_1^{(e)} \\ Y_3^{(e)} - Y_1^{(e)} \\ Z_3^{(e)} - Z_1^{(e)} \end{Bmatrix} = \frac{1}{6.481} \begin{Bmatrix} 5-0 \\ 4-0 \\ 2-1 \end{Bmatrix} = \begin{Bmatrix} 0.7715 \\ 0.6172 \\ 0.1543 \end{Bmatrix}$$

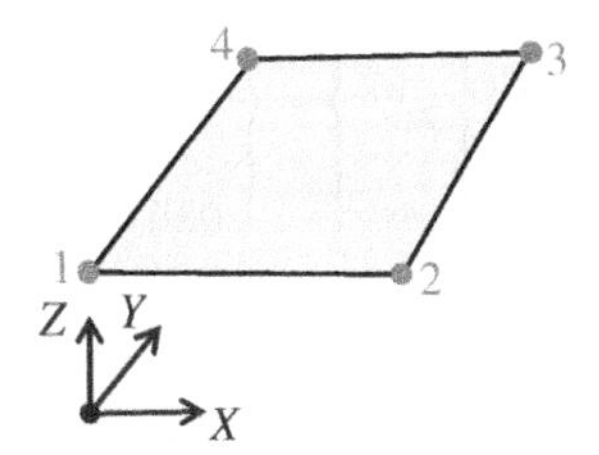

Node	X (in.)	Y (in.)	Z (in.)
1	0	0	1
2	5	0	1
3	5	4	2
4	1	4	2

Figure 14.16 Example quadrilateral shell element.

$$\vec{v}_2^{(e)} = \begin{Bmatrix} v_{2x}^{(e)} \\ v_{2y}^{(e)} \\ v_{2z}^{(e)} \end{Bmatrix} = \frac{1}{\ell_{41}^{(e)}} \begin{Bmatrix} X_4^{(e)} - X_2^{(e)} \\ Y_4^{(e)} - Y_2^{(e)} \\ Z_4^{(e)} - Z_2^{(e)} \end{Bmatrix} = \frac{1}{5.745} \begin{Bmatrix} 1-5 \\ 4-0 \\ 2-1 \end{Bmatrix} = \begin{Bmatrix} -0.6963 \\ 0.6963 \\ 0.1741 \end{Bmatrix}$$

We can also find the local coordinate axis

$$\vec{n}^{(e)} = \frac{\vec{v}_1^{(e)} \times \vec{v}_2^{(e)}}{\left\| \vec{v}_1^{(e)} \times \vec{v}_2^{(e)} \right\|} = \begin{Bmatrix} 0 \\ -0.2425 \\ 0.9701 \end{Bmatrix}$$

Finally, we can establish the other two, in-plane, local coordinate axes of the element, $\vec{t}_1^{(e)}$ and $\vec{t}_2^{(e)}$:

$$\vec{t}_1^{(e)} = \frac{\vec{v}_1^{(e)} - \vec{v}_2^{(e)}}{\left\| \vec{v}_1^{(e)} - \vec{v}_2^{(e)} \right\|} = \begin{Bmatrix} 0.9985 \\ -0.0538 \\ -0.0135 \end{Bmatrix}, \vec{t}_2^{(e)} = \frac{\vec{v}_1^{(e)} + \vec{v}_2^{(e)}}{\left\| \vec{v}_1^{(e)} + \vec{v}_2^{(e)} \right\|} = \begin{Bmatrix} 0.0555 \\ -0.9686 \\ 0.2422 \end{Bmatrix}$$

We now establish the 3 × 3 array:

$$\begin{bmatrix} t_{1x}^{(e)} & t_{1y}^{(e)} & t_{1z}^{(e)} \\ t_{2x}^{(e)} & t_{2y}^{(e)} & t_{2z}^{(e)} \\ n_x^{(e)} & n_y^{(e)} & n_z^{(e)} \end{bmatrix} = \begin{bmatrix} 0.9985 & -0.0538 & -0.0135 \\ 0.0555 & 0.9686 & 0.2422 \\ 0 & -0.2425 & 0.9701 \end{bmatrix}$$

The coordinates of each node I in the local coordinate system of the element can be obtained from:

$$\begin{Bmatrix} x_I'^{(e)} \\ y_I'^{(e)} \\ z_I'^{(e)} \end{Bmatrix} = \begin{bmatrix} t_{1x}^{(e)} & t_{1y}^{(e)} & t_{1z}^{(e)} \\ t_{2x}^{(e)} & t_{2y}^{(e)} & t_{2z}^{(e)} \\ n_x^{(e)} & n_y^{(e)} & n_z^{(e)} \end{bmatrix} \begin{Bmatrix} X_I^{(e)} \\ Y_I^{(e)} \\ Z_I^{(e)} \end{Bmatrix}.$$

By substitution, we have:

$$\begin{Bmatrix} x_1'^{(e)} \\ y_1'^{(e)} \\ z_1'^{(e)} \end{Bmatrix} = \begin{Bmatrix} -0.0135 \\ 0.2422 \\ 0.9701 \end{Bmatrix}, \begin{Bmatrix} x_2'^{(e)} \\ y_2'^{(e)} \\ z_2'^{(e)} \end{Bmatrix} = \begin{Bmatrix} 4.9789 \\ 0.5195 \\ 0.9701 \end{Bmatrix}, \begin{Bmatrix} x_3'^{(e)} \\ y_3'^{(e)} \\ z_3'^{(e)} \end{Bmatrix} = \begin{Bmatrix} 4.7502 \\ 4.6362 \\ 0.9701 \end{Bmatrix}, \begin{Bmatrix} x_4'^{(e)} \\ y_4'^{(e)} \\ z_4'^{(e)} \end{Bmatrix} = \begin{Bmatrix} 0.7563 \\ 4.4144 \\ 0.9701 \end{Bmatrix}$$

It is worth noticing that the value of local z-coordinate is identical for the four-nodal points (this must always be the case for a planar shell element). We can now treat the element as an isoparametric 4Q one, using the values of local x- and y-coordinate for the four nodal points.

The given distributed force vector is expressed in the global coordinate system (because we know that one of the components, along the global Z axis, is nonzero). When we calculate the equivalent nodal force vector, we must have the three components of distributed force (per unit area of the shell) along the local axes of the element (so that the first two components are membrane forces, the third component is a force perpendicular to the plane of the element). The same coordinate transformation equation that applies for nodal displacements and coordinates must be satisfied for the distributed forces:

$$\begin{Bmatrix} p_x'^{(e)} \\ p_y'^{(e)} \\ p_z'^{(e)} \end{Bmatrix} = \begin{bmatrix} t_{1x}^{(e)} & t_{1y}^{(e)} & t_{1z}^{(e)} \\ t_{2x}^{(e)} & t_{2y}^{(e)} & t_{2z}^{(e)} \\ n_x^{(e)} & n_y^{(e)} & n_z^{(e)} \end{bmatrix} \begin{Bmatrix} p_x^{(e)} \\ p_y^{(e)} \\ p_z^{(e)} \end{Bmatrix}$$

$$= \begin{bmatrix} 0.9985 & -0.0538 & -0.0135 \\ 0.0555 & 0.9686 & 0.2422 \\ 0 & -0.2425 & 0.9701 \end{bmatrix} \begin{Bmatrix} 0 \\ 0 \\ -1 \end{Bmatrix} = \begin{Bmatrix} 0.0135 \\ -0.2422 \\ -0.9701 \end{Bmatrix}$$

The five-component distributed force/moment vector in the local coordinate system for the specific element can be written as: $\{p'^{(e)}\} = \left[p_x'^{(e)} \; p_y'^{(e)} \; p_z'^{(e)} \; 0 \; 0\right]^T$, where we have accounted for the fact that, in the specific example, we do not have any distributed moments in our element.

We can also establish the block-matrices for the generalized stress-strain matrix of the element, which is constant:

$$\left[\hat{D}_m\right] = \frac{E \cdot d}{1-\nu^2} \begin{bmatrix} 1 & \nu & 0 \\ \nu & 1 & 0 \\ 0 & 0 & \dfrac{1-\nu}{2} \end{bmatrix} = \begin{bmatrix} 31250 & 6250 & 0 \\ 6250 & 31250 & 0 \\ 0 & 0 & 12500 \end{bmatrix},$$

$$\left[\hat{D}_f\right] = \frac{E \cdot d^3}{12(1-\nu^2)} \begin{bmatrix} 1 & \nu & 0 \\ \nu & 1 & 0 \\ 0 & 0 & \dfrac{1-\nu}{2} \end{bmatrix} = \begin{bmatrix} 93750 & 18750 & 0 \\ 18750 & 93750 & 0 \\ 0 & 0 & 37500 \end{bmatrix}$$

$$\left[\hat{D}_s\right] = \kappa \cdot G \cdot d \begin{bmatrix} 1 & 0 \\ 0 & 1 \end{bmatrix} = \begin{bmatrix} 10417 & 0 \\ 0 & 10417 \end{bmatrix}$$

The generalized stress-strain matrix of the element can be written per Equation (14.4.42):

$$[\hat{D}] = \begin{bmatrix} [\hat{D}_m] & [0] & [0] \\ [0] & [\hat{D}_f] & [0] \\ [0] & [0] & [\hat{D}_s] \end{bmatrix}$$

$$= \begin{bmatrix} \begin{bmatrix} 31250 & 6250 & 0 \\ 6250 & 31250 & 0 \\ 0 & 0 & 12500 \end{bmatrix} & \begin{bmatrix} 0 & 0 & 0 \\ 0 & 0 & 0 \\ 0 & 0 & 0 \end{bmatrix} & \begin{bmatrix} 0 & 0 \\ 0 & 0 \\ 0 & 0 \end{bmatrix} \\ \begin{bmatrix} 0 & 0 & 0 \\ 0 & 0 & 0 \\ 0 & 0 & 0 \end{bmatrix} & \begin{bmatrix} 93750 & 18750 & 0 \\ 18750 & 93750 & 0 \\ 0 & 0 & 37500 \end{bmatrix} & \begin{bmatrix} 0 & 0 \\ 0 & 0 \\ 0 & 0 \end{bmatrix} \\ \begin{bmatrix} 0 & 0 & 0 \\ 0 & 0 & 0 \end{bmatrix} & \begin{bmatrix} 0 & 0 & 0 \\ 0 & 0 & 0 \end{bmatrix} & \begin{bmatrix} 10417 & 0 \\ 0 & 10417 \end{bmatrix} \end{bmatrix}$$

We will conduct in detail the computations for the first integration (Gauss) point of the element, with parametric coordinates $\xi_1 = \frac{1}{\sqrt{3}}, \eta_1 = -\frac{1}{\sqrt{3}}$

$$N_1^{(4Q)} = N_1^{(4Q)}(\xi = \xi_1, \eta = \eta_1) = \frac{1}{4}(1 - \xi_1)(1 - \eta_1) = 0.6220$$

$$N_2^{(4Q)} = N_2^{(4Q)}(\xi = \xi_1, \eta = \eta_1) = \frac{1}{4}(1 + \xi_1)(1 - \eta_1) = 0.1667$$

$$N_3^{(4Q)} = N_3^{(4Q)}(\xi = \xi_1, \eta = \eta_1) = \frac{1}{4}(1 + \xi_1)(1 + \eta_1) = 0.0447$$

$$N_4^{(4Q)} = N_4^{(4Q)}(\xi = \xi_1, \eta = \eta_1) = \frac{1}{4}(1 - \xi_1)(1 + \eta_1) = 0.1667$$

$$\left[N_{,\xi}^{(4Q)}\right]_1 = \begin{bmatrix} \frac{\partial N_1^{(4Q)}}{\partial \xi} & \frac{\partial N_2^{(4Q)}}{\partial \xi} & \frac{\partial N_3^{(4Q)}}{\partial \xi} & \frac{\partial N_4^{(4Q)}}{\partial \xi} \\ \frac{\partial N_1^{(4Q)}}{\partial \eta} & \frac{\partial N_2^{(4Q)}}{\partial \eta} & \frac{\partial N_3^{(4Q)}}{\partial \eta} & \frac{\partial N_4^{(4Q)}}{\partial \eta} \end{bmatrix}_{\substack{\xi = \xi_1 \\ \eta = \eta_1}}$$

$$= \begin{bmatrix} -0.3943 & 0.3943 & 0.1057 & -0.1057 \\ -0.3943 & -0.1057 & 0.1057 & 0.3943 \end{bmatrix}$$

$$[J]_1 = \left[N_{,\xi}^{(4Q)}\right]_1 \left[\left\{x'^{(e)}\right\} \; \left\{y'^{(e)}\right\}\right]$$

$$= \begin{bmatrix} -0.3943 & 0.3943 & 0.1057 & -0.1057 \\ -0.3943 & -0.1057 & 0.1057 & 0.3943 \end{bmatrix} \left[\begin{Bmatrix} -0.0135 \\ 4.9789 \\ 4.7502 \\ 0.7563 \end{Bmatrix} \begin{Bmatrix} 0.2422 \\ 0.5195 \\ 4.6362 \\ 4.4144 \end{Bmatrix} \right]$$

$$= \begin{bmatrix} 2.9307 & 0.1328 \\ 0.2794 & 2.0802 \end{bmatrix}$$

$$J_1 = \det\left([J]_1\right) = 4.9361$$

$$[\tilde{J}]_1 = \left([J]_1\right)^{-1} = \begin{bmatrix} 0.4214 & -0.0269 \\ -0.0566 & 0.4843 \end{bmatrix}$$

$$\left[N_{,x}^{(4Q)}\right] = \begin{bmatrix} \dfrac{\partial N_1^{(4Q)}}{\partial x} & \dfrac{\partial N_2^{(4Q)}}{\partial x} & \dfrac{\partial N_3^{(4Q)}}{\partial x} & \dfrac{\partial N_4^{(4Q)}}{\partial x} \\ \dfrac{\partial N_1^{(4Q)}}{\partial y} & \dfrac{\partial N_2^{(4Q)}}{\partial y} & \dfrac{\partial N_3^{(4Q)}}{\partial y} & \dfrac{\partial N_4^{(4Q)}}{\partial y}. \end{bmatrix}_1 = [\tilde{J}]_1 \left[N_{,\xi}^{(4Q)}\right]_1$$

$$= \begin{bmatrix} -0.1556 & 0.1690 & 0.0417 & -0.0551 \\ -0.1687 & -0.0735 & 0.0452 & 0.1970 \end{bmatrix}$$

We can establish the 5×6 block shape-function arrays for the four nodal points, evaluated at Gauss point 1. For each node I, we have:

$$\left[N_I^{(e)}\right]_1 = \begin{bmatrix} N_I^{(4Q)}(\xi,\eta) & 0 & 0 & 0 & 0 & 0 \\ 0 & N_I^{(4Q)}(\xi,\eta) & 0 & 0 & 0 & 0 \\ 0 & 0 & N_I^{(4Q)}(\xi,\eta) & 0 & 0 & 0 \\ 0 & 0 & 0 & 0 & -N_I^{(4Q)}(\xi,\eta) & 0 \\ 0 & 0 & 0 & N_I^{(4Q)}(\xi,\eta) & 0 & 0 \end{bmatrix}_{\substack{\xi=\xi_1 \\ \eta=\eta_1}}$$

The block shape function array at Gauss point 1 for nodes 1, 2, 3 and 4 is:

$$\left[N_1^{(e)}\right]_1 = \begin{bmatrix} 0.6220 & 0 & 0 & 0 & 0 & 0 \\ 0 & 0.6220 & 0 & 0 & 0 & 0 \\ 0 & 0 & 0.6220 & 0 & 0 & 0 \\ 0 & 0 & 0 & 0 & -0.6220 & 0 \\ 0 & 0 & 0 & 0.6220 & 0 & 0 \end{bmatrix}$$

$$\left[N_2^{(e)}\right]_1 = \begin{bmatrix} 0.1667 & 0 & 0 & 0 & 0 & 0 \\ 0 & 0.1667 & 0 & 0 & 0 & 0 \\ 0 & 0 & 0.1667 & 0 & 0 & 0 \\ 0 & 0 & 0 & 0 & -0.1667 & 0 \\ 0 & 0 & 0 & 0.1667 & 0 & 0 \end{bmatrix}$$

$$\left[N_3^{(e)}\right]_1 = \begin{bmatrix} 0.0447 & 0 & 0 & 0 & 0 & 0 \\ 0 & 0.0447 & 0 & 0 & 0 & 0 \\ 0 & 0 & 0.0447 & 0 & 0 & 0 \\ 0 & 0 & 0 & 0 & -0.0447 & 0 \\ 0 & 0 & 0 & 0.0447 & 0 & 0 \end{bmatrix}$$

$$\left[N_4^{(e)}\right]_1 = \begin{bmatrix} 0.1667 & 0 & 0 & 0 & 0 & 0 \\ 0 & 0.1667 & 0 & 0 & 0 & 0 \\ 0 & 0 & 0.1667 & 0 & 0 & 0 \\ 0 & 0 & 0 & 0 & -0.1667 & 0 \\ 0 & 0 & 0 & 0.1667 & 0 & 0 \end{bmatrix}$$

Given the four block-shape-function arrays at Gauss point 1, we can also obtain the 5×24 shape function array $[N^{(e)}]$ of the shell element, using Equation (14.6.6) for the special case $n = 4$:

$$\left[N^{(e)}\right]_1 = \left[\left[N_1^{(e)}\right]_1 \; \left[N_2^{(e)}\right]_1 \; \left[N_3^{(e)}\right]_1 \; \left[N_4^{(e)}\right]_1\right]$$

We now proceed to find the block strain-displacement arrays for the four nodal points, evaluated at Gauss point 1:

For each node I, we have:

$$\left[B_{mI}^{(e)}\right]_1 = \begin{bmatrix} \dfrac{\partial N_I^{(4Q)}}{\partial x} & 0 & 0 & 0 & 0 & 0 \\ 0 & \dfrac{\partial N_I^{(4Q)}}{\partial y} & 0 & 0 & 0 & 0 \\ \dfrac{\partial N_I^{(4Q)}}{\partial y} & \dfrac{\partial N_I^{(4Q)}}{\partial x} & 0 & 0 & 0 & 0 \end{bmatrix}_1, \quad \left[B_{fI}^{(e)}\right]_1 = \begin{bmatrix} 0 & 0 & 0 & 0 & -\dfrac{\partial N_I^{(4Q)}}{\partial x} & 0 \\ 0 & 0 & 0 & \dfrac{\partial N_I^{(4Q)}}{\partial y} & 0 & 0 \\ 0 & 0 & 0 & \dfrac{\partial N_I^{(4Q)}}{\partial y} & -\dfrac{\partial N_I^{(4Q)}}{\partial x} & 0 \end{bmatrix}_1,$$

$$\left[B_{sI}^{(e)}\right]_1 = \begin{bmatrix} 0 & 0 & \dfrac{\partial N_I^{(4Q)}}{\partial x} & 0 & N_I^{(4Q)} & 0 \\ 0 & 0 & \dfrac{\partial N_I^{(4Q)}}{\partial y} & -N_I^{(4Q)} & 0 & 0 \end{bmatrix}_1$$

The block strain-displacement arrays at Gauss point 1 for node 1 are:

$$\left[B_{ml}^{(e)}\right]_1 = \begin{bmatrix} -01556 & 0 & 0 & 0 & 0 & 0 \\ 0 & -0.1687 & 0 & 0 & 0 & 0 \\ -0.1687 & -0.1556 & 0 & 0 & 0 & 0 \end{bmatrix},$$

$$\left[B_{f1}^{(e)}\right]_1 = \begin{bmatrix} 0 & 0 & 0 & 0 & 0.1556 & 0 \\ 0 & 0 & 0 & -0.1687 & 0 & 0 \\ 0 & 0 & 0 & -0.1556 & 0.1687 & 0 \end{bmatrix}$$

$$\left[B_{s1}^{(e)}\right]_1 = \begin{bmatrix} 0 & 0 & -0.1556 & 0 & 0.6220 & 0 \\ 0 & 0 & -0.1687 & -0.622 & 0 & 0 \end{bmatrix}$$

The strain-displacement arrays at Gauss point 1 for node 2 are:

$$\left[B_{m2}^{(e)}\right]_1 = \begin{bmatrix} 0.1690 & 0 & 0 & 0 & 0 & 0 \\ 0 & -0.0735 & 0 & 0 & 0 & 0 \\ -0.0735 & 0.1690 & 0 & 0 & 0 & 0 \end{bmatrix},$$

$$\left[B_{f2}^{(e)}\right]_1 = \begin{bmatrix} 0 & 0 & 0 & 0 & -0.1690 & 0 \\ 0 & 0 & 0 & -0.0735 & 0 & 0 \\ 0 & 0 & 0 & 0.1690 & 0.0735 & 0 \end{bmatrix}$$

$$\left[B_{s2}^{(e)}\right]_1 = \begin{bmatrix} 0 & 0 & 0.1690 & 0 & 0.167 & 0 \\ 0 & 0 & -0.0735 & -0.1667 & 0 & 0 \end{bmatrix}$$

The same arrays at Gauss point 1 for node 3 are:

$$\left[B_{m3}^{(e)}\right]_1 = \begin{bmatrix} 0.0417 & 0 & 0 & 0 & 0 & 0 \\ 0 & 0.0452 & 0 & 0 & 0 & 0 \\ 0.0452 & 0.0417 & 0 & 0 & 0 & 0 \end{bmatrix},$$

$$\left[B_{f3}^{(e)}\right]_1 = \begin{bmatrix} 0 & 0 & 0 & 0 & -0.0417 & 0 \\ 0 & 0 & 0 & 0.0452 & 0 & 0 \\ 0 & 0 & 0 & 0.0417 & -0.0452 & 0 \end{bmatrix}$$

$$\left[B_{s3}^{(e)}\right]_1 = \begin{bmatrix} 0 & 0 & 0.0417 & 0 & 0.0447 & 0 \\ 0 & 0 & 0.0452 & -0.0447 & 0 & 0 \end{bmatrix}$$

The block strain-displacement arrays at Gauss point 1 for node 4 are:

$$\left[B_{m4}^{(e)}\right]_1 = \begin{bmatrix} -0.0551 & 0 & 0 & 0 & 0 & 0 \\ 0 & 0.1970 & 0 & 0 & 0 & 0 \\ 0.1970 & -0.0551 & 0 & 0 & 0 & 0 \end{bmatrix},$$

$$\left[B_{f4}^{(e)}\right]_1 = \begin{bmatrix} 0 & 0 & 0 & 0 & 0.0551 & 0 \\ 0 & 0 & 0 & 0.1970 & 0 & 0 \\ 0 & 0 & 0 & -0.0551 & -0.1970 & 0 \end{bmatrix}$$

$$\left[B_{s4}^{(e)}\right]_1 = \begin{bmatrix} 0 & 0 & -0.0551 & 0 & 0.1667 & 0 \\ 0 & 0 & 0.1970 & -0.1667 & 0 & 0 \end{bmatrix}$$

We can write the 8×24 generalized strain-displacement matrix of Gauss point 1, $[B^{(e)}]_1$, in block-matrix form in accordance with Equation (14.6.10d), for the special case $n = 4$:

$$\underset{8\times 24}{\left[B^{(e)}\right]_1} = \begin{bmatrix} \begin{bmatrix} \underset{3\times 6}{\left[B_{m1}^{(e)}\right]_1} \\ \underset{3\times 6}{\left[B_{f1}^{(e)}\right]_1} \\ \underset{2\times 6}{\left[B_{s1}^{(e)}\right]_1} \end{bmatrix} & \begin{bmatrix} \underset{3\times 6}{\left[B_{m2}^{(e)}\right]_1} \\ \underset{3\times 6}{\left[B_{f2}^{(e)}\right]_1} \\ \underset{2\times 6}{\left[B_{s2}^{(e)}\right]_1} \end{bmatrix} & \begin{bmatrix} \underset{3\times 6}{\left[B_{m3}^{(e)}\right]_1} \\ \underset{3\times 6}{\left[B_{f3}^{(e)}\right]_1} \\ \underset{2\times 6}{\left[B_{s3}^{(e)}\right]_1} \end{bmatrix} & \begin{bmatrix} \underset{3\times 6}{\left[B_{m4}^{(e)}\right]_1} \\ \underset{3\times 6}{\left[B_{f4}^{(e)}\right]_1} \\ \underset{2\times 6}{\left[B_{s4}^{(e)}\right]_1} \end{bmatrix} \end{bmatrix};$$

for clarity, we have written below each matrix its corresponding dimensions.

The contributions of Gauss point 1 to the stiffness matrix in the local coordinate system, $[k'^{(e)}]$, and to the equivalent nodal force/moment vector, $\{f'^{(e)}\}$, in the local coordinate system, are given by the expressions $\left(\left[B^{(e)}\right]_1\right)^T \left[D^{(e)}\right]_1 \left[B^{(e)}\right]_1 J_1 \cdot W_1$ and $\left(\left[N^{(e)}\right]_1\right)^T \left\{p^{(e)}\right\}_1 J_1 \cdot W_1$, respectively, where $W_1 = 1$ is the weight coefficient of the first Gauss point for a 2×2 quadrature rule.

We use the same exact computations for Gauss points 2 $\left(\xi_2 = 1/\sqrt{3}, \eta_2 = -1/\sqrt{3}\right)$, 3 $\left(\xi_3 = 1/\sqrt{3}, \eta_3 = 1/\sqrt{3}\right)$, and 4 $\left(\xi_4 = -1/\sqrt{3}, \eta_4 = 1/\sqrt{3}\right)$. We can then sum the Gauss point contributions to obtain $[k'^{(e)}]$ and $\{f'^{(e)}\}$:

$$[k'^{(e)}] = ([B^{(e)}]_1)^T [D^{(e)}]_1 [B^{(e)}]_1 J_1 \cdot W_1 + ([B^{(e)}]_2)^T [D^{(e)}]_2 [B^{(e)}]_2 J_2 \cdot W_2 + \left([B^{(e)}]_3\right)^T [D^{(e)}]_3 [B^{(e)}]_3 J_3 \cdot W_3 + ([B^{(e)}]_4)^T [D^{(e)}]_4 [B^{(e)}]_4 J_4 \cdot W_4$$

$$= \left[\begin{array}{cccccccccccccccccccccccc}
12213 & 4005 & 0 & 0 & 0 & 0 & -6641 & -1413 & 0 & 0 & 0 & 0 & -5863 & -4342 & 0 & 0 & 0 & 0 & 291 & 1750 & 0 & 0 & 0 & 0 \\
4005 & 13087 & 0 & 0 & 0 & 0 & 1712 & 2563 & 0 & 0 & 0 & 0 & -4342 & -6283 & 0 & 0 & 0 & 0 & -1375 & -9367 & 0 & 0 & 0 & 0 \\
0 & 0 & 6024 & 7331 & -6762 & 0 & 0 & 0 & -971 & 0 & 0 & 0 & 0 & 0 & -2892 & 0 & 0 & 0 & 0 & 0 & -2161 & 0 & 0 & 0 \\
0 & 0 & 7331 & 61928 & -12016 & 0 & 0 & 0 & 4803 & 7689 & -5137 & 0 & 0 & 0 & -3665 & -18848 & 13027 & 0 & 0 & 0 & -8469 & -28101 & 4126 & 0 \\
0 & 0 & -6762 & -12016 & 59306 & 0 & 0 & 0 & 7436 & 4238 & -19922 & 0 & 0 & 0 & 3381 & 13027 & -17589 & 0 & 0 & 0 & -4055 & -5249 & 873 & 0 \\
0 & 0 & 0 & 0 & 0 & 0 & 0 & 0 & 0 & 0 & 0 & 0 & 0 & 0 & 0 & 0 & 0 & 0 & 0 & 0 & 0 & 0 & 0 & 0 \\
-6641 & 1712 & 0 & 0 & 0 & 0 & 14774 & -5438 & 0 & 0 & 0 & 0 & 291 & -1375 & 0 & 0 & 0 & 0 & -8424 & 5101 & 0 & 0 & 0 & 0 \\
-1413 & 2563 & 0 & 0 & 0 & 0 & -5438 & 15831 & 0 & 0 & 0 & 0 & 1750 & -9367 & 0 & 0 & 0 & 0 & 5101 & -9027 & 0 & 0 & 0 & 0 \\
0 & 0 & -971 & 4803 & 7436 & 0 & 0 & 0 & 7287 & 0 & 0 & 0 & 0 & 0 & -2161 & 0 & 0 & 0 & 0 & 0 & -4155 & 0 & 0 & 0 \\
0 & 0 & 0 & 7689 & 4238 & 0 & 0 & 0 & 0 & 47493 & 16314 & 0 & 0 & 0 & 0 & -28101 & -5249 & 0 & 0 & 0 & 0 & -27081 & -15303 & 0 \\
0 & 0 & 0 & -5137 & -19922 & 0 & 0 & 0 & 0 & 16314 & 44322 & 0 & 0 & 0 & 0 & 4126 & 873 & 0 & 0 & 0 & 0 & -15303 & -25273 & 0 \\
0 & 0 & 0 & 0 & 0 & 0 & 0 & 0 & 0 & 0 & 0 & 0 & 0 & 0 & 0 & 0 & 0 & 0 & 0 & 0 & 0 & 0 & 0 & 0 \\
-5863 & -4342 & 0 & 0 & 0 & 0 & 291 & 1750 & 0 & 0 & 0 & 0 & 13800 & 3921 & 0 & 0 & 0 & 0 & -8228 & -1328 & 0 & 0 & 0 & 0 \\
-4342 & -6283 & 0 & 0 & 0 & 0 & -1375 & -9367 & 0 & 0 & 0 & 0 & 3921 & 14788 & 0 & 0 & 0 & 0 & 1797 & 862 & 0 & 0 & 0 & 0 \\
0 & 0 & -2892 & -3665 & 3381 & 0 & 0 & 0 & -2161 & 0 & 0 & 0 & 0 & 0 & 6807 & 0 & 0 & 0 & 0 & 0 & -1754 & 0 & 0 & 0 \\
0 & 0 & 0 & -18848 & 13027 & 0 & 0 & 0 & 0 & -28101 & 4126 & 0 & 0 & 0 & 0 & 44363 & -11763 & 0 & 0 & 0 & 0 & 2586 & -5390 & 0 \\
0 & 0 & 0 & 13027 & -17589 & 0 & 0 & 0 & 0 & -5249 & 873 & 0 & 0 & 0 & 0 & -11763 & 41401 & 0 & 0 & 0 & 0 & 3985 & -24685 & 0 \\
0 & 0 & 0 & 0 & 0 & 0 & 0 & 0 & 0 & 0 & 0 & 0 & 0 & 0 & 0 & 0 & 0 & 0 & 0 & 0 & 0 & 0 & 0 & 0 \\
291 & -1375 & 0 & 0 & 0 & 0 & -8424 & 5101 & 0 & 0 & 0 & 0 & -8228 & 1797 & 0 & 0 & 0 & 0 & 16362 & -5522 & 0 & 0 & 0 & 0 \\
1750 & -9367 & 0 & 0 & 0 & 0 & 5101 & -9027 & 0 & 0 & 0 & 0 & -1328 & 862 & 0 & 0 & 0 & 0 & -5522 & 17532 & 0 & 0 & 0 & 0 \\
0 & 0 & -2161 & -8469 & -4055 & 0 & 0 & 0 & -4155 & 0 & 0 & 0 & 0 & 0 & -1754 & 0 & 0 & 0 & 0 & 0 & 8070 & 0 & 0 & 0 \\
0 & 0 & 0 & -28101 & -5249 & 0 & 0 & 0 & 0 & -27081 & -15303 & 0 & 0 & 0 & 0 & 2586 & 3985 & 0 & 0 & 0 & 0 & 52596 & 16567 & 0 \\
0 & 0 & 0 & 4126 & 873 & 0 & 0 & 0 & 0 & -15303 & -25273 & 0 & 0 & 0 & 0 & -5390 & -24685 & 0 & 0 & 0 & 0 & 16567 & 49085 & 0 \\
0 & 0 & 0 & 0 & 0 & 0 & 0 & 0 & 0 & 0 & 0 & 0 & 0 & 0 & 0 & 0 & 0 & 0 & 0 & 0 & 0 & 0 & 0 & 0
\end{array}\right]$$

It is worth noticing that the rows and columns of $\left[k'^{(e)}\right]$ corresponding to the nodal rotations about the local z-axis are all zero.

$$\{f'^{(e)}\} = \left(\left[N^{(e)}\right]_1\right)^T \{p^{(e)}\}_1 J_1 \cdot W_1 + \left(\left[N^{(e)}\right]_1\right)^T \{p^{(e)}\}_1 J_1 \cdot W_1 + \left(\left[N^{(e)}\right]_1\right)^T \{p^{(e)}\}_1 J_1 \cdot W_1 + \left(\left[N^{(e)}\right]_1\right)^T \{p^{(e)}\}_1 J_1 \cdot W_1$$

$$= [0.065\ -1.165\ -4.667\ 0.00\ 0.00\ 0.00\ 0.065\ -1.165\ -4.667\ 0.00\ 0.00\ 0.00\ 0.060\ -1.082\ -4.333\ 0.00\ 0.00\ 0.00\ 0.060\ -1.082\ -4.333\ 0.00\ 0.00\ 0.00]^T$$

The 24×24 coordinate transformation array, $[R^{(e)}]$, of the element can be established in block form as:

$$\left[R^{(e)}\right] = \begin{bmatrix} \left[r^{(e)}\right] & [0] & [0] & [0] \\ [0] & \left[r^{(e)}\right] & [0] & [0] \\ [0] & [0] & \left[r^{(e)}\right] & [0] \\ [0] & [0] & [0] & \left[r^{(e)}\right] \end{bmatrix}$$, where each block has dimensions 6×6 and

$$\left[r^{(e)}\right] = \begin{bmatrix} t_{1x}^{(e)} & t_{1y}^{(e)} & t_{1z}^{(e)} & 0 & 0 & 0 \\ t_{2x}^{(e)} & t_{2y}^{(e)} & t_{2z}^{(e)} & 0 & 0 & 0 \\ n_x^{(e)} & n_y^{(e)} & n_z^{(e)} & 0 & 0 & 0 \\ 0 & 0 & 0 & t_{1x}^{(e)} & t_{1y}^{(e)} & t_{1z}^{(e)} \\ 0 & 0 & 0 & t_{2x}^{(e)} & t_{2y}^{(e)} & t_{2z}^{(e)} \\ 0 & 0 & 0 & n_x^{(e)} & n_y^{(e)} & n_z^{(e)} \end{bmatrix}$$

$$= \begin{bmatrix} 0.9985 & -0.0538 & -0.0135 & 0 & 0 & 0 \\ 0.0555 & 0.9686 & 0.2422 & 0 & 0 & 0 \\ 0 & -0.2425 & 0.9701 & 0 & 0 & 0 \\ 0 & 0 & 0 & 0.9985 & -0.0538 & -0.0135 \\ 0 & 0 & 0 & 0.0555 & 0.9686 & 0.2422 \\ 0 & 0 & 0 & 0 & -0.2425 & 0.9701 \end{bmatrix}$$

Finally, we can use Equations (14.8.7) and (14.8.8) to establish the element stiffness matrix and equivalent nodal force/moment vector in the global coordinate system:

$$[k^{(e)}] = [R^{(e)}]^T [k'^{(e)}] [R^{(e)}]$$

$$
= \begin{bmatrix}
12659 & 3909 & 977 & 0 & 0 & 0 & -6596 & -877 & -219 & 0 & 0 & 0 & -6345 & -4209 & -1052 & 0 & 0 & 0 & 282 & 1178 & 294 & 0 & 0 & 0 \\
3909 & 12251 & 1557 & -1684 & 1684 & 421 & 2155 & 2313 & 821 & 0 & 0 & 0 & -4209 & -5629 & -684 & 0 & 0 & 0 & -1854 & -8935 & -1693 & 0 & 0 & 0 \\
977 & 1557 & 6413 & 6737 & -6737 & -1684 & 539 & 821 & -766 & 0 & 0 & 0 & -1052 & -684 & -3063 & 0 & 0 & 0 & -464 & -1693 & -2584 & 0 & 0 & 0 \\
0 & -1684 & 6737 & 60589 & -11726 & -2931 & 0 & -1263 & 5053 & 7554 & -6464 & -1616 & 0 & 842 & -3369 & -17401 & 12628 & 3157 & 0 & 2105 & -8421 & -28075 & 5562 & 1391 \\
0 & 1687 & -6737 & -11726 & 57078 & 14269 & 0 & -1684 & 6737 & 2631 & -18623 & -4656 & 0 & -842 & 3369 & 12628 & -17916 & -4479 & 0 & 842 & -3369 & -3533 & 796 & 199 \\
0 & 421 & -1684 & -2931 & 14269 & 3567 & 0 & -421 & 1684 & 658 & -4656 & -1164 & 0 & -211 & 842 & 3157 & -4479 & -1120 & 0 & 211 & -842 & -883 & 199 & 50 \\
-6596 & 2155 & 539 & 0 & 0 & 0 & 14175 & -5186 & -1297 & 0 & 0 & 0 & 282 & -1854 & -464 & 0 & 0 & 0 & -7861 & 4886 & 1221 & 0 & 0 & 0 \\
-877 & 2313 & 821 & -1263 & -1684 & -421 & -5186 & 15892 & 2151 & 0 & 0 & 0 & 1178 & -8935 & -1693 & 0 & 0 & 0 & 4886 & -9270 & -1279 & 0 & 0 & 0 \\
-219 & 821 & -766 & 5053 & 6737 & 1684 & -1297 & 2151 & 7825 & 0 & 0 & 0 & 294 & -1693 & -2584 & 0 & 0 & 0 & 1221 & -1279 & -4475 & 0 & 0 & 0 \\
0 & 0 & 0 & 7554 & 2631 & 658 & 0 & 0 & 0 & 49290 & 15559 & 3890 & 0 & 0 & 0 & -28075 & -3533 & -883 & 0 & 0 & 0 & -28770 & -14657 & -3664 \\
0 & 0 & 0 & -6464 & -18623 & -4656 & 0 & 0 & 0 & 15559 & 40024 & 10006 & 0 & 0 & 0 & 5562 & 796 & 199 & 0 & 0 & 0 & -14657 & -22192 & -5549 \\
0 & 0 & 0 & -1616 & -4656 & -1164 & 0 & 0 & 0 & 3890 & 10006 & 2501 & 0 & 0 & 0 & 1391 & 199 & 50 & 0 & 0 & 0 & -3664 & -5549 & -1387 \\
-6345 & -4209 & -1052 & 0 & 0 & 0 & 282 & 1178 & 294 & 0 & 0 & 0 & 14238 & 3833 & 958 & 0 & 0 & 0 & -8174 & -802 & -200 & 0 & 0 & 0 \\
-4209 & -5629 & -684 & 842 & -842 & -211 & -1854 & -8935 & -1693 & 0 & 0 & 0 & 3833 & 13907 & 1775 & 0 & 0 & 0 & 2230 & 657 & 603 & 0 & 0 & 0 \\
-1052 & -684 & -3063 & -3369 & 3369 & 842 & -464 & -1693 & -2584 & 0 & 0 & 0 & 958 & 1775 & 7250 & 0 & 0 & 0 & 557 & 603 & -1603 & 0 & 0 & 0 \\
0 & 0 & 0 & -17401 & 12628 & 3157 & 0 & 0 & 0 & -28075 & 5562 & 1391 & 0 & 0 & 0 & 43051 & -11500 & -2875 & 0 & 0 & 0 & 2424 & -6690 & -1672 \\
0 & 0 & 0 & 12628 & -17916 & -4479 & 0 & 0 & 0 & -3533 & 796 & 199 & 0 & 0 & 0 & -11500 & 40201 & 10050 & 0 & 0 & 0 & 2405 & -23080 & -5770 \\
0 & 0 & 0 & 3157 & -4479 & -1120 & 0 & 0 & 0 & -883 & 199 & 50 & 0 & 0 & 0 & -2875 & 10050 & 2513 & 0 & 0 & 0 & 601 & -5770 & -1443 \\
282 & -1854 & -464 & 0 & 0 & 0 & -7861 & 4886 & 1221 & 0 & 0 & 0 & -8174 & 2230 & 557 & 0 & 0 & 0 & 15754 & -5262 & -1315 & 0 & 0 & 0 \\
1178 & -8935 & -1693 & 2105 & 842 & -211 & 4886 & -9270 & -1279 & 0 & 0 & 0 & -802 & 657 & 603 & 0 & 0 & 0 & -5262 & 17548 & 2369 & 0 & 0 & 0 \\
294 & -693 & -2584 & -8421 & -3369 & -842 & 1221 & -1279 & -4475 & 0 & 0 & 0 & -200 & 603 & -1603 & 0 & 0 & 0 & -1315 & 2369 & 862 & 0 & 0 & 0 \\
0 & 0 & 0 & -28075 & -3533 & -883 & 0 & 0 & 0 & -28770 & -14657 & -3664 & 0 & 0 & 0 & 2424 & 2405 & 601 & 0 & 0 & 0 & 54420 & 15785 & 3946 \\
0 & 0 & 0 & 5562 & 796 & 199 & 0 & 0 & 0 & -14657 & -22196 & -5549 & 0 & 0 & 0 & -6690 & -23080 & -5770 & 0 & 0 & 0 & 15785 & 44481 & 11120 \\
0 & 0 & 0 & 1391 & 199 & 50 & 0 & 0 & 0 & -3664 & -5549 & -1387 & 0 & 0 & 0 & -1672 & -5770 & 1443 & 0 & 0 & 0 & 3946 & 11120 & 2780
\end{bmatrix}
$$

We can see that, for the global coordinate system, there are nonzero stiffness terms for all rows and columns of $[k^{(e)}]$.

$$\{f^{(e)}\} = [R^{(e)}]^T \{f'^{(e)}\}$$

$$= [0.00\ 0.00\ -4.810\ 0.00\ 0.00\ 0.00\ 0.00\ 0.00\ -4.810\ 0.00\ 0.00\ 0.00\ 0.00\ 0.00\ -4.467\ 0.00\ 0.00\ 0.00\ 0.00\ 0.00\ -4.467\ 0.00\ 0.00\ 0.00]^T$$

One may observe that $\{f^{(e)}\}$ has nonzero nodal forces along the global Z-axis only. This should come as no surprise, since the distributed loads (expressed in the global coordinate system) that are applied on our element only have a nonzero component along that same axis! ■

14.9 A "Clever" Way to Approximately Satisfy C^1 Continuity Requirements for Thin Shells—The Discrete Kirchhoff Formulation

We will now dedicate a section to the finite element analysis of thin planar shells. As mentioned in Section 14.1, the Kirchhoff-Love (K-L) shell theory, which applies to such structures, can be thought of as an extension (to shells) of the Euler-Bernoulli beam theory. In the K-L theory, the through-thickness shear strain components, γ_{xz} and γ_{yz}, vanish. This implies that *plane shell sections that are normal to the undeformed reference surface remain plane and normal to the reference surface in the deformed configuration.* In terms of the previously derived Equations (14.1.3e, f), the K-L theory leads to satisfaction of the following expressions.

$$\gamma_{yz} = 0 \rightarrow \frac{\partial u_{oz}}{\partial y} - \theta_2 = 0 \rightarrow \theta_2 = \frac{\partial u_{oz}}{\partial y} \tag{14.9.1a}$$

$$\gamma_{zx} = 0 \rightarrow \frac{\partial u_{oz}}{\partial x} - \theta_1 = 0 \rightarrow \theta_1 = \frac{\partial u_{oz}}{\partial x} \tag{14.9.1b}$$

If we plug Equations (14.9.1a,b) into Equations (14.1.4d–f), the curvature (flexural) generalized strains for thin shells are defined as follows.

$$\varphi_{11} = \frac{\partial \theta_1}{\partial x} = \frac{\partial^2 u_{oz}}{\partial x^2} \tag{14.9.2a}$$

$$\varphi_{22} = \frac{\partial \theta_2}{\partial y} = \frac{\partial^2 u_{oz}}{\partial y^2} \tag{14.9.2b}$$

$$\varphi_{12} = \frac{1}{2}\left(\frac{\partial \theta_2}{\partial x} + \frac{\partial \theta_1}{\partial y}\right) = \frac{\partial^2 u_{oz}}{\partial x \partial y} \tag{14.9.2c}$$

The above expressions for curvature components include second-order derivatives of the displacement field u_{oz}. Similar expressions can be established for the virtual curvatures $\bar{\varphi}_{11}$, $\bar{\varphi}_{22}$ and $\bar{\varphi}_{12}$, which will now include second-order derivatives of the virtual normal displacement field w_{oz}:

$$\bar{\varphi}_{11} = \frac{\partial w_{\theta 1}}{\partial x} = \frac{\partial^2 w_{oz}}{\partial x^2} \quad (14.9.3a)$$

$$\bar{\varphi}_{22} = \frac{\partial w_{\theta 2}}{\partial y} = \frac{\partial^2 w_{oz}}{\partial y^2} \quad (14.9.3b)$$

$$\bar{\varphi}_{12} = \frac{1}{2}\left(\frac{\partial w_{\theta 2}}{\partial x} + \frac{\partial w_{\theta 1}}{\partial y}\right) = \frac{\partial^2 w_{oz}}{\partial x \partial y} \quad (14.9.3c)$$

Equations (14.9.3a–c) are a consequence of the kinematic assumptions of the K-L theory, according to which the virtual generalized shear strains $\bar{\gamma}_{zx}$ and $\bar{\gamma}_{yz}$ must vanish:

$$\bar{\gamma}_{zx} = 0 \quad (14.9.4a)$$

$$\bar{\gamma}_{yz} = 0 \quad (14.9.4b)$$

The weak form for shells derived in Box (14.5.1) is also valid for the K-L shell theory. If we take into account Equations (14.9.1a,b) and (14.9.4a,b), we can understand that all the weak-form terms depending on the through-thickness shear strains must vanish! This, in turn, allows us to define a generalized strain vector, $\{\hat{\varepsilon}\} = \left[\varepsilon_{ox}\ \varepsilon_{oy}\ \gamma_{oxy}\ \varphi_{11}\ \varphi_{22}\ 2\varphi_{12}\right]^T$, which only contains the membrane and bending (curvature) components. Similarly, we can use a generalized stress vector $\{\hat{\sigma}\} = \left[\hat{n}_{xx}\ \hat{n}_{yy}\ \hat{n}_{xy}\ \hat{m}_{xx}\ \hat{m}_{yy}\ \hat{m}_{xy}\right]^T$, and a virtual generalized strain vector, $\{\hat{\bar{\varepsilon}}\} = \left[\bar{\varepsilon}_{ox}\ \bar{\varepsilon}_{oy}\ \bar{\gamma}_{oxy}\ \bar{\varphi}_{11}\ \bar{\varphi}_{22}\ 2\bar{\varphi}_{12}\right]^T$. The remainder of this Section will provide an overview of the finite element implementation of the discrete Kirchhoff theory. We will specifically discuss the case of the discrete Kirchhoff quadrilateral (DKQ) element.

Before we move on with the derivations of the shape functions for the specific element, we need to establish a useful lemma in Box 14.9.1.

Box 14.9.1 Lemma—Partial Derivative of a Function along a Specific Direction

The partial derivative of any function $f(x, y)$, with respect to a direction defined by a unit vector $\{n\}$ is given by the following relation.

$$\frac{\partial f}{\partial n} = \left(\vec{\nabla} f\right) \bullet \vec{n} \quad (14.9.5)$$

For a planar problem, if $\vec{n} = \begin{Bmatrix} n_x \\ n_y \end{Bmatrix}$, then

$$\frac{\partial f}{\partial n} = \frac{\partial f}{\partial x} n_x + \frac{\partial f}{\partial y} n_y \quad (14.9.6)$$

We are now going to derive the shape functions for a four-node, DKQ element, as described in Batoz and Ben-Tahar (1982). As a starting point, we will consider an eight-node quadrilateral (8Q) element, whose sides in the physical (Cartesian) space are all straight lines, as shown in Figure 14.17. The four nodal points 1,2,3 and 4 are located at the corners of the quadrilateral. The remaining nodes 5,6,7 and 8 are the midpoints of the four straight sides of the quadrilateral in the physical space.

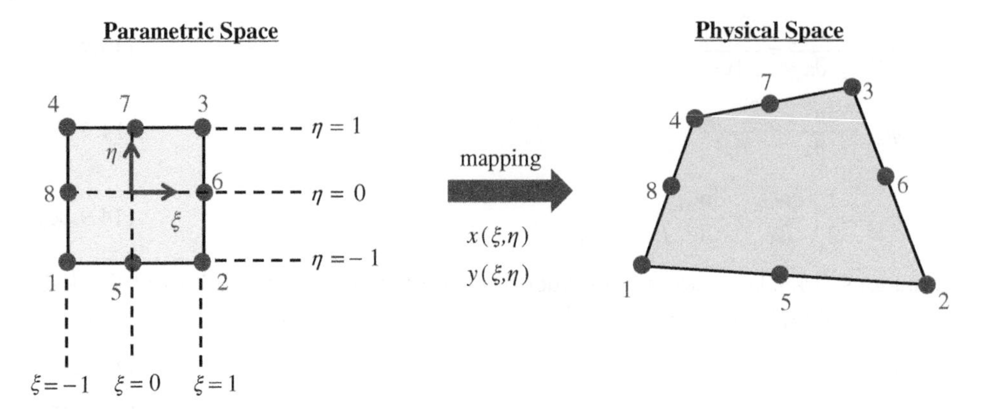

Figure 14.17 Isoparametric mapping for an 8Q element with straight sides in physical space.

For each of the element sides, we need to define the *unit normal outward vector*, $\vec{n}$, and the *unit tangential vector*, $\vec{s}$, as shown in Figure 14.18. The figure also provides the mathematical definition for the two vectors for each side, *ij*. The quantity $\ell_{ij}^{(e)}$ is merely the length of the line joining nodes i and j.

For the element sides, the rotations must be expressed in terms of components along the normal and tangential direction. The rotations about axis n and about axis s are related to the rotations θ_x, θ_y, in accordance with the following coordinate transformation equation (which is a special case of the general coordinate transformation rule in Section 7.7).

$$\begin{Bmatrix} \theta_n \\ \theta_n \end{Bmatrix} = \frac{1}{\ell_{ij}^{(e)}} \begin{bmatrix} -y_{ij}^{(e)} & x_{ij}^{(e)} \\ x_{ij}^{(e)} & y_{ij}^{(e)} \end{bmatrix} \begin{Bmatrix} \theta_x \\ \theta_x \end{Bmatrix} = \frac{1}{\ell_{ij}^{(e)}} \begin{bmatrix} -y_{ij}^{(e)} & -x_{ij}^{(e)} \\ x_{ij}^{(e)} & -y_{ij}^{(e)} \end{bmatrix} \begin{Bmatrix} \theta_2 \\ \theta_1 \end{Bmatrix} \tag{14.9.7}$$

We now come to the weak form for thin shells, which is the same as the expression in Box 14.5.1, except that we must remove the terms associated with the strains γ_{zx}, γ_{yz} and $\bar{\gamma}_{zx}$, $\bar{\gamma}_{yz}$. In accordance with our discussion above, the virtual work terms for flexure will now involve second-order derivatives of kinematic fields. This leads to the requirement for C^1 continuity for a finite element model, and C^1 continuity is impossible to strictly enforce in a conventional, multidimensional approximation with polynomial shape functions. This difficulty is circumvented in the DKQ formulation by an

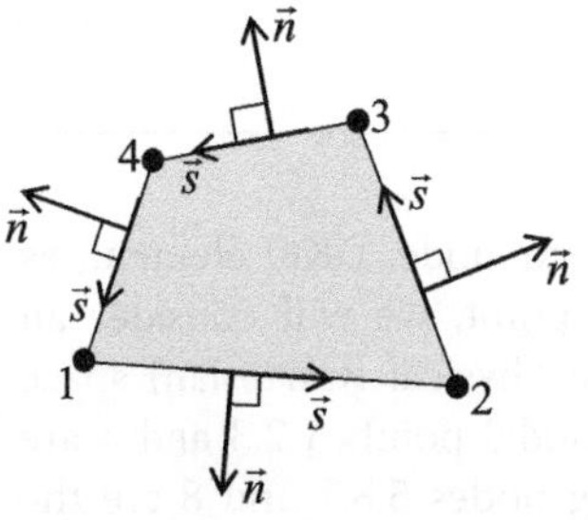

For each side "*ij*" (e.g., side 12):

$$\vec{s} = \frac{1}{\ell_{ij}^{(e)}} \begin{Bmatrix} x_{ij}^{(e)} \\ y_{ij}^{(e)} \end{Bmatrix} = \frac{1}{\ell_{ij}^{(e)}} \begin{Bmatrix} x_j^{(e)} - x_i^{(e)} \\ y_j^{(e)} - y_i^{(e)} \end{Bmatrix}$$

$$\vec{n} = \frac{1}{\ell_{ij}^{(e)}} \begin{Bmatrix} y_{ij}^{(e)} \\ -x_{ij}^{(e)} \end{Bmatrix} = \frac{1}{\ell_{ij}^{(e)}} \begin{Bmatrix} y_j^{(e)} - y_i^{(e)} \\ x_i^{(e)} - x_j^{(e)} \end{Bmatrix}$$

Figure 14.18 Definition of unit normal outward and unit tangential vectors for element sides.

"approximate" enforcement of Equations (14.9.1a, b) and (14.9.4a, b), through stipulation that they only apply at specific locations in an element.

Let us establish the approximation of the DKQ element. We begin by stipulating the approximation for the rotation fields, $\theta_1^{(e)}$ and $\theta_2^{(e)}$, in the interior of the element:

$$\theta_1^{(e)}(\xi,\eta) = -\sum_{i=1}^{8}\left(N_i^{(8Q)}(\xi,\eta)\cdot\theta_{yi}^{(e)}\right) \tag{14.9.8a}$$

$$\theta_2^{(e)}(\xi,\eta) = \sum_{i=1}^{8}\left(N_i^{8Q}(\xi,\eta)\cdot\theta_{xi}^{(e)}\right) \tag{14.9.8b}$$

The eight shape functions in Equations (14.9.8a,b) are those of an 8Q serendipity element, which were presented in Section 6.8 and are also provided here:

$$N_1^{(8Q)}(\xi,\eta) = -\frac{1}{4}(1-\xi)(1-\eta)(1+\xi+\eta) \tag{14.9.9a}$$

$$N_2^{(8Q)}(\xi,\eta) = -\frac{1}{4}(1+\xi)(1-\eta)(1-\xi+\eta) \tag{14.9.9b}$$

$$N_3^{(8Q)}(\xi,\eta) = \frac{1}{4}(1+\xi)(1+\eta)(\xi+\eta-1) \tag{14.9.9c}$$

$$N_4^{(8Q)}(\xi,\eta) = -\frac{1}{4}(1-\xi)(1+\eta)(1+\xi-\eta) \tag{14.9.9d}$$

$$N_5^{(8Q)}(\xi,\eta) = \frac{1}{2}(1-\xi^2)(1-\eta) \tag{14.9.9e}$$

$$N_6^{(8Q)}(\xi,\eta) = \frac{1}{2}(1+\xi)(1-\eta^2) \tag{14.9.9f}$$

$$N_7^{(8Q)}(\xi,\eta) = \frac{1}{2}(1-\xi^2)(1+\eta) \tag{14.9.9g}$$

$$N_8^{(8Q)}(\xi,\eta) = \frac{1}{2}(1-\xi)(1-\eta^2) \tag{14.9.9h}$$

As mentioned above, the intermediate nodes, 5,6,7,8, are the midpoints of the four sides of the quadrilateral element. Based on Figure 14.17, node 5 is located in the middle of side 12, node 6 is located in the middle of side 23, node 7 in the middle of 34, and node 8 in the middle of 41. We are going to stipulate that the following equations apply for the four intermediate nodal points:

$$\theta_{nk} = \left[\frac{\partial u_{oz}}{\partial s}\right]_{node\ k}, \quad \text{for } k = 5,6,7,8 \tag{14.9.10a}$$

$$\theta_{sk} = \frac{1}{2}\left(\theta_{si}+\theta_{sj}\right), \quad \text{for } k = 5,6,7,8 \tag{14.9.10b}$$

Equation (14.9.10a) is nothing other than the K-L hypothesis along the s-direction (i.e., the tangential direction) of each side of the element. On the other hand, Equation (14.9.10b) stipulates that the rotation about the side of the element at the middle is the average of the values at the two end points of the side. The specific equation relates the rotations of an intermediate node k to the corresponding rotations of the corner nodes i and j. The correspondence between a pair of nodes i and j and the node k

is provided in Table 14.9.1. Equation (14.9.10b) is also equivalent to the stipulation that the variation of the rotation in the direction s is linear along the element boundaries.

Next, we stipulate that *the variation of the transverse displacement* u_{oz}, *along each one of the boundary line segments* (i.e., the element sides) is determined with the expressions that we would be using for an Euler-Bernoulli beam element! For example, we can draw the side elevation of side (1-2) of our element, as shown in Figure 14.19, where we have also drawn the nodal transverse displacements and the rotations θ_n of the two end nodal points, 1 and 2. It is worth mentioning that this elevation with the nodal displacements/rotations is "reminiscent" of a planar beam segment! In fact, given the nodal displacements that we have drawn in the figure, and given that s corresponds to the axial coordinate along the beam, one can easily verify that Equation (14.9.10a) for $k = 5$ is nothing else that an Euler-Bernoulli beam statement (as described in Section 13.3) for the location of nodal point 5! Based on this observation, we will stipulate that the approximation of the normal displacement, $u_{oz}(s)$, along side (1-2) is given by the following expression.

$$u_{oz}(s) \approx N_{u1}^{(e)}(s) \cdot u_{z1}^{(e)} + N_{\theta 1}^{(e)}(s) \cdot \theta_{n1}^{(e)} + N_{u2}^{(e)}(s) \cdot u_{z2}^{(e)} + N_{\theta 2}^{(e)}(s) \cdot \theta_{n2}^{(e)} \tag{14.9.11}$$

The shape functions $N_i^{(e)}(s)$ in Equation (14.9.11) are those of a two-dimensional, two-node, Euler-Bernoulli beam element (Section 13.6) with a local coordinate system s-z, as shown in Figure 14.19. Thus, we have to establish the following Hermitian polynomials.

$$N_{u1}^{(e)}(s) = 1 - \frac{3s^2}{\left(\ell_{12}^{(e)}\right)^2} + \frac{2s^3}{\left(\ell_{12}^{(e)}\right)^3} \tag{14.9.12a}$$

Table 14.9.1 Correspondence between pair of nodes i, j and mid-side node k.

i	j	k
1	2	5
2	3	6
3	4	7
4	1	8

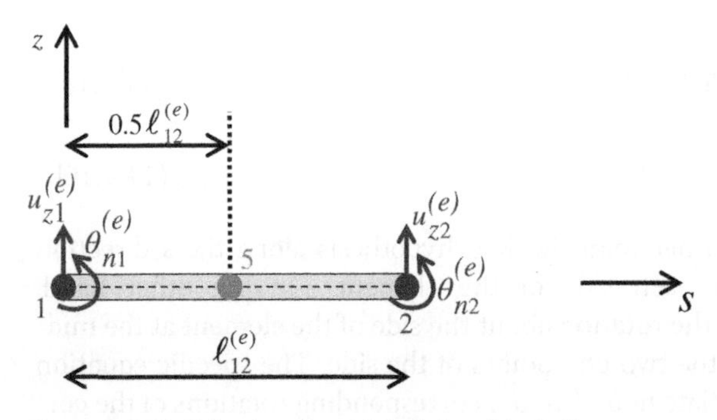

Figure 14.19 Side elevation of shell, for side (1-2).

$$N_{\theta 1}^{(e)}(s) = s - \frac{2s^2}{\ell_{12}^{(e)}} + \frac{s^3}{\left(\ell_{12}^{(e)}\right)^2} \quad (14.9.12b)$$

$$N_{u2}^{(e)}(s) = \frac{3s^2}{\left(\ell_{12}^{(e)}\right)^2} - \frac{2s^3}{\left(\ell_{12}^{(e)}\right)^3} \quad (14.9.12c)$$

$$N_{\theta 2}^{(e)}(s) = -\frac{s^2}{\ell_{12}^{(e)}} + \frac{s^3}{\left(\ell_{12}^{(e)}\right)^2} \quad (14.9.12d)$$

If we plug Equations (14.9.12a–d) into Equation (14.9.11) and then take the first derivative of the resulting expression with respect to the axial local coordinate, s, we obtain:

$$\frac{du_{oz}}{ds} \approx \left(\frac{6s}{\left(\ell_{12}^{(e)}\right)^2} - \frac{6s^2}{\left(\ell_{12}^{(e)}\right)^3}\right) \cdot \left(u_{z2}^{(e)} - u_{z1}^{(e)}\right) + \left(1 - \frac{4s}{\ell_{12}^{(e)}} - \frac{3s^2}{\left(\ell_{12}^{(e)}\right)^2}\right) \cdot \theta_{n1}^{(e)} + \left(-\frac{2s}{\ell_{12}^{(e)}} - \frac{3s^2}{\left(\ell_{12}^{(e)}\right)^2}\right) \cdot \theta_{n2}^{(e)} \quad (14.9.13)$$

Since node 5, $k = 5$, is located at the middle of side (12), we can set $s = 0.5\ell_{12}^{(e)}$ in Equation (14.9.13) to obtain:

$$\left[\frac{\partial u_{oz}}{\partial s}\right]_{node5} = \frac{3}{2\ell_{12}^{(e)}}\left(u_{z2}^{(e)} - u_{z1}^{(e)}\right) - \frac{1}{4}\left(\theta_{n1}^{(e)} + \theta_{n2}^{(e)}\right) \quad (14.9.14)$$

If we plug Equation (14.9.10a) for $k = 5$ into (14.9.14), we obtain:

$$\theta_{n5} = \frac{3}{2\ell_{12}^{(e)}}\left(u_{z2}^{(e)} - u_{z1}^{(e)}\right) - \frac{1}{4}\left(\theta_{n1}^{(e)} + \theta_{n2}^{(e)}\right) \quad (14.9.15)$$

Finally, if we use the first row of the matrix Equation (14.9.7) (i.e., the row giving θ_n), we can express the left-hand side of Equation (14.9.15) in terms of θ_{x5}, θ_{y5}:

$$-\frac{y_{12}^{(e)}}{\ell_{12}^{(e)}}\theta_{x5} + \frac{x_{12}^{(e)}}{\ell_{12}^{(e)}}\theta_{y5} = \frac{3}{2\ell_{12}^{(e)}}\left(u_{z2}^{(e)} - u_{z1}^{(e)}\right) + \frac{y_{12}^{(e)}}{4\ell_{12}^{(e)}}\left(\theta_{x1}^{(e)} + \theta_{x2}^{(e)}\right) - \frac{x_{12}^{(e)}}{4\ell_{12}^{(e)}}\left(\theta_{y1}^{(e)} + \theta_{y2}^{(e)}\right) \quad (14.9.16a)$$

If we plug $k = 5$ into Equation (14.9.10b) and also account for the second row of the matrix Equation (14.9.7) (i.e., the row giving θ_s), we obtain:

$$\frac{x_{12}^{(e)}}{\ell_{12}^{(e)}}\theta_{x5} + \frac{y_{12}^{(e)}}{\ell_{12}^{(e)}}\theta_{y5} = \frac{x_{12}^{(e)}}{2\ell_{12}^{(e)}}\left(\theta_{x1} + \theta_{x2}\right) + \frac{y_{12}^{(e)}}{2\ell_{12}^{(e)}}\left(\theta_{y1} + \theta_{y2}\right) \quad (14.9.16b)$$

Equations (14.9.16a) and (14.9.16b) can be rearranged into a new pair of equations, wherein θ_{x5}, θ_{y5} are expressed in terms of the normal displacements and in-plane rotations of nodes 1 and 2:

$$\theta_{x5} = -\frac{3y_{12}^{(e)}}{2\left(\ell_{12}^{(e)}\right)^2}\left(u_{z2}^{(e)} - u_{z1}^{(e)}\right) + \left[\frac{\left(x_{12}^{(e)}\right)^2}{2\left(\ell_{12}^{(e)}\right)^2} - \frac{\left(y_{12}^{(e)}\right)^2}{4\left(\ell_{12}^{(e)}\right)^2}\right]\left(\theta_{x1}^{(e)} + \theta_{x2}^{(e)}\right) + \frac{3x_{12}^{(e)}y_{12}^{(e)}}{4\left(\ell_{12}^{(e)}\right)^2}\left(\theta_{y1}^{(e)} + \theta_{y2}^{(e)}\right) \tag{14.9.17a}$$

$$\theta_{y5} = \frac{3x_{12}^{(e)}}{2\left(\ell_{12}^{(e)}\right)^2}\left(u_{z2}^{(e)} - u_{z1}^{(e)}\right) + \frac{3x_{12}^{(e)}y_{12}^{(e)}}{4\left(\ell_{12}^{(e)}\right)^2}\left(\theta_{x1}^{(e)} + \theta_{x2}^{(e)}\right) + \left[\frac{\left(y_{12}^{(e)}\right)^2}{2\left(\ell_{12}^{(e)}\right)^2} - \frac{\left(x_{12}^{(e)}\right)^2}{4\left(\ell_{12}^{(e)}\right)^2}\right]\left(\theta_{y1}^{(e)} + \theta_{y2}^{(e)}\right) \tag{14.9.17b}$$

Using the same exact procedure, we can express:

- Rotations θ_{x6} and θ_{y6} in terms of the normal displacements and in-plane rotations at nodes 2 and 3.
- Rotations θ_{x7} and θ_{y7} in terms of the normal displacements and in-plane rotations at nodes 3 and 4.
- Rotations θ_{x8} and θ_{y8} in terms of the normal displacements and in-plane rotations at nodes 4 and 1.

We can eventually obtain the following matrix equation.

$$\begin{Bmatrix} \theta_{x5}^{(e)} \\ \theta_{y5}^{(e)} \\ \theta_{x6}^{(e)} \\ \theta_{y6}^{(e)} \\ \theta_{x7}^{(e)} \\ \theta_{y7}^{(e)} \\ \theta_{x8}^{(e)} \\ \theta_{y8}^{(e)} \end{Bmatrix} = \begin{bmatrix} d_5^{(e)} & -e_5^{(e)} & b_5^{(e)} & -d_5^{(e)} & -e_5^{(e)} & b_5^{(e)} & 0 & 0 & 0 & 0 & 0 & 0 \\ -a_5^{(e)} & b_5^{(e)} & -c_5^{(e)} & a_5^{(e)} & b_5^{(e)} & -c_5^{(e)} & 0 & 0 & 0 & 0 & 0 & 0 \\ 0 & 0 & 0 & d_6^{(e)} & -e_6^{(e)} & b_6^{(e)} & -d_6^{(e)} & -e_6^{(e)} & b_6^{(e)} & 0 & 0 & 0 \\ 0 & 0 & 0 & -a_6^{(e)} & b_6^{(e)} & -c_6^{(e)} & a_6^{(e)} & b_6^{(e)} & -c_6^{(e)} & 0 & 0 & 0 \\ 0 & 0 & 0 & 0 & 0 & 0 & d_7^{(e)} & -e_7^{(e)} & b_7^{(e)} & -d_7^{(e)} & -e_7^{(e)} & b_7^{(e)} \\ 0 & 0 & 0 & 0 & 0 & 0 & -a_7^{(e)} & b_7^{(e)} & -c_7^{(e)} & a_7^{(e)} & b_7^{(e)} & -c_7^{(e)} \\ -d_8^{(e)} & -e_8^{(e)} & b_8^{(e)} & 0 & 0 & 0 & 0 & 0 & 0 & d_8^{(e)} & -e_8^{(e)} & b_8^{(e)} \\ a_8^{(e)} & b_8^{(e)} & -c_8^{(e)} & 0 & 0 & 0 & 0 & 0 & 0 & -a_8^{(e)} & b_8^{(e)} & -c_8^{(e)} \end{bmatrix} \begin{Bmatrix} u_{z1}^{(e)} \\ \theta_{x1}^{(e)} \\ \theta_{y1}^{(e)} \\ u_{z2}^{(e)} \\ \theta_{x2}^{(e)} \\ \theta_{y2}^{(e)} \\ u_{z3}^{(e)} \\ \theta_{x3}^{(e)} \\ \theta_{y3}^{(e)} \\ u_{z4}^{(e)} \\ \theta_{x4}^{(e)} \\ \theta_{y4}^{(e)} \end{Bmatrix} \tag{14.9.18}$$

where

$$a_k = \frac{3x_{ij}^{(e)}}{2\left(\ell_{ij}^{(e)}\right)^2} \tag{14.9.19a}$$

$$b_k = \frac{3x_{ij}^{(e)} y_{ij}^{(e)}}{4\left(\ell_{ij}^{(e)}\right)^2} \tag{14.9.19b}$$

$$c_k = \frac{\left(x_{ij}^{(e)}\right)^2}{4\left(\ell_{ij}^{(e)}\right)^2} - \frac{\left(y_{ij}^{(e)}\right)^2}{2\left(\ell_{ij}^{(e)}\right)^2} \tag{14.9.19c}$$

$$d_k = \frac{3y_{ij}^{(e)}}{2\left(\ell_{ij}^{(e)}\right)^2} \tag{14.9.19d}$$

$$e_k = -\frac{\left(x_{ij}^{(e)}\right)^2}{2\left(\ell_{ij}^{(e)}\right)^2} + \frac{\left(y_{ij}^{(e)}\right)^2}{4\left(\ell_{ij}^{(e)}\right)^2} \tag{14.9.19e}$$

and k is the midpoint of side ij, in accordance with Table 14.9.1.

In summary, if we know the normal displacements and in-plane rotations at the four corner nodes, we also know the in-plane rotations at the intermediate nodes.

If we plug Equation (14.9.18) into Equations (14.9.8a) and (14.9.8b), we obtain:

$$\begin{Bmatrix} \theta_1^{(e)}(\xi,\eta) \\ \theta_2^{(e)}(\xi,\eta) \end{Bmatrix} = \begin{bmatrix} H_1^{1(e)} & H_2^{1(e)} & H_3^{1(e)} & H_4^{1(e)} & H_5^{1(e)} & H_6^{1(e)} & H_7^{1(e)} & H_8^{1(e)} & H_9^{1(e)} & H_{10}^{1(e)} & H_{11}^{1(e)} & H_{12}^{1(e)} \\ H_1^{2(e)} & H_2^{2(e)} & H_3^{2(e)} & H_4^{2(e)} & H_5^{2(e)} & H_6^{2(e)} & H_7^{2(e)} & H_8^{2(e)} & H_9^{2(e)} & H_{10}^{2(e)} & H_{11}^{2(e)} & H_{12}^{2(e)} \end{bmatrix} \left\{ U_f^{(e)} \right\} \tag{14.9.20}$$

where $H_i^{1(e)}(\xi,\eta)$, $H_i^{2(e)}(\xi,\eta)$ constitute modified shape functions accounting for the discrete Kirchhoff kinematic assumptions and $\left\{ U_f^{(e)} \right\}$ is a (12 × 1) column vector, containing the *flexural degrees of freedom of the four corner nodes*:

$$\left\{ U_f^{(e)} \right\} = \left[u_{z1}^{(e)} \; \theta_{x1}^{(e)} \; \theta_{y1}^{(e)} \; u_{z2}^{(e)} \; \theta_{x2}^{(e)} \; \theta_{y2}^{(e)} \; u_{z3}^{(e)} \; \theta_{x3}^{(e)} \; \theta_{y3}^{(e)} \; u_{z4}^{(e)} \; \theta_{x4}^{(e)} \; \theta_{y4}^{(e)} \right]^T \tag{14.9.21}$$

The expressions for the modified shape functions of the first three columns of the shape function array in Equation (14.9.20) (i.e., the shape functions corresponding to the nodal values of corner node 1) can be written as follows.

$$H_1^{1(e)}(\xi,\eta) = a_5 \cdot N_5^{(8Q)}(\xi,\eta) - a_8 \cdot N_8^{(8Q)}(\xi,\eta) \tag{14.9.22a}$$

$$H_2^{1(e)}(\xi,\eta) = b_5 \cdot N_5^{(8Q)}(\xi,\eta) + b_8 \cdot N_8^{(8Q)}(\xi,\eta) \tag{14.9.22b}$$

$$H_3^{1(e)}(\xi,\eta) = N_1^{(8Q)}(\xi,\eta) - c_5 \cdot N_5^{(8Q)}(\xi,\eta) - c_8 \cdot N_8^{(8Q)}(\xi,\eta) \tag{14.9.22c}$$

$$H_2^{1(e)}(\xi,\eta) = d_5 \cdot N_5^{(8Q)}(\xi,\eta) - d_8 \cdot N_5^{(8Q)}(\xi,\eta) \tag{14.9.22d}$$

$$H_2^{2(e)}(\xi,\eta) = -N_1^{(8Q)}(\xi,\eta) + e_5 \cdot N_5^{(8Q)}(\xi,\eta) + e_8 \cdot N_8^{(8Q)}(\xi,\eta) \tag{14.9.22e}$$

$$H_3^{2(e)}(\xi,\eta) = -b_5 \cdot N_5^{(8Q)}(\xi,\eta) - b_8 \cdot N_8^{(8Q)}(\xi,\eta) = -H_2^{1(e)}(\xi,\eta) \tag{14.9.22f}$$

Along the same way, we can obtain the remaining shape functions of Equation (14.9.20), corresponding to corner nodes 2, 3, and 4. We need to keep in mind the following:

- For node 2, we must replace subscript 1 in the right-hand side of Equations (14.9.22c,e) by 2 and subscripts 5 and 8 in Equations (14.9.22a–f) by subscripts 6 and 5, respectively.
- For node 3, we must replace subscript 1 in the right-hand side of Equations (14.9.22c,e) by 3 and subscripts 5 and 8 in Equations (14.9.22a–f) by subscripts 7 and 6, respectively.
- For node 4, we must replace subscript 1 in the right-hand side of Equations (14.9.22c,e) by 4 and subscripts 5 and 8 in Equations (14.9.22a–f) by subscripts 8 and 7, respectively.

Remark 14.9.1: We started our derivations by stipulating the variation of θ_1, θ_2 in the interior of the element, and then stipulated the variation of u_{oz} along the boundary of the element (i.e., along the perimeter of the element). This is an example of a special case of multifield (mixed) element, called *hybrid element,* in which we stipulate the approximation of several field functions (in our case, the rotation fields θ_1 and θ_2) in the interior of the element, while for other field functions (in our case, the displacement field u_{oz}; see Equation (14.9.11)), we stipulate their approximate variation along the boundary. ∎

Remark 14.9.2: The modified shape functions for DKQ elements in Equations (14.9.20) and (14.9.22a–f) only affect the bending (plate) behavior. For the in-plane membrane degrees of freedom, the finite element approximation of Equations (14.6.1a,b) still applies. Furthermore, it is worth mentioning that we no longer establish an approximation for the field u_{oz} in the interior of the element! We can now establish the complete finite element approximation for a four-node, DQK element:

$$\begin{Bmatrix} u_{ox}^{(e)}(\xi,\eta) \\ u_{oy}^{(e)}(\xi,\eta) \\ \theta_1^{(e)}(\xi,\eta) \\ \theta_2^{(e)}(\xi,\eta) \end{Bmatrix} = \left[N_1^{(e)}(\xi,\eta)\right]\left\{U_1^{(e)}\right\} + \left[N_2^{(e)}(\xi,\eta)\right]\left\{U_2^{(e)}\right\} + \ldots + \left[N_4^{(e)}(\xi,\eta)\right]\left\{U_4^{(e)}\right\} \rightarrow$$

$$\rightarrow \begin{Bmatrix} u_{ox}^{(e)}(\xi,\eta) \\ u_{oy}^{(e)}(\xi,\eta) \\ \theta_1^{(e)}(\xi,\eta) \\ \theta_2^{(e)}(\xi,\eta) \end{Bmatrix} = \left[\left[N_1^{(e)}(\xi,\eta)\right]\left[N_2^{(e)}(\xi,\eta)\right]\ldots\left[N_4^{(e)}(\xi,\eta)\right]\right] \begin{Bmatrix} \left\{U_1^{(e)}\right\} \\ \left\{U_2^{(e)}\right\} \\ \vdots \\ \left\{U_4^{(e)}\right\} \end{Bmatrix} \tag{14.9.23}$$

where each 4 × 6 subarray, $\left[N_I^{(e)}(\xi,\eta)\right]$, containing the shape functions for the contribution of node I to the approximate fields, is given by:

$$\left[N_I^{(e)}(\xi,\eta)\right]=\begin{bmatrix} N_I^{(e)}(\xi,\eta) & 0 & 0 & 0 & 0 & 0 \\ 0 & N_I^{(e)}(\xi,\eta) & 0 & 0 & 0 & 0 \\ 0 & 0 & H_{3I-2}^{1(e)}(\xi,\eta) & H_{3I-1}^{1(e)}(\xi,\eta) & H_{3I}^{1(e)}(\xi,\eta) & 0 \\ 0 & 0 & H_{3I-2}^{2(e)}(\xi,\eta) & H_{3I-1}^{2(e)}(\xi,\eta) & H_{3I}^{2(e)}(\xi,\eta) & 0 \end{bmatrix} \quad (14.9.24)$$

∎

We can also formulate the generalized strain-displacement relation for a DKQ element, accounting for the fact that we do not have through-thickness shear strains and by taking appropriate derivatives of the approximate fields. Specifically, we can write:

$$\left\{\hat{\varepsilon}^{(e)}\right\}=\begin{Bmatrix}\left\{\hat{\varepsilon}_m^{(e)}\right\}\\ \left\{\hat{\varepsilon}_f^{(e)}\right\}\end{Bmatrix}=\begin{bmatrix}\begin{bmatrix}\left[B_{m1}^{(e)}(\xi,\eta)\right]\\ \left[B_{f1}^{(e)}(\xi,\eta)\right]\end{bmatrix} & \begin{bmatrix}\left[B_{m1}^{(e)}(\xi,\eta)\right]\\ \left[B_{f1}^{(e)}(\xi,\eta)\right]\end{bmatrix} & \cdots & \begin{bmatrix}\left[B_{m4}^{(e)}(\xi,\eta)\right]\\ \left[B_{f4}^{(e)}(\xi,\eta)\right]\end{bmatrix}\end{bmatrix}\begin{Bmatrix}\left\{U_1^{(e)}\right\}\\ \left\{U_2^{(e)}\right\}\\ \vdots \\ \left\{U_4^{(e)}\right\}\end{Bmatrix}\rightarrow$$

$$\rightarrow\left\{\hat{\varepsilon}^{(e)}\right\}=\left[B^{(e)}(\xi,\eta)\right]\left\{U^{(e)}\right\} \quad (14.9.25)$$

where $\left[B_{mI}^{(e)}(\xi,\eta)\right]$ is the membrane strain-displacement array for node I, given by Equation (14.6.10a), and $\left[B_{fI}^{(e)}(\xi,\eta)\right]$ is the flexural strain-displacement array for node I of a DKQ element, given by:

$$\left[B_{fI}^{(e)}(\xi,\eta)\right]=\begin{bmatrix} 0 & 0 & 0 & \dfrac{\partial H_{3I-2}^{1(e)}}{\partial x} & \dfrac{\partial H_{3I-1}^{1(e)}}{\partial x} & \dfrac{\partial H_{3I}^{1(e)}}{\partial x} \\ 0 & 0 & 0 & \dfrac{\partial H_{3I-2}^{2(e)}}{\partial y} & \dfrac{\partial H_{3I-1}^{2(e)}}{\partial y} & \dfrac{\partial H_{3I}^{2(e)}}{\partial y} \\ 0 & 0 & 0 & \dfrac{\partial H_{3I-2}^{1(e)}}{\partial y}+\dfrac{\partial H_{3I-2}^{2(e)}}{\partial x} & \dfrac{\partial H_{3I-1}^{1(e)}}{\partial y}+\dfrac{\partial H_{3I-1}^{2(e)}}{\partial x} & \dfrac{\partial H_{3I}^{1(e)}}{\partial y}+\dfrac{\partial H_{3I}^{2(e)}}{\partial x} \end{bmatrix} \quad (14.9.26)$$

The $\left[\hat{D}^{(e)}\right]$ array of a DKQ element, giving the generalized stresses from the generalized strains, must only include the terms corresponding to membrane forces and the terms corresponding to flexural moments. In other words, the generalized stress-strain law for a DKQ element attains the following form.

$$\left\{\hat{\sigma}^{(e)}\right\}=\begin{Bmatrix}\left\{\hat{\sigma}_m^{(e)}\right\}\\ \left\{\hat{\sigma}_f^{(e)}\right\}\end{Bmatrix}=\left[\hat{D}^{(e)}\right]\left\{\hat{\varepsilon}^{(e)}\right\} \quad (14.9.27)$$

For the special case of linear isotropic elasticity where the material parameters E and ν do not vary with z, we can use the arrays $\left[\hat{D}_m^{(e)}\right]$ (Equation 14.4.37) and $\left[\hat{D}_f^{(e)}\right]$ (Equation 14.4.29) to write:

$$\left[\hat{D}^{(e)}\right] = \begin{bmatrix} \left[\hat{D}_m^{(e)}\right] & [0] \\ [0] & \left[\hat{D}_f^{(e)}\right] \end{bmatrix} \tag{14.9.28}$$

Remark 14.9.3: The element stiffness matrix $[k^{(e)}]$ of a DKQ element can be obtained using Equation (14.6.13). ▮

Remark 14.9.4: The use of the DKQ shape functions of Equation (14.9.24) for calculation of the equivalent nodal force/moment vector $\{f^{(e)}\}$ may create theoretical difficulties for the quantities associated with plate behavior. To circumvent these difficulties, which will not be discussed here, $\{f^{(e)}\}$ can be obtained using the shape functions and equations of Section 14.6. ▮

Remark 14.9.5: The considerations in Remarks 14.9.3 and 14.9.4 provide the element stiffness matrix and equivalent nodal force/moment vector in the local coordinate system. The transformation equations presented in Section 14.8, which are still valid for DKQ elements, must be used to obtain the corresponding quantities in the global coordinate system. ▮

14.10 Continuum-Based Formulation for Nonplanar (Curved) Shells

The discussion for shells presented so far has been focused on planar (flat) shells. The finite element method can also be used for the analysis of curved shells, which are three-dimensional surfaces. We will rely on a continuum-based (CB) formulation, in accordance with descriptions in Hughes (2000). The CB formulation constitutes an extension (to shells) of the pertinent approach for beams, presented in Sections 13.12 and 13.13. If we go to a location of a shell in the physical space, which corresponds to given values of ξ and η, and draw many points with the same, given ξ and η values but with varying ζ-values, these points define (in the physical space) a curve called the *fiber* of the shell at the specific location. Any surface defined by the points with $\zeta = \zeta_o$ = fixed is called a *lamina*. For nonplanar shells, we need to modify (more accurately, to generalize!) the kinematic assumptions of shell theory. Specifically, we no longer stipulate that the normal stress along one of the physical axes is zero. *Instead, we assume that the normal stresses perpendicular to the lamina plane are zero.* This is the generalization of the statement that $\sigma_{zz} = 0$ for flat shells. *We also stipulate that the change in fiber length can be neglected* when we establish the displacement fields. This kinematic assumption is a generalization of the stipulation that we neglect the effect of thickness changes in planar shells.

We will obtain the formulation of a curved shell using the concept of an underlying continuum element, similar to the procedure provided in Section 13.13 for curved beams. Of course, instead of having the reference curve and top and bottom curves of beams, we will be having a (curved) reference surface and the top and bottom surfaces, as shown in Figure 14.20. Our description will rely on a four-node shell element, although a curved geometry might be better represented using more nodes—for example, eight nodes.

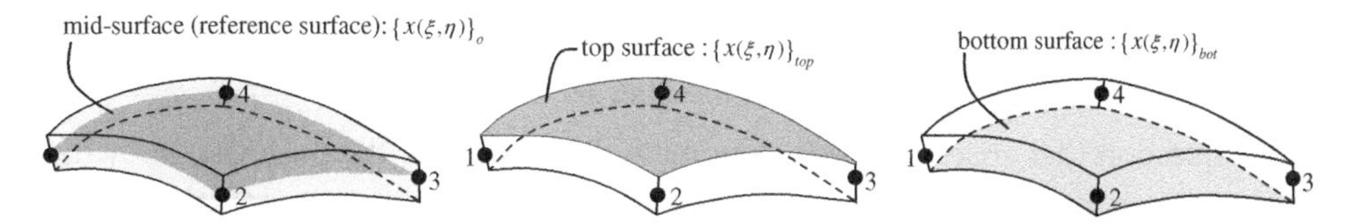

Figure 14.20 Reference, top, and bottom surfaces for curved shell.

Any three-dimensional surface is *parameterized* by two variables. This means that we only need a pair of parametric coordinates ξ, η to uniquely identify the physical coordinates of each point on the surface. The physical location of a point in a three-dimensional surface is established by a set of three Cartesian coordinates, $\{x\} = [x\ y\ z]^T$. As shown in Figure 14.20, the location of each point in the reference surface, the top surface and the bottom surface can be obtained in terms of ξ and η through a coordinate mapping. The element itself will be described by the nodal points that lie on the reference surface, as shown in Figure 14.20. In the derivations described herein, the reference surface will coincide with the mid-surface of the element. For each nodal point, we can define the fiber vector, as shown in Figure 14.21, if we know the physical coordinates of the top and bottom surface, for the same ξ and η values as those of the nodal point. Per Figure 14.21, the fiber vector of each node I is a unit vector parallel to the line that joins the top and bottom surface at the parametric location of node I.

For each node I of the shell element, the components of the unit vector in the fiber direction can be obtained from the following expression.

$$\vec{e}_f^I = \begin{Bmatrix} e_{fx}^I \\ e_{fy}^I \\ e_{fz}^I \end{Bmatrix} = \frac{1}{\sqrt{\left(x_{top}^I - x_{bot}^I\right)^2 + \left(y_{top}^I - y_{bot}^I\right)^2 + \left(z_{top}^I - z_{bot}^I\right)^2}} \begin{Bmatrix} x_{top}^I - x_{bot}^I \\ y_{top}^I - y_{bot}^I \\ z_{top}^I - z_{bot}^I \end{Bmatrix} \tag{14.10.1}$$

For given values of ξ, η, the distance between the top and bottom surfaces is the *thickness, d,* of the element. We now introduce a third parametric coordinate, ζ, which measures the location along the fiber. The value $\zeta = -1$ corresponds to the bottom

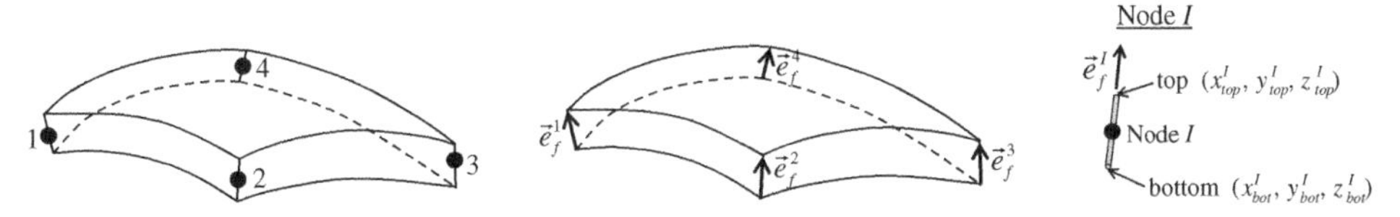

Figure 14.21 Fiber vector for each nodal point.

surface, the value $\zeta = 1$ corresponds to the top surface, and the value $\zeta = 0$ to the mid-surface. The physical coordinates of any point in the interior of the element can now be described by the following expression:

$$\{x(\xi,\eta,\zeta)\} = \{x(\xi,\eta)\}_o + \zeta\left[\frac{d(\xi,\eta)}{2}\vec{e}_f(\xi,\eta)\right] \tag{14.10.2}$$

where $\{x(\xi,\eta)\}_o$ is the vector with the three physical coordinates of the reference surface.

Equation (14.10.2) states that, for given parametric coordinates ξ and η, the location of any point can be found by detecting the location $\{x(\xi,\eta)\}_o$ at the reference surface, and then moving along the direction of the fiber by the amount $\zeta \cdot \dfrac{d(\xi,\eta)}{2}$.

One question that arises is how to determine the product of the thickness, d, by the fiber vector, $\vec{e}_f$, at any location (ξ,η). The answer is: in the very same way that we determine any quantity in the interior of the element—by interpolating between the nodal points through the shape functions! Thus, we can write:

$$d(\xi,\eta)\cdot\vec{e}_f(\xi,\eta) = \sum_{I=1}^{n}\left[N_I^{(e)}(\xi,\eta)\left(d_I\cdot\vec{e}_f^{\,I}\right)\right] \tag{14.10.3a}$$

where d_I is the thickness value at node I. Similarly, the physical coordinates for a location (ξ,η) of the reference surface can be found from the physical coordinates of the four nodal points as follows.

$$\{x(\xi,\eta)\}_o = \sum_{I=1}^{n}\left[N_I^{(e)}(\xi,\eta)\left\{x_I^{(e)}\right\}\right] \tag{14.10.3b}$$

Combining Equations (14.10.2) and (14.10.3a,b) eventually yields the following expression.

$$\{x(\xi,\eta,\zeta)\} = \sum_{I=1}^{n}\left[N_I^{(e)}(\xi,\eta)\left(\left\{x_I^{(e)}\right\} + \zeta\cdot\frac{d_I}{2}\cdot\vec{e}_f^{\,I}\right)\right] \tag{14.10.4}$$

Equation (14.10.4) constitutes a *mapping* from a three-dimensional parametric space (ξ, η, ζ) to the three-dimensional physical space (x, y, z).

The next step is to define the lamina, that is, the tangent plane to the shell reference surface in the element. It is known from considerations of differential geometry (e.g., O'Neill 2006) that, for a curved surface *parameterized* (i.e. described) by two coordinates ξ, η, the tangent plane at any location P on the surface is the plane defined by two tangential unit vectors, $\vec{e}_\xi$ and $\vec{e}_\eta$, schematically shown in Figure 14.22 and defined by the following two expressions.

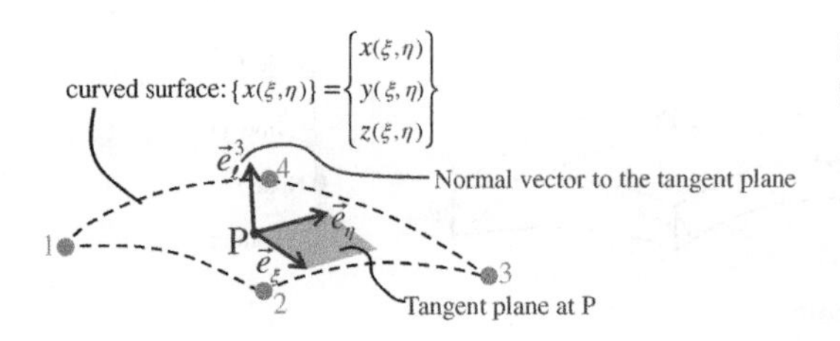

Figure 14.22 Tangent vectors, tangent plane, and normal vector to lamina.

$$\vec{e}_\xi = \frac{\{x_\xi\}}{\|\{x_\xi\}\|} \quad (14.10.5a)$$

$$\vec{e}_\eta = \frac{\{x_\eta\}}{\|\{x_\eta\}\|} \quad (14.10.5b)$$

where

$$\{x_\xi\} = \begin{bmatrix} \frac{\partial x}{\partial \xi} & \frac{\partial y}{\partial \xi} & \frac{\partial z}{\partial \xi} \end{bmatrix}^T \quad (14.10.6a)$$

$$\{x_\eta\} = \begin{bmatrix} \frac{\partial x}{\partial \eta} & \frac{\partial y}{\partial \eta} & \frac{\partial z}{\partial \eta} \end{bmatrix}^T \quad (14.10.6b)$$

Next, we will define a set of orthogonal unit vectors $\vec{e}_\ell^{\,1}, \vec{e}_\ell^{\,2}, \vec{e}_\ell^{\,3}$, describing *the local coordinate axes of a lamina*. We begin by establishing the laminar vector $\vec{e}_\ell^{\,3}$, which, as shown in Figure 14.22, is normal to the tangent plane and can be computed using the following expression.

$$\vec{e}_\ell^{\,3} = \frac{\vec{e}_\xi \times \vec{e}_\eta}{\|\vec{e}_\xi \times \vec{e}_\eta\|} \quad (14.10.7)$$

Next, we define the following two auxiliary vectors.

$$\vec{e}_a = \frac{\vec{e}_\xi + \vec{e}_\eta}{\|\vec{e}_\xi + \vec{e}_\eta\|} \quad (14.10.8a)$$

$$\vec{e}_b = \frac{\vec{e}_\ell^{\,3} \times \vec{e}_a}{\|\vec{e}_\ell^{\,3} \times \vec{e}_a\|} \quad (14.10.8b)$$

Finally, the two *laminar basis unit vectors* $\vec{e}_\ell^{\,1}, \vec{e}_\ell^{\,2}$ can be established using the equations:

$$\vec{e}_\ell^{\,1} = \frac{\sqrt{2}}{2}\left(\vec{e}_a - \vec{e}_b\right) \quad (14.10.9a)$$

$$\vec{e}_\ell^{\,2} = \frac{\sqrt{2}}{2}\left(\vec{e}_a + \vec{e}_b\right) \quad (14.10.9b)$$

Remark 14.10.1: The laminar basis vectors $\vec{e}_\ell^{\,1}, \vec{e}_\ell^{\,2}$ lie in the tangent plane to the curved surface. For a general curved surface, this means that each of these two laminar basis vectors is a linear combination of $\vec{e}_\xi$ and $\vec{e}_\eta$. ∎

We will now establish the considerations of the CB element. Specifically, as shown in Figure 14.23, we will consider an underlying, three-dimensional continuum element, whose nodes are the endpoints of the fibers at the nodal points of the shell element. The figure specifically describes the case where the shell element has four nodes; in this case, the underlying continuum element is an eight-node hexahedral (8H) one. As also shown in the figure, the motion (nodal displacements) of the nodes of the continuum element will

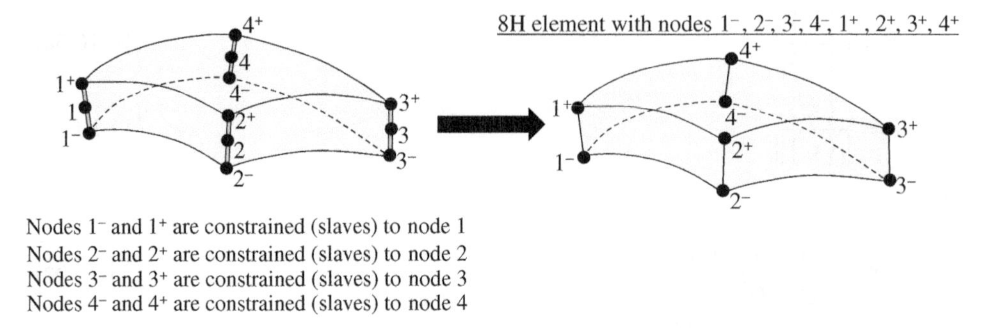

Figure 14.23 Underlying continuum element for four-node, curved shell element.

be *constrained* to the motion (nodal displacements and rotations) of the shell element. This is why the specific formulation is sometimes referred to as a *constrained continuum approach*. The constraints are essentially established by regarding the fibers as rigid bars, connecting the top and the bottom surface of the shell. For each node I ($I = 1, 2, 3, 4$) of the shell element, we have a corresponding pair of nodes, I^- and I^+.

The computations for the underlying continuum element can be conducted in accordance with the standard methodology for three-dimensional isoparametric elements, presented in Chapter 9. The coordinate mapping is described by Equation (14.10.4). The mapping considered for the underlying continuum element is also schematically shown in Figure 14.24.

The displacements of the underlying continuum element can be obtained from the displacements of the nodal points in the shell element, using rigid-body kinematic constraint equations in accordance with Appendix D. If we know the translations and rotations for node I of the shell reference surface, then the translations of the top surface node I^+ can be obtained from the following equation, which is *based on the assumption that node I^+ is connected by a rigid bar* (i.e., the fiber) *to node I* and the position vector of node I^+ with respect to node I is equal to $\frac{d_I}{2}\vec{e}_f^I$:

$$\begin{Bmatrix} u_x \\ u_y \\ u_z \end{Bmatrix}_{I^+} = \begin{Bmatrix} u_x \\ u_y \\ u_z \end{Bmatrix}_I + \frac{d_I}{2}\begin{bmatrix} 0 & e_{fz}^I & -e_{fy}^I \\ -e_{fz}^I & 0 & e_{fx}^I \\ e_{fy}^I & -e_{fx}^I & 0 \end{bmatrix}\begin{Bmatrix} \theta_x \\ \theta_y \\ \theta_z \end{Bmatrix}_I \tag{14.10.10}$$

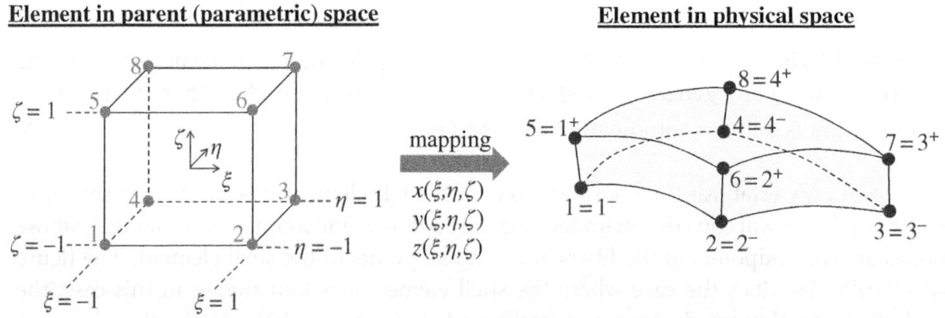

Element in parent (parametric) space | Element in physical space

$\zeta = 1$, $\zeta = -1$, $\eta = 1$, $\eta = -1$, $\xi = -1$, $\xi = 1$

mapping: $x(\xi,\eta,\zeta)$, $y(\xi,\eta,\zeta)$, $z(\xi,\eta,\zeta)$

$8 = 4^+$, $5 = 1^+$, $4 = 4^-$, $7 = 3^+$, $6 = 2^+$, $1 = 1^-$, $3 = 3^-$, $2 = 2^-$

Figure 14.24 Mapping for underlying continuum element of a four-node, curved shell element.

Equation (14.10.10) can also be cast in the following matrix transformation form.

$$\{U_{I^+}\} = [T_{I^+}]\{U_I\} \tag{14.10.11}$$

where

$$\{U_{I^+}\} = \begin{bmatrix} u_{xI^+} & u_{yI^+} & u_{zI^+} \end{bmatrix}^T \tag{14.10.11a}$$

$$\{U_I\} = \begin{bmatrix} u_{xI} & u_{yI} & u_{zI} & \theta_{xI} & \theta_{yI} & \theta_{zI} \end{bmatrix}^T \tag{14.10.11b}$$

and

$$[T_{I^+}] = \begin{bmatrix} 1 & 0 & 0 & 0 & \frac{d_I}{2}e_{fz}^I & -\frac{d_I}{2}e_{fy}^I \\ 0 & 1 & 0 & -\frac{d_I}{2}e_{fz}^I & 0 & \frac{d_I}{2}e_{fx}^I \\ 0 & 0 & 1 & \frac{d_I}{2}e_{fy}^I & -\frac{d_I}{2}e_{fx}^I & 0 \end{bmatrix} \tag{14.10.11c}$$

Similarly, the translations of node I^- of the bottom surface can be obtained as follows.

$$\begin{Bmatrix} u_x \\ u_y \\ u_z \end{Bmatrix}_{I^-} = \begin{Bmatrix} u_x \\ u_y \\ u_z \end{Bmatrix}_I - \frac{d_I}{2}\begin{bmatrix} 0 & e_{fz}^I & -e_{fy}^I \\ -e_{fz}^I & 0 & e_{fx}^I \\ e_{fy}^I & -e_{fx}^I & 0 \end{bmatrix}\begin{Bmatrix} \theta_x \\ \theta_y \\ \theta_z \end{Bmatrix}_I \tag{14.10.12}$$

We can once again write:

$$\{U_{I^-}\} = [T_{I^-}]\{U_I\} \tag{14.10.13}$$

where

$$\{U_{I^-}\} = \begin{bmatrix} u_{xI^-} & u_{yI^-} & u_{zI^-} \end{bmatrix}^T \tag{14.10.14a}$$

and

$$[T_{I^-}] = \begin{bmatrix} 1 & 0 & 0 & 0 & -\frac{d_I}{2}e_{fz}^I & \frac{d_I}{2}e_{fy}^I \\ 0 & 1 & 0 & \frac{d_I}{2}e_{fz}^I & 0 & -\frac{d_I}{2}e_{fx}^I \\ 0 & 0 & 1 & -\frac{d_I}{2}e_{fy}^I & \frac{d_I}{2}e_{fx}^I & 0 \end{bmatrix} \tag{14.10.14b}$$

Next, we define two vectors with nodal values. One vector, $\{U^{(e)}\}$, will contain the nodal displacements and rotations of the shell element (in our case, of the four nodes of the reference surface). The other vector, $\{U^{(8Q)}\}$, contains the nodal displacements of the underlying continuum element. The two vectors are defined by the following block expressions.

$$\{U^{(e)}\} = \begin{bmatrix} \{U_1^{(e)}\} \\ \{U_2^{(e)}\} \\ \{U_3^{(e)}\} \\ \{U_4^{(e)}\} \end{bmatrix} \tag{14.10.15a}$$

$$\left\{U^{(8H)}\right\} = \begin{bmatrix} \left\{U_{1^-}^{(e)}\right\} \\ \left\{U_{2^-}^{(e)}\right\} \\ \left\{U_{3^-}^{(e)}\right\} \\ \left\{U_{4^-}^{(e)}\right\} \\ \left\{U_{1^+}^{(e)}\right\} \\ \left\{U_{2^+}^{(e)}\right\} \\ \left\{U_{3^+}^{(e)}\right\} \\ \left\{U_{4^+}^{(e)}\right\} \end{bmatrix} \tag{14.10.15b}$$

Now, we can combine the transformation equations for all the nodal points of the shell element and of the underlying (8*H*) element into a single constraint matrix equation:

$$\left\{U^{(8H)}\right\} = [T]\left\{U^{(e)}\right\} \tag{14.10.16}$$

The transformation matrix $[T]$ in Equation (14.10.16) can be written in block form as follows:

$$[T] = \begin{bmatrix} [T_{1^-}] & [0] & [0] & [0] \\ [0] & [T_{2^-}] & [0] & [0] \\ [0] & [0] & [T_{3^-}] & [0] \\ [0] & [0] & [0] & [T_{4^-}] \\ [T_{1^+}] & [0] & [0] & [0] \\ [0] & [T_{2^+}] & [0] & [0] \\ [0] & [0] & [T_{3^+}] & [0] \\ [0] & [0] & [0] & [T_{4^+}] \end{bmatrix} \tag{14.10.17}$$

Each of the block matrices in Equation (14.10.17) has dimensions (3×6).

There is one final consideration that we need to make—that is, how to apply the stress-strain law in a CB shell element: the quasi-plane stress condition must be enforced in the laminar coordinate system. Specifically, the normal stress component $\sigma_{33}^{(\ell)}$ along the third laminar local direction must be equal to zero for a curved shell. In accordance with Section 14.4, this allows us to establish a reduced constitutive equation for five strain components. If we write a general linearly elastic stress-strain law for the local laminar coordinate system, we have:

$$\left\{\sigma^{(\ell)}\right\} = \left[D^{(\ell)}\right]\left\{\varepsilon^{(\ell)}\right\} \tag{14.10.18}$$

where

$$\left\{\sigma^{(\ell)}\right\} = \left[\sigma_{11}^{(\ell)} \; \sigma_{22}^{(\ell)} \; \sigma_{33}^{(\ell)} \; \sigma_{12}^{(\ell)} \; \sigma_{23}^{(\ell)} \; \sigma_{31}^{(\ell)}\right]^T \tag{14.10.19a}$$

$$\left\{\varepsilon^{(\ell)}\right\} = \left[\varepsilon_{11}^{(\ell)} \ \varepsilon_{22}^{(\ell)} \ \varepsilon_{33}^{(\ell)} \ 2\varepsilon_{12}^{(\ell)} \ 2\varepsilon_{23}^{(\ell)} \ 2\varepsilon_{31}^{(\ell)}\right]^T \tag{14.10.19b}$$

Just like we did in Section 14.4, we will establish a reduced stress-strain relation in the form:

$$\left\{\tilde{\sigma}^{(\ell)}\right\} = \left[D_{red}^{(\ell)}\right]\left\{\varepsilon^{(\ell)}\right\} \tag{14.10.20}$$

where $\left\{\tilde{\sigma}^{(\ell)}\right\} = \left[\sigma_{11}^{(\ell)} \ \sigma_{22}^{(\ell)} \ \sigma_{12}^{(\ell)} \ \sigma_{23}^{(\ell)} \ \sigma_{31}^{(\ell)}\right]^T$ contains the stress components except for $\sigma_{33}^{(\ell)} = 0$, and $\left[D_{red}^{(\ell)}\right]$ is the material stiffness matrix obtained if we erase the third row of $[D^{(\ell)}]$. We can eventually reach an equation of the form:

$$\left\{\tilde{\sigma}^{(\ell)}\right\} = \left[\tilde{D}^{(\ell)}\right]\left\{\tilde{\varepsilon}^{(\ell)}\right\} \tag{14.10.21}$$

where, per Section 14.4, $\left[\tilde{D}^{(\ell)}\right]$ is a (5 × 5), reduced (*condensed*) material stiffness matrix. We can directly write the following equation for the components of $\left[\tilde{D}^{(\ell)}\right]$.

$$\tilde{D}_{ij}^{(\ell)} = D_{ij}^{(\ell)} - D_{i3}^{(\ell)} \frac{D_{3j}^{(\ell)}}{D_{33}^{(\ell)}} \tag{14.10.22}$$

To use the stress-strain law in the lamina coordinate system, we need to express the strains and the displacements in this system. To this end, a 3 × 3 transformation matrix, $[q]$, is constructed, which allows the transformation of vectors from the global coordinate system to the lamina coordinate system. The rows of matrix $[q]$ are simply equal to the unit vectors of the lamina coordinate system:

$$[q] = \left[\vec{e}_\ell^{\,1} \ \vec{e}_\ell^{\,2} \ \vec{e}_\ell^{\,3}\right]^T \tag{14.10.23}$$

Each displacement component in direction i of the laminar coordinate system is obtained from the displacement vector components in the global coordinate system, through the following transformation equation.

$$u_i^{(l)} = \sum_{m=1}^{3} (q_{im} \cdot u_m) \tag{14.10.24}$$

We can then symbolically write the following expression, giving the nodal displacements of the (8*H*) element in the laminar coordinate system, from the nodal displacements of the same element in the global coordinate system:

$$\left\{U_{(\ell)}^{(8H)}\right\} = \left[Q^{(e)}\right]\left\{U^{(8H)}\right\} \tag{14.10.25}$$

where $\left\{U_{(\ell)}^{(8H)}\right\}$ contains the nodal displacements of the underlying continuum element expressed in the *local lamina* coordinate system, and the transformation matrix $[Q^{(e)}]$ can be written in block form as follows.

$$\left[Q^{(e)}\right]=\begin{bmatrix} [q] & [0] & [0] & [0] & [0] & [0] & [0] & [0] \\ [0] & [q] & [0] & [0] & [0] & [0] & [0] & [0] \\ [0] & [0] & [q] & [0] & [0] & [0] & [0] & [0] \\ [0] & [0] & [0] & [q] & [0] & [0] & [0] & [0] \\ [0] & [0] & [0] & [0] & [q] & [0] & [0] & [0] \\ [0] & [0] & [0] & [0] & [0] & [q] & [0] & [0] \\ [0] & [0] & [0] & [0] & [0] & [0] & [q] & [0] \\ [0] & [0] & [0] & [0] & [0] & [0] & [0] & [q] \end{bmatrix} \tag{14.10.26}$$

Remark 14.10.2: Since the laminar basis vectors generally vary inside a curved shell, the values of $\left\{U_{(\ell)}^{(8H)}\right\}$ are different at each location in the element! ∎

We can now go ahead, and define a matrix strain-displacement relation for the laminar local coordinate system:

$$\left\{\tilde{\varepsilon}^{(\ell)}\right\}=\left[B_{(\ell)}^{(e)}\right]\left\{U_{(\ell)}^{(8H)}\right\} \tag{14.10.27}$$

where the local laminar strain-displacement matrix, $\left[B_{(\ell)}^{(e)}\right]$, is given by the following expression.

$$\left[B_{(\ell)}^{(e)}\right]=\begin{bmatrix} \frac{\partial N_1^{(8H)}}{\partial x_1^{(\ell)}} & 0 & 0 & \frac{\partial N_2^{(8H)}}{\partial x_1^{(\ell)}} & 0 & 0 & \cdots & \frac{\partial N_8^{(8H)}}{\partial x_1^{(\ell)}} & 0 & 0 \\ 0 & \frac{\partial N_1^{(8H)}}{\partial x_2^{(\ell)}} & 0 & 0 & \frac{\partial N_2^{(8H)}}{\partial x_2^{(\ell)}} & 0 & \cdots & 0 & \frac{\partial N_8^{(8H)}}{\partial x_2^{(\ell)}} & 0 \\ \frac{\partial N_1^{(8H)}}{\partial x_2^{(\ell)}} & \frac{\partial N_1^{(8H)}}{\partial x_1^{(\ell)}} & 0 & \frac{\partial N_2^{(8H)}}{\partial x_2^{(\ell)}} & \frac{\partial N_2^{(8H)}}{\partial x_1^{(\ell)}} & 0 & \cdots & \frac{\partial N_8^{(8H)}}{\partial x_2^{(\ell)}} & \frac{\partial N_8^{(8H)}}{\partial x_1^{(\ell)}} & 0 \\ 0 & \frac{\partial N_1^{(8H)}}{\partial x_3^{(\ell)}} & \frac{\partial N_1^{(8H)}}{\partial x_2^{(\ell)}} & 0 & \frac{\partial N_2^{(8H)}}{\partial x_3^{(\ell)}} & \frac{\partial N_2^{(8H)}}{\partial x_2^{(\ell)}} & \cdots & 0 & \frac{\partial N_8^{(8H)}}{\partial x_3^{(\ell)}} & \frac{\partial N_8^{(8H)}}{\partial x_2^{(\ell)}} \\ \frac{\partial N_1^{(8H)}}{\partial x_3^{(\ell)}} & 0 & \frac{\partial N_1^{(8H)}}{\partial x_1^{(\ell)}} & \frac{\partial N_2^{(8H)}}{\partial x_3^{(\ell)}} & 0 & \frac{\partial N_2^{(8H)}}{\partial x_1^{(\ell)}} & \cdots & \frac{\partial N_8^{(8H)}}{\partial x_3^{(\ell)}} & 0 & \frac{\partial N_8^{(8H)}}{\partial x_1^{(\ell)}} \end{bmatrix} \tag{14.10.28}$$

The matrix $\left[B_{(\ell)}^{(e)}\right]$ contains derivatives of the shape functions with respect to the local laminar coordinates. The partial derivative of each shape function $N_a^{(8H)}$ with respect to the laminar coordinate $x_j^{(\ell)}$ can be obtained by means of the chain rule of differentiation, in terms of the corresponding partial derivatives with respect to the global coordinates $x_1 = x$, $x_2 = y$, $x_3 = z$:

$$\frac{\partial N_a^{(8H)}}{\partial x_j^{(\ell)}} = \frac{\partial N_a^{(8H)}}{\partial x_1}\frac{\partial x_1}{\partial x_j^{(\ell)}} + \frac{\partial N_a^{(8H)}}{\partial x_2}\frac{\partial x_2}{\partial x_j^{(\ell)}} + \frac{\partial N_a^{(8H)}}{\partial x_3}\frac{\partial x_3}{\partial x_j^{(\ell)}} \tag{14.10.29}$$

The spatial coordinates (x_1, x_2, x_3) and $\left(x_1^{(\ell)}, x_2^{(\ell)}, x_3^{(\ell)}\right)$ are components of a position vector; thus, they follow the same coordinate transformation rule as the displacements in Equation (14.10.24). We can eventually obtain that the partial derivatives of laminar coordinates with respect to global coordinates are given by:

$$\frac{\partial x_i}{\partial x_j^{(\ell)}} = q_{ji} \tag{14.10.30}$$

If we plug Equation (14.10.30) into Equation (14.10.29), we obtain:

$$\frac{\partial N_a^{(8H)}}{\partial x_j^{(\ell)}} = \frac{\partial N_a^{(8H)}}{\partial x_1}q_{j1} + \frac{\partial N_a^{(8H)}}{\partial x_2}q_{j2} + \frac{\partial N_a^{(8H)}}{\partial x_3}q_{j3} = \sum_{m=1}^{3}\left(\frac{\partial N_a^{(8H)}}{\partial x_m}q_{jm}\right) \tag{14.10.31}$$

We can now establish an algorithm for the calculation of the stiffness matrix and nodal force/moment vector of a continuum-based, curved shell element. This procedure is summarized in Box 14.10.1, for the case that we use Gaussian quadrature with N_g quadrature points for the underlying continuum element.

Box 14.10.1 Calculation of Stiffness Matrix and Nodal Force Vector for Curved, CB Shell Element

1) For each quadrature point g, $g = 1, 2, \ldots, N_g$ of the underlying continuum element:
   a) Calculate the laminar coordinate system, the matrix $[q_g]$, the transformation matrix $\left[Q_g^{(e)}\right]$, the shape function array $\left[N_g^{(8H)}\right]$ and the matrix $\left[B_{(\ell)g}^{(e)}\right]$. The latter is obtained after we have calculated $\left[N_{x,g}^{(8H)}\right]$, containing the derivatives of the shape functions with respect to x, y, z, evaluated at the location of point g, using the procedures described in Section 9.3.1.
   b) Calculate the stiffness contribution of the quadrature point to the stiffness matrix of the continuum element, expressed in the global coordinate system:

$$\left[Q_g^{(e)}\right]^T\left[B_{(\ell)g}^{(e)}\right]^T\left[\tilde{D}_g^{(\ell)}\right]\left[B_{(\ell)g}^{(e)}\right]\left[Q_g^{(e)}\right]J_g W_g$$

   c) Calculate the nodal force contribution of the quadrature point to the equivalent nodal force vector of the continuum element, expressed in the global coordinate system:

$$\left[N_g^{(e)}\right]^T\left\{b_g^{(e)}\right\}J_g W_g$$

2) Calculate the stiffness matrix of the underlying continuum element, by summing the contributions of all the quadrature points. If we have a total of N_g quadrature points, we obtain:

$$\left[k^{(8H)}\right]=\sum_{g=1}^{N_g}\left(\left[Q_g^{(e)}\right]^T\left[B_{(\ell)g}^{(e)}\right]^T\left[\tilde{D}_g^{(\ell)}\right]\left[B_{(\ell)g}^{(e)}\right]\left[Q_g^{(e)}\right]J_gW_g\right)$$

3) Calculate the equivalent nodal force vector due to body forces, $\left\{f_\Omega^{(8H)}\right\}$, of the underlying continuum element, by summing the contributions of all the quadrature points. If we have a total of N_g quadrature points, we obtain:

$$\left\{f_\Omega^{(8H)}\right\}=\sum_{g=1}^{N_g}\left(\left[N_g^{(e)}\right]^T\left\{b_g^{(e)}\right\}J_gW_g\right)$$

4) Calculate the equivalent nodal force vector $\left\{f_\Gamma^{(8H)}\right\}$ due to prescribed tractions (if any) on the boundary surfaces of the underlying continuum element, using the procedure described in Section 9.3.3.
5) Calculate the total equivalent nodal force vector of the underlying continuum element, using the equation: $\left\{f^{(8H)}\right\}=\left\{f_\Omega^{(8H)}\right\}+\left\{f_\Gamma^{(8H)}\right\}$
6) Using the transformation matrix $[T]$, given by Equation (14.10.17), which expresses the relation between the nodal displacements of the continuum element and the displacements and rotations of the nodes in the shell element, obtain the stiffness matrix, $[k^{(e)}]$, and the equivalent nodal force/moment vector $\{f^{(e)}\}$, of the shell element e:

$$\left[k^{(e)}\right]=[T]^T\left[k^{(8H)}\right][T] \qquad \left\{f^{(e)}\right\}=[T]^T\left\{f^{(8H)}\right\}$$

Remark 14.10.3: The algorithm in Box 14.10.1 gives the stiffness matrix and nodal force vector in the global coordinate system. ■

Remark 14.10.4: For a shell element, we commonly state that we arrange the quadrature points in *stacks*. A stack of quadrature points corresponds to a set of quadrature points having the same values of ξ and η. Thus, if we use a conventional $2 \times 2 \times 2$ quadrature rule for the underlying continuum element, as shown in Figure 14.25, we will be having a total of four stacks of quadrature points (each stack having two points, one corresponding to $\zeta = -1/\sqrt{3}$ and the other to $\zeta = 1/\sqrt{3}$). In practice, the quadrature points are numbered by stack. Thus, for Figure 14.25, we would have quadrature stacks 1, 2, 3, and 4, and for each stack we would be having quadrature points 1 and 2. The number of quadrature points per stack is usually an input parameter for a finite element model. For linearly elastic analysis, two-point stacks are expected to give good results. For nonlinear material behavior (e.g., elastoplastic materials, cracking materials, etc.), at least five or six quadrature points per stack are required to obtain accurate results. ■

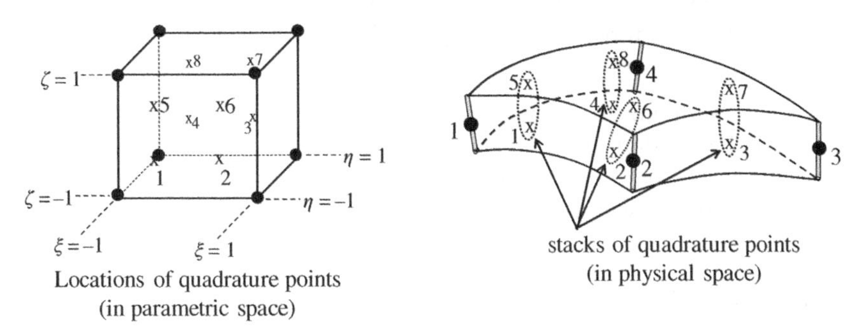

Figure 14.25 Stacks of quadrature points for a curved shell element.

Remark 14.10.5: As a follow-up note to Remark 14.10.4, it is worth mentioning that the analysis of thin shells under flexure with the CB approach presented herein may lead to shear locking, similar to the case described for beam elements in Section 13.14. Selective-reduced integration can be used for the through-thickness shear stiffness terms to address locking issues. All other stiffness terms are calculated with full integration. For the specific case of the four-node CB element shown in Figure 14.25, reduced integration amounts to the use of a single stack of quadrature points (all points in the stack corresponding to $\xi = \eta = 0$), while full integration is conducted using all four stacks shown in the specific figure. ∎

Problems

14.1 We are given the two-shell-element mesh shown in Figure P14.1. The formulation of element 1 is based on the Reissner-Mindlin theory, while element 2 is a discrete Kirchhoff quadrilateral element. Both elements have a thickness of 0.2 m, and are made of linearly elastic, isotropic material with $E = 30$ GPa, $\nu = 0.2$. All translations and rotations of nodes 1 and 2 are fixed to a zero value.

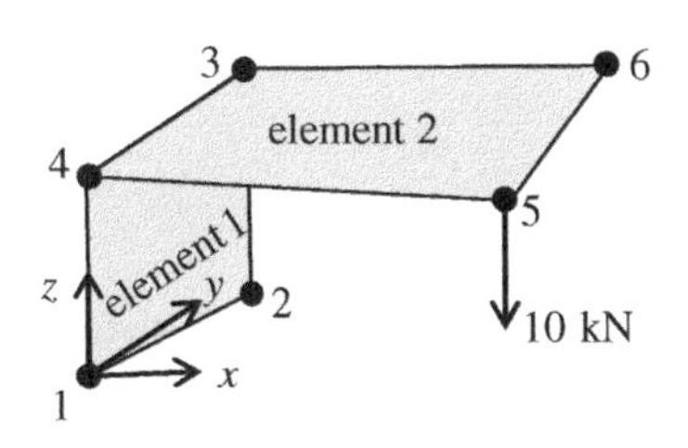

Node	x (m)	y (m)	z (m)
1	0	0	0
2	1	2	−1
3	1	2	4
4	0	0	4
5	5	−1	5
6	4	−1	5

Figure P14.1

a) Obtain the stiffness matrix of each element in the corresponding local coordinate system.
b) Obtain the coordinate transformation array, $[R^{(e)}]$, for each element.

c) Obtain the global stiffness matrix, $[K]$, of the structure.
d) Obtain the nodal displacements and rotations of the model.
e) Calculate the generalized strains and generalized stresses for the quadrature points of each element.

14.2 We are given a continuum-based shell element, for which we have the nodal coordinates of the hexahedral underlying solid element, as shown in Figure P14.2. The material is isotropic, linearly elastic, with $E = 29000$ ksi, $\nu = 0.25$. The nodal coordinates of the shell element are those of the mid-surface.

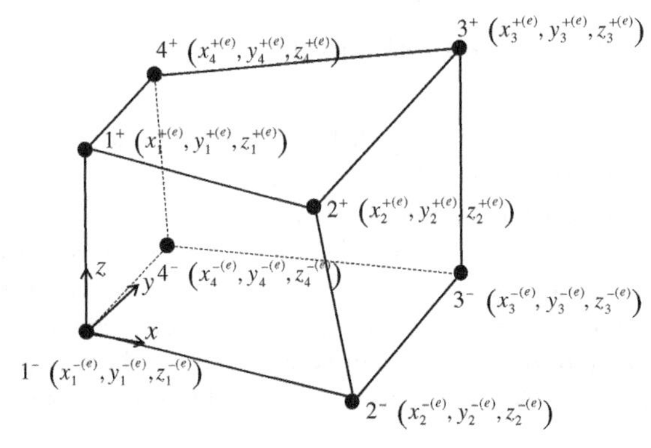

Node	x (ft)	y (ft)	z (ft)
1^-	0	0	0
2^-	5	0	−0.2
3^-	6	5	−0.1
4^-	1	4	0
1^+	0	0	0.8
2^+	4	−1	0.6
3^+	6	5	1
4^+	−1	3.5	0.9

Figure P14.2

f) Obtain the shell thickness values at each nodal location of the shell element (Note: The thickness is not constant!).
g) Obtain the constraint transformation array, giving the nodal displacements of the solid element in terms of the nodal displacements and rotations in the shell element.
h) Obtain the stiffness matrix of the shell element in the global coordinate system.
i) Obtain the stiffness matrix of the shell element in the global coordinate system, using selective-reduced integration in accordance with Remark 14.10.5.

References

Batoz, J.-L., and Ben Tahar, M. (1982). "Evaluation of a New Quadrilateral Thin Plate Bending Element." *International Journal for Numerical Methods in Engineering*, 18 (11), 1655–1677.

Cottrell, A., Hughes, T., and Bazilevs, Y. (2006). *Isogeometric Analysis: Toward Integration of CAD and FEA*. Hoboken, NJ: John Wiley & Sons.

Hughes, T., and Brezzi, F. (1989). "On drilling degrees of freedom." *Computer Methods in Applied Mechanics and Engineering*, 72 (1), 105–121.

Hughes, T. (2000). *The Finite Element Method—Linear Static and Dynamic Finite Element Analysis*, 2nd ed. New York: Dover.

O'Neill, B. (2006). *Differential Geometry*, 2nd ed. Burlington, MA: Academic Press.

15

Finite Elements for Elastodynamics, Structural Dynamics, and Time-Dependent Scalar Field Problems

15.1 Introduction

So far, we have examined the use of the finite element method for various types of *steady-state* physical problems, that is, problems where the applied source terms (heat source, body forces, etc.) and the boundary conditions remain constant with time. In the context of solid mechanics and structural engineering, steady-state problems are called *static problems*, wherein the governing differential equations are force equilibrium equations for an infinitesimally small piece of material. It is important to remember that the concept of *equilibrium* is much more general: equilibrium for any physical system means that all the state variables—that is, the quantities that completely describe the state of the system (as described in Section 1.1)—remain constant with time. For example, in steady-state problems, the temperature field is that corresponding to a state of thermal equilibrium, where the fluxes, temperatures, and so on inside the system do not vary with time.

This chapter will focus on time-dependent problems, where the various terms can change with time. Emphasis will be laid on solid and structural mechanics, wherein time-dependent problems are referred to as *dynamic problems*. In dynamic problems, the applied loadings on a structure and/or the boundary conditions *vary with time*; that is, the applied body forces, tractions and prescribed displacements at essential boundaries are functions of time. Dynamic problems for elastic solid mechanics are called *elastodynamic* problems, while dynamic problems for beams and shells are called *structural dynamics* problems. In dynamic problems, the state variables for a structure (e.g., the displacements) will vary with time. *From a mathematical point of view, the description of dynamic problems introduces time, t, as an extra independent variable, in addition to the spatial variables, x,y,z.* Furthermore, additional kinematic fields, related to the rate of change of field quantities such as displacements, need to be introduced and determined in the problem, to characterize the *process* by which the state of a dynamic system continuously changes. For time-dependent problems *the rate of any field is simply the partial derivative of the field with respect to time.* For solid and structural mechanics, we need to consider the *velocity field and the acceleration field.* The velocity field, $\dot{u}(x,t)$, expresses the rate of change of the displacements with time, and the acceleration field, $\ddot{u}(x,t)$, expresses the rate of change of the velocities with time. The following two expressions apply.

Fundamentals of Finite Element Analysis: Linear Finite Element Analysis, First Edition.
Ioannis Koutromanos, James McClure, and Christopher Roy.

Companion website: www.wiley.com/go/koutromanos/linear

$$\dot{u}(x,t) = \frac{\partial u}{\partial t} \tag{15.1.1a}$$

$$\ddot{u}(x,t) = \frac{\partial \dot{u}}{\partial t} = \frac{\partial}{\partial t}\left(\frac{\partial u}{\partial t}\right) = \frac{\partial^2 u}{\partial t^2} \tag{15.1.1b}$$

We will begin by describing the strong and weak form for the dynamic version of one-dimensional elasticity problems. We will realize that we can quickly obtain the expressions for the strong and weak form of a dynamic problem by simply adding the effect of the *inertial forces* to the distributed loading per unit length. The governing equations for the strong form of dynamic problems, include partial derivatives with respect to time and are called the *differential equations of motion*. The mathematical structure of these equations classifies them as *hyperbolic partial differential equations*.

An additional consideration in dynamics stems from the need to account for *damping* in the material. The term *damping* generally refers to physical mechanisms that *dissipate* part of the energy of a system under dynamic loading. In this chapter, we will describe the damping effect through the constitutive law of the material, by assuming that we have a *viscoelastic* material (which introduces a dependence of the stress on both the strain and on the *strain rate*). This assumption will naturally give rise to a "damping contribution" in the governing equations of both the strong and weak form. After obtaining the strong and weak form for one-dimensional elastodynamics, we will establish the pertinent finite element equations in Section 15.5. The finite element equations, which are based on the same exact approximations that we used in Chapter 3 for one-dimensional static problems, are also called *semi-discrete equations of motion* (because they express the effect of discretization on spatial coordinates, but not time).

Similar considerations for multidimensional elastodynamics will give the strong and weak form by adding an inertial term to the body force. We will discuss this topic (for the most general case of three-dimensional problems) in Sections 15.6 and 15.7. We will also examine the description of structural dynamics, i.e., for dynamic problems in beam and shell structures, in Section 15.8.

Remark 15.1.1: It is worth emphasizing that the equations of motion including inertial forces are a direct consequence of the principle of conservation of linear momentum, as discussed in Remark 1.1.1. The reader is encouraged to revisit the specific discussion before proceeding further with this Chapter. ∎

Although this chapter primarily focuses on elastodynamics and structural dynamics, a discussion of the strong form, weak form, and finite element equations for time-dependent scalar field problems (heat conduction, chemical diffusion, and porous flow) is also provided in Sections 15.10 and 15.11. We will see that the governing equations for these problems include first partial derivatives of the field functions (temperature field, concentration field etc.) with respect to time. The time-dependent scalar-field problems that we will consider are also called *parabolic problems*, a term associated with the mathematical structure of the governing equations. The chapter is concluded with a brief presentation of the *state-space* description of elastodynamics and structural dynamics, which allows us to mathematically describe these problems as parabolic ones.

15.2 Strong Form for One-Dimensional Elastodynamics

We will now examine the behavior of a one-dimensional bar subjected to distributed loadings and tractions, as shown in Figure 15.1. The problem is very similar to the one we considered in Section 2.1 (and Figure 2.1). The only difference is that the displacement field now depends not only on the spatial coordinate x, but also on time, t. We will be mathematically stating this by writing $u(x,t)$ for the displacement field. Similarly, the applied loadings and boundary conditions depend on t. Notice that, for the example case of Figure 15.1, we have a prescribed displacement at the location $x = L$, which in general is a function of time (it can vary with time). Per Figure 15.1, we need to provide (as input) an additional material quantity for solid and structural dynamics, namely, the *(mass) density of the material* $\rho(x)$, which gives the mass per unit volume of the material. We consider the most general case where the density varies along the length of the bar.

The key to obtain the governing differential equations is to isolate and consider the equilibrium of a very small segment of the structure and examine its equilibrium, as shown in Figure 15.2. As seen in the figure, there is an additional force term, the *inertial force*, f_I, associated with the dynamic nature of the loading. In accordance with *D'Alembert's principle* (see Remark 1.1.1), *the inertial force is equal to the product of mass times the opposite of the acceleration.*

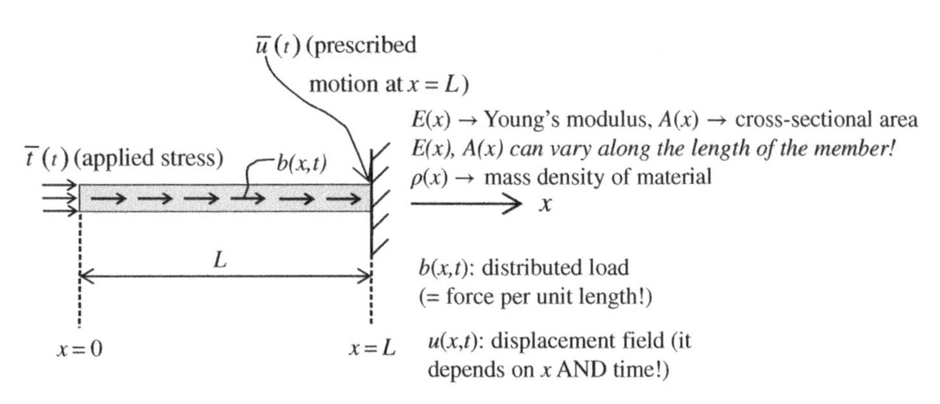

Figure 15.1 One-dimensional bar subjected to time-dependent loading.

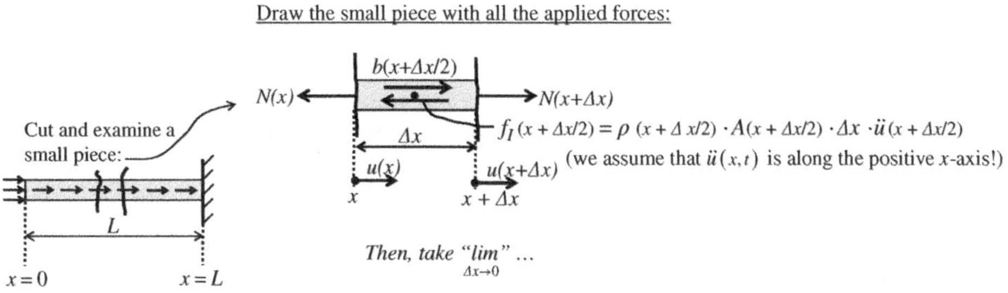

Draw the small piece with all the applied forces:

Cut and examine a small piece:

$b(x+\Delta x/2)$

$N(x)$

$N(x+\Delta x)$

$f_I\,(x + \Delta x/2) = \rho\,(x + \Delta\,x/2)\cdot A(x + \Delta x/2)\cdot \Delta x\cdot \ddot{u}\,(x + \Delta x/2)$

Δx

(we assume that $\ddot{u}(x,t)$ is along the positive x-axis!)

$u(x)$ $\quad u(x+\Delta x)$

x $\quad x + \Delta x$

L

$x = 0$ $\quad x = L$

Then, take "$\lim\limits_{\Delta x \to 0}$" ...

Figure 15.2 Free-body diagram of small segment of 1D elastic structure.

We can now establish the force equilibrium equation in the axial direction, for the free-body diagram of Figure 15.2. In the figure, we assume that, since the piece that we have considered is small, then the mass density, the cross-sectional area and the body force are constant and equal to the corresponding values at the middle of the piece. We can then write the inertial force, $f_I|_{\left(x+\frac{\Delta x}{2}\right)}$, at the center of the piece as follows:

$$f_I|_{\left(x+\frac{\Delta x}{2}\right)} = \rho|_{\left(x+\frac{\Delta x}{2}\right)} \cdot A|_{\left(x+\frac{\Delta x}{2}\right)} \cdot \ddot{u}|_{\left(x+\frac{\Delta x}{2}\right)} \cdot \Delta x \tag{15.2.1}$$

We now write down the force equilibrium equation:

$$N(x+\Delta x) - N(x) + b|_{\left(x+\frac{\Delta x}{2}\right)} \cdot \Delta x - f_I|_{\left(x+\frac{\Delta x}{2}\right)} = 0 \tag{15.2.2}$$

If we plug Equation (15.2.1) into Equation (15.2.2) and then divide by Δx, we obtain:

$$\frac{N(x+\Delta x) - N(x)}{\Delta x} + b|_{\left(x+\frac{\Delta x}{2}\right)} - \rho|_{\left(x+\frac{\Delta x}{2}\right)} \cdot A|_{\left(x+\frac{\Delta x}{2}\right)} \cdot \ddot{u}|_{\left(x+\frac{\Delta x}{2}\right)} = 0 \tag{15.2.3}$$

If we now take the limit as the size of the piece tends to zero, that is, the limit as Δx tends to zero, we obtain:

$$\frac{\partial N}{\partial x} + b(x,t) - \rho(x) \cdot A(x) \cdot \ddot{u}(x,t) = 0 \tag{15.2.4}$$

Finally, if we use the same considerations (constitutive law, strain-displacement equation etc.) as those in Section 2.1, Equation (15.2.2) yields:

$$\frac{\partial}{\partial x}\left(EA\frac{\partial u}{\partial x}\right) + b(x,t) - \rho(x) \cdot A(x) \cdot \ddot{u}(x,t) = 0 \tag{15.2.5}$$

Equation (15.2.5) is the governing partial differential equation of motion for one-dimensional elastodynamics.

Remark 15.2.1: Since we now have two independent variables, x and t, the derivatives with respect to x in the strong form have become partial derivatives with respect to x. ▌

Now, besides the fact that the essential and natural boundary conditions are time-dependent, we need to add some new conditions in the problem, so that it is *well-defined* (i.e., allowing us to obtain a solution): we need to define appropriate *initial conditions*—mathematical expressions which give us the values of the field functions at the instant of time at which *we begin to examine* our system. This instant of time is mathematically taken to correspond to a value of time equal to zero, that is, $t = 0$. For time-dependent problems involving only first derivatives of the field function with respect to time, the required initial conditions must define the value of the field function itself at $t = 0$. For problems like dynamics, which include second derivatives of the field function, the initial conditions must define the values of the field function (i.e., u) and of its first derivative with respect to time (i.e., $\dot{u}$) at $t = 0$. Thus, for dynamic problems, we will be stipulating that we know the *initial displacement field*:

$$u_o(x) = u(x, t=0) \tag{15.2.6a}$$

and the *initial velocity field*:

$$v_o(x) = \dot{u}(x, t = 0) = \left.\frac{\partial u}{\partial t}\right|_{t=0} \tag{15.2.6b}$$

We can finally combine the governing partial differential equations of motion, the initial conditions and boundary conditions, to define the strong form for one-dimensional elastodynamics, provided in Box 15.2.1.

Box 15.2.1 Strong Form for One-Dimensional Elastodynamics

$$\frac{\partial}{\partial x}\left(EA\frac{\partial u}{\partial x}\right) + b(x,t) - \rho(x)\cdot A(x)\cdot \ddot{u}(x,t) = 0, \ 0 < x < L \tag{15.2.7}$$

$$u(x,t) = \bar{u}(t) \ \text{ at } \Gamma_u (\text{ essential boundary conditions}) \tag{15.2.8a}$$

$$\left[n \cdot E\frac{\partial u}{\partial x}\right]_{\Gamma_t} = \bar{t}(t) \ \text{ at } \Gamma_t (\text{ natural boundary conditions}) \tag{15.2.8b}$$

$$\left.\begin{aligned} u(x,t=0) &= u_o(x)\ , \ 0 < x < L \\ \dot{u}(x,t=0) = \left.\frac{\partial u}{\partial t}\right|_{t=0} &= v_o(x)\ , \ 0 < x < L \end{aligned}\right\} \rightarrow \text{initial conditions (IC)} \tag{15.2.9}$$

Remark 15.2.2: Notice an interesting duality: the boundary conditions include values at a given spatial location. These values are generally time-dependent. The initial conditions define values for a given time, and these initial values generally vary with the spatial coordinate x. ■

Remark 15.2.3: The governing equations for time-dependent elasticity include second derivatives with respect to time (correspondingly, the initial conditions involve initial displacement and initial velocity) and up to second derivatives with respect to x. The governing partial differential equation for elasticity problems is called a *hyperbolic equation*. Accordingly, problems governed by hyperbolic equations are called hyperbolic problems. The main physical aspect of hyperbolic problems is that the "information" (e.g., displacements, stresses etc.) propagates in the system with a finite velocity. For elastodynamics, this means that we have *wave propagation* in the interior of the body, and the propagation velocity of the information is the wave speed of the material. The partial differential equations of motion for multi-dimensional elastodynamics and for structural dynamics are also hyperbolic. ■

15.3 Strong Form in the Presence of Material Damping

In the previous section, we established the governing differential equations of motion for one-dimensional elastodynamics, assuming that the material has the same exact linearly elastic behavior as the one presented in Section 2.1 and specifically Equation (2.1.6).

However, it is well known that in reality, *the stresses that develop in a material may depend on the rate of loading* (i.e., how "fast" the strains change) as well. Furthermore, the material has the capability to dissipate part of the energy that is added to the system (the added energy is the work done by the external forces). A simple way to capture this dissipation at the material level and also introduce a dependence on the rate of loading is by replacing the standard "linearly elastic material assumption" (*Hookean material*, i.e., material obeying Hooke's law), schematically described in Figure 15.3a as a linear spring with constant E, with a *viscoelastic material* model, where the stresses depend on both the strains and on the strain rates (i.e. partial derivatives of the strains with time). The simplest case of such a viscoelastic material is the *Kelvin material*, which can be conceptually thought of as a combination of a linear spring and a viscous dashpot, as shown in Figure 15.3b. The dashpot coefficient, η, expresses a linear dependence of the stress on the strain rate.

The stress-strain law for a Kelvin viscoelastic material can be written as:

$$\sigma(x,t) = E(x)\cdot\varepsilon(x,t) + \eta(x)\cdot\dot{\varepsilon}(x,t) \tag{15.3.1}$$

where the *strain rate*, $\dot{\varepsilon}(x,t)$, is simply the partial derivative of the axial strain with respect to time:

$$\dot{\varepsilon}(x,t) = \frac{\partial\varepsilon}{\partial t} = \frac{\partial}{\partial t}\left(\frac{\partial u}{\partial x}\right) = \frac{\partial^2 u}{\partial t\partial x} = \frac{\partial}{\partial x}\left(\frac{\partial u}{\partial t}\right) = \frac{\partial\dot{u}}{\partial x} \tag{15.3.2}$$

Remark 15.3.1: An assumption for damping that establishes a linear dependence of stress on the strain rate is called a *viscous material damping* model. ■

The axial force for a Kelvin material is given by:

$$N(x,t) = A(x)\cdot\sigma(x,t) = A(x)\cdot\left(E(x)\cdot\frac{\partial u}{\partial x} + \eta(x)\cdot\frac{\partial\dot{u}}{\partial x}\right) = E(x)\cdot A(x)\cdot\frac{\partial u}{\partial x} + \eta(x)\cdot A(x)\cdot\frac{\partial\dot{u}}{\partial x} \tag{15.3.3}$$

We can now account for Equation (15.3.3) and establish the *strong form for one-dimensional elastodynamics in the presence of damping*, provided in Box 15.3.1.

We can verify that, if we set $\eta = 0$ in Box 15.3.1, we obtain the strong form of Box 15.2.1 for a linearly elastic material model. This should be the case, since $\eta = 0$ in Figure 15.3b would give the linearly elastic Hookean material of Figure 15.3a.

We can now realize that the only differences in the governing differential equations compared to the case of one-dimensional static elasticity are: (1) the existence of

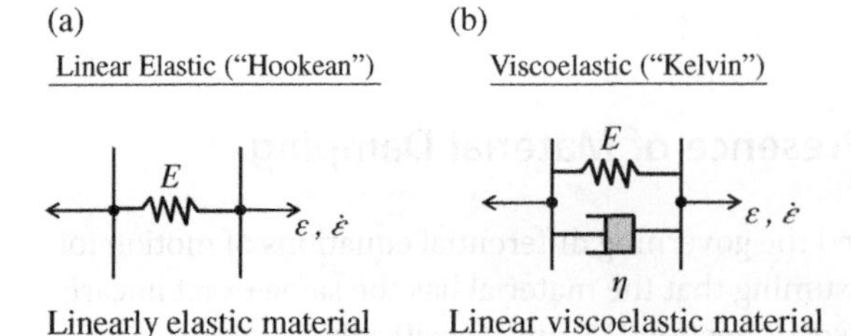

Figure 15.3 Conceptual comparison of elastic Hookean material and Viscoelastic Kelvin material.

Box 15.3.1 Strong form for one-dimensional elastodynamics and viscoelastic (Kelvin) material

$$\frac{\partial}{\partial x}\left(E\cdot A\frac{\partial u}{\partial x}\right)+\frac{\partial}{\partial x}\left(\eta\cdot A\frac{\partial \dot{u}}{\partial x}\right)+b(x,t)-\rho(x)\cdot A(x)\cdot \ddot{u}(x,t)=0,\ \ 0<x<L \quad (15.3.1)$$

$$u(x,t)=\bar{u}(t)\ \text{ at } \Gamma_u(\text{essential boundary conditions}) \quad (15.3.2a)$$

$$[n\cdot\sigma]_{\Gamma_t}=\bar{t}(t)\ \text{ at } \Gamma_t(\text{natural boundary conditions}) \quad (15.3.2b)$$

$$\sigma(x,t)=E(x)\cdot\varepsilon(x,t)+\eta(x)\cdot\dot{\varepsilon}\,(x,t)\ \ (\textit{constitutive law})$$

$$\left.\begin{aligned}&u(x,t=0)=u_0(x)\ ,\ 0<x<L\\ &\dot{u}(x,t=0)=\left.\frac{\partial u}{\partial t}\right|_{t=0}=v_0(x)\ ,\ 0<x<L\end{aligned}\right\}\rightarrow \text{initial conditions (I.C.)} \quad (15.3.2c)$$

distributed inertial forces (which are proportional to the accelerations), (2) the modified expression for the axial force (accounting for the viscous part of the Kelvin material). These are the only modifications required to obtain the equations of motion for dynamic loading, for any type of linearly elastic system!

15.4 Weak Form for One-Dimensional Elastodynamics

The procedure to obtain the weak form for one-dimensional elastodynamics is identical to that followed in Section 2.3 for one-dimensional elasticity (multiply by an arbitrary function $w(x)$, which vanishes on the essential boundary, then render the terms *symmetric* in terms of differentiation with x). The only difference for a dynamic problem described by the strong form of Box 15.3.1 is the presence of the inertial terms and the different constitutive law. The derivation of the weak form for elastodynamics and for a viscoelastic Kelvin material is presented in Box 15.4.1 and will not be proven here in detail; instead, it is left as an exercise.

Box 15.4.1 Weak Form for One-Dimensional Elastodynamics and Viscoelastic (Kelvin) Material

$$\int_0^L w\cdot\rho\cdot A\cdot\ddot{u}\,dx+\int_0^L\frac{\partial w}{\partial x}\eta\cdot A\frac{\partial\dot{u}}{\partial x}dx+\int_0^L\frac{\partial w}{\partial x}EA\frac{\partial u}{\partial x}dx=[w\cdot A]_{\Gamma_t}\cdot\bar{t}(t)+\int_0^L w\cdot b dx \quad (15.4.1)$$

$\forall w$, *with* $w=0$ *at* Γ_u (w vanishes on the essential boundary, Γ_u)

$$u(x,t)=\bar{u}(t)\ \text{ at } \Gamma_u(\text{essential boundary conditions}) \quad (15.4.2)$$

$$\sigma(x,t)=E(x)\cdot\varepsilon(x,t)+\eta(x)\cdot\dot{\varepsilon}\,(x,t)\ \ (\text{constitutive law})$$

$$\left.\begin{aligned}&u(x,t=0)=u_0(x)\ ,\ 0<x<L\\ &\dot{u}(x,t=0)=\left.\frac{\partial u}{\partial t}\right|_{t=0}=v_0(x)\ ,\ 0<x<L\end{aligned}\right\}\rightarrow \text{initial conditions (I.C.)} \quad (15.4.3)$$

Remark 15.4.1: It is worth noticing a significant difference between the weak form for elastodynamics and the weak form for static elasticity problems (the latter was presented in Box 2.3.2). The weak form for elastodynamics must include not only the essential boundary conditions for the actual displacement field, $u(x,t)$, but also the initial conditions. ∎

15.5 Finite Element Approximation and Semi-Discrete Equations of Motion

We now introduce the one-dimensional finite element approximation, in the same exact manner as the one described in Section 3.3 for static one-dimensional elasticity. More specifically, we discretize the domain into N_e elements, each element having n nodes, as schematically shown in Figure 3.1, and introduce a piecewise approximation, in accordance with Section 3.3. We can establish the following expressions for the approximate displacement field in each element, e:

$$u(x,t) \approx u^{(e)}(x,t) = \sum_{j=1}^{n} \left(N_j^{(e)}(x) \cdot u_j^{(e)}(t) \right) = \left[N^{(e)}\right]\left\{U^{(e)}(t)\right\} \tag{15.5.1a}$$

The shape functions, contained in the n-dimensional row vector $[N^{(e)}]$, are functions of the spatial coordinate x alone. The dependence of the approximate displacement field on time stems from the stipulation that the nodal displacements, contained in the n-dimensional vector $\{U^{(e)}(t)\}$ of Equation (15.5.1a), are now functions of time. We can also obtain the approximate expression for the strain field, by taking the partial derivative of the approximate displacements with x:

$$\frac{\partial u}{\partial x} \approx \frac{\partial u^{(e)}}{\partial x} = \sum_{j=1}^{n} \left(\frac{dN_j^{(e)}}{dx} \cdot u_j^{(e)}(t) \right) = \left[B^{(e)}\right]\left\{U^{(e)}(t)\right\} \tag{15.5.1b}$$

The same finite element approximation is established for the virtual displacement field $w(x)$ in Box 15.4.1 and the virtual strain field:

$$w(x) \approx w^{(e)}(x) = \left[N^{(e)}\right]\left\{W^{(e)}\right\} = \left\{W^{(e)}\right\}^T \left[N^{(e)}\right]^T \tag{15.5.2a}$$

$$\frac{\partial w^{(e)}}{\partial x} = \left[B^{(e)}\right]\left\{W^{(e)}\right\} = \left\{W^{(e)}\right\}^T \left[B^{(e)}\right]^T \tag{15.5.2b}$$

Remark 15.5.1: It is worth emphasizing that the shape functions only depend on the spatial coordinates, while the nodal values are now continuous functions of time! The procedure in which we approximate—by means of discretization—only the spatial variation of the field functions is called *semi-discretization.* ∎

The approximate velocity and acceleration fields in each element e can be obtained by taking the first and second partial derivative of the approximate displacement field with time. Accounting for the fact that only the nodal displacements depend on time, we have:

$$\frac{\partial u}{\partial t} = \dot{u}(x,t) \approx \frac{\partial u^{(e)}}{\partial t} = \frac{\partial}{\partial t}\sum_{j=1}^{n}\left(N_j^{(e)}(x)\cdot u_j^{(e)}(t)\right) = \sum_{j=1}^{n}\left(N_j^{(e)}(x)\cdot \dot{u}_j^{(e)}(t)\right) = \left[N^{(e)}\right]\left\{\dot{U}^{(e)}(t)\right\} \tag{15.5.3a}$$

$$\frac{\partial^2 u}{\partial t^2} = \ddot{u}(x,t) \approx \frac{\partial^2 u^{(e)}}{\partial t^2} = \frac{\partial^2}{\partial t^2}\sum_{j=1}^{n}\left(N_j^{(e)}(x)\cdot u_j^{(e)}(t)\right) = \sum_{j=1}^{n}\left(N_j^{(e)}(x)\cdot \ddot{u}_j^{(e)}(t)\right) = \left[N^{(e)}\right]\left\{\ddot{U}^{(e)}(t)\right\} \tag{15.5.3b}$$

Also, the nodal quantities $\{U^{(e)}(t)\}$ and $\{W^{(e)}\}$ of element e can be obtained from the corresponding global vectors, by means of a gather array $[L^{(e)}]$ (as also explained in Section 3.3):

$$\left\{U^{(e)}(t)\right\} = \left[L^{(e)}\right]\{U(t)\} \tag{15.5.4}$$

$$\left\{W^{(e)}\right\} = \left[L^{(e)}\right]\{W\} \tag{15.5.5}$$

We can also establish gather operations for the nodal velocities and nodal accelerations of each element, by taking the first and second partial derivatives of Equation (15.5.4) with respect to time. If we account for the fact that the gather array, $[L^{(e)}]$, is constant and does not change with time, we can write:

$$\left\{\dot{U}^{(e)}(t)\right\} = \left[L^{(e)}\right]\{\dot{U}(t)\} \tag{15.5.6a}$$

$$\left\{\ddot{U}^{(e)}(t)\right\} = \left[L^{(e)}\right]\{\ddot{U}(t)\} \tag{15.5.6b}$$

Finally, we need to remember that the domain integrals can be separated into element contributions. That is, for any domain integral:

$$\int_0^L \ldots dx = \int_{x_1^{(1)}}^{x_2^{(1)}} \ldots dx + \int_{x_1^{(2)}}^{x_2^{(2)}} \ldots dx + \ \ldots \ + \int_{x_1^{(Ne)}}^{x_2^{(Ne)}} \ldots dx \tag{15.5.7}$$

In light of Equation (15.5.7), the integral term in the weak form (Equation 15.4.1) corresponding to the effect of the inertial forces becomes:

$$\begin{aligned}\int_0^L w\cdot\rho\cdot A\cdot\ddot{u}\,dx &= \int_{x_1^{(1)}}^{x_2^{(1)}} w\cdot\rho\cdot A\cdot\ddot{u}\,dx + \int_{x_1^{(2)}}^{x_2^{(2)}} w\cdot\rho\cdot A\cdot\ddot{u}\,dx + \ \ldots \ + \int_{x_1^{(Ne)}}^{x_2^{(Ne)}} w\cdot\rho\cdot A\cdot\ddot{u}\,dx \\ &= \sum_{e=1}^{N_e}\left[\int_{x_1^{(e)}}^{x_2^{(e)}} w\cdot\rho\cdot A\cdot\ddot{u}\,dx\right]\end{aligned} \tag{15.5.8}$$

We can now plug Equations (15.5.2a) and (15.5.3b) in each integral term of the summation in Equation (15.5.8), to obtain the following contribution of each element to the term with the inertial forces.

$$\int_{x_1^{(e)}}^{x_2^{(e)}} w \cdot \rho \cdot A \cdot \ddot{u}\, dx = \int_{x_1^{(e)}}^{x_2^{(e)}} \left(\left\{W^{(e)}\right\}^T \left[N^{(e)}\right]^T \cdot \rho^{(e)} \cdot A^{(e)} \cdot \left[N^{(e)}\right] \left\{\ddot{U}^{(e)}\right\} \right) dx \tag{15.5.9}$$

Since the nodal vectors $\{\ddot{U}^{(e)}\}$ and $\{W^{(e)}\}^T$ do not depend on the variable of integration, x, they can be moved outside of the integrals:

$$\int_{x_1^{(e)}}^{x_2^{(e)}} w \cdot \rho \cdot A \cdot \ddot{u}\, dx = \left\{W^{(e)}\right\}^T \left(\int_{x_1^{(e)}}^{x_2^{(e)}} \left[N^{(e)}\right]^T \cdot \rho^{(e)} \cdot A^{(e)} \cdot \left[N^{(e)}\right] dx \right) \left\{\ddot{U}^{(e)}\right\} = \left\{W^{(e)}\right\}^T \left[m^{(e)}\right] \left\{\ddot{U}^{(e)}\right\} \tag{15.5.10}$$

where the matrix $[m^{(e)}]$ is called the *element mass matrix* and is given by the following expression.

$$\left[m^{(e)}\right] = \int_{x_1^{(e)}}^{x_2^{(e)}} \left[N^{(e)}\right]^T \cdot \rho^{(e)} \cdot A^{(e)} \cdot \left[N^{(e)}\right] dx \tag{15.5.11}$$

If we also account for the gather operations, that is, Equations (15.5.5) and (15.5.6b), Equation (15.5.10) yields:

$$\int_{x_1^{(e)}}^{x_2^{(e)}} w \cdot \rho \cdot A \cdot \ddot{u}\, dx = \{W\}^T \left[L^{(e)}\right]^T \left[m^{(e)}\right] \left[L^{(e)}\right] \{\ddot{U}\} \tag{15.5.12}$$

Plugging Equation (15.5.12) into Equation (15.5.8) finally gives:

$$\int_0^L w \cdot \rho \cdot A \cdot \ddot{u}\, dx = \sum_{e=1}^{N_e} \left(\{W\}^T \left[L^{(e)}\right]^T \left[m^{(e)}\right] \left[L^{(e)}\right] \{\ddot{U}\} \right) \tag{15.5.13}$$

The two global vectors, $\{W\}^T$ and $\{\ddot{U}\}$, are common for all the terms in the summation of Equation (15.5.13) and can be taken outside the summation operation as common factors:

$$\int_0^L w \cdot \rho \cdot A \cdot \ddot{u}\, dx = \{W\}^T \sum_{e=1}^{N_e} \left[\left[L^{(e)}\right]^T \left[m^{(e)}\right] \left[L^{(e)}\right] \right] \{\ddot{U}\} = \{W\}^T [M] \{\ddot{U}\} \tag{15.5.14}$$

where $[M]$ is the *global mass matrix of the structure*, given as an assembly of the element mass matrices:

$$[M] = \sum_{e=1}^{N_e} \left[\left[L^{(e)}\right]^T \left[m^{(e)}\right] \left[L^{(e)}\right] \right] \tag{15.5.15}$$

We now continue with the second integral term in the left-hand side of Equation (15.4.1), which corresponds to the effect of viscous damping:

$$\int_0^L \frac{\partial w}{\partial x}\eta\cdot A\frac{\partial \dot{u}}{\partial x}dx = \int_{x_1^{(1)}}^{x_2^{(1)}} \frac{\partial w}{\partial x}\eta\cdot A\frac{\partial \dot{u}}{\partial x}dx + \int_{x_1^{(2)}}^{x_2^{(2)}} \frac{\partial w}{\partial x}\eta\cdot A\frac{\partial \dot{u}}{\partial x}dx + \ldots + \int_{x_1^{(Ne)}}^{x_2^{(Ne)}} \frac{\partial w}{\partial x}\eta\cdot A\frac{\partial \dot{u}}{\partial x}dx \rightarrow$$

$$\rightarrow \int_0^L \frac{\partial w}{\partial x}\eta\cdot A\frac{\partial \dot{u}}{\partial x}dx = \sum_{e=1}^{N_e}\left[\int_{x_1^{(e)}}^{x_2^{(e)}} \frac{\partial w}{\partial x}\eta\cdot A\frac{\partial \dot{u}}{\partial x}dx\right] \tag{15.5.16}$$

Now, in each element e, we can take the expression for the strain rate, $\frac{\partial \dot{u}}{\partial x}$, by differentiating Equation (15.5.3a) with x:

$$\frac{\partial \dot{u}}{\partial x} \approx \frac{\partial \dot{u}^{(e)}}{\partial x} = \frac{\partial}{\partial x}\left(\left[N^{(e)}\right]\left\{\dot{U}^{(e)}(t)\right\}\right) = \frac{\partial}{\partial x}\sum_{j=1}^{n}\left(N_j^{(e)}(x)\cdot \dot{u}_j^{(e)}(t)\right) \rightarrow$$

$$\rightarrow \frac{\partial \dot{u}}{\partial x} \approx \frac{\partial \dot{u}^{(e)}}{\partial x} = \sum_{j=1}^{n}\left(\frac{\partial N_j^{(e)}}{\partial x}\cdot \dot{u}_j^{(e)}(t)\right) = \left[B^{(e)}\right]\left\{\dot{U}^{(e)}(t)\right\} \tag{15.5.17}$$

Remark 15.5.2: One can observe (as it was probably expected) that the approximate strain rate in each element is obtained as the derivative (with time) of the element strain-displacement matrix relation in the element. ▮

The contribution of each element e in the right-hand side of Equation (15.5.16) can be written in the following form, if we account for Equations (15.5.2b) and (15.5.17):

$$\int_{x_1^{(e)}}^{x_2^{(e)}} \frac{\partial w}{\partial x}\eta\cdot A\frac{\partial \dot{u}}{\partial x}dx = \int_{x_1^{(e)}}^{x_2^{(e)}} \left\{W^{(e)}\right\}^T\left[B^{(e)}\right]^T\cdot\eta^{(e)}\cdot A^{(e)}\cdot\left[B^{(e)}\right]\left\{\dot{U}^{(e)}\right\}dx \tag{15.5.18}$$

Just like we did for the inertial terms, we take the nodal vectors $\{\dot{U}^{(e)}\}$ and $\{W^{(e)}\}^T$ outside of the integral in the right-hand side of Equation (15.5.18):

$$\int_{x_1^{(e)}}^{x_2^{(e)}} \frac{\partial w}{\partial x}\eta\cdot A\frac{\partial \dot{u}}{\partial x}dx = \left\{W^{(e)}\right\}^T\left(\int_{x_1^{(e)}}^{x_2^{(e)}}\left[B^{(e)}\right]^T\cdot\eta^{(e)}\cdot A^{(e)}\cdot\left[B^{(e)}\right]dx\right)\left\{\dot{U}^{(e)}\right\} = \left\{W^{(e)}\right\}^T\left[c^{(e)}\right]\left\{\dot{U}^{(e)}\right\} \tag{15.5.19}$$

where $[c^{(e)}]$ is the *element (viscous) damping matrix*, given by:

$$\left[c^{(e)}\right] = \int_{x_1^{(e)}}^{x_2^{(e)}}\left[B^{(e)}\right]^T\cdot\eta^{(e)}\cdot A^{(e)}\cdot\left[B^{(e)}\right]dx \tag{15.5.20}$$

Remark 15.5.3: Equation (15.5.20) for $[c^{(e)}]$ is only valid for Kelvin viscoelastic materials. ■

Plugging Equation (15.5.19) into Equation (15.5.16) and accounting for Equations (15.5.5) and (15.5.6a) yields:

$$\int_0^L \frac{\partial w}{\partial x}\eta \cdot A \frac{\partial \dot{u}}{\partial x}dx = \sum_{e=1}^{N_e}\left(\left\{W^{(e)}\right\}^T\left[c^{(e)}\right]\left\{\dot{U}^{(e)}\right\}\right) = \sum_{e=1}^{N_e}\left(\{W\}^T\left[L^{(e)}\right]^T\left[c^{(e)}\right]\left[L^{(e)}\right]\{\dot{U}\}\right) \tag{15.5.21}$$

Finally, taking the global vectors, $\{W\}^T$ and $\{\dot{U}\}$, as common factors outside of the summation in Equation (15.5.21) gives:

$$\int_0^L \frac{\partial w}{\partial x}\eta \cdot A \frac{\partial \dot{u}}{\partial x}dx = \{W\}^T\left[\sum_{e=1}^{N_e}\left(\left[L^{(e)}\right]^T\left[c^{(e)}\right]\left[L^{(e)}\right]\right)\right]\{\dot{U}\} = \{W\}^T[C]\{\dot{U}\} \tag{15.5.22a}$$

Where the global damping matrix, $[C]$, if given by:

$$[C] = \sum_{e=1}^{N_e}\left(\left[L^{(e)}\right]^T\left[c^{(e)}\right]\left[L^{(e)}\right]\right) \tag{15.5.22b}$$

In accordance with Equations (3.3.3) to (3.3.10) in Section 3.3, the third term in Equation (15.4.1) is given by the expression:

$$\int_0^L \frac{\partial w}{\partial x}E \cdot A \frac{\partial u}{\partial x}\,dx = \sum_{e=1}^{N_e}\left(\int_{x_1^{(e)}}^{x_2^{(e)}} \frac{\partial w}{\partial x}E \cdot A \frac{\partial u}{\partial x}\,dx\right)$$

$$= \sum_{e=1}^{N_e}\left(\int_{x_1^{(e)}}^{x_2^{(e)}} \left\{W^{(e)}\right\}^T\left[B^{(e)}\right]^T \cdot E^{(e)} \cdot A^{(e)} \cdot \left[B^{(e)}\right]\left\{U^{(e)}\right\}dx\right) \rightarrow$$

$$\rightarrow \int_0^L \frac{\partial w}{\partial x}E \cdot A \frac{\partial u}{\partial x}\,dx = \sum_{e=1}^{N_e}\left(\left\{W^{(e)}\right\}^T\left[k^{(e)}\right]\left\{U^{(e)}\right\}\right) \rightarrow$$

$$\rightarrow \int_0^L \frac{\partial w}{\partial x}E \cdot A \frac{\partial u}{\partial x}\,dx = \sum_{e=1}^{N_e}\left(\{W\}^T\left[L^{(e)}\right]^T\left[k^{(e)}\right]\left[L^{(e)}\right]\{U\}\right) = \{W\}^T[K]\{U\} \tag{15.5.23}$$

where $[K]$ is the global stiffness matrix, given as an assembly of the corresponding element matrices, $[k^{(e)}]$:

$$[K] = \sum_{e=1}^{N_e}\left(\left[L^{(e)}\right]^T\left[k^{(e)}\right]\left[L^{(e)}\right]\right) \tag{15.5.24}$$

$$\left[k^{(e)}\right] = \int_{x_1^{(e)}}^{x_2^{(e)}} \left[B^{(e)}\right]^T \cdot E^{(e)} \cdot A^{(e)} \cdot \left[B^{(e)}\right]dx \tag{15.5.25}$$

Now, the right-hand side of (15.4.1) can also be separated into element contributions; using the same exact considerations as in Equations (3.3.11–14) and (3.3.21b), we can write:

$$[w \cdot A]_{\Gamma_t} \cdot \bar{t}(t) + \int_0^L w \cdot b dx = \sum_{e=1}^{N_e} \left([w \cdot A \cdot \bar{t}(t)]_{\Gamma_t^{(e)}} \right) + \sum_{e=1}^{N_e} \left(\int_{x_1^{(e)}}^{x_2^{(e)}} w \cdot b\, dx \right) = \{W\}^T \{f(t)\} \tag{15.5.26}$$

where the *equivalent nodal force vector*, $\{f(t)\}$, is given through assembly of the corresponding element vectors, $\{f^{(e)}(t)\}$:

$$\{f(t)\} = \sum_{e=1}^{N_e} \left(\left[L^{(e)}\right]^T \left\{f^{(e)}(t)\right\} \right) \tag{15.5.27}$$

where

$$\left\{f^{(e)}(t)\right\} = \left[\left[N^{(e)}\right]^T \cdot A \cdot \bar{t}(t)\right]_{\Gamma_t^{(e)}} + \int_{x_1^{(e)}}^{x_2^{(e)}} \left[N^{(e)}\right]^T \cdot b(x,t)\, dx \tag{15.5.28}$$

Remark 15.5.4: $[K]$, $\{f\}$ have the same definitions as for the static case (the only difference being that $\{f\}$ now changes with time). ■

Thus, after the introduction of the finite element approximation, we can plug Equations (15.5.14), (15.5.22), (15.5.23), and (15.5.26) into the weak form to establish the finite element equations for elastodynamics:

$$\begin{aligned} &\{W\}^T \left[[M]\{\ddot{U}\} + [C]\{\dot{U}\} + [K]\{U\}\right] = \{W\}^T \{f\} \quad \forall \{W\} \Rightarrow \\ &\Rightarrow \{W\}^T \left[[M]\{\ddot{U}\} + [C]\{\dot{U}\} + [K]\{U\} - \{f\}\right] = 0 \ \forall \{W\} \end{aligned} \tag{15.5.29}$$

Since the vector $\{W\}$ in (15.5.29) is arbitrary, the only way to always satisfy Equation (15.5.29) for all the arbitrary possible $\{W\}$ is to have the term in the bracket being equal to a zero vector:

$$[M]\{\ddot{U}\} + [C]\{\dot{U}\} + [K]\{U\} - \{f\} = \{0\} \tag{15.5.30}$$

It is worth emphasizing that all the nodal vectors in Equation (15.5.30) depend on time.

We also need to enforce the initial conditions of the problem. In the context of the finite element equations, these conditions establish initial values for the nodal displacement and velocity vectors. Thus, by slightly rearranging the terms in Equation (15.5.30) and by including initial conditions, we obtain the following expressions:

$$[M]\{\ddot{U}\} + [C]\{\dot{U}\} + [K]\{U\} = \{f\} \tag{15.5.31a}$$

$$\{U\}_{t=0} = \{U_o\} \tag{15.5.31b}$$

$$\{\dot{U}\}_{t=0} = \{V_o\} \tag{15.5.31c}$$

where $\{U_o\}$ and $\{V_o\}$ are the initial nodal displacement and initial nodal velocity vectors, respectively. These two vectors must be given as input in a finite element analysis.

Remark 15.5.5: The nodal displacement vector must also satisfy the essential boundary conditions. In case one of the nodal displacements, for example, U_i, has an applied essential boundary condition, we need to add a reaction force, $r_i(t)$, to the right-hand-side term of the i-th equation. In other words, we need to replace "$f_i(t)$" by "$f_i(t) + r_i(t)$". Of course, the reactions will now be (in general) functions of time! ∎

The obtained equations and initial conditions in Equations (15.5.31a,b,c) and the essential boundary conditions for the nodal displacements constitute the *semi-discrete finite element equations of motion* for one-dimensional elastodynamics. The quantity to be determined using these equations is the nodal displacement vector, $\{U(t)\}$, which is now a continuous function of time. Thus, semi-discretization has now lead to a slightly more complicated problem: instead of having to calculate a vector of constant nodal displacements (which was calculated solving the system of linear equations) for static problems, we need to determine a set of functions of time, that is, the components of $\{U(t)\}$. From a mathematical point of view, the semi-discrete equations of motion (15.5.31a) comprise a *system of coupled differential equations* (since the equations involve derivatives with respect to t). The term *coupled* means that the various components of $\{U(t)\}$ depend on each other: for example, nodal displacements and/or velocities in one node must satisfy equations of equilibrium, which equations depend on displacements and/or velocities at other nodes.

In Sections 15.6–15.8, we will briefly establish the weak form and the semi-discrete equations for multi-dimensional elastodynamics and structural dynamics. The global finite element equations for these cases are still given by Equations (15.5.31a–c). What will change is the expressions giving the element arrays and equivalent force vectors.

15.6 Three-Dimensional Elastodynamics

The considerations of the previous sections pertaining to one-dimensional elastodynamics also apply for the case of multidimensional elastodynamics, that is, analysis of solids subjected to time-dependent loads and boundary conditions. For the general case of a three-dimensional problem, we have a displacement vector field, which varies in both space and time. More specifically, the only difference in the force equilibrium considerations, compared to the case of elastostatics, is the need to account for distributed inertial forces. The direction of the inertial force is opposite to the direction of the accelerations, as shown in Figure 15.4, and their value per unit volume is equal to the product of the density, ρ, of the material, times the acceleration components along the three axes, x, y, and z.

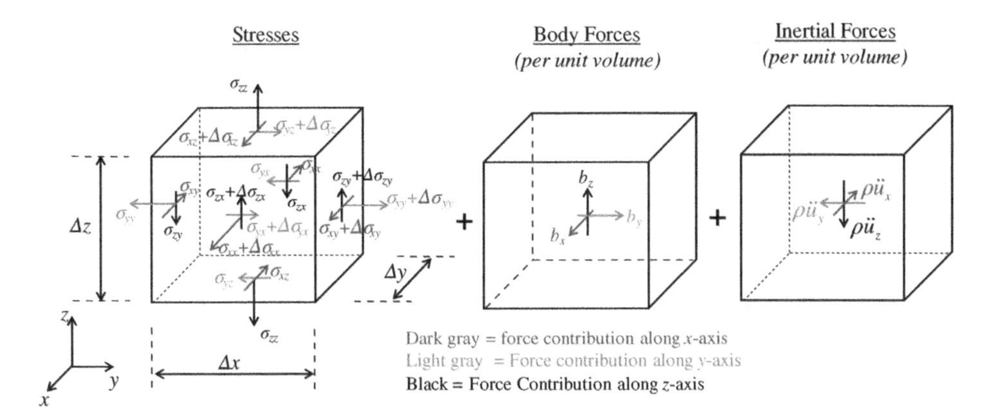

Figure 15.4 Forces at small piece of material in three-dimensional solid under dynamic loading.

To account for viscous damping, a rate dependence of the stress tensor is introduced, by modifying the constitutive equations. Specifically, we will generalize the notion of a Kelvin viscoelastic material of Section 15.3 to write:

$$\{\sigma\} = [D]\{\varepsilon\} + [H]\{\dot{\varepsilon}\} \tag{15.6.1}$$

where $[H]$ is a constant "material viscous damping" matrix, expressing the linear dependence of the stresses $\{\sigma\}$ on the strain-rate vector, $\{\dot{\varepsilon}\}$. The latter vector is established using Voigt notation in accordance with Section (7.4):

$$\{\dot{\varepsilon}\} = \begin{bmatrix} \dot{\varepsilon}_{xx} & \dot{\varepsilon}_{yy} & \dot{\varepsilon}_{zz} & \dot{\gamma}_{xy} = 2\dot{\varepsilon}_{xy} & \dot{\gamma}_{yz} = 2\dot{\varepsilon}_{yz} & \dot{\gamma}_{zx} = 2\dot{\varepsilon}_{zx} \end{bmatrix}^T \tag{15.6.2}$$

The strain-rate components are simply the partial derivatives with time of the corresponding components of the strain tensor.

$$\dot{\varepsilon}_{ij} = \frac{\partial \varepsilon_{ij}}{\partial t} = \frac{\partial}{\partial t}\left[\frac{1}{2}\left(\frac{\partial u_i}{\partial x_j} + \frac{\partial u_j}{\partial x_i}\right)\right] = \frac{1}{2}\left[\frac{\partial}{\partial t}\left(\frac{\partial u_i}{\partial x_j}\right) + \frac{\partial}{\partial t}\left(\frac{\partial u_j}{\partial x_i}\right)\right] = \frac{1}{2}\left(\frac{\partial \dot{u}_i}{\partial x_j} + \frac{\partial \dot{u}_j}{\partial x_i}\right) \tag{15.6.3}$$

Per Equation (15.6.3), the strain-rate components are obtained by applying the strain-displacement equations to the components of the velocity vector (i.e., the rate of the displacement vector).

The strong form for three-dimensional elastodynamics can now be written as presented in Box 15.6.1. The detailed derivation of the strong form is left as an exercise.

Remark 15.6.1: One can verify that the strong form of Box 15.6.1 is identical to that for elastostatics, presented in Box 7.11.1, except that we have to replace the body force $\{b\}$ by a version which includes the inertial effect: $\{\tilde{b}\} = \{b\} - \rho\{\ddot{u}\}$, and also account for the additional dependence of the stresses on the strain rates. ▮

Furthermore, we can also obtain the weak form for three-dimensional elastodynamics, which is presented in Box 15.6.2.

Box 15.6.1 Strong Form for Three-Dimensional Elastodynamics and Viscoelastic (Kelvin) Material

$$\frac{\partial \sigma_{xx}}{\partial x}+\frac{\partial \sigma_{xy}}{\partial y}+\frac{\partial \sigma_{xz}}{\partial z}+b_x=\rho\ddot{u}_x$$
$$\frac{\partial \sigma_{yx}}{\partial x}+\frac{\partial \sigma_{yy}}{\partial y}+\frac{\partial \sigma_{yz}}{\partial z}+b_y=\rho\ddot{u}_y \tag{15.6.4}$$
$$\frac{\partial \sigma_{zx}}{\partial x}+\frac{\partial \sigma_{zy}}{\partial y}+\frac{\partial \sigma_{zz}}{\partial z}+b_z=\rho\ddot{u}_z$$

$$u_x=\bar{u}_x(t)\ on\ \Gamma_{ux},\ u_y=\bar{u}_y(t)\ on\ \Gamma_{uy},\ u_z=\bar{u}_z(t)\ on\ \Gamma_{uz}\ \text{(essential bound. cond.)} \tag{15.6.5a}$$

$$\vec{\sigma}_x\bullet\vec{n}=t_x(t)\ on\ \Gamma_{tx},\ \vec{\sigma}_y\bullet\vec{n}=t_y(t)\ on\ \Gamma_{ty},\ \vec{\sigma}_z\bullet\vec{n}=t_z(t)\ on\ \Gamma_{tz}\ \text{(natural bound. cond.)} \tag{15.6.5b}$$

where $\vec{\sigma}_x=[\sigma_{xx}\ \ \sigma_{xy}\ \ \sigma_{xz}]^T,\vec{\sigma}_y=[\sigma_{yx}\ \ \sigma_{yy}\ \ \sigma_{yz}]^T,\vec{\sigma}_z=[\sigma_{zx}\ \ \sigma_{zy}\ \ \sigma_{zz}]^T$

$$\{\sigma\}=[D]\{\varepsilon\}+[H]\{\dot{\varepsilon}\}\ \text{(constitutive law)} \tag{15.6.6}$$

$$\{u(x,y,z,t=0)\}=\left\{\begin{matrix}u_x(x,y,z,t=0)\\u_y(x,y,z,t=0)\\u_z(x,y,z,t=0)\end{matrix}\right\}=\left\{\left\{\begin{matrix}u_{xo}(x,y,z)\\u_{yo}(x,y,z)\\u_{zo}(x,y,z)\end{matrix}\right\}\ in\ \Omega\right\}\rightarrow\text{(initial displacement field)} \tag{15.6.7a}$$

$$\{\dot{u}(x,y,z,t=0)\}=\left\{\begin{matrix}\dot{u}_x(x,y,z,t=0)\\\dot{u}_y(x,y,z,t=0)\\\dot{u}_z(x,y,z,t=0)\end{matrix}\right\}=\left\{\left\{\begin{matrix}v_{xo}(x,y,z)\\v_{yo}(x,y,z)\\v_{zo}(x,y,z)\end{matrix}\right\}\ in\ \Omega\right\}\rightarrow\text{(initial velocity field)} \tag{15.6.7b}$$

Box 15.6.2 Weak Form for Three-Dimensional Elastodynamics

$$\iiint_\Omega \{w\}^T\rho\{\ddot{u}\}dV+\iiint_\Omega \{\bar{\varepsilon}\}^T[H]\{\dot{\varepsilon}\}dV+\iiint_\Omega \{\bar{\varepsilon}\}^T[D]\{\varepsilon\}dV$$
$$=\iiint_\Omega \{w\}^T\{b\}dV+\iint_{\Gamma_{tx}} w_x t_x\,dS+\iint_{\Gamma_{ty}} w_y t_y\,dS+\iint_{\Gamma_{tz}} w_z t_z\,dS$$

$$\forall\{w\}=[w_x\ \ w_y\ \ w_z]^T\ \text{ with } w_x=0\ \text{ on } \Gamma_{ux}, w_y=0\ \text{ on } \Gamma_{uy}, w_z=0\ \text{ on } \Gamma_{uz}$$

where $\{\ddot{u}\}=[\ddot{u}_x\ \ \ddot{u}_y\ \ \ddot{u}_z]^T$, $\{b\}=[b_x\ \ b_y\ \ b_z]^T$

$$u_x=\bar{u}_x(t)\ on\ \Gamma_{ux},\ u_y=\bar{u}_y(t)\ on\ \Gamma_{uy},\ u_z=\bar{u}_z(t)\ on\ \Gamma_{uz}\ \text{(essential bound. cond.)} \tag{15.6.5a}$$

$$\{\sigma\}=[D]\{\varepsilon\}+[H]\{\dot{\varepsilon}\}\ \text{(constitutive law)} \tag{15.6.6}$$

$$\{u(x,y,z,t=0)\}=\left\{\begin{matrix}u_x(x,y,z,t=0)\\u_y(x,y,z,t=0)\\u_z(x,y,z,t=0)\end{matrix}\right\}=\left\{\left\{\begin{matrix}u_{xo}(x,y,z)\\u_{yo}(x,y,z)\\u_{zo}(x,y,z)\end{matrix}\right\}\ in\ \Omega\right\}\rightarrow\text{(initial displacement field)} \tag{15.6.7a}$$

$$\{\dot{u}(x,y,z,t=0)\}=\left\{\begin{matrix}\dot{u}_x(x,y,z,t=0)\\\dot{u}_y(x,y,z,t=0)\\\dot{u}_z(x,y,z,t=0)\end{matrix}\right\}=\left\{\left\{\begin{matrix}v_{xo}(x,y,z)\\v_{yo}(x,y,z)\\v_{zo}(x,y,z)\end{matrix}\right\}\ in\ \Omega\right\}\rightarrow\text{(initial velocity field)} \tag{15.6.7b}$$

15.7 Semi-Discrete Equations of Motion for Three-Dimensional Elastodynamics

Given the weak form of Box 15.6.2, we can establish a finite element *semi-discretization*, where we discretize the solid in finite elements (see Figure 9.1) and stipulate that the dependence of the displacement field on the spatial coordinates x, y and z is described through a shape function array $[N^{(e)}]$ in each element, e, in accordance with Section 9.2. If we establish the piecewise finite element approximation and plug it into the weak form of Box 15.6.2, we obtain a global system of semi-discrete finite element equations of motion, described with Equations (15.5.31a,b,c). The only difference is in the definition of the element arrays and force vector, which are obtained from the following expressions.

$$\left[m^{(e)}\right] = \iiint_{\Omega^{(e)}} \left[N^{(e)}\right]^T \rho^{(e)} \left[N^{(e)}\right] dV \tag{15.7.1}$$

$$\left[c^{(e)}\right] = \iiint_{\Omega^{(e)}} \left[B^{(e)}\right]^T \left[H^{(e)}\right] \left[B^{(e)}\right] dV \tag{15.7.2}$$

$$\left[k^{(e)}\right] = \iiint_{\Omega^{(e)}} \left[B^{(e)}\right]^T \left[D^{(e)}\right] \left[B^{(e)}\right] dV \tag{15.7.3}$$

$$\begin{aligned}\left\{f^{(e)}(t)\right\} = &\iiint_{\Omega^{(e)}} \left[N^{(e)}\right]^T \{b(t)\}\, dV + \iint_{\Gamma_{tx}^{(e)}} \left[N^{(e)}\right]^T \{t_x(t)\}\, dS \\ &+ \iint_{\Gamma_{ty}^{(e)}} \left[N^{(e)}\right]^T \{t_y(t)\}\, dS + \iint_{\Gamma_{tz}^{(e)}} \left[N^{(e)}\right]^T \{t_z(t)\}\, dS\end{aligned} \tag{15.7.4}$$

The three-dimensional isoparametric finite elements described in Section 9.3 can be used for elastodynamics.

Remark 15.7.1: The same expressions of weak form and semi-discrete equations apply for finite element analysis of two-dimensional elastodynamics. The only difference is that the various vectors and arrays have smaller dimensions; for example, the $[D^{(e)}]$ and $[H^{(e)}]$ arrays will be (3×3). ∎

15.8 Structural Dynamics Problems

This section will describe the equations for structural mechanics (i.e., beam and shell problems) for dynamic loading. We will directly provide the weak form and finite element equations for such problems, without any detailed derivation. Several important aspects of the weak form (e.g., the origin of "rotational inertia" terms), will be explained with some detail. It is worth reiterating that the global, semi-discrete finite element equations of motion (15.5.31a,b,c) still apply for structural dynamics; thus, we will only need to provide the expressions for the element arrays and equivalent nodal force/moment vectors.

15.8.1 Dynamic Beam Problems

The weak form for a beam structure under dynamic loading is described by the expressions provided in Box 15.8.1. This weak form corresponds to most general loading case where, besides having distributed forces p_x and p_y per unit length, we also have a distributed moment, $m(x,t)$ per unit length.

Box 15.8.1 Weak Form for Beam Problems

$$\int_0^L \{w\}^T[\tilde{\rho}]\{\ddot{u}\}^T dx + \int_0^L \{\bar{\varepsilon}\}^T[\hat{H}]\{\dot{\hat{\varepsilon}}\}dx + \int_0^L \{\bar{\varepsilon}\}^T[\hat{D}]\{\hat{\varepsilon}\}dx$$

$$= \int_0^L \{w\}^T\{p\}dx + [w_{ox}F_x]_{\Gamma_{tx}} + \left[w_{oy}F_y\right]_{\Gamma_{ty}} + [\theta_o M_z]_{\Gamma_{t\theta}}$$

$\forall\{w\} = [w_{ox}\ \ w_{oy}\ \ w_\theta]^T$ with $w_{ox}=0$ on Γ_{ux}, $w_{oy}=0$ on Γ_{uy}, $w_\theta = 0$ on $\Gamma_{u\theta}$

$$[\tilde{\rho}] = \begin{bmatrix} \rho A & 0 & 0 \\ 0 & \rho A & 0 \\ 0 & 0 & \rho I \end{bmatrix} \{\ddot{u}\} = \left[\ddot{u}_{ox}\ \ \ddot{u}_{oy}\ \ \ddot{\theta}_o\right]^T, \{p\} = [p_x(x,t)\ \ p_y(x,t)\ \ m(x,t)]^T,$$

$u_{ox} = \bar{u}_{ox}(t)$ on Γ_{ux}, $u_{oy} = \bar{u}_{oy}(t)$ on Γ_{uy}, $\theta_o = \bar{\theta}(t)$ on $\Gamma_{u\theta}$ (essential boundary conditions)

$$\left.\begin{aligned} u_{ox}(x,t=0) &= u_{oxo}(x) \\ u_{oy}(x,t=0) &= u_{oy}(x) \\ \theta_o(x,t=0) &= \theta_{oo}(x) \end{aligned}\right\} \rightarrow \text{initial "displacements"}$$

$$\left.\begin{aligned} \dot{u}_{ox}(x,t=0) &= v_{xo}(x) \\ \dot{u}_{oy}(x,t=0) &= v_{yo}(x) \\ \dot{\theta}_o(x,t=0) &= v_{\theta o}(x) \end{aligned}\right\} \rightarrow \text{initial "velocities"}$$

The weak form of Box (15.8.1) applies for both Euler-Bernoulli and Timoshenko beam theory. The actual expressions giving several of the terms in Box 15.8.1 do depend on whether we have Euler-Bernoulli or Timoshenko beam theory. Specifically, we have:

- For Euler-Bernoulli theory:

$$\theta_o = \frac{\partial u_{oy}}{\partial x} \tag{15.8.1}$$

$$w_\theta = \frac{\partial w_{oy}}{\partial x} \tag{15.8.2}$$

$$[\hat{D}] = \begin{bmatrix} EA & 0 \\ 0 & EI \end{bmatrix} \tag{15.8.3}$$

$$[\hat{H}] = \begin{bmatrix} \eta A & 0 \\ 0 & \eta I \end{bmatrix} \tag{15.8.4}$$

$$\{\hat{\varepsilon}\} = [\varepsilon_o \;\; \varphi]^T = \begin{bmatrix} \frac{\partial u_{ox}}{\partial x} & \frac{\partial^2 u_{oy}}{\partial x^2} \end{bmatrix}^T \tag{15.8.5}$$

$$\{\hat{\bar{\varepsilon}}\} = [\bar{\varepsilon}_o \;\; \bar{\varphi}]^T = \begin{bmatrix} \frac{\partial w_{ox}}{\partial x} & \frac{\partial^2 w_{oy}}{\partial x^2} \end{bmatrix}^T \tag{15.8.6}$$

- For Timoshenko theory:

$$[\hat{D}] = \begin{bmatrix} EA & 0 & 0 \\ 0 & EI & 0 \\ 0 & 0 & \kappa GA \end{bmatrix} \tag{15.8.7}$$

$$[\hat{H}] = \begin{bmatrix} \eta A & 0 & 0 \\ 0 & \eta I & 0 \\ 0 & 0 & \eta_G A \end{bmatrix} \tag{15.8.8}$$

$$\{\hat{\varepsilon}\} = \left[\varepsilon_o \;\; \varphi \;\; \gamma_{xy}\right]^T = \begin{bmatrix} \frac{\partial u_{ox}}{\partial x} & \frac{d\theta_o}{dx} & \frac{du_{oy}}{dx} - \theta_o \end{bmatrix}^T \tag{15.8.9}$$

$$\{\hat{\bar{\varepsilon}}\} = \left[\bar{\varepsilon}_o \;\; \bar{\varphi} \;\; \bar{\gamma}_{xy}\right]^T = \begin{bmatrix} \frac{\partial \bar{u}_{ox}}{\partial x} & \frac{dw_{o\theta}}{dx} & \frac{dw_{oy}}{dx} - w_{o\theta} \end{bmatrix}^T \tag{15.8.10}$$

Remark 15.8.1: An inertial term $\rho I \cdot \ddot{\theta}_o$ in the first integral term of the weak form calculating the moment resultant of the inertial body forces on a section, if we account for the kinematic equation for the axial displacements. Specifically, if we take Equation (13.3.3) and differentiate twice with respect to time, we have:

$$u_x(x,y,t) = u_{ox}(x,t) - y \cdot \theta_o(x,t) \rightarrow \ddot{u}_x(x,y,t) = \ddot{u}_{ox}(x,t) - y \cdot \ddot{\theta}_o(x,t) \tag{15.8.11}$$

We can calculate the inertial force per unit volume in the x-direction (i.e., the axial direction), f_{1x}, and then integrate its moment contribution over the cross-sectional area to obtain the inertial moment per unit length:

$$m_I(x,t) = \iint_A (-y \cdot f_{Ix}) dS = \iint_A (-y \cdot \rho \ddot{u}_x) dS = \iint_A \left[-y \cdot \rho\left(\ddot{u}_{ox} - y \cdot \ddot{\theta}_o\right)\right] dS \tag{15.8.12}$$

which eventually gives:

$$m_I(x,t) = -\ddot{u}_{ox} \cdot \rho \iint_A y dS + \ddot{\theta}_o \cdot \rho \iint_A y^2 dS = \rho \cdot I \cdot \ddot{\theta}_o \tag{15.8.13}$$

where we have accounted for the fact that the density is assumed to be a function of x alone (density constant over a cross-section!) and also that y is the distance from the centroidal axis: $\iint_A y dS = 0$. The quantity $\rho \cdot I$ can be called the rotational mass per unit length of the beam. ■

Remark 15.8.2: The inertial moment per unit length $m_I(x, t)$ must also be accounted for in the moment-related differential equation of equilibrium, in the strong form for beam dynamics. ∎

We will now proceed with the finite element formulation for beam problems. The element stiffness matrix and viscous damping matrix for both beam theories are given by the following expressions.

$$\left[c^{(e)}\right] = \int_{x_1^{(e)}}^{x_2^{(e)}} \left[B^{(e)}\right]^T \left[\hat{H}^{(e)}\right] \left[B^{(e)}\right] dx \tag{15.8.14}$$

$$\left[k^{(e)}\right] = \int_{x_1^{(e)}}^{x_2^{(e)}} \left[B^{(e)}\right]^T \left[\hat{D}^{(e)}\right] \left[B^{(e)}\right] dx \tag{15.8.15}$$

The expressions for the element mass matrix, $[m^{(e)}]$, and for the equivalent nodal force vector, $\{f^{(e)}(t)\}$, depend on which beam theory is used.

- For Euler-Bernoulli theory:

$$\left[m^{(e)}\right] = \int_{x_1^{(e)}}^{x_2^{(e)}} \left[N^{(e)}\right]^T \begin{bmatrix} \rho^{(e)}A^{(e)} & 0 \\ 0 & \rho^{(e)}A^{(e)} \end{bmatrix} \left[N^{(e)}\right] dx + \int_{x_1^{(e)}}^{x_2^{(e)}} \left[N_\theta^{(e)}\right]^T \rho^{(e)} I^{(e)} \left[N_\theta^{(e)}\right] dx \tag{15.8.16}$$

$$\left\{f^{(e)}(t)\right\} = \int_{x_1^{(e)}}^{x_2^{(e)}} \left[N^{(e)}\right]^T \{p\} dx + \left[N^{(e)}\right]^T_{\Gamma_{tx}^{(e)}} \begin{Bmatrix} F_x(t) \\ 0 \end{Bmatrix}_{\Gamma_{tx}^{(e)}} + \left[N^{(e)}\right]^T_{\Gamma_{ty}^{(e)}} \begin{Bmatrix} 0 \\ F_y(t) \end{Bmatrix}_{\Gamma_{ty}^{(e)}} + \left[N_\theta^{(e)}\right]^T_{\Gamma_{t\theta}^{(e)}} M_z(t)_{\Gamma_{t\theta}^{(e)}} \tag{15.8.17}$$

- For Timoshenko theory:

$$\left[m^{(e)}\right] = \int_{x_1^{(e)}}^{x_2^{(e)}} \left[N^{(e)}\right]^T \left[\tilde{\rho}^{(e)}\right] \left[N^{(e)}\right] dx \tag{15.8.18}$$

$$\left\{f^{(e)}\right\} = \int_{x_1^{(e)}}^{x_2^{(e)}} \left[N^{(e)}\right]^T \{p\} dx + \left[N^{(e)}\right]^T_{\Gamma_{tx}^{(e)}} \begin{Bmatrix} F_x(t) \\ 0 \\ 0 \end{Bmatrix} + \left[N^{(e)}\right]^T_{\Gamma_{ty}^{(e)}} \begin{Bmatrix} 0 \\ F_y(t) \\ 0 \end{Bmatrix} + \left[N^{(e)}\right]^T_{\Gamma_{t\theta}^{(e)}} \begin{Bmatrix} 0 \\ 0 \\ M_z(t) \end{Bmatrix} \tag{15.8.19}$$

Remark 15.8.3: In many textbooks, the second term in the right-hand-side of Equation (15.8.16) is neglected, because its contribution is generally smaller than the contribution of the first term. ▮

Remark 15.8.4: The various shape function arrays and the generalized strain-displacement arrays in the above equations have been explained in detail in Sections 13.6 (for Euler-Bernoulli theory) and 13.11 (for Timoshenko theory). ▮

15.8.2 Dynamic Shell Problems

Next, the equations describing dynamic loading of shells will be provided. The focus is on flat shells and on Reissner-Mindlin theory, which have been described in Sections 14.3, 14.4 and 14.5. The weak form of dynamic flat shell problems, provided in Box 15.8.2, includes two additional terms (related to inertial forces and to the rate dependence

Box 15.8.2 Weak form for flat shell problems

$$\int_0^L \{w\}^T [\tilde{\rho}] \{\ddot{u}\}^T dx + \int_0^L \{\bar{\varepsilon}\}^T [\hat{H}] \{\dot{\hat{\varepsilon}}\} dx + \int_0^L \{\bar{\varepsilon}\}^T [\hat{D}] \{\hat{\varepsilon}\} dx = \int_0^L \{w\}^T \{p\} dx + \int_{\Gamma_{tx}} w_{ox} f_x dS$$

$$+ \int_{\Gamma_{ty}} w_{oy} f_y dS + \int_{\Gamma_{tz}} w_{oz} f_z dS + \int_{\Gamma_{t\theta 1}} w_{\theta 1} m_{1\Gamma} dS + \int_{\Gamma_{t\theta 2}} w_{\theta 2} m_{2\Gamma} dS$$

$\forall \{w\} = [w_{ox} \; w_{oy} \; w_{oz} \; w_{\theta 1} \; w_{\theta 2}]^T$

with $w_{ox} = 0$ *on* Γ_{ux}, $w_{oy} = 0$ *on* Γ_{uy}, $w_{oz} = 0$ *on* Γ_{uz}, $w_{\theta 1} = 0$ *on* $\Gamma_{u\theta 1}$, $w_{\theta 2} = 0$ *on* $\Gamma_{u\theta 2}$

$$[\tilde{\rho}] = \begin{bmatrix} \rho \cdot d & 0 & 0 & 0 & 0 \\ 0 & \rho \cdot d & 0 & 0 & 0 \\ 0 & 0 & \rho \cdot d & 0 & 0 \\ 0 & 0 & 0 & \rho \cdot \frac{d^3}{12} & 0 \\ 0 & 0 & 0 & 0 & \rho \cdot \frac{d^3}{12} \end{bmatrix}, \quad \{\ddot{u}\} = \left[\ddot{u}_{ox} \; \ddot{u}_{oy} \; \ddot{u}_{oz} \; \ddot{\theta}_1 \; \ddot{\theta}_2\right]^T,$$

$\{p\} = [p_x \; p_y \; p_z \; m_1 \; m_2]^T$

$u_{ox} = \bar{u}_{ox}(t)$ *on* Γ_{ux}, $u_{oy} = \bar{u}_{oy}(t)$ *on* Γ_{uy}, $u_{oz} = \bar{u}_{oz}(t)$ *on* Γ_{uz}, $\theta_1 = \bar{\theta}_1(t)$ *on* $\Gamma_{u\theta 1}$, $\theta_2 = \bar{\theta}_2(t)$ *on* $\Gamma_{u\theta 2}$
(essential boundary conditions)

$$\begin{Bmatrix} u_{ox} \\ u_{oy} \\ u_{oz} \\ \theta_1 \\ \theta_2 \end{Bmatrix}_{t=0} = \begin{Bmatrix} u_{oxo}(x,y) \\ u_{oyo}(x,y) \\ u_{ozo}(x,y) \\ \theta_{1o}(x,y) \\ \theta_{2o}(x,y) \end{Bmatrix} \text{ (initial "displacements")}, \quad \begin{Bmatrix} \dot{u}_{ox} \\ \dot{u}_{oy} \\ \dot{u}_{oz} \\ \dot{\theta}_1 \\ \dot{\theta}_2 \end{Bmatrix}_{t=0} = \begin{Bmatrix} \dot{u}_{oxo}(x,y) \\ \dot{u}_{oyo}(x,y) \\ \dot{u}_{ozo}(x,y) \\ \dot{\theta}_{1o}(x,y) \\ \dot{\theta}_{2o}(x,y) \end{Bmatrix} \text{ (initial "velocities")}$$

of the stresses compared to the weak form for static shell problems in Box 14.5.1. Additionally, the "body" and boundary loadings are functions of both the spatial coordinates and of time, t.

In Box 15.8.2, we have used the following generalized stress-strain law for a shell, to incorporate the viscous damping effect:

$$\{\hat{\sigma}\} = [\hat{D}]\,\{\hat{\varepsilon}\} + [\hat{H}]\,\{\dot{\hat{\varepsilon}}\} \tag{15.8.20}$$

Remark 15.8.5: For the special case of an isotropic material and assuming that the material parameters do not vary along the thickness of the shell, we can write:

$$\{\hat{\sigma}\} = \begin{Bmatrix} \{\hat{\sigma}_m\} \\ \{\hat{\sigma}_f\} \\ \{\hat{\sigma}_s\} \end{Bmatrix} = \begin{bmatrix} [\hat{D}_m] & [0] & [0] \\ [0] & [\hat{D}_f] & [0] \\ [0] & [0] & [\hat{D}_s] \end{bmatrix} \begin{Bmatrix} \{\hat{\varepsilon}_m\} \\ \{\hat{\varepsilon}_f\} \\ \{\hat{\varepsilon}_s\} \end{Bmatrix} + \begin{bmatrix} [\hat{H}_m] & [0] & [0] \\ [0] & [\hat{H}_f] & [0] \\ [0] & [0] & [\hat{H}_s] \end{bmatrix} \begin{Bmatrix} \{\dot{\hat{\varepsilon}}_m\} \\ \{\dot{\hat{\varepsilon}}_f\} \\ \{\dot{\hat{\varepsilon}}_s\} \end{Bmatrix} \tag{15.8.21}$$

■

Remark 15.8.6: The inertial terms in the weak form which include $\ddot{\theta}_1$, $\ddot{\theta}_2$, i.e., the terms multiplied by components (4,4) and (5,5) of the $[\tilde{\rho}]$ –matrix, stem from the general, three-dimensional expression for virtual work associated with inertial terms, after we introduce the shell kinematic equations for the displacement and virtual displacement fields and integrate over the thickness of the shell.

$$\iint_\Omega \left[\int_{-d/2}^{d/2} \left[w_x(-\rho\ddot{u}_x) + w_y(-\rho\ddot{u}_y) + w_z(-\rho\ddot{u}_z) \right] dz \right] d\Omega =$$

$$= \iint_\Omega \left[\int_{-d/2}^{d/2} \left[-(w_{xo} - z\cdot w_{\theta 1})(\rho\ddot{u}_{xo} - \rho\cdot z\cdot\ddot{\theta}_1) - (w_{yo} - z\cdot w_{\theta 2})(\rho\ddot{u}_{yo} - \rho\cdot z\cdot\ddot{\theta}_2) - w_{zo}(\rho\ddot{u}_{zo}) \right] dz \right] d\Omega =$$

$$= -\iint_\Omega w_{xo}\cdot\rho\ddot{u}_{xo}\left(\int_{-d/2}^{d/2} dz\right) d\Omega - \iint_\Omega w_{yo}\cdot\rho\ddot{u}_{yo}\left(\int_{-d/2}^{d/2} dz\right) d\Omega - \iint_\Omega w_{zo}\cdot\rho\ddot{u}_{zo}\left(\int_{-d/2}^{d/2} dz\right) d\Omega -$$

$$-\iint_\Omega w_{\theta 1}\cdot\rho\ddot{\theta}_1\left(\int_{-d/2}^{d/2} z^2 dz\right) d\Omega - \iint_\Omega w_{\theta 2}\cdot\rho\,\ddot{\ }_2\left(\int_{-d/2}^{d/2} z^2 dz\right) d\Omega +$$

$$+\iint_\Omega w_{xo}\cdot\rho\ddot{\theta}_1\left(\int_{-d/2}^{d/2} zdz\right) d\Omega + \iint_\Omega w_{yo}\cdot\rho\ddot{\theta}_2\left(\int_{-d/2}^{d/2} zdz\right) d\Omega + \iint_\Omega w_{\theta 1}\cdot\rho\ddot{u}_{xo}\left(\int_{-d/2}^{d/2} zdz\right) d\Omega + \iint_\Omega w_{\theta 2}\cdot\rho\ddot{u}_{yo}\left(\int_{-d/2}^{d/2} zdz\right) d\Omega$$

$$= 0, \text{ because } \int_{-d/2}^{d/2} zdz = \left[\frac{z^2}{2}\right]_{-d/2}^{d/2} = 0$$

$$= -\iint_\Omega w_{xo}\cdot p_{xI}\, d\Omega - \iint_\Omega w_{yo}\cdot p_{yI}\, d\Omega - \iint_\Omega w_{zo}\cdot p_{zI}\, d\Omega - \iint_\Omega w_{\theta 1}\cdot m_{1I}\, d\Omega - \iint_\Omega w_{\theta 2}\cdot m_{2I}\, d\Omega$$

where

$$p_{xI} = \rho\ddot{u}_{xo}\left(\int_{-d/2}^{d/2} dz\right) = \rho\cdot d\cdot\ddot{u}_{xo} \tag{15.8.22a}$$

$$p_{yI} = \rho\ddot{u}_{yo}\left(\int_{-d/2}^{d/2} dz\right) = \rho\cdot d\cdot\ddot{u}_{yo} \tag{15.8.22b}$$

$$p_{zI} = \rho \ddot{u}_{zo} \left(\int_{-d/2}^{d/2} dz \right) = \rho \cdot d \cdot \ddot{u}_{zo} \tag{15.8.22c}$$

$$m_{1I} = \rho \ddot{\theta}_1 \left(\int_{-d/2}^{d/2} z^2 dz \right) = \rho \cdot \frac{d^3}{12} \cdot \ddot{\theta}_1 \tag{15.8.22d}$$

and

$$m_{2I} = \rho \ddot{\theta}_2 \left(\int_{-d/2}^{d/2} z^2 dz \right) = \rho \cdot \frac{d^3}{12} \cdot \ddot{\theta}_2 \tag{15.8.22e}$$

■

Finally, the element arrays for finite element analysis of shells under dynamic loading are given by the following equations:

$$\left[m^{(e)}\right] = \iint_{\Omega^{(e)}} \left[N^{(e)}\right]^T \left[\tilde{\rho}^{(e)}\right] \left[N^{(e)}\right] dV \tag{15.8.23}$$

$$\left[c^{(e)}\right] = \iint_{\Omega^{(e)}} \left[B^{(e)}\right]^T \left[\hat{H}^{(e)}\right] \left[B^{(e)}\right] dV \tag{15.8.24}$$

$$\left[k^{(e)}\right] = \iint_{\Omega^{(e)}} \left[B^{(e)}\right]^T \left[\hat{D}^{(e)}\right] \left[B^{(e)}\right] dV \tag{15.8.25}$$

$$\begin{aligned} \left\{f^{(e)}\right\} = \left\{f_\Omega^{(e)}\right\} &+ \int_{\Gamma_{tx}^{(e)}} \left[N^{(e)}\right]^T \{f_x\} dS + \int_{\Gamma_{ty}^{(e)}} \left[N^{(e)}\right]^T \{f_y\} dS + \int_{\Gamma_{tz}^{(e)}} \left[N^{(e)}\right]^T \{f_z\} dS \\ &+ \int_{\Gamma_{t\theta 1}^{(e)}} \left[N^{(e)}\right]^T \{m_{1\Gamma}\} dS + \int_{\Gamma_{t\theta 2}^{(e)}} \left[N^{(e)}\right]^T \{m_{2\Gamma}\} dS \end{aligned} \tag{15.8.26}$$

Just like for the other problems, the element equivalent nodal force vector will generally vary with time.

Remark 15.8.7: The expressions provided in this section for the element arrays and force vectors yield the various quantities in the local coordinate system of a beam or flat shell element. Coordinate transformation is necessary to find the corresponding quantities in the global coordinate system, before assembling the global mass, damping, and stiffness matrices. The transformation equations that we established for the stiffness matrix of a beam element (Equation 13.7.8) and a flat shell element (Equation 14.8.8) can also be used to transform the viscous damping and mass matrices. ■

Remark 15.8.8: Beam and shell dynamic problems can also be formulated as continuum-based (CB) problems, using the considerations of Sections 13.12 and 14.10. In such cases, we can obtain the arrays of an underlying continuum element, using the procedures of Sections 15.6 and 15.7. Then, using the kinematic transformation Equations (13.12.10), (11.12.13) for beams and the equations of Box 14.10.1 for shells, we can obtain the arrays and vectors of the CB beam or shell element in the global coordinate system. Specifically, the constraint transformation equations that were written for the stiffness matrices of continuum-based beam (Equation 13.12.13) or shell elements (Step 6 in Box 14.10.1) are also applicable to the damping and mass matrix for such elements. ∎

15.9 Diagonal (Lumped) Mass Matrices and Mass Lumping Techniques

In previous sections, we obtained expressions for element mass matrices, as element integral terms involving the shape function arrays, for example, Equations (15.5.11) and (15.7.1). Such matrices are called *consistent mass matrices*, because they are consistent with the finite element approximation that we use (they are obtained by plugging the finite element approximation into the weak form of a dynamic problem).

In the early efforts in solid and structural dynamic analysis, the masses would usually be treated as concentrated, or *lumped*, at the nodes. This means that the inertial effect was accounted for through concentrated point masses at the nodes of a model. Computationally, the global mass matrix $[M]$ for a lumped-mass assumption is obtained by adding the mass of each node to the diagonal entries of $[M]$ that correspond to the degrees of freedom of that specific node. For example, in a two-node truss element with a constant cross-section and density along the length, it would intuitively appear meaningful to calculate the total mass of the element, then attach half of that mass as a concentrated point mass to each one of the two nodes. Treating the masses as lumped at the nodes leads to a diagonal mass matrix in an analysis. Mass matrices obtained with this approach are called *lumped mass matrices*. The use of lumped mass matrices simplifies the solution procedure and allows some very efficient algorithms to be developed (especially for nonlinear problems), without significantly affecting the accuracy. It is necessary to establish automated procedures to obtain a lumped mass matrix, since an intuition-based approach—such as the one mentioned above for a two-node truss element—may not always be mathematically valid. This section will be dedicated to such procedures, called *mass lumping techniques*, which lead to diagonal mass matrices $[M]$ in dynamic analysis.

15.9.1 Mass Lumping for Continuum (Solid) Elements

A requirement that needs to be satisfied by any mass lumping technique is that the sum of the lumped masses of each element (for each direction) must equal the total mass of the element. For example, in a three-dimensional hexahedral solid element with eight nodes and a constant density, one must ensure that the sum of all eight diagonal mass quantities corresponding to degrees of freedom in the x-direction is equal to the product of the density times the volume of the element. The same applies for the y- and z-directions.

A common mass lumping procedure used in many programs is the *row-sum technique*, and it is very simple and efficient, especially for low-order elements. For the case of a three-dimensional element with n nodes, the lumped mass for each node I, is given by the following expression.

$$m_{II,x}^{(e)} = m_{II,y}^{(e)} = m_{II,z}^{(e)} = \int_{\Omega^{(e)}} \rho^{(e)} \cdot N_I^{(e)} dV \ , \quad I = 1,2,\ldots,n \tag{15.9.1}$$

where, for example, $m_{II,x}^{(e)}$ is the diagonal term of the lumped matrix corresponding to the x-displacement of node I. All other entries of the mass matrix $[m^{(e)}]$ (besides those in Equation 15.9.1) are set equal to zero. We can cast a computation-oriented algorithm for the row-sum technique, which is summarized in Box 15.9.1.

Box 15.9.1 Row-sum technique for diagonal (lumped) mass matrix $[m^{(e)}]$

For each node I, $I = 1, 2, \ldots, n$:

If we have a three-dimensional element:

$$m_{ij}^{(e)} = \int_{\Omega^{(e)}} \rho^{(e)} \cdot N_I^{(e)} dV \quad \text{if } i=j=3I-2 \ \text{ or } i=j=3I-1 \ \text{ or } i=j=3I \tag{15.9.2a}$$

If we have a two-dimensional element:

$$m_{ij}^{(e)} = \int_{\Omega^{(e)}} \rho^{(e)} \cdot N_I^{(e)} dV \quad \text{if } i=j=2I-1 \ \text{ or } i=j=2I \tag{15.9.2b}$$

If we have a one-dimensional element:

$$m_{ij}^{(e)} = \int_{\Omega^{(e)}} \rho^{(e)} \cdot N_I^{(e)} dV \quad \text{if } i=j=I \tag{15.9.2c}$$

All other entries of the mass matrix: $m_{ij}^{(e)} = 0$

The major issue with the row-sum technique is that it can sometimes give negative diagonal mass terms (and this may prove problematic from both a physical and a computational point of view) for higher-order elements. For example, for an eight-node, serendipity quadrilateral element (8Q element), the row-sum technique will give negative diagonal masses for the nodes in the interior of the element sides.

To resolve potential issues of the row-sum method, a *special lumping technique* can be used instead, to ensure that the diagonal entities of the lumped mass matrix are all positive. In this technique, the concentrated mass of each node I in an element e is given by the following expression:

$$m_I^{(e)} = a \int_{\Omega^{(e)}} \rho^{(e)} \cdot \left(N_I^{(e)}\right)^2 dV \ , \quad I = 1,2,\ldots,n \tag{15.9.3}$$

where

$$a = \frac{\displaystyle\int_{\Omega^{(e)}} \rho^{(e)} dV}{\displaystyle\sum_{I=1}^{n} \left(\int_{\Omega^{(e)}} \rho^{(e)} \cdot \left(N_I^{(e)}\right)^2 dV \right)} \tag{15.9.4}$$

Remark 15.9.1: One can easily verify that the integrand in all the terms of Equations (15.9.3) and (15.9.4) is non-negative, which ensures that no negative lumped masses are obtained. The coefficient a in Equation (15.9.3) is required to ensure that the sum of the lumped nodal masses is equal to the total mass of the element. The numerator in the expression giving a in Equation (15.9.4) is equal to the total element mass. ▮

15.9.2 Mass Lumping for Structural Elements (Beams and Shells)

A separate discussion is deemed necessary for mass lumping in structural elements, which also include rotational degrees of freedom. For such elements, the assumption of lumped nodal masses entails not only zero nondiagonal entities, but also zero diagonal mass terms for the rotational degrees of freedom. The additional assumption of zero rotational mass terms results from the assumption of concentrated masses at the nodes, since such masses do not generate rotational inertia.

For example, in a two-node Euler-Bernoulli beam element, the lumped mass matrix can be obtained through the same techniques presented for solid elements, provided that we calculate the lumped masses using the shape functions for the axial direction, $N_{ux1}^{(e)}(x)$ and $N_{ux2}^{(e)}(x)$ (see Section 13.8). For a Timoshenko beam element, the shape functions for the axial and transverse displacements are identical. These shape functions can be used to obtain the diagonal terms of the lumped mass matrix.

For flat shell elements, we can directly obtain the lumped mass matrix, by setting the terms of rows and columns of $[M]$ corresponding to the rotational degrees of freedom to zero. The diagonal mass terms corresponding to the translational degrees of freedom of the element can then be obtained by using the translational (in-plane) shape functions with either the row-sum or the *special lumping* technique.

For continuum-based beam or shell elements, we can first use a mass lumping technique to obtain a diagonal mass matrix for the underlying continuum element. We can then use the transformation Equation (13.12.13) (for beams) or Step 6 in Box 14.10.1 (for shells) to obtain a mass matrix for the structural element. The final, lumped-mass matrix of the element results by setting all the mass matrix terms in the rows and columns corresponding to rotational degrees of freedom to zero.

15.10 Strong and Weak Form for Time-Dependent Scalar Field (Parabolic) Problems

We will now move on to the analysis of time-dependent scalar field problems. Specifically, the problems of time-dependent heat conduction, mass flow in porous media and chemical diffusion will be examined. We will begin with the discussion of heat conduction, and then we will continue with a discussion of time-dependent flow in porous media and chemical diffusion.

15.10.1 Time-Dependent Heat Conduction

In Sections 5.1 and 5.2, we examined the strong and weak form for steady-state heat conduction problems. The major difference for time-dependent heat conduction (which will introduce additional terms, as we will see) is that the temperature field Θ and all the other quantities (source term, s, and flux vector, $\{q\}$) now additionally depend on time, t. We will focus to the most general case of a three-dimensional problem (although we did not discuss the steady-state three-dimensional case).

The governing equations for time-dependent heat conduction rely on the conservation of energy principle. When we discussed steady-state heat conduction, we stated that the amount of heat that is added to any piece of material is equal to the amount of heat that leaves that piece. In other words, we tacitly assumed that the material does not have the capability to *store energy*. This assumption is going to change now. That is, we are going to allow the material to store energy. Specifically, we will define a fundamental quantity, e, which we will call the *specific internal energy (per unit mass)* of the material.[1] It is simply the amount of energy stored in a piece of material, per unit mass of that piece. If the mass density of the material is ρ, then we can define the internal energy per unit volume as the product $\rho{\cdot}e$. Now, let us focus on a three-dimensional solid, and let us isolate a small, box-shaped region (i.e., a control volume) from the interior of the body, as shown in Figure 15.5. We are going to establish the conservation of energy per unit time. Since

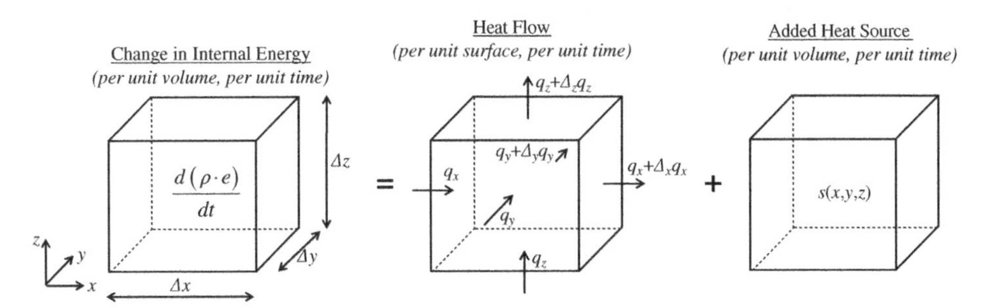

Figure 15.5 Change in energy and energy addition/subtraction in a small piece of material, for multidimensional, time-dependent, heat conduction.

---

1 The specific internal energy e can be thought of as energy stored into the molecules of the material of the solid. In fact, the internal energy is typically defined as a fundamental quantity of a solid in the *FIRST LAW of THERMODYNAMICS*, which constitutes the general statement of conservation of energy.

the box-shaped piece is small, we will assume that the values of the heat source, s, and of the specific energy per unit volume, $\rho \cdot e$, are spatially constant (i.e., for that small piece, they only depend on time and not on spatial coordinates). Now, the rate of change of the internal energy with respect to time will simply be equal to the product of the derivative of $\rho \cdot e$ with time, multiplied by the volume of the piece. In other words, we write:

$$\text{(rate of change of internal energy)} = \frac{d(\rho \cdot e)}{dt} \cdot (\Delta x) \cdot (\Delta y) \cdot (\Delta z) \tag{15.10.1}$$

We can also draw the other energy quantities that enter or leave our piece of material. As shown in Figure 15.5, these quantities stem from heat flow, which is quantified through the components of the heat flux $\{q\}$, and from the heat source.

We can now formulate the energy balance equation per unit time, in accordance with Figure 15.5, as follows:

$$\underbrace{\frac{d(\rho \cdot e)}{dt} \cdot (\Delta x) \cdot (\Delta y) \cdot (\Delta z)}_{\text{change in stored internal energy per unit time}} = \underbrace{q_x \cdot (\Delta y) \cdot (\Delta z) + q_y \cdot (\Delta x) \cdot (\Delta z) + q_z \cdot (\Delta x) \cdot (\Delta y) + s \cdot (\Delta x) \cdot (\Delta y) \cdot (\Delta z)}_{\text{energy added per unit time}}$$
$$\underbrace{- (q_x + \Delta_x q_x) \cdot (\Delta y) \cdot (\Delta z) - (q_y + \Delta_y q_y) \cdot (\Delta x) \cdot (\Delta z) - (q_z + \Delta_z q_z) \cdot (\Delta y) \cdot (\Delta z)}_{\text{energy subtracted per unit time}}$$

We can now proceed by dividing both sides of the above expression by the size of the piece, $(\Delta x)(\Delta y)(\Delta z)$, conduct some mathematical simplifications and then take the limit as the size of the piece tends to zero, i.e. as all three dimensions of the box-shaped piece of Figure 15.5 tend to zero. We obtain:

$$\frac{d(\rho \cdot e)}{dt} - \frac{\partial q_x}{\partial x} - \frac{\partial q_y}{\partial y} - \frac{\partial q_z}{\partial z} + s(x, y, z, t) = 0 \tag{15.10.2}$$

Furthermore, we will assume that the mass density of the material does not change with time, at which case we will have:

$$\rho \frac{de}{dt} - \frac{\partial q_x}{\partial x} - \frac{\partial q_y}{\partial y} - \frac{\partial q_z}{\partial z} + s(x, y, z, t) = 0 \tag{15.10.3}$$

We can also write Equation (15.10.3) as follows:

$$\rho \frac{de}{dt} - \vec{\nabla} \bullet \vec{q} + s(x, y, z, t) = 0 \tag{15.10.4}$$

Furthermore, we will use the constitutive law for heat conduction (i.e., Fourier's law), which was provided in Equation (5.1.11):

$$\vec{q} = -[D]\{\nabla \Theta\} \tag{15.10.5}$$

Finally, we need to somehow treat the first term of Equation (15.10.4), so that it includes the temperature field. The key is to stipulate that the specific internal energy e is a function of the temperature and use the chain rule of differentiation to write:

$$\rho \frac{de}{dt} = \rho \frac{de}{dT} \cdot \frac{dT}{dt} = \rho \cdot c \cdot \dot{\Theta} \tag{15.10.6}$$

where the quantity c is termed the *specific heat capacity* of the material, and is defined by the following expression:

$$c = \frac{de}{dT} \tag{15.10.7}$$

The specific heat capacity is the amount of heat (i.e., energy) required to increase the temperature of a piece of the material with unit mass by one degree. In our discussion here, we will assume that c does not depend on the temperature. This means that the specific internal energy at a location is assumed to be linearly proportional to the temperature at the same location.

Plugging Equations (15.10.5) and (15.10.6) into Equation (15.10.4), and also introducing essential and natural boundary conditions, we can reach the strong form for time-dependent, multidimensional heat conduction, provided in Box 15.10.1.

Box 15.10.1 Strong Form for Time-Dependent Heat Conduction

$$\rho \cdot c \cdot \dot{\Theta} + \vec{\nabla} \bullet ([D]\{\nabla\Theta\}) + s(x,y,z,t) = 0 \tag{15.10.8}$$

$$\Theta = \bar{\Theta}(t) \text{ on } \Gamma_\Theta \tag{15.10.9}$$

$$\vec{q} \bullet \vec{n} = \bar{q}(t) \text{ on } \Gamma_q \tag{15.10.10}$$

$$\Theta(x,y,z,t=0) = \Theta_o(x,y,z) \quad \text{(Initial Condition: Initial Temperature)} \tag{15.10.11}$$

Remark 15.10.1: Box 15.10.1 includes an additional term (the first one) in the left-hand side of Equation (15.10.8), compared to the governing differential equation for steady-state heat conduction. We also now have an initial condition. The governing equations for time-dependent heat conduction include only first derivatives with respect to time; correspondingly, the initial conditions only involve initial temperature, and do not prescribe a value for the initial rate of temperature. ∎

Remark 15.10.2: Partial differential equations such as Equation (15.10.8) are classified as *parabolic equations.* Accordingly, problems governed by parabolic equations are called parabolic problems. As we will see in Section 15.12, the semi-discrete equations of motion for elastodynamics and structural dynamics can be cast in a form which has the mathematical structure of a parabolic problem (but at the same time, the system of that "equivalent parabolic problem" has twice the number of degrees of freedom of the original system!). ∎

The weak form for time-dependent heat conduction can also be obtained with the procedures presented in previous chapters, and in particular in Section 5.2. This form is provided (without a proof) in Box 15.10.2.

Box 15.10.2 Weak Form for Time-Dependent Heat Conduction

$$\iiint_{\Omega} w \cdot \rho \cdot c \cdot \dot{\Theta}\, dV + \iiint_{\Omega} \left(\vec{\nabla} w\right) \bullet [D] \vec{\nabla}\, \Theta\, dV = \iiint_{\Omega} w \cdot s\, dV - \iint_{\Gamma_q} w \cdot \bar{q}\, dS \quad (15.10.12)$$

$\forall w(x,y,z)$ with $w=0$ at Γ_Θ (w vanishes at essential boundary)

$$\Theta = \bar{\Theta}(t) \text{ on } \Gamma_\Theta \quad (15.10.13)$$

$$\Theta(x,y,z,t=0) = \Theta_o(x,y,z) \text{ (Initial Condition: Initial Temperature)} \quad (15.10.14)$$

15.10.2 Time-Dependent Fluid Flow in Porous Media

We will now continue with time-dependent fluid flow in a porous medium. The considerations that we apply are identical to those for heat conduction. The only difference is that we now need to rely on the conservation of mass principle to obtain the strong form. We will also stipulate that the porous medium has the capability to store fluid mass in the pores. We need to define a field θ_w, which is the *fluid content* (i.e., fluid mass) *per unit volume of the porous solid*. The boundary conditions and the constitutive law (i.e., Darcy's law) are the same as those presented for the steady-state case in Section 5.4.2. Specifically, the constitutive law will be written as follows:

$$\{q_w\} = \vec{q}_w = -\frac{\kappa}{\mu}\{\nabla P\} = -\frac{\kappa}{\mu}\vec{\nabla} P \quad (15.10.15)$$

where $\vec{q}_w$ is the fluid flux vector (mass per unit area per unit time) and the other material constants have been defined in Section 5.4.2. The problem will be formulated in terms of the pore fluid pressure field, $P(x,y,z,t)$. We isolate a box-shaped piece of material, as shown in Figure 15.6, and we draw all the mass inflow and outflow. We also consider the rate of change of stored fluid mass in the piece:

$$(\textit{rate of change of fluid mass}) = \frac{d(\theta_w)}{dt} \cdot (\Delta x) \cdot (\Delta y) \cdot (\Delta z) \quad (15.10.16)$$

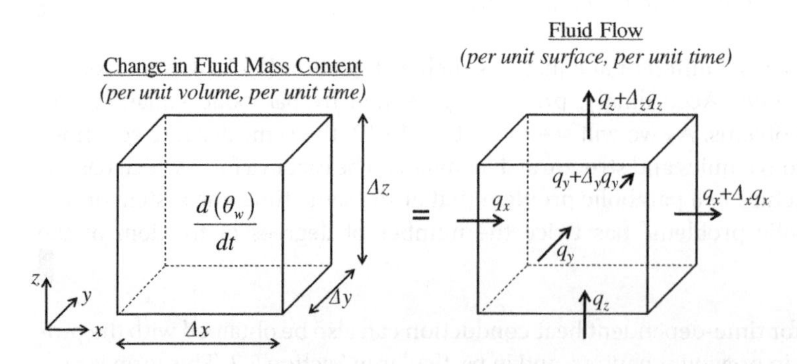

Figure 15.6 Conservation of fluid mass per unit time, for fluid flow in a porous medium.

We can now formulate the mass balance equation per unit time, divide both sides of the resulting expression by $(\Delta x)\cdot(\Delta y)\cdot(\Delta z)$, and then take the limit as (Δx), (Δy), (Δz) all tend to zero. The equation that we finally obtain is the following.

$$\frac{d\theta_w}{dt}-\frac{\partial q_{wx}}{\partial x}-\frac{\partial q_{wy}}{\partial y}-\frac{\partial q_{wz}}{\partial z}=0 \tag{15.10.17}$$

Now, we will transform the first term of Equation (15.10.17) into a version that includes the time derivative of the fluid pressure field. We use the chain rule of differentiation and obtain:

$$\frac{d\theta_w}{dt}=\frac{d\theta_w}{dP}\cdot\frac{dP}{dt}=i_P\cdot\dot{P} \tag{15.10.18}$$

where the term i_P is simply given by:

$$i_P=\frac{d\theta_w}{dP} \tag{15.10.19}$$

Remark 15.10.3: The term i_P will be assumed to be constant in this text. In reality, the relation between fluid content and the fluid pressure is nonlinear. For example, whenever we have moisture (water) flow in porous materials such as concrete, the relation between θ_w (called the *moisture content*) and P (the pressure of the water in the pores) is given by curves that are called *moisture isotherms* (because we typically define a curve for a constant temperature). The interested reader is referred to the textbook by Coussy (2010) for more details on description of porous media and fluid flow. ∎

Plugging Equations (15.10.15) and (15.10.19) into Equation (15.10.17), and including the boundary and initial conditions, yields the strong form for time-dependent flow in porous media, provided in Box 15.10.3.

We can also obtain the weak form (with the procedures presented in previous chapters, and in particular in Section 5.2) for time-dependent porous flow, as summarized in Box 15.10.4.

Box 15.10.3 Strong Form for Time-Dependent Flow in Porous Media

$$i_P\cdot\dot{P}+\vec{\nabla}\bullet\left(\frac{\kappa}{\mu}\vec{\nabla}P\right)=0 \tag{15.10.20}$$

$$P=\bar{P}(t) \text{ on } \Gamma_P \tag{15.10.21}$$

$$\vec{q}_w\bullet\vec{n}=\bar{q}_w(t) \text{ on } \Gamma_{qw} \tag{15.10.22}$$

$$P(x,y,z,t=0)=P_o(x,y,z) \quad \text{(Initial Condition: Initial Pressure)} \tag{15.10.23}$$

Box 15.10.4 Weak Form for Time-Dependent Porous Flow

$$\iiint_{\Omega} w \cdot i_P \cdot \dot{P}\, dV + \iiint_{\Omega} \left(\vec{\nabla} w\right) \bullet \left(\frac{\kappa}{\mu}\vec{\nabla} P\right) dV = -\iint_{\Gamma_q} w \cdot \bar{q}_w \, dS \quad (15.10.24)$$

$\forall w(x,y,z)$ with $w=0$ at Γ_P (w vanishes at essential boundary)

$$P = \bar{P}(t) \text{ on } \Gamma_P \quad (15.10.25)$$

$$P(x,y,z,t=0) = P_o(x,y,z) \quad \text{(Initial Condition: Initial Pressure)} \quad (15.10.26)$$

15.10.3 Time-Dependent Chemical Diffusion

We will complete our consideration of time-dependent scalar-field problems by briefly discussing time-dependent chemical (molecular) diffusion in a medium. Once again, the problem physics are identical to those presented for the steady-state case in Section 5.4.3. The only difference is that we will allow the concentration, C, at each location to change with time. We can then isolate a box-shaped piece of the medium, as shown in Figure 15.7, and then establish the equations describing the conservation of mass per unit time, and finally obtain the strong form provided in Box 15.10.5. We can also obtain the weak form, given in Box 15.10.6.

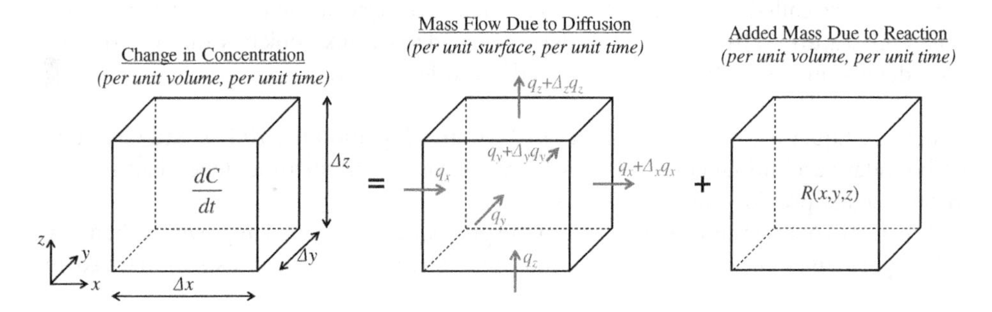

Figure 15.7 Conservation of mass of substance in a medium, for chemical diffusion.

Box 15.10.5 Strong Form for Time-Dependent Chemical (Molecular) Diffusion

$$\dot{C} + \vec{\nabla} \bullet \left(D_C \vec{\nabla} C\right) + R(x,y,z,t) = 0 \quad (15.10.27)$$

$$C = \bar{C}(t) \text{ at } \Gamma_C \text{ (essential boundary condition)} \quad (15.10.28)$$

$$\vec{q}_C \bullet \vec{n} = -\left(D_C \vec{\nabla} C\right) \bullet \vec{n} = \bar{q}_C \text{ at } \Gamma_{qC} \text{ (natural boundary condition)} \quad (15.10.29)$$

$$C(x,y,z,t=0) = C_o(x,y,z) \quad \text{(Initial Condition: Initial Concentration)} \quad (15.10.30)$$

Box 15.10.6 Weak form for time-dependent chemical diffusion

$$\iiint_{\Omega} w \cdot \dot{C}\, dV + \iiint_{\Omega} \left(\vec{\nabla} w\right) \cdot \left(D_C \vec{\nabla} C\right) dV = \iiint_{\Omega} w \cdot R\, dV - \iint_{\Gamma_q} w \cdot \bar{q}_C\, dS \quad (15.10.31)$$

$\forall w(x,y,z)$ with $w = 0$ at Γ_C (w vanishes at essential boundary)

$$C = \bar{C}(t) \text{ on } \Gamma_C \quad (15.10.32)$$

$$C(x,y,z,t=0) = C_o(x,y,z) \quad \text{(Initial Condition: Initial Concentration)} \quad (15.10.33)$$

15.11 Semi-Discrete Finite Element Equations for Scalar Field (Parabolic) Problems

Now, the semi-discrete, finite element equations for time-dependent scalar fields will be provided, without detailed derivations. The conceptual steps to obtain these equations are identical to those in Section 15.5. The global semi-discrete equations for any time-dependent scalar field (parabolic) problem have the following form:

$$[C]\{\dot{X}(t)\} + [K]\{X(t)\} = \{f(t)\} \quad (15.11.1)$$

where $\{X(t)\}$ is the *global nodal solution vector* (temperature for heat conduction, pressure for porous flow, and concentration for chemical diffusion), $\{f(t)\}$ is the equivalent nodal vector and $[C]$, $[K]$ are coefficient arrays. In fact, $[K]$ is the same exact coefficient array obtained for the steady-state problems in Sections 6.1 (heat conduction) and 6.9 (porous flow and chemical diffusion). The global arrays and right-hand-side vector are obtained through assembly of the element contributions. Thus, for a mesh with N_e elements, we can write:

$$[C] = \sum_{e=1}^{Ne} \left(\left[L^{(e)}\right]^T \left[c^{(e)}\right] \left[L^{(e)}\right] \right) \quad (15.11.2)$$

$$[K] = \sum_{e=1}^{Ne} \left(\left[L^{(e)}\right]^T \left[k^{(e)}\right] \left[L^{(e)}\right] \right) \quad (15.11.3)$$

$$\{f(t)\} = \sum_{e=1}^{Ne} \left(\left[L^{(e)}\right]^T \left\{f^{(e)}(t)\right\} \right) \quad (15.11.4)$$

The element arrays $[k^{(e)}]$, $[c^{(e)}]$ and element right-hand-side vectors $\{f^{(e)}(t)\}$ are obtained using Boxes 15.11.1, 15.11.2, and 15.11.3, for heat conduction, porous flow, and chemical diffusion, respectively.

Remark 15.11.1: Lumped matrices $[c^{(e)}]$ can also be defined with the procedures of Section 15.9 and used for finite element analysis of parabolic problems. ∎

Box 15.11.1 Element Arrays and Vectors for Heat Conduction

$$\{X(t)\} = \{\Theta(t)\} \tag{15.11.5}$$

$$\left[c^{(e)}\right] = \iiint_{\Omega^{(e)}} \left[N^{(e)}\right]^T \rho^{(e)} c^{(e)} \left[N^{(e)}\right] dV \tag{15.11.6}$$

$$\left[k^{(e)}\right] = \iiint_{\Omega^{(e)}} \left[B^{(e)}\right]^T \left[D^{(e)}\right] \left[B^{(e)}\right] dV \tag{15.11.7}$$

$$\left\{f^{(e)}(t)\right\} = \iiint_{\Omega^{(e)}} \left[N^{(e)}\right]^T s(t)\, dV - \iint_{\Gamma_q^{(e)}} \left[N^{(e)}\right]^T \cdot \bar{q}(t)\, dS \tag{15.11.8}$$

Box 15.11.2 Element Arrays and Vectors for Fluid Flow in Porous Bodies

$$\{X(t)\} = \{P(t)\} \tag{15.11.9}$$

$$\left[c^{(e)}\right] = \iiint_{\Omega^{(e)}} \left[N^{(e)}\right]^T i_P^{(e)} \left[N^{(e)}\right] dV \tag{15.11.10}$$

$$\left[k^{(e)}\right] = \iiint_{\Omega^{(e)}} \left[B^{(e)}\right]^T \frac{\kappa^{(e)}}{\mu^{(e)}} \left[B^{(e)}\right] dV \tag{15.11.11}$$

$$\left\{f^{(e)}(t)\right\} = -\iint_{\Gamma_{qw}^{(e)}} \left[N^{(e)}\right]^T \cdot \bar{q}_w(t)\, dS \tag{15.11.12}$$

Box 15.11.3 Element Arrays and Vectors for Molecular (Chemical) Diffusion

$$\{X(t)\} = \{C(t)\} \tag{15.11.13}$$

$$\left[c^{(e)}\right] = \iiint_{\Omega^{(e)}} \left[N^{(e)}\right]^T \left[N^{(e)}\right] dV \tag{15.11.14}$$

$$\left[k^{(e)}\right] = \iiint_{\Omega^{(e)}} \left[B^{(e)}\right]^T D_C^{(e)} \left[B^{(e)}\right] dV \tag{15.11.15}$$

$$\left\{f^{(e)}(t)\right\} = \iiint_{\Omega^{(e)}} \left[N^{(e)}\right]^T R(t)\, dV - \iint_{\Gamma_{qc}^{(e)}} \left[N^{(e)}\right]^T \cdot \bar{q}_C(t)\, dS \tag{15.11.16}$$

15.12 Solid and Structural Dynamics as a "Parabolic" Problem: The State-Space Formulation

The algorithms for the solution of parabolic problems (which are discussed in Chapter 16) can be classified as algorithms for systems of ordinary differential equations with respect to time (as explained in, for example, Chapter 9 in Pozrikidis 2009). These algorithms differ from those employed for elastodynamics and structural dynamics, which are examined in Chapter 17. One might wonder whether we can express the semi-discrete equations of motion for elastodynamics and structural dynamics in a form that would seem like that of the semi-discrete equations of a parabolic system, so that we can tackle these cases with the solution algorithms of Chapter 16. The answer is yes, and the means to accomplish this is to use the *state-space formulation*. More specifically, what we do is establish an expanded vector $\{X\}$, combining two response vectors, namely the displacement and velocity vectors:

$$\{X\} = \begin{Bmatrix} \{X_1\} = \{U\} \\ \{X_2\} = \{\dot{U}\} \end{Bmatrix} \tag{15.12.1}$$

The vector $\{X\}$ is called the *state vector* of the dynamic system.

We can easily verify that:

$$\{\dot{X}_1\} = \{X_2\} \rightarrow [I]\{\dot{X}_1\} = [I]\{X_2\} \tag{15.12.2}$$

Now, if we revisit the semi-discrete equations of motion, that is, Equation (15.5.31a), and also account for Equation (15.12.1), we have:

$$[M]\{\ddot{U}\} + [C]\{\dot{U}\} + [K]\{U\} = \{f\} \rightarrow [M]\{\dot{X}_2\} + [C]\{X_2\} + [K]\{X_1\} = \{f\} \tag{15.12.3}$$

We can now combine Equations (15.12.2) and (15.12.3) (*Note:* Each one of them is a system with N degrees of freedom) into a single system with $2N$ degrees of freedom:

$$\begin{bmatrix} [I] & [0] \\ [0] & [M] \end{bmatrix} \begin{Bmatrix} \{\dot{X}_1\} \\ \{\dot{X}_2\} \end{Bmatrix} + \begin{bmatrix} [0] & [I] \\ [K] & [C] \end{bmatrix} \begin{Bmatrix} \{X_1\} \\ \{X_2\} \end{Bmatrix} = \begin{Bmatrix} \{0\} \\ \{f\} \end{Bmatrix} \rightarrow [A]\{\dot{X}\} + [B]\{X\} = \{f_{St-Sp}\} \tag{15.12.4}$$

where

$$[A] = \begin{bmatrix} [I] & [0] \\ [0] & [M] \end{bmatrix} \tag{15.12.5a}$$

$$[B] = \begin{bmatrix} [0] & [I] \\ [K] & [C] \end{bmatrix} \tag{15.12.5b}$$

and

$$\{f_{St-Sp}\} = \begin{Bmatrix} \{0\} \\ \{f\} \end{Bmatrix} \tag{15.12.5c}$$

The initial conditions can also be converted into initial conditions for the state vector:

$$\{X\}_{t=0} = \{X_o\} = \left\{ \begin{array}{l} \{X_{1o}\} = \{U_o\} \\ \{X_{2o}\} = \{V_o\} \end{array} \right\} \text{ (initial state vector)}$$

We can verify that the semi-discrete equations in the state space, that is, Equation (15.12.4), have the same form as those of the semi-discrete, parabolic problem in Equation (15.11.1). Thus, the same types of algorithms as for parabolic problems can be used for the solution of the state-space formulation.

Problems

15.1 We are given the three-element mesh shown in Figure P15.1. The material is isotropic linearly elastic, with $E = 5000$ ksi, $\nu = 0.2$, and $\rho = 2.5 \cdot 10^{-7}$ kip·s^2/in.^{4}. Additionally, there is a concentrated mass of 0.2 kip·s^2/in. at node 8. Finally, there is a compliant (spring) support in the y-direction for node 4, with $k_{sy} = 1000$ kip/in. The material damping matrix is $[H] = 0.001[D]$, where $[D]$ is the material stiffness matrix (the conceptual basis for establishing the damping matrix with this approach is provided in Section 17.2.1).

a) Assemble the global mass, damping and stiffness matrices for the structure.
b) Obtain the arrays $[M_{ff}]$, $[C_{ff}]$, and $[K_{ff}]$, and the vector $\{F_f\}$, corresponding to the unrestrained degrees of freedom.
c) Find $[M_{ff}]$ for the case that we have a lumped-mass matrix based on the row-sum technique.
d) Same as in c, but use the special lumping technique.

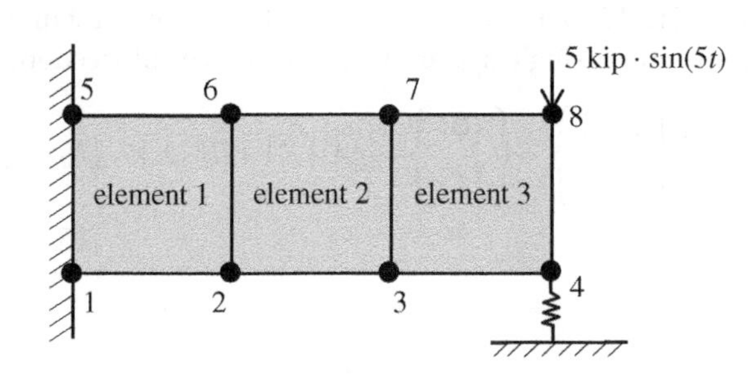

Node "i"	x_i (ft)	y_i (ft)
1	0	0
2	5	0
3	10	0
4	15	0
5	0	5
6	5	5
7	10	5
8	15	5

Figure P15.1

15.2 We are given the two-element mesh shown in Figure P15.2. The same case was considered in Problem 6.10, but we now have a heat source which depends on time. Assume that we have unit thickness. The material is isotropic with $\kappa = 8$ W/(°C·m), $c = 500$ J/kg/K, and $\rho = 8000$ kg /m^3 = 8000 N·s^2 /m^4.

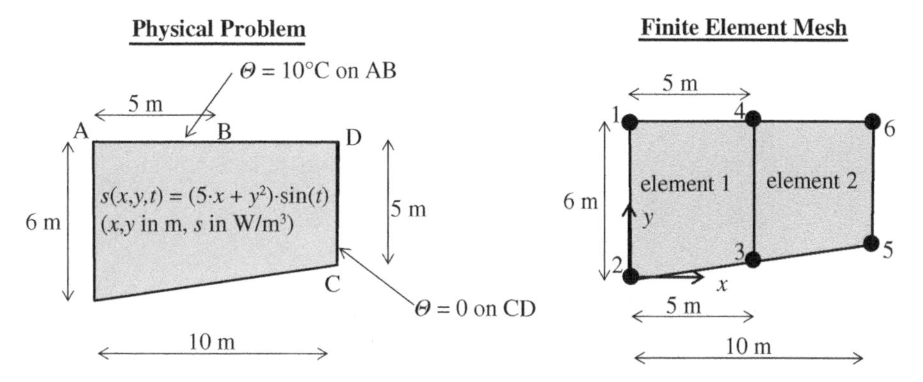

Figure P15.2

a) Obtain the arrays $[C_{ff}]$ and $[K_{ff}]$, and the vector $\{F_f\}$, corresponding to the unrestrained degrees of freedom.
b) Obtain a *lumped* $[C_{ff}]$, using the special lumping technique.

15.3 Establish the semi-discrete equations of the porous flow problem shown in Figure P15.3. The same case was considered (for steady-state conditions) in Problem 6.8. Use $i_P = 1$ for the time-dependent problem.

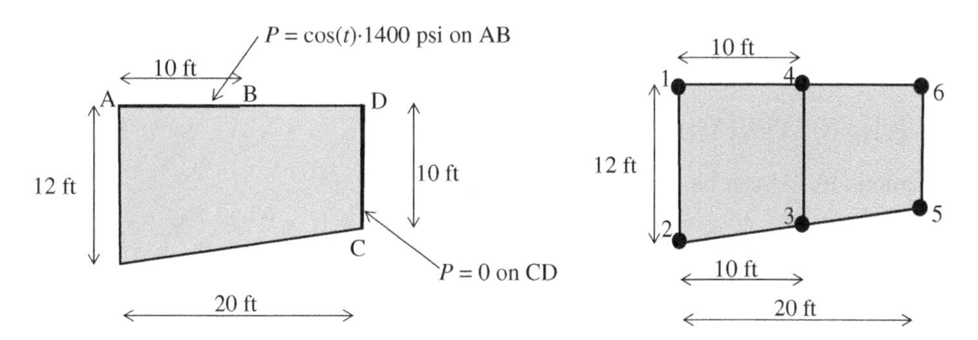

Figure P15.3

15.4 Establish the state-space matrices and semi-discrete equations, for the system examined in Problem 15.1.

References

Pozrikidis, C. (2009). *Numerical Computation in Science and Engineering*, 2nd ed. Oxford: Oxford University Press.

Coussy, O. (2010). *Mechanics and Physics of Porous Solids*. Hoboken, NJ: John Wiley & Sons.

16

Analysis of Time-Dependent Scalar Field (Parabolic) Problems

16.1 Introduction

This chapter is devoted to the finite element analysis of scalar field problems (e.g., heat conduction, chemical diffusion etc.). As mentioned in Section 15.10, time-dependent scalar field problems lead to parabolic equations, where we have first partial derivatives with respect to time. The algorithms presented in this chapter are to determine the evolution (in time) of the global nodal vector $\{X(t)\}$, which will contain the nodal values for the field function. The physical meaning of the vector $\{X(t)\}$ depends on the problem at hand. For example, we may have $\{X(t)\} = \{\Theta(t)\}$ (nodal temperatures) for time-dependent heat conduction, $\{X(t)\} = \{C(t)\}$ (nodal concentrations) for chemical diffusion problems, and so on.

Per Section 15.11, the governing equations for parabolic problems attain the following form.

$$[C]\{\dot{X}(t)\} + [K]\{X(t)\} = \{f(t)\} \tag{16.1.1}$$

Equation (16.1.1) can be recast in a modified form:

$$[C]\{\dot{X}(t)\} = \{R(t)\} \tag{16.1.2}$$

where

$$\{R(t)\} = \{f(t)\} - [K]\{X(t)\} \tag{16.1.3}$$

The "exact" analytical determination of $\{X(t)\}$ as a function of time, also termed the *time history* of $\{X\}$ is extremely difficult, if at all possible. The only exceptions are cases where the right-hand-side vector $\{f(t)\}$ includes simple functions of time. Furthermore, exact solution for the time histories of the nodal values cannot be conducted for non-linear systems.

For this reason, instead of presenting "exact" solution procedures, we will focus on the use of so-called *step-by-step (transient) integration methods*. Such methods do not provide a *closed-form* solution (i.e., an analytical expression $X(t)$, of a quantity as a function of time), but a series of approximate values of the solution X_o, X_1, ..., X_K at specific (discrete) time instants t_o, t_1, ..., t_K, respectively. As we shall see in the next chapter, the

Fundamentals of Finite Element Analysis: Linear Finite Element Analysis, First Edition.
Ioannis Koutromanos, James McClure, and Christopher Roy.

Companion website: www.wiley.com/go/koutromanos/linear

concept of step-by-step methods is also applicable to elastodynamics and structural dynamics.

In step-by-step algorithms, the global system of ordinary differential equations provided in Equation (16.1.1) is converted to an "approximate," discrete form by stipulating that the equation of evolution is satisfied at *discrete time instants*; that is, we enforce the following equation:

$$[C]\{\dot{X}(t_i)\} + [K]\{X(t_i)\} = \{f(t_i)\} \tag{16.1.4}$$

for a set of specific instants t_i, $i = 0, 1, 2, \ldots, N_{step}$

Equation (16.1.4) can be more concisely written as:

$$[C]\{\dot{X}_i\} + [K]\{X_i\} = \{f_i\} \tag{16.1.5}$$

where $\{\dot{X}_i\}$, $\{X_i\}$ and $\{f_i\}$ are the rate of the nodal solution, the nodal solution, and the equivalent right-hand-side vector, respectively, at time $t = t_i$. It is worth noticing that step-by-step methods essentially introduce a "discretization" in time, since we now enforce the governing equations only at discrete time instants $t = t_o, t_1, \ldots, tN_{step}$.

The following sections will be devoted on the various step-by-step algorithms for the solution of time-dependent, scalar field problems. In these algorithms, we will be establishing an equation to find the *updated solution* for each step. In other words, we will be assuming that we know the value of the solution vector $\{X_n\}$, corresponding to some time instant t_n, and we will then be proceeding to obtain the solution for the next step, $\{X_{n+1}\}$, corresponding to some new value of time, t_{n+1}. This is well tailored for the problem at hand: Since we have the initial conditions (i.e., we know the values of $\{X\}$ at $t = t_0 = 0$), we can use the algorithm to find the solution at $t = t_1$, then use the algorithm again to find the solution at $t = t_2$, ... etc. The equations that we will be solving to find the solution of a step (given the solution of the previous step) are called the *update equations*.

The step-by-step algorithms described below are broadly classified into two basic categories, namely, *explicit* and *implicit*. In explicit algorithms, if we know the solution at an instant $t = t_i$, then the solution at the next time instant, that is, at $t = t_{i+1}$, can be directly obtained from known quantities related to t_i. On the other hand, for implicit algorithms, the nodal solution $\{X\}$ and its rate $\{\dot{X}\}$ at time t_{i+1} are coupled, and we need to enforce the governing equation at t_{i+1} to obtain the solution of the next step. While the computations for one step in explicit algorithms are generally faster than those of implicit ones, explicit algorithms introduce limitations to the size of time increment that can be used in the analysis, as explained in Section 16.5.

Before proceeding with the solution algorithms of the following section, we need to remember an important theorem of mathematical analysis, namely, the *mean value theorem*, as applied to the solution vector $\{X\}$ of the parabolic problem:

If we have the vectors $\{X_n\}$ and $\{X_{n+1}\}$, corresponding to the solution at instants t_n and t_{n+1}, then there is some time value t_{int}, with $t_n \le t_{int} \le t_{n+1}$, such that:

$$\{\dot{X}\}\big|_{t=t_{int}} = \frac{1}{\Delta t}(\{X_{n+1}\} - \{X_n\}) \tag{16.1.6}$$

where $\Delta t = t_{n+1} - t_n$ is called the *time-step* of the step-by-step algorithm.

If we premultiply Equation (16.1.6) by [C] and account for Equation (16.1.2), we obtain:

$$[C]\frac{1}{\Delta t}(\{X_{n+1}\}-\{X_n\})=[C]\{\dot{X}\}|_{t=t_{int}}=\{R\}|_{t=t_{int}} \quad (16.1.7)$$

where, by virtue of Equation (16.1.3), we have:
$\{R\}|_{t=t_{int}}=\{R(t=t_{int})\}=\{f(t=t_{int})\}-[K]\{X(t=t_{int})\}$

Equation (16.1.7) does not really help us to obtain the updated solution $\{X_{n+1}\}$, since we do not know what is the actual value of t_{int}! Still, the specific equation facilitates our discussion, in the sense that the popular family of *single-step* algorithms, described in Section 16.2, can be obtained by assuming that $\{R\}|_{t=t_{\text{int}}}$ can be expressed in terms of quantities that we can calculate. For example, we can assume that $\{R\}|_{t=t_{\text{int}}}$ is a linear combination of the vectors $\{R_n\}=\{f_n\}-[K]\{X_n\}$ and $\{R_{n+1}\}=\{f_{n+1}\}-[K]\{X_{n+1}\}$.

16.2 Single-Step Algorithms

This section will focus on *single-step transient algorithms,* which stem from Equation (16.1.7), after assuming the value of $\{R\}|_{t=t_{\text{int}}}$. The following methods can be established, depending on our assumption for $\{R\}|_{t=t_{\text{int}}}$:

- Forward Euler: $\{R\}|_{t=t_{\text{int}}}=\{R_n\}=\{f_n\}-[K]\{X_n\}$
- Backward Euler: $\{R\}|_{t=t_{\text{int}}}=\{R_{n+1}\}=\{f_{n+1}\}-[K]\{X_{n+1}\}$
- Trapezoidal (also called Crank-Nicolson method): $\{R\}|_{t=t_{\text{int}}}=\frac{1}{2}(\{R_n\}+\{R_{n+1}\})$,
- Generalized trapezoidal: $\{R\}|_{t=t_{\text{int}}}=(1-\theta)\{R_n\}+\theta\{R_{n+1}\}$, $0\le\theta\le 1$,
- Midpoint: $\{R\}|_{t=t_{\text{int}}}=\{R_{n+1/2}\}=\{f_{n+1/2}\}-[K]\{X_{n+1/2}\}$, where $\{f_{n+1/2}\}=\{f(t=t_{n+1/2})\}$, $\{X_{n+1/2}\}=\{X(t=t_{n+1/2})\}$, and $t_{n+1/2}=\frac{1}{2}(t_n+t_{n+1})$
- Generalized midpoint: $\{R\}|_{t=t_{\text{int}}}=\{R_{n+\theta}\}=\{f_{n+\theta}\}-[K]\{X_{n+\theta}\}$, where $\{f_{n+\theta}\}=\{f(t=t_{n+\theta})\}$, $\{X_{n+\theta}\}=\{X(t=t_{n+\theta})\}$, and $t_{n+\theta}=(1-\theta)t_n+\theta t_{n+1}$, $0\le\theta\le 1$

Once we have assumed the value of $\{R\}|_{t=t_{\text{int}}}$, we can actually establish the *time-stepping* algorithm, which yields the updated value of the solution, $\{X_{n+1}\}$, if we know the solution of the previous step, $\{X_n\}$. For example, the adoption of the generalized trapezoidal rule leads to the following expression.

$$[C]\frac{1}{\Delta t}(\{X_{n+1}\}-\{X_n\})=(1-\theta)\{R_n\}+\theta\{R_{n+1}\}\rightarrow$$

$$\rightarrow[C]\frac{1}{\Delta t}(\{X_{n+1}\}-\{X_n\})=(1-\theta)\{f_n\}-(1-\theta)[K]\{X_n\}+\theta\{f_{n+1}\}-\theta\cdot[K]\{X_{n+1}\}$$

$$\rightarrow\left(\frac{1}{\Delta t}[C]+\theta\cdot[K]\right)\{X_{n+1}\}=\left(\frac{1}{\Delta t}[C]-(1-\theta)[K]\right)\{X_n\}+(1-\theta)\{f_n\}+\theta\{f_{n+1}\} \quad (16.2.1)$$

If we multiply both sides of Equation (16.2.1) by the time-step, Δt, we obtain:

$$([C]+\theta\cdot\Delta t\cdot[K])\{X_{n+1}\}=([C]-(1-\theta)\cdot\Delta t\cdot[K])\{X_n\}+\Delta t\cdot[(1-\theta)\{f_n\}+\theta\{f_{n+1}\}] \quad (16.2.2)$$

Remark 16.2.1: Setting $\theta = 0$ in the generalized trapezoidal or generalized midpoint method gives the forward Euler method. Similarly, setting $\theta = 1$ gives the backward Euler method. ∎

Remark 16.2.2: For practical reasons, for the cases where $\theta \neq 0$, Equation (16.2.2) can be cast in a computationally more convenient form, which is provided in Hughes (2000). Specifically, we begin by writing the governing equations for time instants t_n and t_{n+1}:

$$[C]\{\dot{X}_n\} + [K]\{X_n\} = \{f_n\} \tag{16.2.3a}$$

$$[C]\{\dot{X}_{n+1}\} + [K]\{X_{n+1}\} = \{f_{n+1}\} \tag{16.2.3b}$$

If we multiply Equation (16.2.3a) by $(1-\theta)$, then multiply Equation (16.2.3b) by θ, then add the two resulting expressions, and also account for Equation (16.1.7), we can obtain:

$$\{\dot{X}\}\big|_{t=t_{int}} = (1-\theta)\{\dot{X}_n\} + \theta\{\dot{X}_{n+1}\} \tag{16.2.4}$$

Then, we can solve Equation (16.2.4) for $\{\dot{X}_{n+1}\}$, to obtain:

$$\{\dot{X}_{n+1}\} = \frac{1}{\theta}\{\dot{X}\}\big|_{t=t_{int}} - \frac{(1-\theta)}{\theta}\{\dot{X}_n\} \tag{16.2.5}$$

Plugging Equation (16.2.5) into Equation (16.2.3b) gives (after some mathematical manipulations) an alternative version of the equation that we can use to find the solution of the new step:

$$\left(\frac{1}{\theta \cdot \Delta t}[C] + [K]\right)\{X_{n+1}\} = \{f_{n+1}\} + [C]\frac{1}{\theta \cdot \Delta t}\{X_n\} + [C]\frac{(1-\theta)}{\theta}\{\dot{X}_n\} \tag{16.2.6}$$

Equation (16.2.6) can also be written as:

$$[\tilde{K}]\{X_{n+1}\} = \{\tilde{f}_{n+1}\} \tag{16.2.7}$$

where

$$[\tilde{K}] = \frac{1}{\theta \cdot \Delta t}[C] + [K] \tag{16.2.8a}$$

and

$$\{\tilde{f}_{n+1}\} = \{f_{n+1}\} + [C]\frac{1}{\theta \cdot \Delta t}\{X_n\} + [C]\frac{(1-\theta)}{\theta}\{\dot{X}_n\} \tag{16.2.8b}$$

The version of the generalized trapezoidal rule given by Equation (16.2.6) is much more convenient for implementation in analysis (especially for nonlinear time-dependent scalar field problems, which are outside the scope of this text). ∎

Example 16.1: Application of Single-Step Algorithms

We have a finite element model of a parabolic problem with two degrees of freedom, $\{X(t)\} = \begin{Bmatrix} X_1(t) \\ X_2(t) \end{Bmatrix}$. The coefficient arrays are $[K] = \begin{bmatrix} 100 & -50 \\ -50 & 50 \end{bmatrix}$, $[C] = \begin{bmatrix} 1000 & 0 \\ 0 & 500 \end{bmatrix}$.

We are given that, for time instant $t_n = 2$ seconds: $\{X_n\} = \begin{Bmatrix} 3 \\ 4 \end{Bmatrix}$, $\{\dot{X}_n\} = \begin{Bmatrix} -0.2 \\ 0.5 \end{Bmatrix}$, and $\{f_n\} = \begin{Bmatrix} 10 \\ 40 \end{Bmatrix}$. We want to determine the values of $\{X_{n+1}\}$ for time $t_{n+1} = 2.1$ seconds, using the forward Euler method and the Crank-Nicolson method. We also know that, at $t = 2.1$ seconds, $\{f_{n+1}\} = \begin{Bmatrix} 0 \\ 50 \end{Bmatrix}$.

We need to use the two methods to conduct a single-step update, for $\Delta t = 2.1 - 2 = 0.1$ seconds. We proceed with the update for the two methods.

Forward Euler Method

1) Find $\{R\}|_{t=t_{mean}} = \{R_n\} = \{f_n\} - [K]\{X_n\} = \begin{Bmatrix} 10 \\ 40 \end{Bmatrix} - \begin{bmatrix} 100 & -50 \\ -50 & 50 \end{bmatrix} \begin{Bmatrix} 3 \\ 4 \end{Bmatrix} = \begin{Bmatrix} -90 \\ -10 \end{Bmatrix}$

2) Find rate $\{\dot{X}\}|_{t=t_{int}} = \{\dot{X}_n\}$ (for forward Euler), solving Equation (16.1.7):

$$[C]\{\dot{X}_n\} = \{R_n\}| \rightarrow \{\dot{X}_n\} = [C]^{-1}\{R_n\} = \begin{Bmatrix} -0.09 \\ -0.02 \end{Bmatrix}$$

3) Set $\{\dot{X}\}|_{t=t_{int}} = \{\dot{X}_n\}$ in Equation (16.1.6), then solve for $\{X_{n+1}\}$:

$$\{X_{n+1}\} = \Delta t \cdot \{\dot{X}\}|_{t=t_{int}} + \{X_n\} = 0.1\begin{Bmatrix} -0.09 \\ -0.02 \end{Bmatrix} + \begin{Bmatrix} 3 \\ 4 \end{Bmatrix} = \begin{Bmatrix} 2.991 \\ 3.998 \end{Bmatrix}$$

Crank-Nicolson Method

We have to use the generalized midpoint rule for $\theta = 0.5$. We can directly use Equation (16.2.2):

$$([C] + \theta \cdot \Delta t \cdot [K])\{X_{n+1}\} = ([C] - (1-\theta) \cdot \Delta t \cdot [K])\{X_n\} + \Delta t \cdot [(1-\theta)\{f_n\} + \theta\{f_{n+1}\}]$$

$$\left(\begin{bmatrix} 1000 & 0 \\ 0 & 500 \end{bmatrix} + 0.5 \cdot 0.1 \cdot \begin{bmatrix} 100 & -50 \\ -50 & 50 \end{bmatrix}\right)\{X_{n+1}\}$$

$$= \left(\begin{bmatrix} 1000 & 0 \\ 0 & 500 \end{bmatrix} - (1-0.5) \cdot 0.1 \cdot \begin{bmatrix} 100 & -50 \\ -50 & 50 \end{bmatrix}\right)\begin{Bmatrix} 3 \\ 4 \end{Bmatrix}$$

$$+ 0.1 \cdot \left[(1-0.5)\begin{Bmatrix} 10 \\ 40 \end{Bmatrix} + 0.5\begin{Bmatrix} 0 \\ 50 \end{Bmatrix}\right] \rightarrow$$

$$\begin{bmatrix} 1005 & -2.5 \\ -2.5 & 502.5 \end{bmatrix}\{X_{n+1}\} = \begin{Bmatrix} 2995.5 \\ 2002 \end{Bmatrix} \rightarrow$$

$$\{X_{n+1}\} = \begin{bmatrix} 1005 & -2.5 \\ -2.5 & 502.5 \end{bmatrix}^{-1} \begin{Bmatrix} 2995.5 \\ 2002 \end{Bmatrix} = \begin{Bmatrix} 2995.5 \\ 2002 \end{Bmatrix}$$ ■

We can now establish a generic algorithm for the use of the generalized trapezoidal method, as summarized in Box 16.2.1. A similar algorithm can be established for other single-step algorithms, such as the generalized midpoint scheme. Several steps in the algorithm of Box 16.2.1 involve division by θ; this means that Box 16.2.1 cannot be used for the forward Euler method (i.e., for $\theta = 0$). A different algorithm, provided in Box 16.2.2, must be used for the forward Euler method.

Box 16.2.1 Generalized Trapezoidal Rule ($\theta \neq 0$) for Step-by-Step Solution of Parabolic Problems

0) Initial conditions: Set initial values vector, $\{X_0\}$

   Then find initial rate of nodal solution, $\{\dot{X}_0\}$, by solving: $[C]\{\dot{X}_0\} = \{f_0\} - [K]\{X_0\}$

For each time-step *n*:

1) Given the solution vector $\{X_n\}$ and rate vector, $\{\dot{X}_n\}$:

   For each element, *e*:

   Find the element nodal field and nodal rate vectors:

$$\left\{X_n^{(e)}\right\} = \left[L^{(e)}\right]\{X_n\},\ \left\{\dot{X}_n^{(e)}\right\} = \left[L^{(e)}\right]\{\dot{X}_n\}$$

   Calculate the element contributions to the right-hand side of Equation (16.2.7):

$$\left\{\tilde{f}_{n+1}^{(e)}\right\} = \left\{f_{n+1}^{(e)}\right\} + [C]\frac{1}{\theta \cdot \Delta t}\left\{X_n^{(e)}\right\} + [C]\frac{(1-\theta)}{\theta}\left\{\dot{X}_n^{(e)}\right\}$$

   Calculate the element contributions to the coefficient array $\left[\tilde{K}\right]$:

$$\left[\tilde{k}^{(e)}\right] = \frac{1}{\theta \cdot \Delta t}\left[c^{(e)}\right] + \left[k^{(e)}\right]$$

2) Assemble the global "effective" right-hand-side vector and the global "effective coefficient matrix" of Equation (16.2.7):

$$\left\{\tilde{f}_{n+1}\right\} = \sum_{e=1}^{Ne}\left(\left[L^{(e)}\right]^T\left\{\tilde{f}_{n+1}^{(e)}\right\}\right)$$

$$\left[\tilde{K}\right] = \sum_{e=1}^{Ne}\left(\left[L^{(e)}\right]^T\left[\tilde{k}^{(e)}\right]\left[L^{(e)}\right]\right)$$

   (note: if Δt is constant, then we can assemble $\left[\tilde{K}\right]$ only once in the analysis!).

3) Update the solution vector (i.e., find the vector values of the new step, $n+1$):

$$\left[\tilde{K}\right]\{X_{n+1}\} = \left\{\tilde{f}_{n+1}\right\} \rightarrow \{X_{n+1}\} = \left[\tilde{K}\right]^{-1}\left\{\tilde{f}_{n+1}\right\}$$

$$\{\dot{X}_{n+1}\} = \frac{1}{\theta}\{\dot{X}\}\big|_{t=t_{int}} - \frac{(1-\theta)}{\theta}\{\dot{X}_n\} = \frac{1}{\theta \cdot \Delta t}(\{X_{n+1}\} - \{X_n\}) - \frac{(1-\theta)}{\theta}\{\dot{X}_n\}$$

4) If we have run all the time-steps we needed to run, terminate the algorithm. Otherwise, set $n \leftarrow n+1$ and go to Step 1.

Box 16.2.2 Forward Euler Method (generalized trapezoidal rule with $\theta = 0$) for Step-By-Step Solution of Parabolic Problems

0) Initial conditions: Set initial values vector, $\{X_0\}$

For each time-step *n*:

1) Given the solution vector $\{X_n\}$:
   For each element, *e*:
   Find the element nodal field vectors:

$$\left\{X_n^{(e)}\right\} = \left[L^{(e)}\right]\{X_n\}$$

   Calculate the element contributions to the global vector $\{R_n\}$:

$$\left\{R_n^{(e)}\right\} = \left\{f_n^{(e)}\right\} - \left[k^{(e)}\right]\left\{X_n^{(e)}\right\}$$

   Calculate the element $[c^{(e)}]$ array.
2) Assemble the global $\{R_n\}$ vector and $[C]$ array:

$$\{R_n\} = \sum_{e=1}^{Ne}\left(\left[L^{(e)}\right]^T\left\{R_{n+1}^{(e)}\right\}\right)$$

$$[C] = \sum_{e=1}^{Ne}\left(\left[L^{(e)}\right]^T\left[c^{(e)}\right]\left[L^{(e)}\right]\right)$$

3) Calculate the rate vector $\left\{\dot{X}_n\right\}$, which, for the forward Euler method, is equal to $\left\{\dot{X}\right\}\big|_{t=t_{int}}$. Using Equations (16.1.6) and (16.1.7):

$$[C]\left\{\dot{X}_n\right\} = \{R_n\} \rightarrow \left\{\dot{X}_n\right\} = [C]^{-1}\{R_n\}$$

4) Find the solution vector $\{X_{n+1}\}$ of the new step: $\{X_{n+1}\} = \{X_n\} + \Delta t \cdot \left\{\dot{X}_n\right\}$.
5) If we have run all the time-steps we needed to run, terminate the algorithm. Otherwise, set $n \leftarrow n + 1$ and go to Step 1.

Example 16.2: Analysis of time-dependent heat conduction

To illustrate some aspects of the physical behavior of time-dependent scalar-field problems (as obtained from a finite element solution), we are now going to analyze the heat conduction problem presented in Figure 16.1a. We have a three-element mesh (all elements have identical dimensions), representing a thin plate with $\kappa = 50\ \mathrm{W \cdot m^{-1} \cdot {}^\circ C^{-1}}$, $c = 500\ \mathrm{J \cdot kg^{-1} \cdot {}^\circ C^{-1}}$, $\rho = 2500\ \mathrm{kg/m^3}$ and a thickness of 0.1 m. We will solve the problem, for the case that the value of the prescribed boundary heat flow, $\bar{q}$, changes linearly with time (from zero to a value of $-4000\ \mathrm{W/m^2}$) up to a time of $t = 1$ second, then retains a constant value of $-4000\ \mathrm{W/m^2}$ throughout the remainder of the analysis, as also shown in Figure 16.1b. We will use the generalized trapezoidal rule (i.e., the algorithm of Box 16.2.1) with $\theta = 0.5$ for the solution of the problem.

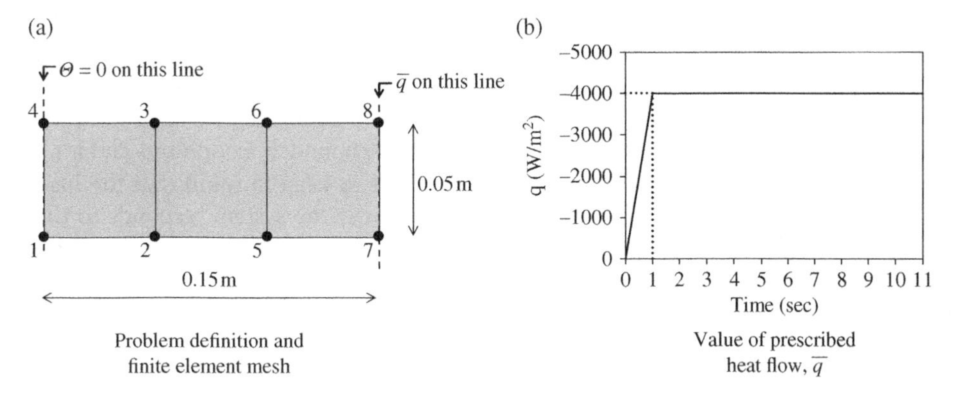

Figure 16.1 Example time-dependent heat conduction problem.

The obtained temperature contour plot at two different time instants ($t = 1$ second and 500 seconds) is presented in Figure 16.2a. The heat flux vector plot is shown in Figure 16.2b (a qualitatively identical heat flux vector plot is obtained at all instants for the specific problem—only the magnitude of the vectors changes). As expected, we have a heat flow from the boundary with the prescribed heat inflow toward the

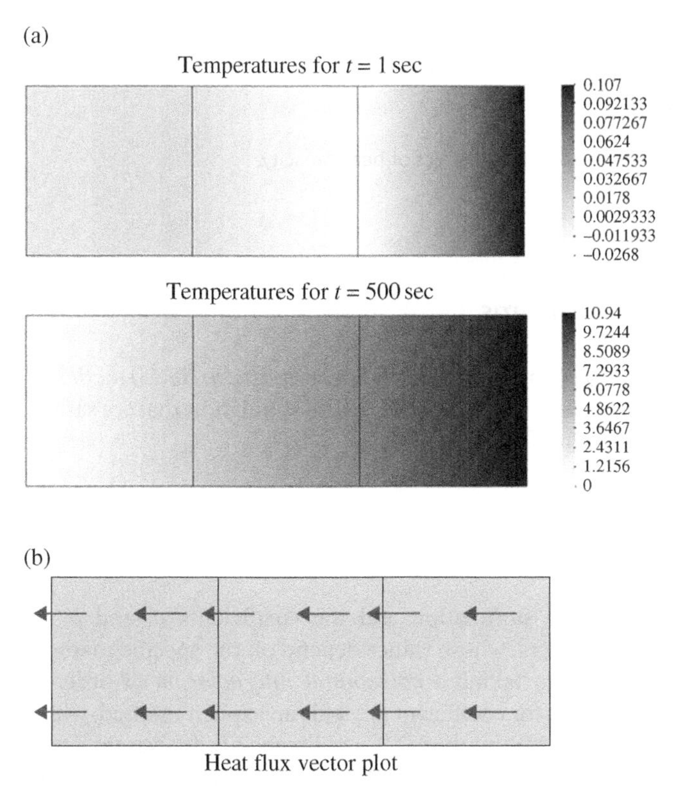

Figure 16.2 Solution for example time-dependent heat conduction problem.

two nodal points, 1 and 4, that have a prescribed temperature equal to zero. Finally, Figure 16.3 shows the change (with time) of the temperature at node 7. It can be seen that the temperature increases, until it becomes equal to 12°C, which is the *steady-state* temperature at node 7 (i.e., the value that would have been obtained if we had solved the problem for steady-state heat conduction, with the same boundary conditions and for a constant $\bar{q} = -400\ W/m^2$)! Additionally, it is important to keep in mind that the heat capacity c is a measure of *thermal inertia* (i.e., how slowly the system responds to the prescribed boundary conditions). To demonstrate this effect, we repeat the analysis for values of $c = 250$ and $c = 100$. As also shown in Figure 16.3, the use of smaller heat capacity yields a system with *lower thermal inertia*, which gives the steady-state solution temperatures faster than the case with $c = 500$. ∎

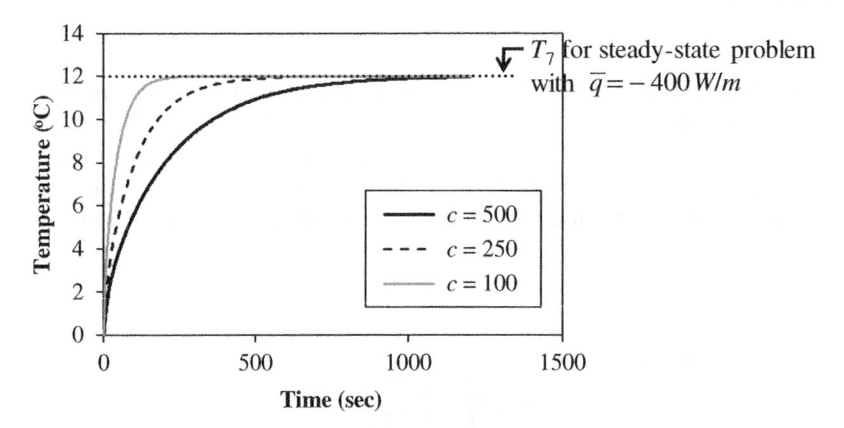

Figure 16.3 Evolution of temperature at node 7 and effect of heat capacity.

16.3 Linear Multistep Algorithms

We will now examine a different type of methods, called *linear multistep (LMS) methods*. In these methods, the time-stepping algorithm is based on the stipulation that we satisfy a polynomial equation of the following form:

$$\sum_{i=0}^{p} (a_i[C]\{X_{n+1-i}\} + \beta_i\{R_{n+1-i}\}) = \{0\} \tag{16.3.1}$$

where the number of terms p in the summation and the coefficients α_i and β_i of Equation (16.3.1) are constant parameters, whose values depend on the specific method used. Equation (16.3.1) is equivalent to taking a *polynomial interpolation* of order p between $(p + 1)$ time-steps. A zero value for coefficient β_0 yields an explicit method, while a nonzero value of β_0 gives implicit LMS methods. We will specifically examine two families of LMS methods, namely, the explicit Adams-Bashforth (AB) and the implicit Adams-Moulton (AM) methods.

16.3.1 Adams-Bashforth (AB) Methods

The p-order, explicit Adams-Bashforth (AB) method uses a (p − 1)-th degree polynomial interpolation through the solution at the steps t_{n-p}, t_{n-p+1}, ..., t_{n-1}, and t_n. We can establish various AB methods with their respective update equations:

- Second-order AB: $[C]\frac{1}{\Delta t}(\{X_{n+1}\}-\{X_n\})=\frac{1}{2}(-\{R_{n-1}\}+3\{R_n\})$
- Third-order AB: $[C]\frac{1}{\Delta t}(\{X_{n+1}\}-\{X_n\})=\frac{1}{12}(5\{R_{n-2}\}-16\{R_{n-1}\}+23\{R_n\})$
- Fourth-order AB:
  $$[C]\frac{1}{\Delta t}(\{X_{n+1}\}-\{X_n\})=\frac{1}{24}(-9\{R_{n-3}\}+37\{R_{n-2}\}-59\{R_{n-1}\}+55\{R_n\})$$
- Fifth-order AB:
  $$[C]\frac{1}{\Delta t}(\{X_{n+1}\}-\{X_n\})=\frac{1}{720}(251\{R_{n-4}\}-1274\{R_{n-3}\}+2616\{R_{n-2}\}-2774\{R_{n-1}\}+1901\{R_n\})$$

16.3.2 Adams-Moulton (AM) Methods

The p-order, implicit Adams-Moulton (AM) method uses a (p − 1)-th degree polynomial interpolation through the solution at the steps t_{n-p+1}, t_{n-p+2}, ..., t_n, and t_{n+1}. As explained in Pozrikidis (2009), the first-order AM method is the backward Euler method, while the second-order AM method is the Crank-Nicolson one. Higher-order AM methods yield the following update equations.

- Third-order AM: $[C]\frac{1}{\Delta t}(\{X_{n+1}\}-\{X_n\})=\frac{1}{12}(-\{R_{n-1}\}+8\{R_n\}+5\{R_{n+1}\})$
- Fourth-order AM: $[C]\frac{1}{\Delta t}(\{X_{n+1}\}-\{X_n\})=\frac{1}{24}(\{R_{n-2}\}-5\{R_{n-1}\}+19\{R_n\}+9\{R_{n+1}\})$
- Fifth-order AM:
  $$[C]\frac{1}{\Delta t}(\{X_{n+1}\}-\{X_n\})=\frac{1}{720}(-19\{R_{n-3}\}+106\{R_{n-2}\}-264\{R_{n-1}\}+646\{R_n\}+251\{R_{n+1}\})$$

Remark 16.3.1: The reason why we use the terms "second order," "third order," etc., will become apparent in Section 16.5.2. ∎

16.4 Predictor-Corrector Algorithms—Runge-Kutta (RK) Methods

Another popular family of methods for transient integration of parabolic problems is that of the Runge-Kutta (RK) methods, which constitute a subset of so-called *predictor-corrector schemes*. In these methods, we initially make one or more "trial" (predictor) calculations for the updated solution. The actual updated solution, $\{X_{n+1}\}$, is finally obtained using a weighted average $\{R\}$ as the right-hand-side vector. The weighted averaging is conducted using the $\{R\}$-vector values of the previous step and of the predictor steps. The final calculation to yield $\{X_{n+1}\}$ is called the *corrector* of the updated solution. From a conceptual point of view, we use the predictor solutions to obtain an understanding of how the solution "wants to change" for our step, then we exploit this understanding for the calculation of the actual updated solution. The RK methods considered herein are all explicit.

- Second-order (RK2): The update equation is:

$$[C]\frac{1}{\Delta t}(\{X_{n+1}\}-\{X_n\})=\left[(1-a)\{R_n\}+a\{R_{tmp}\}\right]$$

We need to find the predictor solution $\{X_{tmp}\}$, obtained from the following equation: $[C]\frac{1}{\kappa\cdot\Delta t}(\{X_{tmp}\}-\{X_n\})=\{R_n\}$, and then calculate:

$$\{R_{tmp}\}=\{f(t=t_n+\kappa\cdot\Delta t)\}-[K]\{X_{tmp}\}$$

The two constants, a and κ, in the update equations of the RK2 method, must satisfy the condition $2a\cdot\kappa=1$.

There are multiple alternative combinations of a and κ that we can use. The standard choice—which yields the so-called *modified Euler method*—corresponds to $a=0.5$, $\kappa=1$. Other options are $a=1$, $\kappa=0.5$ (*midpoint RK2 method*) and $a=3/4$, $\kappa=2/3$ (*Heun method*).

- Third-order (RK3): The update equation is:

$$[C]\frac{1}{\Delta t}(\{X_{n+1}\}-\{X_n\})=\left[(1-a-\beta)\{R_n\}+a\{R_{tmp1}\}+\beta\{R_{tmp2}\}\right]$$

To find $\{R_{tmp1}\}$, $\{R_{tmp2}\}$, we need to conduct the following computations:

a) Find $\{X_{tmp1}\}$: $[C]\frac{1}{\kappa\cdot\Delta t}(\{X_{tmp1}\}-\{X_n\})=\{R_n\}$.

b) Calculate $\{R_{tmp1}\}=\{f(t=t_n+\kappa\cdot\Delta t)\}-[K]\{X_{tmp1}\}$.

c) Find $\{X_{tmp2}\}$: $[C]\frac{1}{\kappa\cdot\Delta t}(\{X_{tmp2}\}-\{X_n\})=\left[(\lambda-\mu)\{R_n\}+\mu\{R_{tmp1}\}\right]$.

d) Then, calculate $\{R_{tmp2}\}=\{f(t=t_n+\lambda\cdot\Delta t)\}-[K]\{X_{tmp2}\}$.

In summary, the RK3 method requires the user to provide the values of five constant parameters, namely, a, β, κ, λ, and μ. As explained in detail in Pozrikidis (2009), the following conditions must be satisfied:

$$\alpha\cdot\kappa+\beta\cdot\lambda=\frac{1}{2},\ \alpha\cdot\kappa^2+\beta\cdot\lambda^2=\frac{1}{3},\ \beta\cdot\kappa\cdot\mu=\frac{1}{6}$$

We can set the value for two of the parameters (the best approach is to set the values for κ and λ) and then obtain the other three parameters, using the three conditions. In the standard RK3 method, we have $\alpha=2/3$, $\beta=1/6$, $\kappa=1/2$, $\lambda=1$, $\mu=2$. An alternative selection of parameters, which yields the *Runge-Kutta-Wray (RKW3)* method, is $\alpha=0$, $\beta=3/4$, $\kappa=8/15$, $\lambda=2/3$, $\mu=5/12$.

- Fourth-order (RK4): The update equation is:

$$[C]\frac{1}{\Delta t}(\{X_{n+1}\}-\{X_n\})=\frac{1}{6}\left(\{R_n\}+2\{R_{tmp1}\}+2\{R_{tmp2}\}+\{R_{tmp3}\}\right)$$

To find $\{R_{tmp1}\}$, $\{R_{tmp2}\}$, $\{R_{tmp3}\}$, we conduct the following sequence of computations:

a) Find $\{X_{tmp1}\}$: $[C]\frac{1}{2\Delta t}(\{X_{tmp1}\}-\{X_n\})=\{R_n\}$.

b) Find $\{R_{tmp1}\}=\{f(t=t_n+\Delta t/2)\}-[K]\{X_{tmp1}\}$.

c) Find $\{X_{tmp2}\}$: $[C]\frac{1}{2\Delta t}\left(\{X_{tmp2}\}-\{X_n\}\right)=\{R_{tmp1}\}$.

d) Find $\{R_{tmp2}\}=\{f(t=t_n+\Delta t/2)\}-[K]\{X_{tmp2}\}$.

e) Find $\{X_{tmp3}\}$: $[C]\frac{1}{\Delta t}\left(\{X_{tmp3}\}-\{X_n\}\right)=\{R_{tmp2}\}$.

f) Find $\{R_{tmp3}\}=\{f(t=t_n+\Delta t)\}-[K]\{X_{tmp3}\}$.

As explained in Pozrikidis (2009), the RK4 method is preferred when the calculation of $\{R\}$ is not very expensive computationally.

Example 16.3: Application of predictor-corrector algorithm
We will now repeat the step update for the case considered in Example 16.1, using the RKW3 method.

i) Find $\{R_n\}=\{f_n\}-[K]\{X_n\}=\begin{Bmatrix}-90\\-10\end{Bmatrix}$ (this value was calculated in Example 16.1).

ii) Find $\{X_{tmp1}\}$: $[C]\frac{1}{\kappa\cdot\Delta t}\left(\{X_{tmp1}\}-\{X_n\}\right)=\{R_n\}\rightarrow\{X_{tmp1}\}=\kappa\cdot\Delta t\cdot[C]^{-1}\{R_n\}+\{X_n\}$

$$\{X_{tmp1}\}=\frac{8}{15}\cdot 0.1\cdot\begin{bmatrix}1000 & 0\\0 & 500\end{bmatrix}^{-1}\begin{Bmatrix}-90\\-10\end{Bmatrix}+\begin{Bmatrix}3\\4\end{Bmatrix}\rightarrow\{X_{tmp1}\}=\begin{Bmatrix}2.9952\\3.9989\end{Bmatrix}.$$

iii) Find $\{f(t=t_n+\kappa\cdot\Delta t)\}$: since we do not know how $\{f\}$ varies between the two steps n and $n+1$, the best we can do is assume that it varies linearly:

$$\{f(t=t_n+\kappa\cdot\Delta t)\}=\{f_n\}+\kappa(\{f_{n+1}\}-\{f_n\})$$
$$=\begin{Bmatrix}10\\40\end{Bmatrix}+\frac{8}{15}\left(\begin{Bmatrix}0\\50\end{Bmatrix}-\begin{Bmatrix}10\\40\end{Bmatrix}\right)=\begin{Bmatrix}4.667\\45.333\end{Bmatrix}.$$

iv) Find $\{R_{tmp1}\}=\{f(t=t_n+\kappa\cdot\Delta t)\}-[K]\{X_{tmp1}\}=$

$$\begin{Bmatrix}4.667\\45.333\end{Bmatrix}-\begin{bmatrix}100 & -50\\-50 & 50\end{bmatrix}\begin{Bmatrix}2.9952\\3.9989\end{Bmatrix}=\begin{Bmatrix}-94.91\\-4.85\end{Bmatrix}.$$

v) Find $\{X_{tmp2}\}$: $[C]\frac{1}{\kappa\cdot\Delta t}\left(\{X_{tmp2}\}-\{X_n\}\right)=\left[(\lambda-\mu)\{R_n\}+\mu\{R_{tmp1}\}\right]$

$$\rightarrow\{X_{tmp2}\}=\kappa\cdot\Delta t\cdot[C]^{-1}\left[(\lambda-\mu)\{R_n\}+\mu\{R_{tmp1}\}\right]+\{X_n\}$$

$$\rightarrow\{X_{tmp2}\}=\frac{8}{15}\cdot 0.1\cdot\begin{bmatrix}1000 & 0\\0 & 500\end{bmatrix}^{-1}\left[\left(\frac{2}{3}-\frac{5}{12}\right)\begin{Bmatrix}-90\\-10\end{Bmatrix}+\frac{5}{12}\begin{Bmatrix}-94.91\\-4.85\end{Bmatrix}\right]+\begin{Bmatrix}3\\4\end{Bmatrix}$$

$$\rightarrow\{X_{tmp2}\}=\begin{Bmatrix}2.9967\\3.9995\end{Bmatrix}.$$

vi) Find $\{f(t=t_n+\lambda\cdot\Delta t)\}=\{f_n\}+\lambda(\{f_{n+1}\}-\{f_n\})=\begin{Bmatrix}10\\40\end{Bmatrix}+\frac{2}{3}\left(\begin{Bmatrix}0\\50\end{Bmatrix}-\begin{Bmatrix}10\\40\end{Bmatrix}\right)$

$$=\begin{Bmatrix}3.333\\46.667\end{Bmatrix}.$$

vii) Calculate:

$$\{R_{tmp2}\} = \{f(t = t_n + \lambda \cdot \Delta t)\} - [K]\{X_{tmp2}\} = \begin{Bmatrix} 3.333 \\ 46.667 \end{Bmatrix} - \begin{bmatrix} 100 & -50 \\ -50 & 50 \end{bmatrix} \begin{Bmatrix} 2.9967 \\ 3.9995 \end{Bmatrix}$$

$$= \begin{Bmatrix} -96.36 \\ -3.47 \end{Bmatrix}.$$

viii) Finally, find the updated solution from the equation:

$$[C]\frac{1}{\Delta t}(\{X_{n+1}\} - \{X_n\}) = [(1-a-\beta)\{R_n\} + a\{R_{tmp1}\} + \beta\{R_{tmp2}\}] \rightarrow$$

$$\rightarrow \{X_{n+1}\} = \Delta t \cdot [C]^{-1}[(1-a-\beta)\{R_n\} + a\{R_{tmp1}\} + \beta\{R_{tmp2}\}] + \{X_n\}$$

$$\rightarrow \{X_{n+1}\} = 0.1 \cdot \begin{bmatrix} 1000 & 0 \\ 0 & 500 \end{bmatrix}^{-1} \left[\left(1 - 0 - \frac{3}{4}\right)\begin{Bmatrix} -90 \\ -10 \end{Bmatrix} + 0 \cdot \begin{Bmatrix} -94.91 \\ -4.85 \end{Bmatrix} + \frac{3}{4}\begin{Bmatrix} -96.36 \\ -3.47 \end{Bmatrix}\right] + \begin{Bmatrix} 3 \\ 4 \end{Bmatrix}$$

$$\rightarrow \{X_{n+1}\} = \begin{Bmatrix} 2.991 \\ 3.999 \end{Bmatrix}$$

■

16.5 Convergence of a Time-Stepping Algorithm

So far, we have considered various time-stepping algorithms, but we have discussed nothing about their accuracy. It is important to establish the process by which we can verify the accuracy and *convergence of a time-stepping algorithm*. To this end, this section will provide an introduction to the concept of stability for such algorithms. Subsequently, we will discuss the notions of error, order of accuracy, consistency, and the conditions for convergence.

16.5.1 Stability of Time-Stepping Algorithms

Let us consider the first-order differential equation (with time) of response for a *single-degree-of-freedom* system, that is, a scalar version of Equation (16.1.1) for a model that only has one degree of freedom:

$$C \cdot \dot{X}(t) + K \cdot X(t) = f(t) \tag{16.5.1}$$

where $C > 0$, $K \geq 0$.

We will focus on the case of *free response*, wherein we do not have any external (right-hand-side) term on the system: $f(t) = 0$. In this case we have:

$$C \cdot \dot{X}(t) + K \cdot X(t) = 0 \tag{16.5.2}$$

This is a *homogeneous differential equation* with time, for which we can obtain the general solution:

$$X(t) = C_h \cdot e^{-\lambda \cdot t} \tag{16.5.3}$$

where C_h is a constant of integration (which can be determined if we have the initial condition for the problem, that is, the value of X for $t = 0$) and

$$\lambda = \frac{K}{C} \tag{16.5.4}$$

Let us consider the scenario where we know the value of the solution at some time instant, $t = t_n$:

$$X_n = X(t = t_n) = C_h \cdot e^{-\lambda \cdot t_n} \tag{16.5.5a}$$

We can write the solution X_{n+1}, which corresponds to a new time $t = t_{n+1} = t_n + \Delta t$, as follows.

$$X_{n+1} = X(t = t_{n+1}) = C_h \cdot e^{-\lambda \cdot t_{n+1}} = C_h \cdot e^{-\lambda \cdot (t_n + \Delta t)} = C_h \cdot e^{-\lambda \cdot t_n} \cdot e^{-\lambda \cdot \Delta t} \tag{16.5.5b}$$

As mentioned in Section 16.1, the quantity Δt is called the *time-step* of our step-by-step algorithm. If we account for Equation (16.5.5a), Equation (16.5.5b) yields:

$$X_{n+1} = e^{-\lambda \cdot \Delta t} \cdot X_n \tag{16.5.6}$$

We can verify that, from the very definition of λ in Equation (16.5.4), we have: $\lambda \geq 0$. This observation, combined with Equation (16.5.6), leads to the realization that the absolute value of the updated solution X_{n+1}, can be either lower than (if $\lambda > 0$) or equal (if $\lambda = 0$) to that of X_n. The capability of an algorithm to reproduce this free response will decide the algorithm's stability. Thus, a time-stepping algorithm will be considered *stable* when, for free response (i.e., with $f = 0$), it satisfies:

$$|X_{n+1}| \leq |X_n| \tag{16.5.7}$$

An algorithm that can always satisfy inequality (16.5.7), no matter what the value of the time-step Δt is, is termed *unconditionally stable.* On the other hand, an algorithm that can only satisfy (16.5.7) for specific values of the time-step Δt, is termed *conditionally stable.*

As an example, let us consider the stability for the generalized trapezoidal rule. We saw in Section 16.2 that the update equation for this algorithm is the following:

$$(C + \theta \cdot \Delta t \cdot K) X_{n+1} = [C - (1-\theta) \cdot \Delta t \cdot K] \cdot X_n + \Delta t \cdot [(1-\theta) f_n + \theta \cdot f_{n+1}]$$

If we divide both sides by C, we obtain:

$$(1 + \theta \cdot \Delta t \cdot \lambda) X_{n+1} = [1 - (1-\theta) \cdot \Delta t \cdot \lambda] \cdot X_n + \frac{\Delta t}{C} \cdot [(1-\theta) f_n + \theta \cdot f_{n+1}]$$

We will now focus on the case where we do not have any "external effect", that is, $f(t) = 0$, which means that the terms f_n and f_{n+1} are both zero. This allows us to express the updated solution, X_{n+1}, in terms of the solution X_n as follows.

$$(1 + \theta \cdot \Delta t \cdot \lambda) X_{n+1} = [1 - (1-\theta) \cdot \Delta t \cdot \lambda] \cdot X_n \rightarrow X_{n+1} = \frac{1 - (1-\theta) \cdot \Delta t \cdot \lambda}{1 + \theta \cdot \Delta t \cdot \lambda} \cdot X_n = A \cdot X_n \tag{16.5.8}$$

The term A in Equation (16.5.8) is called the *amplification factor* of the algorithm.

Now, we need to ensure that inequality (16.5.7) is always satisfied by our algorithm. If we account for Equation (16.5.8), we can directly conclude that stability for an algorithm requires:

$$|A| \leq 1 \tag{16.5.9}$$

where A is the multiplier of X_n in an equation where the only unknown term is the left-hand side, which in turn is equal to X_{n+1}. For the generalized trapezoidal method, plugging Equation (16.5.8) into Inequality (16.5.9) yields:

$$\left|\frac{1-(1-\theta)\cdot\Delta t\cdot\lambda}{1+\theta\cdot\Delta t\cdot\lambda}\right| \leq 1 \rightarrow -1 \leq \frac{1-(1-\theta)\cdot\Delta t\cdot\lambda}{1+\theta\cdot\Delta t\cdot\lambda} \leq 1$$

We now check each of the sides of the obtained inequality separately. We begin with the right-side inequality:

$$\frac{1-(1-\theta)\cdot\Delta t\cdot\lambda}{1+\theta\cdot\Delta t\cdot\lambda} \leq 1 \rightarrow 1-(1-\theta)\cdot\Delta t\cdot\lambda \leq 1+\theta\cdot\Delta t\cdot\lambda \rightarrow -\Delta t\cdot\lambda \leq 0$$

The obtained inequality is always satisfied, since $\Delta t > 0$ and $\lambda > 0$. The left-side inequality gives:

$$-1 \leq \frac{1-(1-\theta)\cdot\Delta t\cdot\lambda}{1+\theta\cdot\Delta t\cdot\lambda} \rightarrow -1-\theta\cdot\Delta t\cdot\lambda \leq 1-(1-\theta)\cdot\Delta t\cdot\lambda \rightarrow (1-2\theta)\Delta t\cdot\lambda \leq 2 \tag{16.5.10}$$

Thus, to ensure the stability of the generalized trapezoidal method, Inequality (16.5.10) must be satisfied. Now, we will further examine the effect of parameter θ on stability:

- If $\theta \geq \frac{1}{2}$, Inequality (16.5.10) is satisfied for any possible value of Δt (because the left-hand side of the inequality is bound to be lower than or equal to zero, while the right-hand side is positive). Thus, the generalized trapezoidal method is unconditionally stable if $\theta \geq \frac{1}{2}$.
- If $\theta < \frac{1}{2}$, inequality (16.5.10) is only satisfied if $\Delta t\cdot\lambda \leq \frac{2}{1-2\theta}$. Thus, for $\theta < \frac{1}{2}$, the generalized trapezoidal method is conditionally stable; for stability, the time-step must satisfy:

$$\Delta t \leq \Delta t_{cr} = \frac{2}{\lambda(1-2\theta)} \tag{16.5.11}$$

where, for a given system, λ is known (by virtue of Equation 16.5.4).

The maximum value, Δt_{cr}, that the time-step Δt is allowed to obtain to satisfy the stability requirements of a conditionally stable algorithm is termed the *critical time-step* of the time-stepping algorithm. Inequality (16.5.11) essentially defines the critical time-step for the generalized trapezoidal rule with $\theta < \frac{1}{2}$.

16.5.2 Error, Order of Accuracy, Consistency, and Convergence

The stability of a step-by-step method is a necessary condition to ensure that a reasonable approximation of the actual function *X(t)* will be obtained through use of a time-stepping algorithm. Now, we need to define an additional measure of "how good" an approximation of $X(t)$ we obtain with the step-by-step scheme. Given that we have presented a number of alternative step-by-step methods that we can use, a "goodness" measure will allow us to quantify the relative accuracy of these methods!

Let us revisit the general problem described by Equation (16.5.2):

$$C \cdot \dot{X}(t) + K \cdot X(t) = 0$$

Using Equation (16.5.4), we can also write Equation (16.5.11) in an equivalent form:

$$\dot{X}(t) = -\lambda \cdot X(t) \tag{16.5.12}$$

We can also differentiate Equation (16.5.12) with time, to obtain:

$$\ddot{X}(t) = -\lambda \cdot \dot{X}(t) \tag{16.5.13}$$

Let us now imagine that we are using some time-stepping method, for a given value of time-step, Δt. We have the solution at step n, X_n, corresponding to time $t = t_n$, and we can use the time-stepping method to find the value X_{n+1}, for the next step, corresponding to time $t = t_{n+1} = t_n + \Delta t$. We will define the *error* of the time-stepping algorithm as follows. If the exact value of the solution at $t = t_{n+1}$ is $X_{n+1}^{(ex)}$, then the error e is given from the following equation.

$$e = X_{n+1}^{(ex)} - X_{n+1} \tag{16.5.14}$$

where

$$X_{n+1}^{(ex)} = X(t = t_{n+1}) \tag{16.5.15}$$

We will now the approach described in the textbook by Hughes (2000), and write the error e in the following form:

$$e = \Delta t \cdot \tau(t) \tag{16.5.16}$$

The term $\tau(t)$ in Equation (16.5.16) is called the *local truncation error*.

We will now demonstrate how we can obtain the *order of accuracy of a time-stepping method*. To provide specific context, we will examine the special case of the generalized trapezoidal rule, but the same exact approach that we will employ can be used for the other step-by-step schemes presented in Sections 16.2, 16.3, and 16.4.

We will first find an expression for $X_{n+1}^{(ex)}$, assuming that we know X_n, which is the solution at $t = t_n$. To this end, we express $X_{n+1}^{(ex)}$ as a Taylor series expansion about X_n:

$$X_{n+1}^{(ex)} = X_n + \Delta t \cdot \dot{X}\big|_{t=t_n} + \frac{(\Delta t)^2}{2} \cdot \ddot{X}\big|_{t=t_n} + \frac{(\Delta t)^3}{6} \cdot \dddot{X}\big|_{t=t_n} + \dots \tag{16.5.17}$$

We also have the expression (16.5.8), which gives us the updated X_{n+1} for our time-stepping scheme:

$$X_{n+1} = \frac{1-(1-\theta)\cdot \Delta t \cdot \lambda}{1+\theta \cdot \Delta t \cdot \lambda} X_n \tag{16.5.8}$$

We will now make some rearrangements to Equation (16.5.8), to make the right-hand side of the equation "as similar as possible" to that of Equation (16.5.17):

$$X_{n+1} = \frac{1-(1-\theta)\cdot\Delta t\cdot\lambda}{1+\theta\cdot\Delta t\cdot\lambda}X_n = \frac{1+\theta\cdot\Delta t\cdot\lambda-\Delta t\cdot\lambda}{1+\theta\cdot\Delta t\cdot\lambda}X_n \rightarrow X_{n+1} = X_n - \frac{\Delta t}{1+\theta\cdot\Delta t\cdot\lambda}\cdot\lambda\cdot X_n \tag{16.5.18}$$

If we set $t = t_n$ in Equation (16.5.12), we have:

$$\dot{X}\big|_{t=t_n} = -\lambda\cdot X(t=t_n) = -\lambda\cdot X_n \tag{16.5.19}$$

Now, if we substitute Equation (16.5.19) into Equation (16.5.18), we obtain:

$$X_{n+1} = X_n + \frac{1}{1+\theta\cdot\Delta t\cdot\lambda}\cdot\Delta t\cdot\dot{X}\big|_{t=t_n} \tag{16.5.20}$$

We now also work with the last term of the right-hand side of Equation (16.5.20). We have:

$$\begin{aligned}\frac{1}{1+\theta\cdot\Delta t\cdot\lambda}\cdot\Delta t\cdot\dot{X}\big|_{t=t_n} &= \frac{1+\theta\cdot\Delta t\cdot\lambda-\theta\cdot\Delta t\cdot\lambda}{1+\theta\cdot\Delta t\cdot\lambda}\cdot\Delta t\cdot\dot{X}\big|_{t=t_n}\\ &= \Delta t\cdot\dot{X}\big|_{t=t_n} - \frac{\theta}{1+\theta\cdot\Delta t\cdot\lambda}\cdot(\Delta t)^2\cdot\lambda\cdot\dot{X}\big|_{t=t_n}\end{aligned} \tag{16.5.21}$$

If we set $t = t_n$ in Equation (16.5.13) and substitute in Equation (16.5.21), we have:

$$\begin{aligned}&\frac{1}{1+\theta\cdot\Delta t\cdot\lambda}\cdot\Delta t\cdot\dot{X}\big|_{t=t_n} = \Delta t\cdot\dot{X}\big|_{t=t_n} + \frac{1}{1+\theta\cdot\Delta t\cdot\lambda}\theta\cdot(\Delta t)^2\cdot\ddot{X}\big|_{t=t_n}\\ &= \Delta t\cdot\dot{X}\big|_{t=t_n} + \frac{1+\theta\cdot\Delta t\cdot\lambda-\theta\cdot\Delta t\cdot\lambda}{1+\theta\cdot\Delta t\cdot\lambda}\theta\cdot(\Delta t)^2\cdot\ddot{X}\big|_{t=t_n}\end{aligned}$$

Rearranging the final term in the obtained expression leads to:

$$\frac{1}{1+\theta\cdot\Delta t\cdot\lambda}\cdot\Delta t\cdot\dot{X}\big|_{t=t_n} = \Delta t\cdot\dot{X}\big|_{t=t_n} + \theta\cdot(\Delta t)^2\cdot\ddot{X}\big|_{t=t_n} - \frac{1}{1+\theta\cdot\Delta t\cdot\lambda}\theta^2\cdot(\Delta t)^3\cdot\lambda\cdot\ddot{X}\big|_{t=t_n} \tag{16.5.22}$$

Finally, if we substitute Equation (16.5.22) into Equation (16.5.20), we reach the following expression.

$$X_{n+1} = X_n + \Delta t\cdot\dot{X}\big|_{t=t_n} + \theta\cdot(\Delta t)^2\cdot\ddot{X}\big|_{t=t_n} - \frac{1}{1+\theta\cdot\Delta t\cdot\lambda}\theta^2\cdot(\Delta t)^3\cdot\lambda\cdot\ddot{X}\big|_{t=t_n} \tag{16.5.23}$$

Notice that we can also modify the last term in the obtained expression; to this end, we differentiate Equation (16.5.13) with time and then set $t = t_n$:

$$-\lambda\cdot\ddot{X}\big|_{t=t_n} = \dddot{X}\big|_{t=t_n} \tag{16.5.24}$$

Now, we can plug Equation (16.5.24) into Equation (16.5.23) and subtract the resulting expression from Equation (16.5.17) to obtain the error e:

$$e = X_{n+1}^{(ex)} - X_{n+1} = \left[\frac{1}{2}-\theta\right]\cdot(\Delta t)^2\cdot\ddot{X}\big|_{t=t_n} + \left[\frac{1}{6}-\frac{\theta^2}{1+\theta\cdot\Delta t\cdot\lambda}\right]\cdot(\Delta t)^3\cdot\dddot{X}\big|_{t=t_n} + \ldots \tag{16.5.25}$$

If we now divide Equation (16.5.25) by the time-step Δt and account for Equation (16.5.16), we obtain the expression for the local truncation error $\tau(t)$:

$$\tau(t) = \left[\frac{1}{2} - \theta\right] \cdot \Delta t \cdot \ddot{X}\big|_{t=t_n} + \left[\frac{1}{6} - \frac{\theta^2}{1 + \theta \cdot \Delta t \cdot \lambda}\right] \cdot (\Delta t)^2 \cdot \dddot{X}\big|_{t=t_n} + \ldots \tag{16.5.26}$$

We can now notice that the expression for $\tau(t)$ in Equation (16.5.26) has several terms, each one of which is proportional to a power of the time-step, Δt. We will only care about the term with the smallest power for Δt. If this power is equal to some number k, we will be saying that the *order of accuracy of the time-stepping method is equal to k*. The reason why we do not care for the terms having a higher power for the time-step (i.e., the *higher-order terms*) in Equation (16.5.26) is that, for sufficiently small values of the time-step, these terms are usually much lower than the term with the smallest power for the time-step, which is called the *leading-order* term. Based on these considerations, we will always write the local error in the following form:

$$\tau(t) = C \cdot (\Delta t)^k + \ldots \tag{16.5.27}$$

where k will be the order of accuracy of the method, and C will be quantity that multiplies the power of Δt in the leading-order term (the exact value of C is not important).

If we account for the definition of order of accuracy and Equation (16.5.26), we can deduce that the order of accuracy for the generalized trapezoidal method is equal to 1 if $\theta \neq \frac{1}{2}$. If $\theta = \frac{1}{2}$, then the order of accuracy of the specific method is equal to 2. The order of accuracy for other methods can also be obtained using the same conceptual procedure.

We need to establish an additional necessary definition: a time-stepping method is called *consistent*, if we have $k > 0$ for the exponent k of the local truncation error in Equation (16.5.27).

We can now proceed and establish the two conditions for a *convergent time-stepping scheme*, in the sense that, as we reduce the value of the time-step Δt, the values of a function of time obtained with the scheme get closer to those of the exact solution. In other words, as Δt tends to zero, the error e defined by Equation (16.5.14) will also tend to zero. The conditions for convergence are provided in Box 16.5.1.

Box 16.5.1 Conditions for Convergence of Time-Stepping Schemes

A time-stepping algorithm is convergent, if it is stable and consistent. Some textbooks in the literature (e.g., Hughes 2000) establish the requirements in the following concise form:

Consistency + Stability = Convergence

Remark 16.5.1: One may now be able to understand that the terms "second-order," "third-order," etc. in Sections 16.3 and 16.4 were referring to the order of accuracy of the various methods. For example, a third-order method corresponds to a local truncation error with $k = 3$ in Equation (16.5.27). ∎

The methods presented in Sections 16.2, 16.3, and 16.4 are all consistent. In fact, the order of accuracy of the various methods is provided by their name. For example, the RK4 method is fourth-order accurate. The stability of the various methods is obtained with a similar rationale as that presented for the generalized trapezoidal rule. However, the exact procedure may be a bit more complicated than that for the generalized trapezoidal method. This is why we will dedicate the remainder of this section to the determination of stability of higher-order methods. We will present a general process for obtaining the critical time-step, using the AB3 method (defined in Section 16.3) as an example case. We begin by writing the update equation, and keeping the solution at step $n+1$ in the left-hand side.

$$(X_{n+1}-X_n) = -\frac{1}{12}\lambda\cdot\Delta t\cdot(5\cdot X_{n-2}-16\cdot X_{n-1}+23\cdot X_n)\rightarrow$$

$$\rightarrow X_{n+1} = \left(1-\frac{23\bar{\lambda}}{12}\right)\cdot X_n + \frac{4\bar{\lambda}}{3}\cdot X_{n-1} + \frac{5\bar{\lambda}}{12}\cdot X_{n-2} \qquad (16.5.28)$$

where we have defined

$$\bar{\lambda} = \lambda\cdot\Delta t \qquad (16.5.29)$$

We now define an *expanded solution vector*, $\{\hat{X}_n\}$, which includes all the components of solution at time-steps prior to step $n+1$, appearing in the time-marching scheme. For the AB3 method, the scheme involves the values X_n, X_{n-1} and X_{n-2} at steps n, $n-1$ and $n-2$. Thus, the expanded vector for the AB3 method is $\{\hat{X}_n\} = [X_n \; X_{n-1} \; X_{n-2}]^T$.

Next, we define the *expanded solution vector of the new step*, $\{\hat{X}_{n+1}\}$, which is obtained by adding 1 to each one of the subscripts in the components of vector $\{\hat{X}_n\}$ (this is equivalent to advancing these terms by one step in time). For the specific case of the AB3 method, we have: $\{\hat{X}_{n+1}\} = [X_{n+1} \; X_{n-1+1} \; X_{n-2+1}]^T = [X_{n+1} \; X_n \; X_{n-1}]^T$. Then, we establish a matrix expression of the form:

$$\{\hat{X}_{n+1}\} = [A]\{\hat{X}_n\} \qquad (16.5.30)$$

The matrix $[A]$ in Equation (16.5.30) will be called the *amplification matrix* of the time-stepping algorithm. For the case of the AB3 method, we have:

$$\{\hat{X}_{n+1}\} = \left\{\begin{matrix} X_{n+1} = \left(1-\frac{23\bar{\lambda}}{12}\right)\cdot X_n + \frac{4\bar{\lambda}}{3}\cdot X_{n-1} - \frac{5\bar{\lambda}}{12}\cdot X_{n-2} \\ X_n = X_n \\ X_{n-1} = X_{n-1} \end{matrix}\right\}$$

$$= \begin{bmatrix} 1-\frac{23\bar{\lambda}}{12} & \frac{4\bar{\lambda}}{3} & -\frac{5\bar{\lambda}}{12} \\ 1 & 0 & 0 \\ 0 & 1 & 0 \end{bmatrix} \begin{Bmatrix} X_n \\ X_{n-1} \\ X_{n-2} \end{Bmatrix} \qquad (16.5.31)$$

From Equations (16.5.30) and (16.5.31), we deduce that the amplification matrix for the AB3 method is given by the following expression.

$$[A] = \begin{bmatrix} 1 - \frac{23\bar{\lambda}}{12} & \frac{4\bar{\lambda}}{3} & -\frac{5\bar{\lambda}}{12} \\ 1 & 0 & 0 \\ 0 & 1 & 0 \end{bmatrix} \tag{16.5.32}$$

Remark 16.5.2: The components of the first row of the amplification matrix $[A]$ will be equal to the factors of the right-hand side of the time-stepping algorithm. For the AB3 method, this is the right-hand side of Equation (16.5.28). All the components of the remaining rows in the amplification array are equal to either 1 or 0. ∎

We will now establish a very important lemma, which is required to determine stability for time-stepping algorithms defined by Equation (16.5.31). This lemma is provided in Box 16.5.2, and relies on the concept of matrix eigenvalues, discussed in Section A.3 of Appendix A. The reader is strongly encouraged to read that section before continuing further in the present chapter.

Box 16.5.2 Stability of Time-Stepping Algorithms

Let us imagine that we have a time-stepping algorithm governed by Equation (16.5.30); that is, $\{\hat{X}_{n+1}\} = [A]\{\hat{X}_n\}$. Let us also define λ_A as the eigenvalue of $[A]$ with the maximum magnitude (for real numbers, the magnitude is equal to the absolute value). The algorithm is stable (in the sense that the magnitude of the vector $\{\hat{X}\}$ does not uncontrollably grow) if $|\lambda_A| < 1$.

Remark 16.5.3: The proof of Equation (16.5.2) can be obtained if we write matrix $[A]$ in the form of Equation (A.3.7):

$$[A] = [\Phi][\Lambda_A][\Phi]^{-1} \tag{A.3.7}$$

where $[\Phi]$ is a matrix whose columns are the eigenvectors of $[A]$, and $[\Lambda_A] = \begin{bmatrix} \lambda_{A1} & 0 & \dots & 0 \\ 0 & \lambda_{A2} & \dots & 0 \\ \vdots & \vdots & \ddots & \vdots \\ 0 & 0 & \dots & \lambda_{Ap} \end{bmatrix}$, where p is the dimension of $[A]$. Let us imagine that we use the algorithm of Equation (16.5.30) for some initial vector $\{\hat{X}_o\}$ in the first step. We have:

$$\{\hat{X}_1\} = [A]\{\hat{X}_o\} \tag{16.5.33}$$

Similarly, for the second step:

$$\{\hat{X}_2\} = [A]\{\hat{X}_1\} \tag{16.5.34}$$

If we plug Equation (16.5.33) into Equation (16.5.34):

$$\{\hat{X}_2\} = [A][A]\{\hat{X}_o\} = [A]^2\{\hat{X}_o\} \tag{16.5.35}$$

By induction, we can verify that for some step n:

$$\{\hat{X}_n\} = [A]^n\{\hat{X}_o\} \tag{16.5.36}$$

If we account for Equation (A.3.7), we have:

$$\{\hat{X}_n\} = \underbrace{[A][A][A]...[A]}_{n \text{ terms } [A]}\{\hat{X}_o\} = \overbrace{[\Phi][\Lambda_A]\underbrace{[\Phi]^{-1}[\Phi]}_{=[I]}[\Lambda_A]\underbrace{[\Phi]^{-1}[\Phi]}_{=[I]}[\Lambda_A]\underbrace{[\Phi]^{-1}...[\Phi]}_{=[I]}[\Lambda_A][\Phi]^{-1}}^{n \text{ terms } [\Phi][\Lambda][\Phi]^{-1}}\{\hat{X}_o\} = [\Phi][\Lambda_A]^n[\Phi]^{-1}\{\hat{X}_o\}$$

Thus, we obtain:

$$\{\hat{X}_n\} = [\Phi][\Lambda_A]^n[\Phi]^{-1}\{\hat{X}_o\} = [\Phi]\begin{bmatrix} (\lambda_{A1})^n & 0 & \dots & 0 \\ 0 & (\lambda_{A2})^n & \dots & 0 \\ \vdots & \vdots & \ddots & \vdots \\ 0 & 0 & \dots & (\lambda_{Ap})^n \end{bmatrix}[\Phi]^{-1}\{\hat{X}_o\} \tag{16.5.37}$$

One can verify from Equation (16.5.37) that, as the step number n increases, the nonzero components of the diagonal matrix are raised to a higher value. If the magnitude of any of the eigenvalues λ_{A1}, λ_{A2}, ..., λ_{Ap} is greater than 1, then the power of that eigenvalue will grow more and more as n increases, leading to a similar growth of $\{\hat{X}_n\}$, and our time-stepping algorithm will be unstable. ■

Using the lemma of Box 16.5.2, we can determine the critical time-step for a time-stepping method as follows: first, we find the value of $\bar{\lambda}$, $\bar{\lambda} = \bar{\lambda}_1$, for which 1 is an eigenvalue of $[A]$. Similarly, we find the value of $\bar{\lambda}$, $\bar{\lambda} = \bar{\lambda}_2$, for which -1 is an eigenvalue of $[A]$. Notice that, by virtue of Equation (16.5.29), $\bar{\lambda}$ cannot be negative (because both λ and Δt are non-negative). Thus, we only keep the positive values of $\bar{\lambda}$ as acceptable. We then define a value $\bar{\lambda}_{max}$ as follows: if both $\bar{\lambda}_1$ and $\bar{\lambda}_2$ are positive, then $\bar{\lambda}_{max}$ is the minimum of the two values. If only one of them is positive, then $\bar{\lambda}_{max}$ will be equal to that positive value. Let us clarify this process by examining the specific case of the AB3 method. We first find the value $\bar{\lambda} = \bar{\lambda}_1$ for which 1 is an eigenvalue of $[A]$. Per Remark A.3.4, this would mean that the determinant of the matrix $[A] - 1 \cdot [I]$ would be zero:

$$\det\left(\begin{bmatrix} 1 - \frac{23\bar{\lambda}}{12} - 1 & \frac{4\bar{\lambda}}{3} & -\frac{5\bar{\lambda}}{12} \\ 1 & 0-1 & 0 \\ 0 & 1 & 0-1 \end{bmatrix}\right) = 0 \rightarrow \det\left(\begin{bmatrix} -\frac{23\bar{\lambda}}{12} & \frac{4\bar{\lambda}}{3} & -\frac{5\bar{\lambda}}{12} \\ 1 & -1 & 0 \\ 0 & 1 & -1 \end{bmatrix}\right) = 0$$

$$\rightarrow -\frac{23\bar{\lambda}}{12} - \frac{4\bar{\lambda}}{3} - \frac{5\bar{\lambda}}{12} = 0 \rightarrow \bar{\lambda} = \bar{\lambda}_1 = 0$$

Similarly, we find $\bar{\lambda} = \bar{\lambda}_2$ for which -1 is an eigenvalue of $[A]$. In this case, the determinant of the matrix $[A]-(-1)\cdot[I]$ would be zero:

$$\det\left(\begin{bmatrix} 1-\dfrac{23\bar{\lambda}}{12}+1 & \dfrac{4\bar{\lambda}}{3} & -\dfrac{5\bar{\lambda}}{12} \\ 1 & 0+1 & 0 \\ 0 & 1 & 0+1 \end{bmatrix}\right) = 0 \rightarrow \det\left(\begin{bmatrix} 2-\dfrac{23\bar{\lambda}}{12} & \dfrac{4\bar{\lambda}}{3} & -\dfrac{5\bar{\lambda}}{12} \\ 1 & 1 & 0 \\ 0 & 1 & 1 \end{bmatrix}\right) = 0$$

$$\rightarrow 2-\frac{23\bar{\lambda}}{12}-\frac{4\bar{\lambda}}{3}-\frac{5\bar{\lambda}}{12} = 0 \rightarrow$$

$$\rightarrow 2-\frac{44\bar{\lambda}}{12} = 0 \rightarrow \bar{\lambda} = \bar{\lambda}_2 = \frac{12}{44}\cdot 2 \approx 0.54545$$

Out of the two values of $\bar{\lambda}$, we only keep the positive value $\bar{\lambda}_2 \approx 0.54545$ as an acceptable one, and set $\bar{\lambda}_{\max} = \bar{\lambda}_2 \approx 0.54545$. After we have obtained $\bar{\lambda}_1$ and $\bar{\lambda}_2$, we find the eigenvalues of the amplification matrix $[A]$, for a test value of $\bar{\lambda}$, $\bar{\lambda} = \bar{\lambda}_{test}$, which is lower than $\bar{\lambda}_{\max}$, that is $\bar{\lambda}_{test} < 0.54545$. For example, let us set $\bar{\lambda}_{test} = 0.5$. We then need to calculate the eigenvalues of the amplification matrix corresponding to $\bar{\lambda} = \bar{\lambda}_{test}$. In the specific case, if one solved for the eigenvalues, they would obtain: $\lambda_{A1} = -0.9239$, $\lambda_{A2} = 0.5700$, $\lambda_{A3} = -0.3956$.

If the magnitude of all the eigenvalues of $[A]$ obtained for $\bar{\lambda}_{test}$ is lower than one, then we conclude that our method is conditionally stable and that the value $\bar{\lambda}_{cr} = \lambda\cdot\Delta t_{cr}$ determines the critical time-step. For the case of the AB3 method, we can verify that all three eigenvalues obtained for $\bar{\lambda}_{test}$ have a magnitude lower than 1, thus, we can write $\bar{\lambda}_{cr} \approx 0.54545$, which means that (if we account for Equation 16.5.29):

For the AB3 method,

$$\Delta t_{cr} \approx \frac{0.54545}{\lambda} \tag{16.5.38}$$

We can now establish a concise algorithm with all the necessary steps to decide the stability of a time-stepping algorithm with an amplification matrix $[A]$. This algorithm is provided in Box 16.5.3.

Box 16.5.3 Process for Establishing Stability of Time-Stepping Algorithms (for $\lambda > 0$)

i) Find the amplification matrix, $[A]$. The values in the first row of $[A]$ will depend on the time-step parameter, $\bar{\lambda} = \lambda\cdot\Delta t$.

ii) Find the values $\bar{\lambda}_1$ and $\bar{\lambda}_2$, for which 1 and −1, respectively, are eigenvalues of $[A]$.

iii) If $\bar{\lambda}_1 \le 0$ and $\bar{\lambda}_2 \le 0$, then the time-stepping algorithm is unconditionally stable.

iv) If at least one of the values $\bar{\lambda}_1$ and $\bar{\lambda}_2$ is positive, then select a value $\bar{\lambda}_{test}$:
- If $\bar{\lambda}_1 > 0$ and $\bar{\lambda}_2 \le 0$, then we must use $\bar{\lambda}_{test} < \bar{\lambda}_1$. Also, set $\bar{\lambda}_{max} = \bar{\lambda}_1$.
- If $\bar{\lambda}_2 > 0$ and $\bar{\lambda}_1 \le 0$, then we must use $\bar{\lambda}_{test} < \bar{\lambda}_2$. Also, set $\bar{\lambda}_{max} = \bar{\lambda}_2$.

- If $\bar{\lambda}_1 > 0$ and $\bar{\lambda}_2 > 0$, then we must use a $\bar{\lambda}_{test}$ that simultaneously satisfies $\bar{\lambda}_{test} < \bar{\lambda}_1$ and $\bar{\lambda}_{test} < \bar{\lambda}_2$. Also, set $\bar{\lambda}_{max} = \min(\bar{\lambda}_1, \bar{\lambda}_2)$.
  Then, we find the matrix $[A_{test}]$ that corresponds to the value $\bar{\lambda}_{test}$, and also find the eigenvalue λ_{Atest} of $[A_{test}]$ with the maximum magnitude (λ_{Atest} may be real or complex).
- IF $|\lambda_{Atest}| \leq 1$, then the time-stepping algorithm is conditionally stable, with $\bar{\lambda}_{cr} = \bar{\lambda}_{max}$.
- IF $|\lambda_{Atest}| > 1$, then the time-stepping algorithm is unconditionally unstable, meaning that it is always unstable, for all positive values of Δt.

Remark 16.5.4: It is interesting to note that, for the Runge-Kutta family of methods, one can obtain a closed-form relation between the eigenvalue of $[A]$ with the maximum magnitude, λ_A, and the time-step parameter $\bar{\lambda}$:

- For RK2: $\lambda_A = 1 + \bar{\lambda} + \frac{\bar{\lambda}^2}{2}$.
- For RK3: $\lambda_A = 1 + \bar{\lambda} + \frac{\bar{\lambda}^2}{2} + \frac{\bar{\lambda}^3}{6}$.
- For RK4: $\lambda_A = 1 + \bar{\lambda} + \frac{\bar{\lambda}^2}{2} + \frac{\bar{\lambda}^3}{6} + \frac{\bar{\lambda}^4}{24}$.

This facilitates the computation of $\bar{\lambda}_{cr}$, without a need to resort to the procedure outlined in Box 16.5.3. ■

Remark 16.5.5: The critical time-step parameter of various linear multistep and predictor-corrector methods is provided in Table 16.1. One may notice that the critical time-step of the implicit Adams-Moulton (AM) methods is much larger than that of the explicit Adams-Bashforth (AB) methods. This pattern is typically observed for time-stepping implicit and explicit schemes. In fact, many implicit schemes used in practice (such as the Backward-Euler method) are unconditionally stable. ■

Table 16.1 Critical Time-Step Parameter, $\bar{\lambda}_{cr}$, for Various Methods.

Method	$\bar{\lambda}_{cr}$
AB3	0.5455
AB4	0.3000
AB5	0.1633
AM3	6.0000
AM4	3.0000
AM5	1.8367
RK2	2.0000
RK3	2.5127
RK4	2.7853

Example 16.4: Accuracy and stability of time-stepping schemes
Let us now examine the case of a single-degree-of-freedom system, governed by Equation (16.5.12), with initial condition $X_o = 1$ and for different values of λ. We will begin by examining the case with $\lambda = 5$. We already know the exact solution of the system, and we examine the results obtained when we use the backward Euler (BE) scheme, for values of time-step equal to 0.1 and 0.01, as shown in Figure 16.4. In the same figure, we provide the results obtained with the RK3 method and a time-step equal to 0.1.

One can verify from Figure 16.4 that the BE method does not give very accurate results when we use $\Delta t = 0.1$. The accuracy becomes much better when we use a time-step $\Delta t = 0.01$. It is worth mentioning that the use of the RK3 method with $\Delta t = 0.1$ gives very good accuracy, identical to that obtained with the BE for a time-step which is 10 times smaller. This happens because the BE method has first-order accuracy, while the RK3 method has third-order accuracy.

We will now try using the forward Euler method, that is, the generalized trapezoidal rule with $\theta = 0$, and the RK3 method, for calculations of a problem with $\lambda = 21$. The results obtained for different values of the time-step with the forward Euler method are presented in Figure 16.5a, while those obtained with the RK3 method are shown in Figure 16.5b. One can see that the use of $\Delta t = 0.1$ gives bad results for both methods. This happens because this value of time-step renders the forward Euler method unstable, while it is also close to the critical time-step of the RK3 method. Instability can occur for low values of time- step, if we have relatively large values of the parameter λ, which correspond to so-called *stiff parabolic problems* (Pozrikidis 2009). ∎

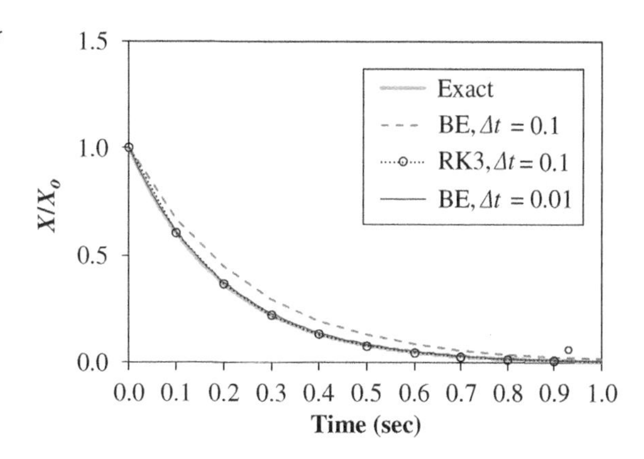

Figure 16.4 Results for backward Euler (BE) method and RK3 method, for different values of the time-step, for example analysis of single-degree-of-freedom system.

16.6 Modal Analysis and Its Use for Determining the Stability for Systems with Many Degrees of Freedom

So far, we have considered the procedure to obtain the stability of step-by-step algorithms in the context of single-degree-of-freedom system response. We now need to discuss how our considerations can be extended to multiple-degree-of-freedom finite element models (i.e., with more than one degrees of freedom). The cornerstone of

(a)

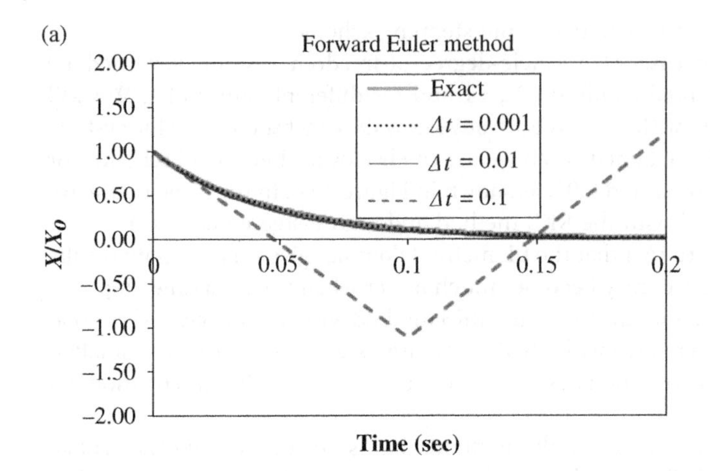

(b)

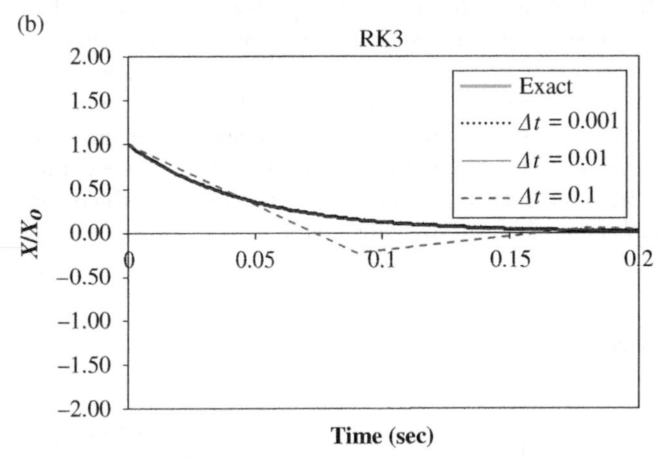

Figure 16.5 Results with forward Euler method and RK3 method, for system with $\lambda = 21$.

our discussion will be *modal analysis*. The response $\{X(t)\}$ of a semidiscretized finite element model of a parabolic problem with N degrees of freedom can be described by establishing the response vectors, $\{X\}$ and $\{\dot{X}\}$, as linear combinations of N vectors $\{\varphi_1\}$, $\{\varphi_2\}$, ..., $\{\varphi_N\}$ which are called the *modal vectors or modes*:

$$\{X(t)\} = z_1(t)\{\varphi_1\} + z_2(t)\{\varphi_2\} + \ldots + z_N(t)\{\varphi_N\} \tag{16.6.1}$$

Note that the scalar coefficients $z_i(t)$ multiplying the modal vectors in Equation (16.6.1) are functions of time. If we differentiate Equation (16.6.1) with time, we obtain an expression giving the rate of $\{X\}$ as a linear combination of the modes:

$$\{\dot{X}(t)\} = \dot{z}_1(t)\{\varphi_1\} + \dot{z}_2(t)\{\varphi_2\} + \ldots + \dot{z}_N(t)\{\varphi_N\} \tag{16.6.2}$$

The modal vectors for a system governed by Equation (16.1.1) can be determined by solving the following *eigenvalue problem*:

$$[K]\{\varphi\} = \lambda[C]\{\varphi\} \rightarrow ([K] - \lambda[C])\{\varphi\} = \{0\} \tag{16.6.3}$$

Equation (16.6.3) obviously has a trivial solution: $\{\varphi\} = \{0\}$, but because we are interested in nonzero solutions, the only way to obtain them is to set (in accordance with Remark A.3.4): $\det([K] - \lambda[C]) = 0$. If we write the complete expression for the specific determinant, we will obtain the *characteristic polynomial*, having N roots. These roots, $\lambda_1, \lambda_2, \ldots, \lambda_N$, are the eigenvalues of the semidiscretized, time-dependent, scalar-field problem described by Equation (16.1.1).

Remark 16.6.1: For the parabolic problems described in Section 15.10 and the respective finite element approximations of Section 15.11, all the eigenvalues are positive. We will also further assume that the N eigenvalues are placed in increasing order, that is:

$$0 \leq \lambda_1 \leq \lambda_2 \leq \ldots \leq \lambda_N \tag{16.6.4}$$ ■

It can be proven that *the modes are orthogonal with respect to the coefficient matrices,* $[C]$ and $[K]$. That is, they satisfy the following equations.

$$\{\varphi_i\}^T [C]\{\varphi_j\} = \begin{cases} 0, & i \neq j \\ \hat{C}_i, & i = j \end{cases} \tag{16.6.4a}$$

$$\{\varphi_i\}^T [K]\{\varphi_j\} = \begin{cases} 0, & i \neq j \\ \hat{K}_i, & i = j \end{cases} \tag{16.6.4b}$$

where

$$\hat{K}_i = \lambda_i \hat{C}_i \tag{16.6.4c}$$

If we take the global semidiscrete equations, $[C]\{\dot{X}\} + [K]\{X\} = \{f\}$, premultiply them by $\{\varphi_i\}^T$, and account for Equations (16.6.1), (16.6.2), (16.6.4a,b), we obtain:

$$\hat{C}_i \cdot \dot{z}_i(t) + \hat{K}_i \cdot z_i(t) = f_i(t) \tag{16.6.5}$$

where

$$f_i(t) = \{\varphi_i\}^T \{f\},\ i = 1, 2, \ldots, N \tag{16.6.6}$$

Equation (16.6.5) is better cast in the following form (obtained if we divide both sides by $\hat{C}_i$):

$$\dot{z}_i(t) + \lambda_i \cdot z_i(t) = \frac{f_i(t)}{\hat{C}_i} \tag{16.6.7}$$

Equation (16.6.7) is very significant, because it is revealing that the solution of Equation (16.1.1) (for a system with N degrees of freedom) is mathematically equivalent to the solution of N single-degree-of-freedom systems, each one governed by Equation (16.6.7). Although we will not discuss modal analysis for time-dependent scalar field problems in further detail, we needed the description provided in this section, to reach the stipulation of Box 16.6.1 for the stability of time-stepping algorithms in the solution of scalar field problems with multiple degrees of freedom.

Box 16.6.1 Stability of Time-Stepping Algorithms for Multi-Degree-of-Freedom Problems

A time-stepping algorithm used for the solution of a problem governed by equation (16.1.1) is stable, if it ensures stability for all the single-degree-of-freedom problems governed by equation (16.6.7). This is equivalent to stating that, for a time-stepping algorithm having a critical parameter $\bar{\lambda}_{cr}$, we must ensure that we satisfy:

$$\lambda_i \cdot \Delta t < \bar{\lambda}_{cr} \tag{16.6.8}$$

for all the eigenvalues of Equation (16.6.3). In accordance with Inequality (16.6.4), this condition is equivalent to stipulating that

$$\lambda_N \cdot \Delta t < \bar{\lambda}_{cr} \tag{16.6.9}$$

In other words, ensuring the stability of the solution requires that the product of the time-step with the maximum eigenvalue given by Equation (16.6.3) must be less than the critical parameter, $\bar{\lambda}_{cr}$, of our algorithm.

Remark 16.6.2: The solution of the eigenvalue problem, Equation (16.6.3), is computationally expensive, especially if all the eigenvalues are to be calculated. Since the stability criterion in Box 16.6.1 only requires the maximum eigenvalue λ_N, the computational cost is relatively small. Still, one could completely avoid working with the eigenvalues of the global system (governed by the arrays $[C]$ and $[K]$), and instead solve small eigenvalue problems for each element to determine the stability. Specifically, the following theorem applies:

"The maximum eigenvalue, λ_N, of a finite element model, does not exceed the maximum eigenvalue obtained from eigenvalue analysis of the individual elements":

$$\lambda_N \le \max_{i,e}\left(\lambda_i^{(e)}\right) \tag{16.6.10}$$

where $\lambda_i^{(e)}$ are the eigenvalues of each element e, obtained from solution of the following eigenvalue problem:

$$\left(\left[k^{(e)}\right] - \lambda^{(e)}\left[c^{(e)}\right]\right)\{\varphi\} = \{0\} \tag{16.6.11}$$

In summary, what we need to do is calculate the eigenvalues of each element, take the maximum eigenvalue for each element, and then use the maximum value among all the element eigenvalues obtained. Then, we use that overall maximum value in the left-hand side of Equation (16.6.9), to determine if we have stability. ∎

Remark 16.6.3: It is important to distinguish the two sets of eigenvalues presented in Sections 16.5 and 16.6. The eigenvalues of matrix $[A]$ in Section 16.5 characterize the time-stepping algorithm, while the eigenvalues of Equation (16.6.3) characterize the physical response of the semidiscretized parabolic problem. ∎

Problems

16.1 Obtain the critical time-step parameter, $\bar{\lambda}_{cr} = \lambda \cdot \Delta t_{cr}$:

a) For the AB5 method.
b) For the AM4 method.

16.2 We want to solve the semidiscrete equations of the porous flow system of Problem 15.3, and obtain the pressure time histories for nodes 2 and 3, from $t = 0$ until $t = 10$.

a) Obtain the time histories using a generalized trapezoidal rule with $\theta = 1$ and $\Delta t = 0.1$.
b) Obtain the time histories using the RK4 method, for $\Delta t = 0.05$.
c) Obtain the time histories using the AM3 method, for $\Delta t = 0.1$.

16.3 Solve the state-space formulation of Problem 15.4, and obtain the time histories of the displacement, velocity and acceleration in the y-direction for node 8, between $t = 0$ sec and $t = 10$ sec, for a generalized trapezoidal rule with $\theta = 1$, and:

a) $\Delta t = 0.5$ sec
b) $\Delta t = 0.2$ sec
c) $\Delta t = 0.5$ sec

Compare the time histories of displacement, velocity and acceleration for the three different values of time-step size.

References

Hughes, T. J. R. (2000). *The Finite Element Method—Linear Static and Dynamic Finite Element Analysis*, 2nd ed. Mineola, NY: Dover.

Pozrikidis, C. (2009). *Numerical Computation in Science and Engineering*. Oxford: Oxford University Press.

17

Solution Procedures for Elastodynamics and Structural Dynamics

17.1 Introduction

In Chapter 15, we obtained the semidiscrete equations of motion for elastodynamics and structural dynamics. A major difference of these equations compared to those of static analysis is that they include three vectors of nodal quantities instead of one, namely, the nodal acceleration vector, $\{\ddot{U}\}$, the nodal velocity vector $\{\dot{U}\}$, and the nodal displacement vector $\{U\}$. These vectors are also termed the *response vectors* for a dynamic system. The semidiscrete equations of motion for dynamics were found to have the following form.

$$[M]\{\ddot{U}(t)\} + [C]\{\dot{U}(t)\} + [K]\{U(t)\} = \{f(t)\} \quad (17.1.1)$$

where $[K]$, $[C]$, and $[M]$ are the global stiffness, damping, and mass matrices, respectively, and $\{f\}$ is the equivalent nodal force vector (which in general varies with time). All the global arrays and vector are obtained by an assembly operation of the individual element contributions.

$$[K] = \sum_{e=1}^{Ne}\left(\left[L^{(e)}\right]^T\left[k^{(e)}\right]\left[L^{(e)}\right]\right) \quad (17.1.2a)$$

$$[C] = \sum_{e=1}^{Ne}\left(\left[L^{(e)}\right]^T\left[c^{(e)}\right]\left[L^{(e)}\right]\right) \quad (17.1.2b)$$

$$[M] = \sum_{e=1}^{Ne}\left(\left[L^{(e)}\right]^T\left[m^{(e)}\right]\left[L^{(e)}\right]\right) \quad (17.1.2c)$$

$$\{f\} = \sum_{e=1}^{Ne}\left(\left[L^{(e)}\right]^T\left\{f^{(e)}\right\}\right) \quad (17.1.2d)$$

We also saw in Chapter 15 how to obtain the element damping and mass matrices for different types of solid and structural elements.

The nodal vectors $\{\ddot{U}\}$, $\{\dot{U}\}$ and $\{U\}$ are all functions of time. This means that the solution of the system will give us the accelerations, velocities and displacements for a period of time that interests us. Textbooks in structural dynamics (e.g., Chopra 2011) usually emphasize the use of *modal analysis*, also called *modal superposition method*. While this method can be used for any linear dynamic system, it will not be the main focus

Fundamentals of Finite Element Analysis: Linear Finite Element Analysis, First Edition.
Ioannis Koutromanos, James McClure, and Christopher Roy.

Companion website: www.wiley.com/go/koutromanos/linear

of this chapter. Instead, the chapter will lay emphasis on *direct time integration* of the equations of motion, where the solution is conducted in a step-by-step fashion with appropriate time-stepping schemes, just like we did for parabolic problems in Chapter 16.

This chapter will proceed based on the assumption that the reader has some knowledge of structural dynamics, especially those of single-degree-of-freedom (SDOF) systems. Detailed derivations and considerations are provided in Chopra (2011). The equations of motion for a SDOF system with mass M, damping coefficient C, and stiffness K, under applied dynamic loading $f(t)$ are:

$$M\cdot\ddot{u}(t)+C\cdot\dot{u}(t)+K\cdot u(t)=f(t) \tag{17.1.3}$$

If we divide both sides of the equation by the mass M, we obtain:

$$\ddot{u}(t)+2\xi\cdot\omega\cdot\dot{u}(t)+\omega^2\cdot u(t)=\frac{f(t)}{M} \tag{17.1.4}$$

where ω is the *natural frequency of vibration* of the system and ξ is the *damping ratio* of the system. The following equations apply.

$$\omega=\sqrt{\frac{K}{M}} \tag{17.1.5a}$$

$$\xi=\frac{C}{2M\omega}=\frac{C}{2\sqrt{M\cdot K}} \tag{17.1.5b}$$

An accompanying quantity to ω is the *natural period*, T, of the system, which is given by the expression:

$$T=\frac{2\pi}{\omega}=2\pi\sqrt{\frac{M}{K}} \tag{17.1.6}$$

The natural period is the period by which the SDOF system will conduct a periodic free vibration, that is, vibration due to initial displacement and/or velocity, without any applied force $f(t)$.

Modal analysis contributes to the understanding of several aspects of direct time integration, such as the stability of step-by-step algorithms (as presented for parabolic problems in Section 16.6 and as will become apparent in Section 17.3). For this reason, we will briefly discuss the modal superposition concept and the accompanying modal analysis procedure before continuing with the description of step-by-step algorithms, which we will also call *transient integrators.*

17.2 Modal Analysis: What Will NOT Be Presented in Detail

Just as we did in Section 16.6 for parabolic (scalar-field) problems, the solution of a discrete dynamic system with N degrees of freedom can be represented by establishing the global response vectors $\{U\}$, $\{\dot{U}\}$ and $\{\ddot{U}\}$ as linear combinations of N modal vectors (or simply modes) $\{\varphi_1\}$, $\{\varphi_2\}$, ..., $\{\varphi_N\}$:

$$\{U(t)\}=z_1(t)\{\varphi_1\}+z_2(t)\{\varphi_2\}+\ldots+z_N(t)\{\varphi_N\} \tag{17.2.1}$$

where each multiplier $z_i(t)$, $i=1, 2, \ldots, N$ is a continuous function of time.

If we take the derivative of Equation (17.2.1) with respect to time, we obtain expressions giving the velocity and acceleration vectors as linear combinations of the modes:

$$\{\dot{U}(t)\} = \dot{z}_1(t)\{\varphi_1\} + \dot{z}_2(t)\{\varphi_2\} + \ldots + \dot{z}_N(t)\{\varphi_N\} \tag{17.2.2}$$

$$\{\ddot{U}(t)\} = \ddot{z}_1(t)\{\varphi_1\} + \ddot{z}_2(t)\{\varphi_2\} + \ldots + \ddot{z}_N(t)\{\varphi_N\} \tag{17.2.3}$$

The modal vectors can be determined by solving the *eigenvalue problem for elastodynamics/structural dynamics*:

$$[K]\{\varphi\} = \omega^2[M]\{\varphi\} \rightarrow ([K] - \omega^2[M])\{\varphi\} = \{0\} \tag{17.2.4}$$

Remark 17.2.1: If we have a lumped, diagonal mass matrix (obtained from use of the procedures in Section 15.9), Equation (17.2.4) can be written as follows.

$$\begin{bmatrix} K_{11}-\omega^2 M_{11} & K_{12} & \cdots & K_{1N} \\ K_{21} & K_{22}-\omega^2 M_{22} & \cdots & K_{2N} \\ \vdots & \vdots & \ddots & \vdots \\ K_{N1} & K_{N2} & \cdots & K_{NN}-\omega^2 M_{NN} \end{bmatrix} \begin{Bmatrix} \varphi_1 \\ \varphi_2 \\ \vdots \\ \varphi_N \end{Bmatrix} = \begin{Bmatrix} 0 \\ 0 \\ \vdots \\ 0 \end{Bmatrix} \tag{17.2.5}$$

■

In accordance with Remark A.3.4, to obtain a nontrivial (nonzero) solution $\{\phi\}$ in Equation (17.2.4), we need to set:

$$det([K] - \omega^2[M]) = 0 \tag{17.2.6}$$

Equation (17.2.6) yields the *characteristic polynomial,* which can be solved for ω^2 to give N values $\omega_1, \omega_2, \ldots, \omega_N$, called the *modal frequencies or the eigen-frequencies or even the natural frequencies* of the dynamic system. For each modal frequency, ω_i, we can find the corresponding modal vector $\{\varphi_i\}$, by solving the following system of equations.

$$([K] - \omega_i^2[M])\{\varphi_i\} = \{0\} \tag{17.2.7}$$

Remark 17.2.2: The only way to solve Equation (17.2.7) for $\{\varphi_i\}$ after we have found ω_i is to prescribe the value of one of the components of $\{\varphi_i\}$, and then determine the remaining $(N - 1)$ components. For example, if we have a diagonal mass matrix and we set the first component of $\{\varphi_i\}$ equal to 1, we will have to solve:

$$\begin{bmatrix} K_{11}-\omega^2 M_{11} & K_{12} & \cdots & K_{1N} \\ K_{21} & K_{22}-\omega^2 M_{22} & \cdots & K_{2N} \\ \vdots & \vdots & \ddots & \vdots \\ K_{N1} & K_{N2} & \cdots & K_{NN}-\omega^2 M_{NN} \end{bmatrix} \begin{Bmatrix} \varphi_{i1}=1 \\ \varphi_{i2} \\ \vdots \\ \varphi_{iN} \end{Bmatrix} = \begin{Bmatrix} 0 \\ 0 \\ \vdots \\ 0 \end{Bmatrix} \tag{17.2.8}$$

Sometimes, setting the first component of $\{\varphi_i\}$ to 1 may not allow us to solve Equation (17.2.8), at which case we should try setting one of the other components, e.g., the second component of $\{\varphi_i\}$, equal to 1, etc. ■

Remark 17.2.3: It can be proven that if a system performs free vibration with $\{u(t)\} = z_i(t)\{\varphi_i\}$, then the motion will be periodic and the period of the vibration will be equal to $T_i = \frac{2\pi}{\omega_i}$. The value T_i is called the *modal period* of mode i. ▮

Solving the eigenvalue problem gives us N *eigenpairs* $(\omega_i, \{\varphi_i\})$, $i = 1, 2, \ldots, N$, with each pair containing an eigen-frequency and the corresponding mode. As explained in Section A.3, every N-dimensional vector can be expressed as a linear combination of the modes. This is what allows us to express the response vectors as linear combinations of the modes per Equations (17.2.1) to (17.2.3)!

It can be proven that *the modes are orthogonal with respect to the mass and stiffness matrices*—that is, they satisfy the following equations.

$$\{\varphi_i\}^T[M]\{\varphi_j\} = \begin{cases} 0, & i \neq j \\ \hat{M}_i, & i = j \end{cases} \tag{17.2.9a}$$

$$\{\varphi_i\}^T[K]\{\varphi_m\} = \begin{cases} 0, & i \neq j \\ \hat{K}_i, & i = j \end{cases} \tag{17.2.9b}$$

where

$$\hat{K}_i = \omega_i^2 \hat{M}_i \tag{17.2.9c}$$

The quantities $\hat{K}_i$, $\hat{M}_i$ are called the *modal stiffness and the modal mass*, respectively, for mode i.

Now, let us consider the equations of motion for an undamped system (i.e., without damping, $[C] = [0]$):

$$[M]\{\ddot{U}\} + [K]\{U\} = \{f\} \tag{17.2.10}$$

If we premultiply Equation (17.2.10) by $\{\varphi_i\}^T$, for some mode i, and account for Equations (17.2.1) to (17.2.3) and (17.2.9a, b), we obtain:

$$\hat{M}_i \cdot \ddot{z}_i(t) + \hat{K}_i \cdot z_i(t) = f_i(t) \tag{17.2.11}$$

where

$$f_i(t) = \{\varphi_i\}^T\{f\}, i = 1, 2, \ldots, N \tag{17.2.12}$$

Thus, if we express the response vectors as linear combinations of the modes, we end up with a set of N SDOF systems to analyze. The response $z_i(t)$ of each SDOF system i, $i = 1, 2, \ldots, N$, is governed by Equation (17.2.11). This is the strength of the modal superposition method for elastodynamics and structural dynamics: It allows us to *decouple* the equations governing the evolution of functions $z_i(t)$.

Remark 17.2.4: For structural dynamics problems (beams or shells), a lumped-mass assumption usually leads to zero diagonal elements in the (diagonal) mass matrix, as explained in Section 15.9.2. These zero elements correspond to the rotational degrees of freedom of the system. In such case, Equation (17.2.6) becomes *ill-defined* (because some eigenfrequencies will be equal to infinity!). This difficulty can be circumvented either by replacing the zero diagonal entries of $[M]$ with small, artificial mass terms or, more commonly, by conducting *static condensation*, in which the equations of motion are expressed in terms of the translational degrees of freedom only (the effect of the rotational degrees of freedom is indirectly accounted for). Let us imagine that we have undamped vibrations and that we have numbered the degrees of freedom (DOFs) in such a fashion, that all the translational DOFs $\{U_t\}$ are first, followed by the rotational DOFs $\{U_r\}$. Assuming that we have a lumped mass matrix that does not include any mass terms for the rotational DOFs, we can partition the global equations of motion as follows:

$$\begin{bmatrix} [M_{tt}] & [0] \\ [0] & [0] \end{bmatrix} \begin{Bmatrix} \{\ddot{U}_t\} \\ \{\ddot{U}_r\} \end{Bmatrix} + \begin{bmatrix} [K_{tt}] & [K_{tr}] \\ [K_{rt}] & [K_{rr}] \end{bmatrix} \begin{Bmatrix} \{U_t\} \\ \{U_r\} \end{Bmatrix} = \begin{Bmatrix} \{f_t\} \\ \{f_r\} \end{Bmatrix} \tag{17.2.13}$$

We essentially have two sets of equations of motion:

$$[M_{tt}]\{\ddot{U}_t\} + [K_{tt}]\{U_t\} + [K_{tr}]\{U_r\} = \{f_t\} \tag{17.2.13a}$$

and

$$[K_{rt}]\{U_t\} + [K_{rr}]\{U_r\} = \{f_r\} \tag{17.2.13b}$$

Assuming that $[K_{rr}]$ is invertible, we can "solve" (17.2.13b) for $\{U_r\}$, to obtain:

$$\{U_r\} = [K_{rr}]^{-1}\{f_r\} - [K_{rr}]^{-1}[K_{rt}]\{U_t\} \tag{17.2.14}$$

Then, if we plug Equation (17.2.14) in equation (17.2.13a), we have:

$$\begin{aligned} &[M_{tt}]\{\ddot{U}_t\} + [K_{tt}]\{U_t\} + [K_{tr}]\left([K_{rr}]^{-1}\{f_r\} - [K_{rr}]^{-1}[K_{rt}]\{U_t\}\right) = \{f_t\} \rightarrow \\ &\rightarrow [M_{tt}]\{\ddot{U}_t\} + [\bar{K}_t]\{U_t\} = \{\bar{f}_t\} \end{aligned} \tag{17.2.15}$$

where

$$[\bar{K}_t] = [K_{tt}] - [K_{tr}][K_{rr}]^{-1}[K_{rt}] \tag{17.2.16}$$

and

$$\{\bar{f}_t\} = \{f_t\} - [K_{tr}][K_{rr}]^{-1}\{f_r\} \tag{17.2.17}$$

Equation (17.2.15) describes the dynamic response through a *condensed system of equations of motion*, corresponding to the N_c translational DOFs $\{U_t\}$, for which all the diagonal entities in the mass matrix are nonzero. This yields a dynamic system for which all eigenfrequencies are finite. After that, we can use the same methods as those presented in this chapter for the solution of the condensed equations of motion. It is important to emphasize that, if we want to use the time-stepping algorithms described in Section 17.3, it is not necessary to conduct static condensation. ▌

17.2.1 Proportional Damping Matrices—Rayleigh Damping Matrix

We have seen that, for a system without damping, the modal superposition method requires the determination of the response of "N" SDOF systems. However, in the presence of damping, the problem may become challenging: the global damping matrix [C] does not generally satisfy any orthogonality conditions like those of the mass and stiffness matrix in Equations (17.2.9a,b)! To obtain a damping matrix which satisfies the orthogonality conditions, allowing us to decouple the equation giving the response of each mode from the equations of the other modes, we need to construct a *proportional* $[C]$-matrix, in the sense that $[C]$ must be equal to a sum of terms, each term being proportional to $[K]$ and/or $[M]$.

The most common type of proportional $[C]$ matrix used in practice is the *Rayleigh damping matrix*, defined from the following expression.

$$[C] = \alpha_0[M] + \alpha_1[K] \tag{17.2.18}$$

where a_0, a_1 are proportionality constants. The part $\alpha_0[M]$ of $[C]$ is called the *mass-proportional* part, while the part $\alpha_1[K]$ is called the *stiffness-proportional* part.

For a Rayleigh proportional damping matrix, we can once again verify that we can separately examine the equations of motion of the "N" SDOF systems of the model. Premultiplying Equation (17.1.1) by $\{\varphi_i\}^T$ and accounting for (17.2.1–3), (17.2.9a, b), (17.2.12), and (17.2.18) gives:

$$\hat{M}_i \cdot \ddot{z}_i(t) + \hat{C}_i \cdot \dot{z}_i(t) + \hat{K}_i \cdot z_i(t) = f_i(t) \tag{17.2.19a}$$

where

$$\hat{C}_i = \{\varphi_i\}^T [C] \{\varphi_i\} = \{\varphi_i\}^T (\alpha_0[M] + \alpha_1[K]) \{\varphi_i\} = \alpha_0 \hat{M}_i + \alpha_1 \hat{K}_i \tag{17.2.19b}$$

If we divide both sides of the Equation (17.2.19a) by $\hat{M}_i$, we obtain:

$$\ddot{z}_i(t) + 2\xi_i\omega_i \cdot \dot{z}_i(t) + \omega_i^2 \cdot z_i(t) = \frac{f_i(t)}{\hat{M}_i}, \; i = 1,2,..,N \tag{17.2.20}$$

where ξ_i is the *modal damping ratio* corresponding to mode i.

It can be proven that, if we want to have a prescribed damping ratio value ξ for two modes, i (with modal frequency ω_i) and "j" (with modal frequency ω_j), then we can uniquely determine the values a_0 and a_1 in terms of the prescribed modal damping ratios ξ_i and ξ_j:

$$\alpha_0 = \frac{2\omega_i\omega_j}{\omega_j^2 - \omega_i^2}\left(\omega_j\xi_i - \omega_i\xi_j\right) \tag{17.2.21a}$$

$$\alpha_1 = \frac{2}{\omega_j^2 - \omega_i^2}\left(\omega_j\xi_j - \omega_i\xi_i\right) \tag{17.2.21b}$$

where ξ_i is the target modal damping ratio for mode i and ξ_j is the target damping ratio for mode j.

Remark 17.2.5: It can be verified that a stiffness-proportional damping matrix is obtained if we establish a Maxwell viscoelastic material (defined by Equation 15.6.1) with $[H] = \alpha_1[D]$ (or $\left[\hat{H}\right] = \alpha_1\left[\hat{D}\right]$ for beam and shell problems). ∎

Remark 17.2.6: Additionally, for elastodynamics, the mass-proportional part of the damping matrix can be obtained if we stipulate (in the considerations of equilibrium of a very small piece of material) that we have an additional vector of body forces (forces per unit volume) which is given by:

$$\{b_d\} = -\alpha_0 \cdot \rho\{\dot{u}\} = \begin{cases} -\alpha_0 \cdot \rho \begin{Bmatrix} \dot{u}_x \\ \dot{u}_y \end{Bmatrix} \; for\ 2D\ problems \\ -\alpha_0 \cdot \rho \begin{Bmatrix} \dot{u}_x \\ \dot{u}_y \\ \dot{u}_z \end{Bmatrix} \; for\ 3D\ problems \end{cases} \tag{17.2.22}$$

■

The construction and use of proportional damping matrices has been emphasized in structural dynamics textbooks, which usually focus on modal analysis. The actual Rayleigh damping matrices are typically determined at the element level. That is, instead of the considerations presented in Remarks 17.2.5 and 17.2.6 (starting with a strong form with modified material laws for a Kelvin solid and with additional body forces depending on velocity), an actual program may calculate the element stiffness matrix $[k^{(e)}]$ and mass matrix $[m^{(e)}]$, then provide the element damping matrix as $[c^{(e)}] = a_0[m^{(e)}] + a_1[k^{(e)}]$, where a_0, a_1 are given as input by the user.

While solution procedures based on modal superposition are invaluable in understanding the nature of the response of elastic systems and even in guiding the selection of the time-step in transient integration (presented in the next section), they are not the most efficient approach for finite element analysis; instead, direct step-by-step integration of Equation (17.1.1) may be preferable. This happens for the following reasons:

1) Many structures examined in practice exhibit nonlinear behavior; thus, the modal superposition (which is only valid for linear systems where the principle of superposition of Section 1.8 applies), cannot be used.
2) Even for linear structures, modal superposition requires the solution of the eigenvalue problem, which can be computationally demanding for large systems.
3) The modal superposition requires the determination of the modal response functions $z_i(t)$, $i = 1, 2, \ldots, N$. This determination still requires the use of step-by-step integration algorithms (with the exception of very simple cases of loading).

The reason why we explained the principles of modal superposition and the related modal analysis method in this section, is that it allows us to think of a multi-degree-of-freedom (MDOF) elastic system with N DOFs as a superposition of N SDOF systems. This will facilitate the discussion of the notions of stability and accuracy of transient algorithms: the discussion can be presented for a SDOF system, but the conclusions will also be valid for each of the N modal SDOF systems that correspond to a finite element model with N DOFs.

17.3 Step-by-Step Algorithms for Direct Integration of Equations of Motion

We will now proceed to establish transient integrators, also called step-by-step algorithms, for approximating the evolution of the response vectors with time. To facilitate our discussion, we will initially formulate these algorithms for analysis of single-degree-of-freedom (SDOF) systems, whose response is described by Equation (17.1.3). The procedure that we will follow is identical to that employed in Chapter 16. More specifically,

Equation (17.1.3) has not yet been discretized with time, in the sense that the displacement, velocity, and acceleration are assumed to be continuous functions of time. The transient solution algorithms do not provide a "closed form" solution (i.e., the response $u(t)$ as a function of time), but a "series" of approximate values of the response u_o, u_1, ..., u_{Ns} at specific (discrete) time instants t_o, t_1, ..., t_{Ns}, respectively. As we shall see below, transient integrators also provide estimates of the values of the velocities and accelerations at the discrete time instants of interest.

In numerical transient algorithms, the equation of motion of a SDOF system:

$$M \cdot \ddot{u}(t) + C \cdot \dot{u}(t) + K \cdot u(t) = f(t) \tag{17.3.1}$$

is converted to an approximate, discrete form, by stipulating that the equation of motion is satisfied at discrete time instants. That is:

$$M \cdot \ddot{u}(t_i) + C \cdot \dot{u}(t_i) + K \cdot u(t_i) = f(t_i) \tag{17.3.2}$$

for a set of specific instants t_i, $i = 0, 1, 2, ..., N_s$. We will imagine that these time instants are spaced apart from each other with a constant *time step* value Δt.

Equation (17.3.2) can be more concisely written as:

$$M \cdot \ddot{u}_i + C \cdot \dot{u}_i + K \cdot u_i = f_i \tag{17.3.3}$$

where $\ddot{u}_i$, $\dot{u}_i$, u_i and f_i are the acceleration, velocity, displacement, and external load (right-hand side), respectively, at time $t = t_i$. A *discretization* in time has now been obtained, since we enforce the equations of motion only at discrete time instants $t = t_o, t_1, ..., t_{Ns}$.

We will formulate various step-by-step algorithms by establishing their update equations. These equations will be using the response values at some step i as input, to give (as output) the response of the next step, $i + 1$. Just like for the case of algorithms for parabolic problems, transient integration algorithms for structural dynamics are classified into two basic categories, namely, *explicit and implicit.* In explicit methods, if we know the solution at an instant $t = t_i$, then the solution at the next time instant, that is, at $t = t_{i+1}$, can be directly obtained from known quantities related to t_i. In such methods, there is usually no need to invert coefficient arrays in the update equations. On the other hand, for implicit methods, the displacement, velocity, and acceleration at time t_{i+1} are coupled, and we need to enforce the equation of motion at t_{i+1} to obtain the updated response quantities. This, in turn, introduces the need to solve a system of linear equations and (for linear systems) "invert" a coefficient array. As mentioned in Section 16.1, the computations for a time-step in explicit algorithms are faster than those of implicit ones; however, explicit algorithms introduce limitations to the size of the time step Δt, which can be used in the analysis.

The strength of transient integrators lies to the fact that the response can be obtained with accuracy, for any type of loading, even for nonlinear systems. In this chapter, we will discuss three of the most popular explicit and implicit methods, namely, the central difference method, the Newmark method, and the Hilber-Hughes-Taylor (alpha) method, in the context of analysis of SDOF systems. In Section 17.4, we will proceed with the use of step-by-step algorithms for MDOF systems.

17.3.1 Explicit Central Difference Method

One of the most popular transient integration algorithms (especially for complicated nonlinear problems) is the central difference method (CDM), which is based on central difference approximations for the derivatives of response quantities with respect to time. The CDM is described in textbooks and programs in two versions:

- Version A (*popular with textbooks*):

$$\dot{u}_i = \frac{u_{i+1} - u_{i-1}}{2\Delta t} \tag{17.3.4}$$

$$\ddot{u}_i = \frac{u_{i+1} - 2u_i + u_{i-1}}{\Delta t^2} \text{ (valid for constant } \Delta t) \tag{17.3.5}$$

- Version B (*popular with many programs*):

$$\dot{u}_{i+1/2} = \frac{u_{i+1} - u_i}{\Delta t} \quad \rightarrow \quad u_{i+1} = u_i + \Delta t \cdot \dot{u}_{i+1/2} \tag{17.3.6a}$$

$$\ddot{u}_i = \frac{\dot{u}_{i+1/2} - \dot{u}_{i-1/2}}{\Delta t} \quad \rightarrow \quad \dot{u}_{i+1/2} = \dot{u}_{i-1/2} + \Delta t \cdot \ddot{u}_i \tag{17.3.6b}$$

In the remainder of this chapter, we will primarily be using version B of the CDM.

Remark 17.3.1: It is worth mentioning that the velocities in version B of the CDM are "staggered" in time compared to the accelerations and displacements (meaning that if the accelerations and displacements are obtained at instants t_i, t_{i+1}, t_{i+2} etc., the velocities will be known in $t_{i-1/2}$, $t_{i+1/2}$, $t_{i+3/2}$ etc. Of course, we can calculate the velocity values at the "correct instants" by using half-step equations:

$$\dot{u}_i = \dot{u}_{i-1/2} + \frac{\Delta t}{2} \cdot \ddot{u}_i = \dot{u}_{i+1/2} - \frac{\Delta t}{2} \cdot \ddot{u}_i, \quad \dot{u}_{i+1} = \dot{u}_{i+1/2} + \frac{\Delta t}{2} \cdot \ddot{u}_{i+1} = \dot{u}_{i+3/2} - \frac{\Delta t}{2} \cdot \ddot{u}_{i+1}, \text{ etc.}$$ ■

Enforcing the equation of motion (17.3.2) at time instant i, and assuming that we are given (from the previous step) the displacement u_i and the velocity $\dot{u}_{i-1/2}$, we can obtain the following equation.

$$M \cdot \ddot{u}_i = f_i - C \cdot \dot{u}_i - K \cdot u_i \tag{17.3.7}$$

Now, we can use a half-step equation to obtain the velocity at step i, $\dot{u}_i$, from $\dot{u}_{i-1/2}$:

$$\dot{u}_i = \dot{u}_{i-1/2} + \ddot{u}_i \frac{\Delta t}{2} \tag{17.3.8}$$

Plugging Equation (17.3.8) into Equation (17.3.7) yields:

$$M \cdot \ddot{u}_i = f_i - C\left(\dot{u}_{i-1/2} + \ddot{u}_i \frac{\Delta t}{2}\right) - K \cdot u_i \quad \rightarrow \quad \left(M + \frac{\Delta t}{2} C\right) \ddot{u}_i = f_i - C \cdot \dot{u}_{i-1/2} - K \cdot u_i \tag{17.3.9}$$

It may be obvious that all the values in the right-hand side of Equation (17.3.9) are known, and we can calculate the acceleration $\ddot{u}_i$. After we find $\ddot{u}_i$, we can use (17.3.6a) to find the updated velocity and then (17.3.6b) to find the updated displacement. After this, we will be able to continue with the next step, t_{i+1}, etc.

Remark 17.3.2: A common approximation in practice for the CDM is to calculate the damping forces using $\dot{u}_{i-1/2}$ instead of $\dot{u}_i$. This introduces a very desirable and efficient algorithm when we have a diagonal, lumped mass matrix (an explanation of why this may be the best choice if we use the CDM is provided in Remark 17.4.2). In this case, Equation (17.3.7) is replaced by the expression:

$$M \cdot \ddot{u}_i = f_i - C \cdot \dot{u}_{i-1/2} - K \cdot u_i \tag{17.3.10}$$

■

Remark 17.3.3: Special considerations are required for the solution at time $t_o = 0$, which is typically the first step that we need to solve for in an analysis. In the first step, we will know the initial conditions, which means we will know the initial displacement, u_0, and the initial velocity, v_0. What we do is use a "half-step" version of Equation (17.3.6b) for $i = 0$:

$$\dot{u}_{0+1/2} = v_0 + \frac{\Delta t}{2} \cdot \ddot{u}_o \tag{17.3.11}$$

where the initial acceleration $\ddot{u}_o$ is obtained by enforcing the equation of motion at time $t_0 = 0$:

$$\ddot{u}_o = \frac{1}{M}(f_0 - K \cdot u_0 - C \cdot \dot{u}_0) \tag{17.3.12}$$

After this point, we can find $u_1 = u_0 + \Delta t \cdot \dot{u}_{0+1/2}$, and then continue, with the update equation for $i = 1$, etc. ■

17.3.2 Newmark Method

A very popular (perhaps the most popular) step-by-step method in research and engineering practice is the Newmark method, named after Nathan Newmark, a professor at the University of Illinois at Urbana-Champaign (Newmark 1959). As a starting point, we assume that we are given the displacement and the velocity at a time instant i ($t = t_i$), and we want to find the updated displacement and velocity for the next time instant, $i + 1$ ($t = t_{i+1}$). As a first step to this end, we state that the acceleration, $\ddot{u}(t)$, varies in accordance with the following expression.

$$\ddot{u}(t) = \ddot{u}_i + \varphi(t) \tag{17.3.13a}$$

Remark 17.3.4: The function $\phi(t)$ in Equation (17.3.13a) determines how the value of the acceleration changes over the time step of interest, that is, during the time interval between t_i and t_{i+1}. ■

Remark 17.3.5: Note that $\ddot{u}_i$ in Equation (17.3.13a) is a constant number, not a function of time (it is the value of acceleration at the specific time t_i). ■

Now, we take the definite integral (with time) of both sides of Equation (17.3.13a), from t_i to some other time, $t > t_i$:

$$\int_{t_i}^{t} \ddot{u}(t)\, dt = \int_{t_i}^{t} \ddot{u}_i\, dt + \int_{t_i}^{t} \varphi(t)\, dt \rightarrow [\dot{u}(t)]_{t_i}^{t} = \ddot{u}_i \cdot (t - t_i) + [\chi(t)]_{t_i}^{t}$$

$$\rightarrow \dot{u}(t) - \dot{u}_i = \ddot{u}_i \cdot (t - t_i) + \chi(t) - \chi(t = t_i)$$

$$\rightarrow \dot{u}(t) = \dot{u}_i + \ddot{u}_i \cdot (t - t_i) + \chi(t) - \chi(t = t_i) \quad (17.3.13b)$$

where $\chi(t)$ is a function of time, such that $\varphi(t) = \frac{d\chi}{dt}$.

If we also take the definite integral (with time) of both sides of Equation (17.3.13b), from $t = t_i$ to some time $t > t_i$, we obtain:

$$\int_{t_i}^{t} \dot{u}(t)\, dt = \dot{u}_i \cdot (t - t_i) + \ddot{u}_i \cdot \frac{(t - t_i)^2}{2} + \int_{t_i}^{t} \chi(t)\, dt - \chi(t = t_i) \cdot (t - t_i)$$

$$\rightarrow [u(t)]_{t_i}^{t} = \dot{u}_i \cdot (t - t_i) + \ddot{u}_i \cdot \frac{(t - t_i)^2}{2} + \int_{t_i}^{t} \chi(t)\, dt - \chi(t = t_i) \cdot (t - t_i) \rightarrow$$

$$\rightarrow u(t) = u_i + \dot{u}_i \cdot (t - t_i) + \ddot{u}_i \cdot \frac{(t - t_i)^2}{2} + \int_{t_i}^{t} \chi(t)\, dt - \chi(t = t_i) \cdot (t - t_i) \quad (17.3.13c)$$

Now, since we are interested in finding the values at $t = t_{i+1}$, we set $t = t_{i+1}$ in Equations (17.3.13b) and (17.3.13c) to obtain:

$$\dot{u}_{i+1} = \dot{u}_i + \Delta t \cdot \ddot{u}_i + \chi(t = t_{i+1}) - \chi(t = t_i) \quad (17.3.14a)$$

$$u_{i+1} = u_i + \Delta t \cdot \dot{u}_i + \frac{\Delta t^2}{2} \ddot{u}_i + \int_{t_i}^{t_{i+1}} \chi(t)\, dt - \chi(t = t_i) \cdot \Delta t \quad (17.3.14b)$$

where $\Delta t = t_{i+1} - t_i$.

The Newmark method introduces the following two approximations in Equations (17.3.14a,b):

$$\int_{t_i}^{t_{i+1}} \chi(t)\, dt - \chi(t = t_i) \cdot \Delta t \approx \beta \cdot \Delta t^2 (\ddot{u}_{i+1} - \ddot{u}_i) \quad (17.3.15a)$$

$$\chi(t = t_{i+1}) - \chi(t = t_i) \approx \gamma \cdot \Delta t (\ddot{u}_{i+1} - \ddot{u}_i) \quad (17.3.15b)$$

where β and γ are user-defined constants, typically selected so that the method has several "desired characteristics", as explained below.

If we account for equations (17.3.15a,b), equations (17.3.14a,b) become:

$$u_{i+1} = u_i + \Delta t \cdot \dot{u}_i + \frac{\Delta t^2}{2} \ddot{u}_i + \beta \cdot \Delta t^2 (\ddot{u}_{i+1} - \ddot{u}_i) \quad (17.3.16a)$$

$$\dot{u}_{i+1} = \dot{u}_i + \Delta t \cdot \ddot{u}_i + \gamma \cdot \Delta t (\ddot{u}_{i+1} - \ddot{u}_i) \quad (17.3.16b)$$

Equations (17.3.16a,b) can be rewritten as follows.

$$u_{i+1} = u_i + \Delta t \cdot \dot{u}_i + \frac{\Delta t^2}{2}[(1-2\beta)\ddot{u}_i + 2\beta \cdot \ddot{u}_{i+1}] \tag{17.3.17a}$$

$$\dot{u}_{i+1} = \dot{u}_i + \Delta t[(1-\gamma)\ddot{u}_i + \gamma \cdot \ddot{u}_{i+1}] \tag{17.3.17b}$$

Equations (17.3.17a,b) constitute the update equations for the Newmark method, which is explicit when $\beta = 0$, otherwise it is implicit. When $\beta = 0$ and $\gamma = \frac{1}{2}$, we obtain a central difference method (which, admittedly, includes slightly different expressions than those in the CDM presented in Section 17.3.1). Two of the most popular implicit Newmark methods are the *constant-average acceleration method*, which is obtained for $\beta = \frac{1}{4}$ and $\gamma = \frac{1}{2}$, and the *linear acceleration method*, which is obtained for $\beta = \frac{1}{6}$ and $\gamma = \frac{1}{2}$. For the implicit Newmark method, we combine Equations (17.3.17a, b) with the equation of motion at $t = t_{i+1}$, to obtain the three response quantities u_{i+1}, $\dot{u}_{i+1}$, and $\ddot{u}_{i+1}$.

Remark 17.3.6: The constant-average acceleration method can also be obtained by stipulating that the acceleration between t_i and t_{i+1} has a constant value, equal to the average of $\ddot{u}_i$ and $\ddot{u}_{i+1}$. The linear acceleration method can be obtained by stipulating that the acceleration between times t_i and t_{i+1} varies linearly with time, and it is equal to $\ddot{u}_i$ at time t_i and equal to $\ddot{u}_{i+1}$ at time t_{i+1}. The variation of the acceleration between t_i and t_{i+1} is governed by the function $\phi(t)$ of Equation (17.3.13a). We can write:

- For the constant average acceleration method: $\varphi(t) = constant = \dfrac{\ddot{u}_{i+1} + \ddot{u}_i}{2}$
- For the linear acceleration method: $\varphi(t) = (\ddot{u}_{i+1} - \ddot{u}_i)\dfrac{(t-t_i)}{\Delta t}$ ■

For linear systems, the implicit Newmark method ($\beta \neq 0$) can be expressed in a "stiffness" form. Specifically, if we solve Equation (17.3.17a) for the acceleration $\ddot{u}_{i+1}$, plug the resulting expression into Equation (17.3.17b), and then plug both obtained expressions into the equation of motion at $t = t_{i+1}$, we can obtain:

$$\widetilde{K} \cdot u_{i+1} = \widetilde{f}_{i+1} \tag{17.3.18}$$

where:

$$\widetilde{K} = \frac{M}{\beta \cdot \Delta t^2} + \frac{\gamma \cdot C}{\beta \cdot \Delta t} + K \tag{17.3.19}$$

and

$$\widetilde{f}_{i+1} = f_{i+1} + \left(\frac{M}{\beta \cdot \Delta t^2} + \frac{\gamma \cdot C}{\beta \cdot \Delta t}\right)u_i + \left[\frac{M}{\beta \cdot \Delta t} + \left(\frac{\gamma}{\beta} - 1\right)C\right]\dot{u}_i + \left[\left(\frac{1}{2\beta} - 1\right)M + \frac{\Delta t}{2}\left(\frac{\gamma}{\beta} - 2\right)C\right]\ddot{u}_i \tag{17.3.20}$$

After we have u_{i+1}, we can use Equations (17.3.17a,b) to also obtain $\dot{u}_{i+1}$ and $\ddot{u}_{i+1}$.

17.3.3 Hilber-Hughes Taylor (HHT or Alpha) Method

A third popular method for the solution of elastodynamics and structural dynamics is the Hilber, Hughes & Taylor (HHT) method, also called the alpha (α) method (Hilber et al.

1977). In the HHT method, the following modified equation of motion needs to be satisfied:

$$M \cdot \ddot{u}_{i+1} + (1+\alpha) \cdot C \cdot \dot{u}_{i+1} - \alpha \cdot C \cdot \dot{u}_i + (1+\alpha) \cdot K \cdot u_{i+1} - \alpha \cdot K \cdot u_i = f_{i+1} \quad (17.3.21)$$

The update of the displacement and velocity relies on the same equations as for the Newmark Method, i.e., Equations (17.3.17a,b). Thus, the only difference compared to the Newmark method is on the type of equation of motion, which must be satisfied; this modified equation includes an additional parameter α, which is used to improve the numerical properties of the solution algorithm. The various parameters in the HHT method should be assigned the following values.

$$-\frac{1}{3} \leq \alpha \leq 0 \quad (17.3.22)$$

$$\gamma = \frac{1}{2}(1-2\alpha) \quad (17.3.23)$$

$$\beta = \frac{1}{4}(1-\alpha)^2 \quad (17.3.24)$$

The Newmark method can be found to be a special case of the HHT method (obtained for $\alpha = 0$).

Example 17.1: Transient step-by-step algorithms
A SDOF system with $K = 100$ kN/m, $M = 10$ Mgr και $\zeta = 5\%$ undergoes forced vibration. We are given that, for time instant $t_i = 3$ seconds: $u_i = 0.10$ m, $\dot{u}_i = 0.50$ m/s, $\ddot{u}_i = 1.262$ m/s^2. We want to determine the displacement, velocity, and acceleration at instant $t = 3.05$ seconds, using the Central Difference Method and the Average Acceleration Method, if the force at $t = 3$ seconds is equal to 24.1 kN, while for $t = 3.05$ seconds the force equals 20 kN.

Based on the input information, we want to use the explicit CDM and the average acceleration method for $\Delta t = 0.05$ second.

We first obtain the viscous damping coefficient, C, for the system:

$$C = 2M \cdot \omega \cdot \zeta = 2\sqrt{K \cdot M} \cdot \zeta = 3.162\, kN \cdot s/m$$

We also know that $f_i = 24.1$ kN, $f_{i+1} = 20$ kN. We proceed with the update for the two methods.

<u>Central Difference Method</u>

We are given $u_i = 0.10$ m, $\dot{u}_i = 0.50$ m/s, $\ddot{u}_i = 1.262$ m/s^2. Note that in the CDM, we need to have the value $\dot{u}_{i+1/2}$, while we are given $\dot{u}_i$ in this example. For this reason, we conduct a half-step calculation to find $\dot{u}_{i+1/2}$:

$$\dot{u}_{i+1/2} = \dot{u}_i + \ddot{u}_i \cdot \frac{\Delta t}{2} \rightarrow \dot{u}_{i+1/2} = 0.50 + 1.262 \cdot \frac{0.05}{2} = 0.53155\, m/s$$

We can now update the displacement:

$$u_{i+1} = u_i + \dot{u}_{i+1/2} \cdot \Delta t \rightarrow u_{i+1} = 0.10 + 0.53155 \cdot 0.05 = 0.126578\, m$$

Using Equation (17.3.8), we can also find the updated acceleration:

$$M \cdot \ddot{u}_{i+1} = f_{i+1} - C \cdot \dot{u}_{i+1/2} - K \cdot u_{i+1} \rightarrow 10 \cdot \ddot{u}_{i+1} = 20 - 3.162 \cdot 0.53155 - 100 \cdot 0.126578$$
$$\rightarrow \ddot{u}_{i+1} = 0.566\, m/s^2$$

Finally, we can use another half-step update, to find $\dot{u}_{i+1}$:

$$\dot{u}_{i+1} = \dot{u}_{i+1/2} + \ddot{u}_{i+1} \cdot \frac{\Delta t}{2} \rightarrow \dot{u}_{i+1} = 0.53155 + 0.566 \cdot \frac{0.05}{2} = 0.5457\, m/s$$

Average Acceleration Method (i.e., Newmark method with $\gamma = 0.5$, $\beta = 0.25$)

First, we use Equation (17.3.19):

$$\widetilde{K} = \frac{M}{\beta \cdot \Delta t^2} + \frac{\gamma \cdot C}{\beta \cdot \Delta t} + K = \frac{10}{0.25 \cdot 0.05^2} + \frac{0.5 \cdot 3.162}{0.25 \cdot 0.05} + 100 = 16226.5\, kN/m$$

We also need to use Equation (17.3.20):

$$\widetilde{f}_{i+1} = f_{i+1} + \left(\frac{M}{\beta \cdot \Delta t^2} + \frac{\gamma \cdot C}{\beta \cdot \Delta t}\right) u_i + \left[\frac{M}{\beta \cdot \Delta t} + \left(\frac{\gamma}{\beta} - 1\right) C\right] \dot{u}_i + \left[\left(\frac{1}{2\beta} - 1\right) M + \frac{\Delta t}{2}\left(\frac{\gamma}{\beta} - 2\right) C\right] \ddot{u}_i$$

$$= 20 + \left(\frac{10}{0.25 \cdot 0.05^2} + \frac{0.5 \cdot 3.162}{0.25 \cdot 0.05}\right) \cdot 0.1 + \left[\frac{10}{0.25 \cdot 0.05} + \left(\frac{0.5}{0.25} - 1\right) \cdot 3.162\right] \cdot 0.5$$

$$+ \left[\left(\frac{1}{2 \times 0.25} - 1\right) \cdot 10 + \frac{0.05}{2}\left(\frac{0.5}{0.25} - 2\right) \cdot 3.162\right] \cdot 1.262 = 20 + 1612.648$$

$$+ 401.581 + 12.62 = 2046.849\, kN$$

Now, we can find the updated displacement through use of Equation (17.3.18):

$$u_{i+1} = \frac{\widetilde{f}_{i+1}}{\widetilde{K}} = \frac{2046.849}{16226.5} = 0.126\, m$$

We can also find the acceleration at the new time, $i+1$, by rearranging Equation (17.3.17a):

$$u_{i+1} = u_i + \Delta t \cdot \dot{u}_i + \frac{\Delta t^2}{2}\left[(1 - 2\beta)\ddot{u}_i + 2\beta \cdot \ddot{u}_{i+1}\right] \rightarrow$$

$$\rightarrow \ddot{u}_{i+1} = \frac{1}{\beta \cdot \Delta t^2}\left[u_{i+1} - u_i - \Delta t \cdot \dot{u}_i - \frac{\Delta t^2}{2}(1 - 2\beta)\ddot{u}_i\right]$$

$$= \frac{1}{0.25 \cdot 0.05^2}\left[0.126 - 0.10 - 0.05 \cdot 0.5 - \frac{0.05^2}{2}(1 - 2 \cdot 0.25)1.262\right] \rightarrow \ddot{u}_{i+1} = 0.338\, m/s^2$$

We can finally find the updated velocity, through use of Equation (17.3.17b):

$$\dot{u}_{i+1} = \dot{u}_i + \Delta t\left[(1 - \gamma)\ddot{u}_i + \gamma \cdot \ddot{u}_{i+1}\right] \rightarrow \dot{u}_{i+1} = 0.50 + 0.05\left[(1 - 0.5)1.262 + 0.5 \cdot 0.338\right] \rightarrow$$

$$\rightarrow \dot{u}_{i+1} = 0.54\, m/s$$

■

17.3.4 Stability and Accuracy of Transient Solution Algorithms

An important parameter to be selected in a dynamic analysis is the time step in the transient integration, that is, the value $\Delta t = t_{i+1} - t_i$, $i = 0, 1, \ldots, N_s$. It is worth mentioning that the integration time step does not have to be constant throughout an analysis, but in practice (and especially for linear systems) it is usually set to a constant value. As explained in Section 16.5, the time step should be such that we satisfy the two requirements for convergence of a step-by-step algorithm, namely, stability and consistency. In this section, we will examine in detail the stability of the central difference method (CDM), then we will briefly discuss the stability of the Newmark and Hilber-Hughes-Taylor (HHT) methods.

To obtain the stability condition for the CDM, we will establish the general amplification matrix (defined in Section 16.5.1) for free-vibration problems, where the external force $f(t)$ is always equal to zero. To this end, we need to consider an expanded solution vector, $\{\hat{X}_i\} = \begin{bmatrix} u_i & \dot{u}_{i-1/2} \end{bmatrix}^T$, which includes the displacement and velocity values. We will determine stability based on the considerations provided in Box 16.5.2. The amplification matrix, $[A]$, for the CDM in solid and structural dynamics, will give the updated vector $\{\hat{X}_{i+1}\} = \begin{bmatrix} u_{i+1} & \dot{u}_{i+1/2} \end{bmatrix}^T$ as the product $[A]\{\hat{X}_i\}$. If we divide Equation (17.3.9) by M and account for Equation (17.1.5b), we can write:

$$(1+\zeta\cdot\omega\cdot\Delta t)\ddot{u}_i = -2\zeta\cdot\omega\cdot\dot{u}_{i-1/2} - \omega^2\cdot u_i \rightarrow \ddot{u}_i = -\frac{2\zeta\cdot\omega}{1+\zeta\cdot\omega\cdot\Delta t}\cdot\dot{u}_{i-1/2} - \frac{\omega^2}{1+\zeta\cdot\omega\cdot\Delta t}\cdot u_i \tag{17.3.25}$$

If we plug Equation (17.3.25) into Equation (17.3.6b), we obtain:

$$\begin{aligned}\dot{u}_{i+1/2} &= \dot{u}_{i-1/2} - \frac{2\zeta\cdot\omega\cdot\Delta t}{1+\zeta\cdot\omega\cdot\Delta t}\cdot\dot{u}_{i-1/2} - \frac{\omega^2\cdot\Delta t}{1+\zeta\cdot\omega\cdot\Delta t}\cdot u_i \\ &= \left(1-\frac{2\zeta\cdot\omega\cdot\Delta t}{1+\zeta\cdot\omega\cdot\Delta t}\right)\cdot\dot{u}_{i-1/2} - \frac{\omega^2\cdot\Delta t}{1+\zeta\cdot\omega\cdot\Delta t}\cdot u_i\end{aligned} \tag{17.3.26}$$

If we now plug Equation (17.3.26) into Equation (17.3.6a), we have:

$$\begin{aligned}u_{i+1} &= u_i + \left(1-\frac{2\zeta\cdot\omega\cdot\Delta t}{1+\zeta\cdot\omega\cdot\Delta t}\right)\cdot\Delta t\cdot\dot{u}_{i-1/2} - \frac{\omega^2\cdot(\Delta t)^2}{1+\zeta\cdot\omega\cdot\Delta t}\cdot u_i \rightarrow \\ \rightarrow u_{i+1} &= \left(1-\frac{\omega^2\cdot(\Delta t)^2}{1+\zeta\cdot\omega\cdot\Delta t}\right)\cdot u_i + \left(1-\frac{2\zeta\cdot\omega\cdot\Delta t}{1+\zeta\cdot\omega\cdot\Delta t}\right)\cdot\Delta t\cdot\dot{u}_{i-1/2}\end{aligned} \tag{17.3.27}$$

We can now combine Equations (17.3.26) and (17.3.27) into a single matrix expression to finally obtain:

$$\begin{Bmatrix} u_{i+1} \\ \dot{u}_{i+1/2} \end{Bmatrix} = \begin{bmatrix} 1-\dfrac{\omega^2\cdot(\Delta t)^2}{1+\zeta\cdot\omega\cdot\Delta t} & \left(1-\dfrac{2\zeta\cdot\omega\cdot\Delta t}{1+\zeta\cdot\omega\cdot\Delta t}\right)\cdot\Delta t \\ -\dfrac{\omega^2\cdot\Delta t}{1+\zeta\cdot\omega\cdot\Delta t} & 1-\dfrac{2\zeta\cdot\omega\cdot\Delta t}{1+\zeta\cdot\omega\cdot\Delta t} \end{bmatrix} \begin{Bmatrix} u_i \\ \dot{u}_{i-1/2} \end{Bmatrix} \tag{17.3.28}$$

Thus, we have established an update equation $\{\hat{X}_{i+1}\} = [A]\{\hat{X}_i\}$ for the CDM, in which the amplification matrix $[A]$ is:

$$[A] = \begin{bmatrix} 1-\dfrac{\bar{\omega}^2}{1+\zeta\cdot\bar{\omega}} & \left(1-\dfrac{2\zeta\cdot\bar{\omega}}{1+\zeta\cdot\bar{\omega}}\right)\cdot\Delta t \\ -\dfrac{\bar{\omega}\cdot\omega}{1+\zeta\cdot\bar{\omega}} & 1-\dfrac{2\zeta\cdot\bar{\omega}}{1+\zeta\cdot\bar{\omega}} \end{bmatrix} \tag{17.3.29}$$

where

$$\bar{\omega} = \omega \cdot \Delta t \tag{17.3.30}$$

Using the procedure in Box 16.5.3 yields the critical value for parameter $\bar{\omega}$:

$$\bar{\omega}_{cr} = \omega \cdot \Delta t_{cr} = 2 \tag{17.3.31}$$

Thus, the CDM is conditionally stable, and the condition for stability in analysis of a single-degree-of-freedom system is:

$$\Delta t \leq \Delta t_{cr} = \frac{2}{\omega} = \frac{T}{\pi} \tag{17.3.32}$$

Obviously, for multiple-degree-of-freedom (MDOF) systems, we need to check the stability by using the maximum value of natural frequency ω that pertains to the system. One may understand that, if the system's eigen-frequencies (defined in Section 17.2) are numbered such that $\omega_1 \leq \omega_2 \leq \ldots \leq \omega_N$, then we merely need to ensure that we satisfy the stability criterion for the N-th mode, since ω_N is the maximum value of eigen-frequency for the dynamic system. The theorem presented in Remark 16.6.2 for parabolic problems can also be extended to elastodynamics and structural dynamics; that is:

"The maximum eigen-frequency of a finite element model does not exceed the maximum eigen-frequency value obtained from eigenvalue analysis of the individual elements":

$$\omega_N \leq \max_{i,e} \left(\omega_i^{(e)} \right) \tag{17.3.33}$$

This means that we can determine the maximum frequency for the individual elements (which is easier because the elements have much less degrees of freedom), and then take the overall maximum to use it for determining stability.

Remark 17.3.7: If Equation (17.3.10) is used instead of (17.3.9), then the amplification matrix for the CDM attains a different form:

$$[A] = \begin{bmatrix} 1-\bar{\omega}^2 & (1-2\zeta \cdot \bar{\omega}) \cdot \Delta t \\ -\bar{\omega} \cdot \omega & 1-2\zeta \cdot \bar{\omega} \end{bmatrix} \tag{17.3.34}$$

This eventually yields:

$$\bar{\omega}_{cr} = 2\left(\sqrt{1+\zeta^2} - \zeta^2 \right) \tag{17.3.35}$$

For $\zeta = 0$, the critical time step from Equation (17.3.35) is identical to that obtained with Equation (17.3.31). One can verify that, for nonzero values of viscous damping coefficient ζ, Equation (17.3.35) yields lower critical time step values than for zero values of viscous damping. This may be surprising, because one could expect that the presence of energy dissipation due to viscous damping would "stabilize" the response. It turns out that this is not the case, and the fact that we use a "staggered-step-velocity" in Equation (17.3.10) leads to a destabilizing effect of viscous damping, as also explained in Belytschko et al. (2014). ∎

Remark 17.3.8: In practice, a slightly different equation is employed for finding the critical time step, Δt_{cr}, of the CDM. Specifically, Δt_{cr} is determined from the following expression.

$$\Delta t_{cr} = F_s \cdot \min_e \left(\frac{\ell^{(e)}}{C^{(e)}} \right) \tag{17.3.36}$$

where, for each element e, $C^{(e)} = \sqrt{\frac{E^{(e)}}{\rho^{(e)}}}$ is the elastic wave propagation speed in the material, F_s is a coefficient (typically set equal to 0.5) and $\ell^{(e)}$ is the element size. For one-dimensional elements, $\ell^{(e)}$ is taken as the element length, for two-dimensional elements, $\ell^{(e)}$ can be taken as the square root of the area, and for three-dimensional elements, $\ell^{(e)}$ can be taken as the cubic root of the volume. ∎

We will now also discuss the stability of the Newmark method. This method is unconditionally stable if $\gamma \geq \frac{1}{2}, \beta \geq \frac{\gamma}{2}$, conditionally stable if $\gamma \geq \frac{1}{2}, \beta < \frac{\gamma}{2}$, and (unconditionally) unstable if $\gamma < \frac{1}{2}$. When the method is conditionally stable, the critical value $\bar{\omega}_{cr} = \omega \cdot \Delta t_{cr}$ is given by the following expression.

$$\bar{\omega}_{cr} = \frac{\xi(\gamma - 1/2) + \left[\gamma/2 - \beta + \xi^2(\gamma - 1/2)^2\right]^{\frac{1}{2}}}{(\gamma/2 - \beta)} \tag{17.3.37}$$

Per Equation (17.3.37), a value of $\gamma = \frac{1}{2}$ makes the critical time step value independent of the damping ratio, ξ.

Remark 17.3.9: The HHT method is unconditionally stable if parameters α, β, and γ are assigned values in accordance with relations (17.3.22), (17.3.23), and (17.3.24). ∎

We will now conclude this section with a discussion about the accuracy and consistency of step-by-step algorithms for elastodynamics and structural dynamics. As mentioned above, consistency is required (in addition to stability) to guarantee convergence. We are reminded here that a transient algorithm is convergent when the numerical solution gets closer and closer to the exact solution as we reduce the time step size, and in the limit, as $\Delta t \to 0$, the two solutions become identical. In other words, the *error* of the algorithm must tend to zero as the time step goes to zero. An important consideration pertains to the order of accuracy of the various methods—a definition of the order or accuracy has been provided in Section 16.5.2. It is important to emphasize that, per the discussion in Section 16.5.2, algorithms with order of accuracy greater than 0 are consistent. We will only briefly discuss the order of accuracy of the various step-by-step methods that we have considered in this chapter. These considerations are provided in Box 17.3.1.

Now, some practical considerations pertaining to accuracy of step-by-step algorithms for elastodynamics/structural dynamics are deemed necessary. As a practical rule to ensure accurate results in such algorithms, the time step in analysis of SDOF systems

Box 17.3.1 Accuracy of Central Difference Method and Newmark Method

- The central difference method is second-order accurate.
- The Newmark method is second-order accurate if $\gamma = \frac{1}{2}$; otherwise, it is first-order accurate.
- The Hilber-Hughes-Taylor method is second-order accurate, if parameters α, β, γ satisfy Equations (17.3.22), (17.3.23), and (17.3.24).

is recommended to be assigned values lower than 0.1 T, preferably around 0.05 T (T being the natural period of the system), to ensure a satisfactory accuracy of the numerical solution. For MDOF systems with a large number of degrees of freedom, N, it is usually impractical to select a time step so that it is lower than, say, $0.1T_N$, T_N being the period of the N-th mode of the system (T_N can be thought of as the "minimum natural period that describes the MDOF system"!). Instead, we usually select Δt so that it is much smaller than the eigen-period of the first few "significant" modes. For this reason, we may sometimes observe *noise* (fast, spurious fluctuations in the time-histories of the solution) due to increased error in the high-mode response. This noise becomes more evident in acceleration time histories. Still, the error in high-mode response may not be a major concern, and in some cases, it may not even have a noticeable effect. This typically happens because the very high modes have very little (if any) participation to the response vectors (meaning the $z(t)$ functions of Equation 17.2.1 for the very high modes have very small, practically zero values).

Now, if we focus on the plot of the time-histories of response quantities, the numerical error in a transient algorithm can be visually manifested in three ways:

i) *Period distortion*, meaning that the solution obtained with the transient integrator is found to correspond to the solution of a system with period slightly different than that of the actual system.
ii) *Artificial damping* (also called numerical dissipation), meaning that the solution algorithm "numerically" dissipates part of the system's energy, and the effect of this numerical dissipation is added to the dissipation due to the actual damping of the system,
iii) *Artificial (or spurious) amplification* of the response under forced vibrations (i.e., when the right-hand side of Equation 17.3.1 is nonzero, $f(t) \neq 0$), which is actually a consequence of the period distortion, as explained in the following Remark.

Remark 17.3.10: Regarding period distortion, it has been established that the CDM usually distorts a period in such a fashion that the system to which the numerical solution corresponds has a slightly lower period than the actual system. This is demonstrated in Example 17.2. The converse applies for the average acceleration method, as shown in Example 17.3. For free vibration problems, the period distortion does not have significant effect on the response quantities. However, for forced vibrations, the period distortion can lead to significant errors in the response histories. To briefly explain why this happens, let us consider the displacement *response spectrum*[1] for a specific dynamic loading history, as shown in Figure 17.1.

In the response spectrum plot, two period values are marked, one corresponding to the actual period of the system, the other to the slightly distorted period obtained with a transient integration algorithm. Using the response spectrum, we can understand that the maximum response for the two systems will differ due to the period distortion. As the same figure suggests (and this applies for most types of loading that we encounter in practice), the difference in maximum response due to period distortion is much greater for the case of very low or zero damping. For sufficiently high damping ratio values (e.g., 10% in Figure 17.1), the difference in maximum response (and in the response history for that matter) is much less significant. ∎

Remark 17.3.11: Artificial damping has the same effect as the actual viscous damping in a system. In other words, it dissipates energy from the system (this artificial dissipation can sometimes be deemed favorable for some types of solution procedures, focusing on static loading, which will not be discussed here). It can be proven that the Newmark method does not include artificial damping when $\gamma = \frac{1}{2}$. If values greater than $\frac{1}{2}$ are assigned to γ, then the Newmark method will introduce artificial damping. Such artificial damping may be desirable, because it can eliminate the noise caused by the increased

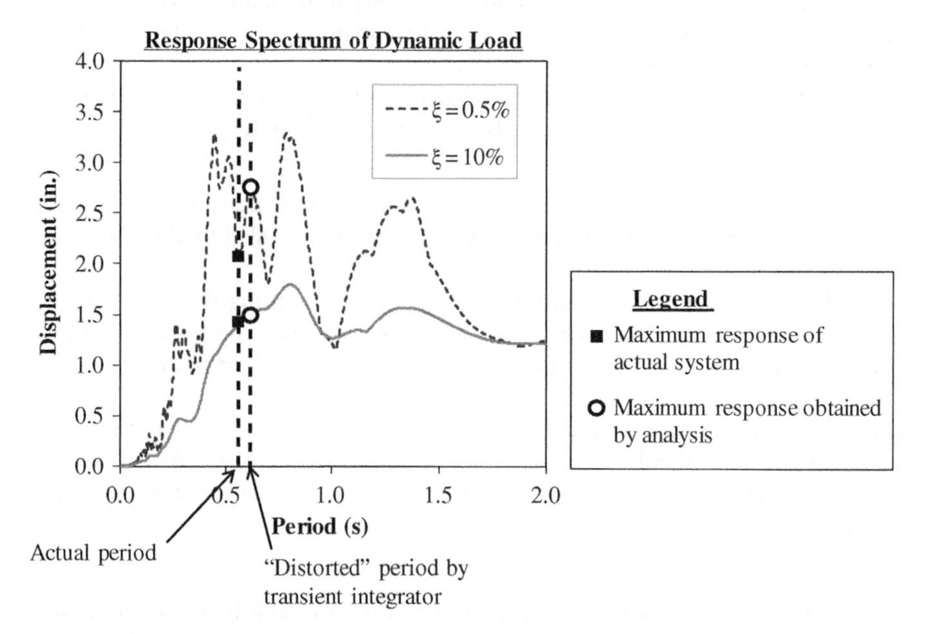

Figure 17.1 Error in the maximum response of a single-degree-of-freedom system, due to period distortion from the numerical step-by-step algorithm.

1 The response spectrum of a dynamic load simply gives the maximum absolute value of a response quantity (displacement, velocity or acceleration) as a function of the natural period of a SDOF system. It is a very efficient tool to quickly understand what level of response will be observed for a wide variety of SDOF systems, if they are subjected to a specific dynamic loading.

numerical error in higher-mode response. However, there is a problem using $\gamma > \frac{1}{2}$ in the Newmark method, because—as explained in Box 17.3.1—this would imply that the order of accuracy of the method degrades from 2 to 1. This is one of the motivations for using the HHT method, which can provide artificial viscous damping for higher modes while also ensuring a second-order accuracy. ■

Example 17.2: Effect of time step for explicit CDM
In this example, we will use the CDM to obtain the displacement history of an undamped SDOF system, with a mass equal to 1 kip·s^2/in. and a stiffness equal to 438kip/in., corresponding to a natural period $T = 0.3$ seconds. The system is undergoing a free vibration with initial displacement $u_o = 1$ in. and initial velocity equal to zero. The exact response time history is given by:

$$u(t) = 1.0 \cdot \cos\left(\frac{2\pi \cdot t}{0.3}\right)$$

The critical time step for this system is found from Eq. (17.3.31) to be equal to 0.095 seconds. The analysis is conducted using three different values of time step, that is, 0.01 seconds 0.05 seconds, and 0.10 seconds. The results obtained for the three different values of time step are compared with the exact solution in Figure 17.2.

It can be seen that for $\Delta t = 0.01$ seconds, the approximate solution is practically identical to the exact one. For $\Delta t = 0.05$ seconds, the approximate solution is stable, but there is an obvious error due to period distortion. This means that the analytically obtained time history for $\Delta t = 0.05$ seconds appears to correspond to a SDOF system with lower period than the actual one. Finally, it is obvious that for $\Delta t = 0.10$ seconds, which is greater than the critical time step, the solution grows uncontrollably (in fact, for $t = 3$ seconds, we have reached values of displacement greater than 300 inches!). The results clearly demonstrate the fact that time step values even slightly greater than Δt_{crit} render the numerical solution completely useless and meaningless. ■

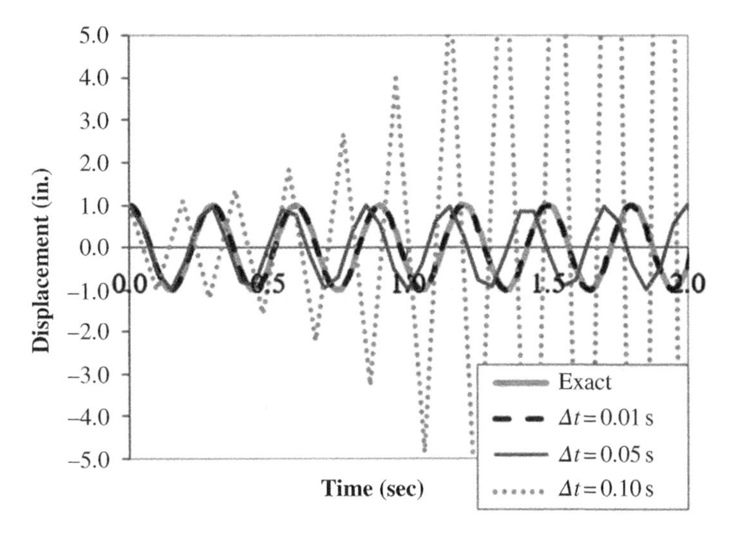

Figure 17.2 Time histories of displacement for example single-degree-of-freedom system, using the central difference method and different values of time step.

Example 17.3: Accuracy of Newmark Method
We will now determine the displacement history of an undamped SDOF system, with a natural period $T = 1$ second, undergoing a free vibration with $u_o = 0.01$ m, $\dot{u}_o = 0$, using the Newmark method ($\gamma = 0.5$ και $\beta = 0.25$) and time step values $\Delta t = 0.02$, 0.10, and 0.20 second.

For the case of a single-degree-of-freedom system with $T = 1$ second, $u_o = 0.01$ m, $\dot{u}_o = 0$, the exact response time history is given by:

$$u(t) = 1.0 \cdot \cos(2\pi \cdot t)$$

The exact displacement time history is compared to the numerically obtained time histories (for different Δt values) in Figure 17.3.

The response in Figure 17.3a is due to free vibration; since $\gamma = 0.5$ (no numerical damping) there should be no error in the obtained amplitude (i.e., the maximum and minimum values of displacement). However, as shown in the figure, several "instants" of maximum displacement have not been "captured" from the transient integration, which only gives the response at instants that are integer multiples of the time step. We can also clearly observe an "increase in the period" of the numerical solutions, as we increase the time step of the transient integration.

Let us now repeat the computation for a time step $\Delta t = 0.02$ s, using $\beta = 0.40$, $\gamma = 0.8$. From Figure 17.3b, we can observe that using $\gamma > 0.5$ leads to numerical damping in our solution, which is attested by the fact that the displacement values obtained by the Newmark method have a decay. ∎

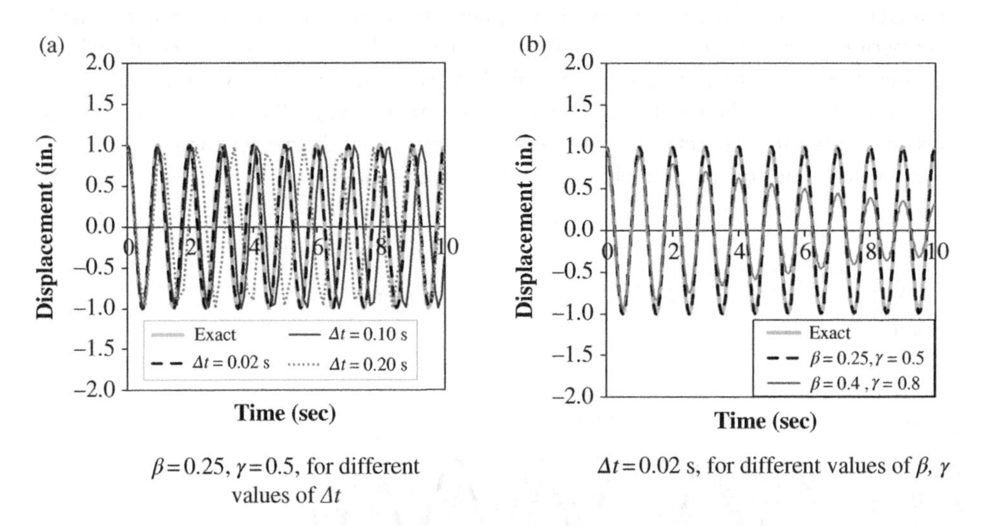

Figure 17.3 Results for example single-degree-of-freedom system, for use of the Newmark method.

17.4 Application of Step-by-Step Algorithms for Discrete Systems with More than One Degrees of Freedom

The transient integration of the semidiscrete equations of motion of any MDOF finite element model can be conducted in the same fashion as for the analysis of SDOF systems presented in the previous section. The only difference is that we have displacement, velocity, acceleration, and force vectors, and mass, damping, and stiffness matrices. This section will describe the use of transient integrators for MDOF systems, starting with the presentation of the explicit central difference method (CDM), and continuing with a description of the Newmark method and the Hilber-Hughes-Taylor (HHT) method.

For the CDM, the calculation of the acceleration vector for each step is conducted using the following equation.

$$[M]\{\ddot{U}\}_i = \{f\}_i - [C]\{\dot{U}\}_{i-1/2} - [K]\{U\}_i \tag{17.4.1}$$

where the vectors $\{\dot{U}\}_{i-1/2}$ and $\{U\}_i$ are known from the calculations of the previous step. The right-hand side of Equation (17.4.1) is sometimes referred to as the *unbalanced force* or *residual* vector, $\{r\}_i$, for time t_i, given by the following expression.

$$\{r\}_i = \{f\}_i - [C]\{\dot{U}\}_{i-1/2} - [K]\{U\}_i \tag{17.4.2}$$

Thus, Equation (17.4.1) can also be written as follows.

$$[M]\{\ddot{U}\}_i = \{r\}_i \tag{17.4.3}$$

Remark 17.4.1: Notice the conceptual meaning of Equation (17.4.3): *Mass times acceleration equals unbalanced force.* Thus, Equation (17.4.3) can be seen as a matrix version of Newton's second law! ∎

After the acceleration vector is obtained from the solution of Equation (17.4.1), the velocity vector and displacement vector of the new step can be obtained from the central difference update expressions:

$$\{\dot{U}\}_{i+1/2} = \{\dot{U}\}_{i-1/2} + \Delta t\{\ddot{U}\}_i \tag{17.4.4}$$

$$\{U\}_{i+1} = \{U_i\} + \Delta t\{\dot{U}\}_{i+1/2} \tag{17.4.5}$$

Then, we can replace i by $i+1$ in Equation (17.4.1) and calculate the new acceleration vector, $\{\ddot{U}_{i+1}\}$.

Remark 17.4.2: A very efficient solution scheme is obtained if a lumped mass matrix is combined with the explicit CDM. In this case, if the discrete dynamic system has N degrees of freedom, then we simply need to establish N scalar update equations, each equation corresponding to one degree of freedom of the system. ∎

The computations for the solution of a dynamic problem using the CDM can be conducted using the algorithm summarized in Box 17.4.1.

If the Newmark method is used, the following equation must be enforced to obtain the updated solution of some step $t = t_{i+1}$:

$$[M]\{\ddot{U}\}_{i+1} + [C]\{\dot{U}\}_{i+1} + [K]\{U\}_{i+1} = \{f\}_{i+1} \tag{17.4.3}$$

Box 17.4.1 Solution of Dynamic Finite Element Problems with Explicit Central Difference Method

0) **Initial conditions:** If we know the initial displacement vector, $\{U\}_0$, and initial velocity vector, $\{V\}_0$, set:

$$\{\dot{U}\}_{0+1/2}=\{V\}_0+\frac{\Delta t}{2}\{\ddot{U}\}_0, \quad \text{where}\{\ddot{U}\}_0=[M]^{-1}(\{f\}_0-[K]\{U\}_0-[C]\{V\}_0)$$

Then find $\{U\}_1=\{U\}_0+\Delta t\{\dot{U}\}_{0+1/2}$

For each time step *i*:

1) Given the displacement vector, $\{U\}_i$, and the velocity vector, $\{\dot{U}\}_{i-1/2}$:

For each element, *e*:

Find the element displacement and velocity vectors:

$$\left\{U^{(e)}\right\}_i=\left[L^{(e)}\right]\{U\}_i, \quad \left\{\dot{U}^{(e)}\right\}_{i-1/2}=\left[L^{(e)}\right]\{\dot{U}\}_{i-1/2}$$

Calculate the element unbalanced force contribution, i.e. the element equivalent nodal force vector minus the sum of the element's damping and stiffness forces:

$$\left\{r^{(e)}\right\}_i=\left\{f^{(e)}\right\}_i-\left[k^{(e)}\right]\left\{U^{(e)}\right\}_i-\left[c^{(e)}\right]\left\{\dot{U}^{(e)}\right\}_{i-1/2}$$

2) Assemble the global unbalanced force vector:

$$\{r\}_i=\sum_{e=1}^{Ne}\left(\left[L^{(e)}\right]^T\left\{r^{(e)}\right\}_i\right)$$

3) Calculate the components of the acceleration vector, $\{\ddot{U}\}_i$: $[M]\{\ddot{U}\}_i=\{r\}_i$

Note: If $[M]$ is diagonal, the update equation becomes:

For each degree of freedom, *I*, calculate:

$\ddot{U}_{Ii}=\frac{1}{M_{II}}r_{Ii}$ (the capitalized subscript refers to degree of freedom, the lowercase one to step)

4) Find the updated displacement and velocity vectors:

$$\{\dot{U}\}_{i+1/2}=\{\dot{U}\}_{i-1/2}+\Delta t\{\ddot{U}\}_i, \quad \{U\}_{i+1}=\{U\}_i+\Delta t\{\dot{U}\}_{i+1/2}$$

5) If we have run all the time steps we needed to run, terminate the algorithm. Otherwise, Set $i \leftarrow i+1$ and go to Step 1.

Equation (17.4.3) must be combined with the update expressions for the displacement vector, $\{U\}_{i+1}$, and velocity vector, $\{\dot{U}\}_{i+1}$:

$$\{U\}_{i+1}=\{U\}_i+\Delta t\{\dot{U}\}_i+\frac{\Delta t^2}{2}\left[(1-2\beta)\{\ddot{U}\}_i+2\beta\{\ddot{U}\}_{i+1}\right] \tag{17.4.4}$$

$$\{\dot{U}\}_{i+1}=\{\dot{U}\}_i+\Delta t\left[(1-\gamma)\{\ddot{U}\}_i+\gamma\{\ddot{U}\}_{i+1}\right] \tag{17.4.5}$$

One can actually combine Equations (17.4.3), (17.4.4), and (17.4.5) to establish a system of equations for the updated displacement vector:

$$\left[\widetilde{K}\right]\{U\}_{i+1} = \left\{\widetilde{f}\right\}_{i+1} \tag{17.4.6}$$

where

$$\left[\widetilde{K}\right] = \frac{1}{\beta \cdot \Delta t^2}[M] + \frac{\gamma}{\beta \cdot \Delta t}[C] + [K] \tag{17.4.7}$$

and

$$\left\{\widetilde{f}\right\}_{i+1} = \{f\}_{i+1} + \left(\frac{1}{\beta \cdot \Delta t^2}[M] + \frac{\gamma}{\beta \cdot \Delta t}[C]\right)\{U\}_i + \left[\frac{1}{\beta \cdot \Delta t}[M] + \left(\frac{\gamma}{\beta} - 1\right)[C]\right]\{\dot{U}\}_i$$
$$+ \left[\left(\frac{1}{2\beta} - 1\right)[M] + \frac{\Delta t}{2}\left(\frac{\gamma}{\beta} - 2\right)[C]\right]\{\ddot{U}\}_i \tag{17.4.8}$$

The algorithm for transient step-by-step solution using Newmark's method is summarized in Box 17.4.2.

Remark 17.4.3: The assembly and inversion of the $\left[\widetilde{K}\right]$-array in Equation (17.4.6) needs to be conducted only once in a linear analysis. To obtain this array, we need to calculate the contribution $\left[\widetilde{k}^{(e)}\right]$ of each element to $\left[\widetilde{K}\right]$ and conduct an assembly operation just like that employed for the stiffness matrix in static analysis. A similar assembly operation is conducted for the vector $\left\{\widetilde{f}\right\}_{i+1}$, which generally changes in a dynamic analysis and thus needs to be assembled at each step. ■

Remark 17.4.4: If the HHT method is used for analysis of a MDOF system, then Equation (17.4.6) is still valid, but the expressions for $\left[\widetilde{K}\right]$ and $\left\{\widetilde{f}\right\}_{i+1}$ differ. Specifically, the following equations must be employed.

$$\left[\widetilde{K}\right] = \frac{1}{\beta \cdot \Delta t^2}[M] + \frac{\gamma(1+a)}{\beta \cdot \Delta t}[C] + (1+a)[K] \tag{17.4.9}$$

and

$$\left\{\widetilde{f}\right\}_{i+1} = \{f\}_{i+1} + \left(\frac{1}{\beta \cdot \Delta t^2}[M] + \frac{\gamma(1+a)}{\beta \cdot \Delta t}[C] + a[K]\right)\{U\}_i$$
$$+ \left[\frac{1}{\beta \cdot \Delta t}[M] + \left(\frac{\gamma}{\beta} + \frac{a \cdot \gamma}{\beta} - 1\right)[C]\right]\{\dot{U}\}_i + \left[\left(\frac{1}{2\beta} - 1\right)[M] + \frac{\Delta t}{2}\left(\frac{\gamma}{\beta} - 2\right)(1+a)[C]\right]\{\ddot{U}\}_i \tag{17.4.10}$$

We can then use the algorithm of Box 17.4.2 for the HHT method, except that we need to appropriately modify the expressions for $\left[\widetilde{k}^{(e)}\right]$ and $\left\{\widetilde{f}^{(e)}\right\}_{i+1}$. ■

Box 17.4.2 Newmark Method for Step-by-step Solution of Equations of Motion

0) Initial conditions: If we know the initial displacement vector, $\{U\}_0$, and the initial velocity vector, $\{V\}_0$, set:

$$\textit{initial displacement}: \ \{U\}_0, \ \textit{initial velocity}: \ \{\dot{U}\}_0 = \{V\}_0$$

Then find initial acceleration, $\{\ddot{U}\}_0$, by solving: $[M]\{\ddot{U}\}_0 = \{f\}_0 - [C]\{\dot{U}\}_0 - [K]\{U\}_0$

For each time step *i*:

1) Given the displacement vector, $\{U\}_i$, the velocity vector, $\{\dot{U}\}_i$, and the acceleration vector, $\{\ddot{U}\}_i$:

   For each element, *e*:

   Find the element displacement, velocity, and acceleration vectors:

$$\left\{U^{(e)}\right\}_i = \left[L^{(e)}\right]\{U\}_i, \ \left\{\dot{U}^{(e)}\right\}_i = \left[L^{(e)}\right]\{\dot{U}\}_i, \ \left\{\ddot{U}^{(e)}\right\}_i = \left[L^{(e)}\right]\{\ddot{U}\}_i$$

Calculate the element contributions to the right-hand side of Equation (17.4.8):

$$\left\{\tilde{f}^{(e)}\right\}_{i+1} = \{f^{(e)}\}_{i+1} + \left(\frac{1}{\beta\cdot\Delta t^2}\left[m^{(e)}\right] + \frac{\gamma}{\beta\cdot\Delta t}\left[c^{(e)}\right]\right)\{U^{(e)}\}_i + \left[\frac{1}{\beta\cdot\Delta t}\left[m^{(e)}\right]\right.$$

$$\left. + \left(\frac{\gamma}{\beta}-1\right)\left[c^{(e)}\right]\right]\left\{\dot{U}^{(e)}\right\}_i + \left[\left(\frac{1}{2\beta}-1\right)\left[m^{(e)}\right] + \frac{\Delta t}{2}\left(\frac{\gamma}{\beta}-2\right)\left[c^{(e)}\right]\right]\left\{\ddot{U}^{(e)}\right\}_i$$

Calculate $\left[\tilde{k}^{(e)}\right] = \frac{1}{\beta\cdot\Delta t^2}\left[m^{(e)}\right] + \frac{\gamma}{\beta\cdot\Delta t}\left[c^{(e)}\right] + \left[k^{(e)}\right]$

2) Assemble the global effective force vector $\left\{\tilde{f}\right\}_{i+1}$ and the global effective stiffness matrix $\left[\tilde{K}\right]$:

$$\left\{\tilde{f}\right\}_{i+1} = \sum_{e=1}^{Ne}\left(\left[L^{(e)}\right]^T\left\{\tilde{f}^{(e)}\right\}_{i+1}\right)$$

$$\left[\tilde{K}\right] = \sum_{e=1}^{Ne}\left(\left[L^{(e)}\right]^T\left[\tilde{k}^{(e)}\right]\left[L^{(e)}\right]\right)$$

(*Note:* If Δt is constant, then we can assemble $\left[\tilde{K}\right]$ only once in an analysis).

3) Update the displacement vector, acceleration vector, and velocity vector (i.e., find the vector values of the new step, $i+1$):

$$\left[\tilde{K}\right]\{U\}_{i+1} = \left\{\tilde{f}\right\}_{i+1} \rightarrow \{U\}_{i+1} = \left[\tilde{K}\right]^{-1}\left\{\tilde{f}\right\}_{i+1}$$

$$\{\ddot{U}\}_{i+1} = \frac{1}{\beta\cdot\Delta t^2}\left(\{U\}_{i+1} - \{U\}_i\right) - \frac{1}{\beta\cdot\Delta t}\{\dot{U}\}_i - \frac{(1-2\beta)}{2\beta}\{\ddot{U}\}_i$$

$$\{\dot{U}\}_{i+1} = \{\dot{U}\}_i + \Delta t\left[(1-\gamma)\{\ddot{U}\}_i + \gamma\{\ddot{U}\}_{i+1}\right]$$

4) If we have run all the time steps we needed to run, terminate the algorithm. Otherwise, Set $i \leftarrow i+1$ and go to Step 1.

Problems

17.1 a) Use the constant-average acceleration method (Newmark method with $\beta = 0.25$, $\gamma = 0.5$) for the system in Problem 15.1, and obtain the time histories of displacement, velocity, and acceleration in the y-direction for node 8, between $t = 0$ and $t = 10$ seconds. Use a time-step $\Delta t = 0.1$ second.

b) Same as in part a, but use the HHT method with $a = -0.3$. Select appropriate values for parameters β and γ to ensure unconditional stability.

17.2 We are given the one-dimensional truss structure shown in Figure P17.1, which has a prescribed displacement at node 4. The material in all three members is linearly elastic, with $E = 29000$ ksi. The density of the material is $\rho = 7.5 \cdot 10^{-7}$ kip·s^2/in.^{4} for all members. The cross-sectional area is equal to 2 in.^{2} for member (1-2), 1.5 in.^{2} for member (2-3), and 1 in.^{2} for member (2-4). The mass matrix is diagonal, obtained from the row-sum technique (Section 15.9.1). The applied displacement at node 4 is zero at $t = 0$, then it increases linearly with time and attains the value shown in Figure P17.1 at $t = 1$ second. After that, the displacement of node 4 remains constant, equal to the value of Figure P17.1. The system does not have any viscous damping.

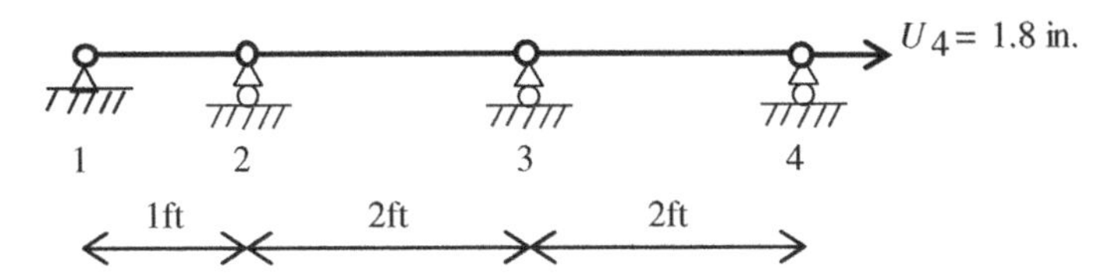

Figure P17.1

a) Obtain the time history of applied horizontal force (reaction) at node 4 and of reaction at node 1, between $t = 0$ and $t = 2$ sec, using the explicit central difference method with $\Delta t = 5 \cdot 10^{-5}$ seconds. Print the requested quantities every 0.005 seconds.

b) Same as in a, but use a time step of 0.005 seconds for the solution. Print the requested quantities for the plot every 0.005 seconds. What do you observe? What explanation do you have for what you observe?

c) Same as in a, but for the case that the displacement of node 4 reaches the target value of Figure P17.1 at $t = 0.05$ seconds. Print the requested quantities for the plot every 0.0005 seconds. How does the shape of the reaction time history at node 1 differ compared to that obtained in a? What explanation do you have for what you observe?

References

Belytschko, T., Liu, W.K., Moran, B., and Elkhodary, K. (2014). *Nonlinear Finite Elements for Continua and Structures*. Hoboken, NJ: John Wiley & Sons.

Chopra, A. (2011). *Dynamics of Structures*, 4th ed. Upper Saddle River, NJ: Prentice-Hall.

Hilber, H., Hughes, T. J. R., and Taylor, R. (1977). "Improved Numerical Dissipation for Time Integration Algorithms in Structural Dynamics." *Earthquake Engineering and Structural Dynamics*, 5, 283–292.

Hughes, T. (2000). *The Finite Element Method—Linear Static and Dynamic Finite Element Analysis*, 2nd ed. New York: Dover.

Newmark, N. (1959). "A Method of Computation for Structural Dynamics." *ASCE Journal of the Engineering Mechanics Division*, 67–94.

18

Verification and Validation for the Finite Element Method

Christopher Roy

18.1 Introduction

Mathematical *models* are used in the finite element method (FEM) to describe the behavior of structural systems. The complexity of the governing equations, which are a set of coupled partial differential or integral equations, generally requires that solutions be obtained using numerical techniques (i.e., *simulations*) through the use of FEM software. Verification and validation provide a means for assessing the credibility and accuracy of mathematical models and their subsequent simulations (see Oberkampf and Roy 2010, Roy and Oberkampf 2016, Roache 2009, AIAA 1998, ASME 2006, and ASME 2009). *Verification* deals with assessing the numerical accuracy of a simulation relative to the true solution to the mathematical model. Verification includes two distinct parts: code verification and solution verification. *Validation*, on the other hand, is the assessment of the accuracy of the mathematical model relative to observations of nature that come in the form of experimental measurements. Verification and validation activities are applicable to a broad range of scientific computing disciplines. As such, the content of this chapter draws heavily from the author's prior book (Oberkampf and Roy 2010) and book chapter (Roy and Oberkampf 2016).

18.2 Code Verification

Code verification ensures that the FEM software is an accurate representation of the underlying governing equations and their solution. It is accomplished by employing appropriate software engineering practices and by using code order of accuracy testing to ensure that there are no mistakes in the computer code or inconsistencies in the chosen FEM algorithm for solving the discretized form of the governing equations. While software engineering is a vast subject unto itself, some aspects that are particularly

Fundamentals of Finite Element Analysis: Linear Finite Element Analysis, First Edition.
Ioannis Koutromanos, James McClure, and Christopher Roy.

Companion website: www.wiley.com/go/koutromanos/linear

useful for FEM software include version control, static analysis, dynamic testing, and regression testing (see Oberkampf and Roy 2010, Pressman 2005, and Sommerville 2004 for more details). For more information on the code verification process, see Oberkampf and Roy 2010, Roache 2009, Knupp and Salari 2003, and Roy 2005.

Before proceeding with a discussion of code verification, it is important to identify what simulation output quantities should be evaluated. Traditionally, finite element methods focused on the energy norm of the discretization error, which was defined in Remark 11.1.4 and can be written as:

$$\left\| e_{en(h)} \right\| = \left[\frac{1}{V_\Omega} \int_\Omega \left\| \vec{\nabla}\, \upsilon_{(h)} - \vec{\nabla}\, \upsilon \right\|^2 d\Omega \right]^{1/2} \tag{18.2.1}$$

where $\upsilon_{(h)}$ is the finite element solution, υ is the exact solution to the governing equations and V_Ω is the total volume of the domain. The energy norm is natural to consider since the FEM can be shown to produce a numerical solution that minimizes the energy norm of the discretization error (Szabo and Babuska 1991). Thus, the energy norm provides a global measure of the overall optimality of the finite element solution. One could also examine error norms (e.g., L_1, L_2, and L_∞ norms) of the dependent variables themselves. For example, the L_2–error norm, $\|e_{(h)}\|_0$ for a FEM solution can be computed in accordance with Section 11.1, as follows.

$$\left\| e_{(h)} \right\|_0 = \left[\frac{1}{V_\Omega} \int_\Omega \left(\upsilon_{(h)} - \upsilon \right)^2 dV \right]^{1/2} \tag{18.2.2}$$

where $\upsilon_{(h)}$ is the FEM solution in the domain Ω and υ is the exact solution to the governing equations. Finally, one should examine the error in any system response quantities (SRQs) that one is interested in predicting.

18.2.1 Order of Accuracy Testing

Various criteria can be used for testing the correctness of software during code verification. However, the most rigorous code verification criterion is the order of accuracy test, where one assesses whether the numerical solutions converge to the exact solution to the governing equations at the expected rate (i.e., the *formal order of accuracy*) for the numerical algorithm. The formal order of accuracy of an algorithm is commonly estimated by performing a truncation error analysis that addresses the convergence of the discrete equations to the original governing equations; however, the order of accuracy of the *truncation error* is generally one order lower than the order of the numerical solution error (i.e., the *discretization error*) on unstructured grids (Katz and Sankaran 2012, Diskin and Thomas 2012). For consistent numerical algorithms, the truncation error will be proportional to the discretization parameters (e.g., spatial element size Δx, time step size Δt) to some exponents that usually correspond to the formal order of accuracy of the discretization scheme.

The *observed order of accuracy* is the actual rate at which the numerical solutions converge to the exact solution to the governing equations with systematic refinement of the mesh and/or time step (Oberkampf and Roy 2010, Roache 2009). Consider a power series expansion of some SRQ in terms of a generic discretization parameter h about the exact solution to the governing equations υ:

$$\upsilon_{(h)} = \upsilon + g_{p_c} h^{p_c} + O\left(h^{p_c+1}\right) \tag{18.2.3}$$

A similar expansion with a discretization parameter that is r times larger (e.g., $r = h_{\text{coarse}} \ / \ h_{\text{fine}} = \Delta x_{\text{coarse}} \ / \ \Delta x_{\text{fine}}$) gives:

$$\upsilon_{(rh)} = \upsilon + g_{p_c} (rh)^{p_c} + O\left(h^{p_c+1}\right) \tag{18.2.4}$$

where r is called the grid refinement factor. Combining these two expressions, neglecting the higher-order terms, and solving for p_c yields an expression for the observed order of accuracy $\hat{p}_c$:

$$\hat{p}_c = \frac{ln\left(\frac{\upsilon_{(rh)} - \upsilon}{\upsilon_{(h)} - \upsilon}\right)}{ln(r)} \tag{18.2.5}$$

where $\upsilon_{(rh)}$ and $\upsilon_{(h)}$ are the coarse- and fine-mesh SRQs, respectively. This expression for the observed order of accuracy requires solutions on two systematically refined mesh levels as well as knowledge of the exact solution to the governing equations υ.

By comparing the observed order of accuracy with the formal order of accuracy, one can *empirically test* if $\hat{p}_c$ approaches the formal order of accuracy with systematic mesh refinement (discussed further in Section 18.2.2). In addition to requiring that there are no mistakes in the software programming or weaknesses in the numerical algorithms that affect the solution, the following conditions are required to pass the order of accuracy test. First, the iterative and round-off errors in the numerical solution must be *significantly less* (we recommend two orders of magnitude smaller) than the fine grid discretization error (i.e., the difference between the fine grid numerical solution and the exact solution to the governing equations). Second, the mesh and time step must be sufficiently small so that the lowest-order terms in the truncation and discretization error expansions dominate the higher-order terms in the power series expansion for the numerical solution (i.e., the numerical solutions must be in the asymptotic convergence range). Third, the mesh and time step must be systematically refined as discussed in the next section.

18.2.2 Systematic Mesh Refinement

Systematic mesh refinement (Oberkampf and Roy 2010) requires that the mesh be refined uniformly by a factor h in each coordinate direction, for example,

$$h = \frac{\Delta x}{\Delta x_{ref}} = \frac{\Delta y}{\Delta y_{ref}} = \frac{\Delta z}{\Delta z_{ref}} \tag{18.2.6}$$

and that the mesh quality be constant or improve with mesh refinement. While it might be advantageous to perform local mesh adaptation to reduce discretization errors, local adaptation is not part of systematic mesh refinement and thus code order of accuracy verification. Ensuring systematic mesh refinement can be challenging for unstructured meshes, especially for those that contain more than one element type (e.g., hexahedral, tetrahedral, and prismatic elements). For structured grids, where grid transformations can be used to transform the grid to a Cartesian computational space, systematic refinement can be ensured by starting with the fine grid and removing every other grid line in each direction, resulting in a grid refinement factor of $r = 2$. The practical drawback to this approach is that each level of refinement requires a factor of 8 increase in elements for three-dimensional problems.

18.2.3 Exact Solutions

Rigorous code order of accuracy testing requires an exact solution to the governing equations. Traditional methods of obtaining exact solutions to partial differential or integral equations are rarely available for the governing equations of mechanics, which may include complicated geometry, non-constant coefficients, complicated submodels (e.g., surface contact, aging, crack propagation), and/or multi-physics coupling. When exact solutions are found for complex governing equations, they often depend on significant simplifications in dimensionality, geometry, or physics, for example.

An alternative to the traditional approach for obtaining exact solutions to governing equations is the *Method of Manufactured Solutions* (MMS) (Roache and Steinberg 1984, Knupp and Salari 2003, Roy et al. 2004, Roy 2005, Roache 2009, and Oberkampf and Roy 2010). The concept behind MMS is to take an original governing equation and modify it by appending an analytic source term so that it satisfies a chosen (usually nonphysical) solution. Consider an original governing equation with dependent variable v written in operator notation as $L(v) + s = 0$, where s is the *source term,* denoting the set of terms in the differential equations that do not depend on the solution v. For heat conduction problems, s is merely the heat source. For elasticity, s is the body force vector. Next, choose an analytic manufactured solution $\hat{v}$ which has nontrivial analytic derivatives. The governing equation is then operated onto the manufactured solution in order to obtain the analytic source term: $s = -L(v)$. The modified governing equation is found by appending this source term to the original governing equation $L(v) + s = 0$, which will be exactly solved by the chosen manufactured solution $\hat{v}$.

Manufactured solutions should be chosen to be analytic functions with smooth derivatives. It is also important to ensure that all of the derivatives of the manufactured solution appearing in the governing equation are nonzero. Trigonometric and exponential functions are recommended since they are smooth and infinitely differentiable. Although the manufactured solutions do not need to be physically realistic when used for code verification, they should be chosen to obey the physical constraints that are embodied in the code (e.g., positive temperatures and energies). Finally, care should be taken that one or two terms in the governing equations do not dominate the other terms.

Example 18.1: Method of manufactured solutions for one-dimensional elasticity
We use a specific example of a one-dimensional elasticity problem to demonstrate the steps for MMS. Let us focus on a one-dimensional elastic bar, shown in Figure 18.1a with $E = 4000$, $A = 100$, and $L = 100$. We could just as easily set "unrealistic values" for these parameters (since, as mentioned above, a manufactured solution does not have to be "physically meaningful"), but let us use these specific "realistic values." We then begin by coming up with an exact, manufactured solution for the actual displacement field, $u(x)$. For example, let us set $\hat{u}(x) = sin\left(\frac{\pi \cdot x}{L}\right)$

We can then readily find the quantity:

$$\frac{d}{dx}\left(A\cdot E\cdot\frac{d\hat{u}}{dx}\right)=\frac{d}{dx}\left[100\cdot 4000\cdot\frac{d}{dx}\left[sin\left(\frac{\pi\cdot x}{100}\right)\right]\right]=-40\pi^2\cdot sin\left(\frac{\pi\cdot x}{100}\right)$$

We know that, per the descriptions in Box 2.2.1, we need to satisfy the following differential equation:

$$\frac{d}{dx}\left(A\cdot E\cdot\frac{du}{dx}\right)+b(x)=0.$$

This allows us to directly find the distributed force $b(x)$ that we need to have so that our manufactured solution $\hat{u}(x)$ satisfies the differential equation of equilibrium:

$$b(x)=-\frac{d}{dx}\left(A\cdot E\cdot\frac{d\hat{u}}{dx}\right)=40\pi^2\cdot sin\left(\frac{\pi\cdot x}{100}\right)$$

This distributed force will be given as input in our finite element analysis to compare the finite element solution to the "exact," manufactured one. Of course, to have a well-defined model for finite element analysis, we also need to define the boundary conditions. It is always preferable to have both essential and natural boundary conditions. For this specific problem, let us use an essential boundary condition, that is, a prescribed displacement, for the left end of the bar ($x=0$), as shown in Figure 18.1a, and a natural boundary condition, that is, a prescribed traction, for the right end of the bar ($x=100$), as shown in the same figure. The prescribed values at the essential and natural boundary will be the displacement and traction, respectively, that we obtain from our manufactured solution. Thus, for the location $x=0$, where we have decided to use an essential boundary condition, we will prescribe a displacement $u(x=0)=40\pi^2\cdot sin\left(\frac{\pi\cdot 0}{100}\right)=0$ at the left end of the bar, and a traction $\left[n\cdot E\frac{du}{dx}\right]_{x=100}=-\left[1\cdot 40\pi\cdot cos\left(\frac{\pi\cdot x}{100}\right)\right]_{x=100}=40\pi$ at the right end of the bar.

We will now proceed to obtain a finite element solution with a five-element mesh, each element having two nodal points. As shown in Figure 18.1b, our approximate solution

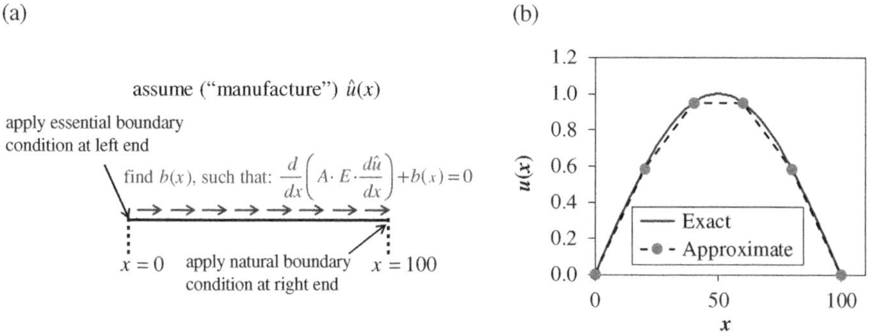

(a) Process for defining the problem corresponding to the manufactured solution

(b) Qualitative comparison of finite element solution with manufactured solution

Figure 18.1 Example use of the method of manufactured solutions for one-dimensional elasticity.

closely matches the exact, manufactured one. If we were to further refine the mesh (e.g., use 10 two-node elements), then the approximate and exact solutions would appear to practically coincide. It is important to emphasize that, for the specific example, we merely rely on a qualitative comparison of the exact and finite element solution and do not pursue a rigorous code verification, because the example is aimed to demonstrate the steps associated with the MMS. ■

Example 18.2: Method of manufactured solutions for a two-dimensional scalar field problem
We will now establish a manufactured solution for a chemical diffusion problem (see Section 5.4) in the rectangular, two-dimensional domain shown in Figure 18.2a.

Let us manufacture the following concentration field: $\hat{C}(x,y) = \left(\frac{x}{6} - \frac{x^2}{18} + \frac{x^3}{216}\right)\left(1 - \frac{y^2}{16}\right)$. The corresponding gradient of the concentration is: $\nabla\hat{C} = \begin{Bmatrix} \partial\hat{C}/\partial x \\ \partial\hat{C}/\partial y \end{Bmatrix} = \begin{Bmatrix} \left(\frac{1}{6} - \frac{x}{9} + \frac{x^2}{72}\right)\left(1 - \frac{y^2}{16}\right) \\ -\left(\frac{x}{6} - \frac{x^2}{18} + \frac{x^3}{216}\right)\cdot\frac{y}{8} \end{Bmatrix}$. Let us also assume that our manufactured solution corresponds to a value $D_C = 1$.

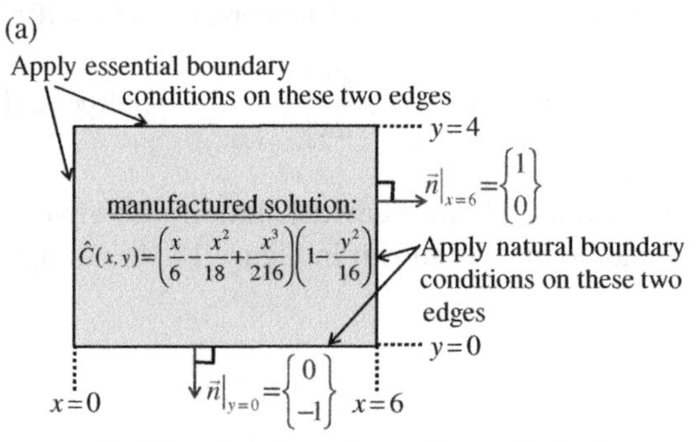

Problem domain and manufactured solution

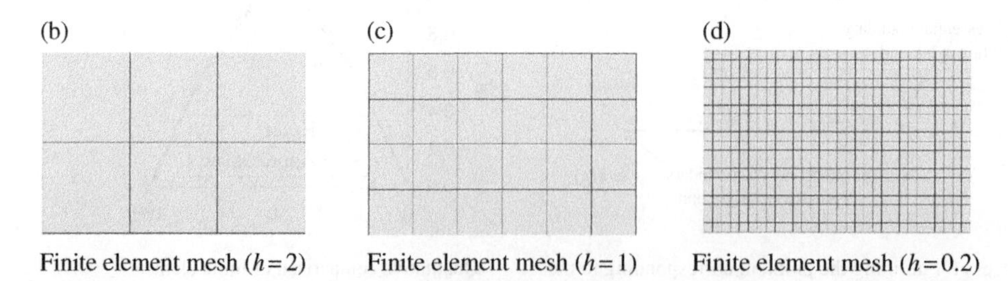

Finite element mesh (h=2) Finite element mesh (h=1) Finite element mesh (h=0.2)

Figure 18.2 Example use of the method of manufactured solutions for two-dimensional scalar field problem.

We can now use the differential Equation (5.4.17) for chemical diffusion (presented in Box 5.4.3) to obtain what should be the reaction term, $R(x,y)$. Thus, we have:

$$R(x,y) = -\nabla\bullet\left(D_C\nabla\hat{C}\right) = -\nabla\bullet\left(\nabla\hat{C}\right) = -\nabla^2\hat{C} = -\frac{\partial^2\hat{C}}{\partial x^2} - \frac{\partial^2\hat{C}}{\partial y^2}$$

$$= -\left(-\frac{1}{9}+\frac{x}{36}\right)\left(1-\frac{y^2}{16}\right)+\frac{1}{8}\cdot\left(\frac{x}{6}-\frac{x^2}{18}+\frac{x^3}{216}\right)$$

We will apply essential boundary conditions at two edges of the domain, and natural boundary conditions at the other two edges, as shown in Figure 18.2a. Let us see what boundary conditions we need to apply:

- For the left vertical edge, where we have $x = 0$: $C(x=0,y) = \left(\frac{0}{6}-\frac{0^2}{18}+\frac{0^3}{216}\right)\left(1-\frac{y^2}{16}\right) = 0$
- For the top horizontal edge, where we have $y = 4$: $C(x,y=4) = \left(\frac{x}{6}-\frac{x^2}{18}+\frac{x^3}{216}\right)\left(1-\frac{4^2}{16}\right) = 0$
- For the right vertical edge, where we have $x = 6$ and $\vec{n}\big|_{x=6} = \begin{Bmatrix}1\\0\end{Bmatrix}$

$$\bar{q}_C\big|_{x=6} = -\left(D_C\vec{\nabla}C\right)\Big|_{x=6}\bullet\vec{n}\big|_{x=6} = -\left(\frac{1}{6}-\frac{x}{9}+\frac{x^2}{72}\right)\left(1-\frac{y^2}{16}\right)\Bigg|_{x=6} = 0$$

- For the bottom horizontal edge, where we have $y = 0$ and $\vec{n}\big|_{x=0} = \begin{Bmatrix}0\\-1\end{Bmatrix}$

$$\bar{q}_C\big|_{y=0} = -\left(D_C\vec{\nabla}C\right)\Big|_{y=0}\bullet\vec{n}\big|_{y=0} = -\left(\frac{x}{6}-\frac{x^2}{18}+\frac{x^3}{216}\right)\cdot\frac{y}{8}\Bigg|_{y=0} = 0$$

We are now going to obtain the finite element solution for three different finite element discretizations, as shown in Figures 18.2b, 18.2c, and 18.2d. Each case will use four-node quadrilateral (4Q) elements. We will select $\Delta x_{ref} = \Delta y_{ref} = \Delta z_{ref} = 1$. The cases presented in Figures 18.2b, 18.2c, and 18.2d correspond to values of discretization parameter h equal to 2, 1, and 0.2, respectively.

We can create the contour plots of the error field, as shown for the three discretizations in Figure 18.3. It can be easily seen that the error reduces as we refine the mesh.

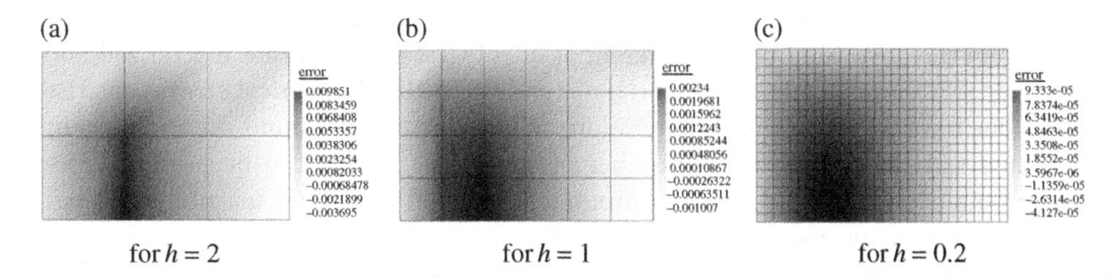

Figure 18.3 Contours of approximation error for example use of the MMS for scalar-field problem, and for different values of the discretization parameter h.

Table 18.1 provides the value of the error norm $||e_{(h)}||_0$ obtained for each discretization case. We can further create a plot of the error norm $||e_{(h)}||_0$ as a function of the discretization parameter h, as shown in Figure 18.4a. The order of accuracy of the method can be verified by creating a logarithmic plot of $||e_{(h)}||_0$ as a function of h. As shown in Figure 18.4b, this plot is a straight line with a slope equal to 2. This should be expected, because the value of the slope is equal to p_c. For 4Q elements, we have a degree of completeness $k = 1$. Plugging this value in Equation (11.1.7a), we can obtain that the error norm $||e_{(h)}||_0$ must be proportional to h^2. This leads to the conclusion that the logarithmic plot of Figure 18.4b must be a straight line with a slope equal to 2. ∎

Table 18.1 Values of Error Norm for the Different Discretization Parameter Values

h	2.0	1.0	0.2
$\|\|e_{(h)}\|\|_0$	$9.68 \cdot 10^{-3}$	$2.37 \cdot 10^{-3}$	$9.46 \cdot 10^{-5}$

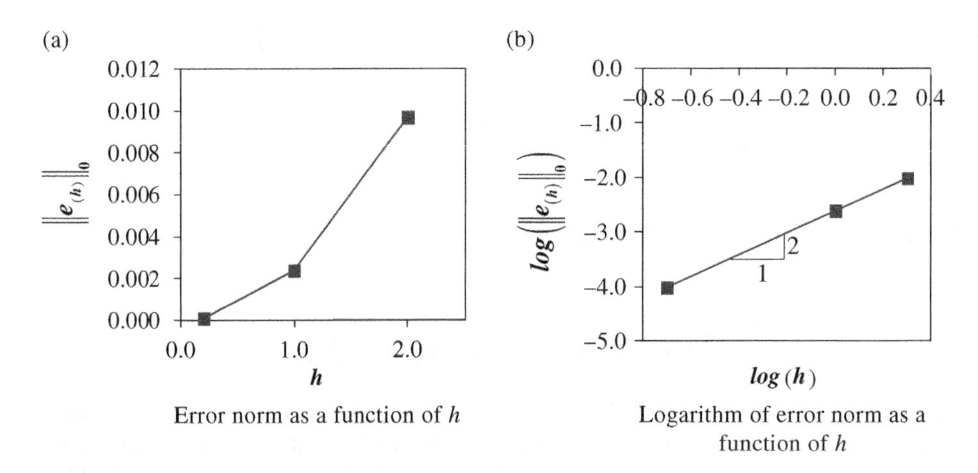

Figure 18.4 Verification of order of accuracy for analysis of manufactured chemical diffusion problem.

18.3 Solution Verification

The main focus of solution verification is the estimation of the numerical errors that occur when the governing partial differential or integral equations are discretized and solved numerically (Roache 2009, Oberkampf and Roy 2010). Numerical errors can arise in FEM due to computer round-off, iteration, and discretization. Round-off errors occur due to the fact that only a finite number of significant figures can be used to store floating-point numbers in a digital computer. Round-off errors are usually small, but can be reduced if necessary by increasing the number of significant figures used in floating point computations (e.g., by changing from single to double precision arithmetic). Iterative convergence errors are present when the global system of finite element equations is solved with iterative solution algorithms (described in detail in Section 19.3).

This type of errors is described in the following subsection. Discretization error arises due to the fact that the spatial domain of the governing equations is decomposed into elements and, for unsteady problems, time is advanced with a finite time step. Iterative and discretization errors are discussed in detail below. Further discussion of all three error types can be found in Oberkampf and Roy 2010.

18.3.1 Iterative Error

Let us assume that we are trying to solve the system of linear equations corresponding to a finite element analysis. We are interested in obtaining an approximate field $\upsilon_{(h)}$. Let us further assume that we are using an iterative solution algorithm for the system of equations, and we have already conducted a total of 'k' iterations for our solution. We can define the *iterative error* as the difference between the approximate solution $\upsilon_{(h)}^{k}$ corresponding to the nodal values of the k-th iteration of solving the linear system of equations, and the actual finite element solution $\upsilon_{(h)}$. We can thus write the following mathematical expression.

$$e_{(h)}^{k} = \upsilon_{(h)}^{k} - \upsilon_{(h)} \tag{18.3.1}$$

where h refers to the discrete solution on a mesh with discretization parameters ($\Delta x, \Delta y$, Δt, etc.) represented collectively by h, $\upsilon_{(h)}^{k}$ is the current iterative solution, and $\upsilon_{(h)}$ is the exact solution to the discrete equations (not to be confused with the exact solution to the governing equations υ).

For stationary iterative methods (e.g., Jacobi, Gauss-Seidel, multigrid) applied to linear systems, iterative convergence is governed by the eigenvalues of the iteration matrix. For linear problems, when the maximum eigenvalue of the iteration matrix is real, the limiting iterative convergence behavior will be monotone. When it is complex, however, the limiting iterative convergence behavior will generally be oscillatory. In these cases, convergence of the iterative method requires that the spectral radius of the iteration matrix be less than unity (Golub and Van Loan 1996). For nonlinear problems, the linearized system is often not solved to convergence, but only solved for a few iterations (sometimes as few as one) before the nonlinear terms are updated, and the convergence behavior is often much more difficult to characterize.

The discrete equations can be symbolically written in the form

$$L_h\left(\upsilon_{(h)}\right) = 0 \tag{18.3.2}$$

where L_h can be thought of as a finite element discretization operator and $\upsilon_{(h)}$ is the finite element solution. The *iterative residual* is found by substituting the current iterative solution $\upsilon_{(h)}^{k}$ into Equation (18.3.2):

$$\mathfrak{R}_{(h)}^{k} = L_h\left(\upsilon_{(h)}^{k}\right) \tag{18.3.3}$$

where $\mathfrak{R}_{(h)}^{k} \to 0$ as $\upsilon_{(h)}^{k} \to \upsilon_{(h)}$. Although monitoring the iterative residuals of each of the discretized equations often serves as an adequate indication as to whether iterative convergence of the solution has been satisfactorily achieved, it does not by itself provide any guidance as to the *size* of the iterative error in the SRQ of interest. Since the iterative residual norms have been shown to follow closely with actual iterative errors for many problems (Oberkampf and Roy 2010, Roy 2005, and Roy et al. 2004), a small number of cases should be sufficient to determine how the iterative errors in the SRQ scale with the iterative residuals for the range of cases of interest.

An example of this procedure is given in Figure 18.5 for laminar viscous flow through a packed bed of spherical particles (Duggirala et al. 2008). The SRQ (f) of interest is the average pressure gradient across the bed, and the desired relative iterative error level in the pressure gradient is 0.01%. The iterative residuals in the conservation of mass and conservation of momentum equations are first converged to 10^{-7} (relative to their initial levels). The value of the pressure gradient at this point ($\hat{f}_{(h)}$) is taken as an approximation to the exact solution to the discrete equations, that is, $\hat{f}_{(h)} \approx f_{(h)}$. The iterative error for all previous iterations is then approximated as $e^k_{(h)} \approx f^k_{(h)} - \hat{f}_{(h)}$. Figure 18.5 shows that to achieve the desired iterative error level in the pressure gradient of 0.01%, the iterative residual norms should be converged down to approximately 10^{-6}. Simulations for similar problems can be expected to require approximately the same level of iterative residual convergence in order to achieve the same iterative error tolerance in the pressure gradient.

18.3.2 Discretization Error

The discretization error is the difference between the exact solution to the discretized equations and the exact solution to the original governing equations. The discretization error is difficult to estimate for practical FEM applications and is often the largest of the numerical error sources. Methods for estimating discretization error can be broadly categorized as either recovery methods or residual-based estimators (Oberkampf and Roy 2010). Recovery methods involve post-processing of the solution(s) and include Richardson extrapolation (Roache 2009, Oberkampf and Roy 2010), order extrapolation

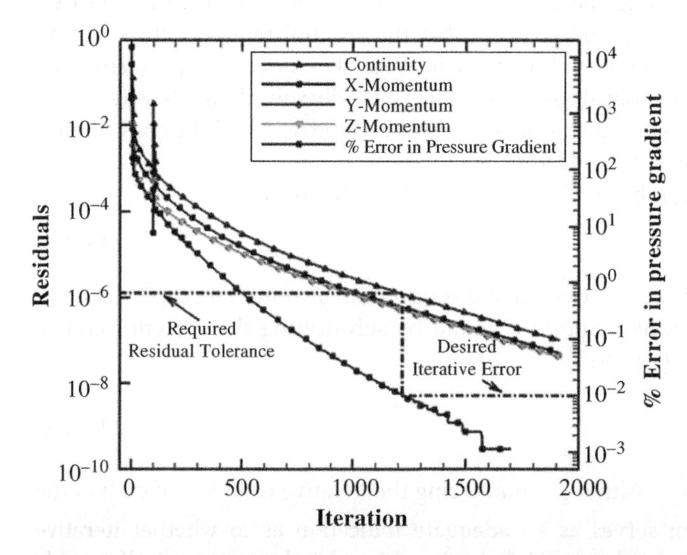

Figure 18.5 Norms of the iterative residuals (left axis) and percentage error in pressure gradient (right axis) for laminar flow through a packed bed Duggirala 2008. Reproduced with permission from American Society of Mechanical Engineers.

(Bank 1996, Oberkampf and Roy 2010), and finite element recovery methods (Zienkiewicz and Zhu 1992, Ainsworth and Oden 2000, Zhang and Naga 2005). The residual-based methods employ additional information about the problem being solved and include discretization error transport equations (e.g., Zhang et al. 2000, Roy 2009, Shih and Williams 2009, Phillips and Roy 2011), defect correction methods (Stetter 1978, Skeel 1986), implicit/explicit residual methods in finite elements (Stewart and Hughes 1998, Ainsworth and Oden 2000, and Cao 2005), and adjoint methods for estimating the error in solution functionals / SRQs (Ainsworth and Oden 2000, Pierce and Giles 2000, Venditti and Darmofal 2000 and 2003). The recovery methods have the drawback of requiring multiple mesh levels (Richardson extrapolation), requiring multiple solutions with different order (order extrapolation), or providing accurate error estimates for only a limited class of problems (finite element recovery methods). The residual methods all require an additional solution to be computed, generally on the same grid, and may provide more accurate error estimates since they use additional information about the problem being solved. As they are the simplest to implement, we will focus our discussion on finite element recovery methods and Richardson extrapolation.

Finite element recovery methods for estimating the discretization error were pioneered by Zienkiewicz and Zhu (1987, 1992). For the standard finite element method, gradient-related quantities (i.e., partial derivatives) of the field function are not continuous across inter-element boundaries. The user of a finite element code is often more interested in gradient quantities such as stresses than the solution itself, so most finite element codes provide for reconstruction of these discontinuous gradients into piece-wise linear gradients using existing finite element infrastructure.

In some cases (see the discussion of superconvergence below), this reconstructed gradient is of a higher order of accuracy than the gradient found in the underlying finite element solution. Recall the definition of the energy norm of the discretization error given in Equation (18.1.1). If the true gradient from the mathematical model is available, then this important error measure can be computed exactly. For the case where a reconstructed gradient is higher-order accurate than the finite element gradient, then it can be used to approximate the true gradient in the energy norm. In addition to providing estimates of the local discretization error in the solution gradients, due to their local nature, recovery methods are also often used as indicators of where solution refinement is needed in adaptive solutions.

In order to justify the use of the recovered gradient in the energy norm, it must in fact be higher-order accurate than the gradient from the finite element solution. This so-called *superconvergence property* can occur when certain regularity conditions on the mesh and the solution are met (Wahlbin 1995) and results in gradients that are up to one order higher in accuracy than the underlying finite element gradients. For finite elements with linear shape functions, the superconvergence points occur at the element centroids, whereas for higher-order finite elements, the location of the superconvergence points depends on the element topology. If the reconstructed gradient is superconvergent, and if certain consistency conditions are met by the gradient reconstruction operator itself, then error estimators based on this recovered gradient can be shown to be asymptotically exact (Ainsworth and Oden 2000). While the superconvergence property appears to be a difficult one to attain for complex finite element applications, the discretization error estimates from some recovery methods tend to be "astonishingly good" for reasons that are not well-understood (Ainsworth and Oden 2000), especially in the case of linear elliptic problems.

Recovery methods have been shown to be most effective when the reconstruction step employs solution gradients rather than solution values. The *superconvergent patch recovery* (SPR) method (Zienkiewicz and Zhu 1992) is the most widely used recovery method in finite element analysis. Assuming the underlying finite element method is of order p_c, the SPR approach is based on a local least squares fitting of the solution gradient values at the superconvergence points using polynomials of degree p_c. The SPR recovery method was found to perform extremely well in an extensive comparison of *a posteriori* finite element error estimators (Babuska et al. 1994). A more recent approach called *polynomial preserving recovery* (PPR) was proposed by Zhang and Naga (2005). In their approach, they use polynomials of degree $p_c + 1$ to fit the solution values at the superconvergence points, then take derivatives of this fit to recover the gradient. Both the SPR and PPR gradient reconstruction methods can be used to obtain error estimates in the global energy norm and in the local solution gradients, but extensions to SRQs must be done heuristically.

Richardson extrapolation uses solutions on two or more systematically refined meshes to estimate the exact solution to the governing equations, which can in turn be used to provide an error estimate for the numerical solutions. Consider the two series expansions for the numerical solution about the exact solution to the governing equations given earlier by Equations (18.2.3) and (18.2.4) for systematically refined meshes with discretization parameter values h and rh, respectively. Assuming for now that the solutions are in the asymptotic range (i.e., that the observed order of accuracy is near the formal order), these two equations can be solved for an estimate of the exact solution to the governing equations by neglecting the higher-order terms. One obtains the Richardson extrapolation estimate for the exact solution υ_{RE} to be

$$\upsilon_{RE} = \upsilon_{(h)} + \frac{\upsilon_{(h)} - \upsilon_{(rh)}}{r^{p_c} - 1} \tag{18.3.4}$$

which is generally a $(p_c + 1)$-order accurate estimate of the exact solution to the governing equations υ. This equation can be used to estimate the discretization error in the fine grid solution, that is, $e_{(h)RE} = \upsilon_{(h)} - \upsilon_{RE}$, resulting in the error estimate:

$$e_{(h)RE} = \frac{\upsilon_{(rh)} - \upsilon_{(h)}}{r^{p_c} - 1}. \tag{18.3.5}$$

Note that in addition to the assumption that both solutions are in the asymptotic range, this error estimate will be accurate only when iterative and round-off errors are much smaller than the fine grid discretization error (100 times smaller is recommended).

Regardless of the approach used for estimating the discretization error, the *reliability* of the discretization error estimate depends on the solution, or solutions, being in the asymptotic mesh convergence range, which is extremely difficult to achieve for complex FEM applications. Verifying that the solutions are in the asymptotic range can be done by computing the observed order of accuracy using numerical solutions on three systematically refined meshes. For systematic refinement by the factor r, one has $h_{fine} = h$, $h_{medium} = rh$, and $h_{coarse} = r^2 h$ and the observed order of accuracy can be found as (Roache 2009):

$$\hat{p}_c = \frac{ln\left(\dfrac{\upsilon_{(r^2h)} - \upsilon_{(rh)}}{\upsilon_{(rh)} - \upsilon_{(h)}}\right)}{ln(r)}. \tag{18.3.6}$$

The case when the grid refinement factor between the fine and medium meshes differs from that between the medium and coarse meshes is more complicated and is addressed in Roache (2009) and Oberkampf and Roy (2010). Note that the observed order of accuracy will only match the formal order when *all three grid levels* are in the asymptotic range. A similar expression for the observed order of accuracy can be derived in terms of error estimates found on two systematically refined meshes (e.g., for use with residual-based error estimators):

$$\hat{p}_c = \frac{ln\left(\frac{e_{(rh)}}{e_{(h)}}\right)}{ln(r)}. \qquad (18.3.7)$$

Example 18.3: Use of Richardson extrapolation to estimate the exact solution for an SRQ

We will now examine the use of Richardson extrapolation, by returning to the analysis of the system considered in Example 18.2. The SRQ of interest will be the value of concentration at the location $x = 2$, $y = 0$. This location constitutes a nodal point for all three mesh cases considered in Example 18.2. We had obtained results for three different values of element size, namely, for $h = 2$, 1, and 0.2. Since we have a manufactured solution, we can directly plug $x = 2$, $y = 0$ into the assumed expression for the exact solution field, $C(x,y)$, to find the exact value of the SRQ of interest:

$$C(x=2,y=0) = \left(\frac{2}{6} - \frac{2^2}{18} + \frac{2^3}{216}\right)\left(1 - \frac{0^2}{16}\right) = 0.14815$$

If we use the solution for $h = 1$, we have: $C_{(h)}(x=2,y=0) = 0.1505$

If we use the solution for $2\,h = 2$, we have: $C_{(2h)}(x=2,y=0) = 0.158$

Since these two cases correspond to a refinement factor $r = \frac{2}{1} = 2$, and since the formal rate of convergence is $p_c = 2$ (we use elements which have a degree of completeness equal to 1), we have:

$$C_{RE} = C_{(h)} + \frac{C_{(h)} - C_{(2h)}}{r^{p_c} - 1} = 0.1505 + \frac{0.1505 - 0.158}{2^2 - 1} = 0.148$$

One can immediately verify that, for the specific problem Richardson extrapolation yielded an estimate of the SRQ of interest, which is very close to the value corresponding to the exact solution of Example 18.2. The Richardson extrapolated value can then be used to estimate the discretization error in the finite element solution for the SRQ. ■

18.4 Numerical Uncertainty

In some cases, when numerical errors can be estimated (both sign and magnitude) with a high degree of confidence, they can be removed from the numerical solution—a process similar to that used for well-characterized bias errors in an experiment. More often, however, the numerical errors are estimated with significantly less certainty, for example, only an estimate of the absolute value of the error can be made. As a result, these error estimates should be treated as numerical uncertainties, with the uncertainty coming from the error estimation process itself (Roache 1994, Oberkampf and Roy 2010, Roy and Oberkampf 2011). One of the simplest methods for converting an error estimate to an uncertainty is to use the magnitude of the error estimate to apply uncertainty bands about the simulation prediction, possibly with an additional factor of safety included. For example, the Richardson extrapolation estimate of discretization error $e_{(h)}$ discussed above can be represented as a numerical uncertainty U_{DE} as

$$U_{\text{DE}} = F_s \left| e_{(h)} \right| \tag{18.4.1}$$

where $F_s \geq 1$ is the factor of safety (this is, in fact, a generalization of Roache's Grid Convergence Index, see Roache 1994, 2009). The resulting interval for the numerical solution, accounting for numerical uncertainties, can be approximated by applying this uncertainty symmetrically about the fine grid solution

$$\upsilon_{(h)} \pm U_{\text{DE}} = \upsilon_{(h)} \pm F_s \left| e_{(h)} \right|. \tag{18.4.2}$$

These concepts are shown graphically in Figure 18.6 with a factor of safety of approximately $F_s = 1.5$. The numerical solution $\upsilon_{(h)}$ has a signed error estimate $e_{(h)}$ as well as an uncertainty band created by taking plus/minus the absolute value of $e_{(h)}$ centered on the numerical solution. The factor of safety is needed because υ_{RE} is only an approximation of υ. In other words, even when the error estimate is reasonably accurate, the true exact solution υ could still be slightly larger or slightly smaller than the estimated exact solution υ_{RE}. When the error estimate $e_{(h)RE}$ is poor, this heuristic approach is designed to still potentially provide conservative numerical uncertainty estimates, depending of course on the chosen factor of safety. It is recommended that this uncertainty be centered about the numerical solution $\upsilon_{(h)}$ rather than the estimated exact solution υ since the latter can lead to erroneous (and possibly physically nonrealizable) values.

When multiple sources of numerical error are present, then a conservative approach is to simply add the numerical uncertainties together (Oberkampf and Roy 2010, Roy and Oberkampf 2011), that is,

$$U_{\text{NUM}} = U_{\text{RO}} + U_{\text{IT}} + U_{\text{DE}}. \tag{18.4.3}$$

where U_{RO} and U_{IT} refer to round-off and iterative error, respectively. Note that Equation (18.4.3) must be applied to *each* SRQ to estimate the total numerical uncertainty on each quantity of interest to the analyst. It is currently an open question as to whether these uncertainties should be characterized probabilistically (Stern et al.

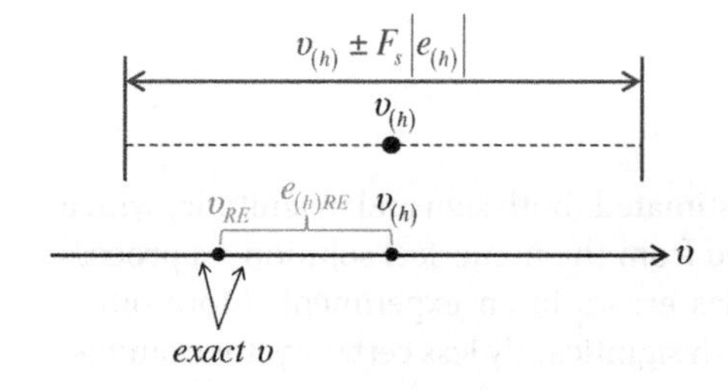

Figure 18.6 Example of converting a discretization error estimate into a numerical uncertainty centered on the numerical solution Roy 2012. Reproduced with permission from Begell House Inc. Publishers.

2001, Phillips and Roy 2013) or in some other fashion (e.g., as intervals) (Roy and Oberkampf 2011, Roy and Balch 2012).

18.5 Sources and Types of Uncertainty

There are many sources of *uncertainty* in FEM, including the model inputs, the form of the model (which embodies all of the assumptions and approximations in the formulation of the model), and poorly characterized numerical approximation errors. These sources of uncertainty can be classified as (1) *aleatory*—the inherent variation in a quantity; (2) *epistemic*—uncertainty due to lack of knowledge; or (3) a mixture of the two (Roy and Oberkampf 2016). Aleatory uncertainty is generally characterized probabilistically by either a probability density function (PDF) or a cumulative distribution function (CDF), the latter being simply the integral of the PDF from minus infinity up to the value of interest (see Figure 18.7). Epistemic uncertainties are usually characterized as uniform probabilities or as intervals (Ferson and Ginzburg 1996). We focus here on aleatory uncertainties in model inputs, which are characterized probabilistically.

There are commonly one or more system outputs (i.e., SRQs) that the analyst is interested in predicting with a FEM code. When uncertain model inputs are aleatory, there are a number of different approaches for propagating this uncertainty through the model. The simplest approach is sampling (e.g., Monte Carlo or Latin Hypercube) where inputs are sampled from their probability distribution and then used to generate a sequence of SRQs; however, sampling methods tend to converge slowly as a function of the number of samples. Other approaches that can be used to propagate aleatory uncertainty include perturbation methods and polynomial chaos (both intrusive and nonintrusive formulations) (Ghanem and Spanos 1991, Xiu 2010). Furthermore, when a response surface approximation of an SRQ as a function of the uncertain model inputs is available, then any nonintrusive method discussed above (including sampling) can be computed efficiently.

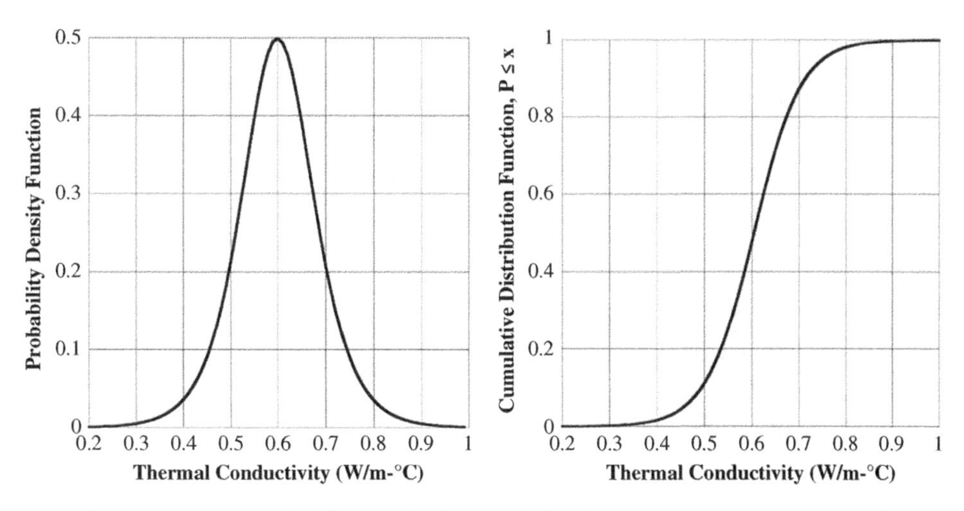

Figure 18.7 Example of a probability density function (PDF, left) and a cumulative distribution function (CDF, right).

18.6 Validation Experiments

A *validation experiment* is an experiment conducted with the primary purpose of assessing the predictive capability of a mathematical model. As such, validation experiments differ from traditional experiments used for exploring a physical phenomenon or obtaining information about a system's performance or reliability. There are six primary guidelines for validation experiments (Oberkampf and Roy 2010). Validation experiments in solid mechanics should:

1) be jointly designed by experimentalists and modelers, with the FEM code used to provide pretest computations of the proposed experiment,
2) be designed to capture the relevant physics and measure all initial conditions, boundary conditions, and other relevant modeling data required by the simulation,
3) strive to emphasize the inherent synergism that is attainable between FEM and experimental approaches,
4) be a blind comparison between FEM and experiment—that is, the experiment should provide all required model inputs and boundary conditions, but not the measured SRQs,
5) be designed to ensure that a hierarchy of SRQs are measured, for example, from globally integrated quantities to local quantities, and
6) be constructed to analyze and estimate the components of random (precision) and systematic (bias) experimental uncertainties in both the SRQs and the model inputs.

Due to the complexity of most structural systems, validation of the models used in FEM can be difficult. The most common approach to model validation is the *validation hierarchy* (Sindir et al. 1996, AIAA 1998), which divides the complex engineering system of interest into an arbitrary number of progressively simpler tiers as shown in Figure 18.8. If the system is divided into three tiers, they are usually referred to as subsystem, benchmark, and unit; additional tiers can be added as necessary. The strategy of the tiered approach encourages assessment of the accuracy of the model at multiple levels

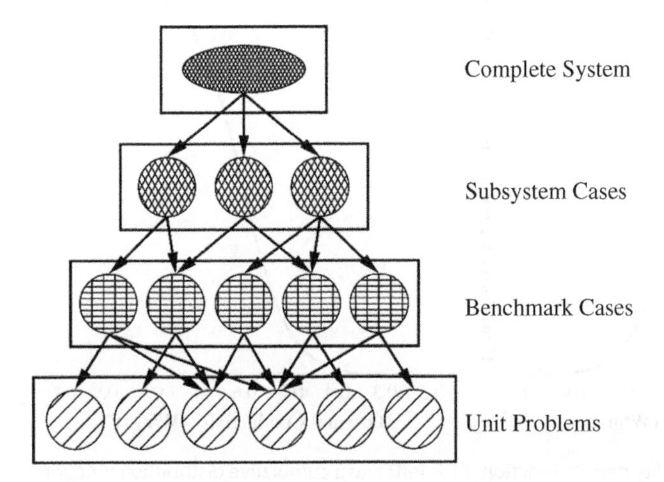

Figure 18.8 Validation hierarchy tiers (AIAA 1998. Reproduced with permission from American).

of complexity and physics coupling. This approach highlights that there is a hierarchy of complexity in real systems and simulations and that the quantity and accuracy of information that is obtained from experiments vary radically over the range of hierarchy. The arrows in Figure 18.8 from a higher tier to a lower tier indicate the (notional) impact of that element on lower tier elements.

The hierarchical view of validation is distinctly a system-oriented perspective. The purpose of the validation hierarchy is to help identify a range of lower level tiers where experiments can be conducted for assessment of model accuracy for simpler systems and physics. Most validation hierarchies differ at the top tiers because the system or subsystems of interest are different, but they may share common elements at the lower tiers. At the top of the validation hierarchy, the complete system tier is composed of actual system hardware at operating conditions, but this tier typically produces very limited measurements of model inputs and outputs along with little or no estimated uncertainties in these measurements. At the bottom of the validation hierarchy, the unit problem tier employs simplified, nonfunctional hardware, simple initial and boundary conditions, and avoids any coupled physics. Validation experiments at the unit problem level should measure all model inputs and outputs, include multiple experimental realizations (i.e., replicate measurements), and provide experimental uncertainty estimates on all quantities. For the common elements at the lower tiers, model validation can have a widespread impact on many systems.

18.7 Validation Metrics

Validation metrics provide a means by which the accuracy of a model can be assessed relative to experimental observations (Ferson et al. 2008, Oberkampf and Roy 2010, Roy and Oberkampf 2011). Liu et al. (2011) recommend that validation metrics provide a quantitative distance-based measure that can be used to quantify modeling error and that they be objective with a given set of simulations and data resulting in a single metric value (i.e., it should not depend on the analyst evaluating the metric, their preferences, or prior assumptions). The field of validation metrics is an area of active research, but for the purposes of this chapter, we focus only on stochastic validation metrics that provide distance-based measures of the agreement/disagreement between the model and experimental data, thus we omit any discussion of approaches such as classical hypothesis testing and Bayesian model comparison. See Oberkampf and Roy (2010) for more details on these alternative approaches.

It is important to draw clear distinctions between the concepts of validation and calibration. While *validation* involves the quantitative assessment of a model relative to experimental data, *calibration* (a.k.a., parameter estimation, parameter optimization, or model updating) instead involves the adjustment of model input parameters to improve agreement with experimental data. For example, if all uncertain model inputs are probabilistic, then Bayesian updating (Hasselman 2001) can be used to update the probability distributions of the model inputs. While calibration is often an important part of the model building and improvement process, it does not in itself provide quantitative estimates of model accuracy. The key difference is that model calibration results in a new model (because model parameters or their distributions are updated), and the new model must still be assessed for accuracy with independent experimental data.

While there are many possible validation metrics, we will focus on one implementation called the area validation metric (Ferson et al. 2008), which is a mathematical metric that provides a quantitative measure of disagreement between a stochastic model output and replicate experimental measurements of the same SRQ. When aleatory uncertainties are present in the model inputs, then propagating these uncertainties through the model produces a CDF of the SRQ. Experimental measurements are then used to construct an empirical CDF of the SRQ by ordering the replicate experimental values from smallest to largest and separating them by equal increments in cumulative probability. The absolute value area between these two CDFs is referred to as the area validation metric d (also called the Minkowski L_1 norm) and is given by:

$$d(F,S_n) = \int_{-\infty}^{\infty} |F(x) - S_n(x)|\,dx \tag{18.7.1}$$

where $F(x)$ is the CDF from the simulation, $S_n(x)$ is the CDF from the experiment, and x is the SRQ. The area validation metric d has the same units as the SRQ and thus provides a measure of the *evidence for disagreement* between the simulation and the experiment (Ferson et al. 2008). Note that the area validation metric represents an epistemic uncertainty because it embodies the bias effect of all of the assumptions and approximations in the formulation of the mathematical model relative to the experimentally measured SRQ. The area validation metric attempts to capture the intrinsic modeling uncertainty and is commonly referred to as *model form uncertainty*.

An example of this area validation metric for a case with only aleatory uncertainties occurring in the model input parameters is given in Figure 18.9. In this figure, the aleatory uncertainties have been propagated through the model (e.g., with a large number of Monte Carlo samples), but only four experimental replicate measurements are available. The stair-steps in the experimental CDF are due to the different values observed in each of the four experimental measurements. The stochastic nature of the measurements can be due to variability of the experimental conditions as well as random measurement uncertainty in the SRQ. The area validation metric can also be computed for cases involving both aleatory and epistemic uncertainty in the model inputs (e.g., see Ferson et al. 2008, Oberkampf and Roy 2010).

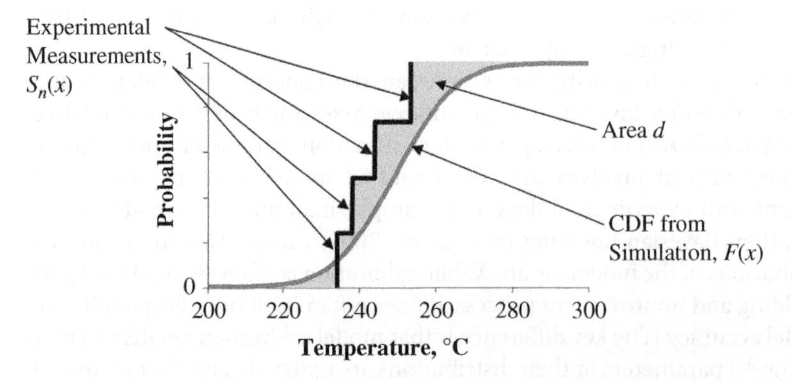

Figure 18.9 Area validation metric example (Ferson 2008. Reproduced with permission from Elsevier).

18.8 Extrapolation of Model Prediction Uncertainty

It is generally too expensive (or even impossible) to obtain experimental data over the entire multidimensional space of model input parameters for the application of interest. As a result, techniques are needed for estimating model prediction uncertainty at conditions where there are no experimental data. Two key contributors to model prediction uncertainty are model form uncertainty and uncertainty due poorly known input parameters at the application conditions. Consider a simple example when there are only two input parameters for the model: α and β (Figure 18.10). The validation domain consists of the set of points in this parameter space where experiments have been conducted and a validation metric has been computed (denoted by a V in the figure). In this common example, the application domain (sometimes referred to as the operating envelope of the system) is larger than the validation domain, although many other set relationships are possible. Thus, one must choose between (1) ignoring the observed inaccuracy in the model, (2) using the flexibility of the model by way of calibrating the model parameters at the validation conditions, (3) extrapolating the validation metric outside of the validation domain, or (4) performing additional validation experiments in the application domain. Figure 18.10 denotes conditions for candidate validation experiments by a C. The key point is that the validation domain is generally not coincident with the application domain; thus, either interpolation or extrapolation of the model form uncertainty to the conditions of interest is needed (see Oberkampf and Roy 2010 or Roy and Oberkampf 2011 for more details on model extrapolation).

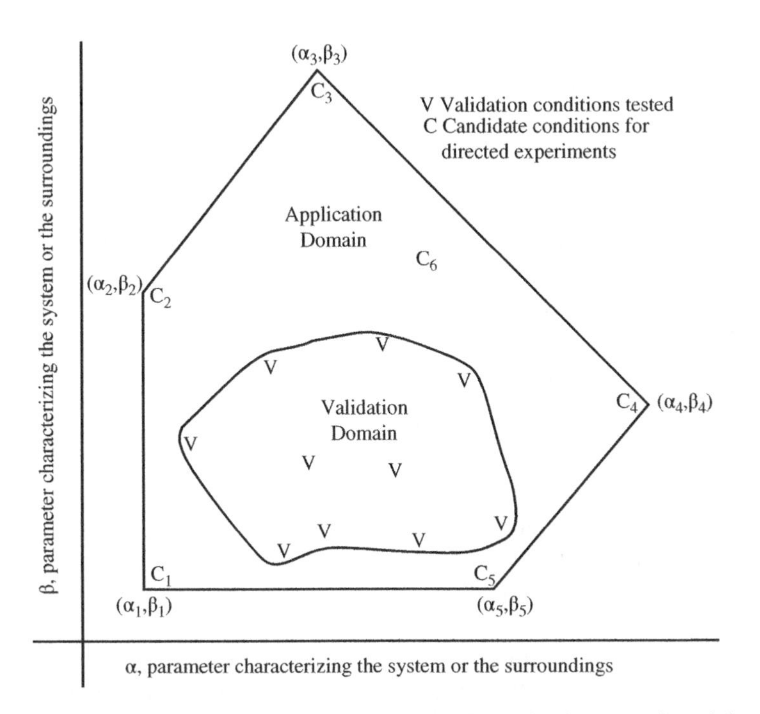

Figure 18.10 Schematic showing a possible relationship between the validation domain and the application domain (reproduced from Oberkampf and Roy 2010).

18.9 Predictive Capability

The total prediction uncertainty should include contributions from the form of the model and those due to the numerical error estimation process. For example, if the predicted SRQ is a single, deterministic value F (i.e., there is no uncertainty in the model inputs), then accounting for the model form and numerical uncertainties would result in the interval $F \pm U_{TOTAL}$, where $U_{TOTAL} = U_{MODEL} + U_{NUM}$. Although uncertainties are not necessarily additive in this way (Ferson and Tucker 2006), this approach estimates the compounding roles of the various sources of modeling and numerical uncertainty. This approach attempts to achieve two goals. First, it is a simple computational procedure to estimate the impact of model form and numerical uncertainty on the SRQs of interest. Second, it explicitly shows the user of the simulation results the magnitude of each contributing uncertainty, thus providing guidance as to where additional resources could be spent to reduce this predictive uncertainty. If the model form uncertainty is large, one may choose to perform additional model validation experiments or to improve/calibrate the model. If numerical uncertainty is large, then the simulations may be run on larger (i.e., finer) meshes. For the case where there is aleatory uncertainty in the model inputs, the resulting CDF for the SRQ could be treated in a similar fashion. See Oberkampf and Roy 2010, Roy and Oberkampf 2016, and Ferson et al. 2008 for more details.

References

AIAA (1998). *Guide for the Verification and Validation of Computational Fluid Dynamics Simulations*, American Institute of Aeronautics and Astronautics, AIAA-G-077–1998, Reston, VA.

Ainsworth, M., and Oden, J. T. (2000). *A Posteriori Error Estimation in Finite Element Analysis*. New York: Wiley Interscience.

ASME (2006). *Guide for Verification and Validation in Computational Solid Mechanics*, American Society of Mechanical Engineers, ASME Standard V&V 10-2006, New York, NY.

ASME (2009). *Standard for Verification and Validation in Computational Fluid Dynamics and Heat Transfer*. American Society of Mechanical Engineers, ASME Standard V&V 20-2009, New York, NY.

Bank, R. E. (1996). "Hierarchical Bases and the Finite Element Method." *Acta Numerica*, 5, 1–45.

Cao, J. (2005). "Application of a Posteriori Error Estimation to Finite Element Simulation of Incompressible Navier-Stokes Flow." *Computers and Fluids*, 34 (8), pp. 972–990.

Diskin, B., and Thomas, J. L. (2012). "Effects of Mesh Regularity on Accuracy of Finite-Volume Schemes," AIAA Paper 2012-0609.

Duggirala, R., Roy, C. J., Saeidi, S. M., Khodadadi, J., Cahela, D., and Tatarchuck B. (2008). "Pressure Drop Predictions for Microfibrous Flows Using CFD." *Journal of Fluids Engineering*, 130 (7), 13 pages.

Ferson, S., and Tucker, W. T. (2006). "Sensitivity in Risk Analyses with Uncertainty Numbers." Sandia National Laboratories Report, SAND2006-2801, Albuquerque, NM.

Ferson, S., and Ginzburg, L. R. (1996). "Different Methods Are Needed to Propagate Ignorance and Variability." *Reliability Engineering and System Safety*, 54, 133–144.

Ferson, S., Oberkampf, W. L., and Ginzburg, L. (2008). "Model Validation and Predictive Capability for the Thermal Challenge Problem." *Computer Methods in Applied Mechanics and Engineering*, 197, 2408–2430.

Ghanem, R., and Spanos, P. (1991). *Stochastic Finite Elements: A Spectral Approach*. New York: Springer Verlag.

Golub, G. H., and C. F. Van Loan (1996). *Matrix Computations*, 3rd ed. Baltimore: The Johns Hopkins University Press.

Hasselman, T. K. (2001). "Quantification of Uncertainty in Structural Dynamic Models." *Journal of Aerospace Engineering*, 14 (4), pp. 158–165.

Katz, A., and Sankaran, V. (2012). "An Efficient Correction Method to Obtain a Formally Third-Order Accurate Flow Solver for Node-Centered Unstructured Grids." *Journal of Scientific Computing*, 51 (2), 375–393.

Knupp, P. M., and K. Salari (2003). *Verification of Computer Codes in Computational Science and Engineering*, K. H. Rosen (ed.), Boca Raton, FL: Chapman and Hall/CRC.

Liu, Y., Chen, W., Arendt, P., and Huang, H.-Z. (2011). "Toward a Better Understanding of Model Validation Metrics." *Journal of Mechanical Design*, 133, pp. 1–13.

Oberkampf, W. L., and Roy, C. J. (2010). *Verification and Validation in Scientific Computing*. Cambridge: Cambridge University Press.

Phillips, T. S., and Roy, C. J. (2013). "A New Extrapolation-Based Uncertainty Estimator for Computational Fluid Dynamics." AIAA Paper 2013-0260.

Phillips, T. S., and Roy, C. J. (2011). "Residual Methods for Discretization Error Estimation." AIAA Paper 2011-3870.

Pierce, N. A., and Giles, M. B. (2000). "Adjoint Recovery of Superconvergent Functionals from PDE Approximations." *SIAM Review*, 42 (2), 247–264.

Pressman, R. S. (2005). *Software Engineering: A Practitioner's Approach*, 6th ed. Boston: McGraw-Hill.

Roache, P. J. (1994). "Perspective: A Method for Uniform Reporting of Grid Refinement Studies." *Journal of Fluids Engineering*, 116, 405–413.

Roache, P. J. (2009). *Fundamentals of Verification and Validation*, Hermosa Publishers, Socorro, New Mexico.

Roache, P. J., and Steinberg S. (1984). "Symbolic Manipulation and Computational Fluid Dynamics." *AIAA Journal*, 22 (10), 1390–1394.

Roy, C. J. (2005). "Review of Code and Solution Verification Procedures for Computational Simulation." *Journal of Computational Physics*, 205, 131–156.

Roy, C. J. (2009). "Strategies for Driving Mesh Adaptation in CFD." AIAA Paper 2009–1302.

Roy, C. J., and Balch, M. S. (2012). "A Holistic Approach to Uncertainty Quantification with Application to Supersonic Nozzle Thrust." *International Journal for Uncertainty Quantification*, 2 (4), 363–381.

Roy, C. J., and Oberkampf, W. L. (2016). "Chapter 44: Verification and Validation in Computational Fluid Dynamics." In R. W. Johnson, ed. *Handbook of Fluid Dynamics*, 2nd ed. Boca Raton, FL: CRC Press.

Roy, C. J., and Oberkampf, W. L. (2011). "A Comprehensive Framework for Verification, Validation, and Uncertainty Quantification in Scientific Computing." *Computer Methods in Applied Mechanics and Engineering*, 200 (25–28), 2131–2144.

Roy, C. J., Nelson, C. C., Smith, T. M., and Ober, C. C. (2004). "Verification of Euler/Navier–Stokes Codes using the Method of Manufactured Solutions." *International Journal for Numerical Methods in Fluids*, 44, 6, 599–620.

Shih, T. I.-P., and B. R. Williams (2009). "Development and Evaluation of an A Posteriori Method for Estimating and Correcting Grid-Induced Errors in Solutions of the Navier-Stokes Equations." AIAA Paper 2009–1499.

Sindir, M. M., Barson, S. L., Chan D. C., and Lin, W. H. (1996). "On the Development and Demonstration of a Code Validation Process for Industrial Applications." AIAA Paper 96-2032.

Skeel, R.D. (1986). "Thirteen Ways to Estimate Global Error." *Numerische Mathematik,* 48, 1–20.

Sommerville, I. (2004). *Software Engineering,* 7th Ed. Harlow, Essex, England: Pearson Education Ltd.

Stern, F., Wilson, R. V., Coleman, H. W., and Paterson, E. G. (2001). "Comprehensive Approach to Verification and Validation of CFD Simulations—Part 1: Methodology and Procedures." *Journal of Fluids Engineering,* 123, 793–802.

Stetter, H. J. (1978). "The Defect Correction Principle and Discretization Methods." *Numerische Mathematik,* 29, 425–443.

Stewart, J. R., and Hughes T. J. R. (1998). "A Tutorial in Elementary Finite Element Error Analysis: A Systematic Presentation of a priori and a posteriori Error Estimates," *Computer Methods in Applied Mechanics and Engineering,* 158 (1–2), 1–22.

Szabo, B. A., and Babuska, I. (1991). *Finite Element Analysis,* New York: John Wiley & Sons.

Venditti, D. A., and Darmofal, D. L. (2000). "Adjoint Error Estimation and Grid Adaptation for Functional Outputs: Application to Quasi-One Dimensional Flow." *Journal of Computational Physics,* 164, 204–227.

Venditti, D. A., and Darmofal, D. L. (2003). "Anisotropic Grid Adaptation for Functional Outputs: Application to Two-Dimensional Viscous Flows." *Journal of Computational Physics,* 187, 22–46.

Xiu, D. (2010). *Numerical Methods for Stochastic Computations: A Spectral Method Approach.* Princeton University Press.

Zhang, X. D., J.-Y. Trepanier, and R. Camarero (2000). "a posteriori Error Estimation for Finite-Volume Solutions of Hyperbolic Conservation Laws," *Computer Methods in Applied Mechanics and Engineering,* Vol. 185, No. 1, pp. 1–19.

Zhang, Z. and A. Naga (2005), "A New Finite Element Gradient Recovery Method: Superconvergence Property," *SIAM Journal of Scientific Computing,* Vol. 26, No. 4, pp. 1192–1213.

Zienkiewicz, O. C., and Zhu, J. Z. (1992). "The Superconvergent Patch Recovery and a posteriori Error Estimates, Part 2: Error Estimates and Adaptivity." *International Journal for Numerical Methods in Engineering,* 33, 1365–1382.

Zienkiewicz, O. C., and Zhu, J. Z. (1987). "A Simple Error Estimator and Adaptive Procedure for Practical Engineering Analysis." *International Journal for Numerical Methods in Engineering,* 24, 337–357.

19

Numerical Solution of Linear Systems of Equations

James McClure

19.1 Introduction

Finite element methods are usually developed to make predictions about physical systems. It is essential for software to be able to describe the physics accurately, and numerical methods are the means to this end. Our interest in particular numerical methods is largely determined by the physics that they allow us to describe. From a philosophical standpoint, there exist a separation between the physics and the numerical details. The numerical details can change, provided that these changes do not compromise the accuracy of the physical description. This provides a useful perspective for software design: when possible, we would like to decouple the physics from the numerical details. In a particular piece of software, some code sections will be very closely tied to the physics. Other code sections will be entirely concerned with the details of the numerical implementation. The solution of linear systems falls into the latter category. In this section, we consider some of the basic strategies that have been developed to solve linear systems of equations numerically.

The finite element method provides the machinery needed to convert partial differential equations into a discrete form that can be solved numerically. Many approaches have been developed to solve the resulting systems of linear equations. These may be divided into two general classes: (1) *direct methods* that directly solve the system of linear equations based on algorithms that recover an exact solution; and (2) *iterative methods* that generate a sequence of approximate solutions to the linear system by applying algorithm that converges to the true solution with numerical errors that can be bounded. In this chapter, we review several basic approaches that can be used to solve linear systems of equations and remark on their application to finite element methods. Specific objectives of this chapter are to provide a practical introduction to (1) several basic methods that can used to solve linear systems of equations; (2) the machinery used to evaluate and compare the computational costs of numerical algorithms; and (3) fundamental aspects of parallel computing that impact the implementation of finite element methods.

Fundamentals of Finite Element Analysis: Linear Finite Element Analysis, First Edition.
Ioannis Koutromanos, James McClure, and Christopher Roy.

Companion website: www.wiley.com/go/koutromanos/linear

19.2 Direct Methods

19.2.1 Gaussian Elimination

Gaussian elimination represents the most basic method used to solve linear systems of equations. The approach should be familiar to anyone who has solved a system of equations by hand. In this case, a solution to the linear system of equations $[A]\{x\}=\{b\}$ is obtained by first converting the system into an upper triangular form, and then using the process of back-substitution to determine the final solution. A generic linear system of N equations can be written as:

$$\begin{bmatrix} A_{11} & A_{12} & \dots & A_{1N} \\ A_{21} & A_{22} & \dots & A_{2N} \\ \vdots & \vdots & \ddots & \vdots \\ A_{N1} & A_{N2} & \dots & A_{NN} \end{bmatrix} \begin{Bmatrix} x_1 \\ x_2 \\ \vdots \\ x_N \end{Bmatrix} = \begin{Bmatrix} b_1 \\ b_2 \\ \vdots \\ b_N \end{Bmatrix} \tag{19.2.1}$$

Gaussian elimination proceeds by successively eliminating the unknowns x_i from equations i through N. This is accomplished by defining a new system of equations $[A^{(\alpha)}]\{x\}=\{b^{(\alpha)}\}$ according to Box 19.2.1, where the index $\alpha=1,2,\ldots,N$

Box 19.2.1 Gaussian Elimination, Part 1

for $\alpha=1,\ldots,N-1$ **do**

$L_{i\alpha} \leftarrow A_{i\alpha}^{(\alpha)}/A_{\alpha\alpha}^{(\alpha)}$

$A_{ij}^{(\alpha+1)} \leftarrow A_{ij}^{(\alpha)} - L_{i\alpha}A_{\alpha j}^{(\alpha)}$

$A_{ij}^{(\alpha+1)} \leftarrow A_{ij}^{(\alpha)} - L_{i\alpha}A_{\alpha j}^{(\alpha)}$

$b_i^{(\alpha+1)} \leftarrow b_i^{(\alpha)} - L_{i\alpha}b_\alpha^{(\alpha)}$

end for

corresponds to the various stages of the solution procedure. In part 1, the unknown x_i has been eliminated from all equations $j>i$. The new system of equations therefore corresponds to

$$\begin{bmatrix} A_{11}^{(N)} & A_{12}^{(N)} & \dots & A_{1N}^{(N)} \\ 0 & A_{22}^{(N)} & \dots & A_{2N}^{(N)} \\ \vdots & \vdots & \ddots & \vdots \\ 0 & 0 & \dots & A_{NN}^{(N)} \end{bmatrix} \begin{Bmatrix} x_1 \\ x_2 \\ \vdots \\ x_N \end{Bmatrix} = \begin{Bmatrix} b_1^{(N)} \\ b_2^{(N)} \\ \vdots \\ b_N^{(N)} \end{Bmatrix} \tag{19.2.2}$$

The resulting system is said to be *upper triangular* because all entries in the lower left triangle are identically zero (i.e., $A_{ij}^{(N)}=0$ for all $i<j$). By convention, this matrix is usually

renamed $[U] \equiv [A^{(N)}]$ to remind us that only the upper right triangle of this matrix has nonzero entries. The solution of this system is straightforward. We note that the last row of $[A^{(N)}]$ has only one nonzero entry, and we can solve for x_N to obtain

$$x_N = b_N^{(N)} / L_{NN}. \tag{19.2.3}$$

With x_N known, we solve for x_{N-1} by using a procedure known as *back substitution*. Using successive back substitution, all unknowns can be determined based on the approach described by Box 19.2.2. The process of back substitution simply solves the upper triangular system for each element in the unknown vector $\{x\}$, completing the process of Gaussian elimination and returning an exact solution to the linear system.

Box 19.2.2 Gaussian Elimination, Part 2 (Back substitution)

$x_N \leftarrow b_N^{(N)} / U_{NN}$
for $\alpha = N-1, N-2, \ldots, 1$ **do**
$\quad x_\alpha \leftarrow \left(b_\alpha^{(N)} - \sum_{\gamma=\alpha+1}^{N} U_{\gamma\alpha} x_\gamma\right) / U_{\alpha\alpha}$
end for

19.2.2 The LU Decomposition

The back-substitution process highlights the simplicity of solving a system of the form $[U]\{x\} = \{d\}$, where $[U]$ is an upper triangular matrix. The LU decomposition is a matrix factorization that exploits this opportunity. We seek a factorization of the matrix A such that $[A] = [L][U]$, where $[L]$ is lower triangular

$$[L] = \begin{bmatrix} 1 & 0 & \ldots & 0 \\ L_{21} & 1 & \ldots & 0 \\ \vdots & \vdots & \ddots & \vdots \\ L_{N1} & L_{N2} & \ldots & 1 \end{bmatrix} \tag{19.2.4}$$

and U is upper triangular

$$[U] = \begin{bmatrix} U_{11} & U_{12} & \ldots & U_{1N} \\ 0 & U_{22} & \ldots & U_{2N} \\ \vdots & \vdots & \ddots & \vdots \\ 0 & 0 & \ldots & U_{NN} \end{bmatrix} \tag{19.2.5}$$

It can be proved that the nonzero coefficients of the matrices $[L]$ and $[U]$ are obtained from Box 19.2.1 (where $[U] = [A^{(N)}]$). Since the diagonal entries of $[L]$ are all one, these do not need to be stored in memory ($L_{ii} = 1$ can be assumed). For memory efficiency, both $[L]$ and $[U]$ can be stored within a single dense matrix data structure $[D]$

$$[D] = \begin{bmatrix} U_{11} & U_{12} & \dots & U_{1N} \\ L_{21} & U_{22} & \dots & U_{2N} \\ \vdots & \vdots & \ddots & \vdots \\ L_{N1} & L_{N2} & \dots & U_{NN} \end{bmatrix} \tag{19.2.6}$$

Taking advantage of this data structure, the matrices L and U can be determined from Algorithm Box 1 can retained and reused. This is especially useful when the system $[A]\{x\} = \{b^k\}$ must be solved for multiple right-hand-side vectors $\{b^k\}$, $k = 1,\dots,M$. To see why this is the useful, we must consider the computational cost associated with Box 19.2.1 and Box 19.2.2.

The computational cost of an algorithm is determined by counting the total number of operations that must be performed. First we consider the costs of computing the matrices L and U in Box 1. To determine $L_{i\alpha}$ a single division operation is performed. At step α, a total of $N - \alpha$ divisions must be performed. Since there are $N - 1$ steps, the total number of divisions is calculated as

$$\text{LU division} = \sum_{\alpha=1}^{N-1} (N - \alpha) = \frac{N(N-1)}{2} \tag{19.2.7}$$

We must also include the cost of determining $[U]$. To update $A_{ij}^{(\alpha+1)} \leftarrow A_{ij}^{(\alpha)} - L_{i\alpha} A_{\alpha j}^{(\alpha)}$, one subtraction operation and one multiplication. Each of these operations must be performed for $i = 1,\dots,N-1$ and $\alpha = 1,\dots,N-1$, but only the nonzero values of $A_{ij}^{(\alpha+1)}$ must be updated. As a result, at step α we must perform $(N-\alpha)^2$ operations each for subtraction and multiplication. The total costs associated with these operations are

$$\text{LU multiply/add} = \sum_{\alpha=1}^{N-1} (N - \alpha)^2 = \frac{N(N-1)(2N-1)}{6} \tag{19.2.8}$$

On the basis of Eqs. 19.2.7 and 19.2.8, it is apparent that the computational cost of computing the matrices $[L]$ and $[U]$ is $\sim \frac{2N^3}{3}$ for large N. Separately, we can determine the cost of updating $\{b\} \rightarrow \{b^{(N)}\}$ to be $N(N-1)$ operations (multiplication and subtraction) and the cost associated with back substitution, which can be determined from Box 19.2.2 as $N^2 - N$ (multiplication and subtraction). Since the cost of the LU factorization is $\sim N^3$, this calculation dominates the total costs of solving the linear system. Conveniently, if $[L]$ and $[U]$ are retained, then we can efficiently solve for many right-hand sides in with N^2 operations.

19.3 Iterative Methods

Computers represent numbers in an inherently approximate way, and are not usually able to generate exact solutions. Numerical calculations are inherently limited by machine precision, and even direct methods are associated with numerical error. Within this context, iterative methods are often attractive because they provide a way to obtain an accurate approximation to the solution of the linear system with lower computational costs. In the previous section, we considered direct solution of the linear system $[A]\{x\} = \{b\}$, with algorithms constructed based on the assumption that the matrix $[A]$ was dense. Direct methods can be quite practical for the solution of relatively small linear systems. However, since the computational cost increases as $\sim N^3$ direct methods become less attractive as the size of the linear system increases. Direct methods are almost never applied to solve the large systems of equations produced by finite element methods for three-dimensional problems. For these systems, iterative methods are the preferred class of solution methods. The computational cost of efficient iterative methods can be dramatically lower than direct methods. In this section, we provide a basic introduction to iterative methods. A rich literature is available on this topic for the interested reader [5, 3].

Instead of solving a linear system $[A]\{x\} = \{b\}$ directly, iterative methods define a procedure to obtain successively better approximations for the solution. The relationship between this hierarchy of approximate solutions can be expressed as

$$\left\{x^{(m+1)}\right\} = [R]\left\{x^{(m)}\right\} + \{c\} \tag{19.3.1}$$

where m is the iteration index, $m = 0, 1, \ldots M$, and the approximation to the solution after m iterations is $\{x^{(m)}\}$. To make use of Equation (19.3.1), an initial guess $\{x^{(0)}\}$ is required. Since we do not know the solution in advance, we must define the matrix $[R]$ and vector $\{c\}$ such that the method will converge to a solution no matter what is provided for $\{x^{(0)}\}$. Convergence implies that eventually the difference between the true solution x and the approximation $\{x^{(m+1)}\}$ will be arbitrarily small. Using Equation (19.3.1) and the fact that convergence requires $\{x\} = [R]\{x\} + \{c\}$, we can write

$$||x^{(m+1)} - x|| \leq ||R|| \cdot ||x^{(m)} - x|| \leq ||R||^{m+1} \cdot ||x^{(m)} - x^{(0)}||. \tag{19.3.2}$$

A criterion for convergence therefore is established for the matrix norm as $||R|| < 1$, which is independent from the initial guess $\{x^{(0)}\}$. Convergence criteria for iterative methods are common topic in numerical analysis texts, and it can be shown that Equation (19.3.1) will converge if the maximum eigenvalue of $[R]$ (also known as the *spectral radius*) is less than 1 (Demmel 1997). In addition to convergence, other practical concerns must be taken into account when choosing iterative methods. First is the computational cost associated with the iterative scheme, including: (1) the cost of each iteration, including the cost of computing $[R]$ and $\{c\}$; and (2) the number of iterations required to obtain a solution at an acceptable level of accuracy, which is determined by the *rate of convergence*). An iterative method is only useful if it allows us to obtain an accurate approximation of $\{x\}$ at lower computational cost than a direct method. The results of the previous section demonstrate that the cost of direct solution is $\mathcal{O}(N^3)$. This provides a practical upper bound on the computational cost of an iterative method. If the cost of an iterative method is higher than the cost of direct solution, we would tend prefer the latter. However, many computationally efficient iterative methods can be constructed. In this section, we provide a basic introduction to the *Jacobi method* (a very basic iterative scheme) and the *Conjugate gradient method* (a scheme that is used very often in practice). It should be noted that this does not provide a full perspective on the possibilities for iterative methods. For the solution of partial differential equations, the most widely used iterative schemes fall within two general classes: *Krylov subspace methods* (Kelley 1995) and *multigrid methods* (Briggs et al. 2000). We refer the interested reader to more comprehensive works that focus on the mathematical construction of these iterative methods and their application to general classes of problems.

19.3.1 The Jacobi Method

Jacobi's method is frequently used to illustrate the basic ideas associated with iterative methods. Based on the notation of the previous section, the matrix A is split into a diagonal component $[D]$ such that $D_{jj} = A_{jj}$ for $j = 1,2,\ldots,N$ and $D_{ij} = 0$ for all $i \neq j$, and the remainder $[R]$ (which is obtained by subtracting $[D]$ from $[A]$). The linear system is then re-expressed as

$$([D] + [R])\{x\} = \{b\} \tag{19.3.3}$$

$$D]\{x\} = \{b\} - [R]\{x\} \tag{19.3.4}$$

$$\{x\} = [D]^{-1}(\{b\} - [R]\{x\}). \tag{19.3.5}$$

To convert this into a useful form, we must develop a way to approximate the unknown value of $\{x\}$ on the right-hand side of this expression. This is done by defining an iterative scheme where the approximate solution $\{x^{(k)}\}$ (at iteration k) is used to predict the subsequent approximation

$$\left\{x^{(k+1)}\right\} = [D]^{-1}\left(\{b\} - [R]\left\{x^{(k)}\right\}\right). \tag{19.3.6}$$

for $k = 0,1,\ldots,M$. Since $[D]$ is a diagonal matrix, its inverse is easily computed and $\{x^{(k)}\}$ can be determined in an iterative fashion based on the initial guess $\{x^{(0)}\}$. One can show that Equation (19.3.6) is in the general form of Equation (19.3.1). Implementing Jacobi's method is straightforward. However, the potential pitfalls Jacobi's method become evident by considering Equation (19.3.6) in the case where any one the diagonal entries of $[A]$ are close to zero. While Jacobi's method provides an intuitive introduction to iterative methods, more sophisticated approaches predominate in the solution of practical problems.

19.3.2 The Conjugate Gradient Method

The conjugate gradient method is among several widely used and important schemes categorized as Krylov subspace methods. Many variants of the conjugate gradient method exist, and the method is an excellent choice for solving linear systems in the case where $[A]$ is symmetric and positive definite (i.e., has positive eigenvalues). While the conjugate gradient method was originally developed as a direct method, it can be used to define an iterative procedure to solve for $\{x^{(k)}\}$ (Kelley 1995). The conjugate gradient method is constructed using a set of conjugate (i.e. orthogonal) vectors $p_1, p_2, \ldots p_N$, where N is the rank of A. Since the vectors form a basis for $\mathbb{R}^N$, any vector in $\mathbb{R}^N$ can be expressed as a linear combination the conjugate vectors. Clearly, this must also include the solution vector

$$\{x\} = \sum_{i=1}^{N} \alpha_i \{p_i\}. \tag{19.3.7}$$

Based on this representation, we consider the challenge of determining (1) a sequence of orthogonal search directions defined by the sequence of conjugate vectors $\{p_1\}$, $\{p_2\}$, ..., $\{p_n\}$; and (2) the projection of the solution onto each search direction, which are given by α_i, $i = 1,2,...,N$. As an iterative method, the usefulness of the conjugate gradient method depends in part on how the conjugate vectors are chosen. For example, if we were able to select $\{p_1\} = \{x\}$, then we would have $\alpha_i = 0$ for all $i \neq 1$. Clearly, this would be the very best choice that could be made. Since we in general will not know $\{x\}$, we make the less ambitious but more realistic goal to carefully choose $\{p_m\}$ carefully with the hope that a good approximation to x can be constructed using only a subset of the conjugate vectors.

Given any initial guess $\{x^{(0)}\}$, we can compute the associated residual vector

$$\{r_0\} = \{b\} - [A]\left\{x^{(0)}\right\}. \tag{19.3.8}$$

The residual $\{r_0\}$ defines a search direction based on the difference between the true solution and the initial approximation $\{x^{(0)}\}$. Therefore, we choose $\{p_0\} = \{r_0\}$. Based on the approach outlined in Section 19.3, we can compute the residual associated with the approximation at iteration m, which is denoted by $\{x^{(m)}\}$ (even with the specific strategy for computing $\{x^{(m)}\}$ still to be determined)

$$\{r_m\} = \{b\} - [A]\left\{x^{(m)}\right\}. \tag{19.3.9}$$

Our previous assertions require that the associated search direction $\{p_m\}$ be orthogonal to all previous search directions $\{p_i\}$, where $i < m$. Since each of these vectors are known at iteration m, this suggests an approach to compute $\{p_m\}$

$$\{p_m\} = \{r_m\} - \sum_{i=0}^{m-1} \frac{\{p_i\}^T[A]\{r_m\}}{\{p_i\}^T[A]\{p_i\}}\{p_i\}. \tag{19.3.10}$$

This effectively constrains the vector $\{p_m\}$ such that it will be orthogonal to all previously used search directions. A mechanism to compute the conjugate vectors is thereby introduced. The coefficients α_m are still to be determined. Given Equation (19.3.7), we can write the linear system as

$$[A]\{x\} = \sum_{i=1}^{N} \alpha_i[A]\{p_i\} = \{b\} \tag{19.3.11}$$

If we multiply both sides of the equation by $\{p_m\}^T$, then we can use the fact that $\{p_m\}$ is conjugate to all vectors $\{p_i\}$ for which $i \neq m$

$$\{p_m\}^T\left(\sum_{i=1}^{N} \alpha_i[A]\{p_i\}\right) = \{p_m\}^T\{b\} \tag{19.3.12}$$

$$\alpha_m\{p_m\}^T[A]\{p_m\} = \{p_m\}^T\{b\}. \tag{19.3.13}$$

Solving for the coefficient α_m follows, with the vector p_m determined by previous steps.

$$\alpha_m = \frac{\{p_m\}^T\{b\}}{\{p_m\}^T[A]\{p_m\}}. \tag{19.3.14}$$

The conjugate gradient algorithm is constructed by synthesizing the information form Equations (19.3.7) to (19.3.14). In the numerical implementation, an error tolerance ε is introduced. When the norm of the residual vector $\{r_m\}$ is less than the tolerance, we conclude that the associated solution $\{x^{(m)}\}$ is sufficiently accurate and the algorithm completes in m iterations. At this point, the iterative method is said to have converged. The basic form of the conjugate gradient method is presented in Box 19.3.1.

Box 19.3.1 Conjugate Gradient Method

$\{r_0\} \leftarrow \{b\} - [A]\{x^{(0)}\}$

$\{p_0\} \leftarrow \{r_0\}$

while $||r_m|| > \varepsilon$ **do**

$$\alpha_m \leftarrow \frac{\{r_m\}^T\{r_m\}}{\{p_m\}^T[A]\{p_m\}}$$

$$\{x^{(m+1)}\} \leftarrow \{x^{(m)}\} + \alpha_m\{p_m\}$$

$$\{r_{m+1}\} \leftarrow \{r_m\} - \alpha_m[A]\{p_m\}$$

$$\beta_m \leftarrow \frac{\{r_{m+1}\}^T\{r_{m+1}\}}{\{r_m\}^T\{r_m\}}$$

$$\{p_{m+1}\} \leftarrow \{r_{m+1}\} + \beta_m\{p_m\}$$

$$m \leftarrow m+1$$

end while

The conjugate gradient method is constructed entirely from basic numerical operations: vector addition, dot product, and matrix-vector multiplication, and multiplication of a vector by a scalar. With the exception of matrix-vector multiplication, the computational costs of these operations are $\mathcal{O}(N)$. However, if the matrix $[A]$ is sparse (which is often the case), then the cost of matrix-vector multiplication can in many cases be performed $\mathcal{O}(N)$ operations. Since the cost of each individual iteration is cheap, the conjugate gradient method has the potential to provide an efficient solution procedure. Of course, the computational costs of the conjugate gradient method depend explicitly on the number of iterations required to compute a sufficiently accurate solution. While the number of iterations required for convergence cannot be determined *a priori*, approaches to accelerate the convergence of iterative methods are widespread. There are in general two ways to accelerate the convergence of the conjugate gradient method. The first is to provide a good initial guess $\{x^{(0)}\}$. The second is to transform the linear system into an alternate form with the same solution, an approach known as a preconditioning. From the mathematical standpoint, preconditioning strategies are applied to construct linear systems with a favorable condition number, which enhances the convergence of iterative methods.

19.4 Parallel Computing and the Finite Element Method

Parallel computing refers to the practice of simultaneously using multiple computing devices to solve a particular computational problem. Such problems may require inter-process communication, or they may allow for many tasks to execute independently. The latter class of problems are often referred to as *embarrassingly parallel*—no special effort is required to get the full performance from the hardware for this class of problems. Academics often find themselves embarrassed when they are able to obtain a result too easily. Fortunately, solving finite element methods in parallel is rarely easy; they can expect to feel very pleased with themselves in the event that they are able to develop an efficient parallel implementation. Modern parallel computers present several challenges that complicate this endeavor. In this section, we provide a basic introduction to parallel computing as it pertains to the finite element method.

Parallel computing has been essential for finite element methods for decades. For many practical problems of interest, laptop and desktop computers are often not sufficient to solve the systems of equations generated from the finite element method. This is particularly true when the method is applied to solve time-dependent three-dimensional PDEs. Large meshes and long simulation times are common for this type of problem. While parallel computing enables simulations that could not otherwise be performed, it also introduces additional complexity that complicates implementation. However, due to the fact that computer architectures evolve rapidly, strategies to target parallel computers must continuously adapt to these changes. An effective software design plan must respect this reality if code is to be maintained over the long term. In particular, it is desirable to avoid the possibility to continuously rewrite code as architectures change. At the same time, tuning code to a particular architecture will often provide significant increases in performance, and is usually essential to obtain the full benefit from an architecture. These two goals are seemingly contradictory, but both must be considered to design effective finite element software. How can this dilemma be resolved? In many cases, software engineers have addressed this problem by organizing architecture-specific code into numerical libraries.

At the simplest level, a computer must process both *instructions* and *data* to produce a result. Instructions contain the operations that are to be performed. Data are the quantities to which the operations are applied. For example, if we add two quantities $A+B$ together, A and B are the data and the addition operation is an instruction. In a simulation, an example of data would be the elements of a matrix or vector that was stored in computer memory. Instructions operate on this data to carry forward the finite element analysis. Previously, we considered the computational costs associated with algorithms developed to solve linear systems based on the number of operations required. For example, consider two vectors $\{a\}$ and $\{b\}$, each of which are length N. If we add the two vectors together to produce a new vector $\{c\}=\{a\}+\{b\}$, the computational cost is $\mathcal{O}(N)$, since N addition operations must be performed to produce the results $\{c\}$. One obvious possibility is that all of these operations are performed sequentially. In this case the time that we must wait for our answer will be proportional to N. However, an alternate possibility is that we may perform some number of these operations in parallel so that we might not wait quite so long to produce a result. This is easy to envision for the operation $\{a\}+\{b\}=\{c\}$. In general, parallel algorithms will rarely be so obvious. In this section, we consider the different models for parallel computing based on the fact there are different ways to apply instructions to data.

19.4.1 Efficiency of Parallel Algorithms

We frequently wish to understand how efficiently we are able to use a computer. In practice, a program is most typically limited by one of three things: the rate that I/O is performed (data read / written to disk); (2) the rate at which data are loaded from computer memory; and (3) the rate at which operations can be performed by the processor. Since data must be loaded from memory to carry out operations on that data, items (2) and (3) are linked in an important way. If a processor can perform operations more rapidly than it can load data, the rate of computation is limited by the memory bandwidth of the device rather than the theoretical peak performance in the operations sense. Since the number of instructions that can be performed depends on the data type, theoretical performance is conventionally reported in terms of the number of floating-point operations per second (FLOPs), which corresponds to the maximum number of single-precision operations (e.g., addition, multiplication) that the processor can perform. The theoretical peak performance of a single processor is determined as $T_{peak} = CI$ from the clock speed C and the number of instructions I that can be performed per clock cycle. For parallel architectures where multiple processors are used together, the theoretical performance is obtained by summing the total of all processors involved. For example, if a processor has 16 cores, each capable of carrying out four instructions per clock cycle with a clock speed of 2.0 GHz, the theoretical peak performance of the processor is 128 gigaflops.

The theoretical performance of the processor represents that maximum performance possible. While there is a firm upper bound on the performance of a code, there is no lower bound. Since poorly written code can perform arbitrarily badly, it is important to assess the actual performance of a code in the context of the theoretical performance to assess how efficiently the code uses the architecture. This performance should be expected to vary from one architecture to the next, and the choice of compiler can also have an impact (due the variety of compiler optimizations that are performed in practice). The efficiency of a code is determined by comparing the number of operations performed by the code during a measured CPU time to the theoretical peak performance:

$$(\text{Efficiency}) = \frac{(\text{Number of operations})}{(\text{CPU time}) \times (\text{Theoretical peak})} \tag{19.4.1}$$

This equation can be used to compute the efficiency of a code for both serial and parallel architectures.

For parallel computations, the concept of parallel efficiency is important and useful. Parallel efficiency describes the speedup obtained by adding additional processors:

$$(\text{Speedup}) = \frac{(\text{Wall time for parallel computation})}{(\text{Wall time for serial computation})} \tag{19.4.2}$$

We obviously expect to be able to solve a problem faster in parallel than we can in serial. Based on the theoretical performance, the speedup will increase proportionately to the number of processors. The parallel efficiency looks like this:

$$(\text{Parallel efficiency}) = \frac{(\text{Speedup})}{(\text{Number of processors})} \tag{19.4.3}$$

Ideal scaling is obtained when the parallel efficiency is one. In rare occasions, the parallel efficiency can be greater than one, which is known as super-linear scaling. This is distinct from the fraction of theoretical peak performance, which provides a strict upper bound. In most situations, the parallel efficiency will be less than one, with the performance degrading as the number of processors is increased. There are two primary ways to evaluate the scaling of parallel codes. Strong scaling studies are performed by solving an identical problem using different numbers of processors, keeping the size of the computational mesh the same as the number of processors is increased. Weak scaling studies are performed by increasing the problem size (e.g., the total size of the mesh) proportionate to the number of processors. Weak scaling studies will generally show better parallel efficiency, but strong scaling studies are generally considered to provide more useful information about a code. Eventually, the efficiency reported in a strong scaling study will degrade based on parallelization overhead and the amount of parallelism available from the selected problem. The trajectory of this degradation provides useful information about the performance of a particular code.

19.4.2 Parallel Architectures

Shared Memory

Shared memory architectures allow multiple processors to access a shared pool of memory, as depicted in Figure 19.1. The predominant shared memory systems that exist today are cache-coherent nonuniform memory access (cc-NUMA) architectures. This refers to two facts:

1) A cache hierarchy exists (usually to hide the costs of memory accesses). Coherence is maintained for all data held within the cache in the event that these data are updated.
2) The costs of accessing data within the memory system are nonuniform.

In many situations, the cost of moving data to the processor can be the bottleneck in a computation. In nearly all situations, data must be moved to the processor arithmetic logic unit (ALU) before an instruction can be performed. *Data locality* refers to the proximity of data to processing, which determines the costs of this data movement on a shared-memory system. The most widely used approach to target shared memory systems is to rely on the OpenMP standard.

Distributed Memory

Algorithms to solve the finite element method in distributed memory are widely used. On a distributed memory computers, each processor is associated with its own memory that is not directly accessible from other processors. As a result, processors must communicate explicitly to share data. High-performance computing systems are often equipped with fast networks to accommodate these communications, as shown in Figure 19.2. In scientific computing, programs often rely on Message Passing Interface (MPI), which provides a standard protocol for communication that has implemented based on a wide range of network technologies. The distributed memory programming model of MPI can also be used to target shared memory systems, and can even be associated with performance advantages due to the resulting data locality. A key consideration for distributed memory implementations is how to distribute the problem to the

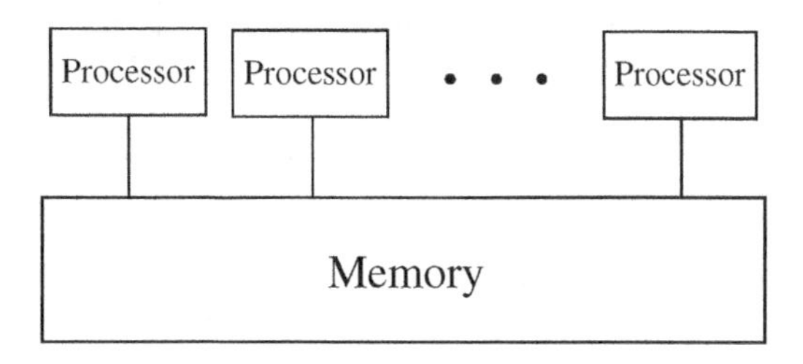

Figure 19.1 Schematic illustration of a shared memory parallel computer where multiple processors are able to access the same memory.

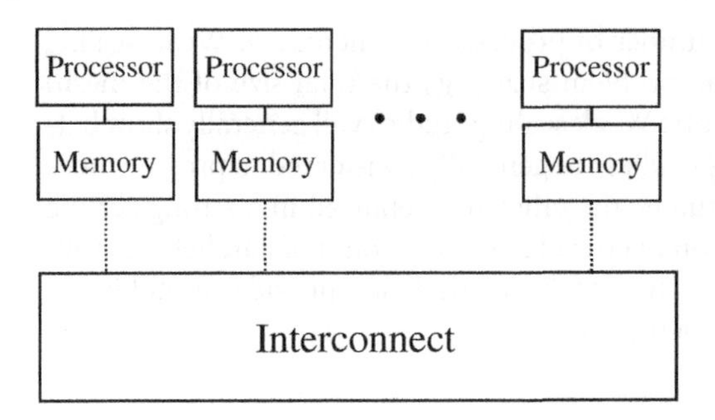

Figure 19.2 Schematic illustration of a distributed memory parallel computer where each processor has its own private memory and rely on an interconnect to exchange data between different processors.

processors. This requires *load balancing* to distribute the computational workload across processors, taking advantage of the local memory for each process, and minimizing the cost of interprocess communications. For finite element methods (an many other problems), load balancing is usually achieved by defining a *domain decomposition* strategy. In the finite element method, a mesh (refer to other section) is generated based on the problem discretization.

Single Instruction Multiple Data (SIMD)

Vectorization has taken on an increased role in current parallel computers. Within an individual processor core, vector instructions can be executed and are required to get the full performance. Architectures such as graphics processing units (GPUs) rely heavily on data parallelism to achieve a high theoretical performance. The enemy of data parallelism are *data dependencies*. In the course of a computation, a single instruction may be applied to many different pieces of data. In Figure 19.3 we consider the addition of two vectors $\{A\} + \{B\} = \{C\}$, where all three vectors have N elements. When we perform vector addition computationally, the vectors $\{A\}$, $\{B\}$ and $\{C\}$ are data, and reside in memory. The only instruction is the addition operation, which must be applied N times to different pieces of data. Furthermore, each individual addition operation is completely independent of all others. Since the operations are independent, they can be executed concurrently by applying a single instruction to multiple pieces of data. Consequently,

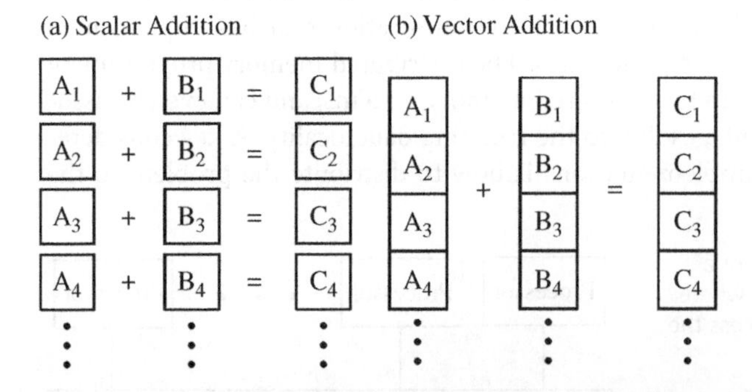

Figure 19.3 Schematic illustrating the execution of vector addition based on (a) a scalar operation in which the array elements are added individually and (b) a vector operation where a single instruction is applied to simultaneously add multiple data elements at the same time.

this type of parallel computation is classified as *single instruction multiple data* (SIMD), and are colloquially known as vector operations.

19.5 Parallel Conjugate Gradient Method

The basic conjugate gradient algorithm can be used to solve the linear system $[A]\{x\} = \{b\}$ for cases where $[A]$ is symmetric and positive definite (see Box 19.3). In this section, we consider we consider a parallel solution of the conjugate gradient method. Here, we consider the specific example of a two-dimensional solution scalar field problem with rectangular elements (see full development of the finite element model from Chapters 5 and 6). An example mesh is shown in Figure 19.4, where a typical ordering of the nodes is presented. A boundary condition constrains the temperature along the left boundary. In general, the structure of $[A]$ is determined by the approximations within the elements as well as the connectivity between elements (e.g., the mesh structure). The matrix $[A]$ represents the full linear system produced when considering all elements within the domain Ω. For the problem considered, $[A]$ has a symmetric, block tridiagonal structure:

$$[A] = \begin{bmatrix} [B] & [C] & [0] & [0] & \cdots & [0]C] \\ [B] & [C] & \ddots & & [0]0] & [C] \\ [B] & [C] & \ddots & [0] & & \\ \vdots & \ddots & \ddots & \ddots & \ddots & [0] \\ \vdots & & \ddots & [C] & [B] & [C]0] \\ \cdots & \cdots & [0] & [C] & [B] & \end{bmatrix}, \tag{19.5.1}$$

where $[B]$ and $[C]$ are square submatrices. For the simple mesh shown in Figure 19.4 the structure of the matrix $[A]$ is determined from the square submatrices $[B]$ and $[C]$

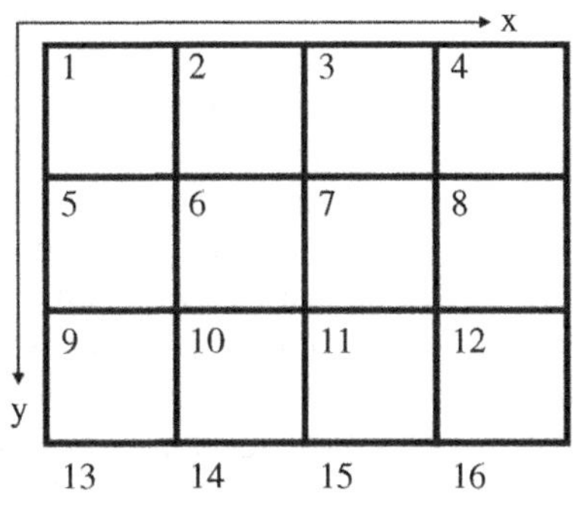

Figure 19.4 Example mesh yielding the linear system of equations given in Equations (19.5.4) and (19.5.5).

$$[B] = \begin{bmatrix} 8 & -1 & 0 & 0 \\ -1 & 8 & -1 & 0 \\ 0 & -1 & 8 & -1 \\ 0 & 0 & -1 & 4 \end{bmatrix}, \tag{19.5.2}$$

and

$$[C] = \begin{bmatrix} -2 & -2 & 0 & 0 \\ -2 & -2 & -2 & 0 \\ 0 & -2 & -2 & -2 \\ 0 & 0 & -2 & -1 \end{bmatrix}. \tag{19.5.3}$$

The small finite element system shown in Figure 19.4 is constructed for heuristic purposes; typical problems of practical interest require finer meshes that yield much larger systems of equations. It is these systems for which parallel computing can be of great aid. In the more general situation where a two-dimensional mesh contains $N_x \times N_y$ nodes, $[B]$ and $[C]$ will be $N_x \times N_x$ block submatrices with the following structure

$$[B] = \begin{bmatrix} 8 & -1 & 0 & \cdots & \cdots & 0 \\ -1 & 8 & -1 & \ddots & & \vdots \\ 0 & -1 & 8 & -1 & \ddots & \vdots \\ \vdots & \ddots & \ddots & \ddots & \ddots & 0 \\ \vdots & & \ddots & -1 & 8 & -1 \\ 0 & \cdots & \cdots & 0 & -1 & 4 \end{bmatrix}, \tag{19.5.4}$$

and

$$[C] = \begin{bmatrix} -2 & -2 & 0 & \cdots & \cdots & 0 \\ -2 & -2 & -2 & \ddots & & \vdots \\ 0 & -2 & -2 & -2 & \ddots & \vdots \\ \vdots & \ddots & \ddots & \ddots & \ddots & 0 \\ \vdots & & \ddots & -2 & -2 & -2 \\ 0 & \cdots & \cdots & 0 & -2 & -1 \end{bmatrix}, \tag{19.5.5}$$

In general, the memory requirement of storing the matrix $[A]$ is $\sim N^2$. However, based on foreknowledge of the structure of $[A]$, we are required only to know $[B]$ and $[C]$. The memory requirements for $[A]$ do not increase with the size of the linear system N. Instead, the memory requirements are driven by the right-hand-side vector $\{b\}$ and the solution $\{x\}$, which require $\mathcal{O}(N)$ memory. The number of operations associated with the conjugate gradient algorithm is also dramatically reduced. For example, consider the cost of multiplying $[A]$ by the vector $\{p\}_m$: instead of requiring $\mathcal{O}(N^2)$ operations, the matrix-vector product can be obtained in $\mathcal{O}(N)$ operations ($\sim 30N/4$). The per-iteration cost of the conjugate algorithm is therefore reduced dramatically.

Distributed memory supercomputers are a frequent target for finite element methods. A strategy for *domain decomposition* is key to develop an efficient parallel implementation for these systems. In the distributed memory model, each processor has its own local private memory space. Domain decomposition is used to distribute the computational work and associated data to the allocated processors. For the finite element method, domain decomposition partitions the computational mesh and assigns a subregion of the mesh to each processor. An example domain decomposition is shown in Figure 19.5. In this case, a two-dimensional mesh is subdivided into four different processors in distributed memory. Each processor is assigned a *process rank*, and receives a subregion of the mesh. The four processors are each responsible for the computations and data associated with the mesh subregion assigned to that process. Since the linear system is coupled as a consequence of the connections between the edges of the elements, there is an advantage to assigning adjacent elements to the same process when possible. Clearly, some adjacent elements will be assigned to different processors, requiring communication to exchange data along the boundary between processors.

The decomposition strategy is applied to distribute the elements of the vector $\{b\}$ to different processors. This, in turn, corresponds with a decomposition of the physical system based on the spatial layout of the mesh, since this layout determines the the coupling between elements (e.g., which leads to the block diagonal or banded structure of the matrix $[A]$). An example of a basic domain decomposition strategy is shown in Figure 19.6. In this case, a rectangular domain is approximated using triangular elements. A rectangular process grid is constructed to distribute the mesh to $P \times Q$ processors, as shown in Figure 19.6. Each processor receives a square mesh subregion of equal size. Communications between processors arise as a consequence of the mesh structure. The coupling between adjacent elements requires information to be shared along the subdomain boundaries. A common strategy is to allocate additional space to store the data, which is known as a *halo*. In Figure 19.6, the active computational region for each processor are shown in gray, and the halo regions are shown in white. Communications between processors are used to fill in the data for the halo region, with MPI being the most widely-used mechanism to carry out this data exchange The costs of moving data across an interconnect can be hidden behind the costs of the computation itself, but this usually requires that a relatively large mesh be retained on each per processor. The challenge of partitioning the computational mesh to different processors can present a

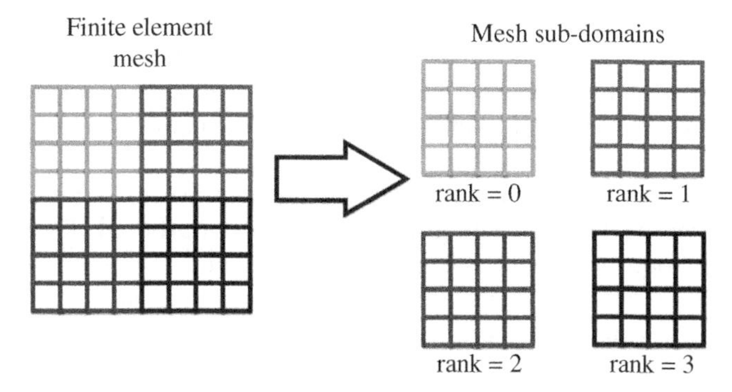

Figure 19.5 Basic domain decomposition in which a finite element mesh is subdivided and distributed between processors in distributed memory.

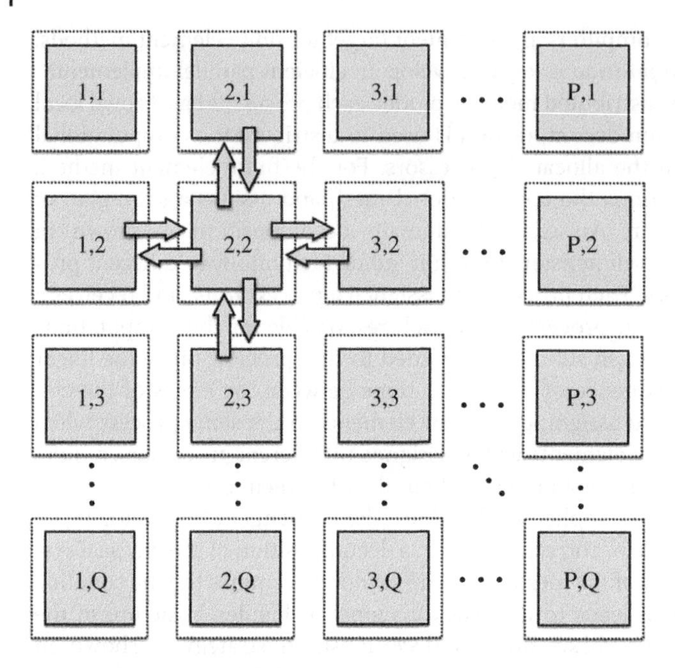

Figure 19.6 Process grid for a basic two-dimensional decomposition of a partial differential equation.

challenge, particularly for more complex element shapes. Graph analysis tools such as Metis and ParMetis are often used to analyze and partition complex meshes for this purpose. The fundamental challenge is in assigning equal work to the various processors while preserving data locality and minimizing communications (Karypis and Kumar 1999).

The advantage of parallel computing is to reduce the overall time-to-solution. A serial processor can perform $\mathcal{O}(N)$ operations in $\mathcal{O}(N)$ time. The goal of parallel processing is to perform computations concurrently such that $\mathcal{O}(N)$ operations can be performed using p processors in $\mathcal{O}(N/p)$ time. Implementations that meet this target are said to exhibit *ideal scaling*. In reality, several factors prevent algorithms implementations from scaling ideally. Any work that cannot be parallelized (e.g., *serial* work) requires the same amount of computational time irrespective of the number of processors p. Imperfect *load balancing* occurs when the amount of work assigned to each processor is not equal, which will also prevent ideal scaling. Finite element algorithms require communications to exchange data, which introduces additional overhead. To achieve an efficient parallel implementation, strategies must be developed to efficiently balance the computational costs while hiding the costs of communication.

Once a strategy to map the data for the problem to distributed memory has been constructed, parallel implementation of the conjugate gradient method can proceed. The parallel algorithm can be understood from four basic operations from linear algebra that are used to develop the algorithm for the conjugate gradient method: (1) vector addition; (2) multiplication of a vector by a scalar; (3) dot product between two vectors; and (4) matrix-vector multiplication. Given the domain decomposition shown in Figure 19.6, the first two operations are trivial. Each processor retains a subset of the vector v_i for $i \in \Omega_p$ vector in local memory. Since the scalar multiplication and the vector addition are entirely local, the serial and parallel algorithm are effectively identical. The dot product of two vectors a and b can be computed by first summing the elements within each subdomain, and then using a *parallel reduction* operation to add the component sums from all processors to obtain the final result in parallel. In MPI, reduction operations can return the final result either to all processors or to a single processor specified by the programmer.

The matrix-vector product $\{v\} = [A]\{y\}$ motivates the domain decomposition shown in Figure 19.6, since this part of the conjugate gradient method introduces a critical burden of communication. Each processor stores locally in memory the data v_i and y_i for $i \in \Omega_p$. However, the computation $v_i = A_{ij}y_j$ can depend on values $y_j \notin \Omega_p$. These values must be supplied by communication from the processors where they reside.

The conjugate gradient algorithm is expressed entirely in terms of basic linear algebra operations (e.g., matrix-vector product, dot product, vector addition, etc.). There are many possible ways to correctly implement these routines in computer code. However, the performance associated with the implementation can vary considerably depending on the underlying computer architecture. While it is straightforward to mix the various aspects of a finite element code together, there are great advantages associated with separating the physics from the numerical details. This is particularly valuable when developing finite element codes for parallel computers, as the difficulty in developing programs increases considerably.

References

Briggs, W. L., Henson, V. E., and McCormick, S. F. (2000). *A Multigrid Tutorial*, 2nd ed. Philadelphia: SIAM.

Demmel, J. W. (1997). *Applied Numerical Linear Algebra*, SIAM, Philadelphia, Pennsylvania, USA.

Karypis, G. and Kumar, V. (1999). "A fast and high quality multilevel scheme for partitioning irregular graphs." *SIAM Journal on Scientific Computing*, 20, 1, 359–392.

Kelley, C. T. (1995). *Iterative Methods for Linear and Non-linear Equations*. Philadelphia: SIAM.

Keyes, D.E., et al. (2013). "Multiphysics Simulations: Challenges and Opportunities." *International Journal of High Performance Computing Applications*, 27 (1), 4–83.

Appendix A

Concise Review of Vector and Matrix Algebra

This Appendix is to provide a very brief summary of vector and matrix algebra, along with some necessary notation and definitions/rules that are used throughout this book.

A.1 Preliminary Definitions

A matrix is a rectangular array of numerical values, arranged in *rows and columns.* For example, Figure A.1 presents a matrix with two rows and three columns.

A matrix with "m" rows and "n" columns is called an $(m \times n)$ matrix. For example, matrix $[M]$ in Figure A.1 is a 2×3 matrix.

Each numerical value of a matrix is called an *element or a component* of the matrix. The component ij of a matrix is the element that belongs to the i-th row and to the j-th column. For example, the component M_{23} of matrix $[M]$ in Figure A.1 is equal to 4.

In the remainder of this Appendix, matrices will be denoted with capital letters, while the elements of matrices will be denoted with lowercase letters.

A *square matrix* is an $N \times N$ matrix, where N is an integer number (in other words: a matrix with the same number of rows and columns).

A *vector* (often called column vector) is a matrix with a single column, i.e., any $(N \times 1)$ matrix. A vector with N components is called an *N-dimensional vector.* For example:

$$\{A\} = \begin{Bmatrix} 1 \\ -1 \\ 0 \end{Bmatrix} \quad \{A\} \text{ is a three-dimensional vector.}$$

We will denote a vector as $\{A\}$ or $\vec{A}$.

The *magnitude* $\left\|\vec{V}\right\|$ *of an N-dimensional vector* $\vec{V}$—also called the *Euclidean norm*—is the square root of the sum of the squares of all the components of $\vec{V}$:

$$\left\|\vec{V}\right\| = \sqrt{(v_1)^2 + (v_2)^2 + \ldots + (v_N)^2} = \sqrt{\sum_{i=1}^{N} \left[(v_1)^2\right]} \tag{A.1.1}$$

Fundamentals of Finite Element Analysis: Linear Finite Element Analysis, First Edition.
Ioannis Koutromanos, James McClure, and Christopher Roy.

Companion website: www.wiley.com/go/koutromanos/linear

Figure A.1 An example matrix with two rows and three columns.

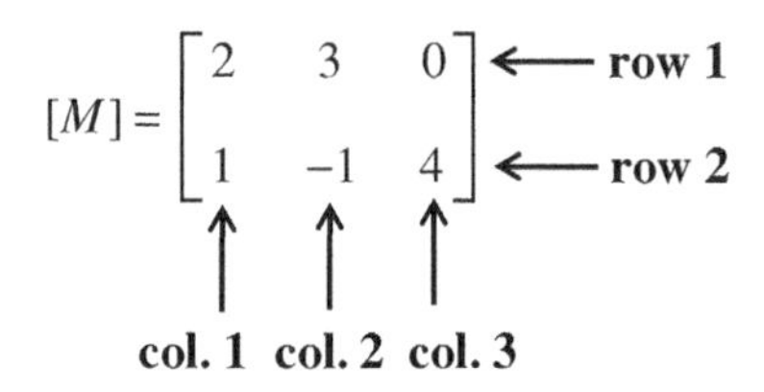

The *transpose of a matrix* is a transformed version of a matrix, in which the columns have been converted to rows and the rows have been converted to columns. We denote the transpose of a matrix using a superscript T.

A.1.1 Matrix Example

For the matrix $[M]$ in Figure A.1, we have the transpose, $[M]^T$:

$$[M]^T = \begin{bmatrix} 2 & 1 \\ 3 & -1 \\ 0 & 4 \end{bmatrix}$$

Note that the first column of $[M]$, which was $\begin{bmatrix} 2 \\ 1 \end{bmatrix}$, has become the first row of $[M]^T$, the second column of $[M]$ has become the second row of $[M]^T$, and so on.

The *transpose of a vector* is a single-row matrix, often called a *row vector*. For example, if we have the vector $\{A\}$ that we defined above, then we can write: $\{A\}^T = [1 \ \ -1 \ \ 0]$.

A (column) vector can be written as the transpose of a row vector. For example, we could write the three-component vector $\{A\}$ as: $\{A\} = [1 \ \ -1 \ \ 0]^T$. This notation is very helpful because it helps us "save" space in written documents. The present text often resorts to this notation for column vectors.

A *symmetric matrix* is a *square* matrix which is equal to its transpose, that is: if $[A]$ is symmetric, then $[A] = [A]^T$. For a symmetric matrix $[A]$, each component a_{ij} is equal to the component a_{ji}.

A *skew-symmetric* or *antisymmetric* matrix is a *square* matrix which is equal to the opposite of its transpose; that is, if $[M]$ is a skew-symmetric matrix, then $[M] = -[M]^T$. For a skew-symmetric matrix $[M]$, the component m_{ij} of $[M]$ is equal to $-m_{ji}$.

A.1.2 Vector Equality

If $\{C\}$ is a M-dimensional vector and $\{D\}$ is another M-dimensional vector (*Note:* The vectors must have the same dimension, and they must *both* be either column or row vectors), then we say that *the two vectors are equal* and write $\{C\} = \{D\}$, if each component, c_i, of $\{C\}$, is equal to the corresponding component, d_i, of $\{D\}$:

- The first component, c_1, of $\{C\}$, is equal to the first component, d_1, of $\{D\}$.
- The second component, c_2, of $\{C\}$, is equal to the second component, d_2, of $\{D\}$.

  ⋮

- The last component, c_M, of $\{C\}$, is equal to the last component, d_M, of $\{D\}$.

We can concisely write:

$$\{C\} = \{D\} \rightarrow c_i = d_i \text{ for every } i = 1, 2, \ldots, M \quad \text{(A.1.2)}$$

Matrix Equality: If $[A]$ is an $M \times N$ matrix and [B] is another $M \times N$ matrix (*Note:* The matrices must have the same dimensions), then we say that $[A]$ *is equal to* $[B]$ and write $[A] = [B]$, if each component, a_{ij}, of $[A]$, is equal to the corresponding component, b_{ij}, of $[B]$. That is:

$$[A] = [B] \rightarrow a_{ij} = b_{ij} \text{ for every } i = 1, 2, \ldots, M, \text{ and every } j = 1, 2, \ldots, N \quad \text{(A.1.3)}$$

A.2 Matrix Mathematical Operations

Sum (or difference) of two vectors or matrices (with the same dimensions): If $[A]$ is an $M{\times}N$ matrix and $[B]$ is another $M{\times}N$ matrix, then *the sum*, $[C]$, of $[A]$ and $[B]$, is denoted as $[C] = [A] + [B]$ and has components, which are given by:

$$c_{ij} = a_{ij} + b_{ij} \quad \text{(A.2.1)}$$

Also, *the difference*, $[D]$, of $[B]$ from $[A]$, is denoted as $[D] = [A] - [B]$ and has components that are given by:

$$d_{ij} = a_{ij} - b_{ij} \quad \text{(A.2.2)}$$

Since vectors are single-column or single-row matrices, *the sum and difference of two row or two column vectors* can be established as a special case of the above definition (with, e.g., $N = 1$).

The scalar product (or dot product or inner product) p_{scalar}, of an N-dimensional vector, $\vec{A} = \{A\} = [a_1 \; a_2 \; \ldots \; a_N]^T$, and another N-dimensional column vector, $\vec{B} = \{B\} = [b_1 \; b_2 \; \ldots \; b_N]^T$ is the following scalar (=single value) quantity:

$$p_{scalar} = \vec{A} \bullet \vec{B} = \{A\}^T\{B\} = \{A\}^T\{B\} = \sum_{i=1}^{N}(a_i \cdot b_i) = a_1 \cdot b_1 + a_2 \cdot b_2 + \ldots + a_N \cdot b_N \quad \text{(A.2.3)}$$

We can also write the following expression:

$$p_{scalar} = \left\|\vec{A}\right\|\left\|\vec{B}\right\| \cdot \cos\left(\vec{A}, \vec{B}\right) \quad \text{(A.2.4)}$$

where $\left\|\vec{A}\right\|$ is the magnitude of $\vec{A}$, $\left\|\vec{B}\right\|$ is the magnitude of $\vec{B}$, and $\cos\left(\vec{A}, \vec{B}\right)$ is the cosine of the angle that the two vectors form if they are thought of as "arrows" in N-dimensional space.

Remark A.2.1: It can be easily shown that, for any vector $\vec{V}$:

$$\vec{V} \bullet \vec{V} = \{V\}^T\{V\} = \left\|\vec{V}\right\|^2 \quad \text{(A.2.5)}$$ ■

Remark A.2.2: Two nonzero vectors $\vec{A} = \{A\}$, and $\vec{B} = \{B\}$ are called *orthogonal*, if $\vec{A} \bullet \vec{B} = \{A\}^T\{B\} = \{A\}^T\{B\} = 0$. This definition stems from Equation (A.2.4), which leads to the conclusion that if the dot product is zero and the magnitude of each vector is nonzero, then we must have $\cos\left(\vec{A}, \vec{B}\right) = 0$, which is geometrically equivalent to stating that the two vectors "form a right (90°) angle with each other." ∎

A.2.1 Exterior Product

The *cross product* (*or vector product*) of two three-dimensional vectors, $\{A\} = \begin{Bmatrix} a_1 \\ a_2 \\ a_3 \end{Bmatrix}$ and $\{B\} = \begin{Bmatrix} b_1 \\ b_2 \\ b_3 \end{Bmatrix}$, is *a vector* $\{C\}$, given by:

$$\vec{C} = \{C\} = \{A\} \times \{B\} = \vec{A} \times \vec{B} = \begin{Bmatrix} a_2 b_3 - a_3 b_2 \\ a_3 b_1 - a_1 b_3 \\ a_1 b_2 - a_2 b_1 \end{Bmatrix} \tag{A.2.6}$$

Remark A.2.3: One can verify that

$$\{A\} \times \{B\} = -\{B\} \times \{A\} \tag{A.2.7}$$ ∎

A.2.2 Product of Two Matrices

If $[A]$ is an $M \times N$ matrix with components $[a_{ij}]$ and $[B]$ is an $N \times P$ matrix with elements $[b_{ij}]$, then the product of the two matrices, $[C] = [A][B]$, is an $M \times P$ matrix (note that the number of columns of the first matrix in the product must be equal to the number of rows of the second matrix, otherwise the product cannot be defined!). We will say for the product $[A][B]$ that we *premultiply* $[B]$ *by* $[A]$. Similarly, for the same product, we will also be saying that we *postmultiply* $[A]$ *by* $[B]$. The element c_{ij} of matrix $[C] = [A][B]$ is the scalar product of the i-th row vector of $[A]$ and the j-th column vector of $[B]$, that is:

$$c_{ij} = \sum_{k=1}^{N} \left(a_{ik} \cdot b_{kj}\right) = a_{i1} \cdot b_{1j} + a_{i2} \cdot b_{2j} + \ldots + a_{iN} \cdot b_{Nj} \tag{A.2.8}$$

Example: $[A] = \begin{bmatrix} 2 & 3 & 0 \\ 1 & -1 & 4 \end{bmatrix}$, $[B] = \begin{bmatrix} 3 & 0 & 1 \\ 1 & 2 & 1 \\ 0 & 1 & 2 \end{bmatrix}$. Notice that $[A]$ is a 2×3 matrix, and $[B]$ is a 3×3 matrix.

Then $[C] = [A][B]$ is a 2×3 matrix, and, for example, the element c_{12} is the scalar product of the first row vector of $[A]$ and the second column vector of $[B]$:

$$c_{12} = \sum_{k=1}^{N} a_{1k} \cdot b_{k2} = a_{11} \cdot b_{12} + a_{12} \cdot b_{22} + a_{13} \cdot b_{32} = 2 \cdot 0 + 3 \cdot 2 + 0 \cdot 1 = 6$$

Remark A.2.4: *For* two *N*-dimensional vectors $\{A\}$ and $\{B\}$, we have: $\{A\}^T\{B\} \neq \{B\}\{A\}^T$ (the left-hand side is a scalar value, the right-hand side is a matrix! The reader can verify this, using, for example, two 2-dimensional vectors). ■

Remark A.2.5: The cross-product of two vectors can be written as a matrix operation:

$$\{A\} \times \{B\} = [M]\{B\} \tag{A.2.9}$$

Where $[M]$ is a skew-symmetric matrix, given by:

$$[M] = \begin{bmatrix} 0 & -a_3 & a_2 \\ a_3 & 0 & -a_1 \\ -a_2 & a_1 & 0 \end{bmatrix} \tag{A.2.10}$$ ■

Remark A.2.6: It is also important to remember the following lemma, which is used a lot throughout this text:

$$\text{if } [C] = [A][B], \text{ then } [C]^T = [B]^T[A]^T \tag{A.2.11}$$ ■

Remark A.2.7: Expressions involving "large" arrays can be partitioned in *block-matrix expressions*, where we treat each of the block matrices as single (scalar) numbers. To explain this concept, let us imagine that we want to obtain the product $[A]\{V\}$, where

$$[A] = \begin{bmatrix} 5 & 0 & 2 & 1 \\ 0 & 3 & 1 & 2 \\ -1 & -1 & 1 & 1 \\ 0 & 0 & 0 & 1 \end{bmatrix} \text{ and } \{V\} = \begin{Bmatrix} 2 \\ -1 \\ 0 \\ 1 \end{Bmatrix}$$

$$\text{We have}: [A]\{V\} = \begin{bmatrix} 5 & 0 & 2 & 1 \\ 0 & 3 & 1 & 2 \\ -1 & -1 & 1 & 1 \\ 0 & 0 & 0 & 1 \end{bmatrix} \begin{Bmatrix} 2 \\ -1 \\ 0 \\ 1 \end{Bmatrix} = \begin{Bmatrix} 11 \\ -1 \\ 0 \\ 1 \end{Bmatrix}$$

We can partition (subdivide) matrix $[A]$ into smaller arrays and the vector $\{V\}$ into smaller vectors, to cast the matrix product in the following form:

$$[A]\{V\} = \begin{bmatrix} 5 & 0 & 2 & 1 \\ 0 & 3 & 1 & 2 \\ -1 & -1 & 1 & 1 \\ 0 & 0 & 0 & 1 \end{bmatrix} \begin{Bmatrix} 2 \\ -1 \\ 0 \\ 1 \end{Bmatrix} = \begin{bmatrix} \begin{bmatrix} 5 & 0 \\ 0 & 3 \end{bmatrix} & \begin{bmatrix} 2 & 1 \\ 1 & 2 \end{bmatrix} \\ \begin{bmatrix} -1 & -1 \\ 0 & 0 \end{bmatrix} & \begin{bmatrix} 1 & 1 \\ - & 1 \end{bmatrix} \end{bmatrix} \begin{Bmatrix} \begin{Bmatrix} 2 \\ -1 \end{Bmatrix} \\ \begin{Bmatrix} 0 \\ 1 \end{Bmatrix} \end{Bmatrix}$$

$$= \begin{bmatrix} [A_{11}] & [A_{12}] \\ [A_{21}] & [A_{22}] \end{bmatrix} \begin{Bmatrix} \{V_1\} \\ \{V_2\} \end{Bmatrix}$$

where $[A_{11}] = \begin{bmatrix} 5 & 0 \\ 0 & 3 \end{bmatrix}$, $[A_{12}] = \begin{bmatrix} 2 & 1 \\ 1 & 2 \end{bmatrix}$, $[A_{21}] = \begin{bmatrix} -1 & -1 \\ 0 & 0 \end{bmatrix}$, $[A_{22}] = \begin{bmatrix} 1 & 1 \\ 0 & 1 \end{bmatrix}$, $\{V_1\} = \begin{Bmatrix} 2 \\ -1 \end{Bmatrix}$, $\{V_2\} = \begin{Bmatrix} 0 \\ 1 \end{Bmatrix}$

Then, we can imagine that we have the following expression: $\begin{bmatrix} A_{11} & A_{12} \\ A_{21} & A_{22} \end{bmatrix} \begin{Bmatrix} V_1 \\ V_2 \end{Bmatrix} = \begin{Bmatrix} A_{11} \cdot V_1 + A_{12} \cdot V_2 \\ A_{21} \cdot V_1 + A_{22} \cdot V_2 \end{Bmatrix}$

This same expression applies for the partitioned block matrices:

Thus:

$$[A]\{V\} = \begin{Bmatrix} [A_{11}]\{V_1\} + [A_{12}]\{V_2\} \\ [A_{21}]\{V_1\} + [A_{22}]\{V_2\} \end{Bmatrix} = \begin{Bmatrix} \begin{bmatrix} 5 & 0 \\ 0 & 3 \end{bmatrix} \begin{Bmatrix} 2 \\ -1 \end{Bmatrix} + \begin{bmatrix} 2 & 1 \\ 1 & 2 \end{bmatrix} \begin{Bmatrix} 0 \\ 1 \end{Bmatrix} \\ \begin{bmatrix} -1 & -1 \\ 0 & 0 \end{bmatrix} \begin{Bmatrix} 2 \\ -1 \end{Bmatrix} + \begin{bmatrix} 1 & 1 \\ 0 & 1 \end{bmatrix} \begin{Bmatrix} 0 \\ 1 \end{Bmatrix} \end{Bmatrix}$$

$$= \begin{Bmatrix} \begin{Bmatrix} 11 \\ -1 \end{Bmatrix} \\ \begin{Bmatrix} 0 \\ 1 \end{Bmatrix} \end{Bmatrix} = \begin{Bmatrix} 11 \\ -1 \\ 0 \\ 1 \end{Bmatrix},$$

which is the same exact result that we obtained when we directly multiplied without any block-partitioning.

The only aspect that we need to be careful about, every time that we conduct block-partitioning of matrix-vector multiplications, is to ensure that the number of columns of $[A_{11}]$ and $[A_{21}]$ must equal the number of components in $\{V_1\}$, and the number of columns of $[A_{12}]$ and $[A_{22}]$ must equal the number of components in $\{V_2\}$. ∎

The *diagonal* of a matrix $[A]$ is the set of all the components a_{ii}, $i = 1, 2, \ldots, N$. That is, the diagonal is defined by the components for which the numbers of row (the first subscript) and column (second subscript) are the same.

If all the components of a matrix below the diagonal are zero, then the matrix is called *upper triangular*. If all components above the diagonal are zero, then the matrix is called *lower triangular*.

An $N \times N$ square matrix, with its diagonal elements equal to one, and all the other elements equal to zero, is called an $N \times N$ *identity matrix* or an N-dimensional identity matrix, denoted in this text as $[I]$:

$$[I] = \begin{bmatrix} 1 & 0 & \ldots & 0 \\ 0 & 1 & \ldots & 0 \\ \vdots & \vdots & \ddots & \vdots \\ 0 & 0 & \ldots & 1 \end{bmatrix} \tag{A.2.12}$$

A.2.3 Inverse of a Square Matrix

If $[M]$ is an $N{\times}N$ matrix, then the inverse $[M]^{-1}$ of $[M]$ is an $N{\times}N$ matrix, such that:

$$[M]^{-1}[M] = [M][M]^{-1} = [I] \tag{A.2.13}$$

Remark A.2.8: A matrix is called *singular*, if it has no inverse. ∎

A.2.4 Orthogonal Matrix

A square matrix $[Q]$ for which $[Q]^{-1} = [Q]^T$ is called an *orthogonal matrix*. That is, the inverse of an orthogonal matrix is equal to its transpose:

$$[Q]^T[Q] = [I] \text{ for an orthogonal matrix } [Q] \tag{A.2.14}$$

A.3 Eigenvalues and Eigenvectors of a Matrix

An important set of quantities related to a matrix are its eigenvalues and eigenvectors. More specifically, let us imagine that we have an $N{\times}N$ square matrix, $[A]$, and we want to find whether there are values λ and nonzero vectors $\{\varphi\}$, which satisfy the equation:

$$[A]\{\varphi\} = \lambda\{\varphi\} \leftrightarrow ([A] - \lambda[I])\{\varphi\} = \{0\} \tag{A.3.1}$$

For any matrix $[A]$, we can find N pairs of values of λ and corresponding vectors $\{\varphi\}$. The values λ_i, $i = 1, 2, \ldots, N$ are called *the eigenvalues* of $[A]$ and the vectors $\{\varphi_i\}$, $i = 1, 2, \ldots, N$ are called *the eigenvectors* of $[A]$. An *eigenpair* is the pair consisting of an eigenvalue λ_i and the corresponding eigenvector $\{\varphi_i\}$. Each eigenpair satisfies the equation:

$$[A]\{\varphi_i\} = \lambda_i\{\varphi_i\} \leftrightarrow ([A] - \lambda_i[I])\{\varphi_i\} = \{0\} \tag{A.3.2}$$

If matrix $[A]$ is symmetric, then all the eigenvalues and eigenvectors are real-valued. The eigenvectors of a matrix are *linearly independent*, and thus they form a *basis* for the N-dimensional space. This means that any N-dimensional vector $\{v\}$ can be written as a linear combination of the eigenvectors:

$$\{v\} = a_1\{\varphi_1\} + a_2\{\varphi_2\} + \ldots + a_N\{\varphi_N\} \tag{A.3.3}$$

where $a_1, a_2, \ldots, a_N$ are real numbers. The term $a_i\{\varphi_i\}$, $i = 1, 2, \ldots, N$, gives us *the projection of* $\{v\}$ *on* $\{\varphi_i\}$ *or simply the component of* $\{v\}$ *along* $\{\varphi_i\}$.

For a symmetric matrix, the eigenvectors satisfy the *orthogonality* condition:

$$\{\varphi_i\}^T\left\{\varphi_j\right\} = 0, \text{ for } i \neq j \tag{A.3.4}$$

This orthogonality condition allows us to write a symmetric matrix $[A]$ as:

$$[A] = [\Phi][\Lambda][\Phi]^T \tag{A.3.5}$$

where $[\Phi]$ is a matrix whose columns are the eigenvectors and $[\Lambda]$ is a *diagonal* matrix whose diagonal entities are the eigenvalues:

$$[\Phi] = [\{\varphi_1\} \ \{\varphi_2\} \ \ldots \ \{\varphi_N\}], \ [\Lambda] = \begin{bmatrix} \lambda_1 & 0 & \ldots & 0 \\ 0 & \lambda_2 & \ldots & 0 \\ \vdots & \vdots & \ddots & \vdots \\ 0 & 0 & \ldots & \lambda_N \end{bmatrix} \tag{A.3.6}$$

Remark A.3.1: Matrix $[\Phi]$ in Equation (A.3.6) is orthogonal: $[\Phi]^T[\Phi] = [\Phi][\Phi]^T = [I]$ ■

For a *nonsymmetric* matrix, we can write a more general expression:

$$[A] = [\Phi][\Lambda][\Phi]^{-1} \tag{A.3.7}$$

We can also write (for a symmetric matrix):

$$[A] = \lambda_1\{\varphi_1\}\{\varphi_1\}^T + \lambda_2\{\varphi_2\}\{\varphi_2\}^T + \ldots + \lambda_N\{\varphi_N\}\{\varphi_N\}^T \tag{A.3.8}$$

The above considerations apply for the so-called *linear eigenvalue* problem. In several cases, we may also come across (and we do come across, as shown in Chapters 16 and 17!) the so-called *generalized eigenvalue problem*:

$$[A]\{\varphi\} = \lambda[B]\{\varphi\} \leftrightarrow ([A] - \lambda[B])\{\varphi\} = \{0\} \tag{A.3.9}$$

Remark A.3.2: A generalized eigenvalue problem can be written as a linear eigenvalue problem, if we premultiply the pertinent Equation (A.3.9) by the inverse of $[B]$ (assuming that this inverse exists):

$$[B]^{-1}([A] - \lambda[B])\{\varphi\} = \{0\} \rightarrow \left([B]^{-1}[A] - \lambda[I]\right)\{\varphi\} = \{0\} \rightarrow \left([\tilde{A}] - \lambda[I]\right)\{\varphi\} = \{0\} \tag{A.3.10}$$

Where

$$[\tilde{A}] = [B]^{-1}[A] \qquad \text{(A.3.11)} \blacksquare$$

Remark A.3.3: In parabolic (scalar field) time-dependent problems, described in Chapters 15 and 16, we have: $[A] = [K]$, $[B] = [C]$. In elastodynamics and structural dynamics, described in Chapters 15 and 17, we have $[A] = [K]$, $[B] = [M]$, $\lambda = \omega^2$. ∎

Remark A.3.4: If we want to find a *nonzero* vector $\{\varphi\}$ satisfying Equation (A.3.1), we need to satisfy:

$$det([A] - \lambda[I]) = 0 \qquad \text{(A.3.12)}$$

That is, the determinant of the matrix $([A] - \lambda[I])$ must be zero. Equation (A.3.12) yields an N-th degree polynomial for λ, called the *characteristic polynomial of matrix* $[A]$. The N eigenvalues are the roots of the characteristic polynomial. ∎

A.4 Rank of a Matrix

The rank of a matrix is a notion of paramount importance in problems where matrix equations arise. We will define the rank here partially relying on eigenvalue-related concepts. The *rank of a matrix* $[A]$, denoted *rank*($[A]$), is the number of linearly independent rows and columns of the matrix. Sometimes, we refer to "row" rank and "column" rank, but these two are equal. *For a square matrix, the rank is equal to the number of nonzero eigenvalues!*

If an $N \times N$ *matrix* $[A]$ is *singular*, then it will have *at least one zero eigenvalue!* This means that *rank* $([A]) < N$. In such case, we say that matrix $[A]$ is *rank-deficient* or that $[A]$ has a rank deficiency[1]!

We know that, if $[A]$ is singular and we are looking for a solution $\{x\}$ to the system of equations:

$$[A]\{x\} = \{b\} \qquad \text{(A.4.1)}$$

where $\{b\}$ is a nonzero vector, then we *cannot* find a solution. The reason is that, if we call $\{\varphi_K\}$ the eigenvector corresponding to a zero eigenvalue ($\lambda_K = 0$), then application of Equation (A.3.1) for $\{\varphi_K\}$ gives us:

$$[A]\{\varphi_K\} = \{0\} \qquad \text{(A.4.2)}$$

which can then lead us to the conclusion that the vector $\{x\}$ in Equation (A.4.1) is *non-unique*. Specifically, if we write $\{x\}$ as a linear combination of the eigenvectors using Equation (A.3.3), we have:

$$\{x\} = a_1\{\varphi_1\} + a_2\{\varphi_2\} + \ldots + a_K\{\varphi_K\} + \ldots + a_N\{\varphi_N\} \qquad \text{(A.4.3)}$$

1 If the rank of a matrix is equal to the maximum possible number (for an $N \times N$ matrix, this means that *rank* $([A]) = N$), then we say that the matrix has *full rank*. Otherwise, the matrix is *rank-deficient*. Note that if we have a non-square, $M \times N$, matrix, then the maximum possible rank is equal to the minimum of M and N.

we can verify that if $\{x\}$ has a nonzero component along the eigenvector $\{\varphi_K\}$, $a_K \neq 0$, then the solution is non-unique: for *any* real value of a_K, the right-hand side of Equation (A.4.1) does not change (due to Equation A.4.2). Thus, the only way for the system of equations (A.4.1) to *have a unique* solution $\{x\}$, is to enforce the condition that the component of $\{x\}$ along $\{\varphi_K\}$ is zero: $a_K = 0$! Such conditions must be enforced for *all* the eigenvectors corresponding to zero eigenvalues! The set of all eigenvectors corresponding to a zero eigenvalue—that is, all the linearly independent vectors satisfying Equation (A.4.2), is called the *nullspace* of matrix $[A]$.

The above mathematical considerations can explain what takes place when we enforce the support conditions to the equilibrium equations in a static matrix analysis (explained in Section B.3 of Appendix B). The global system of equations for such problems is:

$$[K]\{U\} = \{f\} \tag{A.4.4}$$

If we have not enforced support conditions to $[K]$, then $\{U\}$ can contain *rigid-body modes.* Rigid-body mode vectors do not mobilize any resistance for the structure, i.e. they do not lead to any forces at the nodes. This means that each rigid-body mode vector, $\{U_{RBM}\}$, satisfies the following condition.

$$[K]\{U_{RBM}\} = \{0\} \tag{A.4.5}$$

It is worth remembering that for one-dimensional structures we have one rigid-body mode, for two-dimensional structures we have three rigid-body modes, and for three-dimensional structures we have six rigid-body modes. It can easily be verified from Equations (A.4.2) and (A.4.5) that the rigid-body modes are actually eigenvectors of $[K]$ corresponding to zero eigenvalues (i.e., the rigid-body modes are components of the nullspace of $[K]$)! If we do not enforce the constraints of the support conditions in our global stiffness equations, the rank of the global $[K]$ is less than N, which means that $[K]$ is singular. Enforcing the support conditions in our global stiffness equations is mathematically equivalent to stipulating that our global displacement vector, $\{U\}$, does *not* contain any component along the rigid-body mode vectors! This is what happens when we remove rows and columns of $[K]$ and obtain $[K_{ff}]$ (as explained in Section B.3), which has full rank and is invertible. Of course, there are cases (one of them is discussed in Section 11.5) where there may be additional vectors (besides rigid-body-modes), which belong to the nullspace of the matrix $[K]$ and can then create difficulties in an analysis.

Appendix B

Review of Matrix Analysis for Discrete Systems

The finite element analysis of any physical problem is inherently tied to matrix analysis of discrete systems. For this reason, we will review the fundamentals and the methodologies used for the matrix analysis of discrete systems. We will focus on the simple case of truss structures with a constant cross-section. The procedures that we will describe are applicable to the solution of the matrix equations of any finite element formulation.

B.1 Truss Elements

A *truss member (or truss element)* is a line member (i.e., a member where one of the three dimensions, the length, is much larger than the other two and thus the geometry of the member can be represented by a line!), which can only develop *axial resistance*—that is, resistance that is parallel to the line representing the member, also called the *member axis*. In this chapter, we will focus on truss members that have a constant cross-sectional area, $A^{(e)}$, and a constant Young's modulus, $E^{(e)}$, for the linearly elastic material of which the members are made.

The two end points of a truss member are called the nodes or nodal points, and for each one of these nodes we have an associated end (or nodal) force as shown in Figure B.1. It is worth mentioning that the notation for forces is $F_{ij}^{(e)}$, to denote the end force at node i, in the direction j, for member e. Later on in this chapter, we will also examine two-dimensional truss members, where each node has two force components, as shown in Figure B.2.

A truss member with constant cross-sectional area develops a constant axial force, $N^{(e)}$, along its length. For member e, the axial force, $N^{(e)}$, is given by the product of the axial stress, $\sigma^{(e)}$, in each cross-section, times the cross-sectional area, $A^{(e)}$:

$$N^{(e)} = \sigma^{(e)} \cdot A^{(e)} \tag{B.1.1}$$

Fundamentals of Finite Element Analysis: Linear Finite Element Analysis, First Edition.
Ioannis Koutromanos, James McClure, and Christopher Roy.

Companion website: www.wiley.com/go/koutromanos/linear

where $N^{(e)}$ has the same sign as $\sigma^{(e)}$. Both axial stress and axial force are taken positive when they are *tensile*, that is, when the member is being "pulled apart" by the stresses. On the other hand, the end forces are taken positive when they are pointing along the positive x-axis.

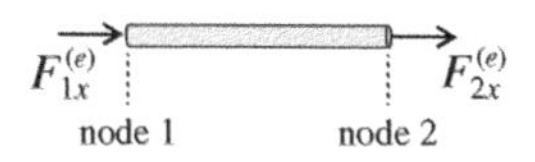

Figure B.1 Truss member.

Now, we are going to obtain a relation between the axial force and the member end forces. To this end, we make a *section* (i.e., a cut) at a location in the interior of the member, as shown in Figure B.3, and take the free-body diagram of the piece to the right of the section:

If we take equilibrium in the x-direction to the right of the section, we have:

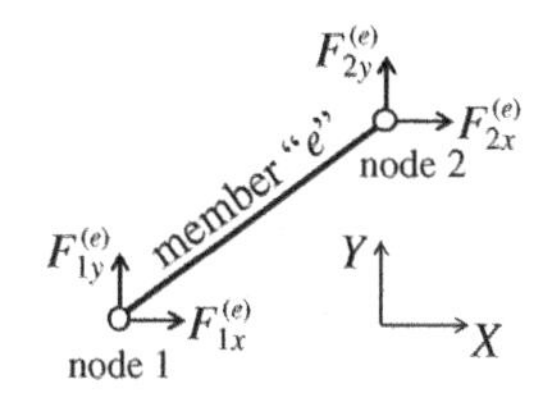

Figure B.2 Two-dimensional truss element.

$$F_{2x}^{(e)} = N^{(e)} \tag{B.1.2}$$

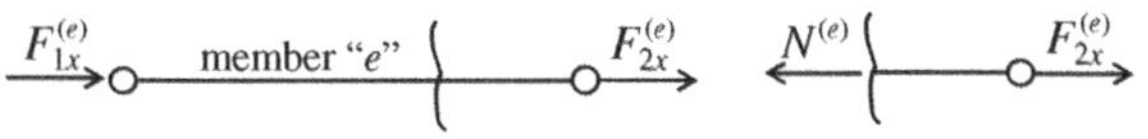

Figure B.3 Relation between right-end (node 2) force and axial force.

The axial stress of the member is given by the product of the modulus of elasticity times the axial strain:

$$\sigma^{(e)} = E^{(e)} \cdot \varepsilon^{(e)} \tag{B.1.3}$$

where

$$\varepsilon^{(e)} = axial\ strain = \frac{\Delta \ell^{(e)}}{\ell^{(e)}} \tag{B.1.4}$$

$\Delta \ell^{(e)}$ is the change in length of the element (taken positive when the element *elongates*), and $\ell^{(e)}$ is the initial length of the member.

Now, we define the element end (nodal) displacements (degrees of freedom, DOF) in the x-direction, as shown in Figure B.4.

The change in length of the element is given by:

$$\Delta \ell^{(e)} = u_{2x}^{(e)} - u_{1x}^{(e)} \tag{B.1.5}$$

If we now combine Equations (B.1.1) to (B.1.5), we obtain:

$$F_{2x}^{(e)} = \frac{E^{(e)} A^{(e)}}{\ell^{(e)}} \left(u_{2x}^{(e)} - u_{1x}^{(e)} \right) = k^{(e)} \left(u_{2x}^{(e)} - u_{1x}^{(e)} \right) \tag{B.1.6}$$

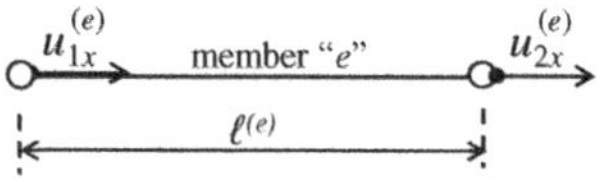

Figure B.4 Nodal displacements for a two-node truss element.

where

$$k^{(e)} = \frac{E^{(e)}A^{(e)}}{\ell^{(e)}} \tag{B.1.7}$$

If we now take force equilibrium of the element as a whole (from Figure B.3), we have:

$$F_{1x}^{(e)} + F_{2x}^{(e)} = 0 \rightarrow F_{1x}^{(e)} = -F_{2x}^{(e)} = \frac{E^{(e)}A^{(e)}}{\ell^{(e)}}\left(u_{1x}^{(e)} - u_{2x}^{(e)}\right) = k^{(e)}\left(u_{1x}^{(e)} - u_{2x}^{(e)}\right) \tag{B.1.8}$$

Equations (B.1.6) and (B.1.8) can be combined into a single matrix expression:

$$\begin{Bmatrix} F_{1x}^{(e)} \\ F_{2x}^{(e)} \end{Bmatrix} = \begin{bmatrix} k^{(e)} & -k^{(e)} \\ -k^{(e)} & k^{(e)} \end{bmatrix} \begin{Bmatrix} u_{1x}^{(e)} \\ u_{2x}^{(e)} \end{Bmatrix} = \frac{E^{(e)}A^{(e)}}{\ell^{(e)}} \begin{bmatrix} 1 & -1 \\ -1 & 1 \end{bmatrix} \begin{Bmatrix} u_{1x}^{(e)} \\ u_{2x}^{(e)} \end{Bmatrix}$$
$$\rightarrow \{F^{(e)}\} = [k^{(e)}]\{U^{(e)}\} \tag{B.1.9}$$

where $\{F^{(e)}\}$ is the element end force vector, $\{U^{(e)}\}$ is the element end displacement vector and $[k^{(e)}]$ is called the *stiffness matrix of the truss element*. The following expressions apply:

$$\{F^{(e)}\} = \begin{Bmatrix} F_{1x}^{(e)} \\ F_{2x}^{(e)} \end{Bmatrix} \tag{B.1.10a}$$

$$\{U^{(e)}\} = \begin{Bmatrix} u_{1x}^{(e)} \\ u_{2x}^{(e)} \end{Bmatrix} \tag{B.1.10b}$$

$$[k^{(e)}] = \begin{bmatrix} k^{(e)} & -k^{(e)} \\ -k^{(e)} & k^{(e)} \end{bmatrix} = \frac{E^{(e)}A^{(e)}}{\ell^{(e)}} \begin{bmatrix} 1 & -1 \\ -1 & 1 \end{bmatrix} \tag{B.1.10c}$$

Remark B.1.1: The stiffness matrix that we obtained has the following properties:

- It has symmetry.
- It is singular (because the member has one rigid-body mode in the axial direction: rigid-body translation, for which $u_1^e = u_2^e$).

To eliminate the singularity, we must restrain the rigid-body modes of motion through appropriate support (restraint) conditions. In one dimension, we need to restrain a single rigid mode (translation along the x-axis), while in two dimensions we need to restrain three (3) rigid-body modes (translation along x-axis, translation along y-axis, and rotation about z-axis). ∎

B.2 One-Dimensional Truss Analysis

We will now proceed to examine *one-dimensional truss analysis*, wherein all the members are aligned with the same axis and we make the tacit assumption that nodes can only translate along the axial direction of the members. To facilitate our descriptions, let us consider an example two-member, one-dimensional truss structure shown in Figure B.5. The structure has a total of three nodal points. One of the nodal points, Node 3, is supported (or restrained), and we know that its displacement is equal to zero. For the other two nodal points, Node 1 and Node 2, we have applied (given) forces, f_1 and f_2, as also shown in Figure B.5. The fact that Node 3 is supported means that we will have an (unknown) support reaction, r_3, at that node, as also shown in Figure B.5.

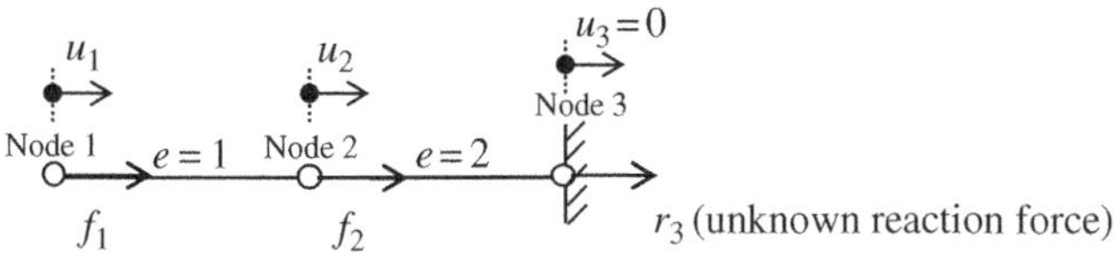

Figure B.5 Example, two-member, one-dimensional truss structure.

For each one of the two truss members, $e = 1$ and $e = 2$, we have the stiffness matrix:

$$\left[k^{(1)}\right] = \frac{E^{(1)}A^{(1)}}{\ell^{(1)}}\begin{bmatrix} 1 & -1 \\ -1 & 1 \end{bmatrix} \tag{B.2.1a}$$

$$\left[k^{(2)}\right] \frac{E^{(2)}A^{(2)}}{\ell^{(2)}}\begin{bmatrix} 1 & -1 \\ -1 & 1 \end{bmatrix} \tag{B.2.1b}$$

We must now somehow *assemble* (i.e., appropriately combine!) these two matrices in a *global (structural) stiffness matrix*, which will express the combined resistance of the entire structure at the locations of the nodal points! This will allow us to formulate the global stiffness equations, ultimately enabling us to solve for the displacements of the unrestrained nodes.

Let us separate the structure into the two members, as shown in Figure B.6, each member drawn with its nodal points and nodal forces. Note that we have two nodal points, 1 and 2, for each one of our members. Of course, in reality, we know that nodes $2_{element\ 1}$ and $1_{element\ 2}$ are the same!

Furthermore, there is a different numbering approach in the global system, where we have nodes 1, 2, and 3. We will now consider equilibrium on nodes 1, 2, and 3 of the structure, by separating these nodes from the rest of the structure, as shown in Figure B.7. As shown in the same figure, we must consider force equilibrium of three nodal forces—namely, the total forces at each one of the three nodal points. Furthermore, we can consider a vector $\{\tilde{F}\}$, expressing the combined contribution of all the elements to the force equilibrium of the nodal points. The vector $\{\tilde{F}\}$ can be further separated to the contributions of the individual elements. For the structure at hand, we will have two such individual element contributions, namely, $\left\{\tilde{F}^{(1)}\right\}$ (contribution from element 1) and $\left\{\tilde{F}^{(2)}\right\}$ (contribution from element 2).

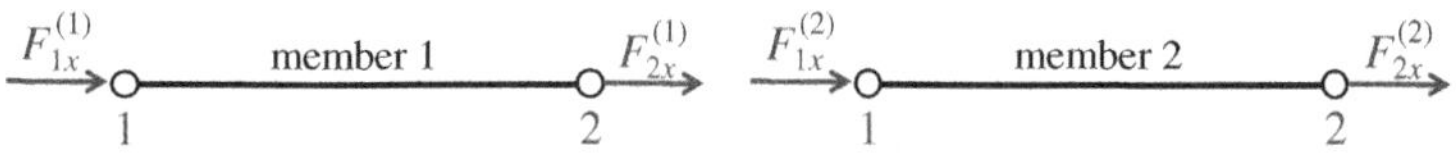

Figure B.6 Separation into two members comprising the structure of Figure B.5.

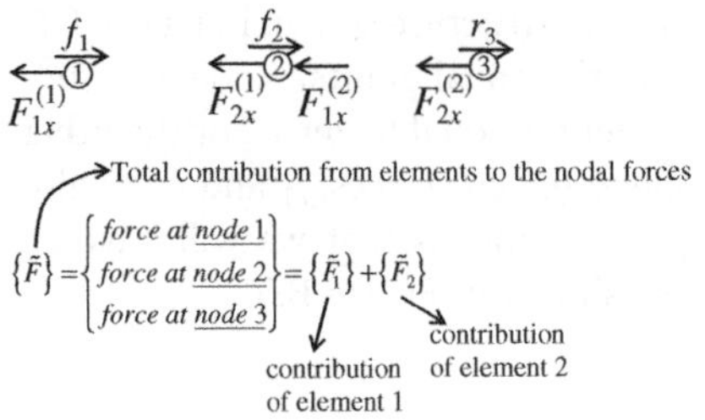

$$\xrightarrow[\textcircled{1}]{f_1} \quad F^{(1)}_{1x} \qquad F^{(1)}_{2x} \xleftarrow{\textcircled{2}} \xrightarrow{f_2} F^{(2)}_{1x} \qquad F^{(2)}_{2x} \xleftarrow{\textcircled{3}} \xrightarrow{r_3}$$

Total contribution from elements to the nodal forces

$$\{\tilde{F}\} = \begin{Bmatrix} \textit{force at node 1} \\ \textit{force at node 2} \\ \textit{force at node 3} \end{Bmatrix} = \{\tilde{F}_1\} + \{\tilde{F}_2\}$$

contribution of element 1; contribution of element 2

Figure B.7 Consideration of force equilibrium at the three nodal points comprising the structure of Figure B.5.

Now, we can write the force equilibrium equations for the nodes in vector form, as shown in Figure B.8:

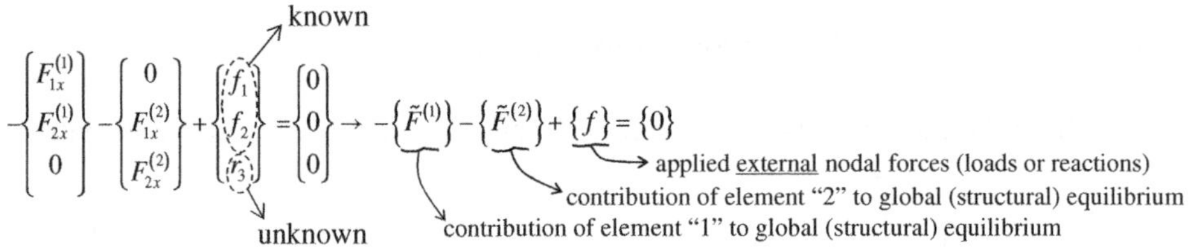

$$-\begin{Bmatrix} F^{(1)}_{1x} \\ F^{(1)}_{2x} \\ 0 \end{Bmatrix} - \begin{Bmatrix} 0 \\ F^{(2)}_{1x} \\ F^{(2)}_{2x} \end{Bmatrix} + \begin{Bmatrix} f_1 \\ f_2 \\ r_3 \end{Bmatrix} = \begin{Bmatrix} 0 \\ 0 \\ 0 \end{Bmatrix} \rightarrow -\{\tilde{F}^{(1)}\} - \{\tilde{F}^{(2)}\} + \{f\} = \{0\}$$

known; unknown

applied external nodal forces (loads or reactions)

contribution of element "2" to global (structural) equilibrium

contribution of element "1" to global (structural) equilibrium

Figure B.8 Equilibrium equation for nodal points of structure.

The element end forces depend on the corresponding end displacements in accordance with the element stiffness equations. Our pursuit is to express the three-component element force vectors $\left\{\tilde{F}^{(1)}\right\}$ and $\left\{\tilde{F}^{(2)}\right\}$ as functions of the three nodal displacements of the entire structure, which are the components of the so-called *global* or *structural nodal displacement vector*, $\{U\}$:

$$\{U\} = [u_1 \;\; u_2 \;\; u_3]^T \tag{B.2.2}$$

We can expand the element stiffness equations to give us a relation between the three-component vector $\left\{\tilde{F}^{(e)}\right\}$ of each element and $\{U\}$. The first step to this end is to understand that the left end point of element (1) coincides with node 1, while the right end point of element (1) coincides with node 2. Thus, we can write for the end displacements of element 1:

$$\left\{U^{(1)}\right\} = \begin{Bmatrix} u^{(1)}_{1x} = u_1 \\ u^{(1)}_{2x} = u_2 \end{Bmatrix} \rightarrow \begin{Bmatrix} u^{(1)}_{1x} \\ u^{(1)}_{2x} \end{Bmatrix} = \begin{Bmatrix} 1\cdot u_1 + 0\cdot u_2 + 0\cdot u_3 \\ 0\cdot u_1 + 1\cdot u_2 + 0\cdot u_3 \end{Bmatrix} \tag{B.2.3}$$

Equation (B.2.3) can now be cast in matrix form as follows.

$$\begin{Bmatrix} u^{(1)}_{1x} \\ u^{(1)}_{2x} \end{Bmatrix} = \begin{bmatrix} 1 & 0 & 0 \\ 0 & 1 & 0 \end{bmatrix} \begin{Bmatrix} u_1 \\ u_2 \\ u_3 \end{Bmatrix} \rightarrow \left\{U^{(1)}\right\} = \left[L^{(1)}\right]\{U\} \tag{B.2.4a}$$

where

$\left[L^{(1)}\right] = \begin{bmatrix} 1 & 0 & 0 \\ 0 & 1 & 0 \end{bmatrix}$ is called the *gather array* of element 1, because it "gathers" appropriate components of the global (structural) displacement vector $\{U\}$ into the element end displacement vector, $\{U^{(1)}\}$.

Similarly, for element 2, we have:

$$\left\{U^{(2)}\right\}=\left\{\begin{matrix} u_{1x}^{(2)}=u_2 \\ u_{2x}^{(2)}=u_3 \end{matrix}\right\} \rightarrow \left\{\begin{matrix} u_{1x}^{(2)} \\ u_{2x}^{(2)} \end{matrix}\right\}=\left\{\begin{matrix} 0\cdot u_1+1\cdot u_2+0\cdot u_3 \\ 0\cdot u_1+0\cdot u_2+1\cdot u_3 \end{matrix}\right\}$$

$$\left\{\begin{matrix} u_{1x}^{(2)} \\ u_{2x}^{(2)} \end{matrix}\right\}=\begin{bmatrix} 0 & 1 & 0 \\ 0 & 0 & 1 \end{bmatrix}\left\{\begin{matrix} u_1 \\ u_2 \\ u_3 \end{matrix}\right\} \rightarrow \left\{U^{(2)}\right\}=\left[L^{(2)}\right]\{U\} \tag{B.2.4b}$$

where $\left[L^{(2)}\right]=\begin{bmatrix} 0 & 1 & 0 \\ 0 & 0 & 1 \end{bmatrix}$ is the gather array of element 2.

In general, for an element e, in a structure with global displacement vector, $\{U\}$, we can write a gather operation for the nodal displacements:

$$\left\{U^{(e)}\right\}=\left[L^{(e)}\right]\{U\} \tag{B.2.4c}$$

Now, we can also establish (with the aid of Figure B.7) a matrix expression giving the element contributions to the structural equilibrium, as functions of the element end force vectors. We begin with the contribution $\left\{\tilde{F}^{(1)}\right\}$ of element 1:

$$\left\{\tilde{F}^{(1)}\right\}=\left\{\begin{matrix} F_{1x}^{(1)} \\ F_{2x}^{(1)} \\ 0 \end{matrix}\right\}=\left\{\begin{matrix} 1\cdot F_{1x}^{(1)}+0\cdot F_{2x}^{(1)} \\ 0\cdot F_{1x}^{(1)}+1\cdot F_{2x}^{(1)} \\ 0\cdot F_{1x}^{(1)}+0\cdot F_{2x}^{(1)} \end{matrix}\right\}=\begin{bmatrix} 1 & 0 \\ 0 & 1 \\ 0 & 0 \end{bmatrix}\left\{\begin{matrix} F_{1x}^{(1)} \\ F_{2x}^{(1)} \end{matrix}\right\}$$

By inspection, we can observe:

$$\left\{\tilde{F}^{(1)}\right\}=\left[L^{(1)}\right]^T\left\{F^{(1)}\right\} \tag{B.2.5a}$$

and the matrix $[L^{(1)}]^T$ is called the *scatter array* of element 1 (because it "scatters" the components of the element end force vector to a larger vector giving the contribution of the element forces to the equilibrium of the entire structure).

Similarly, we can write for element 2:

$$\left\{\tilde{F}^{(2)}\right\}=\left\{\begin{matrix} 0 \\ F_{1x}^{(2)} \\ F_{2x}^{(2)} \end{matrix}\right\}=\left\{\begin{matrix} 0\cdot F_{1x}^{(2)}+0\cdot F_{2x}^{(2)} \\ 1\cdot F_{1x}^{(2)}+0\cdot F_{2x}^{(2)} \\ 0\cdot F_{1x}^{(2)}+1\cdot F_{2x}^{(2)} \end{matrix}\right\}=\begin{bmatrix} 0 & 0 \\ 1 & 0 \\ 0 & 1 \end{bmatrix}\left\{\begin{matrix} F_{1x}^{(2)} \\ F_{2x}^{(2)} \end{matrix}\right\} \rightarrow$$

$$\left\{\tilde{F}^{(2)}\right\}=\left[L^{(2)}\right]^T\left\{F^{(2)}\right\} \tag{B.2.5b}$$

In light of Equations (B.2.5a, b), the equilibrium equation in Figure B.8 for the structure can now be written as:

$$-\left\{\tilde{F}^1\right\}-\left\{\tilde{F}^2\right\}+\{f\}=\{0\} \rightarrow -\left[L^{(1)}\right]^T\{F^1\}-\left[L^{(2)}\right]^T\{F^2\}+\{f\}=\{0\} \tag{B.2.6}$$

Now, we can write the stiffness equations (B.2.1a,b) for the two elements, 1 and 2, and also account for the gather equations, (B.2.4a,b):

$$\left\{F^{(1)}\right\}=\left[k^{(1)}\right]\left\{U^{(1)}\right\}=\left[k^{(1)}\right]\left[L^{(1)}\right]\{U\} \tag{B.2.7a}$$

$$\left\{F^{(2)}\right\} = \left[k^{(2)}\right]\left\{U^{(2)}\right\} = \left[k^{(2)}\right]\left[L^{(2)}\right]\{U\} \tag{B.2.7b}$$

Now, if we plug Equations (B.2.7a, b) into Equation (B.2.6), we obtain:

$$-\left[L^{(1)}\right]^T\left[k^{(1)}\right]\left[L^{(1)}\right]\{u\} - \left[L^{(2)}\right]^T\left[k^{(2)}\right]\left[L^{(2)}\right]\{U\} + \{f\} = \{0\} \rightarrow$$

$$\rightarrow -\left(\left[L^{(1)}\right]^T\left[k^{(1)}\right]\left[L^{(1)}\right] + \left[L^{(2)}\right]^T\left[k^{(2)}\right]\left[L^{(2)}\right]\right)\{U\} + \{f\} = \{0\} \rightarrow -[K]\{U\} + \{f\} = \{0\} \rightarrow$$

$$\rightarrow [K]\{U\} = \{f\} \tag{B.2.8}$$

Where $[K]$ is the *global or structural stiffness matrix* for the structure under consideration, and $\{f\}$ is the global or structural nodal force vector. The global stiffness matrix is obtained from the following expression:

$$[K] = \left[L^{(1)}\right]^T\left[k^{(1)}\right]\left[L^{(1)}\right] + \left[L^{(2)}\right]^T\left[k^{(2)}\right]\left[L^{(2)}\right] = \begin{bmatrix} 1 & 0 \\ 0 & 1 \\ 0 & 0 \end{bmatrix}\begin{bmatrix} k^{(1)} & -k^{(1)} \\ -k^{(1)} & k^{(1)} \end{bmatrix}\begin{bmatrix} 1 & 0 & 0 \\ 0 & 1 & 0 \end{bmatrix}$$

$$+\begin{bmatrix} 0 & 0 \\ 1 & 0 \\ 0 & 1 \end{bmatrix}\begin{bmatrix} k^{(2)} & -k^{(2)} \\ -k^{(2)} & k^{(2)} \end{bmatrix}\begin{bmatrix} 0 & 1 & 0 \\ 0 & 0 & 1 \end{bmatrix} = \begin{bmatrix} k^{(1)} & -k^{(1)} & 0 \\ -k^{(1)} & k^{(1)} & 0 \\ 0 & 0 & 0 \end{bmatrix}$$

$$+\begin{bmatrix} 0 & 0 & 0 \\ 0 & k^{(2)} & -k^{(2)} \\ 0 & -k^{(2)} & k^{(2)} \end{bmatrix} = \begin{bmatrix} k^{(1)} & -k^{(1)} & 0 \\ -k^{(1)} & k^{(1)} + k^{(2)} & -k^{(2)} \\ 0 & -k^{(2)} & k^{(2)} \end{bmatrix}$$

We will now generalize our results for a structure with N members (elements). This generalization is provided in Box B.2.1.

Box B.2.1 Assembly for Global Stiffness Matrix

In general, the global stiffness matrix, $[K]$, of a structure with N_e members (elements) can be obtained as an *assembly* of the individual element contributions:

$$[K] = \sum_{e=1}^{N_e}\left(\left[L^{(e)}\right]^T\left[k^{(e)}\right]\left[L^{(e)}\right]\right) \tag{B.2.9}$$

In other words, for a structure with N_e members, we will have a summation of N_e terms:

$$[K] = \left[L^{(1)}\right]^T\left[k^{(1)}\right]\left[L^{(1)}\right] + \left[L^{(2)}\right]^T\left[k^{(2)}\right]\left[L^{(2)}\right] + \ldots + \left[L^{(N_e)}\right]^T\left[k^{(N_e)}\right]\left[L^{(N_e)}\right]$$

The first term in the summation is the contribution of member 1, the second term in the summation is the contribution of member 2, …, and the $(N_e)^{\text{th}}$ term in the summation is the contribution of member N_e.

B.3 Solving the Global Stiffness Equations of a Discrete System and Postprocessing

Having completed the assembly operation, we have obtained the following system of equations for the nodal displacement vector of the example structure of Figure B.5:

$$\begin{bmatrix} k^{(1)} & -k^{(1)} & 0 \\ -k^{(1)} & k^{(1)}+k^{(2)} & -k^{(2)} \\ 0 & -k^{(2)} & k^{(2)} \end{bmatrix} \begin{Bmatrix} u_1 \\ u_2 \\ u_3 \end{Bmatrix} = \begin{Bmatrix} f_1 \\ f_2 \\ r_3 \end{Bmatrix} \tag{B.3.1}$$

In Equation (B.3.1), we have three unknown quantities, u_1, u_2, r_3 and three known quantities f_1, f_2, and $u_3 = 0$, because node 3 is at the stationary support location.

We can also write Equation (B.3.1) as a set of three equations for the unknown quantities:

$$\begin{bmatrix} k^{(1)} & -k^{(1)} & 0 \\ -k^{(1)} & k^{(1)}+k^{(2)} & -k^{(2)} \\ 0 & -k^{(2)} & k^{(2)} \end{bmatrix} \begin{Bmatrix} u_1 \\ u_2 \\ u_3=0 \end{Bmatrix} = \begin{Bmatrix} f_1 \\ f_2 \\ r_3 \end{Bmatrix} \rightarrow \left\{ \begin{array}{r} k^{(1)}u_1 - k^{(1)}u_2 = f_1 \\ k^{(1)}u_1 - k^{(1)}u_2 - k^{(2)}\cdot 0 = f_2 \\ -k^{(2)}u_2 + k^{(2)}\cdot 0 = r_3 \end{array} \right\}$$

We can keep the set of the first two equations and cast it in matrix form, to obtain a system of equations for u_1 and u_2:

$$\begin{bmatrix} k^{(1)} & -k^{(1)} \\ -k^{(1)} & k^{(1)}+k^{(2)} \end{bmatrix} \begin{Bmatrix} u_1 \\ u_2 \end{Bmatrix} = \begin{Bmatrix} f_1 \\ f_2 \end{Bmatrix} \tag{B.3.2}$$

After we solve for the two unknown nodal displacements, we can proceed with the *postprocessing* stage, which means we calculate other significant quantities, which can only be calculated *after* we have solved and obtained all the nodal displacements!

For example, for each element e, where $e = 1, 2$, we know that $\{U^{(e)}\} = [L^{(e)}]\{U\}$, and can calculate:

- the axial strain:

$$\varepsilon^{(e)} = \frac{u_{2x}^{(e)} - u_{1x}^{(e)}}{\ell^{(e)}} \tag{B.3.3}$$

- the axial stress:

$$\sigma^{(e)} = E^{(e)}\cdot\varepsilon^{(e)} = E^{(e)}\frac{u_{2x}^{(e)} - u_{1x}^{(e)}}{\ell^{(e)}} \tag{B.3.4}$$

- and the axial force

$$N^{(e)} = A^{(e)}\cdot\sigma^{(e)} = A^{(e)}E^{(e)}\frac{u_{2x}^{(e)} - u_{1x}^{(e)}}{\ell^{(e)}} \tag{B.3.5}$$

We can also use the third equation in the global stiffness equations, namely, the equation corresponding to equilibrium of the supported node 3, to obtain the reaction of that node, r_3:

$$r_3 = -k^{(2)}u_2 \tag{B.3.6}$$

Note that, if node 3 had a nonzero, *prescribed* displacement $u_3 = \bar{u}_3$, then we would write the global equilibrium equations in the following form.

$$\begin{bmatrix} k^{(1)} & -k^{(1)} & 0 \\ -k^{(1)} & k^{(1)}+k^{(2)} & -k^{(2)} \\ 0 & -k^{(2)} & k^{(2)} \end{bmatrix} \begin{Bmatrix} u_1 \\ u_2 \\ \bar{u}_3 \end{Bmatrix} = \begin{Bmatrix} f_1 \\ f_2 \\ r_3 \end{Bmatrix} \rightarrow \begin{Bmatrix} k^{(1)}u_1 - k^{(1)}u_2 = f_1 \\ k^{(1)}u_1 - k^{(1)}u_2 - k^{(2)}\cdot\bar{u}_3 = f_2 \\ -k^{(2)}u_2 + k^{(2)}\cdot\bar{u}_3 = r_3 \end{Bmatrix} \quad \text{(B.3.7)}$$

Then, the two first equations will once again give a system for u_1 and u_2:

$$\begin{bmatrix} k^{(1)} & -k^{(1)} \\ -k^{(1)} & k^{(1)}+k^{(2)} \end{bmatrix} \begin{Bmatrix} u_1 \\ u_2 \end{Bmatrix} = \begin{Bmatrix} f_1 \\ f_2 \end{Bmatrix} - \begin{bmatrix} 0 \\ -k^{(2)}\cdot\bar{u}_3 \end{bmatrix} \quad \text{(B.3.8)}$$

Equation (B.3.8) can be solved for u_1 and u_2. After we find u_1 and u_2, we can calculate the reaction at node 3 using the third equation:

$$r_3 = -k^{(2)}u_2 + k^{(2)}\cdot\bar{u}_3 \quad \text{(B.3.9)}$$

This two-stage approach (wherein the first stage involves the calculation of the displacements of the unrestrained nodes) can be generalized for the global system of linear equations for any general discrete system (governed by Equation B.2.8).

The global vector containing the nodal displacements can be partitioned into two parts, namely, the part $\{U_f\}$ containing the (unknown) displacements of the unrestrained (free) nodes and the part $\{U_s\}$ containing the (known) displacements of the restrained (supported) nodes. Then, the global equilibrium (stiffness) equations of the system can be partitioned as follows:

$$\begin{bmatrix} [K_{ff}] & [K_{fs}] \\ [K_{sf}] & [K_{ss}] \end{bmatrix} \begin{Bmatrix} \{U_f\} \\ \{U_s\} \end{Bmatrix} = \begin{Bmatrix} \{f_f\} \\ \{f_s\} \end{Bmatrix} \quad \text{(B.3.10)}$$

where the vectors $\{U_s\}$ and $\{f_f\}$ are *known* (i.e., the displacements of the supported nodes and the applied forces at the free nodes) and vectors $\{U_f\}$ and $\{f_s\}$ are *unknown* (i.e., the displacements of the free nodes and the reactions of the supported nodes).

We can write Equation (B.3.10) as two distinct matrix equations:

$$[K_{ff}]\{U_f\} + [K_{fs}]\{U_s\} = \{f_f\} \quad \text{(B.3.10a)}$$

$$[K_{sf}]\{U_f\} + [K_{ss}]\{U_s\} = \{f_s\} \quad \text{(B.3.10b)}$$

In equation (B.3.10a), we can take all the known terms to the right-hand side to obtain:

$$[K_{ff}]\{U_f\} = \{f_f\} - [K_{fs}]\{U_s\} \rightarrow [K_{ff}]\{U_f\} = \{P_f\} \quad \text{(B.3.11)}$$

where

$$\{P_f\} = \{f_f\} - [K_{fs}]\{U_s\} \quad \text{(B.3.12)}$$

Remark B.3.1: If all the supported displacements are *zero*, then the second term of the right-hand side of Equation (B.3.12) equals zero, and thus we obtain:

$$\text{If}\{U_s\} = 0 \rightarrow [K_{ff}]\{U_f\} = \{f_f\} \quad \text{(B.3.13)}$$ ▮

Equation (B.3.11) can be solved to give us the vector $\{U_f\}$:

$$\text{“}\{U_f\} = [K_{ff}]^{-1}\{P_f\}\text{”} \tag{B.3.14}$$

At any case, the above equations show that in reality, we only need to “invert” the submatrix $[K_{ff}]$ containing the global stiffness terms corresponding to the unrestrained displacements.

Remark B.3.2: The reason why we use *quotation* in Equation (B.3.14) and in the term *invert* is that in reality we do not need to calculate the inverse of $[K_{ff}]$ to solve the system—instead, we use other, computationally more efficient approaches, which are described in Sections 19.2 and 19.3. ▮

After we obtain the values of the free (unrestrained) displacements, $\{U_f\}$, we can use equation (B.3.10b) to calculate the support reactions $\{f_s\}$—that is, the forces corresponding to the supported displacements (one can verify that after obtaining $\{U_f\}$ all the quantities at the left-hand side of B.3.10b are known).

Remark B.3.3: While the global stiffness matrix $[K]$ of a structure is noninvertible (*why?*), it can be proven that the submatrix $[K_{ff}]$ is invertible (and thus, the system of equations for $\{U_f\}$ has a unique solution), *provided that the structure has adequate supports that completely eliminate rigid-body motion modes.*

A consequence of Equation (B.3.11) is that, in reality, we need only assemble the $[K_{ff}]$ subarray of the global stiffness matrix to solve the structure. Actual finite element programs use the *ID* array and *LM* (connectivity) array concepts to assemble the $[K_{ff}]$ array and the $\{P_f\}$ vector. In the next two sections, we will discuss the *ID* and *LM* array concepts. These sections can be skipped, if the reader is not interested in efficient programming of matrix analysis of discrete systems. We will begin by examining (with an example) the assembly of $[K_{ff}]$ for the case where all the displacements at the supports are zero. An advanced interlude is then provided to explain how we can use the *ID* and *LM* array concepts to account for the contributions of nonzero support displacements to $\{P_f\}$.

A second advanced interlude will provide a discussion of how to program the postprocessing stage, that is the calculation of the element stresses and of the support reactions, using the *ID* and *LM* arrays. ▮

B.4 The ID Array Concept (for Equation Assembly)

The assembly of the global stiffness matrix using the element gather/scatter arrays is not a very efficient approach for actual analysis programs, because it requires the computer to keep in memory all the gather/scatter *Boolean arrays* (in a Boolean array, the values of all components are equal to either 0 or 1), which may be a waste of computer memory, especially if we want to analyze a structure with a very large number of elements. To this end, the actual computer implementation of the assembly process is based on the connectivity (*LM*) array and on the *ID* array concepts. In this section, we will discuss the *ID* array, while the next section will describe the use of the *LM* array.

The *ID* array has the same number of components as the global nodal displacement vector, *{U}*, and it gives the unrestrained global displacement to which each component of *{U}* corresponds. The definition of the components of the *ID* array is provided in Box B.4.1.

Box B.4.1 Definition of *ID* Array

If the ith component of $\{U\}$ is equal to the jth component of $\{U_f\}$, then $ID(i) = j$. If the ith component of $\{U\}$ is a restrained (supported or fixed) nodal displacement, then $ID(i) = 0$.

In summary: $ID(jn)$ = unrestrained equation number (= row of $[K_{ff}]$!) of nodal displacement *jn*.

Example B.4.1: Definition of *ID* array
Let us consider the one-dimensional truss structure shown in Figure B.9, wherein the global displacement vector $\{U\}$ contains four components:

$\{U\} = [u_1\ u_2\ u_3\ u_4]^T$. We know that nodes 1 and 4 are restrained, thus, the vector of the unrestrained displacement components will have two entities: $\{U_f\} = \begin{Bmatrix} u_{f1} \\ u_{f2} \end{Bmatrix} = \begin{Bmatrix} u_2 \\ u_3 \end{Bmatrix}$.

If we want to write the *ID* array, we must make the following observations:

- The displacement u_2, which is the second component of the ID array, corresponds to the 1st component of the $\{U_f\}$-vector: $u_2 = u_{f1}$, thus, $ID(2) = 1$.
- The displacement u_3, which is the 3rd component of the ID array, corresponds to the second component of the $\{U_f\}$-vector: $u_3 = u_{f2}$, thus, $ID(3) = 2$.
- The displacements u_1 and u_4 correspond to restrained (i.e., supported) nodes, and they do not correspond to any component of the $\{U_f\}$ vector. For this reason, we mathematically state the condition that u_1 and u_4 correspond to no component of $\{U_f\}$ by setting $ID(1) = 0$ and $ID(4) = 0$.

We will now show, with a schematic (graphics-based) example, how the use of the *ID* array facilitates the assembly of the matrix $[K_{ff}]$ by hand, without a need to resort to the determination and use of the element gather-scatter arrays, $[L^{(e)}]$. In fact, no matrix multiplication will be required for the assembly! A fully automated assembly process, which can be programmed in a computer, will be obtained after we combine the *ID* and *LM* arrays in Section B.5. ∎

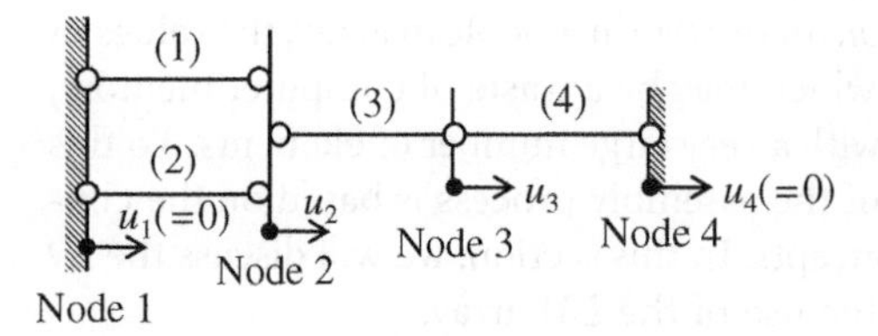

Figure B.9 Example one-dimensional truss structure consisting of four members.

Example B.4.2: Assembly of global stiffness $[K_{ff}]$ with *ID* array

Let us consider the three-member structure shown in Figure B.10. Let us assume that we know the stiffness matrices for the three members, $[k^{(1)}]$, $[k^{(2)}]$, and $[k^{(3)}]$, and want to determine the stiffness matrix $[K_{ff}]$, corresponding to the free (unrestrained) degrees of freedom. Notice that, in the figure, we have also determined the *ID* array for the four nodal displacements of the structure.

To conduct the assembly, we first determine the ID entities of the end nodal displacements (degrees of freedom, or DOF) for each of the three elements:

Element 1 $\Rightarrow$
ID of DOF 1 for element 1 $= ID(1) = 1$
ID of DOF 2 for element 1 $= ID(2) = 2$

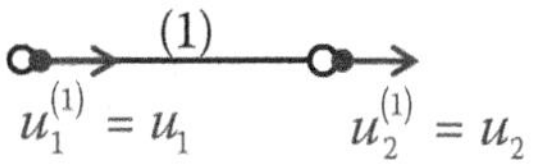

(1) $u_1^{(1)} = u_1$ $u_2^{(1)} = u_2$

Element 2 $\Rightarrow$
ID of DOF 1 for element 2 $= ID(2) = 2$
ID of DOF 2 for element 2 $= ID(3) = 0$

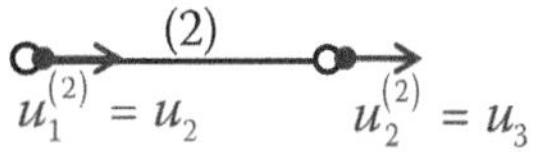

(2) $u_1^{(2)} = u_2$ $u_2^{(2)} = u_3$

Element 3 $\Rightarrow$
ID of DOF 1 for element 3 $= ID(3) = 0$
ID of DOF 2 for element 3 $= ID(4) = 3$

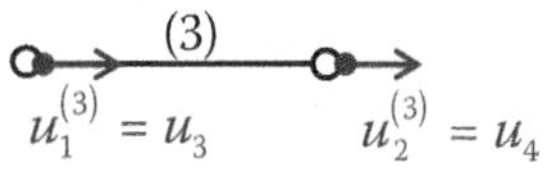

(3) $u_1^{(3)} = u_3$ $u_2^{(3)} = u_4$

After we have determined the *ID* values for the end displacements of each element, we assemble $[K_{ff}]$ by taking the corresponding contributions of each element. The structure in Figure B.10 has three free (unrestrained) displacements, thus, $[K_{ff}]$ will be a (3×3) array. We begin with an empty $[K_{ff}]$, meaning that we write the array with all the values being equal to zero, since we have not yet added the contribution of any element:

$$[k_{ff}] = \begin{bmatrix} 0 & 0 & 0 \\ 0 & 0 & 0 \\ 0 & 0 & 0 \end{bmatrix}$$

We now begin examining the contribution of each element:

Element 1:

We will first show how to assemble the stiffness contribution of element 1 to $[K_{ff}]$. Let us write the stiffness matrix of element 1:

$$\begin{bmatrix} k^{(1)} & -k^{(1)} \\ -k^{(1)} & k^{(1)} \end{bmatrix}$$

Then, we draw next to each row and column of $[k^{(1)}]$ the corresponding *ID* values. Specifically:

ID of DOF 1 for element 1 $= ID(1) = 1 \rightarrow$ Draw the number 1 next to the 1st row and 1st column of $[k^{(1)}]$:

$$\begin{matrix} 1 & & \\ \end{matrix}$$
$$\begin{bmatrix} k^{(1)} & -k^{(1)} \\ -k^{(1)} & k^{(1)} \end{bmatrix} \begin{matrix} 1 \\ \\ \end{matrix}$$

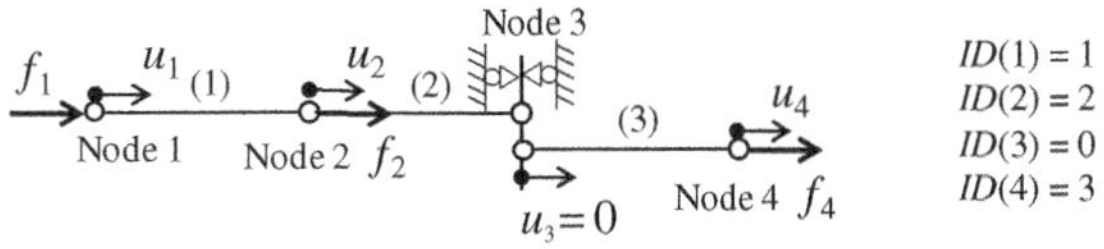

Figure B.10 Three-member structure for explanation of assembly procedure.

$ID(1) = 1$
$ID(2) = 2$
$ID(3) = 0$
$ID(4) = 3$

Also: ID of DOF 2 for element 1 = $ID(2) = 2 \rightarrow$ Draw the number 2 next to the second row and second column of $[k^{(1)}]$:

$$\begin{matrix} 1 & 2 & \\ \end{matrix}$$
$$\begin{bmatrix} k^{(1)} & -k^{(1)} \\ -k^{(1)} & k^{(1)} \end{bmatrix} \begin{matrix} 1 \\ 2 \end{matrix}$$

We now have a number next to all the rows and columns of $[k^{(1)}]$. The numbers that we have drawn next to $[k^{(1)}]$ will guide us in the assembly process for $[K_{ff}]$. The general procedure, shown in Box B.4.2, will be followed. ■

Box B.4.2 Assembling the Stiffness Matrix $[K_{ff}]$ through the *ID* Array

If a value of the stiffness matrix $[k^{(e)}]$ for some element e is on a row next to number $I \neq 0$ and a column next to number $J \neq 0$, then the specific value of the stiffness matrix will contribute (i.e., it must be added!) to the component IJ of $[K_{ff}]$.

If $I = 0$ or $J = 0$, then the specific value of $[k^{(e)}]$ does not contribute at all to $[K_{ff}]$.

Let us apply the procedure in Box B.4.2, for each one of the four components of the stiffness matrix of element 1. The numbers to the right of the two rows will tell us to which row of $[K_{ff}]$ each row of $[k^{(1)}]$ corresponds. Similarly, the numbers above the two columns will tell us to which column of $[K_{ff}]$ each column of $[k^{(1)}]$ corresponds.

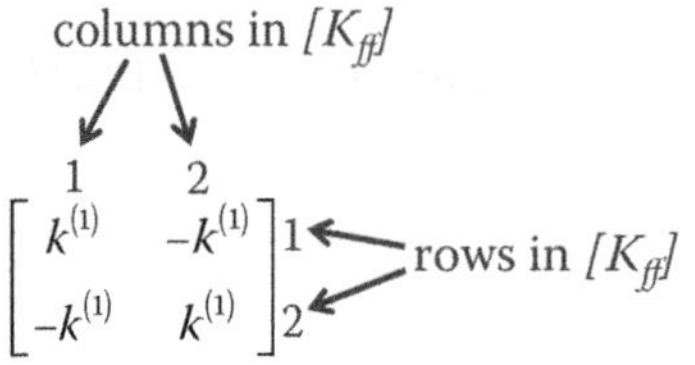

columns in $[K_{ff}]$

$$\begin{matrix} 1 & 2 \end{matrix}$$
$$\begin{bmatrix} k^{(1)} & -k^{(1)} \\ -k^{(1)} & k^{(1)} \end{bmatrix} \begin{matrix} 1 \\ 2 \end{matrix}$$

rows in $[K_{ff}]$

We start with the first component of $[k^{(1)}]$:

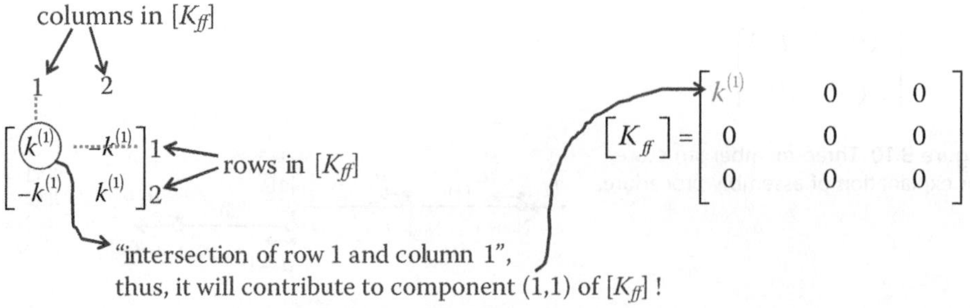

columns in $[K_{ff}]$

$$\begin{matrix} 1 & 2 \end{matrix}$$
$$\begin{bmatrix} k^{(1)} & -k^{(1)} \\ -k^{(1)} & k^{(1)} \end{bmatrix} \begin{matrix} 1 \\ 2 \end{matrix}$$

rows in $[K_{ff}]$

"intersection of row 1 and column 1", thus, it will contribute to component (1,1) of $[K_{ff}]$!

$$[K_{ff}] = \begin{bmatrix} k^{(1)} & 0 & 0 \\ 0 & 0 & 0 \\ 0 & 0 & 0 \end{bmatrix}$$

Similarly, we continue with the other three components of $[k^{(1)}]$:

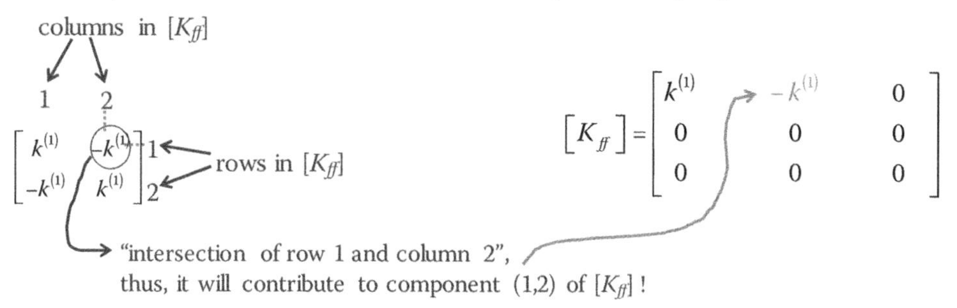

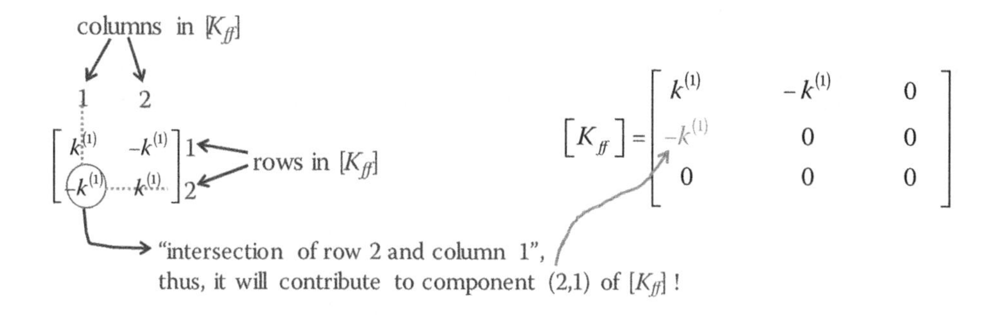

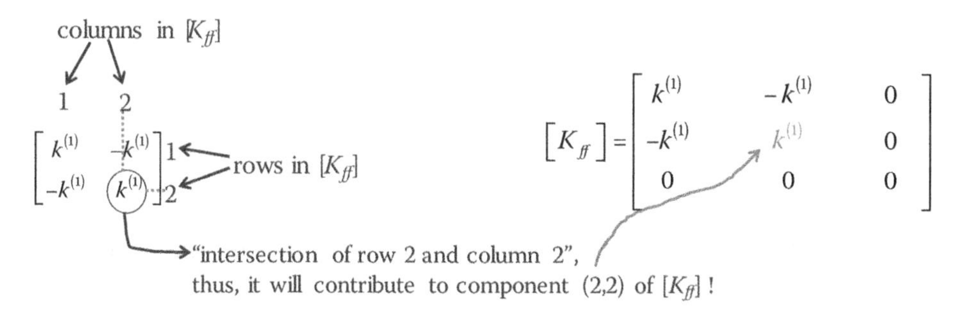

Notice that with the schematically described procedure, we have added all the contributions of element 1 (all four values of this element's stiffness matrix) to $[K_{ff}]$! Thus, we have finished with the contribution of element 1.

We can repeat the same process for the other two elements:

Element 2:

ID of DOF 1 for element 2 = *ID*(2) = 2 → Draw the number 2 next to the first row and first column of $[k^{(2)}]$:

$$\begin{matrix} 2 & \\ \begin{bmatrix} k^{(2)} & -k^{(2)} \\ -k^{(2)} & k^{(2)} \end{bmatrix} & \begin{matrix} 2 \\ \end{matrix} \end{matrix}$$

Also: *ID* of DOF 2 for element 2 = *ID*(3) = 0 → Draw the number 0 next to the second row and second column of $[k^{(2)}]$:

$$\begin{matrix} 2 \quad\quad 0 & \\ \begin{bmatrix} k^{(2)} & -k^{(2)} \\ -k^{(2)} & k^{(2)} \end{bmatrix} & \begin{matrix} 2 \\ 0 \end{matrix} \end{matrix}$$

We can now assemble the contribution of $[k^{(2)}]$ to $[K_{ff}]$. We begin with the first component of $[k^{(2)}]$:

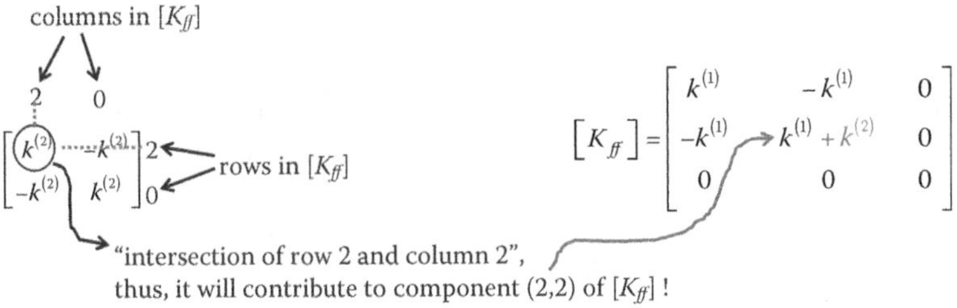

columns in $[K_{ff}]$

$$\begin{matrix} 2 & 0 & \\ \begin{bmatrix} k^{(2)} & -k^{(2)} \\ -k^{(2)} & k^{(2)} \end{bmatrix} & \begin{matrix} 2 \\ 0 \end{matrix} \end{matrix} \quad \text{rows in } [K_{ff}] \qquad [K_{ff}] = \begin{bmatrix} k^{(1)} & -k^{(1)} & 0 \\ -k^{(1)} & k^{(1)} + k^{(2)} & 0 \\ 0 & 0 & 0 \end{bmatrix}$$

"intersection of row 2 and column 2", thus, it will contribute to component (2,2) of $[K_{ff}]$!

In accordance with the descriptions in Box B.4.2, we will establish the following practical rule: "If the *ID* value next to a row or column of an element stiffness matrix is 0, then the components of that row or column will not contribute to $[K_{ff}]$. Thus, we can simply erase all rows and columns which are next to a zero, since the values of these rows and columns will not contribute to $[K_{ff}]$." We can do this with the second row and the second column of $[k^{(2)}]$:

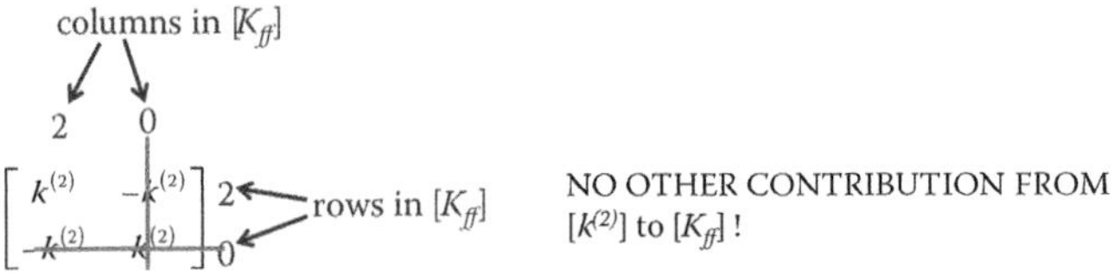

columns in $[K_{ff}]$

$$\begin{matrix} 2 & 0 & \\ \begin{bmatrix} k^{(2)} & -k^{(2)} \\ -k^{(2)} & k^{(2)} \end{bmatrix} & \begin{matrix} 2 \\ 0 \end{matrix} \end{matrix} \quad \text{rows in } [K_{ff}]$$

NO OTHER CONTRIBUTION FROM $[k^{(2)}]$ to $[K_{ff}]$!

Thus, we have no other values for $[k^{(2)}]$ contributing to $[K_{ff}]$.

Element 3:

ID of DOF 1 for element 3 = $ID(3) = 0 \rightarrow$ Draw the number 0 next to the first row and first column of $[k^{(3)}]$:

$$\begin{matrix} 0 & & \\ \begin{bmatrix} k^{(3)} & -k^{(3)} \\ -k^{(3)} & k^{(3)} \end{bmatrix} & \begin{matrix} 0 \\ \ \end{matrix} \end{matrix}$$

Also: *ID* of DOF 2 for element 3 = $ID(4) = 3 \rightarrow$ Draw the number 3 next to the second row and second column of $[k^{(3)}]$:

$$\begin{matrix} 0 \quad 3 & \\ \begin{bmatrix} k^{(3)} & -k^{(3)} \\ -k^{(3)} & k^{(3)} \end{bmatrix} & \begin{matrix} 0 \\ 3 \end{matrix} \end{matrix}$$

We are ready to find the contribution of $[k^{(3)}]$ to $[K_{ff}]$:

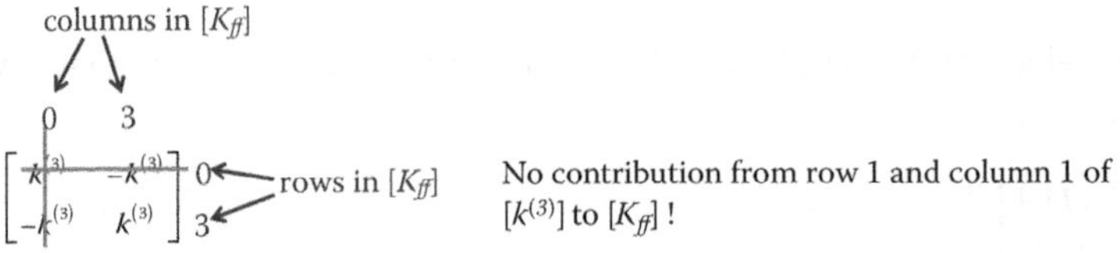

columns in $[K_{ff}]$

$$\begin{matrix} 0 \quad 3 & \\ \begin{bmatrix} k^{(3)} & -k^{(3)} \\ -k^{(3)} & k^{(3)} \end{bmatrix} & \begin{matrix} 0 \\ 3 \end{matrix} \end{matrix} \quad \text{rows in } [K_{ff}]$$

No contribution from row 1 and column 1 of $[k^{(3)}]$ to $[K_{ff}]$!

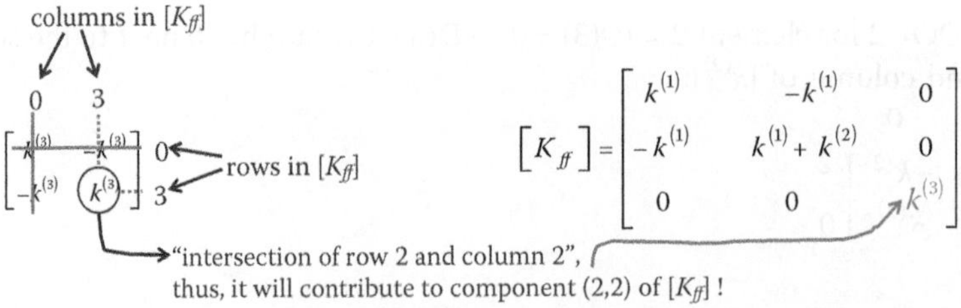

columns in $[K_{ff}]$

$$\begin{matrix} 0 \quad 3 & \\ \begin{bmatrix} k^{(3)} & -k^{(3)} \\ -k^{(3)} & k^{(3)} \end{bmatrix} & \begin{matrix} 0 \\ 3 \end{matrix} \end{matrix} \quad \text{rows in } [K_{ff}] \qquad [K_{ff}] = \begin{bmatrix} k^{(1)} & -k^{(1)} & 0 \\ -k^{(1)} & k^{(1)} + k^{(2)} & 0 \\ 0 & 0 & k^{(3)} \end{bmatrix}$$

"intersection of row 2 and column 2", thus, it will contribute to component (2,2) of $[K_{ff}]$!

We have completed the assembly of $[K_{ff}]$ and the stiffness equations for the unrestrained displacements are:

$$\begin{bmatrix} k^{(1)} & -k^{(1)} & 0 \\ -k^{(1)} & k^{(1)}+k^{(2)} & 0 \\ 0 & 0 & k^{(3)} \end{bmatrix} \begin{Bmatrix} u_1 \\ u_2 \\ u_4 \end{Bmatrix} = \begin{Bmatrix} f_1 \\ f_2 \\ f_4 \end{Bmatrix}$$

Let us find $[K_{ff}]$ with the alternative approach, which is based on assembling $[K]$ from the element stiffness matrices by means of the gather/scatter arrays, $[L^{(e)}]$. Then, we will keep the part of $[K]$ corresponding to the three unrestrained nodal displacements, $\{U_f\} = [u_1 \ u_2 \ u_4]^T$.

For element 1: $$\begin{Bmatrix} u_{1x}^{(1)} \\ u_{2x}^{(1)} \end{Bmatrix} = \begin{bmatrix} 1 & 0 & 0 & 0 \\ 0 & 1 & 0 & 0 \end{bmatrix} \begin{Bmatrix} u_1 \\ u_2 \\ u_3 \\ u_4 \end{Bmatrix}, \text{ thus, } \left[L^{(1)}\right] = \begin{bmatrix} 1 & 0 & 0 & 0 \\ 0 & 1 & 0 & 0 \end{bmatrix}$$

For element 2: $$\begin{Bmatrix} u_{1x}^{(2)} \\ u_{2x}^{(2)} \end{Bmatrix} = \begin{bmatrix} 0 & 1 & 0 & 0 \\ 0 & 0 & 1 & 0 \end{bmatrix} \begin{Bmatrix} u_1 \\ u_2 \\ u_3 \\ u_4 \end{Bmatrix}, \text{ thus, } \left[L^{(2)}\right] = \begin{bmatrix} 0 & 1 & 0 & 0 \\ 0 & 0 & 1 & 0 \end{bmatrix}$$

For element 3: $$\begin{Bmatrix} u_{1x}^{(3)} \\ u_{2x}^{(3)} \end{Bmatrix} = \begin{bmatrix} 0 & 0 & 1 & 0 \\ 0 & 0 & 0 & 1 \end{bmatrix} \begin{Bmatrix} u_1 \\ u_2 \\ u_3 \\ u_4 \end{Bmatrix}, \text{ thus, } \left[L^{(3)}\right] = \begin{bmatrix} 0 & 0 & 1 & 0 \\ 0 & 0 & 0 & 1 \end{bmatrix}$$

We can now assemble the global stiffness matrix $[K]$:

$$[K] = \left[L^{(1)}\right]^T \left[k^{(1)}\right] \left[L^{(1)}\right] + \left[L^{(2)}\right]^T \left[k^{(2)}\right] \left[L^{(2)}\right] + \left[L^{(3)}\right]^T \left[k^{(3)}\right] \left[L^{(3)}\right]$$

$$= \begin{bmatrix} k^{(1)} & -k^{(1)} & 0 & 0 \\ -k^{(1)} & k^{(1)} & 0 & 0 \\ 0 & 0 & 0 & 0 \\ 0 & 0 & 0 & 0 \end{bmatrix} + \begin{bmatrix} 0 & 0 & 0 & 0 \\ 0 & k^{(2)} & -k^{(2)} & 0 \\ 0 & -k^{(2)} & k^{(2)} & 0 \\ 0 & 0 & 0 & 0 \end{bmatrix} + \begin{bmatrix} 0 & 0 & 0 & 0 \\ 0 & 0 & 0 & 0 \\ 0 & 0 & k^{(3)} & -k^{(3)} \\ 0 & 0 & -k^{(3)} & k^{(3)} \end{bmatrix}$$

$$= \begin{bmatrix} k^{(1)} & -k^{(1)} & 0 & 0 \\ -k^{(1)} & k^{(1)}+k^{(2)} & -k^{(2)} & 0 \\ 0 & -k^{(2)} & k^{(2)}+k^{(3)} & -k^{(3)} \\ 0 & 0 & -k^{(3)} & k^{(3)} \end{bmatrix}$$

We can quickly obtain $[K_{ff}]$ from $[K]$ by erasing the row and column corresponding to the fixed (restrained) displacement u_3, i.e. the third row and the third column:

$$\begin{bmatrix} k^{(1)} & -k^{(1)} & 0 & 0 \\ -k^{(1)} & k^{(1)}+k^{(2)} & -k^{(2)} & 0 \\ 0 & -k^{(2)} & k^{(2)}+k^{(3)} & -k^{(3)} \\ 0 & 0 & -k^{(3)} & k^{(3)} \end{bmatrix} \rightarrow [K_{ff}] = \begin{bmatrix} k^{(1)} & -k^{(1)} & 0 \\ -k^{(1)} & k^{(1)}+k^{(2)} & 0 \\ 0 & 0 & k^{(3)} \end{bmatrix}$$

Thus, once again, we obtain the following system for the unrestrained displacements:

$$\begin{bmatrix} k^{(1)} & -k^{(1)} & 0 \\ -k^{(1)} & k^{(1)}+k^{(2)} & 0 \\ 0 & 0 & k^{(3)} \end{bmatrix} \begin{Bmatrix} u_1 \\ u_2 \\ u_4 \end{Bmatrix} = \begin{Bmatrix} f_1 \\ f_2 \\ f_4 \end{Bmatrix}$$

Remark B.4.1: Using the *ID* array and assembling only the global stiffness matrix corresponding to the *free* (unrestrained) degrees of freedom, requires the storage of an $N_f \times N_f$ (N_f = number of free degrees of freedom) array and two vectors with N components each: the nodal displacement vector $\{U\}$ and the nodal force vector $\{F\}$ ($\{F\}$ contains the forces applied to the free degrees of freedom and the reactions of the restrained degrees of freedom). ■

B.5 Fully Automated Assembly: The Connectivity (LM) Array Concept

In the previous section, we saw how we can use the *ID* array to avoid the need to use the element gather/scatter arrays for the assembly of the $[K_{ff}]$ array. The procedure described above using the *ID* array cannot be implemented in a computer program. The reason is that we needed to use inspection and "draw" the *ID* values of each element end displacement next to the rows and columns of the element's stiffness matrix, then find *by inspection* to which location of the $[K_{ff}]$ array each element stiffness term should contribute.

Since an actual computer is not really "endowed" with vision and reason, which would be required to perform the inspection-based procedure that we followed above, we need to devise a second entity describing the data of the problem under consideration. This entity, called the *element connectivity array, LM,* actually eliminates the need to "draw numbers" next to an element's $[k^{(e)}]$ array. The element connectivity array simply tells us to which global displacement component (i.e., to which component of $\{U\}$) do the nodal displacements of each element correspond. The *LM* array has dimensions $N_{el} \times N_{dpe}$, where N_{el} is the number of elements (members) in the structure and N_{dpe} is the number of degrees of freedom that each element has (i.e., the dimension of $[k^{(e)}]$).

For one-dimensional truss analysis, we have $N_{dpe} = 2$; that is, for each element *iel*, *iel* = 1, 2, …, N_{el}, we have two corresponding entities of the *LM* array. These entities provide the following information:

- *LM*(*iel*,1) = To which component of $\{U\}$ does the first end displacement of element *iel* correspond?

- $LM(iel,2)$ = To which component of $\{U\}$ does the second end displacement of element *iel* correspond?

Example B.5.1: Definition of *LM* array
Let us revisit the example structure we had in Figure B.10. We can define the entities of the *LM* array based on the considerations outlined above. All that we need to know is to which global nodal displacement each element nodal displacement corresponds. We can begin with element 1:

$$u_1^{(1)} = u_1 \rightarrow LM(1,1) = 1$$

$$u_2^{(1)} = u_2 \rightarrow LM(1,2) = 2$$

We can repeat the same process for the other two elements:

Element 2: $\begin{array}{l} u_1^{(2)} = u_2 \rightarrow LM(2,1) = 2 \\ u_2^{(2)} = u_3 \rightarrow LM(2,2) = 3 \end{array}$, Element 3: $\begin{array}{l} u_1^{(3)} = u_3 \rightarrow LM(3,1) = 3 \\ u_2^{(3)} = u_4 \rightarrow LM(3,2) = 4 \end{array}$

■

Now, let us revisit the process of the previous section, where we were writing the "ID entries" next to the element stiffness matrices. What we were doing can be summarized as follows: We found, where in $[K_{ff}]$ each component (i,j) of $[k^{(e)}]$ would contribute. This was done by checking the number next to the ith row and the jth column of $[k^{(e)}]$. The number next to the ith row gave us the row of $[K_{ff}]$ to which we had a contribution and the number next to the jth column gave us the column of $[K_{ff}]$ to which we had a contribution. We can easily verify that the number next to the ith row was equal to the entity of the *ID* array to which the ith displacement of element e corresponds. Notice that this number is equal to $(ID(LM(e,i)))$ = the *ID* value of the global displacement to which $u_i^{(e)}$ corresponds. Similarly, the number we drew next to the jth column was equal to $(ID(LM(e,j)))$ = the *ID* value of the global displacement to which $u_j^{(e)}$ corresponds. So, to summarize, the procedure by which we were adding entries of $[k^{(e)}]$ to appropriate locations of $[K_{ff}]$ can be summarized as shown in Box B.5.1.

Box B.5.1 Fully Automated Assembly of $[K_{ff}]$ through the *ID* and *LM* Arrays
For each element e: If $ID(LM(e,i)) = 0$ or $ID(LM(e,j)) = 0$, then the term $k_{ij}^{(e)}$ has no contribution to $[K_{ff}]$. Otherwise, we must add $k_{ij}^{(e)}$ to the location (I, J) of $[K_{ff}]$, where $I = ID(LM(e,i))$ and $J = ID(LM(e,j))$.

The computer implementation of the assembly procedure can be summarized in the pseudo-code[1] provided in Box B.5.2.

1 The term pseudo-code in programming refers to a notation resembling a simplified programming language and used in the conceptual design stage of programs. The readers who are familiar with programming languages such as MATLAB or Fortran will notice a resemblance of the pseudo-code provided here with code in these languages.

Box B.5.2 Pseudo-code for the Assembly of the Global Stiffness Matrix $[K_{ff}]$ = GK:

For each element, *iel*:
Calculate the ($N_{dpe} \times N_{dpe}$) element stiffness matrix, EK

- For each column number, *jc* (*jc* =1, N_{dpe}):
- If ID(LM(*iel*,*jc*)) > 0, THEN
  - For each row number, *ir* (*ir* =1, N_{dpe}):
  - If ID(LM(*iel*,*ir*)) > 0, THEN
  - GK(ID(LM(*iel*,*ir*)),ID(LM(*iel*,*jc*))) = GK(ID(LM(*iel*,*ir*)),ID(LM(*iel*,*jc*))) + EK(*ir*,*jc*)
  - END LOOP for *ir*
- END LOOP for jc

END LOOP for *iel*

B.6 Advanced Interlude—Programming of Assembly When the Restrained Degrees of Freedom Have Nonzero Values

The pseudo-code in Box B.5.2 for the assembly of the GK = $[K_{ff}]$ array corresponding to the free degrees of freedom is applicable to situations where the restrained (fixed) degrees of freedom (DOFs) have zero values. For the most general case that the restrained DOFs have nonzero values, we need to modify the assembly equations to account for the contribution of the nonzero restrained DOFs on the nodal force vector corresponding to the free DOFs (this effect can be observed in the 2×2 matrix Equation B.3.8). The pseudo-code for the assembly of $[K_{ff}]$ and for the nodal force vector $\{P_f\}$ of Equation B.3.12 is provided in Box B.6.1.

Box B.6.1 Pseudo-Code for the GK = $[K_{ff}]$ Array

Pseudo-code for the assembly of the global stiffness matrix $[K_{ff}]$ = GK, and for the corresponding nodal force vector $\{P_f\}$ = F, for the general case that the restrained degrees of freedom may have nonzero values:
For each element, *iel*:
Calculate the ($N_{dpe} \times N_{dpe}$) element stiffness matrix, EK

- For each row number, *ir* (*ir* =1, N_{dpe}):
- If ID(LM(*iel*,*ir*)) > 0, THEN
  - For each column number, *jc* (*jc* =1, N_{dpe}):
    - If ID(LM(*iel*,*jc*)) > 0, THEN
      GK(ID(LM(*iel*,*ir*)),ID(LM(*iel*,*jc*))) = GK(ID(LM(*iel*,*ir*)),ID(LM(*iel*,*jc*))) + EK(*ir*,*jc*)
    - Find element contribution to F:
      If ID(LM(*iel*,*jc*)) = 0, THEN
      FEL(*ir*) = FEL(*ir*) − EK(*ir*,*jc*)∗U(LM(*iel*,*jc*))
  - END LOOP for *jc*
  - Add element contribution to F:
    F(ID(LM(*iel*,*ir*))) = F(ID(LM(*iel*,*ir*))) + FEL(*ir*)
- END LOOP for *ir*

END LOOP for *iel*

B.7 Advanced Interlude 2: Algorithms for Postprocessing

One can also create efficient algorithms for the post-processing stage, where we know the global displacement vector, $\{U\}$, after we have calculated the displacements $\{U_f\}$. Note that the restrained nodal displacements, $\{U_s\}$, are known as input information in an analysis. The post-processing stage involves two types of quantities: (1) quantities such as strains, stresses, and axial forces, which must be locally calculated for each individual element, (2) reactions at the restrained nodes, which must be calculated at a global level, through assembly of element contributions to the reactions!

The post-processing computations at each element e require the gather of the displacements of the element from the global vector $\{U\}$. A direct approach would be to use the gather array $[L^{(e)}]$, in accordance with Equation (B.2.4c). In actual finite element programs, we never establish the gather arrays. Instead, we leverage the *LM* array, similar to the procedures in the previous sections. We will present two algorithms (in pseudo-code form) for one-dimensional truss elements, one algorithm presented in Box B.7.1 for the computation of element strains, stresses and axial forces and another algorithm, presented in Box B.7.2 for the calculation of the nodal reactions for the fixed nodal displacements.

Box B.7.1 Pseudo-code for Printing the Element Strains, ε = EPS, the Element Stresses, σ = SIG, and Axial Forces, N

For each element, *iel*:

We know Young's modulus, E, length, L, and cross-sectional area, A, so:

- Calculate the axial strain, EPS:
  EPS = [U(LM(*iel*,2)) − U(LM(*iel*,1))] / L
- Calculate the axial stress, SIG:
  SIG = E*EPS
- Calculate the axial force, N:
  N = A*SIG
- PRINT IEL, EPS, SIG, N (on the screen or on a file)

END LOOP for *iel*

Box B.7.2 Pseudo-code for Calculating Nodal Reactions

Notice that the global reactions are stored in the components of the global force vector, F, which correspond to the restrained degrees of freedom!

For each element, *iel*:

- Calculate the (N_{dpe} x N_{dpe}) element stiffness matrix, EK
- For each row number, *ir* (*ir* = 1, N_{dpe}):
  - If ID(LM(*iel*,*ir*)) = 0, THEN
  - For each column number, *jc* (*jc* = 1, N_{dpe}):
    F(LM(*iel*,*ir*)) = F(LM(*iel*,*ir*)) + EK(*ir*,*jc*)*U(LM(*iel*,*jc*))
  - END LOOP for *jc*
- END LOOP for *ir*

END LOOP for *iel*

B.8 Two-Dimensional Truss Analysis—Coordinate Transformation Equations

We will now proceed to examine the case of two-dimensional (2D) trusses. In such cases, each of the nodal points in a structure has two nodal displacement and two nodal force components, one component along some *global* coordinate axis, X, and another along a second global coordinate axis, Y, as shown in Figure B.11. The only further consideration for two-dimensional trusses stems from the fact that truss elements only develop resistance (i.e., stiffness) in the direction of their axes, and the axial direction of the various truss elements is not identical. Before we conduct the assembly operation for the global stiffness matrix, we must first establish all the stiffness matrices in the global (or structural) coordinate system.

For each element e, we need to define an angle $\varphi^{(e)}$, shown in Figure B.11, which is equal to the angle by which we need to rotate the global X-axis (in a counterclockwise direction) until it becomes parallel to the member's axis. If we have the coordinates of the element end points in the global coordinate system, that is, $\left(X_1^{(e)}, Y_1^{(e)}\right)$ for nodal point 1 and $\left(X_2^{(e)}, Y_2^{(e)}\right)$ for nodal point 2, then the element length, $\ell^{(e)}$, is given by:

$$\ell^{(e)} = \sqrt{\left(X_2^{(e)} - X_1^{(e)}\right)^2 + \left(Y_2^{(e)} - Y_1^{(e)}\right)^2} = \sqrt{\left(X_{21}^{(e)}\right)^2 + \left(Y_{21}^{(e)}\right)^2} \quad \text{(B.8.1)}$$

and the trigonometric quantities of the angle $\varphi^{(e)}$ are:

$$\cos\varphi^{(e)} = \frac{X_2^{(e)} - X_1^{(e)}}{\ell^{(e)}} = \frac{X_{21}^{(e)}}{\ell^{(e)}} \quad \text{(B.8.2a)}$$

$$\sin\varphi^{(e)} = \frac{Y_2^{(e)} - Y_1^{(e)}}{\ell^{(e)}} = \frac{Y_{21}^{(e)}}{\ell^{(e)}} \quad \text{(B.8.2b)}$$

The stiffness equations for the element can be initially expressed in the *local coordinate system*, shown in Figure B.12, which is defined by the axis parallel to the member and the axis perpendicular to the member. As shown in the same figure, we can also define the two nodal displacements and forces of each node in the local coordinate system.

The local coordinate system is convenient, because it easily allows us to express the member stiffness equations. More specifically, we know that a truss member can only develop forces along its axial direction (i.e., the direction of $x^{(e)}$), and these forces only depend on the axial nodal displacements (i.e., the displacements along $x^{(e)}$). Thus, the previously presented one-dimensional stiffness relations in Equation (B.1.9) can be extended to the local coordinate system of a 2D truss element as follows.

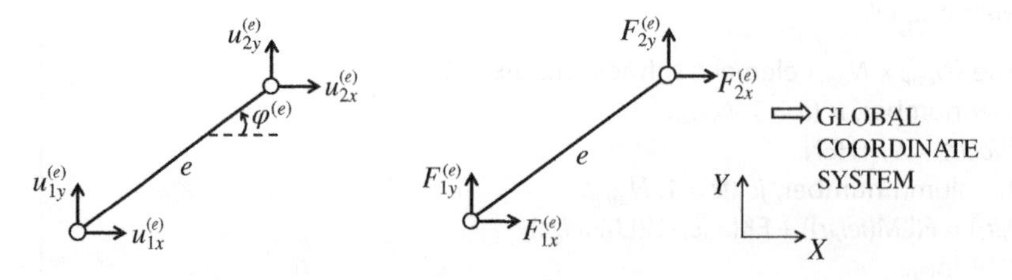

Figure B.11 Two-dimensional truss element.

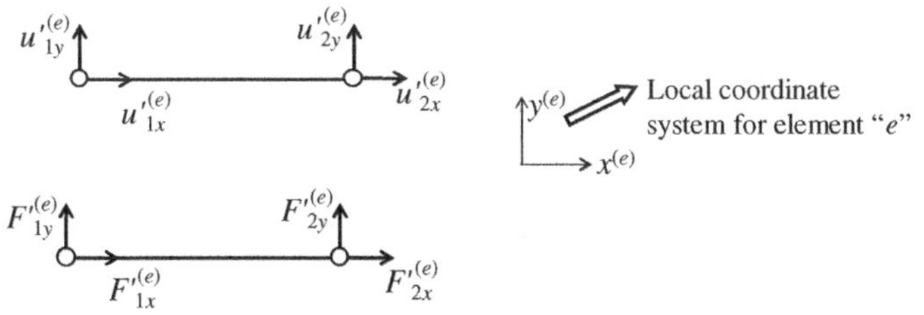

Local coordinate system for element "e"

Figure B.12 Local coordinate system and nodal displacements/forces for 2D truss element.

$$\left\{\begin{array}{c} F'^{(e)}_{1x} = \dfrac{E^{(e)}A^{(e)}}{\ell^{(e)}}\left(u'^{(e)}_{1x} - u'^{(e)}_{2x}\right) \\ F'^{(e)}_{1y} = 0 \\ F'^{(e)}_{2x} = \dfrac{E^{(e)}A^{(e)}}{\ell^{(e)}}\left(u'^{(e)}_{2x} - u'^{(e)}_{1x}\right) \\ F'^{(e)}_{2y} = 0 \end{array}\right\} \rightarrow \begin{Bmatrix} F'^{(e)}_{1x} \\ F'^{(e)}_{1y} \\ F'^{(e)}_{2x} \\ F'^{(e)}_{2y} \end{Bmatrix} = \frac{E^{(e)}A^{(e)}}{\ell^{(e)}} \begin{bmatrix} 1 & 0 & -1 & 0 \\ 0 & 0 & 0 & 0 \\ -1 & 0 & 1 & 0 \\ 0 & 0 & 0 & 0 \end{bmatrix} \begin{Bmatrix} u'^{(e)}_{1x} \\ u'^{(e)}_{1y} \\ u'^{(e)}_{2x} \\ u'^{(e)}_{2y} \end{Bmatrix}$$

We can then write the following equation:

$$\left\{F'^{(e)}\right\} = \left[k'^{(e)}\right]\left\{U'^{(e)}\right\} \tag{B.8.3}$$

where $\{F'^{(e)}\}$ is the element end force vector in the local coordinate system, $\{U'^{(e)}\}$ is the element end displacement vector in the local coordinate system, and $[k'^{(e)}]$ is the *element stiffness matrix in the local coordinate system*. The following expressions apply:

$$\left\{F'^{(e)}\right\} = \left[F'^{(e)}_{1x} \; F'^{(e)}_{1y} \; F'^{(e)}_{2x} \; F'^{(e)}_{2y}\right]^T \tag{B.8.4a}$$

$$\left\{U'^{(e)}\right\} = \left[u'^{(e)}_{1x} \; u'^{(e)}_{1y} \; u'^{(e)}_{2x} \; u'^{(e)}_{2y}\right]^T \tag{B.8.4b}$$

and

$$\left[k'^{(e)}\right] = \frac{E^{(e)}A^{(e)}}{\ell^{(e)}} \begin{bmatrix} 1 & 0 & -1 & 0 \\ 0 & 0 & 0 & 0 \\ -1 & 0 & 1 & 0 \\ 0 & 0 & 0 & 0 \end{bmatrix} \tag{B.8.4c}$$

The global stiffness (equilibrium) equations for the structure must account for all the element force contributions to the nodes. The force contributions from the elements must all be expressed in the same coordinate system (i.e., the global coordinate system)! It is thus necessary to establish *coordinate transformation equations* for the element stiffness matrix. We will first determine the transformation equations for the element end displacements. We will do so for node 1, then continue with node 2. The considerations for node 1 are presented in Figure B.13, for two cases. Case 1 corresponds to the situation where we have a displacement $u^{(e)}_{1X}$ along the global X-axis. We can use geometry, as shown in Figure B.13, to find the displacements of the same nodal point along the local coordinate axes. We have similar considerations in the figure for Case 2, which corresponds to the situation where we have a displacement $u^{(e)}_{1Y}$ along the global Y-axis.

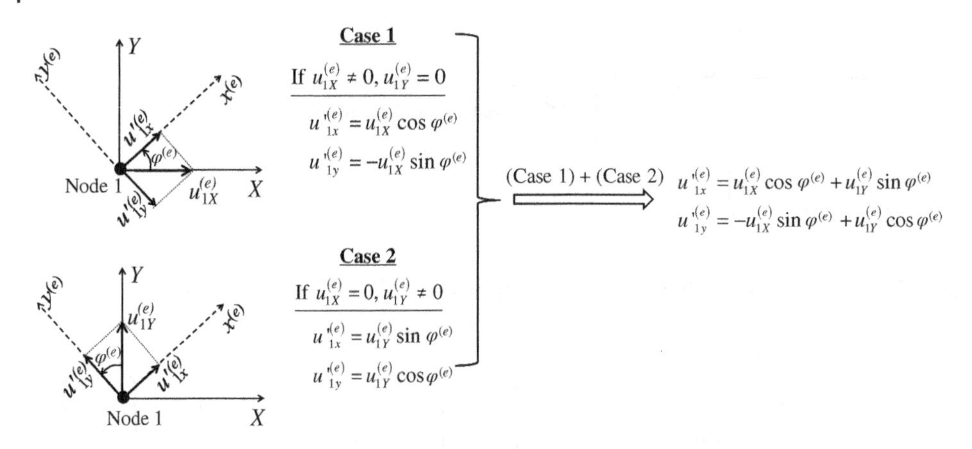

Figure B.13 Coordinate transformation equations for displacements.

If we combine the two equations we obtained from the two cases in Figure B.13 in matrix form, we establish the coordinate transformation equation for the displacements of node 1:

$$\begin{Bmatrix} u'^{(e)}_{1x} \\ u'^{(e)}_{1y} \end{Bmatrix} = \begin{bmatrix} \cos\varphi^{(e)} & \sin\varphi^{(e)} \\ -\sin\varphi^{(e)} & \cos\varphi^{(e)} \end{bmatrix} \begin{Bmatrix} u^{(e)}_{1X} \\ u^{(e)}_{1Y} \end{Bmatrix} \tag{B.8.5a}$$

Similarly, for the displacements of node 2:

$$\begin{Bmatrix} u'^{(e)}_{2x} \\ u'^{(e)}_{2y} \end{Bmatrix} = \begin{bmatrix} \cos\varphi^{(e)} & \sin\varphi^{(e)} \\ -\sin\varphi^{(e)} & \cos\varphi^{(e)} \end{bmatrix} \begin{Bmatrix} u^{(e)}_{2X} \\ u^{(e)}_{2Y} \end{Bmatrix} \tag{B.8.5b}$$

Equations (B.8.5a) and (B.8.5b) can be cast in a single matrix equation, which is the coordinate transformation equation for the nodal displacements of a 2D truss element:

$$\begin{Bmatrix} u'^{(e)}_{1x} \\ u'^{(e)}_{1y} \\ u'^{(e)}_{2x} \\ u'^{(e)}_{2y} \end{Bmatrix} = \begin{bmatrix} \cos\varphi^{(e)} & \sin\varphi^{(e)} & 0 & 0 \\ -\sin\varphi^{(e)} & \cos\varphi^{(e)} & 0 & 0 \\ 0 & 0 & \cos\varphi^{(e)} & \sin\varphi^{(e)} \\ 0 & 0 & -\sin\varphi^{(e)} & \cos\varphi^{(e)} \end{bmatrix} \begin{Bmatrix} u^{(e)}_{1x} \\ u^{(e)}_{1y} \\ u^{(e)}_{2x} \\ u^{(e)}_{2y} \end{Bmatrix} \rightarrow \left\{U'^{(e)}\right\} = \left[R^{(e)}\right]\left\{U^{(e)}\right\} \tag{B.8.5}$$

where $[R^{(e)}]$ is the *coordinate transformation matrix for element e*: If we want to transform a vector expressed in the global coordinate system into a vector expressed in the local coordinate system, we need to premultiply the original vector by $[R^{(e)}]$. The following equation gives the coordinate transformation array for 2D truss elements.

$$\left[R^{(e)}\right] = \begin{bmatrix} \cos\varphi^{(e)} & \sin\varphi^{(e)} & 0 & 0 \\ -\sin\varphi^{(e)} & \cos\varphi^{(e)} & 0 & 0 \\ 0 & 0 & \cos\varphi^{(e)} & \sin\varphi^{(e)} \\ 0 & 0 & -\sin\varphi^{(e)} & \cos\varphi^{(e)} \end{bmatrix} \tag{B.8.6}$$

We will now proceed to obtain a coordinate transformation equation for the nodal forces in the local and the global coordinate systems. We demonstrate the procedure to obtain this expression in Figure B.14. As shown in the figure, we make a cut (section) very close to each of the nodal points, 1 and 2. The element end force components on one side of each cut are drawn in the global coordinate system, while on the other side of the cut they are drawn in the local coordinate system.

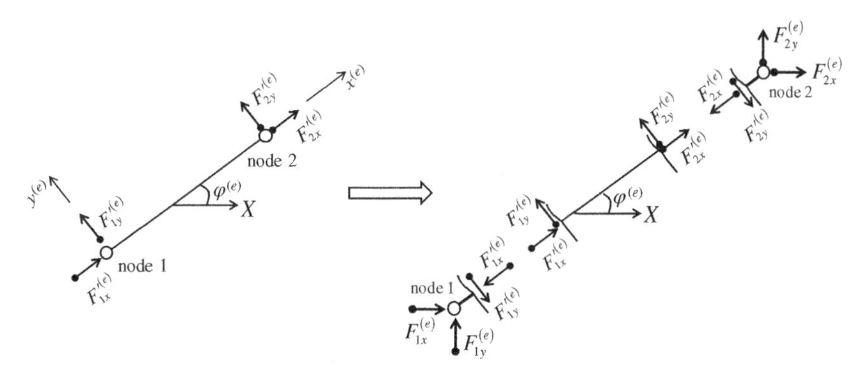

Figure B.14 Schematic procedure to obtain coordinate transformation expression for nodal forces.

We can now establish the force equilibrium equations for the two global directions, X and Y, using the free-body diagrams of the two small pieces next to the nodal points, shown in Figure B.14. The free-body diagram of the small piece next to node 1 is isolated in Figure B.15. A similar diagram can be established for node 2.

Writing down the equilibrium equations for node 1 yields the following two expressions:

$$F_{1x}^{(e)} - F'^{(e)}_{1x}\cos\varphi^{(e)} + F'^{(e)}_{1y}\sin\varphi^{(e)} = 0 \rightarrow F_{1x}^{(e)} = F'^{(e)}_{1x}\cos\varphi^{(e)} - F'^{(e)}_{1y}\sin\varphi^{(e)} \quad \text{(B.8.7a)}$$

$$F_{1y}^{(e)} - F'^{(e)}_{1y}\cos\varphi^{(e)} - F'^{(e)}_{1x}\sin\varphi^{(e)} = 0 \rightarrow F_{1y}^{(e)} = F'^{(e)}_{1x}\sin\varphi^{(e)} + F'^{(e)}_{1y}\cos\varphi^{(e)} \quad \text{(B.8.7b)}$$

We can now continue with the corresponding force equilibrium equations for node 2:

$$F_{2x}^{(e)} - F'^{(e)}_{2x}\cos\varphi^{(e)} + F'^{(e)}_{2y}\sin\varphi^{(e)} = 0 \rightarrow F_{2x}^{(e)} = F'^{(e)}_{2x}\cos\varphi^{(e)} - F'^{(e)}_{2y}\sin\varphi^{(e)} \quad \text{(B.8.7c)}$$

$$F_{2y}^{(e)} - F'^{(e)}_{2y}\cos\varphi^{(e)} - F'^{(e)}_{2x}\sin\varphi^{(e)} = 0 \rightarrow F_{2y}^{(e)} = F'^{(e)}_{2x}\sin\varphi^{(e)} + F'^{(e)}_{2y}\cos\varphi^{(e)} \quad \text{(B.8.7d)}$$

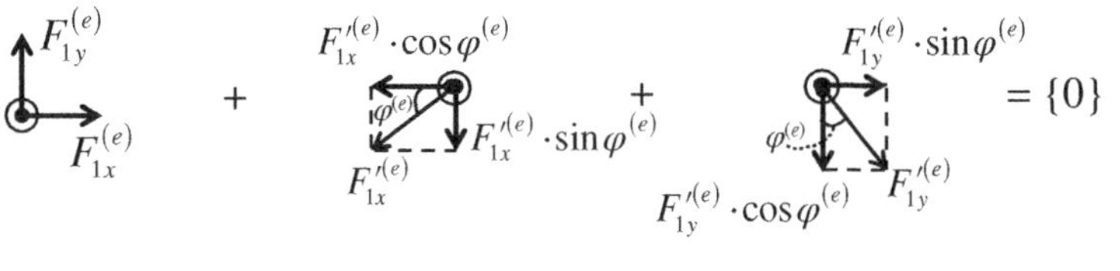

Figure B.15 Local coordinate system and nodal displacements/forces for 2D truss element.

We can collectively cast Equations (B.8.6a–d) to a single matrix one:

$$\begin{Bmatrix} F_{1x}^{(e)} \\ F_{1y}^{(e)} \\ F_{2x}^{(e)} \\ F_{2y}^{(e)} \end{Bmatrix} = \begin{bmatrix} \cos\varphi^{(e)} & -\sin\varphi^{(e)} & 0 & 0 \\ \sin\varphi^{(e)} & \cos\varphi^{(e)} & 0 & 0 \\ 0 & 0 & \cos\varphi^{(e)} & -\sin\varphi^{(e)} \\ 0 & 0 & \sin\varphi^{(e)} & \cos\varphi^{(e)} \end{bmatrix} \begin{Bmatrix} F'^{(e)}_{1x} \\ F'^{(e)}_{1y} \\ F'^{(e)}_{2x} \\ F'^{(e)}_{2y} \end{Bmatrix} \tag{B.8.8}$$

By inspection, we can verify that the following equation applies.

$$\left\{F^{(e)}\right\} = \left[R^{(e)}\right]^T \left\{F'^{(e)}\right\} \tag{B.8.9}$$

where $\{F^{(e)}\}$ is the element nodal force vector in the global coordinate system and $\{F'^{(e)}\}$ is the nodal force vector in the local coordinate system:

$$\left\{F^{(e)}\right\} = \left[F_{1x}^{(e)} \; F_{1y}^{(e)} \; F_{2x}^{(e)} \; F_{2y}^{(e)}\right]^T \tag{B.8.10a}$$

$$\left\{F'^{(e)}\right\} = \left[F'^{(e)}_{1x} \; F'^{(e)}_{1y} \; F'^{(e)}_{2x} \; F'^{(e)}_{2y}\right]^T \tag{B.8.10b}$$

Now, if we plug Equation (B.8.3) into Equation (B.8.9), we obtain:

$$\left\{F^{(e)}\right\} = \left[R^{(e)}\right]^T \left\{F'^{(e)}\right\} = \left[R^{(e)}\right]^T \left[k'^{(e)}\right] \left\{U'^{(e)}\right\}$$

Finally, if we plug Equation (B.8.5) into the obtained expression, we have:

$$\left\{F^{(e)}\right\} = \left(\left[R^{(e)}\right]^T \left[k'^{(e)}\right] \left[R^{(e)}\right]\right) \left\{U^{(e)}\right\} \tag{B.8.11}$$

Equation (B.8.11) constitutes an *element stiffness equation in the global coordinate system*:

$$\left\{F^{(e)}\right\} = \left[k^{(e)}\right] \left\{U^{(e)}\right\} \tag{B.8.12}$$

where the element stiffness matrix in the global coordinate system, $[k^{(e)}]$, is given by the following expression.

$$\left[k^{(e)}\right] = \left[R^{(e)}\right]^T \left[k'^{(e)}\right] \left[R^{(e)}\right] \tag{B.8.13}$$

Remark B.8.1: As shown in Chapters 13 and 14, Equations (B.8.12) and (B.8.13) are always valid for coordinate transformations, provided that we establish the appropriate definition for the element stiffness matrix $[k'^{(e)}]$ in the local system and for the coordinate transformation array $[R^{(e)}]$.

For the special case of a 2D truss element with constant $E^{(e)}$, $A^{(e)}$, the stiffness matrix $[k^{(e)}]$ in the global coordinate system is given by the following expression.

$$\left[k^{(e)}\right] = \frac{E^{(e)}A^{(e)}}{\ell^{(e)}} \begin{bmatrix} \cos^2\varphi^{(e)} & \cos\varphi^{(e)}\cdot\sin\varphi^{(e)} & -\cos^2\varphi^{(e)} & -\cos\varphi^{(e)}\cdot\sin\varphi^{(e)} \\ \cos\varphi^{(e)}\cdot\sin\varphi^{(e)} & \sin^2\varphi^{(e)} & -\cos\varphi^{(e)}\cdot\sin\varphi^{(e)} & -\sin^2\varphi^{(e)} \\ -\cos^2\varphi^{(e)} & -\cos\varphi^{(e)}\cdot\sin\varphi^{(e)} & \cos^2\varphi^{(e)} & \cos\varphi^{(e)}\cdot\sin\varphi^{(e)} \\ -\cos\varphi^{(e)}\cdot\sin\varphi^{(e)} & -\sin^2\varphi^{(e)} & \cos\varphi^{(e)}\cdot\sin\varphi^{(e)} & \sin^2\varphi^{(e)} \end{bmatrix} \tag{B.8.14}$$

After we express the element stiffness matrix in the global coordinate system, we can proceed to conduct the assembly operation for the global stiffness matrix. It is also worth mentioning here that it is customary to establish a different nodal displacement numbering for the four displacement quantities of the element, as shown in Figure B.16.

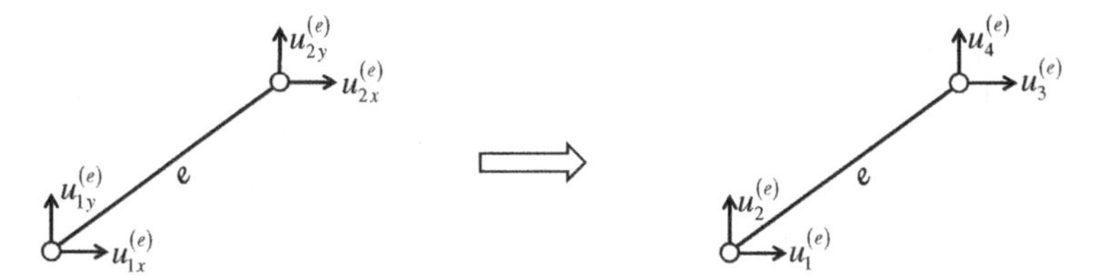

Figure B.16 Alternative numbering for nodal displacement in 2D truss element.

Using the numbering scheme shown in Figure B.16, the assembly procedure proceeds in the same fashion as before, using, for example, the gather/scatter arrays of the elements. In actual analysis programs, the assembly is conducted using the *ID* array and the *LM* array concepts as described in Sections B.4 and B.5. The only difference (compared to what was presented for one-dimensional truss analysis) is that, since each 2D truss element has four nodal displacements, there will be four *LM* array entries for each element *iel*: *LM*(*iel*,1), *LM*(*iel*,2), *LM*(*iel*,3) and *LM*(*iel*,4).

As a reminder, the *ID* and *LM* array give the information provided in Box B.8.1. ■

Box B.8.1 Reminder for *ID* and *LM* Arrays
ID(i1): "to which unrestrained DOF does the *global* DOF *i1* correspond?" *LM(iel,i1)*: "to which global DOF does the DOF *i1* of element *iel* correspond?"

Remark B.8.2: Given that only the axial displacements create stiffness in the local coordinate system, we can cast the coordinate transformation equations in a slightly more compact form, wherein the local stiffness matrix, $[k'^{(e)}]$, is defined as the (2×2) matrix given in Equation (B.1.10c). Similarly, the local displacement vector will only contain displacements that generate deformations (and stresses and axial forces), namely, the two nodal displacements along the local x-axis: $\left\{U'^{(e)}\right\} = \left[u'^{(e)}_{1x} \;\; u'^{(e)}_{2x}\right]^T$. In this case, one can easily verify that the coordinate transformation array, $[R^{(e)}]$ can be obtained by only including the first and third rows from Equation (B.8.5):

$$\left[R^{(e)}\right] = \begin{bmatrix} \cos\varphi^{(e)} & \sin\varphi^{(e)} & 0 & 0 \\ 0 & 0 & \cos\varphi^{(e)} & \sin\varphi^{(e)} \end{bmatrix} \tag{B.8.15}$$

In this case, the transformation Equations (B.8.5) for displacements and (B.8.13) for stiffness remain valid, but now we can work with smaller matrices $[k'^{(e)}]$ (2×2 instead of 4×4) and $[R^{(e)}]$ (2×4 instead of 4×4). ■

Remark B.8.3: An important rule that we need to remember when we have analyses involving coordinate transformations is that the assembly of the global stiffness matrix is conducted after we have transformed all element stiffness matrices in the global coordinate system. Conversely, during the post-processing stage, we will need to transform the nodal displacements of each element in the local coordinate system, after we have applied the gather operation. The reason for having to transform the element nodal displacements in the local system is that post-processing equations, such as Equations (B.3.3), (B.3.4), and (B.3.5) are valid for the nodal displacements in the local coordinate system. ■

Example B.8.1: Two-dimensional truss structure

We are given the structure shown in Figure B.17. The figure also provides the coordinates of the three nodal points. The cross-sectional area of member (1) is $A^{(1)} = 1$ in.^{2}, while that of member (2) is $A^{(2)} = 2$ in.^{2}. Both members are made of a material with $E = 30000$ ksi. We want to find the nodal displacements and the member axial strains, axial stresses and axial forces.

We will first find the stiffness matrix of each element in the corresponding local coordinate system, then transform to the global coordinate system, and then we will assemble the global stiffness matrix. We will then keep $[K_{ff}]$, which corresponds to the free (unrestrained) degrees of freedom, and then solve for the displacements $\{U_f\}$. Finally, we will conduct the post-processing stage, where we will gather the displacements of each element, transform them to the element's local coordinate system, and then find the axial strains, axial stresses, and axial forces of the two members.

The vector with the global degrees of freedom is:

$$\{U\} = \begin{bmatrix} u_1 & u_2 & u_3 & u_4 & u_5 & u_6 \end{bmatrix}^T = \begin{bmatrix} u_{1x} & u_{1y} & u_{2x} & u_{2y} & u_{3x} & u_{3y} \end{bmatrix}^T$$

We can now proceed to obtain the stiffness matrix contributions for the two elements:

Element 1

- Node 1 for element 1 = node 1: $X_1^{(1)} = 0,\ Y_1^{(1)} = 0$
- Node 2 for element 1 = node 2: $X_2^{(1)} = 120\text{ in.},\ Y_2^{(1)} = 0$

$$\ell^{(1)} = \sqrt{\left(X_2^{(1)} - X_1^{(1)}\right)^2 + \left(Y_2^{(1)} - Y_1^{(1)}\right)^2} = 120\text{ in.}$$

$$\cos\varphi^{(1)} = \frac{X_2^{(1)} - X_1^{(1)}}{\ell^{(1)}} = 1,\ \sin\varphi^{(1)} = \frac{Y_2^{(1)} - Y_1^{(1)}}{\ell^{(1)}} = 0$$

$$\left[k'^{(1)}\right] = \frac{E^{(1)}A^{(1)}}{\ell^{(1)}}\begin{bmatrix} 1 & -1 \\ -1 & 1 \end{bmatrix} = \frac{30000\cdot 1}{120}\begin{bmatrix} 1 & -1 \\ -1 & 1 \end{bmatrix} = \begin{bmatrix} 250 & -250 \\ -250 & 250 \end{bmatrix}\text{kip/in.}$$

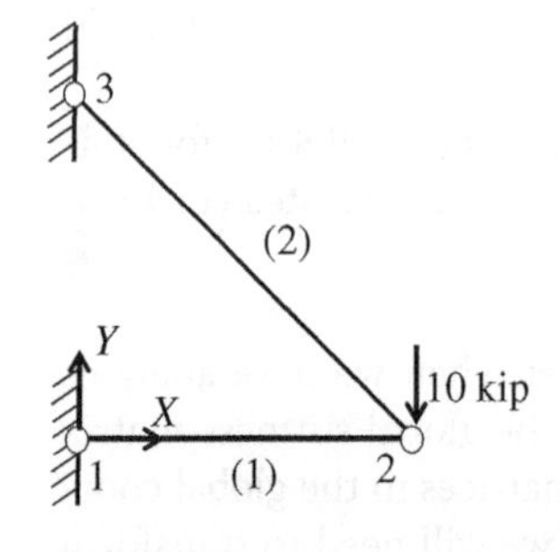

Node	X_i (in.)	Y_i (in.)
1	0	0
2	120	0
3	0	80

Figure B.17 Example truss structure.

Find the coordinate transformation array for element (1):

$$\left[R^{(1)}\right] = \begin{bmatrix} \cos\varphi^{(1)} & \sin\varphi^{(1)} & 0 & 0 \\ 0 & 0 & \cos\varphi^{(1)} & \sin\varphi^{(1)} \end{bmatrix} = \begin{bmatrix} 1 & 0 & 0 & 0 \\ 0 & 0 & 1 & 0 \end{bmatrix}$$

Calculate stiffness matrix of element (1) in the global coordinate system:

$$\left[k^{(1)}\right] = \left[R^{(1)}\right]^T \left[k'^{(1)}\right] \left[R^{(1)}\right] = \begin{bmatrix} 250 & 0 & -250 & 0 \\ 0 & 0 & 0 & 0 \\ -250 & 0 & 250 & 0 \\ 0 & 0 & 0 & 0 \end{bmatrix} \text{kip/in.}$$

Find the gather array, $[L^{(1)}]$, of member (1):

$$\left[L^{(1)}\right] = \begin{bmatrix} 1 & 0 & 0 & 0 & 0 & 0 \\ 0 & 1 & 0 & 0 & 0 & 0 \\ 0 & 0 & 1 & 0 & 0 & 0 \\ 0 & 0 & 0 & 1 & 0 & 0 \end{bmatrix}$$

Element 2

- Node 1 for element 2 = node 2: $X_1^{(2)} = 120\,\text{in.}$, $Y_1^{(2)} = 0$
- Node 2 for element 2 = node 3: $X_2^{(2)} = 0$, $Y_2^{(2)} = 80\,\text{in.}$

$$\ell^{(2)} = \sqrt{\left(X_2^{(2)} - X_1^{(2)}\right)^2 + \left(Y_2^{(2)} - Y_1^{(2)}\right)^2} = 144.22\,\text{in.}$$

$$\cos\varphi^{(2)} = \frac{X_2^{(2)} - X_1^{(2)}}{\ell^{(2)}} = -0.832, \sin\varphi^{(2)} = \frac{Y_2^{(2)} - Y_1^{(2)}}{\ell^{(2)}} = 0.555$$

$$\left[k'^{(2)}\right] = \frac{E^{(2)}A^{(2)}}{\ell^{(2)}} \begin{bmatrix} 1 & -1 \\ -1 & 1 \end{bmatrix} = \frac{30000 \cdot 2}{144.22} \begin{bmatrix} 1 & -1 \\ -1 & 1 \end{bmatrix} = \begin{bmatrix} 416.03 & -416.03 \\ -416.03 & 416.03 \end{bmatrix} \text{kip/in.}$$

Find the coordinate transformation array for element (1):

$$\left[R^{(1)}\right] = \begin{bmatrix} \cos\varphi^{(1)} & \sin\varphi^{(1)} & 0 & 0 \\ 0 & 0 & \cos\varphi^{(1)} & \sin\varphi^{(1)} \end{bmatrix} = \begin{bmatrix} -0.832 & 0.555 & 0 & 0 \\ 0 & 0 & -0.832 & 0.555 \end{bmatrix}$$

Calculate stiffness matrix of element (2) in the global coordinate system:

$$\left[k^{(2)}\right] = \left[R^{(2)}\right]^T \left[k'^{(2)}\right] \left[R^{(2)}\right] = \begin{bmatrix} 288.017 & -192.012 & -288.017 & 192.012 \\ -192.012 & 128.008 & 192.012 & -128.008 \\ -288.017 & 192.012 & 288.017 & -192.012 \\ 192.012 & -128.008 & -192.012 & 128.008 \end{bmatrix} \text{kip/in.}$$

Find the gather array, $[L^{(2)}]$, of member (2):

$$\left[L^{(2)}\right]=\begin{bmatrix}0&0&1&0&0&0\\0&0&0&1&0&0\\0&0&0&0&1&0\\0&0&0&0&0&1\end{bmatrix}$$

Assemble the global stiffness matrix:

$$[K]=\left[L^{(1)}\right]^T\left[k^{(1)}\right]\left[L^{(1)}\right]+\left[L^{(2)}\right]^T\left[k^{(2)}\right]\left[L^{(2)}\right]$$

$$=\begin{bmatrix}250&0&-250&0&0&0\\0&0&0&0&0&0\\-250&0&538.017&-192.012&-288.017&192.012\\0&0&-192.012&128.008&192.012&-128.008\\0&0&-288.017&192.012&288.017&-192.012\\0&0&192.012&-128.008&-192.012&128.008\end{bmatrix}\text{kip/in.}$$

We now know that the only unrestrained degrees of freedom are those of node 2, that is, we can obtain $[K_{ff}]$ by keeping the third and fourth rows and columns of $[K]$:

$$[K]=\begin{bmatrix}250&0&-250&0&0&0\\0&0&0&0&0&0\\-250&0&538.017&-192.012&-288.017&192.012\\0&0&-192.012&128.008&192.012&-128.008\\0&0&-288.017&192.012&288.017&-192.012\\0&0&192.012&-128.008&-192.012&128.008\end{bmatrix}\text{kip/in.}\Rightarrow\left[K_{ff}\right]=\begin{bmatrix}538.017&-192.012\\-192.012&128.008\end{bmatrix}\text{kip/in.}$$

Forces at free DOF: $\begin{matrix}f_3=f_{2x}=0\\f_4=f_{2y}=-10\,kip\end{matrix}$

Thus, we can establish: $\{F_f\}=\begin{Bmatrix}0\\-10\end{Bmatrix}$.

Since all the restrained nodal displacements are zero, $\{U_s\}=\{0\}$, we can solve the following equation to obtain the value of the unrestrained displacements:

$$[K_{ff}]\{U_f\}=\{F_f\}\rightarrow\{U_f\}=[K_{ff}]^{-1}\{F_f\}=\begin{Bmatrix}-0.06\\-0.168\end{Bmatrix}\text{in.}=\begin{Bmatrix}u_{2x}\\u_{2y}\end{Bmatrix}$$

Now, we know the values of the global displacement vector:

$$\{U\}=\left[u_{1x}\ u_{1y}\ u_{2x}\ u_{2y}\ u_{3x}\ u_{3y}\right]^T=[0\ \ 0\ \ -0.06\ \ -0.168\ \ 0\ \ 0]^T$$

We can now proceed with the post-processing stage:

<u>Element 1:</u>

$\{U^{(1)}\} = [L^{(1)}]\{U\}$ (end displacements of member 1 in global coordinate system). We can also obtain the two nodal displacements along the local x-axis of the element:

$$\{U'^{(1)}\} = \begin{Bmatrix} u'^{(1)}_1 \\ u'^{(1)}_2 \end{Bmatrix} = [R^{(1)}]\{U^{(1)}\} = [R^{(1)}][L^{(1)}]\{U\} = \begin{Bmatrix} 0 \\ -0.06 \end{Bmatrix} \text{ in.}$$

(end displacements of member 1 in local coordinate system). Finally, we can apply Equations (B.3.3), (B.3.4), and (B.3.5) to find the axial strain, stress and force of the element:

$$\varepsilon^{(1)} = \frac{u'^{(1)}_{2x} - u'^{(1)}_{1x}}{\ell^{(1)}} = -0.0005, \quad \sigma^{(1)} = E^{(1)}\varepsilon^{(1)} = -15\,\text{ksi}, \quad N^{(1)} = A^{(1)}\sigma^{(1)} = -15\,\text{kip}$$

<u>Element 2:</u>

$$\{U'^{(2)}\} = \begin{Bmatrix} u'^{(2)}_{1x} \\ u'^{(2)}_{2x} \end{Bmatrix} = [R^{(2)}]\{U^{(2)}\} = [R^{(2)}][L^{(2)}]\{U\} = \begin{Bmatrix} -0.04332 \\ 0 \end{Bmatrix} \text{ in.}$$

$$\varepsilon^{(2)} = \frac{u'^{(2)}_{2x} - u'^{(2)}_{1x}}{\ell^{(2)}} = 0.0003, \quad \sigma^{(2)} = E^{(2)}\varepsilon^{(2)} = 9.01\,\text{ksi}, \quad N^{(2)} = A^{(2)}\sigma^{(2)} = 18.02\,\text{kip}$$

As an exercise, the reader can determine how the assembly/postprocessing operations for this example would be conducted using the *ID* and *LM* arrays. ▮

B.9 Extension to Three-Dimensional Truss Analysis

If we have a three-dimensional truss structure with elements having constant cross-sections and constant modulus of elasticity, E, then the stiffness equations for the axial direction are still valid. The only additional consideration is that each nodal point of each

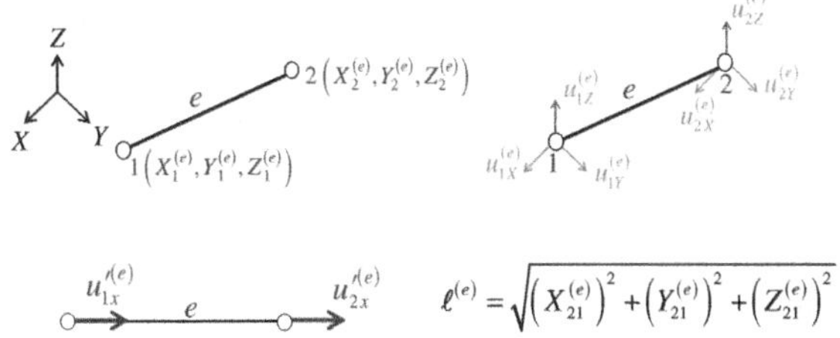

$$\ell^{(e)} = \sqrt{\left(X^{(e)}_{21}\right)^2 + \left(Y^{(e)}_{21}\right)^2 + \left(Z^{(e)}_{21}\right)^2}$$

$$\{U'^{(e)}\} = \begin{Bmatrix} u'^{(e)}_{1x} \\ u'^{(e)}_{2x} \end{Bmatrix} = \underbrace{\begin{bmatrix} \frac{X^{(e)}_{21}}{\ell^{(e)}} & \frac{Y^{(e)}_{21}}{\ell^{(e)}} & \frac{Z^{(e)}_{21}}{\ell^{(e)}} & 0 & 0 & 0 \\ 0 & 0 & 0 & \frac{X^{(e)}_{21}}{\ell^{(e)}} & \frac{Y^{(e)}_{21}}{\ell^{(e)}} & \frac{Z^{(e)}_{21}}{\ell^{(e)}} \end{bmatrix}}_{[R^{(e)}]} \begin{Bmatrix} u^{(e)}_{1x} \\ u^{(e)}_{1y} \\ u^{(e)}_{1z} \\ u^{(e)}_{2x} \\ u^{(e)}_{2y} \\ u^{(e)}_{2z} \end{Bmatrix} = \begin{bmatrix} \frac{X^{(e)}_{21}}{\ell^{(e)}} & \frac{Y^{(e)}_{21}}{\ell^{(e)}} & \frac{Z^{(e)}_{21}}{\ell^{(e)}} & 0 & 0 & 0 \\ 0 & 0 & 0 & \frac{X^{(e)}_{21}}{\ell^{(e)}} & \frac{Y^{(e)}_{21}}{\ell^{(e)}} & \frac{Z^{(e)}_{21}}{\ell^{(e)}} \end{bmatrix} \{U^{(e)}\}$$

Figure B.18 Three-dimensional truss element and coordinate transformation equation.

element will have three displacement components in the global coordinate system, i.e. one displacement along each one of three global axes X, Y, and Z. The coordinate transformation Equations (B.8.5) and (B.8.13) apply. The displacement vector in the local coordinate system is $\left\{U'^{(e)}\right\} = \left[u'^{(e)}_{1x} \;\; u'^{(e)}_{2x}\right]^T$, in accordance with Remark B.8.2. The procedure for a three-dimensional truss element will not be described in detail here. The coordinate transformation equation the element nodal displacements in the local and global coordinate systems are summarized in Figure B.18.

Problems

B.1 We are given the two-dimensional truss structure in Figure PB.1. We are also given the table shown below, which includes the nodal coordinates in the global coordinate system, xy.

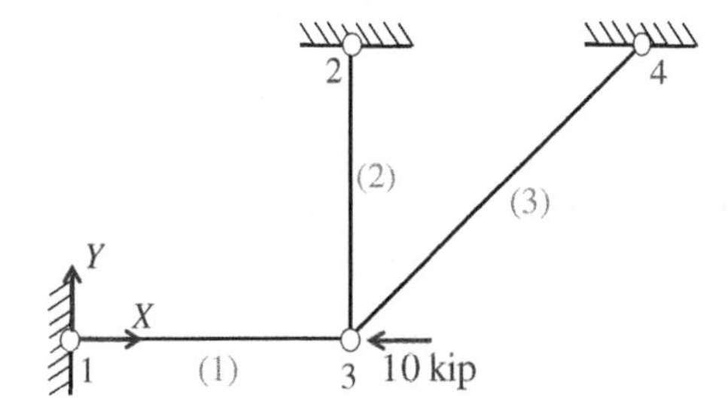

Node	X_i (in.)	Y_i (in.)
1	0	0
2	50	50
3	50	0
4	100	50

Figure PB.1

We are also given that $E = 20{,}000$ ksi, $A = 5$ in.^{2} for all members.

a) Draw and number the structure's global degrees of freedom (DOF).
b) Draw each element with its end DOFs in the local coordinate system, and each element with its end DOFs in the global coordinate system.
c) Determine the coordinate transformation array for each element.
d) Determine the stiffness matrix in the global coordinate system for each element.
e) Determine the gather/scatter array, $[L^{(e)}]$ for each element, and write the equation for the assembly of the global stiffness matrix, $[K]$. (you do *not* need to make calculations!).
f) Using the *ID* array concept, find the global stiffness matrix, $[K_{ff}]$, corresponding to the free (unrestrained) DOF of the structure.
g) Calculate the displacements of the free (unrestrained) DOF.
h) Find the axial stress for each element.
i) We want to use the element gather/scatter array concept ($[L^{(e)}]$) to obtain the global stiffness equations. We want the assembly to give us *only* the global stiffness matrix, $[K_{ff}]$, corresponding to the free (unrestrained) DOF (so far we have seen that using the $[L^{(e)}]$ arrays gives us the global stiffness $[K]$ corresponding to all the DOF, free and restrained). Explain how we need to modify the definition of the $[L^{(e)}]$ arrays to allow us to assemble $[K_{ff}]$. Write the modified versions of the $[L^{(e)}]$ arrays for the three members in the structure at hand, then show that the use of these modified arrays in the assembly equations using $[L^{(e)}]$ gives the same $[K_{ff}]$ that we obtained in part f of this problem.

Appendix C

Minimum Potential Energy for Elasticity—Variational Principles

Chapter 2 presented a procedure to obtain the weak form for one-dimensional problems, using purely mathematical considerations (e.g., integration by parts to render terms "symmetric"). This appendix briefly presents an alternative procedure for obtaining the weak form of the governing equations for elasticity problems, by means of a *variational principle. For the specific problem of elasticity, this variational principle mathematically expresses the minimization of the potential energy for the structure.* The variational principle for one-dimensional elasticity will be presented here, because several books (especially older ones) on the finite element method rely on the calculus of variations for deriving the weak form.

A quantity that can be defined for elasticity problems is the *potential energy* of the body, W. If we know the displacement field, $u(x)$, for a one-dimensional, linearly elastic solid, then the potential energy W is given by the following expression:

$$\text{Given } u(x)\text{, we have: } W(u(x)) = \underbrace{\frac{1}{2}\overbrace{\int_V E\cdot\varepsilon^2 dv}^{\text{volume integral}}}_{W_{int}} - \underbrace{\left(\int_0^L u\cdot b dx + [u\cdot A\cdot \bar{t}]_{\Gamma_t}\right)}_{W_{ext}} \quad \text{(C.1)}$$

where the various quantities in the integrals have been defined in Section 2.1.

The quantity W_{int} in Equation C.1, called the *strain energy* of the linearly elastic body, is the integral of the quantity $\bar{U} = \frac{1}{2}E\cdot\varepsilon^2$ over the *volume, V,* of the bar. The quantity $\bar{U}$ is called the *specific strain energy or the strain energy density.* It is simply the strain energy per unit volume of the linearly elastic bar. A more generic definition, which is applicable to both linear and nonlinear elasticity, is that *the strain energy density gives us the area under the stress-strain* plot if we increase the strain from a zero value to a nonzero value. The graphical meaning of the strain energy density for the specific case of a linearly elastic material and for uniaxial stress is presented in Figure C.1.

A more general mathematical expression, giving the strain energy density as the area under the stress-strain diagram, is:

$$\bar{U} = \int \sigma\, d\varepsilon \quad \text{(C.2)}$$

Fundamentals of Finite Element Analysis: Linear Finite Element Analysis, First Edition.
Ioannis Koutromanos, James McClure, and Christopher Roy.

Companion website: www.wiley.com/go/koutromanos/linear

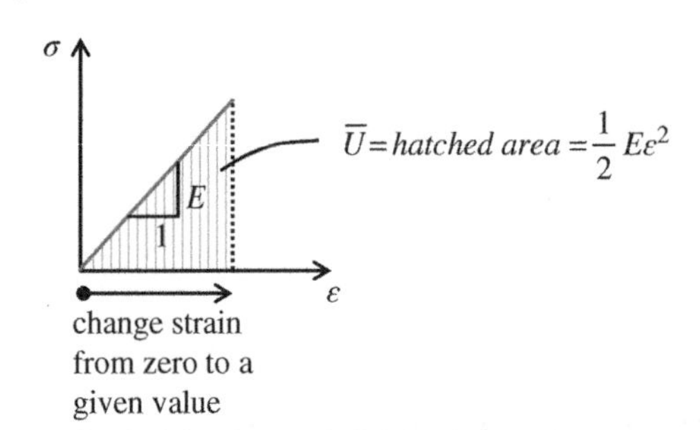

Figure C.1 Graphical definition of strain energy density for linearly elastic, uniaxial stress-strain law.

where the stress σ is expressed as a function of the strain, ε. It is also worth observing that a direct consequence of Equation (C.2) is that:

$$\sigma = \frac{d\bar{U}}{d\varepsilon} \tag{C.3}$$

For a one-dimensional truss bar, we can separate the (three-dimensional) volume integration into two stages, one conducted over the cross-sectional area, A, of each location x, and then integrate the result over the length of the bar. We can also account for the fact that, for one-dimensional elasticity, we have stipulated that E and ε are functions of x alone, and thus they can be taken outside of the integral over the cross-section:

$$W_{\text{int}} = \frac{1}{2}\int_V E\cdot\varepsilon^2 dx = \int_0^L \left(\frac{1}{2}\int_A E\cdot\varepsilon^2 dA\right) dx = \int_0^L \frac{1}{2}E\cdot\varepsilon^2\left(\int_A dA\right) dx = \int_0^L \frac{1}{2}E\cdot A\cdot\varepsilon^2 dx$$

Finally, if we account for the strain-displacement relation, that is, Equation (2.1.5), $\varepsilon = \frac{du}{dx}$, we obtain:

$$W_{\text{int}} = \int_0^L \frac{1}{2}E\cdot A\cdot\left(\frac{du}{dx}\right)^2 dx \tag{C.4}$$

We eventually plug Equation (C.4) into Equation (C.1) to obtain the following expression for the potential energy:

$$W(u(x)) = \int_0^L \frac{1}{2}E\cdot A\cdot\left(\frac{du}{dx}\right)^2 dx - \left(\int_0^L u\cdot b dx + [u\cdot A\cdot \bar{t}]_{\Gamma_t}\right) \tag{C.5}$$

Remark C.1: It is worth mentioning that W is a function of $u(x)$, which in turn is a function of x. Functions like W (i.e., functions of other functions of x) are called *functionals.* ▪

Now, let us imagine an *arbitrary* kinematically admissible axial displacement function, $u_2(x)$, for the bar under consideration, which "slightly" differs from $u(x)$. The term *kinematically admissible* in this appendix implies that a function satisfies the essential boundary conditions and the strain-displacement relation. We can set $u_2(x) - u(x) = \delta u(x)$, as shown in Figure C.2, where $\delta u(x)$ is called *the variation of* $u(x)$. Since $u_2(x)$ "slightly"

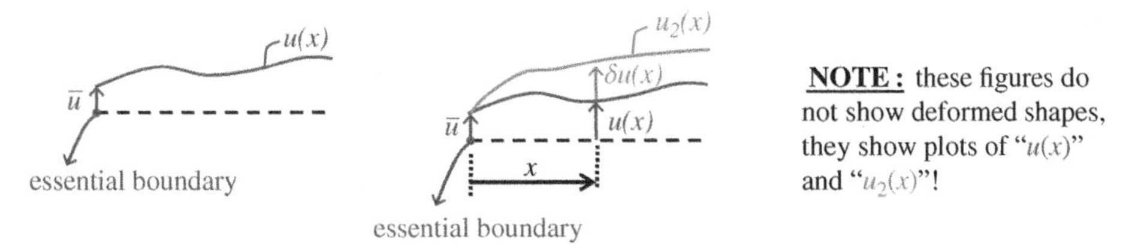

Figure C.2 Kinematically admissible axial displacement field $u(x)$ and variation of displacement field obtained for a "slightly" different kinematically admissible displacement field, $u_2(x)$.

differs from $u(x)$, we can understand that the (arbitrary) variation $\delta u(x)$ must be small, thus we can write $\delta u(x) = \zeta \cdot w(x)$, where ζ is a very small positive number and $w(x)$ is an arbitrary function of x.

We can write the following mathematical expression:

$$u_2(x) = u(x) + \delta u(x) = u(x) + \zeta \cdot w(x), \quad \zeta = \textit{small positive number} \tag{C.6a}$$

We know that both $u(x)$ and $u_2(x)$ must *satisfy the essential boundary conditions*, as schematically shown in Figure C.2. This means that we have:

$$u_2(x)|_{\Gamma_u} = u(x)|_{\Gamma_u} = u, \text{ thus } w(x)|_{\Gamma_u} = 0! \text{ (} w \text{ vanishes on the essential boundary)} \tag{C.6b}$$

Thus, the stipulation that the essential boundary conditions must be satisfied by all possible displacement fields—which are obtained by adding an *arbitrary* variation $\delta u(x) = \zeta \cdot w(x)$ to the actual displacement field—gives Equation (C.6b), that is, that $w(x)$ must vanish at the essential boundary.

The potential energy $W(u_2(x))$ corresponding to $u_2(x)$ is given by:

$$W(u_2(x)) = \int_0^L \frac{1}{2} E \cdot A \cdot \left(\frac{du_2}{dx}\right)^2 dx - \left(\int_0^L u_2 \cdot b dx + [u_2 \cdot A \cdot \bar{t}]_{\Gamma_t}\right) \tag{C.7}$$

Plugging Equation (C.6a) into Equation (C.7) yields:

$$W(u_2(x)) = \int_0^L \frac{1}{2} E \cdot A \cdot \left(\frac{du}{dx} + \zeta \cdot \frac{dw}{dx}\right)^2 dx - \left(\int_0^L (u + \zeta \cdot w) \cdot b dx + [(u + \zeta \cdot w) \cdot A \cdot \bar{t}]_{\Gamma_t}\right)$$

$$= \int_0^L \frac{1}{2} E \cdot A \cdot \left(\frac{du}{dx}\right)^2 dx + \int_0^L \frac{1}{2} E \cdot A \cdot \zeta^2 \left(\frac{dw}{dx}\right)^2 dx + \int_0^L \frac{1}{2} E \cdot A \cdot 2 \frac{du}{dx} \cdot \zeta \frac{dw}{dx} dx$$

$$- \left(\int_0^L (u + \zeta \cdot w) \cdot b dx + [(u + \zeta \cdot w) \cdot A \cdot \bar{t}]_{\Gamma_t}\right)$$

Now, we notice that there is a specific term in the obtained expression, i.e. the term $\int_0^L \frac{1}{2} E \cdot A \cdot \zeta^2 \left(\frac{dw}{dx}\right)^2 dx$, which is multiplied by ζ^2. Given that ζ is a small number, the term multiplied by ζ^2 will be much smaller from the rest and can thus be neglected! Thus, we obtain:

$$W(u_2(x)) = \int_0^L \frac{1}{2} E \cdot A \cdot \left(\frac{du}{dx}\right)^2 dx + \int_0^L \frac{1}{2} E \cdot A \cdot 2 \frac{du}{dx} \cdot \zeta \frac{dw}{dx} dx$$

$$- \left(\int_0^L u \cdot b dx + \int_0^L \zeta \cdot w \cdot b dx + [u \cdot A \cdot \bar{t}]_{\Gamma_t} + [\zeta \cdot w \cdot A \cdot \bar{t}]_{\Gamma_t} \right) \rightarrow$$

$$W(u_2(x)) = \int_0^L \frac{1}{2} E \cdot A \cdot \left(\frac{du}{dx}\right)^2 dx - \left(\int_0^L u \cdot b dx + [u \cdot A \cdot \bar{t}]_{\Gamma_t} \right)$$

$$+ \int_0^L \frac{1}{2} E \cdot A \cdot 2 \frac{du}{dx} \cdot \zeta \frac{dw}{dx} dx - \left(\int_0^L \zeta \cdot w \cdot b dx + [\zeta \cdot w \cdot A \cdot \bar{t}]_{\Gamma_t} \right)$$

If we now account for Equation (C.5), we have:

$$W(u_2(x)) = W(u(x)) + \zeta \cdot \int_0^L E \cdot A \cdot \frac{du}{dx} \cdot \frac{dw}{dx} dx - \zeta \cdot \left(\int_0^L w \cdot b dx + [w \cdot A \cdot \bar{t}]_{\Gamma_t} \right) \tag{C.8}$$

The (first) variation in the potential energy, δW, is simply the change in potential energy due to the variation, $\delta u(x)$, of the displacement field $u(x)$ that we considered. Thus, we can subtract Equation (C.5) from Equation (C.8) to obtain:

$$\delta W = W(u(x) + \delta u(x)) - W(u(x)) = W(u_2(x)) - W(u(x))$$

$$= \zeta \cdot \int_0^L E \cdot A \cdot \frac{du}{dx} \cdot \frac{dw}{dx} dx - \zeta \cdot \left(\int_0^L w \cdot b dx + [w \cdot A \cdot \bar{t}]_{\Gamma_t} \right) \tag{C.9}$$

For the calculus of variations, the variation of a functional has the same significance as the derivative of a function in standard calculus.

Linear elasticity theory stipulates that the actual displacement field $u(x)$ *minimizes* the potential energy, $W(u(x))$. We can alternatively say that the actual displacement field, $u(x)$, is the minimizer of $W(u(x))$.

From our standard calculus, we know that the locations where a function of x has a local minimum or maximum, correspond to a zero derivative of the function with x. Similarly, for the calculus of variations, for functions that minimize or maximize a functional (the minimum or maximum of a functional are called *stationary values* of the functional), the (first) variation in the functional vanishes! Thus, the condition that $W(u(x))$ is a minimum (otherwise: the stationarity condition for W!) is mathematically expressed by the following relation.

$$\delta W = 0 \tag{C.10}$$

Plugging Equation (C.9) into Equation (C.10) eventually yields:

$$\delta W = 0 \rightarrow \zeta \cdot \int_0^L E \cdot A \cdot \frac{du}{dx} \cdot \frac{dw}{dx} dx - \zeta \cdot \left(\int_0^L w \cdot b dx + [w \cdot A \cdot \bar{t}]_{\Gamma_t} \right) = 0 \tag{C.11}$$

Since ζ is positive (and thus nonzero), we can divide both sides of Equation (C.11) by ζ to obtain:

$$\int_0^L E \cdot A \cdot \frac{du}{dx} \cdot \frac{dw}{dx} dx - \left(\int_0^L w \cdot b dx + [w \cdot A \cdot \bar{t}]_{\Gamma_t} \right) = 0$$

$$\rightarrow \int_0^L E \cdot A \cdot \frac{du}{dx} \cdot \frac{dw}{dx} dx = \int_0^L w \cdot b dx + [w \cdot A \cdot \bar{t}]_{\Gamma_t} \tag{C.12}$$

which is the same integral expression as the one that we obtained with the weak form in Box 2.3.2! If we additionally provide the essential boundary condition (which must be added to stipulate that the actual displacement field is kinematically admissible) and the statement for arbitrary $w(x)$, which vanishes on the essential boundary, we in fact obtain the same exact set of expressions as Box 2.3.2. Thus, a variational formulation can provide the weak form for a problem.

Remark C.2: Despite the fact that the variational principle for elasticity can yield the weak form, we do not much rely on calculus of variations in this text. There are three main reasons for this:

1) The calculus of variations includes rather abstract mathematical definitions (e.g., variation, functional, etc.), which might make the understanding of the finite element method a bit hard for readers to follow. This is not so much true with the weak form approach, since we simply multiply by an "arbitrary $w(x)$" and integrate over the domain (and then try to use integration by parts to render some terms "symmetric").
2) For problems other than elasticity, it might be hard to establish the appropriate functional whose variational principle will give the weak form.
3) The fact that we stated that parameter ζ in Equation (C.6a) is very small may lead to the erroneous impression that the weak form is an approximate expression (which of course is NOT the case!).

We still mention variational principles in Chapter 12, when we establish multifield weak forms for elasticity. The reason is to facilitate understanding of pertinent multifield finite element formulations in the literature. These formulations are usually employed in a weak form, which is presented as a Hellinger-Reissner or Hu-Washizu variational principle. ∎

Appendix D

Calculation of Displacement and Force Transformations for Rigid-Body Connections

In this appendix, we examine the case where we have two nodes, 1 and 2, which are connected with a rigid body (e.g., a rigid bar), as shown in Figure D.1. We will prove that the displacements and rotations of one of the nodes, i.e. node 2, can be uniquely defined from the displacements and rotations of the other node, that is, node 1. In other words, *the displacements of node 2 will be constrained to comply with those of node 1.* We will be calling node 1 *the master node* and node 2 *the slave node.* Which node is treated as master and which one is treated as slave is usually a matter of choice in an analysis. What we need to remember is that if we have a set of nodes connected to each other with a rigid body, then we are bound to have a single master node, and all the other nodes in the set will be slave ones.

First of all, if we were to move node 1 by a displacement u_{x1}, while all other displacements and rotations at this node were zero, then node 2 would translate by the same exact displacement, $u_{x2} = u_{x1}$. Also, if we were to move node 1 by a displacement u_{y1}, while all other displacements and rotations at this node were zero, then node 2 would translate by the same exact displacement, $u_{y2} = u_{y1}$. Finally, if we were to move node 1 by a displacement u_{z1}, while all other displacements and rotations at this node were zero, then node 2 would translate by the same exact displacement, $u_{z2} = u_{z1}$.

Now, let us see what happens if we rotate node 1 by an angle, θ_{z1}, while all other displacements and rotations at this node are kept to zero. This scenario is presented in Figure D.2, where we facilitate our discussions by considering the origin of the Cartesian coordinate system at the master node 1. The relative coordinates of node 2 with respect to node 1 are Δx, Δy, and Δz. A rotation about axis z will only give translations in the xy-plane. As shown in Figure D.2, we have the length r_{xy} of the projection of the rigid bar on the xy-plane. The relative coordinates in the xy-plane can be obtained through Figure D.2.

$$\Delta x_{new} = r_{xy}\cos(\theta_o + \theta_{z1}) = r_{xy}\cos\theta_o \cdot \cos\theta_{z1} - r_{xy}\sin\theta_o \cdot \sin\theta_{z1} \tag{D.1a}$$

$$\Delta y_{new} = r_{xy}\sin(\theta_o + \theta_{z1}) = r_{xy}\cos\theta_o \cdot \sin\theta_{z1} + r_{xy}\sin\theta_o \cdot \cos\theta_{z1} \tag{D.1b}$$

Fundamentals of Finite Element Analysis: Linear Finite Element Analysis, First Edition.
Ioannis Koutromanos, James McClure, and Christopher Roy.

Companion website: www.wiley.com/go/koutromanos/linear

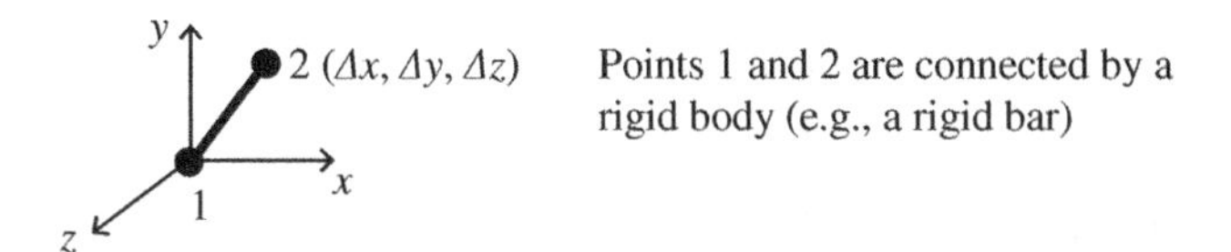

Figure D.1 Nodes interconnected with a rigid body.

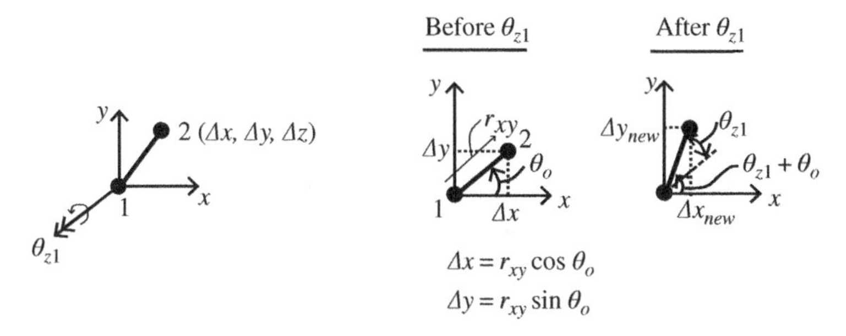

Figure D.2 Effect of rotation of node 1 by an angle θ_{z1}.

We can now calculate the two displacement components due to θ_{z1}. The displacements along the x- and y-direction are the changes in the relative x-coordinate and y-coordinates:

$$\begin{aligned} u_{x2} &= \Delta x_{new} - \Delta x = r_{xy}\cos\theta_o \cdot \cos\theta_{z1} - r_{xy}\sin\theta_o \cdot \sin\theta_{z1} - r_{xy}\cos\theta_o \\ &= r_{xy}\cos\theta_o(\cos\theta_{z1} - 1) - r_{xy}\sin\theta_o \cdot \sin\theta_{z1} \end{aligned} \tag{D.2a}$$

$$\begin{aligned} u_{y2} &= \Delta y_{new} - \Delta y = r_{xy}\cos\theta_o \cdot \sin\theta_{z1} + r_{xy}\sin\theta_o \cdot \cos\theta_{z1} - r_{xy}\sin\theta_o \\ &= r_{xy}\sin\theta_o(\cos\theta_{z1} - 1) + r_{xy}\cos\theta_o \cdot \sin\theta_{z1} \end{aligned} \tag{D.2b}$$

If θ_{z1} is small, then: $\cos\theta_{z1} \approx 1$, $\sin\theta_{z1} \approx \theta_{z1}$. Plugging these two expressions into Equations (D.2a, b), we obtain:

$$u_{x2} \approx -r_{xy}\sin\theta_o \cdot \theta_{z1} = -\Delta y \cdot \theta_{z1} \tag{D.3a}$$

$$u_{y2} \approx r_{xy}\cos\theta_o \cdot \theta_{z1} = \Delta x \cdot \theta_{z1} \tag{D.3b}$$

We apply the same process for the case shown in Figure D.3, where node 1 rotates by an angle, θ_{x1}, while all other displacements and rotations at this node are kept to zero. In this case, we will have translations of node 2 in the yz-plane.

$$\Delta y_{new} = r_{yz}\cos(\theta_o + \theta_{x1}) = r_{yz}\cos\theta_o \cdot \cos\theta_{x1} - r_{yz}\sin\theta_o \cdot \sin\theta_{x1} \tag{D.4a}$$

$$\Delta z_{new} = r_{yz}\sin(\theta_o + \theta_{x1}) = r_{yz}\cos\theta_o \cdot \sin\theta_{x1} + r_{yz}\sin\theta_o \cdot \cos\theta_{x1} \tag{D.4b}$$

$$\begin{aligned} u_{y2} &= \Delta y_{new} - \Delta y = r_{yz}\cos\theta_o \cdot \cos\theta_{x1} - r_{yz}\sin\theta_o \cdot \sin\theta_{x1} - r_{yz}\cos\theta_o \\ &= r_{yz}\cos\theta_o(\cos\theta_{x1} - 1) - r_{yz}\sin\theta_o \cdot \sin\theta_{x1} \end{aligned} \tag{D.5a}$$

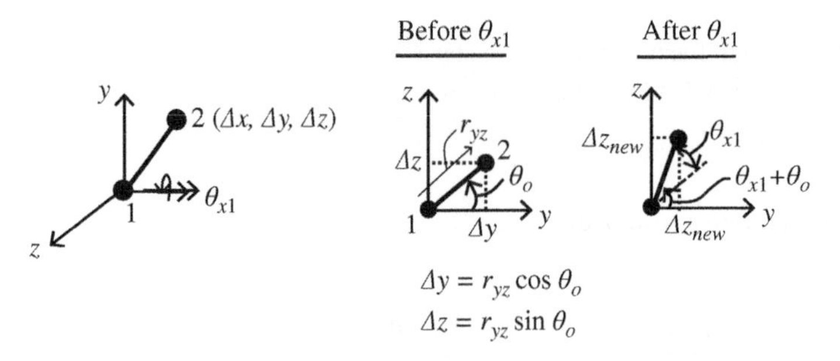

Figure D.3 Effect of rotation of node 1 by an angle θ_{x1}.

$$u_{z2} = \Delta z_{new} - \Delta z = r_{yz} \cos\theta_o \cdot \sin\theta_{x1} + r_{yz} \sin\theta_o \cdot \cos\theta_{x1} - r_{yz} \sin\theta_o$$
$$= r_{yz} \sin\theta_o (\cos\theta_{x1} - 1) + r_{yz} \cos\theta_o \cdot \sin\theta_{x1} \tag{D.5b}$$

For $\cos\theta_{x1} \approx 1$, $\sin\theta_{x1} \approx \theta_{x1}$, we obtain:

$$u_{y2} \approx -r_{yz} \sin\theta_o \cdot \theta_{x1} = -\Delta z \cdot \theta_{x1} \tag{D.6a}$$

$$u_{z2} \approx r_{yz} \cos\theta_o \cdot \theta_{x1} = \Delta y \cdot \theta_{x1} \tag{D.6b}$$

We repeat the same process for the case that we rotate node 1 by an angle θ_{y1}, while all other displacements and rotations at this node are kept to zero. This process will only lead to the translation of node 2 in the zx-plane, as shown in Figure D.4.

$$\Delta z_{new} = r_{zx} \cos(\theta_o + \theta_{y1}) = r_{zx} \cos\theta_o \cdot \cos\theta_{y1} - r_{zx} \sin\theta_o \cdot \sin\theta_{y1} \tag{D.7a}$$

$$\Delta x_{new} = r_{zx} \sin(\theta_o + \theta_{y1}) = r_{zx} \cos\theta_o \cdot \sin\theta_{y1} + r_{zx} \sin\theta_o \cdot \cos\theta_{y1} \tag{D.7b}$$

$$u_{z2} = \Delta z_{new} - \Delta z = r_{zx} \cos\theta_o \cdot \cos\theta_{y1} - r_{zx} \sin\theta_o \cdot \sin\theta_{y1} - r_{zx} \cos\theta_o$$
$$= r_{zx} \cos\theta_o (\cos\theta_{y1} - 1) - r_{zx} \sin\theta_o \cdot \sin\theta_{y1} \tag{D.8a}$$

$$u_{x2} = \Delta x_{new} - \Delta x = r_{zx} \cos\theta_o \cdot \sin\theta_{y1} + r_{zx} \sin\theta_o \cdot \cos\theta_{y1} - r_{zx} \sin\theta_o$$
$$= r_{zx} \sin\theta_o (\cos\theta_{y1} - 1) + r_{zx} \cos\theta_o \cdot \sin\theta_{y1} \tag{D.8b}$$

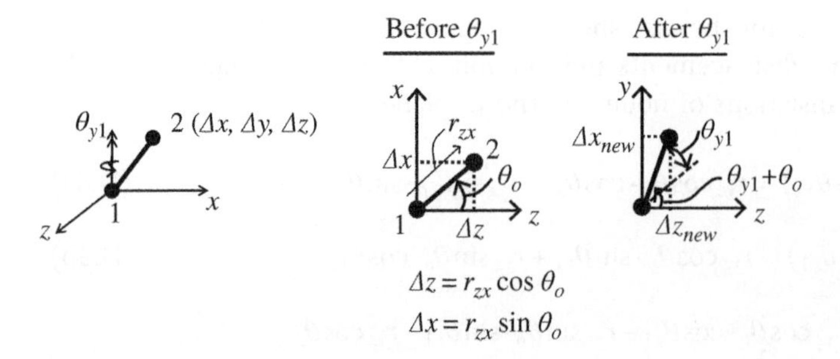

Figure D.4 Effect of rotation of node 1 by an angle θ_{y1}.

Since $\cos\theta_{y1} \approx 1$, $\sin\theta_{y1} \approx \theta_{y1}$, we obtain:

$$u_{z2} \approx -r_{zx}\sin\theta_o \cdot \theta_{y1} = -\Delta x \cdot \theta_{y1} \tag{D.9a}$$

$$u_{x2} \approx r_{zx}\cos\theta_o \cdot \theta_{y1} = \Delta z \cdot \theta_{y1} \tag{D.9b}$$

Now, if we have simultaneous nonzero translations and rotations at node 1, then the displacements at node 2 will be found as follows:

$$\begin{aligned}\text{displacements at node 2} = &\left(\textit{displacements at node 2 due to } u_{x1}\right)\\ &+\left(\textit{displacements at node 2 due to } u_{y1}\right)\\ &+\left(\textit{displacements at node 2 due to } u_{z1}\right)\\ &+\left(\textit{displacements at node 2 due to } \theta_{x1}\right)\\ &+\left(\textit{displacements at node 2 due to } \theta_{y1}\right)\\ &+\left(\textit{displacements at node 2 due to } \theta_{z1}\right)\end{aligned}$$

In accordance with all the above considerations, we can write:

$$\begin{Bmatrix} u_x \\ u_y \\ u_z \end{Bmatrix}_2 = \begin{Bmatrix} u_{x1} + \Delta z \cdot \theta_{y1} - \Delta y \cdot \theta_{z1} \\ u_{y1} + \Delta x \cdot \theta_{z1} - \Delta z \cdot \theta_{x1} \\ u_{z1} + \Delta y \cdot \theta_{x1} - \Delta x \cdot \theta_{y1} \end{Bmatrix} = \begin{Bmatrix} u_{x1} \\ u_{y1} \\ u_{x1} \end{Bmatrix} + \begin{Bmatrix} \Delta z \cdot \theta_{y1} - \Delta y \cdot \theta_{z1} \\ \Delta x \cdot \theta_{z1} - \Delta z \cdot \theta_{x1} \\ \Delta y \cdot \theta_{x1} - \Delta x \cdot \theta_{y1} \end{Bmatrix}$$

$$= \begin{Bmatrix} u_{x1} \\ u_{y1} \\ u_{x1} \end{Bmatrix} + \begin{bmatrix} 0 & \Delta z & -\Delta y \\ -\Delta z & 0 & \Delta x \\ \Delta y & -\Delta x & 0 \end{bmatrix} \begin{Bmatrix} \theta_{x1} \\ \theta_{y1} \\ \theta_{z1} \end{Bmatrix}$$

$$\rightarrow \vec{u}_2 = \vec{u}_1 + \vec{\theta}_1 \times \Delta\vec{r} \tag{D.10}$$

where $\vec{\theta}_1 = \begin{Bmatrix} \theta_{x1} \\ \theta_{y1} \\ \theta_{z1} \end{Bmatrix}$, $\Delta\vec{r} = \begin{Bmatrix} \Delta x \\ \Delta y \\ \Delta z \end{Bmatrix}$ and $\vec{\theta}_1 \times \Delta\vec{r} = \begin{bmatrix} 0 & \Delta z & -\Delta y \\ -\Delta z & 0 & \Delta x \\ \Delta y & -\Delta x & 0 \end{bmatrix} \begin{Bmatrix} \theta_{x1} \\ \theta_{y1} \\ \theta_{z1} \end{Bmatrix}$ is the cross-product (or vector product) of the two vectors, as explained in Section A.2.

Additionally, the rotations at node 2 will equal the corresponding rotations at node 1:

$$\vec{\theta}_2 = \begin{Bmatrix} \theta_{x2} \\ \theta_{y2} \\ \theta_{z2} \end{Bmatrix} = \begin{Bmatrix} \theta_{x1} \\ \theta_{y1} \\ \theta_{z1} \end{Bmatrix} = \vec{\theta}_1 \tag{D.11}$$

We can combine Equations (D.10) and (D.11) to obtain:

$$\begin{Bmatrix} u_x \\ u_y \\ u_z \\ \theta_x \\ \theta_y \\ \theta_z \end{Bmatrix}_2 = \begin{bmatrix} 1 & 0 & 0 & 0 & \Delta z & -\Delta y \\ 0 & 1 & 0 & -\Delta z & 0 & \Delta x \\ 0 & 0 & 1 & \Delta y & -\Delta x & 0 \\ 0 & 0 & 0 & 1 & 0 & 0 \\ 0 & 0 & 0 & 0 & 1 & 0 \\ 0 & 0 & 0 & 0 & 0 & 1 \end{bmatrix} \begin{Bmatrix} u_x \\ u_y \\ u_z \\ \theta_x \\ \theta_y \\ \theta_z \end{Bmatrix}_1 \rightarrow \{U_2\} = [T]\{U_1\} \tag{D.12}$$

where $[T] = \begin{bmatrix} 1 & 0 & 0 & 0 & \Delta z & -\Delta y \\ 0 & 1 & 0 & -\Delta z & 0 & \Delta x \\ 0 & 0 & 1 & \Delta y & -\Delta x & 0 \\ 0 & 0 & 0 & 1 & 0 & 0 \\ 0 & 0 & 0 & 0 & 1 & 0 \\ 0 & 0 & 0 & 0 & 0 & 1 \end{bmatrix}$ is the *constraint transformation array for two nodes connected with a rigid body*, and $\{U_1\}$, $\{U_2\}$ are the nodal displacement/rotation vectors for nodes 1 and 2, respectively.

If we only focus on planar two-dimensional (2D) problems, we can establish the following transformation equation for rigid-body constraints.

$$\begin{Bmatrix} u_x \\ u_y \\ \theta_z \end{Bmatrix}_2 = \begin{bmatrix} 1 & 0 & -\Delta y \\ 0 & 1 & \Delta x \\ 0 & 0 & 1 \end{bmatrix} \begin{Bmatrix} u_x \\ u_y \\ \theta_z \end{Bmatrix}_2 \rightarrow \{U_2\} = [T_{2D}]\{U_1\} \tag{D.13}$$

where $[T_{2D}] = \begin{bmatrix} 1 & 0 & -\Delta y \\ 0 & 1 & \Delta x \\ 0 & 0 & 1 \end{bmatrix}$.

Equation (D.13) accounts for the fact that, for two-dimensional problems, we only allow translations in the xy-plane and rotations about the z-axis. Now, we will obtain a transformation relation for the forces and moments. We will discuss in detail the two-dimensional case. To this end, we consider the situation in Figure D.5, wherein we have two forces and a moment applied at node 2. We are going to transform these forces and moment into an equivalent set of forces and moment at node 1, as also shown in Figure D.5.

From equilibrium considerations, we can obtain the following equation.

$$\begin{Bmatrix} F_{x1} \\ F_{y1} \\ M_{z1} \end{Bmatrix} = \begin{Bmatrix} F_{x2} \\ F_{y2} \\ -\Delta y \cdot F_{x2} + \Delta x \cdot F_{y2} + M_{z2} \end{Bmatrix} = \begin{bmatrix} 1 & 0 & 0 \\ 0 & 1 & 0 \\ -\Delta y & \Delta x & 1 \end{bmatrix} \begin{Bmatrix} F_{x2} \\ F_{y2} \\ M_{z2} \end{Bmatrix}$$

By inspection, we can verify that:

$$\{F_1\} = [T_{2D}]^T \{F_2\} \tag{D.14}$$

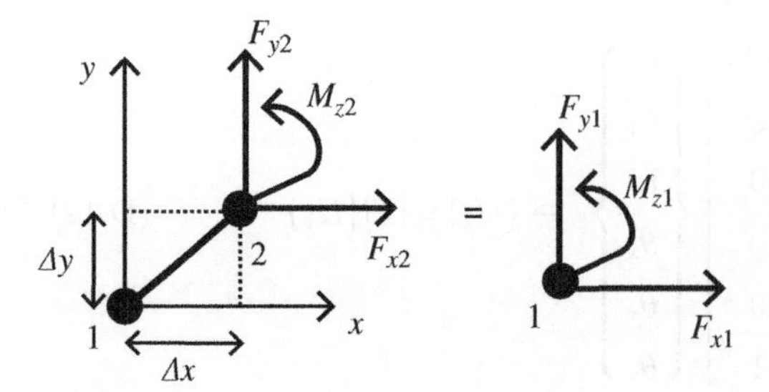

Figure D.5 Transformation of forces and moments at node 2, to a corresponding set at node 1.

In a similar fashion, for three-dimensional problems, we can find that

$$\{F_1\} = [T]^T\{F_2\} \tag{D.15}$$

So, to summarize, for two nodes, 1 and 2, which are parts of a rigid body so that the motion of node 2 is constrained to follow the motion of node 1, we have:

$\{U_2\} = [T]\{U_1\}$ (if we know the displacements of node 1, we can also find the displacements of node 2)

$\{F_1\} = [T]^T\{F_2\}$ (if we know the forces of node 2, we can express them as equivalent forces at node 1).

We can finally summarize the general process to handle the relations of displacements and forces for nodes connected through a rigid body, as summarized in Box D.1.

Box D.1 Transformation Equations for Nodes Connected with a Rigid Body

For any problem where we have master degrees of freedom, $\{U_{master}\}$, and slave degrees of freedom, $\{U_{slave}\}$, due to connection of nodes with a rigid body:

$$\{U_{slave}\} = [T]\{U_{master}\} \tag{D.12}$$

$$\{F_{master}\} = [T]^T\{F_{slave}\} \tag{D.15}$$

where $[T]$ is a transformation matrix expressing the kinematic constraints imposed by the presence of the rigid-body connection.

We will now proceed to obtain a transformation equation for stiffness matrices. If the forces of the slave node can be obtained as the product of a stiffness matrix times the displacements of the same node:

$$\{F_{slave}\} = [k_{slave}]\{U_{slave}\} \tag{D.16}$$

then we can obtain an expression for a stiffness matrix of the master node as well. Using Equation (D.12), we have:

$$\{F_{slave}\} = [k_{slave}]\{U_{slave}\} = [k_{slave}][T]\{U_{master}\} \tag{D.17}$$

Plugging Equation (D.17) into Equation (D.15), we have:

$$\{F_{master}\} = [T]^T\{F_{slave}\} = [T]^T[k_{slave}][T]\{U_{master}\} = \left([T]^T[k_{slave}][T]\right)\{U_{master}\}$$
$$= [k_{master}]\{U_{master}\}$$

where we have obtained the expression for the transformed stiffness matrix, in terms of the master degrees of freedom, $\{U_{master}\}$:

$$[k_{master}] = [T]^T[k_{slave}][T] \tag{D.18}$$

Remark D.1: Equations (D.12), (D.15), and (D.18) are mathematically identical to those obtained for coordinate transformations in Section B.8, if we replace the coordinate transformation matrix $[R^{(e)}]$ with the constraint matrix $[T]$. ∎

Index

Fundamentals of Finite Element Analysis: Linear Finite Element Analysis, First Edition.
Ioannis Koutromanos, James McClure, and Christopher Roy.

Companion website: www.wiley.com/go/koutromanos/linear